USED

Plant Science

Plant Science

AN INTRODUCTION TO WORLD CROPS

THIRD EDITION

Jules Janick
PURDUE UNIVERSITY

Robert W. Schery
THE LAWN INSTITUTE

Frank W. Woods
THE UNIVERSITY OF TENNESSEE

Vernon W. Ruttan
UNIVERSITY OF MINNESOTA

W. H. FREEMAN AND COMPANY
NEW YORK

Project Editor: Patricia Brewer

Manuscript Editor: Gunder Hefta

Designer: Sharon Helen Smith

Production Coordinator: Bill Murdock

Illustration Coordinator: Cheryl Nufer

Compositor: Allservice Phototypesetting Company

Printer and Binder: Arcata Book Group

Cover photograph: An aerial view near Gilroy, California. Copyright Barrie Rokeach 1981.

Part-opening photograph credits
Page 1: Larry Daughters for World Bank Group; page 121: USDA; page 321: Larry Rana/USDA; page 495: Alain Prott/CIRIC for World Bank; page 735: Chicago Board of Trade.

Library of Congress Cataloging in Publication Data

Main entry under title:

Plant science.

　　Includes bibliographies and index.
　　1. Agriculture.　2. Plants, Cultivated.　I. Janick,
Jules, 1931–
SB91.P55 1981　　　　631　　　　81-4897
ISBN 0-7167-1261-X　　　　　AACR2

Printed in the United States of America

4 5 6 7 8 9 0 KP 9 9 8 7 6 5 4 3 2 1 0

Contents

Preface

Some crop production practices are older than written history and form an integral part of human legend and ritual. It is no accident that all people celebrate festivals based on the agricultural calendar. The origins of the plant sciences can be traced to the seventeenth and eighteenth centuries, but they are in many respects truly twentieth-century disciplines. Our purpose is to blend these twin themes of plant science and crop agriculture into a single narrative. The formal disciplines in agriculture (agronomy, horticulture, forestry), botany (systematics, physiology, genetics, ecology), and the social sciences (economics, sociology, political science, history) must be interrelated to provide a full understanding of the contemporary relationship between human beings and plants.

This text was written to give students the scientific, technological, and economic foundations of world crop production. The sciences dealing with plants and the technology and economics of crop production and marketing have become irrevocably entwined. Each contributes relevance and meaning to the other.

This text was planned for use primarily as an introductory, university-level course in plant science, crop science, or economic botany. The entire book can be covered in a two-semester sequence: Parts One to Four in one semester, Parts Five and Six in the second. Under the quarter system various combinations of chapters and parts can be used, the choice depending on the particular objectives of the instructor. For example, Part Five could form the basis for a quarter course in Economic Crops.

Part One, *Plants and People*, introduces the relationship between crops and civilization in both historical and biological terms. Part Two, *Nature of Crop Plants*, is a formal presentation of plant relationships, structure, and development. Part Three, *Plant Environment*, is an introduction to the ecology of crop plants. Part Four, *Strategy of Crop Production*, features technological aspects of agricultural practices. Part Five, *Industry of Plant Agriculture*, surveys individual crop species. Finally Part Six, *The Marketplace*, explores the relation between crops and the economic community.

In this third edition, many modifications reflect the changes in the field that have occurred since the second edition, published in 1974. In addition, Part Six has been reorganized. A major new section, "The Evolution of Farm Commodity Policy," has been added in Chapter 24 to provide perspective on the dramatic changes in United States commodity policy during the 1970s. Two chapters from the earlier editions have been reorganized and integrated into a single new chapter, "World Commodity Markets and Policies" (Chapter 25).

Each of the four authors who collaborated on this book has had specific responsibility for particular chapters: Jules Janick, 1, 3, 7–9, 18, 19; Robert W. Schery, 6, 15, 20–23; Frank W. Woods, 2, 5, 10–14, 16, 17; Vernon W. Ruttan, 4, 24–28. We all, however, bear collective responsibility for the book as a whole.

We wish to thank our families, colleagues, and friends, who have suffered through this endeavor with us. We are pleased to acknowledge the assistance of Arthur H. Westing, who read through the entire manuscript of the first edition, and Laurence N. Skold, who made useful suggestions in various chapters during the preparation of the third edition. We gratefully acknowledge Gunder Hefta, who guided this edition with care and devotion.

February 1981

Jules Janick
Robert W. Schery
Frank W. Woods
Vernon W. Ruttan

Plant Science

One
PLANTS AND PEOPLE

1
Crop Plants and World Affairs

Earth is a planet dominated by plants. The green plant is fundamental to all other life. Were the human species to perish tomorrow, vines would destroy our mighty temples and grass would soon grow in the main streets of the world. In contrast, the disappearance of plants would be accompanied by the disappearance of human beings along with every other animal.

The importance we attribute to any product is usually related to the probability or actuality of a shortage of that product. Thus, those that are plentiful and readily available are often held in low esteem, even though our very existence may depend upon them. The oxygen we breathe, the nutrients we consume, the fuels we burn, and many of the most important materials we use were produced by plants.

Plants have a wide spectrum of uses. The most obvious is for human sustenance. Plants supply us with all of our food, either directly or indirectly as feed for animal intermediaries. They are also used as a source of construction materials and as the raw material in the manufacture of fabrics and paper and such "synthetics" as plastics and rayon. We have come to depend upon many of the complex substances that plants produce—dyes, tannins, waxes, resins, flavorings, medicines, and drugs. Living plants, besides having a direct effect on our ecological position, are used to control erosion by water and wind, to provide a setting for recreation and sports,

as landscape materials, and to satisfy our desire for beautiful objects.

The story of our species is largely a chronicle of struggle for dominion over the environment. The efficiency of this control is thought of in terms of civilization, or culture. To a great extent, controlling the environment means controlling plant life. The failure to provide a high standard of living for the population as a whole in so-called underdeveloped countries is generally associated with an inefficient system of crop production and distribution. Similarly, the decline of developed societies can in many instances be associated with a disruption in the basic system of resource utilization. Crop production, the management of useful plants, is the very basis of our civilization.

CROP PLANTS

In its broadest sense, the term "crop plant" means any plant utilized by people for any purpose. In the restricted but common sense of the term, crop plants are those useful plants that fit economically into the scheme of our work and existence. They are plants that are managed to some degree. This management may involve merely a systematic harvest, as in some types of forestry, or completely artificial culture throughout the entire life of the plant, as in greenhouse fruit production.

The number of species considered to be crop plants depends on the definition employed. For example, in Bailey's *Manual of Cultivated Plants,*° a well-known compilation for the United States and Canada, 5347 species are listed. A worldwide compilation would number between 10,000 and 20,000, and would include many plants that can be classified as crops only under the broadest definition. The number of plants of economic importance is probably between 1000 and 2000 species. Those of major importance in world trade are relatively few, perhaps between 100 and 200. Fifteen species provide the bulk of the world's food crops: rice, wheat, corn, sorghum, and barley; sugarcane and sugar beet; potato, sweetpotato, and cassava; bean, soybean, and peanut; coconut and banana. The course of history has been changed—and continues to be changed—by only a few plants, out of an estimated world total of 350,000 species. People have struggled to obtain them and to cultivate them. These are the plants of legend and literature, of song and sacrifice; they are the plants with which this book is largely concerned.

DEVELOPMENT OF AGRICULTURE

The use of fire and the development of agriculture are the two innovations that form the basis of civilization. Fire is so basic to our existence that it is hard to imagine human culture without it. Our first use of fire not only marked the beginning of social living but eventually gave birth to a whole series of related technologies that otherwise would have been inconceivable. The most important immediate result of the use of fire was the expansion of the food supply, because a great number of foods are inedible, unpalatable, or unsanitary unless they are cooked. The continued development of any society hinges on the availability of an adequate source of food. In primitive societies based on food gathering or hunting, each individual must be totally concerned with the urgencies of securing sustenance. Abundance is temporary and exceptional. The solution to this problem came with the "invention" of a series of complex and

related technologies involving an intimate relation between crop plants and domestic animals—the development of **agriculture.**

Agriculture is a relatively recent innovation when considered in relation to human history, because we have been mere collectors of food for the greatest part of our existence (Fig. 1-1). The first production

Figure 1-1   Hunting and gathering in the Stone Age. Upper Paleolithic cave paintings (ca. 30,000–15,000 B.C.E.) from Castellon, Spain, portray hunting of stags (A) and gathering of honey (B). A close-up of a beautifully wrought cave painting of a bison from Lascaux in southwestern France (C) shows what appears to be a representation of plants. Could this be the first plant illustration? [A and B from C. Singer, E. J. Holmyard, and A. R. Hall (editors), *History of Technology,* vol. 1, Oxford University Press, 1954, courtesy The Metropolitan Museum of Art. C from P. J. Ucko and A. Rosenfeld, *Paleolithic Cave Art,* Weidenfeld & Nicolson, 1967.]

°Liberty H. Bailey, *Manual of Cultivated Plants,* rev. ed. (New York: Macmillan, 1949).

A   B   C

Figure 1-2 Domestication of our crop plants was achieved in prehistory. These Peruvian ceramic jars, which date from 4000 B.C.E., combine human and animal forms with the forms of plant crops: (A) squash; (B) peanut; (C) potato. [From The University Museum, University of Pennsylvania.]

of food by crop cultivation and actual domestication dates back 7000 to 10,000 years, to the Neolithic Age. Agriculture seems to have developed independently at widely separated times in a number of parts of the world. The gradual development of agriculture brought forth a dependable surplus. Such a surplus releases from food production those individuals with skills in other specialties. The development of new specialists is possible only when increased agricultural efficiency allows the exploitation of newly acquired leisure. This is still true. The resultant increases in the standard of living are measured by the degree to which former luxuries become everyday necessities.

The origins of civilization can be traced to the discovery that a plentiful food supply could be assured by the planting of plant parts or seed. Plants that grow rapidly and produce a crop within a season were probably the first cultivated. The technology involved in the cultivation of such long-lived plants as fruit trees is time consuming and complex; it is most likely, therefore, that such crops were simply gathered from the wild until fairly recently. Some crops are still harvested from the wild. The chicle

industry of Central America and the blueberry industry of Maine are dependent on wild plants.

Practically every one of our present-day crop plants (as well as domestic animals) was developed in prehistoric times (Fig. 1-2). This was achieved by means of two distinct processes: **domestication,** the bringing into cultivation or management of some wild species; and **selection,** the differential reproduction of these species. Primitive people displayed uncanny ingenuity in the process of domesticating wild plants and preparing them for food. For example, the cassava, a major source of starch in tropical South America, contains a deadly poison (hydrocyanic acid), but somehow it was long ago learned that the poison could be removed by means of various treatments, especially cooking. Selection often led to the creation of new types. Most of our cultivated crops differ markedly from their wild ancestors, and many are so changed that precise lines of descent have become obscure. The primitive agriculturist was a plant breeder, and a very effective one, without any knowledge of genetics.

Crop plants have accompanied us in our wanderings and migrations. The introduction of species into

Table 1-1 The centers of world production and centers of origin for selected world crops.

Crop	Centers of production	Centers of origin
Cacao	Africa	Brazil
Coffee	Brazil	Abyssinia
Maize	Midwest U.S.	Tropical America
Monterey pine	Australia	California
Pineapple	Hawaii	Brazil
Potato	Eastern Europe	Peru
Wheat	North-central North America	Central Asia

new habitats has been one of the most important features in world agricultural development. Today the centers of production of nearly all agricultural crops are far removed from their centers of origin, as is shown in Table 1-1.

Civilization rests on discoveries made by people little remembered by history. Precisely where the first plants were cultivated is unknown. According to archeological evidence, however, agriculture had its origin 7000 to 8000 years ago, somewhere in the then well-watered highlands of the Indus, Tigris, Euphrates, and Nile rivers. The antecedents of this agriculture undoubtedly date back many more thousands of years. Southeast Asia, with its diverse geography and consequent diversity of vegetation, its mild climate, and its capacity to support a stable population with a fishing/hunting economy, has been proposed as a likely location for the birth of primitive agriculture. This area is especially rich in asexually propagated crop plants, and perhaps the planting of vegetative parts preceded the planting of seed. Primitive agriculture may have originated independently in a number of parts of the world, and developed through the dispersal and divergence of new plant forms in new environments. As agriculture moved into more severe climates, seeding replaced planting of vegetative parts as the dominant technique.

When agriculture came to the Old World there was a movement to river valleys, in which the twin obstacles of drought and flood had to be overcome. Gigantic changes were catalyzed by the innovations required by cereal culture and irrigation. The new technology increased the need for a higher level of social order; great works are required to make rivers work to the benefit of farmers rather than to their detriment. The success of this technology can be measured in terms of the ever-increasing populations that primitive agriculture supported. More than 5000 years ago a stone age culture was transformed into the full-fledged urban civilizations of Egypt and Sumeria.

NEAR EASTERN ORIGINS

The birthplace of many early civilizations lay in the territory now referred to as the Near East. Particularly important was Mesopotamia—"the country between the rivers." The superlative juxtaposition of habitats that exists there supports a diversity of wild species. Such diversity is likely to foster change and to stimulate speedy adaptation under the pressure of early selection. Important plant discoveries and early domestications were presumably made in the lower elevations of the Zagros Mountains and brought down to the fertile Tigris–Euphrates flood plain for full domestication under management. The earliest traces of the transition from food-gathering to food-producing economies are found here. Modest genetic change (an all but inevitable consequence of selection for harvestable types) would encourage rapid adaptation to intensive cultivation in river valleys. It is likely that these first crops, mainly grains, were selected for high yield and valued principally as high-energy foods. Vegetable oils were no doubt secured by gathering fruits from the wild; other requisites for a balanced diet were obtained by catching game. Acorns, pistachio nuts, olives, and hazel nuts abounded in the region.

Figure 1-3 is a map of Mesopotamia and the Fertile Crescent, where so many important civilizations have arisen. The high plateau east of the Zagros Mountains is inhospitable, and probably only the deltas of the mountain valleys were habitable in primitive times. But west of the Zagros are fairly moist

Figure 1-3 Greater Mesopotamia, showing environmental zones, mineral resources, and archeological sites. [After K. V. Flannery, "The Ecology of Early Food Production in Mesopotamia," *Science*, vol. 147, pp. 1247–1255, fig. 2, 12 March 1965. Copyright © by the American Association for the Advancement of Science.]

uplands, below them somewhat drier steppes, and even drier alluvial bottoms below the steppes. Irrigation of the deep, rich soils of the alluvial bottoms would make for very productive agriculture. Ancient settlements were amazingly abundant on the western foothills of the Zagros, indicating that these foohills once had a much greater carrying capacity than seems possible today. Very likely, settlements ranged from essentially permanent ones in the lowlands, to temporary camps of migratory tribes that followed game (and eventually herded livestock after its domestication), to seasonally suitable grazing and browsing land in the mountains. Itinerants may have carried on trade with peoples of the Iranian plateau,

exchanging copper (which was used for making tools and utensils) for food and seasonal accommodations at the more permanent settlements.

Preagricultural subsistence in Mesopotamia relied greatly upon the chase. The wild sheep, goat, pig, gazelle, deer, ass, birds, and small game were all hunted, and fish were caught by various means. Wild fruits and seeds of the vicinity were gathered, too. Wild alfalfa and other legumes, and of course the primitive grains, were widely used. It is not unlikely that a migrating tribe might harvest one crop on the steppes in March, another higher in the foothills in April or May, and still another at higher elevations in summer. A region with so richly varied a natural fare would offer many candidate species for selection and trial for adaptation and storage qualities in new environments.

The first crop plants would not be acceptable today. Wild grains shatter easily, and their tough husks make threshing difficult. Perhaps unconsciously, early farmers overcame these difficulties by selecting grains with a tough stalk and by roasting and grinding husked seeds. The requirement of intensive agriculture brought about directed selection to improve the crop and the preservation of other genetic change. The improved crop, entirely an artifact of human manipulation, and perhaps unable to persist in the wild, is the culmination of a long process involving changes in the ecological relationship between farmers and locally available plants.

As in modern agriculture, early farmers would endeavor to secure the greatest yield with the least labor. Concentrating the crop by sowing seed in areas where weeding and protection against pests were possible would have afforded a definite advantage over collection from wild stands. This practice was established very early both for wheat and for barley. Barley may have become domesticated quite fortuitously, as an impossible-to-eliminate weed in wheat (rye and oats were apparently introduced into European agriculture in this way from the Near East). Certain millets, field peas, lentils, vetches, chickpea, horsebean, and flax were also introduced from the Near East. The wine grape was cultivated at least 6000 years ago and gradually the olive, date, apple, pear, cherry, fig, and other fruits were managed in orchards rather than sought in the wild.

AGRICULTURE AND ANTIQUITY

Mesopotamia, aptly called the cradle of civilization, was to influence all of antiquity. The first urban economy to develop there was based on an agricultural technology that led to temples and priests, granaries and clerks. The creation of a "social surplus" made warfare and slavery economic institutions. The administration of the stored surplus pressed the need for a system of accounts. The solution to this problem came 6000 years ago with the invention of writing, which marks the initiation of civilization in its present-day meaning. The Mesopotamian cultures existed for thousands of years under many different regimes. Their influence, although difficult to define precisely, radiated to Syria and Egypt and perhaps to India and China.

The backbone of Mesopotamian agriculture consisted of crop plants that are still important to the world's food supply: wheat and barley, date and fig, olive and grape. The ancient cultures of Mesopotamia—Sumerian, Babylonian, Assyrian, Chaldean—developed an increasingly complex and integrated agriculture. Their ruins reveal the remnants of irrigated terraces, parks, and gardens. Four thousand years ago irrigation canals lined with burnt brick with asphalt-sealed joints helped keep 10,000 square miles under cultivation to feed 15 million people.

By 700 B.C.E. (Before the Common Era) an Assyrian herbal was compiled that contained the names of more than 900 plants, including 250 vegetable, drug, and oil crops. By the second century B.C.E., a treatise on agriculture consisting of 28 volumes was compiled by Mago, who lived in the ancient city-state of Carthage in North Africa. The books of the old Testament are rich in allusion to the agriculture of this era.

Our knowledge of ancient agricultural development is nowhere greater than in Egypt, where the drifting desert sands have preserved the records of an astonishing era. Although the Nile Valley had supported human activity for at least 20,000 years, its agricultural development is believed to have stimulated the transformations that occurred in the Mediterranean area. The great Egyptian civilization, to which our own culture owes so much, flourished on the agricultural abundance made possible by the constant renewal of fertility from the overflow of the Nile. By 3500 B.C.E. Egypt had established a centralized government with Memphis as its capital, and by 2800 B.C.E. had developed to a level of sophistication capable of such gigantic works as the construction of the pyramids.

The Egyptians were masters at developing techniques of drainage and irrigation. Drainage, the removal of excess water, is a necessity in areas like the

Figure 1-4 Irrigating with a shaduf. The farmer has raised water from a pool with the aid of a clay counterpoise to irrigate a palm garden. From a tomb at Thebes (ca. 1500 B.C.E.). [From C. Singer, E. J. Holmyard, and A. R. Hall (editors), *History of Technology*, vol. 1, Oxford University Press, 1954, courtesy The Metropolitan Museum of Art.]

Nile Valley, in which flooding was an annual event and of concern to municipal as well as agricultural centers. Drainage requires both the development of the slope of the ground and the establishment of efficient conduction systems. Irrigation, the artificial supplying of water to crops, involves the procurement, conduction, and application of water. The problems of drainage and irrigation are intertwined; the Egyptian solution was to construct a series of dikes for storing water and canals that served both purposes. The Egyptians developed techniques for raising water that are still being used today. The most basic invention was the **shaduf,** a suspended container balanced with a counterweight (Fig. 1-4). With this innovation, it was possible for one person to raise 600 gallons of water as much as six feet per day.

Land preparation technologies can be traced through refinements in the simple hoe. The hoe was originally a pointed, crotched stick and was used (as it is today) with a chopping motion. The early plow was merely a hoe drawn by the farmer (later with the help of animals) to scratch the surface of the soil (Fig. 1-5), and is still in use in many parts of the world. Later the plow was improved by reinforcing the parts in contact with the soil with iron and

by more substantial and efficient construction. The modern, or "soil inverting," plow is a more recent European invention that dates back to the eleventh century. The Egyptians employed various cutting tools in harvesting, of which the curved metal sickle was the most refined.

The Egyptians developed the various technologies associated with the culinary arts—the ceramics industry, baking, wine making, and food storage. Their methods of preserving food included fermentation and pickling, drying, smoking, and salting. Many plants were cultivated for fiber, oil, and other "industrial" uses. These included papyrus for paper, date palm for fiber, castor bean for oil, pines for resins. The Egyptians created the first pharmacopoeia, a collection of drug and medicinal plants, and the first spice, perfume, and cosmetic industries.

Along the Nile great formal gardens were created, replete with exotic ornamentals and pools containing fish and water lilies. In orchards the date, grape, fig, lemon, and pomegranate were cultivated. Vegetable gardens included cucumber, artichoke, lentil, garlic, leek, onion, lettuce, mint, endive, chicory, radish, and various melons. (See Fig. 1-6.)

Egyptian civilization persisted for an incredible 35 centuries. It produced not only a remarkable technology but magnificent art and a complex, if bewildering, theology. By the time Egypt had become a Roman province (3 B.C.E.), its influence had already permeated the ancient world—an influence that is still felt today.

In approximately 1000 B.C.E. the introduction of iron and the rise of the Indo-European invaders were to create a period of turbulence in the history of civilization and cause a shift to the Aegean—the bridge between Europe, Asia, and Africa. Phoenician sailors passed on the technological heritage of Mesopotamia and Egypt to the emerging Greek Isles. The agricultural products of the eastern Mediterranean region—wheat, barley, grapes, figs, and olives—were never in abundant supply because of a combination of uncertain rainfall and shallow soil. Consequently, the strength of Greek civilization was destined to be based on sea power and to be nourished by trade and conquest.

Figure 1-5 Development of cultivation technology shows divergent and parallel development in Egypt (A–D) and the New World (E–F).

(A) Treading in seed by sheep (from a tomb at Saqqara, Egypt, ca. 2400 B.C.E.). Herodotus the Greek historian describes the process as he found it in about 450 B.C.E.: "They gather in the fruits of the earth with less labor than any other people . . . for they have not the toil of breaking up the furrow with the plough, nor of hoeing, nor of any other work which all other men must labor at to obtain a crop of corn; but when the river has come of its own accord and irrigated their fields, and having irrigated them has subsided, then each man sows his land and turns his swine into it; and when the seed has been trodden into it by the swine he waits for harvest time; then . . . he gathers it in."

(B) A primitive hoe (*top*) made from a forked branch and more developed form (*below*) with hafted wooden blade (Middle Kingdom, Egypt, 2375–1800 B.C.E.).

(C) Plowing and hoeing (from a tomb at Neni Hasan, ca. 1900 B.C.E.).

(D) Agricultural implements and land reclamation in ancient Egypt (from a tomb at Thebes, ca. 1420 B.C.E.).

(E,F) New World agriculture based on an illustrated calendar by a Peruvian, half Indian and half Spanish, who drew it in the 1580s to present to the King of Spain as part of a treatise on Inca life. Note the primitive hoe (E) and the foot hoe or taclla (F).

[A–D from C. Singer, E. J. Holmyard, and A. R. Hall (editors), *History of Technology*, vol. 1, Oxford University Press, 1954, courtesy The Metropolitan Museum of Art.]

Figure 1-6 Gardening in ancient Egypt. (A) Formal gardens containing doum palms, date palms, acacias, and other trees and shrubs (from a tomb at Thebes, ca. 1450 B.C.E.). Note the statue being towed in the lotus pool toward the pavilion. (B) Picking figs (from a tomb at Beni Hasan, ca. 1900 B.C.E.). (C) Collecting grapes from a round vine arbor and treading to express the juice (from a tomb at Thebes, ca. 1500 B.C.E.). (D) Irrigating a vegetable garden. [A–C from C. Singer, E. J. Holmyard, and A. R. Hall (editors), *History of Technology*, vol. 1, Oxford University Press, 1954, courtesy The Metropolitan Museum of Art. D From M. Gothein, *A History of Garden Art*, vol. 1, Dutton, 1928.]

GREECE AND ROME

The "Greek miracle," the extraordinary achievement wrought some 2500 years ago by an ancient civilization, was the result of a combination of factors that is still difficult to explain. Because our cultural heritage in art, literature, and ethics is in large part traceable to Greek influences, it is easy to overemphasize the technological level of Greek culture (Fig. 1-7). The achievements of the Hellenic world owed much

Figure 1-7 Greek plow (from a black-figured Nikosthenes cup, sixth century B.C.E.). Note the similarity to Egyptian implements which, at that time, had been in use for 1000 years. [From C. Singer, E. J. Holmyard, and A. R. Hall (editors), *History of Technology*, vol. 2, Oxford University Press, 1956, courtesy The Metropolitan Museum of Art.]

to the established technologies of Mesopotamia and Egypt; they were developed when the pyramids were more than a thousand years old. The rise of Greece, and later of Rome, is in some ways comparable to the still later rise of the Huns and the Goths—a victory of barbarism over advanced but tired cultures. The technology of the invader was often inferior to that of the invaded and subdued. Although the Greeks added only incidentally to the practical, their curiosity and their analytical attitude concerning the nature of things were to have profound effects on the course of future technological advance. For example, the Greeks, while contributing little to practical agriculture, made great contributions to the formal study of plants. The science of botany originated in the Greek mind.

Theophrastus of Eresos (372–288 B.C.E.), brilliant student and disciple of Aristotle (whose botanical works are lost), appears to have been among the first to become interested in plants for their own sake. His two treatises, *History of Plants* and *Causes of Plants*, have earned him the appellation "Father of Botany," and influenced botanical thinking until the

seventeenth century. These writings deal with such diverse topics as morphology, classification, seed and vegetative propagation, geographic botany, forestry and horticulture, pharmacology, insect pests, and plant flavors and odors. Theophrastus treated more than 500 kinds of plants, both cultivated and wild. He distinguished between angiosperms and gymnosperms, between monocots and dicots, and discussed the formation of annual rings and methods of collecting resins and pitch. He even discussed the "dusting" of the fruit-forming "female" date palm with flowers of the nonfruiting "male" to effect fruiting. This straightforward account of sexuality in plants was virtually lost for 2000 years.

The Greek genius proved to be incapable of political survival. The intercity struggles and rivalries paved the way for complete destruction by the armies of Macedonia. There are some who trace the downfall of Greece to the effect of increasing populations on declining resources as much as to wars and internal decay. It appears that the agricultural base of Greece was not sufficient to support an ever-growing civilization.

Greek civilization was absorbed by a new breed to the west. The Roman empire, unlike the Greek, was built on a firm and powerful resource base. In contrast to the Greeks, the Romans were extremely interested in the practical aspects of agriculture. Agriculture was a vital part of the economy and of real concern. The Romans' main source of revenue was land taxes; their most important legislation dealt with agrarian planning. Great fortunes were invested in farm land. Rome grew to grandeur on the foundation of a sound and functional agricultural technology. As they conquered, they built a culture that was Greek in origin but Roman in application. Although the Romans can be credited with relatively few original ideas, it must be recognized that they did improve upon what they found. Their enduring trademarks were roads and aqueducts. The Romans were of a modern mind—civilized and urban, yet bound to the land through business and inclination. The combination of their empiricism and their genius for political management produced the largest contiguous empire the world has seen—an empire destined to last a millennium.

Roman agricultural practices are well recorded. The earliest of the agricultural writings in Latin is the terse *De agri cultura* of Marcus Porcius Cato (234–149 B.C.E.), who wrote of the practical aspects of crop and livestock management, and especially of profits—slaves should be fed only the barest minimum. The origins of a rural philosophy are found in his conclusion that farmers are not only the best citizens but make the best soldiers. About a hundred years later a longer, better organized, and more complete work was produced by Marcus Terentius Varro (116–28 B.C.E.). Varro's *De re rustica libri III* (three books on farming) emphasized the dependence of the commonwealth on a sound agriculture. The *Georgics* of Virgil (70–19 B.C.E.), one of the era's literary masterpieces, spoke not only of the details of husbandry but extolled the virtues of rural life. In the second half of the first century A.D. a number of works dealing with agriculture were compiled. These include a twelve-book treatise by Columella and the encyclopedia *Historia naturalis* of Pliny the Elder (A.D. 23–79), a monumental conglomeration of science and ignorance. It is from these and other works that the agrarian history of Rome has been deduced.

During the early period of Roman history (up to 200 B.C.E.) the basic agricultural institution was the village community. Early Rome has been described as a marketplace that served a hamlet of truck gardeners. Individual holdings were small, ranging in size from one to four acres, and intensively managed. (To wish for more was considered dangerous.) But as the Roman state expanded in territory and acquired a slave-labor force through conquest, a larger production unit, the *latifundium*, came into being. These large-scale holdings were carved out of public lands and distributed by the state. The resultant plantation system was to foster the growth of enormous individual fortunes and to encourage graft and corruption, which spread virulently. The abundance of low-cost labor that resulted from the increasing use of captive "barbarians" and the expanding size of individual holdings resulted in a disequilibrium in the social system. The soldier–farmer–citizen lost his place as a stabilizing force in Roman life. Absentee ownership, slavery, and soil exhaustion resulted in ever-declining agricultural productivity. In addition, continual warfare had the end result of reducing the class of small landholders by attrition, only to increase the holdings of the remaining few. Tribute from conquered states in the form of grain and other food products relieved temporary shortages, but in the long run probably contributed to a decline in agricultural productivity on the Italian peninsula.

An alternative system of agricultural production arose about 200 C.E. (Common Era) with the substitution of tenancy for slavery. Rents rather than produce became the return from investment in agriculture. The increase of an indebted tenant class in the rural areas and of a landless, discontented "proletariat" in the cities was to have unhappy consequences. With a decline in land productivity the bond between landowner and tenant deteriorated from a symbiotic to a parasitic relationship and was accompanied by the gradual erosion of freedom of the erstwhile renter. For example, it became a crime for the son of a farmer to leave the farm on which he was reared, and soon "free" men as well as slaves were bound to the land they cultivated. Such legislation was no doubt originally enacted to increase lagging agricultural productivity, but it failed badly in this respect. A declining population in Europe followed the declining productive capacity of the land. The surplus necessary to support cities evaporated, and the decimation of the metropolitan centers was followed by a lowering in the quality of civilization.

The tenant was systematically reduced from a citizen to a serf, and the slave retained his yoke under a new name. The renter who traded his labor for land (and production) was in fact bound to it—originally through the machinery of law but soon even more effectively through force of custom and habit. The resultant manorial system (named after the great houses or manors of the owner) became a stable system and signaled a return to a village economy. It lasted in Europe as a dominant condition for well over a thousand years, and it still exists in many areas of South America. It was the general system of life in the Middle Ages.

The agricultural technology of Rome at its height, although probably inferior to that of Egypt, was not equaled in Europe until the Renaissance. Agricultural writings note grafting and budding, the use of many kinds and cultivars of fruits and vegetables, legume rotation, the use of manure and marl, soil fertility appraisals, and even cold storage of fruit. Mention can be found of a prototype greenhouse (*specularium*) for vegetable forcing, which was constructed of sheets of mica to let in the light.

It was in Rome that ornamental gardening was developed to a high level. The good life was that of a gentleman farmer, his expression of conspicuous consumption being the country estate, or *suburbanum*. It contained fruit orchards and flower gardens. Splendid mansions were enclosed by frescoed walls, statuary, fountains, and pools. The look of luxury was artfully conceived.

The rise and fall of the political fortunes of imperial Rome were paralleled by similar trends in agriculture. Which was the cause and which the effect is not clear; probably they were interrelated. The strain of supporting and defending an overexpanded empire undermined the agricultural foundations, and an unstable and exhausted agriculture reduced the capacity for further growth and development. By the third century C.E. Rome had become an economic, political, and military failure. Nevertheless, the remarkable fact remains that Rome, an amalgamation of diverse peoples and customs, survived for a millennium. The end was rather inglorious: the empire retreated to the villages from which it had been created. Western civilization and agriculture were to take a step backward into ignorance and misfortune.

MEDIEVAL AGRICULTURE

With the decline of Rome and the West, the mantle of technological superiority passed to the Near East and Far East. During the thousand years from 500 to 1500, it was the East that provided the most skilfully crafted products. The rise of Islam after 700 provided a repository for Greek culture and preserved the heritage of antiquity. It is intriguing to note that the Eastern civilizations were at the peak of their achievements when Western Europe was at its low ebb.

During the early decline in technological progress in the West, much of the agricultural art survived in the gardens of monasteries. Gardening became an integral part of monastic life, providing food, wine, decoration, and medicinals. The job of gardener (*Hortularium*, or *Gardinarium*) became a regular monastic office. Fruit varieties and vegetable strains were preserved and improved. The few botanical and agricultural writings of this period were derived from the church, the guardian of literacy, but were for the most part compilations traceable to Roman sources, especially Pliny's *Natural History*. It was to be many centuries before the technology of antiquity was equaled.

The Byzantine empire, the eastern heir of Rome, was to be the filter through which passed techniques and ideas, as well as silks and ivories, from Persia, India, and China. Slowly developing contacts with the East brought many things that were to have a profound influence on the future course of history. Gunpowder, paper making, the compass, and what we know as arabic numerals all come to us from the Far East. In the field of agriculture and crop production many important products of Eastern origin were continually introduced. Such Eastern crop plants as the peach were known to the Chinese before 2000 B.C.E., and had reached the Mediterranean at about the height of Greek civilization. The peach was described in certain writings as a Persian fruit, a misconception that survives in the present botanical name, *Prunus persica*. Sugar cane, native to eastern Asia, although known to classical antiquity, was not commonly used in Europe until it was introduced by the Arabs into Palestine, Sicily, Spain, and the Greek Islands. The soybean is the most recent of a long list of important crop introductions from the Far East. One of the most important inventions brought from the East was an efficient collar that would not choke a plow horse under strenuous pulling. The development of such an efficient harness made possible the utilization of the full power of the horse in agriculture. The padded collar, known to have been used in China from the third century, was unknown in Europe until the ninth century.

The lure of the Orient and its wealth had long been a determining force in Western civilization. It was the search for new routes to the spice-rich East that gave impetus to the exploits of the daring navigators of Italy, Spain, and Portugal. The discovery by

Columbus of the New World in 1492 marks a convenient (if not truly accurate) date to assign to the beginning of the Modern Age.

In the Middle Ages the manor, or estate, was the basic unit of agricultural production and the smallest component of the feudal state. The estate, consisting of small tenant holdings and the lord's land, was characterized by a high degree of self-sufficiency. This very high degree of independence retarded communications and kept progress down to a low level.

Crop growing was based on extensive production in large fields and intensive production in small gardens. The tenant had the use of strips in a number of fields, the right to a share of hay and firewood, and the right to graze his cow on a common pasture (the commons). The usual system of crop production was a **fallow system,** in which part of the land was kept unplanted to restore soil fertility. Fields were rotated among the tenants, a practice that ensured an equitable distribution of the good land but probably removed any incentive to improve the soil.

The present classification of crop agriculture into **agronomy, horticulture,** and **forestry** is medieval in origin, and traces back to the system of land distribution of the manor. The practice of planting extensive areas in grains and forages gave rise to the specialized discipline of crop agriculture now called agronomy. Horticulture developed from the practice of cultivating intensively managed kitchen gardens that provided fruits and "herbs" as well as the increasing degree of ornamental planting on the lord's manor. Forestry developed from the protection of the wild lands that produce timber and support game. The relationship between forestry and game management is an old one.

A decline in feudalism came about with the proliferation and growth of towns and the emergence of strong national states. The re-emergence of towns offered an alternative way of life for the peasant. The rise of the modern state was based on the destruction of power of the manorial lords. Feudalism's breakdown was forewarned by a series of unsuccessful peasant revolts as early as the fifteenth century, yet the complete disappearance of feudalism was slow to come. Although it was gradually eliminated in England in the sixteenth century, serfdom was not legally abolished in many countries until very recently. It was legally abolished in Prussia in 1807, in Russia in 1861, and in Japan in 1874.

The recovery of Europe from the backwardness of the so-called Dark Ages was a slow but continuous process. It came with the increase in trade and economic activity. With the decline in feudal ways and the accumulation of technology (incorporated from Islamic cultures in Spain, old Roman techniques preserved by the monasteries, and the ever-increasing appearance of new processes and inventions), agricultural methods, tools and products improved. The rising population of towns and the expansion of trade and finance also enriched the rural economy. New industries created a market for industrial crops—sugar cane, hemp, flax, vegetable oils, and dyes. The production of wine became a key industry and vineyards began to flourish in favorable regions. Strip farming came to be abandoned in favor of an enclosed system in which individual farmers took continuing responsibility for their own land. Crop rotation and improved plowing techniques became more widespread. Water mills, although known to the East, began to appear in Europe toward the end of the twelfth century. From the eleventh to the fourteenth centuries great dikes (the "Golden Wall") were constructed to hold back the sea from inundating Belgium and Holland. At the end of the Middle Ages the agricultural land of Europe was cleared, drained, and cultivated to provide the foundations for the eventual triumph of European civilization.

As feudalism gave way to trade, a real rise in the standard of living was produced. The diet of the peasant class improved from one consisting mainly of cereals to one that included garden products and meat. By the thirteenth and fourteenth centuries orchards and gardens were common outside of monasteries. The revival of learning known as the Renaissance had a great impact on agriculture, and a horticultural revival spread from Italy to France and then to England (Fig. 1-8). The peak of Renaissance horticulture is to be found in the magnificent gardens of LeNôtre (1613–1700), the most famous being those at the Palace of Versailles, built for the French monarch Louis XIV.

A

B

Figure 1-8 Renaissance gardens. (A) Frantic garden activity as depicted by the famous Flemish artist Pieter Brueghel (1525–1569). (B) An herbal garden from the *Kreuterbuch* of 1587. [A from The Metropolitan Museum of Art, The Harris Brisbane Dick Fund, 1926. All rights reserved, The Metropolitan Museum of Art. B from F. Crisp, *Medieval Gardens*, Hacker Art Books, 1966.]

THE NEW WORLD

The discovery of the New World raised great expectations in Europe for such riches as gold, spices, and new foodstuffs—expectations that were to be fulfilled on all counts. Gold and silver in quantities the world had never seen were to fill the coffers of indebted sovereigns. Pizarro returned to Seville from

Peru with a gold cargo of more than 17 tons, most of which consisted of artistic works melted down into ingots. This was the cargo of but one ship! The foodstuffs were in the form of strange and wonderful plants. The people of the New World had brought under cultivation practically all the indigenous plants we now use: such crops as maize,° potato, tomato, sweetpotato, squash, pumpkin, peanut, kidney bean, Lima bean, cranberry, avocado, Brazil nut, cashew, black walnut, pecan, pineapple, cacao, vanilla, chili, quinine, coca, rubber, and tobacco.

The metal treasure is all gone—hardly a single piece of the golden objects shipped to Spain remains. The gold was not only responsible for the annihilation of the Aztec, Mayan, and Inca empires but led to the ruin of the nations that expropriated it. As Von Hagen has said of the mighty cultures in America, "what was not destroyed by man was overwhelmed by the insults of time." The inheritance of the greed that conquered a continent is the remnant of feudalism that still lingers in many parts of Latin America. But the plants of America have produced, and are still producing, wealth that pales into insignificance all the golden booty (Fig. 1-9).

The exact time that people first came to the Americas is not known. It is generally believed that the first arrivals reached North America at least 40,000 years ago from Asia over a now submerged land bridge across the Bering Strait. It seems that few cultural traditions were carried along; the development of civilization in the Americas was mostly home-grown. Similar needs inspire similar inventions. New World agriculture paralleled many developments in the Old World, although there were important differences. The wheel, for one, was practically unknown, except as a child's toy, and there was no "iron age." When the conquistadores came to the New World, they discovered civilizations in the age of stone. The basic agricultural tool was the digging stick.

Some of the American cultures never developed beyond the Paleolithic mold. For example, the Plains

---

°Maize (*Zea mays*), also "corn," originally "Indian corn" in the United States. The term "corn" originally referred to any cereal grain, especially the seed of wheat, rye, oats, and barley.

Indians of North America never completed the transition to an agricultural economy. The men hunted; only the women planted. Until the horse was introduced into North America by the Spanish, the nomadic Plains Indians were able to hunt buffalo only with great difficulty; the horse gave them ready access to the tremendous herds and completely altered the course of their cultural development.

TVRCICVM
FRVMENTVM.
Türckich torn.

Figure 1-9 The first woodcut of maize, from Leonhard Fuchs's herbal *De historia stirpium*, published in 1542. Fuchs erroneously stated that maize was introduced into Germany by the Turks, hence his name *Turcicum Frumentum*, or "Türckish korn." Maize was unknown to the Old World until after the time of Columbus.

To the south, however, a different situation emerged. By 1000 B.C.E. the great migrations that had filled the continent were over, and the agricultural patterns were formed. Temple-cities began to rise as an agriculture based on the culture of maize bestowed a sufficient surplus of food to free some individuals from laboring in the fields. A series of cultures developed, culminating in the Aztec in Mexico, the Mayan in Central America, and the Inca, which spread from Peru throughout most of the Andes.

In the New World the sequence of developmental steps to a sedentary agriculture has been particularly well worked out for the Tehuacan Valley, some 150 miles south of the site of Mexico City. Maize, one of the world's most important crops, appears to have been brought into cultivation there, along with beans, squash, and other prime food plants. The detailed record unearthed in the Tehuacan Valley is a tribute to the efforts of skilled archeologists.

As in the Fertile Crescent in the Near East, the Tehuacan Valley is ringed by foothills and mountains that provided a wealth of environmental diversity. The valley climate is dry, preserving well the archeological evidence in the middens. Even desiccated feces have been found, providing some direct evidence of diet. The trash accumulations of ancient habitations are sufficiently well preserved to shed light on such details as the customs of the occupants, the seasons of occupancy, and climatic change. Sufficient material for accurate carbon-14 dating is available.

By 7200 B.C.E. some form of primitive agriculture had been in existence for three or four millennia. The valley was sparsely occupied by small, nomadic families or tribes, which changed camps several times yearly and subsisted chiefly on wild plants and hunting. Pollen studies indicate that the valley was sufficiently dry to have been grassland. Its occupants used flint tools, but apparently had no devices for cultivating the soil.

Within the next 2000 years the small nomadic groups gathered at times in base camps, but dispersed to follow the seasons from valley floor to uplands, as did the primitive farmers in the Near East. The first crude planting of seeds seems to have begun by about 5000 B.C.E., with at least squash, chili, and avocado having been domesticated; cotton and maize were probably still gathered from the wild. There were tools for grinding grain, and no doubt primitive digging sticks for planting seeds.

During roughly the next 2000 years, up to about 3400 B.C.E., valley populations increased tenfold and became still more sedentary (although small bands dispersed seasonally for plant collection and hunting). Within another 2000 years, maize, several species of bean, several species of *Amaranthus*, yellow sapote (*Pouteria campechiana*), certain squashes, and perhaps pumpkin were added to the rolls of domesticated plants. Their agriculture was clearly efficient, and their settlements gradually increased in size. But fecal deposits indicate that wild animals and undomesticated plants still accounted for more than half of their diet.

By the last millennium B.C.E., the people of the valley were essentially full-time agriculturists, supporting developing civilizations; one archeological site shows a sequence of 17 consecutive floors. Improved maize was by then a mainstay, and doubtless there were choice selections of the many plants already mentioned. By the beginning of the Common Era wattle-and-daub houses were grouped in villages that sported various ceremonial structures, the occupants being full-time farmers who employed irrigation. New plant introductions included tomato, peanut, lima bean, and guava. Agriculture continued to expand until the Spanish conquest in 1540.

Archeological findings indicate pretty well what the earliest forms of maize were like. A logical explanation of their origin is transformation from wild teosinte (or a teosinte-like ancestor) under the pressure of selection by human beings. The plants were quite unlike modern maize, the seeds not being enclosed by husks and borne on fragile branches of the tassel as well as on ears at high positions on the stalk. Wild maize similar to this was an important food item until the third millennium B.C.E. in the Tehuacan Valley. Evidently it was brought down to the valley floor for cultivation, to an environment too dry for agriculture until irrigated. Planting techniques probably involved nothing more than dropping a seed into a hole poked in the soil with a pointed stick.

From about the fifth millennium B.C.E. until nearly the beginning of the Common Era, a primitive maize was much cultivated. About 1500 B.C.E., improved types began to appear, leading to selection of modern maizes. Substantial change may have come as early as several millennia B.C.E., probably involving introgression, or possibly even hybridization, with races of teosinte and related genera (Mangelsdorf once speculated that *Zea* may have crossed with *Tripsacum dactyloides*). Wild maize apparently became extinct before the Common Era, perhaps genetically "swamped" by cultivated types. Undoubtedly the natural habitat of wild maize had become thoroughly replaced by cultivated fields.

Here, as in the Near East, human cultures gradually adopted techniques found useful by trial and error to reduce need for migration in search of wild plants and animals. New crop originations were profuse in the Tehuacan Valley, but certainly a number of other centers must have domesticated particular species independently. A crop so important to New World civilization as maize undoubtedly gained great renown and became widely spread from its center of domestication. It was grown throughout Mexico and in parts of North America 4000 years ago. Fossil maize pollen more than 2000 years old has been identified in swamp deposits as far north as Virginia.

Agriculture in the New World was developed by the genius of Native Americans for adaptation to diverse conditions. For example, the Aztecs developed unique "floating" gardens in watery environments. Reed baskets eight feet in diameter were filled with earth and set in shallow lake waters to become stationary as roots grew down to anchor them. The Mayans, using the corn stalk as a support for the bean, developed (perhaps unwittingly) a basis for a balanced cropping program and a balanced diet that still sustains the people of the region. The form of agriculture perfected by the Incas was uniquely adapted to their mountain terrain. It was based on the potato, which produced abundantly in high altitudes and could be preserved by dehydration, which required repeated freezing and crushing. Deep narrow valleys were irrigated through the long-distance channeling of streams and the creation of stone-lined reservoirs. Stone terraces, many still in use to this day, prevented erosion. More people were maintained on a higher standard of living 500 years ago than at present in the forbidding Andes region.

THE RISE OF SCIENCE

Philosophy, literally "love of knowledge," denotes the orderly arrangement of information—information dealing with abstract ideas or with the substance of the real world (natural philosophy). Such an orderly arrangement of information can be built up in various ways. Beginning with a reasonable premise, and drawing conclusions (generalizations) in a logical manner is one very attractive method. This method, called **deduction,** was formalized by the Greek philosophers, who formulated rules of logical reasoning designed to extract true inference from generalizations. The flaw in this method of reasoning is that it depends on the correctness of the original premise, which in this method of reasoning is not subjected to test. An alternative method of reasoning is **induction.** In this method generalizations are derived from observations. The generalizations must then be continually tested against new observations and, if possible, against specially devised observations to test the validity of the assumption. Most important, the generalization is considered an *approximation* of the "truth," subject to change, if necessary, to fit any new observations that challenge the validity of the assumption. A generalization, to be useful and valid, must not only fit each and every observation, but must prove useful in prediction. The concept of the testing of generalizations developed a new natural philosophy known as **"science"** (from the Latin "to know"). The word science is now used in place of the older term "philosophy" in reference to the body of knowledge of a subject (for example, plant science). Science is in fact a method of inquiry for the organization of information. **Science** depends upon the **scientific method:** the scientist is one who contributes to this body of information by making generalizations, performing observations, and testing one observation against the other. Experimentation is his most powerful tool. The creative part of science lies in the devising of experiments to test conclusions and the synthesis of generalizations. The discoveries of science are often intuitive or accidental, sometimes removed from actual experimentation except as they are used in subsequent testing. They are a combination of intelligence and insight—from the ability of some to look at old problems in fresh ways. In the creative process, experimentation aids, but is not necessarily a guarantee of, discovery. "Experiment is experience sharpened to a point—useful as a digging stick but not as a divining rod."°

In the ancient world technology gave rise to science; only recently has science been the source of technology. The early contact between science and technology had been only marginal. Philosophers felt little responsibility for making practical use of their findings, and the artisan understood little beyond his "craft mysteries"—the inherited, traditional practices. The concept that science and technology are interrelated, both being concerned with natural phenomena, developed very slowly. The fusion of craft mysteries with the new technology of science led to an industrial society that was to engulf the Western World. The technological superiority of the West over the East became evident only at the close of the Middle Ages and then only in military technology. Conquest and economic intrusion were the signs of the vigor of Western civilization.

The accumulation of information through observation (or experience) and experimentation began to reach significant proportions in the sixteenth century and reached maturity in the awesome discoveries in astronomy and physics of Leonardo Da Vinci (1452–1519), Galileo Galilei (1564–1642), and Isaac Newton (1642–1727). Renewed interest in botanical studies accompanied the revival of learning, as evidenced by the increasing appearance of herbals— descriptive treatises of plant material (Fig. 1-10). The development of an appetite for scholarly works coincided with the European invention, in the middle of the fifteenth century, of printing from movable type (printing from wood blocks was known earlier in China), which had tremendous impact on the spread of ideas.

°H. S. Harrison, "Discovery, Invention, and Diffusion," *in* C. Singer, E. J. Holmyard, and A. R. Hall (editors), *History of Technology*, vol. 1, (London: Oxford University Press, 1954), p. 61.

Figure 1-10 Title page of Thomas Johnston's 1633 edition of John Gerard's *Herball* engraved by John Payne. John Gerard (*lower center*) holds a potato stem in flower and fruit. Homage is paid to the Roman goddess Ceres (growing vegetation) and Pomona (fruit trees) along with the "ancients": Theophrastus of Eresos (ca. 372–288 B.C.E.), "Father of Botany," and Pedanius Dioscorides of Anazarbos, who flourished during the second half of the first century and whose *Re materia medica* was the authoritative work on plant medicinals for 1500 years. The Latin quotation from Genesis 1:29 is translated as "Behold, I have given you every herb bearing seed." [From B. Henrey, *British Botanical and Horticultural Literature Before 1800*, Oxford University Press, 1975.]

Fundamental studies of botanical matters began to appear in print in the seventeenth century, stimulated by the invention of the microscope (1590), just as the invention of the telescope (1608) was to be the generating factor in the advance of astronomy. Robert Hooke (1635–1703), with the aid of the microscope, described "cells" (the name still in use) in cork (the bark of the cork oak) and extended these findings to other plants. The studies of Anton van Leeuwenhoek (1632–1723) in observing and describing protozoa and bacteria are well known. The microscope also influenced the fundamental studies on plant anatomy of Marcello Malpighi (1628–1694) and Nehemiah Grew (1641–1712), research unsurpassed for a century.

The importance of experimentation as a tool in biology increased during the seventeenth and eighteenth centuries. Experimental studies made by the Dutch botanist Rudolph Jacob Camerarius (1665–1721) clearly established the nature of sexual reproduction in plants. The clarification of this basic concept, together with the later hybridization experiments of Joseph Gottlieb Koelreuter (1733–1806), were to lay the foundation for the brilliant experiments on inheritance in the garden pea by the Austrian Abbot Gregor Johann Mendel (1822–1884).

Plant taxonomy was to mature in the eighteenth century. During the more than twenty centuries following Theophrastus, countless lists of plants were developed. The usefulness of these compilations was limited by the lack of a functional and consistent method of classification. This problem became increasingly evident as the number of known plants and plant products increased. For example, the famous medical herbal compiled by the Greek physician Pedanius Dioscorides in the first century lists 600 plants; a later Arabic work, The *Kitah al-Jami* of Abu Muhammed ibn Baitat (1190–1248), lists 1400 plants, foods, and drugs; the *Pinex theatri botanici* (1623), written by Gaspard Bauhin (1560–1624), describes 6000 species! The Swedish physician Carl von Linné (1707–1778), better known as Linnaeus (the Latinized form of his name), developed a workable system of classifying plants and animals. The system he devised is based on structural differences and on similarities between the morphology of reproductive parts, the organs least likely to be influenced by environmental conditions. Linnaeus' fame further rests on the development, with his students, of the binomial system of naming organisms according to genus and species—the system we use today.

The eighteenth century also saw the start of fundamental studies in plant physiology. The Englishman Stephen Hales (1677–1761), greatly influenced by Harvey's discovery of the circulation of blood (1628), was the first to use careful experimental techniques and quantitative methods. He pioneered studies of transpiration and the movement of sap. Another Englishman, Joseph Priestley (1733–1804), the discoverer of oxygen, was the first to recognize the analogy between combustion and respiration. He demonstrated that growing plants "restored" air that had been respired or had candles burnt in it. It was not for many years after this, however, that the significance of his findings came to be understood.

AGRICULTURAL REVOLUTION

The term "agricultural revolution" refers to the transition from medieval to modern practices. The emergence of new agricultural techniques in the seventeenth and eighteenth centuries can be traced to new crops and economic changes. As serfdom disappeared, so did the rights of the peasant to graze livestock on the lord's common field, to hunt and fish on his lands, or to cut wood in his forest. The system in which the cultivated land of the village was divided up into different regions, with each peasant managing one or more strips, gave way to the enclosure of fields to be cultivated in one large-scale operation to increase efficiency. The eventual disruption of rural life was cruel, but it was on these expanded farms that major advances in agricultural technology originated. During the eighteenth century the value of good farmland rose tenfold. The increased productivity was responsible for the growth of the urban population, and in effect allowed the industrial revolution to take place.

The practices of crop agriculture were greatly affected by the acquisition of new foods during the Age of Exploration. Three of the world's great food plants—potato, maize, and rice—entered European agriculture. The potato, introduced into Spain before 1570, was to become one of the most important food and industrial (starch and alcohol) crops in Europe. Introduced into Ireland in the late sixteenth century, it became in time virtually the only source of food for most of the population. When the potato crop failed several years in succession in the mid-nineteenth century as a result of the late-blight disease, the Emerald Isle was depopulated by death and migration. Maize, another New World introduction, became an important crop in southern Europe. Rice, although introduced into Spain in the eighth century, was not cultivated in Italy until the sixteenth century.

The demand for three new beverages: tea (from China), coffee (from Abyssinia via the Moslem world), and cocoa (New World) also gave great importance to sugar as a sweetening agent. Sugarcane was imported to the New World, and the West Indies soon provided the main source of Europe's supply in the seventeenth and eighteenth centuries. Beet sugar did not become important until about 1800. The exploitation and taxation of these crops were to have a profound effect on the body politic.

The problem of insufficient feed for livestock during the winter in the temperate countries of Europe was alleviated by such new winter crops as clovers, which did not deplete soil fertility, and turnips. The new farming systems, which made use of enclosed fields, made it possible to regulate animal populations, and the more intensive rotations led to increased development of animal agriculture. Pasture land that had been inefficient under the old practices tended to decrease. The relation between pasture for sheep and arable land for cereals was kept in balance in England by the changing prices for wool and grain. In the newly developing systems the cattle were fed in stalls from fodder grown in the fields. The system required considerable labor, but was very efficient in land utilization. Moreover, it provided large quantities of manure, which previously had been largely wasted but could now be used to fertilize the soil. Fertilization practices also improved with the use of marl—a clay-like material rich in calcium and magnesium carbonates.

The old fallow system, in which one-third of the land was unplanted to preserve and restore fertility, gave way to improved, "scientific" methods of crop rotations that permitted great increases in production through more efficient utilization of the land. Although these methods were not new in themselves, the idea that crop rotation could be used to bring about several desirable results was new. The advantages of crop rotation (although not entirely understood) were (1) the control of diseases and weeds that tend to become severe problems under con-tinual one-crop plantings, (2) convenience and efficiency, (3) fertility improvements through planting of nitrogen-fixing legumes, (4) erosion control by the use of winter cover crops, and (5) spreading the risk by increasing the number of crops.

The beginnings of mechanization were the preview of a full-fledged revolution that was to change the course of agriculture completely. The development of improved flour-milling techniques heralded the use of mass production for the processing of foodstuffs. The increases in long-distance trade of agricultural crops and products necessitated testing and evaluation procedures, and the resulting establishment of standards led to a rise in the quality of agricultural equipment. Farm machinery was at first relatively simple, and was derived more from the ingenuity of local craftsmen than from specialized knowledge (Fig. 1-11). New plows, seed drills, threshing and harvesting machines, however, soon became increasingly complex. Rising industrialism was to find a real market in agricultural tools, and by 1850 much of the agricultural machinery in Britain was furnished by factories.

NORTH AMERICAN AGRICULTURE

The North American colonists, although lacking wealthy empires to plunder, found rich agricultural resources to develop. Nearest at hand was the forest, which immediately provided two exportable surpluses, timber and fur. Practically the entire Eastern seaboard was continuous forest. While the development of farming necessitated land clearing, the forest was to provide riches in the form of fuel; building materials, posts, rails, furniture, implements, staves, ship frames, and masts; tar, pitch, and turpentine; potash for fertilizer and soap; and hemlock bark for tanning. The early development of colonial manufacturing such as ship building was based on timber resources; the forest provided the chief source of foreign exchange. The "Broad Arrow" policy of England (three blazes cut into the bark of trees repre-

Figure 1-11   Farming in France, 1735. Plowing, broadcast-sowing by hand, and harrowing-in the seed. [From C. Singer, E. J. Holmyard, and A. R. Hall (editors), *History of Technology*, vol. 4, Oxford University Press, 1958.]

sented the broad arrow, symbol of the British Navy), which reserved to the Crown all trees over two feet in diameter to ensure a supply of mast timber, was to contribute to the growing disenchantment between Britain and her American colonies.

Farming practices in North America were developed out of the experiences of the Old World. Agriculture in the first colonies resembled the early agriculture of Europe. The potential of maize was realized immediately, however, and it soon became the dominant grain crop. The rise of tobacco and, soon after, rice as export "cash" crops stimulated the development of slavery, for both crops require high labor inputs. (The decline in the price of tobacco decreased the virulence of slavery; in 1778 Virginia prohibited the importation of slaves.) With Eli Whitney's invention in 1793 of the cotton gin, a machine that could separate the cotton fibers from the seed, cotton became an extremely profitable crop; Whitney's gin did the work of 100 men. Exports increased from 275 bales in 1797 to 36,000 in 1800. King cotton was to establish the plantation system as the basic agricultural pattern in the southern United States. The slavery issue and the future of the plantation system, which depended upon slaves, were eventually to be decided by a civil war.

The young United States was primarily an agricultural nation. In 1800, 90 percent of the working population was engaged in farming. The early agricultural pattern in the United States was strongly affected by the seeming limitlessness of new lands. It was new soils that were most responsible for the unprecedented levels of productivity.

The improvements in European agricultural practice were followed closely in the United States, particularly in the field of mechanization. The mechanical reaper of Cyrus McCormick, invented in Virginia in 1831, was the first to be generally adopted. Labor shortages and high grain prices during the Civil War accelerated the adoption of machines, and the leadership in mechanization was to pass to the United States, a fact that does much to explain the technological progress in American crop production in the twentieth century (Fig. 1-12).

BEGINNINGS OF THE AGRICULTURAL SCIENCES

The changes in agriculture brought into sharp focus many problems that traditional practices could not cope with. In the forefront were the difficulties associated with plant and animal nutrition and disease control. Until the sixteenth century little had been added to man's knowledge of plant nutrition since Aristotle had formulated his hypothesis that plants absorbed nutrients from the soil. Malpighi had mentioned that leaves might have a role in the elaboration of food, and Nehemiah Grew had suggested that light absorption might play a role in this process.

A

Figure 1-12 Development of the reaper. (A) Cyrus McCormick's first reaper, 1831. (B) The 1851 McCormick reaper. (C) The twine binder (1881) reaped and tied sheaves of grain in one operation.

B

C

Various eighteenth-century researchers had worked out some notion of the relation between oxygen, carbon dioxide, and soil nitrogen, but in spite of this the main theories of plant nutrition accepted at the beginning of the nineteenth century were grossly incorrect. It was felt that the carbon source of plants was decomposed soil organic matter ("humus") and that the inorganic salts in the soil were little more than growth stimulants. The beginnings of modern plant nutrition and agricultural chemistry are found in the work of the German chemist Justus von Liebig (1803–1873), who in 1840 presented evidence that contradicted the "humus" theory of plant nutrition. Although he erroneously concluded that plant nitrogen is derived only from the ammonia of the air, he initiated the correct view that plants obtain carbon from carbon dioxide and mineral constituents (potassium, calcium, sulfur, and phosphorus) from the soil. Pasteur's suggestion in 1862 that microorganisms might be involved in soil nitrification was confirmed 15 years later. Studies in plant nutrition advanced at a rapid rate at the end of the nineteenth century, and were paralleled by the growth of the artificial-fertilizer industry.

The understanding of the biologic nature of disease was to have a profound effect on agricultural practices. The studies of early mycologists provided the foundations but did little to eliminate the ancient belief that disease was a scourge descended from the demons. The results of experiments by Matthier du Tillet in 1755 (stimulated by a prize offered by the Academy of Bordeaux) clearly showed the disease-producing potential of the "black dust" from bunted wheat and described treatments of the grain with lye and lime to control the disease. It remained for Benedict Prevost (1755–1819) to demonstrate that the black dust of bunt was in fact spores of the fungus *Tilletia tritici*. Prevost described the fungicidal properties of low concentrations of copper sulfate and its use in seed treatment. These early studies were generally ignored until Louis Pasteur (1822–1895) established that microorganisms are responsible for, rather than engendered by, disease. A rational explanation of the diseases of plants was offered by the German botanist Heinrich Anton De Bary (1831–1888) through his classical work on the late-blight disease of potato. He established the concept that specific plant diseases are caused by specific microorganisms, and defined the concepts of parasitism, infection, and resistance. The control of plant disease as an agricultural practice traces to the discovery in the 1880s that combinations of copper sulfate and lime (Bordeaux mixture) when applied as a spray to grapes would control downy mildew, a ruinous disease in the vineyards of France.

DEVELOPMENT OF AGRICULTURAL EDUCATION AND RESEARCH

Scholarly writings in agriculture, although at first infrequent, appeared continuously in Western writings. The important work *Maison Rustique* (1594), by the Frenchman Charles Estienne, was to appear in many editions, including English translations. One expression of the agricultural revolution in the eighteenth and nineteenth centuries was the increasing appearance of works on agriculture. The emergence of agricultural professionalism was characterized by the appearance of agricultural societies and agricultural publications. The Danish Agricultural Society was founded in 1769; the establishment in 1785 of the Philadelphia Society for the Promotion of Agriculture was soon followed by the formation of many similar societies in the United States. The Royal Agricultural Society in England was incorporated by Royal Charter in 1840. A Board of Agriculture established in England in 1793 was the first of a long and continuous series of governmental agencies to deal with the problems of agriculture.

Formal educational training was another consequence of the increasing stature of agriculture. A school of veterinary medicine was founded in Copenhagen in 1773. Chairs of agriculture were established at the leading universities—Edinburgh, 1790; Oxford (rural economy), 1796; and Yale (agricultural chemistry), 1847. In the early 1800s a number of agricultural institutions were founded in England and on the Continent. The beginnings of the agricultural experiment station trace to the creation in 1834 by John Bennet Lawes (1814–1901), with the later collaboration of Joseph Henry Gilbert (1817–1901), of a private laboratory dedicated to research in agriculture at Rothamsted. This famous institution, which is still in existence, was to be extremely productive. In addition to the early work on fertilizers and fertilization practices, work on experimental technique was to have important consequences for statistics. In 1842 Lawes patented superphosphate, formed by treating phosphate rock with sulfuric acid, and in effect initiated the chemical fertilizer industry.

The increasing interest in agriculture in the United States resulted in the establishment in 1862 of the Department of Agriculture, formerly a division of the Patent Office. In the same year two other famous pieces of agricultural legislation were signed by President Lincoln: the Homestead Act, which provided for staking farms on governmental lands, and the Morrill Act (the "land-grant college act"), which established educational institutions in "agriculture and the mechanical arts." Iowa became the first state to accept the provisions of the Morrill Act. This legislation was to have a tremendous impact on agriculture and technology as well as on the American educational system. Whereas higher education had previously been the privilege of a few, the new institutions established the principles of people's colleges. The state agricultural experiment stations were to develop a vast pool of agricultural information— information on soils, diseases, and insects, and the development of improved cultivars of crop plants and livestock breeds. This body of knowledge provided the basis for the great strides made in American agriculture in the twentieth century.

TWENTIETH-CENTURY TECHNOLOGY

Although the population of the United States is continually increasing, the country is under no immediate internal pressure to increase food supplies, even though its farm force is decreasing and its farm acreage declining (Figs. 15-17 and 27-10). In 1910 each farmer produced for himself and seven others; in 1980 he "supported" more than 60 people. This increase in labor efficiency is the result of improved technology. The magnitude of this increase has been such that the recent agricultural problems have tended to be due to overproduction and low prices. Although the new technology in agriculture had no definite beginnings, its effects became noticeable in the late 1930s, when there was a sharp increase in the rate of technological change (see Chapter 27).

The increase in the United States' agricultural production from 1880 to 1920 was largely a reflection of the increase in expenditures—more land and more labor. Returns per acre remained relatively constant. At the outbreak of World War I the number of people engaged in farming was the highest in history (even though the percentage was declining), as were the number of acres under cultivation and the number of horses and mules (25 million). The replacement of animal power by the gasoline engine in the 1920s was the first great step in the twentieth-century technological revolution. The release of the land formerly used for feed stabilized the theretofore expanding farm force and acreage during the decade following World War I. In the following decades the early pioneering work of the agricultural experiment stations began to pay off in the form of improved agricultural practices. The resurgence of genetic investigations in the quarter century since the rediscovery in 1900 of Mendel's paper on inheritance was translated into new and better crop cultivars. The development of hybrid corn and its rapid adoption (Fig. 27-4) in the corn belt was the most spectacular of these achievements. These improved genetic stocks, in combination with the increased use of inorganic fertilizers, accounted for a large part of the tremendous increase in production necessitated by World War II.

A whole new set of technological advances was initiated in the late 1940s, made possible by the research in the preceding decades. Agricultural chemicals in the form of weed killers, organic fungicides, and insecticides quickly followed the spectacular commercial success of the insecticide DDT and later the broadleaf weed killer 2,4-D. These improvements increased yields per unit area as well as conserved labor. Mechanization increased in the 1950s to include even "chore" tasks. The increasing use of supplemental irrigation in the United States made economical the use of additional fertilizer. The net result has been a steady increase in agricultural efficiency.

In the late 1950s a new trend in research developed when the Federal government decided to support basic research on a very large scale. Although the amount of research funds provided to support applied agricultural research has not diminished, the percentage of funds has declined sharply over the years. In essence a cycle is being repeated: the freedom from the exigencies of agricultural problems has enabled a concentrated effort in the "luxury" areas of research. Present studies of the basic mechanism of plant growth and development cannot but have a profound effect on the future technologies of crop production.

The ecology movement

In the 1960s two approaching crises impinged on world consciousness. One was an emerging world food shortage caused by an unchecked population increase in the tropical world. The second, first brought into prominence by Rachel Carson's *Silent Spring* (1962), was the polluting consequences of technology itself. The two problems were soon shown to be part of a single issue—the distribution of the earth's renewable and nonrenewable resources. Ironically, as we raced to conquer the moon, we discovered the earth. Whereas our plant-dominated planet had traditionally been thought of as "Mother Earth," the source of bounty and plenty, the new concept, that of a "Spaceship Earth," stressed the finiteness of her resources. The new earth was seen as if from the vantage of an observer on the moon: small and isolated, yet beautiful and precious as never before. It soon became apparent that there was indeed a limit to earth's resources and that increasing population acted as a growing divisor of the world's wealth. Worst of all was the realization that the narrow application of technology, geared only to short-term increases in production, was leading at best to a fouling of the nest, and at worst to mass poisoning.

The social consequence was the rise of the ecology movement. Conceived in fear and born in crisis, the movement inspired religious fervor and struck a responsive chord in the United States during the 1960s. Although its message was strident and its thrust sometimes veered uncomfortably toward anti-intellectualism and antiscience, the ecology movement proved itself to be a force to be reckoned with. The banning of DDT, its *cause célèbre*, was the most notable of its victories. The ecology movement spread to the field of social problems as concern for the displaced agricultural worker, the human jetsam of agricultural progress, arose.

The major environmental problems of that decade have not been solved as of this writing (1980), but events of the past few years have led to increased optimism. In the late 1960s a dramatic increase in the production of major food crops (rice, wheat, maize, as well as grain sorghum and potato) occurred in the developing world. A central breakthrough occurred in plant breeding—the creation of high-yielding, fertilizer-responsive cultivars of wheat and rice suitable for the tropics. At the same time more emphasis was placed on the use of agricultural technology in developing countries, such as India, for example. Increased use of technology in rural areas plus the "miracle" grains, then, introduced what is known as the **"green revolution,"** and advanced a number of tropical countries a small step toward self-sufficiency in foodstuffs. Progress in population control has also been encouraging in at least some countries. The clamor over pollution has made agricultural technologists in the United States more aware of the long-range implications of their action. The feeling now exists that technological progress combined with statesmanship can reverse the trend toward environmental deterioration and lead to not only sustained but increased agricultural production.

The energy crisis

In the decade of the 1970s a new shock wave reverberated on world consciousness—the energy crisis. The Arab oil embargo of 1973, coinciding with a quadrupling of crude oil prices, initiated a general economic disorientation. The world, already divided into rich and poor, East and West, free economy and controlled, now was divided once again into oil exporters and oil importers. As oil prices rose, a massive redistribution of wealth and power occurred whose consequences are still unknown and unpredictable.

The energy crisis, like the ecology movement, had a destabilizing effect on agriculture because many of the new technological imports—chiefly fertilizer and pesticides—turned out to be directly related to the cost of energy. Those who had been disturbed about presumed deleterious effects of modern agriculture on the environment were joined by new voices concerned about the high energy subsidy of high-technology agriculture. For the first time progress itself was questioned and continued progress was labeled a dangerous myth. A new concept called **intermediate** or **appropriate technology** was offered as a counterforce. Appropriate technology was defined as a level of technology between primitive and advanced, but one that was at the same time "appropriate to get the job done, to conserve resources, to preserve long-term human values, and to maximize human happiness!"

It is still too early to decide if intermediate technology is merely a nostalgic yearning to return to what is perceived as the simpler ways of the past. To those who have experienced hard times it is difficult to glorify what appears to be an attempt to return to primitive living and subsistence agriculture. It also seems foolhardy, however, to deny the argument of declining resources as energy costs spiral upward. And those who are caught in the frustrations of a complex modern society find it difficult to scoff, without some pangs of self-doubt, at the idea that perhaps "small is beautiful," a motto proposed by E. F. Shumacher, one of the prophets of intermediate technology.

However, it turns out that energy consumption by agriculture is trivial when compared to energy consumption by the rest of modern society. Although the food industry consumes one-sixth of all energy used in the United States, actual food production consumes only a fifth of the food-systems total energy use, or only 3 percent of the U.S. total. It is interest-

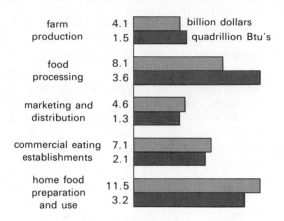

farm production — 4.1 billion dollars, 1.5 quadrillion Btu's
food processing — 8.1, 3.6
marketing and distribution — 4.6, 1.3
commercial eating establishments — 7.1, 2.1
home food preparation and use — 11.5, 3.2

Figure 1-13 Energy used in the U.S. food system in 1975. [Data from USDA.]

ing to note that it takes more than twice as much energy for home food preparation as food production (Fig. 1-13). In this light it is difficult to see how the U.S. farmer can have much of an impact on the energy crisis—and it seems likely that a reduction in food production by conserving energy on the farm may indeed aggravate our current problems.

Recent changes and future trends

One result of the new technology in the United States is the breakdown of the differences that once separated the disciplines of agronomy, horticulture, and forestry. These divisions of plant agriculture had been based on a combination of intensity of agricultural production and tradition. Thus forestry was concerned with extensively managed timber lands, whereas horticulture dealt with intensively managed "gardens." The acceptance of the horticultural concept of intensive care throughout the life of the crop plant and the growing use of mechanization and agricultural chemicals have obscured these differences. Thus, for all crops, there is a growing concern with each phase of development, a decrease in labor input, and at the same time an expanding scale of production—trends that emphasize the increasing role of management and science. On modern farms in the Midwest, for example, there are few differences today between the production of cannery tomatoes and the production of field corn. Pruning, once confined to horticulture, is now an established practice in the production of timber; the nursery end of forestry is a true horticultural operation. Similarly, the large-scale practices of field-crop production are being adapted to the production of fruit, vegetables, and ornamentals. The traditional differences between the disciplines—differences based on species—were never completely satisfactory, and each new crop raised new difficulties and jurisdictional disputes. Tobacco and potatoes, for example, are considered horticultural crops in some locations and agronomic crops in others. None of the tropical crops fit neatly into either category.

A further indication of the growing unity is the speed with which practices initiated in one discipline are quickly adopted by the other. Growth regulators, which have been widely used in horticulture for many years, are now being extensively tested for use on agronomic crops. The use of computer programming, first accepted in forestry, is being adapted to all cropping practices. Tradition remains the main factor separating the three crop disciplines.

A striking feature of the new agriculture is not only the importance of the new discoveries but the increasing rapidity with which these discoveries have been made. No invention comparable to the gin affected cotton production for a century and a half; but the mechanical cotton picker (Fig. 1-14), an innovation of similar magnitude to Whitney's gin—one machine replaces 100 pickers—has completely transformed cotton production, along with such innovations as chemical weed control, defoliants, and hybrid cotton.

The future course of agricultural technology cannot be charted with certainty, but the trend in the United States is unmistakable. The increased devel-

Figure 1-14   The mechanical cotton picker is the most sophisticated present-day farm machine. [Courtesy John Deere Co.]

opment and use of chemical herbicides points to the elimination of most cultivation. Plant growth in a weed-free environment will change many of our present agricultural practices, such as field preparation and planting. Pesticide development and the incorporation into crop species of genetic resistance to pests and diseases can be expected to further reduce losses. Many crops will be improved genetically through the utilization of hybrid vigor. Mechanical harvesting will extend to hard-to-pick fruit crops. The decline in the farm labor force as well as increased yields per unit area will continue.

In addition to the refinement of present agricultural practices, new methods of producing food have been seriously considered since the close of World War II. These vary from the extraction of plant proteins to the culture of algae or microorganisms. Recent stimulation has come from aerospace research on algae as a means of oxygen renewal in space flight. The mass growth of microorganisms is of course a highly developed present-day technology. The end product has a high value per unit weight, as in antibiotic production or the fermentation industries. However, the claims that some new method of food production would quickly arise to solve the world's fast approaching food crisis appear to have been exaggerated or at least premature.

The future course of the plant sciences promises to be an exciting journey. Merely to provide a sustaining diet for the present world population, much less an expanding one, is a herculean task. The goal of proper nutrition for all people and the judicious exploitation of plant life will require the joint effort of creative scientists and wise statesmen. It is an endeavor worthy of the best men and women in our society.

Selected references

Gras, N. S. B., 1940. *A History of Agriculture*, 2nd ed. New York: F. S. Crofts. (Development and history of rural life in Europe and the United States.)

Harland, J. R., 1975. *Crops and Man*. Madison, Wis.: American Society of Agronomy. (A well-written review of crop origins.)

Huxley, A., 1978. *An Illustrated History of Gardening*. New York: Paddington Press. (A profusely illustrated tour-de-force of horticultural history.)

Leonard, J. N., 1973. *The First Farmers*. New York: Time-Life Books. (A well-illustrated popular account of the origins of agriculture.)

Reed, H. S., 1942. *A Short History of the Plant Sciences*. New York: Ronald Press. (An excellent review.)

Sarton, G., 1952, 1959. *A History of Science* (2 vols.). Cambridge, Mass.: Harvard University Press. (The development of science up to the "golden age" of Greece. Only the first two volumes of a projected eight-volume treatment of the history of science were written before the author's death. Consult his monumental *Introduction to the History of Science* for a complete and annotated bibliography.)

Simmonds, N. W., 1976. *Evolution of Crop Plants*. London and New York: Longman. (Origins on a crop-by-crop basis.)

Singer, C., E. J. Holmyard, and A. R. Hall (editors), 1954–1958. *History of Technology* (5 vols.). London: Oxford University Press. (A monumental work on science and technology up to 1900. Agriculture and related fields are well documented.)

Von Hagen, V. W., 1961. *The Ancient Sun Kingdoms of the Americas*. Cleveland: World Publishing. (A popular history with much technological information on the Aztec, Maya, and Inca civilizations.)

Wright, R., 1938. *The Story of Gardening*. Garden City, N.J.: Garden City Publishing. (A popular history of horticulture and gardening.)

2
Energy and Crop Production

Our industrialized society feeds greedily and insatiably on energy made available by burning fossil fuels—coal, petroleum, and natural gas. This stored energy was captured from the sun eons ago by plants whose remains became buried deep in the earth's crust. These fossil fuels are of course limited, and the efficiency with which we use them, which depends on our level of technology, will determine our standard of living for some time to come. Although we obtain some energy from water power and increasing amounts from nuclear reactors, the amounts obtained from these sources are still insignificant on a world basis. The world energy use is equivalent to the consumption of 4.8 billion metric tons of coal annually. Dividing this figure by the earth's estimated population of 4.6 billion gives an energy equivalent of 1.04 metric tons of coal per capita per year.

To compare the energy costs of today's technology with those required for primitive cultures is revealing (Fig. 2-1). In primitive agricultural societies, of which there are many in the world even today, a total of 12 kilocalories of energy is required by each person per day. Compare this with the 243 kilocalories required per day by each member of a medium-income family in the United States, and we find that our standard of living requires 20 times as much energy. Obviously, it is the affluent nations that are creating today's energy problems (and, consequently, today's pollution problems as well).

Each day every human being consumes, on the average, the equivalent of 0.82 kilograms (1.8 pounds) of plant material containing 0.32 kilograms (0.7 pounds) of carbon. This is about 118 kilograms (260 pounds) of carbon per person annually, which amounts to a world total of about 0.5 billion metric tons. Most of this food is produced from cultivated land, but the oceans, other bodies of water, forests, and savannahs are important in developing countries. The primary production of the major plant communities of the world is listed in Table 2-1.

Because of the inefficiencies involved, 4.1 billion metric tons of carbon must be produced each year to supply the amount actually consumed. The difference between the amount consumed and the amount produced is due to a number of factors. On the average only about 20 percent of the cultivated plant is eaten—for example, the grain constitutes only 28 percent of the dry weight of the wheat plant. Fur-

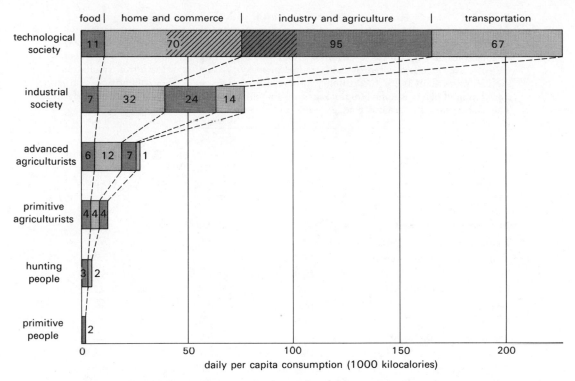

Figure 2-1   Daily consumption of energy per capita was calculated for six stages in human development. Primitive people (East Africa about 1,000,000 years ago), without the use of fire, had only the energy of the food they ate. Hunting people (Europe about 100,000 years ago) had more food and also burned wood for heat and cooking. Primitive agriculturists (Fertile Crescent in 5000 B.C.E.) were growing crops and had gained energy by putting animals to work. Advanced agriculturists (northwestern Europe in A.D. 1400) had some coal for heating, some water power, wind power, and animal transport. Industrial people (in England in 1875) had the steam engine. Today technological people (in the United States) consume an average of 243 kilocalories per person per day, much of it in form of electricity (*hatched area*). Food is divided into plant foods (*far left*) and foods fed to animals. [After E. Cook, "The Flow of Energy in an Industrial Society." Copyright © 1971 by Scientific American, Inc. All rights reserved.]

ther, one-half of the plant material produced on cropland is consumed by animals, and only 3 percent of this stored energy ends up as human food. Losses to pests and diseases account for one-third of the total production. On the average only 1–2 percent, or less, of the total energy from the sun is utilized in fixing the carbon in plants, although some plants, such as corn and sugarcane, are somewhat more efficient. In the end, to produce the 0.36 billion tons of carbon per year needed at present to sustain the human population, 50 to 100 times as much energy is required as is used for all other human purposes.

Basically, agriculture is concerned with the conversion of solar energy to energy usable by people—for food, fiber, and fuel. Sadly, the average technology employed in the world's agriculture is insufficient to meet the immediate needs of our expanding populations. Fortunately, some cultures

Table 2–1 Major plant communities of the earth, with their area, their net primary production, and the amount of carbon they hold in storage. Net primary production is the amount of carbon a plant community provides annually for harvesting or for the support of various consumer organisms, either wild or domesticated. Although only about 30 percent of the earth's surface is covered by land, the net primary production of terrestrial vegetation is slightly more than twice the primary production of the oceans. The quantity of carbon stored in land plants is some 500 times greater than the quantity stored in marine ecosystems. The carbon stored in trees is roughly equal to the carbon in the atmosphere.

Plant community	Area (10^6 sq. km)	Net primary production (10^9 metric tons of carbon per year)	Plant mass (10^9 metric tons of carbon)
Tropical rainforest	17.0	16.8	344.0
Tropical seasonal forest	7.5	5.4	117.0
Temperate evergreen forest	5.0	2.9	79.0
Temperate deciduous forest	7.0	3.8	95.0
Boreal forest	12.0	4.3	108.0
Woodland and shrubland	8.5	2.7	22.0
Savannah	15.0	6.1	27.0
Temperate grassland	9.0	2.4	6.3
Tundra and alpine meadow	8.0	0.5	2.3
Desert scrub	18.0	0.7	5.9
Rock, ice and sand	24.0	0.03	0.2
Cultivated land	14.0	4.1	6.3
Swamp and marsh	2.0	2.7	13.5
Lake and stream	2.0	0.4	0.02
Total continental	149.0	52.8	826.5
Open ocean	332.0	18.7	0.45
Upwelling zones	0.4	0.1	0.004
Continental shelf	26.6	4.3	0.12
Algal bed and reef	0.6	0.7	0.54
Estuaries	1.4	1.0	0.63
Total marine	361.0	24.8	1.74
World total	510.0	77.6	828.0

Source: G. M. Woodwell, "The Carbon Dioxide Question." Copyright © 1977 by Scientific American, Inc. All rights reserved.

have developed intensive techniques so that a few people are able to produce the food required by hundreds of others. To increase agricultural production and efficiency, we need to be concerned with all phases of energy transformation—from the sun to the supper table.

CROP ENERGETICS

The problem of understanding and quantifying the total energy balance at the surface of the earth has been of concern to meterologists for many years. They have become interested in how much of the

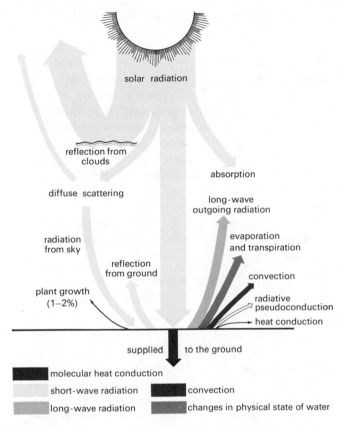

solar radiation

reflection from
clouds

absorption

diffuse scattering

long-wave
outgoing radiation

radiation
from sky

evaporation
and transpiration

reflection
from ground

convection

plant growth
(1–2%)

radiative
pseudoconduction

heat conduction

supplied to the ground

molecular heat conduction

short-wave radiation   convection

long-wave radiation   changes in physical state of water

Figure 2-2 Energy exchange at noon on a summer day. The width of the arrows indicates relative amounts of energy transferred. Note that plant growth accounts for a very small part of the total energy budget. [After R. Geiger, *The Climate Near the Ground,* Harvard University Press, 1950.]

directly to water availability for human needs. Unfortunately, the amount of energy captured by plants is so small that meterologists seldom take it into account. The total amount used in photosynthesis usually amounts to only 1 or 2 percent of the total solar energy input, which is within the limits of computational error. Energy requirements have been calculated for various physical processes that take place in croplands. For example, it has been estimated that about two-thirds of the net radiation falling on vegetation is used in the evaporation and transpiration of water.

In recent years considerable work has been done by biologists who are trying to quantify the productivity of plants. This field of research, called **production ecology,** is concerned with the capture of radiant energy in photosynthesis, its conversion into chemical energy, and its flow through plant and animal communities.

The energy captured in photosynthesis is conveniently represented in part by the total biomass on a unit area of the earth's surface at any given time. **Biomass** merely refers to total living organic matter, usually on an oven-dry-weight basis, but for some purposes it may also include nonliving materials, such as bark, wood, cuticle, and resinous deposits. Even though a crude measure, biomass is useful for making comparisons of different crops and different land areas. The amount of organic matter present is only a partial measure of production, because respiration requires a high proportion of carbohydrates, which are part of the total or gross production.

Photosynthetic efficiency

Every year, an average of 263,000 langleys° of solar energy is received at the outer edge of the earth's atmosphere. Of this, approximately 123,000 langleys of energy is either absorbed or reflected back into the atmosphere by molecules, water vapor, and dust, while the balance of 140,000 langleys actually

sun's radiant energy is used for the evaporation of water, how much is reflected back to the sky, and how much is diverted into various other channels (Fig. 2-2). This accounting process is a matter of "balancing the energy budget" and is a bookkeeping procedure in every sense of the word, although much more complex than the business accountant's double-entry system. Precise and ingenious instrumentation has been devised to measure the energy flux at the surface of the earth and to determine how the total energy budget balances. This objective of meteorology is of more than academic interest, for it relates

°1 langley = 1 gram calorie per square centimeter. One gram calorie is the amount of heat energy necessary to raise the temperature of 1 gram of water by 1° C (specifically from 14.5 to 15.5° C); 1 kilocalorie = 1000 gram calories.

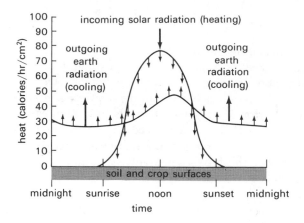

Figure 2-3 Incoming energy is balanced by losses. Gains exceed losses during the day; the reverse is true at night. [After J. E. Newman and B. O. Blair, *Crops and Soils Magazine,* June–July 1964.]

reaches the surface of the earth. In terms of energy values one square meter of the earth's surface will intercept in one day the amount of energy required daily by an active person. This tremendous input of energy is potentially available for use by plants and animals. But solar radiation is not uniformly distributed over the surface of the earth. It varies with cloudiness, the amount of dust in the atmosphere, latitude, altitude, local topography, season, and time of day (Fig. 2-3).

It is interesting to compare the input of solar energy per unit area of land surface at various places on earth. In the United States the average amount of solar radiation received per day is 300 langleys and varies from about 100 to 800 langleys depending on season and region. At the University Experiment Station in Alaska, the total annual input of radiation is about 17,920 langleys during the course of a 64-day frost-free season. In Miami, Florida, where the growing season lasts about 10 months, more than 132,300 langleys of solar energy is available. The great difference in solar input between tropical and temperate regions is not indicative of the actual productivity, however, for many other factors must be taken into account. The effects of air pollution must not be overlooked in heavily populated or industrial areas. Total sunlight reaching the ground in and around

London, for example, is reduced by more than half by pollution and fog.

Of special interest to agriculture is the radiant energy in the visible part of the spectrum. Our biological bookkeeping is concerned mainly with this part of solar radiation, for green plants are able to convert energy from the visible part of the spectrum to chemical energy through the process of photosynthesis. This important reaction has been the subject of intensive study for many years, and in 1961 Dr. Melvin Calvin received a Nobel Prize for his contributions to the elucidation of the physiological processes involved in photosynthesis.

The raw materials and the end products of photosynthesis can be summarized in the following chemical equation:

$$n\text{CO}_2 + 2n\text{H}_2\text{O} + \text{light energy} \longrightarrow$$
$$(\text{CH}_2\text{O})_n + n\text{H}_2\text{O} + n\text{O}_2$$

which merely says, if we assume that $n = 6$, that 6 molecules of carbon dioxide, 12 molecules of water, and sufficient light energy will yield 1 molecule of glucose, 6 molecules of molecular oxygen (O_2), and 6 molecules of water.° This statement does not show the true complexity of the process, for dozens of contributory biochemical processes and energy exchanges take place during the photosynthetic process. See Chapter 8 for a more detailed discussion.

The 673 kilocalories of energy required by this reaction is released as heat when the molecule of glucose is burned by the plant to fuel its life processes (or by the animal that eats the plant). Because of the inefficiencies of the numerous reactions involved in photosynthesis more than 2 or 3 thousand kilocalories

---

° Although photosynthesis usually involves the reduction of carbon dioxide and the splitting of water to form sugars, some primitive organisms carry on variations of this process. Purple sulfur bacteria, for example, use H_2S rather than H_2O and produce elemental sulfur rather than oxygen. The equation for photosynthesis, therefore, can be generalized as follows:

$$n\text{CO}_2 + 2n\text{H}_2\text{X} + \text{light energy} \longrightarrow (\text{CH}_2\text{O})_n + n\text{H}_2\text{O} + 2n\text{X}$$

Here H_2X is a compound that donates electrons; X may be either oxygen or sulfur. Light supplies the energy required to separate hydrogen from its donor.

of energy is required for each molecule of glucose produced. Under special laboratory conditions experiments on the efficiency of the chlorophyll molecule in converting light into chemical energy have shown efficiencies as high as 75 percent, whereas estimates based on short-term experiments with growing plants show efficiencies that range from 15 to 22 percent. (In diffuse light cultures of algae have reached efficiencies of 20–50 percent, but this range drops to 2–6 percent when large tanks are used). Under field conditions the overall efficiency of crops over extended time periods is only a few percent.

The solar energy used by plants is derived only from wavelengths between 390 and 760 nanometers, the portion of the electromagnetic spectrum that we perceive as visible light. When captured by plants the energy is incorporated in the molecular bonds of many different kinds of compounds. Because all chemical bonds are not of equal strength, different compounds may possess different energy levels. Some, such as cellulose, are terminal structural products. Others, including starches and oils, function as intermediate storage compounds and may be converted to other forms, subject to demand by metabolic "messengers." Sugars are not only mobile transfer products, but are used almost immediately in respiration, the biological combustion of energy-rich carbon compounds. In a gross manner, respiration can be characterized as the reverse of the simplified photosynthetic equation. Of the energy released and made available by respiration, part is lost as heat and part is utilized in biosynthesis and chemical work. To a large extent the net amount of energy incorporated into organic matter represents the difference between photosynthesis and respiration. Transfers of energy from one compound to another within the plant are handled by special energy carriers such as adenosine triphosphate (ATP). This flow of energy is a specialized part of the energy concept, and is receiving much attention from biologists.

One important factor that determines the actual photosynthetic efficiency of a leaf is the manner in which the rate of photosynthesis changes with light intensity. The rate of photosynthesis changes with light intensity only up to a certain point. At this intensity it is convenient to say that the leaf is light-saturated.

Although photosynthesis increases with increasing light, it does so at a decreasing rate, and the efficiency of light capture consequently decreases. In fact, maximum efficiency can be obtained only under relatively low light intensities. When intensities are high, relatively more light passes through leaves and is reflected from them. At low intensities a high proportion of light may be absorbed and used. For short periods of time efficiencies of 7–10 percent are possible for some crop plants, but efficiencies of 2–3 percent are probably the most that can be expected over extended periods of time, even if temperature, carbon dioxide, water, and mineral elements are optimal.

Even though less efficient, total production is still greater at high intensities. Although photosynthesis in a single leaf may level off at about 3,000 foot-candles, the rate of photosynthesis in the whole plant may continue to increase up to 10,000 foot-candles because more light reaches the lower leaves, which are shaded.

Light may pass through only a few layers of leaves in low crops but may pass through as many as 15 or 20 layers in tropical forests, where 95 percent of the light may be absorbed before reaching the ground. Sugar beet leaves have a vertical distribution of about a foot, whereas in forests the distribution may extend more than 300 feet. In natural forest stands the amount of light reaching the lower leaves is below the **compensation point**—that light intensity required to maintain a rate of photosynthesis equal to the rate of respiration. Lower branches may actually persist at the partial expense of the upper ones, which supply them with some carbohydrates, and thus such lower branches have been called "negative branches."

Low crops, with leaves in a relatively narrow zone close to the ground, seem to be more efficient producers than forests, in which leaves are spread over a wide range of heights. This is probably because in low crops a smaller percentage of leaves is exposed to light below the compensation point. Moreover, trees have a more extensive transportation system through which the compounds move, and the construction and maintenance of the "plumbing system" and the movement of compounds through it requires much energy.

The growth of plants is a function of the efficiency with which they produce dry matter. This involves the efficiency with which they capture light energy and the efficiency with which they transform it into organic matter. The combined efficiency of these two processes in our most efficient cultivated crops—sugarcane, sugar beet, wheat, and rice—is between 2 and 3 percent. But since only a portion of the crop plant is edible (28 percent of the dry weight of cereals, 55 percent of the root of sugar beet), the net, or overall, efficiency is much lower. Furthermore, there are losses due to diseases, pests, and fires.

Measuring productivity

The fundamental objective of agronomists, horticulturists, and foresters is to increase the efficiency with which solar energy is converted to useful products. Although the products are sold in arbitrary units (pounds of potatoes, bushels of corn, gallons of cider, or cords of wood), much better and more sophisticated units for measuring energy conversion are available. How much air space is there in a bushel of apples? How large is a head of lettuce? It is easy to see why bushels, pecks, and cords are not adequate for scientific use, and not even so for highly intensive agriculture. Furthermore, it is often desirable to measure rates of production of standing crops in the field. What are some of the possibilities for more precise measurement?

Plant productivity can be precisely estimated by measuring either the oxygen released or the carbon dioxide used in photosynthesis. Since the amount of carbon in CO_2 is directly proportional to the amount of carbon fixed in sugars during photosynthesis (Fig. 2-4), productivity can be estimated by the rate of disappearance of CO_2 from its environment. This is a straightforward task in a small growth chamber, but it is difficult in the field, where the apparatus for collecting and measuring gases must be disturbingly complex.

It is also possible to estimate production by determining the amount of chlorophyll present in a given amount of vegetation on a given area of land. To do

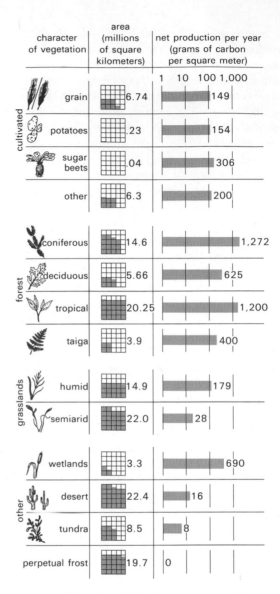

Figure 2-4 The amount of carbon incorporated in organic compounds is a measure of the amount of organic matter produced. Within any given area, cultivated vegetation is less efficient than forest vegetation, but more efficient than for grassland vegetation. Increasing the productivity of the vast areas of desert on the surface of the globe may be our greatest scientific challenge. [From E. S. Deevey, "The Human Population." Copyright © 1970 by Scientific American, Inc. All rights reserved.]

Table 2–2 Yield per acre per year and equivalent caloric values from selected crops produced from 1974 to 1979.

Crop	Average yearly yield (lb/acre)	Kilocalories/lb	Millions of kilocalories/acre	Day equivalents of energy/ person/acre[1]
Potato	5622[2]	1270	7.1	2380
Rice	3819	1460	5.6	1858
Corn	5413	1450	7.9	2616
Soybean	1798	1600	2.9	958
Wheat	1898	1470	2.8	930

[1]At rate of 3000 kilocalories per day per person.
[2]Dry weight (22 percent).

this a sample of leaves of known weight is collected, and the chlorophyll extracted in boiling alcohol. By knowing (1) the total weight of leaves from which the chlorophyll was extracted and (2) the efficiency of chlorophyll in photosynthesis, the total photosynthetic efficiency of vegetation can be estimated. This technique is sometimes used for forest vegetation.

The use of energy to measure productivity gives a degree of precision not heretofore possible. It offers a single unit, the calorie, which is equally useful from the time that light energy is captured by plants until it is incorporated into consumer products. For example, a fairly active man requires approximately 3000 kilocalories of energy each day, a yearly requirement of about 1,100,000 kilocalories. This energy theoretically could be supplied by about 2 metric tons of potatoes (fresh weight) or 340 kilograms (750 pounds) of wheat—a filling if incomplete and uninspiring menu (Table 2-2). The energy unit of measurement is also useful for expressing the production of fiber crops such as wood. Paper mills are already buying wood on a weight basis. Even though few foresters regard boards or sticks of pulpwood as bundles of energy, this is precisely what they are. The specific energy content of a woodpile or a bale of cotton depends largely upon the total amount of cellulose present. Energy levels of crops can be used precisely as expressions of quantity regardless of volume or specific gravity.

The process of estimating energy values in terms of caloric equivalents is a fairly complicated process, and must be done by statistically controlled sampling of the biomass. The energy values of organic materials are determined by burning known quantities of materials under carefully controlled conditions and determining how much heat is given off.

It may well be that many agricultural products that we enjoy today will not be produced on the farm, but will be manufactured chemically from component parts. If people could become conditioned to each and enjoy reconstituted foods and make other such substitutions, agricultural production could be channeled into crops that offer maximum productivity of one or more of the basic requirements of the human diet. For example, certain crops would be grown primarily for their high caloric content, whereas others would be grown for their content of particular organic compounds, such as amino acids. The basic materials, after being synthesized by plants, could be subsequently elaborated into foods with a wide variety of flavors, tastes, and textures, thus satisfying the psychological need for a varied diet.

Programming crop production

We have stressed the capture of light energy by plants. Using the energy concept does not alter the objectives of crop producers, nor would many practices necessarily be changed if the concept were universally adopted, for we have learned through experience to manipulate our crops to make good use of available energy. We will, however, be able to control crop production with much greater precision than is now possible.

Business uses computers to "plan" the most economical and profitable succession of events for particular purposes. One of the important mathematical decision-making devices employed is called linear programming. It has been accepted by American industry and become a routine management tool. But it is only usable because all of the different kinds of materials and efforts involved have a common denominator: Time, materials, and manpower can be expressed in terms of dollars and cents.

By convention, biologists describe the ecological environments of organisms in terms of rainfall, temperature, soil moisture, wind, sunlight, available nutrients, soil conditions, composition of the atmosphere, pollutants, and living organisms. Although these are adequate parameters, they must be used correctly and expressed in comparable terms if they are truly to be understood. The most suitable common denominator for crop production may be energy. Measurements of input and of output can be compared by making use of caloric equivalents. Thus the energy concept will allow more precise analysis, computerized if necessary, in crop production.

In order to maximize the capture and storage of energy, the specialty of crop energetics must take into account not only the standard cultural practices, but many factors that have received little attention in the past, such as plant density and distribution, leaf coloration, and height of individual plants.

Scientists are quantifying the chemistry and physics of plants in terms of their metabolic requirements. The water relations of plants and soils are especially easy to quantify, and much is being done to explain and describe the movement of water in the environment on the basis of energy gradients, using mathematical expressions. Sooner than we realize, most of the biological aspects of crop production will yield to mathematical quantification.

THE ENERGY-FLOW CONCEPT

The function of agriculture is to direct and maximize the capture and concentration of solar energy by plants and to optimize the flow of this energy through the various manufacturing and distribution processes in a manner that provides the greatest possible benefits to society. This point of view gives a common denominator, energy, by which crop production can be measured, and provides an absolute standard for equating the effects of different cultural treatments and practices. Considering crops as traps that can be manipulated to capture, transfer, and store energy provides insight into the means by which we can improve and increase agricultural efficiency and productivity.

Energy capture

There have been many proposals to broaden the base of energy capture. Most of the photosynthesis on the surface of the earth takes place in water, mainly in oceans and seas (see Table 2-1). Only an insignificantly small fraction of this production is used by people for food. By systematic fertilization and management, it might be possible to "farm" certain areas of the seas and oceans to raise their levels of productivity even higher. Can we produce fiber as well as food in this manner?

In some places extending agricultural production to deserts and other areas with unfavorable climates may be feasible. Many desert soils are extremely fertile and only need carefully controlled irrigation to become highly productive.

One obvious method of increasing energy capture is to increase the length of the growing season. Artificially extending the growing season by using greenhouses makes it possible to produce crops where the growing season is very short. But this technique has limited application in today's economy because of the high cost of building and maintaining greenhouses, and it has generally been restricted to the production of flowers and such high-value food crops as tomatoes and grapes. In the future though, solar heating may be used more extensively, even to the point of making possible the production of relatively low-value carbohydrate sources throughout the year.

These proposals require extremely high inputs of technology and capital. The process of energy capture in present agricultural areas offer opportunities for improvement. For example, in many situations of intensive production the factor that limits energy capture is the availability of carbon dioxide. Producers of greenhouse crops can increase yields by releasing CO_2 gas from storage cylinders to "fertilize" the air.

Air movements affect photosynthesis by influencing CO_2 exchange between leaves and the surrounding atmosphere (Fig. 2-5). Thus the orientation of rows with respect to prevailing winds may influence productivity. Where the prevailing winds are negligible to moderate, corn yields can be increased if the rows are oriented at right angles to the wind. The tops are blown back and forth and there is greater turbulence and mixing of air.

Plant morphology also has a relation to energy capture. Such characters as plant type, leaf shape, and branch angle influence light absorption. Rede-

signing crop plants to increase the capture and use of energy will be the future course of the technology of plant improvement.

The energy-flow concept is particularly applicable in forestry. The present production of lumber and pulpwood may well give way to a forest economy based on two commodities: alpha cellulose for structural purposes and degradation products of lignin for stock feed. The primary need, however, will be for molecules of cellulose. Such a change in the objectives of production seems logical. In the form of dissolving pulp, a product of alpha cellulose, wood can be exploded, foamed, extruded, molded, shaped, and combined with other materials to form products that are dimensionally stable, resistant to fire, insects, decay, and rot, and highly adaptable to many uses. Unwoven fabrics are already finding a good market, and it may not be too long before fiberless paper is in use.

Since straight tree trunks and a high degree of structural ability are not important to cellulose production *per se*, tree form would be of no consequence. The form and structure of the forest stand would receive primary consideration. Forests would be designed to make energy capture and storage most efficient. It is probably safe to assume that harvesting and processing techniques would keep pace with forest management and wood utilization, so that even the shrubs, saplings, and suppressed trees of multilayered forests would be used profitably. Under such conditions as these, there might well be a change to a more naturalistic silviculture (tree culture) based on the hypothesis that vegetation types with many layers are photosynthetically more efficient than those with only a single layer.

Structure and function of agricultural ecosystems

All ecosystems have structure and function. *Structure* refers to the relations among producers, users, and decomposers. Strictly speaking, only green plants are producers. However, some ecologists refer to transfers in food form also as a secondary sort of production. For example, animals that eat plants produce animal flesh.

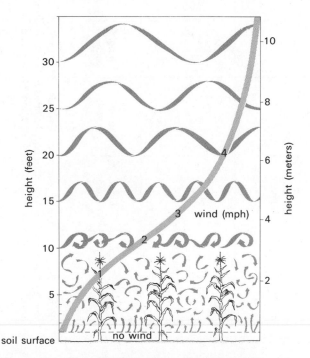

Figure 2-5 The turbulence of wind increases with distance above the soil surface. Crop spacing can increase turbulence at lower elevations, thereby increasing the exchange of carbon dioxide between leaves and the atmosphere. [After J. E. Newman and B. O. Blair, *Crops and Soils Magazine*, June–July 1964.]

structural unit	function	functional unit	pond	forest	grassland	corn	wheat	orchard crops	energy pathways for all ecosystems	organism types common to all ecosystems
producers	production	green plants	algae, phytoplankton	trees, shrubs	grasses, forbs	corn	wheat	apples, peaches, nuts, etc.	heat	green plants
X₁ users	herbivore, symbiosis	animals, nongreen plants	fish (bluegill), snails, zooplankton, bloodworms, water fleas	small rodents, deer, humans, vegetarian birds	rodents, sheep, bison, cattle, prairie dogs	insects, fungi, humans, bacteria, cattle	humans	humans, vegetarian birds, deer, squirrels	heat	arthropods, disease organisms, symbionts, insects, herbivores
X₂ users	predation, symbiosis	animals, nongreen plants	fish (bass), humans, snakes, bloodworms	birds of prey, snakes, weasels, humans, mountain lions	hawks, humans, wolves, vultures, snakes	bacteria, fungi (fermentation products), humans, pets	humans	humans, birds of prey	heat	carnivores, disease organisms, carrion eaters, insects, predators, symbionts, insectivores
X₃ users	predation, symbiosis	animals, nongreen plants	fish (bass), humans	foxes		humans			heat	carnivores, disease organisms, insects, predators, symbionts, insectivores
X₄ users	predation, symbiosis	animals, nongreen plants	fish, humans						heat	carnivores, disease organisms, insects, predators, symbionts, insectivores
decomposers	decay	fungi, bacteria	fungi, bacteria	fungi, bacteria	fungi, bacteria	fungi, bacteria	fungi, bacteria	fungi, bacteria	fungi, bacteria; heat	fungi, bacteria

(Note: "transportation, storage, and processing" appears as spanning notes between the crop columns for the producer, X₁, and X₂ user levels.)

Figure 2-6 All types of ecosystems have similar but not identical structural and functional units. The flow of energy through the different trophic levels is in the same direction, but is complicated, in the care of crop plants, by the energy costs of transportation, storage, and processing.

The structure of an ecosystem may be very complex, such as that of a forest or ocean ecosystem, or it may be fairly simple, as are those of most agricultural ecosystems. In untrammeled nature we seldom find simple systems but are confronted with highly complex ones. To maximize productivity, we try to keep agricultural ecosystems as simple as possible, and the food chains in it as short as possible, thereby decreasing energy losses.

Function refers to the role that each member of the ecosystem plays in its maintenance and development. The primary function of a corn plant in a cornfield is to produce corn. But if the corn is infested with pests, then it also assumes the function of supplying food for the pests, which use it as an energy source. Thus the corn plant may have primary, secondary, tertiary, and even further functions. When one of its functions is undesirable we attempt to block it by using pesticides or some other cultural practice.

Figure 2-6 illustrates the similarities of all ecosystems, all of which share structural and functional units at some of the same levels of complexity. While the producers in a pond ecosystem are algae, they share a position at an identical level with corn in one type of agricultural ecosystem.

Ultimately, the organic structures of all ecosystems can be decomposed to inorganic compounds, for the most part CO_2 and H_2O. In the process of decomposition the minerals bound in organic matter are released back into the environment once again as inorganic matter, a process called **mineralization.**

The energy-efficient farm

The goal of many persons is self-sufficiency, but in today's economy total self-sufficiency is usually impossible. Also, we have learned that a degree of mutual dependency can provide a level of personal comfort and satisfaction that one person, working alone, cannot attain.

The crop-production unit, whether it be a farm or a giant cooperative, is one place in which the energy subsidy from outside sources can be reduced, though few farms can probably ever reach total self-suffi-

ciency. There are waste products at nearly every step in the process of crop production. In years past the disposal of wastes has been a problem, contributing to pollution and to the expense of crop production. But today waste is recognized as a potential asset, convertible to energy and to the nutrition of crops.

Figure 2-7 illustrates some, but not all, of the potential for recycling minerals and energy that would have been wasted in years past. At practically every stage of crop production there are waste organic materials that can be used to generate methane gas, which can be stored in a compact form and used at a later date. Such materials as sawdust and corn cobs can be stored for later use as fuel, but they are bulkier and not so versatile.

Figure 2-8 illustrates the value of auxiliary energy in increasing the percentage of the population available to do work other than that directly connected with food production. People not needed in the agricultural work force can apply their talents, time and energy to writing, music, art, teaching, and other occupations that contribute to the completeness of human society.

INCREASING CROP PRODUCTION

One of the present-day anomalies is that some areas of the world are blessed (?) by agricultural surpluses while others are plagued by persistent and agonizing shortages. There is, of course, a marked difference in the intensity and efficiency of crop-production practices used throughout the world. Consequently, appropriate methods for increasing production will depend on the existing level of technology.

In developing areas the productivity of land can be increased by many agricultural techniques that are now routine in the developed countries. In an area saturated with plants, some environmental factor soon becomes scarce. Light, soil moisture, and nutrients are the factors most often in short supply. These are some of the well-known growth-limiting factors with which agriculturists concern themselves. The technology necessary to identify and overcome

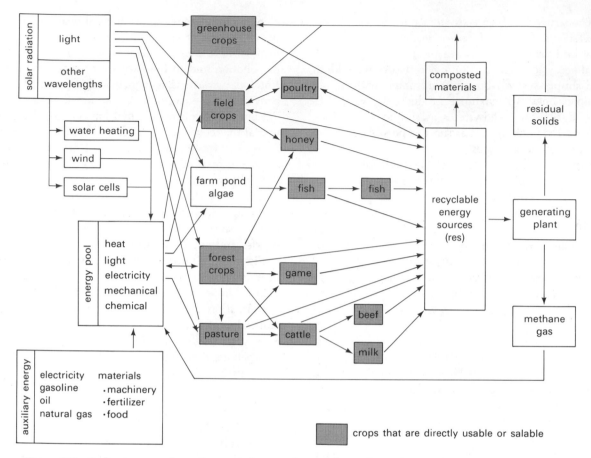

Figure 2-7 Radiant energy from the sun is harnessed to produce plants of many kinds. These primary producers, in turn, provide energy and structural materials for the animals of lower trophic levels. Such a system must receive subsidies from outside the farm, such as fertilizer, to replace the materials that are lost when such a product as poultry, beef, or hay is sold and moved to another location. These subsidies can be reduced, however, if efficient use is made of farm wastes by recycling them to produce fuel and soil amendments.

these limiting factors and to control such crop hazards as pests, disease, and fire is the subject of a large part of this book. This is the direction that agricultural improvement in developing areas must take.

As limiting factors are recognized and corrected it becomes increasingly difficult to obtain further gain. Nevertheless, there is much room for improvement in the developing areas of the world. Fertilization is one of the most powerful tools for increasing crop production, yet in many developing agricultural areas little or no fertilizer is applied. Large increases in efficiency can be obtained with crop improvement through genetics and breeding. The ruinous losses to pests and diseases in developing areas can be prevented with better control methods.

In the developed agricultural areas the immediate, practical problems of production—the amount of fertilizer and the method of applying it, the control

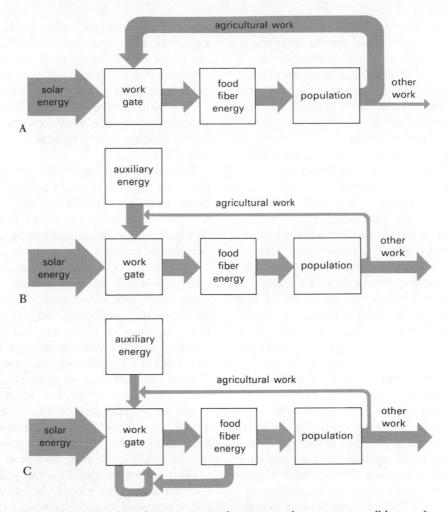

Figure 2-8 The role of auxiliary energy in determining the economic well-being of a society is illustrated by these diagrams. (A) In an economically less developed country, the bulk of the population must be devoted to agriculture in order to support itself at a subsistence level. (B) In an economically more developed and industrial country, auxiliary energy sources "open the gate" to the more efficient utilization of the sun's energy, and the entire population can maintain a higher standard of living. (C) In the future, agriculture must be much less dependent on auxiliary energy, a feat that can be accomplished by making use of products formerly regarded as waste.

of crop pests, the optimum depth to which plow-shares should be set, the length of the rotation period, the choice of weed killers—have received much attention. These types of problems have often been solved on an experience and expediency basis, but they have also been tackled by organized research groups using experimental approaches. Many problems can be now handled routinely by "cook-book solutions." The surplus of farm and forest products produced each year in the United States bears witness to the efficiency of this system.

In the United States most of the small, inefficient farmers and timber growers are being forced, by economic pressures, either to leave their land or to seek supplemental employment in nearby cities. Consequently, the production of farm and forest products has increasingly become the privilege of a relatively few large producers. Even though small applied research projects have been useful in years past in providing immediate answers to many problems, this approach no longers works in advanced agricultural economies. Most of the remaining problems are of a basic nature. To solve them we need a hard core of biological and economic facts. Crop production is taking on a new personality. It is big business, and has become more nearly a science than an art. It is only prudent to assume, in the face of increasing national and world populations, that local overproduction will cease to be a problem. Should international crop sharing become a complete reality, the day of underproduction in the United States is likely to face us sooner than we realize. To prepare for this almost certain eventuality, it is essential that we learn to improve the efficiency of each phase of the energy-flow system of agriculture.

Residue utilization

The first approach to increasing agricultural efficiency should be to use a greater part of current production. This is partly a simple matter of thrift. In some countries straw, cobs, husks, shells, and manure are carefully collected and used either for feed or for fuel. But in other, more prosperous countries they are usually discarded or, at best, composted. In the forests of Europe, even the smallest branches and twigs are carefully collected, while in the United States, tops, bark, slabs, and sawdust are usually burned or left to rot. The cellulose in these discarded parts of trees is potentially as good as that in the wood of the finest firs and pines, and the bark has many unusual and useful properties that have infrequently been exploited. Furthermore, many forest species are totally rejected for use simply because of their small size or poor form.

The extraction of usable fibers from plant residues is a field that holds much promise. Of the many kinds of residues, grain straw, corn stalks, sugarcane bagasse, cottonseed hulls, cotton stems and pods have been utilized to the greatest extent. Strawboard is a commercial product; ground walnut and pecan shells are used as fillers in plastic molding powders; sugarcane bagasse pith is useful as a filler in low-density dynamite. Fine quality papers of the kinds required for cigarettes, fine books, and stationery are made from the fibers of hemp and flax. Sawdust is collected at sawmill sites for manufacture into pulp, an operation that has greatly increased the utilization of wood. Sawdust does not make paper equal to that made from wood chips, but nonetheless it has a number of excellent applications. Someday, a widespread commercial use may be found for the enormous quantities of lignin that are usually dissolved and discarded in the manufacture of paper.

Many industries based on the utilization of discarded plant products could be established. Tops of plants produced mainly for their starchy roots and tubers may be used as sources of protein. Proteins from leaves are now used for animal feed and may someday be extracted for human consumption. Under any circumstances, a comparison of the efficiency of protein production of corn by two different methods is revealing (Fig. 2-9). If 1524 kilograms of corn (3360 pounds, or about 56 bushels) is fed to beef cattle, the end product will be only 32.8 kilograms (72.3 pounds) of meat protein. However, if all of the vegetable protein were extracted from the same 1524 kilograms of corn, the yield would be 117 kilograms (258 pounds). By using the residual starch in the corn to produce a crop of fungi, we can produce another 125 kilograms (276 pounds) of extractable protein. The 242 kilograms (534 pounds) of vegetable protein

corn crop,
3763 kg/ha
(56 bu/acre)

yields

1524 kg
(3360 lb)
of corn

produces   produces   fed to
beef animals
produces

117 kg
(258 lb)
of extracted
protein

914 kg
(2015 lb)
of starch

32.8 kg
(72.3 lb)
of meat
protein

equal to

1051 kg
(2317 lb)
of glucose

fungus
conversion

503 kg
(1109 lb)
of fungal
tissue

contains

+

125 kg
(276 lb)
of
extractable
protein

242 kg
(534 lb)
of vegetable
protein

Figure 2-9 More than seven times as much protein can be produced by the extraction and fungal conversion of corn as by feeding the corn to animals to produce meat protein. However, the technology of producing vegetable protein is complicated, and we have not yet learned how to make vegetable protein as palatable as beefsteak, lamb, and pork roast.

thus produced might not be as tasty as that of the beef, but future generations will undoubtedly consume it in increasing quantities, prepared as a meat substitute. Even today vegetable protein is used to supplement ground meat and reduce its price.

Food processing is an area in which tremendous strides can be made in decreasing waste. Potatoes, which are produced in greater abundance in the United States than any other vegetable, provide a good example. Many potato-processing plants handle more than 450,000 kilograms (nearly a million pounds) of potatoes per day, with the resulting organic pollution equal to that of a city of 300,000 people. Potatoes were once commonly peeled by dipping them in a vat of 16–20 percent lye solution to loosen the skin, and then spraying them with a high-pressure water jet to remove it. In a new process, they are dipped in a 12 percent lye solution, held for 3–5 minutes, heated for 1 minute by infrared lamps, and peeled by rotating rolls with half-inch rubber studs. The waste thus produced is mixed with other organic wastes and fed to steers as 80 percent of their dry ration, practically eliminating potato skins as a source of pollution.

Utilization efficiency

One of the most troublesome problems, to those concerned with supplying the necessities of life to our expanding population, is the inefficiency with which consumers use food and other plant products. Plants usually function with an average efficiency of no more than a few percent in storing the energy that arrives from the sun. Animals and other organisms that feed on the green plants are equally inefficient. And so it goes, with each successive consumer utilizing only a part of the food it consumes. Consumption

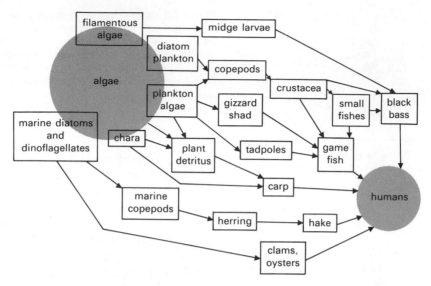

Figure 2-10 An aquatic food web from algae to people. Each time a transfer of energy is made, a 90–95 percent loss occurs. [After E. N. Transeau, H. C. Sampson, and L. H. Tiffany, *Textbook of Botany,* Harper & Row, 1940.]

is, of course, a means of transferring energy and building materials from one organism to another. Such a transfer of materials takes place through a **food chain** or **food web,** depending upon the linearity of transfer, to different kinds of organisms (Fig. 2-10). Eventually, all of the original energy produced by plants, and not dissipated by fire, is depleted by respiration (Fig. 2-11). We are primarily concerned with the food chain from plants to ourselves. Food webs, however, may be extremely complex, and may involve various degrees of predation among lower forms.

It is easy to see that the **ecological efficiency** of the various organisms in the food chain is of the utmost importance. Ecological efficiency may be either a measure of the effectiveness with which (1) solar radiation is captured or (2) the biomass of one organism, either plant or animal, is converted to the biomass of another. Respiration and the production of such indigestible materials as hair, hide, and bone all detract from the efficiency of conversion.

The flow of energy through food chains and food webs is essentially a one-way process. It moves from the photosynthetic plant through an herbivore (plant eater) and then through carnivores until the last of it is used by decay organisms. Each time a transfer of material occurs about 90 percent of the energy is lost and never recovered. It has been estimated that 100,000 kilograms of algae are required to produce 1 kilogram of codfish! All the remaining energy incorporated into the algae is lost in respiration, somewhere in the food chain between the algae and the cod, and even the codfish doesn't last but a few years before it too is reduced into carbon dioxide, oxygen, a few minerals, and heat energy. The heat energy is lost, however, for it cannot be recovered and incorporated directly into the biomass, even though heat from biological combustion does affect the rates of reactions that take place in living organisms (Fig. 2-12).

Suppose, for a moment, that the world population has increased and that the demand for food is so great that every possible means at hand must be applied to avoid starvation. When we eat animals we lengthen the food chain from plant to human; to shorten this chain means less steak and more starch. Shortening the food chain precludes the great loss of energy resulting from each conversion to the next link. In time of shortage the strategy must be to get as close to the primary producers as possible, eliminating the intermediaries. We have eaten primary plant producers in part, or in their entirety, for many thousands of years, and will continue to do so as a matter of choice. In the Orient the diet of most people consists almost exclusively of plant materials, whereas in the United States we have one of the highest rates of meat consumption of any nation.

But the improvement of biological efficiency is not the only way to save energy. In earlier cultures, crops needed to be transported only a short distance from the field to the homestead. The crops were stored in caves, root cellars, rooms, and bins, all of which were close to the preparation center—the stove or the

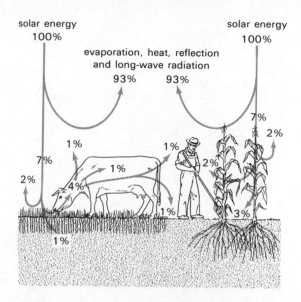

Figure 2-12 Of the solar energy that reaches the earth, about 93 percent returns to the atmosphere. With nutrients and water abundant and with a full cover of leaves, about 7 percent of that solar energy can be converted in photosynthesis by corn; 2 percent goes into respiration, which is required for the growth and maintenance of the crop, and 5 percent goes into the dry matter of the crop. In corn, 3 percent goes into roots, stems, and leaves, which constitute a crop residue that is recycled to the soil or fed to animals, and 2 percent emerges as grain that can be eaten by people. For grass, as much as 4 percent may be consumed by a cow. [From R. S. Loomis, "Agricultural Systems." Copyright © 1976 by Scientific American, Inc. All rights reserved.]

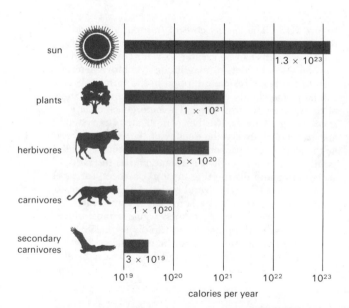

Figure 2-11 The efficiency with which solar energy is used decreases with each step along a food chain. Plants use only a small fraction of the energy that reaches them; herbivores indirectly use only a part of this energy, and carnivores still less. [From L. C. Cole, "The Ecosphere." Copyright © 1958 by Scientific American, Inc. All rights reserved.]

loom. But today, transporting the crop to the consumer in the city must take place in a number of stages, each of which contributes to its marketability and availability (Fig. 2-13). In addition to processing, there must be some provision for storing both the raw materials and their products until they are needed. The energy costs of transportation and storage add to the cost of the end product. Because of the longer chain of events in the preparation of meat for the consumer, the costs of meat production are significantly higher. It is fair to say that meat production is ecologically less efficient not only in the biological production of protein but also in its processing and marketing.

A

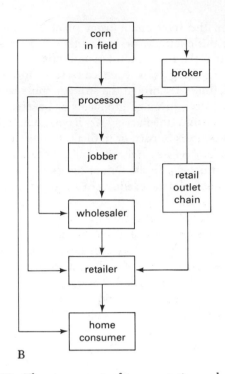

B

Figure 2-13 The energy costs of transportation and storage are great. If corn is consumed in the form of meat, such costs can be greater than for fresh or canned corn by several orders of magnitude. (A) Movement of corn to the home for consumption in the form of beef can be tremendously complicated, involving many stages of transportation and storage. The emergence of large retail outlet chains has further complicated the system and has contributed to the decline of the traditional role of the jobber. The packer, which can be either a small family operation or a giant corporation, has the task of slaughtering and distributing meat as box beef, carcasses, or primal cuts through several outlets. (B) Movement of fresh or canned corn to the home is less complicated. Because sweet corn has a freshness value, it must move to the consumer or the processor as rapidly as possible. Transportation and storage must be minimized. The broker serves only as an agent and may actually be a nonprofit farmers' cooperative.

Let us ignore, for the moment, all matters of dietary custom and consider the most efficient methods of providing a hungry world with food. It has been demonstrated, in a laboratory, that algae can produce a biomass, on a dry-weight basis, of 90 met-ric tons per hectare (about 40 tons per acre) per year. Of this production, 45 metric tons would be protein, 7 metric tons would be fat, and 38 metric tons would be carbohydrates. No present crop produces organic matter so efficiently.

The culture of algae as a crop is possible with a one-celled species called *Chlorella*. Much is known about its physiology and reproductive habits, and it would be a simple matter to get it into production should economic factors become favorable. Perhaps few readers would find soup made from algae as desirable as sirloin, but it is at least as rich in some vitamins and amino acids. One group of people ate such a soup for an extended time and found it to be palatable and nutritious. Powdered algae could probably be incorporated into many of our foods with little apparent change in quality or flavor. If it is not acceptable as human food, the *Chlorella* might be rendered into feed for animals, thereby allowing people to maintain a more conventional diet. Many other kinds of algae are eaten by people all over the world (Fig. 2-14).

In a food crisis even wood can be made edible. By treating wood with a strong acid, molasses can be made that is easily purified for human consumption. How many years would a giant redwood sustain human life if converted into food? Cultivating yeast on wood molasses could produce a complete, well-balanced food. In Germany, during World War II, edible fats were manufactured from coal. This, however, is a complicated process that can be used only when expense is no object.

Woods that cannot be used as structural materials or food could be used for the production of gaseous and liquid hydrocarbons. Even though the costs of conversion are high, some countries do find these techniques economically feasible.

Brazil, a country that must import all of its petroleum, has explored the possibility of using biochemically produced alcohol to extend its gasoline supply. They have found that as much as 10 percent of wood alcohol can be added to gasoline with no special engine adjustment. While the energy content of alcohol is not as great as that of gasoline, burning alcohol is virtually nonpolluting, which is important in many of our smog-bound cities. At present, this mixture, gasohol, is in use throughout the country. The alcohol component is made largely from sugarcane.

In the United States, gasohol is sold in some states but is likely to be more of a novelty than a competi-

Figure 2-14 Intensive agriculture can be conducted even in the oceans. Posts and nets support a crop of edible seaweed, *Porphyra*, in the Inland Sea of Japan. [Courtesy J. H. Ryther.]

tive product for several years. Whether farmers can grow corn for producing fuel alcohol is still in doubt, but as the price of crude oil increases, the economics improve. In years past, crop producers have been charged with the responsibility for producing "food and fiber." In the future this slogan may change to "food, fiber, and fuel."

Intensive agriculture and energy conservation

There have been many pleas from environmentalists that farmers use less fertilizer, which is a prime pollutant of water. This can be done, these environmentalists maintain, by the use of more extensive farming practices—using more land to produce crops so that better use is made of natural fertility. Unfortunately, land suitable for crop production is becoming scarcer, because productive land is being usurped by urban sprawl, superhighways, and shopping centers. It seems inevitable that such losses of farmland will increase at an even greater rate in the future. Farm managers must therefore produce more crops on less land, and this means using even more intensive agricultural practices.

Defining "intensive agriculture" precisely is not easy, even though some of its characteristics may be quickly listed. It includes, but is not limited to (1) closely controlled plant spacing, (2) precise fertility control, (3) mechanization of field cultural practices, (4) insect control, (5) disease control, (6) weed control, (7) irrigation, (8) drainage, (9) burning, (10) the use of genetically improved plant materials, (11) properly timed seeding and harvesting practices, and (12) decreased distances of crop transport. The relative importance of each of these factors is subject to local conditions and constraints. For example, insects may be a first-order problem in some areas for some crops, but not important for the same crops in other areas. Intensive agriculture might best be defined, rather generally, as the application of the optimal balance of farming practices to produce crops with the largest possible net yield of value.

There are many benefits of intensive crop production. For example, fertilizers can be used more effectively if individual crop plants are close together. This, of course, calls for good placement and timing. Chemical weed control can be minimized, because with close spacing crops are better able to compete with weeds. Because of the smaller areas of tilled land, soil losses by wind and water can be minimized by stubble cropping and by alternating crops with strips of grasses or other permanent soil covers. Also, there should be less use of land that is poorly adapted for tillage and difficult to manage.

If less land is used, the energy subsidies of transportation and cultivation can be smaller. There can be less unused residue, because collection will be made easier. Further, irrigation and drainage can be more efficient when land is used intensively, which in turn requires less of the materials and technology of these practices.

In the United States, agriculture is intermediate in intensity. The high degree of reliance on machinery would seem to indicate highly intensive agriculture, but machinery is not the only indicator of intensity. Machinery can produce a better seedbed and a more timely harvest, but yields may be lowered when wide spacings are used to accommodate machines, and the machines may compact the soil and thereby decrease productivity. What a machine-oriented agriculture represents is an optimization of labor.

The most intensive agriculture is practiced in countries that have a small land base, such as Japan. There, the use of sunlight is maximized by the use of large quantities of fertilizer and by the intensive use of human labor. Japan and other land-poor countries serve to show us how close the level of productivity can come to the potential provided by soil, climate, crop selection, and genetic improvement.

The least intensive agriculture can be found in those parts of the world in which some environmental factor is in short supply or in which the people are too poor to pay for such practices as irrigation or fertilization. If the length of time for which some of these agricultural systems have been operating is any measure of success, then some of them are among the most successful in the world. However, their productivity is usually low, and it must be sustained with great care, patience, and planning.

In the end, intensive agriculture may free some land for other uses. The demands of hunters, realtors, engineers, backpackers, campers, conservationists, city planners, and many other special-interest groups will play an important part in shaping agricultural practices of the future, because each group is eager to have more land for its own purposes. Unfortunately, the amount of prime agricultural land available for crop production is more limited now than it ever has been in history.

ENERGY FARMS AND FORESTS

In Europe during the Middle Ages, kings and barons frequently owned vast estates of land, most of it forested. While timber cutting, hunting, and even trespassing were forbidden, they gave their serfs—and others, for a fee—permission to pick up fallen branches to use for firewood, the main fuel for cooking and heating. As a result, the forest floor was nearly always devoid of branches, dead trees, cones, and other detritus. Even though the feudal system broke down, the practice of branch collecting continued until after World War II, when Europe became

Table 2–3 Bioconversion technologies for producing selected energy products from wood.

Product	Technology	Remarks
Electricity	Combustion by spreader Stokers in boiler Furnaces	Developed and in use for capacities up to 55 MW equivalent
	Fluid-bed combustion	Limited use for steam generation with wood-yard waste as fuel
	Charcoal/oil slurry firing	Under development by ERDA
Substitute natural gas	Gasification/methanation	Under intense development for coal. New development activity required for gasification of wood. Current experience at atmospheric pressure using municipal solid waste.
	Anaerobic bacterial digestion	Well-known technology for municipal sewage treatment. Lignin may inhibit bacterial activity.
Medium-Btu fuel gas	Gasification	Same as for substitute natural gas. Product can be a raw material (synthesis gas) for manufacturing methanol and ammonia.
Fuel oil	Pyrolysis	Flash-pyrolysis pilot plant using municipal solid waste to produce fuel oil now in early stages of operation. Moving-bed pyrolysis developed on significant scale by Georgia Institute of Technology. Low-grade oil may require aftertreatment.
Charcoal	Pyrolysis	Coproduct of processing or fuel-oil production. Multiple-hearth furnace process now standard practice for charcoal production
Ammonia	Gasification/hydrogen production/ ammonia synthesis	Gasification same as for substitute natural gas. Carbon monoxide shift to hydrogen and ammonia synthesis widely used and well-developed technology
Methanol	Gasification/methanol synthesis	Gasification same as for substitute natural gas. Methanol synthesis widely-used, and well-developed technology
Ethanol	Cellulose hydrolysis and fermentation	Fermentation step widely used, and well-developed. Industrial experience significant for dilute sulfuric acid as the hydrolytic agent

heavily industrialized. At that time, because of industrial employment, it became economically feasible for most people to buy oil for fuel instead of searching the forest for branches.

In the past, people have used plant agriculture mainly to produce food and fiber. People have always burned plants and plant parts to provide heat for cooking and warmth, but plants have only rarely been grown for this specific purpose. Fallen branches, bagasse, crop residues, and surplus or undesirable trees, as well as otherwise useful trees cut specifically for firewood, have been used for fuel through the ages. Until recent years there has been an abundance of such materials in nearly all parts of the world.

Today we are entering an age of energy awareness, brought on by increasing fuel shortages and rising prices. We are on the threshold of economic feasibility for growing plants to be used as energy sources rather than for food and fiber. With plants as bioconverters of radiant energy to chemical energy, it is increasingly probable that *energy farms,* or *energy forests,* will be used to produce plant biomass as a source of energy (Table 2-3).

As the following table shows, the energy values of plants are not as rich as those of coal.

Energy source	Btu per lb
Bituminous coal	15,000
Slash pine	7,000
Sugarcane	6,500
Alfalfa	6,500
Sycamore	5,800

Despite the fact that their energy values are low, there are many good reasons for using plants and plant products for energy production:

1. They have high location value because of short-distance transportation.
2. They are relatively abundant at many locations.
3. They are renewable resources.
4. They are largely materials that would otherwise be wasted.
5. They are otherwise materials of relatively low value.
6. They can be used in many combinations or mixes.
7. Their sulfur content is very low.
8. Land unsuitable for other purposes can be used to produce them.
9. The land itself is not disrupted (forests).
10. Their residues are often useful as soil amendments.

Because of the bulkiness of such materials, it seems unlikely that they will be shipped in quantity to central locations for energy production. Rather, it would seem most feasible to produce and use the conversion products locally. However, the conversion products themselves, having a higher energy value, might profitably be transported for use elsewhere.

There are many possibilities for using plants as sources of energy. Nobel laureate Melvin Calvin has proposed the mass production of hydrocarbons very much like gasoline from plants of the genus *Euphorbia*. Plants in this genus produce a milky sap, called latex, which contains hydrocarbons of low molecular weight that are similar to those in petroleum. Grown in arid regions not suitable for growing food plants, they might be capable of producing from 25 to 125 barrels of oil per hectare per year. The plants would be harvested by cutting close to the soil, run through a crushing mill to extract the sap, and processed in a refinery very similar to those used for crude oil. Since the plants sprout prolifically from stumps, replanting would be necessary only after about 20 years. Oil produced in this manner would be practically free of sulfur.

Methane is an example of an energy source that can be produced from a variety of farm and forest waste products. The use of small waste digesters for the production of methane for heating and lighting in rural, single-family situations has proved feasible in India, Europe, and the United States. Methane from such digesters is rapidly replacing dried cow dung as the main fuel for heating and cooking in rural India today. In Europe, small digesters were constructed during World War II to produce methane gas as a replacement for fossil fuels, which were in very short supply. The use of such digesters has continued in Europe, particularly in rural, isolated areas of France and Germany, the methane serving as an important supplement to fossil fuels. Simple digesters applicable to single-family use, which can be constructed from used oil drums and plumbing materials for a cost of less than fifty dollars, can produce approximately 0.5 cubic meter (15–20 cubic feet) of methane gas per day, enough to cook three meals a day for two people.

Digesters of the single-family type (Fig. 2-15) and the types used on intensive livestock-production operations, have the potential to produce far more methane per kilogram of digested material than do large municipal-type digesters. This is true because it is possible to monitor closely, and relatively simple to control, environmental conditions within a small digester.

Total dependence on conventional fuels, especially in rural areas, is likely to become a serious handicap in years to come as reserve shortages and increasing

Figure 2-15 An organic digester, using 50 gallon drums for digestion units, can produce significant amounts of gas from waste materials. In India, such units are commonplace, and the gas they produce is used principally for home heating and cooking.

recovery costs continue to push energy prices upward. However, by producing energy from local resources, farmers can be partially freed from dependence on remote sources of increasingly expensive fuel supplies.

Selected references

Asimov, I., 1960. *Life and Energy*. New York: Doubleday. (A broad, popularized treatment of fundamental concepts of energy relations.)

Brown, H., J. Bonner, and J. Weir, 1957. *The Next Hundred Years*. New York: Viking Press. (An exciting and stimulating discussion of the earth's natural resources in relation to human society and technology.)

Buttel, F. H., 1977. *Energy and Agriculture: A Bibliography of Social Science Literature*. Monticello, Ill.: Council of Planning Librarians. (A list of references on the sociological aspects of energy use for crop production.)

Cox, G. W., and M. D. Atkins, 1979. *Agricultural Ecology: An Analysis of World Food Production Systems*. San Francisco: W. H. Freeman and Company.

Fritsch, A. J., L. W. Dujack, and D. A. Jimerson, 1975. *Energy and Food*. Washington, D.C.: Center for Science in the Public Interest. (An analysis of the amounts of energy used for food production and preparation.)

Gates, D. M., and L. E. Papian, 1971. *Atlas of Energy Budgets of Plant Leaves*. New York: Academic Press. (A review of the factors—that is, energy balance—needed to keep a leaf viable, and therefore a crop growing and productive. The book discusses leaf temperatures, transpiration rates, and the like as they relate to such factors as wind speed, relative humidity, radiation, and leaf shape.

Leach, G., 1976. *Energy and Food Production*. Guildford, England: IPC Science and Technology Press. (A short and readable book on the energy analyses of crop production.)

Lockeretz, W., 1977. *Agriculture and Energy*. New York: Academic Press. (A comprehensive treatment of energy use and conservation in crop production.)

Margaleff, R., 1968. *Perspectives in Ecological Theory*. Chicago: University of Chicago Press. (An excellent small book dealing with the energetics of ecosystem structure and management.)

Odum, E. P., 1971. *Fundamentals of Ecology*. Philadelphia: W. B. Saunders. (A textbook on ecology with an excellent section on the energetics of ecosystems.)

Peirotti, A., A. G. Keeler, and A. Fritsch, 1977. *Energy and Food*. Washington, D.C.: Center for Science in the Public Interest. (A comprehensive examination of the sources and inputs of energy used in food production.)

3
Food and Human Needs

Food is any material that provides an organism with energy and nutrients. Green plants utilize the sun's energy to synthesize their food supply directly from atmospheric carbon dioxide and inorganic substances from the soil. (Because soil nutrients do not supply energy, it is not technically correct to refer to them as "plant food.") Some microorganisms have very simple requirements and need only an energy source (such as sugar) and various inorganic nutrients. Animals, however, required their food "ready-made" in the form of highly complex molecules. Their diet may be quite specialized, but they ultimately depend on the energy captured in photosynthesis. Human beings are omnivorous, and obtain their food both directly from plants and from animals.

Insufficient food leads to intolerable consequences. Most of us experience great discomfort if we do not eat once every five or six of our waking hours. A person cut off from all food for 7 to 8 weeks faces certain death. There are also qualitative aspects to food: People who go for extended periods without the proper kind of food suffer irreparable damage; the consequences are especially severe in the young. Further, there is an important relation between proper diet and mental competence and achievement. There is also a cultural dimension to food: All peoples hold strong emotional preferences in their choice of diet. Dietary customs become deeply ingrained and difficult to change. Thus the food we

depend upon for nourishment and sustenance greatly influences our style of living.

Improper diet can lead to starvation, malnutrition, obesity, poisoning, disease, or even death. The problems of food, from production to ultimate use, are the proper concern of all people.

NUTRITION

Nutrition refers to the processes by which living things ingest and assimilate food. Proper growth, maintenance, and functioning of the body are dependent on the choice, amounts, and combination of food materials eaten. The nutritional requirements of human beings are not perfectly known, partly because of the inevitable hazards involved in attempting to use people as test organisms. Many nutrients are required only in minute amounts, and some of these are present in almost all natural foods. Furthermore, precise quantitative determinations are difficult to make because bacteria residing in the intestinal tract may supply some nutrients (such as biotin, folic acid, and vitamin K). Consequently, we must obtain much of our nutritional information from work on other animals (particularly rats), and much more is known about the nutritional requirements of chickens or pigs than of people.

The nutritive value of food must be discussed in terms of chemical and energy units. (Food as an

Table 3–1 Essential human nutrients.

Established as essential	Probably essential	Established as essential	Probably essential
Inorganic		*Organic*	
Water		*Amino acids*	
		Isoleucine	Histidine[3]
Minerals		Leucine	Arginine[3]
Macroelements:		Lysine	
Calcium		Methionine	
Chlorine		Phenylalanine	
Magnesium		Threonine	
Phosphorus		Tryptophan	
Potassium		Valine	
Sodium			
Sulfur		*Fatty acids*	
Microelements:		Linoleic acid	Arachidonic acid
Chromium	Arsenic[1]		
Cobalt	Nickel[2]	*Vitamins*	
Copper	Silicon[2]	Ascorbic acid	Riboflavin
Fluorine	Tin[1]	Biotin	Thiamin
Iodine	Vanadium[2]	Choline[4]	Vitamin A
Iron		Folic acid	Vitamin B_{12}
Manganese		Niacin[5]	Vitamin D[3]
Molybdenum		Pantothenic acid	Vitamin E
Selenium		Pyridoxine	Vitamin K
Zinc			

[1]Possibly essential in animals.
[2]Essentiality not proved but can be assumed from animal studies.
[3]Indicated to be unnecessary in adults in short-term studies but probably necessary for normal growth of children. (Vitamin D required by pregnant women; requirement can be met in children by exposure to sunlight.)
[4]Requirement met under circumstances of adequate dietary methionine.
[5]Requirement may be provided by synthesis from dietary tryptophan.

energy source will be discussed in a later section.) The chemical substances required by human beings include inorganic materials (water and certain elemental minerals) and organic materials (amino acids, fatty acids, and accessory factors popularly grouped as vitamins). Those nutrients that cannot be synthesized from other constituents of the diet are considered **essential** (Table 3-1). A deficiency of any essential nutrient generally results in a specific physiological response (symptom) for each species (see Box 3-1); the most common first symptom in children is decreased growth. When all nutrients but one are present in adequate quantities, growth is usually proportional to the supply of the limiting nutrient. Essential nutrients are usually not consumed in pure form but as components of foods. Although our nutritional requirements can be listed rather specifically in terms of essential nutrients, nutritionists prefer to list them in terms of the broad classification of nutrients into carbohydrates, fats, proteins, vitamins, and minerals.

Box 3–1   *A glossary of some human nutritional diseases*

Beriberi: (From the Singhalese, meaning "I cannot," since persons with severe beriberi cannot move easily.) A condition caused by thiamin (vitamin B_1) deficiency. It is characterized by numbness or tingling of toes and feet, stiffness of ankles, cramping pain in legs, difficulty in walking, and, finally paralysis of legs with atrophy of leg muscles.

Dental caries: A local disease of the teeth (*caries* is derived from the Greek word for "rottenness") that involves the action of specific bacteria in the mouth and the presence of certain substrates (such as sugar) required for bacterial action. Fluorine appears to prevent decay by rendering the structure of tooth enamel more resistant to acid-producing bacteria.

Goiter: A chronic enlargement of the thyroid gland as a result of iodine deficiency.

Iron-deficiency anemia: a condition resulting from low stores of iron, caused by inadequate dietary intake, blood loss, or malabsorption of iron. It is characterized by a reduction of the oxygen-carrying capacity of the blood. Symptoms include paleness of skin, weakness, shortness of breath, lack of appetite, and a general slowing of vital body functions.

Kwashiorkor: A severe clinical condition, occurring most frequently in children 1 to 3 years of age, resulting from a deficiency of protein (and other nutrients) in the face of a relative excess of calories. It is characterized by growth failure, edema, and muscle wasting. There is often a preceding or associated infection, such as diarrhea, respiratory infection, or measles. Other frequently associated changes are reduced pigmentation and easy pluckability of the hair, liver enlargement, dermatosis, and apathy.

Marasmus: A condition occurring mostly in infants (3 to 18 months) as a result of a chronic gross deficiency of calories and an accompanying lack of protein and other nutrients. It is characterized by low body weight, loss of subcutaneous fat, and wasting of muscle tissue. The condition is frequently accompanied by diarrhea.

Obesity: Excessive fatness; or, more strictly defined, body weight in excess of 20 percent more than the ideal weight for a particular body type. It is caused by overeating and inactivity, usually in combination.

Pellagra: A clinical condition caused by niacin deficiency and characterized by tissue damage, chiefly of the skin, gastrointestinal tract, and nerves. The most striking symptom is a reddish skin rash, especially on the face, hands, and feet when exposed to sunlight. Other associated disorders include inflamed membranes in the digestive tract with bloody diarrhea and distressing nervous and mental disturbances (hence "the 3 Ds" of pellagra: dermatitis, diarrhea, and depression or dementia).

Pernicious anemia: A condition caused by insufficient absorption of vitamin B_{12} from the intestine, regardless of food intake, due to a lack of vitamin B_{12}-binding protein in the gut. It is characterized by anemia, degeneration of the spinal cord, and, if untreated, sore tongue, weakness, loss of weight, back pain, tingling of extremities, apathy, and mental and nervous abnormalities. A daily intramuscular injection of B_{12} will restore normal blood count and cause symptoms to disappear.

Rickets: A condition, seen most strikingly in young children, caused by vitamin D deficiency. It is characterized by poor growth and lack of normal development of bones (typically, bow legs) as a result of disturbed calcium and phosphorus metabolism.

Xerophthalmia: Extreme dryness of the eyes caused by a deficiency of vitamin A.

Nutrients

Carbohydrates

The carbohydrates comprise a large group of compounds composed of the elements carbon, hydrogen, and oxygen. Because there are usually two hydrogen atoms for every oxygen atom, they incorrectly appeared to early chemists to be "hydrates of carbon" ($C \cdot H_2O$), and thus the epithet. Carbohydrates make up the bulk of the dry weight of plants. Although there are many kinds, the main plant carbohydrates are sugars, starches, and the various celluloses. Animal products contain only small amounts, with the exception of milk, which contains lactose, a sugar.

These carbon-containing compounds exist in many different forms, each with a molecular architecture of its own. **Sugars** are simple carbohydrates that form more complex materials (polymers), such as starch, cellulose, inulin, and pectins, when they become linked in chainlike fashion. Sucrose, the common table sugar, is composed of two linked simple-sugar molecules: glucose and fructose. Linked chains of hundreds of glucose molecules make up **starch,** a storage form of carbohydrate found in fleshy tubers and roots, and the principal reserve carbohydrate of many plants. Two forms of starch are known: a straight, unbranched chain of glucose molecules linked to each other (amylose) and a branched chain (amylopectin).

Inulin, a polymer of fructose, is the principal reserve carbohydrate of the Jerusalem artichoke (*Helianthus tuberosus*). Inulin has interesting medical uses because, unlike the glucose-containing carbohydrates, it can be metabolized by diabetics.

Cellulose consists of interwoven and interconnected strands of a polymer of glucose. Cellulose is very insoluble and cannot be broken down into sugars by the human digestive system, but it can be broken down by microorganisms. Some plant-eating animals can utilize cellulose because of the fermentation action of microorganisms in their gastrointestinal tract. This is accomplished very efficiently within ruminant animals (cattle, sheep, goats) and those species with a large cecum, such as the horse and the rabbit.

Carbohydrate foods are commonly classified on the basis of digestibility into two components: **crude fiber,** the coarse portion that consists largely of cellulose and related substances, and **nitrogen-free extract,** principally starches and sugars. The crude fiber is not easily digested and yields a lesser proportion of its energy than does the nitrogen-free extract. Although crude fiber is of little direct nutritional use to people, it is now believed to perform important indirect functions, such as stimulating peristaltic action of the intestines. Crude fiber is the basis of many animal rations; the average animal ration is 75 percent carbohydrates.

Carbohydrates are not in themselves essential as a human food. Like other animals, we can be maintained on a carbohydrate-free diet by substituting lipids as an energy source, which the Eskimos do. But carbohydrates in the form of sugars and starches are an inexpensive source of food energy. About half of the calories in the diet of the average American are provided by carbohydrates. In other countries carbohydrates supply an even greater fraction of calories in the diet.

Fats

Fats and fatlike substances are grouped under the general term **lipids.** They are soluble in such fat solvents as ether and chloroform. Fats are esters of long-chain organic acids (known as fatty acids) and glycerol, an alcohol. Fats contain carbon, hydrogen, and oxygen, but, unlike carbohydrates, the proportion of oxygen is very low.

Fats occur in small amounts in all living cells. In plants they generally accumulate in seeds.

Crop	Percent crude fat in seed
Coconut	65
Peanut	40–50
Flax (linseed)	30–35
Soybean	15–20
Cotton seed	15–20
Maize	5

There is an inverse relation between the amount of carbohydrate and the amount of fat in seeds. The amount of fat in the seed may be altered by selection. Long-term selection in maize has produced lines with high (> 14 percent) and low (< 2 percent) oil contents (see Fig. 19-10). Some fruits, such as olives and avocados, are also high in fat.

The differences in fats are related to the kind and arrangement of their fatty acids. Fatty acid molecules differ in their length (usually from 4 to 26 carbons) and in the degree of hydrogen saturation—that is, in the number of double bonds between carbon atoms. The degree of saturation affects their stability and firmness. The distinction between oils and fats is that fats are solid at room temperature, whereas oils are liquid. Oils have large numbers of double bonds; that is, they do not carry all of the hydrogen possible. Plant oils may be changed to fats by the addition of hydrogen (hydrogenation).

Fats are insoluble in water and must be rendered soluble (digested) before they can be utilized by the body: digestion splits the fatty acids from glycerol. The position at which a fatty acid is attached to the glycerol affects digestibility, because a fatty acid attached at mid-position is removed last, if at all. Digestibility depends also on the length of the fatty acid molecule; the short-chain fatty acids are most easily digested.

Fats, the most concentrated energy source, have always been highly valued as food. Some of the unsaturated fatty acids are essential. The most important is linoleic acid, an 18-carbon unsaturated fatty acid with reactive double bonds at the ninth and twelfth positions. Fats aid the absorption of certain fat-soluble vitamins and are partly responsible for the feeling of satiety we experience after eating.

In the United States, fat contributes 40–50 percent of the calories in the average diet; in some less prosperous parts of the world, fat contributes only about 20 percent. It is not clear with whom the advantage lies, for in some individuals the level of cholesterol—a natural fatty material in the blood—increases to high levels with high intakes of saturated fats. High cholesterol levels have been associated with cardiovascular disorders, but the precise relation between fat intake, cholesterol levels, and cardiovascular disease are not at all clear. Much remains to be learned about fat metabolism and its relation to human nutrition and health.

Proteins

Protein (the term applied to proteins collectively) is a major constituent of living material. Next to water, protein is the most abundant substance in our bodies, accounting for about half the dry weight. Protein is the most characteristic constituent of foods of animal origin. On a dry-weight basis, muscle (beefsteak) is 47 percent protein, milk 27 percent, and eggs 50 percent. If cellulose is excluded (and lignin in woody plants), protein also makes up the greatest proportion of the dry weight of most plant cells.

Proteins are complex molecules made up of different combinations of amino acids. Amino acids, the basic unit of proteins, are synthesized in plants (and microorganisms) from carbohydrate fragments and from the nitrogen in ammonium ions (NH_4^+). Although proteins contain much carbon, hydrogen, and oxygen, nitrogen is the characteristic element. Since most proteins average 16 percent nitrogen, the protein content of foods may be estimated by multiplying the nitrogen content by a factor of 6.25. Food protein utilization by human beings is studied by measuring the nitrogen entering and leaving the body. Of the essential elements, sulfur is found in two of the amino acids, methionine and cysteine. Other essential elements may be found in protein in small quantities.

Each of the amino acids contains a carboxyl group (—COOH) and an amino group (— NH_2). Although there are many amino acids, only about twenty are found in protein. Amino acids differ in the number of carbon atoms and attached groups. The amino acids combine at their amino and carboxyl groups to form peptide linkages. A protein can be defined as a group of amino acids held together by peptide linkages. Proteins may be very large. The molecular weight of gliadin, one of the wheat proteins, is 27,500 and that of glutenin, another wheat protein, is in the millions.

There are many kinds of proteins. In plants, proteins are commonly classified on the basis of their origin. Seed proteins are the primary reserve forms. The proteins found in the endosperm of wheat (commonly referred to as gluten) are responsible for the characteristic elastic property of dough. Proteins in plant tissue occur primarily in the cytoplasm and chloroplasts. Nucleoproteins, proteins that complex with nucleic acids, are a major constituent of chromosomes. Enzymes, the major regulatory materials of living organisms, are protein.

Animals require amino acids rather than protein *per se*, and since they cannot synthesize all amino acids, they ultimately depend on plants or microorganisms for those they cannot synthesize. Cattle, sheep, and goats can obtain their amino acids from bacteria that reside in their rumen, and in this way can exist on a diet of purified carbohydrates and simple nitrogen compounds. It is thus possible to produce milk from the cellulose in corn cobs plus urea and ammonium salts as a source of nitrogen. Proteins differ in the kind and proportion of amino acids and differ in nutritive value. High-quality proteins are those that supply a proper balance of essential amino acids. In order for them to be utilized by the body to build proteins, the essential amino acids must be provided together. Rats fed the necessary amino acids in sequence, three hours apart each day, fail to grow.

Proteins are digested when specific enzymes in the gastrointestinal tract break the peptide linkages between the amino acids. The amino acids are later reassembled in the tissues to form new proteins required by the particular species. Amino acids left over are not stored but are stripped of their amino groups by the liver. The remainder of the molecule is used to provide energy or is converted to fat and stored.

Many animal proteins have a greater biological value as human food than proteins from single plant sources (Fig. 3-1). This is because the assortment and proportion of amino acids in animal proteins are similar to those required by the human body. The amino acids that occur in proteins are shown in Figure 3-2. Cereal grains are very low in lysine and moderately low in methionine and tryptophan, whereas proteins from legumes are low in the sulfur-containing amino acids cysteine and methionine. Plant proteins also tend to be less digestible than animal proteins. Although proteins are generally found in lower concentrations in plants than in animals, adequate protein levels can be supplied by the skillful complementation of different plant protein sources. Further, it has been found that the amino acid composition of plant proteins may be improved. The *opaque*–2 gene of maize greatly increases the biological value of corn protein by increasing its lysine and tryptophan content (Fig. 3-3).

Protein is in short supply in the world. It is the most prized and the most expensive of the major nutrients. The world food problem is largely one of insufficient protein. Insufficient protein in the diet of children leads to the deficiency syndrome called **kwashiorkor** ("displaced child"), from its occurrence in Central African infants deprived of their mothers' milk (Box 3-1; Fig. 3-4).

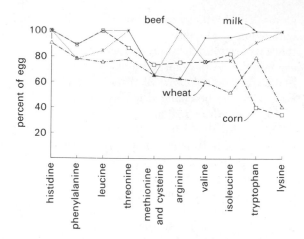

Figure 3-1 Evaluation of various kinds of protein, as compared with egg protein. Each value is the percentage by which the amino acid content of a protein departs from that of whole egg protein, with excesses taken as 100 percent. [After E. W. Crampton and L. E. Harris, *Applied Animal Nutrition*, 2nd ed., W. H. Freeman and Company, copyright © 1968.]

Vitamins

The relationship between diet and disease is very old. That eating liver could cure night blindness was recorded by Hippocrates 24 centuries ago. The use of lime juice by British sailors to prevent scurvy (hence the term "limies") has been traced to a suggestion by James Lind in 1757 that eating fresh fruits and vegetables might help prevent the disease. Early nutritionists found that animals fed on rations of purified forms of known foods sickened and died, yet thrived when small amounts of certain substances, such as water-soluble extracts from yeast or wheat germ, were added.

name	formula	name	formula
alanine	$CH_3-CH-COOH$ $\quad\quad NH_2$	lysine	$H_2N-(CH_2)_4-CH-COOH$ $\quad\quad\quad\quad\quad NH_2$
arginine	$NH_2-C-NH-(CH_2)_3-CH-COOH$ $\quad\quad\parallel\quad\quad\quad\quad\quad NH_2$ $\quad\quad NH$	methionine	$CH_3-S-(CH_2)_2-CH-COOH$ $\quad\quad\quad\quad\quad NH_2$
asparagine	$H_2N-C-CH_2-CH-COOH$ $\quad\quad\parallel\quad\quad\quad NH_2$ $\quad\quad O$	phenylalanine	(phenyl ring)$-CH_2-CH-COOH$ $\quad\quad\quad\quad\quad NH_2$
aspartic acid	$HOOC-CH_2-CH-COOH$ $\quad\quad\quad\quad NH_2$	proline	(pyrrole ring with N–H)$-COOH$
cysteine	$HS-CH_2-CH-COOH$ $\quad\quad\quad NH_2$	serine	$CH_2-CH-COOH$ $OH\quad NH_2$
glutamic acid	$HOOC-(CH_2)_2-CH-COOH$ $\quad\quad\quad\quad\quad NH_2$	threonine	$CH_3-CH-CH-COOH$ $\quad\quad OH\quad NH_2$
glutamine	$H_2N-C-(CH_2)_2-CH-COOH$ $\quad\quad\parallel\quad\quad\quad\quad NH_2$ $\quad\quad O$	tryptophan	(indole ring with N–H)$-CH_2-CH-COOH$ $\quad\quad\quad\quad\quad\quad NH_2$
glycine	CH_2-COOH $\quad NH_2$	tyrosine	$HO-$(phenyl ring)$-CH_2-CH-COOH$ $\quad\quad\quad\quad\quad\quad NH_2$
histidine	(imidazole ring, N, NH)$-CH_2-CH-COOH$ $\quad\quad\quad\quad\quad\quad NH_2$	valine	$CH_3-CH-CH-COOH$ $\quad\quad CH_3\quad NH_2$
isoleucine	$CH_3-CH_2-CH-CH-COOH$ $\quad\quad\quad\quad CH_3\quad NH_2$		
leucine	$CH_3-CH-CH_2-CH-COOH$ $\quad\quad CH_3\quad\quad\quad NH_2$		

Figure 3-2 Structure of the twenty amino acids that occur in proteins. Isoleucine, leucine, lysine, methionine, phenylalanine, threonine, tryptophan, and valine are essential to humans; arginine and histidine are required by children but probably not by adults.

In 1912, Frederick G. Hopkins and Casimir Funk suggested that specific human diseases, such as beri-beri, rickets, and scurvy, were caused by the absence of certain nutritional substances in the diet. These were termed vitamines ("vital amines"), because the first such substance isolated, thiamin (vitamin B_1), was an amine (a compound containing an amino group, ($-NH_2$). When other such essential substances were isolated and analyzed, they proved not to be amines, but the term **vitamin** was retained to refer to any essential growth factor (other than a mineral) required in very small amounts. Many essential vitamins have been discovered since that time. Although the functions of some vitamins are unknown, many have been shown to be coenzymes, which are required for the function of particular enzymes. The water-soluble B vitamins tend to play a part in energy transfer.

Figure 3-3 (A) Kernels on an ear of maize, showing *opaque-2* mutant (*center*) and normal kernels. The *opaque-2* endosperm contains 69 percent more lysine than the normal endosperm. Cross sections through a normal kernel (B) and an *opaque-2* kernel (C) show a reduced horny endosperm associated with the *opaque-2* gene. The superior feed value of *opaque-2* corn (D) is due to its higher protein quality. Because lysine is the limiting amino acid in corn, the *opaque-2* mutant almost doubles the feed value of corn for monogastric animals, such as rats, pigs, and people. [Courtesy Purdue University.]

Figure 3-4 Symptoms of kwashiorkor (protein deficiency) in African children. [Courtesy E. Wilson.]

Letters of the alphabet were first used to describe the mysterious nutritional factors. These letters (A, B, C, D, E, K, and others) have persisted. It was found, however, that some factors actually consisted of more than one substance. The original B factor has been shown to consist of more than a dozen entities. These factors are now designated as specific substances: for example, thiamin (B_1); riboflavin (B_2); pantothenic acid (B_3); three related substances, pyridoxine, pyridoxal, and pyridoxamine (B_6); and niacin (nicotinic acid). Because these commonly occur together, they are referred to as the B-complex vitamins.

Vitamins are classified as fat soluble (A, D, E, and K) or water soluble (B complex and C), and are often described in terms of their deficiency effects. For example, a deficiency of vitamin C results in loose teeth, bleeding gums, a tendency to bruise easily, and extreme weakness; the syndrome is called scurvy.

Nutritional disorders due to vitamin deficiencies are often a result of ignorance as well as economic conditions. For example, vitamin-A deficiency is common in tropical regions even though the vitamin is readily available in native plants that could be, but are not, utilized as food because of dietary preferences. The deficiency may be most marked when food is plentiful, and may actually decrease during severe shortages, when people are forced to consume "unacceptable" but vitamin-A-rich food.

Minerals

Ninety-six percent of the weight of the human body is made up of four elements: oxygen, carbon, hydrogen, and nitrogen. The remainder consists of other essential elements. On the basis of the amounts required, these essential elements (commonly referred to as **minerals**) can be divided into **macroelements** (calcium, phosphorus, potassium, sulfur, sodium, chlorine, magnesium) and **microelements,** or **trace elements** (iron, manganese, copper, iodine, chromium, cobalt, fluorine, molybdenum, selenium, zinc, and perhaps others; see Table 3-1). Of the macroelements, calcium and phosphorus are required in greatest amounts. They make up 70 percent of the total mineral content of the human body and are found mostly in the bones and teeth. Foods vary widely in their content of these elements. Milk, an especially good source, is therefore a very valuable food for children, whose need for calcium and phosphorus is especially great. Inorganic forms of calcium and phosphorus may be added to the rations of animals. Human deficiencies of the remaining macroelements are rare because the mixed diets of most people provide these nutrients in fair quantities.

Of the microelements, iodine is the one most commonly lacking in human diets. Deficiency symptoms, which include an enlarged thyroid gland (goiter), are common in geographic areas in which the diet is low in iodine. Supplementary iodine is usually provided in an inorganic form, as potassium or sodium iodide, and added to common table salt. In general, trace-element deficiencies are more common in livestock and poultry than they are in human beings.

Food as an energy source

Food must supply the fuel for the living machine. The energy requirement depends on the size of the machine and the amount of work performed. The potential chemical energy in a given food is determined by burning it and measuring the amount of heat released. This is done in a device called a "bomb calorimeter." The kilocalorie, commonly known as the Calorie (with a capital C), is the unit used for measuring the energy produced by food oxidized in the body.

For a given quantity of food the number of kilocalories actually utilized by the body is somewhat less than that measured by the bomb-calorimeter method, because some food may not be digested, some may be stored in the body in the form of tissue, and some components (such as protein) are not completely oxidized in the body.

Fats and carbohydrates are metabolized in the body to produce carbon dioxide, water, and heat. Protein, however, leaves residues (such as urea), which are eliminated in the urine.

Nutrient	Kilocalories per gram	Kilocalories utilized by the body
Carbohydrates	4.1	4
Fats	9.5	9
Proteins	9.7	4

Basal metabolism refers to the amount of energy expended (while awake and at rest) of all involuntary vital processes (respiratory mechanisms, the production of body heat, and so on) under uniform environmental conditions. Although basal metabolism varies with age and sex, it tends to be the same for individuals of similar size, differing only by about 5–10 percent. Extreme changes in basal metabolism are then an indication of some pathological metabolic disturbance, such as an overactive or underactive thyroid gland.

The total energy requirement above that required for basal metabolism depends on muscular activity and adjustments to the environmental temperature. Muscular activity requires energy. In addition, the body must eliminate the heat produced from the combustion of food. In an environment at the comfortable temperature of 21°C (70°F), about three-quarters of the heat produced by the human body is eliminated by radiation and conduction and one-quarter by the evaporation of perspiration. More energy is required in cold climates in order to maintain body temperature. During cold weather we can warm ourselves by exercising. Heat can also be produced involuntarily, as it is during the involuntary reflex we call shivering.

Food-energy requirements depend on activity. A man who weighs 66 kilograms (about 145 pounds) and leads a sedentary existence requires approximately 2150–2350 kilocalories per day, but a man of the same weight who is extremely active requires as much as twice that amount. The human requirements for food energy and nutrients are summarized in Table 3-2.

Energy requirements take first call on either food or reserves. Excess energy expenditure over receipts takes place at the expense of stored fat or even body tissues. The short-term response is a loss of weight. The long-term response to a deficit of energy foods leads to reduced vitality, physical exhaustion, and ultimately death. Our need for energy is the basis of the hunger urge. Appetite in most people serves to regulate the balance of energy intake with energy output. When intake exceeds output the body stores energy in the form of fat.

Diet

Nutrients and energy-rich substances may be consumed in many forms. Most of us, however, have a pattern of eating that includes only certain selected foods. Food consumption patterns are a result of many factors—economic, technological, geographical, and cultural. Our diet is inextricably bound up in our pattern of living.

Although sustenance may be derived from a seemingly limitless supply of other life forms, relatively few plants and animals are actually used to satisfy human nutritional requirements. Nevertheless, there are distinct differences in diets. These differences portend important consequences in the quality of life for humankind and for the individual. Food, unlike the air we breathe, is obtained only with a great deal of effort. In many areas of the world food is expensive, and a high percentage of the world's people trade the greater proportion of their productive labor to purchase it. Food, an absolute necessity, comes first.

From a nutritional standpoint, the important components of diet reduce to an accounting of calories, proteins, and vitamins. Dietary differences reflect the way these needs are satisfied. Ultimately, the most important factors that determine the adequacy of a food supply are the availability of suitable land per person and the level of agricultural technology.

The energy food that humankind requires has been made plentiful through the development of agriculture. Certain plants are capable of high energy yield per acre in convenient edible form, including the grains (especially wheat, rice, and corn), and potato, cassava, and sugarcane. All great civilizations have been able to cultivate at least one of these im-

portant energy producers in abundance. These foods are versatile and are readily transformed to such products as bread, spaghetti, tortillas, and rice cakes.

Our demand for protein can also be satisfied in a great many ways. Although the choice is seemingly dependent on many factors, there is one underlying economic principle: the further along the food chain from plants, the higher is the cost of the food. Only where there is a land surplus, and sufficient technology to exploit it, can people afford the luxury of converting grain into meat.

The use of animal sources of protein has always been very attractive. There is a firm biological basis for this because animal proteins are complete; that is, they contain all eight or nine essential amino acids in amounts fairly close to our needs. The beneficial qualities of meat also include minerals and vitamins. This is reflected in the widespread human preference for meat, especially in combination with fat. In this combination, the protein and energy are supplied together. A few cultures exist entirely on animal protein and fat; for example, the Eskimo's diet consists chiefly of fish and blubber, and the staples in the diet of the Masai in Africa are the blood and milk of cattle.

Animals, however, are inefficient converters of feed to meat. Only 3 percent of the total energy in plants is converted to potential human food. Not all of the plant is consumed by animals; of the part that is consumed, some is "walked off," or wasted. Moreover, much of the animal is not consumed. As a result, when meat is expensive, people have shifted to eating that plant food formerly used as feed. In highly developed societies the convertibility of feed into meat proteins provides an ever-present variety of foods, one's choice being governed by price. If meat is scarce, high prices merely mean a shift to plant protein. Where such an abundance of food is available to feed animals, there is seldom any real threat of famine.

In parts of the world in which the pressure of human population on land resources is great, animal protein satisfies only a small fraction of the human requirements: the plant food produced is virtually all consumed by people. When shortages occur there is no other immediately available supply of food. Un-

Table 3-2 Example of a human dietary standard—Recommended Daily Dietary Allowances.

		Weight		Height						Water-soluble vitamins					
	Age (years)	(kg)	(lb)	(cm)	(in)	Energy (kcal)	Protein (g)	Ascorbic acid (mg)	Folacin[1] (μg)	Niacin[2] (mg)	Riboflavin (mg)	Thiamin (mg)	Vitamin B$_6$ (mg)	Vitamin B$_{12}$ (μg)
Infants	0.0–0.5	6	14	60	24	kg × 117	kg × 2.2	35	50	5	0.4	0.3	0.3	0.3
	0.5–1.0	9	20	71	28	kg × 108	kg × 2.0	35	50	8	0.6	0.5	0.4	0.3
Children	1–3	13	28	86	34	1300	23	40	100	9	0.8	0.7	0.6	1.0
	4–6	20	44	110	44	1800	30	40	200	12	1.1	0.9	0.9	1.5
	7–10	30	66	135	54	2400	36	40	300	16	1.2	1.2	1.2	2.0
Males	11–14	44	97	158	63	2800	44	45	400	18	1.5	1.4	1.6	3.0
	15–18	61	134	172	69	3000	54	45	400	20	1.8	1.5	2.0	3.0
	19–22	67	147	172	69	3000	54	45	400	20	1.8	1.5	2.0	3.0
	23–50	70	154	172	69	2700	56	45	400	18	1.6	1.4	2.0	3.0
	51+	70	154	172	69	2400	56	45	400	16	1.5	1.2	2.0	3.0
Females	11–14	44	97	155	62	2400	44	45	400	16	1.3	1.2	1.6	3.0
	15–18	54	119	162	65	2100	48	45	400	14	1.4	1.1	2.0	3.0
	19–22	58	128	162	65	2100	46	45	400	14	1.4	1.1	2.0	3.0
	23–50	58	128	162	65	2000	46	45	400	13	1.2	1.0	2.0	3.0
	51+	58	128	162	65	1800	46	45	400	12	1.1	1.0	2.0	3.0
Pregnant						+300	+30	60	800	+2	+0.3	+0.3	2.5	4.0
Lactating						+500	+20	80	600	+4	+0.5	+0.3	2.5	4.0

Source: Reproduced with permission of the National Academy of Sciences from NAS–NRC (U.S.), *Recommended Dietary Allowances* (9th ed.), National Academy of Sciences–National Research Council, Washington, D.C. (1980).

Note: The allowances are intended to provide for individual variations among most normal persons as they live in the United States under usual environmental stresses. Diets should be based on a variety of common foods in order to provide other nutrients for which human requirements have been less well defined.

[1]The folacin allowances refer to dietary sources as determined by *Lactobacillus casei* assay. Pure forms of folacin may be effective in doses less than one-fourth of the recommended dietary allowance.

der such circumstances the specter of famine and starvation can be expected to recur.

The special value ascribed to animal protein is reflected in the many cultural attitudes about the consumption of animal flesh, in contrast with the few about the consumption of plants. This appears to be manifested in religious prohibitions of various types. For example, Catholics still observe a number of meatless days as penance. Most of the complex dietary laws in Judaism govern the eating of animal products, specifying the species and parts of the animal that may be eaten and the kinds of food that may be combined (mixing of milk and meat is prohibited). Devout Moslems and Jews do not eat pork.

Hindus do not eat cattle, for they regard them as sacred, and some sects abstain from meat in any form. All of us scoff at the "taboos" of others, but strongly defend our own "preferences." In the United States many will not eat rabbit, and most draw the line at horsemeat or squirrel.

FOOD STORAGE AND PROCESSING

Although the demand for food is continuous, its production is not. The growth of many crops conforms to a seasonal cycle and the harvest becomes available

	Age (years)	Fat-soluble vitamins				Minerals					
		Vitamin A activity (RE)[3]	(IU)	Vitamin D (IU)	Vitamin E activity (IU)[5]	Calcium (mg)	Phosphorus (mg)	Iodine (μg)	Iron (mg)	Magnesium (mg)	Zinc (mg)
Infants	0.0–0.5	120[4]	1400	400	4	360	240	35	10	60	3
	0.5–1.0	400	2000	400	5	540	400	45	15	70	5
Children	1–3	400	2000	400	7	800	800	60	15	150	10
	4–6	500	2500	400	9	800	800	80	10	200	10
	7–10	700	3300	400	10	800	800	110	10	250	10
Males	11–14	1000	5000	400	12	1200	1200	130	18	350	15
	15–18	1000	5000	400	15	1200	1200	150	18	400	15
	19–22	1000	5000	400	15	800	800	140	10	350	15
	23–50	1000	5000		15	800	800	130	10	350	15
	51+	1000	5000		15	800	800	110	10	350	15
Females	11–14	800	4000	400	12	1200	1200	115	18	300	15
	15–18	800	4000	400	12	1200	1200	115	18	300	15
	19–22	800	4000	400	12	800	800	100	18	300	15
	23–50	800	4000		12	800	800	100	18	300	15
	51+	800	4000		12	800	800	80	10	300	15
Pregnant		1000	5000	400	15	1200	1200	125	18+[6]	450	20
Lactating		1200	6000	400	15	1200	1200	150	18	450	25

[2]Although allowances are expressed as niacin, it is recognized that on the average 1 mg of niacin is derived from each 60 mg of dietary tryptophan.
[3]Retinol equivalents.
[4]Assumed to be all as retinol in milk during the first six months of life. All subsequent intakes are assumed to be half as retinol and half as β-carotene when calculated from international units. As retinol equivalents, three-fourths are as retinol and one-fourth as β-carotene.
[5]Total vitamin E activity, estimated to be 80 percent as α-tocopherol and 20 percent other tocopherols.
[6]This increased requirement cannot be met by ordinary diets; therefore, the use of supplemental iron is recommended.

from particular regions only once each year. In practically all areas there is a period during which crops will not grow—winter in temperate regions and periods of drought or excessive moisture in the tropics. In order to insure a steady supply without recourse to nomadism, it is essential that food be preserved.

In order to extend the supply of harvested foodstuff a number of destructive forces must be circumvented. These include natural physiological deterioration, decay by microorganisms, and losses to pests, such as rodents and insects. The methods of food preservation are many, and the choice of method depends on the product, its use, the destructive agen-

cies, dietary preferences, and the length of time the product must be preserved.

Acceptable quality of some products is a transient and ephemeral thing. The cells of a fresh apple are actually still alive when it is eaten; thus the preservation of some foods involves extending the living state. For the long-term storage of most food crops, however, life must be terminated and microbial action prevented. Edibility is maintained even though the processes involved materially alter the food from its original condition. Often this results in a new product with utility beyond its preservation function—for example, milk can be made into cheese, tomatoes into ketchup.

Storage of perishables

Plant products, such as fresh fruits and vegetables, and the great bulk of animal products—meat, milk, fish—have a usable period measured in days. Living plant products deteriorate largely through respiration and excessive loss of water. Animal food products are subject to rapid microbial decay. Short-term storage without extensive changes in form by processing may be accomplished by controlling the environment—specifically, by controlling temperature, humidity, and oxygen level.

Temperature is the critical factor in storage. It affects the respiration rate of living commodities and the growth rate of the microbial population. The lower limit must be determined for each commodity. Storage temperature is usually kept above the point at which the commodity freezes. For example, apples store best at about −1° to 0°C (30° to 32°F). Some tropical fruits, however, are injured at temperatures below 10°C (50°F). The evolution of heat by respiration in living materials must be considered in determining the proper storage temperature. Animal carcasses and such products as milk are originally at high temperatures—37° to 40°C (98° to 104°F)—and must be rapidly cooled. Year-round consumption of fresh fruits and vegetables and animal products requires a high level of technology, which is manifested in today's refrigerated transport and storage equipment.

Humidity affects the perishability of many plant products. In general, low humidity results in desiccation and wilting. High humidity, however, favors the development of decay, especially if temperatures are too high.

In addition to being temperature-dependent, respiration is also directly affected by the amounts of oxygen and carbon dioxide in the storage atmosphere. Reducing the amount of oxygen slows the respiration rate. Controlling the amount of oxygen in order to retard respiration is known as **modified atmosphere storage.** In a gas-tight storage space filled with freshly harvested living produce, oxygen is consumed and equal concentrations of carbon dioxide released. By controlling the introduction of air, oxygen levels may be kept low enough to reduce respira-

tion and yet not low enough to permit anaerobic respiration, which results in the formation of alcohol. This has been accomplished quite simply with the use of polyethylene film liners, which are differentially permeable to carbon dioxide and oxygen. Oxygen may be removed rapidly by flushing the storage space with nitrogen. The combination of low temperature and modified atmosphere makes it possible to keep fresh fruit for periods of months.

Food processing

Relatively long-term preservation of food may be achieved by physical and chemical processes that sterilize the food or render it incapable of supporting the growth of microorganisms. These processes include drying, canning, freezing, the use of preservatives, and irradiation. Food processing has become one of the major agricultural industries.

Drying

Drying is the most ancient method of food preservation. Removing water from plant or animal tissues results in a highly concentrated material of enduring quality. The natural deterioration of the product by respiration is stopped because of enzyme inactivation, and the lack of free water protects the dried product from decay by microorganisms. Drying may be achieved naturally (**sun drying**) or artificially (**dehydration**), by passing warm air over the product. Freeze-dehydration, the drying of quick-frozen food in a high vacuum, holds promise of increased usefulness because the quality of the reconstituted product is much higher than that achieved by ordinary dehydration.

Drying is the natural means of preservation for grains. They are harvested when nearly dry, and consequently can be stored for extended periods of time. Drying is also useful for many other products; much fruit is preserved by drying, including grapes, plums, apricots, dates, figs, bananas, peaches, apples, and pears. Meat and fish can also be preserved in this manner. In tropical areas sun-dried meat (**charqui** or "jerked" meat) is widely used. For these products salt may be used in combination with the drying process.

Dried, concentrated, processed foods such as flour, protein supplements, powdered milk, and powdered eggs can be stored for long periods if protected from rodents and insects. The storage life of dried materials is extended at cool temperatures. High humidities encourage the growth of molds.

Drying, the most ancient of processing methods, appears to be the technique of the future. It has many advantages: by removing water it reduces weight and, in some foods, bulk. Food preservatives are not required for the process, and packaging problems are minimal. Storage life is long, and nutrients are virtually unaffected. Loss of quality, the limiting factor in dried foods, is being overcome by advances in technology.

Canning

Canning consists of heat-sterilizing food and sealing it in an airtight container. The process destroys both the pathogenic and the food-spoilage microorganisms and inactivates the enzymes that would otherwise cause decomposition during storage. The sealed container prevents reinfection after the food has been sterilized and prevents gaseous exchange.

The application of sufficient heat to sterilize food and inactivate enzymes results in alteration of color, flavor, texture, and nutritive value. Although the transformations are similar to those that occur in the normal cooking process, the treatment must be more extreme for many foods. Consequently, loss of quality is one of the limiting factors of the canning process. Low-acid foods require relatively severe heat treatment, since they can support growth of *Clostridium botulinum*, a bacterium that causes severe, often fatal, food poisoning: a millionth of a gram of the toxin produced by this organism will cause death. Thus, all foods capable of sustaining growth of this organism are processed on the assumption that the organism is present.

Although highly efficient systems have made the cost of canning reasonable in highly developed economies, canned foods are still out of the reach of a large part of the world's population. In addition, the finished product is bulky and heavy, and, unless proper precautions are taken, not suitable for extended storage under all conditions. At storage temperatures above 49°C (120°F) certain heat-loving bacteria not ordinarily killed by the sterilization process will continue to grow. In humid regions, and especially in coastal areas where salt concentrations are high, storage life is limited by corrosion of the metal containers. Nevertheless, in developed countries, canning is still the most important processing method, largely because of convenience.

Freezing

Freezing protects food from spoiling because most microorganisms cannot grow at temperatures below 0°C (32°F). The freezing process stops most enzymatic activity and is not in itself destructive to nutrients. Some products, however, require heat treatment before freezing, in order to inactivate enzymes that affect flavor and color during storage.

The basic principle in freezing food is the speedy removal of heat. This is achieved by cold air blasts, direct immersion in a cooling medium, contact with refrigerated plates, or liquid air, nitrogen, or carbon dioxide. Freezing in still air is the slowest method. The rate of freezing has a large effect on quality. In slow freezing relatively large ice crystals develop and fracture the tissue cells, resulting in the loss of cellular fluid and an undesirable soft texture when the food is thawed. If freezing occurs rapidly, the crystals formed are about one-hundredth the size of those formed in slow freezing. Because these crystals are tightly packed, fewer cells are ruptured. In quick freezing the refrigerant is kept at temperatures of −29° to −40°C (−20° to −40°F), and the water in food is completely frozen in thirty minutes or less.

The success of freezing as a method of preservation depends on the continuous application of the process. Thus the method is only suitable in places where people have their own refrigeration equipment. Because of the high quality of frozen foods compared to canned foods, freezing has taken over the market for many products. For example, frozen concentrate juice now accounts for the greatest use of the orange crop in the United States.

Preservatives

There are many materials that extend the usable life of foods, including salt, sugar, vinegar, spices, and a number of synthetic chemicals.

Common salt, because of its action in controlling bacterial growth, is widely used for preservation, especially for fish, meat, and meat products. "Curing" meats with salt is an ancient art—the term "corned" is derived from an old word for coarsely granulated salt. In addition, nitrate and nitrite salts, and sugar and other materials, may be added for color and flavor. Recently it has been found that nitrites in the presence of amines (NH_2R, where R is an organic radical such as — CH_3), when exposed to high temperatures such as occur during frying, give rise to toxic compounds called nitrosamines. Some of these are potent carcinogens. As a result, the levels of nitrites permissible in foods have been greatly reduced.

Smoking, often used in combination with curing for adding flavor, also acts as a meat preservative. Sugar, like salt, preserves foods by reducing the availability of free water. Acid fruits concentrated to at least 69 percent solids are commonly preserved with sugar in the form of jelly (concentrated juice), jam (concentrated fruit pieces) or preserves (concentrated whole fruit), butter (a semisolid product), and glacéed, or candied, fruits.

Many preservative materials are obtained naturally through **fermentation,** the anaerobic microbial decomposition of carbohydrates. Fermentation may be accomplished by a number of different organisms. Among the variety of end products are carbon dioxide and water (complete oxidation), acids (partial oxidation), and alcohols (alcoholic fermentation). The buildup of some of these products to sufficient concentrations creates unfavorable conditions for microorganisms, including the ones responsible for the fermentation. In addition, fermentation may impart certain flavors and secondary effects that are regarded as desirable. Fermentation may be controlled by conditions that favor the growth of one type of organism through the regulation of pH, oxygen, temperature, and the use of salt (pickling refers to fermentation in combination with salting) and other curing materials.

Fermentation plays an important role in processing many kinds of food. Cheese is a fermentation product; fermented vegetables make highly prized relishes. The fermentation of grain or fruits yields the alcoholic beverages: beer, wine, and spirits. Ensilage, fermented fodder, is an important animal feed.

A number of chemicals are used to retard deterioration, usually in conjunction with other methods of preservation. These include inorganic agents, such as sulfur dioxide and chlorine, and organic agents, such as benzoic acid, certain fatty acids, and sorbic acid. Ethylene, propylene oxide, and various antibiotics have been used experimentally, but they have very little chance of gaining approval because their reaction products may be toxic, or in the case of antibiotics, there is a fear that growth of antibiotic-resistant strains of microorganisms will be promoted. Chemical preservatives and other food additives have a legitimate place in food processing. They are unacceptable, however, if they are used to deceive or if there is any associated health hazard.

Ionizing radiation

Radiation has, at present, a limited use in preserving some foodstuffs. Gamma radiation extends the storage life of perishables by inhibiting growth processes, such as sprouting in potatoes. It is also used to decontaminate food by killing insects or parasites. There is evidence that radiation changes the flavor, texture, appearance, and nutritive value of food, as do other forms of processing, such as heat treatment. Although there is no direct evidence of toxic effects, the radiation of foodstuffs for preservation purposes must be considered as only experimental. Food preservation by irradiation has not lived up to early expectations.

PROVIDING FOR WORLD FOOD NEEDS

A large proportion of the world population—estimates run as high as half—is underfed or malnourished (Fig. 3-5). More distressing, the situation is de-

high energy
high protein

high energy
minimum protein

low energy
minimum protein

low energy
low protein

Figure 3-5 Nutritional status of the world's population. The land areas are distorted to indicate the numbers of people in each area. Note that the malnourished constitute more than half the world's population. [From H. W. Mattson, *International Science and Technology*, December 1965.]

teriorating as it is inextricably bound up with the "population explosion" (see Chapter 4). Although feeding the present world population of 4.5 billion people seems difficult (but conceivable, even to the unduly pessimistic), the problem of nourishing a population of 6 billion (to be reached by the year 2000, according to most projections) is staggering. Because of the present age distribution and other factors, the world's population will surely continue to rise in the next several decades. At best, only the rate of increase can be slowed. The population must at some point decline, however, either because of its impossible demands on resources or because we will have found some way to control our numbers—by regulating births or regulating deaths! The solution to better nutrition for the world's inhabitants must therefore involve both population control and increased food supply. The upshot of failure will be incomprehensible misery.

The suggestions for increasing world food production may be simply summarized: more land, new foods, greater efficiency. The simplicity of the suggestions, however, obscures the serious problems that

each entails. The potentials and limitations of each must be carefully considered.

Developing productive lands

A map of the world's population is in large part a map of the world's productive areas. People have already filled to overflowing most of those areas easily capable of supporting a large population. This is not to imply that all the usable land is gone, but the large areas of land presently unfarmed are generally unsuited for agriculture unless modern technology can be employed on a massive scale to remedy their deficiencies. Among these deficiencies are low soil fertility, excess or insufficient rainfall, and unsuitable terrain and climate. Making such lands agriculturally productive is very expensive and in many places economically prohibitive. For example, the recent cost of opening new lands in the western United States was about one or two thousand dollars per acre, much more than the value of the crops for many years to come. If production is the main concern, these capital resources could, in general, have been used more efficiently to increase the yield from land already cultivated. Furthermore, because the unused lands of the world are largely outside of populated areas, the problems of distribution would be very great. The exploitation of new lands (and new seas?) does not seem to provide the key to the world's present food problem.

New foods

The suggestion that some new food will appear to alleviate human problems is a recurring and wistful theme—from ambrosia to heavenly manna. In contrast to the change taking place everywhere around us, little or no change seems to be taking place in our diet; our present foods have been with us interminably long. Many people are quick to point out that more efficient food sources are available and need only be exploited (not for those of us who are quite content with our present fare, but for others). Their arguments appear persuasive at first hearing. The culture of microorganisms, including such aquatic algae as *Chlorella*, or the extraction of protein from forages, or the fermentation of petroleum fractions all seem to be more efficient—at least on theoretical grounds—than our present food sources. The culture of microorganisms requires little space, is independent of climate, and takes place extremely rapidly. The use of fish proteins to enrich less nourishing local products such as flour and meal, instead of being "wasted" on animal feed supplements—or worse, on fertilizers—seems reasonable. The substance 1,3-butanediol, a cheap industrial chemical derived from natural gas, can be produced in very large quantities, is completely digestible, and is a good energy source.

Unfortunately, because of shortages of time and capital, such processes do not seem to be ready solutions to the world's immediate requirements. There are many special problems inherent in the adaptation of these advanced technologies, in addition to those of acceptance. The enthusiasm that developed to grow bacteria on petroleum fractions to produce the "single-cell protein" resulted in the construction of pilot plants, but the dream faded when crude-oil prices spiraled upward in 1973. Further, single-cell protein is high in nucleic acids, which present special problems in many diets. The process of extracting protein from leaves or algae requires much energy, and the resulting protein is expensive. Fish protein has a special problem: the U.S. Food and Drug Administration considered it "contaminated," which prevented its use in the United States and made it politically unacceptable to encourage its use in other countries. The substance 1,3-butanediol, though edible, has an extremely objectionable odor. Finally the fear of toxic substances necessitates extensive and expensive testing procedures, and this has dampened enthusiasm for all new food products.

Even if the processes were entirely practicable, the acceptance of the new foods would remain a serious problem. New foods more readily find acceptance in highly developed countries, where they are least needed. In the United States, for example, soybean products are being substituted for a whole spectrum of dairy products (butter, whipped cream, ice cream), and soy protein is beginning to find a place in the market as a binder and extender for meats. Synthetic chicken and ham are being marketed. Vitamin enrichment of such foods as bread, milk, breakfast foods, and fruit drinks is commonplace and often legally required. New foods typically become acceptable when they mimic highly acceptable familiar products. The cost of educating populations even about *acceptable* new products is difficult at best, as any of our large food-processing firms will verify. It is a certainty that any preparation of algae (or bacteria, or yeast) will not be consumed as some green, bitter slurry, but rather to fortify familiar products, very likely different ones in different parts of the world, or as livestock feed. And if the new foods come from outside of the food-deficit area, the related costs of packaging, distribution, and development would probably make them too expensive in the area where they are needed most.

Improving agricultural practices

Remarkable increases in productivity have been obtained merely by applying new techniques to present-day agricultural systems and using land presently under cultivation. Employment of these techniques in developing areas offers the most acceptable means of increasing the world's food production in the short run. Food production should be increased in precisely those areas in which the food will be consumed.

Of course, the economic difficulties in this approach cannot be ignored. It has been estimated, for example, that the capacity of the chemical-fertilizer industry must double if productivity is to double. Similarly, a great expansion of the food-preservation and storage industries is essential. The accomplishment of these ends requires heavy investments of capital. Further, and contrary to what is generally believed, the developments that have increased production in temperate regions are not directly transferable to tropical regions, where most of the world's food problems exist. For example, such improvements as genetic gain can be spectacular, but cultivars must be developed in large part for specific areas of adaptation—a time-consuming and expensive process. The most important requirement is "pay-off" technology—that is, technology that produces immediate gains. This is absolutely necessary because the subsistence-level farmer has essentially no surplus capital to risk. Results must be guaranteed. This seemingly obvious fact is too often overlooked or little appreciated. This kind of technology, if not developed in the area of application, must at least be refined and adapted for use in the area of application. Thus, technology and technologists must be improved together. Still further, agriculture is an evolving system in which one improvement depends on another. New pesticides, to be efficient, require efficient application equipment. Increased production requires added storage capacity. Agricultural development cannot be conducted piecemeal.

It has been said that a developing country can more easily obtain a modern steel industry than a modern system of agriculture. This is because of the diffuse nature of the agricultural industry—spread over large areas and involving whole populations. Improving the agriculture of a developing country involves improving the whole structure of society. Thus, the development of a modern agricultural system depends as much on improving education and communication channels as it does on technology. Whole philosophies must be created to change the concept of agriculture from an extractive, feudal system to an investment-business enterprise. We will face extinction if we shirk our obligation to project, to foresee, to take measure of the calamities that are fast approaching.

Selected references

Bogert, J., G. M. Briggs, and D. Howes Calloway, 1979. *Nutrition and Physical Fitness*, 10th ed. Philadelphia: W. B. Saunders. (The principles of human nutrition.)

Borgstrom, G., 1968. *Principles of Food Science* (2 vols.). New York: Macmillan. (An introduction to food science: volume 1 covers food preservation; volume 2 covers food chemistry, microbiology, marketing, and utilization.)

Desrosier, N. W., 1977. *The Technology of Food Preservation*, 4th ed. Westport, Conn.: Avi Publishing. (A general textbook.)

Hamilton, E. M., and E. Whitney, 1979. *Nutrition: Concepts and Controversies*. St. Paul, Minn.: West Publishing. (A text exploring food issues and problems.)

Hoff, J. E., and J. Janick (editors), 1973. *Food: Readings from* Scientific American. San Francisco: W. H. Freeman and Company. (A collection of brief but stimulating articles on nutrition and food production.)

Kretchmer, N., and W. van B. Robertson (editors), 1978. *Human Nutrition: Readings from* Scientific American. San Francisco: W. H. Freeman and Company. (More *Scientific American* articles with the accent on food.)

Lloyd, L. E., B. E. McDonald, and E. W. Crampton, 1978. *Fundamentals of Nutrition*, 2nd ed. San Francisco: W. H. Freeman and Company. (An integrated treatment of nutrition and metabolism.)

National Academy of Sciences, 1979. *Recommended Dietary Allowances*, 9th ed. (The basis of U.S. dietary allowances.)

U.S. Department of Agriculture, 1959. *Food* (USDA Yearbook, 1959). (The story of food, written for the layman.)

Wohl, M. G., and R. S. Goodhart (editors), 1968. *Modern Nutrition in Health and Disease*, 4th ed. Philadelphia: Lea & Febiger. (An advanced treatise.)

4
World Population

More than one million identified species of animals and plants inhabit the earth. Of these only *Homo sapiens* can willfully control and modify the environment. Because of this ability, we now dominate the earth to an extent that has not been achieved by any other species. This domination is relatively recent in human history. It has given rise to a new dimension in population growth with which we are now only beginning to cope.

The evolutionary process has endowed all species of plants and animals with a reproductive potential that, if unchecked, would overpopulate the earth within a few generations. This reproductive potential has been and still is controlled by diseases, limitation of food supply, and competition among species in the struggle for existence. No species has yet been able to free itself from the biological regulation of growth in large numbers over any significant time period. We cannot yet claim to be an exception to this rule. Our failure to achieve an optimal balance between food supply and rate of population growth reflects limitations of social organization rather than technical feasibility.

THE POPULATION EXPLOSION

The length of time that people have inhabited the earth is now estimated to be between one and two million years, although some estimates indicate an even earlier origin. The total human population probably had not reached one-quarter billion until approximately 2000 years ago (Fig. 4-1). World population did not reach one-half billion until the middle of the seventeenth century. Only two more centuries were required to again double the world population to slightly more than one billion. The time required to add an additional half billion has successively shortened. At the rate of growth now being achieved, during the fourth quarter of the twentieth century less than ten years is required to add one-half billion people to the world's population.

The "population explosion" of the last several decades has given rise to two sharply opposed viewpoints regarding its implication for human welfare. On the one hand there is a small group of alarmists, often termed "neo-Malthusians" after Thomas Robert Malthus, the famous clergyman-economist who,

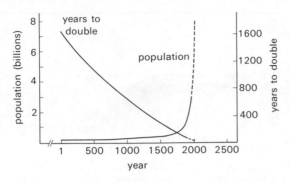

year	population (billions)	number of years to double
1	0.25 (?)	1,650 (?)
1650	0.50	200
1850	1.1	80
1930	2.0	45
1975	4.0	35
2010	8.0	?

Figure 4-1 Estimated population of the world and the number of years required for it to double. [After H. F. Dorn, "World Population Growth," in P. M. Hauser, *The Population Dilemma*, © 1963 by the American Assembly, Columbia University. Reprinted with permission of Prentice-Hall, Inc., Englewood Cliffs, N.J.]

around 1800, first drew attention to the economic implications of rapid population growth. The neo-Malthusians have coined such phrases as "population bomb" to express their concern about the implication of a continued growth of population at the present rate. They would pose population policy as the major issue facing humankind in the next several decades.

At the other extreme are the "technocrats," who have faith that the possible application of science and technology to the production of crops and other raw materials will make possible a rising level of living for most of humanity for the indefinite future. Under these conditions they would expect a slowly declining birth rate to solve the population problem.

A careful sifting of the evidence seems to reveal that advancing science and technology could permit an adequate life for the world's present population, as well as for a growing population over the next several decades. This does not imply, however, that matters of population policy are not of immediate significance. Indeed, to cast the problem in terms of whether science and technology can meet the requirements for food, fiber, and other forms of energy implied by current or anticipated population growth is highly misleading.

Current population growth rates in countries with high "medieval" birth rates and low "modern" death rates are close to 3 percent per year. Under these circumstances a high rate of investment in agricultural development and in educational and health facilities and services is required simply to prevent declines in individual levels of consumption and welfare. The resource demands of rapid population growth can probably be met, but only at the expense of qualitative improvements in the health, nutrition, education, and other dimensions of human welfare.

It would appear that the warnings and the concerns of both the neo-Malthusians and the technocrats are relevant to current population policy. Science and technology must be applied to the fullest extent possible if improved standards of living are to be achieved for the people who are now alive. At the same time, the rising population growth rates of the last several decades must be reversed if economic development is to be a continuing process for most regions of the earth.

The analysis needed to support this conclusion will be discussed in greater detail in the rest of this chapter. Attention will first be given to the principles of demographic changes and then to the problem of expanding crop production.

POPULATION DYNAMICS

Inferences drawn from world population growth rates are likely to be quite unrealistic for any particular country or region in the immediate future. The world is not a single economic unit. The earth's resources and human technological knowledge are not reservoirs available to all. Population growth rates vary widely among countries and regions. In general,

Box 4–1 *Birth and death rates, 1770–1980*

In the industrialized nations in the last century, the declining death rate was accompanied by a decline in the birth rate. From a rate of about 40 per 1000 in 1875, the birth rate in developed countries went down to about 20 per 1000 by 1970. Demographers refer to this gradual shift in both birth and death rates as the "demographic transition."

The difference between the birth rate and the death rate is a measure of how fast the population is increasing. For developed countries, this difference is about 10 per 1000 population, or about 1 percent. This means that, despite going through the demographic transition, developed countries continue to grow in population.

In a large number of developing countries the birth rate continues to be close to 40 per 1000, while the death rate has fallen to about 15 per 1000. This results in an increase in population of 25 per 1000 each year, or a 2.5 percent population growth rate.

A number of developing countries have experienced significant declines in birth rates in the 1960s and 1970s. In Cuba and Taiwan, for example, the birth rate is now approaching 20 per 1000.

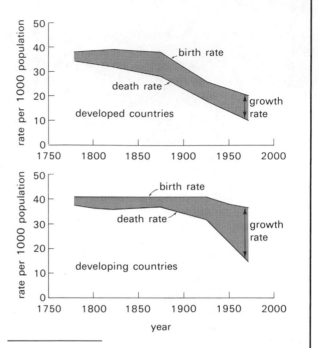

[After Population Reference Bureau, Inc., *The World Population Dilemma*, Columbia Books, 1972.]

technological knowledge is expanding the availability of resources most rapidly in exactly those regions or countries in which population growth rates are lowest.

In Europe, and in the USSR, North America, and Oceania, population growth rates are typically low— 1 percent per year or less, and declining. In most countries of Africa, Asia, Central America, and South America—South Africa and Japan being exceptions—population growth rates range from 2 percent to slightly more than 3 percent per year. Most of the countries with relatively low population growth rates have experienced a transition from high to low population growth rates within relatively recent history. This transition to relatively low population growth rates, based on both low birth rates and low death rates rather than high birth rates and high death

rates, is referred to as the **demographic transformation** (Box 4-1).

Will the rapidly growing population of Asia, Africa, and Latin America go through the same demographic transition that has occurred in Western Europe, North America, and Oceania? A purely biological theory of population would imply a negative answer to this question.

Malthusian population growth

In a population system in which growth rates are determined only by biological or "natural" factors, population growth would be determined by (1) the biological capacity of women to produce children (fecundity); (2) female survival rates through reproductive ages under optimum conditions; and (3) the natural ecological conditions that affect food availability and health and reduce the actual number of children born (fertility) to below the biological maximum and reduce the average length of life (mortality) to below the natural optimum.

The best-known biological or natural theory of population growth was that attributed to Thomas Robert Malthus (1766–1834), an English clergyman-economist. Malthus concluded from the historical and scientific evidence of his time that food production had a tendency to increase at an arithmetic rate (1, 2, 3, 4, . . .) while human population increased at a geometric rate (1, 2, 4, 8, . . .) over time. He concluded that since human population growth is ultimately limited by the amount of food the world can produce, the rate of human population growth would of necessity be brought into balance with the rate of growth of food production only by the natural checks of famine and pestilence.

The Malthusian theory of population growth was not, however, a completely biological theory. Malthus noted that population growth could also be limited by such factors as delayed marriage (restraint) or prostitution (vice). He did not regard these factors as having much practical significance, nor did he visualize the potential impact of technological change, resulting from advances in the natural sciences, on our capacity to increase the production of food and fiber. Malthus visualized a pattern of population growth and decline like that shown in Figure 4-2. A similar pattern was suggested in a recent study by Meadows and associates in *The Limits to Growth*.

There have also been other biological or natural theories of population growth. Some have emphasized the negative effects of population density or complexity of life on the capacity to reproduce. One that gained widespread attention in the 1950s was put forth by the Brazilian biologist Josué de Castro. He suggested in *The Geography of Hunger*° that improved diets, particularly those that are high in protein, may impair the ability to reproduce. Other biologists have suggested that undernutrition of nurs-

---

°Josué de Castro, *The Geography of Hunger* (Boston: Little, Brown, 1952).

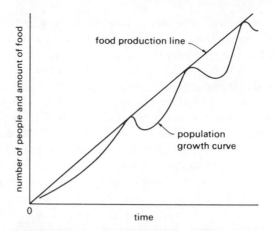

Figure 4-2 Diagram of Malthus's theory of population growth. [After M. M. Snodgrass and L. T. Wallace, *Agriculture, Economics, and Growth*, Fig. 5–3, Meredith Publishing Company, copyright © 1964. Reprinted with permission of Appleton-Century-Crofts.]

ing mothers delays the period of potential conception. Recent field studies in Bangladesh and Guatemala provide fairly convincing proof, however, that fecundity is not significantly affected by undernutrition.

Sociological explanation of population growth

The demographic transformation in Western Europe, which began in about 1875 when birth rates were in the middle and high 30s (per 1000 population per year), continued until the late 1930s when birth rates had declined to between 15 and 20; this decline in birth rates is generally taken as a refutation of Malthus's hypothesis. Today it is generally accepted that the rate of population growth, even in the high-growth-rate countries, is more a reflection of social and cultural customs and institutions than of natural or biological factors. Population growth is thought of as a social rather than a natural phenomenon. Explanations for differences in rates of human population growth are sought in terms of sociological or institutional influences, such as religion and law;

technological influences, such as medical services and public health; economic factors, such as income and the rising economic value of women's time; and cultural factors, such as attitudes toward the role of women in society.

These factors affect the rate of population growth through their separate effects on the birth rate and on the death rate (see Box 4-1). During the Western demographic revolution the factors that tended to decrease death rates were more influential than the factors that tended to decrease birth rates. Better control of disease by means of improved housing, better food and water supplies, adoption of sanitary measures, advances in preventive medicine, and, more recently, development of insecticides and antibiotics have all contributed to reducing the death rate. Furthermore, rapid adoption of these advances for the purpose of reducing infant mortality or prolonging average life span has not provoked any serious moral or social conflicts.

The adoption of practices intended to reduce birth rates required much more complex changes in personal and social motivation. Religious beliefs and social values that had developed under conditions of high mortality—for example, perpetuation of large families—were often in conflict with new values resulting from conditions in which improvement in welfare and opportunities could be gained by maintaining small families. Only after the new values were widely accepted did the control of family size through some combination of delayed marriage and contraceptive practice become extensive enough to reduce birth rates.

The lag in the decline in birth rates behind the decline in death rates has resulted in exceptionally rapid population growth during the middle years of the demographic transformation in most countries. Even the seemingly modest change from a three-child family to a two-child family can have a tremendous impact when the change occurs over several generations. In the United States an average of three children per family would, for example, result in a population of about 320 million in the year 2000 and almost one billion by 2070 (Fig. 4-3). An average of two children per family would, in contrast, result in a national population of about 270 million in 2000

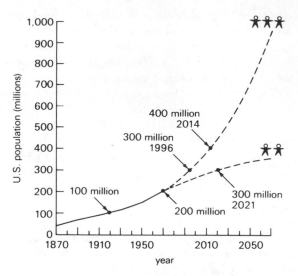

Figure 4-3 Population growth and family size in the United States. [After Commission on Population Growth and The American Future, *Population and the American Future*, The New American Library, 1972.]

and a leveling off of the population at approximately 350 million by 2070. The pattern of fluctuations in fertility rates in the United States throughout the past several decades indicates that considerable caution should be exercised in attempts to forecast the fertility behavior of American families. Recent data suggest, however, that the U.S. fertility rate is becoming one that is consistent with a stationary, or nearly stationary, population (Fig. 4-4).

A neo-Mathusian perspective

The continuum described above was once regarded as a transition through which all nations would pass. All nations and people seemed to be moving through the demographic transformation, although at different rates. Developments during the 1950s and 1960s did not, however, appear to support this optimistic view. There appeared to be an increasing polarization into high-income, developed nations, in which the population increases slowly, and low-income, developing nations, in which the population increases rapidly.

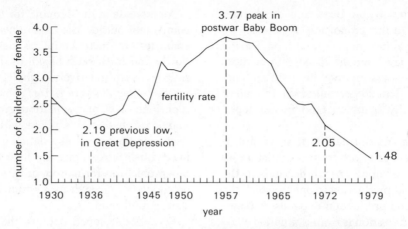

Figure 4-4 Fluctuation in the U.S. fertility rate, 1930–1979.

This polarization was related to the extremely rapid decline in infant mortality and to an increase in average longevity. For example, the death rate among the Moslem population in Algeria declined more in the eight years between 1946 and 1956 than the death rate in Sweden did during the period 1775 to 1875. The result is that population growth rates in the least developed countries exceeded those in the most developed countries even during their periods of most rapid population growth.

In Algeria, Mexico, Pakistan, and a number of other developing countries, for example, the population growth rate continues to exceed 3 percent per year. Rates of more than 3 percent per year are entirely outside the range of historical experience in Western countries. In the West, population growth rates have rarely exceeded 2 percent. The high population growth rates in some developing countries led some popular critics to propose a "life-boat ethic"— implying that the wealthier countries should let the poor countries with high population growth rates drown in their own fecundity.

Since the mid-1960s, there has been a pervasive decline in both fertility and population growth rates in a number of developing countries. The decline has been most apparent in Asia and Latin America. In Colombia, for example, the birth rate declined from 46 in 1964 to 33 in 1975; and in Thailand the birth rate declined from 43 in 1960 to 36 in 1976. The population growth rate in Africa has not yet been significantly altered. By the late 1970s, however, there was less panic about the "population bomb" and more concern about effective ways of meeting world food requirements.

POPULATION AND FOOD

In a Malthusian economy, the relation between population growth and food production is relatively simple: the rate of increase of food production limits the rate of population growth. Wars and epidemics or the discovery of new land resources might ease the food situation for a time, but population growth would ultimately be limited by the difficulty of increasing food production as cultivation extended to the less productive soils and as intensity of labor use increased. In the Malthusian system, food production is the independent variable and population growth the dependent variable.

In a world in which countries are passing through the several stages of the demographic revolution, the relation between population and food production is much more complex than in a simple Malthusian system. As a nation undergoes the transition from high birth and death rates to low birth and death rates there is a decrease in the proportion of national resources, or of personal income, spent on food. During the early stages of development 75 percent or more of the income of a population may be spent on food. By comparison, less than 25 percent of the income of consumers is typically spent on food in developed countries of today.

The demographic transformation, if successful, is accompanied by a transition from a situation in which the rate of population growth depends on the rate of growth in food production to one in which the rate of growth of food production depends on population growth. This transition is from a situation characterized by the "pressure of population on food supplies" to one characterized by the "pressure of food supplies on population." In economies like those of Bangladesh and Nepal, population growth may still be limited by growth in food production (plus food imports). In the United States the growth of food production is restricted by the growth of population (plus food exports).

During the intermediate stages of the demographic transformation, the population-food relation is even more complex. During the initial stages of the transformation, death rates decline more rapidly than birth rates. This alone would result in a rise in the food needs. If per capita income is also rising, this adds further to the demand for food. As incomes rise from levels of $100 to $200 per capita per year, much of the increase in income is spent in increasing the quantity and the quality of food consumed. The result is a rise in the demand for food that is substantially greater than the increase in population.

Before death rates begin to decline substantially, population and the demand for food may both be increasing at little more than 1.0–1.5 percent per year. As death rates continue to decline and birth rates remain high, the rate of population growth may rise to more than 3 percent per year, as in Pakistan, most of black Africa, and a number of countries in Latin America. If rapid population growth is accompanied by an increase in per capita income of 2–3 percent per year, the demand for food may increase by 5–6 percent per year.

An increase in the demand for food in this range is completely outside the experience of the developed countries. In these countries death rates declined slowly and birth rates began to decline while death rates were still relatively high. The question that remains to be answered, for countries experiencing such high rates of population growth and food demand, is whether this phase will be followed by a decline in the rate of population growth as a result of lower birth rates, along the lines of a "classical" demographic transformation, or by a decline in the rate of population growth as a result of a new "Malthusian" equilibrium.

For less developed nations, the rate of growth of domestic crop production has clearly been behind the rate of growth in demand during most of the period since World War II (Fig. 4-5). The projections to 1990 indicate that if these countries are to escape the Malthusian "trap" and proceed along the path of demographic transformation, either the rate of growth of food production must expand more rapidly or the rate of population growth must decline. Recently, a few countries have begun to experience changes. An agricultural "green revolution" has resulted in more rapid growth of grain production in Colombia, India, the Philippines, and a few other countries since the mid-1960s. The rate of population growth has begun to decline in an increasing number of developing countries—even some such as India and Indonesia, where the prospect for a decline in population growth rates seemed completely unrealistic in the mid-1960s. By and large, however, the institutional capacity for the dissemination of knowledge and materials, needed to speed the rate of growth of food production or reduce the rate of population growth, is still inadequate.

POPULATION POLICY

None of the problems of population growth can be solved without sustained increases in both agricultural and industrial production. The fundamental problems that must be solved to achieve sustained increases in agricultural production are dealt with throughout this book.

Figure 4-5 Low-income, food-deficit developing-market economies: production and consumption of major staples is shown for the period 1960–1975 and projected to 1990. The deficit is smallest if consumption remains at the 1975 per capita level, and the deficit is greatest if consumption increases rapidly because of high income-growth. [After *Food Needs of Developing Countries: Projections of Production and Consumption to 1990,* International Food Policy Research Institute, 1977.]

It is also clear that population growth itself represents an obstacle to the achievement of sustained growth in agricultural and industrial production. High rates of population growth impose heavy demands on technical progress and capital formation simply to maintain the existing per capita rates at which food, fiber, and other forms of energy are consumed. Rapid population growth is, therefore, competitive with improvements in the quality of human existence.

There must, therefore, be a dual approach to the problems of population. One involves technological, social, and economic development to improve life for the people who are now alive. The other approach is demographic. There must be declines in birth rates if

economic development is to become a continuing process. The problem, then, is not whether to speed up the rate of technological development or to reduce birth rates. Rather, for each country, the problem is to determine what combination of policies should be followed to achieve agricultural development and reduce population growth.

Adoption of an effective population policy is complicated by a number of social and biological factors. The social problem centers primarily on the ethical issues related to the use of contraceptive devices and the development of motivation for the control of family size. The biological problem centers on the development of low-cost, effective, socially acceptable contraceptive devices or techniques.

Ethical problem of population control

Most of the world's major religions have no generally accepted or firmly enforced objections to birth control. Individual Moslems, Hindus, Buddhists, and Confucians may be found to favor or oppose birth control, but there is no "official" theological or ethical rationalization of either position.

The Roman Catholic position does represent a more difficult theological obstacle to population control. The Catholic position does not oppose population control as such. Rather, it prescribes the means by which fertility may be regulated. Catholic doctrines on marriage and procreation have been significantly modified during the twentieth century. Periodic abstinence as a method of regulating births has been endorsed. There has also been increased emphasis on personal and social values in marriage. Some observers anticipate further change in Catholic doctrine that will facilitate the development of policies designed to lower the birth rate in Catholic families and in predominantly Roman Catholic nations.

Social factors

Social inertia has, in the past, been at least as important an obstacle to the decline of birth rates as have religious and ethical values. The changes in the status of women, aspiration for social mobility, and the growth of the social and economic importance of formal education have contributed to declines in birth rates in the Catholic and non-Catholic countries of Europe alike. Where motivation to control family size was strong, birth rates began to decline even before the development of modern contraceptive devices.

Among the important social, or economic, factors leading to the decrease in family size in the higher-income countries is the rising value of female labor. As employment opportunities for women have risen, paid employment has become increasingly competitive with bearing and raising children as a demand on women's time. Concurrently, the material worth of children, at least in large numbers, has lessened.

Urbanization has reduced their economic value in production, and the development of social security and other institutional support for the elderly has reduced their importance as a source of security in old age.

For thousands of years people have sought reliable means of avoiding unwanted children by contraception—preventing the male sperm from fertilizing the female egg during sexual intercourse, or in other ways preventing the fetus from forming. Progress has not been as rapid in this field of medical science as in many others, partly because less effort has been devoted to methods of birth control than to improving the methods of extending life by deferring death. Furthermore, until very recently the available methods of birth control were not well adapted to meet the needs of poor, illiterate couples lacking either privacy or adequate sanitary facilities.

The principal methods of contraception include the use of jellies and foams to kill the sperm; the wearing of condoms and diaphragms to keep the sperm from reaching the egg; the ingestion of oral contraceptives by the woman to make her temporarily infertile by chemical means; and the insertion of a plastic or metal device (an IUD, or Intra-Uterine Device) into the uterus to make the woman temporarily infertile by physical means (Fig. 4-6). In addition to contraception, surgical sterilization has achieved growing popularity, particularly among mature adults. The usual operation is a vasectomy for the man or a tubal ligation for the woman. Both are virtually 100 percent effective. Abortion is another surgical method of reducing unwanted births. It tends to be used most extensively where contraceptives are not easily available.

All of the methods discussed above have a place, depending on a particular country's cultural traditions and stage of medical development. Also, among individuals, variation in birth-control methods is substantial, the one employed often depending on the age of the woman and her parity status—that is, the number of children she has borne (Fig. 4-7). Within individual countries the transition to lower birth rates also involves a fairly complex relation among live births, induced abortion, and contraception (Box 4-2).

Figure 4-6 In the Indian village of Deoli, a family-planning fieldworker from the Central Family Planning Institute in New Delhi discusses with village women the use of contraceptive devices. The small white object she is showing them is an IUD. [United Nations photograph.]

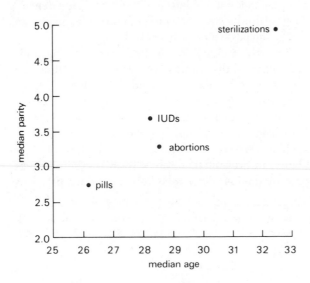

Figure 4-7 The choice of birth-control method is greatly influenced by the age and parity status of the woman (the number of children she has borne). These data have been obtained from postpartum family-planning programs and hospitals throughout the world. [After *Population Program Assistance*, Agency for International Development, December 1971.]

Box 4–2 *The transition to lower fertility: interrelation of births, contraceptive practice, and abortion*

The pattern of induced abortion during the demographic transition seems fairly constant and relatively uninfluenced by abortion law. In most instances, the transition to lower fertility rates involves an initial increase in the incidence of abortion and creates pressure for liberalization of abortion laws. If modern contraceptives are made widely available at the same time that abortion is legalized, countries undergoing this demographic transition can avoid long-term dependence on abortion as the primary method of birth prevention. However, if family planning education and service programs receive minimal public support, abortion may continue to account for an abnormally large share of births avoided long after lower birth rates have been achieved.

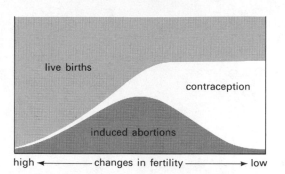

[Adapted from M. Requena, "Chilean program of abortion control and fertility planning: present situation and forecast for the next decade," Centro Latinamericano de Demografia (CELADE), 1969.]

Population programs

There has been widespread interest in the development of population policies. Both the Republic of China (Taiwan) and the People's Republic of China have active programs to reduce birth rates. The Republic of Korea (South Korea) has a strong government-backed program to spread family planning. Japan has achieved one of the world's lowest birth rates. India and Indonesia have initiated strong birth-control programs. Tunisia and Egypt have both developed programs to spread family planning. Even in Latin America there is widespread concern about the "population explosion," and a few Latin American countries make contraceptive services available in government hospitals.

The fertility control efforts in Taiwan were one of the first effective efforts in a poor country. The results have been particularly interesting because they have been related to a well-planned research program designed to provide information on the efficiency and acceptability of different methods and devices.* In Taiwan, a preprogram survey indicated that most women were having more children than they wanted and that they approved of the idea of limiting family size. Not only were they aware that limiting family size would improve the welfare of their families, but they believed the number of children should not be left to "fate" or "providence." Their attitudes were, in general, more advanced than many officials believed them to be.

Similarly, a study in the Philippines in the mid-1960s showed that rural wives would like to limit the size of their families. The study also showed, however, that the wives did not know that effective methods were available. In spite of their lack of knowledge, a very high percentage of the women studied were trying desperately, to the best of their ability, to prevent themselves from becoming pregnant by the use of the relatively ineffective practices

---

*Bernard Berelson and Ronald Freedman, "A Study in Fertility Control," *Scientific American*, vol. 210, no. 5, May 1964, pp. 29–37.

available to them. Between 1965 and 1975 the Philippines has achieved a decline in the birth rate from 44 to 36. Indonesia now seems well on the way to repeating the Taiwan and Philippine experience.

Prospects for the future

There is every reason to expect that birth rates in many of the newly developing countries can be reduced more rapidly than might have been predicted from the earlier experiences of Western Europe. Possibilities appear even more favorable than they did in the 1960s. Religious, ideological, and nationalistic disputes about birth control have subsided; governments are increasingly aware of the need to reduce the birth rate; much of the public is already anxious to control its childbearing; and there are new methods of contraception that are cheap, safe, and effective for the large majority of the childbearing population.

It is now possible to think that the world will—before 2000—be well on the way toward finding positive solutions to its problem of excess reproduction and population growth. This optimism is based on a number of favorable developments.

1. The concept of fertility control, and its individual and social benefits, has been broadly disseminated in an increasingly receptive world.

2. Many countries have removed or modified laws that restricted the availability and use of the most effective means of birth control.

3. Resources available for the development of population and family-planning programs have increased rapidly.

4. Key breakthroughs in fertility-control technology have occurred or now seem feasible.

5. Many countries are developing increasingly effective family-planning programs.

6. Recent census and other demographic data indicate a more favorable world population figure than has been projected from earlier data.

The optimism about future population growth should not be allowed to obscure the fact that the increase in world population over the next several decades will be very large. In 1980, world population was slightly more than 4.5 billion people. By 2000, it will almost certainly be in the neighborhood of 6 billion.

In spite of the apparent success of family planning programs in some countries, there is still a good deal of controversy about their effectiveness. Critics argue that socioeconomic development is the primary source of fertility decline and that successful family-planning programs are little more than a reflection of more basic changes in socioeconomic development. Socioeconomic development is, however, a slow process. Family planning and related health and nutrition programs are among the few instruments available to governments in poor countries to reduce fertility.

If population growth rates are to decline fast enough to allow savings and capital to be diverted from simply meeting food needs to improving the quality of life, positive government programs will be required. At least part of the resources that would otherwise be devoted to expanding national agricultural and industrial potential could be productively devoted to the extension of public education and health services directed to reducing the birth rate.

Selected references

Commission on Population Growth and the American Future, 1972. *Population and the American Future*. Bergenfield, N.J.: New American Library (Signet). (An authoritative analysis, by a commission proposed by President Nixon and authorized by the Congress, of population trends and of the implications of future population growth for the United States.)

Demeny, P., 1979. *On the End of the Population Explosion* (Population Council Working Paper No. 30, March 1979). New York: The Population Council. (The author challenges the analysis employed by Tsui and Bogue in deriving their optimistic perspective on population trends.)

Ehrlich, P. R., A. H. Ehrlich, and J. R. Holdren, 1977. *Ecoscience: Population, Resources, Environment*. San

Francisco: W. H. Freeman and Company. (A textbook of environmental science that provides an ecological perspective on population growth in relation to resource constraints.)

Mauldin, W. P., and B. Berelson, 1978. Conditions of Fertility Decline in Developing Countries, 1965–1975. *Studies in Family Planning*, 9:90–147. (A definitive review and analysis of the sources of fertility change during the period from 1965 to 1975. The results indicate that the level of "modernization" has a substantial relation to fertility decline and that family-planning programs have a significant independent effect over and above the effect of socioeconomic factors.)

Meadows, D. H., D. L. Meadows, J. Randers, and W. W. Behrens III, 1972. *The Limits to Growth*. New York: Universe Books. (A neo-Malthusian exposition of the relation between resources, technology, and population growth. Utilizes systems-simulation methodology and rhetoric to project a world population crisis during the second half of the twenty-first century.)

Tsui, A. O., and D. J. Bogue, 1978. *Declining World Fertility: Trends, Causes, Implications*. Washington, D.C.: Population Reference Bureau. (An optimistic perspective on world population trends. The authors place major emphasis on the effectiveness of family-planning programs in reducing fertility.)

5
Agriculture, Pollution, and the Environment

Pollution is a general term applied to the presence of undesirable substances in the environment. It may be caused by substances introduced into soil, air, or water, or to natural substances modified so that they become undesirable. Pollutants may also be biological in nature, such as pests or pathogens, or forms of unwanted energy, such as waste heat, noise, light, and radiation. The term has also been expanded to include visual or esthetic insults to the environment. Although everyone agrees that pollution is undesirable, there is considerable argument over who is responsible for it and what can be done about it. In this chapter, we will consider pollution as it affects agriculture and the environment, and the ways in which it relates to present-day circumstances. It is a particularly important subject now, since minimizing pollution and maximizing production has become a legal issue as well as a moral one.

MOTHER EARTH VERSUS SPACESHIP EARTH

Early people usually spent their lifetime, averaging perhaps 20 or 25 years, within a day's walking distance of their birthplace. They did not need to go farther. Living in the rough comforts of generations past, a cave or some other well-concealed spot, they secured their food by hunting and gathering. But as the population of the world increased, it became necessary to live in small, less mobile groups for mutual aid and protection. The hunting and foraging parties of these groups would deplete the food in their immediate vicinities and were then obliged to range increasingly greater distances to secure sustenance. With the development of agriculture, the proportion of food obtained from cultivation increased. Through time, and countless trials and errors, people learned to husband the soil and to respect it (Fig. 5-1).

Thus, **Mother Earth** became the great provider for the cultivator, too, although her bounty was often grudgingly given. Throughout the centuries, the reverence for "nature" as a sustainer of life has endured. Even today we assign the properties of life to soil itself and describe it in terms we would use for living things—for example, "beautiful," "productive," "dead," "mature," "old," and "young." The forces of nature were thus personified and deified, and we worship them still.

With the passage of time, our sense of direct dependence on the land became obscured and our reverence for the soil and the environment faded.

Figure 5-1 People and nature can coexist in harmony and peace, creating an environment in which there is abundance for all. [Courtesy Tennessee Valley Authority.]

Today, one person can produce food for 50 people, and a city dweller may never see a field of corn planted or a wheat field harvested. For many years planners and engineers considered the supply of rich tillable land unlimited, and they covered our most productive soils with cement and asphalt. The millions of people who work in cities, and then escape each day to the more pleasant surroundings of suburbia, tempted developers to transform the countryside into a plethora of look-alike housing developments and trailer parks. Even the laws in many parts of the United States discourage the use of farmland for crop production: increasingly more land is being taxed according to its "highest use," which means its greatest

economical potential for residential, industrial, or other commercial exploitation, and this makes it economically infeasible to use the land to raise agricultural or forest crops. These materialistic and impersonal values we have placed on the soil that sustains us must ultimately change if this nation's traditional self-sufficiency in agricultural production is to endure.

Until the 1960s, little public concern was expressed over the prevailing methods of waste disposal. Refuse in general—including the wastes from factories, paper mills, canneries, power plants, and city sewers, as well as those directly from agriculture itself—was assumed to be harmless, until contrary evidence was

Figure 5-2   When conservation measures are neglected, destruction often ensues.
[Courtesy Tennessee Valley Authority.]

produced by some regulatory body. The rationale that allowed this permissive attitude to develop in the United States had its roots in the early history of this nation at a time when resources seemed almost unlimited and pressures to industrialize and develop a greater national economic base dominated nearly all other aspects of life. Most people welcomed new industry, regardless of what damage it did to the countryside, and disregarded the evidence that streams were becoming polluted and vast areas of land disfigured and ruined (Fig. 5-2). Unfortunately, the same robber-baron philosophy still prevails today, and many developing nations have decided to ignore pollution in their efforts to industrialize as

rapidly as possible. We can hardly censure them, for they are only following our example.

To be sure, many people still pay verbal homage to Mother Earth, but most continue to assume that her resources are infinite: she is expected to be forever tolerant of the abuse of her erring and spoiled children. Although concern for the environment had long been championed by eloquent and forceful exponents (such as Henry Thoreau, Gifford Pinchot, and John Muir), their combined voices never really touched the conscience of America, whose heroes were frontiersmen rather than Indians, cowboys rather than farmers, and industrialists rather than craftsmen. However, in the early 1960s an awareness

Figure 5-3 The finite nature of spaceship earth is most apparent when it is viewed from the moon. [Courtesy NASA.]

of environmental issues emerged in both the scientific community and in society generally; we awakened to a growing apprehension that much of what we cherish might be lost forever. This revolution in thinking, known as the environmental or ecology movement, found an especially receptive audience in the youth, a generation at least partially jaded by abundance. A series of dramatic events and findings quickly unfolded: the publication of Rachel Carson's *Silent Spring;* the military use of defoliants; the association of DDT with the potential extinction of certain species of birds; and the linkage of smoking and cancer. The astronauts hurtled through space to the moon, and rediscovered the earth; and the rest of us, too, saw it anew as a lonely, tiny craft, with limited resources: Spaceship Earth (Fig. 5-3). Environmental pollution became a cause with which all could identify.

The concern for environmental quality has come from two major sources. The first is the increased demand for the environmental assimilation of residues derived from growth in commodity production and consumption, and from the energy production and transportation services associated with this growth. The second, and related, source is the rising demand for outdoor "retreats" blessed by such amenities as the absence of pollution and congestion. The increased competition between the demand for environmental services for disposal of residues and that for untouched natural areas resulted in the 1960s in a heightened awareness of environmental problems in industrial countries.

At the same time, it is useful to emphasize certain considerations that have frequently been ignored in the heat of the debate generated by the new awareness of environmental issues. First, it is well to remember that throughout history people have been continuously challenged by the dual problem of (1) how to provide themselves with adequate sustenance and (2) how to manage the production and disposal of residues or, in less elegant language, garbage. Failure to make balanced progress along both fronts has at times imposed serious constraints on society's growth and development. Even in poor communities, whose people are living relatively close to nature, use of energy to dispose of residues is directly competitive with the use of energy to provide for sustenance.

It is also worth remembering that the capacity of a society to solve either the problem of sustenance or the problems posed by the production of residues is inversely related to the rate of population growth and positively related to the society's capacity for innovation in science and technology and in social institutions. Continued technological advance is essential for further advances in both the materials and esthetic dimensions of consumption. The fundamental significance of advances in science and technology is that they permit the substitution of knowledge for resources. Scientific and technical effort must now be redirected toward reduction of environmental stress.

Technology, however, its potential benefits notwithstanding, was one of the first culprits to be identified as a major polluter, and those who did so assumed the status of heroes. Agriculture, the self-proclaimed friend of the soil, was not to be spared as a "special case," and its technological aids were considered to be as harmful as those of industry. Although we may indeed be able to ascribe pollution to a few major sources, and although all of us may identify with the cause of a "clean environment," it is also clear that each of us contributes to its degradation. We pollute every time we drive our automobile, wash our clothes, or turn on the light. If cars, detergents, the power industries, or agricultural production practices are among the causes of pollution, then so are we who make use of them.

AGRICULTURAL PRACTICES AND THE ENVIRONMENT

Modern agriculture is presently identified both as an object of pollution and as a polluter itself. This dual role is particularly ambiguous, however, because of the continuing fluctuations in the interrelations between agriculture, the chemical revolution, and the population explosion. The growth of cities and the increase of population have resulted in the continued encroachment of asphalt and steel on farmland, even as the numbers of mouths to feed have increased. Cities and their industrial complexes have preempted still more valuable space for landfills for their refuse, and noxious gases are constantly spewed from smokestacks, burning dumps, and automobiles. However, just as the urbanite relies on new artificial fibers and medicines, so has agriculture turned to the chemical industries for mineral fertilizers and for pesticides to control weeds, fungi, and insects. As a consequence of its search for increased efficiency, agriculture, too, has helped to pollute the environment with pesticide residues and leached nutrients. Nevertheless, as we seek solutions, the basic purpose of agriculture—to produce food and fiber—must be clearly kept in mind: to eliminate pollution and at the same time to create a food shortage will not better the human condition.

That agriculture contributes to pollution in various ways (Fig. 5-4) can be most clearly illustrated by descriptions of two cropping situations—corn production practices in the Midwest and forest clearcutting.

Corn production

In the Midwest, approximately 120 days are needed to grow a crop of corn from the time the seed is planted. During the other 245 days of the year, the soil is bare and subject to erosion, since it is not protected by a mantle of vegetation. In some parts of the world a second crop is grown that will protect the soil, but in the climate of the Corn Belt of the United States there is no respite from winter.

Corn is generally planted on a bare site. As the seedbed is prepared for planting, dust is sent flying into the air to be blown away by the wind. Soil is lost in a similar manner while the crop is growing and during the fallow period between crops.

To grow corn most efficiently one must apply pesticides, starting before or at the same time the seed is being planted. These chemical materials can infiltrate into the environment by scattering and blowing, vaporization, leaching and adsorption onto soil particles, and incorporation into the crop and crop residues. The fate of these pesticides is now con-

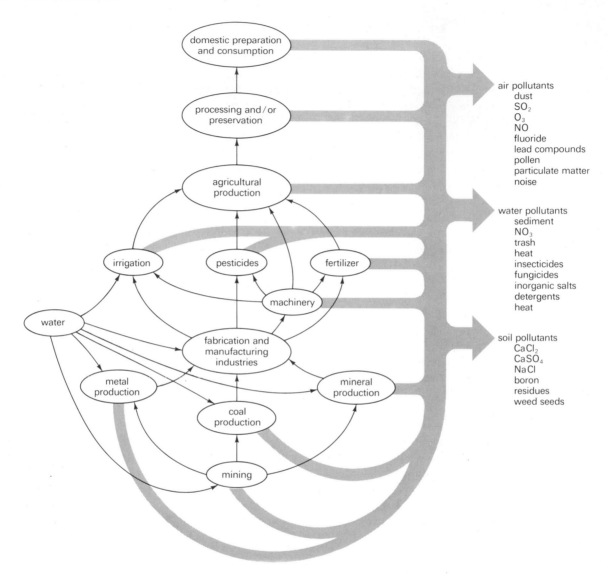

Figure 5-4 At every stage of modern food production, pollution results. From the mining of metals to make machinery with which to work the fields, to the ultimate disposal of human wastes, food production creates a complex of pollution problems that overlap, feedback, and loop, so that the exact sources of pollution are not always clear. This diagram is much too simple to indicate the complexity and immensity of the situation. The transportation system, which carries materials from one place to another in the production network, alone creates a staggering amount of pollution that must also be accounted for.

sidered to be one of agriculture's major problems. They will eventually break down chemically or biologically, although some chemicals may take many years to do so. However, if the amounts applied are greater than the amounts broken down in a given time, an accumulation will result (Fig. 5-5).

Unless the fields are contour plowed, rainfall drains almost immediately into waterways and

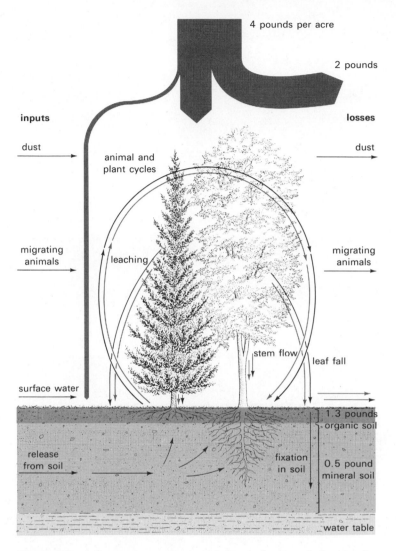

Figure 5-5   A forest community is an integrated array of plants and animals that accumulates and reuses nutrients in cycles. DDT and other substances introduced into such a community are also cycled. Three years after a New Brunswick forest had been sprayed with DDT (2 pounds per acre in the course of seven years), the DDT residues were as shown in the figure (in pounds per acre). [From G. M. Woodwell, "Toxic Substances and Ecological Cycles." Copyright © 1967 by Scientific American, Inc. All rights reserved.]

streams. Hence, runoff frequently contributes to floods and soil erosion. A conservative estimate of annual soil loss by erosion from a well-managed cornfield with little slope is approximately 18 metric tons per hectare (8 tons per acre). Since the total weight of soil in the upper 15 centimeters (6 inches) of soil is approximately 2200 metric tons per hectare (a little less than 1000 tons per acre), such a loss is

Figure 5-6 Obliteration of lakes and other kinds of water impoundments is a process that starts at the edge of the water. (A) A few bog-adapted conifers rise in a forest of hardwoods. (B) Next, the debris of shallow-water plants turns the lake margin into marsh that is gradually invaded by mosses and bog plants, such bog-adapted bushes and trees as blueberry and willow, and additional conifers. (C) Eventually the lake, however deep, is almost entirely filled with silt from its tributaries and with plant debris. (D) In the final stage, the last central bog grows up into forest. [From C. F. Powers and A. Robertson, "The Aging Great Lakes." Copyright © 1966 by Scientific American, Inc. All rights reserved.]

hardly significant in any one year. However, the annual loss is commonly four times as great, and it is sometimes greater by a factor of 10. Unfortunately, most of the soil that is lost goes directly into streams and ponds, creating sedimentation problems by silting (Fig. 5-6).

If corn production in the U.S. Corn Belt is to be profitable, grain yields of about 5 metric tons per hectare (nearly 100 bushels per acre) are necessary. If these yields are to be obtained, the nutrients that have been removed by the growing crop must be replaced, the usual method being the application of mineral fertilizers. However, those nutrients that are soluble in water, or that do not become tightly affixed to soil particles, are lost in the soil-water runoff, thereby creating the possibility of excess aquatic vegetation from an overenrichment of nutrients in streams and lakes, and eventually fish kills caused by the depletion of dissolved oxygen by the vegetation. This process is called **eutrophication.**

Although phosphorus is a cause of water deterioration through eutrophication, the amount lost from the soil to water because of agricultural practices is probably insignificant, because phosphate ions are tightly bound to soil particles. However, nitrogen loss from the soil to water is much greater, because nitrates are highly soluble and are readily leached from soil. Consequently these compounds may be the main elements contributing to eutrophication of many lakes and farm ponds. A critical problem can also occur when nitrogen compounds find their way into underground aquifers, because excess nitrate (NO_3^-) or nitrite (NO_2^-), when ingested, may be hazardous, especially to the young.

Box 5-1   *The nitrogen-ozone connection*

Ozone (O_3), a form of oxygen, is a pollutant in the lower atmosphere (see p. 99), but its presence in minute concentrations (less than 1 part per 100,000) in the upper atmosphere is essential to the life forms on earth. Ozone is formed in the upper atmosphere when atmospheric oxygen (O_2) absorbs ultraviolet radiation. Ozone itself absorbs ultraviolet radiation, and the absorption processes shield life on earth from exposure to the potentially lethal radiation.

The scientist who suggested that nitrogen oxide exhausts from supersonic aircraft may harm the earth's protective ozone layer, Dr. Harold Johnston of the University of California, has more recently said that the use of nitrogen fertilizers may produce the same effect. Much of the nitrogen fertilizer applied to crops is used by plants, and a large part of it is removed from the fields when crops are harvested. However, all of the nitrogen fertilizer that is not harvested with the crop is eventually denitrified. In this process nitrogen is reduced to form a gas, nitrous oxide (N_2O), which eventually finds its way into the upper atmosphere. In the lower atmosphere this gas is inert, but in the upper atmosphere it becomes a catalyst for the destruction of ozone. The reactions are as follows:

$$N_2O \text{ (formed in soil)} + O_2 \rightarrow NO$$
$$NO + O_3 \rightarrow NO_2 + O_2$$
$$O_3 + \text{radiant energy} \rightarrow O_2 + O$$
$$NO_2 + O \rightarrow NO + O_2$$

There are many complications in Johnston's estimate, and not all scientists agree with him. Whereas he has said that 15 percent of the ozone layer may be destroyed in a century, others have estimated only a "few" percent. All estimates are in the same order of magnitude, however, and there seems to be no doubt that a decrease, which occurs normally even without fertilization, will take place.

The worst effect of a decrease in ozone would be an increase in skin cancer, because more high-intensity ultraviolet radiation would reach the earth's surface. A National Academy of Sciences Report suggests that a 5 percent depletion of ozone could produce an additional 8000 cases of skin cancer per year among the white population of the United States alone. Other estimates are as high as 200,000 to 600,000 cases.

The solution will undoubtedly be politically complicated. Fertilizers are far more important than supersonic planes, which produce the same effect. Yet the American public is often loath to give up its luxuries even when the forfeiture is logical.

Corn is a row crop and, as usually managed, requires intensive cultivation; however a new system, called "minimum tillage," has been developed for the production of corn and certain other crops that reduces erosion but relies on greater use of herbicides (Fig. 5-7). Tractors, used to cultivate corn, produce exhaust fumes containing hydrocarbons. These fumes are not a major source of pollution in rural surroundings, but they must nevertheless be counted as a contributing factor.

Smut, a common fungus that attacks corn, releases spores that can cause severe allergies in some persons. Similarly, pollen from ragweed and other weeds growing at the uncultivated edges of fields is a cause of asthmatic attacks. Corn pollen is irritating to some, but is usually no more than a local nuisance to those working directly with the crop.

Under modern systems of production, most crop residues are returned to the soil, but some are burned in the field. Until recently, before the large-scale adoption of picker-shellers, corn cobs were harvested along with the grain and were later disposed of by burning. Thus burning contributed, and still does to some degree, to air pollution.

Clear-cutting in forestry

Clear-cutting is a practice in which all the merchantable timber in a given area is removed at one time, usually leaving the soil essentially treeless. Once the forest canopy is removed by clear-cutting, precipitation falls directly on the soil and a greater proportion of the water runs off immediately. Many of the environmental problems that result from corn production are also created by clear-cutting, though they are not as severe. Natural fertility is lost. Pesticides are introduced into the ecosystem. Machinery that is used to harvest the crop releases hydrocarbons into the atmosphere. The humus in the topsoil is exposed to the sun so that it decomposes faster, thus reducing the overall quality of the soil. Smoke from fires used to burn brush piles pollutes the atmosphere. Esthetically, a clear-cut is nearly always displeasing. One can rationalize all of these objections by saying that clear-cutting is necessary to grow certain kinds of trees that are especially light demand-

Figure 5-7 By growing such crops as corn in soil with minimum tillage (so that the soil has a perennial cover of sod), soil losses are negligible. This system of production will be used much more in the future. [Courtesy Ohio Agriculture Research and Development Center.]

ing, especially in their juvenile stages, and that it has a minimal negative effect in certain environments. However, this is a weak defense against the onslaught of preservationists, many of whom would prefer that no trees at all be cut.

Agriculture as a polluter

As indicated in the discussions of corn production and clear-cutting practices, agriculture does indeed contribute to the deterioration of the environment, as does industry. In general, those specific practices most likely to cause pollution are soil erosion, the improper use of plant nutrients, and pesticide usage.

Erosion has long been an agricultural problem, and soil conservation has always been the byword of progressive agriculture. Long before it became fashionable to wave the banner of environmentalism, agriculture had been concerned with this problem, because long-range maintenance of productivity was in agriculture's obvious self-interest. Consequently, a discussion of soil and soil conservation is necessarily included in any basic coverage of agricultural practices (see Chapters 12 and 16).

In the past, the excessive use of plant nutrients has usually been considered an agricultural problem only in relatively isolated situations. These include arid areas where the penalty of overfertilization is severe, leading to the salinization of water supplies and the accumulation of salts in the soil. Overfertilization of greenhouse soils has similarly generated an appreciation of the salt-accumulation problem that can result from this practice. However it is fair to say that in the past underfertilization has received the greatest attention, and the problems of excess nutrients in humid regions have not been considered except for economic purposes: that is, overfertilization has always been unprofitable, and it has been cautioned against only for that reason.

The pesticide problem appears to have been underestimated, even though some problems caused by overuse were recognized long ago. The discovery of the new "miracle" pesticides, such as DDT, seemed to provide a welcome solution because the concentrations in which they were applied were infinitesimally smaller than that of lead arsenate, for example, and

their potential for polluting was largely overlooked in agriculture. Today, even after the banning of DDT, the solution to the problem of pesticide pollution is still not clearly apparent. We now find that some of the replacements, the so-called "hard" or persistent pesticides, are quite dangerous, particularly for those who apply them. It also appears that the toxic effects of such substances as DDT may have been exaggerated. Many have felt that the banning of this chemical was more an emotional response than a rational one, the result of a need to identify a scapegoat that would be symbolic of the general problem of environmental pollution.

Agriculture has never relied exclusively on chemical control of pests. Genetic resistance to pests has been a key objective in crop-breeding programs, and biological control of pests is not at all a new agricultural practice.

Finally, it would be folly to attempt to grow food and fiber crops at a "zero" level of pollution. The cost of "organically grown" foods is approximately twice that of food produced by conventional practices. We could eliminate farm machinery and return to primitive ways, but it would require a mass movement back to the land, a chilling prospect to all but the romantic. In other words, overfastidiousness is not the solution. This is not to imply that we should not use conservation practices with sound ecological bases. We must! But we must determine how we can feasibly produce sufficient amounts of food, given the economic realities *and* the necessity to safeguard public health and welfare. Our goal must be not only sustained but increased yield, achieved with minimum damage to the environment.

INDUSTRIALIZATION AND AGRICULTURE

As a result of the trend toward urbanization and industrialization, the threat to agriculture, primarily from air pollution has increased. Figure 5-8 is a diagram of the global circulation patterns of certain atmospheric pollutants.

Air pollution is not only esthetically offensive but also a genuine health hazard to animals and people. It can be extremely injurious to vegetation, so that

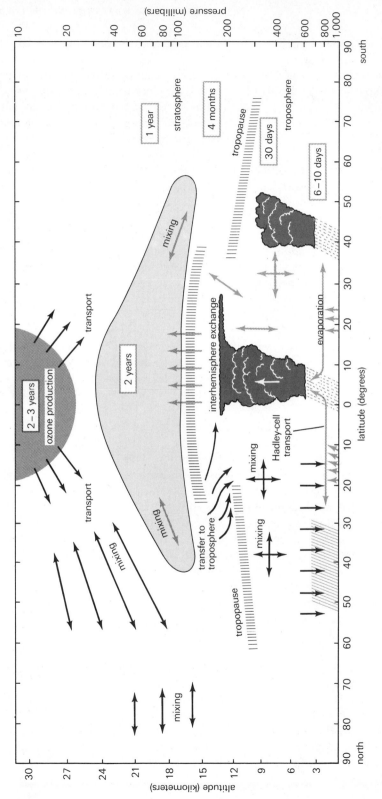

Figure 5-8 Atmospheric pollutants follow global circulation patterns. The numbers in the boxes indicate the periods of residence in different layers of the atmosphere for aerosols, one of the forms in which pesticides are applied to crop plants. [From R. E. Newell, "The Global Circulation of Atmospheric Pollutants." Copyright © 1971 by Scientific American, Inc. All rights reserved.]

agricultural industries located around population centers, such as market gardens, are especially endangered. Air pollution is often harmful to greenery in urban parks, medians of avenues, and residential plantings. Plants are injured by particulate matter as well as by such gases as sulfur dioxide and ozone.

The principal sources of air pollution are automobiles (60%), industries (19%), electric-power generating plants (12.5%), space heating (6%), and refuse disposal (3%). In the United States, these sources generate about 125 million tons of air pollutants annually. The responsible agents are mainly carbon monoxide, oxides of sulfur, ozone, hydrocarbons, and particulate matter. As a result of the urban concentration of the sources of these pollutants, the air in urban areas is, on the average, three times as polluted as the air in rural areas, but the gradient may be 50-fold or greater.

The principal air pollutants that are toxic to plants are discussed below.

Ozone

Ozone was recognized as a major phytotoxic pollutant in 1958. Since then, ozone damage has been observed on ornamental, agronomic, and forest crops of many kinds.

Ozone is most abundant in urban environments where there are large quantities of the chemicals necessary for its formation. Automobile engines are the chief source. Nitric oxide (NO) is formed in the engine, whence it passes into the atmosphere to oxidize quickly, forming nitrogen dioxide (NO_2). NO_2 reacts with other atmospheric components and releases O, which combines with molecular oxygen (O_2) in the atmosphere to form ozone (O_3).

Whereas 1–2 parts per hundred million is a normal level of ozone in the atmosphere, the daily maximum in the Los Angeles basin has reached 100 parts per hundred million. Daily peaks frequently reach 30 parts per hundred million in the summer and fall. But ozone reacts quickly, once in the atmosphere, so that photochemically produced ozone disappears soon after it ceases to be produced.

Ozone has various effects on plants (Fig. 5-9). Photosynthesis is decreased as long as plants are exposed, and growth is thus impaired. Respiration may be increased, resulting in a reduction of carbohydrates. The stomata, the openings that permit gas exchange, close when ozone is in the atmosphere, limiting gas exchange and all other processes that depend on exchange of CO_2 and O_2 between the plant and the atmosphere.

Ozone injury first presents a shiny, oily appearance on upper leaf surfaces. If the exposure is sublethal and is soon halted, the leaf will return to normal. In tissues that have been killed, the shiny areas take on a dull, gray water-soaked appearance. The groups of cells subsequently turn white, giving the leaves a stippled appearance. In monocotyledons, such as corn, wheat, rye and oats, there is a chlorotic stipple between the veins. Leaf bronzing and early senescence of the leaves occur in some woody ornamentals.

Some species are so sensitive that they need no urban source to produce symptoms of ozone damage. White pine (*Pinus strobus*) is one such plant, being subject to a disorder called white pine needle blight or emergence tip burn. Stipple is an ozone-induced disorder of grapes that begins with small brown or black lesions on young leaves, ultimately leading to bronzing and defoliation.

Sulfur dioxide

Sulfur dioxide (SO_2) is released from burning coal. Power plants, refineries, and home furnaces especially contribute to the SO_2 content of the air. Additional SO_2 is cast into the air when certain ores of nickel, copper, and iron are smelted.

This compound can kill vegetation and threaten the health of people and animals. It enters leaves through their stomata, passing into the intercellular spaces of the mesophyll where it is absorbed on the wet cell walls. The mechanism by which it damages plants is still a matter of speculation, but one theory suggests that it interferes with protein synthesis. Sulfur damage can also result if accumulations are great enough to be directly toxic to cells: it causes cells to shrink and collapse, and chloroplasts to disintegrate.

Figure 5-9 Ozone has various effects on plants. (A) Grape leaves develop small lesions, which are distributed interveinally over the entire leaf surface. (B) On tomatoes, injury is concentrated on the terminal leaves, which become yellow near the tip and die gradually as the damage extends from the tip toward the base. [Courtesy USDA, EPA, and Walter Kender.]

The external symptoms of SO_2 damage are varied (Fig. 5-10). Alfalfa and red clover develop white spots on leaves, whereas white pine tips lose color. Both the spots and the pine tips will become green again in a few days if the source of SO_2 is removed. If not, they become necrotic (the affected tissues die).

Studies with alfalfa, barley, and wheat have shown that crop yield is directly proportional to the degree of leaf necrosis. The vertical growth of white pine, an especially sensitive species, may be depressed by SO_2. Other trees that can be seriously retarded include Douglas-fir and lodgepole pine.

One of the major examples of SO_2 damage to an ecological system occurred in the Copper Basin of Tennessee. In the early days of smelting, the ore was placed on large pyres of logs which were ignited. The copper melted and flowed to the bottom of the fire, where it solidified and was picked out after the ashes had cooled—a crude but effective smelting operation. Not only were trees cut to build the pyres, but others were killed by the large amounts of SO_2 released in the smelting operation. More than 2800 hectares (about 7000 acres) of land were completely denuded. Similar disasters have occurred in British Columbia and Montana. Even today the SO_2 emis-

Figure 5-10 Sulfur dioxide fumes have damaged the white birch leaf on the left. The leaf on the right is from a tree grown in filtered air. [Courtesy USDA.]

sion from modern copper and sulfuric acid plants restricts vegetation growth. Because all species of plants are not equally affected, the constant presence of SO_2, even at low concentrations, changes the composition of native vegetation.

Fluoride

When ores containing such minerals as mica, hornblende, and cryolite are smelted, fluoride escapes, as it does when fluoride compounds are manufactured or used for catalysts and fluxes. Either gases or particulate forms may be released.

Fluoride enters plants by diffusing through the stomata of the leaves into intercellular spaces, where it is absorbed into the cells and transported through vascular tissues to the tips and margins of leaves, and it then accumulates in chloroplasts, cell walls, nuclei, and cytoplasm. Although plants may normally have a fluoride content of from 1–20 parts per million, those in industrial areas may contain 100–1000 parts per million. A few plants can accumulate large amounts of fluoride even in areas that have no industrial pollution. For example, tea plants may contain as much as 1300 parts per million without apparent damage.

The exact processes through which fluorides injure plants are not known, but they seem to interfere with enzymes that affect many metabolic processes. It is thought, too, that fluoride may precipitate the calcium within the plant, thus causing a calcium deficiency.

Fluoride causes mottling and necrosis at the tips and margins of leaves of sensitive broad-leaved species. On grasses, tips are killed, and stippling or mottling is found along the margins. On conifers, necrosis starts at the tips of the current year's needles.

The fruits of some species are extremely sensitive, even more so than the leaves. In peaches, fluoride cause the soft-suture disease, a premature ripening along one or both sides of the suture line. This area overripens or rots before the rest of the fruit matures.

Fluoride pollution does not always damage plants. Conifers, aspen, grapes, and oranges trees may actually increase in growth upon exposure to small doses of atmospheric fluoride. But if the amounts are large, crop production is decreased and natural vegetation may be changed in composition.

Figure 5-11 Petunia plants react very strongly to smog, which has become a common component of the atmospheric environment of most cities. [Courtesy California Statewide Air Pollution Research Center.]

Smog

The term **smog** was coined in London in 1905 by combining the words *smoke* and *fog*. Los Angeles, the smog "capital" of the world, established an Air Pollution Control Board in the late 1940s to study the problem and to prescribe a course of action leading to its control. Research has shown that smog contains dusts, oxides of nitrogen, hydrocarbons, ozones, sulfur dioxide, aldehydes, and numerous other compounds. The components of smog most damaging to plants seem to be oxidants, which include ozone, nitrogen oxides, and peroxides (Fig. 5-11).

Dr. Aarie Haagen-Smit, a pioneer in the study of the atmosphere in and around Los Angeles, found that the hydrocarbons and nitrogen oxides from automobile exhausts react in a normal atmosphere in the presence of sunlight to yield toxic gases, which cause plant damage and eye irritation. The end products from this series of reactions are a group of compounds collectively called peroxyacyl nitrate (PAN). PAN and ozone (photochemical oxidants) cause great damage to plants (Fig. 5-12). Even if exposure is as slight as 5 parts per million for 10 minutes, some sensitive species are damaged (Fig. 5-13). PAN

Figure 5-12 These pinto-bean plants were exposed to air containing PAN (at a concentration of 5 parts per million) for ten minutes. The plant on the left was in the light during the exposure while the plant on the right was in the dark. The variation between the two illustrates the photochemical nature of the phytotoxic effects of PAN. [Courtesy EPA.]

Figure 5-13 'Cherry Belle' radishes grown in unfiltered air (left) and filtered air (right). The damage to the plant on the left was caused by various oxidants. [Courtesy USDA.]

causes several types of injury to plants. Glazing and bronzing occur on spinach, lettuce, endive, and sugar beets.

Other gaseous pollutants

Additional gases that have caused air-pollution problems include ammonia, chlorine, and other gases that have been accidentally spilled into the atmosphere and have caused local problems. Even ethylene, a natural plant product, will injure plants if it is present in too great a quantity (Fig. 5-14).

Suspended particles

Another type of pollution injurious to plants, suspended particles or dust (often referred to as "particulate matter"), can be introduced into the air by smelters, smokestacks, mining operations, cement plants, quarries, strip mines, mineral-processing operations, and other industries, as well as by various agricultural practices.

Dust, when it is most damaging, can sandblast seedlings. The epidermis of cotton, tobacco, and even young pine seedlings can be eroded away at the soil surface by wind-blown particles. Such damage oc-

Figure 5-14 Ethylene can injure greenhouse crops, such as the orchids shown here. [Courtesy California Statewide Air Pollution Research Center.]

curs frequently in the coastal plain of the southeastern United States.

When particulate matter coats leaves, photosynthesis can be impaired. Finer particles may obstruct stomatal openings so that gas exchange is inhibited. When airborne particles settle on some crops, such as lettuce and celery, the quality of these vegetables may be impaired, as may that of fruits and ornamentals.

Perhaps the most subtle of all damage caused by dust may be indirect: dust particles can carry pesticides. Even particles that have been airborne for as much as 1600 kilometers (1000 miles) have been found to have a coating of pesticides. This means, then, that the spraying practices of a corn farmer in Kansas could have a direct effect on, for example, grape production in New York.

Another kind of potentially damaging particle comes from the smoke produced by controlled burning. Foresters have for a long time burned ground cover and low-shrub vegetation of certain forest types. These creeping ground fires are used for several purposes: (1) fuel reduction, to prevent the accumulation of dangerous amounts of brush that could fuel a serious forest fire; (2) destruction of pathogen spores (such as the fungus that causes the brown-spot needle blight); and (3) improvement of visibility. Unfortunately, such controlled burns produce much objectionable smoke (Fig. 5-15). However, an alternative solution for the residue problem is not yet apparent.

Dusts from various sources, aerosols, and other kinds of particulate matter, including smoke, all cause injury to shade trees.

SOLUTIONS TO AGRICULTURE'S POLLUTION PROBLEMS

Many approaches are necessary to solve the problems of pollution—to reduce agriculture's own contribution as well as to alleviate the ill effects of pollution on agriculture. Many of the cures are already well known, such as soil-conservation practices. Others are new, because many of the pollutants they are meant to control, such as mercury, were not previously a problem, or the problems they caused remained

Figure 5-15 Controlled burns have been used for many years in routine forest management. Designed to reduce ground cover and prevent litter accumulation, they are "cool" fires that are managed so that they spread into the wind. Consequently, much objectional smoke is generated. [Courtesy St. Regis Paper Company.]

unrecognized or underestimated (Fig. 5-16). It should also be remembered that there is almost always a trade-off between alternative approaches to environmental problems. For example, the use of fertilizer and pesticides has prevented much soil erosion. Corn acreage has remained at about 40 million hectares (100 million acres) since the mid-1920s in the United States, while output has more than doubled. If yield increases had not occurred, it would have been necessary to push corn production onto the more fragile margins—the steeper slopes and the areas with poorer soil. Thus, the control of one pollution problem, soil erosion, is achieved in part at the expense of chemical pollution.

Improving agricultural practices

Because many of the solutions for pollution have a negative short-term economic value, there is no immediate profit incentive, which is the usual motivat-

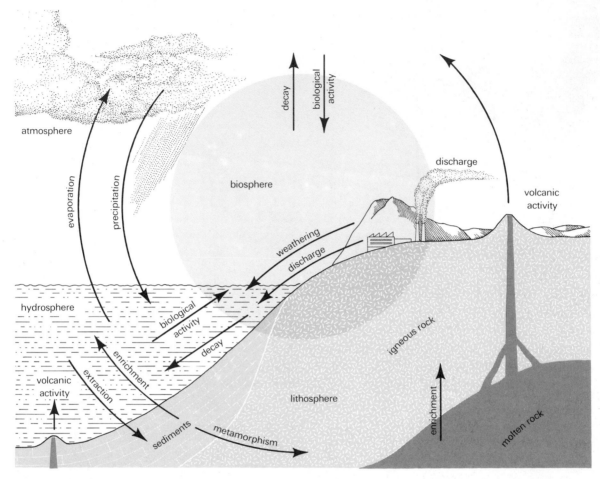

Figure 5-16 Mercury is present in the lithosphere, the hydrosphere, and the atmosphere in trace amounts; however, it becomes concentrated in organisms by means of various biological processes. [From L. J. Goldwater, "Mercury in the Environment." Copyright © 1971 by Scientific American, Inc. All rights reserved.]

ing force behind capital investment and action. The following consideration of the possibilities for cleaning up agricultural pollution will indicate the complexity of the task.

Increased use of agriculture residues

Rather than to discard waste products from agriculture, it would be infinitely wiser to find uses for them. Attempts have been made, but the primary motivation has usually not been the reduction of pollution but rather the possibility of increasing profits.

The incentive to "sell" a material that has posed an expensive disposal problem is very strong.

Organic waste from processing plants may be used in the future as a soil additive. Experimental trials have already begun and feasible technology is known. However, if these wastes are to be successfully used, soils must be porous enough to accept the large volumes of water that carry the solid waste material, and transportation costs from factory to field must not be prohibitive.

Disposal of waste with a high cellulose content is especially difficult. Cellulose wastes of many kinds could be converted to protein if they were broken down into sugars and fermented in the presence of nitrogen and phosphorus. Although not a profitable process at this time, it could become so in the future. Termites can use wood as an energy source because microorganisms in their intestine break down the cellulose into sugars. In a controlled environment, which would favor growth of such microorganisms, raw sawdust could be biologically converted to produce sugars. Acid hydrolysis is another available technique for transforming cellulose to sugars.

The amount of utilizable wastes from food and food processing in the United States is nearly 50,000 metric tons. Under controlled conditions the potential yield of methane gas from the breakdown of these wastes is almost 30,000 cubic meters (1 trillion cubic feet) each day—hardly an amount of energy to be ignored. There are several methods by which this can be done, but one of the most promising is a microbiological two-step fermentation. Acids, alcohols, and aldehydes are produced in the first step; methane and carbon dioxide are among the products in the second step. Although pilot plants for this process exist, problems of garbage sorting and transportation remain to be solved.

Many other uses of agricultural wastes are technically feasible. Citrus wastes are now commercially processed into a nutritious feed. Paper can be made from rice and cereal straw. Uses of residues from forestry operations have been discussed in Chapter 2.

Water and soil conservation

Water conservation is essential at all stages of crop production. This problem is extensively discussed in Chapters 13 and 17. Industry uses much less water than agriculture, but it still requires vast amounts of fresh, clean water for cooling and processing, and returns most of it, frequently in a highly polluted condition, to its original source. In contrast, most of the water that is used for agriculture is returned into the atmosphere in a "pure form" by transpiration and evaporation, though some of the remainder, which is returned to the original source, is polluted with dissolved salts and particulate matter. The key to better water conservation lies in more diligent application of known water and soil conservation practices.

Soil conservation, discussed at length in Chapters 12 and 16, cannot be separated from water management (Fig. 5-17). The sediment from surface erosion constitutes, by far, the greatest volume of waste from forestry and farming operations.

In Chapter 16, the use of mineral fertilizers is discussed in detail. The pollution problems that have resulted from the practice of fertilization have, in recent years, been largely due to overuse and misuse. Three important principles of fertilizer use have emerged: (1) apply fertilizers only if and when they are needed; (2) avoid overuse; (3) place them precisely for immediate uptake by the plants. Adherence to these principles, rather than the use of "shotgun" methods, reduces the cost of fertilization as well as the attendant pollution problems. The proper use of fertilizer materials requires an exact understanding of what nutrients are required and when they are needed.

Reducing pesticide hazards

Minimizing the hazards associated with pesticide usage is a complicated problem without easy solutions. In assessing the effects of pesticides, consideration must be given to the protection of the applicator and the consumer of the crop, and the possible consequences of those residues that may pass into the ecosystem to become worldwide contaminants. Still, if pesticides were completely eliminated, the alternatives would be severe crop losses or failures, increased filth from pest infestation, and problems resulting from disease-associated toxins (mycotoxins)—all creating, of course, totally unacceptable problems for the human population.

Feasible solutions include a variety of measures: more precise pesticide monitoring, greater use of nonpersistent and biodegradable materials, and increased efficiency. Finally the reliance on chemicals alone must be decreased. This goal may be accomplished by integrating other systems of pest control with the use of chemical pesticides.

Pesticides have undergone several stages of development. The early insecticides consisted of quite simple and crude chemical poisons, such as nicotine sulfate for use against sucking insects, lead arsenate to kill chewing insects, and kerosene, which was

Figure 5-17 Even the most severe soil erosion can be corrected. (A) The owner of this gullied land in Buncombe County, North Carolina, planted the area with pine and locust seedlings provided by the Tennessee Valley Authority in the late 1930s. (B) The same tract nearly 20 years later is heavily forested, and soil erosion has long since been halted. The Tennessee Valley Authority assists local forestry and agricultural agencies in reforestation for timber production and watershed protection. [Courtesy TVA.]

spread on surfaces of ponds to destroy the larvae of mosquitos and other insects. For these materials to be effective, large amounts were required, and consequently serious problems occurred, not the least of which was the inefficacy of control in many instances. The so-called second-generation insecticides, first used in World War II, were fantastically potent in very minute amounts. DDT, the first of the chlorinated hydrocarbons, can be credited with staving off the plague during the period of reoccupation in southern Europe and has been used throughout the world for malaria control. Many other "miracle" insecticides followed: for example, parathion and malathion. Three major problems soon became obvious: (1) many of the insecticides were broad spectrum killers, eliminating both harmful and beneficial organisms; (2) the insects against which these chemicals were directed soon developed resistance to them, which greatly reduced the efficacy of the chemicals; and (3) insecticide residues began to accumulate in the environment and to be concentrated by passing through food chains (Fig. 5-18).

We now understand that we must require pesticides to be selective as well as safer. Either they must be **biodegradable** so that they can be readily decomposed by the action of microorganisms in the environment into which they are introduced, or they must be chemically degradable. A "third generation" of insecticides, which would overcome present deficiencies, is presently being developed. Some achieve selectivity through their effects on specific phases of insect development. For example, materials that mimic the juvenile hormone of specific insects prevent larvae from maturing, or render adult forms permanently sterile. One juvenile hormone, juvabione, has been found to kill bugs in the family Pyrrhocoridae, which includes several extremely destructive cotton pests in Asia.

But the problem does not end with the selection of a pesticide that seeks out only a specific target. It has been calculated that only one pesticide molecule out of every million released into the environment "hits" its intended target. More effective placement would greatly reduce the minimum amount required. For example, some species of insects produce chemical attractants (pheromones) that attract insects of the

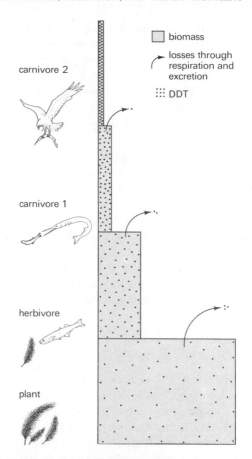

Figure 5-18 As "biomass," or living material, is transferred from one link to another along a food chain, usually more than half of it is consumed in respiration or is excreted; the remainder forms new biomass. However, the losses of DDT residues along the chain (*curved arrows*) are small in proportion to the amount that is transferred from one link to the next. For this reason, high concentrations of DDT residues occur in carnivores. [From G. M. Woodwell, "Toxic Substances and Ecological Cycles." Copyright © 1967 by Scientific American, Inc. All rights reserved.]

opposite sex over great distances. The isolation and synthesis of certain of these sex attractants has made it possible to use them as lures to achieve more effective destruction with very reduced quantities of the active insecticide.

Other notable alternatives to chemical control are the various "biological methods." These include the use of genetic resistance and the introduction and control of natural predators, diseases, or genetic defects in the pests themselves. The principles of biological control are not new (see Chapter 18 and 19). Although biological control is not adaptable to all situations, the use of these methods must increase, and a more stringent regulation of agricultural chemicals must be put into effect.

Perhaps the best hope for effective and safe pest control lies in a combination of methods that depend on minimal use of chemical pesticides coupled with other physical or biological methods. This **integrated control** appears to be more realistic from an ecological point of view for it relies on a number of complimentary methods. It assumes that control, though perhaps not perfect, will be at least sufficient to prevent serious loss, and it assures that the cure will not be more hazardous than the malady.

Protecting plants from pollution injury

The obvious solution to the ill effects of industrial pollution on agriculture is to reduce pollution at the source. Unfortunately, although many give verbal support to such a solution, few would willingly pay for its consequences. There is no evading the fact that the reduction of pollution must increase the cost of the products we all use, whether it be the cost of buying or running our automobiles or that of turning on the light. But completely eliminating pollution is neither a realistic nor a necessary goal. The aim must be to reduce pollution to a level that will not cause deterioration of our environment in the course of an indefinite period of time. This goal is essential, not only for our own health and well-being but also for the health of the plant life upon which our survival depends. Materials that pollute the air, water, and soil tend in various ways to have adverse effects on plant growth, productivity, and quality.

Although plants may themselves help to detoxify the environment, this effect appears to be minor and therefore provides no excuse for us to relent in our efforts to reduce pollution. In fact the evidence is clear that, in many areas, present levels of pollution are highly toxic to plant life. Growing plants in the

filtered environments of greenhouses on a large scale is not a practical possibility. Although a number of studies have indicated that such environmental factors as light and nutrition affect plant reactions to pollutant injury, this does not seem to offer a viable method of control. There do, however, appear to be genetic differences in the ability of plants to react to many of the common pollutants.

At present, we know of two mechanisms of pollution "resistance" in plants: (1) physiological systems that allow the plant to tolerate excesses of pollutants, (2) structural and physiological mechanisms that tend to prevent pollutants from entering the plant. The second system is the better understood. Airborne pollutants enter the leaves largely through their stomata. Resistance to ozone injury has been related to the sensitivity of guard cells, which close the stomata in the presence of ozone, just as drought resistance in many species is a function of stomatal response to water stress: closure of stomata by these cells prevents pollutants from entering the plants just as it prevents excessive water loss by preventing transpiration. Significant differences in guard-cell sensitivity occur between species and within species.

In areas of high pollution, one can preferentially plant those species that resist pollution injury through their inherent protective mechanisms. For example, deciduous trees are much more resistant to air pollution than evergreen trees. Pollution injury in Germany is almost entirely restricted to conifers, and in the highly industrialized regions of that country foresters are in the process of replacing spruces and firs with hardwoods and such resistant conifers as Austrian pine and Corsican pine. Even within the same stands, it is possible to find variation from tree to tree in resistance and susceptibility. Eastern white pine, which is native to the eastern United States and Canada, is highly variable in its resistance to oxidants and to coal smoke and gas from power plants. Propagation from cuttings from particularly resistant trees has made it possible to produce plant stock for areas in which these pollutants are common.

Similar differences in susceptibility have been found in lines and cultivators of such plants as grape, onion, tomato, and corn. The precise basis for these variations in resistance is unknown, but selecting plants with genetic resistance appears to be a viable possibility in the continuing efforts to combat pollution injury.

Recently, evidence that pollution resistance can be obtained by chemical treatment has been discovered. Butanedioic acid mono-(2,2-dimethylhydrazide) (daminozide) applied to petunias provides them with resistance to ozone injury. Various materials that affect stomatal apertures are also known. This new area of research offers another means for the amelioration of the effects of pollution on plants.

POLLUTION AND THE LAW

The National Environmental Policy Act (NEPA) of 1969 places the responsibility for assurance that any Federal activity will not be significantly detrimental to the quality of the human environment directly on the Environmental Protection Agency. Furthermore, in some states, the courts have decided that private individuals and industries also will be subject to the provisions of similar laws established by the governments of those states. Actions that take place in one state and that affect the environment of people in other states will be subject to federal scrutiny too. For example, if pesticide sprayed in one state drifts into another, federal action could be invoked.

Agribusiness will be subject to such laws as NEPA. The provisions of this act call for the presentation of an Environmental Statement, popularly called an **Impact Statement,** prior to the initiation of new federal projects. The statements must include the following items:

1. A description of the proposed action, including information and technical data adequate to permit a careful assessment of its environmental impact by agencies, institutions, and individuals.

2. A description of the probable impact of the proposed action on the environment, including its ecosystems and their constituent organisms, such as wildlife, fish, and marine life. Both primary and secondary significant consequences for the environment should be included.

3. A description of any probable adverse environmental effects that cannot be avoided.

4. A listing of alternatives to the proposed action. Sufficient analysis of such alternatives, their costs, and their impact on the environment should accompany the proposed action through the review process in order not to foreclose prematurely any options that might have less detrimental effects.

5. A discussion of the relation between local short-term human uses of the environment and the maintenance and enhancement of its long-term productivity.

6. A listing of any irreversible and irretrievable commitments of resources that would be involved in the proposed action.

7. A discussion of the problems and objectives raised by Federal, state, and local agencies and by private organizations and individuals in the review process, and proposals for the disposition of those issues.

The statement is routinely sent to various agencies, organizations, and individuals for review, and to others on request. If one is not satisfied that the environment will remain unharmed, court action in the form of an injunction may be taken that can delay construction or other action until the statement is satisfactory to all concerned.

NEPA applies particularly to projects with large potential impacts on the environment. Dams, power plants, and canals are clearly within the scope of the law. It is reasonable to expect that, in the future, projects having less potential for impact on the environment will also be subject to scrutiny. Even individual crop production units, such as farms, may be reviewed, and their owners may then be required to implement the best possible conservation practices, a measure that would increase the cost of crop production.

In 1972, the Federal Environmental Pesticide Control Act was passed. This Act will have a strong effect on American agriculture because it provides that private citizens who wish to apply restricted-use pesticides must be certified and have a permit even to purchase them. It also delegates the responsibility for a plan to carry out the provisions of the Act to the individual states. The 1972 Act also specifies a list of pesticides for which no permit need be obtained.

The laws on the use of pesticides are complex, and they are constantly changing. Some pesticides, such as DDT, sodium fluoroacetate, and organic mercury fungicides have been permanently banned. Others, including parathion, paraquat, and 2,4-D, are presently restricted to use by persons with an appropriate permit. A few, such as endrin, are banned from use only on food crops. Those that present very little or no danger to people, such as pyrethrum, malathion, and rotenone, are unrestricted. It seems likely that additional changes in legislation and in administrative rulings will be necessary in order to achieve effective safety standards in the marketing of pesticides and at the same time to continue to provide incentives for the development and marketing of more effective pesticides.

ECOLOGY AND CHANGE

Perhaps the greatest ecological principle is that change is inevitable. Whether or not we inflict ourselves and our artifacts on an ecosystem, that ecosystem is constantly changing. A mature forest ecosystem, such as a deciduous forest of beech and maple, one of the most stable of all kinds of vegetative communities, undergoes fluctuations, compensations, and counterbalances that take place in a dynamic equilibrium. The components are constantly changing in intensity, numbers, kinds, and complexity.

The second great important principle is that change is directional—that is, always toward greater complexity, stability, and maturity. In plant succession, various kinds of organisms give way to succeeding ones until, at last, the organisms and the physical environment are in dynamic harmony.

When we inflict ourselves on a successionally mature and stable forest by cutting trees or by clearing the whole forest for pasture or for cropland, an earlier successional stage of plant life occurs. In a cornfield, our weed problems are not caused by the trees and shrubs of the mature forest that once occu-

pied the cornfield and that will eventually occupy it again, but by invading herbaceous weeds of an earlier successional stage that are once again able to compete for soil moisture and nutrients.

Nature does not necessarily give us trouble when we manipulate it wisely, but our actions must be planned, so that we take into account the checks and balances that have evolved in the course of time. Common sense dictates that we should try to predict the consequences of our actions. Realistic assessments of what the consequences might be, if a dam is built, the soil plowed, pesticides are applied, or a forest is cut, are completely within the capabilities of the human mind. If we fail to consider ecological principles in producing our crops, then we can expect more and even greater disasters than in the past.

THE POLITICS OF POLLUTION

"Public health is purchasable. Within natural limitations any community can determine its own death rate." So said Herman Biggs, the State Commissioner of Health for New York, in 1914. Even though his words were unduly optimistic at a time when communicable disease was rampant, it reflected the fundamental truth that a community or a nation can do much to determine its own destiny.

Today this motto is more realistic. Medicine and biology have controlled many of the diseases that were serious problems in the past. Yellow fever, cholera, malaria, typhus, diphtheria—all those scourges of the past promise to disappear in time, as smallpox did in the 1970s. The black plague, which is estimated to have killed three-quarters of the population of Europe and Asia in the fourteenth century, is now only a textbook curiosity.

Now, however, we have other kinds of diseases—those caused by air and water pollution. They affect our food sources as well as our bodies. These problems, even though multiplying at a constant rate, are of a different type, for while they may not always affect our eventual life span, they usually do affect our immediate well-being. Solutions to the problems can be only temporary, because our populations, communities and industries are constantly changing and increasing in size.

Although business leaders, politicians, ministers, and newspaper editors may try to move the public to take action against pollution, a high degree of public impatience is usually required before any action is actually taken. No matter what the objective facts may be, air pollution will continue to be a problem until community action is galvanized by the subjective perception that it is harmful, unpleasant, and inconvenient. Few people understand scientific measurements of suspended particles, gases, and residues in terms of illness, lowered crop production, and impaired quality of life. Indeed, some people who have grown up in areas with severe air pollution regard it as normal.

After twenty years of intensive study, British authorities have concluded that atmospheric SO_2 has no effects on human health. Yet, they failed to consider its effect on crop plants, which are harmed by SO_2 (in combination with ozone, which is ever-present) at levels that are hundreds of times lower than those proven to be directly harmful to humans. It seems that immediate personal comfort and prosperity predominate over eventual human needs, and that it will require a major catastrophe, in the form of a food shortage, to provoke the public into action. Our economy is based on attaining more speed, more power, and more immediate wealth. Personal well-being and job security have been our guidelines. We must overcome our reluctance to view the world as a life-supporting ecosystem.

CHEMICALS IN FOOD: POLLUTION VERSUS PROTECTION

In the fall of 1959, cranberry growers in the United States had harvested their largest crop in ten years. There were large stocks of canned cranberries on grocers' shelves as well as fresh ones at the produce counters. Everyone, it seemed, would have cranberries that Thanksgiving and Christmas.

On November 9, only about two weeks before Thanksgiving, the Secretary of the Department of Health, Education, and Welfare, Arthur Flemming, held his regular Monday press conference. He told the public that a chemical, aminotriazole, used to kill weeds in certain cranberry fields, had also been found to induce cancerous growths in the thyroid glands of rats. He told the public that, if they could not be sure that their cranberries had not been contaminated with aminotriazole, they should not eat them in any form. As a result, grocers removed them from their shelves and restaurateurs dropped from their menus.

Then, in March 1960, the Food and Drug Administration announced that the general public had been confused by ". . . misleading information from various sources," and was therefore uncertain about the magnitude and the significance of the problem of chemicals in food. Even black jellybeans seemed to be unsafe to eat. People wondered: Is our food safe? Must chemicals be added to it? The answer is yes, to both questions, but not without evidence and reason.

Aminotriazole is a chemical that, applied selectively to weeds, stops chlorophyll formation, prevents carbohydrates from being formed, and causes the weeds to die. By feeding rats from 10 to 100 times the amount of this chemical found on a cranberry bush daily for a period of 70 to 100 weeks, thyroid carcinomas were induced. This means that each of the test rats that developed cancer had received from 700 to 10,000 times as much of the chemical as was found on an entire cranberry bush—fruits, leaves, stems, and all. Is this the kind of evidence we need for making decisions about which foods to include in our diets?

Common table salt, or sodium chloride (NaCl), is essential to life, yet it could cause death if taken in large quantities. Even glucose has caused death when injected into animals as a 15-percent solution, yet glucose is needed to sustain life. Chickens that were fed a diet of bran and hen's eggs developed cancer of the ovary more frequently, and at an earlier age, than chickens that were fed a normal diet. How many other dangerous substances could be found in the foods we eat every day, if there were sufficient time and labor to make the tests? Even lettuce and carrots might have to be eliminated from our diets after careful investigation because of the large amount of carotene (provitamin A) they contain. We would also be forced to ban those foods and drugs that cause some people to have allergic reactions, such as tomatoes and fish. Even penicillin might be banned, for it has been known to produce severe allergic reactions in some people, and even death.

There is a difference between *hazard* and *toxicity,* but it is a difference that many food-testing agencies have not always recognized. Many of the foods that we eat every day contain substances that are toxic to some degree but that do not present a hazard to health. The philosophical approach to *zero hazard-tolerance* for all persons is neither feasible nor possible.

Then, too, there is the possibility that unscrupulous people might add harmful chemicals to food in order to maximize their profit, a practice we can punish but not prevent. A fish dealer in Philadelphia, to cite one example, wanted to sell a batch of spoiling fish, so he dosed the fish liberally with sodium nitrate to retard the spoilage and sent them out to markets along the Atlantic coast. As a result, 150 people became ill with nitrate poisoning and a three-year-old child died of it. Sodium nitrate, or Chile saltpeter, has been used since the Middle Ages to fix the color of meat and, in small quantities, is fairly innocuous. In large quantities, it causes vasodilation and a subsequent drop in blood pressure, which can be fatal. Think of the army of inspectors that would be required to protect us from all incidents of this sort.

Merely because a chemical occurs naturally in foods does not imply that it poses no danger. Yucca (*Manihot esculenta*) is a plant with an enlarged starchy rhizome that is eaten in the tropics throughout the world. Yet the skin of the rhizome contains a deadly poison, potassium cyanide. The rhizome is safe to eat only if it has been peeled or if it has been cooked until the toxin is leached out or destroyed. Nature has not provided us with a poison-free environment.

A hamburger, a bowl of spinach, nonnutritive sweeteners, mold inhibitors, and the other foods and additives we ingest in one way or another are all basically chemicals, even though the foods have considerably more complexity than most of the chemicals we add to them. Who could write out the chemical structure of a T-bone steak, crepe suzettes, a western omelet, an Italian salad, or a pizza? Obviously, most of the chemicals in these items are not hazardous, at least to many of us. But some of us cannot even consume certain animal proteins without an allergic reaction.

The most important reason for intentionally adding chemicals to foods is to enhance their storage and keeping properties. For example, wheat flour normally contains the spores of a bacterium called *Bacillus mesentericus,* which are able to survive the high temperatures of baking. During storage, the bacteria multiply and alter the flour, which causes it to produce a ropy dough, making the bread inedible. Small amounts of sorbic acid or sodium propionate retard the growth of molds and bacteria, enabling the foods to which they are added to withstand longer periods of storage. In canned foods, especially those that are basic rather than acidic, chemicals are added to inhibit the growth of *Clostridium botulinum,* an anaerobic bacterium that produces what is perhaps the most devastating toxin known.

With increased numbers of women joining the labor force, and an increased desire to spend less time in the preparation of meals, there will be greater emphasis on precooked, prepackaged, ready-to-eat foods. This will require more extensive use of food additives and other methods of food preservation. We must continually search for new methods of food preservation, even while we test the chemicals we now use to determine whether they pose any hazard to health. The results of our tests must be true and valid, and the manner in which we apply the results must be logical and reasonable. It would be just as unreasonable to ban a chemical because it has caused allergic reactions in 10 out of 10,000 mice when added to their diets for a period of two years starting immediately after weaning as it would be to ban it because it has an unpronounceable name, such as cyclopentanoperhydrophenanthrene (a widely used food additive, otherwise known as vitamin D).

ECOLOGY AND AGRICULTURAL PERFORMANCE

Ecology is the study of the interactions of organisms with one another and with their environments. In practice, this study is tempered by subjective consid-

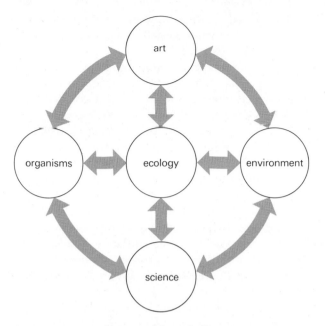

Figure 5-19 Ecology is neither wholly a science nor an art; rather, it is a way of considering all of the relations between organisms and their environments in both space and time. The judgments that must be made regarding the interplay of organisms and environments require perspectives ranging from the objectivity of science to the subjectivity of art.

erations (Fig. 5-19). The task of the ecologist is to investigate, understand, and explain such phenomena as forest succession, the growth of cities, and the conditions that must be maintained for sustained crop yield. The ecologist does not necessarily make economic decisions on the maximization of profits, nor is he expected to provide a commentary on the benefits of his thoughts to society. Rather his role is to understand the functioning of ecosystems and to predict what will happen when they become unbalanced.

Human disturbances of the balance of nature are not limited to the twentieth century. In the course of the past millennia, we have caused countless ecological disasters through our various efforts to circumvent or thwart nature. A few examples illustrate this point.

In the United States, the Great Plains region has always been subject to periodic drought. As long as the plains were covered with short, sod-forming native grasses, the dry spells did little damage. But in 1889, when the Oklahoma territory was opened for homesteading, the number of settlers in that territory increased from a few thousand to 60,000, and more and more acres of protective grass were plowed under to make room for crops. This took place during a period of abundant rainfall, and by 1900, the rich fertile soils of Oklahoma supported 390,000 people. In 1910 a dry spell came, but by 1914 the soils were, once again, back in full production. In 1924 there came a prolonged drought, and the parched topsoil of Oklahoma was blown away by the wind, some of it as far as the Atlantic Ocean. Great hardship befell the people of Oklahoma, resulting in the migration of thousands of them to the West Coast, a phenomenon documented in John Steinbeck's moving epic *The Grapes of Wrath*.

The Sahel is a broad expanse of land extending across Africa from Senegal to Ethiopia, between the Sahara Desert to the north and the tropical forest to the south. In the Middle Ages it was the home of the vast empires of the Mali, Ghana, and Songhai peoples. Until recently, it was culturally and economically one of the most important regions of Africa, largely owing to the remarkable adaptation of its nomadic herders to a semiarid environment.

For hundreds of years the nomads herded their cattle as far south in the dry season as they could go without reaching the habitat of the dreaded tsetse fly, the carrier of sleeping sickness. There, their cattle were allowed to graze stubble in the fields of the sedentary farmers of the region in exchange for manure. They also exchanged manure for millet, one of their staple crops.

When the rains came, grass came up once again and the herds moved northward. The migration continued north as long as the grass ahead looked greener than that at hand, until they reached the northern edge of the Sahelian rain belt. When that grass was eaten the herds started south once more, grazing on grass that had grown since they had passed some months before. The tribal chiefs worked out the routes and rates of travel, as well as how long a herd might stay at a given well. In this manner, food and water was provided to all with the least possible danger from disease and from conflict with other tribes.

But today, much of the Sahel is an unproductive desert. Western medicine has increased the birth and survival rates, so it has been necessary to increase the herds of cattle to support the growing population. As a result, overgrazing has damaged the land. Cotton and peanuts were introduced by the French to earn the best possible rates of exchange, but using 5-year rather than 20-year rotations. Consequently, the crops depleted the soils and productivity decreased. The process of *desertification* was thus started.

By 1960 many small countries in the region had declared independence, and the traditional migrations of the nomads were halted by political boundaries. As a result, they were forced to graze their cattle in a smaller area, further depleting the resources of the land they occupied.

About 300 meters (1000 feet) below the Sahel there are vast reservoirs of water that can be tapped by wells. Thousands of wells were dug, costing as much as $200,000 each. During the recent drought, thousands of dead and dying animals from hundreds of miles away clustered around the wells creating miniature deserts of 15 to 20 hectares (about 35 to 50 acres) by their incessant overgrazing and trampling of the soil. The animals depleted the forage around the wells and died of starvation while bloated with water. At the present time there seems to be neither a social nor an ecological solution to the problems of the Sahel.

In the Middle East, engineers had long contemplated building a great dam across the Nile to impound water from the spring rains for irrigation during the hot, dry summers. The Aswan Dam, completed in 1971, was built to fulfill this objective as well as to provide hydroelectric power. Water storage was begun in 1964, in the partially completed dam, and a river that had flowed freely for thousands of years was suddenly trapped and harnessed.

Some of the anticipated benefits of the dam have materialized. Power is being produced, and higher yields of cotton, grain, fruits, and vegetables have been obtained in some locations. But at the same time, many problems have resulted. The sediment that was formerly washed out to sea carried mineral nutrients with it that were important to the food web of offshore life. Consequently, the annual catch of sardines decreased by 18,000 tons. Each year the flooding waters carried away salts that would otherwise have accumulated in the soil. Once the flooding stopped, salts began to accumulate and soil salinity is now a threat to productivity. Flooding also carried the silt and its nutrients into the fields downstream from the dam. Now that the nutrients are no longer being added to the soil naturally, chemical fertilizers must be added. The shorelines of the Nile delta are being eroded now that sediments are no longer being deposited. Other kinds of problems have arisen too. During the dry periods, a snail fluke called *Bilharzia*, which can cause a serious human disease known as schistosomiasis, was kept under natural control. After the irrigation canals were built, that natural annual control ceased to exist, and the fluke, which spreads easily to people, now infects 80 percent of those people who work in the fields that are irrigated by water from the canals.

In parts of India today, overgrazing by goats is destroying ground cover while goatherds, using sickles with 20-foot handles, cut the limbs from trees so that their animals can eat the foliage. The trees then die and the unprotected soil blows away, so that the sites are no longer productive.

Each of these case histories is an example of *ecological backlash*—an unforeseen detrimental consequence of a modification of the environment that cancels out, or overcompensates for, the long-term net gains that were expected. Failure to foresee such consequences makes for poor conservation.

In plant agriculture, conservation means *wise use*, the aim being to insure a sustained or increased yield of useful plant products while maintaining an environment that is compatible with esthetic needs, recreation, wildlife, and all other uses to which the land might reasonably be put (Fig. 5-20). It implies planning and foresight and demands the application of ecological principles. The preceding examples of land misuse are the result of faulty planning caused primarily by a misunderstanding of these principles.

Figure 5-20 In the early days of American forestry, the primary emphasis was on fire protection. Today, foresters are expected to provide not only wood, but recreation opportunities, a continuous supply of good water, wildlife, and pleasant surroundings. [Courtesy U.S. Forest Service.]

RATIONAL CHOICES

Throughout history, people have feared, condemned, and destroyed or banished not only other human beings but also various animals, plants, and objects—even ideas and concepts—that it was thought might possibly cause injury or discomfort. Today we still launch inquisitions and attacks on people, and even on various inanimate objects. For example, the banning of DDT would appear to have been the result of

this kind of decision making. The evidence indicates that DDT, used properly, can make the quality of life better for people throughout the world. Used improperly, it may be a true hazard in the environment (see Fig. 5-5). A rational approach to this problem would be to control and restrict its use to situations in which it will not create an environmental hazard. However, the emotional approach—to ban it completely because it poses a potential threat—has prevailed. If the same approach were applied to other compounds and products, we might very well forbid the use of such everyday items as sugar. Eaten in too great a quantity, sugar will cause obesity, cardiovascular problems, and shortened life spans. Yet, most of us would never think of eliminating sugar from our diet, although wise use restricts its intake. Every age has had its witch to attack, and even an enlightened age such as ours has not denied itself this pleasure. DDT has been one of the objects offered for our ritual sacrifices.

In societies that are underdeveloped or that have few resources, pollution is not a problem. When Eskimos kill a seal, practically every part of the animal is used for food, clothing, or tools; no part that has a potential use is discarded. Even if it were, the population density is so low that it is unlikely that the waste would cause a serious problem. When an Indian in the Amazonian jungle grows a crop of manioc and bananas, the detritus from the unused parts of the plants does not accumulate to create a pollution problem because it decays rapidly in the heat and humidity of the jungle. In such societies, a bottle or a can, once emptied of its original contents, is likely to be coveted and used again and again.

Unfortunately the same cannot be said of our own enlightened and highly developed society. Our monumental garbage dumps are dwarfed only by the problems we encounter in trying to find locations for new and even grander ones. Our roadsides have become convenient dumping places for unwanted cans, bottles, paper, garbage, and discarded car bodies. An advanced society or community "learns" to perpetuate itself by recycling the materials essential for continued growth—wasting little, reusing most.

Frequently, pollution is a problem of concentration. The wastes that a herd of beef cattle produce on an open range or pasture usually present no problems; rather, they actually contribute to the maintenance of the ecosystem. The same herd confined in a feed lot next to Kansas City, however, will cause very serious problems! Similarly, a small portable sawmill operating in a farmer's woodlot for four to six weeks

before it is moved away does not produce a pile of sawdust so large that it cannot be leveled and scattered about with a farm tractor—or even be used as a soil amendment, with the addition of appropriate mineral fertilizers. But a large sawmill operating for many years in a single location will create serious problems of sawdust and bark disposal. In years past the sawdust was burned, either to help supply the power requirements of the mill, or in a special disposal furnace. Today, sawdust is increasingly used in paper or as a filler in many products. Even bark, a by-product once thought to be absolutely without positive economic value, is being widely used as a mulch for ornamental plants.

Still, we cannot escape the fact that we must convert enough land, produce enough fertilizers, cut enough timber, mine enough minerals, plow enough land, process enough food, and dispose of enough waste to feed, house, and clothe a population that could nearly double within the next 30 years. Our problems have only begun! If we wish only to maintain our present standards of living, production still must be substantially increased. If we wish to continue to increase our living standards, then production must expand even more. Consumers must ultimately make the decision about what is to happen, for it is they who will pay the costs. Good conservation practices, in which high quality is maintained while pollution is abated, will increase prices for everyone, and we must be prepared to pay.

It is folly to believe that completely unguided individual action will result in greater food production and a cleaner environment simultaneously. Past events suggest that human beings are naturally aggressive, acquisitive, and selfish, and controllable only by mutual coercion. Given the freedom of choice, we dump wastes in rivers because it is cheap and easy. We stop only if our laws say that we must do so. Thus, our laws must be so structured that individuals and groups are motivated to protect the quality of the environment. The regulations must be feasible, enforceable, and enforced.

Selected references

Benarde, M. A., 1970. *Our Precarious Habitat*. New York: W. W. Norton. (This book contains material on the problems of pollution and food contamination as they affect contemporary food production.)

Dasmann, R. F., 1976. *Environmental Conservation*. New York: Wiley. (A textbook that brings pollution into context as an environmental factor.)

Environmental Protection Agency, 1974. *Workshop on Agricultural Non-Point Source Water Pollution Control*, Washington, D.C.: Environmental Protection Agency. (Non-point sources of pollution as an agricultural problem.)

Hellman, H., 1972. *Feeding the World of the Future*. New York: M. Evans. (A short, interesting book about human food production and the problems resulting therefrom.)

Hobson, P. N., and A. M. Robertson, 1977. *Waste Treatment in Agriculture*. London: Applied Science Publishers. (Some of the problems of agricultural pollution and their solutions.)

Jacobson, J. S., and A. C. Hill (editors), 1970. *Recognition of Air Pollution Injury to Vegetation: A Pictorial Atlas* (APCS Informative Report No. 1). Pittsburgh: Air Pollution Control Association. (An excellent collection of photographs showing most types of air-pollution injury to plants.

Kepler, K., D. Carlson, and W. T. Pitts, 1978. *Pollution Control Manual For Irrigated Agriculture*. Springfield, Va.: Environmental Protection Agency. (A manual for the control of pollution that results from irrigation practices.)

Lindahl, K. C., 1972. *Conservation for Survival*. New York: William Morrow. (A textbook of conservation that discusses pollution as an environmental factor.)

Loehr, R. D. (editor), 1977. *Food, Fertilizer, and Agricultural Residues.* Ann Arbor: Ann Arbor Science Publishers. (A comprehensive book on most aspects of agricultural waste management.)

Ludington, D. C. (editor), 1971. *Agricultural Wastes: Principles and Guidelines for Practical Solutions*. Syracuse, N.Y.: Cornell University Conference on Agricultural Waste Management. (A collection of papers dealing with disposal of agricultural wastes as well as with alternative uses.)

McEwan, F. L., and G. R. Stephenson, 1979. *The Use and Significance of Pesticides in the Environment*. New York: Wiley. (A book that considers reasons for pesticide use: the potential for good or harm, their movement and penetration into the environment, their effects on target and nontarget organisms, and legislative controls.)

Ormrod, D. P., 1978. *Pollution in Horticulture*. New York: Elsevier North-Holland. (A presentation relating the nature of pollution problems in horticulture to factors that must be considered for their solutions.)

Tinsley, I. J., 1979. *Chemical Concepts in Pollution Behavior*. New York: Wiley. (A treatment of the distribution of chemicals in the environment based on their properties.)

U.S. Department of Agriculture, 1971. *National Handbook of Conservation Practices*. (A handbook of conservation practices that includes many normal crop-production activities.)

Wadleigh, C. H., 1968. *Wastes in Relation to Agriculture and Forestry* (USDA Miscellaneous Publication No. 1065). (A valuable book on the problems of pollution—those affecting agriculture and those created by it.)

Ware, G. W., 1978. *The Pesticide Book*. San Francisco: W. H. Freeman and Company. (A readable introduction to the chemistry and uses of pesticides, with a discussion of their hazards and the laws that control them.)

White-Stevens, R. (editor), 1971. *Pesticides in the Environment*. (2 vols). New York: Marcel Dekker. (A comprehensive collection of papers on the chemistry, analysis, toxicology, and pest resistance of most pesticides in current use.)

Two
NATURE OF CROP PLANTS

6
Origin and Classification

People and the plants upon which they are dependent are both products of evolutionary development. During the first billion or so years, life forms sustained themselves mainly on what happened to be at hand. Not until the last few tens of thousands of years has there existed any organism both dexterous enough and intelligent enough to oversee other forms of life for its own benefit—not until certain especially advanced primates, early members of the human species, slowly learned to manage the environment sufficiently to maintain for their own use certain of nature's wealth of species. In the past few thousand years the tempo of change and exploitation has increased exponentially, as human beings, with their bent for technology, set about literally to alter the face of the earth and to populate it everywhere. How has this been possible? What is the story behind the hundreds of kinds of plants that help form the basis upon which modern societies rest? And how do modern people cope with the need for ever more useful and efficient plant types to sustain the tremendous superstructure of civilization? Evolution and genetics help clarify the origin. Classification supplies the inventory.

THE EVOLUTIONARY SEQUENCE

Although intelligent speculations can be offered, no one can be sure just when, why, or how life first started. Indeed, it is impossible to characterize life exactly, impossible to distinguish it clearly in some of its simpler and more ancient forms from the nonliving. Two chief criteria characterize earthly life: the presence of carbon, typically in large organic molecules, and reproduction or self-duplication—the ability of life to make new life from the nonliving substances in its milieu.

Simple forms of life, probably quite similar to bacteria, existed two to three billion years ago (Fig. 6-1). Their remains have been found in rocks that old, and there is supplementary evidence of their influence on their ancient environment—for example, abundantly oxidized beds of iron in what seems otherwise to have been a reducing environment. This fairly elaborate form of life must itself have evolved over a long period of time in which simple molecules linked to form progressively larger molecules, some of which were able, eventually, to duplicate themselves by assembling appropriate ingredients found in the primordial sea and atmosphere. Even in space, amino compounds (compounds containing $-NH_2$ groups), which are essential to life as we know it, are known to occur. The newly formed earth must have contained such simple ingredients as methane, ammonia, and carbon dioxide. Laboratory tests show that under certain conditions of heat and radiation these compounds can combine into large molecules. The Russian biochemist A. I. Oparin hypothesized a sort of "colloidal broth" for the beginnings of life, and Oparin's followers have elaborated quite plausible explanations of how early life may have started.

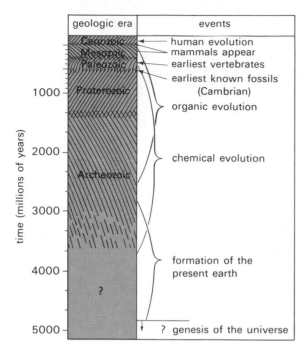

Figure 6-1 Evolutionary time scale.

Although circumstantial evidence indicates that life was created but once, and that from this creation has arisen the whole complexity of the modern living world, life could have been created many times; we cannot be sure. But it is likely that, once created, life so altered the environment that suitable conditions for our hypothesized beginnings no longer existed. For example, it is very likely that the early atmosphere contained little if any free oxygen; it is probable that green plants, through photosynthesis, have altered an original atmosphere rich in carbon dioxide to one quite impoverished of it, and instead rich in oxygen, by tying up the carbon in various biomass products, both living (such as forests) and dead (such as coal and petroleum). Oxygen, a relatively reactive gas, is inimical to the reducing atmosphere seemingly necessary for the slow buildup of compounds leading toward the first forms of life. And with life came such organisms as today's bacteria, pirating for their sustenance large molecules as a source of energy ("food"). No longer could free organic compounds exist for any length of time, awaiting chance concat-

enation in an organic soup; instead they would be immediately utilized by life itself. Though it is possible in the laboratory to assemble proteinaceous molecules into something at least very akin to life, it is unlikely under present-day natural conditions that there would be either time enough or the proper "mix."

All of this, of course, is quite speculative, and in a sense incidental to our story. But from some such beginning have come the myriad of life forms with which classification concerns itself. How new organisms arise and evolve are equally of intellectual concern and essential to providing ever more useful forms for our own utilitarian purposes.

Origin through natural selection

The evidence that evolution has occurred is overwhelming, yet the mechanisms by which it proceeds are still only partly clarified. Not until the publication of Charles Darwin's *Origin of Species* (1859) was any scientifically credible theory formulated. Alfred Russel Wallace, at the opposite end of the earth (Darwin in England, Wallace in Malaysia), arrived independently at roughly the same conclusions at about the same time. But Darwin's work remains the landmark; it is Darwin who is credited with ushering in one of biology's momentous concepts—evolution by means of natural selection.

In essence, natural selection—popularly, "survival of the fittest"—views all life as continuous back in time; all present forms are closely or remotely related, depending upon how long ago and with what rapidity they diverged from a common ancestor. Life is a huge "tree," in which the present forms are represented only by the tips of the many branches; back in time these interconnect. What are now discrete forms arose from similar forms. Only those best adapted to the environment of their times survived.

Organisms are capable of reproducing far beyond the capacity of the environment to sustain all of their offspring. Only a few of the best adapted progeny will survive the competition and yield future generations. Progeny vary, although often only subtly. Those variations that best equip an organism to survive and reproduce its kind are the ones that will be perpetuated by natural selection. Thus did nature, over billions of years, gradually evolve the wide array of organisms inhabiting the modern environment.

Through the ages continents have risen and subsided, seas have come and gone, glaciers have formed and melted, desert has superseded rain forest, and vice versa. With unceasing change there is certainly opportunity for the survival of new forms better adapted to changed conditions. Thus, groups that were once great are now extinct, like the dinosaurs that dominated the earth for many millions of years. A relative upstart, *Homo sapiens,* now has its turn, and, like all other life before, must abide by nature's one pervasive rule—living organisms must come to terms with their environment. To fail means extinction, and millions of extinctions mark the trail from early life to the present. Extinction can come from overaggressiveness that exhausts the resources upon which an organism depends, as well as from inability to meet the challenge of change. Masters of their surroundings to a degree that no other organisms have ever been, people still must prove that they can harmonize their procreation and exploitation with their finite environment.

Origin through crop breeding

Changes in the genetic material (mutations) form the basis for variation in living things. Mutations result from the alteration or displacement of the genes. Along with the considerable variety of gene combinations, mutations are the grist for the mill of natural selection. Evolution can be thought of as occurring by changes in "populations" of genes, from which new harmonious combinations are isolated.

It is evident that mutations that confer some increase in fitness might have a good chance of becoming fixed, at least in a small population. Most mutations, however, tend to be disadvantageous. Low adaptiveness of a mutant tends to lead to its extinction, whereas high adaptiveness of a mutant tends to lead to disappearance of the ancestral type. High adaptiveness in some environments and low adaptiveness in others—a more usual condition—

leads to selection of independent populations. The more progeny that survive, the greater the frequency of their particular genes. This in essence is the mechanism operative in natural selection. Similarly, selection of certain progeny for perpetuation has always been the basic method of the plant or animal breeder.

Most animal life is dependent upon the seed plants, which evolved some hundreds of millions of years ago apparently from fern-like ancestors. People are relatively recent, perhaps only 2 or 3 million years on the scene. Through the ages people learned to save some of the wild seed they found, and to sow this to ensure future supplies. Such was the beginning of agriculture, and the emergence of primitive people from a life of constant roaming in search of sustenance. As was seen in Chapter 1, only when people became sedentary and began living in relatively permanent communities could civilization begin. Even in the most complex of modern societies, the domesticated seed plants and the animals that feed upon them are still the foundation of human existence.

Human alterations of wild plant populations began with domestication—the first step toward crop breeding. It is a marvel that the art could be so cleverly practiced before the dawn of history, and without genetic understanding. The present-day improvement of crops is effected by genetic manipulation in addition to selection. Variable populations of known potentiality are created from whatever useful forms may be selected (see Chapter 19). In the future it is theoretically possible that the breeder will be able to control which gametes survive and which do not.

The plants (and animals) that primitive people found useful were available just as natural selection delivered them. To this day we still use many wild plants, unchanged from the way they occur in nature. Familiar examples are persimmon, hickory, black walnut, Brazil nut, rubber, the various medicinal plants, and the many trees used for lumber, pulp, and firewood.

For most of today's crop plants, however, wild plants were but the starting point for further selection and alteration. In short, people have superimposed upon natural selection their will and their ability to exercise control over their environment in order to hasten the evolution of forms they want. People, rather than nature, now decide which forms will survive and be perpetuated. By undertaking selection, unconsciously or knowingly and with purpose, people have caused the evolution of plants to take place far more rapidly than is normal in nature. As a matter of fact, so complete have been the evolutionary changes that ancestral forms of many domesticated plants are not even known. Such plants that occur only under cultivation—cabbage and corn are examples—are known as **cultigens.** Improvements in yield and quality, achieved through plant breeding, make possible a modern and efficient agriculture.

PLANT SYSTEMATICS

The results of evolution—the plants and animals sharing the earth with us today—have been classified into discrete and identifiable categories termed **taxa** (*singular* **taxon**). Some taxa include members that are only physiologically distinguishable; others manifest unique, clearly visible structural differences. An example of the former might be two soybean varieties that look quite alike but are differentially hardy; an example of the latter might be the giant sequoia, which has few close relatives.

Science must concern itself with some sort of method of classifying life forms. One logical way of arranging the plant kingdom is according to evolutionary relationships. Very closely related taxa are grouped as subunits of the same species; most distantly related and distinctive taxa fall into groups of higher rank. The name attached to an organism not only identifies it, but relates it to all others.

The basic taxon used in classifying living organisms is the **species** (singular and plural). This was presumed by early botanists to be the unit of creation, the distinctive entity. The results of Darwin's concepts and, later on, the results of genetic studies have cast some doubt on that, but the species is still the universally accepted typification that is defined and visualized for practical classification. Whether a species is an arbitary unit on a long intergrading continuum or whether evolution takes place in quantum-like jumps to make discrete units of appreciable magnitude can be argued pro and con. For our purposes a species is best visualized as a population of related and interbreeding forms, which under normal circumstances remain reasonably distinct from similar populations that constitute other species.

Naming plants

To identify and discuss important plants and animals, some name is needed. Orderly trade and commerce depend upon widely recognized names for such familiar plants as bean, maize, white oak, and so on. Our primitive ancestors undoubtedly gave names to the wild plants and animals that they depended upon for food. In fact it appears that primitive peoples, such as the American Indians and the early Micronesians, were much better acquainted with plants and more adept at knowing them by name than is the average highly educated and "civilized" person of today.

Plant identification has been an evolutionary process in itself, starting perhaps with simple names designating use, and ending today in highly codified systems so abstruse as to be intelligible only to taxonomic specialists. (**Taxonomists** are scholars concerned with properly identifying and naming life forms according to internationally accepted rules of procedure, and in the process surmising evolutionary or natural relationships.) As early as the Middle Ages it was recognized that a universal scholarly language must be used that could be understood by learned men everywhere, no matter what their native tongue. The important treatises were written in Latin, and to this day Latin remains the accepted language for the definitive or scientific naming of organisms. Latin is a dead language, unchanging, understandable to scholars everywhere, hence eminently suitable for such universal communication.

The internationally accepted rules for naming plants prescribe that each plant be distinguished by two primary identifying names in Latin form—a more inclusive **generic** one (genus) and a more definitive **specific epithet** (species)—which together serve to identify the organism. (This method of naming organisms is often called the **binomial system.**) The original description of the plant must be given in Latin, too.

The rules further prescribe that (with a few expressly indicated exceptions) the earliest properly published name after 1753 (the date of the publication of Linnaeus's *Species Plantarum*) be the one accepted to identify the plant. If two authors unbeknownst to each other describe the same species and assign it different names, the earliest published name has priority and is the correct designation. Even if a species is placed in an incorrect genus and is later transferred to another, the original specific epithet must be retained if it was the earliest to properly identify the species. The author of a plant name is often listed after the specific name; the names of well-known taxonomists are indicated merely by an initial or an abbreviation. Thus L. stands for Linnaeus, as in the Latin name for wheat: *Triticum aestivum* L.

Keep in mind, however, that even Latin naming according to accepted rules has shortcomings. Sometimes botanists don't agree on what constitutes a species or, as mentioned, were unaware of prior publication. Critical study often reveals that two or more names (**synonyms**) apply to the same species. Thus the Latin names in this book won't necessarily agree with every author's choice. Technically, only one name can be valid, but until the plant group is thoroughly studied by a competent taxonomist, one can't be sure which name applies to exactly what entity. Nevertheless, the combination of common name and Latin name should certainly make clear enough the identity of the plants discussed in this book.

Species may be subdivided into **subspecies** or **varieties,** for additional identification. The names of botanical subspecies are also written in Latin. Popular names given to **cultivars** or **clones** ("agricultural varieties") are sometimes added, but they are not Latinized—for example, *Rosa foetida* 'Austrian Copper.' *Rosa* is the generic name, *foetida* the specific epithet, and 'Austrian Copper' the name of the cultivar. (See Fig. 6-2.)

Until publication of Linnaeus's *Species Plantarum,* the name of a plant was not limited to two primary words, the genus and species. The medieval herbals had employed whole phrases or sentences, **polynomials,** often several lines long. As more and more kinds of plants were discovered, the discoverers simply added additional Latin adjectives to distin-

Figure 6-2   The three plants shown are all members of the same species, but they are distinguishable cultivars. All are wheat, *Triticum aestivum,* but the Turkish cultivar on the left ('Floranza') is obviously less thrifty and poorer yielding than the Mexican cultivars in the middle and on the right ('Lerma Rojo' and 'Sonora'). [Courtesy AID, U.S. Department of State.]

guish new plants from their relatives. The times were certainly ripe for the simpler binomial system finally established by the students of Linnaeus.

Linnaeus is regarded as the father of modern taxonomy. As ingenious as he was he nevertheless had two centuries of the accumulated experience of many herbalists to build upon. Of Linnaeus's predecessors, Vallerius Cordus (1515–1544) in particular stands out as one of those whose work foreshadowed that of Linnaeus. The descriptions and organization in Cordus's *Historia Stirpium* (published posthumously) have a modern ring to them. With the rise in world exploration, during Linnaeus's lifetime, he had unparalleled opportunity to systematize the then expanding biological world. During Linnaeus's time botanists and zoologists were chiefly concerned with gathering and naming new forms, an occupation at which Linnaeus was without peer. So great was his renown, and so widely accepted his new system (based chiefly upon the reproductive structures of plants) that explorers the world over sent collections to him and rightly regarded him as the top authority of his time. Few people can match Linnaeus in the quantity of species named, or in the intuition with which he systematized their cataloguing.

Classifying plants

Naming plants as they are discovered is essential for identification. But equally important is the systematic pigeonholing of them for orderly reference—classification. There are many ways in which plants can be classified. An obvious way is to designate them according to their use by man: food, fuel, lumber, feed for grazing stock, and so on. They might be categorized according to where they grow: in the sea, in fresh water, on swampy ground, on high ground, on the prairie, in the forest, in the jungle, on the tundra, and so on. Or they could be grouped according to general appearance and stature: leafy herbs, shrubs or bushes, vines, trees, and so forth. We actually group some plants by a combination of appearance and use; examples are vegetables, fruits, roots, and nuts. These groupings are not very precise, but are adequate for certain everyday purposes. They are not adequate, however, for communication between people of different cultures: one culture's fruit is another culture's vegetable. Rather, the traditional basis for botanical classification has been physical appearance (morphology), especially of the reproductive (flower) structures (which are not so prone to change in response to the environment, and

thus ordinarily reflect natural relationships more accurately than do vegetative characteristics, many of which alter rapidly to adapt to the environment—for example, all desert plants develop water-conserving forms whether or not they are related).

DESCRIPTIVE CLASSIFICATIONS

Growth habit

The practice of distinguishing plants as annuals, biennials, or perennials goes back at least to the early Greeks. **Annuals** complete their growth cycle in a single growing season and are perpetuated by seed. The major crop plants of the world are annuals, including all of the grains and many of the legumes, such as soybean, pea, and bean. **Biennials** are plants that take two growing seasons to complete their growth cycle. Usually during the first year the plant accumulates food reserves in storage organs, and during the second produces flowers and seed. Some of the root crops, such as beets, carrots, and parsnips, are biennials, as are such vegetables as onion and cabbage and a few of the ornamentals, such as hollyhock. By and large, biennials are of secondary importance compared to annuals. **Perennials** are plants that continue growing indefinitely. Some die back to the ground each winter, but revive from the roots the following spring; this habit is typical of most forage grasses and alfalfa, plus some garden plants, such as rhubarb. Most perennials simply add new growth each year, with little loss or destruction of aboveground parts. Some species—especially tropical forms, such as tomato—are perennial in one climate, but are cultivated as annuals in another. The same is true of cotton: tropical forms are often grown as perennials, whereas temperate forms are usually annuals.

Structure and form

Herbaceous plants are soft and succulent, with little or no secondary tissue. They may be annuals, biennials, or perennials, but herbaceous perennials usually die back to the ground each winter in temperate climates. **Woody** plants, in contrast, are those that develop secondary stem tissues and abundant xylem. Herbaceous plants and plant parts are much more likely to be edible than woody ones; woody plants are generally more suitable for building materials or as a source of such substances as cellulose.

Herbs are of course herbaceous plants, but in a limited popular sense an herb is a small aromatic species primarily used for scent or flavoring. **Vines** are trailing or climbing plants that are not sufficiently woody to support themselves (and many are entirely herbaceous). Large vines with somewhat woody stems, particularly of the tropics, are **lianas.** The wild grapevine of temperate climates would be called a liana, if we were accustomed to using that term in temperate regions. **Shrubs** are low, woody plants with stems that are rigid and strong enough to sustain considerable erect growth; shrubs typically have several stems of essentially equal size and rank. **Trees** are plants that are abundantly woody, usually have a single stem or trunk, and typically grow quite tall.

Leaf retention

Although almost all plants drop their leaves sooner or later, those that do so annually, and are leafless for extended periods, usually in winter, are said to be **deciduous.** Many familiar hardwood trees, such as oaks and maples, are of this type. Plants that maintain green leaves throughout the year are said to be **evergreen.** Needle trees, such as the pines and spruces, are of this type, as are certain broadleaf plants (Dicotyledonae), such as holly and rhododendron.

Climatic adaptation

Classification by climatic adaptation is essentially self-explanatory. **Tropical** plants grow in warm climates where freezing rarely if ever occurs. Most shed their leaves once a year in response to changing season, but the period during which old leaves are replaced by new ones is relatively brief, and many are essentially "evergreen." **Temperate** plants grow where there is a marked winter season, typically with considerable freezing. Their climatic range is great—

from the arctic, where the growing season (summer) is brief and the winter severe, to approximately the latitude of northern Florida, where the growing season is prolonged and winter is mild. The climate in such places as Florida is often termed **subtropical.** Climates in the colder part of the temperate zone are often referred to as **boreal.** Most temperate species not only withstand cold well, but require winter weather for full vigor and flowering. Kentucky bluegrass, for example, will not set seed unless freezing temperature combines with short day length to trigger this response.

Related is plant hardiness and tenderness. Plants are termed **hardy** if they will survive the winter climate of a particular region. Plants termed **tender** will not survive winter without special protection. Whether a plant is hardy or tender depends on where one wishes to grow the plant. This is true of many fruit trees and ornamental shrubs. The terms "hardy" and "tender" have also been used with reference to a plant's ability to tolerate lack of water or endure other stresses. In general, the diversity of plants is greatest in humid, tropical climates, and decreases progressively toward the arctic and antarctic regions or toward drier environments. Relatively few kinds of plants are hardy enough to withstand the effects of freezing and desiccation that plants are subjected to in arctic climates.

Usefulness

The distinctions between useful, useless, and harmful plants were among the first that primitive people had to make. We can suppose that any species faces extinction when any considerable portion of its environment is toxic. By the same token, evolutionarily successful humankind is harmfully affected by very few plants or plant parts. Of course, almost anything used immoderately can be discomforting, or even lethal. But in general only a few kinds of bacteria, some mushrooms, and very occasional seeds or parts from higher plants are deadly **poisonous.** Discomfort can come from a number of stinging or irritating plants, such as nettles and poison ivy. And of course many plants and parts of plants are **inedible,** mostly because they have an unappealing taste, little nutritive value, or are too tough and woody even to chew.

Plants can be useful in many ways. Those that provide utilizable substance are **crops.** Those providing esthetic satisfaction are **ornamentals,** or ornamental crops. **Forages** are crops that animals graze upon.

Classifications of plants according to the edible parts are often based more upon custom than botanical accuracy, and the delimitations are vague. Horticulturists divide edible garden plants into **fruits** and **vegetables.** Botanically, fruits are seed-bearing structures derived from flowers. As the word is popularly used, "fruits" are pulpy, juicy, and tasteful, and generally, but not always, come from woody perennials, often trees or shrubs—for example, apple, peach, orange, and quince. Strawberry and banana plants, although herbaceous, are grouped by horticulturists as fruit plants. Most herbaceous plants that have edible parts (including, confusingly, the botanical fruit) are termed vegetables. Common examples of vegetables include lettuce (edible leaf), asparagus (edible stem), beet (edible root), cauliflower (edible flower), tomato (edible fruit), and pea (edible seed). No precise distinctions can be made between "fruit" and "vegetable." The edible portions of grasses are the dry, one-seeded fruits ("seeds") called **grains**—such as rice, wheat, oats, barley, millet, and sorghum. **Root crops** are those with swollen roots or underground stems modified for the storage of sugars or starches (tubers). Examples are beet, sweetpotato, manioc (roots), and potato (tuber). There are many other general classifications based on plant parts such as **nuts, berries, bulbs,** and **corms.**

THE MODERN CLASSIFICATION

A number of attempts were made before Linnaeus's time to catalog new plants methodically as they were discovered. With Linnaeus came the great practical step forward. Later botanists all strived for logical order in arranging species. They recognized that general appearance was not fundamental, and that completely unrelated species might have similar characteristics as a result of response to a given environment. For example, many desert plants are waxy or fleshy, and many rain-forest plants are tall trees, no matter what their botanical affinities. General morphology (shape, form) is essential to the description of a species, but the reproductive structures

(flowers) give a truer, more constant reflection of broad relationships. The basic flower structure of related species is relatively constant, and less subject to environmental variation than vegetative structures.

As noted earlier, Darwin's *Origin of Species* provided breathtaking illumination of the evolutionary processes. Reasons for degrees of relationship suddenly became clear. Ever since, taxonomists have considered one of their main obligations to be the decipherment of "natural relationships"—that is, the presumed evolutionary background of the groups they are studying. Because of the immensity of the plant kingdom, intense study of this kind has been undertaken only for relatively few plant groups. Many taxonomists spend a lifetime specializing in a certain group, such as a family, but the great majority of plants have not received such attention.

It is remarkable that the early botanists, who had little more than morphology and a bit of geography to go on, were for the most part intuitively correct in their grouping of plants according to natural relationships. It is apparent now that most ancestral life forms have become extinct, and that the "missing links" between related groups are not likely to be found except as occasional fossils. The gradual accumulation of many lines of evidence make it possible to piece together evolutionary sequences such as the one roughly outlined in Figure 6-3.

We will regard systems adopted since Linnaeus as "modern." Linnaeus's species concept has remained intact, although many groupings of plants have been changed according to various authors' beliefs about what constitutes a natural relationship. Surprisingly, there is good general agreement between systems conceived by eighteenth and nineteenth century botanists and those of more recent date. As taxonomists acquire better tools for deciphering relationships (for example, chromosome counts, serologial and chromatographic tests, improved chemical analysis, and computerized investigations), one would expect that taxonomic changes would be based upon a more universally accepted concept of just what constitutes a natural relationship. However, evidence from these approaches generally confirms the process of evolution and provides valuable support for the traditional morphological and paleontological ap-

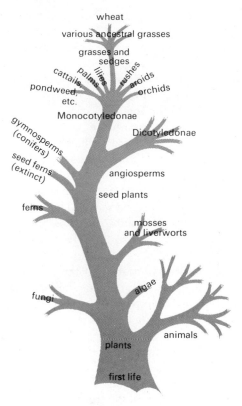

Figure 6-3 Hypothetical evolutionary sequence leading to wheat.

edness on the gene level. Extracted DNA from one taxon is inoculated with radioactively labeled DNA from another. Recombination occurs in proportion to the similarity of the two kinds of DNA. Comparisons have been made between the DNA from several higher organisms—people, monkeys, mice, hamsters, guinea pigs, rabbits, cattle, and salmon. The results show that human DNA has much in common with that of monkeys and little in common with that of the salmon, and that the hamster is more closely related to the mouse than to the guinea pig. This same sort of comparison is possible with such crops as wheat, rye, oats, and their various wild relatives.

More recently, chemical investigation has been used to analyze complex biochemical compounds. Chromatography has proved to be a powerful tool of biochemical taxonomy. It is presumed that similarity of compounds indicates similar biochemical pathways under the control of genes. Chemical analysis has been effective in elucidating relationships in *Eucalyptus* (by essential oils), *Pinus* (by terpenes), *Senecio* and *Veratrum* (alkaloids), various crucifers (isothiocyanates), and many legumes (various organic compounds). Often the results of these techniques support or cast doubt on long-accepted relationships established on the basis of morphology alone. For example, betacyanins occur in all families of the order Centrospermae except the family Caryophyllaceae. Is the family Caryophyllaceae improperly included in the order Centrospermae?

Judging from the results of such studies, it is safe to predict that much of the structure of the system of classifying the plant kingdom will remain essentially as it is now, differing little from its organization according to natural relationships as primarily perfected by German taxonomists of the late nineteenth century. Surely the Linnaean concept of species in a genus will be retained, and so will the hierarchy now so well recognized: genera (plural of genus) in a family, families in an order, orders in a class, classes in a division (the "phylum" of the zoologist). Various "sub" designations (subclass, suborder, and so on), or other in-between groupings ("section," "tribe"), can always be employed as convenience dictates. Just how one may care to delimit and by what name identify these categories is often a matter of preference.

proaches. By and large their usefulness is greatest within a relatively small group, such as a family or a genus. With good microscopes and modern techniques for easily staining cell nuclei, chromosomes have been counted and catalogued for a great many species. The information is especially helpful in suggesting hybrid relationships and polyploid series (see Chapter 19).

Serums made from plant extracts have been used to test affinities between plants. A protein extract from one plant is injected into an animal to "immunize" it to the extract. The degree of precipitation in the reaction between the antiserum obtained from the animal and the extract of another plant is presumably an index of the similarity of the two plants, and hence of their relationship. A similar technique consists in comparing the genetic material, DNA, from different taxa to obtain a measurement of relat-

Because the heyday of plant collection and initial classification coincided with the tremendous surge of world exploration in the nineteenth century, many of the important type specimens (the preserved plant from which the original description was derived) are ordered in the great herbaria of the world according to a system that originated in Germany before the turn of the century and was brought to perfection in the monumental work of Engler and Prantl, *Die Natürlichen Pflanzenfamilien.* Although its details are not necessarily up to date, there is obvious convenience to maintaining the Engler-Prantl classification as a standard. As an example, the pigeonholing of wheat, *Triticum aestivum,* would go as follows:

Kingdom—Plantae
    Division (or Phylum)—Spermatophyta (seed plants)
        Class—Angiospermae (seeds in fruit)
            Subclass—Monocotyledonae (monocots—seeds with a single seed leaf)
                Order—Graminales (grass-like families)
                    Family—Graminae or Graminaceae (the grasses)
                        Genus—*Triticum* (the wheats)
                            Species—*aestivum* (bread wheat)

A reading upward from the smallest taxon efficiently categorizes the plant in the overall scheme of things. It is wheat, in the grass family, related to sedges (also in the Graminales) and more distantly to the palms (Monocotyledonae), and of course even more distantly to other seed plants. If this were drawn as a "family tree," the ancestral genealogy would look something like Figure 6-3.

Major and minor taxa

The categories of division, class, order, and family are referred to as **major taxa** and from genus down as **minor taxa.** The family, a group consisting of related genera, is a distinction that most botanists can agree upon. Certain families stand out as mainstays of modern agriculture. Perhaps the most important of all is the grass family, **Graminae** (also sometimes called Poaceae), the source of most of the important grains, staff of life for the world. Without the easily stored and easily handled grains, human feeding would be exceedingly cumbersome. Almost equally as important as grain in many parts of the world is grass for the grazing of livestock. Sugarcane, too, is a grass, as are corn and sorghum. And in many parts of the world various bamboos, which are also grasses, are used for structural purposes. Everywhere grasses protect and build the soil.

Perhaps next in importance is the pea family, **Leguminosae** (or Fabaceae), which includes beans, peanuts, and such nutritious forages as clover, vetch, and alfalfa. In the tropics, trees of the family Leguminosae are utilized for their fine timber as well as their fruit (pods and seeds). The Leguminosae live in symbiotic relationship with certain bacteria; nodules on the roots house these bacteria, and they trap gaseous nitrogen and convert it to nitrate. Nitrate is an essential nutrient that the legumes of the world had long been producing before there was ever a fertilizer industry. With usable nitrogen so readily available, it is not surprising that legumes are generally rich in protein. Many tropical populations get along largely without animal protein by substituting proteinaceous beans instead.

Members of the palm family, **Palmae** (or Arecaceae), important in the tropics, yield food from their fruits (coconut, date), structural materials from their stems and leaves, and various waxes and oils from their leaves, fruits, and seeds. The rose family (**Rosaceae**) not only contributes many ornamentals, but is important for many familiar fruits (apple, pear, peach, apricot, plum, cherry, strawberry, raspberry, and others). The melon family (**Cucurbitaceae**) was perhaps more important to primitive civilizations than it is to modern ones, but it is still prized as the source of squash, pumpkins, gourds, watermelons, muskmelons, and cucumbers. The mustard family, **Cruciferae** (or Brassicaceae), yields the cole plants (the cabbage group, including broccoli, cauliflower, Brussels sprouts, and so on), and, along with the celery family, **Umbelliferae** (or Apiaceae), is the source of many oil-bearing seeds. The pine family (**Pinaceae**) is, of course, an outstanding source of forest products.

The minor taxa are those below the genus. The genus is usually a group small enough for intensive study. The generic concept is an ancient one; most genera were established before the concepts of evolution were known and are based primarily on morphological distinctions. As a result, many well-known genera differ considerably in complexity and require extensive revision. This is especially true for those of economic importance, such as *Sorghum*, a genus whose genetics have been muddled through breeding and selection.

The species, the basic unit of all taxonomy, is also subject to subdivision according to various morphological criteria. The traditional subdivisions of the species are, in decreasing order of complexity, **subspecies, variety,** and **race** (form). But the definitions for these terms are vague and largely a matter of individual preference. Consequently modern taxonomy, making greater use of genetics and ecology, has developed new terms with more precise definitions that are subject to experimental verification. These include the following:

Comparium: Integrity maintained entirely by genetic barriers; often equivalent to a traditional genus.

Cenospecies: Integrity maintained by genetic barriers reinforced by ecological barriers; often equivalent to a traditional genus or species.

Ecospecies: Integrity maintained by ecological barriers reinforced by genetic barriers; often equivalent to a traditional species.

Ecotype: Integrity maintained entirely by ecological barriers; often equivalent to a traditional subspecies, variety, or race.

Identifying higher plants

Until one has mastered taxonomy sufficiently to categorize most plants (at least to the family level), usually on the basis of floral similarity, the theory behind natural classification is fine, but of limited practical value. In order to facilitate identification, **keys** (also called "analytical keys," or "dichotomous keys") are utilized, and these are of value to both the uninitiated and the expert in determining the identity of a particular plant. The principle of these keys is the selection between alternatives (usually only two choices are given at each characteristic), and one continues progressively into finer subdivisions of family, genus, species, or subspecies.

A key to all of the plants in the world would of course be so complex, cumbersome, and time-consuming to use that any effort at identification would be completely frustrated. Rather, most keys are limited to the plants of certain geographical areas, or to certain categories of the flora, as are, for example, *Spring Flora of Missouri, Illustrated Guide to Trees and Shrubs* [of the Northeastern United States], or *West Virginia Grasses*, in which the subject coverage is restricted to manageable proportions.

To use a key effectively, one ordinarily needs a specimen of the plant being identified that exhibits both vegetative and floral characteristics (for example, a twig with leaves and flowers). Especially if the plant is a large one, the specimen should be accompanied by information indicating size, habitat, growth habit, flower color, fruit characteristics (if available), and so on. Ordinarily it is impossible to "key out," or identify, a native plant unless a flowering (or fruiting) specimen can be obtained, although some attempts have been made to devise keys for such flora as lawn weeds or small groupings of cultivated plants that are based entirely on stem and foliage characteristics. A magnifying lens or binocular microscope will probably be needed to determine such characters as the type of pubescence (hairiness), or the position and shape of stamens and pistil (and similar features determined through dissection of a flower)—characteristics often utilized for making final distinctions.

Dichotomous keys typically identify the two (occasionally more) alternatives to a characteristic by the same code letter or number, with subsequent alternatives (after a choice has been made) following sequentially, and usually indented. A very simple example follows:

```
A—colored flowers, not white
    B—tree ........................... taxon 1
    BB—herb ......................... taxon 2
AA—white flowers
    B—leaves pubescent (hairy)
        C—stamens 4 ................. taxon 3
        CC—stamens 8 ............... taxon 4
    BB—leaves glabrous (smooth) ....... taxon 5
```

If the specimen has white flowers, alternative AA is chosen and all subheadings under A are passed over; if its leaves are pubescent, alternative B (under AA) is selected; then if eight rather than four stamens occur in the flowers, alternative CC (under AA, B) is chosen. If no further alternatives are listed the name of the taxon is given, in this example "taxon 4." Sometimes rather than being sequentially indented alternatives are progressively listed by number, as follows:

1. Trees or shrubs 2
1. Herbaceous 4

2. Leaves scale-like . . . . taxon A
2. Leaves flat and expanded 3

3. Leaves opposite . . . . taxon B
3. Leaves alternate . . . . taxon C

4. Leaf margins serrate
   (toothed) 5
4. Leaf margins entire
   (not toothed). . . . . taxon D

If the specimen is an herb with serrate leaves the progression is 1, 4, 5, and so on.

Because many identification keys are designed for professional use, their terminology may be unintelligible outside of botanical circles; to employ them effectively requires mastery of the terms that botanists use to describe plant form precisely. The brief key that follows is taken from the *Flora of Panama*, for the identification of trees of the genus *Albizzia* (members of the family Leguminosae) growing there.

a. Leaflets small and narrow (5–11 mm. long and about 2 mm. wide); numerous (9–50 pairs per pinna).
  b. Leaflets less than 30 pair per pinna, pubescent; mature peduncles 2–3 cm. long; flowers relatively large (calyx about 2.5 mm. long) 1. A. CARBONARIA
  bb. Leaflets usually 30 or more pairs per pinna, glabrous; peduncles less than 1.5 cm. long; flowers small (calyx about 1 mm. long) . . . . 2. A. CARIBAEA
aa. Leaflets larger and broader (mostly 15–40 mm. long and 10–20 mm. wide), few or several (2–9 pairs per pinna).
  b. Leaflets tapered and bluntly acute apically; flowers sessile
     3. A. ADINOCEPHALA

  bb. Leaflets obtuse or rounded to emarginate apically; flowers pedicellate.
    c. Flowers all alike, short-pedicellate (pedicels mostly about 2 mm. long); mature leaflets, young twigs, and legume sparingly pubescent or glabrous. . . . . 4. A. LEBBECK
    cc. Flowers dimorphic, all except center one long-pedicellate (pedicels 8–15 mm. long); mature leaflets, young twigs, and legume tomentose to moderately pubescent . . . . 5. A. LONGEPEDATA

One must know what "pinna," "peduncle," "sessile," "pedicellate," "acute," "obtuse," "glabrous," and "tomentose" signify to use even so simple a key as this. The lengthier key for determining the genera of the family Leguminosae in the same publication (by which one determines, to begin with, whether the specimen at hand is a species of *Albizzia* or of some other genus) requires additional expertise with taxonomic language; and if one did not have enough experience to know that the plant at hand is a member of the family Leguminosae, the use of a comprehensive manual to determine its family would be even more complicated, and probably a task for someone with taxonomic training.

The cultivated variety, or cultivar

In agricultural terminology, the term "variety" has been used to refer to a named group of plants within a particular cultivated species that is distinguished by a character or group of characters. Examples are the 'Bartlett' pear, 'Kenya Farmer' wheat, 'Ranger' alfalfa, and 'U.S. 13' corn. To avoid confusion with the botanical variety, the term **cultivar** (a contraction of *culti*vated *vari*ety) has been suggested as the preferred term. The cultivar is the taxon that is most often of relevance to the individual crop producer.

The term cultivar can refer to any one of several entities, often depending on the method of propagation employed for the particular crop. Thus with reference to crops that are generally asexually propagated, such as potato and various fruits, the term means a particular **clone,** the product of a single individual. For example, the names 'Bartlett' pear,

'Russet Burbank' potato, and 'Senga Sengana' strawberry refer to unique genotypes perpetuated by vegetative propagation. They are often referred to as clonal cultivars. Many improved clonal cultivars are the result of mutations (known as sports) that produce some advantageous change, such as the red pigment in the skin of certain sports of the 'Delicious' apple. In a few plants clonal propagation can be accomplished by means of seed that does not develop sexually and is identical with the mother plant—a condition known as apomixis. Kentucky bluegrass (*Poa pratensis*), a favorite for lawns as well as a prime pasture species, is largely apomictic.

With reference to self-pollinated plants, such as wheat or tomato, the term "cultivar" usually means a particular inbred or pure line that breeds true naturally. These cultivars are maintained by avoiding contamination and mixing. In cross-pollinated crops, such as alfalfa, the term usually refers to a population of plants distinguished on some morphological or physiological basis and maintained by selection and isolation. These are referred to as open-pollinated cultivars. Finally, the term "hybrid cultivar" may refer to a particular combination of inbred lines. Thus it can be seen that the precise meaning of the term cultivar depends on the particular crop in question. The improvement of crop cultivars is the goal of plant breeding.

AN OVERALL VIEW

Although the most important domestications were complete before recorded history, people are still selecting new plants for their use, and improving old ones. The breeder of a particular crop usually works with subdivisions of lesser rank than the genus. By and large, species of different genera will ordinarily not hybridize, but many species naturally cross with others of the same genus. Indeed, this seems to be one of nature's means for creating new combinations for natural selection. Plant breeders select from the progeny of their crosses those individuals having some particular advantage. Useful selections become the cultivars of modern agriculture (Fig. 6-4).

As improved cultivars are created, older ones are abandoned. Scientists have begun to wonder—as uni-

Figure 6-4 A hybrid millet bred in India. Repeated crossings of cultivars often yield advantageous new cultivars, quickly "proved" under the intense selection pressure of the modern agricultural system. No matter how you classify such a "hybrid" (it is usually given a cultivar name in addition to its specific name), it is usually quite distinct from the wild ancestors bearing the same specific name. [Courtesy AID, U.S. Department of State.]

form, high-yielding cultivars are now being pressed upon "underdeveloped" parts of the world, in place of traditional, locally adapted types—whether we are not losing germ plasm that might prove immensely valuable in some future age under different circumstances. By allowing wild plants to become severely decimated, even extinct, and by not perpetuating "obsolete" cultivars, this could happen. Efforts are now being undertaken, perhaps just in the nick of time, to preserve a few bits and snatches of natural plant habitat ("living museums"), and a Federal Seed Storage Laboratory (germ-plasm bank) has been created at Fort Collins, Colorado, to house stocks of older cultivars that might someday prove of use to future plant breeders.

As pedantic and uninspiring as classification may seem to those not engaged in its research, it is apparent that some understanding of it and the terminology that goes with it is fundamental to discussions involving numerous plants. Classification is essential in order to identify—that is, to "key out"—plants; to determine whether they are known or unknown. The system is by no means perfect, even for the more easily categorized wild plants. Applied to modern cultivars and hybrids, it does not always work well, and there is not always agreement about how to categorize such cultivars as those that originate through crosses between species (often such cultivars are given a formal name that includes the symbol ✕, as in *Magnolia* ✕ *soulangiana*, the Saucer Magnolia, which is a hybrid between *M. heptapeta* and *M. quinquepeta*). For lower organisms, such as bacteria, morphology offers little to go on (even with the aid of the best microscopes), and consequently we must rely upon physiological reactions, chemical by-products, and colony formation. Nevertheless, a plant's name should bring to mind diverse information about that plant. In this sense the name of a plant is a summary of it; it is an epitomization of its genetics, physiology, uses, economic importance, and so on. Regarded in this way, plant classification becomes one of the broadest and most dynamic fields of study within the larger field of botany.

Selected references

Bailey, L. H., 1949. *Manual of Cultivated Plants*, rev. ed. New York: Macmillan. (Classic compendium for identifying cultivated plants, especially those having ornamental value, by the outstanding horticultural taxonomist of the century. For reference rather than reading, but worth examining to become aware of how plants are practically catalogued and how keys are used for their identification. The book contains comprehensive glossaries of Latin names, technical terms, and taxonomic authorities. The manual follows the classical concept of Engler and Prantl.)

Bailey Hortorium Staff and guest authorities, 1976. *Hortus Third*. New York: Macmillan. (The horticulturists' "Bible," assembled at Cornell University. An encyclopedia alphabetical by genus [species alphabetical under the generic name] with family characterizations and occasional subject headings included. An authoritative source of information about cultivated plants, botanical authors, botanical terms, and common names.)

Bendz, G., and J. Santesson (editors), 1974. *Chemistry in Botanical Classification*. New York: Academic Press. (After several general papers on chemical taxonomy, various authors explore specific evolutionary indicators and specific plant groups in greater detail. A comprehensive review of chemical assistance to traditional morphological taxonomy.)

Berlin, B., D. E. Breedlove, and P. H. Raven, 1974. *Principles of Tzeltal Plant Classification*. New York: Academic Press. (This interesting study is indicative of ways preliterate society develops botanical classification. Folk designations according to use are first discussed, followed by more sophisticated means of identifying and designating plant species.)

Blunt, W., 1971. *The Compleat Naturalist: A Life of Linnaeus*. New York: Viking. (A readable biography of Linnaeus from his boyhood to his attainment of recognition as the leading botanical taxonomist; instructive of Linnaeus's skill in adapting the prevailing knowledge of his time.)

de Candolle, Alphonse, 1886. *Origin of Cultivated Plants* (translation of *L'Origine de Plantes Cultivées*, Paris, 1833). Reprinted 1959. New York: Hafner. (The classic historical work on cultivated plants, utilizing many avenues of evidence to predict where and how the major crop species were brought into cultivation.)

Dobzhansky, T., F. J. Ayala, G. L. Stebbins, and J. W. Valentine, 1977. *Evolution*. San Francisco: W. H. Freeman and Company. (Covers the theory of evolution, considering many modern avenues of research. Reviews natural selection, speciation, phylogenies, and related topics.)

Eldredge, N., and J. Cracraft, 1980. *Phylogenetic Patterns and the Evolutionary Process*. New York: Columbia University Press. (Comparative biologists examine the ordering of the earth's biota and the evolutionary theory that explains how that order arises. Interrelationships among taxa, speciation, macroevolution, and construction of classifications are discussed.)

Folsome, C. E., 1979. *The Origin of Life: A Warm Little Pond*. San Francisco: W. H. Freeman and Company. (Speculations on "how it all began.")

Gerard, J., 1633. *The Herbal; or General History of Plants* (revision by Thomas Johnson of *The Herball or Generall Historie of Plantes*, 1597). Reprinted 1977. New York: Dover. (This historically important treatise, written in the quaint English of the time, is beautifully illustrated with woodcuts. Exciting to leaf through, it is an example of early interest in, and classification of, plants prior to Linnaeus.)

Grant, V., 1971. *Plant Speciation*. New York: Columbia University Press. (This volume contains many examples of the evolutionary development of species, and it is concerned with the difficulties of adequately defining what a species is both taxonomically and biologically.)

Harborne, J. B. (editor), 1978. *Biochemical Aspects of Plant and Animal Coevolution*. New York: Academic Press. (The editor is well known for his editorial organization of volumes dealing with plant chemotaxonomy, of which this presentation affords an extension by examining the coevolution of plants and animals. More specific is *Chemotaxonomy of the Leguminosae*, which he coedited with Boulter and Turner in 1971, with the same publisher.)

Harborne, J. B. (editor), 1970. *Phytochemical Phylogeny*. New York: Academic Press. (Surveys the impact of phytochemical research on our concepts of the origin of plant life and the evolution of modern plants. At the 1969 meeting out of which this volume grew, various authors presented papers on subjects ranging from palaeobotany through chemical composition of plant parts and secretions to discussions of the environment and enzyme evolution in plants.)

Hawkes, J. G. (editor), 1968. *Chemotaxonomy and Serotaxonomy*. New York and London: Academic Press. (Discussion of serological and biochemical techniques for establishing "chemical" systematics. Proceedings of a symposium on the subject held at the University of Birmingham.)

Heywood, V. H. (editor), 1968. *Modern Methods in Plant Taxonomy*. New York and London: Academic Press. (A series of review papers given at a conference of the Botanical Society of the British Isles and the Linnean Society of London in 1967, resulting in a comprehensive review of the "state of the art" in plant taxonomy as of that date. Chemosystematics and numerical taxonomy are among the topics treated.)

Heywood, V. H. (editor), 1973. *Taxonomy and Ecology*. New York: Academic Press. (A mixture of presentations at an international symposium, with a summary by Heywood entitled "Ecological Data in Practical Taxonomy." The presentations deal both with animals and with plants; some go into great detail, others are of a general nature. More for browsing than as an integrated presentation.)

Jardine, N., and R. Sibson, 1971. *Mathematical Taxonomy*. New York: Wiley. (A review of taxonomic methodology, drawing upon widespread European references, and including a section that discusses the principles of biological classification in a numerical context.)

Lawrence, G. H. M., 1951. *Taxonomy of Vascular Plants*. New York: Macmillan. (A standard text in systematic botany. Part I covers both the theoretical and practical aspects of classification, nomenclature, and identification. Part II describes all tracheophyte families known to be native to or introduced into the United States. The appendix contains a detailed glossary of botanical terms.)

Miller, S. L., and L. E. Orgel, 1974. *The Origins of Life on Earth*. Englewood Cliffs, N.J.: Prentice-Hall. (Additional ruminations on the chemical events that led to organic evolution.)

Oparin, Aleksandr Ivanovich, 1964. *The Chemical Origin of Life* (translation by Ann Synge of the Russian edition of 1936). Springfield, Ill.: Charles C. Thomas. (Tedious for thorough reading, but worth scanning as a landmark that ushered in modern concepts about the creation of life.)

Ponnamperuma, C., 1972. *The Origins of Life*. New York: Dutton. (Discussions suited to beginning students on the relation between living things and their cosmic environment. Stresses "chemical evolution" and the antiquity of life.)

Porter, C. L., 1967. *Taxonomy of Flowering Plants*, 2nd ed. San Francisco: W. H. Freeman and Company. (An excellent elementary text.)

Radford, A. E., W. C. Dickison, J. R. Massey, and C. R. Bell, 1974. *Vascular Plant Systematics*. New York: Harper and Row. (This is a practical book, with an extensive glossary and instructional information. A reader not acquainted with taxonomic terminology may be "overwhelmed," as if reading a dictionary rather than an intriguing story. The book is ambitious and comprehensive, carefully worked and well executed; perhaps best suited as a reference and for systematists.)

Reed, C. A. (editor), 1978. *Origins of Agriculture*. The Hague: Mouton. (A group of specialists—predominantly archaeologists, but some botanists, zoologists, and others—zero in on the origins of agriculture and geographical centers of origin. The book may be overly detailed except for the specialist, but it serves to provide varying viewpoints useful for background.)

Scagel, R. F., G. E. Rouse, J. R. Stein, R. J. Bandoni, W. B. Schofield, and T. M. C. Taylor, 1965. *An Evolutionary Survey of the Plant Kingdom*. Belmont, Calif.: Wadsworth. (A thoroughgoing survey of the plant kingdom, integrating information from the fields of physiology, cytogenetics, and so on; a compendium of information suitable for students planning advanced study in the subject area.)

Simmonds, N. W. (editor), 1976. *Evolution of Crop Plants*. New York: Longmans. (Although not dealing directly with systematics or classification, this book assembles a wealth of information by numerous authors concerning 86 crop species. Genetic background is given, and a smattering of cultivational history. An additional 81 crops receive "capsule" discussion.)

Sneath, P. H. A., and R. R. Sokal, 1973. *Numerical Taxonomy: The Principles and Practice of Numerical Classification*. San Francisco: W. H. Freeman and Company. (These advocates of the "new" approach scorn intuition in assessing relationships, and rely instead upon mass analysis of as many characteristics as possible by computer. Counter to traditional taxonomic methodology, this objective approach has caused furor within the field in recent years.)

Stebbins, G. L., 1971. *Chromosomal Evolution in Higher Plants*. Reading, Mass.: Addison-Wesley. (A review integrating molecular biology, cytology, and population studies of plants. Of interest for concepts of evolutionary mechanisms, although slanted more to the expert than to the beginner.)

Stebbins, G. L., 1974. *Flowering Plants: Evolution Above the Species Level*. Cambridge, Mass.: Harvard University Press. (A leading evolutionary theorist interprets the phylogeny of flowering plants in the light of modern genetics and other botanical fields. Stebbins notes that natural selection is broadly operative at the subspecific level, but at higher evolutionary levels emphasis is on reproductive characteristics. The book offers exciting insights into evolutionary theory rather than practical delimitation of taxa.)

Swain, T. (editor), 1963. *Chemical Plant Taxonomy*. London: Academic Press. (A thorough consideration of chemical and allied techniques for determining relationships among plants.)

Valentine, D. H. (editor), 1972. *Taxonomy, Phytogeography and Evolution*. New York: Academic Press. (Biosystematists participating in a 1971 symposium in England delivered the papers included in this volume, which explore the relationship between geographical distribution and evolution in higher plants. With more recent evidence accumulating on past climates and geological changes, taxonomists and geneticists examine patterns of evolution and taxonomic background.)

7
Structure and Function

Flowering plants exhibit tremendous diversity in size and structure. Plant organs—stem, leaf, root, flower, fruit, and seed—show extreme modification; crop plants especially have unusual features. Although many structures are superficially very different (for example, the aerial root of the orchid and the swollen root of the sweetpotato), they can be shown to be functionally and morphologically related. The present discussion must necessarily strike a balance between generalizations stressing the similarities of structure and function among different plants and specific comments on the differences that distinguish individual crop plants.

THE CELL AND ITS COMPONENTS

The structural unit of plants—as well as of animals, fungi, bacteria, and other living things—is the **cell.** Cytology, the study of cells, is concerned with the organization, structure, and function of cells. The concept that the cell is the basic unit of all living things is still accepted by biologists as dogma. A complex multicellular organism is an integrated collec-

tion of living and nonliving cells. The high degree of synchronization and coordination of the cells in the organism creates an entity that is in effect greater than the sum of its parts.

Plant cells vary in shape from spherical, polyhedral, and ameboidal to tubular. Most cells are between 0.025 millimeter and 0.25 millimeter in size (0.001–0.1 inch), but some fibers (long, tubular cells) are as much as 60 centimeters (2 feet) long. The concept of the typical, or generalized, plant cell is a useful one (Fig. 7-1). The most prominent cell components visible under the light microscope are the densely staining nucleus, and the more or less rigid cell wall enclosing the cytoplasm. The cytoplasm contains a number of structural bodies, or organelles, such as plastids, mitochondria, and vacuoles, and various other entities, such as crystals, starch grains, and oil droplets. A distinction is often made between living and nonliving substances in the cell, but such distinctions have little meaning.

The **cytoplasm** (less precisely termed protoplasm) is an extremely complex substance, both physically and chemically. It is composed of 85 to 90 percent water (by fresh weight); the remaining 10 to 15

Figure 7-1 Three views of a plant cell and its component parts. Recent electron microscope studies have produced new understanding of the ultrastructure of the cell. [A after K. Esau, *Plant Anatomy*, Wiley, 1953. C after J. Brachet, "The Living Cell." Copyright © 1961 by Scientific American, Inc. All rights reserved.]

percent consists of organic and inorganic substances that are either dissolved (salts and carbohydrates) or in the colloidal state (proteins and fats). The physiological activity of the cell occurs in the cytoplasm. Although the living state appears to be dependent on the cytoplasm remaining within the structural confines of a cell, various so-called vital processes (such as photosynthesis) have recently been found to be possible outside of the cell. The concept of what constitutes a vital process changes as biological techniques increase in sophistication.

Surrounding the cytoplasm is the **plasma membrane,** which lies against the cell wall. The cell wall is permeable to nearly all solutes and solvents, whereas the plasma membrane is semipermeable. The plasma membrane is composed of lipoproteins, which account for its elasticity and high permeability to fatty substances. Although it is ultramicroscopic in thickness, the plasma membrane can be defined in plasmolyzed (partially dehydrated) cells. The electron microscope has revealed that the plasma membrane is not a mere boundary to the cell but that it has structural definition. The cell appears to be permeated by internal membranes that, along with the external membrane, are thought to constitute a system called the **endoplasmic reticulum.**

The **nucleus,** a dense, usually spheroidal body, is located within the cytoplasm. The close affinity of the nuclear material for many dyes makes it the most conspicuous feature of a stained cell. The nucleus is, in effect, the control center of the cell, since it contains the **chromosomes** (*chromo,* colored; *soma,* body). The chromosomes contain deoxyribonucleic acid (DNA) associated with protein. The DNA contains genetic information, somewhat analogous to the magnetic tape in an electronic computer. The information is relayed thence to the cytoplasm in the form of RNA (ribonucleic acid), a substance similar to DNA. The information contained in the DNA thereby affects the machinery of the cell through the control of protein synthesis. The actual site of protein synthesis is in small particles in the cytoplasm called **ribosomes.** The information contained in the DNA provides the basis for the physiological function of the cell and determines the metabolic and morphological features of the organism. It is also a major part of the hereditary bridge between generations.

The **cell wall,** one of the structures that distinguishes plant cells from animal cells, is usually thought of as a deposit, or secretion, of the cytoplasm. The plant cell wall is composed of three basic groups of constituents: cellulose, lignin, and pectin compounds. **Cellulose** and a number of closely related compounds, such as hemicellulose, make up the firm layer of the cell wall. These compounds are, basically, unbranched long-chain units (polysaccharides) formed from the simple sugar glucose (see p. 58). Hemicellulose is related to cellulose but contains sugar and nonsugar components. Cellulose and its derivatives are highly combustible materials. **Lignin** consists of a complex network of alcohols that impregnates the cellulose framework of certain plant cells. The deposition of lignin (lignification) hardens the cellulose walls into an inelastic and enduring material resistant to microbial decomposition. Lignins cause yellowing in paper, and so must be dissolved from the wood pulp used in making high-quality pa-

pers. **Pectin compounds** are water-soluble polymers of galacturonic acid that form sols and gels with water. The familiar substance pectin is used as a solidifying agent for jams and jellies.

The cell wall is laid down in distinct layers, and its thickness varies greatly with the age and type of cell. Three regions of the cell wall are generally distinguished:

1. The **middle lamella,** a pectinous material associated with the intercellular substance. The slimy nature of rotted fruit is due to the dissolving of the middle lamella by fungal organisms.

2. The **primary cell wall,** the first wall formed in the developing cell. It is composed largely of cellulose and pectic compounds, but closely related substances and noncellulose compounds may also be present, and it may be lignified. The primary cell wall is the wall of dividing and growing cells; in many cells it is the only wall.

3. The **secondary cell wall,** which is laid down inside the primary cell wall after the cell has ceased to enlarge. It appears to have a mechanical function, and is similar in structure to the primary wall, although it contains more cellulose, and often contains some lignin.

The cell wall in plants is not continuous. It appears to be pierced by cytoplasmic strands (**plasmodesmata**) that provide a living connection between cells. Furthermore, thin areas called **pits** occur in the walls of some cells.

Plastids are specialized, disc-shaped bodies contained in the cytoplasm, and are peculiar to plant cells. They are classified on the basis of the presence or absence of pigment into leucoplasts (colorless) or chromoplasts (colored). Leucoplasts occur in mature cells that are not exposed to light, and some types are involved in the storage of starch. Of the colored plastids, those containing chlorophyll (**chloroplasts**) are the most significant, for they are the complete

Figure 7-2   An electron micrograph of a chloroplast from a tobacco leaf. Note granum (*G*) and starch grain (*S*). [Courtesy W. M. Laetsch.]

structural and functional unit of photosynthesis, the process by which carbon dioxide and water are transformed into sugars and other carbon-containing compounds. Actively photosynthesizing chloroplasts contain grains of starch, which is the form in which they temporarily store the sugars that they produce.

There are about 20 to 100 chloroplasts in each chlorophyllous cell of a green leaf. Electron micrographs have revealed the internal structure of chloroplasts (Fig. 7-2). Structural units that resemble stacked coins, called grana, contain the chlorophyll and are the receptors of light. The actual transformation of carbon dioxide to carbon-containing compounds occurs in the surrounding material, called stroma.

Chloroplasts have a kind of independent existence in the cell. Although they are under the influence of nuclear genes, the chloroplasts are self-replicating, and they apparently arise only from pre-existing plastids.

Mitochondria are the cell's power plants. They appear as small, dense granules under the light microscope, but electron microscopy has revealed that they have a complex, involuted internal structure. the mitochondria are made up of proteins and phospholipids. All of the known functions of mitochondria are related to enzymatic activity connected with oxidative metabolism. This activity occurs through the formation of the energy-carrying substance called adenosine triphosphate (ATP).

Vacuoles are membrane-lined cavities located within the cytoplasm that are filled with a watery substance known as the cell sap. The cell sap may contain a number of dissolved materials—salts, pigments, and various organic metabolic constituents. In actively dividing cells the numerous vacuoles are small, whereas in mature cells the vacuoles coalesce into one large vacuole that occupies the center of the cell, pushing the cytoplasm and the nucleus next to the cell wall.

A number of complex materials may be found in the cytoplasm and vacuoles; among these are crystals, starch grains, oil droplets, silica, resins, gums, alkaloids, and many organic substances. Many of these compounds are reserve or waste products of the cell.

TISSUES AND TISSUE SYSTEMS

Although the plant ultimately originates from a single cell (the fertilized egg), the marvels of cell division and differentiation produce an organism composed of many different kinds of cells—different structurally and physiologically. It is this difference in cell morphology and cell arrangement that results in the complex variation among different plants and within an individual plant.

Plants are made up of groups of similar types of cells in a definite, organized pattern. Continuous, organized masses of similar cells are known as **tissues.** Tissues can be classified in several ways, but no single system of classification is universally used. The following system retains many of the customary botanical terms.

Meristematic tissue: actively dividing, undifferentiated cells

Permanent tissues: nondividing, differentiated cells
    Simple tissue: composed of one type of cell
        Parenchyma: simple, thin-walled cells
        Collenchyma: thick-walled, supporting tissue
        Sclerenchyma: thick-walled, highly specialized elements
    Complex tissue: composed of more than one type of cell
        Xylem: water-conducting tissue
        Phloem: food-conducting tissue

Meristematic tissue

Meristematic tissue is composed of cells actively or potentially involved in cell division and growth. The meristem not only perpetuates the formation of new tissue but perpetuates itself. Since many so-called permanent tissues may, under proper stimulation, assume meristematic activity, no strict line of

Figure 7-3   Diagrammatic longitudinal section of a grass plant, indicating the location of the meristems. These shaded areas are the youngest parts of the plant. [After A. J. Eames and L. H. MacDaniels, *An Introduction to Plant Anatomy*, McGraw-Hill, 1947.]

demarcation exists between meristematic and permanent tissue.

Meristems are located in various portions of the plant (Fig. 7-3). Those at the tips of shoots and roots are known as apical meristems. The apical meristem of the shoot is also known as the growing point. The increase in girth of woody stems results from the growth of lateral meristems, collectively referred to as the **cambium.** The meristematic regions of grasses become isolated near the nodes, and are called intercalary meristems. Thus the mowing of lawns does not interfere with the growing portion of the grass plant. Tissues differentiated from apical meristems are referred to as **primary tissues.** Others, especially tissues formed from the cambium, are **secondary tissues.**

Although there are exceptions, meristematic cells are usually small, roughly spherical to brick-shaped, with thin walls and inconspicuous vacuoles. In sections prepared for microscopic examination, they appear darkly stained, owing to the relatively small amount of cytoplasm in relation to the size of the nucleus.

Simple tissue

Permanent tissues are derived directly from meristems. They are referred to as simple tissues when they are composed of only one type of cell (Fig. 7-4).

Parenchyma is unspecialized tissue. It makes up a large proportion of many plant organs, such as the fleshy parts of fruits, roots, and tubers, and is the seat of activities that depend on living protoplasm, such as photosynthesis, assimilation, storage, and secretion. It is made up of relatively undifferentiated, thin-walled, living cells that are capable of growth and differentiation.

Collenchyma tissue is characterized by elongated cells with unevenly thickened primary walls. Collenchyma cells, which commonly occur in strands or continuous cylinders, function largely as mechanical support in early growth. The strands of the outer edge of celery petioles are made up of collenchyma cells.

Sclerenchyma cells are thick-walled, have small cavities, and occur individually, in small groups, or in continuous masses. They are often heavily lignified. Sclerenchyma cells, in contrast to parenchyma and collenchyma, are nonliving when mature. When these cells are long and tapered, they are usually referred to as fibers. Fibers tend to interlace and interlock to form sheets of tissues. Short or irregularly shaped sclerenchyma cells are called sclereids. Clusters of sclereids, or "stone cells," are responsible for the gritty texture of the pear fruit. In masses, sclereids produce the hardness of walnut shells and of peach and cherry pits.

In common usage, the term "fiber" refers to threadlike strands of any substance. The fibers of commerce may be derived from plants (flax, hemp), animals (silk, wool), or minerals (asbestos). Many plant fibers are derived from the tough, flexible strands that occur in stems, leaves, or roots. Such fibers are generally composed largely of scleren-

Figure 7-4 Simple tissues. [After A. J. Eames and L. H. MacDaniels, *An Introduction to Plant Anatomy*, McGraw-Hill, 1947.]

chyma cells, although other tissues such as vascular bundles may be involved. Some well-known fibers are morphologically unrelated to sclerenchyma; cotton "fibers" are in fact epidermal hairs of the seed coat and are almost pure cellulose. Plant fibers may be classified, on the basis of their origin, as bast fibers (phloem, pericycle, or cortex), wood fibers, vascular bundle fibers, or surface fibers (seed hairs).

Complex tissue

Complex tissue is composed of combinations of simple and specialized cells and tissues. The two major types of complex tissues, xylem and phloem, are continuous throughout the plant; in addition to supporting the plant, they function in the conduction and storage of food and water.

Xylem

The principal water-conducting tissue of woody and herbaceous plants is **xylem,** an enduring tissue that consists of living and nonliving cells (Fig. 7-5). The basic element of xylem is a specialized nonliving cell called the **tracheid.** These are typically angular, elongated, tapered cells having walls that are hard and usually lignified although not especially thick. When mature, the cell is devoid of contents and is adapted for its principal function of water conduction, although the rigid walls also function in support. Water and its dissolved contents move readily through the empty tracheids, flowing from cell to cell through the numerous pits. The tracheid cell walls are often unevenly thickened or sculptured in a spiral fashion, which makes them easy to distinguish in longitudinal section.

vessel element

ray cell

tracheid

fiber

parenchyma cell

Figure 7-5 Cells of the xylem.

In the relatively primitive condition the tracheids constituted the principal cell type of xylem as in gymnosperms, but the xylem has evolved into a more complex and specialized tissue in which specialized fibers have taken over the role of structural support and in which efficient specialized series of cells (tracheae), called vessels, function in water conduction as in angiosperms. **Vessels** are formed from meristematic cells that are lined up end to end and whose end walls have been dissolved. The cells may be very large in diameter, and the vessels may be many feet long. In highly specialized xylem, thin ribbonlike structures of parenchyma tissue, called medullary rays, run inward toward the center of the stem at right angles to the vessels. These rays serve for lateral conduction of water and for food storage.

Primary xylem is formed by differentiation of the apical meristems of root and shoot. In perennial woody plants, secondary xylem is also formed from the cambium. The differential seasonal growth of xylem produces the familiar annual rings of trees. Spring wood consists of larger cells with thinner walls and appears lighter, or less dense, than summer wood.

Wood and lumber

All of the xylem of vascular plants—including the veins of leaves or the stringlike vascular bundles of palms and bamboo—is properly called wood. The name xylem is in fact derived from *xulon,* the Greek word for wood. The lumberman's terms "softwood" and "hardwood" have little relation to actual hardness but refer to the two great classes of trees, gymnosperms and angiosperms (specifically, the dicotyledons), that provide lumber. The wood of gymnosperms (such as pine, spruce, and larch) is called softwood, or nonporous wood; the wood of angiosperms (such as oak, mahogany, and teak) is called hardwood, or porous wood (Fig. 7-6). In hardwoods, water conduction is carried on in vessels, some of which are large enough to be seen with the naked eye. In softwoods, water conduction is carried on in tracheids, which also provide support. Many softwoods contain resin, either in very thin tubes (resin ducts) parallel to the tracheids or in short cells like those of the redwood of California.

Figure 7-6   Block diagrams of the cambium and secondary xylem of (A) white cedar (*Thuja occidentalis*), a gymnosperm (softwood), and (B) tulip tree (*Liriodendron tulipifera*), an angiosperm (hardwood). [From K. Esau, Plant Anatomy, Wiley, 1953.]

In general, the central portion of the trunk of a tree is darker than the outer portion. The darker center is called heartwood and the lighter outer portion is called sapwood. The change from sapwood to heartwood occurs when tannins, gums, resins, and other substances accumulate in aging xylem cells, which is associated with the cessation of their functioning as liquid-conducting elements.

The various odors of different woods are caused by aromatic gums, resins, and essential oils that accumulate in the parenchyma cells. The grain of wood is due to the orientation of the structural elements. Interesting patterns may be seen in cut wood, owing to irregularities of the grain or the orientation of the grain in relation to the sawed or sliced surface. Knots are the basal ends of branches that become embedded in the thickening trunk. Wood's anatomical components and arrangement are characteristic for each species, and therefore can be used as a means of identification.

The remarkable properties of wood—its strength and lightness, its elastic properties, and its relative softness—are a consequence of the physical nature of the wood material and its anatomical structure. Such factors as density, moisture content, and defects influence the mechanical properties of lumber.

Phloem

The basic components of the **phloem** are series of specialized cells called **sieve elements.** It is through these elongated living cells with thin cellulose walls that the food-conducting function of the phloem is carried out. Upon maturity the nucleus of the sieve-tube element disappears. In the angiosperms, however, specialized parenchyma cells called **companion cells** are in intimate association with the sieve cells (Fig. 7-7). In addition to these, fibers and sclereids may also be present. The fibers of hemp and flax are derived from phloem tissue.

parenchyma cell

ray cell

sieve tube element
and companion cell

fiber

Figure 7-7   Cells of the phloem.

The phloem, like the xylem, is formed both by the apical meristem and by the cambium. The phloem, however, is not enduring, and the old phloem gets crushed and largely obliterated in woody stems as they grow. Young, functional phloem is protected by special meristematic tissues (cork cambium) that produce parenchymatous tissue. The phloem, the corky tissue, and the other incidental tissues constitute **bark.**

ANATOMICAL REGIONS

Plant tissues can be classified in terms of structure and function into anatomical regions. These include the vascular system, cortex, and epidermis (Fig. 7-8). These regions are interrelated, and it is sometimes hard to decide where one ends and the other begins. The pith, pericycle, endodermis, and secretory glands are components of one or more of these regions.

The vascular system

The vascular system consists principally of the xylem and phloem tissues. It serves primarily as the conduction system of the plant, but because it may also serve in support it may be compared to both the circulatory and skeletal systems of animals. Although xylem and phloem are usually found in association, there are differences in their structional relations to each other. Typically, the vascular system forms a continuous cylinder in the stem, in which the outer portion is phloem and the inner portion is xylem, which surrounds an area of parenchymatous tissue known as the pith. In some plants—for example, the monocotyledons—the vascular system is discontinuous, and appears as a series of strands in longitudinal section, and as bundles in cross section (Fig. 7-9). In

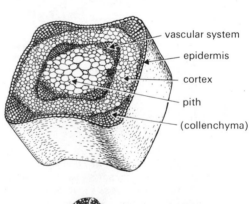

vascular system

epidermis

cortex

pith

(collenchyma)

Figure 7-8   Anatomical regions of the mint stem.

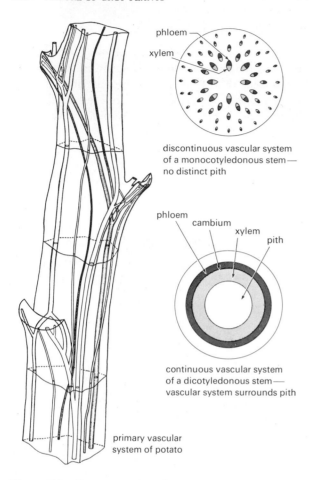

phloem

xylem

discontinuous vascular system
of a monocotyledonous stem—
no distinct pith

phloem

cambium

xylem

pith

continuous vascular system
of a dicotyledonous stem—
vascular system surrounds pith

primary vascular
system of potato

Figure 7-9 Discontinuous and continuous arrangements
of the vascular system. The primary vascular system of
the potato stem initially appears as separate bundles but
becomes embedded in secondary tissue as the stem
matures. The vascular system of the mature stem is
continuous. [After A. J. Eames and L. H. MacDaniels, *An
Introduction to Plant Anatomy*, McGraw-Hill, 1947.]

other plants the primary vascular system is discontinuous and forms a series of strands, and the secondary vascular system forms a continuous cylinder. In some plants the vascular system makes up the greater part of the root, the amount of pith being minimal.

In roots, the vascular system is separated from the cortex by specialized tissues called the pericycle and the endodermis. The **pericycle** encircles the vascular system. It is composed of parenchymatous tissue and

is the source of the branch roots and stems that arise from the root. The **endodermis** is commonly a single sheet of cells separating the vascular system from the cortex; it is not absolutely clear whether it is a part of the vascular system or of the cortex. It appears to have a protective function. The pericycle and endodermis are usually absent in the stem.

The development of the shoot is accompanied by the expansion and elongation of the vascular system. The part of the vascular system that is derived from the apical meristem is called the primary vascular system. When further increase is derived from the cambium, the resultant growth is referred to as the secondary vascular system.

In gymnosperms and dicotyledons the secondary vascular system is derived from the cambium, a cylinder between the xylem and the phloem. The cambium produces secondary xylem toward the inside and secondary phloem toward the outside. Not all plants have a cambium; in those herbaceous plants that do not, the increase in diameter that occurs during growth is the result of enlargement of cells of the cortex and vascular system.

Monocotyledonous plants generally lack secondary growth, but they may produce very large stems, as do the palms, by a special thickening meristem. The thickening meristem usually ceases activity close to the apex of the plant, but in some monocots it remains active and produces secondary growth, and in fact is referred to as cambium.

Cortex

The cortex is the region between the vascular system and the epidermis. It is made up of primary tissues, predominately parenchyma. In old woody stems, the formation of cork in the cortex, with the subsequent disintegration of the outer areas, tends to obliterate the cortex as a distinct area. Cork is formed when mature tissue becomes meristematic and forms cells with walls that contain a waxy substance called suberin. This corky protective sheath that replaces the epidermis when it is killed or sloughed away is known as the periderm. Callus tissue, which is formed as a result of wounding, may also become corky. The cork industry depends on the large amounts of this tissue produced by the cork oak (*Quercus suber*).

Differentiated portions of the periderm that form ruptured, rough areas are referred to as **lenticels.** These are particularly conspicuous in the bark of young woody stems of some species. The "dots" on apple skin are lenticels. This loose arrangement of cells tends to tear the epidermis, and fruit lenticels become subject to the entrance of decay organisms and act as points of water loss. Lenticels do, however, permit further gas exchange.

Epidermis

The epidermis is a continuous cell layer that envelops the plant. Its main functions include mechanical protection, water conservation, gaseous exchange (in shoots), water absorption (in roots), and other functions that predominate in other regions, such as secretion. Except in older stems and roots, where the epidermis may be destroyed, it sheaths the entire plant. The structure of the epidermis varies, and it may be composed of different kinds of primary tissue. The epidermal cells located a short distance from the root tips form tubelike extensions called root hairs, which function in the absorption of water and inorganic nutrients. Hairs are also found in epidermal cells of the shoot, and may be of complex multicellular structure. Cotton fibers are extremely long epidermal hairs on the seed. The velvety feel of rose petals is due to the uneven surface of their upper epidermal cells. Guard cells, which surround **stomata,** openings in the leaves that permit gaseous exchange, are modified epidermal cells.

A significant feature of epidermal cells is the **cuticle,** a waxy layer that appears on the exposed surface of the cell. It consists of a substance called **cutin,** a waxy material that acts as a protective covering to prevent the desiccation of inner tissues. It is particularly noticeable in many fruits—for example, apple, nectarine, and cherry, where its accumulation results in a blush that can be polished to a high gloss. Carnauba wax, which is used to make a number of consumer products, including polishes for wood, is derived from the cutin from the leaves of a native Brazilian palm, *Copernicia cerifera.*

Secretory glands

A number of morphologically unrelated structures— some very simple, others quite complex—secrete or excrete complex metabolic products. These substances are often exuded in viscous or liquid form. Many of them are of economic importance—for example, resin, rubber, mucilages, and gums.

Secretory functions are carried out by many multicelled, hairlike epidermal appendages called trichomes. Moreover, complex secretory structures called glands develop from subepidermal as well as epidermal tissue in various parts of the plant body. The fragrance of flowers is produced in glands called nectaries. Some specialized glands secrete essential oils (see p. 169). In citrus fruits such glands occur in the peel, as can be verified by squeezing a portion of the rind next to a flame. Rupturing the glands releases the volatile and inflammable essential oils. Gums, mucilages, and resins are formed in ducts and canals, intercellular spaces between groups of specialized cells. Often these substances are secreted in response to wounding, as in the stem of peach and cherry.

Latex, a milky, viscous substance containing various materials, particularly gums, is found in some angiosperm families. It may be formed in ordinary parenchyma or may be formed in a complex series of branching tubes called laticifers. They are not considered distinct tissue but are associated with other tissues, commonly the phloem. Latex is released under pressure when these cells are punctured. Rubber is a constituent of some latex-forming plants. The content varies greatly, but in the rubber tree (*Hevea brasiliensis*) 40 to 50 percent of the latex is rubber. The function of the latex duct network is not clear, but it is felt that it probably serves as an excretory or waste storage system for the plant and plays some role in wound healing.

MORPHOLOGICAL STRUCTURES

The flowering plant consists of two basic parts: the root, the portion that is normally underground, and the shoot, the portion that is normally above ground. The shoot is made up of stems and leaves, although

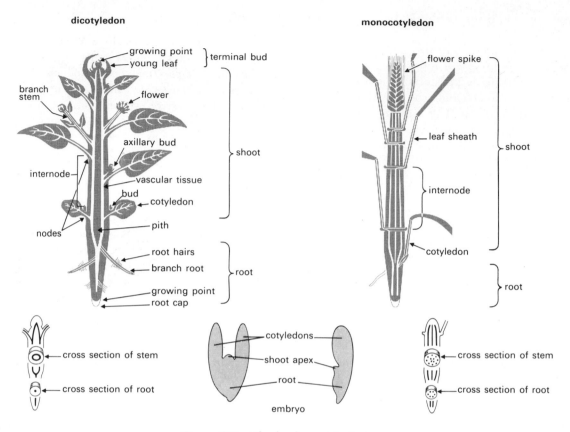

Figure 7-10   The fundamental plant parts.

these structures are not easily distinguished in all plants. Leaves arise from enlarged portions of the stem called nodes. The flower may be thought of as a specialized stem with leaves adapted for reproductive functions. Buds are miniature leafy or flowering stems. The fundamental parts of monocots and dicots are presented in Figure 7-10.

The root

The root, although visually inconspicuous, is a major component of the plant, both in terms of function and absolute bulk. It usually consists of perhaps one-third of the dry weight of the entire plant body. The root is structurally adapted for its major functions of absorption, anchorage, and storage. Owing to its complex branching, which occurs irregularly rather than at nodes as in stems, and to its tip area of root hairs, the root presents a very large surface in intimate contact with the soil. The process of absorption of water and nutrients in most plants is carried out mainly in the root hairs, which are constantly renewed by new growth. In most woody plants and in some herbaceous plants, absorption is further increased by mycorrhizae—the symbiotic association of certain fungi with the tissues of the roots. The growth of the plant is in part limited by the extent of its underground expansion. This vast network of roots anchors the plant and supports the superstructure of food-producing leaves. The older roots also serve as a storage organ for elaborated foodstuffs, as in the sweetpotato and the sugar beet.

The original seedling root, or primary root, generates the root system of the plant by forming various branching patterns (Fig. 7-11). When the primary root becomes the main root of the plant, the network is referred to as a taproot system, as in walnut, carrot, and alfalfa. In many plants, however, the primary root ceases growth when the plant is still young, and the root system is taken over by new roots that grow adventitiously out of the primary root, or, as in many of the grasses, out of the stem, forming a fibrous root system. In addition, many taprooted plants form an upper network of fibrous feeder roots, as does apple. This permits deep anchorage and a more reliable water supply while providing absorptive capacity at the more fertile upper layers of the soil. A fibrous root system may be induced artifically by destroying the taproot. This is accomplished by transplanting or undercutting, and it is a standard practice in nurseries that produce shrubs and trees. The plants so treated develop a fibrous root system concentrated in a "ball" below the plant. This permits even relatively large plants to be transplanted successfully.

In general, plants having a fibrous root system are shallow-rooted in comparison with taprooted plants.

Figure 7-11   Root systems.

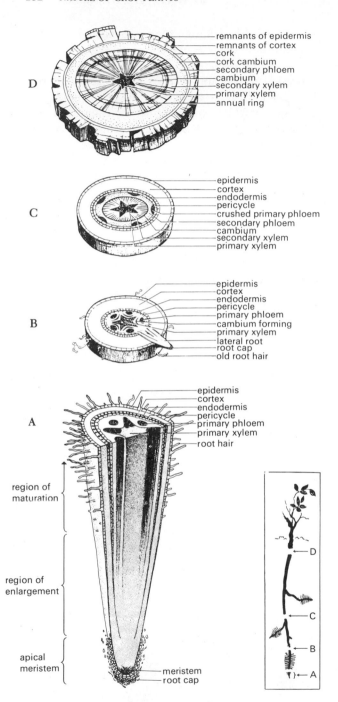

D

- remnants of epidermis
- remnants of cortex
- cork
- cork cambium
- secondary phloem
- cambium
- secondary xylem
- primary xylem
- annual ring

C

- epidermis
- cortex
- endodermis
- pericycle
- crushed primary phloem
- secondary phloem
- cambium
- secondary xylem
- primary xylem

B

- epidermis
- cortex
- endodermis
- pericycle
- primary phloem
- cambium forming
- primary xylem
- lateral root
- root cap
- old root hair

A

- epidermis
- cortex
- endodermis
- pericycle
- primary phloem
- primary xylem
- root hair

region of maturation

region of enlargement

apical meristem

- meristem
- root cap

Figure 7-12 Anatomical structure of the root. [From W. H. Muller, *Botany: A Functional Approach.* Reprinted with permission of The Macmillan Company. Copyright © 1963 by Walter H. Muller.]

Shallow-rooted plants will, of course, be more subject to drought and will show quicker response to variations in fertilizer. Shallow-rooted species depend mostly on the rain that falls during the growing season; deep-rooted species depend mostly on cumulative annual rainfall.

The anatomical structure of the root is shown in Figure 7-12. The arrangement of its vascular system differs from that of the stem. Except in monocots, the root lacks pith. The pericycle and the endodermis surround the vascular system. A cambium serves to increase the girth of roots, just as it does in the stem in some plants.

The root displays considerable structural variation. The names for various roots are descriptive of their functional adaptations—for example, storage, aquatic, brace, aerial, holdfast. Many plants are grown as crops because of their root modifications. Roots of certain species become swollen and fleshy as a result of the storage of food in the form of starches and sugars. Some of these storage roots, such as the carrot, cassava, and sweetpotato, are edible. The ability of some roots—for example, those of the sweetpotato and the dahlia—to form adventitious shoot buds renders them important in propagation.

The shoot

The shoot has been described as a "central axis with appendages." The "central axis"—that is, the stem—supports the food-producing leaves and connects them with the water- and nutrient-gathering roots. The stem also is a storage organ, and in many plants its structure is greatly modified for this function. Because they contain chloroplasts, young green stems also have a small role in food production. Plants assume extremely varied forms, ranging from single upright shoots, as in the date palm, to prostrate branched "creepers." It is the structure and growth pattern of the stem that determines the form of the plant. Basic structural and anatomical features of herbaceous and woody stems are shown in Figure 7-13.

The upright growth of plants that have one active growing point and a rigid stem is considered the basic one, and our descriptive terms are used to differ-

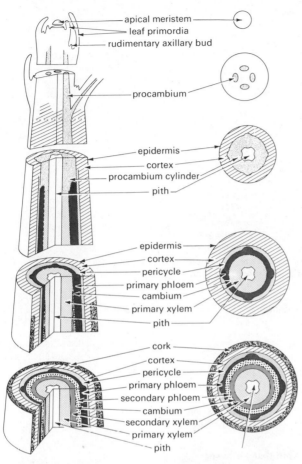

apical meristem

leaf primordia

rudimentary axillary bud

procambium

epidermis

cortex

procambium cylinder

pith

epidermis

cortex

pericycle

primary phloem

cambium

primary xylem

pith

cork

cortex

pericycle

primary phloem

secondary phloem

cambium

secondary xylem

primary xylem

pith

Figure 7-13 Diagrammatic sections through a stem.
[After R. M. Holman and W. W. Robbins, *A Textbook of General Botany*, Wiley, 1939.]

entiate other growth patterns. Typical shrubby or bushy growth is due to the absence of a main trunk or central leader. Plants with this pattern of growth are characterized by a number of erect or semierect stems, none of which dominates. The distinguishing feature is form rather than size. Similarly, slender and flexible stems that cannot support themselves in an erect position are known as vines or creepers. Vines will trail unless an independent support is available to permit them to grow upright. Some vines are herbaceous (morning glory, pea), and others are woody (grape), in which case they are technically called lianas.

Buds

The stem is divided into matur
ing regions in which growth and
place. The "embryonic stem" is
All buds do not grow actively;
of arrested development, or become dormant, but are nevertheless potential sources of further growth. Overwintering is one of the main functions of buds. Although buds may be embedded in the stem tissues, and relatively inconspicuous, they may become quite structured. The form, structure, and arrangement of buds prove to be useful guides in describing woody plants, even when the leaves are absent, as in winter. Growth may originate from a single terminal bud or from lateral buds, which occur in the leaf axis. In addition, buds may be formed in internodal regions of the stem, in leaves, or even in roots, often as a result of injury. These are called **adventitious** buds.

Buds may produce leaves, flowers, or both, and are referred to as leaf, flower, and mixed buds, respectively. When more than one bud is present at a leaf axis, all but the central or basal bud are called accessory buds. The arrangement or topology of buds on a stem (their phyllotaxy) corresponds to the leaf arrangement. If leaves are opposed to each other at the same level, the leaf (and bud) arrangement is said to be opposite or whorled. When single leaves grow at different levels, they are arranged in a spiral and are said to be alternate. The spiral pattern of leaf arrangement is varied and is expressed as a fraction (½, ⅓, ⅖, ⅜), where the numerator is the number of turns to get to a leaf directly above another and the denominator is the number of buds passed. Phyllotaxy has some taxonomic significance, since it is often the same throughout a genus and often even throughout a whole family.

Stem modifications

The stem may be greatly modified from a basically upright cylindrical structure (Fig. 7-14). Some of these alterations may seem quite bizarre; yet upon close analysis (particularly of their development) these modifications can be shown to be basically stemlike in structure; that is, they have nodes, leaves,

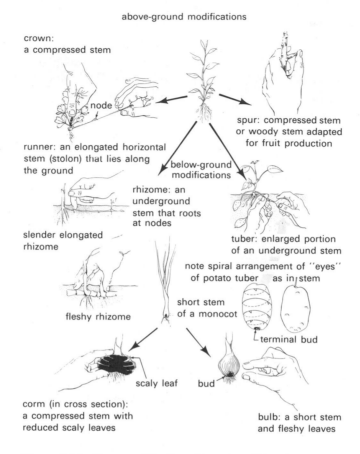

above-ground modifications

crown:
a compressed stem

node

spur: compressed stem
or woody stem adapted
for fruit production

runner: an elongated horizontal
stem (stolon) that lies along
the ground

below-ground
modifications

rhizome: an
underground
stem that roots
at nodes

slender elongated
rhizome

tuber: enlarged portion
of an underground stem

note spiral arrangement of "eyes"
of potato tuber   as in stem

short stem
of a monocot

fleshy rhizome

terminal bud

scaly leaf   bud

corm (in cross section):
a compressed stem with
reduced scaly leaves

bulb: a short stem
and fleshy leaves

Figure 7-14 Stem modifications. Note that all have
nodes and leaflike structures.

or similar scalelike structures, and they function in transport or storage. Stem modification can be divided into above-ground forms (crowns, stolons, spurs) and below-ground forms (bulbs, corms, rhizomes, tubers). Since many stem modifications contain large amounts of stored food, they are especially significant in propagation, and, in potatoes, are important as a source of food.

Above-ground modifications

The crown of a plant is, in general, that portion just above and just below ground level. (The forester's term "crown" refers to the branched top of a tree.) This portion of the plant may become greatly enlarged, as in baldcypress (*Taxodium distichum*). Crowns of some plants that appear to be stemless may be thought of as "compressed" stems. The structure of the strawberry crown can be clearly seen by artificially elongating it through treatment with the plant-growth regulating substances known as gibberellins. Leaves and flowers arise from the crown by buds, as they do in ordinary stems. Fleshy buds from crowns may produce new plants, referred to as crown divisions. The crown may be modified into a food storage organ, as it is in asparagus.

Short, many-sided, horizontal branches that grow out of the crown and bear fleshy buds or leafy rosettes are referred to as offsets, slips, suckers, pips, and so on. These stem modifications, which can be collectively termed **offshoots,** are important for propagation.

Stolons are stems that grow horizontally along the ground. A runner is a stolon with long internodes that originates at the base or crown of the plant. At some of the nodes of a stolon, roots and shoots develop. A well-known example of a plant with runners is the strawberry.

Spurs are stems of woody plants whose longitudinal growth is greatly restricted. They are characterized by greatly shortened internodes, and they appear laterally on branches. In mature fruit trees, such as apple, pear and quince, flowering is largely confined to spurs. Spurs are not irrevocably static, and they may revert to normal stem growth even after many years of fruiting.

Below-ground modifications

Bulbs appear as compressed modifications of the shoot, and consist of a short, flattened, or disc-shaped stem surrounded by fleshy, leaflike structures called scales. They may enclose shoot or flower buds. Bulbs and corms are found only in some monocotyledonous plants. The scales, filled with stored food, may be continuous, and form a series of concentric layers surrounding a growing point, as in the onion or tulip (tunicate bulbs), or may form a more or less random attachment to a small portion of stem, as in the Easter lily (scaly bulbs). Bulbs commonly grow under the ground or at ground level, although bulblike structures (bulbils) may be formed on aerial stems, as they are in some lily species, or even in association with flower parts, as in the onion.

Corms are short, fleshy, underground stems having few nodes. The corm is stem; the few rudimentary leaves are nonfleshy. The gladiolus and crocus are propagated by corms.

Rhizomes are horizontal, underground stems. They may be compressed and fleshy, as in irises, or slender with elongated internodes, as in turfgrasses (for example, Bermuda, bluegrass). Normally, roots and shoots develop from the nodal regions. Such weeds as quackgrass and Canadian thistle are particularly insidious because they spread so rapidly, owing to their natural propagation by rhizomes.

Tubers are greatly enlarged, fleshy port[...] rhizomes. They are typically noncylindrical. [...] word "tuber" is derived from a Latin word meani[...] lump.) The edible portion of the potato plant is a[...] tuber. The "eyes" arranged in a spiral around the tuber are the buds. Each eye consists of a rudimentary leaf scar and a depressed cluster of buds.

The leaf

Leaves are the principal photosynthetic organs of higher plants. The leaf blade is usually a flat appendage of the stem arranged in such a way as to present a large surface for the efficient absorption of light energy, and it is usually attached to the stem by a stalk or petiole. Leaflike outgrowths of the petioles, known as stipules, may be present. Leaves may contain many secretory structures, such as glands or ducts. In many plants, particularly those native to the humid tropics, water can be discharged through small openings in the epidermis, called hydathodes; they are often situated at the leaf tip.

The anatomical structure of the leaf is shown in Figure 7-15. Note that its vascular system consists of a branching network of veins. The veining is typi-

Figure 7-15 Structure of a leaf. [After A. J. Eames and L. H. MacDaniels, *An Introduction to Plant Anatomy,* McGraw-Hill, 1947.]

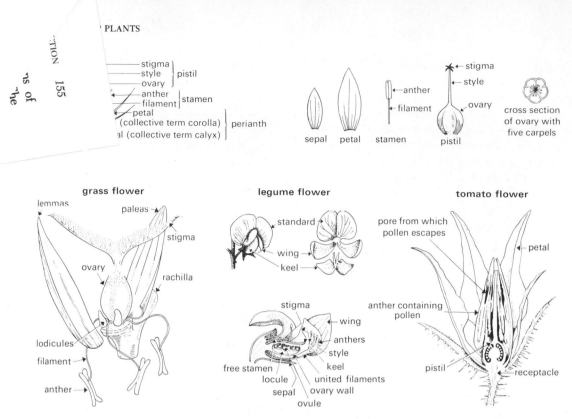

Figure 7-16 Structure of flowers.

cally netlike in dicots and parallel in monocots. The leaf blade, although commonly bilaterally symmetrical, is not radially symmetrical, since it has distinct upper and lower sides. Beneath the upper epidermal layer, which is often characterized by heavy deposits of cutin, lie series of elongated, closely packed palisade cells, which are particularly rich in chloroplasts. The irregular, loose arrangement of cells beneath the palisade cells results in a spongelike region (spongy mesophyll) that provides the air space necessary for gaseous exchange in photosynthesis and transpiration. Depending on the species, stomata occur on the lower, upper, or both upper and lower epidermis.

Leaves of plants vary in shape from the flat, thin blades just described to the needlelike leaves of conifers or the fleshy leaves in the jade plant (*Crassula argentea*). The tendrils of peas are modified leaflets attached to the leaf. Leaves are the edible portions of many plants, such as lettuce, cabbage, and spinach, and are the most prominent feature of some plants, such as grasses.

The flower

The flower shows great variety in structure, composition, and size. The principal flower parts (Fig. 7-16) are as follows:

Sepals (collectively, the **calyx**) enclose the flower when it is in bud. They are usually small, green, leaflike structures below the petals.

Petals (collectively, the **corolla**) are the most conspicuous parts of most flowers. They may be highly colored, but are usually not green. The extremely large, showy flowers of many cultivated ornamentals are the result of rigorous selection for this character.

Stamens, the "male" reproductive structures, are composed of pollen-bearing **anthers** supported by a **filament.** When the pollen is mature, it is discharged through pores or through the ruptured anther wall. Nectaries, secretory structures that produce a viscous, sugary substance, and often a perfume, are most commonly found at the base of stamens.

Pistils (made up of one or more **carpels**), the "female" reproductive structures, are usually modified into an ovule-bearing base (or **ovary**) supporting an elongated region (or **style**) whose expanding tip (or surface) is called the **stigma.** The ovule develops into the seed. The mature ovary becomes the fruit.

The petals and sepals of the flower, as well as the reproductive parts (the stamens and pistils) are essentially modified leaves. That the stamens are modified leaves can be clearly seen in the stamenoids or "extra" petals of the cultivated rose. These flower parts are borne on an enlarged portion of the flower-supporting stem, called the **receptacle.**

Flowers composed of sepals, petals, stamens, and pistils are referred to as **complete. Incomplete** flowers lack one or more of these parts. For example, they may lack stamens (**pistillate,** or "female" flowers) or pistils (**staminate,** or "male" flowers). Those that contain both stamens and pistils (perfect, bisex-

ual, or hermaphroditic flowers) are referred to as **perfect** even if they lack calyx or corolla and are thus incomplete.

Similarly, plants may be referred to as staminate, pistillate, or perfect on the basis of the type of flowers they bear. When separate staminate and pistillate flowers occur on the same plant, as in maize, the sex type is **monoecious.** Species in which staminate and pistillate flowers are borne on different plants are referred to as **dioecious** (date palm, papaya, spinach, asparagus, hemp). Other combinations of flower types also occur. For example, muskmelons have perfect and staminate flowers on the same plant; this sex type is referred to as **andromonoecious.**

There are many ways in which the flowers may be situated and arranged on the plant. The term **inflorescence** refers to a flower cluster. Some of the common types of inflorescence are diagrammed in Figure 7-17.

Figure 7-17 Common inflorescences.

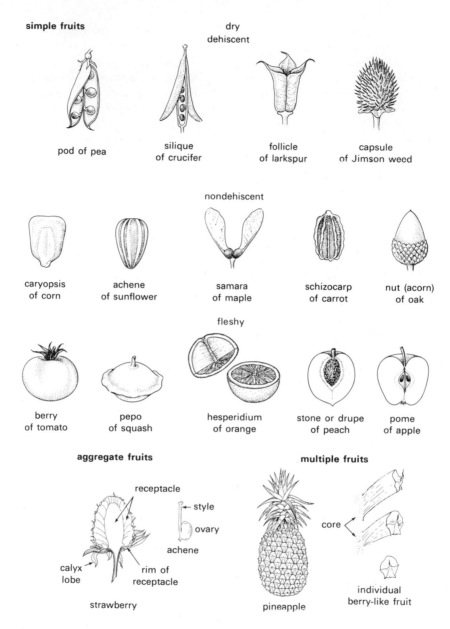

simple fruits

dry
dehiscent

pod of pea

silique
of crucifer

follicle
of larkspur

capsule
of Jimson weed

nondehiscent

caryopsis
of corn

achene
of sunflower

samara
of maple

schizocarp
of carrot

nut (acorn)
of oak

fleshy

berry
of tomato

pepo
of squash

hesperidium
of orange

stone or drupe
of peach

pome
of apple

aggregate fruits

receptacle

style

ovary

achene

calyx
lobe

rim of
receptacle

strawberry

multiple fruits

core

individual
berry-like fruit

pineapple

Figure 7-18 Various types of fruits. Simple fruits are made up of a single ovary. Aggregate fruits are made up of many pistils on a common receptacle. In the strawberry, the seedlike structures are achenes—small, dry, indehiscent, one-seeded fruits. Multiple fruits are derived from closely clustered flowers. The individual fruitlets in the pineapple are berrylike. [From J. Janick, *Horticultural Science*, 3rd ed., W. H. Freeman and Company, copyright © 1979. After R. M. Holman and W. W. Robbins, *A Textbook of General Botany*, Wiley, 1939.]

The fruit

The botanical term **fruit** refers to the mature ovary and other flower parts associated with it. Thus, it may include the receptacle as well as withered remnants of the petals, sepals, stamens, and stylar portions of the pistil. It would also include any seeds contained in the ovary.

The structure of the fruit is related to the structure of the flower. Fruits are classified according to the number of ovaries incorporated in the structure (for example, as simple, aggregate, or multiple fruits), the nature and structure of the ovary wall, whether they split when ripe (dehisce), and the way in which the seed is attached to the ovary (Fig. 7-18).

Simple fruits

The majority of flowering plants have fruits composed of a single ovary, and are referred to as **simple fruits.** In the mature fruit (when the enclosed seed is fully developed), the ovary wall may be **fleshy** (composed of large portions of living succulent parenchyma) or **dry** (made up of nonliving sclerenchyma cells with lignified or suberized walls).

The ovary wall, or pericarp, is composed of three distinct layers. From outer to inner layer, these are the exocarp, mesocarp, and endocarp. When the entire pericarp of simple fruits is fleshy, the fruit is referred to as a berry. The tomato, grape, and pepper are berry fruits. The muskmelon is a berry (specifically, a type of berry called a **pepo**) with a hard rind made up of exocarp and receptacle tissue. Citrus fruit is also a berry, of a type called a **hesperidium,** in which the rind is made up of exocarp and mesocarp; the edible juicy portion is endocarp.

Simple fleshy fruits having a stony endocarp are known as **drupe,** or stone, fruits (peach, cherry, plum, olive). The skin of these fruits is the exocarp; the fleshy, edible portion is the mesocarp. Simple fleshy fruits in which the inner portion of the pericarp forms a dry, papery "core" are known as **pomes** (apple, pear, quince).

Dry fruits may be either **dehiscent** or **indehiscent.** Dehiscent fruits (those that split open at maturity) usually have many seeds, and they are classified by the manner in which dehiscence occurs and the number of carpels from which they develop. For example, the pod (legume) of leguminous plants, such as the pea or bean, develops from a single carpel and separates along both sutures. Among the other types of dehiscent fruits are the **silique** (crucifers), **follicle** (larkspur), and **capsule** (Jimson weed). Indehiscent fruits (those that do not split open at maturity) generally bear one seed. When the pericarp lies close to, but separated from, the seed, the fruit is called an **achene** (sunflower). When the paricarp becomes fused to the seed, the fruit is called a **caryopsis.** The caryopsis is the type of fruit found in most grasses, including maize and wheat. **Nuts** are characterized by a hardened pericarp. Other indehiscent fruits include the **samara** (maple), in which the pericarp forms a winged appendage, and the **schizocarp** (carrot), in which the fruit divides into two or more one-seeded indehiscent parts.

Aggregate fruits

Aggregate fruits are derived from flowers that have many pistils on a common receptacle. The individual fruits of the aggregate are drupes in the blackberry and achenes in the strawberry. The fleshy, edible portion of the strawberry is the receptacle.

Multiple fruits

The multiple fruit is derived from many separate but closely clustered flowers. Familiar examples of multiple fruits are the pineapple, fig, and mulberry. The beet "seed" is a multiple fruit.

The seed

A seed is in essence a miniature plant (the embryo) in an arrested state of development. Most seeds contain a built-in food supply (the orchid seed is an exception). Structurally, the true, or botanical, seed is a mature ovule. The agricultural definition of a seed refers to the unit that is planted, regardless of its structure (Fig. 7-19), and includes true seeds (alfalfa, cucumber); one-seeded dry indehiscent fruits, such as the caryopsis in grasses and the achene in lettuce; and multiple-seeded fruits, in which additional accessory structures are present, as in beet and New Zealand spinach.

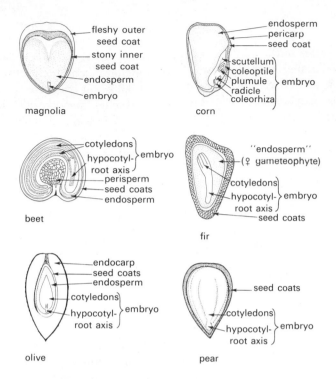

Figure 7-19 Structure of seeds and one-seeded fruits. [After D. E. Kester and H. T Hartmann, *Plant Propagation: Principles and Practices*, 3rd ed., Prentice-Hall, 1975.]

The seed contains a miniature plant, or embryo, which usually develops from the union of generative cells or gametes. The details of fertilization and seed development will be discussed in Chapter 9, "Reproduction and Propagation."

The fully developed embryo commonly consists of a hypocotyl–root axis, a transition zone between the rudimentary shoot and root. On the upper end it contains one or more specialized seed leaves called **cotyledons.** (In grasses the shieldlike cotyledon is called the **scutellum.**) The embryonic shoot bud is called the **plumule** (Latin for little feather). On the lower end of the axis there may be an embryonic root called the **radicle.** In grasses a protective sheath, or **coleoptile,** covers the plumule and a similar sheath, the **coleorhiza,** covers the radicle.

The food stored in seeds is present as carbohydrates, fats, and proteins, one of these forms usually predominating. Seeds are thus a rich source of food as well as of fats and oils for industrial purposes. The reserves in cereals and legumes provide a large share of the world's food supply for people and animals. In some seeds this stored food is derived from a tissue called the endosperm, which is formed as a result of the fertilization process. In mature seeds of some plants the endosperm constitutes a specialized region, as in maize; in others it is absorbed by the developing embryo, in which case the cotyledons serve as the food-storage organ, as in beans and walnuts.

Seeds vary greatly in size, form, and shape. Most plants can be identified by their seeds alone. In addition, great variation exists within some species. Ex-

Figure 7-20   Seed germination and seedling morphology in onion (a monocot) and castor bean (a dicot). [From A. S. Foster and E. M. Gifford, *Comparative Morphology of Vascular Plants*, 2nd ed., W. H. Freeman and Company, copyright © 1974.]

amples of such variation are the presence or absence of spines (spinach), color variation (beans), and differences in the chemical composition of stored food (sugary versus starchy maize).

Seed germination refers to the change from arrested development to active growth. The subsequent seedling stage, the interval during which the young plant becomes dependent on its own food-manufacturing structures, is diagrammed in Figure 7-20.

Selected references

Bold, H. C., 1973. *Morphology of Plants,* 3rd ed. New York: Harper and Row. (A basic text on the subject.)

Eames, A. J., and L. H. MacDaniels, 1947. *An Introduction to Plant Anatomy.* New York: McGraw-Hill. (A well-known textbook on the anatomy of vascular plants.)

Esau, K., 1965. *Plant Anatomy,* 2nd ed. New York: Wiley. (An authoritative treatise and advanced textbook on the anatomy of seed plants.)

Esau, K., 1977. *Anatomy of Seed Plants,* 2nd ed. New York: Wiley. (A comprehensive advanced treatise.)

Foster, A. S., and E. M. Gifford, Jr., 1974. *Comparative Morphology of Vascular Plants,* 2nd ed. San Francisco: W. H. Freeman and Company. (A physiogenetic approach to plant morphology.)

Heyward, H. E., 1938. *The Structure of Economic Plants.* New York: Macmillan. (This book is especially noted for its treatment of particular species of important crop families, including maize, wheat, onion, hemp, beet, radish, alfalfa, pea, flax, cotton, celery, sweetpotato, potato, tomato, squash, and lettuce.)

Weier, T. E., C. R. Stocking, and M. G. Barbour, 1974. *Botany: An Introduction to Plant Biology,* 5th ed. New York: Wiley. (An excellent introductory textbook.)

8
Growth and Development

The life of an organism is characterized by systematic increases in bulk and complexity. These events can be discussed in terms of the interrelated processes of growth and development. **Growth,** in a restricted sense, refers to an irreversible increase in size, reflecting a net increase in protoplasm. **Development** infers differentiation, a higher order of change that involves anatomical and physiological specialization and organization.

PLANT GROWTH PROCESSES

Growth in plants can be measured in many ways; the most straightforward is by measuring the increase in dry weight. The increase in the weight of an organ (such as a fruit) or an entire plant may result from increases in either the number or the size of cells, or both. The growth of multicellular organisms is based on increases in cell number provided by cell division. The mechanical aspects of cell division will be discussed in more detail in Chapter 9.

The increase in protoplasm is brought about through a series of events in which water, carbon dioxide, and inorganic salts are transformed into living material. This involves the production of carbohydrates (**photosynthesis**), the elaboration and degradation of complex molecules (**metabolism**), and the uptake and movement of water and nutrients (**absorption** and **translocation**). The required chemical energy is provided by **respiration.** These physiological processes are functions of individual cells and of multicellular organisms; although these processes are interrelated, it is useful to classify them into separate systems.

Photosynthesis

Photosynthesis is the process in which carbon dioxide and water are transformed, in the presence of light, into carbon-containing, energy-rich, organic compounds. This conversion of light energy into chemical energy is perhaps the most significant of the life processes. With few exceptions all of the organic matter in living things is ultimately provided through this sequence of biochemical reactions.

Photosynthesis takes place primarily in the presence of two pigments, chlorophyll a and chlorophyll b, and, except under laboratory conditions, it takes place only in the chloroplasts of living cells. Photosynthesis is a complex series of integrated processes that can be stated in abbreviated form by the following chemical reaction:

$$nCO_2 + 2nH_2O + \text{light energy} \rightarrow$$
$$(CH_2O)_n + nH_2O + nO_2$$

The series of photosynthetic reactions can be grouped into a light phase (the reactions that require light) and a dark phase (the reactions that do not require light).

The first step in this series is independent of temperature, and consists in the trapping of light energy, which accomplishes the cleavage of the water molecule into hydrogen and oxygen (**photolysis**). The oxygen is released as gaseous molecular oxygen, and the hydrogen is trapped by a hydrogen acceptor, nicotinamide adenine dinucleotide phosphate (NADP). Thus, the liberation of oxygen in photosynthesis is independent of the synthesis of carbohydrates. This step has been referred to as the **Hill reaction.** The trapping of light energy by the conversion of adenosine diphosphate (ADP) to adenosine triphosphate (ATP) occurs in a process known as **photophosphorylation.** The combination of Hill reaction and phosphorylation is known as the **light phase** of photosynthesis.

The conversion of light energy to chemical energy is achieved by the formation of energy carriers such as ATP and $NADPH_2$. ATP has been termed "energy currency" and $NADPH_2$ as "reducing power." ATP is involved in energy transfers in many vital processes of the cell. It is formed from ADP through the addition of a third phosphate group. The captured light energy stored in the third phosphate bond becomes available following the conversion of ATP back to ADP. NADP accepts electrons and hydrogen atoms produced in photolysis and transfers these to other compounds. Indeed the crucial reactions of photosynthesis are the conversion of ADP to ATP and the reduction of NADP to $NADPH_2$.

The **dark phase** of photosynthesis (also called the **Calvin cycle**) is greatly affected by temperature and has been shown to be independent of light. Essentially the hydrogen atoms from water are transferred by the hydrogen-carrying acceptor ($NADPH_2$) to a low-energy organic acid to produce, with the help of the energy of the ATP, a "carbohydrate" of higher energy, from which sugars are formed.

This reduction reaction—that is, the addition of electrons and hydrogen atoms to carbon dioxide— results in the formation of sugar units. The precise pathways by which carbon dioxide is synthesized to sugar are now almost completely known. A substance formed early in this synthesis has been identified as a 3-carbon phosphorus-containing compound, phosphoglyceric acid, two molecules of which eventually give rise to a single 6-carbon sugar and, finally, to starch grains in the chloroplasts, as shown in Figure 8-1.

Photosynthetic efficiency (the rate of net photosynthesis) is equal to the rate of "gross" photosynthesis less any photosynthate that may be lost through respiration:

Net photosynthesis

= Gross photosynthesis — Respiration

Recent studies have indicated that there are two important categories of plants in regard to net photosynthetic rate (Table 8-1). One group, the most typical, fixes carbon from ribulose diphosphate into a 3-carbon acid (phosphoglyceric acid) utilizing the enzyme ribulose-diphosphate carboxylase. Thus, this type of photosynthesis is referred to as the C_3 type. The second group of plants, which has a much higher rate of net photosynthesis, includes tropical grasses, sugar cane, and species of *Atriplex* and *Amaranthus* (these two genera include many fast-growing weeds). In this group, CO_2 is first fixed into phosphoenolpyruvate (PEP) to yield 4-carbon acids (such as oxaloacetic, malic, and aspartic acid) and involves the enzyme PEP carboxylase; hence it is known as the C_4 type of photosynthesis. C_4 plants carry out both C_3 and C_4 photosynthesis, whereas C_3 plants lack the C_4 pathway. In addition, there is a variant of the C_4 photosynthesis referred to as the crassulacean acid metabolism (CAM) type that is characteristic of plants adapted to arid regions.

Although the C_4 pathway requires slightly more energy than the C_3 pathway, this requirement is offset by other features. The most important advantage is the apparent absence of a type of light-dependent respiration (hence the term **photorespiration**) that is linked to the photosynthetic cycle in C_3 plants. Photorespiration lowers the apparent efficiency of CO_2 assimilation in C_3 plants. The apparent absence of photorespiration in C_4 plants occurs because the photosynthetic affinity for CO_2 is so high that photorespiratory CO_2 is recycled. The process is different

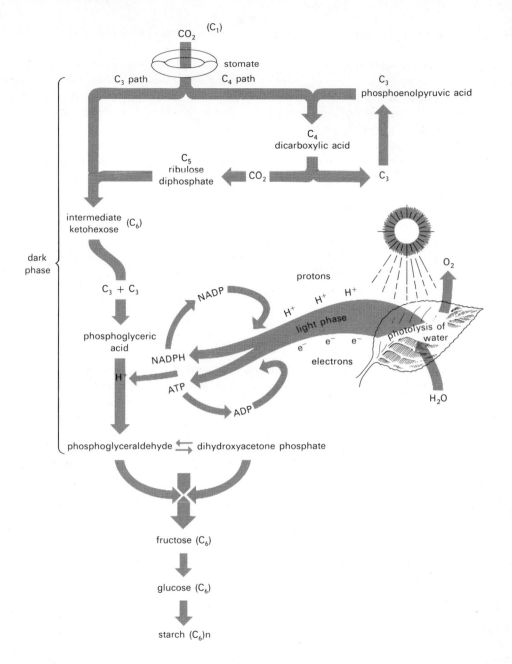

Figure 8-1   Photosynthesis. Note the C_3 and C_4 pathways. [From J. Janick, *Horticultural Science*, 3rd ed., W. H. Freeman and Company, copyright © 1979.]

Table 8-1 Comparative rates of maximum photosynthesis.

Type of plant	Maximum photosynthesis (mg CO_2 fixed per sq. dm of leaf surface per hr)
C_3 photosynthesis	
Slow growing perennials (desert species, orchids)	1–10
Evergreen woody plants	5–15
Deciduous woody plants	15–30
Rapidly growing annuals (wheat, soybean, sugar beet, sunflower)	20–50
C_4 photosynthesis	
Tropical grasses, sugarcane, corn, *Amaranthus, Atriplex*	50–90

Source: Jarvis and Jarvis, *Plant Physiol.* 17:645–666, 1964.

from normal ("dark") respiration, which is independent of light. (See the discussion under "Respiration," p. 165.) The occurrence of photorespiration was difficult to detect because it is hard to distinguish the CO_2 fixed in photosynthesis from the CO_2 given off by photorespiration.

The difference in photorespiration between the C_3 and C_4 systems can be demonstrated by enclosing plants in a sealed, illuminated container. As photosynthesis increases, CO_2 is taken up by the plant and fixed, and the effect is to lower the amount normally present in the air. For C_3 plants, a steady state of CO_2 concentration in the air (known as the **compensation point**) is reached when there remains about 50 parts per million at 25°C (77°F). This amount is as large as it is because photorespiration releases CO_2. With a C_4 plant such as corn, which lacks photorespiration, the amount of CO_2 in such a sealed container diminishes to less than 10 parts per million. A soybean plant (C_3) enclosed with a corn plant (C_4) will die because the corn exhausts the CO_2.

The rate of photorespiration increases with temperature faster than gross photosynthesis. Thus many

C_3 plants are nonproductive at high temperatures (25–35°C, or 77–95°F), whereas C_4 plants such as tropical grasses increase in productivity at these higher temperatures.

Other specific adaptations are associated with the C_4 system. These include enhanced translocation of photosynthetic products, specialized leaf anatomy and chloroplast structure, and greater ratio of chlorophyll *a* to chlorophyll *b*.

Respiration

Energy is required to run the machinery of the cell. The energy incorporated in the chemical bonds of the sugars formed from photosynthesis cannot, of course, be harnessed by the cell from high-temperature combustion, but must be provided at low and constant temperatures in delicately controlled reactions. Respiration, the process of obtaining energy from organic material, is accomplished with great efficiency in the cell. It is in a superficial sense the reverse process of photosynthesis:

$$C_6H_{12}O_6 + 6O_2 \rightarrow 6H_2O + 6CO_2 + \text{energy}$$

The captured energy of light is released from the low-temperature oxidation of (removal of hydrogen from) sugars. Although a small part does appear as heat, the useful energy is channeled into chemical work, initially in the formation of high-energy phosphates, and later in the synthesis of organic materials required in growth and development (Fig. 8-2).

The biologic combustion of sugar is accomplished through an extremely complicated series of reactions involving many specific enzyme systems and energy carriers. The steps from 6-carbon sugar to carbon dioxide involve the transformation of phosphorylated sugars to 3-carbon pyruvic acid ($CH_3COCOOH$). This step, known as **glycolysis,** is common to many organisms. Organisms that do not require oxygen (**anaerobes**) transform pyruvic acid to alcohol or lactic acid. Organisms that do require oxygen (**aerobes**) transform pyruvic acid to water and carbon dioxide. This involves the participation of a number of plant acids in the cyclic series of steps known as the citric

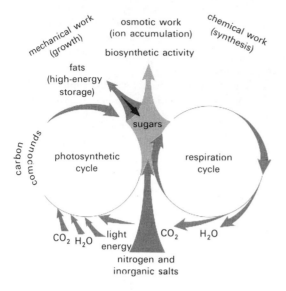

Figure 8-2   Energy cycle in green plants.

acid cycle, or (in honor of its discoverer) the **Krebs cycle.**

The rate of respiration depends on many factors. It is highest in rapidly growing tissues and lowest in dormant tissues. The rate of respiration is greatly influenced by temperature, and approximately doubles for each 10°C (18°F) rise over a range of 4–36°C (40–97°F). Other factors influence the respiration rate, including the availability of oxygen and carbohydrates and the age and condition of cells and tissues. Respiration is a common feature of all living material, and consequently the detection of evolved carbon dioxide is often utilized as a test for life.

Metabolism

All of the various materials produced in the plant are ultimately derived from the carbon compounds produced by photosynthesis and from the inorganic nutrients and water absorbed from the soil (Fig. 8-3). The synthesis (anabolism) and degradation (catabolism) of these organic materials is known as **metabolism.** The degradation of sugars and fats and the release of energy in respiration are examples of catabolic metabolism. The step-by-step elucidation of the

pathways in plant metabolism comprises some of the brightest chapters in biochemistry.

Plants are cultivated for the complex molecules they synthesize; carbohydrates, proteins, and lipids are the compounds of most concern because they are the major constituents of food (see Chapter 3). The metabolism of carbohydrates has already been dealt with briefly in the various discussions of photosynthesis. Proteins are second in abundance to carbohydrates as plant constituents (Table 8-2). An understanding of the interrelations between carbohydrate metabolism and nitrogen metabolism is necessary to the comprehension of plant development.

Green plants can survive in an inorganic world. They are autotrophic for nitrogen as well as carbon. Except for the nitrogen-fixing legumes and a few other plants that depend on their bacterial partners, plants are dependent on the nitrate and ammonium ions in the soil solution as a source of nitrogen. Since microorganisms in the soil rapidly oxidize the reduced forms of nitrogen, the most common source of nitrogen taken up by plants is the nitrate ion, NO_3^-. Before combining with organic acids to form amino acids, nitrate is reduced through a series of enzymatic reactions. The reduced nitrogen groups —NH_2 and NH_3 combine with the carbon frameworks formed during the oxidation of sugars to form amino acids.°

Amino acids join to form proteins through a complex series of reactions regulated by the nucleic acids present in the cell. The primary sites of protein synthesis in plants are the tissues in which new cells are being formed, such as tips of stems and roots, buds, cambium, and developing storage organs. The green leaf is also an important site of protein synthesis, since both the carbon frameworks and the inorganic nitrogen necessary for the formation of amino acids are readily available. Leaf protein is in a continuous state of turnover. When leaves are placed in the

---

°The ammonia (NH_3) initially combines with the organic acid α-ketoglutaric acid to form the amino acid glutamic acid, an energy-requiring process. Other amino acids are constructed via the transfer of the amino group (—NH_2) from this amino acid to other organic acids from the metabolic pool (a process known as transamination).

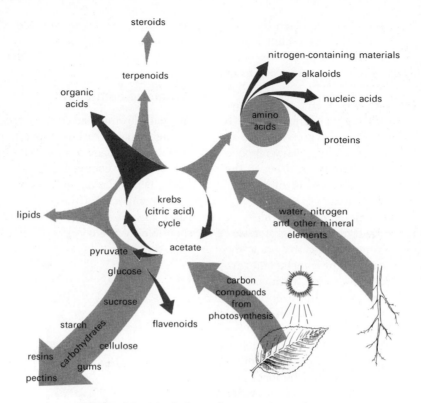

Figure 8-3 Metabolic pathways in green plants.

Table 8–2 Nutrient constitutents of certain plant parts.

		Nutrients as percent of dry matter			
Plant part	Dry matter (%)	Crude protein[1]	Fat	Carbohydrate	Minerals
Cabbage leaf	15.8	16.5	2.5	62.0	19.0
Beet root	16.4	9.8	0.6	82.9	6.7
Potato tuber	21.2	10.0	0.5	84.0	5.2
Wheat seed	89.5	14.7	2.1	81.0	2.1
Soybean seed	90.0	42.1	20.0	32.8	5.1

[1]Includes protein and nitrogen-containing compounds.

dark, are excised from the plant, or reach the state of senescence, protein breaks down and soluble nitrogen compounds are formed that can be reincorporated into amino acids in other leaves or organs of the plant.

Because the nitrogen metabolism and the carbohydrate metabolism of the plant are closely linked, both processes are markedly affected by gross changes in the environment of the plant, such as heavy nitrogen fertilization, reduced light intensity, or drought; the effects on these processes are reflected in altered patterns of growth and development.

A great number of organic compounds, many of great complexity, are formed in metabolic processes. (Hundreds of different compounds contribute to the flavor of coffee alone!) Some of these metabolites are commonly referred to as essential oils, pigments, vitamins, and so forth. These terms, however, are not specific in the chemical sense and often refer to mixtures of materials, many of them unrelated. For example, pigments—substances that preferentially absorb certain wavelengths of visible light—differ structurally and chemically. Not all substances produced in the various metabolic pathways are, so far as is known, essential to the plant. Such substances are characteristically produced only by certain plants or plant groups. Many of them (for example, rubber and menthol) are highly prized and constitute the prime reason for the cultivation of the species from which they are derived.

The following brief sections deal with the various categories of metabolic substances other than carbohydrates, proteins, and lipids.

Organic acids and alcohols

Plants manufacture a great variety of organic acids. A number of them participate in metabolic cycles like those involved in the biological combustion of sugar to carbon dioxide. Organic acids are found in the vascular sap and commonly accumulate in certain plant organs, usually the fruit (citric acid in lemons, malic acid in apples).

Alcohols occur only in very low concentrations as free alcohols, and are generally found combined with organic acids as esters. Fats are esters of the alcohol glyerol and various fatty acids. The characteristic

fruit odors and flavors are due to a combination of the volatile organic acids, esters, and other compounds such as ketones and aldehydes.

Aromatic compounds

Aromatic compounds, in contrast to aliphatic or "straight-chain" carbon compounds, have structures containing at least one benzene ring. Many, but not all, of these compounds have very characteristic odors, hence the name "aromatic." The simple phenolic compounds (those containing one benzene ring), such as vanillin and methyl salicylate, are responsible for the odors of vanilla and wintergreen.

vanillin   methyl salicylate

The volatile phenolic compounds (specifically, the phenylpropane derivatives) are responsible for the characteristic flavors and odors of cinnamon, clover, and parsley.

phenylpropane skeleton

Two commercially important plant constituents, lignins and tannins, are mixtures of complex aromatic and carbohydrate materials. Lignins have already been discussed briefly as constituents of the cell wall (p. 141). Tannins have the property of precipitating proteins and are used in tanning—the transformation of animal hides into leather. Tannins are extracted from a number of plants—the word "tan" comes from the Latin *tannum*, the bark of the oak tree, which has been used for centuries in the production of leather. Unripe persimmons are high in tannin, as any person unwise enough to sample one will discover. The extreme astringent taste is due to protein precipitation on the tongue.

Flavonoids are a group of aromatic compounds characterized by two substituted benzene rings connected by a 3-carbon chain and an oxygen bridge.

flavonoid skeleton

Sugars are commonly attached to flavonoids to form glycosides. Among the flavonoids are many common pigments, including the anthocyanins and flavones. Anthocyanins are responsible for the reds and blues of many fruits and flowers; flavones, for the yellow of lemon. The insecticide rotenone is related to the flavonoids.

Terpenoids and steroids

Such compounds as essential oils, steroids, alkaloids, and various pigments are related in the chemical sense to fused units of the 5-carbon substance isoprene. Two units of isoprene may form a terpene unit, the basic structure of a number of important plant constituents called terpenoids.

isoprene skeleton

myrcene
an open-chain terpene

menthol
a cyclic terpene

The essential oils are mixtures of volatile, highly aromatic substances that have distinctive odors and flavors. They are formed in specialized glands, ducts, or cells in various plant parts. Although called "oils," these compounds are not lipids. An example is menthol, which is the major constituent of both peppermint oil and turpentine. The odors, flavors, and other properties of essential oils make them economically important for a variety of uses.

Resins—gummy exudates of many plants—are terpenoids consisting of three to six isoprene units. Resins are often associated with essential oils (and have the ability to harden as the oils evaporate). Resin synthesis occurs near wounded tissue, and the resins serve to retard water loss and entry of microorganisms. The ability of plants to produce resins is lowered if the general vigor of the plant is poor.

Steroids are complex cyclic terpene compounds composed of eight units of isoprene (tetraterpenoids). Although their function in plants is not well understood, it is known that many of them have important metabolic effects in animals. Cortisone, sex hormones, and vitamin D are steroids.

The carotenoids are complex tetraterpenes— yellow to red pigments that occur in many different kinds of tissue. Common members of this group include carotene (composed exclusively of carbon and hydrogen) and xanthophyll (composed of oxygen in addition to carbon and hydrogen). The most widespread carotenoid is the orange pigment β-carotene, which when split in the digestive tract of animals gives rise to two molecules of vitamin A.

Rubber is a high-molecular-weight terpenoid containing 3000 to 6000 isoprene units. Although it is produced in many dicotyledonous plants, only a few produce enough for commercial purposes. The most important is the rubber tree, *Hevea brasiliensis*. The rubber is extracted from latex, a milky, sticky liquid exuded by glandular structures associated with the phloem.

Nonprotein nitrogen compounds

Nitrogen is found in a wide variety of compounds in addition to amino acids and proteins. These include the nitrogen bases and a heterogeneous group of compounds called alkaloids. Heterocyclic rings con-

nitrogenous bases

alkaloids

Figure 8-4 Cyclic nitrogen compounds in plants.

taining both carbon and nitrogen are a common structural form (Fig. 8-4).

The nucleoproteins, the source of the genetic material, are composed of proteins and nucleic acids. Nucleic acids are high-molecular-weight polymers of nucleotides, each formed from three constituents: a sugar (specifically ribose or deoxyribose), phosphoric acid, and a nitrogen base having the structure of either a purine or pyrimidine ring. In deoxyribonucleic acid (DNA), the genetic material in chromosomes, there are two purines (adenine and guanine) and two pyrimidines (cytosine and thymine). Thus there are four nucleotides, depending on the specific base involved.

The sequence in which the four nucleotides are arranged spells out a message that is decipherable by the cell. This genetic code specifies the sequence of amino acids in protein. The specificity of an enzyme is determined by its sequence of amino acids (which determines its structural configuration); thus, DNA controls the destiny of the cell. The gene, in essence a functional portion of the DNA molecule, contains the genetic information passed from generation to generation through the gametes.

Alkaloids, as their name indicates, are organic bases. (Although the nitrogen bases discussed above fit into the broad definition of alkaloids, they are usually treated separately.) The function of many alkaloids in plants is obscure. Nevertheless, the many effects that these substances have on human physiology render them of extreme pharmacological interest. A list of alkaloids derived from plant sources, along with their medicinal properties or other uses, is shown in Table 8-3. It is of interest to note that several popular beverages—coffee, tea, cocoa, and maté—all contain stimulating alkaloids.

A number of extremely important pigments are complex alkaloids called porphyrins. Chlorophyll is a magnesium-containing porphyrin. The cytochromes are iron-containing porphyrins that function in energy transfer. Hemoglobin, the oxygen carrier in animal blood, is also a porphyrin.

Nutrient absorption and translocation

Although chemical analysis of plant cells indicates the presence of many different elements, only 16 have been shown to be essential. The most abundant elements, carbon, hydrogen, and oxygen, are derived largely from carbon dioxide and water. The other 13 (iron, potassium, calcium, magnesium, nitrogen,

Table 8–3 Origin and uses of some plant alkaloids.

Alkaloid	Common plant source	Medicinal and other uses
Atropine	*Datura stramonium, Atropa belladonna, Duboisia* spp.	Relaxant of gastrointestinal tract, affects parasympathetic nervous system
Caffeine	*Thea sinensis, Coffea arabica, Ilex paraguayensis, Cola acuminata*	Stimulant of central nervous system
Cocaine	*Erythroxylon coca*	Surface anesthetic, stimulant of central nervous system
Colchicine	*Colchicum autumnale*	Induces chromosome doubling, relieves symptoms of gout
Emetine	*Cephaelis ipecacuanha*	Emetic, amoebicide
Ephedrine	*Ephedra gerardiana*	Stimulant of central nervous system, relieves nasal congestion
Hydrastine	*Hydrastis canadensis*	Antihemorrhagic
Morphine	*Papaver somniferum*	Analgesic
Nicotine	*Nicotiana tabacum*	Insecticide
Pelletierine	*Punica granatum*	Vermifuge (tapeworm)
Pilocarpine	*Pilocarpus microphyllus*	Diaphoretic (sweat inducer)
Quinine	*Cinchona* spp.	Antimalarial, cardiac depressant
Reserpine	*Rauwolfia serpentina*	Tranquilizer, sedative
Strychnine	*Strychnos nux-vomica*	Poison, stimulant of central nervous system
Tubocurarine	*Chondodendron tomentosum*	Muscle relaxant
Theobromine	*Theobroma cacao*	Diuretic stimulant

phosphorus, sulfur, manganese, boron, zinc, copper, molybdenum, and chlorine) are derived ultimately from the soil in the form of inorganic salts. Plant growth is dependent on the availability of the essential nutrients. Since nutrients and water are ultimately supplied to the cell from the soil, the study of plant nutrition is largely concerned with the biology and chemistry of the soil (see Chapters 12 and 16).

With respect to absorption, the cell can be considered as a mass of protoplasm surrounded by a differentially permeable membrane that permits passage of water and inorganic salts but restrains the passage of most large complex molecules, such as sucrose. Molecules move through a selectively permeable membrane by **diffusion.** The movement of water through such a membrane is referred to as **osmosis,** and involves diffusion as well as bulk flow due to hydrostatic pressure differences. The osmotic movement of molecules can be demonstrated in nonliving closed systems by immersing a differentially permeable membrane that contains sugar water into a solution of pure water (or one with a lesser amount of

sugar). The water moves from the solution of dilute or no solute concentration (pure water or dilute sugar water) to the solution of high solute concentration (sugar solution). Living cells, however, are able to accumulate certain ions in a manner unaccounted for by diffusion. The cell appears to act as a metabolic pump. This process, known as active uptake, requires energy, which is supplied by respiration. The ability of molecules to move in and out of plant cells is related to their size, their lipid solubility, and their ionic charge; membrane permeability is affected by the ionic concentration of the nutrient medium. Monovalent ions (K^+, Na^+, Cl^-) appear to increase the permeability of membranes, whereas polyvalent cations (Ca^{++} and Mg^{++}) decrease membrane permeability. Furthermore, different ions interact in their effect on membrane permeability.

Translocation may be defined as the movement of inorganic or organic solutes from one part of the plant to another. The transport of water and solutes in and out of single cells and simple multicellular plants is accomplished largely by diffusion. In higher

Figure 8-5 A schematic diagram of the translocation of water and elaborated sugars in the plant. The upward movement of water through the xylem is due to tension on the continuous water column in the plant. This tension, produced by the evaporation of water from the leaf (transpiration), is transmitted to the absorbing cells of the root. Sugars synthesized in the leaves move through the sieve tubes of the phloem. Phloem transport is a pressure flow brought about by a high osmotic concentration in the leaf cells and a low concentration in the receiving cells. A model of this system, called the osmometer, is shown at lower left. [After J. Bonner and A. W. Galston, *Principles of Plant Physiology*, W. H. Freeman and Company, copyright © 1952.]

plants, however, this conduction of solutes is carried out largely in distinct tissue systems. Physiological specialization in multicellular plants is made possible because of the rapid, large-scale transport of substances within the plant. This movement is largely a two way stream, in which water and its dissolved contents move up from the roots through the xylem, and synthesized sugars move out of the leaves to other parts of the plant through the phloem (Fig. 8-5). There is, however, some movement of minerals in the phloem, and the xylem of woody stems functions in the upward movement of organic compounds, especially at certain seasons of the year.

The upward movement of water and solutes in the xylem of higher plants is related in part to **transpiration,** the evaporative loss of water vapor from the leaves through the numerous stomatal openings. As water is lost by the cells, a low water potential draws the water from the xylem elements, which form large numbers of continuous tubes from roots to leaves. Thus, the tension is transmitted through the entire column to the root cells and results in increased water absorption. The rate of transpiration is affected by the number of stomata and the degree to which they are opened, as well as by such environmental factors as temperature and vapor pressure of water in the atmosphere, which affect the rate of water evaporation. The opening of the stomata is a mechanical process regulated by the turgidity of the guard cells (see p. 149).

The movement of sugars occurs principally in the phloem. Phloem transport appears to be accomplished by increased osmotic concentration in the leaf mesophyll cells brought about by the high concentrations of dissolved photosynthates. These sugars then move into the sieve tubes of the phloem by a process that is not clearly understood. The resultant sugar gradient appears to result in a flow, and other substances are swept along the presumed sieve-tube stream. The sugars are utilized in the receiving cells through respiration, growth, or storage processes. There is also evidence of lateral transport between xylem and phloem.

PLANT DEVELOPMENT

The plant is more than the sum of its physiological processes. The orderly cycle of development that the whole plant undergoes involves complex patterns of change in cells, tissues, and organs. This cycle begins with seed germination and progresses through juvenility, maturity, and flowering. Upon fruiting the essential cycle of plant development is completed. In perennials, the plant is ready to recycle after a period of quiescence. In annuals and biennials, fruiting is a signal to the organism to enter the final phases of plant growth—senescence and death (Fig. 8-6).

Differentiation

The problem of differentiation is one of the great themes in biology. There are two levels of complexity to the problem. The first concerns development of individual cells. Unicellular organisms are capable of undergoing complex transformations with no other apparent internal stimulus than their own genetic makeup. The differentiation of individual cells must involve some systematic turning on and off of the multiplicity of complex instructions potentially receivable from the genetic material. Differentiation in multicellular organisms comes about as a result of

differential growth within and among cells. This is accomplished in an orderly and systematic way, with the mitotic process in cell division insuring the genetic continuity of the cells. The differentiation of genetically identical cells in multicellular organisms is indeed a mystifying process. The way in which this is accomplished appears to depend on the interaction between the cell's genetically controlled processes and their external environment. In a multicellular organism the environment of one part of a cell may be quite different from that of another part. Investigations of tissue and organ differentiation have shown that many of these differences involve the interaction of substances produced by different parts of the plant. For example, to artificially culture roots in tomato, not only must root meristem be present, but also thiamin and pyridoxine (familiar as vitamins in the B complex), which are normally provided by the leaves.

In higher organisms particular cells take over the control of differentiation through the action of "chemical messengers." Naturally occurring organic substances that affect growth and development in very low concentrations and whose action may be involved in sites far removed from their origin are known as **hormones.** A great many such substances occur in plants. Minute amounts (as low as one part per billion) exert measurable physiological effects. The term "growth regulators" has been coined to include all naturally occurring and synthetically copied, or created, substances that affect growth and development. They may be either inhibitors or promotors of growth. Sometimes one compound can have either effect, depending upon its concentration.

Growth regulators

From continuing research in the chemical control systems affecting differentiation in plants, a number of important groups of hormonelike substances have been defined: **auxins, gibberellins, cytokinins, abscisic acid,** and **ethylene.** These substances are so grouped on the basis of structural and physiological similarities, but their functions overlap and interact. There are also undoubtedly other groups. The basis

174

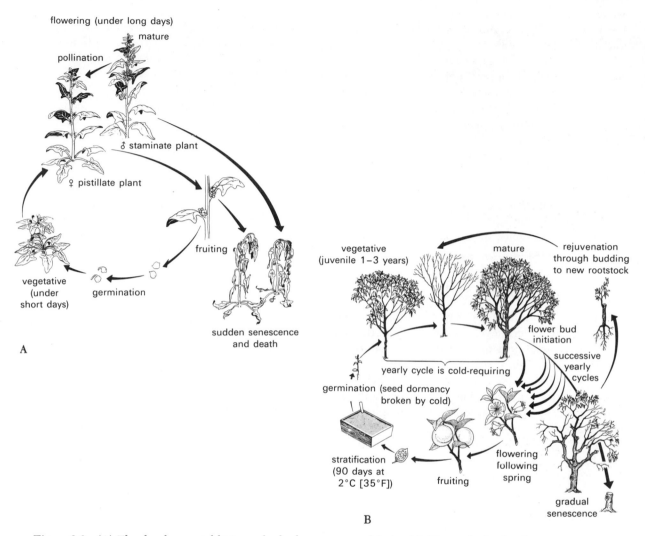

Figure 8-6 (A) The developmental history of a herbaceous annual (spinach): The seed of spinach germinates quickly under proper environmental conditions for growth. If the seedling is grown under short days it forms a distinct vegetative stage (rosette). Under long days the stem elongates to form a seedstalk and initiates flowers. Since spinach is normally dioecious, staminate and pistillate flowers form on separate plants. Soon after the staminate plants flower, and soon after the pistillate plants fruit, they undergo rapid senescence and die. (B) The developmental history of a woody perennial (peach): The peach seed germinates only after a period of exposure to cold. The juvenile stage lasts about 2 to 3 years. The yearly cycles must be interrupted by a period of cold. Usually flower buds are initiated in the third or fourth year and open in the following spring. The plant may live for 40 to 50 years, and undergoes a very gradual deterioration. The plant may be rejuvenated, however, from buds close to the base of the tree. If vegetative growth is continually budded onto new rootstocks the "original" plant remains, in effect, immortal. [From J. Janick, *Horticultural Science*, 3rd ed., W. H. Freeman and Company, copyright © 1979.]

for the chemical control of many distinct growth patterns, such as flowering, remains unknown. In addition, many naturally occurring compounds known to inhibit growth and developmental processes fit no well-defined group.

Auxins

The auxins, the first class of growth regulators to be discovered, have received considerable attention in plant physiology. Auxins are growth-promoting plant hormones. Cell elongation, the simplest example of anatomical differentiation, is directly affected by auxin concentration. The mode of action here appears to involve alterations in the plasticity of the cell wall. This fundamental property of auxin has been used in assaying its activity. The most basic test for activity consists of measuring the elongation of dark-grown oat coleoptile sections. Similar tests assay the rate of curvature of longitudinally halved stems in response to auxin application, as in the split-pea test. The most common natural auxin is indoleacetic acid (IAA).

indoleacetic acid (IAA)

Some very fundamental growth responses have been shown to be controlled by auxins. Phototropism, the bending toward light of the stem apex (growing point) can be explained as the result of differential cell elongation caused by the accumulation of auxin on the darkened side of the meristem. The cells on that side elongate and cause the stem to bend toward the light. Similarly, auxins greatly affect growth patterns in the plant. For example, **apical dominance**—the inhibition by the growing point of the growth of dormant buds below it—appears to be a function of auxin distribution. Auxins are produced in greatest abundance in a vigorously growing stem apex, and high concentrations emanating therefrom have been

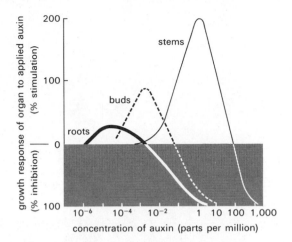

Figure 8-7   The effect of auxin concentration on the growth of roots, buds, and stems. [From L. Machlis and J. G. Torrey, *Plants in Action*, W. H. Freeman and Company, copyright © 1956.]

shown to inhibit bud break. Removal of the auxin-producing stem tip increases lateral bud break and subsequent branching, usually directly below the cut. Thus the form of many plants can be changed by the manipulation of apical dominance through pruning.

Auxins are at present the best understood of the many substances affecting plant development. They are formed in the stem and root apices, whence they move to the rest of the plant. Their distribution, however, is not uniform. The resultant concentration of auxins in various parts of the plant has been correlated with inhibition and stimulation of growth (Fig. 8-7) as well as with differentiation of organs and tissues. Such processes as cell enlargement, leaf and organ abscission, apical dominance, and fruit set and growth have been shown to be influenced by auxins. Auxins have been associated with flowering and sex expression in some plants. Research on auxins has had a deep impact in agriculture. They have been used as herbicides, rooting promoters, and as materials to control the setting, thinning, and dropping of fruit.

The role of auxins has been associated with so many diverse growth systems in plants that it has at times been tempting to suggest that they must

untreated
bush bean

gibberellin-treated
bush bean

untreated
dwarf corn

gibberellin-treated
dwarf corn

Figure 8-8   Gibberellin treatment stimulates normal growth in dwarf plants. In beans 20 µg changes the bush to a vine habit; in corn 60 µg causes a change from dwarf to normal growth.

behave as "master" hormones affecting growth and differentiation, but the importance ascribed to any particular developmental material is influenced by the intensity of study that it receives. Growth and development appear to depend on interactions of many factors. The conclusion is now inescapable that auxins, though important substances, constitute only one of many significant groups of materials involved in differentiation and growth.

Gibberellins

The gibberellins are a group of at least 45 closely related, naturally occurring terpenoid compounds. They were discovered through studies of excessive growth of rice occurring in response to a fungal disease. The gibberellins affect cell enlargement and cell division in subapical meristems. The most startling effect is the stimulation of growth in many compact plant types. Minute applications transform bush beans to pole beans or dwarf corn to normal corn (Fig. 8-8). This effect is utilized as a biological

assay. In addition to the dwarf-reversing response, gibberellins have a wide array of effects in many developmental processes, particularly those controlled by temperature and light (photoperiod), including seed and plant dormancy, germination, seed-stalk development, and fruit development. Barley is treated with gibberellic acid in the malting stages to increase the enzymatic content of malt. The use of gibberellins has had impressive agricultural applications, particularly in grape production. In Japan, seedlessness is induced in grapes, and in California the 'Thompson Seedless' cultivar ('Sultanina') shows a remarkable increase in berry size after gibberellin treatment (Fig. 8-9).

gibberellic acid (GA₃)

Figure 8-9 Effect of gibberellic acid (GA) on 'Thompson Seedless' grapes: upper left, control; upper right, 5 ppm GA; lower left, 20 ppm GA; lower right, 50 ppm GA. [Courtesy R. J. Weaver.]

Cytokinins

The term cytokinins (or simply kinins) has been applied to a group of chemical substances that have a decisive influence in the stimulation of cell division.

Many cytokinins are purines. Kinetin, the original cell-division stimulant isolated from herring-sperm DNA, has been identified as 6-furfurylaminopurine. The synthetic cytokinin benzyladenine, or BA (6-benzylaminopurine) stimulates shoot initiation in woody plants, including conifers.

6-furfurylaminopurine (Kinetin)

zeatin

6-(γ,γ-dimethylallylamino)purine (2iP)

6-benzylamino purine (BA)

6-(benzylamino)-9-(2-tetrahydropyranyl)-9*H*-purine (PBA)

Figure 8-10 Effect of various levels of a cytokinin (kinetin) and an auxin (naphthaleneacetic acid, or NAA) on organogenesis in tissue culture. Leaf discs of salpiglossis were cultured for 40 days. [From J. Janick, *Horticultural Science*, 3rd ed., W. H. Freeman and Company, copyright © 1979.]

Many cytokinins have been detected in research involving tissue culture. Vascular strands laid on top of cultured pith tissue stimulate differentiation in otherwise nondividing cells, an effect also induced by adenine and kinetin. Cytokinins and auxins interact to affect differentiation (Fig. 8-10): high auxin and low cytokinin gives rise to root development; low auxin and high cytokinin gives rise to bud development; equal amounts result in undifferentiated growth. Cytokinins are found in abundance in fruit and seeds (for example, corn endosperm, coconut milk), and are probably important in promoting growth and differentiation of the embryo. Tumor cells are also rich in cytokinins.

Cytokinins affect such diverse physiological mechanisms as leaf growth, light response, and aging. Recently, cytokinins and other plant hormones have been found to be active in regulating protein synthesis, possibly by turning gene transcription on and off or by acting through transfer RNA to control translation of the gene product.

Ethylene

Although it has been known for many years that ethylene has a number of striking effects on plant growth and development, only recently has it been considered to be a regulatory hormone. Thus, it has been a standard practice to achieve ripening in banana through the introduction of ethylene, and ethylene (as well as such closely related compounds as acetylene from calcium carbide) has been used to obtain uniform flowering in pineapple. Ethylene has enormous effects on dark-grown seedlings, causing swelling and disorientation; such effects were used as a method of assay before the days of gas chromatography. Ethylene also influences cell division; thus, tomatoes grown in high concentrations of ethylene show extensive rooting up and down the stem. It is wise not to store budwood or hardwood cuttings in places where apples are stored: apples produce ethylene as they ripen, and ethylene will cause splitting and bark peeling. In the past decade it has been clearly shown that the natural or "endogenous" ethylene produced by plants influences (1) the natural course of development in etiolated seedlings, (2) abscission, (3) floral initiation (in some plants), and (4) fruit ripening. Ethephon, or (2-chloroethyl)phosphonic acid, a material that is transformed into ethylene in the plant, has a wide array of interesting effects, including the induction of uniform ripening in pineapple and tomato, sex conversion, induction of pistillate flowers in androecious (staminate) plants of cucumbers, and the stimulation of latex flow in rubber trees.

H₂C=CH₂

ethylene

Abscisic acid

Abscisic acid (ABA), referred to at one time as "dormin" and "abscisin II," is a natural inhibitory compound that affects bud and seed dormancy and leaf abscission. Similar synthetic substances have been reported. It has a wide range of effects: in addition to promoting dormancy in buds and seeds and accelerating leaf abscission, it also promotes flowering in some short-day plants. The dormancy response may be through an effect of RNA and protein synthesis. Some effects of ABA seem to be reversed by gibberellins (barley seed germination) or cytokinins (stomatal closure).

abscisic acid

Inhibitors

Natural as well as synthetic inhibitory substances are often placed together as a diverse class of growth regulatory materials (Fig. 8-11). Natural inhibitors help control such processes as seed germination, shoot growth, and dormancy. Several synthetic inhibitors have found important agricultural applications. Maleic hydrazide has been effective in preventing the sprouting of onions and potatoes. A number of materials that inhibit the natural formation of gibberellins by the plant act to produce dwarfing. CBBP and chlormequat show promise for use in reducing the height of many ornamental flowering plants without unduly interfering with flowering time or flower size. A gibberellin suppressor, daminozide, in addition to dwarfing, affects fruit maturity, hardiness, and many other plant processes.

Vegetative physiology*

Germination

Germination includes all the sequential steps from the time the seed first imbibes water until the seedling is self-sustaining. After the imbibition of water the complex reserve substances in the seed are enzymatically converted to simple soluble forms that are readily translocated to the embryonic plant. Here some substances undergo respiratory breakdown and release energy; others are utilized in the synthesis of compounds needed for growth. The seed must, of course, be supplied with ample quantities of oxygen and water to satisfy the respiratory requirements and to provide a medium for enzymatic activity and synthesis. The temperature of the environment must be such that the biochemical processes of degradation and synthesis can operate. Although light is not usually required, depending on the wavelength, it can either trigger or inhibit the germination process in some plants (for example, lettuce).

Seed dormancy

Seed that is viable yet fails to germinate in the presence of normally favorable environmental conditions

---

*The section on vegetative physiology is based on the chapter "Plant Development," by C. E. Hess, in J. Janick, *Horticultural Science*, 3rd ed., W. H. Freeman and Company, 1979.

naturally occurring inhibitors

benzoic acid   coumarin   chlorogenic acid

synthetic inhibitors

butanedioic acid mono-(2,2-dimethylhydrazide)
(daminozide)

(2-chloroethyl)trimethylammonium chloride (CCC; cycocel)

1,2-dihydro-3,6-pyridazinedione (MH; maleic hydrazide)

2,4-dichlorobenzyltributylphosphonium chloride (CBBP)

(5-hydroxycarvacryl)trimethylammonium chloride,
1-piperidinecarboxylate (AMO 1618)

methyl-2-chloro-9-hydroxyfluorene-9-carboxylate
(IT 3456, a morphactin)

Figure 8-11 Structural formulas of some naturally occurring and synthetic inhibitors. [From R. J. Weaver, *Plant Growth Substances in Agriculture*, W. H. Freeman and Company, copyright © 1972.]

is said to be dormant. The cause of dormancy may be physical or physiological (Fig. 8-12).

Physical dormancy takes the form of structural limitations to germination, such as hard, impervious seed coats that provide a mechanical barrier to the entrance of either water or, in some plants, oxygen. The pea family provides the greatest number of examples of this type of dormancy. Some hard seeds resist water imbibition by means of a "one-way valve" in the hilum, the scar formed at the point of attachment of the seed to the pod. Although the seed may lose moisture in a dry environment, a moist environment effectively closes the valve. Under natural conditions these seeds do not germinate until soil microorganisms or weathering have sufficiently weakened the seed coat to permit the entrance of water. Thus setting the stage for germination may require a number of years rather than only a small

Some seeds proceed without hindrance provided they have adequate moisture, proper temperature and oxygen; some need light as well.

Internal physiological barriers are overcome in many seeds by low temperature.

Inhibitors in fruits or seed coats may block germination. They may be removed by leaching and by microorganisms.

Physical barriers, such as an impervious seed coat, prevent entrance of water or oxygen. They may be removed naturally through weathering and bacterial action or by treatment with caustic substances.

Figure 8-12   Germination is the route from seed to seedling. Physical or physiological dormancy may prevent germination even though favorable environmental conditions for growth are present.

part of one season. Seeds having an impervious coat may be artificially worn or weakened in order that germination can take place uniformly and without delay (see Chapter 9).

Physiological dormancy may be due to a number of mechanisms. These commonly involve growth-regulating systems of inhibitors or promotors. Inhibiting substances that block the germination process may be present in the flesh of the fruit, in the seed coat, or even in the endosperm of the seed. Germination of seed within a tomato is a rare occurrence, yet as the flesh of the tomato is removed and the seeds are rinsed, germination takes place without delay. This inhibition of germination is due not to the low pH or high osmotic values of the tomato flesh but to a specific chemical entity. Although the inhibitor in tomato fruit has not been isolated and characterized, there is evidence that it belongs to a group of compounds known as the unsaturated lactones. A naturally occurring member of this group is coumarin (see Fig. 8-11), a substance responsible for the aroma of freshly cut hay. The mode of action of such substances as coumarin consists in blocking or activating enzymes essential to the germination process. For example, it has been shown that the activity of α- and β-amylases is blocked in the presence of a germination inhibitor. The amylases are essential for the hydrolysis (splitting) of starch into simple soluble forms.

A number of chemical materials may function as germination inhibitors. They are removed naturally from the fruit and seed coats through leaching by rain and degradation by microorganisms.

Physiological dormancy may also be due to internal factors, such as embryo immaturity. In the seeds of some species the embryo consists of only a few undifferentiated cells even when the fruit is ripe; the seed of the American holly (*Ilex opaca*) is an example. Therefore, an **afterripening,** process—maturation of the seed after harvest—must occur, during which the embryo differentiates at the expense of the endosperm. In the seeds of other species, particularly woody species of the temperate zone, the causes of internal dormancy are more complex. In addition to the inhibitors that block germination, a germination stimulator is also required. To induce production of the stimulator, the seed must be subjected to a period of **cold stratification**—storage under moist, aerobic conditions at temperatures usually above freezing but below 10°C (50°F). A period of about 4 to 6 weeks at 5°C (41°F) is usually required to promote germination of seeds that exhibit this kind of dormancy.

Germination of seed may be blocked by more than one form of dormancy. Such **double dormancy** is characteristic of several members of the pea family. In the redbuds (*Cercis* species) both physical and physiological blocks to germination are present, the former being an impervious seed coat and the latter an internal dormancy that is broken by exposure to cold. In order for seeds with double dormancy to germinate, the physical barrier must be removed before the physiological barrier. If a seed coat is impervious, water essential for biochemical reactions cannot enter, and cold treatments administered to remove inhibitors or to promote the synthesis of stimulators will not be effective. In nature, seeds with double dormancy often require two years for germination. The first year is required for soil micro-organisms and weathering to remove the physical barrier. During the winter of the second year internal dormancy is broken, and germination occurs the following spring. Physiological dormancy in some seeds may be broken by artificial treatments that either interfere with, dilute, or remove inhibitors. In others, a missing growth regulator, such as gibberellin, will break dormancy.

Dormancy of seeds is a biological mechanism that provides protection against premature germination when environmental conditions may not be favorable for seedling growth. Seeds of many woody plants of the temperature zone will not germinate during the late fall, but are internally dormant until the following spring. Weed species possess, almost without exception, one or more of the major types of dormancy. In contrast, the seeds of most crop plants are not blocked by either physical or physiological forms of dormancy. Dormancy mechanisms have been eliminated through selection for rapid germination.

Juvenility

A seed is considered germinated when it has produced a plant that, under proper environmental conditions, is potentially capable of continuous and uninterrupted growth. From the time this stage is reached until the first flower primordium is initiated, the plant is considered to be in the vegetative phase of growth. If during the vegetative phase the plant cannot be made to flower, regardless of the environmental conditions imposed, it is said to be **juvenile.** Juvenility and maturity, however, are relative terms. In many species these growth phases blend into each

other. The end of the juvenile period is indicated when the plant responds to flower-inducing stimuli.

The juvenile phase is characterized by the most rapid rate of growth of the overall organism and, in some plants, by distinct morphological and physiological features. The juvenile phase varies in length from one to two months for annuals to a period of a few years for most fruit trees. Some plants, such as bamboo, require scores of years. Among the morphological features that are associated with juvenility, and that are lost or altered at maturity, are the presence of thorns (*Citrus*), leaf lobing (cotton) or the lack of lobes (*Philodendron*), and the angle of the branches with respect to the main axis of the plant (spruce). The geotropic response varies with developmental stage in some plants. In its juvenile phase of growth, English ivy (*Hedera helix*) is a trailing vine, climbing only with support; in the mature stage of development, the plant grows upright (Fig. 8-13). The ability of juvenile plants to initiate adventitious roots readily is a common feature of many species. This ability decreases or is lost entirely in mature forms.

In some plants it can be shown that there is a gradual change from juvenility to maturity in the same plant. The basal portions can remain juvenile even though the shoot extremities have reached maturity. This feature is utilized in the propagation of certain apple clones. In order to obtain shoots that root easily

Figure 8-13 Morphological changes associated with growth phases in English ivy (*Hedera helix*). In the juvenile stage, the leaves are lobed and growth is horizontal. When the plant becomes mature, the leaves are entire and shoots grow upright and bear flowers.

(a characteristic of juvenility), growth is continually forced from the basal portion of the tree by pruning.

Juvenility can be experimentally terminated by chemical or physical treatments, but many species resist the change from the mature to the juvenile stage. Intergrafting juvenile and mature wood has provided some insight into the physiology of this phenomenon. The termination of juvenility may be explained as an aging of the growing point. This may be the result of dilution or exhaustion of some substance responsible for the juvenile condition. If a mature branch of English ivy is grafted onto a juvenile stock, the mature branch will develop juvenile shoots at first, but after a few weeks growth the juvenility gradually disappears and the branch again becomes mature. This sort of result lends support to the assumption that new shoots on the mature scion first utilize some "juvenile factors" in the stock and gradually mature when such factors become exhausted. No compound has yet been shown to be solely responsible for juvenility in plants, but gibberellins have been implicated as among the factors that maintain juvenility in English ivy.

Bud dormancy and rest

Vegetative growth is not a continuous process but one that is interrupted by periods of arrested development. Dormancy may affect the whole plant or be restricted to certain parts, particularly seeds and buds. In the potato, dormancy is typically limited to vegetative buds, whereas in coffee, dormancy is limited to flower buds.

One type of arrested growth is brought about by unfavorable environment. For example, the growth of bluegrass, a common turf species, ceases when water is lacking, and top growth may die under continued drought. When conditions are more favorable, growth resumes from underground rhizomes. Many plants similarly survive periods of temperature extremes by undergoing a period of quiescence or rest. In tropical areas the periodic flushing and resting of vegetation has been correlated with moisture, light, and temperature. In addition, it has been suggested that internal factors, such as the stage of development or the age of the plant, influence the rhythm of growth. In the banana each stalk or pseudostem stops growing after flowering.

In **bud dormancy,** as in seed dormancy, growth is temporarily suspended even though all the external conditions normally required for growth are present. For example, woody plants of the temperate climates develop vegetative buds at each node throughout the growing season. The lack of growth of these lateral buds after formation is initially an expression of apical dominance, controlled by auxin distribution. At this time, these buds can be induced to grow by removing the growth-inhibiting apical meristem. With the onset of autumn, however, the buds of many species become truly dormant and will not grow, even if the plant is pruned and moved to a warm greenhouse, unless they have been exposed to a suitable period of cold weather. The degree of dormancy varies not only among species, but also among buds on the same plant. The flower buds of many trees, such as peach, cherry, and apple, require less chilling than do the vegetative buds. This is why the flower emerges before the leaves in the spring. In *Forsythia* the cold requirement is so minimal that it may be seen to flower in the fall after a brief period of cold weather.

The physiological basis of dormancy seems to be the accumulation and disappearance of growth inhibitors and promotors. In woody plants of the temperate climates, the onset of dormancy is conditioned by short day length and low temperatures. Bud dormancy in these plants is probably internally regulated by the formation of growth-inhibiting substances. It is broken naturally by cold temperatures. This cold reaction is localized. If an isolated stem of a dormant plant is cold-treated while the remainder of the plant is kept at a warm temperature, only the buds on the treated stem will grow.

The period of cold treatment required to break dormancy not only varies with the species but is sensitive to selection within species. For example, peaches have cold requirements that vary from 350 to 1200 hours at temperatures below 7°C (45°F). Cultivars having low chilling requirements are selected for areas in which the periods of cold weather are likely to be brief. If cultivars having high chilling requirements are grown in mild-winter regions they will leaf out poorly or not at all in the spring. This type of dormancy prevents the production of most temperate fruit crops in subtropical regions.

The survival value of dormancy is clear. A physiological mechanism that prevents growth is a biological check on the perversity of weather. If woody plants lacked internally imposed cold dormancy they might initiate growth under favorable periods in the late fall or winter only to succumb to succeeding severe weather.

The chemical control of dormancy holds promise of important application. A means of extending dormancy would protect the plant against growth during unsuitable periods, such as times of frost. Similarly, a means of breaking dormancy would permit precise control of crop operations. For example, the application of ethylene and related compounds to dormant potato tubers overcomes dormancy and ensures prompt, rapid growth.

Reproductive physiology

A plant is mature when it is potentially capable of reproduction. Maturity is characteristic by abundant and general flower development, although other physiological and morphological changes may occur. The mature plant, although capable, may not necessarily flower. The environment to which the plant is exposed at the time of maturity determines whether the plant will reach this ultimate expression of the mature state.

Flowering

Flowering represents a wide spectrum of physiological and morphological events. The first event, the most critical and perhaps least understood, is the transformation of the vegetable stem primordia into floral primordia (Fig. 8-14). At this time subtle biochemical changes take place that dramatically alter the pattern of differentiation from leaf, bud, and stem tissue to the tissues that make up the reproductive organs—pistil and stamens—and the accessory flower parts—petals and sepals. Among the events that follow initiation are development of the individual floral parts, floral maturation, and anthesis.

Once a meristem has been signaled to change from the vegetative to the reproductive state, microscopic changes in its configuration become apparent. Growth of the central portion is reduced or inhibited, and the meristem becomes flattened in contrast to the conical shape characteristic of the vegetative condition. Next, small protuberances develop in a spiral or whorl arrangement around the meristem. Although this phase of reproductive differentiation is quite similar to vegetative differentiation, a basic difference does exist in that there is no elongation of axis between the successive floral primordia as there usually is between leaf primordia. In most plants the transformation from the vegetative to the reproductive state is irreversible, and the floral parts will continue developing until **anthesis**—the time at which the flower is fully open—even though environmental conditions change. By the time anthesis takes place, the pollen and the embryo sac are completely developed, and the plant is prepared for fruiting, the next major step in its development.

Floral initiation and development have been studied primarily through manipulations of the plant's environment, particularly nutrition, light, and temperature. The rapid advances in the knowledge of the purely nutritive aspects of plant physiology in the latter part of the nineteenth century created an atmosphere in which it was believed that perhaps all aspects of plant growth could be explained or regulated by an alteration or adjustment of a plant's nutrition. In 1918 E. J. Kraus and H. R. Kraybill proposed that the development of flowering was regulated by the **carbohydrate : nitrogen ratio** within the plant. When tomato plants were grown under conditions favoring photosynthesis, and at the same time were supplied with an abundance of nitrogen fertilizers, vegetative growth was lush and flowering was reduced. But when the nitrogen supply was reduced while photosynthesis was maintained at a high level, vegetative growth was reduced, and flowering was abundant. With the combination of low nitrogen and low photosynthesis both vegetative growth and flowering were reduced. This concept of nutritional control of flower development was readily accepted, but the results of extensive investigations indicated that many plants flower over an extremely wide

Figure 8-14 Morphological changes associated with the transition from vegetative to reproductive buds in Easter lily. (A) Vegetative meristem surrounded by leaf primordia in late October. (B) The prefloral meristem becomes domelike in early January. (C) Early reproductive stage in mid-January showing floral primordium regions. Each primordium differentiates into a bract and a floral bud. (D) Reproductive meristem showing five floral buds, late January. (E,F) Reproductive meristem showing continued development of floral buds. [From A. A. De Hertogh, H. P. Rasmussen, and N. Blakely, *J. Amer. Soc. Hort. Sci.*, 101:463–471, 1976.]

185

range of carbohydrate : nitrogen ratios. In view of our present appreciation of the tremendous physiological effects of minute quantities of growth-regulating substances, it is not difficult to understand why the gross ratios of total carbon compounds to total nitrogen compounds does not provide a consistent indication of the physiological condition of the plant in relation to flower initiation and development.

It can be demonstrated that the presence of new growth has an inhibitory effect upon floral initiation. Tomato plants flower abundantly with the continued removal of new leaves, even when only the cotyledons are left for photosynthesis. From these results it is tempting to suggest that the reason nitrogen reduced flowering in the tomato is only incidentally associated with the carbohydrate : nitrogen ratio, but rather that the stimulation of new growth by nitrogen inhibits flower initiation. In support of this concept is the observation that almost any means by which growth can be reduced, such as bending a branch for an upright to a downward position, or applying such growth retardants as daminozide, results in increased floral initiation.

Photoperiodic effect

The discovery of photoperiodism—a growth response to the length of light and dark periods—by W. W. Garner and H. A. Allard, scientists of the United States Department of Agriculture, completely changed the concept of flower initiation. Decreasing the daily exposure to light during the summer induced a fall-flowering cultivar of tobacco to bloom profusely, a feat unattainable with all other environmental changes. It was subsequently demonstrated that the length of the dark period rather than that of the light period is the critical factor in the photoperiod response. Interrupting the dark period with light reverses the effect of a long night, whereas interrupting the light period with darkness has no effect. Nevertheless, "day length" remains the conventional term for describing the photoperiodic response. As the results of investigations accumulated, it became apparent that a majority of plants fell into one of three categories: short-day, long-day, and day-neutral plants. Short-day plants require a dark period exceeding some critical length in order to flower, and cannot flower under continual illumination. These include spring- and fall-flowering tobacco, poinsettia, chrysanthemum, and various tropical species. Long-day plants are inhibited from flowering when the dark period exceeds some critical length, and they can flower under continual illumination. Included in this group are many of the summer-flowering plants of the temperate zones, such as radish, lettuce, spinach, potato, and winter barley. Day-neutral plants apparently can initiate flowers under any night length. These include the dandelion, buckwheat, and many tropical plants that either flower the year round or, if they do not, can be shown to be affected by other environmental conditions. With the continued study of plant response to day length, other categories have been added. For example, some plants are nonobligate; that is, they will flower regardless of the day length but will flower earlier or more profusely when the day is either long or short. In still another category are the plants that flower only after an alteration of day lengths; these are known as long-day, short-day plants (or short-day, long-day plants). Such plants require first an exposure to long days and then to a period of short days, or vice versa.

Investigations relating to photoperiodism have clearly demonstrated the hormonal control of flower initiation. The mature or newly expanded leaf is the perceptor of changes in day length. In some plants the leaves need only to be exposed to one light–dark cycle of the proper day length to cause flower initiation, although most require several to many cycles. Once the leaves have received the photoperiodic message, they produce a substance, or a precursor of a hypothetical substance, called **florigen.** Unfortunately, however, florigen remains only a name, for it has proven to be one of the most elusive of all plant-growth substances. Its transport from the leaves to the growing point can be demonstrated indirectly up and down stems, across graft unions, and from one plant to another, but the substance has not yet been isolated. Other evidence from grafting experiments indicates that the flowering stimulus is identical in long-day and short-day plants. A major advance that may lead to the identification of florigen is the isolation from leaves of the pigment system called **phytochrome,** which specifically receives the external photoperiodic message. This system was discovered by H. A. Borthwick and Sterling B. Hendricks, scientists of the United States Department of Agriculture, and first reported in 1952. Phytochrome is a pigment present in small amounts in all plants. Under short-day conditions an interruption of the dark period with light (specifically red light) causes an intramolecular shift in phytochrome that brings about flowering in long-day plants and prevents flowering in short-day plants. The reason for this differential response is not known. The effect of red light (light with a wavelength of about 660 nanometers) on the structure and thus the activity of phytochrome can be reversed rapidly by exposure to far red light (about 730 nanometers). (See Fig. 8-15.) In addition, this shift is also influenced by the length of the dark period and by temperature. This may explain the light–temperature relation in photoperiodism and may explain why the light break is most effective at the middle of the dark period.

The characterization of phytochrome is still being attempted. It is a protein with chromophoric (pigment) groups. The molecular weight of phytochrome is estimated to be multiples of 60,000 with three chromophores per molecule. The chromophore is a bile pigment closely similar but not identical to the chromophore of phycocyanin, an algal chromoprotein.

Phytochrome appears to have many effects upon growth and development. It has been shown to be responsible for the promotion and inhibition of germination in seeds of some plants. In lettuce, seed germination can be inhibited or promoted by alternating exposure to red and far red light. Although all seeds contain phytochrome, those of many plants do not require this "light signal" to germinate. It probably is involved in other photoperiodically controlled processes, such as dormancy and tuber and bulb formation.

Although the discovery of phytochrome has made a very significant contribution toward explaining the

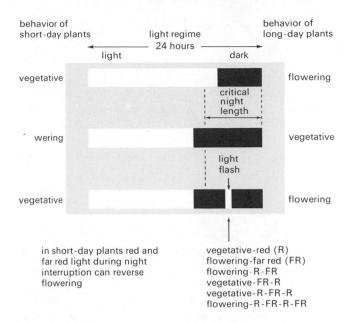

Figure 8-15   The length of the dark period is the critical factor in the photoperiodic control of flowering. Short-day plants flower under periods of long night; long-day plants flower under periods of short nights. Interrupting the dark period with light prevents flowering in short-day plants and permits flowering in long-day plants. The flowering effect can be reversed by substituting red and far red light during night interruption. [After A. W. Galston, *The Life of the Green Plant*, Prentice-Hall, 1961.]

first step in the mechanism of flower initiation, much experimentation must be done before the substances specifically causing the transition of the apical meristems from the vegetative to the flowering condition can be isolated and their mode of action determined.

Temperature effects

Photoperiodism does not have exclusive control over flowering. Temperature, for example, has both indirect and direct effects. It can influence flowering indirectly by modifying the plant response to a given photoperiod. Thus, a poinsettia will initiate and develop flowers in 65 days when grown under short days at 21°C (70°F). When grown under the same day length at 15.5°C (60°F), however, 85 days are required before initiation occurs. A more striking example is that of the strawberry. At temperatures above 19.5°C (67°F), the June-bearing strawberry behaves as a short-day plant and will not initiate flowers in day lengths longer than 12 hours. At temperatures below 19.5°C (67°F), its response is that of a day-neutral plant, and the plant will initiate flowers even in continuous illumination.

The direct effect of low temperature upon flowering was recognized more than a hundred years ago when an American, J. H. Klippart, described a process for "converting" winter wheat to spring wheat. (Winter wheat is planted in the winter and produces a crop the following year; spring wheat produces a crop in the same year it is planted.) The process consisted essentially of reproducing the sequence that occurs in the field. The winter wheat is partially germinated and further development is prevented by maintaining the seed at a temperature near 0°C (32°F). The treated seed, when planted in the spring, produces a crop in the same year. The elimination of the normal biennial habit of the winter wheat has been accomplished by satisfying the cold requirement. Klippart's research received little attention until his work was repeated by the Russian agronomist T. Lysenko,° who in 1928 coined the term **vernalization** to describe the effect upon future flowering behavior of subjecting a plant to low temperatures during one or more of the developmental phases of growth. The low-temperature effect may be obtained by treating the whole plant or part of the plant. In the grains only the moistened seed need be treated. Sometimes a vernalized plant—that is, a plant that has received a cold treatment—can be grafted onto a nonvernalized plant and both will flower. The implication is that some substance has passed from the vernalized plant, across the graft union, to the nonvernalized plant. A similar experiment has been conducted with annual and biennial

---

°Lysenko later became notorious for his unorthodox (now discredited) views on genetics and plant breeding.

forms of the same plant, such as henbane. The annual form requires no cold treatment to flower and induces flowering in the biennial form when grafted to it. Apparently the annual form produced the flowering stimulus without a cold treatment. The difference between spring and winter forms of wheat has been shown to be controlled by a single gene. Although the flower-inducing substance produced during vernalization has not been isolated, a clue to its identity is found in the fact that the cold requirement of some plants (dicotyledonous biennials, but not the cereal grasses) can be partially or completely replaced with gibberellins.

Further evidence that temperature can have a profound effect upon flowering is provided in the phenomenon of **devernalization.** Some plants exposed to a low temperature for a period of time sufficient to initiate flowering (usually at least six weeks) will revert to the original nonflowering condition if exposed to a high temperature. Onion growers take advantage of the devernalization on a commercial scale. Onion sets (small bulbs produced by crowding) are stored during the winter at temperatures near freezing to retard spoilage. The onion sets are vernalized at this temperature; and if they were planted directly from cold storage, they would quickly flower and bulbs would not be formed. Therefore, the onion sets are exposed to temperatures above 27°C (80°F) for 2 to 3 weeks before planting. The devernalized sets form bulbs and do not flower.

Although photoperiodism and vernalization have many similarities and are definitely interrelated, the stimuli produced in the two responses to environment do no seem to be identical. It is possible to separate the effect of vernalization from those of photoperiodism. For example, some biennial plants when exposed for the required time to low temperatures still will not flower unless they are subjected to the proper day length, usually a long day length. Similarly, when gibberellins are substituted for the cold treatment, flower initiation will not occur in some plants unless the day length is correct.

In addition to these temperature effects, it has been shown that the alternation of warm and cool temperatures influences flowering (as well as vegetative growth). This phenomenon, called **thermoperiodism,** has been utilized in the culture of greenhouse tomatoes. The tomato initiates the most flowers when grown at about 27°C (80°F) during the day and 17–20°C (63–68°F) during the night.

Fruit development

Fruit development can be divided into four phases: (1) initiation of the fruit tissue, (2) prepollination development, (3) postpollination growth, and (4) ripening, maturation, and senescence. The increase in fruit size in the prepollination phase of development is primarily the result of cell division. After pollination, cell enlargement is responsible for most of the increase in size. However, in some large-fruited plants (for example, watermelon and squash) cell division continues for some time after pollination, the final size being a consequence both of an increase in cell number and an increase in cell size. The stimuli and nutrients for prepollination growth are supplied primarily by the main body of the plant. In plants with perfect flowers, the stamen primordia are often differentiated before the ovary primordium, and have been shown to be a source of growth stimulus. Early surgical removal of the immature stamens in the flower bud can adversely affect the growth of the ovary. Extraction of the unripe anthers reveals the presence of large amounts of auxins.

Pollination

One of the most critical events in the growth and development of a fruit is pollination. Pollination has at least two separate and independent functions. The first is the initiation of the physiological processes that culminate in **fruit set** (more precisely, the inhibition of fruit or flower abscission); the second is to provide the male gametes for fertilization. The separateness of these two functions can be demonstrated in orchids by the use of dead pollen, which causes some fruit to set and some growth to occur but does not cause fertilization. The fruit-setting function of pollen can be substituted for by the application of auxin (Fig. 8-16). The effectiveness of water extracts

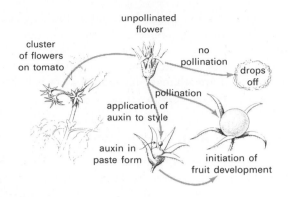

Figure 8-16   The role of pollen in preventing abscission of the flower may be replaced by the application of auxin. [From J. Bonner and A. W. Galston, *Principles of Plant Physiology*, W. H. Freeman and Company, copyright © 1952.]

of pollen in inducing fruit set has led to the postulation that the pollen contains an auxin. But the minute amount of auxin that might be present in the pollen that lands on a stigma cannot account for the auxin response obtained. Instead, it seems that the pollen contributes either an enzyme that converts precursors present in the stigma to an auxin or provides a synergist that renders effective the auxin already present.

Even though pollination takes place and fruit set is obtained, fertilization is not absolutely assured. Sometimes pollen does not germinate, or, if it does, the pollen tube may burst in the style. The germination of pollen is dependent upon the presence of a medium of the proper osmotic concentration (as shown by the effect of various sugar concentrations upon germination) and is stimulated by the presence of certain inorganic substances, such as manganese sulfate, calcium, and boron. In the lily it has been shown that the highest concentration of boron occurs in the style and stigma.

If the growth of the pollen tube is very slow, the style or even the entire flower may abscise. This may be artificially prevented, however, by the application of an auxin, such as α-naphthalene acetamide. When fruit set and growth are obtained without fertilization the fruit is said to be **parthenocarpic.** Although the fruits of some plants have no seeds, they may develop seedlike structures as do some seedless varieties of holly, orange, and grape. The effect of pollination on fruit set and the influence of the resultant seed on fruit growth make pollination a crucial phase in the production of many fruit crops.

Postfertilization development

After fertilization, the plant enters a phase of physiological activity that is second in intensity only to germination. The developing fruit no longer depends primarily upon the parent plant for a source of growth stimuli but instead receives them from developing seed within the fruit. The role of the seed can be empirically demonstrated by observing that misshapen fruits result from uneven distribution of developed seeds. In many fruits, a direct correlation exists between either weight or length and seed number. This effect of the seed on fruit development is mediated through chemical substances. For example, extracts of immature seeds can stimulate growth of unpollinated tomatoes. Furthermore, it is possible to correlate various physiological events in the development of a fruit with the presence of growth substances. It has been demonstrated that the auxin levels reach a low at the time of flower drop, and particularly in fruit trees during the natural abscission of partially developed fruit that occurs when a heavy crop has been set. The growth of the strawberry receptacle (the fleshy part) has been shown to be controlled by the achenes—one-seeded fruits scattered over its surface. The achenes contain high concentrations of auxins, whereas the receptacle tissue contains little. The addition of an auxin (indoleacetic acid) dramatically stimulates the growth of an acheneless receptacle. Indoleacetic acid has been demonstrated to be present in very high concentrations in immature corn kernels. More recent studies indicate that the embryo and the endosperm of the developing seeds are the sites of many substances that promote fruit growth, such as several auxins, gibberellinlike compounds, and cytokinins.

Although the control center of fruit growth is located in the seed, the raw materials for fruit development are supplied by the plant. Thus, the nutrients and moisture available to the plant directly affect fruit size. It has been calculated that at least forty leaves on a mature apple tree are required to support the growth of one apple. If the forty-to-one ratio is substantially reduced by an abnormally high fruit set, the quality and size of the individual fruit is greatly reduced. Therefore, it is a common orchard practice to reduce the number of fruit by thinning either the flowers or the partially developed fruit.

Fruit ripening

A final, dramatic, physiological event, the **climacteric,** marks the end of maturation and the beginning of senescence of fruits in some species. It is characterized by a marked and sudden rise in the respiration rate of a fruit prior to senescence. The respiration rate then returns to a level equal to or below the rate that existed prior to the climacteric. The climacteric is characterized not only by the abrupt increase in the production of carbon dioxide, but also by qualitative changes related to ripening, such as changes in pigmentation. The transition from green to yellow in certain cultivars of apple, pear, and banana takes place during or immediately following the climacteric. The peak of acceptability, or "edible ripeness," of pears coincides with the peak of the climacteric; maximum acceptability in the apple, banana, and avocado is reached immediately after the climacteric. A marked increase in the susceptibility of fruits to fungal invasion follows the climacteric.

Temperature has a profound effect upon maturation and the climacteric. For example, in the apple maximum respiratory activity is five to six times as high at $23°C$ ($73°F$) as at $2°C$ ($36°F$), and it takes twenty-five times as long to reach the climacteric at the lower temperature. However, the total amount of CO_2 liberated during the time between harvest and the end of storage life is approximately the same for both temperatures, equivalent to 16 to 20 percent of the reserve carbohydrates initially present in the fruit.

It is becoming clear that fruit ripening, like other phases of plant development, is a DNA-controlled process. A rise in RNA occurs in climacteric fruits, followed by increases in enzyme proteins such as the synthetases, hydrolases, and oxidases. Therefore, the climacteric is associated with both degradative and synthetic reactions. Ethylene and perhaps other volatiles appear to have key functions in the ripening process, and ethylene has been referred to as a fruit-ripening hormone. It is true that ethylene can accelerate fruit ripening; treatments to remove ethylene from the tissues of preclimacteric fruit can delay ripening. How ethylene exerts its effect on the initiation of ripening is not yet clear.

Senescence

Senescence is the phase of plant growth that begins at full maturity and ends in death. The erosive processes that accompany aging are the most baffling and least understood aspects of plant life. Senescence in plants may be partial or complete (Fig. 8-17). **Partial senescence** refers to the deterioration and death of plant organs, such as leaves, stems, fruits, and flowers. The death and abscission of cotyledons in

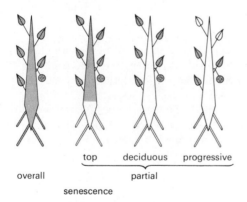

Figure 8-17   Patterns of plant senescence. The dead portions of the plant are indicated by shading. [After A. C. Leopold, *Plant Growth and Development*, McGraw-Hill, 1964.]

bean plants, the death of two-year-old raspberry canes, and the death of the entire shoot of tulip in early summer are all examples of partial senescence. **Overall senescence** refers to the aging and death of the entire plant, except for the seeds. The termination of the life cycle of true annuals and biennials is often sudden and dramatic. For example, after fruiting, whole fields of wheat die in a synchronized pattern during the growing season. In contrast, the senescence of such perennial plants as the apple or peach appears as a gradual erosion of growth and viability. In addition, perennials can be rejuvenated. Mature apple trees may be revitalized by severe pruning and fertilization or by encouraging the growth of adventitious buds at the base of the tree. The senescence of annual plants after flowering and fruiting is irreversible. Spinach plants kept vegetative under short days will not senesce, but, after flower induction, death is a certainty.

The earlier explanations for the senescence of annual plants depended upon a depletion hypothesis. It was suggested that during flowering and fruiting essential metabolites are drained from the main plant and accumulate in the fruit and seed. By the time the fruit is mature the plant is depleted to the extent that further growth is impossible, and death rapidly follows. It can be shown that the removal of fruits can significantly postpone senescence, whereas the greater the number of flowers or fruits left on the plant, the more rapidly senescence occurs. The depletion concept of senescence, however, is an oversimplification of a complex process. Death is delayed, but not averted, by the removal of fruits and flowers. Unpollinated pistillate plants and staminate plants of a dioecious species such as spinach also senesce. Apparently, the biological signal that causes an annual plant to senesce is associated in some way with the flowering process and is merely accentuated through fruiting. The relationship between the senescing signal and flowering is unfortunately not at all clear at present.

Selected references

Annual Reviews, 1950–. *Annual Review of Plant Physiology*. Palo Alto, Calif.: Annual Reviews. (Recent advances in plant physiology are brought up to date in timely, technical reviews. Published each year since 1950.)

Audus, L. J., 1972. *Plant Growth Substances*, 3rd ed. New York: Barnes & Noble. (An advanced treatise on the chemistry and physiology of growth regulating materials.)

Bonner, J., and J. E. Varner (editors), 1976. *Plant Biochemistry*. New York: Academic Press. (An advanced treatise.)

Galston, A. W., and P. J. Davies, 1970. *Control Mechanisms in Plant Development*. Englewood Cliffs, N.J.: Prentice-Hall. (A brief review of plant developmental processes written for the beginning student.)

Leopold, A. C., and P. E. Kriedemann, 1975. *Plant Growth and Development*, 2nd ed. New York: McGraw-Hill. (An excellent text on plant physiology.)

Mayer, A. N., and A. Poljakoff-Mayber, 1974. *The Germination of Seeds*, 2nd ed. New York: Pergamon Press. (An up-to-date work on the physiology, ecology, and environmental biology of seeds.)

Rabinowitch, E., and Govindjee, 1969. *Photosynthesis*. New York: Wiley. (A broad introduction to the physiology of photosynthesis and the enzymatic processes associated with it.)

Salisbury, F. B., 1963. *The Flowering Process*. New York: Pergamon Press. (A biological framework for understanding conversion from the vegetative to the reproductive state in higher plants.)

Salisbury, F. B., and Ross, C., 1978. *Plant Physiology*, 2nd ed. Belmont, Calif.: Wadsworth. (An excellent modern text in plant physiology.)

Smith, H., 1975. *Phytochrome and Photomorphogenesis: An Introduction to the Photocontrol of Plant Development*. New York: McGraw-Hill. (Photobiology of plants.)

Thimann, K. V., 1977. *Hormone Action in the Whole Life of Plants*. Amherst, Mass.: University of Massachusetts Press. (An advanced text on phytohormones and plant growth and development.)

Torrey, J. G., 1967. *Development in Flowering Plants*. New York: Macmillan. (The ontogeny of plant development.)

Vince-Prue, D., 1975. *Photoperiodism in Plants*. London: McGraw-Hill. (An advanced treatise.)

Wareing, P. F., and I. D. J. Phillips, 1970. *The Control of Growth and Differentiation in Plants*. New York: Pergamon Press. (An introduction to the processes underlying and controlling development in plants.)

Weaver, R. J., 1972. *Plant Growth Substances in Agriculture*. San Francisco: W. H. Freeman and Company. (A thorough and comprehensive review of growth regulators from an agricultural point of view.)

Wightman, F., and G. Setterfield (editors), 1968. *Biochemistry and Physiology of Plant Growth Substances*. Ottawa: Runge Press. (Proceedings of the 6th International Conference on Plant Growth Substances held at Carleton University, Ottawa, July 24–29, 1967.)

Wilkins, M. B., 1969. *Physiology of Plant Growth and Development*. New York: McGraw-Hill. (A text on advanced plant physiology and plant biochemistry containing contributions from eighteen specialists in selected areas of whole-plant physiology.)

Zimmerman, R. H. (editor), 1976. *Symposium on Juvenility in Woody Perennials* (Acta Horticulturae 56). International Society for Horticultural Science. (Recent advances in juvenility research.)

Zelitch, I., 1971. *Photosynthesis, Photorespiration, and Plant Productivity*. New York: Academic Press. (A modern treatment of photosynthesis and its effects on plant productivity.)

9
Reproduction and Propagation

Reproduction refers to the sequence of events involved in the perpetuation and multiplication of cells and organisms. There are two essentially different types of reproduction in plants: **sexual,** the increase of plants through seeds formed from the union of gametes; and **asexual,** the increase of plants through ordinary cell division and differentiation. The basis of asexual reproduction is the ability of many plants to regenerate missing parts. **Propagation,** the utilization of reproductive processes in the controlled perpetuation of plants, is in a real sense the basis of agriculture.

REPRODUCTION

Reproduction involves a controlled self-duplication of the life-controlling mechanism. A self-duplicating machine is an intriguing and complex concept. Although people have devised devices that can duplicate themselves, the living organism certainly remains the most elegant and the most fascinating system. We appear to be on the verge of understanding the mystery of reproduction.

Replication of living systems

The principle behind the mechanism by which living material replicates itself is somewhat analogous to that of a typesetting machine (for example, a linotype). The machine arranges the type in certain sequence to transmit information—that is, produce the printed page. New type is continually being formed from a mold or template of the original type. In this analogy the reproducing element is type; in living systems the reproducing elements are the nitrogen bases in DNA.

In the English language, 26 letters of the alphabet produce a dictionary of tens of thousands of words. The same results can be achieved with the two symbols of the Morse code. The four symbols of the genetic alphabet, or code, are the four nitrogen bases that, along with one sugar group and one phosphate group each, make up the **nucleotides** (or "building blocks") of DNA. These bases are adenine, guanine, thymine, and cytosine (see p. 170). If two symbols can express 26 letters, thousands of words, and thus an infinite series of messages, is it not conceivable that four bases can make up the code by which the information for maintaining living matter is transmitted? The way in which this is accomplished is a subject of intense study. The general theory is that the sequence of bases in DNA directs the sequence of amino acids in the synthesis of protein, specifically enzymes. The specificity of an enzyme is related to its amino acid sequence, and it is not unreasonable to assume that enzymes are responsible for the biochemical control of living systems.

The replication of DNA appears to be accomplished by a system quite analogous to the template system of the type machine. The template model,

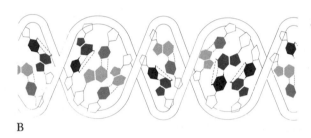

Figure 9-1 (A) Portion of the DNA molecule, shown untwisted. The molecule consists of a long double chain of nucleotides made up of phosphate-linked deoxyribose sugar groups and a side group of one of the four nitrogen bases. Hydrogen bonds (*broken lines*) link pairs of bases to form the double chain. The bases are always paired as shown, although the sequence varies. (B) The double-helix form of the DNA molecule. [After V. G. Allfrey and A. E. Mirsky, "How Cells Make Molecules." Copyright © 1961 by Scientific American, Inc. All rights reserved.]

ration of the bases, adenine is always paired with, and opposite to, thymine; and guanine is always paired with, and opposite to, cytosine. Thus if the sequence of one strand is given, the sequence of the opposite strand is fixed. Each strand, then, is the "template" for the other.

The theory of DNA replication follows nicely from its structure. The separation of the strands of DNA is apparently followed by a realignment of complementary bases on each strand (Fig. 9-2).

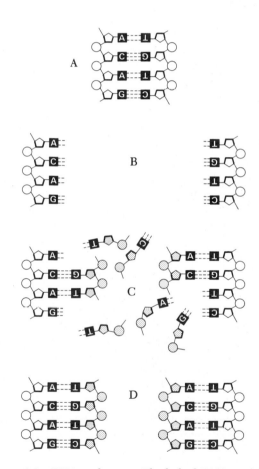

Figure 9-2 DNA replication. The linked DNA strands (A) separate in the region undergoing replication (B). Free nucleotides (indicated by shading) pair with their appropriate partners (C), forming two complete DNA molecules (D). [After F. H. C. Crick, "Nucleic Acids." Copyright © 1957 by Scientific American, Inc. All rights reserved.]

however, is based on chemical bonding rather than physical impression. Chemical analysis of DNA shows that, although the relative quantity of the four nitrogen bases varies, the amount of adenine is always the same as that of thymine, and the amount of guanine is always the same as that of cytosine. This suggests that adenine and thymine, and guanine and cytosine, occur in pairs. The significance of this becomes apparent when the structure of DNA is considered (Fig. 9-1). DNA normally occurs in double strands, and each strand is connected by linkages between the nitrogen bases. But because of the configu-

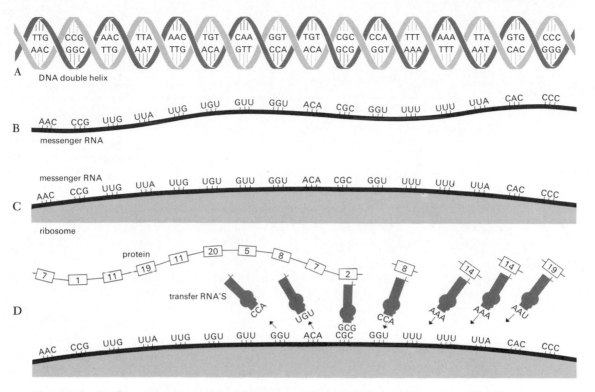

Figure 9-3 A schematic representation of protein synthesis as related to the genetic code. The complementary code, here derived from the dark-lettered strand of DNA (A), is transferred into RNA (B) and moves to the ribosome (C), the site of protein synthesis. In protein formation (D), amino acids (*numbered rectangles*) are presumably carried to proper sites on messenger RNA by transfer RNA and are linked to form protein. [After M. W. Nirenberg, "The Genetic Code." Copyright © 1963 by Scientific American, Inc. All rights reserved.]

The structure of DNA accounts for both its ability to replicate and synthesize. The actual synthesis of protein is carried out not in the nucleus but in the cytoplasm. The information from the nucleus must be transferred to the sites of synthesis in the cytoplasm, the ribosomes. The information is carried by a substance called ribonucleic acid or, RNA. (The components of RNA are almost identical to those of DNA except that the sugar portion of the molecule is slightly different—ribose takes the place of deoxyribose, and the nitrogen base thymine is replaced by a closely related base called uracil.) The particular form of RNA that carries the genetic instructions, in the form of a complementary copy of the DNA series

of bases, is called messenger RNA. Another form of RNA, transfer RNA, presumably brings the amino acids to the ribosomes to construct protein (Fig. 9-3).

Protein synthesis can be achieved outside of the living organism by placing together the component parts: ribosomes (from bacteria), an energy-generating system, amino acids, and RNA. Very elegant experiments involving the direction of protein synthesis have been possible through the use of artificial RNA. In this way, the genetic code has been deciphered; that is to say, the sequence of bases in RNA has been associated with the incorporation of a particular amino acid into protein. The message has been shown to be in the form of a three-letter code;

Table 9–1 The genetic-code dictionary.[1]

First Base	Second Base				Third Base
	U	C	A	G	
U (Uracil)	Phenylalanine	Serine	Tyrosine	Cysteine	U
	Phenylalanine	Serine	Tyrosine	Cysteine	C
	Leucine	Serine	*Nonsense*	*Nonsense*	A
	Leucine	Serine	*Nonsense*	Tryptophan	G
C (Cytosine)	Leucine	Proline	Histidine	Argenine	U
	Leucine	Proline	Histidine	Argenine	C
	Leucine	Proline	Glutamine	Argenine	A
	Leucine	Proline	Glutamine	Argenine	G
A (Adenine)	Isoleucine	Threonine	Asparagine	Serine	U
	Isoleucine	Threonine	Asparagine	Serine	C
	Isoleucine	Threonine	Lysine	Argenine	A
	Methionine	Threonine	Lysine	Argenine	G
G (Guanine)	Valine	Alanine	Asparagine	Glycine	U
	Valine	Alanine	Asparagine	Glycine	C
	Valine	Alanine	Glutamine	Glycine	A
	Valine	Alanine	Glutamine	Glycine	G

Source: This code is derived from the work of M. Nirenberg and others.
[1]The three bases of RNA (remember that uracil replaces the thymine of DNA) specify the amino acids indicated. The *nonsense* combinations act as "periods" to stop polypeptide synthesis.

that is, a sequence of three bases directs each amino acid. For example, RNA composed only of uracil (UUU) incorporates the amino acid phenylalanine; the sequence UUA incorporates leucine. The four-letter alphabet and base sequence is shown in Table 9-1. The sequence of amino acids in the polypeptide (protein unit) of the enzyme is related to the three-letter code sequence of DNA. Thus the four-letter alphabet represented by the four nitrogen bases is used to make three-letter "words," each corresponding to one of the approximately twenty amino acids. Each sentence spells out the amino acid sequence for one protein. This "sentence" is the gene!

Genes and chromosomes

The complete set of instructions is separated into chromosomes, linear structures within the nucleus. A chromosome is in essence a very long molecule of DNA. A specific function-controlling sequence of nucleotides within that molecule of DNA is a gene. Our understanding of the molecular, or "fine structure," of the gene is derived from studies that began in the late 1940s. A tremendous amount of information on the transmission and action of genes has accumulated from work on a wide spectrum of organisms, including crop plants. These investigations trace back to the beginnings of genetics, conveniently marked by the rediscovery in 1900 of Mendel's 1865 paper on inheritance. The theory that the units of inheritance, the genes, are particulate bodies on chromosomes and that they are transmitted from cell to cell and from generation to generation is the contribution of "classical" genetics and was clearly established by 1925.

Studies of many organisms have indicated that genes function through the regulation of chemical events. Genes control each step in the biochemical pathways of metabolism. Our understanding of the gene's role in metabolism dates back to flower pigmentation studies of the 1920s. It was found, for example, that two anthocyanin pigments that are different shades of red are chemically distinguished by a single hydroxyl group. The addition of this single hydroxyl group changes the color of the plant's petals from scarlet to deep red (see Fig. 9-4). This biochemical step is gene controlled. Thus, the petals of a plant that contain a gene (C) controlling the transformation of pelargonidin to cyanidin will be deep red. A plant having an altered version of the gene (c), one that is incapable of adding a hydroxyl group, will have scarlet red petals. Alternate forms of a particular gene are referred to as alleles (C and c are alleles). The change in the gene's internal structure, which gives rise to new alleles, is known as mutation. **Gene mutations** are changes in the genetic information (changes in the sequence of the nitrogen bases in DNA), and are ultimately responsible for the inher-

ent variation in all living things.

Only one set of instructions is necessary for the proper functioning of many organisms. That is, only one set of chromosomes is present in each cell, and there is one gene for each function. There are obvious disadvantages to such a system. Malfunctioning of any part of the essential instructions sacrifices the whole organism. Thus, a system has evolved in which the set of genetic instructions occurs in pairs. Higher plants and animals normally have two complete sets of chromosomes in each cell.

The characteristic number of chromosomes in the somatic cells of higher plants is known as the $2n$, the somatic or diploid number (for example $2n = 12$ in spinach and rice, 14 in barley and red clover, 16 in onion and peach, 18 in cabbage and sugar beet, 20 in corn and soybean, 22 in watermelon and bean, and 24 in tomato and tulip). The reproductive cells, or **gametes,** however, receive only one chromosome from each pair, the haploid number (n). Fertilization subsequently restores the diploid number in the zygote, the fertilized egg.

As each pair of genes may be present in more than one form (for example, C and c), a particular pair of alleles may be present in any one of three combinations: CC, Cc, or cc. A plant containing two identical genes, CC or cc, is **homozygous** for that allele. When the alleles are different, as with Cc, the plant is **heterozygous.** In a plant containing the heterozygous pair Cc the gametes will contain either C or c.

What is the difference in outward expression of

pelargonidin
(scarlet color, as in scarlet asters or pelargoniums)

gene *C* enzyme

cyanidin
(deep red color, as in cranberries or deep red roses)

Figure 9-4   The conversion of pelargonidin to cyanidin involves the addition of a single hydroxyl group and is controlled by a single gene.

the gene (**phenotype**) when the genetic constitution (**genotype**) is *CC*, *Cc*, or *cc?* With reference to our example of petal color, if the allele *c* is completely nonfunctional, as regards hydroxylation of pelargonidin, two alleles (*cc*) should not be any more efficient. That this is so has been shown experimentally, for flowers of *cc* plants are scarlet, and chemical analysis of their petals yields no cyanidin. The difference between plants that are homozygous (*CC*) and those that are heterozygous (*Cc*) cannot be predicted. The allele *C* may be efficient enough to produce sufficient enzyme such that plants with only one functioning allele (*Cc*) cannot be distinguished phenotypically from those having two functioning alleles (*CC*). When this is the case it would appear that the allele *C* dominates *c*. The condition in which heterozygous plants *Cc* are indistinguishable from homozygous *CC* plants is termed **dominance**. In genetic terminology *C* is **dominant**, *c* **recessive**. If the heterozygote *Cc* is intermediate in phenotype between the two homozygous types *CC* and *cc*, dominance is said to be incomplete.

Cell division

The basic difference between sexual and asexual reproduction lies in the way the chromosomes are distributed during cell division—**mitosis** and **meiosis** (Figs. 9-5 and 9-6).

Mitosis

Mitosis is the type of cell division that occurs in growth. It is a synchronized division in which both the chromosomes and the cell divide (Fig. 9-5). The chromosomes split longitudinally, each half then duplicating its missing half from raw materials available in the nucleus and moving to opposite ends of the cell. The two resulting daughter cells thus receive exactly the same number and kind of chromosomes. The division of the cell distributes more or less equally the other constituents of the cell.

The differentiation of cells into tissues and organs ordinarily is not related to any chromosomal difference. That each cell contains all the necessary genetic material implies that any cell holds the potential to give rise to the entire organism. The production of an entire plant from a single carrot parenchyma cell has been achieved. The formation of shoot buds from roots, or of roots from leaves, is a potential result of the genetic continuity made possible by mitosis. The differential patterns of growth would indicate that the instructions carried in the chromosome can be turned on and off. This system itself is no doubt also under the control of the genes.

Meiosis

Meiosis is a specialized sequence of two cell divisions that occurs prior to the formation of gametes. Meiotic divisions reduce the number of chromosomes by half (Fig. 9-6). In somatic (vegetative) cells the two chromosomes of a pair (homologues) are morphologically similar and contain the same kind of genes, although the two members of any given gene pair may not be identical because of some mutation. The combination of different alleles, as well as different genes, is responsible for the genetic variability between living things. The sexual process is one mechanism that provides for the reassortment and recombination of genetic factors, a feature that permits organisms to adapt through time to an ever-changing environment. The reassortment of genetic factors between and within chromosomes is accomplished by meiosis; the recombination is accomplished by fertilization.

Meiosis is basically a series of two divisions in which the cells divide twice but the chromosomes duplicate themselves only once. This results in four cells, each having the haploid number—that is, half the number of chromosomes present in the somatic cells. Each of these four cells may give rise to a gamete. Fertilization, the fusion of two gametes, subsequently restores the diploid number.

The way in which meiosis differs from mitosis becomes apparent at the first division. The chromosomes divide longitudinally, as in mitosis (into half chromosomes called chromatids), but the homologous visibly doubled chromosomes pair (synapse). During synapsis there may be an actual exchange

A

B

C

D

Figure 9-5 Mitosis in the California coastal peony. The vegetative cells of this species have 10 chromosomes ($2n = 10$). Although mitosis is continuous, the process is broken down into a number of stages for descriptive purposes. (A) Prophase: the coiled linear structure of the chromosomes becomes distinct. (B) Metaphase: the chromosomes line up in the equatorial plate; they have reduplicated, and appear visibly doubled. (C) Anaphase: the chromosomes separate and approach the poles of the cell. (D) Telophase: the contracted chromosomes are pressed close together at each end of the cell. A wall is subsequently formed across the cell, forming two "daughter" cells with the same number and kind of chromosomes as existed in the original cell. [Courtesy M. S. Walters and S. W. Brown.]

Figure 9-6 Meiosis in pepper (*Capsicum*). This species has 24 chromosomes in the vegetative cells ($2n = 24$). Note that there are two divisions. (A) In the prophase of the first division, the chromosomes reduplicate and pair. A visibly doubled chromosome pair can be seen in the bottom of the cell. (B) At metaphase of the first division, the 12 chromosome pairs line up on the equatorial plate (face view). Each consists of 2 doubled chromosomes (4 chromatids). (C) Anaphase of the first division. (D) Telophase of the first division. (E) Metaphase of the second division: one plate is a face view, the other is a side view. (F) Telophase of the second division. Walls will form 4 pollen grains, each containing 12 chromosomes, half as many as the original cell.

(crossover) of segments between chromatids of homologous chromosomes. The precise way in which this occurs is still not clearly understood. The attraction between the synapsed chromosomes changes to repulsion and each doubled chromosome moves to opposite poles of the cell without separation of the chromatids. The resultant shuffling of chromosomes occurs at random. In the second division, which immediately follows the first, the chromatids separate.

The most obvious effect of meiosis is the reduction in chromosome number from the diploid to the haploid number necessary to accommodate the fusion of gametes at fertilization and to maintain a uniform chromosome number from generation to generation. This is accomplished by the synchronization of two cell divisions and one chromosome duplication. (The chromosomes duplicate in the first division and separate in the second.) The less obvious but equally important effect in the long run is the reassortment of the genetic material in the gametes. The reassortment involves not only chromosomes, but chromosome segments that result from crossovers.

Life cycle of plants

The life cycle of plants consists of a series of developmental events involving an alternation of diploid (**sporophytic**) and haploid (**gametophytic**) stages, as is illustrated by the example shown in Figure 9-7.

The haploid stage is produced by the sporophyte through a meiotic division. The resulting haploid cells undergo mitotic divisions and produce distinct entities—gametophytes. The fusion of the male and female gametes restores the diploid condition and gives rise to the embryo from which the sporophyte is produced. In the seed plants, the sporophytic ($2n$) stage, or generation, is the dominant, independent one, and the gametophytic (n) generation is short

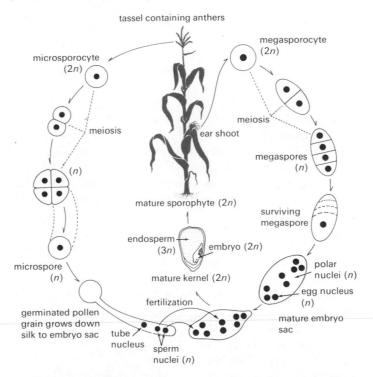

Figure 9-7   The life cycle of corn, illustrating formation of pollen and embryo sac. [From A. M. Srb, R. D. Owen, and R. S. Edgar, *General Genetics*, W. H. Freeman and Company, copyright © 1965.]

and dependent, or parasitic, on the sporophyte. In more primitive plants, such as the ferns, these two generations alternate, each in turn leading an independent existence. In plants that are even more primitive, such as the mosses, the gametophytic generation is independent and of long duration, and the sporophytic generation is transitory and dependent.

In higher plants the male gametophyte, or **pollen spore,** is produced in anthers and usually consists of two sperm nuclei (the gametes) and a tube nucleus. The female gametophyte is produced in a specialized region of the ovary, called the ovule. Although there are various patterns, the developed female gametophyte commonly consists of eight nuclei and is known as the **embryo sac.** One nucleus is the egg; two polar nuclei eventually contribute to the formation of the endosperm, a tissue that supports the young embryo. The function of the five remaining nuclei is obscure.

Pollination

Pollination is the transfer of pollen from the anther to the stigma. The transfer of pollen within the same flower (or any flower on the same plant or clone) is known as **self-pollination;** the transfer of pollen to a flower on another plant (which is likely to have a different genetic makeup) is **cross-pollination.** Self-pollination may be accomplished by gravity or by actual contact of the shedding anther with the sticky surface of the stigma. In cross-pollination wind and insects are the important agents of pollen transfer. Most plants both self- and cross-pollinate naturally, with the proportion of each mode depending upon functional or structural features of the flower or upon genetic incompatibility. Plants may be grouped by their usual method of pollination as **self-pollinated** (cross pollination less than about 4 to 5 percent), **cross-pollinated** (cross-pollination predominant), and **self- and cross-pollinated.** The mode of pollination has important consequences to the plant's genetic structure (see Chapter 19). The likelihood of reproducing a particular plant exactly depends on its natural method of pollination. Seed propagation duplicates plants that are highly self-

pollinated because such plants tend to be homozygous. Since cross-pollinated plants are highly heterozygous, they can be duplicated exactly only by asexual methods. Nevertheless, a high degree of uniformity in some characters may be achieved in the seed propagation of cross-pollinated plants by constant selection.

Natural self-pollination is achieved through functional and structural features of the flower. Perfect flowers—those that contain both stamens and pistils—lend themselves to self-pollination. In the violet, the pollen sheds before the flower is open; in the tomato the pistil grows through a sheath of anthers.

Cross-pollination is brought about in many different ways. Stamens and pistils may occur in separate flowers on the same plant, as in corn and cucumbers (in which case the plant is termed **monoecious**), or staminate and pistillate flowers may be borne as in spinach, asparagus, hemp, hops, and date palm (in which case the species is termed **dioecious**). However, many perfect-flowered plants cross-pollinate. This is achieved by anatomical or physiological features of the flower that prevent self-pollination (selfing). For example, the differential maturation of stamens or pistils will prevent natural selfing. The structural features of the flower that insure cross-pollination are often adapted to pollen transfer by insects. Among the special adaptations that aid in insect pollination are petal color, odor, and the presence of nectar. In some plants an intimate interdependence exists between structural features of the flower and insect pollination (Fig. 9-8).

Incompatibility (self-sterility) is a physiological mechanism that prevents self-fertilization. A genetic factor (or factors) serves to prevent pollen tubes produced by a plant from growing in the style of the same plant. Incompatibility factors prevent self-pollination in such crop plants as alfalfa, cabbage, tobacco, and apple.

Fertilization

After a pollen grain lands on the stigma of the pistil, it absorbs water and such substances as sugars and forms a pollen tube, which literally grows down the

Figure 9-8 Flower structure and pollinating mechanism in alfalfa. Pollen is deposited on the bee when the staminal column (pistil and stamens) is tripped by the bee landing on the flower. [After J. M. Poehlman, *Breeding Field Crops*, Holt, Rinehart & Winston, 1959.]

style to the embryo sac. The pollen tube penetrates the embryo sac, where one male gamete unites with the egg to form the **zygote.** After mitotic division the zygote becomes the **embryo** of the resultant seed. In angiosperms the other male gamete fuses with the two polar nuclei and forms the **endosperm.** This complete process is referred to as **double fertilization.** The endosperm of gymnosperms is formed from mitotic divisions in the gametophyte.

Seed maturation

Fertilization initiates rapid growth of the ovary and subsequent development of the seed. Usually the ovary will not develop unless it contains viable and growing seed. Common exceptions (**parthenocarpy**) are seedless grapes and oranges, in which the developing seed breaks down at an early stage of development.

In the developing seed the development of the embryo is preceded by endosperm growth. When rapid growth of the embryo begins, it does so at the expense of the endosperm. The amount of endosperm at maturity varies with different plants. The endosperm may be completely absorbed at seed maturity, in which case stored food may be in the embryo itself, as in the cotyledons of bean seeds.

Seed development is completed with the formation of hardened integuments, or seed coats. In one-seeded fruits, these integuments include maternal ovarian tissue.

PLANT PROPAGATION

The basic objectives of plant propagation are to achieve an increase in number and to preserve the essential characteristics of the plant. Propagation may be achieved sexually by seed or asexually by utilizing specialized vegetative structures or employing such artificial techniques as grafting, layering, and cutting.

Seed propagation

Seed is the most common means of propagating self-pollinated plants, and is extensively used for many cross-pollinated plants. It is often the only possible or practical method of propagation. There are many advantages to propagating by seed. It is usually the most inexpensive method of propagation. Seeds also offer a convenient means of storing plants over long periods of time. When kept dry and cool, seeds normally remain viable from harvest to the following planting season, and under suitable storage conditions seeds of many species can be kept in viable condition for many years. Seed propagation also makes it possible to start "disease-free" plants. This is especially important with respect to virus diseases, since it is difficult to obtain virus-free tissues for propagation from a virus-infected plant, whereas the seeds of such plants are rarely infected. The disadvantages of seed propagation are the genetic segregation in heterozygous plants and the long time required by some plants to grow from seed to maturity. Potatoes, for example, not only do not breed true from seed but do not produce large tubers the first year. These disadvantages are overcome by vegetative propagation.

Seed technology

The practice of saving seed to plant the following year has developed into a very specialized technology. Seed technology involves all the steps necessary to assure the production and utilization of high-quality seed (seed with high viability, freedom from disease, purity, and trueness to type)—specialized growing and harvesting techniques, cleaning and separation, moisture regulation, and a number of distinct processes for improving the viability and germination powers of the seed or the subsequent performance of the crop. Examples include delinting of cotton seed, separation of multigerm seed such as beet, stratification and scarification of dormant or hard seeds, inoculation of seeds of nitrogen-fixing legumes with bacterial cultures to improve crop performance, and chemical treatment to control seed- or soil-borne diseases. Finally, the marketing of seeds involves exacting storage, packaging, labeling, and transporting procedures. The seed is a crucial stage in the entire agricultural cycle. The regulation and control of the movement and labeling of seed and the establishment of seed-improvement and certification agencies to effect the distribution of improved seed are associated with a highly developed agriculture.

The location of seed-producing area varies with the crop. When the seed or the associated fruit is the commercial part of the crop, as it is in grains, seed is usually produced within the crop-producing areas, in which case seed- and crop-production practices differ little. But when the seed is not the usable part of the plant (as in root crops, forage crops, many vegetables, and flowers), seed production tends to be concentrated in areas separate from crop-production areas and develops into a very specialized practice. The use of distinct seed-producing areas may be imposed by specific flowering requirements, such as cold induction or photoperiod. For example, seeds of many biennial vegetables, such as cabbage and carrots, cannot be produced in tropical areas, as they require cold to flower. Among other requirements that must be taken into account for economical seed production are such environmental factors as low moisture at harvest, which permits proper seed maturation and the reduction of diseases that attack the seed. Isolation is often necessary to prevent genetic contamination in cross-pollinating species. Relatively little tree and shrub seed is grown commercially. Forest tree seed is usually collected from natural stands. The intensification of forest tree plantings, however, has encouraged the development of seed orchards consisting of groves of selected, genetically superior trees intensively managed to produce large

seed crops. Seed to produce rootstocks for fruit trees is often obtained as a by-product in the fruit-processing industries.

The harvesting of many seed crops includes cutting the plant (reaping) and separating the seed from the plant (threshing). Modern combines reap, thresh, and partially clean seed in one operation. Although many seed crops are mechanically harvested, a considerable amount of hand labor is still used on crops whose seed is dispersed easily (shatters) or is produced so gradually that there may be mature seed pods, flowers, and flower buds on a given plant at the same time. In some plants, such as carnation, the entire plant is cut and placed on sheets of canvas or in windows to dry and is then threshed. In many crops the seed must be extracted from the fruits and then cleaned. Seed is removed from dried pods or seed heads by milling. The separation of the seed from fleshy fruits, such as the tomato, is accomplished through fermentation of the macerated pulp.

A great number of special machines have been developed to separate and clean seed through the utilization of differences in the physical characteristics of the seeds.

Size

Separation of seed by size is usually accomplished by the use of screens of different sizes and shapes. Small, light fragments of detritus are removed at the same time by passing an air stream through the seed as it passes from one screen to another. The air-and-screen machines are basic processing apparatus, and practically all seeds are first processed in this manner. The machines vary from small two-screen farm models to large industrial types. Seeds whose widths equal their thicknesses can be separated by length with disc separators. A wheel covered with indentations is passed through a mass of seed, and each indentation picks up a seed. The size and shape of the indentation are determined by the seeds being cleaned.

Specific gravity

Seed separation can be achieved by taking advantage of differences in specific gravity. It is thus possible not only to separate different kinds of seed, or remove foreign material, but to achieve a quality separation of the heavy seed of the same crop. The separation of a mixture of grain and chaff by blowing away the light fragments or tossing the mixture in the air is a very ancient technique. One ingenious machine, the specific-gravity separator, functions by layering a mixture of seed by the use of an air column over a vibrating platform. The different layers can be separated by a combination of deck motion and gravity.

Aspirators are used to separate materials according to their terminal velocity, the speed at which a material falls through an air stream. The terminal velocity is affected by size, shape, and specific gravity.

Surface characteristics

A number of machines have been designed to take advantage of differences in the surface characteristics of seed, such as texture, shape, and color. For example, rollers covered with a velvetlike material lift out rough seeds (such as dodder) and leave smooth seeds behind. Seed with a rough, gelatinous, or granular surface can be separated from smooth seed on the basis of differential retention of a fine iron powder applied to the moistened seed. The actual separation is achieved over a magnetized metal drum. Differences in roundess are exploited by passing seed down a spiral incline. Rounder seed falls faster and is forced out of the channel by centrifugal force; less round and slower rolling seed slides down the spiral. Seed of different colors can be separated by machines equipped with photoelectric cells. These machines are also used to grade such seed crops as coffee.

Seed storage

Seed is living material capable of surviving in a more or less metabolically suspended state. The storage life of seeds varies with the species and with environmental conditions. When storing seed for extended periods it is necessary to avoid conditions favorable for respiration and enzymatic action in general by controlling moisture, temperature, and oxygen availability.

Low moisture content is critical for extended stor-

age life of most seed. Because seeds frequently contain 16–20 percent moisture when harvested, drying is necessary to retain maximum viability. There are exceptions, however; the seeds of sugar maple and citrus species, for example, lose their viability when dried. Some seeds dry naturally in the field except under conditions of high humidity. Others may be dried artificially by means of forced air, which is often heated to increase its drying power. Because the moisture content of seed fluctuates with the humidity of the surrounding atmosphere, the storage humidity is very important. Most seeds retain their high viability with a low relative humidity; a relative humidity of 65 percent or more is generally unsafe for seed storage.

The range of optimum temperatures for long-term seed storage is -18 to $0°$ C (0 to $32°$ F).[*] Moisture and temperature are interrelated; if either is high the other must be low to insure seed viability (Table 9-2). For most seeds, however, a temperature between 0 and $10°$ C (32 and $50°$ F) and a relative humidity of 50 to 60 percent is adequate to maintain full viability for at least one year.

The storage life of seed can be extended by controlling the storage atmosphere. Reducing the oxygen content and increasing the CO_2 content decreases respiration. Seeds stored in sealed containers build up a controlled atmosphere naturally through respiration. The atmosphere in sealed containers can be replaced with an inert gas such as nitrogen, which prolongs the storage life of some seeds.

The long-term storage of agricultural seeds offers a number of important economic advantages. Most important, perhaps, is that it permits adjustment of supply and demand. In such high-value speciality crops as hybrid cauliflower it is often most practical to produce more than a year's supply of seed in one season. This also permits the crop to be tested before the seed is sold. Long-term storage is necessary for such purposes as breeding and experimentation. There are also biological advantages to maintain-

[*]Seeds of an arctic tundra lupine, uncovered after being frozen for 10,000 years, not only germinated but produced normal, healthy plants.

Table 9–2 Effect of moisture content and storage temperature on viability of cotton seed stored in sealed containers.

Approximate moisture content (%)	Percent germination after storage for following number years							
	1	2	3	$5\frac{1}{2}$	7	$13\frac{1}{2}$	15	19
Stored at 0.5° C (33° F)								
7	87	87	87	90	94	90	91	92
9	92	87	93	89	92	93	91	92
11	89	88	91	79	89	88	93	89
13	90	87	86	87	92	91	72	51
14	88	90	85	61	34	10	0	—
Stored at 21° C (70° F)								
7	93	91	90	84	89	85	73	51
9	87	91	82	81	59	0	—	—
11	86	89	68	1	—	—	—	—
13	72	23	3	—	—	—	—	—
14	17	0	—	—	—	—	—	—
Stored at air temperature (uncontrolled)								
7	87	90	83	—	88	—	0	—
9	91	88	84	—	—	—	0	—
11	85	69	18	—	—	—	—	—
13	49	0	—	—	—	—	—	—
14	0	—	—	—	—	—	—	—
Stored at 32.2° C (90° F)								
7	86	86	59	0	—	—	—	—
9	50	33	0	—	—	—	—	—
11	21	0	—	—	—	—	—	—
13	0	—	—	—	—	—	—	—
14	0	—	—	—	—	—	—	—

Source: After D. M. Simpson, *Agron. J.* **49:**608,1957.

ing a supply of old seed. For example, certain seed-transmitted pathogens (for example, bacterial-wilt organism in tomato) have a short storage life, and the disease can be controlled by using old seed.

Germination

Germination—the series of events leading from dormant seed to growing seedling—is dependent upon seed viability, suitable environmental conditions, and, in some crops, the breaking of dormancy.

The germinating seed and young seedlings are vulnerable to certain diseases, and may need protection.

Seed viability

Seed viability refers to the percentage of seed that will complete germination, the speed of germination, and the resulting vigor of newly germinated seedlings. The viability of seed lots can be determined by standardized testing procedures. Probably the most significant measure of viability is the germination percentage—the percentage of seed of the spe-

A

C

B

Figure 9-9   Seed testing: (A) A vacuum seed counter picks up 100 seeds and places them on a germination medium. (B) Germination rooms with complete environmental control at Indiana State Seed Laboratory. (C) Germination of corn seed being tested by the "rolled towel" technique. One hundred seeds are placed on moist paper toweling, which is sealed in wax paper, rolled, and stored under controlled temperature. After seven days, the percentage of germinated seeds is determined. [Courtesy J. C. Allen & Son.]

cies tested that produces normal seedlings under optimum germinating conditions (Fig. 9-9). Germinating tests are usually performed on moistened absorbent paper under rigidly controlled environmental conditions. The length of the test varies with the rate at which different species germinate. Perhaps the greatest problem is in distinguishing between dormant and nonviable seed. Seed dormancy must be overcome to obtain a reliable test. A rapid chemical test with tetrazolium (2,3,5-triphenyltetrazolium chloride) makes possible a rough prediction of viability in nongerminating dormant seed. Living cells turn red, but nonliving cells do not.

Breaking dormancy

The breaking of dormancy and the creation of a suitable environment are necessary to initiate the germination process for some species. The treatments depend on the type of dormancy (see Chapter 8). They include scarification, stratification, embryo culture, and various combinations of these treatments, with suitable environmental control. The germination of seed that contains an impervious seed coat may be promoted by **scarification**—the alteration of the seed coat to render it permeable to gases and water. This is accomplished by a number of techniques, mechanical methods involving abrasive action being the most common. The action of hot water, 77–100° C (170–212° F), is effective for the seed of honey locust and various other species of the pea family. Some seed can be scarified by the corrosive action of sulfuric acid.

The seed of some plants requires dry storage for a period of days or months before it will germinate. The physiological basis of this type of dormancy is not clear but seems to depend on the time required for the dissipation of volatile inhibitors. The after-ripening of some seeds requires a period of moist, aerobic storage known as **stratification.** Cold-stratification—the afterripening of dormant embryos by storing them at high moisture and low temperature—is a prerequisite for the uniform germination of many temperate-zone species, such as apple and pear. The cold-stratification of apple and pear seed involves storing the moist, aerated seed at around 2° C (26° F). The germination percentage increases with time in these species until the third month of treatment. The stratification medium consists of moist sterilized soil, sand, peat, or such substances as vermiculite. Warm-stratification—moist storage above approximately 7° C (45° F)—promotes germination in some species as a result of microbial decomposition of the seed covering. Seeds that possess double dormancy, such as those of viburnum, are first warm-stratified and then cold-stratified. In some seeds, such as those of redbud, a combination of scarification and cold-stratification is used.

Environmental factors affecting germination

The germination of seed that does not require after-ripening, or of seed that has had this requirement satisfied, depends upon external environmental factors: water, favorable temperature, oxygen, and sometimes light.

Water is an essential requirement for germination. The amount of water required varies somewhat with different species. For example, celery requires that soil moisture be near field capacity, whereas tomato will germinate with soil moisture just above the permanent wilting point. For most seed, overwet conditions are harmful, since they prevent aeration and promote conditions favorable for disease.

The effect of temperature upon germination varies with the species, and is related somewhat to the temperature requirements for optimum growth of the mature plant. In general, the germination rate increases with increasing temperature, although the highest germination percentage may be obtained at a relatively low temperature. For grasses, an alternating temperature is usually more favorable than a constant temperature.

Because of its critical role in respiration, oxygen is necessary for seed germination in all plants except some water-loving species, such as rice. The maintenance of proper drainage and tilth in seedbeds promotes rapid germination, largely as a result of good aeration. The effect of light in stimulating or inhibiting the germination of some seed, is a reversible red–far-red phenomenon (p. 187). The action of certain salts has been shown to influence germination. At concentrations of 0.1 to 0.2 percent, potassium nitrate will increase germination in a number of plants, and is used in seed testing. Germination is inhibited, however, by high salt concentrations caused by excessive fertilization.

Embryo culture

The artificial culture of the embryo is used to facilitate seed germination in certain species. For example, the embryo in many early-ripening peach cultivars (for example, 'Mayflower') is not sufficiently mature to germinate when the fruit is ripe. This problem, a serious impediment to the breeding of early-ripening peaches, can be overcome, however, by excising the embryo from the pit and culturing it under aseptic conditions in media providing certain nutrients.

Tukey's solution for culturing mature and
relatively immature embryos

Stock chemical	Amounts (grams/liter H_2O)
KCl	7.5
$CaSO_4$	1.9
$MgSO_4$	1.9
$Ca_3(PO_4)_2$	1.9
$Fe_3(PO_4)_2$	1.9
KNO_3	1.5

The routine germination of orchid seed also involves culture in artificial media. These seeds, which are almost microscopic, consist of a simple, undifferentiated embryo with no reserve food.

Disease control

Disease is a constant threat to the germinating seed. This is especially true for seed that must be stratified or that requires an extensive period of time to germinate. The control of diseases is an integral part of the technology of seed propagation.

Damping-off, the major disease of germinating seeds, can be caused by a number of fungi, mainly species of *Pythium, Rhizoctonia,* and *Phytophthora.* The disease is expressed either by the failure of the seedling to emerge or by the death of the seedling shortly after emergence. A common symptom is the girdling of young seedling stems at the soil surface. Damping-off usually occurs only in young succulent seedlings during or shortly after germination, but older plants may be affected by severe attacks, which may occur either in the field or in greenhouse plantings.

Protection from damping-off and other seedling diseases requires both the direct control of the causative organism and the regulation of environmental conditions such that they favor the rapid growth of the plant rather than the growth of the pathogens. A number of seed treatments are available, either to eliminate the organisms from the seed or to provide protection to the seedling when it is growing in contaminated soil. These consist in coating the seed with a suitable fungicide. A common seed treatment to eliminate surface contaminants consists of a 5-minute dip in a 0.5 percent solution of sodium hypochlorite (the active ingredient in ordinary household bleach). The treatment of seeds in hot water—50° C (122° F) for 15 to 30 minutes—has been used for the control of such seed-borne diseases as loose smut of wheat when the organism (*Ustilago tritici*) is present inside the seed. Such treatment must be precise, however, or the seed may be seriously injured.

Soil may be treated by applying fungicides to the upper surface or by applying heat. Raising the soil temperature to 82° C (180° F) for 30 minutes (pasteurization) is always recommended for potting soils to control weeds and nematodes, as well as the organisms that cause damping-off. Complete sterilization and desiccation of soil interferes with nutrient availability and should be avoided. Sphagnum moss has proved satisfactory as a germination and stratification medium for some seeds because inhibitors and low pH prevent the growth of many of the organisms that cause damping-off.

The temperatures most favorable to the fungi responsible for damping-off are 21 to 29° C (70 to 85° F). Thus, damping-off tends to be severe when the cool season crops are germinated at temperatures that are too high. For best control, germination temperatures should be optimum for the crop. Good viable seed and rapid seedling growth are important. Many of the fungi responsible for damping-off are water loving, and are encouraged in wet, poorly drained soils. Cloudy weather and periods of poor drying encourage damping-off disease.

Vegetative propagation

Asexual reproduction, the basis for vegetative propagation, enables plants to regenerate tissues and parts. In many plants vegetative propagation is a completely natural process; in others it is more or less artificial. The methods of vegetative propagation are many, and the choice of method depends on the plant and the objectives of the propagator. The advantages of vegetative propagation are readily apparent. Heterozygous plants may be perpetuated without alteration. In addition, vegetative propagation may be easier and faster than seed propagation, as seed-dormancy problems may be completely eliminated and the juvenile state reduced. Vegetative propagation also makes it possible to perpetuate seedless clones—for example, 'Washington Navel' orange, 'Gros Michel' banana, and 'Thompson Seedless' grape.

The various methods of vegetative propagation are summarized as follows:

1. Utilization of apomictic seed (citrus)
2. Utilization of specialized vegetative structures:

    Runner (strawberry)

    Bulb (tulip)

    Corm (gladiolus)

    Rhizome (iris)

    Offshoot (banana)

    Stem tuber (potato)

    Tuberous root (sweetpotato)
3. Induction of adventitious roots or shoots:

    Layering—regeneration from vegetative part while still attached to the plant

    Cutting—regeneration from vegetative part detached from the plant
4. Grafting—the joining of plant parts by means of tissue regeneration
5. Tissue culture

Utilization of apomictic seed

Apomixis is the development of seeds without fertilization. It is therefore a form of vegetative reproduction. The most significant type of apomixis is that in which the complete meiotic cycle is eliminated. The seed is formed directly from a diploid cell, which may be either the nonreduced megaspore mother cell or some cell from the maternal ovular tissue, such as the nucellus. As a result of apomixis, a heterozygous cross-pollinating plant will appear to breed true.

Although apomixis is widespread within the plant kingdom, it is not a common means of asexual propagation. It is utilized in the propagation of Kentucky bluegrass, citrus, and mango. These species, however, are only partially apomictic, and seed will be derived from both the sexual and asexual processes.

Virus diseases are not readily transmitted through seed, including apomictic seed. Apomictic seed is therefore utilized to free virus-infected clones of citrus. These "nucellar" clones, being asexually produced, will retain all of the characteristics of the parent plant, but will be free of virus infection. Apomictic seedlings, like ordinary seedlings, show juvenile characteristics.

Utilization of specialized vegetative structures

The natural increase of many plants is achieved through specialized vegetative structures. These modified roots or stems are often also food storage organs (bulbs, corms, tubers), although some function only for natural vegetative increase (runners). These organs enable the plant to survive adverse conditions or to give the plant a means of spreading. They renew the plant through adventitious roots and shoots and are commonly utilized by people as a means of propagation. The propagating process is referred to as **separation** when the structures divide naturally, and as **division** when they must be cut.

Stem modifications

Bulbs are shortened stems with thick, fleshy leaf scales. In addition to developing at the central growing point, bulbs may also develop from buds at the axils of the leaf scales. These buds form miniature bulbs (**bulblets**), which are called **offsets** when grown to full size.

Bulbs are common in the lily family (Liliaceae). The development of bulbs from initiation to flowering size takes a single season in the onion, but most bulbs, like those of the daffodil and hyacinth, continue to grow from the center, becoming larger each year while continually producing new offsets. The asexual propagation of bulb-forming plants is commonly achieved through the development of scale buds. The individual scales, the offsets, or the enlarging mature bulb itself may be used in propagation. The bulb of hyacinth is commonly wounded to encourage adventitious bulblets to form. The bulblets develop into usable size in 2 to 4 years.

Corms, although they resemble bulbs, do not contain fleshy leaves, but are a solid-stem structure containing nodes and internodes. The gladiolus, crocus, and water chestnut (*Eleocharis tuberosa*) are examples of corm-forming plants. In large, mature corms, one or more of the upper buds develop into flowering shoots. The corm is expended by the plant in flower production, and the base of the shoot forms a new corm above the old. At season's end, one or more new corms may develop in this manner. **Cormels,** or miniature corms, are fleshy buds that develop between the old and new corms. Cormels do not increase in size when planted, but produce larger corms from the base of the new stem axis. These require 1 to 2 years to reach flowering size. This is the usual method of propagating corm-producing plants. Corms may also be increased by division, but this is not commonly practiced because of disease problems.

Runners are specialized aerial stems that develop from the leaf axils at the base or crown of some plants. They provide a means of natural increase and spread. The commercial propagation of strawberries is through runner-plant production. Leaf clusters, which root easily, are formed at the second node of the runner, and may in turn produce new runners. Runnering is photoperiod-sensitive, being commonly initiated when the day length is 12 hours or longer and temperatures are above 10° C (50° F). The yield of runner plants per mother plant varies with the variety but averages somewhere between 20 and 30; under optimum conditions it may be as high as 200. Some species of strawberry are nonrunning, and many of the ever-bearing cultivars usually form relatively few runners. Nonrunning plants may be vegetatively propagated by crown divisions, but the increase in plants is much lower.

Horizontally orientated stems growing underground are called **rhizomes.** Rhizomes contain nodes and internodes of various length, and readily produce adventitious roots. Rhizomes may be thick and fleshy (iris) or slender and elongated (Kentucky bluegrass). Growth proceeds from the terminal bud or through lateral shoots. In many plants the older portion of the rhizomes dies out. If new growth proceeds from branching, the new plants eventually separate. Rhizomatous plants are easily propagated by cutting the rhizome into pieces containing a vegetative bud.

Tubers are fleshy portions of rhizomes. The potato, the best known example, is propagated by planting either the whole tuber or pieces containing at least one bud cluster, or "eye." If the whole tuber is planted, the terminal eye commonly inhibits the other buds, but this apical dominance is destroyed when the tuber is cut. Commonly, the "seed pieces" are kept at about 30 to 60 grams (1 to 2 ounces) to provide sufficient food for the young plant. The seed pieces may be cured (stored in a warm place) to allow the cut surface to heal. This is accomplished by active cell division at the cut surface and the formation of a waxy substance called suberin. Chemical treatments may be used to prevent disease.

In many plants lateral shoots (**offshoots**) develop from the stem, which when rooted serve to duplicate the plant. They may also be called offsets, suckers, or crown divisions, depending on the species. Propagation of plants that produce offshoots is easily made by division.

Root modifications

Roots as well as stems may be structurally modified to propagative and food-storage organs. Fleshy, swollen roots that store food materials are known as **tuberous roots.** Shoot buds are readily formed adventitiously. The sweetpotato is commonly propagated from the formation of rooted adventitious shoots called slips. Tuberous roots of some species may contain shoot buds at the "stem end" as part of their structure. In the dahlia, the roots are divided, but each tuberous root must incorporate a bud from the crown; in the tuberous begonia, the primary taproot develops into an enlarged tuberous root that can be propagated by division if each section contains a bud.

Induction of adventitious roots and shoots

The regeneration of structural parts in the propagation of many plants is accomplished by the artificial induction of adventitious roots and shoots. The process is called **layerage** (or layering) when induction of new roots or shoots takes place while the original vegetative part (the layer) is attached to the plant, and **cuttage** when induction takes place after the vegetative part (the cutting) has been detached from the plant. These two processes are part of the same phenomenon—the ability of vegetative plant parts to develop into a complete plant.

Layerage often occurs naturally. In the black raspberry, the drooping stem tips tend to root when in contact with the soil. Because the regenerated stem is still attached to and nourished by the parent plant, the timing and techniques of layerage are not as critical as in cuttage. Layerage is thus a very simple and effective means of propagation. Rooting may be facilitated by such practices as wounding, girdling, etiolation, and disorientation of the stem, which affects the movement and accumulation of carbohydrates and auxins.

Cuttage is one of the most important means of vegetative propagation. The term "cutting" refers to any detached vegetative plant part that can be expected to regenerate the missing part (or parts) to form a complete plant. Cuttings are commonly classified by plant part (root, stem, leaf, bud).

Anatomical basis of adventitious roots and shoots

The formation of adventitious roots can be divided into two phases: (1) initiation, which is characterized by cell division and the differentiation of certain cells into a root initial; and (2) growth, in which the root initial expands by a combination of cell division and elongation. Although the two processes usually occur in rapid sequence, in some plants, such as willows, the time between initiation and development is well separated.

Root initials are formed adjacent to vascular tissue. In herbaceous plants, which lack a cambium, the root initials are formed near the vascular bundles close to the phloem. Thus, roots appear in rows along the stem, corresponding to the major vascular bundles. In woody plants, initiation commonly occurs in the phloem tissue, usually at a point corresponding to the entrance of a vascular ray. Adventitious roots and shoots from leaf cuttings both commonly originate in secondary meristematic tissues—cells that have differentiated but subsequently resume meristematic activity.

Adventitious roots and shoots may be derived from different tissues; for example, in African violet leaf cuttings the roots are initiated from cells between the vascular bundles, whereas shoots are initiated from cells of the epidermis or cortex.

Physiological basis of rooting

The ability of a stem to root is a variable character, depending on the plant, its age, its environmental history, and the type of treatment it receives. Some insight into the physiological basis of rooting has been developed from studies on closely related **easy-to-root** and **difficult-to-root** plants. The ability of a stem to root has been shown to be due to an interaction of inherent factors present in the stem cells, as well as to transportable substances produced in leaves and buds. Some of these transportable substances are auxins, carbohydrates, nitrogenous compounds, vitamins, and others that have not been identified. Substances that interact with auxins to induce rooting are called **rooting cofactors.** In addition, such environmental factors as light, temperature, humidity, and oxygen play an important role in the process. The physiological factors involved in rooting are only beginning to be understood; it is still not possible to induce rooting easily in many plants—for example, in blue spruce, rubber tree, and oak.

The presence of auxins is closely associated with adventitious rooting of stem cuttings, although the precise relations are not clear. Rooting of stems appears to be triggered by the accumulation of auxins at the base of a cutting. The increase in rooting achieved by applying indoleacetic acid or some other auxin supports this concept. It is certain, however, that auxins are not solely responsible, for rooting of many difficult-to-root cuttings is not improved by the application of an auxin alone. Many specific compounds are known that either stimulate rooting (such as catechol) or inhibit it (such as 5-fluorouracil).

The presence of leaves and buds exerts a powerful influence on the rooting of stem cuttings. In many plants the effect of buds is due primarily to their role as a source of auxins, whereas the rooting stimulation provided by leaves can be shown to be due to additional transportable cofactors, which complement both carbohydrate and auxins (Fig. 9-10). Rooting is also affected by the nutritional status of the plant. In general, carbohydrates stimulate rooting.

The accumulation of auxins as well as carbohydrates explains in part the effectiveness of ringing and wounding in stimulating rooting. Callusing of the wounded surface also increases the efficiency of water absorption.

The effectiveness of stem rooting varies with the stage of development and the age of the plant, the type and location of stem, and the time of year. Owing to the great variation between species, precise conclusions concerning the relation of these factors to rooting cannot be made. In general, greater rooting ability is associated with the juvenile stage of growth. Such plants as English ivy, apple, and many conifers become very difficult to root when they reach the mature stage. Mature, difficult-to-root plants may be made easy-to-root by a reversion to the juvenile stage. Generally, adventitious shoots from the base of the mature plants tend to assume juvenile characteristics. In mature plants that become difficult to root, adventitious shoots may be induced by severe pruning. A form of layering called **stooling** maintains the juvenile stage of growth by repeated pruning to induce the development of adventitious shoots at the base of the plant.

The ability of a stem to root is also affected by its position on the plant; lateral shoots tend to root better than terminal shoots. Vegetative shoots also tend to root better than flowering shoots. These differences are probably related in part to the auxin level and the amount of stored food in the shoot.

Cuttings vary in their ability to root, depending upon the type of stem tissue from which they are derived. Cuttings may be made from succulent, nonlignified growth (**softwood cuttings**) or from wood up to several years old (**hardwood cuttings**). Although almost all types of cuttings of easy-to-root plants root readily, softwood cuttings taken from deciduous, woody plants in the spring or summer generally root more easily than hardwood cuttings taken in the winter. Dormant hardwood cuttings are used when possible, however, because of the ease of shipment and handling. Dormant cuttings must be stored until the shoot's rest period is broken, although rooting is less affected by dormancy than is shoot development. The time at which softwood cuttings should be taken varies greatly with the species. In the azalea, softwood cuttings root best in the early spring; in other broad-leafed evergreens, the optimum time for rooting may be from spring to late fall.

Environmental factors affecting rooting

The environmental factors that have significant effects on the rooting of layers and cuttings are humidity, temperature, light, and the nature of the rooting medium.

Humidity is very important in the induction of rooting. The death of the stem as a result of dessication before rooting is achieved is one of the primary causes of failure in propagation by means of

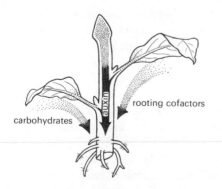

Figure 9-10   The rooting of a cutting is dependent on an auxin, carbohydrates, and other rooting cofactors. The cofactors interact with the auxin to trigger rooting.

cuttings. The lack of roots prevents sufficient water intake, although the intact leaves and new shoot growth continue to lose water by transpiration. Leaves or portions of the leaf may be removed to prevent excess transpiration, but this practice is not desirable because the presence of leaves encourages rooting. The repeated application of water mist maintains high humidity and also reduces leaf temperature by maintaining a water film on the leaf. This allows a high light intensity to be employed in order that photosynthesis will not be reduced. The use of automatic controls to produce an intermittent mist is desirable because excess water may be harmful to many plants.

The use of bottom heat to maintain the temperature of the rooting medium at about 24° C (75° F) facilitates rooting by stimulating cell division in the rooting area. The aerial portion may be kept cool to reduce transpiration and respiration. Daytime air temperatures of 21 to 27° C (70 to 80° F) are optimum for the rooting of most species.

Light in itself appears to inhibit (or conversely, the lack of light encourages) root initiation. The effect of light on rooting varies with the plant and with the method of propagation. Softwood and herbaceous cuttings indirectly respond to light because of its role in the synthesis of carbohydrates. Deciduous hardwood cuttings that contain sufficient stored food, and to which artificial auxins can be supplied, root best in the dark. Root promotion may be achieved through the use of opaque coverings that etiolate the stem. The reason the absence of light favors root initiation in stem tissues is not clear. Etiolation probably affects the accumulation of auxins and other substances that are unstable in light.

The rooting medium must provide sufficient moisture and oxygen and must be relatively disease free. It is not necessary that it be a source of nutrients until a root system has been established. The rooting medium may, however, have an effect on the percentage of cuttings rooted and on the type of roots formed. Various mixes containing soil, sand, peat, vermiculite (expanded mica), and perlite (frothy volcanic glass) have been widely used. Perlite used alone or in combinations with peat moss has proven especially effective because of its good water-holding properties, drainage, and freedom from root-rotting fungi. Sand or water alone may be satisfactory for some easy-to-root cuttings. When water is used alone, improved results are achieved with aeration.

Grafting

Grafting involves the joining together of plant parts by means of tissue regeneration, in which the resulting combination of parts achieves physical union to grow as a single plant (Fig. 9-11). (Natural grafts may form as a result of the close intertwining of roots or stems.) The part of the combination that provides the root is called the **stock;** the added piece is called the **scion.** When the graft combination is made up of more than two parts, the middle piece is referred to as an **interstock.** When the scion consists of a single bud, the process is referred to as **budding** (Fig. 9-11).

There are two basic kinds of grafts: approach grafts and detached-scion grafts. In the **approach graft** the scion and the stock are each connected to a growing root system. In the **detached-scion graft,** the most common method, only the stock supplies roots, since the scion is severed from its parent plant. Approach grafting is used when it is difficult to obtain a union by the ordinary procedures. The root connection to the scion acts as a "nurse" to the scion until union is achieved, at which time the scion is severed from its own roots.

Grafting has many uses in addition to propagation. The interaction of two or more genetic components in a graft combination may affect growth (dwarfing or invigoration), disease resistance, and hardiness. Other uses include cultivar change (topworking) and repair (inarching, bridging, and bracing).

Graft union

The **graft union** is the basis of grafting. It is formed from the intermingling and interlocking of callus tissue produced by the stock and scion cambia in response to wounding. The cambium is continuous in perennial woody dicots. Monocots with a diffused cambium apparently cannot be grafted. Callus tissue is composed of parenchymatous cells. Under the influence of the cambium, the callus tissue differentiates to form new cambial tissue. This new cambium

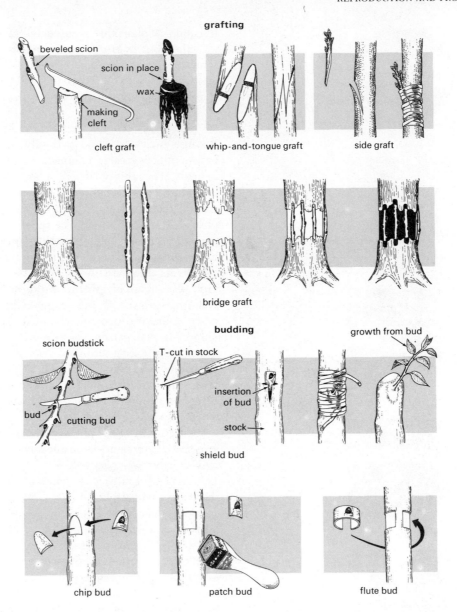

Figure 9-11 Budding and grafting.

differentiates to form xylem and phloem to form a living, growing connection between stock and scion (Fig. 9-12). The basic technique of grafting consists in placing the cambial tissue of stock and scion in intimate association such that the resulting callus tissues produced by stock and scion interlock to form a continuous connection. A snug fit is often obtained by utilizing the tension of a split stock and/or scion. Tape, rubber strips, and nails may also be used to achieve tight contact.

The graft union provides vascular continuity, and offers little if any barrier to translocation in xylem and phloem or to the movement of hormones or viruses. Consequently, grafting is used in virus iden-

Figure 9-12 Developmental sequence during the healing of a cleft-graft union. The graft union is formed from the redifferentiation of the callus tissue under the influence of the stock cambium. [After D. E. Kester and H. T. Hartmann, *Plant Propagation: Principles and Practices*, 3rd ed., © 1975, reprinted by permission of Prentice-Hall, Inc., Englewood Cliffs, N.J.]

tification. A plant that is suspected of having a virus infection, but does not show obvious morphological symptoms, is grafted to a related plant that is especially sensitive to the virus and will show the symptoms of infection; this process is called **indexing.**

Incompatibility

Because plants have no antibody mechanism they have a much greater tolerance to grafting than do animals. The ability of two plants to form a successful graft combination is related in part to their natural relationship and thus to their structural and physiological similarities. The inherent failure of two plants to form a successful union—**graft incompatibility**—may be due both to structural and physiological effects. Incompatibility may be expressed in a high percentage of grafting failures: poor, weak, or abnormal growth of the scion; overgrowths at the graft union; or poor mechanical strength of the union, which in extreme cases results in a clean break at the graft. Incompatibility may be manifested immediately or it may be delayed for several years. In some cases incompatibility has been traced to a virus contributed by one of the graft components, which was itself virus tolerant. Such incompatibility is due to the virus sensitivity of the other component. If the sensitive component is the rootstock, the entire tree may be adversely affected.

In general, grafting of plants of the same species results in compatible unions. Grafts made between species of the same genera or species of closely related genera vary in their degree of compatibility. For example, most pears (*Pyrus communis*) may be successfully grafted to quince stock (*Cydonia oblonga*), but the reverse combination, quince scion on pear stock, is not successful. Incompatibility may be bridged by an interstock that is compatible with both components.

Factors influencing grafting success

In addition to the inherent compatibility of the plants being grafted, there are a number of factors that affect successful "take" in grafting. Skillful grafting or budding techniques are, of course, necessary. Success is then dependent upon environmental factors that promote callus formation. In general, callus formation is optimal at about 27–29°C (80–85°F). Grafts of dormant material are often stored in a warm, moist place for a week or two to stimulate callus formation.

It is very important that high moisture be maintained to prevent dessication of the delicate callus tissue. Special waxes are available that consist of various formulations of resins, beeswax, paraffin, and linseed oil. Bench grafts should be plunged in warm, moist peat to prevent dessication during the period of healing. The use of plastic films has proved successful in conserving moisture. In some plants (for example, the grape), oxygen has been shown to be required for callus formation, hence waxing should not be used.

The percentage of "take" in grafting is often improved if the stock is in a vigorous state of growth. The scion, however, should be dormant to prevent premature growth and subsequent dessication. Growing trees are often grafted with dormant scions from refrigerated storage. In summer budding, irrigation should be supplied before budding in order to invigorate the stock. When grafted in late summer, the leaf buds of most wood plants of the temperate zone remain naturally dormant until the following spring.

Tissue culture

Tissue culture (or **in-vitro culture**) refers to the aseptic growth of cells, tissues, or organs in artificial media. Although the culture of plant cells and tissues has long been a tool of the physiologist, these techniques are now increasingly used as a means of rapid plant propagation. Plant propagation by tissue culture (also known as **micropropagation**) involves a sequence of steps, each of which may necessitate a specific set of conditions. Three distinct steps are usually involved: (1) establishment of an aseptic culture, (2) multiplication of the propagule, and (3) preparation and establishment of the propagule for an independent existence by hardening and acclimation.

The first step requires surgical removal of a suitable tissue or organ. While almost every plant organ has been successfully cultured, the most suitable material often differs from species to species. Shoot tips are extensively used, but sections of stems, leaves, petals, and cotyledons have also been cultured, as have shoot apical meristems (the actual growing point) and nucellar tissues.

The probability of isolating tissues free from contamination by microbes and viruses is inversely proportional to the size of the **explant** (the piece of tissue taken). Explants taken from the true shoot meristem consist only of the apical meristem dome, which is less than 0.1 millimeter in height. These are usually too small for use in practical propagation, but they are used whenever it is especially important to establish a tissue culture entirely free of pathogens, including viruses and viroids. However, shoot tip pieces as tall as 1 millimeter have been used successfully in the establishment of pathogen-free cultures. These shoot tips establish more quickly and growth is faster.

Although such environmental factors as light can be critical to the success of a tissue culture, growth and differentiation of the explant is dependent primarily on the culture medium. Various formulas have been developed that make use of combinations of inorganic salts, sugars, vitamins, amino acids, purine bases, such organic complexes as coconut "water" or yeast extract, and growth regulators (Table 9-3). One of the key factors controlling morphogenesis is the auxin–cytokinin balance. A high cytokinin:auxin ratio promotes shoot formation while a high auxin:cytokinin ratio promotes root formation (see Fig. 8-10). Species and tissues differ in the critical balance required as well as in their reactions to different auxins and cytokinins. A critical mass of tissue is usually required before differentiation can proceed.

Actual propagation after the explant is established can be achieved by one of three pathways: (1) asexual formation of embryos (**embryogenesis**), (2) proliferation of axillary shoots, and (3) formation of adventitious shoots.

Embryogenesis

Embryogenesis cannot yet be induced in all species, and its applicability to commercial propagation is confined to relatively few plant families. This process

Table 9–3 Inorganic constituents of various media for culture of tissues, organs, and embryos.

Compound	Concentration (mg/liter)			
	Knop medium	White medium	Heller medium	Murashige and Skoog medium[1]
KNO_3	200	80		1,900
$Ca(NO_3)_2 \cdot 4H_2O$	800	200		
$NaNO_3$			600	
NH_4NO_3				1,650
KH_2PO_4	200			170
$NaH_2PO_4 \cdot 4H_2O$		17	125	
Na_2SO_4		200		
$MgSO_4 \cdot 7H_2O$	200	360	250	370
KCl		65	750	
$CaCl_2 \cdot 2H_2O$			75	440

Source: D. E. Kester and H. T. Hartmann, *Plant Propagation: Principles and Practices*, 3rd ed., Prentice-Hall, 1975.
[1]Also contains 22.3 mg $MnSO_4 \cdot 4H_2O$, 6.2 mg H_3BO_3, 8.6 mg $ZnSO_4 \cdot 4H_2O$, 0.83 mg KI, 0.25 mg $Na_2MoO_4 \cdot 2H_2O$, 0.025 mg $CuSO_4 \cdot 5H_2O$, and 0.025 mg $CoCl_2 \cdot 6H_2O$. Also add 37.3 mg Na_2 EDTA and 27.8 mg $FeSO_4 \cdot 7H_2O$ (as 5 ml/liter of stock solution containing 5.57 g $FeSO_4 \cdot 7H_2O$ and 7.45 g Na_2 EDTA per liter of water).

may proceed directly from the cultured explant or indirectly from a **callus intermediary,** unorganized growth induced from the cut surface of plant tissues. In certain plants, pollen grains (or microspores) have been induced to proliferate embryos directly, or from callus growth, to yield haploid plants.

During this remarkable process, embryo development proceeds through small globular, heart-shaped, and torpedo-shaped stages and mimics the development of normal embryos from true seeds. The asexually formed embryos may be produced in very large numbers: literally hundreds of embryos may be formed from the relatively small volume of tissue in a culture tube. Each embryo is potentially a separate plant. Embryogenesis was first observed in carrot, but it also occurs in other plants, such as *Kalanchoë,* asparagus, poppy, and citrus (Fig. 9-13).

The clonal increase of the members of some orchid genera has proven very efficient with tissue-culture techniques utilizing a system that appears to be related to embryogenesis. This is potentially very valuable because the usual method of vegetative propagation of orchids (by offshoots) is very slow. In some genera (*Cymbidium, Cattleya,* and *Dendrobium*) the culture of shoot tips results in the development of bulblets that resemble protocorms (orchid plants in the "embryonic" stage between germination and developing seedlings). These protocormlike bulblets may be multiplied by cutting them into quarters. Vegetative propagules of orchid produced in this manner are called **mericlones.** Vegetative increase may be extremely rapid, with the number of propagules quadrupling every 10 days, but such rapid propagation is not possible with all orchids. Species of *Vanda* and *Phaleonopsis* do not form such protocormlike bodies.

Proliferation of axillary buds

Enhancement of the development of axillary buds has wide applicability to tissue-culture propagation. Although the process is the slowest of the three, the tissues are under the normal control of the apical meristem and are not likely to produce genetically deviant plants—especially not those of the sort caused by such chromosome irregularities as chromosome doubling. Shoot tips are usually proliferated

Figure 9-13 *In vitro* embryogenesis in orange from nucellus explants. (A) Embryos and embryogenic callus. (B) Cross section of callus showing cluster of developing proembryos. States of embryo development: (C) Globular, elongated, heart-shaped; and (D) cotyledonary. Note a globular embryo bud at the base of the most advanced stage. [From J. Kochba and P. Spiegel-Roy, *HortScience* 12:110–114, 1977.]

A

B

Figure 9-14 (A) Proliferation of axillary shoots in carnation after 5 weeks of growth in a liquid medium containing 2.0 mg/liter kinetin and 0.02 mg/liter NAA with a single replacement of the medium after 3 weeks. (B) Close-up showing single shoots and clumps. [From E. D. Earle and R. W. Langhans, *HortScience* 10:608–610, 1975.]

with high cytokinin:auxin ratios; 2iP, kinetin, and BA are the cytokinins usually used. The proliferated shoot must be separated and induced to form roots in order for the propagation function to be complete. Carnations (Fig. 9-14) and many tropical foliage plants are typically propagated by axillary-bud proliferation.

Formation of adventitious shoots

The formation of shoots from callus (or other tissues, such as leaf discs) is a very rapid way to propagate certain plants. However, the process of forming adventitious organs (**organogenesis**) is associated with a higher incidence of genetically aberrant plants than

Figure 9-15   Callus culture of geranium. (A) Undifferentiated callus. (B) Various stages in callus differentiation. (C) Shoot development. (D) Root development. (E) A plantlet 2 inches long derived from callus culture. [Courtesy A. C. Hildebrandt.]

A

B

C

D

E

is the proliferation of axillary buds. Although the process is controlled by the cytokinin-auxin balance, organogenesis has yet to be demonstrated in many plants, especially many woody species. Intense effort is now being applied to the problem of developing organogenetic propagation methods for all horticulturally important species. The induction of roots from callus without the induction of shoots is common, and root initiation appears to be a physiologically simpler process than shoot initiation. Of course, when shoots are induced they must be rooted to complete the propagation cycle. Boston fern and geraniums (Fig. 9-15) are among the species propagated by inducing the formation of adventitious shoots.

Plant tissue culture has had a significant impact on clonal multiplication of floricultural crops, including foliage plants and potted plants. The potential rate of increase per year may be a million-fold greater than with conventional methods. Other uses of tissue culture include the production, isolation, and maintenance of disease-free plants and the storage of germplasm.

Seed and plant certification

Seed or plant certification is part of a distribution system designed to preserve and maintain the identity and quality of crop varieties. A new or improved variety may literally start with a handful of seed or a single vegetatively propagated plant, although many tons of seed or millions of plants may be ultimately required in a single growing season. In the multiplication and subsequent distribution, some control must be exercised to assure that the essential characteristics of the variety are not lost by contamination, mix-up, or faulty production practices. This control is best handled through an organization, association, or agency with legal or official status.

The certifying agency obtains a new or improved

strain of crop plant identified by a distinct cultivar name from the originator. The original seed is known as breeder, or basic, seed. Because in many crops the first multiplication of breeder seed will not be sufficient to supply the needs of a grower, a series of multiplicative generations is necessary. Although each generation removed from breeder's seed increases the chance of alteration, the most damaging alterations are those that occur early. Consequently, standards must be highest in the early steps of multiplication. The certifying agency assumes the responsibility for each step, although the actual production may be contracted to private growers. The certifying agency provides supervision, inspection, laboratory services, and sets and maintains production standards. It receives a payment for its services through membership and inspection fees. Fees may be based on the number of fields, acres, or bushels, or all three. The certification tag specifies varietal identification and the year for which it is certified. Certified seed, like all other seed in the United States, is subject to seed-control laws that required detailed labeling specifying germination percentage, mechanical purity, amount of weed seed, origin, and moisture content. The labeling usually also includes the date the seed was produced and the name or code number of the grower who produced it. The first, second, and third generations removed from the original breeder's seed are identified in various ways, usually as foundation, registered, and certified seed, respectively. Foundation and registered seed is used for further seed production. Certified seed is used for commercial crop production.

Standards of certification for vegetatively propagated crops are concerned primarily with diseases, particularly virus disease. Foundation plants, for example, are checked each year by suitable indexing procedures to insure freedom from virus infection. In the United States the propagation of potatoes, strawberries, and citrus is covered by plant-certification programs.

The success of these programs is well documented. Ultimately, everyone—the breeder, the seed producer, the grower, and the eventual consumer— profits from a well-executed program.

Selected references

Gamborg, O. L., and L. R. Welter (editors), 1975. *Plant Tissue Culture Methods*. Saskatoon, Sask.: Prairie Regional Laboratory, National Research Council of Canada. (A useful manual of laboratory methods in tissue culture.)

Hawthorn, L. R., and L. H. Pollard, 1954. *Vegetable and Flower Seed Production*. New York: Blakiston. (The technology of seed production.)

Kester, D. E., and H. T. Hartmann, 1975. *Plant Propagation: Principles and Practices*, 3rd ed. Englewood Cliffs, N.J.: Prentice-Hall. (The most valuable and authoritative book on the subject.)

Kozlowski, T. T. (editor), 1972. *Seed Biology* (3 vols.). New York: Academic Press. (A series of monographs on the physiology, ecology, and pathology of seeds.)

Mayer, A. M., and A. Poljakoff-Mayber, 1975. *The Germination of Seeds*, 2nd ed. New York: Pergamon Press. (A monograph on the physiology of seed germination.)

Murashige, T., 1977. Plant propagation through tissue cultures. *Annual Review of Plant Physiology*, 25:136–166. (An up-to-date review of micropropagation.)

Strickberger, M., 1976. *Genetics*, 2nd ed. New York: Macmillan. (A standard text of basic genetics.)

Thomas, E., and M. R. Davey, 1975. *From Single Cells to Plants*. New York: Springer-Verlag. (An outline of tissue-culture procedures.)

U.S. Department of Agriculture, 1961. *Seeds* (USDA Yearbook, 1961). (The story of seeds.)

U.S. Forest Service, 1974. *Seeds of Woody Plants in the United States* (USDA Agricultural Handbook 450). (A manual on all aspects of the seeds of woody plants and their proper handling.)

Three
PLANT ENVIRONMENT

10
Light

SOLAR RADIATION

The sun is the earth's source of energy. The sun's energy is transferred across 150 million kilometers (93 million miles) of space in the form of radiation. Solar radiation has two phases, one electric and one magnetic, and unless deflected travels in a straight line at 299,792.8 kilometers per second (about 186,000 miles per second). It can gain or lose energy, and its effects on matter can be mathematically predicted. For some purposes it is best construed as bundles of energy called **photons.** For other purposes it is best interpreted as moving in **waves,** and is commonly described in terms of its wavelength, just as sound is described in terms of its pitch. But in spite of all the effort that has been spent in study and experimentation, the true nature of solar radiation eludes science.

The transfer of energy is accomplished by radiation, conduction, and convection. **Radiation** refers to an organized flow of energy through space. It travels in the form of electromagnetic waves. When absorbed by a surface, radiant energy can be transformed into heat energy, which produces an increase in temperature. In **conduction,** energy flows through material from warmer to cooler places. The transfer of heat through the soil is an example of conduction. The ability of a substance to conduct heat is called its conductivity, a property that varies with the material (Table 10-1). The movement of energy by **convection** involves the mass movement of energized particles. For example, warmed air or water rises because its density changes. Still another phenomenon that concerns us greatly is **reflection.** Heat energy as well as light can be reflected from a surface; a sheet of polished metal reflects both heat and light, but not with equal efficiency.

Table 10–1 Heat conductivity of various substances.

Substance	Conductivity (cal cm-sec-°C)
Silver	1.0
Iron	0.1
Water	0.0013
Dry soil	0.0003
Sawdust	0.0001
Air	0.00005

The quantity and quality (spectral composition) of radiation depend on the temperature of the radiating body. The higher the temperature, the greater the rate of radiation and the greater the proportion of short-wavelength (high-frequency) radiation. Solar radiant energy received at the edge of the earth's atmosphere has a great frequency range—from about one-millionth of a nanometer to 3 billion nanometers or longer. However most is short-wave radiation in the visible range (390–770 nanometers) and near-visible ultraviolet and infrared radiation (mostly between 290 and 390 nanometers in the ultraviolet and between 770 and 4000 nanometers in the infrared). The radiant energy occurring as visible light is but a small fraction of the frequency range of the electromagnetic spectrum (Fig. 10-1).

Short-wave solar energy from the sun is absorbed at the earth's surface and transformed into heat. The earth then becomes a radiating body at a lower temperature, which averages 14° C (57° F). The energy radiated by the earth is entirely in the long-wave (low-frequency) range.

Many things happen to solar radiation as it traverses the earth's atmosphere. This is readily apparent in cloudy weather, although not so evident on bright and sunny days. Noticeable changes in the solar spectrum include a shift in peak energies toward the red. Blue light is scattered, producing the blue color of the sky.

The atmosphere in the first 3000 meters (about 10,000 feet) above sea level absorbs about 20 percent of the total radiation received, but on a dark cloudy day only 5 to 10 percent of the total radiation may reach the surface of the earth. A high percentage of solar radiation is deflected by scattering caused by molecular nitrogen, oxygen, water, and other gases.

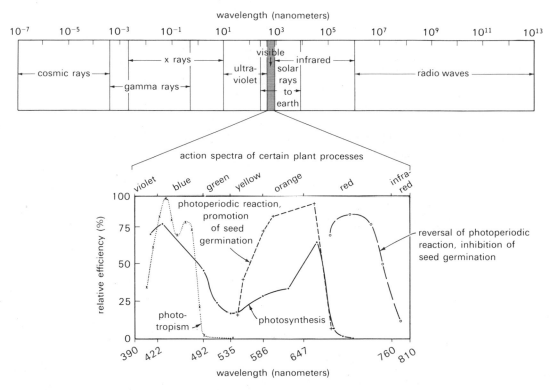

Figure 10-1   The electromagnetic spectrum and the action spectra of certain plant processes. [After L. Machlis and J. G. Torrey, *Plants in Action*, W. H. Freeman and Company, copyright © 1959.]

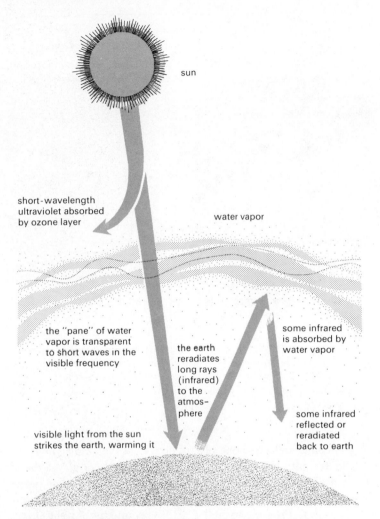

sun

short-wavelength
ultraviolet absorbed
by ozone layer

water vapor

the "pane" of water
vapor is transparent
to short waves in the
visible frequency

the earth
reradiates
long rays
(infrared)
to the
atmos-
phere

some infrared
is absorbed by
water vapor

some infrared
reflected or
reradiated
back to earth

visible light from the sun
strikes the earth, warming it

Figure 10-2 The "greenhouse effect" of the earth's atmosphere
is confined to the zone within which there are enough particles
to intercept a significant proportion of the radiation that passes
through it. [After A. W. Galston, *Life of the Green Plant*,
Prentice-Hall, 1961.]

Ozone absorbs much ultraviolet and cosmic radiation, so that very little radiation shorter than 290 nanometers reaches the earth. The ozone layer in the atmosphere is one of the most effective filters in nature. It also absorbs and reradiates long-wave radiation from the earth. Water vapor absorbs only about 14 percent of the incoming short-wave radiation, but absorbs 85 percent of the earth's long-wave radiation. This tends to keep the surface temperature of the earth much higher than it otherwise would be. The earth's atmosphere thus acts as a pane of glass, transparent to the sun's shortwaves, but partially opaque to the earth's long waves; hence the name **greenhouse effect** for this phenomenon (Fig. 10-2).

>500   400-499   300-399   200-299   100-199   <100
calories/cm²/day

Figure 10-3   The mean annual distribution of solar radiation on the surface of the earth. [After D. Ashbel, "Solar Radiation," in *Third International Geophysical Year*, Hebrew University, 1961.]

Radiation absorbed by plants and soil takes one of several routes. Some is conducted into the ground as heat energy. A large part is used in the evaporation of water. A smaller amount is transferred by convection through the layers of air next to the soil. A small but extremely important part, 1 or 2 percent, is captured by photochemical processes. Eventually, however, all is lost back into the atmosphere as long-wave radiation.

THE DISTRIBUTION OF SOLAR ENERGY

Although visible light is but a small fraction of the total range of the electromagnetic spectrum, it is the most important source of energy in today's world, for part of the radiant energy in this special range is converted by plants through photosynthesis to chemical energy. In addition, both shorter and

longer wavelengths have a profound influence on plant growth. Some of the solar energy captured by plants in past ages has been converted into coal, oil, and gas, and can be released today by burning.

The amount of light energy that any portion of the earth receives is related to factors that affect the intensity and duration of incoming radiation (Fig. 10-3). The intensity of intercepted radiation depends on the angle at which the solar rays penetrate the earth's atmosphere. Water vapor and, to a smaller extent, air and dust diffuse, reflect, and absorb radiant energy.

The solar radiation received by the surface of the earth is greatly affected by latitude. Because the earth is spherical, the sun's rays strike the polar regions at an oblique angle. At latitudes near the poles the sun's rays are distributed over a larger surface of the earth and travel longer through the atmosphere than they do at the equator, where the sun's rays strike at right angles to the surface. At the equinox

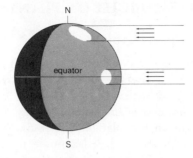

Figure 10-4 The effect of latitude on the dispersion of solar energy over the earth's surface during the equinox. [After W. E. Reifsnyder and H. W. Lull, "Radiant Energy in Relation to Forests," in *USDA Tech. Bull.* 1344, 1965.]

the sun's rays must penetrate an air mass at the poles equivalent to 45 air masses at the equator (Fig. 10-4).

We are all familiar with seasonal differences in light. The duration of exposure to radiation changes seasonally, as a consequence of variation in day length and cloud cover. Seasonal differences in light intensity occur because radiation passes through a thicker layer of atmosphere in winter than in summer. Since atmospheric gases filter out some parts of the spectrum more effectively than others, there are also seasonal differences in the distribution of wavelengths.

The thickness of the air layer through which radiation must pass before reaching the earth changes with altitude. The higher the altitude, the more intense the radiation, particularly of the shorter wavelengths. Some types of radiation, such as cosmic rays, are more abundant at higher elevations and may result in higher mutation rates. Mountain climbers must take special precautions to protect exposed parts of their bodies from the intense ultraviolet radiation at higher elevations; the danger of sunburn is much greater at high altitudes than at sea level.

Local topography also determines how much radiation reaches the earth's surface. In the northern hemisphere, a cool, shady, northern slope may receive only one or two hours of direct radiation during the entire day, and less if it is in a cove or ravine. Slopes that face southward may be droughty as a result of the intense radiation they receive. The pat-

terns of vegetation in mountains develop partly in response to the amount of light the slopes receive.

Vegetation has an influence on the amount of radiation that filters through to the soil surface. The Black Forest in southern Germany, whose dark and mysterious interior was so dreaded by travelers and merchants of the Middle Ages, was so named because of the dense growth of spruce and fir trees, which intercept a high proportion of solar radiation.

Finally, dust and smoke, a special problem in and around cities, may intercept large amounts of radiation. The smog in Los Angeles, a prime example, is composed primarily of gases from automobile exhausts, but industries also contribute their share. In addition to being a light screen, smog also has toxic effects on plants and people.

MEASURING RADIANT ENERGY

There are many units in which the energy of solar radiation can be measured, depending on the objective. For engineering purposes energy is often expressed in British thermal units (Btu). One Btu is the amount of heat energy required to raise the temperature of 1 pound of water by 1° F. In the metric system, the gram-calorie is the amount of heat required to raise the temperature of 1 gram of water by 1° C. One Btu is equivalent to 252 gram-calories. A langley is equivalent to 1 gram-calorie per square centimeter.

Radiant energy can also be measured in terms of wavelength. The intensity of visible light can be measured in foot-candles or lumens. In the English system, a foot-candle was originally the amount of illumination shed by a standard candle on a square foot of surface with a curvature radius of 12 inches from the candle. In recent years, however, more reproducible standards have been developed. In the metric system, the lux, equal to 1 lumen per square meter, is the unit of measurement.

For many purposes, the foot-candle has gained wide acceptance. Full sunlight at sea level in the middle of a clear day may reach an intensity of about 10,000 foot-candles. On a bright night the moon provides about one-fiftieth of a foot-candle. About 20 foot-candles is required for comfortable reading. Although it is convenient to express the light requirements of plants in terms of foot-candles, this unit is not quite precise because it is based on the sensitivity of the human eye. When measuring energy required for photosynthesis, the quantum is the most appropriate unit. A **quantum** is the amount of energy delivered by one photon.

Many kinds of instruments have been devised for measuring solar radiation. Most are useful for measuring a particular narrow range of energies. Instruments useful for measuring all wavelengths of solar radiation are called **radiometers.** They are generally sensitive in the range from 300 to 100,000 nanometers. **Net radiometers** contain two sensitive elements, one facing the sun and the other facing the earth. Long-wave radiation from the earth is subtracted from incoming radiation to give net values.

One of the standard instruments used for measuring direct incoming radiation from the sun is the Eppley normal incidence pyrheliometer. It is designed so that a blackened surface containing 15 thermocouples intercepts a narrow beam of solar radiation and transforms it to heat energy. This heat energy causes the flow of a small current that is measured by a meter.

It is frequently desirable to measure the radiation impinging on a hemispherical surface, for this is the geometry to which a plant is exposed. An Eppley pyrheliometer is one of the instruments used for this purpose. It has a 16-junction thermopile that is exposed to all radiation passing through a hemispherical glass bulb 3 inches in diameter. Other instruments that measure radiation are based on either the differential bending of metals or on the evaporation of liquids.

The intensity of visible light may be measured by various techniques and devices, such as photocells, exposure meters, photographic paper, and extinction meters. Unfortunately the exact wavelengths measured by most techniques are usually unknown. Spectral radiometers, which measure the spectral composition of radiation, have many uses. We can expect the future development of meters that will be sensitive to the wavelengths and intensities that plants use in photosynthesis.

EFFECTS OF LIGHT ON PLANTS

Light and plant growth

Light has tremendous effects on the growth processes of plants. In the absence of light plants will grow until their food reserves are exhausted, but their growth in darkness is abnormal and they develop a syndrome known as etiolation (Fig. 10-5). They have whitish, spindly stems, their leaves are not fully expanded, their root systems develop poorly, and all of their tissues are highly succulent. Internodal elongation is extreme in etiolated plants, and they are frequently not strong enough to remain erect.

Normal plant growth requires light from the entire range of the visible spectrum because the various light reactions of plants (photosynthesis, phototropism, photoperiodism, and so on) depend on photochemical reactions carried on by specific pigment systems that respond to various wavelengths in this

Figure 10-5 Etiolated (*left*) and light-grown (*right*) bean seedlings. [From J. Bonner and A. W. Galston, *Principles of Plant Physiology*, W. H. Freeman and Company, copyright © 1952.]

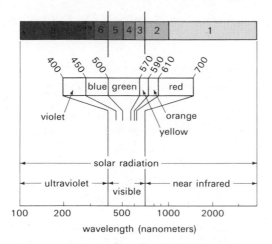

Figure 10-6   Wavelength of solar radiation and spectral bands as defined by the Dutch Committee on Plant Irradiation. *Band 1,* > 1000 nanometers: no specific effects of this radiation on plants are known; to whatever degree it is absorbed by the plant, it is transformed into heat without interference in biochemical processes. *Band 2,* 700–1000 nanometers: the region of specific elongating effect on plants. *Band 3,* 610–700 nanometers: strongest absorption of chlorophyll and strongest photosynthetic activity; in many cases, the region of strongest photoperiodic activity. *Band 4,* 510–610 nanometers: low photosynthetic effectiveness. *Band 5,* 400–510 nanometers: strong chlorophyll absorption and absorption by yellow pigments, and strong photosynthetic activity in the blue-violet. *Band 6,* 315–400 nanometers: fluorescence and strong photographic action. *Band 7,* 280–315 nanometers: antirachitic and germicidal action. *Band 8,* < 280 nanometers: below the atmospheric transmittance limit for sunshine. [After W. E. Reifsnyder and H. W. Lull, "Radiant Energy in Relation to Forests," in *USDA Tech. Bull. 1344,* 1965.]

part of the spectrum. The Dutch Committee on Plant Irradiation has described several bands of radiation according to the biological effectiveness of certain wavelengths, disregarding the vague references to color (Fig. 10-6). Variation in light quality does differentially affect many physiological processes; the light quality over the surface of the earth, however, is relatively uniform. Thus the critical aspects of light as an environmental factor are light intensity and light duration.

For a photoreaction to occur, the energy of light must be transferred to the electrons of atoms. When photons of light hit a light-absorbing molecule or atom (such as chlorophyll), an electron moves away from the nucleus to an orbit further out. When this happens, the atom is said to be "excited." It is then at a higher energy state and capable of delivering its energy for incorporation into chemical bonds. Recent research in photosynthesis has shown that the high-energy electrons in excited chlorophyll molecules actually escape from the molecule. They are accepted momentarily by an iron-containing carrier, such as ferredoxin, which in turn reduces the nucleotide NADP to $NADPH_2$, which then participates in the photophosphorylation of adenosine diphosphate to adenosine triphosphate. The electron from the chlorophyll molecule is replaced by an electron from another source, presumably a hydrogen ion.

It has been found that corn, sugarcane, sorghum, and some species of tropical grasses are photosynthetically much more efficient than might be expected. Two Australian scientists, M. D. Hatch and C. R. Slack, have found that these species carry on certain aspects of the photosynthetic process in a manner unlike most other plants (refer to p. 163 for a discussion of C_4 photosynthesis). Photorespiration, a deterrent to productivity, is nearly absent in C_4 plants because the plants lack the necessary enzymes! Particularly at high light intensities, photorespiration can result in large depressions in productivity. It is exciting to contemplate the possiblities of increasing crop production by experimentally or genetically repressing photorespiration in economic crops of many kinds.

Light has many direct and indirect influences on plants. Light provides the energy for the synthesis of carbon compounds. It also has a profound effect on many physiological processes. Germination, flowering, dormancy, and plant movements are directly affected by light in various plant species. Light affects differentiation of many tissues and organs. It also is a source of heat, although most of the heat received by the plant is from reradiation of long wavelengths outside of the visible range.

Plant growth depends on the amount of carbon fixed during photosynthesis. Neither photosynthesis nor chloroplast development can take place without light. Photosynthesis is most effective at 655 nanometers (red), although there is a secondary peak at 440 nanometers (blue). These peaks agree very well with the light absorption spectra of chlorophyll (Fig. 10-7). Chlorophyll formation, however, is greatest at around 445 nanometers (blue) and next greatest at 620–660 nanometers (orange through red-orange). Thus overall plant growth is usually much less in the green wavelength than in either the blue–violet or orange–red. Photosynthesis and energy have been discussed in Chapter 2; photosynthesis as a growth process has been discussed in Chapter 8.

Carbohydrate accumulation is associated with high light intensity. Therefore many growth processes such as flowering, fruiting, and seed production are indirectly affected by light. Root growth is also affected, and occurs after the photosynthate requirements of the top portions of the plant have been met. The resistance of plants to diseases and insects, as well as stress brought on by drought and high temperature, is also influenced by the vigor and reserves of plants.

Light affects plant movement or phototropism. Phototropic curvatures are affected most strongly by light at the blue end of the spectrum. Yellow pigments—possibly carotenoids, probably flavonoids—absorb the wavelengths effective in phototropism. The familiar twisting of sunflowers to "follow" the sun is due to the effect of light on auxins (see p. 175). Usually auxins move down uniformly along the stem, but light may penetrate and either result in destruction of auxins or migration of auxins away from the side exposed to light. As a result elongation of the stem takes place at a faster rate on the side away from the light. The reaction is produced by light of very low intensity. Even a small amount of weak light is often sufficient to prevent the elongation of etiolated seedlings.

Light indirectly influences the opening of stomata, thereby determining in part the rates of gas exchange between leaves and their environment. Stomatal opening is affected by blue light. Light operates through its effects on the acidity of guard cells, which regulate the conversion of starch to sugar. When sugar concentration is high, water osmotically diffuses into the guard cells, causing them to swell and bend away from each other, widening the stomatal openings. With darkness, acidity is changed again, sugars are converted to starch, and the process reverses.

Light influences germination (Fig. 10-8) and flowering. In both processes the pigment phytochrome seems to be the seat of activity. Phytochrome is influenced by the red and far-red part of the spectrum. Lettuce seeds kept in darkness or exposed to far-red wavelengths will not germinate, but germination will proceed if the seeds are exposed to shorter wavelengths from the red end of the spectrum. This is a reversible process, and it requires very little energy. Exposure of loblolly pine seed to light lasting as little as 0.005 second stimulates germination.

The effect of day length and night length (photoperiodism) on flowering has been discussed in more detail on page 186. Photoperiodism also affects other

Figure 10-7 The maximum apparent rate of photosynthesis indicated by the rate of CO_2 absorption by plants is the average of light absorption by chlorophylls a and b. [After O. F. Curtis and D. G. Clark, *Introduction to Plant Physiology*, McGraw-Hill, 1950.]

Figure 10-8 The relative germination of slash pine (*Pinus caribaea*) seeds in darkness and when exposed to 16 hours of light daily. [Courtesy U.S. Forest Service.]

developmental processes, such as plant dormancy, tuber and bulb formation, and sex expression.

The chemical composition of plants is affected by light. The brilliant red coloration of deciduous trees in autumn is a function of high light intensities. Cool bright conditions favor the conversion of starch to sugar, which in turn is then available for synthesis of red anthocyanins. During cold nights chlorophyll is decomposed, exposing the more stable yellow carotenoid pigments. The variations in fall coloration of different species is due to differences in pigment composition. In most plants anthocyanin formation is stimulated by light; in some, anthocyanin formation is light-dependent.

Light promotes structural differentiation. The internal structure of plants growing in bright sun is likely to be different from those of plants in shade. Sungrown plants usually have fewer chloroplasts, more layers of palisade cells, smaller cells in the leaves, a thicker cuticle, and pubescence on the lower surfaces of leaves.

Light and the plant community

Not all plants use light with equal efficiency. Many deciduous plants can persist and grow in shade that would be insufficient for conifers because they are more efficient photosynthetically. Some weeds use light with extreme efficiency and grow to maturity in the shade of crop plants. Efficiency also changes

with the stage of development. For example, the internal structure of the leaves of juvenile pine seedlings is arranged much more effectively for capturing light and carrying on photosynthesis than is that of the mature leaves of older plants.

Many principles of crop production are based on the relative light requirements of different species. This knowledge is considered part of the professional expertise of those who produce the crops.

Some species show a number of structural adaptations to light. Adaptations that promote light capture include thin cuticles on leaves, several layers of chlorophyllous palisade cells, and high efficiency in using low intensities of reflected light. Trees are well adapted structurally for capturing and fixing light energy because they have large numbers of leaves that are scattered over a wide range of height. In the canopy formed by a stand of trees, both the sides as well as the tops of at least some trees receive direct light, thereby increasing their efficiency. Plants that prosper best in bright sun and even require it for survival are called **heliophytes** (sun plants). Most plants that require intense light are photosynthetically inefficient. Several species of plants, including turkey oak, manzanita, and prickly lettuce, have leaves that are vertically rather than horizontally oriented. Vertically oriented leaves are less exposed to the direct rays of the sun, and subject to smaller transpiration stresses than horizontal leaves. Plants with this characteristic are nearly always found in hot, exposed habitats.

In most locations important crops are not part of the natural vegetation but exist only by virtue of the special care they receive. Many commercially important species of pines must be cultivated and manipulated vigorously to maintain them as a component of forest stands. If left to propagate themselves, they will be succeeded by native vegetation, which is more tolerant of shade and better adapted to the local environment. In turn, each successive vegetation type will be succeeded by still another, until a shade-tolerant species that reproduces and grows in its own shade predominates. Other factors—water, nutrients, and temperature—also influence successional relationships.

THE CONTROL OF LIGHT

Light control has an important role in agriculture. Techniques presently used to obtain maximum utilization of light involve the choice of location, the modification of plant density and distribution, and the use of supplemental illumination or shade.

Geographic location and light

The world distribution of many plant species is greatly determined by their photoperiodic response. The duration and intensity of light both depend on the climatic or geographic regimes. In the tropics the day length approaches 12 hours throughout the year, whereas in polar regions day length varies from 0 to 24 hours. It is useless to attempt to grow many kinds of plants for flowers or fruit in the tropics, for they will remain vegetative indefinitely. The total amount of light energy received by a plant is also affected by local topographic features and by local atmospheric conditions, which determine the number of sunny versus cloudy days during the growing season. Light greatly affects quality in many crops. The bright, clear summer days in the state of Washington favor the production of high-quality, attractive apples. Similarly, grape production is intensive on sunny slopes that receive high amounts of reflected light from rivers.

Plant density and distribution

The control of density is a basic method for controlling the amount of light received by plants. Plant density may be modified by plant spacing, pruning, and training.

The control of plant population is one of the simplest ways to control the amount of light received by individual plants. When plants are grown close together mutual shading prevents direct light from reaching all but the tops.

A good example of plant-density modification is found in forest management. Approximately two to five times more seedlings are planted than are expected to grow to full maturity. This practice helps to suppress weed species by fully occupying the site with the desired species while the trees are small, and it promotes the early natural death and shedding of lower branches (self-pruning). But every 5 to 10 years or so a number of trees must be removed to permit full crown development in the remaining trees.

For row crops, spacing within and between rows determines density. Usually the amount of space between rows is determined by the nature of the cultivation equipment. The trend toward narrow rows in order to make maximum use of available light has encouraged the manufacture of specialized equipment. The use of chemical weed control may eliminate the need for row cropping in many cases.

Density has a tremendous effect on plant performance, largely because of light competition. In general, high yields per unit area are achieved by high populations because maximum use is made of light early in plant development. Eventually however, the performance of individual plants decreases because of competition for light and for other growth factors. Plants respond by decreasing in size or in the number of their parts. Optimum density is dictated by the economic factors involved in optimizing returns. The yield value is a function of the yield per plant, the number of plants per unit area, the yield quality, the price, and production costs. When quality is not a factor (as for example in production of cellulose for dissolving pulp) yield per unit area continually increases, although at a decreasing rate with increasing population. But when quality is a factor, as in the yield of timber, optimum yield often occurs only at a particular population density.

The distribution of plants is also a factor in efficient light utilization. In general, equidistant plant spacing is more efficient than any other spacing because the onset of light competition is delayed (Fig. 10-9).

The orientation of rows can be utilized to make maximum use of light. Crops planted in east–west rows utilize light more efficiently than those planted in north–south rows, because the latter shade each other. In most situations, however, the direction of the row is usually governed by the prevailing slope of the land or by convenience.

Competition for light between parts of the same

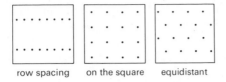

row spacing   on the square   equidistant

Figure 10-9 Plant-distribution patterns.

plant can be modified by pruning and training. **Training** refers to the orientation of the plant in space. Fruit trees may be trained to wires to make maximum use of available light. A new training system for grapes that results in greater exposure to light, called the Geneva double curtain system, has increased yields 40 to 90 percent over those obtained by the standard (Kniffen) system. **Pruning** refers to the judicious removal of plant parts. Pruning may be performed for a great number of reasons. For example, forest trees may be pruned to produce high-quality knot-free lumber. Fruit trees may be pruned for structural strength, and ornamentals for esthetic reasons. One of the main considerations in pruning most fruit trees is to control competition for light within the plant to affect crop size and quality. High yields of extremely small, poorly colored fruit may be economically useless. Pruning is performed to open up the tree so that light may reach normally shaded parts. Flowers and fruit may also be removed by a number of very sophisticated techniques in order to adjust the fruit load to the light-absorbing capacity of foliage.

Finally, weed control can be considered in part a practice to eliminate competition for light between crop and weed. This is discussed further in Chapter 18.

Shading

In some cases it may be necessary to reduce the amount of light that plants receive. Shading is important in many nursery operations to reduce light and temperature and to lower moisture requirements. In permanent lath houses, shade is determined by the spacing between the laths. In recent years plastic screen has been manufactured with various meshes that allow for the passage of different amounts of light. It is thus possible to change the intensity of light throughout the course of a growing season by changing the screen.

Day length may be shortened by shading with black opaque cloth. This is a regular practice to induce or delay flowering in many florist crops.

Tobacco for cigar wrappers is grown under a cover of shade to promote the growth of large, thin, undamaged leaves. This is achieved by placing "tobacco cloth," which is a kind of strong cheese cloth, on frames 2.5 to 3 meters (8 to 10 feet) above the plants. Entire fields of tobacco are grown in this manner, with the cloth suspended high enough to permit cultivation within the enclosure by conventional tractors.

Shading is used to inhibit pigment development in crops in which lack of color is an important factor in "quality." In England celery is "blanched" by mounding the base of the crop with soil to produce pure white (etiolated) stalks.

Supplemental illumination

Supplemental illumination can be used either to satisfy a crop's photosynthetic requirement or to extend day length in photoperiod control. Because of the high energy requirements of photosynthesis and the cost of power, it is not economically feasible to use supplemental light to increase photosynthesis except in special cases in which the plants are especially valuable, such as in seedling production and the production of high-value florist crops. Extension of the photoperiod by illumination is economical on a commercial basis because of the low light intensity required. An increase in photoperiod usually is achieved by extending the day length to about 17 or 18 hours. The same effect can often be achieved by interrupting the dark period for about 3 hours. In terms of power, altering the dark period is more efficient than extending day length. This effect can be made even more efficient by the use of brief light flashes (4-second flashes at 1-minute intervals).

The control of day length by utilizing supplemental illumination and shading has become a standard practice in florist-crop production, particularly in greenhouses. In this way plants can be induced to flower "out of season," and production can be synchronized more closely with market demand. The artificial lengthening of short days (as in winter), or interruption of the dark period, promotes flowering of long-day plants and prevents flowering of short-day plants. (Similarly, under naturally long days, shading to reduce day length prevents flowering of long-day plants and promotes flowering of short-day plants.) The manipulation of photoperiod is standard practice in growing chrysanthemums, the most important florist crop in the United States. Alteration of photoperiod is a valuable tool for the breeder who may wish to cross plants that normally do not flower simultaneously.

Artificial light sources differ greatly in their spectral distribution. Tungsten lamps, which emit light from filaments heated to extremely high temperatures (about 2600° C, or 4648° F), produce a continuous spectrum from blue to infrared. The radiation within the visibile spectrum lies mainly in the red and far-red regions, although the greater part of the overall emission is in the invisible infrared. Fluorescent lamps emit light from both low-pressure mercury vapor and fluorescent powder. Their emission spectrum contains both the continuous spectrum from the fluorescent material and the line spectrum of the mercury vapor. Light from ordinary fluorescent lamps is very low in red light and deficient in far-red light. This is why fluorescent lamps are cool. Different types of lamps vary in their spectral distribution. Special lamps are available that produce light rich in red.

Because of the energy lost in heat (the infrared radiation), tungsten lamps are rather inefficient for promoting plant growth. Only about 5 percent of the energy input is transformed into the light range required by plants, as compared to the more than 15 percent produced by fluorescent lamps. Consequently fluorescent lamps are better for providing the high energy required for photosynthesis. They have not been widely used in commercial greenhouses, because of the high cost of ballasts and installation. Tungsten lamps have proven very efficient for extending the day length because the red light so important in the photoperiodic response is proportionately high.

A curve that shows what proportion of the total energy emitted by a certain light source is contributed by each wavelength is called a spectral-energy distribution curve, or an SED curve. The SED curve is a useful tool when selecting a light source for a particular task. Figure 10-10 shows SED curves for three separate light sources, the sun, a cool-white fluorescent lamp, and an incandescent (tungsten-

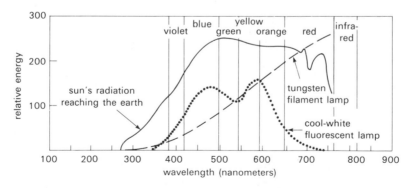

Figure 10-10 Spectral-energy distribution curves for tungsten-filament and cool-white fluorescent lamps compared with the spectrum of sunlight reaching the earth. The weak line spectrum of the mercury discharge from the fluorescent lamp is not shown. The figure offers no quantitative comparison of the energy outputs of the sources. [After General Electric Company.]

filament) lamp. The main differences between sunlight and light from electrical sources are in the amount of light from each wavelength band.

Combinations of light sources are used to achieve satisfactory growth with completely artificial light in sophisticated experimental growth chambers called **phytotrons.** Tungsten and fluorescent lights complement each other to produce a spectrum close to that of sunlight. Special cooling must be provided to dispose of the high heat output.

A still better source of artificial light for plant growth, and one that is widely used in greenhouses and other types of plant-growing rooms, is a fluorescent lamp available under the trade name Gro-Lux. This lamp was designed to provide specifically those wavelengths required by plants for chlorophyll synthesis and for synthesis of carbohydrates. Even though Gro-Lux lamps do not precisely approximate the wavelengths that plants require (Fig. 10-11), they are far better than others, and they eliminate the need for using two different kinds of lamps in plant-growth structures. In addition to being more efficient

Figure 10-12   This plant shop uses plant stands with four plant-growth lamps per shelf. The special lamps not only produce vigorous, healthy plants, but have been installed in a manner so as to enhance their esthetic appeal to potential buyers. [Courtesy GTE Lighting Products.]

Figure 10-11   Comparison of the action spectra of two principal plant photochemical reactions with the spectral-energy distribution curve for the Gro-Lux lamp. (A) The SED curve of the Gro-Lux fluorescent lamp. (B) The photosynthesis curve. (C) The chlorophyll-synthesis curve. [Courtesy GTE Lighting Products.]

for promoting plant growth, they require less power than conventional lamps. By converting from 150-watt incandescent lamps to 40-watt Gro-Lux lamps, electrical energy consumption in one commercial outlet was cut from 48,000 to 12,800 watts. This is an important contribution to energy conservation. An equally important energy savings can be made by using Gro-Lux lamps in merchandising situations (Fig. 10-12). These lamps are relatively cool, while rooms lighted sufficiently for plant growth with tungsten lamps may have to be artificially cooled, even in winter.

Selected references

Bickford, E. D., and S. Dunn, 1972. *Lighting for Plant Growth*. Kent, Ohio: The Kent State University Press. (A reference manual for scientists and commercial producers working with the control of light in greenhouses or plant-growth chambers.)

Canham, A. E., 1966. *Artificial Light in Horticulture*. Eindhoven, The Netherlands: Uitgeversmaatschappij Centrex. (A treatise for the horticulturist on the nature and use of artificial light in plant production.)

Clayton, R. K., 1970–1971. *Light and Living Matter: A Guide to the Study of Photobiology* (2 vols). New York: McGraw-Hill. (The physical and biological aspects of light are considered.)

Evans, L. T., 1975. *Daylength and the Flowering of Plants*. Menlo Park, Calif.: W. A. Benjamin. (A short and interesting history and discussion of the effect of daylength on plants written for the nonphysiologist.)

Pietro, A. S., F. A. Greer, and T. W. Army (editors), 1967. *Harvesting the Sun*. New York: Academic Press. (A scientific treatment on many aspects of photosynthesis.)

Reifsnyder, W. E., and H. W. Lull, 1965. *Radiant Energy in Relation to Forests* (USDA Technical Bulletin No. 1344). (An introduction to radiant energy.)

Ruechardt, E., 1960. *Light, Visible and Invisible*. Ann Arbor: University of Michigan Press. (A useful reference on the physical nature of light.)

Terrien, J., G. Truffant, and J. Carles, 1957. *Light, Vegetation, and Chlorophyll* (translated by M. E. Thompson). London: Hutchinson. (A short, interesting volume dealing with the effects of light on plants.)

Treshow, M., 1970. *Environment and Plant Response*. New York: McGraw-Hill. (A readable book on most aspects of the plant environment, including light.)

Vince-Prue, D., 1975. *Photoperiodism in Plants*. London: McGraw-Hill. (A comprehensive treatment of photoperiodism and its effects on plant metabolism and growth.)

Zalenskii, O. V., 1977. *Photosynthesis and Solar Energy Utilization*. New Delhi: Amerind Publishing. (A series of monographs dealing with photosynthesis from the biochemical to the community level.)

11
Heat

All matter is composed of atoms that are in a state of vibration that depends upon their relative heat. The temperature of a substance is a measure of the relative speed with which its atoms are vibrating. If they are vibrating fast, the temperature is high. Theoretically, at absolute zero ($-273°$ C) all vibration ceases and atoms at that temperature are absolutely still. This low temperature is never found in nature, however, and scientists have not been able to produce it in the laboratory, although they have approached it very closely.

EFFECTS OF HEAT

Physical and chemical processes are controlled by temperature, and these processes in turn control the biological reactions that take place in plants. For example, temperature determines the *diffusion rates* of gases and liquids in plants. As temperature decreases, the viscosity of water increases. The same is true of gases, because the kinetic energy of carbon dioxide, oxygen, and other substances and compounds changes with temperature.

The *solubility* of various substances is temperature dependent. Carbon dioxide is perhaps twice as soluble in cold water as it is in warm water. The inverse

is true of most solids; sugar is much more soluble in warm water than in cold water.

The *rate of reaction* is affected by temperature; usually, the higher the temperature, the faster the reaction. Thus temperature has a very marked and important effect on respiration. However, the relationship of temperature to the biochemical reactions that occur in plants is seldom directly proportional because of certain complicating factors. For example, the end products produced, such as sugars, may accumulate and block further reactions. In some reactions the availability of raw materials may be a limiting factor.

It is useful to think in terms of the Q_{10} of reaction velocities. The symbol Q_{10} refers to the rate of change in reaction activity that results from a change in temperature. The rates of uncatalyzed chemical reactions increase approximately 2.4 times for each $10°$ C ($18°$ F) rise in temperature. Of course, this implies a physical system in which limiting processes are minimal. For the overall growth process the Q_{10} of chemical reactions is more likely to be only 1.2 to 1.3, because there are many factors that slow the rates of chemical reactions in cells and living systems.

Temperature affects the *stability* of enzyme systems. At optimal temperatures, enzyme systems function well and remain stable for long periods of time. At colder temperatures, they remain stable but are nonfunctional, while at high temperatures they completely break down. An enzyme system that is stable at 20° C (68° F) may be active for only a half hour at 30° C (86° F) and for only a few seconds at 38° C (100° F).

The *equilibrium* of various kinds of systems and compounds is a function of temperature. For example, the balance between sugars, starches, and fats is altered when temperature changes. During the fall season, sugars in some species decrease in quantity, whereas starches and fats increase (some plants store food mainly as fats, others as starches). When spring arrives, however, there is a change from both starch and fats to sugars, which are translocated to actively growing parts of plants.

Since temperature has strong effects on physiological and biochemical reactions in plants, it also determines the rates of various plant functions, such as the absorption of mineral elements and water. Not only is the *viscosity* of water greater at low temperatures, but cytoplasmic membranes through which water must pass seem to be less permeable. Photosynthesis is notably slower at low temperatures, and consequently the growth rate is slower. Temperature also affects the rate of cytoplasmic streaming within individual cells.

Plant growth in relation to temperature has been of great interest to many scientists. For example, roots of most plants grow more or less continuously if temperatures are suitable; root growth stops only when temperatures become too cold or too hot. Roots of apple trees start growing at 7.2° C (46.7° F), reach a maximum growth rate at 18.3° C (64.9° F), and stop growing at 32.2° C (89.9° F). Other temperate-zone species that have been investigated are similar in this respect. Near Durham, North Carolina, the roots of loblolly pine will continue to grow all winter if soil temperatures are high enough. Bluegrass roots grow, even if slowly, whenever soil is not frozen.

In view of its profound effects on the biochemical reactions of the plant, the critical effects of temperature on many developmental processes are not un- expected. Among these processes are flowering, sex expression, seed-stalk formation, dormancy of seed and plant, as well as various maturation processes (see Chapter 8).

TEMPERATURE MEASUREMENT

The temperature of a substance may be expressed in terms of several different scales. The Fahrenheit scale (°F) is in common use in the United States, but the Celsius scale (°C) is used in most other parts of the world and in the scientific community. However, because it is based on a system in which 0 is the freezing point and 100 the boiling point, the Celsius scale is coming into much greater use in the United States, as the metric system is being increasingly used for all types of measurements. For some scientific purposes, the Kelvin (K) scale, a scale whose zero point is absolute zero, is used. The gram-calorie is the amount of heat energy required to raise the temperature of 1 gram of water from 14.5° C to 15.5° C.

Temperatures can be measured by various devices. The most common are those that are based on the expansion of liquids, solids, or gases when they are heated. As temperature increases, molecules move faster and require more space. Mercury and alcohol are the liquids most commonly used in inexpensive household and fever thermometers. Liquid-in-glass thermometers, though relatively simple, can be made very precise.

When a high degree of precision in temperature measurement is required, thermocouples are frequently the first choice. A **thermocouple,** or thermoelectric couple, consists of two pieces of dissimilar metals in contact. As the point of contact is heated or cooled, the electrical potential between the two metals changes. Since the change in potential is proportional to the change in temperature, a thermocouple can be calibrated to measure temperature. Thermocouples are easily made and comparatively inexpensive.

Another device for measuring temperature is based on the change in electrical conductivity that metals undergo when their temperatures change. Certain metal alloys have been invented in which the change in conductivity is very great for very small changes in temperature. These metals have been incorporated into devices called **thermistors.** Thermistors, some of which are as small as a pinhead, have found many applications in many kinds of electrical circuits as well as for temperature measurement.

The melting point of a solid is sometimes used to measure temperature. By exposing a series of waxes or metallic alloys that melt at different temperatures, it is possible to tell how high a temperature has been reached. This method has been used to regulate kiln temperatures. A metallic tablet placed between two electrical contacts inside the furnace will melt and shut off the flow of electrical current when the desired temperature is attained. Specially manufactured tablets have also been used for determining maximum temperatures of forest fires and in other situations where it is not possible to use instruments. The main shortcoming of the tablets is that they do not tell how long the temperature remained at the level indicated by the last tablet to melt.

Since the long-wave infrared radiation emitted by objects is a function of their surface temperature, it is possible to use this radiation to measure surface temperature. Instruments that measure long-wave radiation, called **infrared radiometers,** operate on a number of principles. They are generally devised so that they "sense" long-wave radiation emitted by an object, compare it with a standard within the instrument, and give a meter reading that indicates the object's temperature. Such instruments, however, divulge nothing about internal temperatures.

Unusual methods for measuring temperature include the use of sonic frequencies and the refractive index of air. However, these techniques are not in wide use nor is it likely that they will be for biological purposes in the near future.

At times it may be desirable to measure the total amount of heat in a particular environment over a period of time. Such a measurement is called a **heat summation.** If a sterile solution of a disaccharide (compound sugar), such as sucrose, is placed for a period of time in a vial with a given amount of a sugar-digesting enzyme, such as invertase, the inversion of disaccharides to monosaccharides (simple su-

gar) will be proportional to the sum of the heat to which the vial was exposed (provided that temperatures do not rise above the point of diminishing returns). This also gives a measure of potential biological activity, which depends on the amount of heat available.

Other devices for determining temperatures make use of the transmissivity of sound at different temperatures. Such sophisticated techniques have relatively little use in biological research today, but they may become important in the future.

CLIMATE AND TEMPERATURE

In discussing temperature and other climatic elements, it is nearly always necessary to designate the space or area to which they apply. Several expressions for doing this have evolved. The term **macroclimate** refers to the climate of a relatively large part of the earth's surface. It implies a regimen of weather, and is measured by a standardized weather station, which may be 40 to 80 kilometers (25 to 50 miles) distant from some of the locations it represents climatically. Macroclimate is discussed further in Chapter 14.

At the other extreme, **microclimate** refers to the weather conditions around a plant, a leaf, or even a stomate. It expresses the climate of an extremely small volume of space and is not necessarily representative of any place other than the one where measurements are made.

Still another expression, **local climate,** is useful for indicating the climate of a volume of space intermediate in size between that represented by macroclimate and microclimate. Local climate may refer to the weather conditions within a stand of trees, a patch of peas, or a meadow; it is measured by a standard weather-instrument shelter placed strategically to give the best representation of gross weather conditions within the field or other unit of vegetation.

German foresters have an interesting and frequently used expression called **trunk space.** The trunk space, of course, is the space that could be occupied by the trunks of trees, and is usually considered to extend from the soil up to the lower part of the crown. While trunk space is not a scientific term, it is useful insofar as it provides a reference point for estimating local temperatures in forests.

The temperature at any point on the earth's surface depends on the geographic ordinates of latitude and altitude, season and time of day, and the mediating influence of microclimate. The major factor that determines temperature is the amount of solar radiation received, which depends upon both the intensity and duration of solar radiation. The duration of the solar radiation is affected both by the day length and by the variable effects of cloud cover. As cloudiness increases, the amount of solar radiation decreases.

Variation in environmental temperature

Temperatures over the earth's surface range from $-71°$ C ($-91°$ F) at Verkhoyansk, Siberia, to $58°$ C ($136°$ F) at Azizia, Libya. Mean annual temperatures range from an estimated $-30°$ C ($-22°$ F) at the South Pole at an elevation of 2440 meters (8000 feet) to a record $30°$ C ($86°$ F) at Massawa, Ethiopia. In annual crops the important climatic elements are the mean and extreme temperatures and the length of the growing season. Perennial plants are affected by temperature throughout the year. Seasonal variation and average temperature must both be considered in relation to plant growth.

As a general rule, variations in environmental temperatures are greatest near the interface between the soil and the air. Solar radiation is generally intercepted by the surface of the earth with a relatively high efficiency, and so we might expect to find the highest temperatures at this point. At the surface of the soil it is not uncommon for night and day temperatures to differ by $38°$ C ($100°$ F) during the course of a day. Soil temperatures were measured in Florida by covering the bulb of a mercury thermometer with a thin layer of soil less than 1 millimeter (about 0.03 inch) deep. Temperatures varied from $29.3°$ C ($74°$ F) to $74°$ C ($165°$ F) during a single 24-hour period. The variation decreases above and below the soil (Fig. 11-1). At elevations of several hundred feet, temperatures usually range neither as high nor as low as they do at the soil surface. The decrease in temperature with increase in elevation averages $1.4°$ C ($2.6°$F) for every 300 meters (1000 feet) approximately 1000 times the latitudinal rate of temperature change. The reason for this is that much of the thermal energy of the atmosphere is radiated from the earth's surface. Moreover, because the lower tropospheric air contains more water vapor and dust, it is a more efficient absorber of terrestrial radiation; this accounts for the existence of snow-capped mountains on the equator in many parts of the world.

The greater the distance from the earth's surface, the more constant the temperature. At great distances from the earth, temperatures are constantly cold because there is very little matter to absorb radiant energy. Similarly, at soil depths of 7 to 9 meters (about 23 to 30 feet), temperatures may not change more than a degree or two during the course of an entire year. At depths of 3 to 3.5 meters (about 10 to 12 feet), there are likely to be seasonal variations in temperature, but soils at these depths do not undergo daily fluctuations.

Soil temperatures vary much less than air temperatures. Soil beneath the surface is not warmed immediately by the radiation incident on the surface, and a temperature lag occurs because heat must be conducted through the soil. In the spring, soil temperatures may not rise enough for germination to occur for several months after air temperatures have risen enough to support plant growth. Conversely, once soils have been warmed by the hot summer sun, they may retain their heat and delay the dormancy of roots even after air temperatures have become quite low.

Factors that regulate environmental temperatures

Of all the factors that determine soil temperatures, meteorological events and conditions are the most important. The amount of solar radiation striking the soil surface is a primary consideration because it determines the maximum potential input of heat. Va-

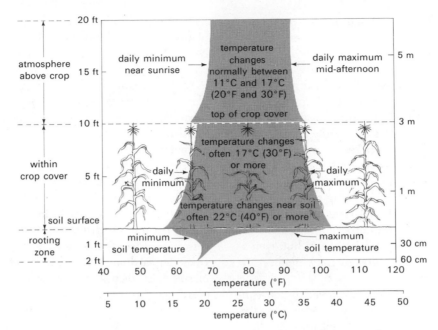

Figure 11-1 Temperatures around crop plants are subject to very regular variations, which decrease both with altitude and with soil depth. [After J. E. Newman and B. O. Blair, *Crops and Soils Magazine,* June–July 1964.]

por, clouds, dust, the angle of the sun's incidence, and many other factors are important influences on soil temperature.

A snow cover on the ground can reflect a high proportion of radiation back into the atmosphere. But snow also insulates the soil, with the result that soil temperatures under a heavy snow are frequently higher than those in an area without a snow cover.

Soil texture and condition influence heat input, heat retention, and heat loss. Soils with a high proportion of fine materials conduct heat faster than coarse-textured soils. In fine-textured soil, particles are in intimate contact with each other and the transfer of heat from one soil particle to another by conduction is relatively efficient. Moisture content is important too, for water has a much greater conductivity than air; the greater the moisture content, the greater the heat conductivity.

Organic matter, unless it is wet, has a very low specific heat, and neither absorbs nor retains much heat. Soil with a high content of organic matter is likely to be a very poor conductor of heat. In some soils the organic content of the surface may be so great that it acts as a mulch, thereby stabilizing the subsurface soil temperatures. Surface color of the soil is also important. Dark surfaces absorb a much higher proportion of radiation than light surfaces.

Soils with a heavy litter are likely to have more constant temperatures than those with no litter. The litter prevents convection losses, stops incoming radiation, and precludes the loss of a large part of the long-wave radiation.

Vegetative cover has a marked effect on soil temperature. During the early part of the growing season, when crops are small and have relatively little foliage, a high percentage of the incident radiation reaches the soil. This is beneficial in the spring of the year because the soil is cold and must be warmed before crop growth can proceed at the maximum rate. By midsummer, plants are large enough so that their tops form a continuous canopy and shade the soil.

Vegetation of any kind reduces the range of temperatures because it reflects heat back into the atmosphere (Fig. 11-1). At the same time, it absorbs heat from the soil and prevents losses into the atmosphere. As a general rule, vegetation raises the average minimum temperature of the environment while lowering the average maximum temperature. The magnitude of this effect will vary according to the size and density of the vegetation. Forest vegetation is usually much more effective in diminishing the effects of solar radiation than grasses or row crops, but even a very sparse cover of short grasses may modify air temperatures greatly.

Still another way in which forest vegetation can change air temperature is to decrease wind velocity and permit the buildup of heat within the plant cover. This has an opposite effect from some of the factors mentioned previously. It should be obvious that any vegetational ecosystem is a complex mass of interacting forces.

CRITICAL TEMPERATURES

Cardinal temperatures are those at which the rates of vital processes are increased or decreased to the point that the survival, growth, or reproductive capacity of a plant is critically impaired. The cardinal maximum temperature for plant life is about 54° C (130° F), and the cardinal minimum temperature for growth is about 5° C (41° F). This means that most plants will not live if temperatures exceed 54° C and that they will not grow at temperatures below about 5° C. The sustained maximum cardinal temperature for growth of temperate-zone plants seems to vary from about 32.2° C to 33.3° C (90° F to 92° F), but the sustained cardinal minimum for life is more obscure. For some tropical plants the cardinal minimum may be a few degrees above freezing, whereas some subarctic plants may survive temperatures as low as −50° C (−58° F) with no apparent ill ef-

fects. There is a wide variation in temperate-zone plants, too. Experimentally, stem sections of mulberry (*Morus alba*), a woody plant, have survived temperatures of −196° C (−321° F) for 160 days.

At temperatures of 4.5–5.5° C (40–42° F) most plants cease to grow. There are exceptions—some subarctic species continue to function normally at temperatures as low as 1° C (about 34° F). The rates of metabolism are slow at these temperatures, but most plants that live in such cold, rigorous climates are small; some species that grow at high altitudes may require 150 years or more to reach a height of 50 centimeters (less than 20 inches).

The concept of cardinal temperatures for life processes is, of course, unsatisfactory if not qualified. Does the cardinal maximum for life refer to protoplasm, cells, tissues, organs, or entire plants? Obviously, many plants live and even grow when exposed to occasional temperatures that exceed 71° C (160° F). Individual living cells of the plant would be killed by these temperatures, but they are protected by characteristics that prevent internal temperatures from being lethal, such as heat-resistant bark and coloration that reflects radiation. Entire plants, then, are obviously unsatisfactory as indicators of temperature effects on protoplasm, and protoplasmic temperatures are extremely difficult to measure.

The cardinal temperatures of different plant parts differ considerably. Flower buds of temperate-zone plants are much more sensitive to low temperatures than vegetative buds. Different vegetative stages vary in sensitivity, too, for young leaves are more sensitive both to high and to low temperatures than mature leaves.

Time is also a factor. A plant that can survive exposure to very high temperatures for a brief time may die when exposed to lower temperatures for an extended period.

The humidity of the environment is important in determining the lethality of temperature. When humidity is low, plants and other organisms can generally tolerate much higher temperatures than when humidity is high. This is partly because organisms dissipate heat as long-wave radiation and by convection more efficiently when humidity is low.

High-temperature injury

There are several fundamentally different ways in which heat can injure plants. Perhaps the most readily apparent way is by directly killing protoplasm. When the temperature of protoplasm reaches about 54° C (130° F), it coagulates in the same manner that the white of an egg solidifies when cooked. Most plants are provided with a measure of protection against lethal internal temperatures by thick bark or large heat-radiating surfaces such as those of leaves.

Direct heat injury is most apparent when particular areas of plants have been injured. **Sunscald,** a condition in which cambium is killed by high temperature, occurs frequently on species of trees with thin bark. If an apple tree is severely pruned, solar radiation may raise bark temperatures and kill cambial tissues beneath bark previously shielded by leaves. The resultant injury may provide a portal of entry for disease-causing microorganisms. Forest species are also subject to sunscald, which results in a deterioration of wood quality. Birch, a thin-barked species, may suffer severe sunscald if trees in a dense stand are thinned heavily.

High temperature is also harmful through its desiccating effect. When temperatures are high, the vapor-pressure gradient between the insides of leaves and the outside environment is large, and transpiration takes place at a very fast rate. Wilting is the first indication of desiccation. It is possible for a plant to wilt and still recover if it has not lost too much water. Tomato plants wilt practically every day in the middle of the summer in some climates, but at night their water balance is restored and they become turgid and firm again. Each day this process is repeated, and it continues as long as soil water is abundant enough to replace the losses. Such wilting is called **incipient wilting.** When a plant fails to recover after water is made available, it is said to be **permanently wilted.**

Another kind of heat injury occurs when plants become hot and respire at an abnormally high rate. When the rate of respiration is such that the photosynthate is being used faster than it is being made, the plant is in a state of **metabolic imbalance.** When

temperature is so high that respiration exceeds photosynthesis over a period of time the plant is "operating in the red." To reverse this and obtain a net increase in photosynthate, temperature must be lowered. Of course, a plant is normally in a state of some kind of imbalance most of the time. During the day, when photosynthesis is proceeding, carbohydrates are manufactured faster than they are used. At night, carbohydrates are not made but are used in respiration and various growth processes. When the ratio of carbohydrate synthesis to breakdown is less than 1 over a period of several days the imbalance may become dangerous to the plant. If a plant continues to use carbohydrates faster than they are manufactured over a long period of time, it will eventually deplete its reserves and die. Metabolic imbalance can be brought about by disease organisms, insects that eat plant parts, and other factors that either limit the accumulation of reserve foods or destroy photosynthetic tissues.

High temperatures may have indirect effects on plants, too. When timber stands of birch in the Northeast are thinned, soil temperatures may increase one or two degrees Fahrenheit because of the increased amount of radiation reaching the ground. This small increase in temperature is apparently enough to stimulate a fungus that is always present in the soil to become pathogenic and cause a disease called "birch dieback."

Low-temperature injury

Freeze damage

Cold-temperature injury to plants is often thought of as simple freezing and tissue death. However, there are several mechanisms through which cold temperatures can injure plants, depending upon the nature of the cold weather and the condition of the plants. The simplest kind of cold-temperature injury is freezing of the living matter of the cells. When protoplasm freezes, its structure becomes disorganized, proteins precipitate, and death results. But this kind of freezing injury takes place only when temperatures drop to subfreezing levels rather quickly.

If temperatures slowly fall below freezing over a period of eight or ten hours, or gradually over a period of days, freeze injury is of quite a different nature. As temperatures decrease, the intercellular water between cells is the first to freeze. This water is not a part of protoplasm, and its sugar content is usually quite low. Upon freezing, its diffusion pressure decreases to practically zero while the diffusion pressure of the unfrozen water inside cells remain relatively high. As a result, water moves out of the cell protoplasm, where diffusion pressure is high, into the intracellular spaces, where the diffusion pressure is low. Once there, it freezes and its vapor pressure drops. This process continues until much of the water in the protoplasm of cells has moved out of the cells and into the intercellular spaces, causing the proteins of the protoplasm to clump together and precipitate. Thus, this injury is a form of cell dehydration.

Still another kind of cold-weather injury is desiccation. If the weather has been cold for several months, as in late winter, the soil will have had a chance to reach a temperature equilibrium with the air, so that both air temperatures and soil temperatures are low. Sometimes air temperatures become unseasonably warm, with the result that the tops of evergreen plants may start transpiring and lose large quantities of water, even though growth processes are not taking place. If the weather remains warm for a few days, transpiration will exhaust the supplies of moisture that are immediately available to the plant, and in the absence of continued root growth and water absorption the plant becomes desiccated. When soil moisture becomes frozen and unavailable, this results in **winter burn**—a very serious and costly form of damage common in ornamental shrubs that have not been planted in the ecological niche to which they are best adapted. Winter burn results in the death of the most tender parts of leaves. Only the tips of needles turn brown on conifers, but broad-leaved evergreens may become uniformly brown.

Cold-temperature injury usually occurs in winter, and is seldom of importance in relation to annual crops, which are not usually exposed to winter temperatures. Winter temperatures should not harm woody perennials either, if selections have been made carefully and with full consideration of the extreme lows to which temperatures can fall every 25 to 50 years. In northwestern Florida in the nineteenth century, thousands of acres were planted to oranges, which were completely destroyed by a short period of unusually cold weather during the winter of 1894.

Glaze damage (Fig. 11-2) can occur when relative humidity (see p. 300) is high and temperatures drop below freezing. The weight of the ice coating on tree limbs causes them to break off.

Other kinds of problems may arise because of low temperatures. Trees may split and crack like a rifle

Figure 11-2 Glaze damage to sugar maple trees. Limbs broken off by the weight of the ice coating provide portals of entry for disease organisms. [Courtesy USDA.]

shot when temperatures fall very low. In Canada and northern Europe, a fungus disease called "snow blight" sometimes attacks coniferous seedlings growing in nurseries while they are covered by snow. In Sweden the fungus does not grow unless 40–60 centimeters (about 16–24 inches) of snow remains on the ground throughout the winter. In temperate parts of the world, seedlings planted in the fall may be pushed out of the soil by **frost heaving.** Frost heaving results when water that has accumulated in cracks, or in holes in which the soil is packed loosely, freezes and expands upward in a column and out of the hole.

Frost

Cold injury is a common hazard in temperate and subtropical climates as a result of spring or fall frost. **Frost** is a thin layer of ice crystals deposited on soil and plant surfaces when temperatures drop below freezing. Two types of weather conditions produce freezing temperatures: (1) rapid radiational cooling, which results in a **radiational frost,** and (2) the introduction of a cold air mass with a temperature below 0° C (32° F), which produces an **air-mass freeze.** The former commonly occurs when skies are clear and calm; the latter, however, usually occurs when skies are overcast and windy. Both occur when temperatures are at or below the freezing point. Because of its local nature, frost may occur when the so-called official temperature (taken nearby and usually at a height of six feet) is above freezing.

The earth radiates energy, as do all bodies. At night, when the earth receives no solar radiation, there is a net loss of heat. Frost occurs when the loss of heat from the ground permits the temperature to drop to, or below, the freezing point. The **white frost** (hoarfrost) commonly seen in the morning results from frozen dew. Its occurrence depends on the **dewpoint**—the temperature at which relative humidity reaches 100 percent. When the air temperature is below the dewpoint but above 0° C (32° F), water vapor condenses in the form of dew. White frost occurs when the air temperature is below the dewpoint and below the freezing point. If the humidity is low, frost damage may occur when the air temperature is below freezing but above the dewpoint. This is known as a **black frost,** because the only indication of it comes when the vegetation turns dark due to cold injury.

Frost injury is potentially extremely destructive to perennial fruit crops that flower in the early spring because flower parts are extremely sensitive to freezing. It is also a problem in tender vegetable transplants.

Meteorological conditions that are especially favorable for frost formation are those conducive to rapid and prolonged surface cooling. These occur on cloudless, dry, calm nights when long-wave radiation from plant and soil surfaces is likely to be at a maximum. When atmospheric humidity is high or when the sky is cloudy, water particles intercept radiation and reflect it back to the earth. The absence of wind leaves the coldest air undisturbed. Hazardous conditions also form in **frost pockets**—depressions into which cold air "drains" from the surrounding higher ground and accumulates. A field or an orchard may have low areas that require special protection during times of forest danger.

Frost may also occur when there is a **temperature inversion**—a situation in which a cold air mass is trapped below a warmer air mass. Normally, temperature decreases with height. Thermal inversions are sometimes very local in extent, but most frequently take place over large areas on windless, overcast nights. As the sun warms the soil during the early morning, conditions that are conducive to inversions are altered.

The methods used to prevent frost involve either the reduction of heat lost through radiation or the addition of heat. The loss of heat by radiation may be reduced by the use of fogging, hot caps, or cold frames. Heat may be obtained from the earth by improving the conduction characteristics of the soil, or it may be obtained from the air by disturbing the temperature inversion. Heat may be added directly by using heaters, and indirectly by means of sprinkler irrigation or techniques that increase the absorption of solar radiation. These techniques will be discussed later in this chapter.

Chilling injury

Some plants are found to be sensitive to temperatures slightly above freezing. **Chilling injury** affects the peanut, velvet bean, sweetpotato, various cucurbits, and many tropical plants. Chilling injury can also interfere with the ripening of such fruits as tomato and banana. Chilling injury affects appearance and may result in actual breakdown of plant tissues.

COLD HARDINESS

Cold hardiness refers to the ability of plants to withstand cold injury. Cold hardiness is a variable characteristic that differs greatly with species and with seasonal change. For example, many plants that survive extreme winter freezing may be severely injured by spring frost. The natural change in hardiness in woody plants is related to temperature, water availability, and day length. The onset of cooler weather and short days in the fall brings about dormancy and a physiological toughening in some plants.

Cold resistance of the whole plant develops in a number of steps. As the end of the growing season approaches, the rate of growth slows down and carbohydrates begin to accumulate. Reserve foods accumulate during this period partly because they are not being utilized in growth and respiration. They are translocated to other plant parts, particularly roots, where they remain over the winter as insoluble fats and starches (although some fraction may be used in respiration). Also during this period, water in the individual cells moves from the protoplasm to the vacuoles. These changes in water relations and in the availability of certain carbohydrates promote stability of the protoplasmic proteins, which have a higher proportion of bound water during winter than during the summer. There seems to be a surface effect on protoplasmic colloids, so that they do not clump or precipitate so readily. During the onset of winter the concentrations of colloidal and dissolved materials become much greater than during the summer, which may contribute to cold-temperature resistance.

Hardening in its broad sense refers to processes that increase the ability of a plant to survive the impact of unfavorable environmental stress. In its more restricted meaning, hardening usually refers to processes that enable plants to withstand cold injury, just as the term hardiness usually refers specifically to cold hardiness.

There are several ways to condition plants for cold weather. One of the simplest techniques is to withhold water and fertilizer, especially nitrogen, toward the end of the growing season. Nitrogen stimulates vegetative growth, and if applied late in the season may cause plants to continue to grow until late in the season rather than **harden off** prior to winter. Succulent green shoots, stimulated by the addition of nitrogen, are not frost hardy, but are quickly killed as temperatures approach freezing. On the other hand, additions of phosphorous and potassium seem to increase frost hardiness, but even these must not be applied too late in the growing season.

Promoting plant vigor during the growing season increases the supply of carbohydrates. Plants with high levels of reserve carbohydrates are better able to withstand cold weather than those in poor condition. Respiration continues to some extent throughout the winter, and there must be a supply of carbohydrates to maintain life processes during this period. Some grape varieties overbear, and if the excess fruit is not removed, the plants are subject to winter injury because their carbohydrate reserves have been depleted in fruiting.

As a general rule, resistance to cold injury increases with age until senescence. When exotic ornamental shrubs and other plants are introduced into a region, seeds are not planted directly in the field. Rather, they are planted in nurseries and carefully sheltered and protected for several years to get them past the juvenile stage. After they have ceased to be tender young plants and are "hardened" to some extent, they survive much better than if they are planted directly in the field. Some species of ornamentals must be 6 to 8 years old before they will survive field planting.

The hardening of transplants may be achieved by any treatment that materially checks new growth. This may be accomplished by gradually exposing plants to cold, by withholding moisture, or by a combination of these two treatments. In general, about 10 days of hardening is sufficient for most plants. This treatment is designed to produce a stocky, toughened plant in contrast to a soft, tender, "leggy" plant. Hardened plants are often darker green in color, and hardened crucifers have an increased waxy covering of the leaves. The induced cold resistance brought about by the hardening treatment in such cool-season crops as cabbage or celery may be considerable. Unhardened cabbage plants show injury at $-2.2°$ C (28° F), whereas hardened plants can withstand temperatures as low as $-5.6°$ C (22° F). In warm-season crops, such as tomato, the degree of cold hardiness imposed is slight, but hardening inures the plant to transplanting shock and hastens plant establishment under adverse conditions. The reduced growth rate enables the plant to withstand desiccation until the root system becomes established. Tender plants, which respire rapidly, have little chance of survival under warm, windy, dry conditions following transplanting. The cessation of growth under the hardening treatment, however, may severely interfere with subsequent performance; thus care must be taken to avoid overhardening. Under ideal transplanting conditions hardening may not be necessary.

HEAT UNITS

The harvest date of some annual crops can be changed by planting earlier or later. But a change in the date of planting does not always cause an equal change in the harvest date. For some crops the time required to reach a harvestable stage is a function of the total heat received. When temperatures are below the minimum required for growth, as may be the case if seeds are planted early, an earlier planting date will not necessarily affect the date of maturity.

For some crops the time required to reach the harvestable stage may be expressed in terms of temperature-time values called **heat units** by calculating time in relation to temperature above a certain minimum. If the minimum temperature for growth of a particular crop is 10° C (50° F), then a day with a mean temperature of 15.6° C (60° F) would provide 5.6 Celsius degree-days of heat units (or 10 Fahrenheit degree-days). A day with an average

temperature of 5° C (41° F) would provide zero degree-days of heat units. The harvest date can be ascertained by an accounting of accumulated heat units. In Wisconsin it has been shown that between 665 and 695 Celsius degree-days (about 1200 to 1250 Fahrenheit degree-days) are required for the 'Alaska' cultivar of pea to mature if planted in early spring. Under increasing spring temperatures it takes longer to accumulate heat units in the early part of the season than later. Assuming that all temperatures above a certain minimum have similar effects on growth, there would be a decreasing interval between planting and harvest dates as the season progresses.

The heat-unit system has a number of limitations. For example, soil temperatures more accurately indicate conditions for early growth than do air temperatures. Differences in day-night temperature shifts, day-length effects, as well as the differential effect of temperature on various stages of plant growth also affect results. In addition, all temperatures above a minimum may not have a similar effect on growth, but may act exponentially, approximately doubling many physiological processes with every 10° C (18° F) rise in temperature. The precise determination of harvest date by the accumulation of temperature data depends upon a knowledge of the general climate of an area and upon experience with harvest dates based upon planting dates for each crop. The system works rather well for some crops, not so well for others.

TEMPERATURE CONTROL

The control of environmental temperatures around plants can be extremely important in crop production. Environmental temperatures may be controlled in several ways.

Choice of site

The choice of site is often a way of controlling temperature. Sunward slopes intercept more radiation than shady slopes or level ground. Selecting these warm slopes for crop production extends the growing season in cool climates. Site selection is very important for frost protection. In order to escape the dam-

A Except in inversion, air temperature decreases with height. (Darker tones indicate cooler air.)

B Temperature inversion sets in when cool air is trapped under a warm layer. The normal temperature gradient is reversed in the inversion layer.

Figure 11-3 (A) A normal air-temperature gradient. (B) Thermal belt formed on a hillside as the result of a temperature inversion. [After R. Geiger, *Climate Near the Ground*, Harvard University Press, 1957.]

aging effects of frost, orchard sites subject to the accumulation of cold air must be avoided (Fig. 11-3).

Proximity to large bodies of water has a stabilizing effect on air temperature. Frost protection afforded by large bodies of water is due to the high specific heat of water as compared to that of the land; water absorbs and gives up heat slowly. Solar radiation penetrates water more deeply than land, and the continuous internal movement of water results in heat distribution throughout the water mass. Thus large bodies of water become "heat reservoirs" in the fall and "heat sinks" in the spring. In the winter and early spring the influence of large bodies of water keeps temperatures moderately low and prevents premature plant growth. In the late spring they may provide enough heat to prevent frost; in the fall they tend to delay the advent of frost. This temperature lag is felt mainly along the windward side of large bodies of water.

The fruit industry of the northeastern United States is localized in counties bordering the Great Lakes to take advantage of the ameliorating effects of large bodies of water. Crops frequently escape extreme low-temperature winter injury as well as crippling spring frosts in these locations. The effect of warm ocean currents is also important. England and Labrador are in the same latitude, but England is warmer because of the influence of the Gulf Stream. The southern coast of Alaska is warmed by the Japanese currents, and commercial truck gardens near the coast produce many vegetables that cannot be grown inland.

Soil mulch

Soil temperatures are frequently controlled by mulches, which are ground coverings of various kinds that reflect intense solar radiation in summer and prevent loss of heat from the soil in winter. In northern Florida two inches of pine-straw (pine-needle) mulch decreases surface soil temperatures in summer by 25° C (45° F). In more extreme but less common situations, mulches may depress surface temperatures by as much as 33° C (60° F). Plant litter or such by-products as leaves, straw, pine needles, wood chips, and hay are good insulating materials and therefore make good mulches. Not only the cost of the mulching material but the cost of application must be considered. Some materials are toxic, and others may lower soil pH excessively; consequently care must be exercised in choosing a mulch. Organic mulches have the added advantage of preventing erosion, conserving moisture, inhibiting weed growth, and contributing plant nutrients after decomposing. Because of the importance of mulches in the production of such crops as pineapples and strawberries, it is economical to use polyethylene or paper as mulching materials.

Soil temperatures can also be controlled by spreading a thin layer of sand on the ground. The sand reflects light and also serves as an insulating soil mulch.

When plant litter is burned the soil is blackened and more radiation is absorbed. Mean surface temperatures in areas that have been burned may be several degrees higher than in unburned areas. An interesting experiment was done in Siberia to determine the effect carbon black might have on hastening the melting of snow by raising the surface temperature. Carbon black spread over the snow at the rate of 4.5 metric tons per hectare (4000 pounds per acre) resulted in the earlier maturation of the crop that had been planted.

Irrigation

Irrigation can be used for temperature control as well as for supplying moisture. Water used in irrigation is relatively warm compared to the surface temperature of soils in fall, winter, and spring. When pumped into a field, it has a warming effect. Moreover, after a field has been thoroughly wetted, heat is more readily transferred from lower depths to the surface because saturated soil conducts heat better than dry soil.

Water serves not only as a vehicle for the transfer of heat but functions in other ways to control temperature. The formation of ice is accompanied by the release of large amounts of heat. The heat released helps maintain plants at the freezing temperature of water for as long as water is being frozen. Of course, once all the water is frozen, temperatures will fall again. If this principle is to be used water must be continually applied until the air temperature is above freezing. Spray irrigation is commonly used to protect strawberries from frost.

Structures

Structures of various kinds have long been used to control temperature. **Cold frames**—low, enclosed beds covered with a removable sash of either glass or plastic—are used to extend the growing season so that young plants can be started before the growing season begins. Temperatures are maintained at a relatively high level because radiant energy passes through the transparent top and is absorbed by the soil. It is lost from the soil as long-wave radiation, which is prevented from leaving the enclosure by the glass or plastic cover. As a result, heat builds up in the cold frame during the day and warms the soil, which releases its heat by convection gradually during the night and keeps plants warm.

Hotbeds are cold frames with supplemental heat, which may be supplied in a number of different ways. Today, electrical heating cables, steam pipes, and hot-water pipes are buried in the soil. In the past, the supplemental heat was supplied by the respiration of microorganisms, which decomposed a layer of organic matter (usually manure) under the plants.

Greenhouses are essentially large hotbeds in which heat is provided usually by steam (Fig. 11-4). Temperatures in winter are controlled automatically. Summer temperatures are regulated by shading, by spraying the tops with water, by pulling air in through wet pads (and thus cooling it by evaporation), and by refrigeration. Glass surfaces are sometimes painted or speckled with an easily removable white paint or whitewash to reflect a part of the solar radiation. Greenhouse maintenance is a highly specialized operation, requiring a knowledge of plant materials and practical mechanics.

Inexpensive plastics, such as polyethylene film, are now widely used for greenhouses in place of glass. They are usually attached to a frame of lumber or metal tubing. The flexible plastics used for such structures usually last only a single season before replacement is necessary. They cannot be used in areas that are very windy or subject to frequent hail. Rigid plastic sheets are also widely used in place of glass.

Cones of translucent paper or plastic, called "hot caps," are sometimes placed over the tops of individual plants as they are growing in the field. They are usually used for crops that are very tender and must be transplanted (Fig. 11-5).

The use of portable, tentlike glass sashes (cloches) over individual plants has long been common in European market gardens to facilitate early vegetable production, but the technique has fallen into

A

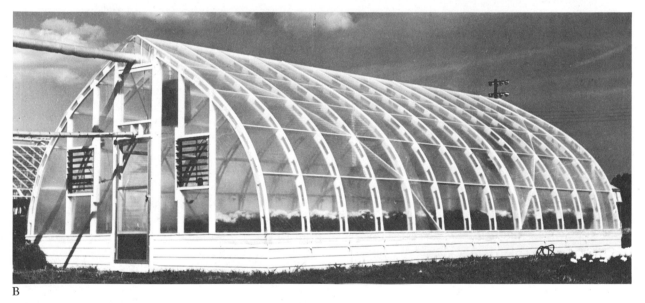

B

Figure 11-4 Greenhouses are structures covered with glass (A) or plastic (B) for temperature control in plant growth. In the glass greenhouse, fan-and-pad cooling is used as an economical system for lowering summer greenhouse temperatures. Cooling is achieved by the evaporation of circulated water through excelsior or some other material with a high surface area per unit volume. Fans opposite the cooling pads draw the cooled air across the greenhouse. The plastic greenhouse is a relatively inexpensive structure. The polyethylene sheeting is removed in the spring when temperatures get too high and is replaced in the fall. The frame can then be covered with shade cloth and the structure converted to a shade house. During the winter, temperature control is achieved with ventilators and steam carried in from the greenhouse range. [A, courtesy Acme Engineering and Manufacturing Corporation, Muskagee, Oklahoma; B, courtesy P. H. Massey, Jr.]

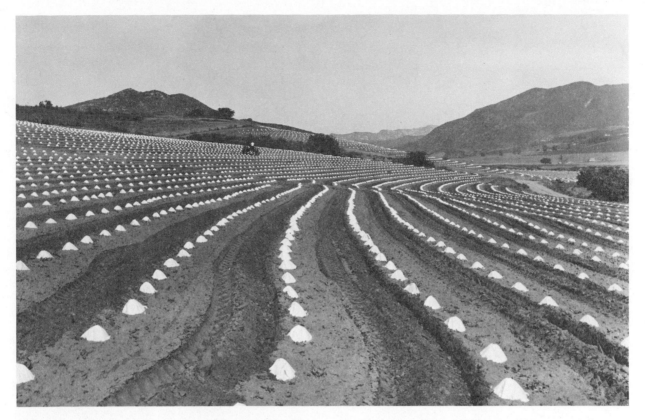

Figure 11-5 Hot caps protect early tomatoes in California's San Luis Rey district. The hot cap is made of a translucent paper and acts as a miniature greenhouse. [Courtesy USDA.]

disuse because of the tremendous amount of labor required. However the principle has been revived on a large scale in some areas, especially in Israel for winter vegetable production, with the innovation of plastic tunnels (Fig. 11-6). Tunnels are made from sheets of polyethylene laid down mechanically, usually over wire hoops. Soil may be used to seal the sides of the plastic. The higher temperature within the tunnel encourages early production. A number of systems dissipate built-up heat as temperatures increase with the approach of summer. In one system the plastic may be temporarily slipped off the hoops during periods of warm weather. In others, the plastic is first perforated for ventilation and then slit in increasingly more places at intervals as temperatures rise. Plastic tunnel and plastic mulch may be com-

bined; in some systems the plastic tunnel is converted to a plastic mulch during the growing season.

In England, it is possible to grow peach trees if they are located near a masonry wall, which will absorb radiant heat and release it gradually during the night. Thus, many English "kitchen gardens" have a welcome addition, otherwise unavailable because of the cool climate.

Although most climate-control structures are intended to increase the temperature of the plant environment, lath houses, or shade houses, are used to reduce temperatures. This kind of structure provides partial shade by admitting a fraction of the radiation that strikes the structure. Evenly spaced wooden laths or metal slats were originally used for this purpose. Today, however, plastic screening, which al-

Figure 11-6 Plastic row covers for winter cucumber production in southern California. [Courtesy Bernarr Hall.]

lows a known percentage of radiation to pass through it, is commercially available. Thus by changing the screening, the amount of shading may be controlled.

Wind machines

Air itself is a reservoir of heat that can sometimes be used to advantage. A large part of the heat of a plant is normally transferred to the air by convection, or vice versa. On calm, clear nights temperature inversions may occur. The air temperature 15 meters (about 50 feet) above the ground may be more than 2.5° C (4.5° F) warmer than that in the vicinity of the crop plants. By using an 85- to 100-horsepower wind machine during the night, a local turbulence can be created to cause air to mix and thereby to increase temperatures in the vicinity of the crop plants by a few degrees (Fig. 11-7). Changes as small as this can be critical for fruit crops during the flowering period. The wind machines merely stir up the air and make the heat in the warm upper layer available to plants. On cold, windy nights, wind machines give no protection, but during calm, clear nights, when temperature inversions occur, they can be extremely helpful. Wind machines are usually permanently mounted on towers at a level 6 to 7.5 meters (about 20 to 25 feet) above the soil surface.

Figure 11-7 Wind machines mix the warm upper air with the cool air around these citrus trees when temperature inversions exist. [Courtesy Food Machinery and Chemical Corp.]

Their elevation is important, because low areas are more subject to the accumulation of cold air than are high ones.

Smudge pots and heaters

In regions in which citrus crops are subject to occasional frost, oil smudges and heat may be used. If orange growers have a warning from the U.S. Weather Bureau that there will be a frost on any particular night, they prepare smudge pots or heaters by filling them with enough oil to burn throughout the anticipated danger period and place them in orchards in strategic locations (Fig. 11-8). Before the cold front moves in, the devices are lighted. The radiant fraction of the thermal output of heaters and smudge pots provides the most protection, although part of the convection heat is also useful, particularly under conditions of good inversion when wind machines can redirect some of the warm layer overhead

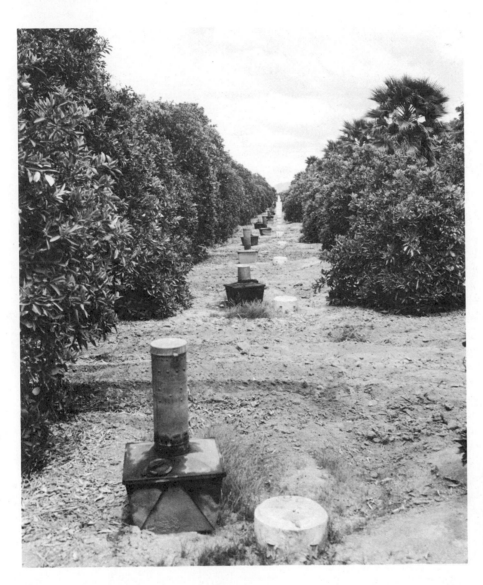

Figure 11-8 Heaters placed at regular intervals between the rows of trees in a citrus orchard are now used instead of smudge pots to protect the trees and fruit from frost. [Courtesy USDA.]

back into the orchard. Although the smoke from smudge pots may be of some value in acting as a radiation "blanket" if atmospheric conditions permit its accumulation in an inversion ceiling during the night, this benefit will be more than lost if more than one night of protection against radiation frost is required, for the smoke will reduce the incoming solar radiation in the succeeding day. "Smudging" is illegal in the citrus-growing areas of California and Washington because of the air pollution that it causes. It seems probable that, in the near future, smudging will not be allowed any place in the world. Environment-conscious people are working to outlaw such sources of air pollution.

Various kinds of heaters have been used to ward off cold temperatures in high-value fruit crops. A recent innovation is the use of solid-fuel candles. Made of solid petroleum wax, they are about eight inches in diameter and burn with a low flame for eight hours. Two such heaters beneath a grapefruit tree will raise the average air temperature within the canopy of the tree by about 4° C (7.2° F).

Manipulation of vegetation

Vegetation itself can be manipulated to control temperatures within crops. Anything that prevents the accumulation of radiation during the day will increase frost. Vegetation that shades the soil will reduce the amount of heat stored in the day. Thus a sodded or mulched area is more likely to become frosted than one under "clean cultivation." The necessity for controlling frost is one of the main reasons why peaches are not grown under permanent sod.

If plants are widely spaced, more solar radiation is admitted to the soil. Plants themselves are relatively poor at storing heat, but soil materials and water are good storage "containers." The low vegetation beneath a forest canopy may be extremely effective in reducing the solar radiation reaching the soil, for shrub covers of rhododendron, laurel, and blueberries may form a very tight layer that admits very little radiation. In both forest and field crops, the density of the canopy can be regulated by careful spacing. If the canopy is too dense after planting,

then some plants may be removed to make the spacing more appropriate. In Norway and Sweden, forests are frequently thinned to admit radiation through the canopy to the soil surface in early spring. By doing this, the growing season of conifers can be effectively extended.

Although it seems impractical at this time, it is quite possible that in the future entire fields may be warmed by subsurface systems that rely on nuclear fission as a source of heat. Water-conducting pipes of the heating system could be laid beneath the soil to carry heat in the same manner as it is carried through the floors of some houses.

Selected references

American Society for Horticultural Science, 1970. Cold hardiness, dormancy, and freeze protection of fruit crops (a symposium). *HortScience* 5:401–432. (A series of papers on the cold-hardiness of plants and techniques for protecting plants from low-temperature injury.)

Cooper, A. J., 1973. *Root Temperature and Plant Growth.* Commonwealth Agricultural Bureaus. East Malling, Maidstone Kent, England. (A small, readable book dealing with the effects of root temperature on various aspects of plant metabolism and growth.)

Levitt, J., 1956. *The Hardiness of Plants.* New York: Academic Press. (A short but thorough monograph dealing with the hardiness of plants.)

Prosser, C. L. (editor), 1967. *Molecular Mechanisms of Temperature Adaptation.* Baltimore: Horn-Shafer. (A collection of monographs dealing with the adaptation of plants and animals to heat and cold.)

Rauner, Y. L., 1977. *Heat Balance of The Plant Cover.* New Delhi: Amerind Publishing. (A monograph dealing with the heat-exchange meteorology of plant cover.)

Weber, R. L., 1941. *Temperature Measurement and Control.* Philadelphia: Blakiston. (The nature of heat and its measurement.)

Weiser, C. J., 1970. Cold resistance and injury in woody plants. *Science* 169:1269–1278. (A review of the physiology of low-temperature injury.)

Wijk, W. R. van (editor), 1963. *Physics of Plant Environment.* New York: Wiley. (A collection of technical papers on the reactions of plants to their environment.)

12
Soil

Soil is different things to different people. To most of us it is something to be washed from clothes and swept out the door. To the geologist soil is the unconsolidated, weathered part of the earth's mantle, an insignificant, small fraction of its total volume. To the civil engineer it is the medium for the support of structures and a construction material. But to the agriculturist, soil is that part of the earth's crust in which the roots of plants grow. It is a vital, living component of the environment—a component that can be manipulated to affect crop performance. When soil is misused, crops become less productive; when handled with due consideration for its biological and physical nature, it can continue to yield crops throughout countless generations of cultivation and use.

Soil has three primary functions in sustaining plant life:

1. It supplies mineral elements, serving both as a medium of exchange and as a place of storage.

2. It supplies water and serves as a storage reservoir.

3. It serves as a medium within which the roots of terrestrial plants, as well as those of many aquatic plants, anchor themselves.

One might ask whether soil is really necessary for plant growth. **Hydroponics**—the soilless cultivation of plants with nutrient solutions alone or with sand or gravel—is practiced today on a limited scale. The United States Armed Services use hydroponic gardens to provide fresh vegetables for personnel on isolated islands in the Pacific Ocean where soil is not available. Even though it is possible to maintain a limited agriculture without soil, the massive production of plant materials that the world requires cannot be accomplished economically without soil under present economic conditions.

SOIL SYSTEMS

For the purposes of crop production, soil must be considered to be a delicate balance of interwoven and interacting systems: (1) inorganic minerals, (2) organic matter, (3) soil organisms, (4) soil atmosphere, and (5) soil water.

Although it is convenient to consider each of these systems individually, they are neither separate nor independent in nature. To change one of them results in a change in all. The nature of a soil is greatly modified when one system dominates. For example, a high water table produces a kind of soil that has particular and definite characteristics, no matter where it is found in the world.

Soil minerals

The inorganic constituents of the soil—soil minerals—are ultimately derived from the parent materials, which are the underlying chemically weathered minerals or organic matter that is the basis of soil. Soil minerals may be **original,** relatively unchanged from the parent minerals (such as quartz in the form of sand), or **secondary,** formed by the weathering of less resistant minerals (such as clays). The amount of inorganic material varies greatly, from more than 99 percent of the weight of sandy and clayey soils, to as little as 1 percent in some organic soils. The inorganic component of soils consists of a mixture of particles that differ in size, in composition, and in physical and chemical properties.

Soil texture

The mineral particles of the soil can be arranged according to size from very coarse to very fine. The term **soil texture** refers to the size of the individual mineral particles. Soil particles have been classed according to size, as shown in Table 12-1. Textural designations are also used to describe soils (Fig. 12-1). The main ones are sand, silt, clay, and loam. The nontechnical terms "lightness" and "heaviness" refer to soil texture. "Heavy soils" are high in clay and other fine particles; "light soils" are low in clay and high in sand and other coarse particles.

The coarse materials such as sands and gravels are usually composed of many small particles cemented together either chemically or by a matrix material. These are bound relatively firmly and present only a

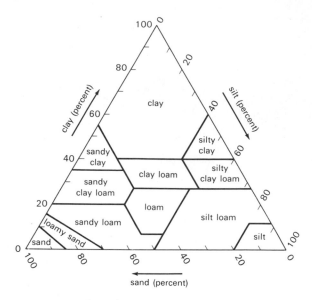

Figure 12-1 The soil-texture triangle shows the relative percentages of sand, silt, and clay in each textural class. (In the United States, the term "loam" refers to a soil with more or less equal proportions of sand, silt, and clay. As used in Great Britain, the term refers to a soil high in organic matter, a "mellow" soil.) [Courtesy USDA.]

Table 12–1 Classification of soil particles.[1]

Particle	Diameter (mm)
Coarse sand	0.5–1
Medium sand	0.25–0.5
Fine sand	0.1–0.25
Very fine sand	0.05–0.1
Silt	0.002–0.05
Clay	<0.002

Source: Soil Taxonomy, USDA Handbook No. 436, 1975.
[1]The International Classification (Atterberg) System refers only to soil particles under 2 mm:

Coarse sand	0.2–2 mm
Fine sand	0.02–0.2
Silt	0.002–0.02
Clay	<0.002

single outer surface. The physical and chemical properties of these coarse materials do not differ greatly from those of their parent materials.

Silt particles, which are smaller than sand particles, are more or less unweathered, but their surfaces are coated with a clayey matter. The properties of silt are therefore somewhat intermediate between those of sand and clay.

The clays, the smallest of the soil particles, show distinct chemical and physical properties. Clays are colloidal°—they are viscous and gelatinous when moist but hard and cohesive when dry. Their structure can only be seen with an electron microscope. Clays are composed of particles called **micelles,** which are formed from the parent materials by a crystallization process; they are not merely finely di-

---

° A colloid is a small, insoluble, nondiffusable particle larger than a molecule but small enough to remain suspended in a fluid medium without settling. Most soils contain organic colloidal particles as well as the inorganic colloidal particles of clays.

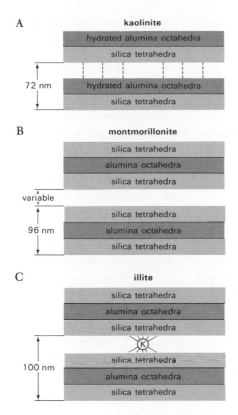

A   kaolinite
hydrated alumina octahedra
silica tetrahedra
72 nm
hydrated alumina octahedra
silica tetrahedra

B   montmorillonite
silica tetrahedra
alumina octahedra
silica tetrahedra
variable
96 nm
silica tetrahedra
alumina octahedra
silica tetrahedra

C   illite
silica tetrahedra
alumina octahedra
silica tetrahedra
K
100 nm
silica tetrahedra
alumina octahedra
silica tetrahedra

Figure 12-2   The crystalline structure of clay. (A) The simplest clay mineral is called kaolinite. The micelles of kaolinite consist of two different layers, one of silica and the other of alumina. Kaolinitic clay micelles are relatively large and are bound together tightly. The distance between the alumina and silica layers is rigidly fixed and does not increase when water is adsorbed on the surfaces of the individual micelles. The internal space is not available for surface reactions. Such clays do not shrink greatly when dry or expand much when hydrated. (B) Montmorillonite is more complex, and each micelle consists of an alumina layer sandwiched between two layers of silica. These clays are not bound together tightly, and they swell when wet because the hygroscopic surfaces between layers adsorb water and force the layers apart. All surfaces can adsorb water and minerals. (C) The micelles of illite consist of an alumina layer between silica layers, and adjacent layers are held together by potassium atoms. Surface adsorption can take place on the surfaces between crystals to a limited degree. Illite is less abundant than kaolinite or montmorillonite.

vided rock. The micelles are sheetlike (laminar), with internal as well as external surfaces, and tend to be held together by chemical linkages or ions between the plates. Their tremendous surface area relative to their volume is one of their most significant features.

The structure of clay micelles may be complex (Fig. 12-2). Clay particles are negatively charged and thus attract, retain, and exchange positively charged ions. The adsorbed water on clay particles acts both as a lubricant and as a binding agent. Clay platelets behave like a stack of wet poker chips. This to a large degree explains the plasticity of clay. Wet clay soils low in organic matter and low in weakly hydrated cations, such as calcium, become sticky or puddled. The unique chemical and physical properties of clays are responsible for most of the important properties of soils.

Soil texture affects the retention of water and the rate of water infiltration. Coarse soils permit the rapid infiltration and percolation of water, so that there is no surface runoff even after a heavy rain. In contrast, clay soils are so finely textured that very little water penetrates to lower levels, especially after the surface clays become wet and expand. Coarse soils, however, are incapable of retaining large amounts of water. The range of capacity for holding moisture available for plant growth among soils of different textures is given in the following tabulation:

Soil type	Water capacity (millimeters per meter of soil)
Very coarse texture—very coarse sands	33–63
Coarse texture—coarse sands, fine sands, and loamy sands	63–104
Moderately coarse texture—sandy loams and fine sandy loams	104–146
Medium texture—very fine sandy loams, loams, and silt loams	125–192
Moderately fine texture—clay loams, silty clay loams, and sandy clay loams	146–208
Fine texture—sandy clays, silty clays, and clays	133–208
Peats and mucks	167–250

Soil structure

Soil structure refers to the gross arrangement of the soil particles into aggregates. A soil may have either a simple or a compound structure. Sands and gravels, examples of soils with a **simple structure**, have very little cohesion, plasticity, and consistency (the resistance of the particles in the soil to separation). Simple-structured soils are usually composed of materials that are relatively resistant to weathering, such as quartz sand. They are also said to have a **single-grain structure.**

Most agricultural soils have a **compound structure;** their particles aggregate, or stick together. Several distinct sizes of compound structures are recognized.

Structural type	Aggregate diameter (millimeters)
Columnar	>250
Blocky	5–25
Granular	3–5
Crumbly	1–3
Massive	Completely puddled or compacted

Soil structure develops when small colloidal soil particles clump together or **flocculate** into granules. Granulation is promoted by freezing and thawing, the disruptive action of plant roots, the mixing effects of soil fauna, the expansion and contraction of water films, and the presence of a network of fungal hyphae. However, unless the granules are stabilized by coatings of organic matter or by their own electrochemical properties, they will coalesce into clods.

Good soil structure is very important for agricultural soils. Highly granulated soils are well aerated and have a high water-holding capacity because of the increased volume of the soil pore space. The **pore space** of soil is occupied by water and air in varying proportions, the soil acting as a huge sponge. The total pore space of soil, which is typically about 50 percent of the total volume, is not as important as the size of the individual pores. Clay soils have more total pore space than sandy soils, but because of the small size of the pores in clay soils, air and water move through them slowly. When the small pores of clay soils become filled with water, aeration is greatly limited. Large pore spaces become filled and are drained by gravity, whereas small pores absorb and retain water by capillary action. Capillary water is of the utmost importance to the plant: it is the soil solution most used by plants.

The crumbly nature of good agricultural soils depends on soil texture and on the percentage of humus (decomposed, stable organic matter). Clay soils low in organic matter typically have poor structure. In order to maintain good compound structure in clay soils, they must be carefully managed. If worked when too wet, the structure may be damaged. (Fig. 12-3). When clods are exposed they become dry, hard, and difficult to work back into the soil. In heavy soils it is necessary to add organic matter to maintain good structure. In sandy soils, where structure is not as critical, it is necessary to add organic matter to increase its water- and nutrient-holding capacities.

Figure 12-3   Puddled clay soils may form large cracks when they dry out. This condition indicates an unproductive soil that has been poorly managed. [Courtesy USDA.]

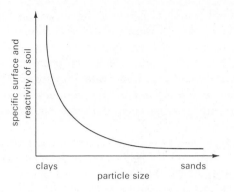

Figure 12-4 The reactivity of soils depends on the total surface area of the soil particles per unit volume of soil. The total surface area varies inversely with particle size. [Courtesy USDA.]

Exchange capacity

The capacity of a soil to retain and exchange such cations as H^+, Ca^{2+}, Mg^{2+}, and K^+, is called its **exchange capacity.** Exchange capacity, a measure of the chemical reactivity of the soil, varies inversely with particle size (Fig. 12-4). Because a given volume of small soil particles has much more surface area than an equal volume of large particles, fine soils accumulate and retain many times more cations than do coarse soils. The soil's colloidal particles, clay and humus, are negatively charged and attract cations. When these particles in a soil are saturated with hydrogen ions, the soil has a strong acid reaction. Soil reaction and pH are discussed more fully in Chapter 16.

The exchange capacity of the soil's colloidal particles is of tremendous importance. Nutrients that would otherwise be lost by leaching are held in reserve and when exchanged become available to the plant.

The process of cation exchange is not a random one. Cations differ in their ability to replace one another; if present in equal amounts,

H^+ replaces Ca^{2+} replaces Mg^{2+}
replaces K^+ replaces Na^+.

If one cation is added in large amounts it may replace another by sheer force of number (mass ac-

tion). This is largely what occurs with the addition of fertilizer.

The release of hydrogen ions in soils tends to promote the exchange of cations, making them available to plants. Hydrogen ions are made available by the dissociation of carbonic acid formed from the carbon dioxide released by roots and from the even larger amount released by the enormous population of microorganisms in respiration and in the decomposition of organic matter:

$$CO_2 + H_2O \rightleftharpoons H_2CO_3 \rightleftharpoons H^+ + HCO_3^-$$

The cations are replenished by the decomposition of the inorganic fraction of the soil, by the degradation of organic materials, and by the application of fertilizer. The cations in a productive soil are in equilibrium between the soil particle, the soil solution, the soil microorganisms, and the plant.

The cation exchange capacity of a soil is expressed in milliequivalents (meq), of H^+; that is, the number of milligrams of H^+ that will combine with 100 grams of dry soil. The exchange capacity of a soil depends on the percentage of humus it contains and on the percentage and composition of its clays. Clays differ markedly in their ability to exchange cations. Kaolinite has an exchange capacity of 10 meq/100 grams, whereas montmorillonite clay has an exchange capacity of about 100 meq/100 grams. The exchange capacity of humus ranges from 150 to 300 meq/100 grams. The ranges of exchange capacity for various soils are given in Table 12-2.

Table 12–2 Ranges of cation exchange capacity for various soil types.

Soil type	Cation exchange capacity (meq/100 g)
Sands	2–4
Sandy loams	2–17
Loams	7–16
Silt loams	9–26
Clay and clay loams	4–60
Organic soils	50–300

Source: Adapted from N. C. Brady, *Nature and Properties of Soils*, Macmillan, 1974.

The ability of a soil to supply mineral ions for absorption by plants is a measure of its **fertility.** It is quite possible for soils to contain relatively large amounts of minerals and yet be infertile because the ions are unavailable to plants. In a region with moderate to large amounts of rainfall, simple soils are usually exceedingly low in fertility because minerals are readily lost by leaching. Such crops as peanuts, which are grown in sands, must be fertilized frequently and carefully.

Soils may be fertile without being **productive.** Some desert soils have an exceedingly high natural fertility but must be irrigated to render them capable of producing crops. A fertile soil can be productive only when moisture, temperature, and other environmental factors are in good balance. The aim of soil management must be to make maximum use of the productive capacity of the soil.

Soil organic matter

Soil organic matter is the fraction that is derived from living organisms. Most apparent is the **litter** on the surface. It is composed of undecayed leaves, branches, reproductive parts, and other residues from the top parts of plants. **Duff** refers to partially decayed litter. It is frequently matted together with fungal mycelia, and in this condition it is called **leaf mold.** Duff forms when soil is moist enough to supply the water essential for microbial activity and when litter is thick enough to prevent evaporative water losses. Leaf mold is an important component of forest soils, but it is never found in cultivated agricultural soils. Plant roots and the excreta, sheddings, and bodies of soil organisms, although not as apparent, also contribute to soil organic matter.

The upper layers of the soil are often high in an organic fraction called **humus.** Humus is relatively resistant to further breakdown and decomposition. Humus, unlike mineral colloids, is noncrystalline. The primary sources of humus in the upper soil layers are surface litter and plant roots. As surface litter is broken down by mechanical action and decomposed into fine particles by microorganisms, it is washed down into the soil, where it becomes a part of the soil complex. The washing action can only affect the surface of most soils. The decomposition of dead roots provides organic matter throughout the upper soil. Prairie soils, which receive small amounts of precipitation, have a low rate of biological decomposition, so that organic matter from dead grass roots accumulates to make them dark in color, friable, and extremely fertile.

Perhaps one of the most important contributions of organic matter to soil is its water-holding capacity. Organic matter acts like a sponge: it can absorb large amounts of water relative to its weight. Because it is porous, it also readily permits the infiltration of water. Organic matter is also a source of mineral elements, which are made available when it decomposes. The decomposition by bacteria, fungi, and other organisms of organic matter to form water and carbon dioxide with the release of minerals is called **mineralization.** This is an important aspect of **chemical cycling** in stands of vegetation. Chemical cycling consists of (1) the absorption of minerals through roots and their incorporation into chemical compounds of various kinds in plants, (2) the death of plants and plant parts, (3) and the decomposition of the plant material and the release of its contained minerals into the soil. The minerals are then reabsorbed and recycled. The high adsorptive capacity of organic matter is also important in the retention and exchange of mineral cations. When organic matter decomposes or when fertilizers are added to the soil, the mineral elements made available are subject to leaching. Organic matter may retain large quantities of minerals and thus prevent their losses from the soil. Organic matter may account for as much as 90 percent of the absorptive and adsorptive capacities of sandy soils.

Organic matter helps maintain the structure of cultivated soils. Finely divided organic matter covers mineral particles and keeps them from sticking together. Clay soils with an appropriate quantity of organic matter are less inclined to be sticky and are more readily cultivated. Crop growers say that soils that are friable and easy to cultivate have good **tilth.**

Soils are frequently classified according to their content of organic matter. One such classification is as follows:

Soil name	Percent organic matter
Mineral	0–10
Muck	10–40
Peat	40–100

Although most agricultural soils are mineral soils, mucks and peats may also be rendered extremely productive under proper management. Mucks and peats develop when organic matter from living plants is covered by water and does not decompose. Once these soils are exposed to air by cultivation and drainage, bacteria and fungi begin to decompose them, with the result that the soil surface is actually lowered. In the Florida Everglades, drained and cultivated peat soils actually subsided nearly 2 meters (about 6 feet) during 40 years of cultivation (Fig. 12-5).

Soil organisms

Soil organisms play an important part in soil development. In addition to the roots of higher plants, the soil is inhabited by a wide variety of plant and animal life (Fig. 12-6). In fact, a soil does not usually develop until the inorganic material is "invaded" by various kinds of organisms. The total weight of soil organisms (excluding higher plants) in the upper 30 centimeters (about 1 foot) of fertile agricultural soils is impressive, as much as 7000 kilograms per hectare (about 6250 pounds per acre), which is equivalent to the weight of 20 to 30 marketable hogs (Table 12-3). This is, however, only about 0.1 percent of the weight of the soil that the organisms occupy.

Higher plants must be considered the principal soil organisms. Roots of trees and other higher plants penetrate crevices in rock, expand them, and split the rocks by the tremendous forces they exert when growing, thereby performing an important function in the continuing process of soil formation. Roots exude many kinds of organic acids and other substances that hasten the solution of soil minerals and make them available to plants. Living roots give off CO_2, which raises the carbonic acid content of the soil solution and increases the rate at which soil minerals dissolve. When leaves fall, or when plants die, their organic matter becomes incorporated into the soil, contributing to its fertility. Channels remaining after dead roots have decayed serve as pathways for the movement of soil water.

Bacteria are present in all soils. Estimates of the weight of living and dead bacteria in forest soils have been as great as 6300 kilograms per hectare (about 5600 pounds per acre), whereas fertile agricultural soils typically contain less than 20 percent of this.

Figure 12-5 The subsidence of drained organic soils in Florida over a period of 40 years. Gravity drainage began in 1914. It was necessary to install pumps for drainage in 1927 in order to keep surface soils dry enough for cultivation. [Courtesy USDA.]

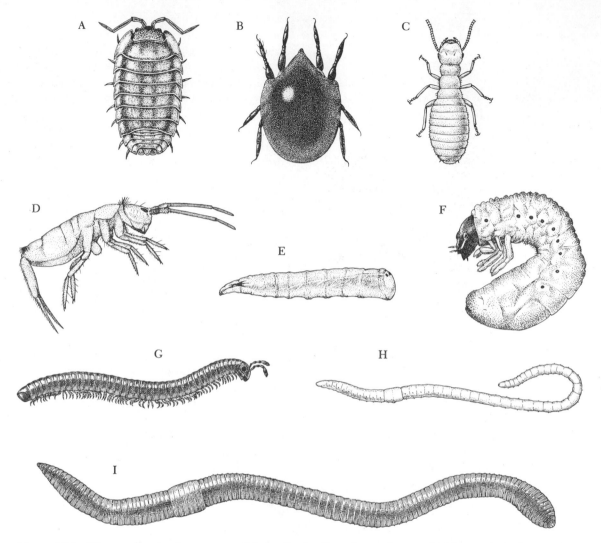

Figure 12-6 Nine small animals are responsible for the fertility of most forest soils. They include the wood louse (A), the oribatid mite (B), the termite (C), the springtail (D), the fly larva (E), the beetle larva (F), the millipede (G), the enchytraeid worm (H), and the earthworm (I). [From C. A. Edwards, "Soil Pollutants and Soil Animals." Copyright © 1969 by Scientific American, Inc. All rights reserved.]

Bacteria in the soil decompose organic matter produced by plants (and other organisms), releasing the minerals they contain, which then become available for another cycle of plant growth. Some kinds of bacteria, such as those in the genus *Azotobacter*, convert molecular nitrogen from the atmosphere into nitrogenous compounds that can be used by plants. Other species live in symbiotic association with the roots of certain kinds of plants and perform the same function. **Nitrogen-fixing bacteria** supply nitrogen to the plant while the plant supplies carbohydrates to the bacteria. Some bacteria play roles in soil formation and degradation, but these processes are poorly understood.

Table 12–3 Average weight of organisms in the upper 30 centimeters of soil (in kilograms/hectare).

Organism	Low	High
Bacteria	560	1120
Fungi	1680	2240
Actinomyces	895	1680
Protozoa	225	450
Algae	225	335
Nematodes	28	55
Other worms and insects	895	1120
Total	4508	7000

Source: O. N. Allen, *Experiments in Soil Bacteriology,* Burgess, 1957.

Not all soil bacteria are helpful. Many species are **pathogenic,** causing crop diseases and large economic losses. Other harmful types oxidize ferrous iron to the slightly soluble ferric form, thereby contributing to the formation of soil horizons called **hardpans,** which are extremely rich in iron and quite hard. Hardpans may prevent the drainage of excess water and inhibit root penetration.

Fungi perform many of the same functions that bacteria carry out in decomposing organic matter and in cycling nutrients. Many soil-borne fungi are also pathogenic. **Saprophytic fungi** decompose the dead organic matter of soils. When temperature, moisture, oxygen, and soil acidity are in the right balance, these fungi decompose protein and cellulose, lignin, and other carbohydrates, changing them to humus. Even the bark of trees, which is highly resistant, is decomposed by fungi.

Many species of fungi live in such close symbiotic association with roots (particularly of woody plants) that organic and inorganic materials pass back and forth. Such an association is known as a **mycorrhiza.** Water and minerals move from the fungi to the roots, and carbohydrates and other organic materials move in the other direction. Some mycorrhizal fungi actually penetrate the interior of the roots; others remain on the outside. The importance of these fungi is becoming increasingly apparent. Most species of

pines cannot survive without them, and there is increasing evidence that they may be associated with the roots of practically all plant species.

Most types of algae are aquatic, but several kinds, mainly the blue-green algae (division Cyanophyta) and the green algae (division Chlorophyta), are also found in the soil, where they hasten the weathering of soil minerals. Algae normally occur near the soil surface, as they require light for photosynthesis. **Lichens** are made up of an alga and a fungus living in a symbiotic relationship. They occupy a special niche in the process of soil formation. They become established on rocks, which they very slowly dissolve, and also serve as traps for dust, which accumulates and may form the soil in which other plants become established. Organic acids are leached from living and dead lichens, increasing the rate of solution of the rocks to which they are attached. In making possible the growth of other plants, lichens hasten their own demise, for they soon become covered with litter and die.

Arthropods (phylum Arthropoda) are among the more obvious soil fauna (see Table 18-1). Into this large phylum fall the crayfish (important in poorly drained areas of the southeastern United States) mites, ants, centipedes, millipedes, sow bugs, insects, and numerous other species that physically cultivate the soil and contribute to its organic matter when they die. They are numerous and have a short life span and thus contribute to the soil hundreds of kilograms of organic matter per acre.

Ants, in particular, cultivate great amounts of soil. In parts of Germany, the red wood ant has been introduced into many forests to compensate for centuries of soil degradation due to timber cutting. Ants also loosen and aerate the soil and destroy many kinds of insects that are harmful to plants. Many ants and termites are severe pests in tropical areas. The large termite mounds in Brazil seriously interfere with cultivation. The leaf-cutting ant is one of the most destructive of tropical pests. Mites and myriapods play an important role in the breakdown and decomposition of forest litter.

The segmented worms (phylum Annelida), commonly called earthworms, and round nonsegmented eelworms, or nematodes (class Nematoda of the phylum Aschelminthes), are both important in agricultural soils. An early study of earthworms by the famous biologist Charles Darwin concluded that about 0.25 to 0.5 centimeters (0.1 to 0.2 inches) of subsurface soil may be brought to the surface each year by earthworms and deposited as castings. Nematodes are economically important because some of them are pathogenic to crop plants and others distribute parasitic fungi and increase the extent of root-rot diseases. Some species kill bark beetles and other insect pests; some give rise to serious disorders in mammals and birds.

A number of vertebrate animals, including many species of burrowing mammals, must be considered part of the soil fauna. Their effect is mainly one of cultivation, although their excrement contributes organic matter to the soil. Some, however, also destroy crops. Badgers, gophers, moles, voles, foxes, shrews, mice, ground squirrels, woodchucks, and even some species of birds, such as woodcocks, frequent the soil for all or a part of their lives. When present in great numbers they may become serious pests. Moles, for example, do considerable damage to turf grasses and certain farm crops. Beavers, through their dam building over the millennia, have had a profound impact on the forest soils of boreal areas.

Soil atmosphere

The soil atmosphere exists in the pore spaces that are not filled with water. These pores contain the same gases as the atmosphere above the ground, but in different proportions. The soil atmosphere is not necessarily a continuous system, for there may be isolated, unconnected pore spaces.

The humidity of the soil atmosphere is near 100 percent much of the time. The carbon dioxide content of the soil is greater than that of the air above it because of the decomposition of organic matter, and it increases with depth because carbon dioxide diffuses very slowly into the air above the soil. Conversely, the oxygen content of the soil atmosphere is less than that of the air above the soil, and it decreases with depth. Oxygen in the soil is used in respiration carried on by roots and microorganisms, and is slowly replaced by diffusion from the atmosphere.

As water is added to a soil, the air in the soil is squeezed out. Consequently, plant roots may be deprived of oxygen in flooded or very wet soils. For some plants even a few days of flooding may be disastrous, especially during the growing season.

Perhaps one of the most characteristic features of the soil atmosphere is its variability. Because the distribution of dead roots, living roots, and microorganisms and the structure of the soil are not uniform, the rates of gas production and rates of exchange with the atmosphere are not equal, and show great seasonal variation. Conditions in cultivated agricultural soils are likely to be much more uniform than in forest soils.

Soil moisture

Soil moisture includes free water as well as capillary water, hygroscopic water, and water vapor. Soil moisture is discussed at length in Chapters 13 and 17.

SOIL ORIGINS

Soil is formed as a result of all the interacting forces that affect the parent rock materials, air and water movement of particles, and the composition and fate of living organisms that inhabit it. Although much has been written about the fascinating subject of soil genesis, only a few of the most important contributing factors can be considered in this brief discussion.

Parent materials

The **parent materials** of soils are the rocks or unconsolidated materials from which the mineral components of soils have been derived. The parent materials determine the physical and chemical properties of soils. The wide variety of soils found within relatively small areas frequently reflects a diversity of parent materials. **Residual materials** are formed by the weathering of rocks that have not been moved from the place where they were formed. **Transported materials,** which are often more important than residual materials, are those that have been moved to the place of soil formation by wind, water, gravity, ice,

Table 12–4 Representative types of rocks of different origins.

Igneous	Sedimentary	Metamorphic
Basalt	Chalk	Gneiss (from granite)
Granite	Conglomerate	Marble (from limestone or chalk)
Obsidian	Limestone	Quartzite (from sandstone)
Pumice	Sandstone	Schist (from micaceous shales)
Rhyolite	Shale	Slate (from shale)

or a combination of these forces. When the Pleistocene glaciers were growing in size and moving southward in the Northern Hemisphere, great quantities of rock and other materials were scooped up on the front edges of the ice sheets and were deposited at great distances from their sources when the ice melted. Locally, landslides deposit unconsolidated materials downslope, which later become the **colluvial** (gravity deposited) parent materials of soils. Soils deposited in the bottomlands along rivers are called **alluvial.**

The three main kinds of rocks are igneous, sedimentary, and metamorphic (see Table 12-4). **Igneous** rocks are formed when molten lava or magma solidifies. **Sedimentary** rocks are formed when sediments that have accumulated in water or have been deposited from the air are compressed and cemented into rocks. **Metamorphic** rocks are formed when igneous, sedimentary, or previously metamorphosed rocks are altered and recrystallized by heat or pressure, or occasionally infiltrated by solutions of minerals that subsequently precipitate.

Weathering and soil formation

Chemical processes

Many changes take place in the weathering of parent materials, both on the surfaces of rocks and within them. The main chemical processes are hydrolysis, carbonation, oxidation, and hydration.

Hydrolysis is a decomposition process in which water is one of the reacting agents. It can be illustrated by the hydrolysis of orthoclase feldspar (potassium aluminum silicate) by carbonic acid and water to produce kaolinite, an important clay mineral:

$$2\ KAlSi_3O_8 + H_2CO_3 + H_2O \rightarrow$$
$$Al_2Si_2O_5\ (OH)_4 + K_2CO_3 + 4\ SiO_2$$

Carbonation may be illustrated by the decomposition of calcite (calcium carbonate) to calcium bicarbonate:

$$CaCO_3 + CO_2 + H_2O \rightarrow Ca(HCO_3)_2$$

Calcium bicarbonate is highly soluble and is readily leached from soils.

Oxidation involves the loss of electrons to a receptor, which is frequently oxygen. Iron is a familiar substance that is easily oxidized. Iron pyrite is an important soil mineral that is oxidized to yield a hydrated iron oxide and sulfuric acid:

$$4FeS_2 + 10H_2O + 15O_2 \rightarrow 4FeO(OH) + 8H_2SO_4$$

Hydration is the combination of molecular water with a compound. In the hydrated form it may be more subject to weathering processes. The mineral anhydrite (calcium sulfate) is hydrated to gypsum (hydrous calcium sulfate) by the addition of water:

$$CaSO_4 + 2H_2O \rightarrow CaSO_4 \cdot 2H_2O$$

Temperature

Temperature is important in the weathering of parent materials because changes in temperature cause differential expansion and contraction of minerals and rocks. Since rocks are aggregates of minerals, each of which has a different coefficient of expansion, changes in temperature result in internal stresses that cause cracks. Even large boulders can be rent asunder by relatively small changes in temperature.

In rocks exposed to solar radiation, a very large temperature gradient is established within the first few millimeters beneath the surface. When this happens, the hot outer surfaces may expand and flake off (**exfoliate**). American Indians and members of other native cultures used the same principle to fashion arrowheads by repeatedly dropping water on heated flint on the spot where a chip was to be removed.

Low temperatures cause water to freeze in the cracks of rocks and split them apart. When water freezes it exerts a force of about 1465 metric tons per square meter (150 tons per square foot). Freezing is probably the main weathering agent at high latitudes and altitudes. The rubble formed by freezing has characteristic sharp edges.

The dissociation of water molecules into hydrogen (H^+) and hydroxyl (OH^-) ions is much greater at high temperatures than at low temperatures. Since H^+ ions are particularly effective in dissolving minerals, the rate of weathering is faster when temperatures are higher. If soil is frozen or very cold throughout the year, bedrock may be weathered only a few centimeters deep. On the other hand, weathering in the tropics may extend to a depth of 50 meters (more than 160 feet).

Rainfall

Soil moisture is obtained in most areas through rainfall, although fog and dew are important in some parts of the world. The pounding of raindrops over thousands of years plays an important role in the weathering of exposed rocks and in the erosion of weathered materials. Such effects are minor, however, when compared to the effects of soil moisture in leaching dissolved minerals and providing a medium in which soil chemicals react to form complex compounds of many kinds. And these effects in turn are minor when raindrop joins raindrop to form streams, which erode and grind away mountains of material, redistributing them hither and yon.

Rainfall is frequently the most important climatic element in soil development. In regions of moderate to high rainfall, soil development is likely to be greater than in those where rainfall is sparse. Desert soils are usually poorly developed, yet, in adjacent areas with adequate rainfall to sustain plant growth, soils originating from identical parent materials may be well developed and productive.

Topography

Topography frequently plays an important part in soil development. If the land is flat, low, and poorly drained, water may accumulate and drain so slowly that biological activity will be arrested for long periods. Under such conditions, soil material, both inorganic and organic, may remain in a state of preservation until the land is drained. When topography is very steep, parent materials may be subject to erosion by water, wind, and ice to such an extent that soil material is removed from the site as soon as it is weathered from the parent material. The most favorable topographic situation for soil development is one in which the slope is steep enough to carry off excess water, but not so steep as to permit the removal of weathered materials that lack the protection of a vegetative cover. In addition, for the development of "normal soil" the water table should be below the weathered parent material so that roots can occupy it to the greatest possible extent.

Time

The factors that contribute to the development of soil are straightforward. They fall into a discrete pattern of events. Yet, unless the time element in soil formation is recognized, the significance of the events that take place cannot be truly appreciated. A mature soil is not formed in a decade, or even ten decades, but over thousands of years. Soil development and vegetational succession are related processes. When vegetational development reaches a **climax condition,** in which the vegetation type is in balance with its environment and is relatively unchanging, the soil will also have reached full development and will be in balance with its environment. The soil is then said to be **mature.** There is a large variation in the absolute length of time required for complete maturation. In the tundra of the far north, cold weather precludes rapid biological and physical weathering; consequently, five to ten thousand years might be required to form a soil only a few inches deep. In the tropics, however, the much faster rates of reaction can lead to the development of soils that may be five to ten feet deep in a much shorter period of time.

The age of a soil is determined by the degree to which it has developed toward maturity. It may be chronologically young (in terms of absolute years) and yet may be mature in terms of its zonal structure—the soil profile.

The soil profile

In excavating the soil to any appreciable depth, it is necessary to dig through distinct zones called **horizons.** The morphology of these horizons is called the **soil profile.** Each soil type has a distinctive profile that is determined by the particular combination of factors that produced the soil. By the time most forest and prairie soils are mature they have developed three major horizons, designated (from the top down) A, B, and C. The underlying bedrock (usually but not always the true parent material) is called the R horizon.

The A horizon is the zone of leaching (eluviation). Roots, bacteria, fungi, and small animals (for example, earthworms and nematodes) are most abundant in this horizon. It is poor in soluble substances and has lost some clay and some iron and aluminum oxides.

The B horizon is the zone of accumulation (illuviation). It is less abundant in living matter. It is higher in clay and in iron and aluminum oxides, and is thus stickier when wet and harder when dry.

The C horizon consists of weathered rock material, often true parent material. When hardpans are present (as in prairie soils and the soils of wheat- and corn-raising regions) they are a part of this horizon.

The R horizon is the underlying consolidated bedrock, such as limestone, granite, or sandstone, from which the overlying horizons were formed.

Often it is desirable to subdivide the A or B horizons further. They are then each divided from top to bottom into three strata: A_1, A_2, and A_3; B_1, B_2, and B_3. The A_1 stratum is often dark in color, owing to its high content of organic matter. The middle stratum of a horizon (A_2 or B_2) is always the typical one for that horizon; the others tend to grade into their neighbors. The layer of partially decomposed duff immediately above the A horizon can be referred to as the A_0 horizon, and the fresh litter above that as the A_{00} horizon. Not all profiles show complete development, and soils that have been used for agricultural purposes for long periods of time are especially likely to be lacking in one or more of the uppermost horizons. Their most prominent feature is the plow horizon—the horizon that has been repeatedly disrupted by plowing, disking, and harrowing.

Soils with well-developed profiles are called **zonal,** or **normal soils.** Zonal soils are found where the factors of soil formation are conducive to profile development. **Azonal soils** have no distinct profile and are frequently alluvial (water deposited) or colluvial (gravity deposited) soils of recent origin. **Intrazonal soils** are those that are limited in extent and are under the control of a local factor, such as high salt content or poor drainage, that does not permit the development normally associated with zonal soils.

Some soils have been **truncated.** That is, they consist of only a portion of the original profile, the upper portion having been lost through erosion (Fig. 12-7). Most agricultural soils are truncated remains of the original soil that developed under natural vegetation, and some are completely new, having been formed by the addition of soil or organic matter from another location in order to increase productivity. At times, truncated soils are "buried" under soil deposited by wind or water (Fig. 12-8).

Major soil groups

Differences in soil-forming processes lead to different zonal soils. The major soil groups are based on climatic differences. In the humid regions three broad groups have been recognized: tundra, podzolic, and lateritic soils. In the arid and semiarid regions specific soils have developed in response to low rainfall. These include the chernozems, or black soils, and various desert soils.

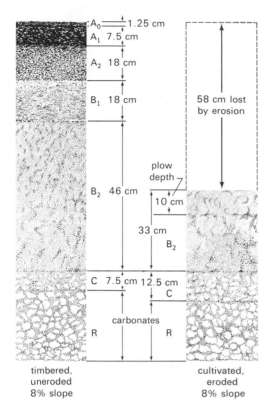

Figure 12-7 Uncontrolled erosion can result in the loss of productive topsoil, as shown by these profiles taken in Miami silt loam on adjoining fields. The profile at the left was taken on a field still covered by virgin timber; the profile on the right was taken on a cleared field that had been farmed for 50 to 75 years. [Data from USDA.]

Figure 12-8 Silt was deposited on a flood-plain soil, and a second soil developed from the silt. The soil whose former surface shows at the 8-foot level is called a buried soil. [Courtesy USDA.]

The **tundra soils** are formed in cold climates, where biological activity is minimal. Organic matter accumulates on the surface, and the deeper horizons remain frozen throughout the year (permafrost). In winter, tundra soils are completely frozen. Their horizons are only poorly differentiated. These soils are of little agricultural significance.

In cold, humid climates a particular combination of soil-forming processes, collectively called **podzolization,** produces **podzolic soils. Podzols,** the most extreme type, have a deep accumulation of litter and humus, a strongly acid reaction, and an upper horizon from which iron and aluminum oxides (the so-called sesquioxides) have been leached. (The name "podzol" is derived from the Russian and means "ash beneath," with reference to the color of the A_2 stratum.) Fungi are the main soil-forming organisms. True podzol soils are found at the higher latitudes and coincide with the boreal forest. The mild summers and long, cold winters in these latitudes permit the accumulation of the litter and humus necessary

SOIL 271

for their formation. Even in temperate climates, however, and sometimes under special conditions in the subtropics, podzolization is a factor. The widespread forests of the temperate zone grow predominantly on podzolic soils. They are potentially good agricultural soils, topography permitting, but they are easily degraded by poor cropping practices.

In the humid tropics and subtropics the combination of soil-forming processes is collectively called **laterization** and forms **lateritic soils. Laterites,** the extreme type, have a very shallow accumulation of litter and humus, a neutral soil reaction, and an accumulation of iron and aluminum oxides in the upper horizons. Bacteria are the primary soil-forming organism. Laterites are typical of tropical rain forests, and are characterized by extremely intensive chemical weathering. Even the most resistant compounds, such as silica, are broken down and leached by the heavy precipitation in the tropical rain forests. These soils are so deeply weathered that some geologists regard laterization as a geologic rather than a soil-forming process. The name "laterite" is derived from the Latin word *later*, which means brick; these soils are widely used as building materials.

Some soils in climates with low precipitation accumulate carbonates on and near the surface. This process, called calcification, tends to produce very alkaline soils, which are characteristic of deserts and steppes.

Podzolization and laterization are compared in Table 12-5. Figure 12-9 shows the relation of all three soil-forming processes.

Table 12–5 Comparison of podzolization and laterization in soil formation under optimal conditions.

Variable	Podzolization	Laterization
Climate	Cold	Warm to hot
Natural vegetation	Conifers	Tropical rain forest
Most active soil-forming microorganisms	Fungi	Bacteria
Raw humus	Abundant	Sparse
Soil reaction	Acid (pH 4–4.5)	Nearly neutral (pH 6.5–7)
Earthworms	Scarce	Abundant
Solubility of silica	Low	High
Iron and aluminum	Leached out of A horizon	Accumulate in A horizon
Hydrolysis	Weak	Intense

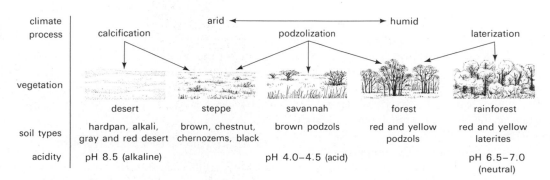

Figure 12-9 The various soil-forming processes are not discrete. They form a continuum, so that some soils are the result of a combination of processes, while others are more clearly a function of a single process. [Courtesy USDA.]

Chernozems are formed in areas that receive 40 to 60 centimeters (15 to 25 inches) of rain per year. The natural vegetation is grass. The upper horizon is dark due to the accumulation of organic matter. Chernozem means "black soil." In the temperate zones these are among the most fertile agricultural soils, and produce much of the world's grain.

Desert soils are only slightly weathered and leached because of low rainfall. Nutrients other than nitrogen are present in moderate or large quantities. As a result, under irrigation and proper management these soils may become very productive.

SOIL CLASSIFICATION

Soils may be classified on the basis of genetic origin—that is, their parent materials—or on the basis of morphology. The National Cooperative Soil Survey initiated a new system called the United States Comprehensive System of Soil Classification, which contains 8000 soil series and about 80,000 soil types and phases. This system, which took 15 years to devise, does not alter the basic content or form of the soil survey report that soil conservationists and land use planners have used in the past. Neither does it require major changes in the technical guides for engineers, biologists, and other users of soil survey information. It does, however, permit a higher degree of accuracy in classifying soils, permit greater efficiency in mapping, and allow the classification of soils that for various reasons could not be included in other systems.

The classification system used previously was developed and published in 1936 by C. F. Marbut, who was chief of the U.S. Soil Survey at that time. A drastic modification, outlined in the 1938 Yearbook of Agriculture (*Soils and Men*) was used until 1965. The Comprehensive System differs from all other systems in two important respects: (1) the nomenclature of the higher categories is entirely new, and (2) definitions of the classes are much more quantitative and specific.

The new system emphasizes the properties of soils as they exist today. Previous classifications were based on assumptions of what the soil was when the country was settled, even though cultivation and erosion might have changed them greatly. It was always difficult to get agreement about the genesis of many soils. Thus, the new classification is much more utilitarian than the older ones.

A new language was created for the higher categories of classification so that there would be no confusion with names or words formerly in use. Soil scientists selected the diagnostic properties for each class, and classicists from the University of Illinois and the University of Ghent proposed Greek and Latin names for them.

In the new system there are 10 **orders** of soils, each defined by a few distinguishing characteristics (Table 12-6; Fig. 12-10). As an example, one of the orders is designated "Oxisol," the *oxi* referring to oxide and *sol* to soil. These are the soils of tropical regions; they contain large amounts of iron and aluminum oxides.

Subcategories include **suborders, great groups, subgroups, families, series,** and **soil types.**

Suborders are named according to the distinguishing features of the horizons that develop. Thus, an "Aquox" is a wet soil, gray or bluish in color, with large amounts of oxides.

Great groups define the soils still more specifically, whereas suborders may make reference to minor properties that are also found in great groups. Families are based on properties important to the growth of plants.

The soil series is a collection of individual soils that are essentially uniform with respect to differentiating characteristics, including color, zonal development, and the depth limit within which each horizon is found. The series is classified according to type, the lowest category of classification based on texture.

Fortunately, the crop grower will still be able to call his soil by the same name as used before the advent of the new system. As a matter of fact, he need not even know that an old familiar soil such as "Lackland sand" is also an "ultic quarzopsamment!"

Land-capability classification

The capability classification of land has been widely used in the United States. In this system, arable soils are grouped according to their potential productivity

Table 12–6 Soil orders according to the comprehensive system of classification.

Order	Formative syllable	Derivation	Meaning	Diagnostic features	Older equivalents
1. Entisol	ent	Meaningless syllable	Recent soil	Very weak or no profile development	Regosols, lithosols, alluvial, some low humic gley
2. Vertisol	ert	Latin: *verto*, turn	Inverted soil	Self-mulching; expanding lattice clays; subhumid to arid climates	Grumusol, regur, black cotton, tropical black clays, smonitza, some alluvial
3. Inceptisol	ept	Latin: *inceptum*, beginning	Young soil	Weak profile development but no strong illuvial horizon; cambic horizon present	Brown forest, subarctic brown forest, tundra, ando, and some lithosols, regosols, and humic gley
4. Aridisol	id	Latin: *aridus*, dry	Arid soil	Soils of arid regions; often have natric, calcic, gypsic, or salic horizons	Desert, red desert, sierozem, reddish brown, solonchak, some regosols, and lithosols
5. Mollisol	oll	Latin: *mollis*, soft	Soft soil	Thick, dark A_1 stratum; usually develops under grassy vegetation	Chernozem, brunizem (prairie), chestnut, reddish prairie, some humic gley, rendzinas, brown, reddish chestnut, and brown forest soils
6. Spodosol	od	Greek: *spodos*, wood ash	Ashy soil (podzol)	Illuvial horizon shows accumulation of iron and organic colloids; weak to strongly cemented hardpan	Podzols, brown podzolic, groundwater podzols
7. Alfisol	alf	Meaningless syllable	Aluminum–iron soil (pedalfer)	Argillic horizon of relatively high base saturation (>35%); usually under boreal or deciduous broad-leaf forest	Noncalcic brown, gray-wooded; many planosols, some half-bog soils
8. Ultisol	ult	Latin: *ultimus*, last	Ultimate (of leaching) soil	Argillic horizon of low base saturation (<35%); plinthite often present; humid climate; usually forest or savanna vegetation	Red-yellow podzolic, reddish-brown lateritic, rubrozem, some gley and groundwater laterites
9. Oxisol	ox	French: *oxide*, oxide	Oxide soil	Argillic horizon very high in iron and aluminum oxides	Latosols, and most groundwater laterites
10. Histosol	ist	Greek: *histos*, tissue	Tissue (organic) soil	Organic surface horizon (>30% organic matter) more than 6 inches thick	Bog and some half-bog soils

Source: Soil Taxonomy, USDA Soil Conservation Service. Agriculture Handbook 436, 1975.

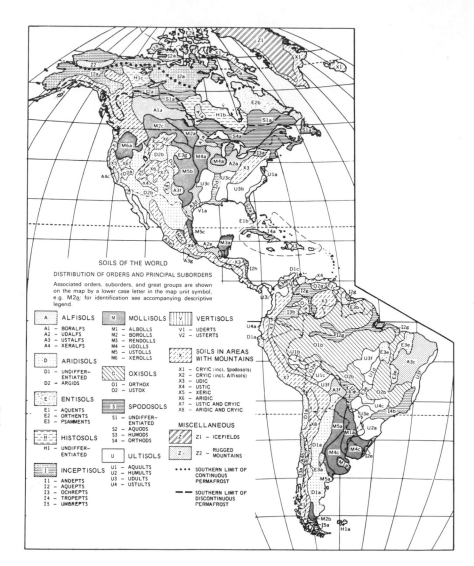

Figure 12-10 Soils of the world: the distribution of the orders and principal suborders. [Courtesy USDA.]

and their limitations for crop production without special treatment. Nonarable soils are grouped in eight classes according to their risk of damage and their potential for sustained production. The first four classes are suitable for cultivation if good management practices are followed: Classes V through VII are not suited for cultivation, but can be used for grazing and forest management. Class VIII land is useful only for recreation, wildlife, and watersheds. Details of the classes are as follows:

Class I soils are potentially the most productive of all soils. They are nearly level, and are not particularly susceptible to erosion. They are deep and well drained, with a good nutrient supply and water-holding capacity, and have all of the other physical and chemical characterisics required for intensive crop production. They can be used for the production of cultivated crops, pasture, range, forests, and wildlife. Irrigated soils may be placed in Class I if water is supplied by a permanent irrigation system.

Class II soils have some limitations. They may have gentle slopes, moderate susceptibility to erosion, a soil depth that is less than ideal, slightly unfavorable soil structure, wetness that can be corrected by drainage, slight climatic limitations, moderate salinity or alkalinity, or other slight limitations. Farm operators have a smaller choice of crops to produce on these soils than they do on Class I soils. Class II soils can be used for crops, pasture, range, forests, and wildlife.

Class III soils have severe limitations that reduce the choice of plants, frequently requiring the use of special conservation practices. They may have moderately steep slopes, high susceptibility to wind or water erosion, overflows that may result in crop damage, only slightly permeable subsoil, a shallow depth to rock or hardpan, which restricts rooting, poor fertility, and moderate salinity or alkalinity, either all in combination or singly. They may be used for cultivated crops, pasture, forest, range, or wildlife. Management of these soils must be very good if they are to remain productive.

Class IV soils have very severe limitations. They may be on steep erodible sites, have a history of erosion, be too shallow, be excessively wet, be severely alkaline, have a poor water-holding capacity, or be subject to damaging overflows. Although these soils may be used for crops, pasture, forests, range, or wildlife, extremely careful soil conservation practices must be applied. It is not advisable to use them for crops that require extensive cultivation.

Class V soils have little or no susceptibility to erosion, but they have other limitations such that they may be used only for pasture, range, forests, or wildlife production. They may be bottomland soils subject to frequent overflow, nearly level soils in areas where the growing season is short and limits the production of cultivated crops, level soils that are rocky, or ponded areas in which drainage is not feasible.

Class VI soils have severe limitations. They may be steep, highly erodible or eroded, stony, shallow, wet, subject to overflow, saline, or alkaline, or they may have a poor water-holding capacity. They can be used for pasture, range, forest, or wildlife. Special practices, such as water control by contour ditches, water spreading, or drainage are usually necessary. Some of these soils are adapted to special crops such as sodded orchards and blueberries.

Class VII soils have such severe limitations that their use is restricted to grazing, forest, or wildlife. They are more severely restricted than Class VI soils because it is impossible to compensate for or to correct one or more of their continuing limitations. They may be too steep, stony, erodible, wet, alkaline, saline, or deficient in some other manner.

Class VIII soils are useful only for recreation, wildlife habitat, watersheds, or aesthetic purposes. They cannot be managed for crops, grasses, or trees. Uncorrectable limitations may include erosion or erosion hazard, wetness, stoniness, salinity or alkalinity, or poor water-holding capacity.

Land-capability classification can be made more specific by means of subclasses, which specify the limitations of the soils in a given class. Four subclasses are recognized:

Subclass e soils have experienced severe erosion, or have a potential erosion hazard.

Subclass w includes soils for which excess water is the dominant limitation. Poor drainage, overflow, and high water tables are likely to be problems.

Subclass c consists of soils for which temperature and lack of precipitation are likely to be problems. This subclass contains soils for which climate is the only limitation.

Subclass s includes soils having characteristics that restrict rooting. These soils may be shallow, stony, alkaline, or saline, they may have low fertility that is hard to correct, or they may have a low water-holding capacity.

The capability classification of land has been used in the past solely for agricultural purposes. However, land-use planners and municipal officials are now beginning to realize that this system can be applied in many other fields as well, such as in planning the location of new subdivisions. For example, a shallow soil with an impermeable layer close to the surface would preclude the use of septic tanks and drainage fields.

SOIL CONSERVATION

Conservation means "careful use." The careful use of our soil resource is an obligation we have not only to ourselves but to future generations as well. This obligation has not always been acknowledged or acted upon. As early as 1894 Assistant Secretary of Agriculture Charles W. Dabney wrote, "Thousands of acres of land in this country are abandoned every year because the surface has been washed and gullied beyond the possibility of profitable cultivation" (Fig. 12-11). Unfortunately, nearly a century later, this statement holds true on an even larger scale!

Soil conservation has many meanings. For crop producers it implies a careful and intensive series of management practices involving plants and soils. In forests, soil conservation requires the use of logging practices that will not permit erosion, the regulation of grazing, and protection from uncontrolled fire. To others less concerned with growing crops, it may mean the stabilization either of sand dunes along a

Figure 12-11 Uncontrolled erosion eventually results in the complete loss of the soil resource, accompanied by poverty and misery. This tragedy has been commonplace in many parts of the world. [Courtesy USDA.]

beach or of steep slopes by planting appropriate plant species.

The basis of agriculture is a layer of topsoil that averages less than 18 centimeters (7 inches) in depth over the earth's surface. This soil cannot be recklessly exploited indefinitely; it must be preserved and refurbished. Although the reduction of productive capacity through the loss of fertility and structure is considerable, the most serious problem is erosion. Nutrients can be artifically supplied, but the loss of topsoil cannot be so quickly or easily remedied. The loss of soil due to wind and water is a national problem. It contributes to silt-clogged rivers, alternate drought and flood, dust bowls, and poverty.

Erosion of the soil is a natural process influenced by climate, topography, and the nature of the soil itself. Where permanent and undisturbed plant cover exists, erosion is more or less gradual and in equilibrium with soil-forming processes. Accelerated erosion comes about in the absence of plant cover. Areas that are unable to support a permanent plant cover as a result of climate or topography undergo large-scale erosion of the kind that carved the Grand Canyon. The accelerated soil erosion brought about by agricultural cultivation or overgrazing comes about largely through the action of water in humid climates and of wind in arid climates.

The maintenance of vegetative cover is basic to soil management. Vegetative cover retards erosion by cushioning the beating force of the rain (Fig. 12-12), increasing the absorptive capacity of the soil, and holding the soil against both water and wind. The soil cover increases the infiltration of water through the soil by preventing the clogging of the soil pores by fine surface particles. The techniques used for increasing soil cover include the increased use of sod culture, proper rotation, cover cropping, and mulching.

Water erodes the soil by literally carrying it away. The carrying power of water increases with its speed and volume. The volume of excess water depends upon the amount of rainfall and the rate at which it is absorbed by the soil. The speed with which this water moves is directly related to the slope of the

Figure 12-12 A stroboscopic photograph has caught the impact of a raindrop on moist soil, bursting upward and outward and carrying soil with it. [Courtesy USDA.]

land and the amount of cover. Any technique that either increases absorption or reduces the speed of the runoff will help prevent soil erosion.

The absorptive capacity of soil may be increased by deep plowing, by increasing the amount of organic matter in the soil, or by increasing drainage. Thus the burning or removal of organic matter is a poor conservation practice. Where natural drainage is poor, tiling may be necessary to remove water and provide air. In some regions hardpans have to be periodically broken up with plows cutting as deep as 2 meters (more than 6 feet).

The control of erosion by reducing the speed of runoff may be accomplished in a number of ways. Most basic is contour tillage, in which plowing, cultivation, and the direction of the "row" follows the contour rather than the slope of the land (Fig. 12-13). This affects the speed and power the surface water attains and thus the ability of the tilled soil to absorb water. The use of intertillage or strip cropping, which alternates strips of sod and row crops planted along the contour, helps to slow runoff by interposing barriers with high absorptive capacity. The alternation of sod with row cropping serves to achieve the benefits of rotation. On steep slopes where greater amounts of surface water must be accommodated,

the use of **waterways**—permanently sodded areas—facilitates water removal and minimizes erosion.

Where contour cultivation and strip cropping are not sufficient to check erosion, terraces constructed on the contour must be used. **Terracing,** an ancient practice, consists in cutting a slope into a number of level areas. Terraces appear as giant steps on the hillside. Although the steps of ancient terraces were made of stone, modern terraces are made by building low rounded ridges of earth across the sloping hillside. Terracing slows down the speed of surface runoff, and, although it is designed primarily to prevent erosion, it facilitates the storage of available water. Thus terracing is an important practice in areas of low rainfall and where the scarcity of arable land necessitates the exploitation of steep hillsides.

Some of the erosion caused by wind, especially on open prairies or plains, may be checked by planting windbreaks, one or more rows of trees or shrubs planted at right angles to the prevailing winds. The effectiveness of a windbreak is local and is related to the thickness and the height of the trees. The maintenance of a permanent plant cover in conjunction with windbreaks effectively reduces wind erosion where it is a problem. On organic soils, rows of small grain may be used as temporary windbreaks to protect seedlings.

Cropping practices almost always result in the loss of soil, much more so for row crops such as potatoes and corn than for crops that provide a permanent soil cover. Until recently, soil losses from sloping cropland could be estimated only in relation to losses from level cropland. Scientists at the University of Tennessee, using data from experimental plots, devised an equation, which applies directly to Tennessee agricultural land, for predicting soil losses for a given field under different management practices. The following components make up the equation:

A = *average annual soil loss*, in tons per acre, over the course of a period of years. It may not be valid for any one particular year.

R = *rainfall factor*, the erosion potential in a given locality. This factor expresses the intensity of rainfall and its kinetic energy.

K = *soil erodibility factor*, a reflection of soil struc-

Figure 12-13   Water was held on this 6 percent slope for 36 hours after a 10-centimeter (4-inch) rain without washing out the corn. Nearby fields planted in straight rows up and down the slopes had to be replanted. [Courtesy USDA.]

ture, texture, organic matter, permeability, depth, and all of the other characteristics that influence the rate of soil erosion.

LS = length-and-steepness-of-slope factor, an extremely important consideration: longer and steeper slopes have more erosion.

C = cropping-management factor, the ratio of soil loss under a specific management system to losses under continuous fallow.

P = conservation-practice factor, a value for conservation practices such as contour plowing and terracing, which reduce soil losses.

T = soil-loss tolerance, the estimated average annual soil loss that can be tolerated in tons per acre if sustained economical crop production is to be achieved.

In the equation, *A* is computed (by multiplying the factors *R*, *K*, *LS*, *C*, and *P*) and then compared with the *T* value. One acre-inch of soil weighs about 150 tons. If the average soil-loss is 5 tons per acre per year, then 30 years would be required to erode 1 acre-inch of soil from a field that is being cropped. If this rate of loss is too high for a particular situation, then either additional conservation must be initiated or cropping practices must be changed.

"NO-TILL" AND "MINIMUM-TILL" AGRICULTURE

The first humans to plant a crop merely punched a hole in the soil with a sharp-pointed stick, dropped in a few seeds, and waited for their crop to grow. They were practicing no-till agriculture. *Minimum till* is a variation in which tillage is avoided in every possible operation.

It may seem strange that today we "moderns" are trying to use the same basic system to conserve soil, soil moisture, and energy—and are finding that it works. The dust storms in the winter of 1977 would have been less severe if no-till systems had been in wide use in the drought-stricken western part of the United States.

There are many variations of no-till agriculture: (1) alternating strips of meadow and no-till corn, sometimes called *sod cropping;* (2) planting in last year's crop residues without cultivation called *stubble cropping;* (3) following corn and fall-seeded small grain with a second crop of soybeans without cultivation, called *double cropping;* (4) broadcast seeding and fertilization by aircraft for a totally undisturbed soil; (5) interseeding legumes into grass pastures; and (6) interplanting corn with small-grain cover crops.

These are only a few of the possible variants of the system, which was once limited only by imagination. Although no-till agriculture is practiced today primarily for ethical reasons, it will be practiced in the future largely for economic ones: Agricultural engineers have estimated that conventional land preparation and planting requires 51.5 to 65.5 liters of fuel per hectare (5.5 to 7 gallons per acre), but no-till agriculture requires only 2.8 to 22.5 liters of fuel per hectare (0.3 to 2.4 gallons per acre).

No-till agriculture is not universally applicable. Soils with low pH and those with low phosphorus cannot be used. The method cannot be practiced where perennial weeds (such as Johnson grass, nutsedge, and bermuda grass) are established. On the other hand, large-seeded broadleaf weeds, such as cocklebur, may be reduced by the elimination of tillage. Damage by above-ground insects may be more common with no-till agriculture. Some diseases, such as gray leaf spot of corn, have been intensified, but a few, including stalk rot of sorghum, have been reduced. Appropriate pesticides can help control these problems.

It is safe to say that no-till and minimum till are in their infancy in modern agriculture. We will see much more of them in the future.

FUTURE DILEMMAS

The costs of further gains in yield are becoming greater with each passing year. The costs of capital and energy for crop production are increased by every ton of topsoil that is washed or blown away, by each reservoir filled with silt, and every hectare claimed by desert sand. The cost-multipliers of crop production—social organization and technological sophistication—will also continue to increase as the natural productivity of our crop ecosystems is impaired.

Human-caused stresses on ecosystems, whether they are on hillsides in Venezuela or Pakistan, or on the rangelands of Botswana or Argentina, reduce the ability of the ecosystems to resist "normal" climatic extremes. With mistreatment of ecosystems, troublesome periods of low rainfall become periods of famine, and serious floods became disasters. In one month, a layer of fertile topsoil that took hundreds of years to accumulate can be lost because of human mismanagement.

In the United States, the productivity of the Dust Bowl was restored because tremendous technical and financial resources were available. We were also able to purchase food abroad during the lag period while productivity was being rebuilt, thus avoiding prolonged shortages and hardships. But most countries whose agricultural ecosystems are threatened lack the necessary technical skills of agronomy, forestry, horticulture, range management and engineering, which must be simultaneously brought to bear on problems that threaten food production. Social and political differences add to the problems. Those who are barely making a living on degenerating land cannot be expected to plant trees, leave land fallow, and sacrifice livestock if they are not provided with alternative food sources. In the words of a newly elected president of one African country, "We cannot afford conservation, we are not rich enough." Thus, the problem seems to be ever-increasing.

Soil conservation must become a matter of ethics and law for crop producers the world over, just as honest weights and measures and the humane treatment of livestock have become matters of ethical concern and legal prescription. To make conservation a vital factor, it may be necesary to sacrifice large short-term gains for smaller profits that accumulate over a long period. In the end, soil conservation implies acceptance (and perhaps pursuit) of a level of biological productivity that can be sustained indefinitely. Soil conservation often provides immediate benefits in terms of plant performance and must be considered a requisite of sound soil management.

Soil has sometimes been called "our most important heritage." Without a doubt, it is the backbone of agricultural productivity, and the greatness of nations can be measured in terms of their skill and persistence in producing agricultural products. The developing nations will do well to heed the lessons of history, and see that their agricultural base, which is circumscribed by their climates and soils, is carefully developed and protected.

Selected references

Allison, F. E., 1973. *Soil Organic Matter and its Role in Crop Production*. New York: Elsevier. (A comprehensive treatment of all aspects of soil organic matter in crop production.)

Barnes, K. K., W. M. Carleton, H. M. Taylor, R. I. Throckmorton, and G. E. Vandenerg, 1971. *Compaction of Agricultural Soils*. The American Society of Agricultural Engineers, Ann Arbor. (A review of the current state of knowledge about soil compaction and a discussion of those areas in which research can be most profitable.)

Baver, L. D., 1972. *Soil Physics*. New York: Wiley. (A widely used text.)

Black, C. A., 1968. *Soil–Plant Relationships*. New York: Wiley. (An outstanding and comprehensive text that treats soil as a medium for plant growth.)

Cruickshank, J. G., 1972. *Soil Geography*. New York: Wiley. (A short introduction to the distribution and formation of soils.)

Donahue, R. L., R. W. Miller, and J. C. Schickluna, 1977. *Soils*. Englewood Cliffs, N.J.: Prentice-Hall. (An excellent elementary text for the study of soils.)

Foth, H. D., 1978. *Fundamentals of Soil Science*. New York: Wiley. (A textbook for students who are beginning the study of soils.)

National Research Council, Committee on Tropical Soils, 1972. *Soils of the Humid Tropics*. Washington, D.C.: National Academy of Sciences. (A collection of 12 papers concerning problems in the management of tropical soils.)

Newton, W., J. R. Postgate, and C. Rodriguez-Barrueco (editors), 1977. *Recent Developments in Nitrogen Fixation*. New York: Academic Press. (Papers on practically all aspects of nitrogen fixation.)

Postgate, J. R. (editor), 1971. *The Chemistry and Biochemistry of Nitrogen Fixation*, New York: Plenum Press. (A technical overview of an often overlooked but fundamentally important aspect of biology—a comprehensive survey of a widely divergent field.)

Smith, L. P., and B. D. Trafford, 1976. *Climate and Drainage*. London: Her Majesty's Stationery Office. (A bulletin dealing with the design and planning of soil-drainage systems for agriculture.)

Taylor, S. A., and G. L. Ashcroft, 1972. *Physical Edaphology*. San Francisco: W. H. Freeman and Company. (A textbook on soils that also deals with the practical aspects of irrigation, fertilization, plant– water relations, and soil behavior under cultivation.)

Trudgill, S. T., 1977. *Soil and Vegetation Systems*. London: Oxford University Press. (A book for undergraduates who need a broad understanding of the relations between soil and vegetation.)

U.S. Department of Agriculture, 1957. *Soil* (USDA Yearbook, 1957). (This is an updated version of the famous 1938 yearbook, *Soils and Men*.)

U.S. Department of Agriculture, Soil Survey staff, 1975. *Soil Taxonomy* (USDA Handbook 436). (Contains the most recent approach to the comprehensive classification of world soils.)

13
Water

Consider, for a moment, a substance that is often taken largely for granted—water. We drink it, breathe its vapors, cook with it, and bathe ourselves with it. In the form of perspiration, it cools our bodies. Water can extinguish fire, dissolve both rocks and sugar, and is transparent. When frozen, unlike most substances, water becomes lighter and floats. In short, this common substance has rather uncommon properties.

Water comes closer than any other liquid to being a universal solvent. Given time, it can dissolve practically anything on earth. This is biologically important, because practically all of the substances of which organisms are composed are either immersed or dissolved in it. In effect, people and plants are columns of water that serve as the media for both passive and active metabolic transfer of materials. The distribution of water throughout plants provides pathways of movement from the tips of the roots to the topmost leaves of even the highest redwoods.

PROPERTIES OF WATER

Once the temperature of water has reached the boiling point, it will rise no further until the water changes from the liquid to the gaseous state. But be-

fore the change in state occurs the water will absorb 539.3 calories per gram. The energy required to change water from the liquid to the gaseous state is called its **latent heat of vaporization** (Fig. 13-1). The uniquely high latent heat of vaporization of water adds greatly to its stabilizing effect on surface temperatures of the earth.

A similar absorption of heat at a constant temperature takes place when ice melts. The energy required to melt ice is called the **latent heat of fusion.** After the temperature of ice has risen to 0°C, it will absorb 79.7 calories per gram before melting. Conversely, 79.7 calories of heat is released when one gram of water changes to ice. Few substances require this much energy to change from the liquid to the solid state or vice versa.

The **specific heat** of a substance is the amount of heat in calories required to change the temperature of one gram of the substance by 1°C. As a consequence of the definition of the gram-calorie, the specific heat of water is 1 calorie per gram °C. Water has an exceptionally high specific heat, as is shown in the table below. Few other substances found in nature require so much heat to change its temperature so slightly. It is easy to see why water has a stabilizing effect on the temperature of organisms. Imagine how

Figure 13-1   The latent heats of water. One gram of water changes from ice to liquid at a constant temperature of 0°C with the absorption of 79.7 calories. It changes from liquid to gas at a constant temperature of 100°C with the absorption of 539.4 calories. [After K. S. Davis and J. A. Day, *Water, the Mirror of Science,* copyright © 1961 by Educational Services, Inc. and Heinemann Educational Books, Ltd.]

rapidly the temperature of an organism, either plant or animal, might change if 90 percent of the weight of its living parts were not made up of water.

Substance	Specific heat (cal/gram °C)
Water	1.00
Paraffin	0.69
Wood	0.42
Cellulose	0.37
Chalk	0.21
Glass	0.19
Granite	0.19
Quartz	0.19
Iron	0.11
Copper	0.09
Zinc	0.09

Water is also a good conductor of heat. It is not as good a conductor as some other compounds, but nonetheless it ranks high in this respect.

One of the most remarkable things about water is that, although it contracts as its temperature de-creases, just as other substances do when cooled, it undergoes a rapid and abrupt expansion as its temperature decreases from 4 to 0°C. Molecules of crystallized water assume a polar orientation with each other, a condition that requires more space than when they are in the liquid state. This is why ice floats.

Water has the greatest surface tension of all common liquids. Surface tension is the force of cohension that exists between the molecules in the surface layer of a liquid. It is this cohesive force that causes liquids to pull into a round shape as they drop. Molecules of water cohere to each other so strongly that a force of approximately 95,275 kilograms (210,000 pounds) would be required to pull apart a column of perfectly pure water having a diameter of 25.4 millimeters (1 inch). Of course, there is no pure water on earth, and so this is only a theoretical value. Even so, experiments have demonstrated that nearly 910 kilograms (about 2000 pounds) of tension is required to pull apart a column of ordinary water. It has a tensile strength approximating that of steel!

Not only is water cohesive, it is also strongly adhesive, meaning that it adheres to other substances with

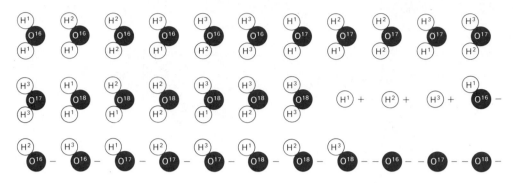

Figure 13-2 Water is not a simple substance, but a complex mixture of 33 different substances. [From A. M. Buswell and W. H. Rodebush, "Water." Copyright © 1956 by Scientific American, Inc. All rights reserved.]

which it comes in contact. Among the few substances that water will not wet are fats and waxes; this is why water forms droplets, rather than a film, on the surfaces of cutinized leaves.

The transparency of water is of immense importance, because visible radiation can pass through it and through the interiors of leaves and other organs. If water were not transparent photosynthesis would be confined to a very narrow layer on the surfaces on leaves. But because it is transparent, cells deep inside plant tissues can carry on photosynthesis.

Because water is relatively inert it serves as a medium in which practically every chemical reaction in the plant takes place, without itself taking part in the reaction. In this manner it serves as a catalyst. (The photolysis of water in photosynthesis is one important exception.)

Although most people don't realize it, water decreases in viscosity as its temperature increases. Water is 36 percent as viscous at 40°C (104°F) as it is at 0°C (32°F).

Temperature (°C)	Relative viscosity (%)
0	100
10	73
20	56
30	45
40	36

Pure water is a good electrical insulator. Of course, pure water is never found in nature, and the substances that are dissolved in it frequently make it an excellent conductor.

Most people think of water as a simple, homogenous substance—H_2O. In fact water is not simply H_2O but a mixture of 33 different substances. Eighteen are combinations of the three isotopes of hydrogen and the three of oxygen. The three hydrogen isotopes are ordinary hydrogen (^1H), deuterium (^2H), and tritium (^3H). The three oxygen isotopes are ordinary oxygen (^{16}O), oxygen-17 (^{17}O), and oxygen-18 (^{18}O). The remaining substances are various ions (Fig. 13-2). Water is indeed a complex substance with many properties and characteristics.

THE HYDROLOGIC CYCLE

The earth's water circulates through a system that extends from 16 kilometers (about 10 miles) up in the atmosphere to about 0.8 kilometer (0.5 mile) below the surface of the oceans. Within this vertical range of space, the waters of the earth are constantly moving and circulating. The **hydrologic cycle** (Fig. 13-3) has neither a beginning nor an end, and all the waters of the world are subject to perpetual recycling.

The hydrologic cycle is sustained by solar energy. The sun supplies the energy that evaporates water from the surface films of plants, from the soil, and

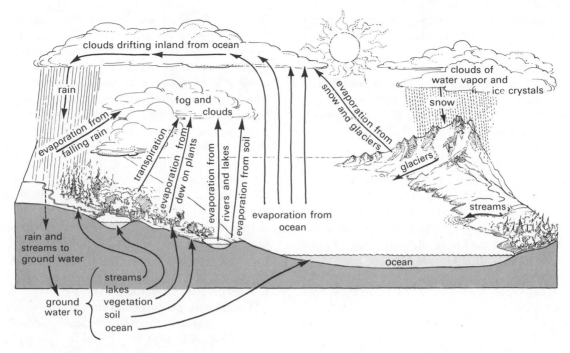

Figure 13-3   The hydrologic cycle has neither beginning nor end. It is as old as the earth and as relentless as time. [From J. Gilluly, A. C. Waters, and A. O. Woodford, *Principles of Geology*, 4th ed. W. H. Freeman and Company, copyright © 1975.]

from the waters of rivers, lakes, and oceans. Although most atmospheric moisture comes from the oceans, transpiration accounts for a large fraction in areas of abundant vegetation. The combination of evaporation and transpiration is called **evapotranspiration.** Among the various phenomena that cause air to rise are the frontal sliding of masses of warm air over wedges of cold air; the upward squeezing of air between two relatively cold air masses (convective circulation); and orographic lifting, the result of rising air being forced upward over mountains. Precipitation completes the hydrologic cycle.

The movement of moisture-laden air masses in the atmosphere is influenced by gravity, centrifugal forces caused by the rotation of the earth, and atmospheric temperature gradients. Differential heating of the earth's surface directs the movement of moisture-laden air. Because solar radiation is not as intense near the poles as it is near the equator, the

general circulation of air near the earth is toward the equator. After it is heated at the equator, it rises and passes at high altitudes over the cooled air to the poles, where it is cooled, drops, and repeats the cycle. Since the atmosphere always contains moisture, and since water has a very high specific heat, great transfers of energy always accompany the movement of air masses.

The temperature of the oceans and other large bodies of water remains relatively constant because of the high specific heat of water. Land masses, however, have a comparatively low specific heat, and tend to cool rapidly in water and heat rapidly in summer. Consequently, low-pressure air masses generally form over the land in the summer, and high-pressure masses over the oceans. The hurricanes that originate in the Gulf of Mexico and the Atlantic Ocean and sweep through the central and southeastern United States in the summer are good examples of this phenomenon. In winter, the reverse occurs. The resultant pressure differentials cause air to move from areas of high pressure to areas of low pressure.

Movements of air masses are also influenced by seasonal and diurnal variation in radiation. Differences between night and day and between the seasons are eventually averaged out, but in the process they strongly influence the direction and speed of the hydrologic cycle.

We can do relatively little to alter the hydrologic cycle. Building dams, modifying vegetation, tapping groundwater via deep wells, and other practices can produce economically significant local changes, but we cannot make major interruptions, and it is unlikely that we will want to in the near future. Before this can be done with safety, we need to know much more about the effects it might have on the ecology of the **biosphere**—that part of the earth's environment in which organisms live.

Unfortunately, our desire to increase the supply of water available for agricultural and industrial use has resulted in the exhaustion of local water supplies, with the resultant "death" of some communities.

ATMOSPHERIC MOISTURE

Forms of atmospheric moisture

There are several forms of atmospheric moisture, including vapor, rain, snow, sleet, hail, and dew. Atmospheric moisture is important to plants in two basically different ways. As precipitation, it supplies moisture to soil and permits the uptake of both water and nutrients by plants. As water vapor, it decreases the loss of water from plants (and soil) by transpiration (and evaporation). These two aspects of the importance of moisture have consequences that bear directly on the fundamental life processes of plants.

Rain forms when the condensation of atmospheric moisture takes place at temperatures above the freezing point. **Snow** forms when atmospheric moisture condenses at temperatures below the freezing point. When rain falls through a layer of air with temperatures below the freezing point, **sleet** is formed; but if air currents carry the frozen particles back up into the frigid layer of air, so that they accumulate a thicker covering of ice before falling again, **hail** is formed. Under certain atmospheric conditions, hail particles may accumulate enough layers of ice to become as large as baseballs. Needless to say, such hailstones can cause considerable damage to plants.

Dew forms at night when atmospheric moisture condenses on cold surfaces. It can form only when the sky is clear enough for the soil and plants to lose enough heat such that their temperature falls below the dew point. Clouds, haze, and dust prevent the loss of heat at night, when the energy balance of the earth is normally negative. Dew cannot, of course, supply all the moisture for the growth of most plants, but it has great potential as a supplemental source of water for crop plants in semiarid areas, such as Israel, where the total amount of rainfall amounts to only a few centimeters (about one inch) per year.

A number of plant species are able to use the moisture from fog and mist. In the South, Spanish moss (*Tillandsia usneoides* sp.) grows as an epiphyte (a plant that grows on other plants but receives no sustenance from them) on the branches of several tree species and even on telephone lines. It is never found in areas that are not subject to mist and high relative humidity. The Torrey pine and several other plants of California grow on very dry, poor soils near the ocean, where fog from the ocean supplies moisture.

Fogs and mists are low clouds. They do not settle, and so they cannot be measured by conventional rain gauges. Even so, the vegetation that develops in areas subject to frequent fogs may be far richer and more abundant than might be expected from precipitation records. Campers on Mount Mitchell, in North Carolina, will vouch for the fact that when a fog rolls in and is intercepted by the foliage of fir trees, the resultant drip of water to the ground beneath is equivalent to a heavy rain.

Measuring atmospheric moisture

There are several ways to measure humidity. Perhaps the most commonly used one is the thermodynamic technique. The most commonly used instrument for this purpose is the **psychrometer.** A psychrometer usually consists of two thermometers, one of which has a wettable wick around the bulb. The two thermometers, fastened side by side, are spun so that a stream of air passes over their bulbs. The thermometer without a wick will indicate the air temperature, and the other will indicate a lower temperature because the water evaporating from the wick causes a heat loss. When the temperature of the water in the wick has been lowered until the vapor pressure of the water remaining is equal to the vapor pressure of the atmosphere, no more water will be lost and the temperature of the wet-bulb thermometer will remain constant. The relative humidity can be obtained from the temperatures indicated by the two thermometers through the use of appropriate calculations or tables.

Another kind of device for measuring humidity is called a **hygrometer.** Hygrometers contain substances that are affected by the moisture content of the environment. The elements of hygrometers are made from hairs and from thin animal membranes that stretch when they become moist. Both the hairs and the membranes are connected, through a series of levers, to a chart or a needle, calibrated to indicate relative humidity.

Also useful are such chemicals as cobalt chloride, which changes color as humidity changes. One of the first techniques used for estimating transpiration consisted of putting a small filter paper, which had been saturated with cobalt chloride and dried, in contact with the leaf surface. As water diffused through the stomata, the color of the cobalt chloride changed from blue to pink. The rate of change and the intensity of color was thought to indicate the internal water conditions of the plant.

Optical techniques also have some value. Certain wavelengths of radiation are absorbed by water to a much greater degree than others. By measuring the absorption of radiation in these wavelengths with a special cell, it is possible to determine how much water vapor is in the air at any time.

Humidity can also be measured electronically with the use of **resistance psychrometers.** This is done by coating a nonconducting glass or plastic element with lithium chloride or some other hygroscopic salt, and measuring the electrical resistance of the material between two electrodes with a meter. The salt will adsorb more water at higher humidities so that electrical resistance between electrodes decreases as atmospheric moisture increases.

Liquid precipitation can be measured in a number of different ways, but by far the most common is to use rain gauges. Rain gauges may be simple buckets, fashioned from cans of various sizes, or they may be much more sophisticated devices that record intensity as well as time of rainfall. In the United States, receptacles with an opening 8 inches in diameter are most frequently used; this size of opening is practically regarded as standard.

Recording rain gauges, which usually weigh the water that falls into the bucket, are best when measurements of intensity and time of precipitation are required. They are nearly always supplemented, however, with simple collecting gauges, which usually give a more precise measurement of total rainfall.

At times, it may be desirable to measure rainfall in remote locations, as in deserts and mountains, where daily or even weekly access to the gauges is not possible. In such locations as these, a can is placed in an appropriate location with a layer of mineral oil in the bottom. The mineral oil, being less dense than water, rises to the top of the accumulated rainfall and prevents evaporation. This technique can be used for stations visited as infrequently as once every six months. Snow gauges, important at high elevations and in cold climates, are frequently made of rain gauges with slatted wooden baffles added on the outside to still the wind so that it drops its burden of snow into the gauge.

It is frequently desirable to determine the amount of rainfall that has been intercepted by the canopy of a forest or by the vegetative cover of some other crop. For this purpose, two sets of rain gauges are used, one placed under the canopy at ground level, and the other placed either above or to one side of the crop. The difference in amounts of precipitation measured in the two cans represents the amount of precipitation intercepted by the vegetation. However, one source of error in such measurements lies in the fact that some of the intercepted precipitation finds its way to the soil by flowing down trunks or stems. This can be counteracted by the construction of troughs around the trees or other plants to catch the water going down the stem, whereupon it is retained and measured.

There are several devices used to estimate evaporation. The simplest consists of nothing more than an open pan of water, frequently about 1 meter in diameter, placed in an exposed location. Evaporation is measured merely by noting changes in the water level. This device has been used often in various parts of the world. A more sophisticated device is the **atmometer,** which consists of a porous clay bulb connected by a water-filled tube to a reservoir of water. As solar radiation strikes the outside of the clay bulb and heats it, the water in the tube evaporates and is replaced by water that moves up from the reservoir. Still another device, the **evaporimeter,** consists of a wick wetted by water from a reservoir. As water evaporates from the wick, it is replaced by water from the reservoir. The amount of water lost indicates the relative rate of evaporation.

SOIL MOISTURE

Forms of soil moisture

Water is found in several forms in the soil. Let us imagine a bucket full of soil, to which enough water has been added to saturate the soil and form a puddle on the surface. If the bucket were equipped with a removable bottom that could be replaced by a screen, the water that would drain out first would be **free water.** It is not bound to the soil but percolates through it and drains through the screen at the bottom under the influence of gravity. Most of the water remaining in the soil is called **capillary water** (Fig. 13-4), because it is held in capillary pores by its own surface tension and adhesive properties. Capillary and free water can be withdrawn from the soil and used by plants. Although capillary water cannot move through the soil, it often is the most important component of soil moisture available to plants.

The rate at which plants extract capillary water from soil is largely a function of the abundance of roots, root hairs, and mycorrhizae, and, to some extent, of mineral concentration of the soil water. Since

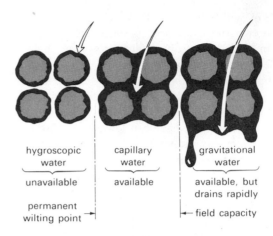

Figure 13-4 Capillary water is the most important in sustaining plant life. Most free water (gravitational water) drains away before it can be used, whereas hygroscopic water is bound too tightly to soil particles to be removed by plants. [After J. Bonner and A. W. Galston, *Principles of Plant Physiology,* W. H. Freeman and Company, copyright © 1952.]

Figure 13-5 The average moisture-extraction pattern for plants growing in a soil without restrictive layers and with an adequate supply of available moisture throughout the root zone. [Courtesy Soil Conservation Service, *National Engineering Handbook,* 1964.]

there are many more roots in the surface soil, water is extracted at a faster rate, and more completely, than from soils at lower depths (Fig. 13-5). The rate of extraction from lower soil zones does increase, however, in times of drought. It is generally believed that the lower soil zones supply the moisture that sustains some trees and other deep-rooted crops during such critical times.

Still another form of soil moisture is called **hygroscopic water.** Hygroscopic water adheres tightly to soil particles and is not normally available to plants. During periods of drought, however, it may evaporate and form water vapor, so that some hygroscopic water may become available, even though the amount may be small. Water vapor does not constitute a large fraction of the moisture in most soils, but in some, especially in sandy, porous soils, it may move from one part of the soil to another—for example, from a low horizon to a higher horizon. Vapor transfer of moisture in the soil is poorly understood.

Water is also present in the soil as **chemically combined water.** The chemical compounds with which water is combined are said to be hydrated. Moisture in this form is bound tightly and is unavailable for use by plants.

When all free water has drained from a soil, it is said to be at **field capacity.** The significance of this term is that it takes into account the physical nature of the soil. In terms of absolute amounts of moisture, field capacity differs with soil type. The field capacity of a heavy clay soil may be 30 percent, whereas it is only 3 percent in a sand. Various direct and indirect methods have been devised for measuring soil moisture in terms of percentage of field capacity.

The water content of soils can also be expressed in terms of the availability of the water to plants. After capillary water has been depleted, moisture is said to have reached the **wilting point,** because plants lose turgor. Actually, it is more nearly correct to say **wilting range,** because there is a small variation in water availability to different plants. The percentage of water depends upon the type of soil and is relatively independent of the species of plant growing in it.

Movement of soil moisture

A number of factors influence **infiltration,** the movement of water into soil, and **percolation,** the movement of water through soil. One of the most important factors is the soil moisture content at the time precipitation falls. If the soil is dry, much more water can be absorbed than if it is wet, provided there are no barriers to absorption. In the spring of the year in temperate regions, the soil is usually saturated and cannot absorb and store additional water, so that runoff is likely to cause floods. Not only is there less pore space through which water can move when the soil is wet, but the soil particles of colloidal size are swollen by hygroscopic water, further reducing pore space available for water storage.

Vegetation has an important influence on infiltration of water. Layers of vegetation, even a single leaf, can break the speed of rain drops and slow them down so that they do not erode the soil (Fig. 13-6).

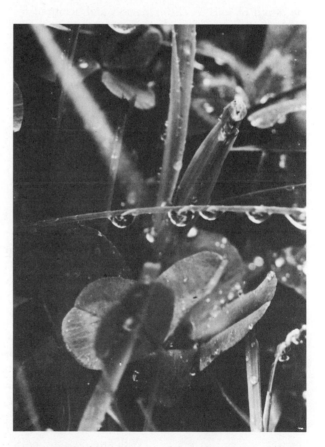

Figure 13-6 Plants protect soil from heavy rain. When a large drop strikes a leaf or stem it bursts into many little drops, some of which evaporate while others coalesce and drop short distances to the soil. [Courtesy USDA.]

Surface litter cushions the impact of rain drops and keeps them from beating directly against the soil and causing raindrop erosion, a significant factor in the deterioration of clay soils. Surface litter also forms myriads of small dams that retain water and permit it to infiltrate the soil. Vegetation also cools the soil by intercepting solar radiation, prevents evaporation, and decreases surface runoff by absorbing water. It slows down the wind and prevents the fine soil particles from blowing away, especially the organic particles, which are comparatively light in weight.

A layer of vegetation maintains the humidity and keeps the surface from drying up completely. As soil dries out, complex compounds that are hard to wet are formed on the surface. Vegetation also increases the depth of soils. As litter decays and accumulates on the surface, a part of the partially decomposed humus is washed down into the soil. Most important, roots decay and form channels through which water can move. These channels provide a rapid access to the subsoil both for precipitation and for roots of succeeding plants, thereby increasing the moisture-storage capacity of the soil.

The use of heavy equipment and even trampling by animals and people, as on paths and roads, will compact surface soil so tightly that water will not infiltrate. Sandy soils do not compact because their particles do not normally stick together. However, for clay soils, in which the fine soil particles can cling to each other, even rain may cause surface soils to puddle and thereby decrease infiltration.

The inwash of fine material may affect infiltration and percolation. If the inwashed material is composed of sticks, stones, rocks, and large particles, it has little effect on the soil that it covers. But if it is fine material such as clay, bits of organic debris, or fine sand, then it is likely to clog the small pores of the surface and prevent infiltration.

Measuring soil moisture

There are several ways of expressing the amount of water in soil. Experienced persons can approximate moisture content rather closely by "feeling" the soil—that is, by rubbing a small amount of soil from the root zone between the hands to determine its consistency (Table 13-1).

The simplest way to express soil moisture quantitatively is as a percentage of the dry weight of the soil from which it has been extracted. To determine moisture content gravimetrically, soil samples from the field are usually placed in cans with tight lids, weighed, oven-dried until their weight remains constant, and reweighed to determine how much water

Table 13–1 The determination of moisture in medium- to fine-textured soils by feel. (With sandy soil, the balls are more friable and fragile throughout the whole range.)

Amount of moisture	Feel	Field capacity (%)
Dry	Powder dry	0
Low (critical)	Crumbly, will not form a ball	<25
Fair (usual time to irrigate)	Forms a ball, but will crumble upon being tossed several times	25–50
Good	Forms a ball that will remain intact after being tossed five times; will stick slightly with pressure	50–75
Excellent	Forms a durable ball and is pliable; sticks readily; a sizable chunk will stick to the thumb after soil is squeezed firmly	75–100
Too wet	With firm pressure, some water can be squeezed from the ball	>100

Source: W. Strong, *Sprinkler Irrigation Manual*, Wright Rain, Ringwood, England, 1956.

Figure 13-7 As the moisture content of soil increases, the soil water is less strongly bound. Note that at the wilting point and at field capacity a clay soil may contain four times as much water as a sand. [After L. D. Baver, *Soil Physics*, 3rd ed., Wiley, 1956.]

was lost. The data obtained in this fashion can be used to compute the percentage of soil moisture on either a dry-weight or a wet-weight basis. This method does not, of course, take soil structure into account. To do this requires determining the water content for a known volume of soil.

It may also be desirable to express soil moisture in terms of the tension with which it is bound to the soil (Fig. 13-7). For example, free water is held with very low tension. Capillary water is bound more tightly, but not extremely so. Hygroscopic water is bound very tightly, and the tension required to remove it is high. Tension may be expressed in atmospheres of pressure or in one of several other units. The greatest advantage to using tension to express soil moisture is that the critical values, wilting coefficient and field capacity, are the same for all soils, even though the total amount of extractable water may be vastly different.

An apparatus called a **tensiometer** measures the tension with which soil water is held in the soil (Fig. 13-8). When the tensiometer is filled with water and placed in moist soil, water is extracted from a porous

ceramic cup until it comes into equilibrium with the soil water. The tension in the ceramic cup is read from a manometer or a vacuum gauge.

Moisture content can be estimated by measuring the electrical conductivity of a volume of soil. Conductivity is not reliable, however, because the soil texture varies greatly from one spot to another, soil minerals are not distributed evenly, and it is difficult to repeat or duplicate measurements.

A modern technique for measuring the moisture content of soils is to use the thermalization and scattering of fast neutrons. Fast neutrons have an energy of about 2 million electron volts, but after colliding elastically with the hydrogen in water, the speed of the neutrons comes into thermoequilibrium with the speed of the molecules composing the material that they are in, and become slow, or thermal, neutrons. Thermal neutrons have energies of about 0.025 electron volt. If a source of fast neutrons is placed in the soil, the number of fast neutrons that will be converted to thermal neutrons is a function of the amount of water in the soil, and they can easily be detected by a thermal neutron counter. This tech-

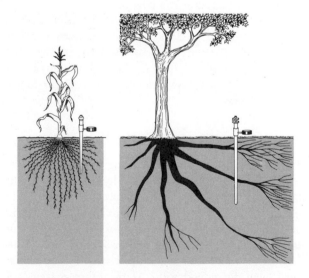

Figure 13-8 Tensiometers are placed in the vicinity of roots to measure the tension with which water is held in soil.

nique has been found so useful in recent years that a number of devices for this technique have become commercially available. Its major drawback in the field is that it does not distinguish between water in the soil, water in the roots, and hydrogen in soil organic matter.

An interesting but seldom used technique is based on the ability of water to attenuate or absorb gamma radiation. This method can be used very easily with small containers, such as pots. A radiation source is placed on one side of the pot and a detector on the other side. The amount of moisture present at any given time is an inverse function of the amount of radiation that can penetrate the pot.

Many investigators have used Bouyoucos blocks to measure soil moisture in experimental plots. These blocks, invented by scientist J. V. Bouyoucos, are of porous ceramic clay with two electrodes embedded in each, a fixed distance apart. They are "buried" in the soil in locations where soil moisture is to be measured with wires going to the surface from the electrodes. When a moisture measurement is desired, an electrical current is applied by means of the exposed wires, and the amount of current flowing through the moist blocks is measured. The greater the soil moisture, the greater the flow of current.

Dry soil is a poor conductor of heat. But as water content increases, so does thermal conductivity. Thus thermal conductivity can be used to estimate the amount of moisture present. This technique requires precise calibration, however, and so far has not found wide acceptance.

PLANTS AND WATER USE

Plants may be classified according to their water requirements. At one extreme are **hydrophytes,** plants that grow in water. Hydrophytes are frequently characterized by having a low oxygen requirement and poorly developed root systems. **Mesophytes,** intermediate in their water requirements, have well-developed root systems and are characteristic of mature, well-developed soils. At the other extreme of the moisture scale are **xerophytes,** which grow in dry habitats and have undergone various adaptations that enable them to grow and reproduce in very droughty situations.

Drought

Drought may be described as a period in which a scarcity of soil moisture is limiting to the growth of plants. The precise definition of drought is elusive; none of those in use satisfies everyone. Drought has been referred to as "a period in which the soil contains little or no moisture." How much is little or no moisture? Some clay soils may contain very little moisture in relation to the amount they are capable of holding, and yet may have a hundred times more than sandy soils that are at field capacity. Another definition of drought is "a period following 14 consecutive days without rainfall." This definition also has limitations because, although some soils may be deficient in moisture after 14 days, others can go for 50 to 90 days and still provide sufficient moisture for plant growth. "Dry-farming" techniques are based on the supposition that there will be very little rainfall during most of an entire growing season. Drought has also been defined as "a period of 21 days or more during which rainfall is 30 percent of normal rainfall for the period."

All of these definitions, though useful in various specific, isolated circumstances, are of little use in generalized descriptions of drought. Perhaps the concept of drought is best understood by using the term **drought day**—a 24-hour period during which soil moisture is limiting to the growth of plants. This definition, proposed by the American soil scientist C. H. M. Van Bavel, is useful for a number of crops, for it provides a reproducible criterion if soil moisture tension is used as a measure of water availability to plants. Obviously, it is not possible to measure moisture content on a 24-hour basis in every area where data are needed, so the drought-day criterion is based on a single measurement made at some time during the day. The time of day that the measurement is made must also be standardized, because there may be substantial variation in the moisture content of a soil during the course of a single day.

Plants respond to drought in different ways. There are many adaptations that enable plants to survive under drought conditions. Usually drought survival is determined by a complex of several factors rather than a single one. Desert plants are especially interesting in this regard, because they have several mechanisms for living through extended rainless periods.

Some desert plants persist and reproduce in areas with little rainfall by a drought-escaping mechanism. Many small plants are able to germinate, grow to maturity, flower, and produce seed within a period of several weeks. Such plants are invariably small, and are sometimes called "belly plants" by ecologists because to observe them closely it is necessary to assume the prone position. The spring rains that frequently occur even in the driest deserts are sufficient to provide these plants with enough moisture to complete their life cycle. The rest of the year, they persist as seeds, and they may remain in this state for years until there is sufficient rainfall to allow them to germinate.

Tolerance of desiccation is well demonstrated by creosote bush (*Larrea tridentata*). Leaves of this species are able to withstand desiccation to about 50 percent of their normal weight. Some of its leaves are less tolerant of drought than others, however, and these may be shed during long periods without rainfall. During extended droughts the remaining leaves may turn brown and appear to be dead. But when precipitation occurs they turn bright green and reassume their normal function.

Some plants seem to rely exclusively on deep root systems that are continually in contact with moist subsoil in order to withstand long periods without precipitation. Creosote bush, a very deep-rooted plant, never grows in soils with a hardpan, since a hardpan precludes deep root development. Even during droughts alfalfa produces relatively high yields of forage because of its deep root system, whereas the growth of shallow-rooted cereal crops and grasses is greatly retarded. It is always a surprise to find that a plant has roots that extend into the soil to a depth of a hundred feet or more, but this is not uncommon in desert climates. Except when landslides occur, one rarely has the opportunity to observe roots at great depths. Such opportunities are usually the fringe benefits of canal projects, mining operations, or well excavations.

The amount of surface area on roots influences the response of plants to drought. Plants that have large roots with few ramifications have little absorbing surface area and extract less water than those with highly divided, fine root systems. Roots that have a thick suberized surface layer absorb less moisture than those that do not.

The growth potential of roots is a significant feature of drought response. If a plant has ample supplies of food reserves, roots continue to grow through drought periods and invade new supplies of capillary soil moisture that would otherwise be unavailable. This is one of the indirect ways in which light can have an effect on drought resistance.

Drought resistance in certain species is influenced by the presence of mycorrhizal fungi. The fungal hyphae, in intimate association with the root and in contact with soil, absorb water otherwise unavailable to the plant. In the conifers, mycorrhizae function in lieu of root hairs.

Since most of the water extracted from soil by plants is lost through transpiration, leaf modifications affect drought resistance. Plants whose leaves have fewer and smaller stomata are less subject to water losses. A thick cuticle will prevent the loss of water through epidermal cells. Some plants have a thick covering of hairs on the lower surfaces. When hairs are present, the movement of air across them is slowed and humidity is more likely to build up around the stomata. As a result, the vapor pressure gradient between the inside and the outside of leaves is reduced.

Stomatal action is much more rapid in some species than in others. Extreme sensitivity in responding to mild drought by closing results in the conservation of water that would otherwise be lost. This is not only true of desert plants, but also of species of eastern deciduous forest trees. In some species, especially conifers, the great thicknesses of leaf veins per unit of leaf area make the leaves stiff and decrease the susceptibility of cells to shrinking during periods of drought. If cells do not shrink, then protoplasm is less likely to coagulate. Grasses have leaves that curl or fold during periods of drought, causing less surface area to be exposed directly to wind and radiation. This mechanism will not stop transpiration, but it slows transpiration considerably during periods of drought. Finally, cactuses and certain other desert plants dispense with leaves altogether, while chloroplasts in their stems take over the job of photosynthesis.

Excess moisture

Excess soil water affects both plants and soils. Crops with a high requirement for oxygen, such as wheat, cannot be grown on poorly drained soil. Soils that are waterlogged usually have a high CO_2 content. The carbon dioxide released by respiration does not move into the upper atmosphere, because water in the soil pores restricts diffusion. At the same time, soil oxygen is used but is not replaced. The increased CO_2 content that accompanies flooding decreases the permeability of the cytoplasmic membranes of root cells and inhibits the uptake of water. As a result, it is possible for living plants to undergo **physiological drought** while their roots are completely submerged in water. Winter flooding is usually less serious than summer flooding, because roots neither grow as fast nor absorb as much water during cold weather as during warm weather, and some plants are dormant.

After the oxygen in a waterlogged soil has been used, anaerobic decomposition of organic matter by soil organisms results in the production of marsh gas (methane), methyl compounds, and aldehydes. In addition, several kinds of ions are reduced and may accumulate to toxic levels.

Oxidized state	Reduced state
Ferric ions	Ferrous ions
Manganic ions	Manganous ions
Nitrate ions	Ammonia, nitrous oxide, nitrogen
Sulfate ions	Hydrogen sulfide and sulfide ions

High water tables restrict root development. In forested areas with a water table 45 to 60 centimeters (18 to 24 inches) below the surface, roots of many species fully occupy the surface but penetrate below the water table only a few centimeters. As a result, root systems may grow in an "inverted umbrella" configuration, which makes trees highly susceptible to being blown down by strong winds.

Poor aeration is directly related to the incidence and severity of some diseases. Plants weakened by poor aeration are much more susceptible to diseases caused by soil-borne pathogens.

Wet soils are slow to warm in the spring, with the result that the growing season for plants may actually be shorter than on adjacent dry soils. Five times the amount of heat energy is required to raise the temperature of water by $1°C$ ($1.8°F$) than is required to raise the temperature of an equivalent weight of dry mineral soil by the same amount. In one study it was found that the heat input required to evaporate water from a saturated soil would raise the temperature of a soil with optimal moisture content by more than $5°C$ ($9°F$).

Soil structure is adversely affected by water saturation, with the result that root development is restricted. In constantly flooded land the lack of periodic wetting and drying also prevents granulation.

WATER RESOURCES AND HUMAN AFFAIRS

About five thousand years ago, people living in the Indus Valley of India designed and built systems that supplied water for irrigation, swimming pools, baths, and other uses. At about the same time, the first dam was built in Egypt to store water. In ancient Babylonia, King Hammurabi oversaw the construction of a network of irrigation canals and enforced laws providing for their upkeep. The Romans' far-flung empire was possible because every bastion and outpost, no matter how remote, was always supplied with water (Fig. 13-9). Later Mohammed said that "No one can refuse surplus water to a thirsty person without sinning against Allah and against man."

Today, matters have changed only in magnitude. People still build dams, make laws providing for the equitable use of water resources, and wage battles both legal and military to determine who will avail themselves of limited water supplies. Today, the dams and the battles are bigger, and so are the problems. Furthermore, the technology that we employ today for the manipulation of water resources differs from that of years past only in its degree of precision, not much in methodology. New ideas relating to methods of water-resource development and allocation have not been put into practice for thousands of years.

Figure 13-9   In Roman times, supplying good drinking water in sufficient quantities to the large towns and cities of Auranitis (now part of Syria and Jordan) was already an acute problem. These remains of an aqueduct were part of the Roman water system. The aqueduct carried water from a spring in the north to the city of Jerash, a distance of about 100 kilometers. [Courtesy Food and Agriculture Organization of the United Nations.]

Is it possible that radically new methods of trapping and moving water may be found? Could large amounts of water be moved from one place to another as clouds, rather than through canals and pipes? Can atmospheric moisture be condensed by cloud seeding with regularity and in a magnitude large enough to provide water for cities of millions of people? Can fresh water trapped as ice at the poles be towed behind barges to places where it is needed? Can seawater be economically desalinated on a large scale? New and startling developments in the wise manipulation of the hydrologic cycle are necessary if the water requirements of future generations are to increase, as predicted, and still be met.

Agriculture, too, has its particular problems relating to water availability. In the past, at least in the United States, the issue of sufficient water for agricultural purposes has received relatively little attention. Nevertheless, the problem continues to become more and more urgent, and consequently we can expect that in the future it will be necessary to make use of new concepts in our attempts to meet the increasing demand for water. Perhaps the prayers and pleas of many persons will be answered by the use of slightly saline water, obtained at least in part from the ocean, for irrigation.

Selected references

Crafts, A. S., H. B. Currier, and C. R. Stocking, 1949. *Water in the Physiology of Plants*. New York: Ronald Press. (A monograph on plant and water relations.)

Davis, K. S., and J. A. Day, 1961. *Water, the Mirror of Science*. Garden City, N.Y.: Doubleday. (This is a short reference book on the physical and chemical nature of water.)

Fox, C. S., 1951. *Water, a Study of Its Properties, Its Constitution, Its Circulation on the Earth, and Its Utilization by Man*. London: Technical Press. (A popular treatment of the subject.)

Frank, B., 1950. *Water, Land, and People*. New York: Knopf. (A discussion of the results of misuse of water resources.)

Hexem, R. W., and E. O. Heady, 1978. *Water Production Functions for Agriculture*. Ames, Iowa: Iowa State University Press. (A consideration of the estimation and application of water production functions based on controlled experiments in the western United States.)

Hockensmith, R. D. (editor), 1960. *Water and Agriculture*. Washington, D.C.: American Association for the Advancement of Science. (Proceedings of a symposium dealing with the role of water in agricultural production.)

Kozlowski, T. T., 1964. *Water Metabolism in Plants*. New York: Harper and Row. (A short book on the function and uses of water in plants.)

Lange, O. L., L. Kappen, and E. D. Schulze, 1976. *Water and Plant Life*. Berlin: Springer-Verlag. (A comprehensive series of papers dealing with the water metabolism of plants and why it influences vegetation patterns.)

14
Crops and Climates

Crop geography deals with all of the factors related to the distribution and cultivation of crop plants. In addition to the physical and biotic environment, it must also be concerned with the social, economic, political, technological, and historical forces that shape cropping practices over the entire world. By its very nature crop geography must treat large areas and long periods of time. Even though crop geography deals solely with crop plants, its scope is broad because of the interplay of social customs and historical events with crops and cropping patterns.

Crop ecology is concerned with the relations between crop plants and their physical and biological environments, frequently with no regard for economics. Its scope ranges from continental land areas to the immediate environment of a single stoma. It deals with the same environmental factors and relations that concern plant ecology, but the focus is on crops. Today, the differences between plant ecology and crop ecology have become small, because many of the exclusive domains of the traditional plant ecologists—virgin forests, tundra, and deserts—are being critically examined to determine whether they will serve a productive function in our economic structure. It seems probable that in the future there will be little or no distinction between plant ecology

and crop ecology, because even the trees of wilderness areas and the plankton of oceans will be considered as crops to be manipulated for production.

Several examples of the complex interplay of factors involved in crop geography are in order. Historical forces have a great effect. During and shortly after the American Civil War (1861–1865) millions of acres of cropland in the southeastern United States were abandoned because of a labor shortage. A high proportion of these fields were naturally seeded by pines, which reached a peak of sawlog merchantability in the period 1920–1940. Cotton never entirely recovered its place in the regional economy in the period after the war, and as a result the agricultural base of the region changed from cotton to forest products. Tobacco production, however, did not shift because tobacco has an exceptionally high value per acre, and because the soils of the region are favorable for certain types of tobacco highly valued for the manufacture of cigarettes.

Cultural factors and personal preference also play a role in the choice of crops. Wheat recently shipped to a nation in the throes of a famine was rejected by all but the starving because they preferred rice. Even if the climate and soils of such a country were well adapted to wheat production, several generations

297

would be required for its people to develop a taste for wheat.

Technological forces are also important. In the tropics a form of agriculture is practiced in which all merchantable trees in an area are cut and removed and those remaining are felled and burned, the soil being enriched by the ashes. Several successive crops, such as hill rice, are grown on the land, until it becomes too unproductive to justify further cultivation. It is then either allowed to revert to forest or is planted to a desirable timber species such as teak (*Tectona grandis*). The repetition of this process over many years has resulted in vast changes in the vegetative cover of the landscape. This pattern of **shifting agriculture** (see Chapter 15) is considered by some to be one of the major agricultural problems of the tropics.

CLIMATE

Since climate is the principal factor that controls crop distribution, an examination of the major climates of the world will reveal the patterns that help establish the distribution of crops. In order to interpret descriptive data, it is first necessary to understand how daily weather data are compiled.

Temperature

Temperature, the simplest weather element to measure, is probably more frequently used than any other kind of data. Temperature and precipitation records are often the only ones that have been collected at many stations throughout the world, which limits the number of ways weather data can be universally expressed.

The **absolute maximum** and **absolute minimum temperatures** are the highest and lowest temperatures reached, if even for only a second, in the course of a year. They are usually accompanied by date and sometimes by duration.

The **yearly mean temperature** is the arithmetic average of the daily mean temperatures (an average of the daily maxima and the minima) over the course of a year. Mean temperatures of winter months are sometimes contrasted with mean temperatures of summer months for climatic comparisons.

The **mean maximum** and **mean minimum temperatures** are averages of the daily maxima and minima over a given period of time.

Deviation from normal is used, as it is with reference to precipitation, to express deviation from the average of the mean temperatures on any particular date. The United States National Weather Service regularly announces this information, deriving it from a "normalized" curve based on averages for a number of years.

In a limited number of locations **soil temperatures** are recorded. However, these data are frequently collected for ecological studies in locations under sods or common field grasses that may be far removed from crop fields.

Precipitation

The amount of precipitation is expressed in many ways, but perhaps most often as **average annual precipitation** (Fig. 14-1). There is a limit, however, to the time period to which an average value may reasonably be applied. Even though data for several hundreds of years are available at some stations in England, the entire period is not used because over this much time a true climatic change may have occurred. Therefore, in order for an average value to be meaningful under today's climatic regime, it should be derived from data accumulated during the past 25 to 35 years. **Mean winter precipitation** and **mean summer precipitation** are frequently used when describing the climate of a region. When precipitation is in the form of snow, the total depth of snow is given along with a factor that can be used to convert the snow to inches of water. A 1:10 ratio of water to snow is sometimes used when actual conversion data are not available.

Growing-season rainfall is an expression frequently used for agricultural purposes, but because the growing season is not the same length at all latitudes, such data are not always directly comparable. The growing season is considered to be the average

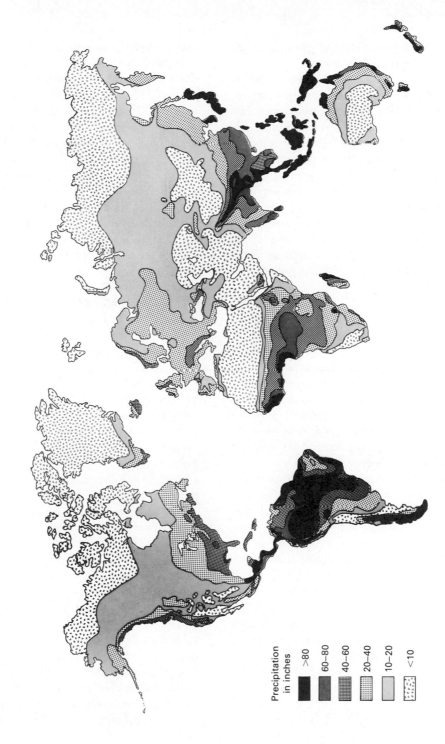

Figure 14-1 Average annual precipitation in the world. [From *A Graphic Summary of World Agriculture* (USDA Miscellaneous Publication 705), 1964.]

Precipitation
in inches

>80
60–80
40–60
20–40
10–20
<10

length of the period between the last spring frost and the first fall frost, the **frost-free season.**

Graphic techniques, showing **average monthly precipitation,** are useful for making rapid surveys. The **total monthly precipitation** is sometimes used as a general index of drought conditions, especially during summer months. When it falls below the necessary minimum determined for a particular area, drought conditions are considered to prevail.

The **deviation from normal precipitation** is an expression frequently used by weather reporters. This value (and its equivalent for temperature) may be useful when studying growth performance of native plants. **Normal precipitation** is defined as the average precipitation for any particular period.

Humidity

Atmospheric moisture may be described in a number of ways. **Relative humidity,** the most frequently used unit of measure, is the actual quantity of water vapor in the atmosphere expressed as a percentage of the maximum amount that the air is capable of holding at a given temperature. The amount of water vapor that the atmosphere can hold increases with temperature. Thus the **drying power** of air, which is proportional to the water-vapor deficit below saturation, is related to relative humidity and temperature. At high temperatures small differences in the relative humidity represent large differences in drying power; at low temperatures, differences in relative humidity represent smaller differences in drying power. **Absolute humidity** is the mass of water vapor per unit volume of air, usually expressed as grams of water per cubic meter of air. **Specific humidity** is the weight of the water vapor per unit weight of air (including the weight of the water vapor). Thus absolute humidity is a function of volume, and specific humidity is a function of weight.

Light

Unfortunately, the amount of light is not generally expressed directly in weather data. Even actual sunshine data are not generally available, and so **poten-**

tial sunlight (sunrise to sunset) obtained from tables calculated for different latitudes is generally used. Such tables do not provide a good quantitative estimate of light because of the variable nature of clouds and fog. The total hours of sunshine each day can be recorded by an instrument called a Stokes-Campbell sunshine duration recorder, which concentrates the sun's rays and burns a line in a strip of paper.

The length of the daily photoperiod varies greatly at different latitudes. Natural vegetation has responded to latitudinal variations by adapting to photoperiod. Thus many plants cannot be successfully moved from one latitude to another even though other environmental factors are compatible.

THE KÖPPEN CLASSIFICATION OF CLIMATES

The climatic classifications of the greatest agricultural value are those based on the interactions of temperature and precipitation. The most widely known and used system was devised by the Austrian geographer Wladimir Köppen. It is based on temperature, precipitation, seasonal characteristics, and the fact that natural vegetation is the best available expression of the climate of a region. A distinctive feature of the **Köppen system** is its use of symbolic terms to designate climatic types. The various climates are described by a code consisting of letters, each of which has a precise meaning.

Köppen identified five basic climates: *A,* tropical rainy; *B,* dry; *C,* humid, mild-winter temperate; *D,* humid, severe-winter temperate; *E,* polar. Each is subdivided to describe different subclimates, which are denoted by a combination of capital and small letters. The capital letters *S* (steppe) and *W* (desert) subdivide the *B,* or dry, climates. Similarly *T* (tundra) and *F* (icecap) subdivide the *E,* or polar, climates. Small letters further differentiate climates. Some of these are indicated in the following list, in which the basic climates to which the small-letter codes pertain are given in the column to the right. (A partial listing of the major and minor categories is given in Table 14-1 and in Fig. 14-2.)

Table 14–1 Selected climatic types according to Köppen.

Symbol	Major climate	Chief characteristic	Subclimates (incomplete)
A	Tropical rainy	Coolest month above 18° C (64.4° F)	*Af*, tropical rainforest *Am*, monsoon rainforest *Aw*, tropical savannah
B	Dry	Evaporation exceeds precipitation	*BS*, steppe *BSh*, tropical and subtropical steppe *BSk*, middle latitude steppe *BW*, desert *BWh*, tropical and subtropical desert
C	Humid, mild-winter temperate	Coldest month between 18° C (64.4° F) and 0° C (32° F)	*Cs*, Mediterranean (dry-summer subtropical) *Csa*, hot-summer Mediterranean *Csb*, cool-summer Mediterranean *Ca*, humid subtropical (warm summer) *Caw*, with dry winter *Caf*, with no dry season *Cb, Cc*, marine west coast (cool summer)
D	Humid, severe-winter temperate	Coldest month below 0° C (32° F); warmest above 10° C (50° F)	*Da*, humid continental, warm summer *Daw*, with dry winter *Daf*, with no dry season *Db*, humid continental, cool summer *Dc, Dd*, subarctic
E	Polar	Warmest month below 10° C (50° F)	*ET*, tundra *EF*, icecap

a = Warmest month above 22°C (71.6°F)	*C, D*	
b = Warmest month below 22°C (71.6°F)	*C, D*	
c = Warmest month below 22°C (71.6°F); less than 4 months below 10°C (50°F)	*C, D*	
d = Coldest month below −38°C (−36.4°F)	*D*	
f = No distinct dry season	*A, C, D*	
h = Hot; average annual temperature above 18°C (64.4°F)	*B*	
k = Cold; average annual temperature below 18°C (64.4°F)	*B*	
m = Monsoon (short dry season)	*A*	
n = Frequent fog	*B*	
s = Dry season in summer	*C*	
w = Dry season in winter	*A, C, D*	

Critics have expressed the opinion that the Köppen classification is based on too few kinds of data, and that boundaries between the various climatic regions are too arbitrary. But in spite of these objections this system has gained widespread recognition and use. Its simplicity and general adherence to vegetational zones has made it the basis for many revisions and other classifications.

Tropical rainy climates (A)

Tropical rainy climates are characterized by the absence of winter; there is no month with an average temperature less than 18°C (64.4°F). This type of climate prevails over about 36 percent of the total surface of the earth and over about 20 percent of its land surface. Rainfall is usually abundant, seldom less than 75 centimeters (about 30 inches) per year. The principal climatic types in the humid tropics are distinguished by the amount and distribution of rainfall.

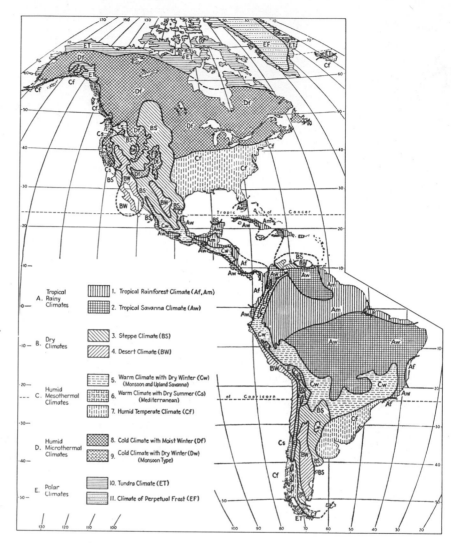

Figure 14-2 Climates of the earth. [From G. T. Trewartha, *An Introduction to Climate,* McGraw-Hill, 1954.]

Tropical
A. Rainy
Climates

IIII 1. Tropical Rainforest Climate (Af, Am)

2. Tropical Savanna Climate (Aw)

B. Dry
Climates

3. Steppe Climate (BS)

4. Desert Climate (BW)

C. Humid
Mesothermal
Climates

5. Warm Climate with Dry Winter (Cw) (Monsoon and Upland Savanna)

6. Warm Climate with Dry Summer (Cs) (Mediterranean)

7. Humid Temperate Climate (Cf)

D. Humid
Microthermal
Climates

8. Cold Climate with Moist Winter (Df)

9. Cold Climate with Dry Winter (Dw) (Monsoon Type)

E. Polar
Climates

10. Tundra Climate (ET)

11. Climate of Perpetual Frost (EF)

Tropical rainforest (*Af*)

Tropical rainforest is found in parts of Africa, Central America, Brazil, Madagascar, the Philippines, and elsewhere. Its climate is characterized by uniformly high temperatures and heavy, more or less uniformly distributed precipitation. Even though temperatures may be more moderate than those in higher latitudes, the high humidity makes the heat oppressive. Violent thunderstorms occur daily at the same hour and with clocklike precision during certain parts of the year.

Plants that grow in tropical rainforests are called **megatherms,** and they require continuous high temperatures with abundant precipitation. The vegetation of tropical rainforests is vigorous and luxuriant. The number of plant species is large. The trees are tall and broad-leaved, held together as a unit by

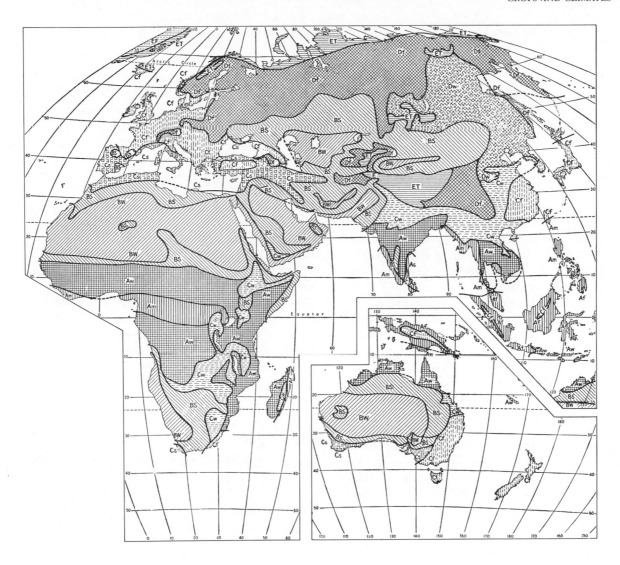

vines and lianas. During World War II, when camp-sites were cleared in the midst of tropical rainforests, a few trees were nearly always left for shade, but this practice was discontinued because high winds blew many of the unsupported trees down.

Soils of the rainforest are deeply weathered be-cause high temperatures intensify chemical action. When cultivated they are quickly exhausted unless heavily fertilized. Rainforest soils may also lose their porosity under cultivation, in which case poor aera-tion becomes a problem.

Monsoon rainforest (*Am*)

The monsoon rainforest is characterized by a long season of fairly evenly distributed rainfall, which supports rainforest vegetation, and a short dry sea-son. The classic monsoon climate is found in India, Indochina, Burma, and the eastern Amazon valley.

Tropical savannah (*Aw*)

The climate of tropical savannahs is transitional between dry climates and those of tropical rainforests. There is less yearly rainfall than in rainforests and it is less evenly distributed throughout the year. Temperatures are much like those of the rainforest climate (*Af*), but yearly ranges are greater. There are distinct cool-dry and hot-wet seasons. Savannah climates are found in South America, Africa, and Australia.

The natural vegetation of tropical savannahs consists typically of tall, coarse grass punctuated with clumps of trees, giving a parklike appearance. Bunch grass, which may reach heights of 12 feet, predominates. The trees are always widely spaced and are leafless during the dry season.

Savannah soils are lateritic. They are low in organic matter, poor in fertility, but porous and easily cultivated. In dry savannah regions, the soils resemble prairie soils.

Owing primarily to the difficulty of weeding and keeping land clear so that crops can be established, the major agricultural exports of tropical rainy (*A*) climates are products of woody plants—trees, shrubs, and canes. Woody perennial plants, which need be replaced only periodically are much less costly to establish than annual crops. The culture of field crops is limited. Important crops include cacao, banana, sugarcane, coffee, rubber, black pepper, vanilla, and sisal.

Sugarcane (*Saccharum* spp.) requires abundant sunshine, 150–165 centimeters (about 60–65 inches) of rain annually, and an ample supply of mineral nutrients. Climates in which there is bright sun and frequent short periods of heavy precipitation are best. Although strictly a tropical plant, sugarcane is cultivated in some parts of the world that are on the margins of regions with tropical climates. In Louisiana, where the frost-free season lasts about 250 days, it is cut in an immature stage.

Cacao (*Theobroma cacao*) beans, used for cocoa and chocolate, grow in pods 15–25 centimeters (6–10 inches) in length and 7.5–12.5 centimeters (3–5 inches) in diameter. The trees grow up to 12 meters (40 feet) tall, bear fruit for as long as 50 years, and have two harvests per year (one more abundant than the other) in well-adapted regions. *Af* climates are best, for cacao requires at least 200 centimeters (about 80 inches) of rainfall. Average annual temperatures should be 24–27°C (75–81°F).

Bananas and plantains (*Musa* spp.) are adapted to *Af* and *Aw* climates. They require at least 190 centimeters (75 inches) of water annually, a temperature of 24°C (75°F), and fertile, well drained soil.

Coffee (*Coffea* spp.) is indigenous to Africa, but has been introduced throughout the world wherever climate permits. It is better adapted to *Aw* climates than to *Af* climates, and is grown from sea level to about 2100 meters (nearly 7000 feet). Some species, however, such as *Coffea arabica,* can be grown at higher altitudes. The mean annual temperature in most coffee-growing regions is at least 20°C (68°F) or higher; coffee is not grown in regions with a mean minimum of less than 13°C (55.4°F). From 115 to 180 centimeters (about 45 to 70 inches) of rainfall is required annually, and good soil drainage is a requisite.

Dry climates (*B*)

The main features of dry climates are low humidity, scanty and unreliable rainfall, intense radiation, and violent winds. Desert (*BW*) and steppe (*BS*) are the principal subtypes.

Potential evapotranspiration (transpiration and evaporation from land and water surfaces) exceeds precipitation in dry climates. In general, the drier the climate, the more uncertain the rainfall. Since dryness is defined as a function of potential evapotranspiration, it is difficult to associate a particular temperature with dryness. For example, Paris has a cool marine climate but receives only about 58 centimeters (23 inches) of rainfall per year; it is in the midst of forest vegetation typical of humid climates. In the United States temperatures are much higher in areas with such low rainfall, and no forests grow in regions with less than 64 centimeters (about 25 inches) of precipitation per year.

Desert (*BW*)

Annual rainfall in most deserts is usually less than 13 centimeters (about 5 inches). In Cairo, rainfall averages less than 5 centimeters (2 inches), whereas in some parts of the world, such as Calama, Chile, there is little evidence to show that rain has ever fallen. When rain does fall in desert climates, it is likely to create local torrential floods.

Desert vegetation is notable by its scarcity. When present, it is highly seasonal in character, bursting out after rainfall and rapidly completing its life cycle or yearly growth (if perennial) in a period of about a month, before available soil moisture is depleted. The mechanisms by which plants survive drought are discussed in Chapter 13. Wide spacing is characteristic of desert vegetation, with bare soil between shrubs and bunch grasses (Fig. 14-3).

Deserts of the middle latitudes tend to be cooler than those of lower latitudes, and frequently have a partial plant cover. The meager vegetation has slight value as food for livestock, but has been exploited for this purpose in practically every part of the world. In contrast to the highly leached soils of wet climates, desert soils are often highly fertile and may become productive when irrigated, but they require careful management. They are highly variable, however, and tend to be alkaline. The accumulation of salts (salinity) is a recurrent problem.

Steppe (*BS*)

Steppe climates are semiarid. Where rainfall is greatest, short grasses prevail; elsewhere, bunch grasses dominate the vegetation. The steppes of the South Africa veldt (grassland in which there may also be scattered shrubs) once supported vast herds of wild animals. It has recently been established that the native animals produce more meat for human consumption than the beef cattle that have been introduced.

Steppe soils are dark in color and have no well developed structure. Those of tropical regions are generally less well developed than those of higher latitudes. Although cultivation is easy, there is not

A

B

Figure 14-3   (A) Desert vegetation is always widely spaced. In some areas the spacing is so regular that the desert shrubs appear to be planted. (B) In contrast, green fescue growing in more humid climates forms a dense ground cover. [A courtesy U.S. Atomic Energy Commission; B courtesy USDA.]

enough rainfall to support a well-developed agriculture unless irrigation is provided.

Wheat (*Triticum* spp.), one of the world's principal grain crops, is well adapted to steppe climates, although it grows well in many other climates. Climate determines the type of wheat that must be grown. High-protein (hard) wheats are grown in *Bs* climates; the starchy (soft) wheats are most frequently grown in *Db* climates. There is not a month of the year during which wheat is not harvested somewhere in the world.

Humid, mild-winter temperate climates (C)

Humid, mild-winter climates are characterized by seasonal variations in temperature. Plant dormancy is usually caused by cold rather than drought. Weather conditions in *C* climates are likely to be changeable, owing to conflicting masses of air from tropical and polar sources. Cyclonic storms are frequent. Regions with *C* climates produce a wide variety of crops, including forage grasses, olives, cotton, grapes, tobacco, cereal grains, and vegetables. There are three principal climatic subtypes: dry-summer subtropical, or Mediterranean (*Cs*), humid subtropical (*Ca*), and marine west coat (*Cb*).

Dry-summer subtropical (*Cs*)

Dry-summer subtropical climates are characterized by hot, dry summers, mild winters with rainfall, and brilliant sunshine throughout the year. They are found on the borders of the Mediterranean Sea, in central Chile, in South Africa, in southern Australia, and in central and coastal California.

Because of their delightfully mild winters and bright sunny weather, these are among the best-known climates, even though they are found in only 2 percent of the world's land area.

Average winter temperatures are typically 4.5–10°C (40–50°F) and average summer temperatures are 21–27°C (70–80°F). Rainfall averages 38–64 centimeters (15–25 inches), less than enough to support a forest vegetation of large trees with a closed canopy. The effective precipitation is high, since the bulk of the rain comes in the winter. Summers are generally hot and dry.

Soils are varied because of the diversity of parent materials. In the Mediterranean region a distinctive soil, known as **terra rosa** (red soil), has developed on the hard limestone bed rock. Soils tend to be easily leached when cultivated. Soils in humid regions with abundant vegetation tend to be poor, whereas those in regions with less rainfall are often very fertile. At both extremes of precipitation, productivity of crop plants may be poor.

Vegetation in this climate is variable and falls into complex patterns. No single type dominates, although trees are always present except in local areas where soil type or fire is the dominant ecological factor. Glades and open woodland constitute the tree vegetation. Shrubs and trees are usually dwarfed and have small, thick, glossy leaves that prevent excessive transpiration. The cork oak (*Quercus suber*) is characteristic, as is the olive. With irrigation, it is possible to grow an extremely wide variety of crop plants.

The fruit crops of *Cs* climates include citrus, fig, date, apricot, olive, grape, and plum. The dry heat of summer is ideal for fruit drying; the fruit is merely laid out in the sun. California vegetables grown under irrigation are shipped throughout the United States and Canada because of their high quality and winter availability.

Humid subtropical (*Ca*)

Humid subtropical climates have abundant precipitation, which may be concentrated in the summer, and are usually found on the eastern sides of continents. Every continent has regions dominated by this type of climate, which has a high potential for the production of agricultural products. There is nearly always enough precipitation to support crops that mature in late summer or in early fall. Summer temperatures average from 24°C to 26.5°C (75°F to 80°F), and winter temperatures average from 4.5°C to (40 to 55°F). There is a wide difference between day and night temperatures, sometimes as much as 20°C (36°F).

Precipitation ranges from 75 to 165 centimeters (30 to 65 inches) per year. There is abundant snow, but in lower latitudes it may remain on the ground for no more than a day. Violent typhoons sometimes occur near land masses, and late summer rains and hailstorms may destroy nearly ripened crops.

Rice (*Oryza sativa*) is the major cereal crop of this climate. It is an annual grass that germinates and grows in an aquatic environment and has specialized tissues through which oxygen moves to roots. Although grown over a range of temperatures, 20–21°C (68°–70°F) is best for grain formation. Rice can be grown in a wide variety of soils, the main requirement being that the soil must hold water for a considerable period of time.

Cotton (*Gossypium* spp.) is used throughout the world. Some species of the genus are native to the New World and others to the Old World. They grow in a wide range of climates, but do best where there are at least 200 frost-free days, an average growing-season temperature of at least 15–17°C (59–63°F), and light soils.

Tobacco (*Nicotiana* spp.) makes very rapid growth during a short season and quickly exhausts the soils of its mineral elements. In some areas it is the main agricultural crop.

Marine west coast (*Cb* and *Cc*)

Marine west coast climates are usually found in relatively narrow strips along the west coasts of continents in the middle latitudes. Characteristic air masses move in from the oceans, resulting in cool summers and moderate winters. Mean summer temperatures are seldom higher than 20°C (68°F). Mean winter temperatures are above freezing, although killing frosts are not unusual. Frequently the transition from a marine climate to a severely cold one (as in Scandinavia) or from a marine climate to a very dry one (as in California) may take place within a mile or two.

In some places there is as much as 380 centimeters (150 inches) of precipitation per year but in others no more than 50 centimeters (20 inches), which nevertheless may be adequate for crop growth because of

cool temperatures. Snow is usually abundant only at high elevations. Always accompanying the precipitation is a high degree of cloudiness. Fog and mist are common, and the sun may not be seen for several weeks.

Soils are quite variable and generally deep. For the most part they are capable of good agricultural production.

Evergreen forests are predominant in some areas, deciduous ones in others. In North America, gigantic redwoods that grow to more than 3 meters (10 feet) in diameter occupy the foggy western slopes of California's coastal ranges. In Chile, the forests are of broad-leaved evergreen trees. Evergreen shrubs, such as heather, gorse, and juniper, take the place of trees in areas that have been subjected to severe glacial scouring, as in Wales and Scotland. In Europe, the dominant trees are deciduous.

These are rich climates for such cool-season crops as apple, pear, strawberry, peas, and lettuce, but too cool for growing many warm-season crops outdoors, with the result that greenhouses are used in Holland and England to grow such crops as tomato and cucumber.

Humid, severe-winter temperate climates (*D*)

There are two main subdivisions of *D* climates that are of agricultural importance, both distinguished by cold, snowy winters and wide annual ranges in temperature. Because their controlling climatic forces originate over large land masses, they are frequently called continental climates. Only in the Northern Hemisphere, and there only in North America and in Eurasia, are there land masses large enough for their development. The climates of Canada and the Soviet Union are typical.

Soils at the northern limits of *D* climates are podzols with a heavy cover of litter. They are poor for agricultural purposes, but they support dense stands of conifers. Soils of the deciduous forests have podzolic characteristics, but generally they are better than those further north. Grasslands have fertile, dark, prairie soils. Their general use for agricultural production testifies to their desirability. The grasses are tall and originally supported large herds of animals. Today, there are few places where relicts of this vegetation in North America have been preserved. The forests were originally maple and mixed oak and hickory in the southern parts, with conifers along the northern edges.

Humid continental with warm summers (*Da*)

Conditions in *Da* climates are excellent for the growth of certain crops. Mean temperatures may be as much as 30°C (86°F), and highs of 38.5°C (105°F) have been recorded. There may be as many as 200 frost-free days. The climate of the north-central United States is an excellent example of this type.

Most rainfall occurs in the summer months. In some years rainfall is sparse, resulting in drought. Snows are occasionally extremely deep, but winter precipitation does not usually exceed 13 centimeters (about 5 inches), whether it is rain or its equivalent in melted snow. Snow cover prevents losses of heat from the earth during the winter, with the result that the soil warms faster than might be expected on the basis of air temperature measurements.

Corn (*Zea mays*) has reached its greatest development in *Da* climates, but it also does well in some of the *Db*, *Ca*, and *Cb* climates. A native of the American tropics, by 1492 it was grown throughout North America by the Indians. In the corn belt of the United States the mean summer temperature is 14.5°C (58°F) or higher. Temperature requirements for the germination of corn are 10°C (50°F). The most critical period for soil moisture is shortly before and after the silking time. Corn is a major crop on every continent that has large areas of *Da* climate.

Humid continental with cool summers (*Db*)

Regions with the *Db* climate lie to the north of those with the *Da* type, and are primarily wheat-production regions. Although summers are a few degrees cooler than those of the hot-summer type, winters are 5–15°C (9–27°F) colder. The long summer days in higher latitudes compensate to some extent for the colder temperatures. But in all, the climate is dominated by winter. In the parts of Europe subject to this climate, winter temperatures are tempered somewhat by oceanic air masses. The frost-free season may be no longer than 50 days.

Precipitation is highly variable. There is, however, considerable snow, and the snow cover may last for more than 120 days. Crops grown in *Db* climates include potato, pea, sweet corn, turnip, carrot, cabbage, and other plants that need only a short growing season.

Polar climates (*E*)

Polar climates are agriculturally unimportant. The presence of the sun above the horizon for six months, its absence for an equal period, the bitter cold, and the perpetual snows or icecaps (in *EF* climates) may signal high adventure for those who wish to test their courage, endurance, and skill, but they hold no appeal for the agriculturist.

There are, however, interesting possibilities for storing surplus agricultural produce (if it ever again exists) in snowpacks and ice that are not subject to lateral movement. The only cost would be that of transportation.

Not all polar climates are characterized by permanent snow. In the warmer tundra (*ET*), which is found over a large area in the Northern Hemisphere, there is a brief growing season and meager vegetative cover, even though the soil may be permanently frozen to a depth of a few inches to a few feet below the surface, a condition called **permafrost.** Tundra vegetation is basically of two types: one a carpet of moss and lichens; the other a mixture of mosses, sedges, flowing herbs, and low woody shrubs. These shrubs (often trees in warmer climes) are dwarfed and stunted, have distorted and crooked stems, and are known as elfinwood, or **krummholtz.** The soil in which tundra vegetation grows may be no more than a few centimeters deep—an environmental characteristic shared by plants that grow at high elevations.

THE THORNTHWAITE CLASSIFICATION OF CLIMATES

The classification of climates proposed by C. W. Thornthwaite resembles Köppen's insofar as combinations of letters are used to designate individual climates and plant response is used to integrate the climatic elements. Three factors enter into Thornthwaite's description of climatic types—precipitation effectiveness, seasonal distribution of precipitation, and temperature efficiency.

Precipitation effectiveness is represented by a P-E ratio, in which monthly precipitation is divided by monthly evaporation. The P-E index is the summation of monthly P-E ratios. There are five humidity provinces (designated by capital letters), each associated with a particular vegetation type:

Letter	Humidity province	Vegetation	P-E index
A	Wet	Rainforest	> 127
B	Humid	Forest	64–127
C	Subhumid	Grassland	32–63
D	Semiarid	Steppe	16–31
E	Arid	Desert	< 16

The humidity provinces are subdivided into subtypes (designated by lower-case letters) on the basis of seasonal distribution:

Letter	Distribution of precipitation
r	Abundant in all seasons
s	Sparse in summer
w	Sparse in winter
d	Sparse in all seasons

The **temperature efficiency** (T-E) index is obtained by summing the monthly mean Fahrenheit temperatures, subtracting 32°F, and dividing by 4. There are six temperature provinces, each designated by a capital letter marked with a prime:

Letter	Temperature province	T-E index
A′	Tropical	> 128
B′	Mesothermal	64–128
C′	Microthermal	32–64
D′	Taiga	16–32
E′	Tundra	0–16
F′	Perpetual frost	< 0

Of the 120 possible combinations, only 32 are recognized as actual climatic types. Figure 14-4 illustrates the relations between P-E and T-E indices and the natural vegetation. The world distribution of climates according to the Thornthwaite classification is shown in Figure 14-5.

CROP GEOGRAPHY

The distribution of natural vegetation depends on (1) environment, (2) plant response, (3) plant migrations, and (4) the evolution of floras and climax vegetation. These all have somewhat different meanings

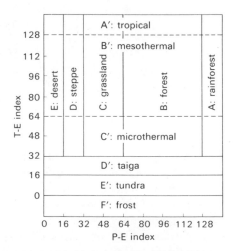

Figure 14-4 Temperature and humidity provinces, based on natural vegetation. [After C. W. Thornthwaite, *Geographical Review* 21:649, 1931.]

Figure 14-5 Climates of the earth. [From G. T. Trewartha, *An Introduction to Climate*, McGraw-Hill, 1954.]

for crops. For example, the purposeful introduction of exotic species by people is quite a different matter from the natural development of floras by migrations that occur over tens of thousands of years. There is, however, a remarkable coincidence between the occurrence of certain types of natural vegetation throughout the world and the occurrence of agricultural regions (Fig. 14-6). The principal factors that influence the geographical distribution of crops are discussed in the following sections.

Climatic control

Plant geographers have always acknowledged the relation between the climate of a region and its natural vegetation. Climatic zonation coincides in a broad and general way with **climax vegetation,** the ultimate type that can develop under a particular local pattern of soil parent material, topography, temperature, precipitation, and other meterological conditions. As a result, natural vegetation has been used as a kind of meteorological "instrument" to define boundaries in the classification of climate, with the result that savannah, steppe, and rainforest are names for climates as well as vegetation types. Climate is also a primary controlling factor in the growth of crops. The effects of climate on crops can, of course, be altered. Inadequate or excessive moisture may be compensated for by irrigation or drainage. Low fertility may be overcome through the application of fertilizers. In greenhouse production

practically all climatic influences can be modified. But fertilization, irrigation, and other cultural practices can be used only to the extent that they yield sufficient return to the grower. It is expensive to modify natural climatic influences.

In 1840 Justus Liebig formulated a hypothesis that plant growth is limited by the nutrient available to it in least quantity. This concept, now popularly known as the **law of limiting factors,** can be expanded to include all environmental factors that contribute to

Figure 14-6 Natural vegetation of the world (*above*) compared with approximate cropland area (*facing page*). [From *A Graphic Summary of World Agriculture*, USDA Miscellaneous Publication 705, 1964.]

Low latitude forest
Tropical rain
Lighter tropical
Scrub and thorn

Middle latitude forest
Mediterranean scrub
Coniferous
Broadleaf and mixed broadleaf–coniferous

Grasslands
Savanna
Prairie
Steppe

Desert and waste
Desert
Tundra
Ice cap
Undifferentiated highlands

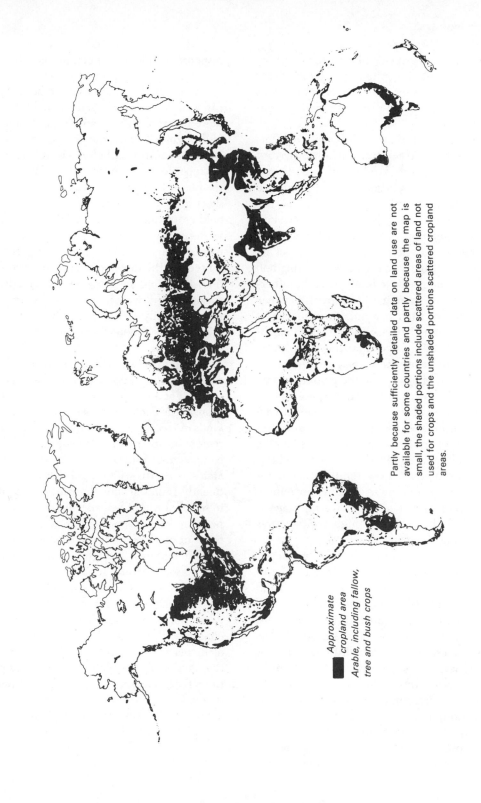

Approximate
cropland area
Arable, including fallow,
tree and bush crops

Partly because sufficiently detailed data on land use are not available for some countries and partly because the map is small, the shaded portions include scattered areas of land not used for crops and the unshaded portions scattered cropland areas.

[313]

growth. It implies that growth, like a chain with its proverbial weakest link, is limited by the factor that is least favorable. However, all sorts of interactions and compensations come into play with environmental factors.

Although plant growth is affected by subtle differences in climate, the extremes produce the greatest effects. Thus, the climatic extremes of a particular region may be more important in maintaining a vegetation type than the average. An occasional severe period of drought may preclude the development of forest vegetation even though the average annual precipitation is adequate. Cotton cannot be grown in regions that have heavy rainfall during boll formation, even though the climate may be otherwise suitable. Climatic extremes have great influence on the distribution of such perennial crops as fruits (Table 14-2). Although citrus can be grown in northern Florida and southern Alabama, the hard freeze that can be expected every few years precludes the establishment of an industry in these areas. Some tropical crops, such as cacao and bananas, are injured by low temperatures above freezing. The relation between temperature and the growth of temperate-zone crops may be quite complex. For example, areas of peach culture are bracketed by high and low temperature requirements. Low temperature is required for dormancy, but injury follows extremely low winter temperatures, and the flowering stage is very sensitive to frost. Relatively high temperatures are required during the growing season.

The boundaries of agricultural regions are determined by climatic factors. In the Northern Hemisphere the limits are determined by low rainfall and low temperature. In Africa low rainfall is limiting both in the south and in the north. In South America the boundaries are set by low temperatures in the south, the harsh conditions of high altitudes and poor soil near the equator, and low rainfall in the northeast and along parts of the Pacific coast. In Australia intensive production is confined to coastal areas that have adequate rainfall.

The action of any single environmental factor is affected by all other environmental factors; that is, the environment is **holocoenotic.** The physiological responses of plants are conditioned by the collective action of the environment.

The stress that one environmental factor imposes can sometimes be relieved by making another factor more favorable. For example, the debilitating effects of drought may be somewhat less severe in crops that were well fertilized before the drought began.

Opposite climatic types may be compensatory. In the far north, long days may partially compensate for short growing seasons and permit the cultivation of field crops that otherwise could not be grown. In the western United States, roughly 100 to 150 centimeters (about 45 to 60 inches) of precipitation is required for the development of mature forest vegetation, including such species as Douglas-fir. Yet this species grows well in England, where precipitation may be only 60–100 centimeters (about 25–40 inches), but where heavy mists and fogs reduce transpiration.

The compensation of one environmental factor for another is most common near the edge of a species' range. For example, spruce (*Picea*) and fir (*Abies*) grow in the cooler climates of high latitudes, but both kinds of trees grow at high elevations far south of the region of their best development. Thus elevation compensates for latitude. In the southern United States, eastern white pine (*Pinus strobus*) is frequently found at low elevations far south of regions in which it might be expected on the basis of its temperature requirements, but it is always found in coves or on slopes that are sheltered by natural topographic features. Topography can therefore compensate, to some extent, for latitude.

Climate exerts a major influence on natural vegetation through its effect on soil development. Natural vegetation progresses through successional stages in concert with the development of the soil. If climatic elements or the parent material are adverse to either the physical or the biological processes of soil formation, there can be no progress to more advanced stages.

Soil may become the limiting factor for some species even when all other climatic factors are favorable. For example, soils composed mainly of quartz sand, which is extremely resistant to weathering, can

Table 14–2 Classification of common fruit crops by temperature requirements.[1]

		Temperate	
Tropical	Subtropical	Mild winter	Severe winter
coconut banana cacao mango pineapple papaya			
	coffee date fig avocado		
	citrus olive pomegranate		
		almond blackberry grape (European) persimmon (Japanese) quince	
		peach cherry apricot (blossoms tender) strawberry ⎱ (very hardy with blueberry ⎰ snow cover) raspberry cranberry	
			pear plum grape (American)
			currant apple
Sensitive to low temperatures	Slightly frost tolerant	Tender	Winter-hardy
Not cold requiring		Cold requiring	

[1]Variation in tolerance depends to a large extent on species, variety, plant part, and stage of growth.

never develop into deep, rich soils such as those in regions where limestone is the parent material. Limestone is dissolved by carbonic acid, which is a natural component of rainwater and also forms from decomposition of organic matter. Some species of trees require deep soils that allow the development of extensive root systems. Sandy soils may not have enough water-holding capacity for the production of some crops, even though rainfall is abundant. Conversely, if the water table is too high, normal root development may be restricted. Soils that hold water on the surface make good rice fields but will not support corn. In some soils a deficiency or excess of minerals can be the limiting factor. Nevertheless, soil must be considered of secondary importance in controlling the distribution of crops. If soil conditions are not extreme they can often be altered to render the soil productive. If the demand for a particular crop warrants the expense, the soil may be physically and chemically altered. Fertilization, irrigation, subsoiling, drainage, and fumigation are the most common means used to alter soils. These will be discussed in greater detail in Chapters 16 and 17.

Environmental gradients

Environmental factors vary along gradients from one area or region to another. Probably the most extreme example of an environmental gradient is the change in temperature from the equator to the poles. Striking temperature gradients commonly occur where topography causes cold air to drain from smooth, gentle slopes and accumulate at the bottom. Temperature, humidity, and precipitation also vary with distance from bodies of water. Precipitation declines from 115 centimeters (about 45 inches) or more per year in the deciduous forests of the eastern United States to less than 25 centimeters (about 10 inches) on the Great Plains.

Environmental gradients are usually not steep except where topography or some other factor intensifies them. The temperature gradient from the coast to the Piedmont in the southeastern United States is very gradual. Similarly, the precipitation gradient between the deciduous forest and the Great Plains is gentle. Soil moisture shows steeper gradients, for even within a few meters moisture conditions may change greatly. The sharp changes in environmental conditions where topography is irregular are troublesome for crop producers, who frequently wish to establish large production units that are managed uniformly. Large-scale agriculture is difficult in mountainous areas.

The environment is constantly changing, although the changes occur only very gradually. The Wisconsin glacial stage, the most recent advance of the ice sheet that once covered a large part of North America, ended about 12 thousand years ago. The climate has since shown a warming trend, and the extent of glacial ice in high mountain valleys has been observed to decrease markedly during the past 50 years. Other major climatic changes have resulted in a change in vegetation from spruce and fir to such plants as oak, hickory, and grasses, which can tolerate drier climates.

Because climatic change is gradual, many crop growers would maintain that it is not realistic to take it into account, but within the large climatic changes and cycles there are smaller changes that can be witnessed within a generation. For example, the United States National Weather Service has abandoned the practice of using the total weather records available for a station to determine mean and average values, and now uses a shorter period of time, 25 to 30 years. However, in the tropics, an even shorter period may be used because of the more nearly constant weather conditions.

Although environmental changes are gradual, there is often a lag in the kinds of natural vegetation that occupy an area. Plant migration does not always immediately follow site availability. This means that some species may be successfully introduced into regions in which they are not native because the rate of migration has not kept apace of climatic change. Slash pine (*Pinus eliottii*), native to Florida and the Gulf coast, is a notable example; this species has been planted as far north as North Carolina, where it is sufficiently productive to justify planting for timber.

Species range and tolerance

The **tolerance** of a species is represented by the range of climatic and soil conditions within which it can exist and reproduce. The expression **ecological amplitude** is sometimes used to describe this range of conditions. There is great variation in tolerance. Some species can exist only in a very narrow range, whereas others can tolerate a broad spectrum of environmental conditions.

In the ecological literature many theories of tolerance have been proposed. Most are concerned with the historical development of natural vegetation and the function that time has served. V. E. Shelford's generalized law of tolerance is of great interest and value to ecologists and geographers, and is defined in terms applicable to crops. Shelford's concepts can be stated as follows:

1. Species with a wide range of tolerance are likely to be widely distributed.

2. Some species may have a wide range of tolerance for one environmental factor and a narrow range for another.

3. When one environmental factor is limiting, the range of tolerance to other factors is likely to be decreased.

4. The range of tolerance is likely to be narrowest during the period of reproduction.

Tolerance has a genetic base. Environment can influence the growth and development of organisms only within genetically imposed limits. Species with a wide range of genetic diversity are more likely to respond favorably to new and changed environments than to those with a narrow genetic base. The evolutionary forces that determine the range of species involve differentiation and speciation. Long-term forces apply only incidentally to crop plants, because people have imposed evolutionary change through artificial hybridization and other kinds of genetic manipulation. Adaptability becomes one of the main goals of modern plant breeding (see Chapter 19). The relatively slow migration of natural species, hampered in many cases by natural topographical barriers, has given way to widespread and rapid introduction of exotic species.

Plants in different stages of growth may have different tolerances. Plants, like other organisms, change in structure as they develop. The developmental phases respond differently to environmental factors. The limits of tolerance are often narrow in the seedling and juvenile stage. For example, the stem of a newly generated long-leaf pine seedling has no protective bark and may be killed quickly by temperatures of 60°C (140°F) or less. The same tree, when 10 years old or more, may tolerate temperatures approaching the boiling point for a few minutes, because by that time a thick layer of bark will

have developed. Tobacco plants must be protected when young, but after a few weeks they grow well under field conditions. As a seedling, the roots of baldcypress (*Taxodium distichum*) must not be submerged, but later in its life the water level of the habitat in which it grows is often up to the trunk.

In practically all crops, the range of tolerance during the flowering period is relatively limited. Excessively hot dry weather during flowering of corn can cause crop failure if pollen is released before silking occurs, and it can even kill pollen. Similarly, cold wet weather during apple bloom adversely affects pollination because bee activity is reduced.

Because the geographic ranges of species are tolerance-limited, much effort has gone into a worldwide search for areas with similar climates in order to weigh the prospects of successfully introducing crop species from one area to another. Areas with similar climates are called **homoclimes,** or with reference to agricultural production, **agroclimatic analogues.** The crop ecologist M. Y. Nuttonson has studied agroclimatic analogues of the United States using mean temperatures, dates of killing frost, average precipitation, precipitation effectiveness, and other types of climatological data. Beer Menuha, Israel, and Calexico, California, were found to have comparable climates, and are nearly thermal equivalents. Irrigated barley is the most important cereal crop in Calexico because it tolerates soil salinity and grows under a wide variety of soil conditions. Since the same conditions prevail in Beer Menuha, barley might be expected to grow profitably there if equivalent cultural techniques are used.

Biotic factors

The members of the biologic community, in their roles as parasite, predator, symbiont, competitor, or vector of pollination and propagation, strongly influence the distribution of natural plant communities. Consider the role of the vast herds of animals in helping to maintain the great South African veldt.

Biological factors also have great significance on the distribution of crops. The crop producer works in situations in which the natural biological equilibria are disturbed. Under these conditions the biotic factors of greatest concern are populations of crop pests. The inability of the crop producer to exert sufficient control over a predator or disease can virtually eliminate a crop species. Thus Panama disease has eliminated commercial banana production from large areas of Central America. A number of diseases of the *Hevea* rubber tree have had a great effect in moving the industry from the native habitat of *Hevea* in Brazil to Southeast Asia. The bacterial disease fire-blight has eliminated commercial pear production in most of the warm humid areas of the United States.

Crop pests are frequently controlled directly by chemical or other cultural means. Control through genetic manipulation of the crop plants is also widely used to produce new varieties with genetic resistance. The adversaries are often biologically resourceful, however, and the occurrence of new races of some pests has made the incorporation of genetic resistance only a temporary stopgap technique. **Biological control,** the manipulation of biotic factors in the control of plant and animal pests, has recently received a great deal of attention (see Chapter 18).

Other biotic factors have influenced crop distribution. Many crops depend on biological vectors for pollination. The production of smyrna fig in California was unsuccessful until the *Blastophaga* wasp was introduced for pollination. The value of many legumes as crops depends on their association with particular nitrogen-fixing bacteria. Many legumes proved ineffectual in tropical areas until specific forms of bacterial symbionts were introduced.

Introduction and adaptation

The term "natural migration" infers that a species has been transported to a new area without human assistance, become established, and by successfully reproducing has become a permanent part of the natural flora. This process is slow and evolutionary. In contrast, the successful dispersal of crop plants, which leads to the permanent incorporation of a particular species into the agriculture of an area, can be a rapid, explosive, revolutionary process.

The dispersal of crop plants typically results from deliberate importation. The widespread dispersal of plant species from the Old and New Worlds during the Age of Exploration took place within a century. At present, plant expeditions to uncover new crop plants and new forms of old ones is a continuing endeavor.

Introduced crop plants do not necessarily have to be able to reproduce in the new area. Many crop industries depend on seed produced outside the area in which the crop is produced. This is often true of plants grown for some part other than the fruit.

Low tolerance in critical periods may be compensated for by cultural practices. The adaptability of crop plants depends heavily on people; most of our crops could not survive on their own in most locations. Corn does not live anywhere except as a crop plant.

Finally, adaptability is encouraged through genetic manipulation. For example, for a long time cabbage would not reproduce itself in the subtropical climate of Brazil, and seed had to be imported each year. But recently in an abandoned field in the state of Sao Paulo, some variants responded to slight chilling and flowered. Selection soon produced types adapted to local seed production, called *repolho loco*, or "crazy cabbage." The tomato is constantly being altered by breeding to maintain adaptability to new locations (it is grown from the equator to as far north as Fort Norman, Canada, at 65° north latitude), to resist local pests, and to adapt to new systems of management such as mechanical harvesting. New strains of wheat must be continually produced to compensate for mutations that occur in the stem-rust fungus (*Puccinia graminis*), a destructive pest of wheat.

In some crop plants it is desirable to prevent change in genetic structure. When crops are particularly adapted to human needs, and where asexual propagation is possible, particular genotypes may be kept intact for long periods of time. The 'Bartlett' pear, the main variety grown in the United States, Canada, and France, originated in England before 1770.

The accidental introduction of destructive weeds has been costly to crop growers. Russian thistle (*Salsola kali*), Japanese honeysuckle (*Lonicera japonica*), and witchweed (*Striga asiatica*) are examples. The migration of fungi and bacteria also causes serious problems. Fifty years after the fungus *Endothia parasitica*, which causes chestnut blight, was introduced to the New York Botanical Garden on an imported chestnut, it had virtually eliminated the native chestnut from the eastern United States. Governmental agencies maintain large and costly systems of quarantine and inspection in an attempt to prevent the introduction of unwanted species. In the United States even individual states have also found it necessary to maintain inspection stations at their borders.

Economic and cultural factors

Crops are a part of human economic and social life. The distribution of crop plants is thus affected by many factors in addition to biological ones. The agriculture of any nation strongly reflects that nation's level of economic development and its standard of living. On a local level, such factors as land costs, availability of labor, distance to market, and transportation facilities frequently override biological considerations of climate and soil suitability. Finally, governmental policy in the form of taxes, subsidies, and tariff barriers become important considerations. Even though a crop may be well adapted to a particular climatic regime, there must be a demand for it by consumers and a profit incentive for producers. Countries in which agriculture is organized in such a manner that production incentives are dampened by collective farming or by absentee landlord systems, for example, have found it extremely difficult to achieve rapid growth in agricultural production. The economics of crop production are discussed in Part Six.

The preferences of people in particular countries and regions for certain kinds of agricultural products has developed over countless generations. Rice is the staple carbohydrate in the Far East; wheat occupies this niche in the West. It would be difficult to reverse the use of these products even if such a change were technologically desirable. Food habits often reflect religious attitudes. Observant Hindus refrain from eating beef, even though India probably has the largest population of cattle of any country in the world. Although preferences are frequently dictated by local productive potential and capacity, they develop into habits or biases that are difficult to change rapidly. Human mobility has increased greatly in the past several decades, resulting in the introduction of new foods and other plant products, but for most of the world's peoples significant dietary changes are unlikely.

THE FUTURE

Agriculture occupies more people than any other trade in the world. Even so, only 8 to 9 percent of the total land surface of the earth is under cultivation—approximately 1 to 1.2 billion hectares (2.5 to 3 billion acres). Most of this cultivated land is in the middle latitudes, between 30° and 60° north and south of the equator. Four countries—the United States, the Soviet Union, China, and India—have a fourth of this land. The distribution of land that is suitable for cultivation is primarily determined by the nature of the physical environment—climate, soil, and land forms.

Vast portions of the earth do not now meet the requirements for crop production. In addition, probably no more than one-half to two-thirds of the available land is used for agriculture during any one year. Although no one knows how much more cropland can be added to that already under cultivation, there is widespread agreement that we have not yet reached the limit. But practically all of the best land is now under cultivation. Advances in agricultural technology are required to make profitable use of the great areas in the tropics and in the deserts that are now agriculturally unproductive. The expansion of the frontiers of present-day agriculture will be more than a challenge for future scientific pioneers.

Selected references

Anderson, J. R., 1970. *A Geography of Agriculture.* Dubuque, Iowa: William C. Brown. (World-wide systems of agricultural production and their geographic patterns.)

Banerjee, B., 1969. *Essays on Agricultural Geography.* Calcutta: Nabasakti Press. (A climatic approach to farming in India.)

Gregor, H. F., 1970. *Geography of Agriculture*. Engelwood Cliffs, N.J.: Prentice-Hall. (An examination of the roles of spatial organization, environment, culture, and history on agriculture.)

Grigg, D. B., 1974. *The Agricultural Systems of the World*. London: Cambridge University Press. (A description of the chief characteristics of the major agricultural regions of the world and how they came into being.)

Jackson, I. J., 1977. *Climate, Water and Agriculture in the Tropics*. New York: Longman Group. (An examination of water availability, characteristics, and use by agricultural plants in the tropics.)

Morgan, W. B., and R. J. C. Munton, 1971. *Agricultural Geography*. New York: St. Martin's Press. (A book dealing with the problems of agricultural location.)

Parry, M. L., 1978. *Climatic Change, Agriculture and Settlement*. Hamden, Conn.: William Dawson & Sons. (The role of climatic change in agricultural history.)

Tarrant, J. R. 1974. *Agricultural Geography*. New York: Wiley. (A study of data availability and techniques of analysis of agricultural production.)

Taylor, J. A., 1972. *Weather Forecasting For Agriculture and Industry*. Rutherford, N.J.: Fairleigh Dickinson University Press. (A volume dealing with the use and value of weather forecasts in agriculture.)

Wheeler, K. E., A. M. Ladley, and F. G. Leong, 1970. *Studies in Agricultural Geography*. Melbourne, Australia: F. W. Cheshire. (A short book dealing with the geographic distribution of different cropping methods and the underlying reasons.)

Williams, C. N., and K. T. Joseph, 1970. *Climate, Soil and Crop Production in the Humid Tropics*. New York: Oxford University Press. (A text dealing with many aspects of crop production in the tropics, including moisture, heat, light, soils, and cultural practices.)

Wilsie, C. P., 1962, *Crop Adaptation and Distribution*. San Francisco: W. H. Freeman and Company. (An excellent textbook or reference dealing with the environmental biology of crop plants.)

Four
STRATEGY OF
CROP PRODUCTION

15
Cropping Systems and Practices

The patterns of different cropping systems are highly variable—as diverse as are the crops themselves and the climates, habitats, levels of mechanization, and human customs under which crop production is undertaken. Depending on the part of the world, certain considerations will weigh more heavily than others. In one place primitive methods are appropriate, in another the system is highly mechanized; in some areas, growth factors are paramount, and in others economic realities are of greater concern. Before reviewing specific cropping systems, let us consider whether there are any general principles underlying the often complex ways in which crops are grown and harvested. Certainly cropping practices enjoy historical continuity; they must exhibit adaptations to ecological realities in order to remain viable.

Primitive harvesting undoubtedly consisted of little more than casual collecting of wild seeds, fruits, and other plant products. The hunting of game supported nomadic tribesmen, but was incidental in increasingly sedentary societies on the highroad toward civilization. Sedentary tribes came to rely chiefly upon the nearby abundance of food that domestication and cropping provided. Thus since the earliest days of agriculture two general "cropping systems" have been in operation: one centering on grazing and browsing animals (the animals harvest the vegetation, and people, secondarily, the milk and meat—a

system still basic in some parts of the world, such as East Africa where it is practiced by the Masai), and the direct harvest by people of the crops they sow and tend (though sharing some of their produce with livestock). The latter system forms the basis for most of today's agriculture and horticulture.

Forages for livestock are discussed in Chapter 20. The rumen or "second stomach" of cattle enables them to digest cellulosic materials indigestible by people and carnivores. The rumen is populated by a flora of microorganisms called flagellates, which symbiotically make possible the digestion of forages and fodders useless to other animals.

Grasses and legumes are the major products used for livestock feed, and various management procedures conforming to climate, soil, and other environmental considerations are followed. Grazing is an especially effective way of utilizing "secondary" land, which is too steep, rocky, dry, or otherwise unsuitable for economical growing of direct-harvest crops. Even today the system of grazing varies from merely turning out half-domesticated livestock onto open range, to cultivating pastures that are as meticulously organized for maximum production as is any agricultural pursuit (with appropriate combinations of grasses and legumes, properly fertilized, irrigated, and otherwise managed). Open-range herding is really not very different in principle than was the

Box 15-1   *Sobering observations*

This chapter opens with some generalizations about cropping systems, provides some perspective on modern practices, and ends with a review of "agribusiness"—a capitalistic society's response to modern needs. But in some circles, the claim that American agriculture is the envy of the world, a grand accomplishment, is met with only muted applause.

Indeed, some people suggest that intensive agriculture means "paying for oil with soil"—that is, that soil contamination, soil deterioration, and loss of soil to erosion are the consequences of the system's incessant demands. Here's what the editor of *Garden* (vol. 1, no. 3, July/Aug. 1977) offers as an aside to an article by Wendell Berry entitled "Let 'em Eat the Future." It epitomizes this rising concern:

> Species impoverishment is inherent in conventional agriculture. Out of the quarter-million extant species of flowering plants and the 2,000-odd species that have ever been used in agriculture, agribusiness deals with only a dozen or so, resulting in vast and dangerous monocultures, all taking up the same nutrients and the same amounts of water and subject to the same diseases. This sinister state of affairs we call American know-how.

Nor is American agriculture's extravagant use of energy exemplary. Obviously, it would not be possible, worldwide, to expend the 750 liters of fuel per hectare (about 80 gallons per acre) used for growing corn in the United States. Measured in terms of input of energy versus output, this is slovenly husbandry, far less efficient than that of the uneducated New Guinea tribesman, and not even close to "sustained yield"!

taking of wild game, whereas advanced animal husbandry is systematized to yield as much as 400 kilograms or more of high-quality beef to the hectare annually. In practice, of course, livestock raising is usually combined with direct crop production; in primitive societies the pigs may be allowed to root in the garden on occasion, whereas in technically advanced operations the grain, hay, and silage may be delivered to the livestock.

A cropping system must relate to the climate or biome (Fig. 15-1) in a general way. Mean annual temperature and mean annual precipitation are major determinants of climate. Notice in Figure 15-1 that certain types of natural vegetation result from the interaction of temperature and rainfall, and these in turn are indicative of the cropping practices suited to the region. We will see in this chapter that swidden (slash-and-burn) and plantation agriculture are especially suitable for tropical forest habitat, diversified farming for the deciduous forest and moist grasslands, ranching and dry-land cropping for the drier grasslands, irrigation agriculture for the deserts, tree cropping for the coniferous forests, and grazing for the alpine or arctic meadows.

Whatever the cropping systems that evolved in each of these dissimilar habitats, undoubtedly each underwent long evolution by trial and error, and depended upon domestications made from the wild there or nearby.

During such long evolutionary periods even climate may change (as almost surely it seems to have done in northern Africa), leading to still newer cropping innovations. Or people may impose change on older cropping systems, as occurred with the development of the Mexican wheats by Norman Borlaug of the International Center for Improvement of Maize and Wheat (CIMMYT) in Mexico and of the new rice cultivars by the International Rice Research Institute (IRRI) in the Philippines. Such new cultivars usually call for greater human control of the environment (such as implementation with fertilizer), and the growing season is likely to be different

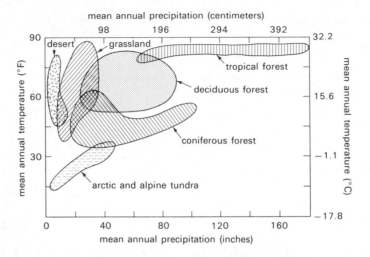

Figure 15-1 Six biomes of the world ecosystem characterized by mean annual temperature and precipitation. [Courtesy National Science Foundation; from *Science* 175:47, 1972, copyright © 1972 by the American Association for the Advancement of Science.]

than that for traditional cultivars (and thus upsetting to established habits). Although changes like these seem trivial on the surface, the influence can be profound on delicately balanced economic and political systems geared to other considerations. For example, the new cultivar may favor well-to-do and knowledgeable growers (who are able to finance the purchase of fertilizer and to use it intelligently) over peasant farmers (who may have to sacrifice their independence as landholders for economic reasons, with consequent, and possibly widespread, political dissatisfaction).

On the whole, the ecosystems (the ecological relationships between all elements of the environment) of the tropics are more complex and more diverse, and allow for many more alternative paths of energy flow, than those of temperate and boreal climates. A tropical ecosystem is therefore usually more resilient than a nontropical one, and able to absorb more abuse with quicker recovery if the deviation is not too drastic. By the same token, monoculture is a greater departure from the norm in the tropics than it is in temperate climates, and pests are more numerous. Lateritic tropical soils, naturally deficient in

organic matter, withstand the drastic imposition of cultivation poorly. Conversely, of course, the ecosystems of colder, harsher environments are less complex, contain fewer species (in larger populations), and show less stability and resistance to the impact of a cropping system imposed upon it.

Any cropping system implies some manipulation of the ecosystem. This ranges from merely taking care not to exhaust a resource completely (as do the Australian Bushmen and the people of remote tribes in Africa and South America), to nearly complete environmental control, in which moisture, fertility, pests, and even temperature are carefully managed (as in the intensive growing of strawberries and fresh vegetables in California, in which even the soil is sterilized before planting).

Considering the relative richness of tropical ecosystems, it is perhaps inevitable that tropical cropping would have gravitated more toward multiple species than has the agriculture of such climates as the United States and Europe, taking advantage of the greater abundance of ecological niches. Simpler systems are appropriate to the harsher environments of temperate climates, more rigidly controlled by

Table 15–1 Areas and nations ranked in order of pesticide usage per acre and in order of yields of major crops.

Area or nation	Pesticide use			Yields		
	Milliliters/hectare	Ounces/acre	Rank	Kilograms/hectare	Lb/acre	Rank
Japan	11,250	154.0	1	5,482	4,890	1
Europe	1,950	26.7	2	3,430	3,060	2
United States	1,557	21.3	3	2,600	2,320	3
Latin America	227	3.1	4	1,973	1,760	4
Oceania	205	2.8	5	1,569	1,400	5
India	153	2.1	6	818	730	7
Africa	131	1.8	7	1,211	1,080	6

Source: Adapted from W. B. Ennis, Jr., *Weed Sci.* 19:632, 1971.

technological means (examples of such systems being the extensive monoculture of corn, wheat, annual cotton, and so on). Attempts to impose a simplified cropping system on tropical habitats (say, the plantation system, which will be discussed), have, on the whole, not met with persistent success, and today there seems to be a trend toward smaller, individual plantings supplemented by multicrop door-yard gardens.

Yet it is impossible to dispute the productivity of cropping systems that utilize the maximum density of highly selected individual plants, even though such a system may be naturally unstable and require close technological supervision. Notice in Table 15-1, in which designation of pesticide usage indicates intensive cropping, that total yields per unit of land area are greatly favored by this highly technical approach. However, it must be recognized that such intensive cropping, if it is to be effective, is heavily dependent on outside resources—capital (which provides the many forms of labor-saving equipment), energy (especially electricity and gasoline to power machinery), and knowledge (the technological background gained from an advanced educational system, as well as a familiarity with such supplementary aids as pesticides).

Simplification of the ecosystem by means of monoculture presents a constantly changing pattern. Advantage is taken of new regulative practices to improve yields and especially the efficiency of human labor. Production at a 1920 rate on even the finest

Corn Belt soils would be unremunerative today. At that time, plowing was done with a single moldboard plow pulled by a mule or a span of horses, the farmer handling the plow. Readying the seedbed and planting were similarly tedious. With open-pollinated corn or any other crop, seed for the next planting was usually from a previous harvest on the farm; even under ideal conditions such strains were incapable of yields half that of modern hybrids. Weed control was solely by tillage, and required additional passes over the field, which were very costly in labor (and doubtless caused some damage to crop roots). Because of the need for row space along which a horse could pass, plant populations per acre were considerably lower than at present. Harvest required much hand labor, field shocking, and multiple gathering or threshing operations.

Yet, this was a beginning for mechanized monoculture. In America farming has gradually evolved into the highly mechanized systems discussed later in this chapter. Multiplow tractor-drawn equipment has replaced the independent moldboard plow, and, with some models, one pass completes seedbed preparation, planting, fertilizing and pesticide application. Herbicides eliminate the need for subsequent weed-control cultivation. Large plant populations in narrow rows are harvested by a field combine, and on-the-farm cribs with grain-on-the-cob are mostly a thing of the past. Yields per acre have increased fourfold, and yields per hour of human labor have increased tenfold. As we will see, the end is not in sight, as more and more environmental variables are brought under control.

Most efficient, in terms of low input per unit of yield, would be "primitive" forms of gardening, still practiced in such remote locations as New Guinea, parts of central Africa, and the Amazon basin. In these regions all ecological niches are utilized, and the result is a mixed planting that looks quite untidy to someone from the temperate zone who is accustomed to neatly kept fields. In taking maximum advantage of sunlight, space, and the generally abundant rainfall, growers in these areas use a thin canopy of trees as a crop overstory, which shelters an understory of bush and herbaceous plants, many of which avail themselves of underground space, too, thereby yielding comestible rootcrops and tubers as well as the fruits and greens found above the ground. Such a cropping system is quite balanced, and requires minimum maintenance once it is established. The shade and crop competition help restrain weeds, and regrowth occurs rapidly to compensate for portions harvested. Moderate attention to weeding and the maintenance of improvised fences to keep pigs or other livestock out (until appropriate occasions when they might undertake harvest for themselves) are the chief demands upon the custodians.

Tropical systems utilize vegetatively propagated crops much more than do temperate systems. Most root and tree crops are propagated by division or from cuttings, and even herbaceous plants, if not perennial in the tropical climate, are more conveniently started from vegetative slips rather than from seed. Quite the opposite is true for temperate-zone agriculture, especially where agriculture is based primarily upon grains and legumes massively planted by mechanized means.

Diversified subsistence gardens in the tropics may continue to be productive for a number of years in favorable locations. Less diversified plantings exhaust soil fertility within a few years, and must be reestablished on "virgin" ground elsewhere—the so-called shifting agriculture, also called the slash-and-burn, or swidden, system of cropping. A forest that has been felled and burned for manioc or maize is a typical example. Once the protective canopy of trees is destroyed, ecological change is rapid and drastic;

soil fertility is quickly exhausted, and sun-loving weeds and other pests often become impossible to control. Although conservative of inputs, the swidden system is extravagant of land use, and suited only to lightly populated regions, which are now quickly vanishing from the face of the earth. When population pressure necessitates continuous cropping of the same soil and the use of seeded plants to increase the yield, problems quickly arise in the tropics. Many lateritic soils become hard as brick and unusable when subjected to cultivation for a long period.

The greatest advancements in tropical cropping have not come in the dense rainforest regions, however, but in habitats where rainy and "dry" seasons alternate. Apparently the contrasting ecological regimens that result provide a rich range of habitat, well suited to origination and domestication of special crops and cropping systems. It appears that many crops (especially the starchy root crops) originated in climates of this sort, where the monotony of rainforest is interrupted by a dry season that opens new opportunities.

Intensive cropping is quite demanding, and best confined to good soil and level land. A crop of some 5 metric tons of grain per hectare will deplete the land of about 100 kilograms of nitrogen, 20 kilograms of phosphorus, 25 kilograms of potassium, 21 kilograms of magnesium, and 9 kilograms of sulfur. If crop fields are to remain productive for intensive use, it is clear that nutrients must be replaced continuously, and to do so in adequate amounts is possible only in those parts of the world where fertilizer is readily available. Although in some places food needs could no longer be met were cropping systems to be limited to the "natural" methods characteristic of subsistence agriculture, it appears that intensive cropping can be continuous if measures are undertaken to replace nutrients, and if the organic content of the soil is maintained at adequate levels through the incorporation of plant residues. Admittedly, however, intensive cropping is too recent a practice to provide the experience (which can only be gained in many decades) necessary for making an evaluation of the permanence of the system, unlike some other cropping systems that have been carried on in certain areas since the dawn of civilization (such as the

Figure 15-2 Rice growing in Southeast Asia is more intensive than it may seem. Although taking advantage of abundant, inexpensive labor, it increasingly uses highly bred cultivars, fertilizers, and pest controls. Here, in Burma, flooding the paddies necessitates hand transplanting from the nurseries, but the trade-off in weed control is worthwhile under prevailing circumstances. [Courtesy Food and Agricultural Organization of the United Nations.]

paddy farming in parts of the Orient—see Figure 15-2—or the terrace agriculture and irrigation systems of the Near East and the South American Andes).

It is quite evident that in many parts of the world misuse of the land has caused the destruction of civilizations that once existed there. The tendency of lateritic soils to become unworkable when exposed to cultivation has been mentioned. Perhaps an even more significant cause of soil depletion is erosion in hilly country, and on overgrazed or overbrowsed lands (like those on which feral goats range in the Near East and Oceania). Sardis, in southern Turkey, is an example of a once wealthy community now virtually abandoned. Lands of the eastern Mediterranean, now denuded and bare (and unfit for any cropping unless accorded expensive renovation), were once covered with magnificent forests (such as the cedars of Lebanon). Although soils respond differently in susceptibility to erosion, depending upon such factors as cultivation, slope, rainfall, and so on, a study in Puerto Rico shows how serious erosion can

be: nearly 48 metric tons per hectare of Pandura sandy loam exposed by clean cultivation were lost in a single simulated storm. If a cropping system is to be enduring, soil protection must certainly be afforded. Unfortunately, scant attention is given to soil conservation in many parts of the world, especially in some of the remote areas whose inhabitants have little if any modern agricultural knowledge and where more primitive cropping practices prevail.

THE MODERN CONTEXT

In general, crop producers aspire to the greatest possible yield commensurate with reasonable cost. The balance struck varies with location and occasion. In relatively prosperous parts of the world where population pressure is not yet excessive, as in the United States, the cost of labor is generally the prime consideration, more important than total production and even yields. It is not economical to farm those types of land, those kinds of crops, or those inconvenient locations that require extra hours of human labor for limited additional production. In other, more crowded parts of the world, such as the heavily populated Orient, labor is far less important than is total production. There each square foot of productive land will be farmed, even at the expense of many hours of tedious hand labor. Given aids to increase growth, such as fertilizer, yields per hectare may be greater than under highly mechanized, massive farming in America, where detailed attention to the land is not practiced because it is too costly.

Goals and methods will continue to change, depending upon economic conditions of the future and upon the ability of people to limit the size of the world population. The day is already at hand when American agriculture is more concerned with increasing production rather than managing the surpluses that had plagued the agricultural economy for some decades. In Japan, where trained labor is becoming more costly, a trend toward greater mechanization in agriculture can be expected. The cultural system best adapted to any given part of the world depends largely upon the human ecology of the place.

One thing can be foreseen: there will come a time when a ceiling on production is imposed beyond which it is impossible to go, practically. No matter how rich the soil is made, and no matter how well the crops are tended, space, sunlight, and crop potentiality will limit further increase. Maximum possible yields will have been achieved, because the crops will have been grown on the richest possible soil under the most suitable conditions obtainable. Somewhat before this ceiling is reached, the law of diminishing returns will have made impractical further increments of possible increase. The cost of practices needed for this last measure of return will be impossible to justify. Most of the world, however, is a long way from reaching such a ceiling, and the increased use of fertilizers alone should greatly extend the world's capacity to produce crops. Fortunately, this gives civilization some margin of time to institute population control and to reduce rapidly mounting demands on natural resources. It is sobering to note that even with all the improved cultural practices developed since World War II, and the consequent increased yields, the world is no better fed today than it was decades ago.

The quality of production must be assessed, too. How long can the United States, for example, continue to enjoy a basic meat diet, an extravagant source of protein in terms of the resources needed to produce it? Will not the time come when a much greater part of our dietary protein must be gained directly from plants? And what cropping practice, what type of fertilization, will yield suitable chemical quality in each foodstuff, even if at the expense of some yield? Cropping practices of the future must take this into consideration. Moreover, people will have to determine not only the best means of growing a crop, but which kind of crop best fits the economic situation of the moment.

Complicated interrelations are involved in agricultural policy decisions. Should such crops as soybean, which is relatively low in yield but high in protein, be grown in preference to crops that are higher in yield but lower in protein, such as grains and root crops? What about select sugarcanes, in which the yield of dry substance is powerfully increased at the expense of nitrogen content? What combination of such crops provides the best balance? Maximum use is made of photosynthesis with the sugarcanes, but certainly the protein is deficient. Broad cultural systems under maximum population pressure will have to provide a daily intake of about 2500 kilocalories. The ratio of protein to carbohydrates plus fat should ideally be about 1 to 11. Rice is one staple crop that matches good yield with this protein quality. Various analyses made to evaluate such factors as yield, protein content, and utilization of straw show that rice is capable of sustaining 25 percent more persons per unit of area than wheat. One can hypothesize sustaining even greater populations by using a high-yielding sugar to ferment proteinaceous yeast. Of course some help will come from the sea; for some time there has been talk of "farming" the marine plants of the continental shelves. Let us hope, however, that people learn to regulate their numbers before turning to yeast and algae just to stay alive. The world will not be a happy place in which to live if it becomes necessary to sustain upwards of 20,000 persons per square mile, though this is theoretically possible with good crop land.

As we enter the latter decades of the twentieth century, resource problems become serious in the face of exploding world populations. It is apparent that exploitation of fossil fuels cannot continue indefinitely at the prodigious rate to which we have become accustomed. At best, the cost of energy will be increasingly dear, and fuels will become progressively difficult to obtain. The implications for mechanized cropping are profound. Perhaps the world will find it necessary gradually to shift back toward the ancient systems, based upon a current sunlight budget (that is, at its simplest something like the New Guinea tribesmen's way of life, mentioned on page 327).

Already the advanced agricultural countries are at a point of no return, with an insufficient land base remaining to nurture the livestock that would be necessary for a return to animal power instead of petroleum-fueled machines. Food shortages would certainly result, even in areas of present surplus, were horses, mules, bullocks, and similar draft animals, all needing food and space perhaps amounting to as much as one-third of the land they work, to be substituted for machines powered by fossil fuels.

The cost of fertilizer will be continuously increasing, particularly for nitrogen compounds heretofore produced in processes involving relatively inexpensive petroleum inputs. Therefore, greater and greater emphasis will no doubt be given to nitrifying systems, such as are presently found in naturally evolved symbioses in rice paddies, and to the use of *Rhizobium*-nodulated legumes in crop rotations.

It may also become necessary to introduce a new type of cropping, that of biomass production. This, at its simplest, involves maximum growth of vegetation (whether native plants, weeds, or crops) to trap energy from sunlight. The biomass so produced might be used in any of several ways: for direct consumption by ruminants, for fermentation to derive such products as alcohol, for chemical manipulation into useful compounds, for fuel, and so on. As this is written, biomass production is still an ill-defined and incompletely investigated area, but it will doubtless involve many of the cropping systems mentioned in this chapter. Already there has been some talk of coppicing—that is, cutting all trees and woody vegetation on a piece of land and then harvesting the stump sprouts periodically. Biomass production by this method would probably be only about one-tenth as productive as natural biomass production by a mature forest at its height of productivity. (The coastal coniferous forests of the Pacific Northwest, among the lushest in the world, are estimated to produce 4000 metric tons of biomass per hectare annually; deciduous woodlands east of the Mississippi may yield 400–500 metric tons per hectare.)

Still another problem that is increasing worldwide is the shortage of fuel wood. Heretofore this has been a matter of only local importance, but now it has dire implications for cropping systems. In many parts of the world, virtually all available woody materials are harvested and salvaged from wood and field, effectively baring the land and preventing the regeneration of forests (Fig. 15-3). Ecosystems are seriously degraded. Nepal provides an example of this: with increasing population, once forested lands have been

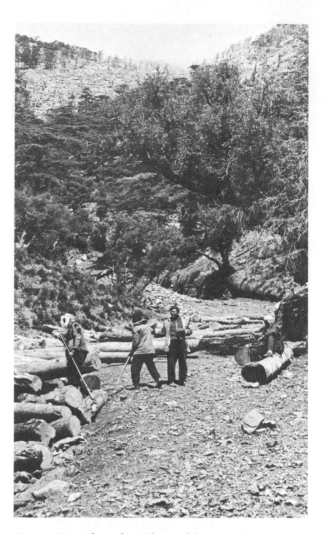

Figure 15-3 Throughout the world, many of the last remaining forests are being exhausted, as here in timber-poor eastern Algeria. Denuded forest land will erode, and, especially in marginal climates, it is unlikely that the forests will regenerate. [Courtesy Food and Agricultural Organization of the United Nations.]

cleared farther and farther up the mountain slopes to make room for terraced crops. Fuel wood, the only source of fuel other than dried cattle dung, thus becomes more and more remote, so that a day's journey by foot may be required to fetch a bundle of fagots that once could be gathered just beyond the dooryard. Denuded land erodes in spite of diligent terracing, and much fertile soil ends up in the river valleys of India. In a similar fashion, the caatinga forests of dry northeastern Brazil have been overharvested to provide fuel for sustaining the electric utilities in the larger coastal cities. And East African savannah forests have been cut both for fuel and for grazing land making a countryside naturally well suited to tree crops overtaxed and unproductive. Such demands for fuel wood cause drastic disturbance, in the sense of diverting much land from its best ecological use (or its proper cropping system).

TECHNIQUES

In the general discussion so far we have alluded to a number of situations and practices of rather fundamental importance. Some can be traced back to the beginnings of agriculture, and are worldwide today. Let's examine them more closely, keeping in mind that people are rather flexible, omnivorous animals that can adapt to a wide range of situations. They can, for example, interrupt the food chain at almost any level, to consume the more energy-efficient vegetable products in place of the more costly animal products, a practice generally followed under stringent circumstances. Also, people can substitute high yielding carbohydrate crops (such as sugarcane and cereals) for proteinaceous crops (mostly legumes, which must sacrifice some energy for nitrogen metabolism to make protein).

Possessing a degree of thoughtfulness, people may make their choices in light of what they expect the future to bring. At present, uneducated peasants are being offered improved "green revolution" crops, which, however, must be fertilized and otherwise given extraordinary care in order for them to fulfill their potential. It is no wonder that the simple farmer often chooses a lower-risk system, one that evolved locally over a long period of time, utilizing low-yielding but trustworthy crops and practices with which he is thoroughly familiar. The alternative would be to become a "vassal" of the more complicated system, neither well understood by him nor financially feasible (unless he is unusually wealthy).

In the tropics, especially, social rather than technological factors are often more important in determining the nature of the agroecosystem.

Obviously, any cropping system is limited by the broad parameters that define the ecology of the area—temperature, length of growing season, adequacy of rainfall, soil quality, topography, and so on. Seldom are all natural conditions favorable, but innovative people have learned to overcome many deficiencies. This is especially true under heavy population pressure, when land is scarce and costly and highly productive agriculture is needed. Various inputs can be provided to help optimize photosynthetic potential. Thus the intensive agriculturists of Japan will alleviate nutrient limitations by fertilization, reduce pest losses by various techniques (including the generous use of pesticides), forestall drought through irrigation, and so on.

In contrast to the intensive agriculture of advanced economies is subsistence agriculture—small-scale, self-sufficient production usually employing heterogeneous germplasm (landrace strains rather than select cultivars), and utilizing little in the way of modern inputs (such as chemical fertilizers, pesticides, and machinery). Subsistence agriculture has been ubiquitous since time immemorial, and it still provides sustenance for half the world's population. One cannot deny its appropriateness for many cultures, and one must recognize its virtues under less sophisticated circumstances. Subsistence agriculture is in actuality a finely tuned, steady-state means of food production that has evolved through trial and error over a long period. Of course it yields little in the way of cash crops with which to barter for manufactured items, which today are often considered necessities; but neither is it dependent upon uncontrollable and unintelligible inputs from the outside world. It cannot sustain the large populations that intensive agriculture can, but should technology falter or civilization wane (as may be the case during revolution, with drastic exhaustion of a sustaining resource, or when a natural disaster occurs), it can be the more reliable way of life (Fig. 15-4).

Each age has had its advanced cropping systems appropriate to the times, in the search for more

Figure 15-4 Traditional subsistence agriculture, such as here in Sri Lanka, utilizes few outside inputs, and sacrifices some net output to the use of draft animal power (in place of complicated, often undependable motorized equipment). As energy resources per capita diminish, will this be the prime methodology of the future? [Courtesy Food and Agricultural Organization of the United Nations.]

abundance and relief from laborious tasks. Among the practices that have been utilized in many places in many different cropping systems are land manipulations, crop introductions and selections, intercropping and multiple cropping, fertilization, and crop protection.

Land manipulations

Since the earliest days of crop production, people have reshaped the terrain in order to irrigate, to provide better drainage, and to conserve the soil. Only recently has it become apparent that raised beds or fields (such as the chinampas of Central America) were not a pre-Columbian rarity in the lowlands, but commonplace. These provided suitably drained and

aerated soil in lowland places that would otherwise be sodden. When one flies the earth in a modern aircraft, it becomes apparent that terracing—the making of contoured banks or ridges on slopes to make them cultivable—is practically ubiquitous. Terraces conserve water and soil in all parts of the world (Fig. 15-5). Likewise land has been engineered for irrigation since the earliest days of farming, a fact well demonstrated by archeological evidence from the Near East, the Andes, and East Asia. Mostly this was done by creating gently sloping sluiceways for gravitational movement of water. Terracing and irrigation have together been especially important in creation of tropical rice paddies, which sustain much of the world's population (Fig. 15-6).

Figure 15-5 Land terraced by hand to conserve the thin soil in an area of subsistence farming of mountainous Java. [Courtesy Food and Agricultural Organization of the United Nations.]

Figure 15-6 A water lift ingeniously constructed with split bamboo, for irrigating rice paddies in China. The vanes of the wheel are activated by the current, and the bamboo scoops carry water to the trough at the top. [Courtesy Food and Agricultural Organization of the United Nations.]

Crop introductions and selections

Ancient people perforce selected and introduced plants in order to achieve their domestication. Species suited to the locality were chosen, and within each species selective pressure was brought to bear upon various advantageous traits (such as greater productivity, longer bearing season, preferred size and shape, and so on). As we have noted, conditions dictated that, in general, annual crops adapted to a short growing season became the mainstays of temperate cropping, while perennial, everbearing crops were more often chosen for tropical environments, where carry-over and storage are more difficult.

Intercropping and multiple cropping

Intercropping, whereby different crops are interplanted, is characteristic of subsistence agriculture. The practice has much to recommend it, because it spreads the risk of a crop failure or devastation by insects or disease over two or more crops. Its disadvantage is the inconvenience it causes by making it impossible to obtain the efficiencies of scale that can be enjoyed with a single crop. Intercropping (often termed mixed cropping, multiple cropping, or polyculture) is being more frequently recommended for tropical regions, where the trend is away from large plantations and toward individual holdings. Double cropping (or triple cropping, in some instances) refers to successive croppings within the same year. This is appropriate for fertile land in a climate with a lengthy growing season, for it increases the productivity of the land proportionally. Often, however, repeated cropping is thwarted by marginal growing conditions, seasonal changes, a build-up of pests, nutrient exhaustion, or other difficulties.

The increase in productivity that came with flood control in the Nile Valley, which eliminated the "wasted" time when the fields were under water, will be discussed later in the chapter. Double and triple cropping is now possible on these fertile lands, although only time will tell whether ecological change—most likely involving fertility loss and pest build-up—will exact its toll. It is interesting that depictions in the tombs of ancient Egypt indicate that seed was planted just before the flood season, so that a silt topdressing (as well as abundant soil moisture) was immediately present for crop encouragement as the waters receded. New methodologies, such as the use of fertilizers, can doubtless be taught to the Egyptian farmer, but it remains to be seen whether the economics of the age-old natural system can be profitably substituted for.

Fertilization

Since ancient times it has been realized that adding nutrients to the soil is a valuable cropping practice (Fig. 15-7). Manures, guano, and (especially in tropi-

Sonora 64

tillering under   tillering under   tillering under
high fertility   medium fertility   low fertility

Figure 15-7 The difference fertilization can make in crop yield, in Turkey. Wheat plants like the one on the left, which was heavily fertilized, will bear much more abundantly than those like the one on the right, which was underfertilized. [Courtesy Agency for International Development.]

cal environments) composted vegetation have been extensively used over the ages. Perhaps second only to a lack of moisture, a lack of nutrients imposes a serious limitation on the photosynthetic potential of a crop plant. The technically advanced agriculture of today takes advantage of soil and tissue tests to zero in on nutrient needs. The nutrients are generally supplied from economical synthetic sources, in proper balance, and in easily handled formulations. Incautious fertilization, however, can induce disturbances and imbalances, cause changes in pH (which, in turn, can affect micronutrient availability), and introduce unwanted elements (such as cadmium or other heavy metals).

Crop protection

People, of course, are not the only consumers of their crops. Especially troublesome competitive consumers are arthropods (primarily insects) and pathogenic microorganisms, but even vertebrate animals often feast on cultivated fields. Weeds, too, are a major problem, competing with crop plants for space and nutrients. Crop predation is especially acute under tropical conditions, and many agriculturists believe that it is more the cause for the deterioration of productivity often seen on farmland in the tropics than is soil depletion. The principal tactics for controlling pests have been the selection of resistant cultivars, the utilization of cultural controls (such as rotational cropping, cultivation, timing of planting, and fallowing), and the institution of biological controls (introducing predators or competitors of the pests). Only recently has the application of pesticides become commonplace, and only in advanced agricultural economies.

At first, almost miraculous results were achieved with the use of pesticides, with small quantities of "novel" molecules sufficing to fell weeds and insects often quite selectively. But all is not well on the pesticide front these days. In addition to newly found hazards to health, the tremendous amounts of pesticides that have been used in some areas have tended to alter ecosystems. Especially with insects, but also to some extent with pathogens and weeds, resistance quickly builds up in the pest so that heavier doses of more toxic chemicals are called for. Eventually the cropping system becomes overly dependent upon pesticides, which become less and less functional and more and more harmful to the environment. Cotton growing is a notorious case; the crop as generally cultivated in the United States requires innumerable treatments. With public sentiment turning against the extravagant use of pesticides, integrated pest management (IPM) is gaining favor. The idea, sensible in principle, is to utilize cultural and natural controls to the greatest extent practicable, reserving pesticides only for emergencies and especially devastating invasions.

Other innovations

The trend toward mechanization has been underway since the first draft animal was hitched to a plow to replace manual cultivation with a digging stick. The remarkable extent to which fossil fuel has been used to power the innumerable labor-saving devices commonplace in advanced agricultural societies is commented upon on p. 350. Similarly, human inventiveness has created irrigation systems far more effective than the gravity-powered systems of old (Fig. 15-8). Hormonal chemicals, usually synthesized, can today regulate plant growth. Pheromones mimic insect sex attractants, luring harmful insects to poison or traps. Synthetic hormones that disrupt growth are used to prevent insects from fulfilling their normal reproductive cycles. No-till and mini-till systems have been developed, in which herbicides are used to control vegetative growth other than that of the planted crop.

SYSTEMS

There is no end to the schemes that might be employed for raising and harvesting a crop. But certain systems adapt better than others to certain circum-

Figure 15-8 Technology steps in, increasing yields through irrigation. Here, in Nalagonda, southern India, a well has been dug and a motorized pump has been installed, resulting in six-fold increase in rice productivity. [Courtesy Agency for International Development.]

stances. Although ill-defined and with many variations, the following procedures cover most of the cropping methods people undertake to gain their livelihood from the land in all parts of the world.

Shifting agriculture (swidden cropping)

Shifting agriculture is a slash-and-burn-and-abandon process employed mainly by primitive cultures in sparsely populated regions. The **milpa** of the

American tropics and the **ladang** system of the Orient are examples. The storehouse for vegetational wealth in the tropics is lush topgrowth, not soil. Shifting agriculture endeavors to release these stores for down-to-earth cropping, making but temporary use of stubborn lateritic soils.

Forest trees are either girdled or felled. The dead vegetation is burned during the dry season (or sometimes left to decay and deteriorate through termite action). The resulting open ground is automatically fertilized with wood ashes. After only a few crops, fertility is typically exhausted. The clearing is then abandoned, and the process is repeated elsewhere. On unfavorable terrain, much soil is lost through erosion.

Cropping entails the planting of seed or, more commonly, vegetative parts as soon after burning as the soil softens with the onset of the rainy season. Under primitive circumstances, planting is done simply by dibbling holes in the soft soil with a planting stick and dropping in the seed or the vegetative part. Some weeds will volunteer in the crop, and hand pulling may be needed to aid the crop. Brush or grasses typically overrun the clearing within a very few years.

In a region as unpopulated as the Amazon Valley has been until recently, slash-and-burn agriculture was a superior system, all things considered. It provided a readily available and effective means of fertilization, from the ashes produced by burning. However, yields typically decrease by about 50 percent with each successive planting, although this could be ameliorated somewhat if the abundant natural vegetation were cut and composted, and then hoed into the soil. As development proceeds in tropical areas, some problems of nutrient deficiency can be corrected with complete fertilizers, but composting is still needed to conserve soil structure. As modern cropping methods have been introduced, trace-element deficiencies, particularly of magnesium, have often proved a limiting factor.

Obviously, shifting agriculture makes extravagant use of land and is wasteful of trees. Societies depending upon it must either keep on the move or extend operations ever farther from the central city as nearby lands become exhausted. Frequent clearing of the same land degrades the vegetation into an unmanageable tangle of brush. Maybe this is why empires based upon shifting agriculture have not endured as well as those established on valley flood plains.

Tree horticulture

Fruits, derived mainly from woody plants, are important crops. They were gathered from the wild by primitive people, and today they help to support the economies of many parts of the world. Tree crops, mostly fruits and nuts, are particularly suited to tropical cultural systems, and, as was noted earlier in this chapter, they allow for a more stable cropping system there than does swidden cropping. In addition to fruits, woody plants are grown for tea, maté, wax, rubber, quinine, and cork, often in plantations (which will be discussed subsequently). In temperate climates, orchards—counterpart to the tropical plantation—yield a variety of fruits and nuts and even such decorative materials as holly.

The tropical climate is generally better suited to tree growth than to the production of the annual crops that people living in temperate climates have come to regard as agricultural mainstays. The lateritic soils of the tropics become hard and refractive; powered equipment for cultivation is expensive and hard to procure and maintain; efficient means of mass handling and storage are generally lacking. But tree groves accompanied by family gardens support individual families and small villages well. They are enduring and harmonize with the tropical forest environment. Thus, as settlements become too mature for shifting agriculture, villagers turn to tending groves of coconut and other palm fruits, breadfruit, banana, citrus, avocado, and so on. This kind of tree horticulture is supplemented by diversified home garden patches (Table 15-2), and usually some domesticated animals are used to help keep down the prolific tropical growth and to supply manure for the plantings. Fishing and hunting often yield additional protein.

In some systems trees are cultivated as a source of

Table 15-2 Some of the many species typically cultivated in a dooryard garden in Central America (Yucatan).

Scientific name	Common name (English)
Annona cherimola	cherimoya
Brosimum alicastrum	breadnut
Byrsonima crassifolia	craboo
Capsicum annuum	chili
Carica papaya	papaya
Citrus aurantiifolia	lime
C. aurantium	sour orange
C. reticulata	mandarine orange
C. sinensis	sweet orange
Cocos nucifera	coconut
Cordia dodecandra	
Diospyros digyna	black sapote
Mangifera indica	mango
Mentha piperita var. *citrata*	bergamot mint
Persea americana	avocado
Psidium guajava	guava
Sabal yapa	guano palm
Sechium edule	chayote
Tagetes erecta	marigold

Source: Adapted from C. E. Smith and M. L. Cameron, *Econ. Bot.* 31:99, 1977.

fertilizer. The tops are constantly cut back to the trunk, the vegetation being used to fertilize cleared ground, usually by burning but sometimes by composting.

Plantation agriculture

A successful agricultural method where large sections of land have been cleared is the **plantation system,** hallmark of tropical agriculture in many regions. A simple example might be the growing of plantains in Ghana. Locally selected cultivars are perpetuated by vegetative "suckers," after felling of forest and burning of brush in the dry season. Reasonable attention is given to mulching, irrigation, propping of fruit bunches, intercropping with vegetables, spraying for

disease and pests, timing of harvest, and so on. Only where an enlightened supervision practices sanitation, selection of disease-free suckers, and fertilization, can a planting endure for many years and thereby avoid the swidden system of moving on to new ground after only a few seasons.

A good plantation combines the better features of shifting agriculture with those of horticulture. Under trained managers, often sent from agriculturally advanced parts of the world, plantations generally utilize the land efficiently (Fig. 15-9). The plantation system requires responsible management and the capital resources to supply the necessary machinery and facilities and to open up marketing opportunities. Although there is no doubt that many foreign-controlled plantations have exploited native populations unfairly, it is regrettable that the rise of nationalism and anticolonialism has tended "to throw

Figure 15-9 A tea plantation in East Africa. Adequate capitalization, improved cultivars, and trained supervision combined to give high yields and economical production. [Courtesy Ivan Massar for World Bank.]

out the baby with the bath water." The rubber plantations of Sumatra and Malaysia, the cacao plantations of Africa and Central America, the nutmeg orchards of Grenada, and the sugar and coffee plantations throughout the tropics probably represent the soundest ways of utilizing the land under prevailing circumstances. The native populations of these parts of the world often lack agricultural training and so must be supervised—a situation that must be changed if world food demands are to be met. There should be some means of combining social enlightenment with the efficiences of large-scale operations, especially in the less developed tropical countries where small holdings are not well managed.

Orcharding

Although the Orient and the tropics still rely a great deal upon backyard plantings of a few fruit trees, in technologically advanced countries commercial orcharding has evolved into a large-scale specialization (Fig. 15-10). The blemished fruits typical of backyard plantings no longer find a market in the United States. Select varieties with proper keeping qualities, exact spray schedules, and harvesting standards, typical of professional orcharding, have taken their place. It is not unexpected that production centers have shifted to particularly favorable climates—in the United States mostly to the coastal Northwest and the Great Lakes region. Europe has seen similar specialization—for example, the heavy plantings of pears in the Rhone Valley, and of both apples and pears in the Po and other valleys of northern Italy. Southern Australia, South Africa, and southern Argentina also produce considerable fruit crops.

Eastern Asia is particularly well suited to fruit production, having climatic conditions equal to other fruit-producing parts of the world, plus the blessing (for fruit production) of an unlimited labor supply. Skillful gardening rather than extensive agriculture has an ancient tradition in Asia. It is said that there are more than 500 species of native Chinese plants that produce edible fruits, of which at least 120 species are in cultivation. These include apple, pear, persimmon, jujube, peach, grape, mandarin orange, sweet orange, pineapple, banana, longans, and litchi nut. Peach cultivars, such as 'Chinese Cling,' reknowned as a parent of the 'Belle of Georgia' and 'Elberta' cultivars, have come from China, and this area with its long history of civilization (during which many selections and adaptations were made) promises a wealth of untapped breeding potential still.

In the United States and Europe, apple, citrus, pear, peach, plum, apricot, cherry, pecan, walnut, and grape are among the principal orchard species. Throughout the tropics, coconut and banana are extremely important; and in special locations such as Hawaii, pineapple is a major crop. Lesser fruits grown in the tropics include mango, papaya, date, fig, avocado, olive, and pomegranate; in temperate climates, almond, persimmon, blueberry, chestnut and various bramble species.

Each type of orchard, each soil and climatic zone, demands its own cropping practices. In general, select planting stock is usually grown in a nursery, through budding, grafting, and rooting of cuttings (see Chapter 9). Saplings are transplanted to the orchard, where they are spaced geometrically to accommodate eventual tree size and the equipment used to tend the orchard. In the United States, trees are generally spaced 6–14 meters (20–45 feet) apart, but the trend is toward high density plantings in new orchards. In Europe, where land space is at a premium, dwarfed "hedgerow" orcharding is very common, especially with apple and pear. There are obvious efficiencies in harvesting fruit that grows low enough to be reached from the ground or from a moving platform. In northern Europe, trees are often trained to wire trellises or are maintained as dwarfs either by pruning or by using size-controlling rootstocks. In Italy and France, plantings range from spaced trees to narrow "fruiting walls." 'Golden Delicious,' an apple cultivar popular in the United States, responds well to hedgerow growing and is now extensively planted in Europe under this system.

Few orchard trees demand as fertile a soil as do annual crops, but, instead, thrive best in deep, well-drained locations. Their root system, much more extensive than that of annuals, is more continuously

Figure 15-10   A portion of land owned by the Dixie Orchard Company, Vincennes, Indiana. This large enterprise manages more than 400 hectares of apples and peaches. [Courtesy U.S. Soil Conservation Service.]

active but does not cross-transfer nutrients well from one side of the plant to the other. Fruit orchards often do well on light soils, and tolerate wide ranges in pH. But they may be susceptible to shortages of minor elements.

Location is important; blossom loss from freezing can be prevented by planting orchard trees in equable climates leeward of bodies of water (which warm later in spring and cool more slowly in autumn, allowing late blooming and ripening), and on rises rather than in depressions (where cold air settles, causing frost).

Various soil-management schemes are practiced in the orchard, ranging from clean cultivation (particularly when the planting is young) to the maintenance of sod or legume covers to hold and build the soil. With modern herbicides, sod and weeds can be restrained without cultivation, resulting in less competition for water and fertilizer and the elimination of cultivation injury to surface feeder roots.

The modern orchard requires elaborate maintenance to produce fruit of marketable quality. As many as a dozen pesticide sprayings may be required during the growing season. Seasonal pruning is needed to regulate shape and to contain the trees for convenient harvesting and maximum fruit production. Fertilization (and irrigation, if required) must match local soil requirements.

Cultivars must be properly selected to insure pollination and fruit set (although this can be implemented with chemical sprays). Many fruits are self-sterile and must cross-pollinate, sometimes with an entirely different cultivar. This entails appropriately spaced interplantings and, for some crops, introduction of pollinating insects (such as honeybees). Moreover, some fruits set well only in alternate years. Cross-pollination is generally necessary for a good crop of apples; many pears, sweet cherries, and plums are likewise self-unfruitful. Peaches, nectarines, apricots, and sour cherries self-pollinate adequately.

Growth-regulating materials have found great application in modern orcharding. In such temperate fruit crops as apples, pears, and peaches, **chemical thinning** is commonly practiced to adjust the crop load. This is necessary in order to control fruit size and to prevent alternate bearing and winter injury—problems that arise from overbearing. Chemical thinning materials may be either caustic substances (such as dinitro compounds) that burn off the blossoms or auxin derivatives (particularly naphthalene-acetic -acid) that bring about embryo abortion. Auxins are also used to prevent premature fruit drop in the fall. In grape production, gibberellins have been used to induce seedlessness and particularly to increase fruit set and fruit size in seedless cultivars. The chemical SADH (succinic acid-2,2-dimethylhydrazide, also known as daminozide) has interesting effects—dwarfing, the induction of earlier production, intensified fruit color, and improved storage qualities.

Added to the orchardist's tasks may be such things as nematode control and protection against freezing (as in citrus orchards in Florida), the proper timing of the harvest (in bananas, for keeping during shipment, and in rubber, for maximum yield), and perhaps above all the securing of seasonal hand labor for harvesting. Fruits that can be shaken from a tree to fall on nets or canvas without damage are increasingly being harvested mechanically (with padded devices that vibrate the tree or bush).

Flood-plain irrigation

Most remunerative of all cultural systems in the tropics and subtropics is flood-plain cropping with irrigation, a system that had its beginnings in Mesopotamia, the Nile Valley, and Tehuacan. Flood-plain irrigation supports dense, sedentary populations along the Nile and throughout the Orient, where individual paddies are created by terracing and intensive land management. Practical skills and much hand labor are needed, with the result that, by and large, flood-plain irrigation has been most successful where the people are peaceful, intelligent, and dexterous. The system has been singularly successful in Southeast Asia.

Irrigated desert

Although stream borders in desert valleys have always been among the world's most productive agricultural locales, it remained for modern technology to make practicable the massive, intensive, highly remunerative farming seen today in such naturally unlikely spots as the interior valleys of California, southern Arizona, and the high plains from Colorado into Texas. Irrigation projects that transport water for many miles are the prime means for turning "worthless" desert into productive agricultural land. Well-managed and scientifically irrigated holdings in California and Arizona can produce twice as much cotton as the famed cotton lands of the Southeast. The same is true with such forages as alfalfa, in comparison to yields in the Northeast with natural rainfall (and shorter growing season).

Where irrigation is not possible through diversion from streams, increased drilling capabilities and huge pumps enable well-capitalized farmers to draw groundwater from as much as 600 meters (nearly 2000 feet) below the surface (the water table, unfortunately, is thereby being lowered in many parts of the country at an alarming rate). **Trickle irrigation** in which water is dripped to the soil directly below the crop plants slowly and in precise quantities, is finding increasing approval as a practice that is more conservative of water than is flood irrigation or sprinkling. (Apparently, trickle irrigation also permits the use of somewhat more saline water, which is typically found in arid regions).

In arid regions, which are naturally inhospitable to people, irrigation permits very remunerative yields under highly technical and intensive farming methods. Some of the world's most specialized large-scale agricultural operations have been established on level desert. Desert soils may not be as good as prairie soils, but their deficiencies can be compensated for by fertilization and proper irrigation. How involved these operations can become is exemplified by the efforts of the Kern County Land Company in California to trap seasonal runoff for underground storage, to be pumped out later for irrigation. Soil treatments that will maximize surface infiltration are assiduously sought, and ponding fields are built merely as insoak reservoirs to get mountain stream runoff into the water table. Almost no water runs "uselessly" to the sea any more in southern California.

Even the huge flow of the Nile is being appreciably drawn down for expanded irrigation of the deserts of lower Egypt, now that the Aswan High Dam is in operation. Expansion in food production more than sufficient to keep pace with the burgeoning Egyptian population has resulted, perhaps more because triple cropping (three crops per year on the same field) is possible now that the annual spring floods are controlled. There is a trade-off, however: the fertility value of the silt that was once deposited by the spring floods has been sacrificed, and commercial fertilizer must increasingly be used (and this mostly by uneducated farmers not accustomed to fertilizer usage, in a system lacking the infrastructure for agricultural supplies or sophisticated marketing means).

The huge outlays generally required for large-scale irrigation farming in the United States are reminiscent of the management practices of plantation agriculture in the tropics. Small, self-sufficient home farms are seldom seen. Many of the large land holdings are owned by absentees and managed by hired professionals. Even with private operations the "farmer" acts more like a "businessman" (not uncommonly commuting hither and yon in his private airplane) than the overalled cultivator of the soil so dear to the American political ethic. These owners and managers are just as interested in credit, national money policies, the stock market, and the Washington political scene as are the executives of industrial corporations. In groups or associations they may even control marketing outlets and have connections overseas. This is probably but a foretaste of what most agriculture will be like eventually in advanced countries.

Diversified general farming

The farming tradition familiar to Western civilization in temperate climates—the system of diversified general farming—embraces most of the temperate-zone crops, planted, where they are well adapted and economically remunerative, by the landholders themselves. This sort of cropping developed in climates where rainfall is sufficient to maintain forest vegetation. Until quite recently Europe and North America have been particularly dependent upon farming of what was formerly forest land. By and large a very flexible farming program is possible on land of this sort. Though the soils are not particularly fertile by nature, they are quite responsive to fertilization practices and proper handling. Much of the best cotton, soybean, corn, and dairy-pasture lands, so important in the eastern United States, are of this type.

The colonist regarded the virgin temperate forest as somewhat inimical to his economic well-being. Trees were to be utilized, of course, for timber, fuel, furniture, and so on. But the forest seemed limitless, and the main task at hand was to cut down the trees and grub out the stumps, so that land could be cultivated and cropped. In North America even the Indians had made a few clearings in the forest, where they planted corn and other crops. Gradually the hand felling of trees and manual hoeing of soil, which were characteristic of the early Virginia colonies, changed to soil cultivation with moldboard plows drawn by oxen or horses, and then to today's use of elaborate farm machinery. This evolution is at the heart of American economic progress, and will be discussed in more detail under "Mechanization."

General farming allows great self-sufficiency. New World colonists or European peasants could provide themselves with life's chief essentials, and perhaps a bit extra for barter. It was a system that encouraged individualism, so dear to the American political image. Gradually farming of the humid temperate lands has given way to greater specialization, although diversification and crop rotation are not yet things of the past. Some once-forested lands prove particularly suitable for vineyards or orchards—especially in equable climates near bodies of water, such as in western Michigan and from northern Ohio to upstate New York. Cotton was the basis of a one-crop economy in the upper South a century ago, where, after a good many fields were nearly ruined, diversification eventually returned (tobacco, beans, peanuts, corn, small grains, orchards, hay, and forage). Once-forested lands in Ohio and Indiana are today among the best in the Corn Belt, where corn is rotated with soybeans and occasionally with small grains or pasture. It is hard to imagine that the state of Ohio, once almost solid forest, has within a century and a half been transformed into a man-made "prairie" with scarcely a virgin stand of trees to be found anywhere in its tens-of-thousands of square miles.

Prairie

The temperate prairies, the modern world's bread-basket, were among the last of the major vegetational complexes to be harnessed. The harsh, windy environment provided little in the way of fuel and shelter. Until sturdy, powered implements became available the plowing down of thick sod was a back-breaking chore. It is not surprising that, in the colonization of America, cultural systems patterned after those of the Old World carried settlement easily to the boundaries of the prairies, where migration paused until new techniques and attitudes (built around wheat cropping) could replace the diversified farming tradition long familiar in the East.

Prairie life breeds cooperation rather than individualism. The very nature of the environment calls for large acreages devoted to but one or a few well-adapted crops. Neighbors, though more distant, are also more cherished—the sole social outlet in a culturally barren environment. And the lifeline for existence itself is a tenuous road or railroad leading to the industrialized parts of the world, from which come the tools and supplies not locally obtainable. With an existence that is anything but self-sufficient, being dependent upon distant markets, there is real concern about other parts of the world, and a cooperative rather than isolationist political outlook develops.

Modern transportation and powered equipment have completed destruction of the native prairie even more quickly than the eastern forests were felled. Only small pieces of the original prairie remain, although it once covered hundreds of thousands of square miles in the United States, from Illinois to the Rocky Mountains and from Texas to Canada. Much of this has been replanted to domesticated grasses and legumes for hay and forage, and even more to such cultivated crops as wheat and grain sorghum. Unfortunately, at times prairie lands that were too dry for cropping have been stripped of their protective sod cover, causing a loss of soil to wind storms. Other prairie lands have been so overgrazed that grassland has become degraded to brush and desert.

Dryland farming

The western reaches of grassland in the United States embrace the drier short-grass prairie, where a system of alternating crop with fallow permits acceptable yields under minimum rainfall. Alternate strips, or fields, are sowed (usually to wheat) and left unplanted (but cultivated to control weeds). Such rain as does fall soaks into the fallow ground and is not dissipated by plant growth. Moisture "stored" in this fashion for one year is usually sufficient to sustain a crop the second, after which the process is repeated. Dryland farming is a human adaptation to the severest part of the prairie environment, and requires a hardy breed of agriculturist to withstand the adversity of dry cycles for the prospect of bounty in relatively rainy ones. Soils are good, and dryland farming is increasingly giving way to irrigation as facilities become available.

Burning and grazing

The natural vegetation in many parts of the world is grass, and its utilization for grazing (originally by wild animals, later by livestock) can support a moderately concentrated population. Where climate is seasonally too dry for forest or parkland, prairie persists, but even there it can be argued that periodic burning is required to keep brush and grass in balance. This is particularly true in tropical savannahs or in forest clearings, which are burned annually to prevent the incursion of trees. Often savannahs are burned merely to get rid of lignified old leafage, which is unpalatable to stock and contains little protein; the animals prefer tender new sprouts. Dispute is endless over whether such burning causes long-term deterioration, which is likely if burning is accompanied by overgrazing. But certainly burning is essential in some regions to prevent reversion to forest. Tough grasses that have invaded tropical clearings are often intentionally destroyed by controlled burning just as new growth starts; food reserves in underground parts are low then, and the grasses are easily weakened.

Browsing

Agriculturists considering various cropping practices might take a hint from wildlife species, among which browsing (the nibbling of woody vegetation as opposed to the grazing of herbaceous ground cover) is commonplace. In some respects the planting and maintenance of low-growing shrubs and trees requires less effort and is more conservative of energy than is establishment and maintenance of pasture. Some tests have shown that a greater quantity of digestible protein is produced per hectare by browsing than by the grazing of forbs and grasses.

Goats and camels in particular are noteworthy browsers among domesticated livestock. Often they are simply turned loose to consume wild vegetation. With reasonable attention to husbandry, it should

prove possible to combine the browsing of certain livestock species with some commercial yield (as of fruits or nuts, and perhaps a modicum of fuel-wood) from shrubs and trees. Plantings of jojoba (*Simmondsia chinensis*) might be very productive in this respect. The species could be a source of agricultural income in arid climates where few other crops are adapted: controlled browsing might be combined with at least some remuneration from harvesting the seeds, which contain a valuable liquid wax.

Forest management

Forests have only recently come to be regarded as crop rather than as natural bounty to be freely taken. With much of the world's virgin timber already cut, there is now concern not only for quantity but for quality also. Save for a few remaining wild parts of the world, such as the Amazon valley and the remote subarctic regions, forests are today more or less managed. In some cases their management is just as intensive as that of cultivated crops, and includes specialized techniques for planting and replanting as well as harvesting.

Figure 15-11 Early-day logging in the United States (this in Shasta National Forest, California) with bullock teams. This mode of operation is still practiced in remote forest regions, such as the Gran Chaco of Paraguay, Tierra del Fuego in Argentina, and parts of Africa and Asia. [Courtesy U.S. Forest Service.]

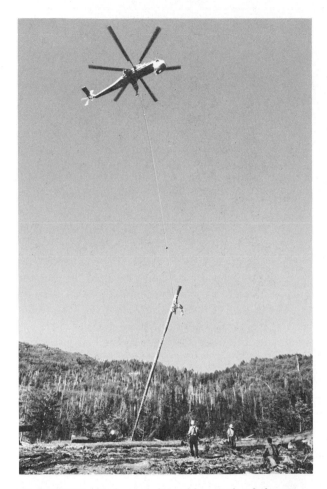

Figure 15-12   Modern yarding of logs with a helicopter in California. [Courtesy U.S. Forest Service.]

Compared to the systems used for annual crops, however, and even orchards, forest cropping systems are still in their developmental stages. Over most of the tropics and much of the temperate zone, management is minimal. Shifting agriculture continues, and in many places trees are cut for timber, pulp, and firewood much as they were by early settlers. But technically advanced societies have come to realize that forests are not unlimited. There is little hope that improved forestry practices will be imposed over most of the tropical woodlands in the near future; rather, high-grading (removal of only the best trees of the best species, leaving inferior types to re-

produce) will continue, as will wanton deforestation for fuel and new cropland. But more intelligent use of temperate forests is at hand, both with regard to their use as a source of raw materials and as an esthetic and recreational resource.

Forest management has many problems that are peculiar to it. Most obvious is the long time required for tree growth. Up to the present, forestry has depended upon old-growth timber, some of it thousands of years old. New growth generally takes 15 to 150 years to reach merchantable size for timber and significantly less time for pulp, depending upon the part of the world. It is apparent that any mistakes made in selecting new planting stock, or in getting it started, will be perpetuated for decades. Besides causing a loss of time, mistakes of this kind cannot be corrected without the additional cost of cutting useless stock and replanting again.

In many parts of the world, forestry relies for its seed mainly upon native trees. It is assumed that natural selection will have provided a suitable genotype for any given area. A specialized forest seed industry has now been developed, whereby seed from years of abundance is properly stored for use in leaner times. In most temperate countries where seedling culture is practical, a 5- to 7-year supply of seeds of commercially important forest trees is always on hand.

Ecological studies are identifying more clearly the environmental factors that favor individual forest species. The modern forester has some information on the silvicultural practices favoring pine, mixed conifers, hardwoods, and so on. Temperature, thermoperiodicity, light, moisture, and soil condition all interact to determine the suitability of a forest site. Of course, these are in turn affected by elevation, slope, and so on. Temperate-forest yields (stem-wood volume) are about 700 to 900 cubic meters per hectare per year (10,000 to 13,000 cubic feet per acre per year), the yield decreasing with increasing dryness or elevation. Evergreen forest is more productive at elevations of 1500 meters (about 5000 feet) or higher or at latitudes north of about 45°. An example of the change in (and overlap of) dominant forest species as the environment alters is shown in Figure 15-13. Ponderosa pine and Douglas-fir, for example, are adapted to the warmer and drier forest belts of the

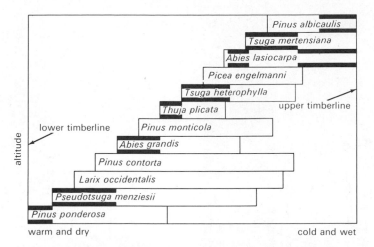

Figure 15-13   The usual order in which coniferous trees are encountered with increasing altitude in eastern Washington and northern Idaho. The horizontal bars indicate the upper and lower limits of the species relative to the climatic gradient. The heavy lines indicate that portion of a species' altitudinal range in which it can maintain a self-reproducing population in the face of intense competition. [After R. Daubenmire, "Vegetation: Identification of Typal Communities," *Science* 151:291–298, 1966. Copyright © 1966 by the American Association for the Advancement of Science.]

Pacific Northwest, and are gradually replaced by other species as the climate becomes colder and wetter with increasing elevation.

Armed with this sort of information, the forest managers can base their decisions on measurements rather than on guess work. Still, there are many aspects of tree growth about which little information has been gathered. Compared to annual crops, trees are peculiar in the tremendous amounts of energy-rich materials they are able to incorporate into their tissues. Just how might the trapped chemical elements be most efficiently replaced? Recirculation through leaf fall is very much related to temperature. Tests in the Great Smoky Mountains show that fallen leaves decompose only about 35 percent in the first year at 1500 meters (about 5000 feet) but 46 percent at 300 meters (about 1000 feet). In the tropics recirculation may take only a few weeks or months. Measurements of energy exchange for individual species and stands are very scanty. Whether one species can trap more energy than another or whether uneven-aged stands can trap more energy than even-aged stands is not known, nor are we certain how many times land can be clear-cut and still yield a vigorous new stand of high-quality timber. Forestry is still quite a way from attaining completely scientific cropping.

Although the ultimate aim in cropping forests is the same as with annual crops—profitable return on investment—the business outlook has to be adjusted to long-term objectives. It is impossible to see fifty years into the future, which is the time needed in some areas to mature a new tree crop. A logical approach under these circumstances is to cut one-fiftieth of the holdings yearly; after fifty years the first parcel would be ready for harvesting again. This is the principle behind **sustained yield** (Fig. 15–14). It is also possible to manage forest made up of trees of diverse ages on a sustained-yield basis by cutting only a portion of the volume during each cutting

Figure 15-14 A carefully managed pine planting, characteristic of sustained-yield operations in the southeastern United States. [Courtesy U.S. Forest Service.]

block cutting
Trees like the Douglas-fir need full sunlight for proper growth. Also, their seeds are carried a good distance by the wind.

seed-tree cutting
Southern pines also need sunlight, and their seeds are carried by the wind. Isolated trees are relatively windfirm.

selective cutting
Young red spruce will grow in the shade of larger trees. Thus trees of all ages and sizes are found growing side by side.

cycle. In practice, 100 percent sustained yield may be difficult to achieve in any but the largest holdings, although it is a practicable ideal to strive for in temperate forest, in which there are essentially solid stands of one or a few species. In the tropics, where there is a tremendous diversity of species, ecological complications are more likely to occur. For one thing, tropical forests do not lend themselves to clear-cutting. Rather, selective cutting is necessary—one tree here, another there—in order to procure a uniform product for economical processing, handling, and marketing. However, in some tropical forests, notably in the lower Amazon valley, attention is now being directed to harvesting all tree species for paper pulp.

In many temperate forests, clear-cutting of timber over extensive areas has given way to sequential harvesting in smaller blocks. This practice leaves adjacent stands of seed trees, which under favorable circumstances will recolonize the cut area within a few years (Fig. 15-15). The size and treatment for cleared

For these reasons the accepted way of harvesting Douglas-fir is to cut all marketable trees in a block of 100 acres or less. Ample standing timber is left between blocks.

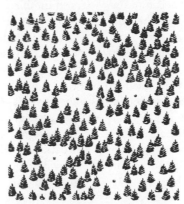

Seeds blow in from the surrounding forest and new trees appear. The new stand is thinned periodically to provide growing space and to remove undesirable trees.

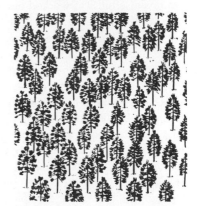

In about 40 years the first cut-over block is mature enough to provide seed for the neighboring block. After a few more decades the first block can be harvested again.

It is therefore possible to remove all trees from a large area except for 4 or 5 per acre, which act as seed trees. These remain until the new seedlings are established.

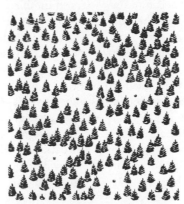

When the seedlings have grown enough so that they are reasonably safe from fire—5 to 10 years—the seed trees can be removed in their turn.

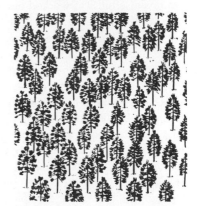

After about 20 years the new stand of pine should be thinned to prevent crowding. Under favorable conditions the pines can be harvested 30 years after original cutting.

In harvesting red spruce the most mature and marketable trees are selected for cutting. (In this picture they are the trees shown slightly darker than the others.)

After the marketable trees have been removed their neighbors provide seed for the new trees that will replace them. In the meantime other trees are maturing.

In 10 to 20 years these trees are ready to be harvested in their turn. If carefully harvested, a forest of red spruce can provide a continuous supply of marketable trees.

Figure 15-15 Harvesting systems in forest management. [After St. Regis Paper Co.]

blocks vary with local conditions; whether woodland should be clear-cut or selectively cut involves such considerations as whether regenerating seedlings demand full sunlight or are of a type that prosper under partial shade, and whether slope is such as to cause significant soil erosion when all plant cover is removed. In areas where insufficient seed trees persist or are of types that do not recolonize naturally, replanting of the forest is necessary. The courts in 1974–1975 in the "Monongahela decision," ruled that a long-disregarded law, the Organic Act of 1897, compels selective harvests (bans clear-cutting) in the national forests of Appalachia. This was a question of considerable contention during the mid-1970s until a compromise bill was worked out and passed by Congress as the National Forest Act of 1976. The new law did provide for some management as even-aged stands, but in general tended to preserve the conservation thrust of the 1897 Organic Act.

Various degrees of success have been obtained in reforesting. Where huge areas have been devastated, such as in the Tillamook burn of Oregon in 1933, the least advantageous sites have regenerated poorly even after a period of more than 40 years. Neither direct seeding nor planting has been very successful in many parts of the burn, perhaps because soils exposed by the forest fire have been altered and eroded, and perhaps for other ecological causes.

Replanting cut-over forest is now commonplace, and in the future this practice will largely replace the hit-or-miss and time-consuming practice of natural regeneration. Tree cropping takes advantage of mechanization, too, although not to the same extent as in field cropping. Some seed is scattered by airplane and by helicopter, usually at the rate of about 0.5 kilogram per hectare (nearly 0.5 pound per acre) for Douglas-fir in the Northwest. Mechanically pulled devices are used for planting nursery seedlings on fairly level land with few rocks. In rough, rocky country, hand planting is still necessary, although expensive. Seedlings of Douglas-fir can be grown in the nursery for a penny or two each, but hand planting them in forest sites may cost as much as $400 per hectare (more than $160 per acre), quite a sizable investment.

Tree seed is still collected mostly from the wild. The difficulties are formidable, what with most pine and fir cones growing at the tips of branches 15 to 75 meters (about 50 to 250 feet) above the ground. Much seed is gathered from felled trees after logging. Sometimes cones cut by squirrels are appropriated; a bushel or more may be cached by a squirrel in one hideaway. Such a system allows little directed genetic selection (although weak seedlings are eliminated in the nursery) and no planned cross-breeding. Tree-breeding possibilities should be tremendous once suitable techniques are more widely developed.

The tree-seed industry is just beginning to come into its own. Cones of coniferous trees are oven-dried and tumbled to free the seed. A temperature of about 49°C (120°F) is best for some pines, whereas 32°C (90°F) is better for fir. Seed is dewinged by brushing or friction, and the residues are blown away. It is typically stored at subfreezing temperatures. At these low temperatures viability is retained for 5 to 10 years. Most tree seeds must be cold-stratified for good germination. Seed planted into the nursery in autumn is cold-stratified naturally by winter weather. Seed for spring planting is typically soaked in water for most of a day and then held at temperatures slightly above freezing for a number of weeks. It will then germinate in 7 to 10 days. Nursery seedlings are tended in much the same way as orchard trees; they are watered, fertilized, weeded, and usually cultivated for one to four years (Fig. 15-16). In the southern United States seedlings are sufficiently mature for field planting in one year.

Of course, direct seeding is much simpler and more economical than nursery operations. It is likely, too, that the root system of a seedling sprouting in place is more adequate than one that has been transplanted once or twice. Direct seeding is generally the only practicable means of replanting large and inaccessible areas; it is also the only means of replanting species that do not transplant well, such as longleaf pine. The comparatively heavy seeds of hardwoods are often more successfully seeded directly than are those of softwoods (conifers). Potential loss is great with direct seeding. Birds and small mammals may eat the seed, and weeds may smother the seedling.

Figure 15-16 Slash pine seedbeds in Georgia. This nursery grows about 30 million seedlings per year. Increased selection of germ plasm has a tremendous impact upon eventual production. [Courtesy U.S. Forest Service.]

Rodents are sometimes poisoned with treated grain prior to seeding. Where terrain permits the use of tillage equipment, seed is liable to be buried. Coatings are applied to the seed (pelleted seed) to repel animals, insects, and birds.

A helicopter or aircraft can seed as much as 1200 hectares (nearly 3000 acres) in a single day. Broadcast seeding from the ground, with hand or tractor-mounted seeders, is tedious but generally more precise and quite satisfactory for small tracts. Seeding from air or ground is more effective if the soil has been mechanically prepared beforehand to the extent that the terrain allows. Mechanized row drilling is even possible on favorable sites.

Large forest-products companies, today accounting for most of the pulp and lumber produced in the United States, are as elaborately organized as are the gigantic field-crop farms discussed earlier. Virtually all of the large lumber and plywood producers now use computers for linear programming in operations

analysis to streamline their planning, production, and sales. Lumber itself may be graded electronically. Raw materials—logs—are thus more efficiently turned into wood products. Logs are no longer inexpensive; they must be fully utilized for profitable operations. The trend is to use all types of timber cut, and programing can indicate quickly whether a certain log of a certain type at a certain time should be used for lumber or plywood. Operations research may suggest that lower-grade logs are satisfactory for plywood sheeting, thus freeing higher-grade logs for premium sale for other uses; or that spruce can be substituted for white fir in interior plys, thus saving costs (largely because spruce has a shorter drying time than fir and because its use relieves a bottleneck at the kiln and speeds the product to market). Or it may take into account glue-spreading capacity and market demand to determine whether plywood should be 3-ply or 5-ply. Computerized predictions can be made for markets months and years ahead, to better adjust product mix to needs. One of these days, no doubt, the processing of forest products will be scientifically controlled from raw material to finished product, as in the processing of farm products. The small, inefficient forestry operation is already as obsolete as the small, inefficient farm.

MODERN CROPPING INNOVATIONS

Mechanization

Mechanization has brought significant new opportunities to plant cropping, especially for seeded crops in temperate climates. Mechanical harvesting is now widespread, and is employed not only for grains but for tree and bush fruits, root crops, leafy vegetables and even crops as perishable as ripe tomatoes. The efficiency in terms of human labor that mechanization brings is nowhere more evident than in North America, where today one farmer produces sufficient food for about 56 non-farming people—a situation in striking contrast to the nearly universal participation in farming found in primitive societies (and in the United States during the colonial period). Mechanization, the biological engineering that provides tailor-made cultivars, and all of the accouterments for crop care have not only had a quantitative effect but have on the whole improved quality as well.

The trend of industrialized societies is to use more capital and less labor. The pressure of labor costs encourages extensive and often rapid mechanization. Since it often proves impossible to hire efficient hand labor in the United States at any wage, much less low wages, it is no wonder that in two years the picking of canning tomatoes in California changed from 100 percent hand picking to 85 percent mechanized picking. Even though a picking machine may cost upwards of $20,000 with its accessories, it still pays for itself in a reasonably short time.

The increase in number of certain farm machines in the two decades between 1940 and 1960 was phenomenal (see Fig. 27-5). There are now nearly a million corn-picking units in operation, most of them put into use since World War II. Sprinkler irrigation has risen fifty-fold in the same interval, with only about 40,000 hectares (about 100,000 acres) irrigated in 1945 compared with more than 2 million hectares (more than 5 million acres) in 1965. Twelve million horses and mules gave way to 5 million tractors within one generation in the United States, and the trend is for tractors to get ever bigger and more powerful. Some units are essentially a self-propelled chassis with power take-offs to which special attachments are hitched for a wide variety of operations. There are cultivators that barely scuff the topsoil, others that chisel deep into the subsoil; rotary hoes that chop weeds, and tractors that pull up to a dozen plows; spray machines that arch over tall crops, and others that carry spray booms 30 meters wide.

Harvesting machines are able to determine the readiness of head lettuce for picking, to cut asparagus under photoelectric control, to shake tree fruit onto cushioned catchers, to vacuum-sweep pecan and tung fruits from the ground, to dig potatoes and sort the tubers while traveling at speeds of more than 3 kilometers per hour, to collect pineapples on conveyers that are many meters wide, and so on. A multitude of devices for separation, cleaning, and sorting of agricultural products minimize the drudgery of yesteryear. Many crops are prepared for harvest with chemical sprays to control ripeness and abscission. Indeed, it seems that some piece of equipment has been invented or adapted for almost every once-laborious agricultural operation. The situation is nicely summarized in Figure 15-17.

The effects wrought by this mechanical revolution are truly astounding. The greater output per hour of human labor was accomplished by substituting tractors for horses and by putting into use about a million each of pickup balers, corn picker-shellers, and grain combines (Fig. 15-18). Cotton is now picked mechanically throughout most of the cotton belt. Airplanes are used to sow rice and to spread pesticides. There are mechanical tobacco harvesters, motors to lift hay and move grain. Machines limb, top, cut, and handle entire trees. Gadgets shake fruit and nuts from trees, prevent frost, grind and blend feeds, dry grain and

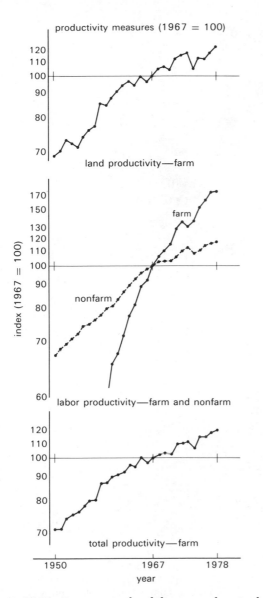

Figure 15-17 Increases in the ability to produce in the United States during the last three decades have been achieved by fewer farm workers and more equipment. (Value for 1967 = 100.)

hay. There is equipment to place just the right amount of fertilizer in just the right soil location, to spray the exact row width with herbicide, to meter seed exactly 12 rows at a pass. For a long time, metering was accomplished by slotted planter plates that accepted seed of a particular size and shape, thus necessitating uniform seed. Now plateless planters can utilize the more economical seed not standardized for size, and are less likely to crack the seed coats. Furrow openers can work through crop residues left on the soil surface to provide an organic soil mulch, with precise depth control even at a speed of 8 kilometers per hour. No wonder corn yields have risen from about 2.3 metric tons per hectare (about 34 bushels per acre) in 1945 to more than 6.7 metric tons per hectare (on well-run farms), with yields of over 13 metric tons per hectare (about 200 bushels per acre) in sight (Fig. 15-19). It took about 350 hours of human labor to produce some 7 metric tons of corn in 1800, only 34 hours by 1950, 3.5 hours in 1960, and less than 1 hour in 1966. Tobacco yields have doubled since 1945, wheat yields gone up 80 percent, and soybeans 30 percent. Today's agribusiness uses capital and brains to increase production and profit.

Such a degree of mechanization is possible only under a fortunate combination of industrial capacity, capital, and technical knowledge. It could hardly be operative today in much of Latin America, Africa, or the heavily populated parts of the Orient (where human labor is often much less expensive than machines). It is ironic that acceptance of agricultural innovation is greatest and fastest where agriculture is already profitable and progressive. It is most adaptable to large farms in sophisticated societies. Even within the United States it is more easily adapted to the prosperous farms of Iowa than to the hills of Appalachia (for a variety of reasons). The fruits of mechanization have been little enjoyed in most tropical countries, except to a small degree where plantation agriculture is practiced. Even today much of the world's crop land is tediously scratched by oxen pulling a forked log as a plow. But, as was noted early in this chapter, this may be the wave of the future what with high energy costs and petroleum shortages!

A

B

Figure 15-18 Mechanized grain harvesting: The invention of the combine, a unitized mechanism for harvesting grain and freeing it of chaff and detritus, has had a tremendous impact on agriculture. A single machine substitutes for hand-cutting, shocking, flailing (threshing), winnowing, and associated procedures of primitive agriculture. (A) In the self-propelled wheat combine an adjustable reel holds grain-bearing stalks against the basal cutter bar; the auger guides cut stems to a central beater and eventually to the rasp-bar cylinder that rotates within the open-grate concave (just above the front wheels). Ninety percent of the grain separation occurs at the concave; the grain is abraded free from the seed head by the cylinder, and falls through open slots in the concave to conveyor belts below. Additional mechanisms in the upper rear part of the combine further agitate stems and chaff to retrieve remaining grain, the spent straw being issued from the rear of the machine. Screening mechanisms in the lower rear part of the machine clean the grain further by siftings and air blasts before it is elevated to the grain tank (behind the operator). The clean grain can be unloaded through the unloading auger (behind the operator) directly into trucks drawn alongside the combine. (B) The corn combine, similar in principle, has a gathering mechanism designed to accommodate rows of corn. Ears are freed from stems by a snap-bar mechanism (instead of cutting the whole stalk). The rasp-bar cylinder and concave are designed to shell the corn from the cob. Grain is retrieved and cleaned as in the wheat combine. [Courtesy John Deere Co.]

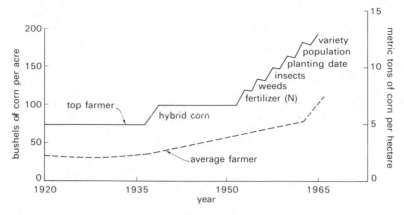

Figure 15-19   Corn-yield trends in Illinois. By successively incorporating improved cultural practices in their cropping system, the most productive farmers achieve dramatic increases in yield in relatively short periods of time. They can be expected to adopt any technological improvement coming to their attention if it provides even small economic return or increased efficiency. [Courtesy J. W. Pendelton.]

Agribusiness—the new system

In agriculturally advanced parts of the world, such as Japan, the United States, Canada, and parts of Europe, the simple farmer of yesteryear is fast becoming a manager of crop production, drawing upon chemistry, biology, engineering, economics, and many other disciplines. The modern farmer runs a rural "factory" producing proteins, fats, and carbohydrates. The modern farmer's goal is the same as that of any person in business—to maximize return. Thus in these parts of the world the trend in farm operations will be toward a high-yield system, drawing upon many technical resources. If soybeans are planted, they will be a select variety, for which a special fertilizer is compounded in keeping with local soil needs (as determined by test). Pre-emergence weed control will be provided. Planting machinery will adjust spacing and row width for maximum utilization of sunlight. Chemical growth regulators may even control shape of the plants, prevent lodging, and defoliate plants before harvest. Irrigation will probably be used, perhaps automatically applied as soil moisture reaches a predetermined critical level. Pesticides will be accurately and efficiently applied

at exactly the correct season (before diseases or insects get out of hand). All of these activities may very well have to be programed by specialists who never work in the fields.

Specialized services make such systems click. Much of American agriculture is already beyond general servicing by a county agent, trying to cover all facets with demonstrations, tours, and bulletins. Experts in many fields are needed, operating as a team. Diverse inputs are demanded that must be correct, timely, and sophisticated. The system is concerned with the whole sequence of steps dealing with the organism from growth to distribution and marketing. The agribusiness manager must be aware of government programs, transportation problems, unionization and financial trends, as well as of new cultivars, cultural techniques, and pest hazards.

The less well-developed parts of the world cannot support such a sophisticated agribusiness, and in heavily populated lands mechanization to relieve labor (as opposed to increasing production) is hardly the answer. What many parts of Africa, Asia, and Latin America need at this stage are simple, practical agricultural techniques developed through local research, not involved investigations or computer controls. Agriculturists from advanced countries may provide guidance, but should not assume that techniques developed for the temperate zone can necessarily be applied elsewhere.

In agriculturally advanced regions the trend is toward fewer but larger farms. Most production may eventually come from incorporated agribusinesses rather than individual holdings too poorly capitalized to operate efficiently. It has been predicted that as few as 500,000 farmers could provide the necessary production for the United States. The trend is unmistakable: from 1954 to 1964 the size of the average farm increased from around 100 hectares (about 250 acres) to 144 hectares (about 356 acres), each farm representing an investment of hundreds of thousands of dollars. The number of farms decreased from 5 million to 3 million, of which 1.5 million provide 90 percent of the total output. During that decade 1.25 million farmers left the farm, and total farm acreage decreased about 3 percent; yet crop production increased remarkably. The farmers themselves now work less strenuously and put in fewer hours.

Agribusiness has immense social consequences. For example, migrant workers work seasonally and may be unemployed in the off-season, causing grave social problems. And what of the political consequences, as fewer and fewer votes come from the farm and more and more come from the city? Farm programs will no doubt change, and the interests of the non-farmer will be given greater consideration.

Present techniques and trends

Improvements in efficiency are eagerly sought in agriculturally advanced parts of the world. Amid the plethora of improvements in product and practice, several trends seem to be shaping up.

Minimum tillage is increasingly practiced, its usefulness varying from soil to soil and from crop to crop (Fig. 15-20). It is a marked departure from the traditional idea that fields are best thoroughly cultivated. Soil is better protected from erosion and is less compacted by machinery. Crop yield is usually equally as good as under conventional cultivation, yet with an economy of operation (at least one plowing or disking is avoided). Equipment now in use can prepare the soil, apply the fertilizer, and

Figure 15-20 "Minimum tillage" implies reduced soil cultivation and separate planting operations. Here wheel-track planting creates two distinct soil zones, the row zone and the interrow zone. The row zone is smooth and firm, providing good soil-to-seed contact for fast germination. The interrow zone is left rough to catch and absorb rain faster and to reduce germination of weed seeds. The planting machine shown here is equipped with attachments for the application of dry fertilizer, insecticides, and liquid herbicides. [Courtesy John Deere Co.]

plant the seed in one pass over the field. Extreme examples of minimum tillage involve relatively little soil disturbance of any kind. For example, a sod blanket may be killed chemically, left as a mulch, and a row crop planted in slits cut into the dead sod (the no-till system).

Vegetables and small fruits can sometimes be grown with no postplanting tillage at all. Weeds are controlled entirely by herbicides. Solid plantings of course take much better advantage of space, since cultivating equipment need not pass between rows. With some crops this may facilitate harvesting, and certainly surface feeder roots are not physically disturbed. Furthermore, no buried weed seed is brought to the surface where it can sprout. Feasibility, of course, depends upon local conditions—for example, whether the soil is of a type that allows adequate water infiltration without cultivation. In any event, several root crops, including potatoes, have been grown in Britain without postplanting tillage. Yields of potatoes, parsnips, carrots, gooseberries, broad beans, and Brussels sprouts have all been close to or slightly better than yields where cultivation is practiced. The cost of herbicide treatment must be weighed against the cost of cultivation for a particular area.

Another trend is toward higher plant populations, to take maximum advantage of sunlight. Increased fertilization and irrigation are needed to accommodate the more densely planted fields. Dense plant populations often shade weeds and protect soil somewhat more adequately from erosion. In some crops nowadays seeds are even being oriented during planting so that leaf position (as in corn, which has leaves in a single plane) is crosswise, not overlapping and partly shading other plants in the row. Dense plantings are being used effectively with cotton, grain sorghum, sugar beets, soybeans, and corn just about as fast as new equipment to fit the narrower rows becomes available. The narrower rows may give an additional 20 percent yield if fertility and moisture are adequate.

Corn populations denser than 60,000 plants per hectare (about 24,000 plants per acre) are generally impractical; excessive crowding can cause lodging and undeveloped ears. Simple arithmetic shows that more symmetrical spacing is had with more closely spaced rows. On traditional 40 inch rows (the old standard, devised to accommodate the mule), the distance between seeds in the row must be about 15 to 18 centimeters (6 to 7 inches) to achieve a population of nearly 60,000 plants per hectare. But a 75-centimeter (30-inch) row allows about 20 to 23 centimeters (8 to 9 inches) between plants, with less crowding in the direction of the row. Perhaps even larger yields would be obtained if approximately 40 centimeters (about 16 inches) were allowed between rows and plants, which would be possible if herbicides were made to substitute for cultivation.

Where cropland is abundant, and high yield with fewer plants is more important than yield per hectare, such as those areas in which planting limitations are in force, plant yield can often be increased by skip-row planting. A row or two of crop is sowed, then a row or two left fallow. The plants then receive light from the sides as well as from above, resulting in more photosynthesis and higher yields per plant. Even greater growth can sometimes be stimulated by placing reflecting boards in the vacant rows, which, like mirrors, direct more sunlight to the sides and bases of the plants.

There is increasing interest in **field processing** of many crops to avoid unnecessary haulage and storing. Preliminary treatment of hay crops in the field ("conditioning," a partial chopping to facilitate drying) is increasingly practiced, as is automatic heat-drying of forage (often as mechanically formed pellets) before storage and marketing. Combines clean up grain in the field, casting straw and other residues back on the land. Corn is increasingly shelled in the field, too. Inexpensive procedures for oxygen-free storage of silage are being developed, as are simplified feed-lot operations.

The following are a few examples of present trends in the production of various crops.

Hay and forage

Field pelleting or wafering of hay and forages is increasing as more dependable and economical machinery becomes available. It is common in dry climates, as in the alfalfa fields of central California. Field conditioning of hay before baling is widespread everywhere. Improvements have been made too in mechanized bale handling and in gathering green chop (see Box 15-2).

Corn

The trend toward denser plant populations in corn production has already been mentioned. The distance between rows is increasingly being reduced from the conventional 40 inches (1 meter), and corn is increasingly being harvested with a four-row cornhead on a combine, which can handle 350 liters (about 10 bushels) per minute. Field shelling is well accepted, now that drying systems and high-moisture storage facilities are more available.

Cotton

In cotton production, narrower rows and planting to assure an even stand without thinning are on the increase. Herbicidal weed control, chemical defoliation prior to picking, and machine harvesting are standard today in efficient cotton farming. Hybrid cotton is under development.

Box 15–2   *Planning high TDN yields for livestock*

The TDN (total digestible nutrients) available to livestock from feed and forage is the key to profitable animal husbandry in technically advanced lands. A high TDN yield is best gained from select cultivars, of course. Yet there are options, and systems must be chosen in view of such local conditions as moisture availability, length of growing season, soil type, slope, and so on. On good land that has been progressively cropped, 11 metric tons of TDN per hectare (about 5 tons of TDN per acre) can be obtained either from 56 metric tons of corn silage, 15 metric tons of corn, or 22 metric tons of alfalfa hay. Which route is best? Or if a meadow can be harvested for silage instead of for hay, to yield an additional 25 percent of TDN, does the added expense justify this choice?

On poorer lands, especially on slopes unsuited to cultivation, yields may amount to only a bit more than 3 metric tons of TDN per hectare (about 1.5 tons per acre). This type of land may be most efficiently exploited by letting livestock graze rather than by trying to produce the 18 metric tons of silage or 4.5 metric tons of grain needed to match this TDN yield. Cost of inputs (fertilizer and capitalization of technology) for high yields on poor land may be so great that intensive cropping may not be the most efficient use! Urging a highly responsive forage such as alfalfa (each metric ton of which removes about 50 kilograms of lime, 6 kilograms of phosphate, and 22 kilograms of potash from the soil) to maximum production on poor soil might require doubling of inputs for an increase of only 50 percent in output! Adding grass to alfalfa may reduce TDN yields, but may still be the wisest course for protection of soil on slopes.

Obviously, management is a vital factor. Proper timing of fertilization is important, not only to provide maximum yields when needed, but to control the botanical composition of the pasture. Choice of cultivar is significant. For example, white clover may prove to be better on a shallow soil (if moisture is not limiting) than deeper-rooted alfalfa or red clover, which are ordinarily more productive. Where moisture is insufficient for maximum corn yields, sorghum may be preferable, or, on stony hills perhaps tall fescue forage. In the southern United States, Coastal bermudagrass, which provides a high TDN yield on good soil, would be inferior to bahiagrass on poor land.

Maximum advantage is not taken of TDN opportunities if a pasture is undergrazed. Yet rotational grazing to allow recovery from overgrazing can increase TDN yields by 50 percent. If corn is harvested in the milk stage it may yield only two-thirds as much energy as silage feed made from the same corn allowed to mature to the dent stage. Yet, delaying alfalfa harvest can impair nutritive quality so much that 50 percent additional feed grain may be required for equivalent animal weight gain. But too early (or too frequent) cutting of alfalfa can hurt the stand and increase the weeds!

Obviously, managing modern forage production intelligently requires no little technical knowledge. Profitable farming is not for the untrained.

Vegetables

Typically, root crops such as beets and leaf crops such as lettuce are densely planted in rows and then thinned. Not only is machinery now available for automatic thinning, but techniques are being rapidly developed for properly metering seed during planting so that thinning is not necessary. Plastic or resinous mulches are often used to assure a better stand. Systems for mechanical harvesting are used in practically all vegetable crops.

In tomato fields it is now possible to fumigate the soil (with ethylene dibromide) and prefertilize with anhydrous ammonia at the same time. Raised beds are formed and planted, and a pre-emergence herbicide and starter fertilizer are drilled in by the same machine. Seeding is metered for a plant population as high as 124,000 plants per hectare (about 50,000 plants per acre). During growth, insecticide and fungicide sprays control insects, mites, and disease. A mechanical harvester cuts spreading tomato vines at root level, lifts them, and shakes the tomatoes into a bin. The old vine is discarded, while a small crew sorts out green tomatoes and tosses them back onto the field. The harvested fruit is handled in bulk for dispatch to processing facilities.

To some extent, high-value greenhouse crops are being fertilized with carbon dioxide, which is vital for photosynthesis but is sometimes limiting to plant growth because it is in scant supply (only 0.07 percent or less is present in the air), as well as with optimal amounts of mineral nutrients.

CONCLUSION

Cropping systems vary with crop, time and location, and level of technology. They have evolved with the development of human society. This development can be considered to have proceeded in three stages. The first, the most primitive, consisted in **gathering**

from the wild. This was the basis of pre-agricultural societies, and it is still practiced in many primitive parts of the world. Many of the plant products used in advanced countries are still gathered from the wild: chicle, Brazil nuts, blueberries, gum arabic, and occasionally rubber. The next stage, the **management of natural stands,** was the basis of early orcharding. Much of the world's forest and range land is still handled in this manner, as is the herding of livestock on open range. Finally, the most advanced stage, **cultivation,** involves management of the plant throughout its life cycle. It usually depends on the introduction of exotic species and complex techniques of soil manipulation, environmental control, and alteration of plant growth processes. The intensity of management varies greatly. The scale of operation ranges from extensive outdoor farming to concentrated cropping in small gardenlike areas. It

---

Box 15–3 *Overtones to advanced agriculture*

Cropping systems, in all of their manifestations, are profoundly disturbing to natural ecosystems—to habitats as they were before they were occupied by people. Some of the consequences of "progress" to human ecology are seldom taken into account. Involved overall is the ethical question of human concern for humankind. When the Third World proclaims its right to unlimited population increase (as it did at the Bucharest conference) and in the next breath accuses "developed" countries of irresponsibility should the masses not be adequately fed, confrontation is unavoidable. We are faced with the age-old ethical dilemma of whether short-term evil (inadequate sustenance) leading to long-term good (population in balance) is justified, or whether short-term good (assistance programs) leading to long-term evil (mass misery) is preferable.

As the United States Congress debated the bill entitled *The Self-Reliant Development and International Food Assistance Reform Act of 1979,* here's how New Directions (a nonprofit organization devoted to leadership in pressing world affairs), one of the instigators of the bill,

sized up the situation. New Directions felt that the dependency of many poor nations obstructs their economic progress. "This is particularly true in the area of food. Our massive food aid program is growing every year, while agriculture in the Third World stagnates. American food dumped on Third World markets depresses the prices of local agricultural products to the point where many stop farming. At the same time, the increased demand for our farm products helps to drive up food prices for Americans" (*New Directions,* vol. 1, no. 2, April 1979).

G. K. Myrdal, in reviewing development projects undertaken in various parts of the world, had this to say about not taking human carrying capacity into consideration: "But seldom is provision made to hold populations at the new levels that land can support. In consequence, the land deteriorates, deserts spread or become barren, and a greater number of people end up worse off than they were before development of the area took place. One can question whether international development agencies should continue to play this losing game."

extends to the artificial environments found in greenhouse production, and to the synthetic factory systems of the pharmaceutical and fermentation industries.

Outdoor cultivation systems usually start with some degree of land and soil preparation. In areas new to agriculture this may entail land clearing, adjusting land slope to prevent erosion and washouts, and the creation of irrigation and drainage systems. It usually entails adding soil amendments (such as lime and fertilizer) and making various physical changes in the soil to create conditions favorable to tender seedlings.

The procedures used to increase yield comprise a wide spectrum of processes to modify the plant environment and to control plant growth processes. Growth control can be achieved by such ancient techniques as training and pruning, grafting, and breeding, or by such modern methods as the application of growth-regulating materials. Regulation of plant physiology is limited only by the extent of our understanding of growth and development—the more complete our knowledge, the more sophisticated the control. It becomes possible to affect not only growth itself but such developmental processes as flowering and fruiting. Advances in this area have been a by-product of basic physiological studies. The field was given great impetus with the discovery of substances that are toxic to plants in very small amounts—the modern herbicides. A number of materials are known that dwarf plants, set or remove fruit, and induce rooting and flowering; many are routinely used in intensive horticultural operations. It is very likely that future changes in cropping patterns will involve chemical manipulation of plant growth processes, which may well usher in revolutionary methods of crop production.

Cropping practices also extend to market preparation of the plant product and to storage and distribution. The path of agricultural products from the growing plant to the ultimate user is complex. **Marketing,** the activities that direct the flow of goods from producer to consumer, often becomes an integral part of cropping systems. This will be discussed further in Part Six.

Selected references

Abelson, P. H. (editor), 1976. *Food: Politics, Economics, Nutrition and Research*, Washington, D.C.: American Association for the Advancement of Science. (A thorough, stimulating review of factors and problems bearing upon world-wide food production. Janzen's "Tropical Agroecosystems," pp. 103–110, and Steinhart and Steinhart's "Energy Use in the U.S. Food System," pp. 33–42, are especially provocative regarding cropping systems.)

Chapman, S. R., and L. P. Carter, 1976. *Crop Production: Principles and Practices*. San Francisco: W. H. Freeman and Company. (An all-around general text discussing the major farm crops of today as grown under advanced farming conditions.)

Childers, N. F., 1976. *Modern Fruit Science* (7th ed.). New Brunswick, N.J.: Rutgers University Press. (A popular text on orchard management with special emphasis on temperate tree fruits.)

Cox, G. W., and M. D. Atkins, 1979. *Agricultural Ecology: An Analysis of World Food Production Systems*. San Francisco: W. H. Freeman and Company. (The book discusses ways in which people manipulate ecosystems and, in so doing, impinge on various "cropping systems." The discussion of roles of varying natural conditions makes crop ecology understandable.)

Donkin, R. A., 1979. *Agricultural Terracing in the Aboriginal New World*. Tucson: University of Arizona Press. (A long overlooked subject is treated in some depth in this paperback. The fundamental importance of water-control systems was apparent even to primitive societies, which used terracing and channeling in various forms.)

Duckham, A. N., and G. B. Masefield, 1970. *Farming Systems of the World*. London: Chatto and Windus. (A worthwhile "textbook" on comparative agriculture: it relates farming systems to ecological, economic, and sociological factors, using case studies from various parts of the world.)

Eddowes, M., 1971. *Crop Technology*. London: Hutchinson. (A crisp, to-the-point presentation of cropping fundamentals and practices in the United Kingdom.)

Folley, R. R. W., 1973. *Intensive Crop Economics*. New York: American Elsevier. (The author explores the application of horticultural techniques to agriculture as a more intensive means of production. An overpopulated world would seem in need of the increased food production that should result.)

Grigg, D. B., 1974. *The Agricultural Systems of The World: An Evolutionary Approach.* New York: Cambridge University Press. (Early chapters discuss the early history of agriculture. Other chapters cover shifting agriculture, paddy rice, pastoral nomadism, mixed (Western) farming, Mediterranean agriculture, dairying, ranching, plantations, and large-scale grain production.)

Harrison, P. D., and B. L. Turner (editors), 1978. *Pre-Hispanic Maya Agriculture.* Albuquerque: University of New Mexico Press. (A series of seventeen papers explores the diversity of Mayan crop-production methods, chiefly from 300 to 900 A.D., with some comparison with similar Eastern cultures. Apparently, slash-and-burn methods were less important than previously thought, and strongly supplemented by tree cropping and—in the lowlands—the use of raised fields.)

Harwood, R. R., 1979. *Small Farm Development: Understanding and Improving Farming Systems in the Humid Tropics.* Boulder, Col.: Westview Press. (Harwood recommends the blending of traditional and modern technology for smaller farms in less-developed parts of the world. The book is brief, with Part I analyzing small farm conditions, Part II offering some cropping suggestions and efficient use of resources.)

Hillel, D. (editor), 1972. *Optimizing the Soil Physical Environment Towards Greater Crop Yields.* New York: Academic Press. (The proceedings of a special invitational symposium held in Israel, dealing with a wide range of topics that influence crop growth and production yields.)

Janick, J. 1979. *Horticultural Science,* 3rd ed. San Francisco: W. H. Freeman and Company. (An introduction to the scientific and technological bases of modern horticulture.)

Kepner, R. A., R. Bainer, and E. L. Barger, 1978. *Principles of Farm Machinery,* 3rd ed., Westport, Conn.: Avi. (An engineering slant on the use of diversified agricultural equipment, including power dynamics. This edition has adopted metric units.)

Loomis, R. S., 1976. Agricultural systems, *Scientific American* 235(3):98–105. (Well-illustrated, understandably written summary of the many considerations that must prevail in determining a cropping system.)

Martin, J. H., and W. H. Leonard, 1976. *Principles of Field Crop Production,* 3rd ed. New York: Macmillan. (A comprehensive reference on field crops, especially cereals and forages but including various rootcrops and less important species.)

Meggers, B. J., 1971. *Amazonia: Man and Culture in a Counterfeit Paradise.* Chicago: Aldine-Atherton. (Discusses the relationship between tropical forest environments and human societies, suggesting that the tropical rainforest is an unsuitable environment for development and persistence of complex societies, intense agriculture, and appreciable population density.)

Nayer, N. M. (editor), 1974. *Proceedings of the First National Symposium on Plantation Crops.* Kerala, India: Bangalore Press. (With tropical agriculture long something of a stepchild to the profession, it is interesting to have emphasis placed upon production of plantation crops. Nearly fifty papers discuss various aspects of the growing of such crops as coconut, tea, coffee, black pepper, ginger, cardamon, clove, rubber, arecanut, cashews, coriander, and chili pepper.)

Ochse, J. J., M. J. Soule, Jr., M. J. Dijkman, and C. Wehlburg, 1961. *Tropical and Subtropical Agriculture* (2 vols.). New York: Macmillan. (A voluminous work dealing with tropical agriculture. The first volume deals with general cultural practices, the second with crops.)

Olehaf, R. C., 1978. *Organic Agriculture: Economic and Ecological Comparisons with Conventional Methods.* Montclair, N.J.: Allanheld, Osmun. (Although favoring the organic approach, Olehaf's treatment of the subject is objective and without the sermonizing characteristic of many "organic gardeners." Long-term advantages would seem to be the strength of organic agriculture, and the holistic approach of a biodynamic soil–crop–livestock system is worth considering. The economics of farming according to organic principles are considered, and are shown not to be currently competitive in most places.)

Ruddle, K., 1974. *The Kukpa Cultivation System: A Study of Shifting Cultivation in Colombia and Venezuela.* Berkeley and Los Angeles: University of California Press. (Shifting cultivation accommodates human needs, but is disappearing as acculturation proceeds. It is not inherently wasteful, and can be adaptive. Much information is given on the tribes in question, including the plants they cultivate.)

Ruthenberg, H., 1971. *Farming Systems in the Tropics.* London: Oxford University Press. (A traditional way of looking at farming methods from the special discipline viewpoints, while covering a subject area generally unfamiliar to temperate-zone people.)

Smith, D. M. 1962. *The Practice of Silviculture*, 7th ed. New York: Wiley. (A widely used textbook on practices and techniques of North American forestry.)

Smith, H. P., and L. H. Wilkes, 1976. *Farm Machinery and Equipment*, 6th ed. New York: McGraw-Hill. (Almost all kinds of farm machinery are illustrated, and their fundamental functions are discussed. Included are such newer developments as minimum tillage, seed-metering, and flaming. A comprehensive book, organized by crop.)

Spedding, C. R. W., 1975. *The Biology of Agricultural Systems*. New York: Academic Press. (The purposes and fundamentals behind cropping are discussed more than specific practices. The interaction of plants and animals, and of agriculture with society, are taken up. The book advocates a systems approach to the use of land.)

Thomas, W. L., Jr. (editor), 1956. *Man's Role in Changing the Face of the Earth*. Chicago: University of Chicago Press. (A collection of analyses by eminent authorities relating to the drastic alteration of world ecology as a result of human progress toward a modern state of civilization.)

Thorne, M. D. (editor), 1977. *Agronomists and Food: Contributions and Challenges*. Madison, Wisc.: American Society of Agronomy. (Brings ecology into the discussions, puts special emphasis on wheat, and makes predictions for the future. The book provides uneven subject coverage in line with the authors' interests, but it is a valuable source of additional bibliographic information.)

Westwood, M. N., 1978. *Temperate-Zone Pomology*. San Francisco: W. H. Freeman and Company. (An up-to-date treatment of fruit and nut production in temperate climates. Includes discussions of the small fruits, such as strawberries and grapes, as well as the tree fruits and nuts.)

16
Crop Nutrition

The control of plant nutrition is one of the foundations of modern agriculture. Although the influence of soil on plant growth is recorded in ancient writings, the role of inorganic minerals was unknown until relatively recently. The first experimental studies on plant nutrition were conducted by a Belgian chemist, J. B. van Helmont (1577–1644). He planted a willow in a weighed quantity of soil and found that after five years growth, the tree increased in weight by 153 pounds while the weight of the soil decreased only two ounces! He erroneously (but understandably) assumed that the increase in the plant's weight was all derived from the water added to the soil, not realizing the role played by atmospheric CO_2. Today we know that the two ounces of material removed from the soil consisted mainly of various inorganic minerals that are essential to plant life and growth. This concept, however, was not established until the early nineteenth century, when the Swiss physicist Théodore de Saussure showed that the ashes of plants contained various mineral elements and that plant nutrition is dependent upon the uptake of nitrates and mineral elements from the soil. In 1860, Julius Sachs and W. Knop determined that certain of those mineral elements—calcium, magnesium, potassium, and nitrogen—are required by plants in relatively large quantities. The nutrition of crops is heavily dependent upon the chemical and physical properties of the soil, which influence its ability to hold and provide water and nutrients for plant growth.

FACTORS INFLUENCING SOIL FERTILITY

The fertility of a soil is only indirectly related to the chemical composition of its primary inorganic minerals. Often the most important factor is the form in which the nutrients exist in the soil. The availability of nutrients depends upon many factors, among which are the solubility of the nutrients, soil pH, the cation-exchange capacity of the soil, soil texture, and the amount of organic matter present.

Plant nutrients

About 90 percent of the entire weight of a living herbaceous plant is water; the remaining 10 percent, the dry matter, consists mainly of three elements: carbon, hydrogen, and oxygen. A small but important fraction of the dry matter consists of other elements that are indispensable for growth. Although soil may supply a large number of elements, only 13 (in addition to carbon, hydrogen, and oxygen) have been proven to be absolutely necessary—that is, **essential** for higher plant life and growth.

These 13 essential elements are divided into two categories on the basis of the abundance with which they are required by plants. **Major elements** are required in relatively large amounts, usually expressed as parts per hundred (percent) per unit of dry matter. These include nitrogen, phosphorus, potassium, calcium, magnesium, and sulfur. **Minor elements** are required in very small quantities, usually expressed in parts per million (ppm) per unit of dry matter. These include iron, boron, manganese, zinc, molybdenum, copper, and chlorine. The essential elements will be discussed individually later in this chapter.

Some elements are not required by higher plants but are nearly always present. These may be necessary to animals and certain microorganisms. Similarly, some algae and fungi do not require calcium, an essential element for higher plants.

Cobalt is found in higher plants, but apparently serves no function. Yet animals that do not have a supply of cobalt suffer from a vitamin B_{12} deficiency. Blue-green algae have been shown to require cobalt. The addition of 1 nanogram (a billionth of a gram) of cobalt to a culture solution in which the fresh-water blue-green alga *Nostoc* was grown increased the yield of dry weight.

Vanadium is apparently necessary for the growth of certain bacteria and algae, but it is not required by higher plants.

Sodium is not absolutely required, but it is found in all plant materials in quantities large enough to detect. It is, however, required by animals. Sodium can partially substitute for some plant needs otherwise satisfied by potassium.

Iodine is required by animals and has been found either in or associated with organic compounds in plants. It has not, however, been found to be essential for higher plants.

Fluorine is frequently found in plants, even though it has not been proven to be essential. But it is absolutely necessary for the development of animal skeletons.

Silicon and **aluminum** are found in most plants in comparatively large quantities, yet they have never been proven to be necessary to sustain life. Silicon, however, apparently imparts structural strength to a great many plants. Scouring rushes were given their name because their silicon content is so great that they were used by early settlers in America for scrubbing pots and pans. The leaves of many tropical plants also have an extremely high content of silicon.

The qualitative determination of which elements plants require has been achieved by growing plants without soil in solutions of inorganic salts. The combination of inorganic salts that supports normal growth makes up a **complete nutrient solution.** Plants may be grown in the solutions alone, if aeration is provided, or in quartz sand wetted with the solution. Whether a particular element is essential is determined by comparing long-term growth in a complete solution with growth in a solution lacking that element. Special precautions are required, because some elements are required in minute quantities. The container, the sand, the chemicals, the distilled water, and even the atmosphere may contain traces of the element under test.

The compositions of two standard nutrient solutions are shown in Table 16-1. Either solution must be supplemented by other essential elements, such as boron and manganese. Frequently these will be supplied in sufficient quantities by contamination when chemicals are used that are not of high-grade analytical quality.

Foliar analyses are frequently useful in determining nutrient requirements from the mineral composition of plants. Leaf samples collected from plants growing under a wide variety of conditions are analyzed chemically and spectrographically. If other environmental factors are not limiting, growth rate can be related to the content of each element, and response curves can be developed—the more favorable the concentration, the better the growth. Assumptions based on foliar analysis must be made with great care until standards have been developed for each crop and each element.

It should not come as a surprise that China has long been deeply involved in conservation and min-

Table 16–1 Composition of two nutrient solutions.

Constituents	Approximate amount	
	ounces per 25 gallons of water	grams per 100 liters of water
Solution 1		
Potassium phosphate (monobasic)	0.5	15
Potassium nitrate	2	60
Calcium nitrate	3	90
Magnesium sulfate	1.5	45
Solution 2		
Ammonium phosphate	0.5	15
Potassium nitrate	2.5	75
Calcium nitrate	2.5	75
Magnesium sulfate	1.5	45

Source: D. R. Hoagland and D. I. Arnon, *Calif. Agr. Exp. Sta. Circ.* 347, Berkeley, 1950.

eral cycling, for only 11 percent of its land can support crops. Night soil, or human excrement, has been used to fertilize crops for hundreds of years. Another recycling technique involves feeding the leaves of cultivated mulberry trees to silkworms; silk is harvested for cloth and the excrement and debris are collected and put in fish ponds to feed fish. Carp nearly 1 meter long are grown in such ponds for human consumption. Silt from the pond is then used as a fertilizer for the mulberry trees, so the cycle starts over again. This is a small but appropriate example of how mineral cycles can be maintained under intensive cultivation for producing several important crops.

Soil reaction

Soil reaction, the degree of acidity or alkalinity of a soil, is determined by the relative number of hydrogen ions (H^+) and hydroxyl ions (OH^-) in the soil solution. A predominance of hydrogen ions makes the soil acid; a predominance of hydroxyl ions makes it alkaline.

Expressing the concentration of these ions in the usual chemical terms results in decimal fractions that are difficult to use. A Danish biochemist, S. P. L. Sorenson, devised a system that is extensively used today for expressing acidity and alkalinity in terms of pH, the logarithm of the reciprocal of the hydrogen ion concentration in moles per liter (Table 16-2).

$$pH = \log \frac{1}{H^+}$$

When pH is less than 7, the soil reaction is acid; when it is greater than 7, the soil is alkaline; at pH 7, the soil is neutral. Table 16-2 gives the concentration in moles per liter of H^+ and OH^- for pH values of 0 to 14. Note that the molar concentration of H^+ times the molar concentration of OH^- equals a constant of 10^{-14}. Since the scale is logarithmic, a soil of pH 5 has 10 times as many hydrogen ions as soil of pH 6. Likewise, a soil of pH 10 has 100 times as many hydroxyl ions as a soil of pH 8.

Hydrogen ions exist in soils in many chemical combinations. Those in the soil solution are in balance with those adsorbed on the surfaces of soil particles. The combination of these two sources of hydrogen ions is called the **total acidity** of a soil. Comparatively few hydrogen ions are in solution, as compared with those adsorbed on soil surfaces. Hy-

Table 16-2 The concentration of H^+ and OH^- with varying pH.

pH	Soil reaction		H^+ concentration (moles/liter)[1]	OH^- concentration (moles/liter)[2]	Reaction of common substances
0			10^{-0}	10^{-14}	
1			10^{-1}	10^{-13}	
2		very strong	10^{-2}	10^{-12}	
3		strong	10^{-3}	10^{-11}	Lemon juice
4	Acidity	moderate	10^{-4}	10^{-10}	Orange juice
5		slight	10^{-5}	10^{-9}	
6			10^{-6}	10^{-8}	Milk
7		neutral	10^{-7}	10^{-7}	Pure water
8		slight	10^{-8}	10^{-6}	Seawater
9		moderate	10^{-9}	10^{-5}	Soap solution
10	Alkalinity	strong	10^{-10}	10^{-4}	
11			10^{-11}	10^{-3}	
12		very strong	10^{-12}	10^{-2}	
13			10^{-13}	10^{-1}	
14			10^{-14}	10^{-0}	

(soil range brackets around pH 2–12)

[1] 1 mole of H^+ = 1 g

[2] 1 mole of OH^- = 17 g

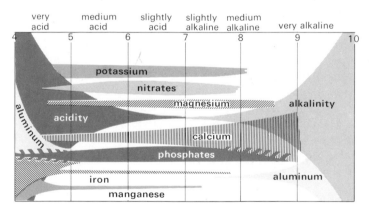

Figure 16-1 Nutrient availability at different pH ranges. [Courtesy Virginia Polytechnic Institute and State University.]

drogen ions are released from these surfaces as fast as they are removed from the soil solution, so that the acidity of the solution is not changed greatly. This resistance to change of acidity is called **buffering.** Most soil solutions are highly buffered.

Either an abnormally high soil pH (above 9) or a low pH (below 4) is toxic to plant roots, but within this range the direct effect of acidity on most plants is not great. However, the indirect effects of acidity and alkalinity on soils are tremendous. Soil pH affects nutrient availability (Fig. 16-1). For example, iron, manganese, and copper become less available to plants as alkalinity increases. The chlorotic condition (chlorosis) found in some plants grown in soils of high pH is often a result of iron deficiency resulting from the formation of insoluble iron compounds. Soil organisms, especially bacteria, are also affected by pH. For example, highly acidic soils are not favorable for the growth of the bacteria that fix atmospheric nitrogen and help rot plant residues.

Crop plants vary in their response to pH (Fig. 16-2). Most crops perform well with a pH between 6.0 and 6.5. Acid-loving plants, many of them in the family Ericaceae (rhododendrons, azaleas, cranberries, and blueberries) perform best at low pH (4.5 to 6.0). Soil reaction may be used to control soil disease in crops that prove less sensitive than the pathogen. Thus potatoes may be grown at pH 5.2 to control scab, a disease caused by a fungus not adapted to acid soils. But potatoes perform equally well at a higher pH in "disease-free" soil.

Measuring soil pH

Soil reaction may be determined approximately by the use of chemically treated papers such as litmus papers, which change color in response to different pH values. The pH of dilute soil-water systems provides an acceptable approximation of soil acidity. The device usually used for measuring acidity is called a glass electrode pH meter. It has two electrodes, usually glass and calomel (mercurous chloride), which are placed in a soil and water mixture. The glass electrode consists of a thin glass bulb filled with a potassium chloride solution. An electrical potential, which is proportional to the hydrogen ion concentration of the solution being tested, develops across the glass membrane and is registered on a meter that is read directly in units from 0 to 14. The calomel electrode is a reference for measuring the potential of the glass electrode.

Regulation of soil reaction

The reaction of soils is due to the interaction of climate with the soil components. Soil pH depends on whether the clay particles are charged with hydrogen ions or mineral cations (Fig. 16-3). In general,

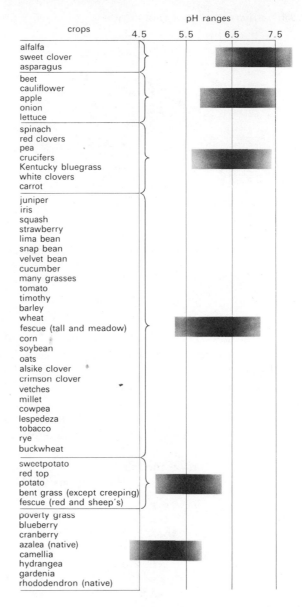

Figure 16-2 Suitable pH ranges for certain crop plants. [Adapted, with permission of the publisher, from *The Nature and Properties of Soils*, 5th ed., by T. L. Lyon, H. O. Buckman, and N. C. Brady, copyright © 1952 by the Macmillan Co.]

acid soils are common where the precipitation is high enough to leach appreciable amounts of exchangeable bases from the surface layers. The ions may be removed through absorption by plants, leaching, and

clay particle charged
with H⁺ ions

ionizes weakly
in solution

clay particle
charged with
Na⁺ ions

hydrolysis in water with
the formation of a
weakly alkaline solution

Figure 16-3 The soil reaction depends on whether the clay particles are charged with hydrogen ions or mineral cations. [From J. Bonner and A. W. Galston, *Principles of Plant Physiology*, W. H. Freeman and Company, copyright © 1952.]

other processes. The soil colloids become charged with hydrogen ions made continually available by the dissociation of carbonic acid formed from the dissolved carbon dioxide released by living roots in respiration, and from the biological decay of carbohydrates (see p. 294). Alkaline soils generally occur in arid regions where there is a higher degree of cation accumulation.

Soil reaction can be modified. Soil can be made more alkaline; that is, the soil pH can be increased by adding basic cations such as calcium, magnesium, sodium, or potassium. Calcium is the most economical cation for increasing pH, and it also has other beneficial effects. The addition of calcium or compounds of calcium and magnesium to reduce the acidity of the soil is known as **liming.** Although the term "lime" refers specifically to calcium oxide (CaO), it is used in the agricultural sense to include such limestone-derived materials as oxides, hydroxides, carbonates, and silicates of calcium or both calcium and magnesium. Liming may significantly improve the performance of crops grown in acid soils. The amount of lime required depends upon the degree of pH change desired, the exchange capacity of the soil, the amount of precipitation, and the liming material and its physical form. When adding lime, enough must be used to neutralize the hydrogen ions that are released from colloid surfaces as well as those that are in solution; that is, the buffering effect of the soil must be overcome. Figure 16-4 illustrates the reactions that take place when an acid soil is limed.

Soil may be made more acid by adding hydrogen ions to the soil colloid. This is done by adding substances that tend to produce strong acids in the soil. Some nitrogen fertilizers increase soil acidity, but sulfur is by far the most effective substance. In warm, moist, well-aerated soils, bacterial action converts sulfur to sulfuric acid.

acid soil + lime ⟶ neutral soil

H⁺   hydrogen ion in soil solution (active acidity)

H⁺   replaceable hydrogen ion (potential acidity)

Ca²⁺   calcium ion in soil solution

Ca²⁺   replaceable calcium ion

Mg²⁺   replaceable magnesium ion

Al³⁺   replaceable aluminum ion

K⁺   replaceable potassium ion

Al(OH)₃   aluminum hydroxide

Figure 16-4 Cation-exchange relations when an acid soil is limed. [After USDA, Yearbook of Agriculture, 1957.]

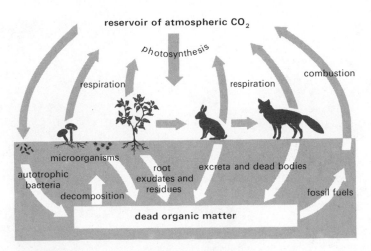

Figure 16-5 The cycle of carbon between organic and inorganic forms can be regulated by cultural manipulations. In natural ecosystems, such as the one illustrated, an imbalance will occur if one of the factors is disturbed.

Soil texture has an influence on the ease with which pH can be changed. Clay soils are more difficult to neutralize than silt or sandy soils because they have more surface area for adsorbing, holding, and supplying hydrogen ions.

Organic matter

Organic matter, formed by the decomposition of plant and animal residues, affects the physical condition and fertility of the soil. Once organic matter becomes a part of the soil, it begins to decay and is soon converted to water, carbon dioxide, and a few mineral elements (Fig. 16-5). Some fractions, such as lignins, waxes, fats, and some proteinaceous materials, resist decomposition and, through complex biochemical processes, form a dark, noncrystalline, colloidal substance called **humus.** Humus has adsorptive and absorptive properties for nutrients and moisture that are even higher than those of clay. Thus small amounts of humus greatly affect the structural and nutritive properties of soil.

The accumulation of organic matter reaches an equilibrium in undisturbed soil. Because the organic matter of the soil is largely under the control of bio-

logic and climatic forces, one must differentiate between a temporary and a long-term increase in organic matter. Unfortunately, it is easy to decrease the organic matter of a soil, but relatively difficult to increase it. The major loss of organic matter comes about from increased oxidation as a result of cultivation and from crop removal. The rate of decomposition is affected by aeration, temperature, nitrogen availability, soil moisture, soil reaction, and the nature of the decomposing organisms.

Organic matter affects the physical condition of the soil in a number of ways. It improves the water-holding capacity of the soil because it acts as a sponge to absorb and hold water that would otherwise percolate through the soil and become unavailable for plant growth. Organic matter also increases soil aeration in this manner. Organic matter coats the mineral particles and gives a crumbly structure to soils, which makes them easier to cultivate.

Organic matter also supplies nutrients, which are released and made available for plant growth as the organic matter decomposes. The decomposition of plant and animal material is accomplished by enzymatic digestion carried out by soil microorganisms. The decomposition of simple carbohydrates (starches

Figure 16-6   This buckwheat crop, planted between orchard trees, will be plowed under at the end of the growing season to provide "green manure" for the trees. [Courtesy USDA.]

and sugars) is a fairly rapid process and results in the release of carbon dioxide in the soil. Water-soluble proteins are decomposed readily to amino acids and then to available ammonium compounds. Ammonium compounds under the action of certain "nitrifying" bacteria are transformed to nitrates, in which form they are available to plants. In addition to nitrogen, organic matter supplies other essential elements.

Because of its high exchange capacity, humus also acts as a cation retainer. Particles of organic matter are negatively charged and attract positively charged ions to their surfaces. Thus organic matter adsorbs and retains cations, such as calcium and magnesium, which would be subject to loss by leaching. It can absorb far more cations per unit weight than clay (see Table 12-2). Because organic matter absorbs large amounts of cations it greatly increases the buffering capacity of soil.

The organic-matter content of a soil may be increased through the extended use of a legume or grass sod. The organic matter produced as a result of root disintegration is protected from excessive oxidation by the constant cover and lack of cultivation. Organic matter can be built up by rotating row crops with legume sods. The plowing under of a growing crop as a **green manure** temporarily increases the organic content of the soil (Fig. 16-6), but this practice cannot be expected to increase the organic matter content for long. The buildup of high populations of microorganisms may actually reduce the net organic matter content of the soil through an unexplained breakdown of the more resistant organic fraction. The rapid breakdown of green manure releases bound nutrients, but minerals that have a high carbon content in relation to their nitrogen content, such as millet or sudan grass, may create a temporary nitrogen shortage because available nitrogen is incorporated into the bodies of decomposing microorganisms. This is called **biological absorption**. It can be avoided either by waiting for the carbohydrates and other organic matter to decompose thoroughly before planting a crop or by adding enough nitrogen to compensate for what is "absorbed" by the bacteria. Sweet clover and other legumes that have a high nitrogen content in proportion to carbon compounds are rapidly decomposed by microorganisms and release a steady supply of nitrogen. Farm manures may be used to supplement soil organic matter, but collecting, storing, and applying them and incorporating them into the soil is often an expensive procedure. Manure is also valuable for its nutrients.

Fertilizer and fertilization

A fertilizer is a material that supplies nutrients to plants. Fertilizers are usually applied to the soil, but some may also be applied to the leaves or stems. Carbon dioxide added to the air in greenhouses may also be considered as fertilizer.

Fertilizers that supply nitrogen, phosphorus, and potassium are referred to as **complete**. The grade (or analysis) of fertilizers is the percentage by weight of nitrogen (expressed as elemental N), phosphorus (expressed as P_2O_5), and potassium (expressed as K_2O). Phosphorus and potassium are not expressed in their elemental form for historical reasons.° An 80-pound bag of a 5-15-30 fertilizer will contain 4 pounds of N, the amount of phosphorus in 12 pounds of P_2O_5, and the amount of potassium in 24 pounds of K_2O. The fertilizer ratio is the analysis expressed in terms of the lowest common denominator. A 10-10-10 analysis

---

°To change oxide analyses to elements and vice versa, multiply the quantity of the oxide or element by the following conversion factors:

Element	Oxide to element	Element to oxide
Phosphorus	.43	2.33
Potassium	.83	1.20

has a 1:1:1 ratio. Fertilizers are referred to as "low analysis" when the amount of available nutrients is less than 30 percent and "high analysis" when it is 30 percent or more. Because of lower weight and less handling, high-analysis fertilizers tend to be less expensive per unit of nutrient than low-analysis fertilizers, but they require more precision in their application. In the United States the analysis of nitrogen, phosphorus, and potassium must be stated on the container, although many fertilizers contain, or consist entirely of, other plant nutrients.

Fertilizers may be classified as **natural organic** or **chemical**. In times past, most fertilizers were waste organic materials, such as manure, crop residues, blood and tankage, or fish scraps. Chemical fertilizers, such as ammonium nitrate and superphosphate, are synthesized from inorganic minerals. Nitrogen fertilizers can be synthesized using nitrogen directly from the air. Recently, a number of forms of nitrogen-containing organic compounds, such as urea, $CO(NH_2)_2$, and cyanamid, $NCNH_2$, have been synthesized. Although they are organic in the chemical sense, they are not necessarily derived from living systems. The nutrients in natural organic fertilizers and some synthetic organics undergo gradual chemical transformations to available forms to provide a means of extending the period of nutrient availability. "Urea-forms" are combinations of urea and formaldehyde—the greater the proportion of formaldehyde, the lower the solubility of urea, and hence the slower the "release." Other types of slow-release materials are composed according to certain specifications for particle size, chemical composition, or solubility. A special slow-release system has recently been developed by encapsulating fertilizer pellets with a plastic coat. The coat acts as a membrane to release the fertilizer salts gradually. These materials are comparatively expensive per unit weight of nutrient, but they have special uses, such as for container-grown nursery stock, because they reduce labor costs.

In order to minimize the cost of fertilization, agricultural practice strives to provide only enough fertilizer to supplement available soil nutrients and to raise the quantity of nutrients to the level actually required by the crop. The nutrient requirements of crop plants are determined by correlating plant response with the mineral content of plant tissue and soil. Unfortunately, the total nutrient content of a soil does not usually give a true picture of nutrient availability. The availability of nutrients is related to the cation-exchange capacity of the soil, soil reaction, and organic cycles. Some biological assays of soil have been made by utilizing the response of particularly sensitive plants or microorganisms. Quick chemical tests have been developed, although these are often not very precise. In many plants, severe shortages of certain nutrients produce characteristic **deficiency symptoms,** the identification of which can often be used as a diagnostic technique. The good agriculturist, however, will not permit nutrient shortages to become this severe.

The relation between nutrient level and plant performance varies with the species and the nutrient in question. Nevertheless, there is a general response of plants to nutrient level. **Deficiency levels** produce definite symptoms of nutrient starvation. At somewhat higher levels, deficiency symptoms are not present, and plants may respond to fertilization by showing increases in yield and performance. At levels that produce no obvious response to fertilizer, the plant may continue to show an increasing level of nutrient absorption termed **luxury consumption.** At abnormally high or **toxic levels,** growth is retarded and death may even occur. These levels differ for each nutrient. Boron may be toxic to certain plants if their tissues contain as little as one part per million.

The level of crop response to fertilizer is related in part to the **productive capacity** of the soil. Crops grown on soils of low productive capacity show maximum response at a lower level of fertility than those grown on soils of high productive capacity. Productive capacity depends on long-term availability of nutrients and on soil condition. Owing to the forces that establish an equilibrium between the soil and the soil solution, optimum fertility is usually not achieved in one quick step. When large amounts of fertilizer are placed on soils of low productive capacity, much of it is wasted by leaching, by being tied up in unavailable forms, or by unequal distribution throughout the soil. Continued application of fertil-

izer at the level of optimum plant response appears to increase the productive capacity of the soil, ultimately raising its yield potential.

Fertilizer may be solid, liquid, or gaseous. Most are solids and are applied to the soil. Fertilizer may be dissolved in irrigation water or dissolved fertilizer may be applied to the foliage. Nitrogen may be added to the soil in the form of ammonia gas (NH_3), since it is heavier than air and dissolves quickly in the soil moisture.

Proper placement and timing are important factors in fertilization. Plant response, the prevention of injury, and convenience and economy of application must be considered. To be effective, fertilizer must be applied where and when the plant needs it. Single yearly applications may be insufficient for some nutrients and unnecessary for others. Concentrated, highly soluble fertilizers cannot be applied to growing plants, especially when young, because of salt injury. In perennials, or in long-season annuals, nutrient availability may be more efficiently controlled by making repeated applications throughout the season. This is especially true with nitrogen fertilization, as excess amounts may be wasted because of leaching, and the timing of applications may be critical to plant response.

There are various methods of fertilizer placement. **Broadcast application** refers to the scattering of material uniformly over the soil surface, usually before the crop is planned. Surface applications of this kind may not be as effective as a **top dressing**—placement of the fertilizer directly over the growing crop. When plants are subject to injury (fertilizer burn), fertilizer can be put next to the plants as a **side dressing.** Side-dressed applications of fertilizer are often made along with a cultivation and are thus mixed into the soil. Fertilizer can also be placed in continuous strips between rows (**band placement**) or it can be dropped behind the plow in the bottom of furrows (**plow-sole placement**).

Fertilizer may be applied along with mechanical transplanting either as a band of solid fertilizer under the plant or dissolved in the supplemental transplanting water as a **starter solution.** In **foliar application,** the fertilizer is absorbed immediately and plant response may be evident within a day or two, but because there is little residual effect, applications must be more frequent than soil applications. Foliar application of nitrogen compounds and minor elements such as manganese and boron has proved practical for some crops.

The timing of fertilizer placement may be critical. In perennial fruit crops, such as apple and peach, excess nitrogen is responsible for poor fruit color and soft fruit as well as for undesirable vegetative growth that occurs late in the season and leaves the plant vulnerable to winter injury. Consequently, nitrogen is usually applied only once, early in the spring, so that any excess will have been used up by late summer. Many plants show a response to spring application of nitrogen because the concentration of nitrogen compounds in cold or wet, poorly drained soils is low due to a lack of nitrification by aerobic bacteria and to excess **denitrification** (the conversion of nitrate to gaseous forms) by anaerobic bacteria. In many crops, nitrogen is applied several times during the growing season because it is subject to leaching and conversion to gaseous or other unavailable forms. Some nutrients that are relatively immobile in the soil, such as phosphorus, are best applied before the crop is sown so that they can be worked into the soil and be evenly distributed by plowing.

BIOGEOCHEMICAL CYCLES

Every ecosystem, agricultural or otherwise, has a *nutrient pool*. This expression refers to the sum total of nutrients found in an ecosystem, at least a part of which are available for crop growth. It includes nutrients in and on the soil as well as those incorporated in undecomposed parent material but which are readily usable by plants. It also includes the minerals incorporated into the tissues of organisms and which will become available for use by other organisms following death. Nutrients in both plants and animals are part of the total pool.

In a mature forest there is a delicate balance between the minerals that are lost from the ecosystem by various processes and those that are added. On any one day, or year, the inputs and outputs will not balance, but they will over a period of years. Without this equilibrium the stability of a mature ecosystem would not be possible. In ecosystems that are subject to sustained, heavy rainfall, such as tropical rainforests, practically all of the nutrient pool is bound in biomass, which includes both living and dead organic matter. Under these conditions mineralization does not take place in the usual manner, for before dead organs can decompose they are invaded by mycorrhizal fungi which absorb their minerals and recycle them directly into other organisms. Were it not for these fungal ghouls, the heavy rains would leach and erode minerals completely out of the ecosystem. This is exactly what happens to soils of tropical forests when the vegetation is cut and cleared for planting food crops such as manioc, and has led to the necessity for various systems of shifting agriculture (see Chapter 15).

Were the mineral cycles of an ecosystem without loss or gain, we would say that it is *closed*. In nature, there can be no such thing because all ecosystems are subject to losses as well as gains from other ecosystems. Rather we strive to keep them *closed* to the greatest degree possible.

But since we have already found that we must use intensive crop production to feed the hungry peoples of the world, we must be prepared to accept some losses in soil productivity due to cultivation and its attendant erosion (ecosystem losses), and be prepared to compensate for them. Even if there were no losses of minerals by this route, those that are incorporated into crops that are harvested and removed would make some kind of adjustments necessary if the ecosystem is to remain productive. In this case, the adjustment must be mineral fertilization.

With the exceptions of mineral fertilization and crop harvest, the inter- and intrasystem functions of agricultural and "natural" ecosystems are similar with respect to pathways and processes (Figure 16-7). However, their balances may be vastly different. Learning how to minimize intersystem processes is of great importance to crop producers. Economics has always dictated minimizing inputs, and more recently a growing concern for environmental issues has dictated minimizing outputs as well.

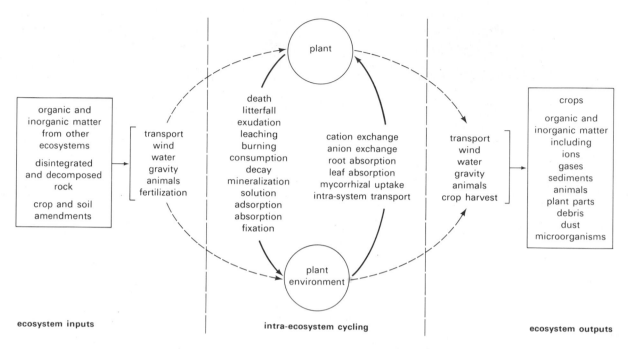

Figure 16-7 Generalized mineral cycle of an agricultural ecosystem—pathways and processes.

ESSENTIAL ELEMENTS IN PLANT NUTRITION

In the ensuing discussion, all the essential elements are treated in turn. Each element is discussed in relation to its function in the plant, deficiency symptoms, soil relations, and fertilization.

Nitrogen

Nitrogen (N_2) is an inert, odorless, tasteless, colorless gas that makes up about 77.5 percent of the air by weight and about 78 percent by volume. A hectare of soil may contain from 1 to 10 metric tons of nitrogen, whereas the air over it holds about 78,000 metric tons. Because atmospheric nitrogen cannot be used by plants, it must be "fixed" in a soluble inorganic form before it can be absorbed and assimilated by plant tissues.

Functions in plants

Amino acids (building blocks of protein) and many other plant substances, including purines, vitamins, and alkaloids, are composed in part of nitrogen. Chlorophyll molecules have one magnesium atom and four nitrogen atoms. Nitrogen accounts for only 1 or 2 percent of the dry weight of a plant, but nitrogen-containing compounds make up perhaps 25 percent of the dry weight.

Nitrogen excesses delay maturity and fruiting of some crops by promoting vegetative growth. Severe nitrogen deficiencies can also delay crop maturity and reduce yields. High-nitrogen plants may be less fruitful than low-nitrogen plants.

Nitrogen causes plants to grow rapidly, resulting in a high proportion of succulent, fleshy plant tissue in contrast to stiff fibrous tissue. As a result, herbaceous plants with a high nitrogen content are often blown

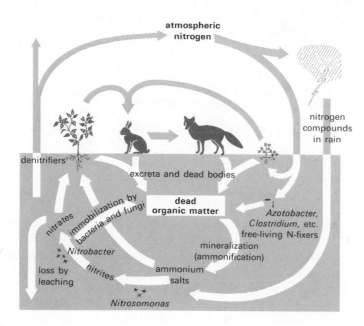

atmospheric
nitrogen

nitrogen
compounds
in rain

denitrifiers

excreta and dead bodies

nitrates

immobilization by
bacteria and fungi

dead
organic matter

Azotobacter,
Clostridium, etc.
free-living N-fixers

Nitrobacter

mineralization
(ammonification)

loss by
leaching

nitrites

ammonium
salts

Nitrosomonas

Figure 16-8   The nitrogen cycle.

down (lodged). Freezing injury is also associated with a high nitrogen content of plant material.

Deficiency symptoms

Leaves of nitrogen-deficient plants are usually very light green, but may be yellow or red because when chlorophyll is deficient the color of other pigments shows through. Leaves of such plants are also small. The lower leaves are usually the first to show discoloration, and they may turn yellow and drop off before the topmost leaves have lost their intense green color. Individual branches may die, and the entire plant is stunted.

Soil relations

Nitrogen is found in both organic and inorganic compounds in the soil. Soil nitrogen is most abundant in climatic regions that favor the accumulation of organic matter, such as those of grasslands. Mineralization of organic nitrogen—the change from an organic to an inorganic form by the decomposition of organic matter—must take place before nitrogen can be absorbed and used again by plants.

The forms of nitrogen most usable by plants are the ammonium (NH_4^+) and nitrate (NO_3^-) ions. Ammonium nitrate is widely used as a fertilizer because it is composed of both ions. The nitrite ion (NO_2^-) can be utilized by the plant, but it tends to be unstable and toxic in large amounts. The transformation of nitrogen-containing compounds to available forms is a part of the **nitrogen cycle** (Fig. 16-8). Nitrogen changes from gaseous molecule to protein and back largely through biological pathways.

Biological nitrogen fixation—the transformation of atmospheric nitrogen into forms available to plants by organisms—is accomplished largely by certain types of bacteria (Table 16-3). The most efficient of these are symbiotic; that is, they convert atmospheric nitrogen to combined forms, but only in association with the roots of certain plants. Symbiotic nitrogen-fixing bacteria are commonly associated with crop plants of the pea family (legumes) with nodular deformations on their roots, which were once thought to be a disease condition (Figure 16-9).

Table 16–3 Nitrogen-fixing bacteria.

Type	Genus	Requirements
Symbiotic	*Rhizobium*	Carbohydrates, inhibited by nitrates and ammonium
Nonsymbiotic Anaerobic	*Clostridium*	Carbohydrates, inhibited by nitrates
Aerobic	*Azotobacter*	Calcium, traces of molybdenum

Rhizobium, an important genus of bacteria associated with legumes, will continue to fix nitrogen for a time even on severed roots. Seeds of leguminous crops are often "inoculated" prior to planting by coating them with a slurry or a powder containing spores of *Rhizobium*.

Prior to World War II, legumes were used extensively to provide nitrogen for crop growth. During the war, however, the production of explosives required vast amounts of nitrogen for the production of smokeless powder. As a result, scientists developed methods for chemically fixing nitrogen to supplement the nitrates that had previously been mined. Following the war, commercially synthesized nitrogen compounds were relatively inexpensive, and so the use of leguminous "green manure" crops was largely abandoned.

Today's energy-conscious world requires ever increasing quantities of nitrogen compounds, and sources are being sought that will make fewer demands on our dwindling energy supplies. For example, the commerical synthesis of fixed nitrogen in the United States requires nearly 3 percent of our annual consumption of natural gas.

In addition to the energy problem, there are other objections to the use of nitrogenous fertilizers, no matter what the source. Transportation, storage, and application costs are high, and losses to the environment are great. Crops use only about 50 percent of applied nitrogenous fertilizer. While a part of the 50 percent is denitrified and goes back into the atmosphere, much of it escapes and is responsible for the eutrophication of water impoundments and the contamination of water supplies. Because the product of biological fixation is ammonia, the problem of accumulating excess nitrates in soil and soil water is decreased through the use of nitrogen-fixing organisms. Nitrogen that is fixed in nodules of legumes is transported directly to the site of use in the plant and is therefore less subject to loss.

There are several types of nitrogen-fixing organisms:

Figure 16-9 Nitrogen-fixing bacteria inhabit the nodules of soybean roots in a symbiotic relationship. [Courtesy J. C. Allen & Son.]

1. nitrogen-fixing bacteria that form nodules on the roots of legumes;

2. nitrogen-fixing bacteria that form nodules on the roots of other plants, such as alder and New Jersey tea (*Ceanothus*), most of which are not normally crop plants;

3. nitrogen-fixing algae associated with higher plants; and

4. free-living nitrogen-fixing organisms that are able to provide their own nitrogen requirements, including blue-green algae and some bacteria.

One answer to the problem of providing an adequate supply of soil nitrogen is to return to the use of leguminous cover crops that add nitrogen to the soil system through biological fixation. This is usually done through crop rotations, which, unfortunately, may cause the total production of the food crop—corn, potatoes, or whatever—to be decreased.

A second approach could be to use nitrogen-fixing organisms that live symbiotically with non-crop plants, but in the same areas and at the same time that crops are being produced. The nitrogen-fixing blue-green alga *Anabaena azollae* lives symbiotically with a very small floating aquatic fern (*Azolla* sp.) in the rice fields of Southeast Asia. Each leaf of the fern has a cavity that holds algal cells. The nitrogen that "leaks" into the water from this association fertilizes the rice that is growing in the same field. When W. K. Talley and D. W. Rains, from the University of California, Davis, plowed in one crop of *Azolla* as a "green manure" and grew another fern crop in the water with the rice crop, the yield of rice was increased by 200 percent.

In recent years nitrogen-fixing bacteria that do not form root nodules have been found growing in association with some grasses, including corn and wheat. For example, Bahia grass (*Paspalum notatum*) has been found to harbor a nitrogen-fixing bacterium under the mucilaginous sheath of its root surfaces. *Digitaria decumbens*, a crab grass, receives nitrogen from the bacterium *Spirillum lipoferrum,* which lives in the root cortex. What a breakthrough it would be to find a nitrogen-fixer that could live sym-

biotically with corn, wheat, or any other of the worlds major grain crops!

Still another approach would be to incorporate nitrogen-fixing genes from bacteria or algae directly into crop plants using recombinant-DNA techniques. It is technically possible to introduce the genes responsible for nitrogen fixation, called *nif* genes, into such crop plants as bananas and potatoes. This, however, carries a potential danger because of the possibility that the genes could also become incorporated into pathogens or weeds. Hence this work must proceed cautiously and with full awareness of the problems that might result.

Atmospheric fixation of nitrogen results when lightning causes nitrogen and oxygen to unite. As much as 56 kilograms of elemental nitrogen per hectare per year (about 50 pounds per acre per year) may be added to soil in this manner in regions with high thunderstorm activity, but 2 or 3 kilograms per hectare is the usual annual average.

The breakdown of the complex proteins of organic material into simpler forms, such as amino acids, is accomplished largely by bacterial action. The nitrogen released by this process is only available after the death and disintegration of the bacteria responsible for the decay. Soil microorganisms have the first call on nutrients. This is especially true for materials with a carbon-to-nitrogen ratio (by weight) greater than 10:1. The subsequent breakdown of amino acids and similar nitrogenous materials to forms of nitrogen available to plants takes place by transformations referred to as **ammonification** and **nitrification.**

The bacteria involved in nitrification are autotrophic and aerobic; that is, they do not require organic nutrition, but they do require oxygen. Thus they are greatly affected by soil aeration, temperature, and moisture.

Amino acids
from protein $\xrightarrow[\substack{\text{(many bacteria} \\ \text{and other} \\ \text{soil organisms)}}]{}$ NH_4^+
degradation   (ammonium ion)

NH_4^+ $\xrightarrow[\substack{\textit{Nitrosococcus} \\ \textit{Nitrosomonas}}]{}$ NO_2^- (nitrite ion) $\xrightarrow[\textit{Nitrobacter}]{}$ NO_3^- (nitrate ion)

Losses from soil

Nitrogen is lost from the soil through erosion, leaching, crop removal, and denitrification. Moreover, nitrate is extremely soluble in water, and thus large amounts are lost because of leaching. When rainwater percolates down through the soil, the highly soluble nitrates are carried with it. Losses are greater from coarse soils than from fine soils. Losses are also greater from fallow soils than from those with a vegetative cover; consequently, a winter "cover crop" may greatly reduce losses due to leaching in humid regions.

Losses due to crop removal are unavoidable. Most agricultural crops remove 28–56 kilograms of nitrogen per hectare (25–50 pounds per acre) during a single growing season, which is 1 to 2 percent of the amount present in various forms.

Denitrification can account for significant losses. Gases are evolved through the bacterial reduction of nitrate to nitrogen gas (N_2), nitrous oxide (N_2O), and nitric oxide (NO). Denitrification is carried out under anaerobic conditions; thus improper aeration can result in the loss of available nitrogen. In compost piles, losses of nitrate may be large. For crops grown in wet soils, such as rice, denitrification may be critical. Still other losses are due to the escape of ammonia gas into the atmosphere.

Fertilization

Nitrogen fertilizers may be either organic or chemical. Organic matter has been used for fertilizer for thousands of years. Legume crops have long been used to add nitrogen to soil, often supplemented by tankage and other organic wastes.

Sodium nitrate, imported from Chile, was the first widely used source of inorganic nitrogen. It is still used, but to a smaller extent than in the years before nitrogen could be extracted from air. Nitrogen was first removed from the air for conversion to forms usable as fertilizers on a commercial basis in 1898.

Ammonium sulfate contains 20.5 percent nitrogen. It is called an **acid-forming** fertilizer because the sulfate can be transformed to sulfuric acid. It is usually obtained as a by-product of coke ovens, but it is also made from ammonia. **Anhydrous ammonia** (dry ammonia) is gaseous in the range of temperatures at which plants grow, but it becomes a liquid when cooled and put under pressure. Kept in refrigerated and pressurized tanks until used, it is injected 5 to 6 inches into the soil and covered immediately to prevent its escape before it becomes fixed. It is also available in solution, in which form it is injected into the soil or added to irrigation water. **Ammonium nitrate** contains 33.5 percent nitrogen and is usually used in the form of pellets. It is one of the most widely used of nitrogen fertilizers. **Ammonium phosphates,** available in both the monoammonium and diammonium forms, are used for making mixed fertilizers. **Magnesium ammonium phosphate** is convenient to use because it is relatively insoluble and will not "burn" plants, even when used in fairly heavy applications. **Urea,** which is synthesized on a massive scale, is an excellent fertilizer because of its high nitrogen content (45 percent). Urea can be applied either to the soil or to the foliage.

Phosphorus

Phosphorus, a member of the nitrogen group of elements, reacts vigorously with oxygen in the atmosphere and must be stored under water. Never found free in nature, it was long ago thought to have mystical properties.

Functions in plants

Phosphorus is found in plants in many forms. In seeds it is stored as phytins (salts of myoinositol hexaphosphoric acid). During germination phosphorus is "mobilized" (converted to a form in which it can be translocated) and becomes a part of the new proteins formed during growth. Nucleoproteins are composed in part of phosphorus. There are many phosphorous compounds involved in metabolic transfer processes; one of the most important is **adenosine triphosphate** (ATP). Surprisingly, only about 0.2 percent of the total dry weight of a plant is phosphorus.

Deficiency symptoms

Phosphorus-deficient plants frequently have purplish leaves, stems, and branches. Maturity is retarded and growth is generally slow. Yields of fruits and seeds are usually poor; fruit often drops prematurely.

Phosphorous deficiency in tobacco results in stunted growth and small, very dark green leaves forming a rosette. Legumes have a high requirement for phosphorus, and seed yields may be greatly reduced if phosphorus is deficient. Oranges from phosphorus-deficient trees have a coarse skin, sour juice, and poor shape.

Soil relations

Phosphorus is released into soils mainly by the weathering of rocks that contain the mineral apatite, $Ca_5F(PO_4)_3$. It occurs naturally as calcium phosphate, iron and aluminum phosphates, and as organic phosphorus. A soil may have an abundance of phosphorus and yet have little available for plant nutrition if soil conditions are unfavorable.

The availability of phosphorus to plants is partially a function of soil reaction. In acid soils (pH 4 and less) insoluble iron and aluminum phosphates form, whereas in highly alkaline soils (pH 8.5 and higher) calcium phosphates form that are equally insoluble.

Drought usually causes a decrease in phosphorus availability that lasts long after the immediate effects of soil desiccation. During the process of dehydration, the concentration of solutes in the soil solution increases, and large, complex mineral crystals form that have less surface area per unit volume and are less readily dissolved than small crystals.

When phosphorus is applied to soil, it seldom moves very far from the point of application because of its numerous complex reactions with clay and organic matter. Consequently, it is usually placed close to the crop row rather than broadcast, because band application reduces surface contact with soil particles. Even when applied in a water-soluble form, phosphorus usually diffuses no more than an inch or so from the point of application before it is converted to an insoluble compound (within 2 or 3 days).

Losses from soil

Losses of phosphorus are usually due to the removal of crop plants and to its becoming unavailable in the soil. It is not volatile; neither is it readily leached. The phosphorus cycle is illustrated in Figure 16-10.

Fertilization

Phosphate rock, which consists largely of apatite, is treated with sulfuric acid to form **superphosphate** according to this reaction:

$$2Ca_5F(PO_4)_3 + 7H_2SO_4 + 3H_2O \rightarrow$$
$$3CaH_4(PO_4)_2H_2O + 7CaSO_4 + 2HF$$

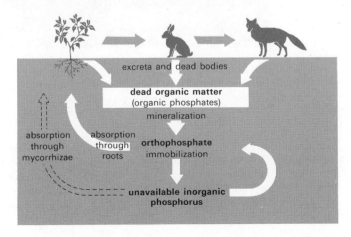

Figure 16-10 The phosphorus cycle.

Superphosphate contains about 8.6 percent available phosphorus. **Triple superphosphate,** which is made by treating phosphate rock with phosphoric acid, contains about 20 percent elemental phosphorus. Other phosphatic fertilizers are **ammonium phosphates, calcium metaphosphate,** and finally ground **phosphate rock.** Phosphate rock contains a high percentage of phosphorus, but it is not always readily available to plants.

Because of the high phosphorus requirements of seedlings, it is important that adequate quantities be made available close to the seed or transplant (Fig. 16-11). Thus phosphorus applications are often made as bands of fertilizer under the seed. The use of a starter solution, water containing a liberal amount of phosphorus (about 1500 parts per million), is recommended for many transplants. The phosphorus is supplied in the form of soluble phosphate salts, such as monoammonium or diammonium phosphate and monopotassium phosphate. A popular starter solution contains 1 kilogram of a mixture of diammonium phosphate and monopotassium phosphate (10–52–17 analysis) per 140 liters of water (or about 3 pounds per 50 gallons). When an extensive root system is established, plant requirements are met by lower concentrations of phosphorus.

Potassium

Potassium is a soft white metal that can be cut with a knife. It is so reactive that it must be stored under oil or in sealed tubes because it will react with oxygen, water, and carbon dioxide. Potassium in the form of potassium carbonate is commonly called **potash.**

Functions in plants

Potassium was first discovered to be essential for plant growth in 1866. At that time it was found that oats would not produce flowers unless potassium were present.

Potassium is necessary for the formation of sugars and starches, for the synthesis of proteins, and for cell division. It also neutralizes organic acids and regulates the activity of other mineral nutrients in plants. It activates certain enzymes, helps to adjust water

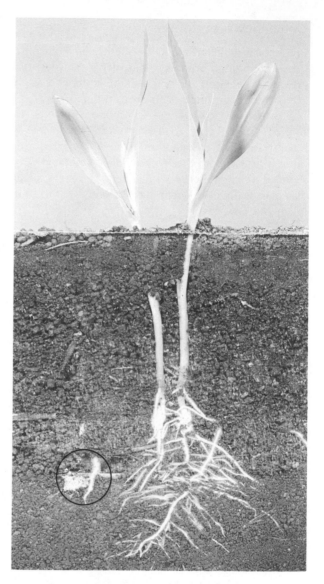

Figure 16-11   Proper placement of "starter" fertilizer for corn is made close to the seed. [Courtesy Howard Knaus.]

relations, and promotes the growth of young plants. It improves the rigidity of straw and helps prevent lodging, increases the oil content of fruits grown for oil, contributes to cold hardiness, and is associated with enhanced flavor and color of some fruit and vegetable crops. Even though it is thought to be essential for the formation of carbohydrates and is somehow involved in many other processes, it is never found as a part of organic compounds. About 1 percent of the dry weight of a plant is potassium.

Deficiency symptoms

Plants deficient in potassium produce low yields even before abnormal external colorations or other symptoms can be observed. Potassium-deficient leaves are usually mottled, spotted, or curled, the older leaves showing these symptoms first. Leaves may appear "scorched" or "burned" along the margins and on the tips, with dead areas falling out and leaving ragged edges. Corn plants with insufficient potassium are usually streaked with yellow and yellowish green. Tobacco leaves become mottled and chlorotic, developing dead spots in the centers of mottled areas.

Potassium-deficient plants nearly always have poorly developed root systems and are easily blown over. Carbohydrate synthesis is so greatly impaired that there is not enough to supply both the tops and the roots. Tops usually have greater priority, partly because they are nearer to the place of synthesis, and the roots get less than they need for optimum growth.

Soil relations

Most potassium in soils is derived from the minerals muscovite, biotite, orthoclase, and microcline. The first two are micas and the second two are feldspars. These minerals are only slightly soluble and are usually found as particles of large size. A small amount of the potassium in soil solutions comes from soluble salts, such as potassium chloride, which are readily available to plants. Under normal conditions there is a balance between highly soluble and slightly soluble forms, and when potassium is added to the soil in the form of soluble salts at least a part of it is "fixed" and rendered unavailable to plants.

Losses from soil

Much of the potassium applied to soils is removed by crops. Since most of the potassium utilized by plants is in the leaves, crops grown for their foliage, such as forage crops, remove much more than those grown for other parts, such as tubers or fruits. Potassium is not readily leached from most soils. In one study it was found that potassium moved downward through the soil profile by less than 1 meter (39 inches) over a 16-year period. Because of its great mobility in plants, potassium can be leached from living leaves by rain.

Potassium is more likely to be deficient in the upper soil layers than in the lower layers, mainly because it is removed from the upper layers by plants. With increasing depth it becomes more abundant and more uniformly distributed. But in soils that have been highly leached, the surface may contain more than lower depths due to reentry following leaching from organic matter on the surface. Potassium in the form of potassium carbonate (potash) leaches readily out of wood ashes.

Fertilization

About 95 percent of all potassium fertilizer is applied in the form of **potassium chloride** (KCl). **Potassium sulfate** (K_2SO_4) and **potash magnesia** ($K_2SO_2 \cdot MgSO_4$) account for most of the remaining 5 percent. Potassium is usually mined as potassium chloride, ground, purified by flotation in a slurry, dried, and screened to the desired size. Cow manure contains about 4 kilograms of potassium per metric ton (wet weight), which is about 8 pounds per short ton.

Since there is a dynamic balance between the amounts of potassium present in different forms in soils, the choice of fertilizer must be based on the natural ability of soil to supply potassium to crop plants. Soils with a large supply of unweathered potassium minerals usually need little or no additional potassium. Its rate of withdrawal from the exchangeable phase is determined largely by the rate at which it is absorbed by plant roots.

For some plants (oats, wheat, cotton, tomatoes, and alfalfa) sodium seems to serve as a semisubstitute for potassium, producing a crop response when the potassium supply is deficient. Some plants (beets, turnips, and cabbage) respond to sodium even if potassium is abundant. The successful replacement of potassium by sodium in water-culture experiments has been as great as 80 percent (sugar beets). Most other plants show no response to sodium.

Calcium

Calcium is the most abundant alkaline earth metal. When unexposed to the atmosphere it is white and silvery, but it rapidly tarnishes when allowed to combine with the oxygen and nitrogen of the air. In the pure state, it is malleable and tends to have a crystalline structure.

Functions in plants

Calcium is fixed in cell walls as a calcium salt of the pectic compounds of the middle lamella. It is necessary for cell growth and division in apical meristems and elsewhere. Calcium has complex relations with other elements. Without calcium some species are unable to assimilate nitrogen.

Calcium is extremely immobile in plants. Thus the older leaves may have large reserves while younger leaves are deficient, although calcium oxalate crystals have been observed to disappear from older leaves during periods of deficiency, indicating some degree of translocation. Plants contain 0.2 to 0.3 percent calcium by dry weight.

Deficiency symptoms

Plants deficient in calcium may be green, but they have greatly deformed terminal leaves and branches, and their root growth is impaired. Beans (*Phaseolus* spp.) deficient in calcium turn black. Calcium deficiencies in corn are manifested by gelatinous tips of the innermost leaves, which stick together when they dry. Beet leaves turn pale green around the margins. Dead spots develop near the midrib of potato leaves and gradually form a circular pattern in the middle of the leaf. Roots of many species turn black and die. A number of fruit disorders in apple and pear that often appear in storage have been associated with calcium deficiency. See Figure 16-12 for other symptoms.

Soil relations

Calcium is derived from many minerals, but calcite and dolomite are the main sources. Calcite and dolomite, the principal mineral components of limestone, readily decompose and are quickly leached from the surface soils of humid regions. In areas where rainfall is less abundant, they accumulate in lower horizons and may form a hard, white, calcareous horizon.

Not only is calcium necessary to plants, but it has a strong influence on the absorption of ions by soil particles and on the relations of other elements to plant

A   B   C

Figure 16-12 Calcium-deficiency symptoms: (A) Cavity spot in carrot; (B) internal browning in Brussels sprouts; (C) blossom-end rot in tomato fruits. [Courtesy D. N. Maynard.]

growth. Lime (calcium oxide, CaO) is frequently applied to soil before application of fertilizers containing other elements in order to reduce soil acidity and to reduce phosphorus fixation. When soil acidity is reduced, magnesium and molybdenum are changed to forms that are more available to plants, with the result that additions of these elements may be unnecessary. Excess liming, however, can cause problems, because manganese and iron are less soluble at higher pH values. Calcium promotes good soil structure by increasing aggregation, and it encourages certain microorganisms, especially the nitrifying and nitrogen-fixing bacteria.

Losses from soil

Calcium is lost from soils mainly by leaching but also by crop removal and erosion. Soils derived principally from limestone materials may actually be deficient in calcium if leaching has carried the calcium below the root zone. Carbon dioxide in rainwater (carbonic acid) reacts with calcium carbonate and other clay minerals to form calcium bicarbonate, which is highly soluble. Losses of calcium sulfate and calcium nitrate by leaching are also significant. Losses by crop removal may be significant in such crops as alfalfa, which requires 140–168 kilograms per hectare (125 to 150 pounds per acre) of elemental calcium for the production of 11 metric tons per hectare (about 5 tons per acre) of alfalfa hay.

Fertilization

Most often calcium is added to the soil in the form of limestone with a high calcium content. Calcium and magnesium carbonates are found over the entire United States and are easily quarried and crushed. The finer it is ground, the more readily it is available to plants. **Agricultural meal,** 95 percent of which will pass through an 8-mesh screen sieve, is a commonly used form of limestone fertilizer. **Marl,** a sedimentary rock composed of precipitated calcium carbonate, is common in some regions and is used because it is immediately available and inexpensive. As a general rule, the calcium in two cubic yards (about 1.5 cubic meters) of marl is equal to that in a ton of limestone. **Oyster shells** and **shell rock** are used in areas where they are readily available. The shell rock is usually pulverized, but oyster shells are sometimes used without being crushed. **Basic slag,** a by-product of iron smelting, contains calcium silicate, calcium oxide, magnesium oxide, and phosphoric oxide, and is frequently used for liming. It must be ground before application. **Burnt lime** (CaO), or unslaked lime, that has been treated with water is called **slaked lime** ($Ca(OH)_2$). Other names for this material are hydrated lime, caustic lime, and agricultural hydrate. It has some objectionable characteristics and is hard to handle, but it reacts very rapidly with acid soils. **Gypsum** (calcium sulfate) is used where it is readily available.

Magnesium

Magnesium is a shiny silvery metal closely related to calcium. It reacts very slowly with water, and its oxide has a melting point of 2800°C (5072°F). The pure metal burns rapidly once ignited, and it is a good conductor of heat.

Functions in plants

Magnesium is an essential part of the chlorophyll molecule and is also necessary for the formation of amino acids and vitamins. Another important function of magnesium is to buffer organic acids in plants. It is essential to the formation of fats, the germination of seed, and the synthesis of sugar, in which processes it functions largely as a coenzyme. On the average plants contain about 0.2 percent magnesium by dry weight.

Deficiency symptoms

Plants deficient in magnesium are usually chlorotic and may develop brilliant colors. In severe cases, leaves become yellow. Chlorosis is frequently most evident between the veins of older leaves, starting at the tips and moving inward (Fig. 16-13). Leaves may also droop.

Figure 16-13 Progressive effects of magnesium deficiency in potatoes. The leaf at left is normal; severity of deficiency increases to right. [Courtesy USDA.]

Soil relations

Magnesium is found in soils that include the minerals biotite, chlorite, dolomite, magnesite, olivine, and serpentine, as well as in lake brines. After being weathered from mineral sources it is adsorbed onto the surfaces of clay and organic particles. Magnesium is leached much more readily than calcium in lower soil horizons. Magnesium and calcium are similar in their soil relations. Both cations are found in many of the same rocks and minerals, and they behave similarly in response to soil reaction and water content.

Losses from soil

Magnesium is lost by leaching, plant removal, and erosion. Deficiencies frequently occur in sandy soils exposed to moderate or heavy rainfall, where leaching is intense. Losses of magnesium through transformations to unavailable forms take place in acidic peats and mucks as well as in poorly drained soils.

Fertilization with other cations may displace magnesium and can result in deficiencies, because magnesium moves readily to the lower soil horizons.

Fertilization

Magnesium is generally added to soils when lime is added. Finely ground **dolomitic limestone** is frequently used to correct deficiencies. When magnesium deficiency is acute, foliar sprays of epsom salts ($MgSO_4$), magnesium nitrate, and magnesium chloride are applied. Sprays produce an immediate response, frequently within 24 hours, but their residual effect is of short duration; consequently sprays cannot be considered a complete substitute for soil fertilization.

Sulfur

Sulfur was used medicinally as early as 1000 B.C.E. It is found in regions where volcanoes have been active and in association with salt domes along the Gulf

Coast of North America. Although far less abundant than oxygen, it is in the same group of elements.

Functions in plants

Sulfur is an important component of several amino acids, such as methionine and cysteine. The vitamins thiamin and biotin also contain sulfur. Sulfur compounds impart flavor to cruciferous plants and pungency to the onion. In general, plants contain perhaps 0.2 percent sulfur by dry weight.

Deficiency symptoms

The symptoms of sulfur deficiency are somewhat similar to those of nitrogen deficiency. Plants deficient in sulfur tend to be light green because sulfur, although not an essential part of the chlorophyll molecule, is necessary for its synthesis (Fig. 16-14).

Soil relations

Sulfur occurs in rocks mainly as iron sulfide. Upon weathering it rapidly changes to a soluble sulfate and subsequently becomes incorporated into organic compounds. Small amounts are absorbed by clay particles. Microorganisms break down the organic compounds and convert them to sulfates. Calcium sulfate

Figure 16-14   Sulfur deficiency in peach; severity of deficiency increases to right. [Courtesy National Plant Food Institute.]

seems to accumulate near the surfaces of arid soils; in humid regions it is leached to lower horizons. The sulfur cycle is illustrated in Figure 16-15.

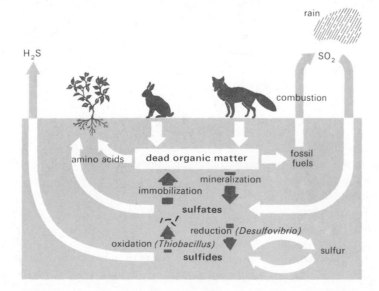

Figure 16-15   The sulfur cycle.

In some regions a large part of the sulfur content of soils originates as sulfur dioxide (SO_2), which is washed from coal smoke in the air by rain, forming sulfurous acid (H_2SO_3). Sulfur dioxide toxicity to plants is an increasing problem in urban and industrial areas, particularly in the vicinity of smelters.

Losses from soil

Sulfur is lost by erosion, leaching, and crop removal. Erosion causes the greatest loss because a large part of the sulfur in soils is in the organic fraction of the upper horizons. Leaching may account for sulfur losses of up to 67 kilograms per hectare (about 60 pounds per acre) per year.

Fertilization

Native sulfur is most frequently mined by drilling into underground deposits and pumping hot steam into them to melt the sulfur and force it to the surface under pressure. It is then shipped, still in the liquid stage, to plants where sulfuric acid is made. (Sulfuric acid is also made by burning iron pyrite, a by-product of certain mining operations.) Sulfuric acid is used to make the many sulfur compounds (such as ammonium sulfate, calcium sulfate, and superphosphate) used in fertilizers.

Iron

Iron, the most widely used metal in the world, belongs to the group of elements that includes cobalt and nickel, and the three metals have many similar properties.

Functions in plants

Iron functions as a catalyst in the synthesis of chlorophyll; much of the iron in plants is found in the chloroplasts. It is involved in the formation of many compounds, even though it is not present in their structure. Proteinaceous iron-porphyrin compounds serve as oxygen carriers and "oxygen activators." Iron is a component of certain important enzymes—for example, the oxidases and peroxidases. It is essential to the normal development of the young growing parts of plants, and it accumulates in nodes. On the average, plants contain about 0.1 percent iron by dry weight.

Deficiency symptoms

Iron deficiency in its early stages is manifested by a patternless paling in leaf color, and in its later stages by yellowing of leaves (Fig. 16-16). Veins are not affected at first, but eventually even the larger veins turn yellow. Grasses develop leaf patterns with alternate rows of green and white.

Soil relations

Among the most common iron-bearing minerals are hornblende, chlorite, and biotite. During weathering the iron in such minerals is changed from the ferrous (Fe^{2+}) to the ferric (Fe^{3+}) state. In regions with ample rainfall, low temperatures, and an accumulation of litter, iron is dissolved and leached to lower horizons, where it accumulates. In regions where rainfall is abundant, temperatures high, and organic cover sparse, iron is converted to a hydrated form and is retained in the horizon in which it weathered from rock. The degree of hydration also affects the soil color, which varies from yellow to red.

Under highly alkaline conditions, iron is rendered unavailable to plants by the formation of insoluble ferrous compounds. As a result the total amount of iron in soils may be a very poor indicator of available iron. When soils are alkaline it may be abundant but unavailable to plants.

Losses from soil

Iron may be lost by crop removal, leaching, and erosion. But because iron is required only in minute amounts, losses are likely to be less significant than those of other elements. Availability is of prime concern.

Fertilization

Most soils contain sufficient iron for plant growth. The main problem is to make it available by proper soil management. This may sometimes be done by

normal leaves

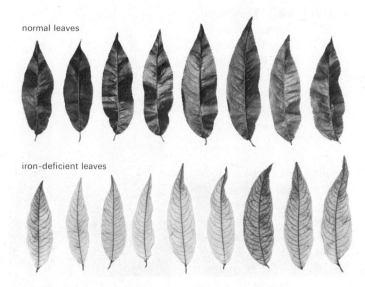

iron-deficient leaves

Figure 16-16   Iron deficiency in peach. [Courtesy National Plant Food Institute.]

controlling soil moisture, for ferrous iron accumulates in wet or water-logged soils. If soil reaction is too high, acidity may be increased by applications of sulfuric acid or ammonium sulfate.

Iron may be applied in several different ways. It is most frequently applied in an inorganic form directly to the soil, but in this form it often becomes bound in heavily limed or alkaline soils. Iron can be applied to fruit trees by drilling a hole into the trunk and inserting a gelatin capsule filled with an iron salt.

Perhaps the most effective way to remedy iron chlorosis is to use iron **chelates.** Iron chelates are compounds that bind iron at two or more positions within their structures. The iron ion is held in such a way that it stays in solution and cannot precipitate unless ion exchange takes place or unless the chelate is precipitated by phosphate. Although chelates are expensive to produce, they are widely used in foliar sprays for ornamental shrubs and other high-value crops. For nursery crops they are frequently applied directly to soils. Citric acid is an example of a compound that acts as a chelating agent. Chelates are not toxic to plants, applications of 1120 kilograms per hectare (1000 pounds per acre) having been tried without ill effects. Chelates are decomposed by mi-

croorganisms, and hence the iron they make available becomes bound again. The effect of small applications of certain chelates on the mineral metabolism of soybeans suggests that they have favorable physiological effects in addition to their capacity to chelate iron in the soil.

Boron

Boron is not an abundant element, but large deposits of hydrated sodium borate ($Na_2B_4O_7 \cdot 10H_2O$), called **borax,** exist in the Mojave Desert of California. Boron is used in the manufacture of glass, soap, enamel, and pottery, as an antiseptic, and in the alloys of machinery where high resistance to abrasion is necessary.

Functions in plants

Boron affects flowering, pollen germination, fruiting, cell division, nitrogen metabolism, salt absorption, water relations and the movement of hormones. It prevents the excessive polymerization of sugars at sites of sugar synthesis. Altogether, at least sixteen functions of boron in plants have been described. Plants contain only a few parts per million of boron, and their requirements for it are small. They must absorb boron throughout their lives because it is not readily transported in the vascular system. Boron uptake seems to be closely related to calcium uptake. Plants with an inadequate supply of calcium have a low tolerance for boron; when there is an excessive supply of calcium, boron requirements are high.

Deficiency symptoms

In the early stages of boron deficiency, terminal buds die and lateral branches begin to grow. Lateral buds then die, to be superseded by the growth of new ones, with the result that a rosette of branches may form at the ends of twigs. This seems to be due to a disintegration of meristematic tissue. Leaves may become thickened, curled, and brittle. Fruits, tubers, and roots may become discolored, cracked (Fig. 16-17), and flecked with brown spots.

Soil relations

Borates of calcium, magnesium, and sodium occur naturally in soils. Once released from rocks by weathering processes, they are utilized by plants and various microorganisms. Upon the death of the organisms the borates are oxidized to inorganic forms. Borates may be leached from soils at any time while they are in an inorganic form. Leaching of borates takes place much more slowly from clay soils than from light-textured soils.

Since most boron compounds are soluble only when the pH is low, soil acidity affects the availability of boron. Liming may therefore result in boron deficiencies due to the fixation or binding of boron on soil colloids at higher pH ranges. Clay and organic soils fix boron in relatively large amounts, whereas sandy soils fix very little. Drying of organic matter also results in fixation. Thus, irrigation increases boron availability.

Losses from soil

Boron is most frequently deficient in humid regions where soils are acidic. Every state east of the Mississippi River has areas that are deficient in this ele-

A

B

Figure 16-17 Boron deficiency in pear fruits (A) and broccoli plants (B). [Courtesy National Plant Food Institute and D. B. Maynard, University of Massachusetts.]

ment to various degrees. Alfalfa has an unusually high requirement for boron. Vegetable crops also deplete soil boron rapidly. When forest trees are fertilized their accelerated growth causes a rapid depletion of boron.

Fertilization

The boron used in fertilizers comes primarily from deposits in the Mojave Desert and from the brine of Searles Lake in California. Borax is the form that has most often been used for agricultural purposes. Colemanite (a form of calcium borate) is used on sandy soils because it does not leach as rapidly as borax.

Boron is frequently a contaminant in various kinds of fertilizers. Lime contains about 30 grams of boron per metric ton (about 1 ounce of boron per ton). Barnyard manure contains some boron, and superphosphate also contains appreciable amounts. Thus, for crops with small boron requirements, such materials may provide adequate amounts for crop growth.

Boron compounds have frequently been used as weed killers. Liberal application of borax to the soil usually results in the death of plants whose roots extend into the treated area. Borate slurries were used for the suppression of forest fires by aerial "bombing" from tanker planes until it was found that plants would not grow in these areas for many years.

Manganese

Manganese is a reddish-gray metal somewhat like iron, which it resembles in many ways. It is found in many minerals that also contain iron. The pure metal has few commercial uses, but is used in many alloys.

Functions in plants

Manganese is required for the synthesis of chlorophyll. It also helps to control some of the complex oxidation–reduction systems that involve iron, glutathione, and ascorbic acid, where it functions as a coenzyme. Plants contain about 300 parts per million of manganese.

Deficiency symptoms

Manganese deficiencies in plants are readily apparent unless obscured by other symptoms. Young leaves may show a network of green veins on a light green background of intervenal tissue, a pattern similar to that caused by iron deficiency (Fig. 16-18). If the deficiency is not corrected, light green parts become white, and leaves are shed. Brownish, black, or grayish spots appear next to veins in the leaves of some plants, giving rise to such names as gray speck, white streak, dry spot, and marsh spot.

Soil relations

Among the minerals that contain manganese are manganite, hausmannite, olivine, biotite, and garnet. In acid soils manganese is likely to be abundantly

Figure 16-18 Five-month-old tung seedling showing symptoms of manganese deficiency. The lower leaves are normal because they developed before the manganese in the seed was exhausted. [Courtesy USDA.]

available, whereas in neutral or alkaline soils plants often show deficiency symptoms even though the total amount present is great. Manganous manganese (Mn^{2+}) is readily available, but the manganic form (Mn^{3+}) is not. In highly acid soils manganese may be available to plants to the extent that it produces toxicity.

Fertilization

Plants require very small amounts of manganese, and soil acidification generally cures deficiency problems. When the manganese content of a soil is too low, manganese sulfate is usually applied either to the soil or to the foliage of the crop.

Zinc

Zinc was one of the first metals to be produced from ores. Pure zinc has a white metallic luster when polished, but it soon tarnishes. It becomes malleable at 100°C (212°F).

Functions in plants

Zinc is required in small amounts for auxin formation, internodal elongation, chloroplast formation, and starch formation. Zinc-deficient plants have abnormal roots with enlarged cells and unusual arrangements of root hairs. Tannin, fats, and calcium oxalate crystals tend to accumulate in the root tissues of zinc-deficient plants. Legumes require zinc for seed formation. Monterey pines (*Pinus radiata*) planted in zinc-deficient soils in Australia grew poorly except along fence rows where rainwater washed enough zinc oxide from the galvanized wire fencing to supply their needs. In general, plants contain only a few parts per million of zinc.

Deficiency symptoms

Zinc-deficient plants may have mottled leaves with irregular chlorotic areas (citrus), bronzed and rosetted leaves (pecan), dwarfed and rosetted leaves (apple, cherry, and various other fruits), chlorotic interveinal leaf tissue (various field crops), or rusty

Figure 16-19   Leaf of a zinc-deficient corn plant. [Courtesy A. Ohlrogge.]

brown flecks on leaves (barley and other grasses). Because zinc deficiency leads to iron deficiency, the usual symptoms of both deficiencies are similar. Zinc-deficient corn is illustrated in Figure 16-19.

Soil relations

Zinc is found as a trace constituent of soils that contain biotite, magnetite, and hornblende. After being weathered from minerals it is adsorbed by particles of clay and organic matter. It accumulates in surface soils after having been released by decomposing leaves and other plant materials, and deficiencies are likely to occur when the surface soil is eroded or otherwise removed.

Zinc is least available in the pH range from 5.5 to 7.0. At lower pH values it becomes more soluble, but at higher values the formation of insoluble calcium zincate is favored, and it may therefore be less available to plants in some alkaline soils.

Zinc toxicity can result when soils are acidified to increase the availability of other elements and when zinc fertilization is continued for a long period.

Fertilization

Zinc deficiencies can usually be corrected by foliar applications of zinc sulfate. Zinc sulfate is also mixed with fertilizers applied to soils. Galvanized nails are sometimes driven directly into the trunks and branches of deficient trees. In recent years the use of chelated zinc compounds has become common. These compounds are applied both to foliage and to soil.

Molybdenum

Molybdenum is a silvery white metal. Most of the large, commercially exploited deposits of this element are in Norway, Colorado, and New Mexico.

Functions in plants

Molybdenum is a component of the enzyme system that reduces nitrates to ammonia. When molybdenum is absent, the synthesis of proteins is blocked and plant growth ceases. The root-nodule bacteria of legumes also require it: when molybdenum is absent, the nodules fail to develop. It also seems to be required for vitamin synthesis. Seeds are frequently not "filled" when molybdenum is lacking. Plants generally contain about 1 part per million or less of molybdenum.

Deficiency symptoms

Plants lacking sufficient molybdenum may become nitrogen-deficient. Leaves of deficient plants are frequently pale green and have rolled or cupped margins. Yellow spots appear on the leaves of some species. Potato tubers may fail to develop, and the glumes of grain do not fill out. Leaves of crucifers become narrow, hence the name "whiptail" for this symptom of deficiency.

Soil relations

Two minerals that contain molybdenum are molybdenite (MoS_2) and powellite ($CaMoO_4$). Molybdenum is sometimes an accessory element in olivine and in hydrous iron and aluminum oxides and other clay minerals. After it weathers from minerals, molybdenum is only slightly soluble. It is fixed much more strongly in acid soils than in alkaline soils. Liming, rather than fertilization, is usually the most suitable corrective measure.

Losses from soil

Molybdenum deficiencies may result either because the element is present only in small amounts in the soil or because it is unavailable to plants. Because it is required only in extremely small amounts, removal by plants is not likely to be a problem.

Fertilization

Molybdenum is usually added to soil at the rate of about 140 grams per hectare (2 ounces per acre). If even this small amount is added for a number of consecutive years it may accumulate to toxic levels. It is usually applied either by coating seeds with sodium molybdate powder or by spraying plants with a solution of it.

Copper

Pure copper is malleable, flexible, and a very good conductor of electricity. Almost all countries produce some copper, but the largest deposits are in the United States. Rich deposits also occur in Peru and Zimbabwe-Rhodesia.

Functions in plants

Although copper is an essential nutrient, it is highly toxic to plants when it is too abundant. It is concentrated in the small absorbing roots of plants, and it plays a part in nitrogen metabolism. It is a component of several enzymes and may be a part of the enzyme systems that utilize carbohydrates and proteins. The functions of copper in plants are not entirely understood. In general, plants contain a few parts per million of copper.

Deficiency symptoms

The ends of copper-deficient shoots are first to be affected: they die back from the tips, and terminal leaves develop dead spots and brown areas. Rosettes of leaves may precede death of the twigs. Plants that have no branches, such as cabbage, become chlorotic, and leaves fail to form. Deficient corn may become gray at the tips or develop striped leaves (Fig. 16-20). Unlike the symptoms caused by deficiencies of iron, zinc, and many other elements, the symptoms caused by copper deficiency are neither clear-cut nor consistent.

Soil relations

Clayey and loamy mineral soils usually have a copper content of 10 to 200 parts per million; since copper is required by plants only in small quantities, this is generally sufficient. Copper is weathered from cuprite, malachite, azurite, and other minerals, after which it is present in the soil as phosphates, hydrox-

Figure 16-20 Copper deficiency in corn. [Courtesy National Plant Food Institute.]

ides, and carbonates. It is bound tightly in organic-matter complexes, with the result that plants in organic soils are likely to be deficient in copper. Most copper is held in the lattices of soil colloids with such tenacity that it cannot be displaced easily by other soil cations in the pH range 7 to 8.

Losses from soil

Copper is not readily lost from the soil. It is rather commonly present in a form in which it is unavailable to plants. Removal by crops is negligible, since plants rarely contain more than 25 parts per million, usually less.

Fertilization

Copper is usually added to soils as copper sulfate ($CuSO_4 \cdot 5H_2O$) and copper oxide (CuO) at the rate of 1–2.25 kilograms per hectare (about 1–2 pounds per acre). It should be added only when deficiencies are evident, because yearly additions may accumulate to produce toxicity.

Chlorine

Chlorine is a heavy, greenish-yellow gas that is highly toxic to plants and animals. It is highly reactive, however, and does not occur naturally in the free state.

Functions in plants

Chlorine is the most recent element to be included in the "essential" category (1954). It is usually present in plants in small quantities (perhaps 0.1 percent by dry weight), and it takes part in the photosynthetic reactions in which oxygen is liberated. At one time it was thought to be necessary for the translocation of carbohydrates, a role that has not been substantiated. Experimental evidence on the functions of chlorine is extremely difficult to obtain because of its high solubility and its ubiquitous distribution in soil, water, and atmosphere. An environment absolutely free of chlorine is extremely difficult to produce.

Deficiency symptoms

Deficiency symptoms of chlorine include wilting, stubby roots with excessive branching, chlorosis, and bronzing. Even the odors of some plants are decreased. These conditions only develop, however, when chlorine is deliberately withheld, with great effort, from nutrient solutions in which plants are grown.

Losses from soil

Chlorine is usually found in highly soluble forms, and thus it is subject to losses by leaching. Removal by plants can account for the loss of a certain amount of chlorine, but in nature it is probably never a serious factor.

Fertilization

Chlorine is added to the soil by rainfall in amounts sufficient to satisfy plant needs. It is never added intentionally by fertilization, although all commer-cial fertilizers are "contaminated" by chlorine. Some plants obtain enough chlorine from the air to prevent deficiency symptoms.

WORLD PLANT NUTRITION

The "agricultural revolution" resulted in more sophisticated methods of cultivation, water management, and fertilization, making possible great increases in the production of food, which were necessary to support the world's expanded population (Fig. 16-21). Fertilization may be the primary key to the problem of feeding even larger future populations; the world is certainly far from having reached its capacity to produce food. It has been estimated that the food requirements of the increasing world population can be met for the next hundred years if the productivity of all existing arable lands can be increased to equal the current productivity of the arable land of Europe. The productivity of European agriculture is approximately twice that of the world average, due in part to the greater use of chemical fertilizers. The productivity of arable lands in Japan is about twice that of the European average,

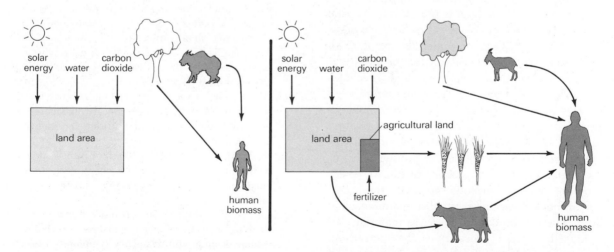

Figure 16-21 Before the "agricultural revolution," hunting, fishing, and food gathering provided sustenance for about 10 million people. The 10 percent of the land now being cropped provides the primary support for a population that approaches 4 billion. [From L. R. Brown, "Human Food Production as a Process in the Biosphere." Copyright © 1970 by Scientific American, Inc. All rights reserved.]

which indicates that doubling the productivity of arable lands is by no means an impossible task. One metric ton of plant nutrients (nitrogen, phosphorus, and potassium) can produce an average increase of about 10 metric tons of basic food. Ten metric tons of basic food will supply 55,000 kilocalories per day for a year, enough to feed 22 more people if the average world caloric intake is considered to be about 2500 kilocalories per day.

Those countries whose people are well fed (3000 kilocalories per day) are advanced technologically. Asia has 54 percent of the world's population but only 16 percent of the earth's agriculturally productive land, and the average diet consists of no more than 1800 kilocalories per day. In India, even if the amount of arable land were increased by 25 percent, there would still not be enough land to feed the population at present productivity levels. But if the rate of productivity per unit of land area could be increased to equal that of Europe, then the crisis could be delayed.

The demand for fertilizers is growing at such a rate that by the year 2000, an estimated 180 million metric tons of the major fertilizer nutrients—nitrogen, phosphorus, and potassium—will be needed annually to replace those removed by cropping. This is 15 times more fertilizer than is presently used. The increase must come from chemical rather than natural organic sources. Organic materials are of more value for secondary products than for fertilizer, and their bulk makes them expensive and precludes widespread use by large crop-producing units.

But the problem should not be oversimplified. Projecting fertilizer needs is not the same as producing the materials. Even though the world has tremendous resources and facilities for producing nitrogen, phosphorus, and potassium fertilizers, there must be an economic incentive to invite the necessary investment of capital to establish manufacturing facilities. There is no doubt that these materials can be produced; what is necessary is the motivation, technology, and capital. Furthermore, farmers need special incentives to use fertilizers. Unfortunately, a tax on fertilizer is used to produce revenue in some countries that can ill-afford to discourage good farming practices. Much of the agricultural output in such

countries is the result of subsistence farming on small holdings. Since their products have no commerical value, farmers have little incentive to produce more. The wise use of fertilizer is a key factor in agricultural development. It would seem that great social and economic changes must accompany technological change if a balanced, self-sustaining world agriculture is to be attained.

Selected references

Cooke, G. W., 1975. *Fertilizing For Maximum Yield*. London: Granada. (A book on nutrition and fertilization of various types of crops for the practicioner.)

Dinauer, R. D. (editor), 1968. *Changing Patterns in Fertilizer Use*. Madison, Wisc.: Soil Science Society of America. (The proceedings of an important symposium dealing with many aspects of fertilizers and their uses for different crops.)

Hillel, D. (editor), 1972. *Optimizing the Soil Physical Environment toward Greater Crop Yields*. New York: Academic Press. (Proceedings of an international panel.)

Olson, R. A. (editor), 1971. *Fertilizer Technology and Use*, 2nd ed. Madison, Wisc.: Soil Science Society of America. (A collection of advanced monographs on the economics, production, and use of fertilizer materials.)

Pearson, R. W. (editor), 1967. *Soil Acidity and Liming* (Agronomy Monograph No. 12). Madison, Wisc.: American Society of Agronomy. (A collection of papers dealing with the soil chemistry of liming and the attendant crop responses.)

Pillai, K. M., 1967. *Crop Nutrition*. Bombay: Asia Publishing House. (A short presentation of fundamental and practical concepts of plant nutrition for students of agriculture.)

Samish, R. M. (editor), 1971. *Recent Advances in Plant Nutrition*. New York: Gordon and Breach. (A compendium of short, informative papers on crop nutrition.)

Sprague, H. B. (editor), 1964. *Hunger Signs in Crops*. Washington, D.C.: National Fertilizer Association. (A volume dealing with deficiency symptoms in selected crops.)

Tisdale, S. L., and W. L. Nelson, 1966. *Soil Fertility and Fertilizers*, 2nd ed. New York: Macmillan. (This college text covers the fundamental concepts of soil fertility and fertilizer manufacture.)

17
Water Management

The chapters of Genesis, which describe the life of Israel's forefathers 4000 years ago in Canaan, abound in references to water and wells. To survive, these people had to practice good water conservation. Even a modern engineer with the best tools would find it difficult, if not impossible, to improve on the methods of water conservation that were used 2400 years ago by the Nabataean inhabitants of Israel's Negev Desert. The Nabataeans collected the runoff from uncultivated slopes and routed it to cultivated areas. In order to do so, they built elaborate diversion and drop structures, terraces, guide walls, storage impoundments, and irrigation systems for basins. They also practiced dry-farming. Practically all of the systems and structures they developed later fell into disrepair because of war.

Today, water management still involves interrupting and manipulating various stages in the hydrologic cycle to make more water available, to remove it where there is an excess, and to improve the quality when necessary. Although water is also managed for industrial, recreational, and other purposes, this discussion will be principally concerned with water management in relation to crop production (Fig.

17-1). This involves **irrigation,** the addition of supplemental water; **drainage,** the removal of excess water; and **conservation,** the protection and wise use of our water resources.

IRRIGATION

Irrigation to provide supplemental water for crop growth was a highly developed art 4000 years ago. Paintings and sculptures in the tombs of the ancient pharaohs depict slaves lifting water to pour on crop fields. One of the earliest aids to irrigation was the **shaduf**—a long pole balanced on an upright, with a bucket on one end and a counterweight on the other (Fig. 1-4). Water could be raised nearly 2 meters (about 6 feet) with this device, and even more if several shadufs were used in series.

The **sakieh,** or chain of pots, was developed later and utilized animal power. It consists of a beam attached to a toothed wheel that engages a vertical wheel, which, in turn, drives an endless rope fitted with buckets. The water in the buckets is spilled into a small flume as they tilt when passing over the vertical wheel. This is similar to the pedal-powered

Figure 17-1   The efficient use of water necessitates multiple use. Storage impoundments can provide water for recreation, crop irrigation, human consumption, industrial use, and many other purposes. [Courtesy USDA.]

Figure 17-2   A Persian water wheel powered by a man's legs. [Courtesy Food and Agriculture Organization of the United Nations.]

Persian wheel, which is still in use in some parts of the world (Fig. 17-2). The water screw, invented by Archimedes about 200 B.C.E., could lift water about 4.5 meters (about 15 feet).

Many of the ancient devices are still used today in some parts of the world. It is interesting to see the modern touch that has been given the sakieh, with tin cans replacing leather buckets as bails. The suction pump, driven by wind, steam, or electricity, is a modern contribution to irrigation. An interesting pump called the hydraulic ram operates by water power (Fig. 17-3). Today, water is usually extracted from the soil by electric or gasoline pumps, even where wells are still dug as they have been for thousands of years (Fig. 17-4).

In present irrigation practice, supplemental water may be applied in three ways. In surface irrigation, water is distributed over the surface of the soil.

Figure 17-3 The hydraulic ram is an interesting pump that uses water power to move water to a greater height. The principle of operation is as follows: water moving down a supply pipe closes an escape valve (*E*) controlled by a spring. The momentum of the moving water causes a ramming action that forces water through a supply valve (*S*) at the base of an air chamber. The force of the water compresses the air in the chamber. When the air pressure equals the pressure of the moving water in the supply pipe, the supply valve closes and the trapped water is forced out of the delivery pipe to a higher elevation. Immediately after the supply valve is closed, there is a rebound or backward flow of water that causes the escape valve to open. Then the water starts to move again down the supply pipe, builds up momentum, and the cycle repeats. The efficiency of this device may be as high as 75 percent under ideal conditions. The hydraulic ram, with only two moving parts, can operate for years without attention. It is especially valuable for providing irrigation where power is unavailable, and it should have wider use in underdeveloped areas.

Figure 17-4 Water from this newly dug well in eastern Algeria is pumped to the surface to irrigate a cereal crop. [Photograph by H. Hull courtesy Food and Agriculture Organization of the United Nations.]

Sprinkler irrigation is the application of water under pressure as simulated rain. Subirrigation, the distribution of water to soil below the surface, provides moisture to crops by upward capillary action. Each system is adapted to a particular cropping situation.

For agricultural purposes water is measured by volume and by rate of flow. Volumes may be given as gallons, liters, cubic feet, cubic meters, acre-inches, and acre-feet. An **acre-inch** of water is the amount of water that will cover one acre to a depth of one inch, and is approximately 27,000 gallons (about 102 cubic meters); an **acre-foot** will cover one acre to a depth of one foot, and amounts to 43,560 cubic feet or neary 326,000 gallons (about 1233 cubic meters). Flow rates are usually given in terms of cubic feet or liters per second, gallons or liters per minute, acre-inches or cubic meters per hour, or acre-feet or cubic meters per day.

Surface irrigation

Water is applied by surface irrigation in arid and semiarid regions. When the surface of a field cannot be contoured to prevent ponding, provisions must be made in advance to drain the lower places. Conversely, when fields are extremely flat, low dikes help to retain water.

Water is usually carried to the place of use in open ditches at a slow velocity in order to prevent erosion. Pipelines and flumes are sometimes used when water is scarce, since they eliminate losses due to seepage and evaporation when water is moved long distances. When ditches are used for this purpose they are frequently lined with asphalt or concrete to eliminate seepage loss (Fig. 17-5). Earth cement, made of cement and soil, can be used to line irrigation ditches. It is inexpensive and can be mixed "in place," but it is not durable. In recent years, plastic has been used for lining ditches, but plastic linings have a comparatively short life. If possible, it is helpful to cover the banks and bottoms of ditches with a permanent grass sod. Small portable gasoline pumps and light, easily moved surface pipes are used to carry water from ponds and ditches to fields.

The distribution of water from ditches is accomplished by various control structures or siphons. The flow of water must be carefully controlled to prevent erosion. **Division boxes** are used to divide a stream into two or more smaller streams. **Checks,** dams of adjustable height for controlling the depth of water at turnouts, are installed at places where water is delivered to the area to be irrigated. When water is transported by pipeline it is sometimes helpful to use a **soaker,** a canvas hose or perforated pipe that distributes the water over a large surface area and prevents erosion.

Figure 17-5   Irrigation canals lined with concrete have no seepage losses. [Photograph by C. A. Weber courtesy USDA Soil Conservation Service.]

Figure 17-6 Rice culture in Nepal. Terracing permits the recycling of water for flood irrigation at successively lower levels. [Courtesy USDA.]

Surface irrigation may be accomplished in a number of ways. In **flood irrigation,** water is allowed to cover the entire surface of the field in a continuous sheet. It must be deep enough and remain long enough to replenish the root zone with water. Provisions must be made to carry away excess water (water that remains on the surface after the soil has reached field capacity) through drainage ditches. Flood irrigation is frequently used for crops that form a complete ground cover, such as permanent pasture and rice. **Surface flooding** is sometimes carried out by building small retaining walls to restrict water movement from one part of a field to another.

Contour flooding is practiced where slopes are steep. A series of dikes are formed along the contour of the land at regular intervals. Each section of the field is then flooded, resulting in a series of shallow impoundments, each 15 to 30 centimeters (6 to 12 inches) above the other. **Contour terraces** are variants of contour furrows (Fig. 17-6).

Figure 17-7 This crop of pinto beans is being furrow-irrigated with water from a feeder canal lined with concrete. [Photograph by H. Bryan courtesy USDA Soil Conservation Service.]

Furrow irrigation is one of the most common methods used for crops (Figs. 17-7 and 17-8). Water runs down the furrows between plant rows (but not

Figure 17-8 Because of the wide spacing of the vines, vineyards lend themselves to furrow irrigation. [Courtesy U.S. Forest Service.]

over them) and moves to all parts of the soil by capillary action or gravity. It is an expensive technique, because constant supervision is required to prevent furrow streams from uniting and forming large channels, but it can be justified for truck crops, orchards, and vineyards if soils are not so coarse that seepage is excessive. Contour furrows can be used effectively when slopes do not exceed 10–15 percent (Fig. 17-9).

Corrugation irrigation is the application of water in small furrows that run downslope from a **head ditch.** The furrows guide rather than carry water. It is adapted to crops on sloping soils that absorb water slowly.

The main advantage of surface irrigation is that under ideal conditions no source of energy is required for the final distribution of water. But since this system depends on gravity flow, it is inefficient in that more water is required than is actually delivered to the crop plants. The chief disadvantage lies in the possible degradation of soil structure. Heavy soils become puddled under the heavy load of water and are poorly aerated, tending to bake and crack when they dry, and they form hard clods when broken by a plow.

Figure 17-9 Contour furrows can be used for irrigation, as for this potato crop, if slopes are carefully controlled. [Courtesy USDA Soil Conservation Service.]

Sprinkler irrigation

Sprinkler irrigation was first used in the United States in about 1900. Sprinkler irrigation systems formerly consisted of permanent overhead sprinklers that were used only in intensive cropping operations such as market gardens. Today, the use of light weight, easily movable aluminum pipe with rapid coupling devices has cut the labor cost and brought the practice into general use.

Water may be applied through sprinkler heads or perforated pipe. **Sprinkler heads,** the most popular, may be used with permanent buried lines or temporary portable lines (Fig. 17-10). Water is delivered through sprinkler heads mounted on riser pipes. Because each sprinkler head applies water to a circular area, distribution is not uniform. There should be an overlap of a fourth to a half of the wetted circles, and the distances between lines of sprayers should not exceed 70 percent of the diameter of the area covered by one sprinkler.

Several types of sprinkler heads are available. **Stationary heads** apply water to small areas and at low rates; usually no attempt is made to obtain overlap. **Revolving heads** may have either one or two nozzles. A single nozzle that emits a large water stream is used when wind is likely to deflect a small stream. When double nozzles are used, one nozzle throws water to the outer part of the circle while a smaller one delivers water close to the sprinkler head.

Traveling heads are mounted on carriages that are propelled at a slow rate by the pressure of the water delivered to the sprinkler. "Big guns" mounted in this manner can throw water great distances, revolving as they move from one part of a field to another.

"Rounds" of crop land may be irrigated by lengths of irrigation tubing and sprays mounted on multiple sets of wheels (Fig. 17-11). A well at the center of each field delivers water to the irrigation pipeline, which travels in a circle at a predetermined rate of speed. Corners of such fields are not watered, hence the term *rounds*, which was coined by farmers in the Midwest.

Perforated pipe sprinklers deliver water through small holes drilled along the length of pipes mounted

Figure 17-10 Sprinkler heads may be permanent or movable. This one was designed to propel itself, using water power, along a predetermined line. [Courtesy Reinke Manufacturing Co. Inc., Deshler, Nebraska.]

on permanent supports. A water-driven motor causes the pipe to oscillate and throw water alternately to each side. Strips about 12 meters (40 feet) wide can be watered in this way. This kind of system requires a water pressure of approximately 1.75 kilograms per square centimeter (nearly 25 pounds per square inch), whereas sprinkler-head systems require as little as 1 kilogram per square centimeter (less than 15 pounds per square inch).

There are many advantages to sprinkler irrigation. Water can be applied precisely and uniformly to all kinds of soils and crops in the exact amount required. As a result, drainage problems in sprinkler irrigation are usually small. Erosion need not be a problem even on steep slopes. Land preparation is not necessary, and labor costs are smaller than for other methods. Small sources of water can be used efficiently from ponds and catchment areas and applied to higher areas. Sprinklers can sometimes be used to reduce crop damage during periods of frost. Water-

Figure 17-11 (A) Close-up view of big waterguns mounted on carriages that traverse a circular route. (B) Big guns mounted in series. (C) Circular crop fields irrigated by the revolving irrigation systems. [Courtesy Valmont Industries, Inc., Valley, Nebraska.]

soluble fertilizers can be applied by sprinklers, giving very uniform distribution.

There are certain disadvantages to sprinkler systems. The initial cost and power requirements are high. Water must be clean, otherwise it will clog the spray equipment. In hot, dry regions, evaporation may consume a significant amount of the water while it is in the air. Tight soils with slow rates of infiltration are difficult to irrigate in hot, windy weather. But in spite of these drawbacks, sprinkler irrigation can be adapted to practically any kind of crop.

Irrigation has been a traditional practice in semiarid regions. Within recent years, however, portable sprinkler irrigation systems have become a part of farm equipment even in regions that receive more than 125 centimeters (about 50 inches) of precipitation annually. Because occasional periods of drought in such regions can cause the loss of a high-value crop, the cost of owning an irrigation system is often justified even though it may not be used every year.

Drip irrigation

About 40 years ago in Israel, Symcha Blass, an engineer, observed that a tree near a leaking faucet was more vigorous than other trees that were growing nearby but that received no water from the faulty faucet. Blass, recognizing that the tree was thriving on a very small amount of water, conceived the idea of an irrigation system in which water would be applied to crop plants drop by drop. Today, this is the fastest growing method of irrigation in the world.

Drip irrigation, also called trickle, high-frequency, or daily-flow irrigation, conserves water by restricting application to the immediate root zone. Water is applied frequently, but very slowly and in small amounts so that little is lost by evaporation or to portions of the field not occupied by plant roots. (Fig. 17-12). Water is generally applied at a rate of 4 to 8 liters (about 1 to 2 gallons) per hour under low pressure (about 1 kilogram per square centimeter, or less than 15 pounds per square inch). Since moisture

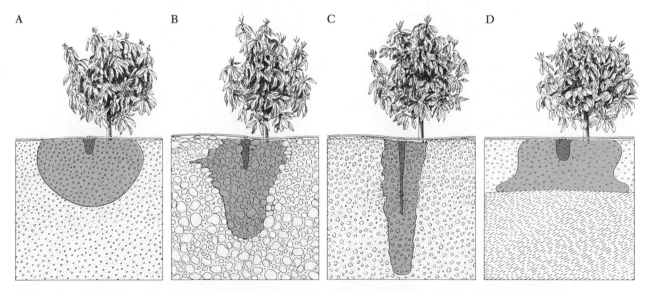

Figure 17-12   The wet zone around the roots of a tree or a plant irrigated by the drip method varies in cross section according to the soil. In each example here an emitter is positioned just to the left of the tree, and the wettest zone (*dark color*) is directly under the emitter. (A) In well-prepared, fine-grained soil capillary forces predominate, and the profile of the wet zone (*light color*) approaches the shape of a hemisphere. (B) If the soil is poorly prepared, lumps and clods interfere with capillary action, and the profile is influenced more by gravity. (C) In soil that is quite coarse, gravitational forces predominate. (D) A layer of hardpan, being fairly impermeable, restricts the water's downward movement and improves its lateral movement. [From K. Shoji, "Drip Irrigation." Copyright © 1977 by Scientific American, Inc. All rights reserved.]

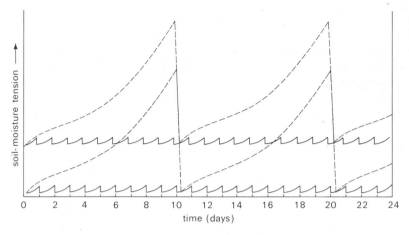

Figure 17-13   Soil-moisture tension, a measure of the force by which water is held in the soil and thus of the effort required of a plant to extract water, is compared for drip irrigation (*solid curves*) and sprinkler or furrow irrigation (*broken curves*). In drip irrigation, water is applied daily. The soil-moisture tension rises slightly during the period of from 12 to 18 hours between applications. With sprinkler or furrow irrigation, which is more likely to be done at longer time intervals, the soil-moisture tension builds up considerably between applications. The upper set of curves shown here represents irrigation with saline water; the lower set, irrigation with relatively fresh water. [From K. Shoji, "Drip Irrigation." Copyright © 1977 by Scientific American, Inc. All rights reserved.]

stresses tend to equalize throughout a plant, as little as 25 percent of the root system of an individual plant need be wetted (Fig. 17-13).

The delivery system consists of a network of water-conducting plastic tubes, with each line terminating in an "emitter" or leak at each plant (Fig. 17-14). Trickle irrigation first proved useful for watering individual containers in greenhouses, where each tube ends in a valve that may be turned on and off and is weighted to keep it in place. The system is flexible because couplings allow fast and simple connections. Many types of emitters can be individually regulated. A popular type of emitter is a very thin tube or pathway through a nozzle (with an internal diameter of about 0.9 millimeter) whose length regulates the pressure—the longer the tube, the slower the rate of delivery. In this way the delivery of water to plants located at different distances from the water source can be equalized.

Water conservation has long been a major concern for crop producers. Even in the 1860s farmers in Germany laid clay pipes about 0.8 meter (about 2.5 feet) below the soil surface in an effort to use the same system for both irrigation and drainage as the water table rose and fell during the year. Then in the 1930s crop managers in Australia invented a system for irrigating peach orchards. They made holes in 5-centimeter pipes so that water could be applied to the peach trees without wetting the soil surface of the entire orchard.

When Blass conceived the idea of drip irrigation, the materials required to build a system were available but too expensive for general application. But the plastics industry developed greatly after World War II, and the required materials became available at prices that were economically justifiable. The method was introduced to the United States in the early 1960s and was used initially by greenhouse op-

Figure 17-14 This nozzlelike emitter is designed to reduce water pressure by causing the water to move along an elongated helical path. With low pressure, the emission hole can be larger, so that its tendency to become clogged is diminished. An emitter of this design is preset at the factory to emit 1, 1.5, or 2 gallons of water per hour. The hook at the right attaches the emitter to the drip-irrigation pipe at a right angle; water enters the conical tip from the center of the pipe. [From K. Shoji, "Drip Irrigation." Copyright © 1977 by Scientific American, Inc. All rights reserved.]

erators. It was first used for orchard and row crops in California in 1968.

Drip irrigation is attractive economically in many agricultural situations because it provides more benefits than any other method of irrigation. The principal benefit is that drip irrigation supplies plants with the exact amount of water they need. Exactly enough water is delivered to the root zone of a plant to replenish the amount consumed in evapotranspiration (the water evaporated from the soil and transpired by the plant), with an additional amount to leach salts from the soil zone close to the roots. This procedure is far more efficient than wetting an entire field or supplying water through furrows.

In addition to requiring less water, there are a number of other advantages to drip irrigation. Compared with other systems, the delivery system is very light and can easily be moved by one person (Fig. 17-15). Soluble fertilizers can be added to the water before it is delivered, and the use of these amendments is thereby rendered very efficient. Water is not wasted between rows nor is it lost to deep percolation. Thus, deep underground water supplies are less likely to be contaminated. Some pesticides can be applied in the irrigation water. Because delivery is in a closed system under pressure, the system has been used on slopes of 50 to 60 percent without problems of erosion. When irrigation must be done with saline water, salt buildup is controlled by continuous leaching. Salts are pushed out to the periphery of the root profile by an advancing front of water coming from the emitter. The roots are able to take up water freely from the middle of the wet zone, where soil-moisture tension is low and the salt level remains nearly the same as it is in the irrigation water (Fig. 17-16).

Since the soil surface stays dry, especially between plants, the crop field is essentially free of weeds. Weeds can be a problem in the wet areas near emitters, and selective herbicides applied through the irrigation water may soon be developed to control them.

Some of the problems of drip irrigation include the high costs of installation and difficulties in keeping equipment operational. Bacterial slime may build up in the plastic tubes and small soil particles may block the small openings. Also, root distribution of perennial plants may be altered to accommodate the

Figure 17-15 A drip-irrigation system in an experimental plot at the University of Hawaii. In the photograph at the left, narrow plastic tubes deliver water to individual lettuce seedlings by means of tiny holes made at appropriate intervals in the tubes. The wet spots on the soil indicate the precision with which the water is applied. In the photograph at the right, the lettuce plants in a nearby plot are shown after growing under drip irrigation for four weeks. In some systems, more complex emitters are installed in the pipes. [Courtesy I-pai Wu, University of Hawaii at Manoa.]

changed pattern of soil-water distribution, with a temporary loss of productivity. Sediment can block the small openings, and the low water pressures used in such systems are frequently insufficient to unclog them. In some areas, ants seeking water cut the tubes open, and rodents use the plastic delivery system to sharpen their teeth. But additives in water to decrease bacterial growth and in plastics to deter animal populations should solve these problems. Finally,

drip irrigation cannot be used for some closely-grown crops, such as cereal grains, alfalfa, and hay, because they would require such a large amount of tubing that the system would be uneconomical.

Most of the world's crop producers irrigate by the same methods that were used 5000 years ago—by flooding and furrows. These methods are only 30 to 60 percent efficient, so that much (usually most) of the water is not used by plants. In addition, they

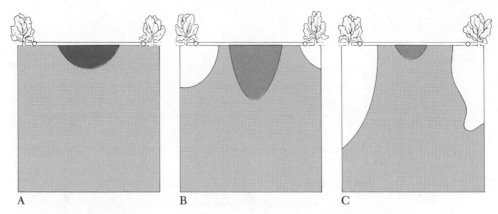

Figure 17-16   Buildup of salt in the root zone of plants irrigated with drip irrigation is depicted for three conditions in which plants are separated by 1 meter and each plant has an emitter next to it. Four degrees of salt buildup, ranging from insignificant to severe, are represented by the spectrum from white through shades of gray to black. If the amount of water applied during irrigation is less than the amount lost through evaporation and transpiration (A), a high concentration of salt builds up in the area between the plants, and the entire root zone contains a significant amount of salt. In the other two conditions depicted, irrigation equals (B) and exceeds (C) evapotranspiration. [From K. Shoji, "Drip Irrigation." Copyright © 1977 by Scientific American, Inc. All rights reserved.]

require considerable preparation of the areas to be irrigated, and they may produce a soil structure that is not favorable to crop growth. In areas where water is abundant and labor not costly, flooding and furrow irrigation may have an economic advantage. But where water is in short supply, labor costs are high, and power is expensive, drip irrigation will likely have the advantage (Fig. 17-17). In 1976, California alone had 15 percent of the worldwide total of land area being irrigated by this system. But even some countries with tropical, high-rainfall climates are using it for such crops as bananas, oil palms, and vegetables (Table 17-1).

Subirrigation

Subirrigation is the application of water beneath the surface of the soil. The objective is to create an artificial water table from which water can move upward by capillary action. Usually the artificial water table is maintained between 30 and 75 centimeters (about 12 to 30 inches) beneath the soil surface.

Ditches are most frequently used to distribute water because they are the least expensive. They have an additional advantage in that the water in them can be seen and the water table can be regulated with relative ease. **Tile drains** can also be used, but they usually cost several hundred dollars an acre to install and are thus the most expensive means of distributing the water. **Mole drains** (discussed under drainage) are useful only in organic soils and have a relatively short life.

Subirrigation is possible only where the topography is absolutely level. Furthermore, there must be a barrier against the irrigation pipe to prevent the loss of water down through the soil by percolation. The barrier may be a water table that is not high enough to provide adequate moisture to the root zone or it may be an impervious soil horizon.

Successful subirrigation in the United States has been carried out in such diverse locations as the flatwoods of Florida, the San Luis valley of Colorado, the Sacramento–San Joaquin Delta in central California, and in the organic soils of Michigan.

Figure 17-17 Piping patterns employed in drip irrigation are shown in a stylized arrangement that emphasizes how pipes and emitters are set up differently for different crops. In an orchard where the trees are small and close together, the pipes can be laid parallel to the rows of trees with numerous orifices or emitters. With large trees, the pipe is looped around each tree, again with several emitters. For a row crop, the pattern may be one pipe per row or one pipe in every other row. When a crop is closely planted, as alfalfa and cereal grains are, drip irrigation may not be the best way to irrigate, since the quantity of pipe required may make the method uneconomical. If drip irrigation is used with such a crop, the pipes and emitters are spaced evenly. Sequencing valves make it possible to water different areas in sequence. Tensiometers measure the moisture in the soil so that the delivery of water can be scheduled efficiently. [After K. Shoji, "Drip Irrigation." Copyright © 1977 by Scientific American, Inc. All rights reserved.]

Determining irrigation requirements

Irrigation, which is capable of yielding enormous benefits, may be wasteful and even harmful if it is applied incorrectly. The determination of when to irrigate and how much water to supply in relation to economic crop production is a crucial part of irrigation technology. In a sense this requires that an accounting system be established that will determine whether the available soil moisture will meet crop needs in spite of losses through evaporation, transpiration, runoff, and percolation. The net deficiency not compensated for by precipitation may be made up by irrigation. The timing of application in relation to the growth stages of the plant can be critical. In beans, for example, moisture stress during flowering and pod formation seriously depresses yield as a result of flower abscission and ovule abortion.

Table 17–1   Extent of drip irrigation in 1975.

Country	Area in 1975		Crops
	Hectares	Acres	
Argentina	152	375	Vineyard plants
Australia	17,337	42,840	Deciduous fruit, vegetables, avocados, bananas, citrus fruit, nursery plants, nuts
Brazil	368	909	Deciduous fruit, vegetables, nursery plants, nuts
Canada	405	1,000	Deciduous fruit, vegetables, nursery plants, ornamentals
Cyprus	405	1,000	Vegetables, vines, bananas, citrus fruit
Costa Rica	10	25	Oil palms (experimental)
France	1,243	3,050	Deciduous fruit, flowers, greenhouse plants
Honduras	2	5	Bananas, palms
India	20	50	Deciduous fruit, vegetables
Iran	800	2,000	Citrus fruit, vegetables
Israel	10,118	25,000	Deciduous fruit, vegetables, avocados, bananas, citrus fruit, greenhouse plants, ornamentals
Japan	4	10	Citrus fruit, grapes, vegetables
Mexico	5,595	13,826	Deciduous fruit, citrus fruit, row crops, olives
Martinique	405	1,000	Bananas, avocados, sugarcane
New Zealand	1,085	2,682	Deciduous fruit, vegetables, citrus fruit, avocados, greenhouse plants
Panama	0.8	2	Deciduous fruit, vegetables
Puerto Rico	73	181	Mangoes, cherries, plantains
South Africa	7,285	18,000	Deciduous fruit, vegetables, greenhouse flowers
United Kingdom	1,619	4,000	Deciduous fruit, vegetables, greenhouse flowers
United States	54,115	133,717	Deciduous fruit, vegetables, citrus fruit, sugarcane, nuts, greenhouse plants, nursery plants, ornamentals, row crops, avocados
West Africa (Senegal)	405	1,000	Vegetables

Source: K. Shoji, "Drip Irrigation," Copyright © 1977 by Scientific American, Inc. All rights reserved. Data from C. D. Gustafson, University of California, San Diego.

There are two approaches to determining irrigation requirements. One consists in the actual measurement of soil moisture and subsequent calculation of available moisture. The other consists in calculating water availability from meteorological data.

Irrigation is most effective if applied before soil moisture becomes limiting to plant. As a rule of thumb, water should be applied when 60 percent of the available water (that is, 60 percent of the water between field capacity and the wilting point) in the root zone has been depleted or when the water tension in the zone of rapid water removal exceeds 4 atmospheres. To determine when irrigation is required, it is first necessary to determine how much water is already in the soil. The most commonly used techniques are discussed in Chapter 13. From these measurements, the availability of water to plants can be determined, as can the amount of irrigation water required in the root zone.

Meteorological and climatological data offer powerful aids for measuring the amount of available water; the procedure involves the calculation of **consumptive use**—the water lost by evaporation and transpiration—which is probably the best index of irrigation requirements. Consumptive use varies with a great number of factors: temperature, hours of sunshine, humidity, wind, amount of plant cover, the stage of plant development, and available moisture.

High rates of water consumption are associated with a high percentage of plant cover and with hot, dry, windy conditions. Crops differ in their water requirement largely in relation to their ground covering ability and, to some extent, to the depth of their roots, even though deep-rooted plants obtain the greatest percentage of their moisture from the upper part of the root zone. Because optimum plant performance depends upon an adequate, uninterrupted supply of water, the plant's peak requirements must be considered. Crops have the highest water requirements in the fruiting or seed-forming periods.

The amount of available moisture may be much less than the total precipitation. Owing to soil evaporation and the slow rate of infiltration, showers that provide less than 0.6 centimeter (about 0.25 inch) during the hot summer days may contribute very little to the available soil moisture. On the other hand, a high proportion of water from heavy precipitation may be lost through runoff. The effectiveness of precipitation therefore depends upon the rate at which it falls, as well as upon the amount, in relation to temperature and the absorptive capacity of the soil.

It has been possible to determine satisfactory indices of consumptive use for particular areas by using monthly averages of mean temperature and hours of sunshine. Empirically derived constants are available for adjusting these values for different crops. Irrigation requirements can then be estimated by a bookkeeping procedure. The moisture content of the soil at the end of any time period is computed by sub-

tracting the amount of moisture lost by evapotranspiration from the amount present at the beginning, and adding the amount of available precipitation during this period. The generalized equation for this relation is:

$$\text{Final soil moisture} = \text{Soil moisture (at start)}$$
$$- \text{Evapotranspiration} + \text{Precipitation} - \text{Runoff}$$

If the maximum possible soil-moisture storage capacity is known, it is a simple matter to compute the soil moisture content at any given time and thus determine whether irrigation is necessary. Computer programs, which require no more than the accumulation of daily temperature and rainfall data, have been developed to make daily estimates of the status of soil moisture.

Because all of the irrigation water applied is not available to the crop, the amount applied must be based on **irrigation efficiency,** the percentage of irrigation water applied in relation to the amount that becomes available for consumptive use. Water should be applied to bring the soil up to field capacity at a depth commensurate with the bulk of the feeder-root system. The rate must be consistent with the absorptive properties of the soil. The frequency of irrigation is determined in part by soil texture—the coarser the soil, the more frequent must be the irrigation (Fig. 17-18). The amount of water that can be efficiently utilized is also related to the fertility of the soil. The maximum benefits of irrigation depend upon a readily available nitrogen supply.

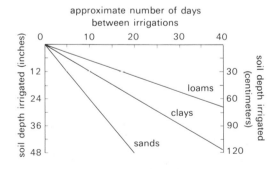

Figure 17-18 Irrigation intervals as influenced by soil texture and the depth in the root zone at which water is used at the rate of 2.5 centimeters (about 1 inch) per week. [Courtesy USDA.]

DRAINAGE

In 1823, Professor W. H. Keating accompanied the Long Expedition to the source of the Minnesota River and described the land south and west of what is now Chicago.

> From Chicago to a place where we forded the Des Plaines River the country presents a low, flat, and swampy prairie, very thickly covered with high grass, aquatic plants, and among others the wild rice. The latter occurs principally in places which are under water; its blades floating on the surface of the fluid like those of the young domestic plant. The whole of this tract is overflowed during the spring, and canoes pass in every direction across the prairie.

Today, there are more miles of drainage ditches in this region than of public highways. The productive fields, numerous good roads, and excellent farm buildings in this area testify to the advantages of

drainage when soil and climatic conditions are favorable for crop production. Drainage has made possible the tillage of approximately 12 million hectares (about 30 million acres) of land in the United States, and has increased the productivity of about 16 million additional hectares (about 40 million acres).

Water tables that are too high for crop production may exist for more than one reason. Impervious layers may prevent percolation of the water down through the soil. Soils along rivers may be flooded yearly during seasons that are critical to crop production. The surfaces of soils near the ocean, which may have an elevation of no more than a meter or so above sea level, may be within a few centimeters of the groundwater level. Even some irrigated lands must be drained periodically to flush out salts that have accumulated in the surface soil layers (Fig. 17-19). The valley of the Tigris and Euphrates rivers in ancient Mesopotamia has returned to desert, largely

Figure 17-19   Salt damage to a carrot crop in California's Coachella Valley. Salt is a by-product of irrigation in deserts; the naturally salty land does not receive enough rain to wash it away. Land can be reclaimed by leaching to flush the salts from the soil. [Photograph by E. E. Hertzog courtesy U.S. Bureau of Reclamation.]

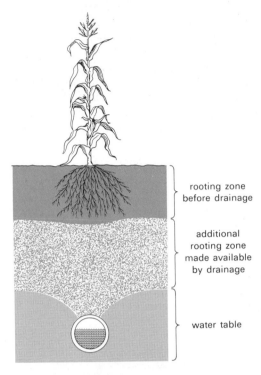

Figure 17-20 Additional volume can be added to the root zone by proper drainage.

because of the accumulation of salts in the plow layer, the zone of surface soil that is subject to disturbance by cultivating equipment.

The function of drainage is to lower the water table, enabling roots to make use of deeper soil horizons and the nutrients they contain (Fig. 17-20). This results in many benefits. Crop yields and quality are increased, and more kinds of crops can be grown on the same land. Soil aeration, microbial action, and fertilizer efficiency are all improved by drainage. At the same time, drought prevention is better, and the growing and harvesting seasons are both extended.

History of drainage

The history of land drainage goes back to the ancient Egyptians and Babylonians. The real beginning of the practice is obscured in antiquity, but extensive embankments and irrigation works for the control of

water in the interest of agriculture were in use more than 2000 years ago. In the second century B.C.E., Cato gave the first specific directions for drainage. Ditches nearly 1 meter (about 3 feet) wide at the top, about 1.25 meters (4 feet) deep, and about 30 centimeters (1 foot) wide at the bottom were dug. Stones were laid in the bottom of the ditches, but in the absence of stone, green willow rods or even twigs tied in bundles were substituted. This method was used for 200 years without change.

In the first century A.D., open ditches were used in heavy soils, but where the soil was loose the tops of the ditches were usually covered. Palladius and Columella, authorities on Roman drainage methods, set standards that were used without improvement for more than 1000 years.

In the 1600s the next advance took place. Trenches 1.25 meters (4 feet) deep were dug and half filled with stone. The excavated soil was then packed solidly above them (Fig. 17-21A). The obvious disadvantage of the stone-filled ditch was that water had to flow in the spaces between the stones, which were quickly filled with silt. As a result, various attempts were made to form a continuous opening along the bottom of the trench, one of which was to lay a row of ordinary clayware roofing tiles along the bottom to form a tunnel, before adding stones. This was so successful that flat tiles were made especially for the purpose. Illustrations in books written at that time show that they were used singly, or in pairs to make a triangular opening, or sometimes even in groups of four to form a diamond-shaped opening (Fig. 17-21B–D). To this day, clay field-drain pipes are still called "tiles."

It was only a matter of time before the flat clay slab was bent into a U shape before firing to provide a more stable opening along the bottom of the trench (Fig. 17-21E). Unfortunately, it was not completely satisfactory because the edges of the tile tended to sink into the bottom of the trench. Various methods were tried to combat this problem. Of all the methods used, a U-shaped tile placed on a flat sole tile was the most satisfactory and was generally adopted (Fig. 17-21F).

With the advent of the extrusion press, it became possible to make the U shape and the flat sole in one

Figure 17-21   A graphic history of drainage. [Courtesy Certain-Teed/Daymond Co.]

piece, producing a **D**-shaped tile. In time the center opening was made oval or round because of improved hydraulic properties, even though the flat base remained for some time before the introduction of the modern round pipe (Fig. 17-21G–I). Clay tiles are still used extensively today and modern methods of handling have helped keep them economical for land drainage. Some manufacturers produce tile with holes along their length to allow water to flow into them more freely (Fig. 17-21J).

Plastic drainage tubing (Fig. 17-21K) was introduced into the United States from Europe in 1966. Its light weight, great strength, and easy handling characteristics, coupled with new installation methods, have completely revolutionized the industry. A 75-meter coil (about 250 feet) of 10-centimeter (4-inch) tubing weighs only 34 kilograms (75 pounds), so that a tractor-trailer can carry 5400 meters of tubing (about 18,000 feet). Easy to install, it is not affected by soil acids, by alkalis, or even by severe freezing. In North America today, about 80 percent of the drainage systems being installed consist of plastic tubing, in part because it can be used in many kinds of drainage patterns (Fig. 17-22).

Surface drainage

Several kinds of ditches are used to drain the surfaces of waterlogged soils. **Field drains,** which are merely ditches as much as 0.5 meter (about 18 inches) deep, collect water from the row crops and land surfaces and carry it to a larger drain. They are usually about 120 meters (about 400 feet) apart on heavy soils, but they may be as much as 300 meters (about 1000 feet) apart on lighter soils. **Lateral drains,** also called **collecting ditches,** receive water from field

double main pattern

herringbone pattern

parallel pattern

Figure 17-22 Plastic drainage pipes can be used to create many drainage patterns. [Courtesy Certain-Teed/Daymond Co.]

random pattern with the main following a natural gully

random pattern—used in rolling terrain and frequently seen in irrigated areas

drains. They are usually about 30 centimeters (1 foot) deeper than the field drains and may have side slopes with a low angle so that they can be easily crossed with motorized farm equipment. **Main drains** collect water from laterals and carry it to drainage outlets.

Topography usually dictates the type of drainage system used. When topography is uniform, it is desirable to use a **parallel system** of surface drains. Field drains are usually spaced uniformly and are at right angles to collecting drains.

So-called **random systems** are used when topography is not uniform and when only some areas require drainage. Each problem area must be considered separately and enough field drains installed to provide the necessary outlets; collecting drains must be so placed that they do not interfere with the overall farming plan.

A number of surface-drainage systems have been adopted in various parts of the world. In Louisiana, one known simply as the **sugarcane system** has been used for sugarcane for more than a century. Split, parallel ditches are dug in the direction of the slope at intervals of 30 to 75 meters (about 100 to 250 feet). Quarter drains carry water from crop rows to split ditches and then to cross ditches that empty into canals. **Rice drainage systems** function to remove water in excess of that required to flood the paddy fields.

Bedding, one of the earliest forms of drainage, consists of turning furrows to form high ridges that are flanked by ditches to carry away excess water. Early English systems used beds 1 meter or more (3 to 4 feet) higher than the bottoms of drainage furrows. Along the coastal plain of the southeastern United States, beds are commonly constructed that are only 10–20 centimeters (about 4–8 inches) high.

Interception drains are sometimes used to catch unwanted water from a higher slope and route it to a place of disposal. Interception ditches are also used to prevent water from flowing across fields during periods of heavy precipitation.

Subsurface drainage

Where soil conditions are suitable, drains installed below ground are used to remove excess water from high-value agricultural land. Buried below the surface, they occupy no arable area, as would open ditches. Such drains also eliminate open-ditch maintenance problems, such as weed control. Water enters the drains through holes in the pipes or cracks between the individual tiles and is carried away by gravity. Although expensive to install, tile drains have proven their value in many places.

Drains can be installed in many patterns, including **parallel line, herringbone, random,** and **double main** (see Fig. 17-22). If necessary, tiles can be laid to intercept seepage water that follows the upper surface of an impervious subsoil. **Lateral drains** remove free water, **submains** collect water from a group of laterals, and **main drains** carry water to the outlets.

Drains must be installed at depths great enough to lower the water table by the desired amount. They should be large enough and laid in rows close enough together to remove excess water in a specified length of time and to provide the required amount of protection from waterlogging. Ten-centimeter (4-inch) tubes are satisfactory in mineral soils with small amounts of fine silt and sand, whereas 12.5-centimeter (5-inch) drains are the minimum generally recommended for heavy clay soils. Fifteen-centimeter (six-inch) drains must be used for peat and muck soils.

It may be necessary to lay the drains on boards if the soil in the trench bottom is not stable. In some cases, better alignment and easier placement can be achieved by connecting tiles with short sheet metal tubes slipped over their ends.

The installation of plastic drains is rapid and efficient. One worker, using a tractor-drawn tubing-burying machine and a laser-controlled leveling device, can install several miles of drain in a single day (Fig. 17-23).

Mole drains are temporary drains made by pulling a bullet-shaped tube through the soil by tractor at the desired drainage depth, usually no less than 75 centimeters (about 30 inches). Since they are no more

Figure 17-23 It is now possible for one person, using a laser-controlled leveling device, to install several miles of plastic drain tube in a day. [Courtesy Ditch Witch.]

than holes in the ground, they must be renewed every two or three years, although in Great Britain and New Zealand mole drains have lasted as long as 10 to 15 years. They can be installed at less than one-tenth the cost of tile drainage.

Drainage of forest soils

Forest drainage presents problems because of the magnitude of individual projects, sometimes involving as much as 18,000 hectares (about 45,000 acres). Fortunately, tree species grown in drained areas are usually fairly resistant to flooding, and there is usually no need for extremely rapid rates of water removal.

Ditches for forest drainage are usually large (Fig. 17-24). Even the **shallow drains** are 60–90 centimeters (about 2–3 feet) deep, and the **lateral drains** that they empty into must be at least twice that deep. **Main drains,** which carry the water to **drainage canals,** are sometimes 3 meters (about 10 feet) deep and require the use of draglines for their construction. In dense forest lands, however, rows of dynamite are sometimes exploded to open up as much as a quarter of a mile of ditch at one time.

Figure 17-24   Drainage can greatly increase forest productivity. The cleared strip adjacent to the drainage ditch can be used as a logging road and for access to fight fires.

Many problems are likely to accompany the drainage of large areas of organic forest soils. When drained, they frequently became dry and are highly subject to fire. Subsurface fires have been known to smolder and burn in the organic soils of the coastal plain of the southeastern United States for many months, and occasionally for more than a year. Such fires are extremely harmful because they are almost impossible to extinguish. They burn out the root systems of plants and make regeneration by sprouting impossible. Because of the rapid rate of decomposition of dried organic soils, it is usually necessary to maintain water-table control with a series of **water gates** rather than to permit complete, free gravitational drainage. Roads built on spoil banks dug from drainage ditches may subside because of the action of microorganisms in destroying the drained organic matter.

Drainage may also have severe environmental effects. When a forest is drained, for example, a series of events may be triggered that may be catastrophic on a local basis, changing the entire character of wetlands (Fig. 17-25). The drainage of swamps for crop production has aroused much opposition from environmentalists in recent years, and will no doubt continue to do so for years to come.

WATER CONSERVATION

Water is of national concern. For a nation to prosper, an abundant source of high-quality water must be available for agricultural and industrial use as well as for human consumption, sanitation, and recreation. The misuse of water resources leads to alternate flood and drought, and to pollution—problems that affect us all.

Water conservation implies the best possible use of water. It may involve large programs to control flooding, to develop hydroelectric power, and to facilitate navigation. Now that cloud seeding seems to be a practical reality when atmospheric conditions are favorable, every phase of the hydrologic cycle is subject to manipulation. Such programs require national effort. Water conservation also involves the control of water resources on a smaller scale. It must be part of the water management of every individual enterprise.

Watersheds

Water supply is closely related to economic development, because water resources must keep pace with other resources. The basic unit of water supply for any particular area is the **watershed**, or catchment area. A watershed is merely a topographic unit or drainage basin from which all precipitation finds its way to a single outlet, called its confluence. A watershed may be no more than a few acres in size, or many minor watersheds may be considered together as part of a vast overall watershed that may occupy an area of thousands of square kilometers. Consider the watershed for the Mississippi River. This includes the entire central part of the United States from the Rockies to the Appalachians. On the other hand, very small watersheds may be of extreme importance for local communities. The problems of maintaining a constant flow of water through the confluence of the watershed are nearly the same, whether they are

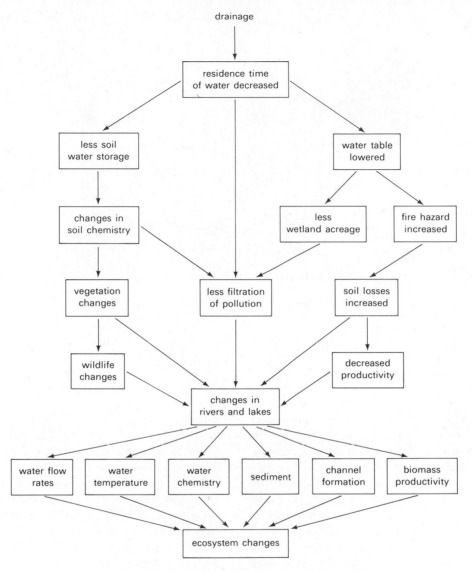

Figure 17-25   The environmental impact of drainage may extend far beyond its immediate effect of decreasing the residence time of water.

large or small. They differ enormously, however, in magnitude, and the problems are compounded by political and jurisdictional disputes.

Water leaves the watershed as a liquid in several different ways. After a period of precipitation, **surface runoff** is the first to be lost. Surface runoff is water that never enters the soil but runs directly over the surface into streams, as it would over a rock. Surface runoff is greatest immediately following storms of high intensity and after soil has become saturated with water to the extent that additional water cannot infiltrate the soil (Fig. 17-26).

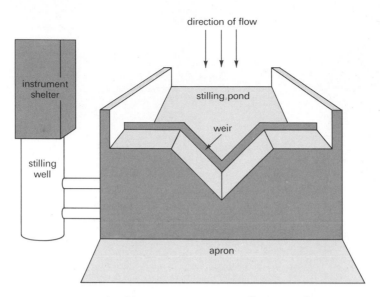

Figure 17-26   A damlike structure permits collection and measurement of surface water that issues from a watershed as it passes over the V-shaped weir. The well stills the water, the level of which is continuously recorded. [From F. H. Bormann and G. E. Likens, "The Nutrient Cycles of an Ecosystem." Copyright © 1970 by Scientific American, Inc. All rights reserved.]

Water may also leave the watershed as **subsurface runoff,** which infiltrates and percolates through the soil pores. Subsurface runoff is largely free (gravitational) water and does not take as long to flow into streams as **retained water.** Water in **retention storage** is poorly understood, but a number of studies indicate that it is held in capillary pores and may be "pulled" downward through soil pores due to tension from below. Water in retention determines the regularity of flow of springs, which in turn allow small tributary creeks to flow the year round.

The plant life of a watershed is extremely important in maintaining the flow of water from streams, for the rate of transpiration largely determines how much water is lost into the atmosphere. There is a common belief that forests increase the water yields of watersheds. This is far from the truth, however, because trees have extensive and deep root systems that can withdraw great amounts of water from the soil that cannot be subsequently used for consumption or industrial purposes. Forests are beneficial,

however, because they not only provide a more even year-round supply but stabilize soil, provide such products as wood, and make good recreation areas. Most important is the fact that the quality of water from forested watersheds is much better than water that flows from denuded areas. Water from forested areas is less difficult to purify for human consumption. In many parts of the world, there must be two water systems: one to supply highly purified water for consumption, and the other to supply less purified water for washing, irrigation, and industrial purposes.

Forests also improve the seasonal distribution of water from the watershed. If one wishes to obtain as much water as possible from a particular area of land, one can do no better than to pave it with asphalt or concrete. By doing this, one stops all losses due to transpiration, and a very high percentage of the water from precipitation will flow through the mouth of the watershed. Of course, this is frequently a highly undesirable condition, because under these circumstances the water must be stored in impoundments and reservoirs until it is needed. In some parts of the world there is no alternative to impoundment. Forests, on the other hand, permit the gradual infiltration of water through the soil and allow it to seep through the subterranean channels at a slow rate, so that it is released in the streams over long periods. It is apparent, therefore, that watersheds are not only areas for trapping water, but are also important for filtration and as storage repositories.

Conserving precipitation

Preventing immediate runoff and subsequent loss of precipitation is a major means of increasing available water. In agricultural practice precipitation is conserved by terraces constructed on the contour (Fig. 17-27). Terracing slows down surface runoff and thereby increases the time for infiltration. Although terracing may be designed primarily to prevent erosion, it also facilitates water storage (Fig. 17-28).

Figure 17-27   Contour plowing and terracing for planting citrus to conserve moisture and prevent erosion. [Photograph by W. A. Garnett.]

Figure 17-28 These steep backslope terraces are coupled with underground tile to provide a maximum reduction in soil and water loss. On a similar, unterraced, straight-row field, approximately 10 times as much water runoff and 100 times as much soil loss occurred. [Courtesy Cal J. Ward, Concord, Nebraska.]

Since precipitation does not infiltrate compacted soil, rapid runoff may result in flash floods. The regulation of grazing in rangelands, where the trampling of soil by animals causes compaction, is truly a water-conservation practice in addition to its function in conserving soil and vegetation.

Rotation cropping, the practice of alternating crops, conserves water as well as soil. Usually a row crop grown for one year is followed by one or more years of grasses and legumes, during which more organic matter is incorporated into the soil and infiltration of water is increased. In a typical study carried out in Wisconsin, continuous cropping of corn permitted 19 percent of the rainfall to run off, whereas when clover was rotated with corn, only 4 percent of the rainfall was lost.

Spectacular increases in watershed yields have been obtained by appropriate timber harvesting practices. In North Carolina, a 13.4-hectare (33-acre) watershed was completely cut over, and in the first year water yield at the confluence was increased by 65 percent. In another watershed, where timber was cut only along the stream channels, the water yield increased by 10 percent. In Colorado, streamflow was increased by 31 percent by the complete harvest (clear-cutting) of merchantable lodgepole pines, and 19 percent by harvesting methods in which one-third of the trees were left to provide seed for future crops. When trees are removed, not only are transpirational losses decreased but less snow is intercepted by leaves and evaporated or sublimated back into the atmosphere. Snow storage on the forest floor is increased and more water infiltrates the soil when the snow melts. Snow accumulation can also be increased along slopes by strategic placement of fences or vegetation.

Water storage

The first dam of record was built in Egypt 5000 years ago. It was 108 meters (about 355 feet) long and 12 meters (about 40 feet) high and stored water for drinking and irrigation. Today, dams are commonplace throughout the world (Fig. 17-29), and there are thousands of dams in the United States alone. Some impound water surfaces of less than 0.5

Figure 17-29   Water impoundment for surface irrigation in a mountainous area of Minas Gerais, Brazil. [Courtesy Hans Mann.]

hectare and may supply water to irrigate only a single field. But Lake Mead, in Nevada and Arizona, has a surface area of 592.6 square kilometers (228.8 square miles), stores 39.8 billion cubic meters (32,300,000 acre-feet) of water, and supplies irrigation water to several states.

Because evaporation from the surface of a body of water can result in great losses of water (particularly serious when the reservoir is in an arid region with low atmospheric humidity), various materials are used to prevent evaporation. These compounds, called surfactants, form a monomolecular layer over the surface of the water and prevent a large part of the evaporative loss. Cetyl alcohol was found to reduce evaporation from a pond by as much as 50 percent on a calm day, although savings are smaller in windy weather because the monomolecular layer is constantly being broken. Hexadecanol, docosanol, and other long-chain fatty alcohols have also been found useful, but they are subject to bacterial decomposition and must be replenished periodically. Surfactants may also be helpful when applied to soil surfaces. They increase the efficiency with which plants use water, decrease evaporation losses from the soil, and increase the rate of water infiltration into the soil. Hexadecanol, when used as a soil treatment, has been found to remain effective for as long as 14 months.

One of the most recent developments in water storage is the use of excavated basins lined with plastic to prevent seepage losses (Fig. 17-30). Since the plastic sheets may be sealed at any edge, there is virtually no limit to the size of such storage basins. Some storage basins of this sort are as large as 50 hectares (nearly 125 acres).

Refilling underground storage reservoirs

A large part of the water used for irrigation comes from natural underground storage reservoirs. In California, for example, 60 percent of the irrigation water is from underground sources. There are several methods by which groundwater supplies may be replenished.

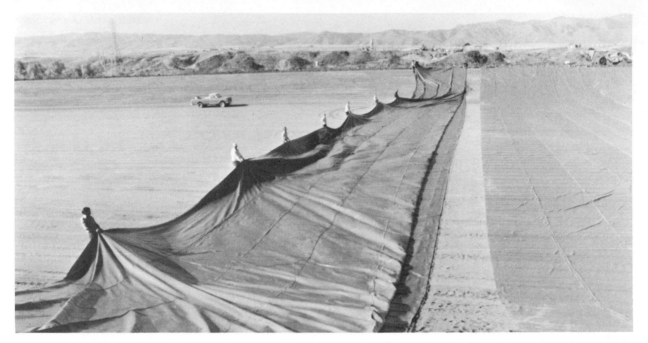

Figure 17-30 An accordion-pleated Watersaver membrane is spread to full panel width for final positioning in a water-storage basin. [Courtesy Watersaver Company, Inc.]

Figure 17-31 By using canals and ditches to "rearrange" the flow of water and make it available to as large an area as possible, erosion is avoided and the best use is made of water resources. This technique is called water-spreading. [Courtesy U.S. Department of the Interior, Bureau of Land Management.]

Water spreading is the flooding of long narrow basins that surround production farm land (Fig. 17-31). The artificially perched water tables spread out under the unflooded crop land, giving the water a better chance to infiltrate. This technique can also be used to recharge groundwater supplies for pumping.

Flooding allows water to move slowly over the land in a thin sheet and to percolate down to the water-bearing strata (Fig. 17-32). Dikes are sometimes used to retain water until it moves into the soil.

In **shaft recharge,** an appropriate number of shafts or wells are drilled to permit the water to move directly to the natural underground reservoir, bypassing obstacles. Deep wells are also used in this manner, but they must be deep enough to reach good water-conducting strata (Fig. 17-33).

Replenishment irrigation is the application of excess water to crop land during winter by means of the normal irrigation system. An extra benefit of this system is the control of weeds, most of which do not survive flooding.

Fallow

The term "fallow" refers to the practice of keeping the land bare of vegetation by cultivation, and also to the land kept bare by this practice. Its main benefit is the conservation of soil moisture. Fallow, more than any other single practice, has made intensive agriculture possible on the Great Plains of the United States and on the Canadian prairie. This ancient practice is most advantageously used in regions with winter rainfall and dry summers. Fallow is recommended where total annual precipitation is less than 40 centimeters (about 16 inches) in the northern United States and in Canada and where it is less than 50 centimeters (about 20 inches) in more southerly and warm regions.

The fallow period between crops is generally 21 months. After harvesting in the late summer or fall, weeds are destroyed by a light surface cultivation. Winter precipitation recharges the soil. A crop is not grown the following year, but the soil is lightly

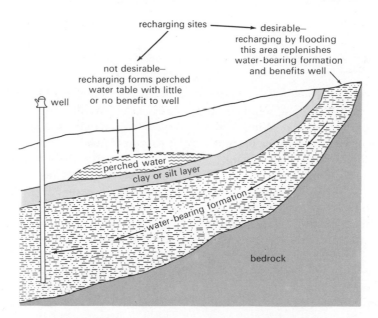

Figure 17-32 Stratigraphy governs the replenishment of groundwater supplies by flooding. The water-bearing formation (aquifer) can be recharged only if there is not an impermeable layer above it. [Courtesy USDA.]

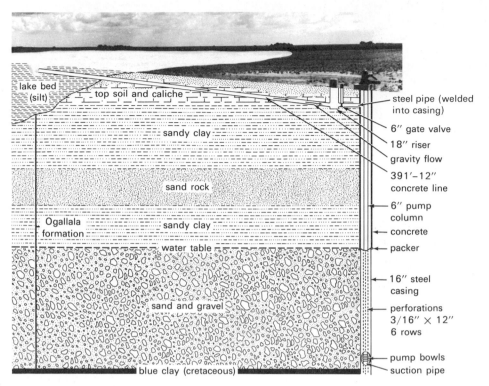

Figure 17-33   Artificial recharge wells may be a mile or more deep. [Courtesy USDA.]

worked to kill weeds during the growing season. After the second winter's precipitation, soil moisture is abundant enough to produce a crop. Winter cultivation, by breaking the surface crust, prevents runoff of water and wind erosion of the smooth surface.

Fallow also permits the accumulation of nitrate. In the fall after a crop is harvested, the stubble, roots, and other organic residues are plowed into the soil. Decomposition of the organic matter releases nitrates and other minerals that remain in the soil if they are not utilized by weeds. Although bacterial denitrification accounts for some nitrogen loss, it is much less than the loss due to weeds.

Herbicides are sometimes used to keep fields in a fallow condition, but supplemental cultivation is often necessary. Chemical weed control is especially useful when erosion is a problem, and it is desirable

to leave as much stubble and organic matter on the surface as possible. Stubble standing upright permits snow to drop to the soil rather than be swept away or to accumulate at windbreaks (Fig. 17-34).

The excessive use of fallow as a dry-farming technique has led to many problems: wind and water erosion, the eventual depletion of soil minerals, and a decrease in the content of organic matter. Nevertheless, the technique permits the growth of crops in many areas that would otherwise be unproductive. In the future, the increased use of fertilizers and herbicides will offset some of the disadvantages of fallow.

An effective 3-year crop rotation scheme, in which reduced tillage and herbicides are coordinated, was developed for use in the central Great Plains as a direct result of research carried out by G. Wicks and D. Smika. Called "ecofallow," it incorporates two fal-

Figure 17-34 Stubble mulching is a common practice in dry-farming areas. Stubble helps snow to form an even blanket rather than to pile up in drifts. Consequently, when the snow melts, the water is distributed evenly over the field. [Courtesy USDA.]

40–50 percent greater than those obtained using conventional systems. The conservation benefits of ecofallow were found to be largely due to wheat residue and stubble. They include:

1. increased movement of precipitation into the soil,
2. less water erosion,
3. less runoff,
4. less sedimentation,
5. less evaporation,
6. greater soil-moisture storage,
7. less blowoff of snow,
8. reduced wind velocity at the soil surface, and
9. reduced soil temperatures.

It is hoped that experimentation will show that this list of benefits can be duplicated by other combinations of crops.

Using dew

The Nabataeans, who lived in the center of the Negev Desert from about the fourth century B.C.E., made agricultural use of dew. Flint stones were

low periods, the only cultivation being the seed-planting operation. Figure 17-35 illustrates the plan of action. Using ecofallow, wheat yields were 8–10 percent greater and corn and sorghum yields were

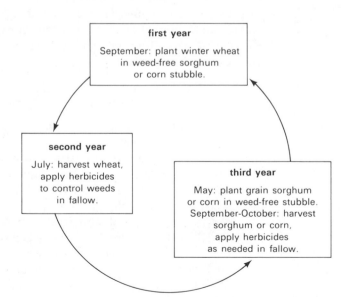

Figure 17-35 The ecofallow method. This method was developed for producing wheat and sorghum in the central Great Plains.

stacked in such a manner that deposited dew dripped to the ground below to help water a single plant, such as a grapevine. Piles of such stones can still be seen on the hillsides.

At least some dew is directly absorbed by leaf surfaces. Young leaves are less cutinized and absorb more water than older leaves. Dew water runs off leaves to the ground and increases soil moisture. It may be possible to increase the deposition of this moisture by breeding for modified plant forms that permit the accumulated droplets of dew to run down the stem.

In an experiment conducted in Ohio, it was determined that an amount of moisture equivalent to 23.1 centimeters (9.1 inches) of rain was gained by the formation of dew over a 6-month period. This was 20 percent of the total water available to the plants! In California, ponderosa pine can survive on soils below the wilting point, apparently because of the ability of the trees to absorb and utilize dew.

Cultural practices

Choice of crops

The amount of water used by a crop during a growing season may determine where it is grown (Table 17-2). If fruit trees require 53 centimeters (21 inches)

Table 17-2 Water requirements of irrigated crops in relation to the requirements of alfalfa.

Crop	Relative seasonal consumptive use in relation to alfalfa
Alfalfa	1.00
Pasture	0.90
Clover	0.90
Sugar beet	0.82
Citrus (fallow)	0.70
Corn	0.65
Potato (winter)	0.50
Tomato	0.48
Bean	0.42
Sorghum	0.38
Potato (spring)	0.30

of water during the growing season, then they obviously cannot be grown in a region that normally receives only 28 centimeters (11 inches) of rainfall unless artificial irrigation can provide the other 25 centimeters (10 inches).

Differences in the sensitivity of crops to moisture stress can be utilized to make maximum use of available moisture. When subjected to identical drought conditions, corn may die while sorghum survives by "waiting" for the drought to be relieved. Rows of specialized cells in the leaves of many grasses lose turgor as a result of a reduction in moisture and the leaves respond by rolling up and curling inward, thus concealing their stomata and reducing transpiration. Corn has relatively rigid leaf tissues and is less able to withstand extreme moisture stresses than sorghum. Sorghum is thus a more logical choice than corn as a cash crop in a semiarid area.

Weed control

Although weeds have many undesirable effects (see Chapter 18), perhaps the most harmful is their use of soil moisture that would otherwise be available to crop plants. It has been estimated that natural vegetation may consume twice as much water as field crops. Cattails growing in the hot sun along a mile of irrigation canal may use as much water as 3.25 hectares (about 8 acres) of alfalfa. Weed control becomes an integral part of water conservation, especially where water may be limited.

Structural modifications and mulches

A variety of artificial measures for increasing agricultural yield are presently in use in different parts of the world. One example of a structural modification is an integrated system of both rigid and air-supported greenhouses, in which the environments are strictly controlled, that provide power, water, and food for the people of Abu Dhabi, an Arabian Gulf shiekdom (Fig. 17-36). Fresh water is produced from sea water by desalting facilities, which harness the heat from small engine-driven generators. Heating and cooling is achieved with sea water. Humidity is very high in the greenhouses, so that moisture is

Figure 17-36   Abu Dhabians and their camels stroll by controlled-environment greenhouses, which use seawater for heating, cooling, and irrigation. [Courtesy Carl O. Hodges.]

conserved. The yield per unit area is often higher than that from field production. Plant-growing structures are frequently shaded or cooled with a water spray to decrease transpirational losses.

Soil mulches also have a great effect in conserving soil moisture (see Chapter 11). Experimental studies in Alabama have shown that the use of black plastic mulch increases yields of corn, partly as a result of the increased availability of soil moisture.

Limiting transpiration

Crop growers have long sought a method of reducing water losses by plants. Some losses due to transpiration seem to be essential in helping to cool leaves, but at least a part of the water transpired merely moves through a relatively open system at rates dictated by solar radiation. One method of reducing such losses would be to apply a substance to leaves that would cause their stomata to close. Most chemicals that do

this also kill the leaves, but compounds have recently been discovered that temporarily close stomatal apertures without harming the plant. For example, the glyceryl ester of decenylsuccinic acid has proven effective, as have related substances, but field methods of application must still be developed. In the future it may be possible to decrease irrigation requirements considerably by adding such chemicals via sprinkler irrigation systems or by applying them aerially to large acreages of unirrigated crops.

WATER MANAGEMENT AND WATER RESOURCES

All life depends upon water. It usually comes as a shock to find that plants require water in such large quantities, mainly because they lose so much in transpiration. An herbaceous plant will transpire perhaps twice its weight in water each day. A single corn plant in full leaf may require 30 liters (about 8

atmosphere

annual precipitation 100

stream flow 29

forests and browse vegetation 16

nonirrigated land 71

farm crop and pasture 23

noneconomic vegetation 32

concentrated supply

mined from aquifers .1

stream flow not withdrawn 22

27

ocean reservoir

irrigation

3.35

industrial use

3.35

1.35

municipal

3.30

.6

.55

consumed loss

2

.05

.05

Figure 17-37 The hydrologic cycle of the United States is typical of highly developed nations. Irrigation and manufacturing uses require about the same amount of water. However, most of the water applied through irrigation returns to the atmosphere through evapotranspiration. The numbers in the figure refer to the relative quantities of water involved in the various parts of the hydrologic cycle. [From R. Revelle, "Water." Copyright © 1963 by Scientific American, Inc. All rights reserved.]

gallons) of water each week. A stand of hardwood may transpire 28,000 to 37,000 liters of water per hectare per day (about 3000 to 4000 gallons per acre per day). A single tree requires perhaps 990 liters of water to produce 1 kilogram of wood (about 250 gallons per pound of wood). The figures required to compute the water requirements of the world's crops are astronomical.

Water must also be regarded as an economic commodity. Costs vary from about $25 to $50 for 30 centimeters of depth per hectare ($10 to $20 per acre-foot) in the United States, and this is cheap for the volume acquired; but the cost of water can be a major expense in crop production because so much is required. For example, if all of the water necessary for crop production had to be obtained at this price, the water needed to produce an economic forest crop would cost $87 to $100 per hectare ($35 to $40 per acre) each year! Given the cash value of forest crops, it would be impossible to make a profit. Many field crops would be uneconomical for the same reason.

Human biological processes require approximately 6 liters (about 1.5 gallons) of water per person per day. People in primitive cultures can satisfy their needs for minimum comfort with 10 to 20 liters (about 3 to 5 gallons) per day, for such uses as washing and cooking, and may have trouble obtaining even this small amount. But people living in a modern technological society require tremendous amounts of water. In 1960 the average daily use of water for all purposes in the United States was about 5675 liters (about 1500 gallons) per capita. An estimate for 1980 is 8700 liters (about 2300 gallons) per day for every man, woman, and child. Truly, the degree to which water is consumed is one measure of the technological development of a nation (Fig. 17-37).

The power of nations is, in part, reflected in their ability to supply water to their peoples. Wherever the Roman empire flourished, it saw to the building and maintenance of vast systems of aqueducts for irrigation and human use. Ruins of these structures can be seen today in Spain, North Africa, and elsewhere. The water-distribution systems failed, or were destroyed, along with the empire that built them.

Water is becoming increasingly precious. Cities, states, and even nations are contesting each other, militarily and legally, for the privilege of using water from particular watersheds, lakes, rivers, and even wells. Water management is of urgent necessity to world health and world peace.

In this chapter, we have dealt with some of the contemporary water-management techniques used principally by crop growers. Managing water for crop growth differs from managing water for human and industrial use, largely in cost.

There is a fantastically large amount of water in the world. Much of it is potable without further treatment, and a large proportion of it is immediately usable for industrial purposes. The problem is that it is frequently not present where it is most needed. Transportation costs depend on the size of the operation. It has been suggested that, on a large scale, it may be cheaper to bring water from Baffin Bay to Tierra del Fuego than to convert seawater to fresh water at the tip of South America. The same applies to waste disposal.

A partial answer to the problem of obtaining water for crop irrigation lies in multiple use (Fig. 17-38). Australia provides an excellent example. The great inland desert of that continent is bordered on the east by the Snowy Mountains. Rivers discharge water from these mountains to the southeast into Bass Strait. The Snowy Mountain–Tumut project has reversed three of the rivers, sending them west through tunnels into artificial lakes. The water then drops, in several stages, through powerhouses. At the end of its route, the water is used to stabilize the flow of the Murray and Murrumbidgee rivers and to supply water to their arid valleys.

The ice cap of the Arctic Ocean is probably the largest body of unused fresh water in the northern part of North America. It might not be an unsurmountable task to build tunnels and aqueducts to take a part of its melt as far south as Mexico.

Cloud seeding is in its infancy. The technological possibilities in this area are limited only by the political, social, and legal consequences that will emerge when the technique is completely developed.

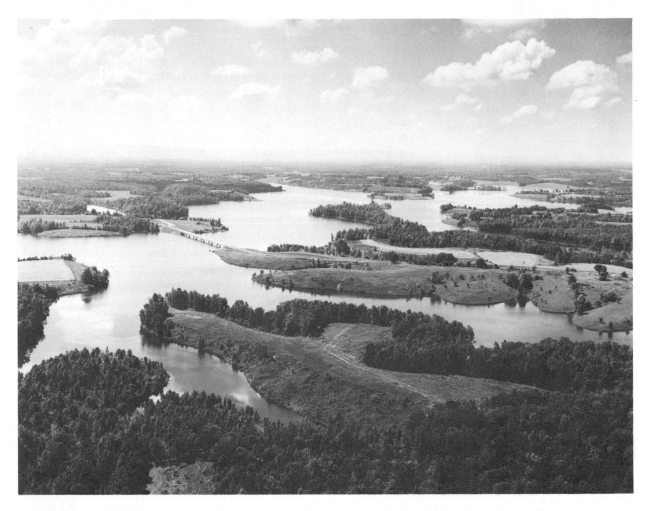

Figure 17-38   Tims Ford Dam and Reservoir in central Tennessee. This is the first major Tennessee Valley Authority project planned specifically to provide opportunities for recreation, hunting, fishing, water supply, water-quality control, and shoreline development, as well as to serve the traditional purposes of flood control and power generation. [Courtesy Tennessee Valley Authority.]

Solar stills, desalinization of seawater, steam stills powered by nuclear energy, electrodialysis, compression distillation, freezing, high-pressure diffusion, and other techniques can be used to convert "impure" water to usable water. Further, brackish water that may kill some crops may safely be used to irrigate others that have a greater tolerance of salts. Crop growers of the future may need to use all of these techniques.

Selected references

Clark, C., 1966. *The Economics of Irrigation*. New York: Pergamon Press. (The applied economics of irrigation, including costs and potential economic returns.)

Colman, E. A., 1953. *Vegetation and Watershed Management*. New York: Ronald Press. (A widely used reference on watershed management.)

Frederick, K. D., 1975. *Water Management and Agricul-*

tural Development: A Case Study of the Cuyo Region of Argentina. Baltimore: The Johns Hopkins University Press. (A case history of water problems and development in the Cuyo region of Argentina.)

Hagan, R. M., H. R. Haise, and T. W. Edminster (editors), 1967. *Irrigation of Agricultural Lands* (Agronomy Series No. 11). Madison, Wisc.: American Society of Agronomy. (A monograph that treats irrigation from the standpoints of engineering and crop production.)

Humlum, J., 1969. *Water Development and Water Planning in the Southwestern United States*. Copenhagen: Kulturgeografisk Institute. (A popular presentation of water problems and water development in an arid part of the United States.)

Israelsen, O. W., and V. E. Hansen, 1962. *Irrigation Principles and Practices*. New York: Wiley. (A good reference book on irrigation practices.)

Koelzer, V. A. (editor), 1974. *Water Policies on U.S. Irrigated Agriculture*. Fort Collins, Colo.: Environmental Resources Center, Colorado State University. (A synthesis of policies of water resource problems and the economic and social consequences of water development.)

La Mer, V. K., 1962. *Retardation of Evaporation by Monolayers*. New York: Academic Press. (Papers on the use of surfactants to reduce evaporation.)

Luthin, J. N. (editor), 1957. *Drainage of Agricultural Lands*. Madison, Wisc.: American Society of Agronomy. (This is an excellent series of essays on practically all of the many aspects of agricultural drainage.)

Roe, H. B., 1950. *Moisture Requirements in Agriculture*. New York: McGraw-Hill. (A comprehensive manual of agricultural water management, both theoretical and practical.)

U.S. Department of Agriculture, 1955. *Water* (USDA Yearbook, 1955). (A thorough coverage of water in relation to agriculture.)

U.S. Department of Agriculture, Soil Conservation Service, 1973. *Drainage of Agricultural Land*. Port Washington, N.Y.: Water Information Center. (Deals with all types of drainage systems and drainage problems in the temperate zone, including the use of dikes and pumps.)

Wright Rain Ltd., 1956. *Sprinkler Irrigation Manual*. Ringwood, Hants., Eng.: Wright Rain. (A compilation by English workers.)

18
Crop Hazards

Worms have destroyed half the wheat, and the hippopotami have eaten the rest; there are swarms of rats in the fields, the grasshoppers alight there, the cattle devour, the little birds pilfer; and if the farmer loses sight for an instant of what remains on the ground, it is carried off by robbers. . . .
—a scribe of ancient Egypt

The successful cultivation of crops involves more than the proper combination of environmental factors. There are many natural shocks to which plants, like people, fall heir. These include pests and various nonbiological factors. Of the latter, heat, frost, drought, flood, and nutrient deficiencies have already been discussed, and only fire, hail, and wind damage will be reviewed in this chapter.

PLANT PESTS

Plant pests include all life forms destructive to plants—the whole biological spectrum is represented. Pests may be grouped under the broad terms *pathogen, predator,* and *weed;* their respective injurious effects are disease, damage, and competition. The toll exacted from world agriculture by plant pests is heavy; field and storage losses are estimated to be as high as 35 percent. In the United States the yearly loss in farm crops alone is about $8 billion. Moreover, the annual cost of pest control to farmers in the United States is $450 million for insects, $250 million for diseases, and $2.5 billion for weeds! Some pests have such awesome destructive power that they are capable of destroying entire crops. Others, though less spectacular, are equally damaging in the long run; they continually erode our abundance while their damage passes more or less unnoticed. Everyone eventually shares the cost of these losses.

Injury

Broadly defined, the term **disease**—literally, "not at ease"—refers to any injurious abnormality. As generally used, however, the term refers to the injurious effects caused by some infectious agent (the **pathogen**) living in intimate association with its host for extended periods of time. This restrictive definition of disease applies nicely to the detrimental effects in plants caused by many microorganisms. The relation in which one living organism derives nourishment at the expense of the other is known as **parasitism.** Some parasites have evolved to such an extent as to be **obligate:** they can survive only on the living tissue of their host. Pathogens, with rare exceptions, are injurious parasites; all parasites, however, are not necessarily injurious. Organisms that live on dead organic matter are called **saprophytes.** Many pathogens exist largely as saprophytes. The host range of some pathogens may be extremely large and may include many different species; that of others may be very specific, limited even to particular cultivars within a species.

The harmful effects that insects and larger animals such as rodents or birds have on plants are often referred to as damage rather than disease because of the kind of association and the time factor, and the term **predator** is used rather than pathogen. However, the distinctions between predator and pathogen, disease and damage, are arbitrary. Many insects, for example, live in intimate association with certain plants for extended periods and are pathogens under the strictest definition. The distinction between predation and parasitism is somewhat clearer when confined to the animal kingdom. Predators are usually larger than their prey, which they destroy outright. Parasites are usually smaller than their host, at whose expense they obtain room and board. Predators live on capital, parasites on income. In the United States, the study of the adverse effects of viruses, fungi, and bacteria on plants has been pre-empted by the field of plant pathology, and the effects are considered diseases. The harmful effects of insects and mites are a specialty of entomology and are referred to as damage. (Jurisdiction over nematodes is claimed by both, rodents and birds by neither.) Any malaise of unknown origin is also the concern of the pathologist. Disorders shown to be caused by climatic or environmental conditions, the so-called physiological disorders, become the concern of physiologists. In the ensuing discussion no special distinction will be made between pathogen and predator or between disease and damage.

The injurious effects of weeds are usually due to direct competition for essential growth factors. A few, however, are also parasitic—for example, dodder, mistletoe, and witchweed—and some secrete or leave a harmful residue in the soil.

The association between plants and pathogens and predators is part of the interrelation and interdependence of living things. The resulting injury should not be regarded as intrinsically evil or particularly unusual. To do so would be to give an egocentric and anthropomorphic interpretation of natural phenomena—an interpretation that is patently false.

Symptoms

The visible responses of plants to injury are known as **symptoms,** which, together with evidence of the particular agent (**signs**), permit diagnosis (Fig. 18-1). Although the basic plant responses to injury are limited, the many changes that take place in different tissues and organs often permit accurate diagnosis from symptoms alone. Death (necrosis) may affect the entire plant, or it may be limited to specific organs (leaves, branches, flowers, fruits) or to small areas that appear as spots or holes. Decline may be gradual and incomplete. For example, chloroplast breakdown appears as yellowing or mottling but does not necessarily result in the death of the plant. Another plant response is reduction in growth rate, which may be general or restricted, and may result in stunting, dwarfing, or incomplete differentiation. In contrast, injury may increase growth abnormally to produce enlargements of organs, tissues, or cells, or tumorlike protuberances called galls. Wilting or loss of turgidity may be due to mechanical obstruction of water flow in the vascular tissue, root injury, or increased water loss.

Disease cycle

The **disease cycle** consists of the sequential changes that take place in both pathogen and host. It involves the life history, the morphological forms and stages, of host and pathogen. There is considerable variation among the life histories of pathogens. For example, some lower organisms do not have a sexual stage but exist as a single continuously repeating form (many bacteria), and others reproduce asexually by means of spores (some fungi). Complex life histories involve diverse developmental stages, including a number of asexual cycles and even alternations of hosts in different stages.

The life history is often adjusted to the seasonal cycle. In temperate areas, nonmigratory organisms

bitter rot

cedar rust

Botryosphaeria rot

black pox

fly speck and sooty blotch

blotch

blister spot

powdery mildew

scab

Figure 18-1   Symptoms of fungal diseases in apple. [Courtesy Purdue University.]

either die during the winter or survive the winter in a stage of inactivity or dormancy. The life history of an organism may involve a number of different morphological stages. The sequence of stages initiated with the advent of the growing season are called primary cycles, and those initiated subsequently are called secondary cycles. In areas where the change in seasons is not pronounced, a secondary cycle may continue uninterrupted.

A disease cycle may be divided into a **pathogenic** phase, in which the organism remains associated with the living tissues of the plant, and an **independent** phase, in which it may be saprophytic, dormant, or pathogenic to another plant. The relative lengths of these two phases vary greatly. For example, the independent phase of plant viruses usually lasts only while they are being transported from plant to plant via insects, mites, or fungi. Some fungi, such as *Fusarium*, live saprophytically as a part of the soil flora and become pathogenic with the introduction of a suitable host plant.

The pathogenic phase may be subdivided into a

number of stages. **Inoculation** consists of the transfer of some form of the pathogen (the inoculum) to the plant. Pathogens may travel short distances under their own power (for example, as free-swimming spores) or they may be transported long distances by wind or water or by some vector, principally insects and people. The **incubation** period begins when the pathogen enters the plant and ends when the plant reacts—the beginning of the infection stage. Inoculation and incubation are often important stages during which to attempt to control the pathogen, for by the time **infection** is manifested the pathogen is least vulnerable to control measures and serious damage has been done.

It is impossible to generalize on the relation between the disease cycle and the life history of a particular pathogen or predator. For example, the primary cycle of the fungus *Venturia inaequalis*, which causes the apple-scab disease, begins with sexual spores (ascospores) produced by the fungus, after it has overwintered saprophytically on dead apple leaves. These are carried by air currents to young apple leaves, where they germinate to begin the primary infection. The secondary cycles begin when asexual spores (conidia) are produced during primary infection. The life history of this pathogen takes a whole year, and includes several asexual cycles (Fig. 18-2). In contrast, the life history of the codling moth, a serious insect pest of several fruits, particularly apple, may recur two or three times within a single year, depending upon the latitude. The larval stage is the damaging one.

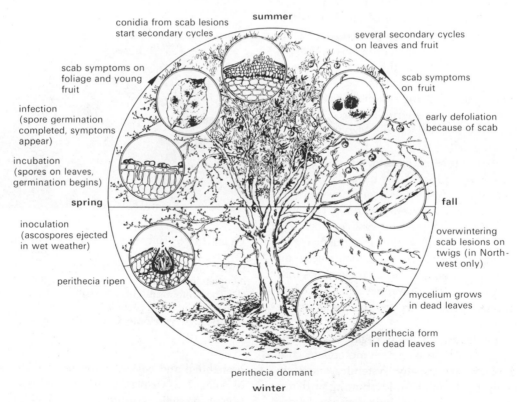

Figure 18-2 The disease cycle of apple scab, caused by *Venturia inaequalis*. [From L. Pyenson, *Elements of Plant Protection*, Wiley, 1951.]

Pathogen and predator

Viruses

Viruses are small infectious particles consisting of a core of nucleic acid surrounded by a protein sheath (Fig. 18-3). Their small size (in the neighborhood of 100 nanometers, or about 4 millionths of an inch) makes them visible only with the electron microscope. Virus particles may be short or long rigid rods, flexible threads, or "spherical" (actually 20-sided) forms. Viruses are obligate parasites; they reproduce only in the living cells of the host, but they may be removed from the organism and remain capable of causing infection. Some viruses remain active in extracted plant juices for many months; the tobacco-mosaic virus remains active for years in dried material. Some viruses can be crystallized and still cause the disease when reintroduced into the plant.

The question of whether viruses are living is a matter of terminology or philosophy. If the living system is defined as a self-duplicating entity capable of reconstructing itself from different component parts, then viruses can be considered alive. But viruses are not complete, living systems, because they are unable to generate the energy required for their multiplication and must exploit the enzyme systems of infected, living cells.

The genetic structure of certain viruses that infect bacteria (bacteriophages) has been determined. These viruses mutate and show genetic recombination that is akin to sexual reproduction. Thus the development of various strains is possible. Their genetic material is arranged in linear sequence, and linkage maps (see Chapter 19) can be constructed on the basis of the recombination of characteristics that affect the expression of symptoms. The complete RNA structures of only a very few bacterial viruses are known.

Viruses cause diseases of many organisms, including many important crop plants. Symptoms of viral diseases of plants are usually systemic, and they range from death, through mild stunting with reduced quality and yield, to no obvious effects. Symptoms may include small lesions, a spotted or mosaic pattern of green and yellowish areas, yellowing, stunting, leaf-curling and edge-crinkling, excessive

A

B

Figure 18-3   Plant viruses. (A) Portions of tobacco-mosaic virus particles. The spiral arrangement of the protein sheath is visible in this electron micrograph. The complete virus particle measures 15 × 300 nanometers. (B) The protein subunits are visible in this micrograph of turnip-mosaic virus particles 300 nanometers in diameter. [Courtesy R. W. Horne, Cambridge, England.]

branching, and color-striping or total disappearance of color. The virus is usually named by the description of the major symptom coupled with the name of the host plant in which the disease was first described. Thus tobacco-mosaic virus retains its name when it infects tomatoes. Viruses are identified by a combination of symptoms, serological tests (with the use of antibodies from warm-blooded animals, usually rabbits), electron microscopy, and chemical analysis.

The effect of viruses on the plant depends upon the sensitivity of the host cultivar and the strain of the virus. Beet curly-top, a virus disease that also affects tomatoes, results in the quick death of the plant. Some viruses produce no obvious symptoms but still cause considerable economic loss by reducing yields and performance. Infection by a combination of viruses generally complicates and increases the severity of symptoms.

Viruses are commonly insect-transmitted, many of them by aphids. A few viruses are soil-borne, and some of these are transmitted by nematodes or by fungi. Some viruses are transmitted from one generation to the next through the seed. But even when a virus is seed-transmitted, only a fraction of the seedlings become infected. Some highly infectious mosaic viruses may be transmitted by people touching healthy leaves after they have touched infected leaves. In laboratory assays, viruses are transmitted by rubbing infectious sap on healthy susceptible leaves. Viruses are commonly transmitted through a graft union, and some pass from plant to plant through dodder, a parasitic plant.

The vegetative propagation of virus-infected plants also propagates the virus. The only satisfactory way of maintaining virus-free stocks of vegetatively propagated plants is by the perpetuation of plants that are determined to be free of the virus. Maintenance of virus-free plant stock is achieved by isolation, roguing of infected plants, and control of insect vectors. As a rule, once a plant contains a virus, little can be done to get rid of it, although heat treatment of the plant is an effective way of inactivating some viruses, and excised shoot tips (particularly if they are from rapidly growing plants) may be free of virus.

Viroids

A number of virus-like particles implicated as the cause of plant diseases have resisted numerous attempts at identification by conventional techniques. Recently the infectious materials causing potato spindle-tuber disease and citrus-exocortis disease have been shown to be low-molecular-weight RNA (about one-tenth the size of the RNA in the smallest known virus). These infectious agents appear to represent a new class of pathogens now referred to as **viroids,** or **infectious** or **pathogenic RNA.** Viroids have also been implicated as the causal agent of at least one animal disease, scrapie, a kind of dysentery suffered by sheep.

Mycoplasmas

For many years a number of strange organisms have been known to exist that have the property of passing through filters that trap bacteria (as do viruses), but that live on inorganic media (as do bacteria). In the late 1960s the discovery was made that these organisms, now called **mycoplasmas,** are also responsible for causing plant diseases. Mycoplasmas are now implicated in a number of plant diseases formerly thought to be caused by viruses, particularly in a group called "yellows" diseases, which are principally spread by leafhoppers. Examples of plant diseases known to be caused by mycoplasmas include aster yellows, mulberry dwarf disease, and potato witches' broom.

The size of mycoplasmas is variable. When grown in culture they produce long filaments resembling fungal hyphae, hence their name, which means "fungal-form." However, the filaments may break up into very small, round cells known as **elementary bodies,** ranging in size from 125 to 250 nanometers. These are the particles that pass through bacterial filters. From the elementary bodies mycoplasmas may form either branching filaments or "large cells" (500 nanometers) that approach the dimensions of bacteria.

Mycoplasmas, unlike viruses, are definitely living organisms. They contain enzyme catalysts and energy systems, as well as genetic information in the form of DNA and RNA. Although surrounded by a membrane, they have no exterior cell wall as do most bacteria. (Some biologists consider mycoplasmas to be simply small bacteria that lack cell walls.) The lack of a cell wall explains their apparent resistance to penicillin, which works on the bacterial cell wall. However, mycoplasmas are inhibited by tetracycline

A

B   C

Figure 18-4 Fire blight, which is caused by the bacterium *Erwinia amylovora*, is one of the most devastating diseases of pear and apple. (A) Profuse production of bacterial strands on stem and leaf petioles of 'Bartlett' pear. The aerial strands may be smooth or beaded. (B) The strands contain large numbers of pathogenic bacteria embedded in a matrix. The strands are disseminated by wind, and they appear to play an important role in fire-blight epidemics. (C) Fire-blight symptoms on a 4-year-old pear tree. [A and B, courtesy Harry L. Keil, USDA; C, courtesy Purdue University.]

compounds (such as chlorotetracycline), and many suspected viral diseases have been proved, through the application of these chemicals, to be caused by mycoplasmas.

Bacteria

Many crop diseases are caused by bacteria, primitive one-celled organisms that (if the mycoplasmas are considered among them) are the smallest of living things. About 180 species of bacteria belonging to seven genera are known to be plant pathogens. Bacteria are able to enter plants only through natural openings, such as the stomata and lenticels, or through wounds. Insects are important agents in the transmission of bacteria.

One of the most serious bacterial diseases—and the one first shown to be caused by bacteria—is fire blight, a disease of apple and pear caused by *Erwinia amylovora* (Fig. 18-4). Insects may disseminate these bacteria, which enter the plant either through the nectar-producing glands in the flower, causing death of the blossom (blossom blight), or through shoot terminals (shoot blight). The bacterial pathogen is also spread from infected parts to other parts of the tree by rain. The organism survives the winter in older bark lesions called cankers. The disease is extremely serious in pear and has confined commercial production in the United States to the Pacific states and the Great Lakes area. Even in these locations, however, careful control measures must be used, such as constant removal of blighted wood, the use of antibiotic sprays, and the avoidance of rapid, succulent growth.

Symptoms of bacterial diseases include the death of tissues (angular leafspot of cotton) and the formation of galls (crown gall of apples). The soft rots common in storage diseases are associated with pectin-dissolving enzymes produced by bacteria. The wilting symptom of some bacterial diseases (Stewart's wilt of corn) are a result of vascular disturbances, specifically, a "plugging" of the vascular system by masses of the bacteria.

Fungi

Fungi cause most plant diseases. Except for certain primitive types, they are characterized by a branching, threadlike (mycelial) growth. Fungi, which were once considered to be plants but are now commonly placed in a separate kingdom of their own, do not have chlorophyll and hence depend ultimately on green plants for their food. They may be saprophytic or parasitic, and many are both. Fungi reproduce by spores, which may be mitotically (asexually) or meiotically (sexually) produced. The life cycles of most fungi are quite complicated, comprising many different stages and forms.

Fungi fall into three, large, well-defined classes (or divisions, in some systems of classification): **Phycomycetes, Ascomycetes,** and **Basidiomycetes.** Because the distinctions between these groups are made in part on the basis of differences in the reproductive organs of fungi when they are in the sexual (perfect) stage of their life cycle, those whose sexual stage is now known (and many apparently never form a sexual stage) are sometimes lumped in an arbitrary group called **Fungi imperfecti.**

Members of the class Phycomycetes are primitive fungi whose charactistic feature is the absence of mycelial cross walls. Because many of the phycomycetes are aquatic and rather like algae, they are popularly known as water molds. The spores of aquatic forms are motile. Examples of common crop diseases caused by phycomycetes are downy mildews (many host plants including grape and various grasses), root rots and damping-off (caused by a number of species in the genera *Pythium* and *Phytophthora*), and the infamous late blight of potato and tomato, caused by *Phytophthora infestans*.

Members of the class Ascomycetes are distinguished by a specialized sac (**ascus**) that contains the sexual spores. Comprising more than 25,000 species of great variety, the ascomycetes range from pure saprophytes to obligate parasites. They are responsible for the largest number of plant diseases: powdery mildews on cereals, fruits, and many other crops; Dutch elm disease; ergot of grains; brown rot of peaches and plums; black spot of roses; and apple scab. Also among the ascomycetes, however, are truffles, which, being highly prized as food, are imported into the United States from Europe to grace the table of the gourmet.

Members of the class Basidiomycetes, the "higher fungi," produce a specialized sexual spore-forming structure called a **basidium.** They include the smuts, rusts, and fleshy fungi, such as mushrooms, and are almost as numerous as the ascomycetes. Rusts (red) and smuts (black) are so named because of the color of spores produced on the host plant. They are among the most destructive scourges of grasses, especially cereal crops. The rust fungi are all obligately parasitic. Some require two unrelated plants for the completion of their extremely complex life history. For example, *Puccinia graminis*, which causes wheat rust, requires barberry for its sexual stage; *Cronartium ribicola*, which causes the white-pine blister rust, alternates between the pine and species of *Ribes* (wild and cultivated currant and gooseberry); stages of the cedar–apple rust organism, *Gymnosporangium juniperi-virginianae*, infect both hosts. The smut fungi are facultative saprophytes; they may subsist as saprophytes but they do not ordinarily complete their life history away from the host. Smuts cause diseases of corn, oats, barley, wheat, and on-

A

B

C

D

Figure 18-5 Symptoms of fungal diseases of corn: (A) Stalk rot caused by *Diplodia maydis*. (B) Stalk rot caused by *Pythium aphanidermatum*. (C) Crazy top caused by *Sclerophthora macrospora*. (D) Smut caused by *Ustilago maydis*. [Courtesy A. J. Ullstrup.]

ions. The fleshy fungi are important as wood rotters, and some species are cultivated. A strain of the common field mushroom, *Agaricus campestris*, is the favorite of the United States' mushroom industry.

Fungi spread in a number of ways. Spores are commonly produced in enormous numbers and are dispersed by air currents, water, and insects. Mycelia growing saprophytically spread soil-borne fungi. Most fungi penetrate directly into plant tissue, although wounds and natural openings greatly facilitate entry for some.

Symptoms of fungal diseases are extremely varied. The diseases may affect the entire plant or just parts of it. Plant-pathogenic fungi generally cause localized injury, although some are responsible for vascular wilts. The presence of visible (to the unaided eye) forms of the fungus, such as mycelial growth, masses of spores, or fruiting bodies, is usually an integral part of disease expression and often serves as an invaluable aid in identification (Fig. 18-5).

Nematodes

Nematodes (Fig. 18-6) are roundworms of the class Nematoda of the phylum Aschelminthes (not to be confused with the segmented earthworms of the phylum Annelida). Although the majority of nematodes are saprophytes and represent a significant portion of the soil fauna, many species are parasitic on plants and animals, including people. The importance of nematodes in causing plant injury is becoming increasingly apparent.

Most nematodes are tiny and many are actually microscopic. The majority of nematodes that attack plants generally feed on the roots. They may feed superficially or be partially or completely embedded in the root tissues. A few species feed on aerial portions (for example, the seed-gall nematode of bentgrass). Some are quite specialized and attack only a few species of plants; others have a wide host range.

There are two common types of plant-attacking nematodes. Those belonging to the genus *Meloidogyne* (root-knot nematodes) produce gall-like overgrowths on roots. These readily observable symptoms allow positive identification of nematode injury and are used as a criterion in the inspection of planting

Figure 18-6 Nematodes of two genera representing different morphological forms. (A) The swollen female of *Meloidogyne* (root-knot nematode). (B) The wormlike female of *Tylenchorhynchus* (stunt nematode). [Photograph by H. H. Lyon courtesy W. F. Mai.]

stock. The infected roots afford access to bacterial and fungal pathogens. The root tissues, especially the vascular tissues, become disorganized; giant multinucleate cells are often formed. The above-ground symptoms appear as excessive wilting and weak yellow growth similar to that resulting from drought injury. Other plant-parasitic nematodes, such as those of the genus *Pratylenchus* (meadow nematodes), do not form galls. Both the root damage and the above-ground symptoms resemble those of root rot. Owing to the diagnostic difficulty, meadow nematode may be easily transported in infected planting stock.

Insects and mites

Insects and mites are members of the phylum Arthropoda. The arthropods (literally "jointed legged") include the invertebrate animals having an external skeleton; paired, jointed limbs; and a segmented, bilaterally symmetrical body. The Arthropoda is a very large group, containing about three-quarters of all known animal species, of which 90 percent belong to one class, Insecta, the true insects. The insects have three pairs of legs, and almost all are winged at a certain stage (although the wings of some are rudimentary or degenerate). They have three body regions—head, thorax, and abdomen. Almost 700,000 species of insects are known, and the estimated total number of species in the world ranges from 2 to 10 million. Mites, the other important group of plant pests in the phylum Arthropoda, belong to the class Arachnida, which also includes the spiders. Arachnids have four pairs of legs, are wingless, and, unlike the insects, do not have a separate head. A brief classification of the Arthropoda, showing the main orders of insects that attack plants, is given in Table 18-1. In the following discussion, the term "insects," when used nontechnically, will include the mites.

Our struggle against crop-damaging insects is a continuous one. The battle lines are not clearly drawn, for we are aided by the intense competition between insect species; predatory insects must be regarded as beneficial. Furthermore, if we were rid of all insects, we might be worse off, since many are necessary for the pollination of crop plants.

Insects have many built-in advantages. These include their small size, which makes them difficult to find; their ability to fly; their extremely rapid rate of reproduction; and their specialized structural adaptions, which enable different species to exist in practically any location and damage any plant species. The division of the life cycle of certain insects into separate stages (**complete metamorphosis**) is of tremendous advantage to them in that it permits specialized structural adaptation for feeding and reproduction. The life cycle of an insect with complete metamorphosis consists of four stages: (1) egg; (2) larva, a feeding stage; (3) pupa, a quiescent stage in which the larva is transformed into the adult form;

Table 18–1   Partial classification of arthropods, including major orders of insects that attack plants.

Class and order	Examples	Typical mouth parts	
		Larvae	Adults
Chilopoda	Centipedes		
Diplopoda	Millipedes		
Crustacea	Crayfish, lobsters, crabs		
Arachnida	Scorpions, mites, ticks, spiders		
Insecta	Insects		
Orders with gradual metamorphosis			
Orthoptera	Grasshoppers, crickets		Chewing
Thysanoptera	Thrips		Rasping-sucking
Hemiptera	True bugs		Piercing-sucking
Orders with complete metamorphosis			
Hymenoptera	Bees, ants, wasps	Chewing	Chewing-lapping
Coleoptera	Beetles	Chewing	Chewing
Lepidoptera	Butterflies, moths	Chewing	Syphoning
Diptera	Two-winged flies and mosquitos	Chewing	Piercing-sucking Sponging

and (4) adult, the reproductive stage. Some insects, such as grasshoppers, have a gradual or **incomplete metamorphosis,** in which the physical changes from egg to adult are gradual. The intermediate stages are known as nymphs.

The larval stage of insects is often the most injurious to plants; thus, many of the descriptive names given to insect pests (tomato hornworm, apple maggot) refer to this stage. There is generally no obvious resemblance to the adult stage (for example, caterpillar and butterfly). The ordinary terms for the larval stages of the insect orders are not specific. The term **maggot** generally refers to the larval stage of a fly (order Diptera) that tunnels within leaves. **Caterpillar** is the larval stage of a butterfly or moth (order Lepidoptera). The name **grub** is used with reference to some of the large soil-borne larvae of beetles (order Coleoptera), although the name is loosely applied to any soil-borne larva. The larvae of "click" beetles are called **wireworms.** The larval and adult forms of beetles that infest grains and seeds are called **weevils.** The name **slug** is applied to any slimy larva, specifically to larvae of bees, ants, and wasps (order Hymenoptera). (Shell-less molluscs that are related to snails are also known as slugs.) Moth or beetle larvae that tunnel within roots or stems are known as **borers.** The term "worm" is only incorrectly used to refer to insect larvae.

Insects and mites injure plants directly in their attempts to secure food and shelter and indirectly by spreading plant pathogens. Moreover, their decomposed bodies or excreta contaminate plant products. They exhibit two basic feeding patterns, each associated with specialized mouth parts. **Chewing insects** tear, bite, or "lap" portions of the plants; **sucking insects** pierce or rasp the plant and suck or sponge up the sap, or often passively allow the pressure of the sap to fill them up.

Chewing insects, adults or larvae, may eat their way through the plant, riddling it with holes and tunnels. Leaves may be skeletonized by insects that do not eat the tougher, vascular portions; the entire plant may be devoured by the less selective. Injury caused by chewing insects that feed externally is seldom confused with anything else (Fig. 18-7). It is

A

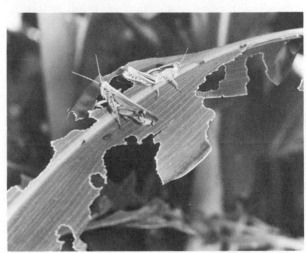

B

Figure 18-7   Insect damage. (A) Cabbage looper on cabbage. (B) Grasshopper on corn. [Courtesy J. C. Allen & Son.]

more difficult to diagnose damage caused by chewing insects that girdle the plant or those that feed internally or on roots. The internal feeders gain entrance to the plant from eggs deposited in the plant tissue or by eating their way in soon after being hatched on or near the surface of the plant. The symptoms of the peach-tree borer are typical of the symptoms caused by internal feeders—namely, weakened, devitalized growth of the tree and yellowing of the foliage. A gummy exudate may be observed where the borer has wounded the trunk. Peach or plum trees infested with borers often die in a few years.

Injury by sucking insects (aphids, scale insects, leafhoppers, and the plant bugs) results in distinctly different symptoms, the most common being curling, stunting, and deforming of plant parts, usually stem terminals. Spotting, yellowing, and a glazed appearance of the leaves are also common symptoms. Red spider mites have sucking mouth parts and produce similar symptoms as well as webs. Mites are difficult to see without a magnifying glass, and by the time webs are formed damage may be extremely severe.

Overgrowths and galls, the result of abnormal growth of plant tissue, are symptoms either of the feeding of chewing or sucking insects or merely of the presence of insect eggs deposited in the tissue. Such galls, suggestive of cancerous growth, may be quite elaborate and structured. Some appear to do little harm; others are obviously quite injurious.

Birds and rodents

Among the vertebrates, or backboned animals, birds and rodents are the greatest crop pests. Rats and mice are serious field pests and often are responsible for large storage losses. In India the diet of an estimated 2.4 billion rats, over four times the human population, has been calculated at 26 million tons of grain, almost one-fifth of India's present production and enough to feed 135 million people. The prevention of this loss would transform India from a nation with a food deficit to one with a food surplus. Mice can be destructive orchard pests. They feed on living trees, which may be completely girdled in winter and early spring. Apple trees are particularly susceptible to injury, and unless prompt action is taken by repair grafting, large trees may be killed outright.

Birds rival rodents as plant pests. Billions of weaver birds (red-billed quelea), one of the most numerous and destructive birds in the world, threaten grain production in Africa. Birds may be quite trou-

blesome to fruit crops, especially cherries and grapes; even a few pecked grapes permit the introduction of diseases and other pests, and in effect spoil the whole cluster.

Weeds

A weed may be defined as any undesirable plant. More typically the term refers to certain naturally occurring aggressive plants injurious to agriculture. Crop losses caused by weeds are usually a direct result of competition with weeds for light, water, and mineral nutrients, as well as the lowering of crop quality because of contamination of the grain by weed seeds. Unchecked noxious weeds tend to dominate crop plants. Weeds also cause crop losses indirectly by harboring insect pests. They present a safety hazard on roadsides and railroad rights-of-way, and they may clog irrigation ditches and streams. Some weeds adversely affect the health and comfort of people and livestock. A few are poisonous or irritating (poison ivy, white snake root), and the pollen of some (such as ragweed) is a source of misery to millions of allergy sufferers. Moreover, weeds are unsightly, as any homeowner will testify. The annual cost of weed control in the United States exceeds that of all other types of pest control combined.

The fierce competition offered by certain weeds is due to a combination of their high reproductive capacity and their vigorous, exuberant growth. Their destructive power is a function of sheer number. An average of 20,000 seeds per plant—varying from 250 (wild oat) to several millions (tumbling pigweed)—was reported in a study of 181 annual, biennial, and perennial weed species. The seeds of many are structurally adapted for dispersal by wind, water, or animals. For example, many weeds have hard seed coats and remain viable when passed through the digestive tract of animals. People have unwittingly become one of the chief disseminators of weed seeds through the shipment of crop seeds and plants. Seed numbers alone do not account for their effectiveness. Many crops are prolific seed producers, yet do not become really weedlike. The high reproductive capacity of weeds is due to their extended seed viability and delayed germination brought about by dormancy. The failure of weed seed to germinate may be due to natural (see Chapter 8) or induced dormancy. Induced dormancy is brought about by environmental conditions that prevent germination. Seeds buried deep in the soil by tillage may lack either sufficient oxygen or, in some plants, the light stimulus necessary for germination. The viability of weed seed buried in the soil is usually much greater than that of the seeds of most crop plants. In one famous "buried seed" experiment, the seeds of more than half of 20 species germinated after 25 years and showed some viability after 90 years. The combination of dormancy, extended seed viability, and high seed production complicates weed control. If weed seeds would all germinate at once, control might be accomplished by a rigorous and intensive eradication program. The high weed-seed populations of agricultural soils make weed control a continuing and integral part of crop culture.

Many weeds reproduce vegetatively as well as by seed. Some of the most pernicious weeds reproduce in this way: Johnson grass and quack grass by rhizomes; nut sedge by corms; wild morning-glory by roots; and wild garlic by bulblets. Vegetative reproduction by roots or underground stem modifications makes control by cultivation particularly difficult.

The competition between weeds and crops adversely affects both, but weeds often win out. The exact physiological basis of the growth advantages of weeds is not clear, but it includes rapid germination and seedling growth and a deeply penetrating yet fibrous root system. Furthermore, weeds often possess a natural resistance to many of the pests that plague crops. Hardiness is undoubtedly another factor that benefits weeds.

The luxuriant growth that is characteristic of weeds might be a trait that has been sacrificed in crop plants during the process of domestication, just as seed dormancy has been lost in most seed-propagated crops. A more reasonable explanation is that weeds must be uniquely adapted to their surroundings in order to survive, whereas crop plants may be grown in locations far removed from the areas to which they are specifically adapted. Weed species differ far more dramatically than crop species across the United States.

Plant-protection methods

The control of plant pests can be considered as an agricultural discipline called **plant protection.** A highly specialized and rapidly changing technology has been developed around the control of particular pests (microorganisms, nematodes, insects, weeds), and those who have participated in this development have organized themselves into groups, complete with society affiliations, special nomenclatures, and scholarly journals. Although the techniques of control vary with pest and crop, the basic approach is either to interfere with some stage in the life cycle of the particular pest or to protect the host plant. The most successful treatments are preventive rather than curative; they aim to deter injury rather than cure plants that are already affected. All methods depend upon an intimate knowledge of the biological agents, their natural history and ecology, and, especially, their specialized relation to the crop. Successful control often depends on the proper use of many different methods (Fig. 18-8). The present-day coordinated control of pests is referred to as **integrated pest management**.

Cultural methods

Cultural practices used to reduce the effective population of a pest include the elimination of diseased or infested plants or seeds (**roguing**), the cutting out of infected plant parts (**surgery**), or the removal of plant debris that may harbor the pest (**sanitation**). The reduction of pest populations may be achieved by alternately planting crops unacceptable to the pest (**rotation**). Weed populations can be reduced by rotations including well-adapted crops, such as alfalfa and silage corn, that may outcompete weeds. Often it is the kind of tillage used in the rotation that brings weeds under control. For example, cultivation of row crops may reduce the populations of grassy weeds in subsequent small-grain plantings.

Plants, unlike animals, do not have an antibody mechanism to help them resist disease and thus cannot be made immune by vaccines. Nevertheless, the physiology of the plant can be altered to affect the plant's ability either to resist invasion by the patho-

Figure 18-8 Onion-seed fields occupy land for 6 to 14 months, depending on whether they are seed-planted or bulb-planted, and the onions compete poorly with weeds. Onion-seed producers must contend with three weed populations: winter, spring, and summer annuals. Control is achieved by (1) early spring cultivation, (2) midspring cultivation with an application of a pre-emergence herbicide, and (3) an early-summer application of herbicide when fragile seed stalks prevent cultivation. (A) An unweeded onion-seed field. (B) A field with nearly complete weed control. [From P. J. Torell and D. F. Franklin, *HortScience* 8(5):419–420, 1973.]

gen or to overcome its deleterious effects. For example, the vascular wilt of maple, caused by the fungus *Verticillium alboatrum,* can be compensated for by applications of fertilizer, which will cause the plant literally to outgrow the pathogen. The reverse technique is utilized in fire blight of pears, in which infection and the growth of the bacterium causing the disease are extremely rapid in fast-growing succulent shoots. Slowing down the growth of the tree by eliminating excessive nitrogen fertilization or by avoiding extensive pruning is one means of control. Application of inorganic nutrients gives protection in some instances. For example, club root of cabbage appears to diminish in severity when the ratio of calcium to potassium in the soil is decreased.

Pest populations may be reduced by methods that render the environment unfavorable, including such diverse practices as draining land, flooding, pruning to reduce foliage density and to increase the rate of drying, and varying the temperature and humidity. Potato scab is commonly controlled by lowering soil pH.

Many serious plant pests are introduced from other areas. Although separation of the pathogen and the plant cannot be legislated, safeguards against the spread of pests between neighboring countries and states may be enforced by **quarantines**—restrictions governing the importation of certain plants or plant products. Many countries prohibit the shipment of uninspected plants.

Physical control

Physical methods may be used to protect the plant against intrusion or to eliminate the pest entirely. The erection of physical barriers to protect crops is among the oldest of treatments. These range from the traditional garden fence to selective screening, such as guards to protect tree trunks against mice and screen-houses to eliminate insects or birds. In Japan, individual fruits of many types are enclosed in bags to protect them from insect injury.

Traps have been used to control insects. Attractants of various types are used to lure them into lethal solutions. These include "black light" (ultraviolet light of wavelengths between 340 and 380 nanometers), various sounds, and chemicals that mimic sex attractants (pheromes). Trapping is recommended in some areas to reduce the population of such birds as the starling.

Physical techniques are obviously difficult to use against microorganisms. When their presence becomes known through symptoms it is difficult to destroy them physically without also damaging the crop. Perhaps the most successful method is the use of heat. Hot-water treatment is used for destroying seed-borne or plant-borne pathogens—for example, loose smut, a fungal disease of wheat; black rot, a bacterial disease of crucifers; and nematodes on dormant strawberry plants. Heat treatment has also been effective in inactivating some viruses. Steaming of potting soils is a common greenhouse practice for eliminating many soil pests, including fungi, nematodes, and weed seeds.

Physical control is still widely used for weed control. The pulling or grubbing of weeds is the simplest and oldest method. The hoe is still the basic farm implement in many parts of the world. Various mechanical devices have been developed to automate this process. The principle is to cut out, chop up, or cover up weeds and thereby destroy them. Weeds propagated by roots are difficult to control by cultivation and may actually be dispersed in this way.

Cultivation and tillage, the loosening or breaking up of the soil, are such widespread agricultural practices that benefits in addition to the control of weeds have been ascribed to them. Many experiments indicate, however, that the primary benefit of cultivation is weed control. The other advantages, such as increased soil aeration, or conservation of soil moisture through the formation of a soil or dust mulch, may be counteracted by the destructive effects of inadvertent root pruning in the upper and most productive soil layer. In addition, extensive cultivation also contributes to considerable soil erosion.

Other physical means of destroying weeds are mulching and fire. Black polyethylene film has been useful as a mulch to control weeds, but cost limits its use to high-value crops. Flame throwers have been adapted to control weeds in such crops as cotton and onions.

Chemical control

Pesticide is the generic term for all chemicals used to control pests. As the name infers, they are usually toxic to some stage of the pest. Included under the legal definition of pesticide are **repellents**—compounds that may not be actively poisonous but that make the crop plant unattractive to animal predators by virtue of their odor, taste, or other physical properties.

Selectivity

A selective pesticide is one that kills certain organisms and does not harm others. Selectivity, however, is a relative concept, and it depends to a large extent on the interaction of a number of factors: dosage, timing, the method of application, the chemical and physical properties of the material applied, and the genetic and physiological state of the organisms involved. Nonselective pesticides kill indiscriminately. In general, most pesticides are selective to some degree. Thus pesticides are commonly classed according to the organisms they affect—for example, bactericide, fungicide, nematicide, miticide, insecticide, herbicide, rodenticide, and so on.

Many pesticides are selective within a broad group of organisms. For example, insecticides may be classified by their action as stomach or contact poisons. **Stomach poisons** are usually used against chewing insects. They are ingested along with the chewed plant material and cause death when they reach the stomach. Since sucking insects feed on the sap and ingest no external plant material, they are seldom affected by stomach poisons. They must be controlled by **contact poisons,** which kill by penetrating the insect's body directly or by entering through breathing or sensory pores.

Herbicides, chemicals that kill plants, are absorbed by the plant and cause death by producing some toxic reaction. Those that kill only the area of the plant actually covered (for example, dinitro compounds, oils, and many arsenicals) are called **contact herbicides;** others that are translocated within the plant (for example, 2,4-D) are called **noncontact herbicides,** or **translocated herbicides.** Selectivity may be brought about by directing contact herbicides away from the crop plant (positional tolerance) or by taking advantage of inherent morphological differences between the crop plant and the weed that will prevent the crop plant from being killed along with the weed. These include the amount and type of waxy cuticle, the orientation and shape of the leaves, and the location of the growing point. Selectivity may also be achieved by controlling the dose. In high concentrates 2,4-D will seriously damage or kill all plants, whereas in low doses it can be effectively used to kill selectively (Fig. 18-9). Selectivity may be achieved in still another way—by exploiting physiological differences between plants. The mechanisms of physiological selectivity are largely unknown; in fact, the precise way in which most herbicides kill is obscure. Some interfere with enzyme systems; others disturb the metabolism of the plant. Physiological selectivity thus reflects metabolic differences between plant species. Selectivity of herbicides has been related in some plants to differences in cell-membrane permeability and in others to differences in the ability of the plants to translocate the herbicide. Physiological selectivity may change drastically with the stage of growth (Fig. 18-10). Many herbicides are most effective during seed germination and early seedling growth.

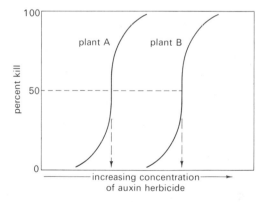

Figure 18-9   Selectivity of 2,4-D is a function of dose. Arrows on mortality curves of two species indicate relative concentration required to kill 50 percent of the plants. [From A. C. Leopold, *Auxin and Plant Growth*, University of California Press, 1955.]

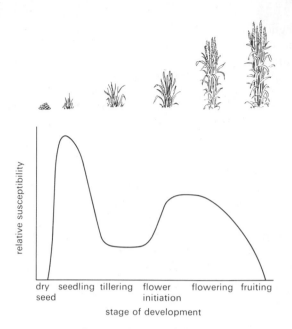

Figure 18-10   Changes of susceptibility of grains to auxin herbicides at various stages of development. [From A. C. Leopold, *Auxin and Plant Growth*, University of California Press, 1955.]

Chemical and physical properties

A great number of compounds, inorganic and organic, natural and synthetic, are pesticidal (Table 18-2). At present, pesticides become outmoded very quickly, as new or improved substances are continually being released by a vigorous and expanding industry.

The earliest fungicides and insecticides were inorganic salts of such metals as copper, mercury, lead, and arsenic; many of these are still useful. Natural organic compounds came into prominence in the nineteenth century. These include the insecticides derived from plant sources—nicotine, pyrethrum, and rotenone. The modern pesticide industry is based on synthetic organic materials. This industry was given great impetus by two widely successful materials whose names are today common terms—DDT and 2,4-D.

The insecticidal value of DDT (dichlorodiphenyl-

Table 18–2 Classification of some important pesticides.

Class of materials	Specific examples	Major type of action
Inorganic	ammonium sulfamate	Herbicide
	copper sulfate	Fungicide, bactericide
	lime sulfur	Fungicide
	sodium arsenite	Herbicide
	sulfuric acid	Herbicide
Organic		
Natural compounds and derivatives		
antibiotics	streptomycin, cycloheximide	Bactericide, fungicide
nicotine	nicotine sulfate	Insecticide
pyrethrum		Insecticide
rotenone		
oils	Stoddard solvent, diesel oil, highly refined oils	Herbicide, insecticide
Synthetics		
benzene compounds	chlorothalonil	Fungicide
benzimidazoles	benomyl, thiophanate	Fungicide
carbamates	maneb, zineb	Fungicide
	carbaryl	Insecticide
carbon tetrachloride		Fumigant
chlorinated aliphatic acids	TCA, dalapon	Herbicide
chlorinated hydrocarbons	DDT, methoxychlor	Insecticide
	DD	Nematicide
formamidines	galecron	Insecticide
heterocyclic compounds	captan, folpet	Fungicide
organophosphates	malathion	Insecticide, miticide
pyrethoids	resmethrin, permethrin	Insecticide
quinones	dichlone	Fungicides
triazines	atrazine, simazine	Herbicide
ureas	monuron, diuron	Herbicide
Miscellaneous	anticoagulants (Warfarin)	Rodenticide
	DCPA (Dacthal)	Herbicide
	trifluraline	Herbicide
	methyl bromide	Soil sterilant
Biologicals		
bacteria	*Bacillus thuringiensis*	Insecticide
viruses		Insecticide

trichloroethane) was discovered in 1939, although the material had been known 65 years previously. It acts on the sensory nerves of insects and is effective at very low concentrations. Introduced by the Geigy Company, a Swiss dye firm, it was manufactured in great quantities in the United States during World War II and was credited with control of the malaria mosquito in many parts of the world. The phenomenal results attending its use led to the investigation of a great number of similarly constructed compounds or analogs, which led in turn to many other chlorinated hydrocarbon insecticides—methoxychlor, BHC (benzene hexachloride), chlordane, aldrin, dieldrin, and many others.

DDT is an extremely persistent and fat-soluble molecule that "magnifies" as it passes along food chains (refer to Fig. 5-18). This magnification aroused great concern and resulted in the application of strong pressure to reduce the indiscriminate use of DDT and other chlorinated hydrocarbons. In the United States, DDT was restricted for most uses in 1972. Many agriculturists have felt that the "banning" of DDT was an emotional issue, because a careful analysis of its action had demonstrated that, when properly used, this chemical posed no threat to environment, wildlife, or people. Despite the ban on most agricultural uses, DDT is still available for public health and quarantine uses.

The origin of the "broad-leaf weed killer" 2,4-D can be traced back to research on the herbicidal action of auxins in the early 1940s. The research effort was stimulated by the war, and most of the early studies were shrouded in military secrecy. The precise way in which minute quantities of auxinlike phenoxy herbicides cause death is still not fully known, but it is believed that these materials cause a rapid depletion of energy carriers and the accumulation of toxic metabolites. The phenoxy herbicides have a close structural relation to each other and to indoleacetic acid, as is shown in Figure 18-11. However, small alterations in the molecular structure of these materials greatly change their herbicidal and physical properties. For example, the amine form of 2,4-D is much less volatile than the ester form, and is thus less subject to drifting. Both have a short residual life in the soil and are apparently not gener-

ally toxic to animals. The selectivity of 2,4-D at low dosages is responsible for its main agricultural use— killing broad-leaved plants without affecting grasses. Since it was discovered that 2,4-D is a successful selective herbicide, research has uncovered many others. For example, the chlorinated acetic acids, such as TCA (trichloroacetic acid) and dalapon (2,2-dichloropropionic acid), selectively kill monocots without harming dicots, and thus control grassy weeds in broad-leaved crops.

Pesticides may be applied in various forms and ways. Efficient application requires uniform coverage at a controlled rate. Because it is seldom practical to apply chemicals in their pure form, they are usually distributed in dilute form in an inert carrier. If the carrier is a solid (talc, clay, or diatomaceous earth), the pesticide is applied as a dust. Dusts have advantages of lightness, and they are particularly suitable for application by airplane or helicopter, but they are difficult to apply uniformly and they do not persist for long. Moreover, some materials (for example, oils) cannot be applied in this manner. These disadvantages may be overcome by the use of a liquid carrier, usually water, in which the chemical is dissolved, suspended, or emulsified. The material is applied under pressure in droplets of various sizes, but usually in the form of a spray (Fig. 18-12). The disadvantages of the weight and bulk of the water carrier can be overcome by using high concentrations of the active material and achieving dispersal

2,4-dichlorophenoxyacetic acid (2,4-D)

2,4,5-trichlorophenoxyacetic acid (2,4,5-T)

methylchlorophenoxyacetic acid (MCPA)

indoleacetic acid (IAA)

Figure 18-11 Structures of auxinlike herbicides compared with structure of indoleacetic acid.

Figure 18-12   The drop-size spectrum.

by a blast of air (Fig. 18-13). On heavily cutinized plant surfaces, sprays form droplets instead of a continuous film. This irregular dispersal may be overcome by adding wetting agents—chemicals that break the surface tension. Other additives, known as "stickers," are used to improve the adherence and persistence of pesticides.

Volatile pesticides active as gases are known as **fumigants.** Fumigants are usually stored under pressure as liquids; in aerosol bombs the active ingredient is dispersed in a gaseous carrier. Some fumigants are solid, and volatilize when heated or moistened. For maximum efficiency, fumigants must be applied in an enclosed system, such as a storage bin or a greenhouse. Fumigants are finding increasing usefulness in soil treatment. To prevent the fumigant from escaping, the surface of the soil may be sealed by covering it with plastic film or a layer of water, or merely by compacting the upper layer.

Pesticides may be **systemic** or **nonsystemic** in their action. Systemic pesticides are absorbed by the plant and translocated within it, rendering the plant itself toxic or repellant to the pest. Their use is limited in edible crops unless they break down within

Figure 18-13   Orchard sprayer. A blast of air is used as a carrier for highly concentrated sprays. [Courtesy Food Machinery and Chemical Corp.]

Figure 18-14   Results of chemical weed control in peanut: left, untreated; right, EPTC applied as a preplanting soil-incorporated treatment has given excellent control of nut sedge and all other broad-leafed weeds and weedy grasses. [Courtesy USDA.]

the plant before consumption. The action of non-systemic pesticides takes place on the surface of the plant, although they may be absorbed to some degree.

Timing of application

The effectiveness of chemical control usually depends on the timing of application. A certain pesticide may not be equally toxic to all forms of a particular pest, nor is this necessary; usually one stage or another is especially vulnerable, which is why timing is important. The weak link may be a germinating fungal spore, the young larval stage of an insect, or the insect vector of a virus disease. For most diseases, chemical control must be applied before symptom expression. For example, it is difficult to kill fungi after they have entered the plant, but control may be achieved by materials that either kill spores or prevent them from germinating on the plant.

Timing is especially important in weed control. Selectivity may be achieved for fumigants and other nonselective herbicides when they are applied before the crop is planted (**preplanting treatment**). Selective preplanting herbicides are also available that attack germinating weed seeds without harming crop seeds or transplants. Applications made after the crop is planted but before it has emerged from the soil are known as **pre-emergence treatments** (Fig. 18-14). The action of many of these materials is restricted to the soil surface. To be effective, pre-emergence materials must have good coverage and be relatively resistant to leaching. Soil moisture is often critical, for water may be required to activate the herbicide, although an excess may cause leaching. Pre-emergence herbicides are physiologically selective between weeds and crops. Selectivity may be enhanced by the early germination of weed seed. The germination of crop seed is usually slightly delayed because of the depth of planting and the time required for the seed to imbibe water. Applications of herbicides to the growing crop are referred to as **post-emergence** treatments. Selectivity must be positional (from directed applications) or physiological.

Spray injury and residues

The beneficial effects of chemicals must be balanced against their potential hazards. To provide for the safe and proper use of these materials, restrictions are imposed, especially when they are used in connection with foodstuffs. Care must be taken to ensure that the cure is less injurious than the disease! Some chemicals leave soil residues that may interfere with plant growth and future land use. Their eventual disappearance may result from vaporization, chemical breakdown, biological decomposition, leaching, or adsorption onto soil colloids. Some pesticides are phytotoxic and can reduce yields and quality. Of greatest concern, however, is the possibility that the pesticide or its harmful breakdown products may persist in the plant product or in animals that feed on the plants.

The potential hazard that pesticidal residues represent to consumers is a topic that has been widely discussed. Every substance, whether it be food, drug, cosmetic, or pesticide, has a finite hazard. All chemicals have some dose at which they are toxic, and some dose at which they are nontoxic. Poison is a relative term: *dosis sola facit venenum*—the dose makes the poison. Many common "poisons"—arsenic, for example—may have a stimulating effect at very low dosages and may, in fact, have medicinal properties. As dosage increases they become injurious and ultimately toxic. Some substances have no noticeable effect at low concentrations but tend to accumulate in certain parts of the body. The problems associated with these substances are real. Any undue risks to the consumer, as well as to the applicator, cannot be tolerated.

The Environmental Protection Agency of the United States, the governmental agency that regulates the use of pesticides and other chemicals, sets tolerance levels. The tolerance level, often conditioned by a 100-fold safety factor, determines the maximum detectable amount, zero or otherwise, that is considered safe. Tolerance levels are determined through toxicity studies on animals. Pesticides are an essential ingredient in agriculture, and, like many substances, they can be used or misused. When used with good judgment, they are a blessing to humankind.

Biological control

Plant pests may be also controlled through the manipulation of biological factors. Biological control may be achieved by directing the natural competition between organisms—for example, by introducing natural parasites or predators of a pest (Fig. 18-15) or incorporating natural resistance to a pest into a host crop. There are other subtle ways in which biological factors may be manipulated to af-

A

B

Figure 18-15   Insects parasitize other insects. (A) Cocoons of braconid parasites emerging from the body of a tomato hornworm. The cocoons contain the pupal stage of a parasitic wasp that lays its eggs on the body of the hornworm. (B) The potato-beetle killer attacking a larva of the Mexican bean beetle. [Courtesy J. C. Allen & Son.]

fect the development of the pest. For example, the screwworm, a severe pest of livestock, has been controlled by releasing into the population large numbers of males rendered sterile by irradiation. Biological control is an attractive measure, because once put into effect it takes place without human influence, and chemical hazards are eliminated.

Natural predators

An example of successful biological control by the introduction of natural predators was the elimination of the threat posed by the cottony-cushion scale, a serious insect pest of citrus in California. This was accomplished by Albert Koebele, who in 1888 introduced from Australia the vedalia beetle, which feeds upon the eggs and larvae of the scale insect. The scale was kept under control until the widespread use of DDT in the late 1940s reduced populations of the beetle and upset the biological balance. A more recent example of biological control is the use of the spores of *Bacillus thuringiensis* to control caterpillars; the spores are applied as a spray.

Biological control has also been used to control weeds. Prickly pear (*Opuntia* sp.) was introduced as an ornamental into Australia prior to 1839 by the early colonists. By 1925, due to the absence of natural predators, more than 24 million hectares (about 60 million acres) of land was infected with the cactus, and the infestation was increasing at the rate of about 400 thousand hectares (about 1 million acres) annually. Ten years after the introduction of the moth borer *Cactoblastis cactorum* from Argentina, control was almost complete, and although there were successive waves of regrowth, they were of diminishing proportions.

The introduction of predators is best suited to the control of pests that lack natural predators, pathogens, or other enemies. For introduced predators to be successful they must thrive in the new habitat and yet not become destructive to agricultural crops. Thus only predators that are highly selective in their feeding habits can be imported. Because of the large areas involved and the inherent problems, this method of control must be placed under the auspices of some national agency rather than left to the discretion of individual growers.

Genetic resistance

The incorporation of genetic resistance, the innate ability of a plant to resist the injurious effects of pathogens or predators, is an ideal method of control (Fig. 18-16). Resistance varies from nil (**susceptibility**), to partial, to complete (**immunity**). **Tolerance** is a type of resistance in which the plant suffers infection and some injury, but is able to live with it.

Many plants are known to have resistance to viruses, bacteria, fungi, or nematodes. These plants owe their resistance to structural features or biochemical effects that either prevent or discourage intrusion and persistence of the particular pest. Resistance may be complex or simple. Some plants have resistance to whole groups of pathogens; others have only a specific resistance to a particular species or race. The combination of resistance and quality is one of the main objects of the plant breeder (see Chapter 19). Where pathogen and plant are closely adapted to each other (as with obligate parasites), a close relation exists between the genetic resistance of the plant and the genetic ability of the pathogen to violate or overcome this resistance. The emergence of a new race of the pathogen by mutation or introduction is one of the major problems the plant breeder faces, and the seeming impossibility of incorporating long-term genetic resistance to some diseases has inspired the search for tolerance or broad kinds of resistance.

Biological balance

The biological balance that exists in the natural environment cannot be ignored in plant protection. The upsetting of this equilibrium by environmental fluctuations is a perfectly natural phenomenon. Agriculture, by its very nature, disturbs the natural biological balance, but it does not operate independently of it. For example, often when one fungicide supercedes another, fungi that were once minor pathogens may begin to cause large-scale damage. Furthermore, the pathogen is not a static factor. New, resistant varieties develop, and new pesticides may act as screening devices for the selection of resistant types. Thus many insects—the common housefly and many others—have developed resistance to DDT since its introduction. Similarly, the develop-

A

B

C

D

Figure 18-16 Genetic resistance and pest control. (A) Susceptibility and resistance to the beta race of anthracnose, caused by *Colletotrichum lindemuthianum,* in beans: 'Red Kidney' (left) is susceptible; 'Michelite' (right) is resistant. (B) Larvae of the corn earworm cause less feeding damage and develop slower on the resistant experimental hybrid sweet corn 'M4399' (left) than on the susceptible 'Iobelle Hybrid' (right). (C) Muskmelon plants exposed for 8 days to mass infestation by the melon aphid, *Aphis gossypii.* The plant on the left is resistant to the aphids. (D) Adult aphids on such resistant plants (left) grew slowly and were stunted as compared with aphids on susceptible plants (right). [A, courtesy USDA; B, courtesy E. V. Wann, USDA; C, D, courtesy G. W. Bohn, A. N. Kishaba, and H. H. Toba.]

ment of rust-resistant plant varieties soon becomes negated by the formation of new, virulent races of the rust fungus.

FIRE

The combustion, or burning, of wood is a reaction between carbohydrate fuels and oxygen; the reaction produces heat energy, carbon dioxide, and water (the whitish color of smoke is due to water vapor). The chemical equation for the reaction is the same as that for respiration, but combustion takes place more rapidly and at a higher temperature. Both combustion and respiration are forms of oxidation. The energy accumulated through many years of photosynthesis may be rapidly dissipated by fire. Combustion requires three elements: heat, fuel, and oxygen. The release of stored energy in combustion starts a chain reaction that continues as long as the fuel and oxygen are in adequate supply. Fire-suppression techniques strive to break one link in this chain of events.

Forest fire

Fire, especially in forests, is one of the destructive forces that affect crops. A **forest fire** is defined as an unenclosed, freely spreading fire in a forest environment. It is a wildfire, in contrast to the **controlled fire** used as a forest-management tool.

Probably the most important factor governing the severity of forest fire is weather: hot, dry weather lowers the moisture content of fuels. Once a fire has started, wind is extremely critical because it influences the oxygen supply and the rate of spread. Forest type and forest conditions are also important factors. For example, fire is more likely to occur in conifers than in hardwoods; slash-covered or brushy areas are especially hazardous because the rate at which combustion consumes fuel is proportional to fuel volume and surface area.

A small fire, with a rate of energy release of approximately 3.5 million joules per second per meter of fire front (nearly 1000 British thermal units per second per foot of fire front) behaves as a two-dimensional surface. High intensity, or blowup, fires releasing 70 million to 105 million joules per second

per meter of fire front (roughly 20 thousand to 30 thousand British thermal units per second per foot of front) also have a vertical dimension (Fig. 18-17). These "three-dimensional" fires are comparable to thunderstorms.

Some fires are caused by lightning; others are caused by people. Those caused by people may be accidental or intentional (incendiary). The causes of fire in the United States show large regional differences (Table 18-3). In the Rocky Mountains more than 70 percent of the fires are caused by lightning, whereas incendiary fires amount to only about 1 percent. In contrast, more than 40 percent of the fires in the South are incendiary, whereas lightning causes only 1 percent. Obviously, fire-prevention techniques must vary from region to region.

Fire control

Of all the potential hazards to forests, fire poses the greatest threat. Consequently, fire control must be an integral part of forest management. Fire control has several dimensions:

Fire prevention is largely a sociological problem. True, many fires are started by sparks from faulty equipment, lightning, and other causes, but 90 percent of them are started by people. More education is probably needed to reduce the number of fires caused by people. The prevention of fires caused by lightning seems to be out of our hands.

Hazard reduction is the elimination of situations in which fire is likely to occur. Each year tens of thousands of acres are lightly burned intentionally in order to prevent litter fuel from accumulating to dangerous proportions. Firebreaks and fire lines are constructed to prevent the spread of fire in case one starts. Safety devices around logging operations and railroads help reduce the fire hazard.

Preparation for suppression includes all the activities that must be carried out in advance of a fire in order to be ready to fight it. This includes organizing fire crews, maintaining equipment in a state of readiness, and ensuring the availability of extra labor for emergencies.

Detection has become a specialized activity, for the earlier a fire is spotted, the sooner suppression

A   B

Figure 18-17   Large blow-up fires extend far into the sky. These extremely dangerous fires can cause total destruction. [A, courtesy *Los Angeles Herald Examiner;* B, courtesy U.S. Forest Service.]

Table 18-3 Causes of forest fire, area burned, and acres protected in the United States by forest regions, 1935–1954.

Region and period	Cause of fires (percent)								Av. annual no. fires	Av. annual area protected (acres × 1000)	Av. annual area burned (acres)	Av. no. fires per million acres protected	Av. annual acres burned per million acres protected
	Lightning	Railroads	Campers	Smokers	Debris burning	Incendiary	Lumbering	Misc.					
Rocky Mountain (10 states)													
1935–1939	67.8	2.0	7.7	14.1	3.4	1.2	0.7	3.1	5,554	97,612	105,452	57	1,080
1940–1944	71.6	4.8	4.3	10.0	3.0	0.7	0.6	5.0	6,118	121,289	326,029	50	2,688
1945–1949	70.0	3.7	5.6	10.3	3.3	0.8	0.9	5.4	5,232	159,275	220,485	33	1,384
1950–1954	68.7	2.0	7.2	10.0	4.0	1.0	1.0	6.1	5,336	163,566	160,692	33	982
Pacific (3 states)													
1935–1939	30.2	3.1	7.1	25.8	9.5	11.6	2.7	10.0	8,783	78,615	658,863	112	8,381
1940–1944	34.6	5.4	4.0	23.9	10.3	5.9	2.3	13.6	7,747	94,517	519,028	82	5,491
1945–1949	30.2	3.7	5.3	24.0	11.2	6.0	3.7	15.9	7,287	103,470	417,780	70	4,038
1950–1954	33.7	3.2	6.5	20.8	10.3	4.2	5.2	16.1	7,129	100,024	342,458	71	3,424
North Central (9 states)													
1935–1939	2.9	6.7	5.5	32.6	26.1	14.6	1.2	10.4	7,326	64,097	273,881	114	4,273
1940–1944	1.3	8.4	3.4	29.3	25.7	19.5	1.5	10.9	6,647	69,699	263,175	93	3,776
1945–1949	1.5	13.7	3.4	24.7	25.3	18.0	0.9	12.5	8,374	80,228	287,445	104	3,583
1950–1954	1.0	6.3	2.7	19.2	36.7	20.1	1.0	13.6	9,422	83,278	276,320	113	3,318
Southern (11 states)													
1935–1939	1.2	2.5	5.8	22.7	12.8	43.0	2.8	9.2	36,698	80,314	1,594,933	457	19,859
1940–1944	1.1	3.7	4.4	20.8	15.1	44.3	2.4	8.2	42,299	90,717	2,137,341	466	23,561
1945–1949	0.8	3.5	3.2	18.9	15.0	45.0	2.3	11.3	39,859	107,999	1,320,583	369	12,228
1950–1954	1.5	1.9	3.4	16.1	19.6	44.9	2.5	10.1	77,162	140,740	2,599,998	548	18,474
Eastern (14 states)													
1935–1939	0.7	9.3	10.2	39.6	19.4	10.5	0.8	9.5	16,161	80,374	332,346	201	4,136
1940–1944	1.0	12.3	7.5	35.0	22.0	9.4	0.8	12.0	17,618	86,079	488,978	205	5,681
1945–1949	1.0	11.4	7.1	33.1	21.4	8.5	1.0	16.5	12,946	93,115	257,353	139	2,764
1950–1954	1.4	5.6	5.8	33.2	25.9	11.3	0.9	15.9	15,096	95,734	594,485	158	6,210
Total U.S.													
1935–1939	9.7	4.4	7.0	27.0	14.5	26.4	2.0	9.0	74,522	400,985	2,965,475	186	7,395
1940–1944	9.7	6.3	4.9	24.1	16.1	27.6	1.8	9.6	80,427	462,301	3,533,551	174	7,643
1945–1949	8.8	6.1	4.3	21.9	16.1	28.5	1.9	12.4	73,775	544,089	2,703,715	136	4,969
1950–1954	6.6	2.8	4.0	18.6	20.5	33.8	2.3	11.4	114,146	583,354	3,953,411	196	6,777

Source: K. P. Davis, *Forest Fire*, McGraw-Hill, 1959.

activities can begin. In the past, detection was accomplished visually—flame at night, smoke by day—using fixed lookouts or moving air or ground patrols, but these methods are hampered by weather conditions that decrease visibility. Infrared sensing offers a partial solution to this problem, and, coupled with remote television stations, it may provide a much more sensitive detection system in all kinds of weather than has ever been available before.

Suppression involves all the work of extinguishing a fire after it has been discovered. Stopping the spread of fire may be achieved by smothering it with water, chemicals, or other materials or by providing a break in its fuel supply. Firebreaks can be made by plowing a fire line near a strip free of combustible materials or by backfiring—setting a fire some distance from the wildfire. Fighting fire with fire is spectacular and sometimes dangerous. The use of chemicals or water (when available) to make the fuels unsuitable for combustion accomplishes the same effect. Airplanes are beginning to be used routinely to drop chemicals on fires.

Particular fire-fighting tactics that depend on the wind, moisture, topography, and special local conditions are beyond the scope of this text. It is axiomatic that fires are best controlled at an early stage. Blow-ups, large "three-dimensional" conflagrations that extend far up into the sky, are most difficult to control, and present great danger to timber, wildlife resources, and to human life.

HAIL

The heaviest and largest unit of precipitation existing in solid form is hail. It is a product of vigorous convection currents that occur in the warm season. If a hailstone is cut in half, one can see concentric layers of differing density and opacity that were formed by successive ascents and descents in these convection currents.

Hail begins as rain, but rather than falling to the ground, the drops are carried upward by rapidly ascending updrafts of warm, moist air, into regions where temperatures are below freezing. They freeze, acquire a coat of frost and snow, and become hailstones. Eventually they enter a weaker convection current and fall, accumulating a coat of water, which also freezes. This process may be repeated many times, with the hailstones accumulating an additional coating of ice each time. Ultimately, they will become so large that convection currents are no longer strong enough to force them upward. Hailstones 12.5 centimeters (about five inches) in diameter and weighing nearly 0.7 kilogram (about 1.5 pounds) have been recorded in Nebraska.

Millions of dollars are lost yearly due to hail damage, principally through the shredding of leaves (Fig. 18-18) or damage to fruit. For crops in which leaf quality is a factor, such as tobacco, a single hailstorm may result in the loss of an entire crop. Losses in fruit crops can be ruinous. For corn, productivity seems to be reduced by approximately the percentage of damage to the leaves.

Figure 18-18 Hail damage to corn. [Courtesy J. C. Allen & Son.]

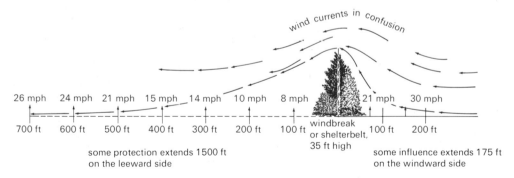

26 mph 24 mph 21 mph 15 mph 14 mph 10 mph 8 mph   21 mph 30 mph

700 ft 600 ft 500 ft 400 ft 300 ft 200 ft 100 ft   windbreak   100 ft 200 ft
or shelterbelt,
35 ft high

some protection extends 1500 ft   some influence extends 175 ft
on the leeward side   on the windward side

Figure 18-19 Tree plantings act as windbreaks to reduce wind velocity and thus protect crops as well as homes and livestock from cold and hot winds. Shelterbelts consisting of 5- to 10-row strips of trees help to control wind erosion, contribute substantially to the saving of moisture, and protect crops from damage by high winds. [Courtesy USDA.]

Efforts have been made to prevent the formation of hail by seeding clouds with silver iodide, which causes moisture to be precipitated earlier, thus preventing hail formation.

WIND DAMAGE

The harmful effects of wind are most apparent through its mechanical effects. Although wind damage is usually local, some regions suffer more than others. For example, tornadoes occur in a belt from Texas to Michigan, and seasonal hurricanes are a hazard along the Gulf and Atlantic coasts of eastern North America. Even winds of 50 to 65 kilometers per hour (about 30 to 40 miles per hour) may cause extensive damage when accompanied by hail, sleet, and rain. Winter winds may cause desiccation and browning of evergreen ornamental shrubs.

Wind can cause serious erosion problems, as it did in the "dust bowl" during the 1930s, where dry-farming practices left soil without the protection of a vegetative cover. Another harmful effect is the transport of propagules of weed plants, insect pests, and spores of pathogenic fungi.

Hot, dry winds can have a debilitating effect on plants by upsetting their internal water balance. When this happens, photosynthesis is decreased and crop production is lowered. Excessive wind at flowering time can evaporate the secretions from stigmas and result in reduced seed set. Windbreaks are an important and necessary adjunct to farming practices in the central United States and elsewhere (Fig. 18-19). The benefits of windbreaks in increasing crop yields have been firmly established.

In areas where wind is a hazard, sensitive crop plants, such as tobacco, should not be grown. Plants native to regions subject to intense, high winds frequently have structural modifications that protect them against wind damage. Hurricanes seldom uproot mangroves growing in the Everglades of Florida, for their stiltlike roots anchor them firmly. In general, close spacing provides protection against wind damage because the plants then form a barrier against the wind and are "forced" to provide mutual support for each other. In forest management it is sometimes necessary to use a cropping system that does not leave trees vulnerable to wind damage. Thus complete rather than partial harvesting is practiced in many parts of the world where trees are especially susceptible to such injury.

Selected references

Anderson, W. P., 1977. *Weed Science: Principles*. St. Paul, Minn.: West Publishing. (A broad perspective of weeds and weed control.)

Ashton, F. M., and A. S. Crafts, 1973. *Mode of Action of Herbicides*. New York: Wiley. (An advanced treatise.)

Chapman, R. F., 1969. *The Insects: Structure and Function*. New York: American Elsevier. (A text on morphology and physiology.)

Crafts, A. S., 1975. *Modern Weed Control*. Berkeley and Los Angeles: University of California Press. (A comprehensive text and manual on the principles of weed control.)

Davis, K. P., 1959. *Forest Fire: Control and Use*. New York: McGraw-Hill. (An authoritative text, with much technical information.)

Horsfall, J. G., and A. E. Dimond (editors), 1959, 1960. *Plant Pathology: An Advanced Treatise* (3 vols.). New York: Academic Press. (A good, detailed work on the science of plant pathology.)

Kasasian, L., 1971. *Weed Control in the Tropics*. New York: Leonard Hill. (A guide and elementary text.)

Kenaga, C. B., 1974. *Principles of Phytopathology*, 2nd ed. Lafayette, Ind.: Balt. (A general introductory text on plant pathology.)

Klingman, G. C., F. M. Ashton, and L. J. Noordhoff, 1975. *Weed Science: Principles and Practices*. New York: Wiley. (An excellent general textbook.)

Metcalf, C. L., W. P. Flint, and R. L. Metcalf, 1962. *Destructive and Useful Insects: Their Habits and Control*, 14th ed. New York: McGraw-Hill. (The bible for economic entomology.)

National Research Council, 1972. *Pest Control Strategies for the Future*. Washington, D.C.: National Academy of Sciences. (Papers on integrated pest control presented at a symposium sponsored by the Agricultural Board.)

National Research Council, 1968–1970. *Principles of Plant and Animal Pest Control* (6 vols.). Washington, D.C.: National Academy of Sciences. (The six volumes concern, respectively, plant disease, weed control, insect-pest management, nematodes, vertebrate pests, and the effects of pesticides on fruit and vegetable physiology. The technical reports are especially useful in offering guidance in resource management.)

Pfadt, R. E. (editor), 1971. *Fundamentals of Applied Entomology*, 2nd ed. New York: Macmillan. (Each chapter was contributed by a specialist.)

Roberts, D. A., and C. W. Boothroyd, 1972. *Fundamentals of Plant Pathology*. San Francisco: W. H. Freeman and Company. (This fine phytopathology text presents specific diseases in groups according to their physiological impact upon plant welfare.)

Roberts, D. A., 1978. *Fundamentals of Plant-Pest Control*. San Francisco: W. H. Freeman and Company. (A comprehensive introduction to the nature and control of plant pests, stressing an interdisciplinary approach.)

U.S. Department of Agriculture, 1952. *Insects* (USDA Yearbook, 1952). (This and the following two publications are broad, popular treatments of aspects of crop protection.)

U.S. Department of Agriculture, 1953. *Plant Diseases* (USDA Yearbook, 1953).

U.S. Department of Agriculture, 1966. *Protecting Our Food* (USDA Yearbook, 1966).

Walker, J. C., 1969. *Plant Pathology*, 3rd ed. New York: McGraw-Hill. (An excellent introduction, with an organization based on causal agents. There are many other textbooks whose organization is based on crops or groups of crops.)

Ware, G. W., 1978. *The Pesticide Book*. San Francisco: W. H. Freeman and Company. (A readable introduction to the chemistry and uses of pesticides, with a discussion of their hazards and the laws that control them.)

Webster, J. M. (editor), 1972. *Economic Nematology*. New York: Academic Press. (A reference text to the important nematode pests of the world's major crops.)

Weed Science Society of America, 1974. *Weed Science Society of America Herbicide Handbook*, 3rd ed. Champaign, Ill. (A list of herbicides, desiccants, plant-growth regulators, various adjuvants, and chemicals of interest to weed scientists.)

Wiley, R. W., and S. B. Heath, 1969. The quantitative relationships between plant population and crop yield. In A. G. Norman (editor), *Advances in Agronomy*, vol. 21, pp. 281–321. New York: Academic Press. (An excellent review that examines the usefulness and biological validity of the different mathematical equations that have been proposed to describe the relationships between plant population and crop yield.)

19
Crop Improvement

Within a given environment, crop improvement may be achieved through superior heredity. **Plant breeding,** the systematic improvement of plants, is an innovation of this century. Genetics, the science of heredity, has placed it on a firm theoretical basis. Plant breeding has become a specialized technology and is responsible for a large share of the current progress in crop production.

Since human needs and standards change, the job of the plant breeder is a continuous one. For example, efficient mechanical harvesting becomes possible only with plants that bear the great bulk of their crop at a given time; it further requires that plants be structurally adapted to the harvesting machine. The raw materials for the necessary changes may be found in the tremendous variation that exists in cultivated plants and related wild species. The incorporation of alterations into crop plants adapted to specific geographical areas demands not only a knowledge of the theoretical basis of heredity, but also the arts and skills necessary for the discovery and perpetuation of small but fundamentally important differences in plant material.

THE GENETIC BASIS OF PLANT BREEDING

Variation

Variety is more than the spice of life; it is the very essence of it. The variation found among crop plants, as among all living things, ranges from the obvious (sugar beet versus sugar maple) to the almost imperceptible differences between two wheat plants growing side by side.

The variation encountered in biological material is measurable. Just as the general performance of a group can be described by a single statistic called the **arithmetic mean,** the amount of variation in an array of observations can be expressed by a comparable statistic called the **standard deviation,** a measure of the average deviation from the mean.

The variation found in biological material (for example, among individual plants in a cotton field) is the result of genetic and environmental components. The genetic component is the result of differences in

461

the inheritied makeup of the organism. It is due largely to differences in genes, the main units of inheritance that control development. The sum total of the genetic endowment of an individual is referred to as the **genotype.** Differences between similar plants grown in identical environments must be due to genotypic differences.

The environmental component of variation is the result of nongenetic factors that influence the expression of the genotype. In our hypothetical cotton field, these factors may include differences in soil fertility and in shading, or any of a number of random events such as insect injury or hail damage. The effects of environmental variation can be observed directly in organisms of identical genetic makeup.

The outward appearance of an organism is known as its **phenotype.** The phenotype may be thought of as the interaction product of genotype with environment. The range of environmental variation is enormous in living things, but its boundaries are set by the genotype. It is difficult, for example, to conceive of any environmental condition that could transform sugar beets into sugar maples.

The variation found in a population of living organisms may be influenced in varying proportions by environmental and genetic factors. The percentage of the total variation due to genetic factors is known as **heritability.** In corn, the heritability of yield and plant height has been estimated to be 25 percent and 75 percent, respectively. This means that in segregating populations, 25 percent of the variation in yield and 75 percent of the variation in height is attributable to genetic factors. Consequently, it is much easier to select for plant height than for yield. When heritability is low, then the relative influence of environment is high, and the quickest way to maximize efficiency is to improve the environment.

The question of whether genetic or environmental variation is the more important has little meaning. The pertinent problems are to find out (1) which genotype is best suited to a particular set of environmental conditions; and (2) what combination of environmental variables will permit the optimum expression of a given genotype.

The fundamental discovery that the genotype is inherited in the form of discrete units belongs to the Austrian Abbot Gregor Mendel, whose experiments on the transmission of seven traits in the garden pea were published in 1865. Environmental variation is not transmitted in this manner. This was firmly established in 1903 by the Danish geneticist and breeder Wilhelm Johannsen, whose studies dealt with the transmission of seed size in beans. Johannsen demonstrated that the inheritance of phenotypic variation is possible only when it is a result of genetic differences.

Genetic factors form the basis for plant improvement through breeding. The unique problem of the breeder is to make the plant better adapted to human needs by bringing about a favorable combination of genes.

Mutation

The genetic variation in living things is ultimately caused by differences in the kind, structure, and amount of genetic material (see pp. 193–198). The source of this variation is found in sudden heritable changes called **mutations.** Changes that affect the gene itself are called **point mutations,** and they result from alterations that affect the internal makeup or arrangement of the functional gene. The rate of these mutations varies from gene to gene. True point mutations may also revert to the original condition (**back mutation**).

Extragenic changes are also included under the broad category of mutation. These may be due to chromosomal aberrations that alter the structural arrangements of the gene-carrying chromosomes. Chromosomal aberrations are of several types:

Aberrations involving one chromosome
   deficiency: segment deleted
   duplication: segment repeated
   inversion: segment reversed
Aberration involving more than one chromosome
   translocation: exchange of segments

Another type of mutation results from a change in chromosome number (**heteroploidy**). The chromosomes in the somatic cells of the higher plants are composed of sets called **genomes.** Each genome contains a particular number of chromosomes. Although most plants have two genomes, the diploid number, a wide variation in number is possible. Variation in

Table 19-1 Genomic constitution of wheats and their relatives.

Species	Genome formula	Common name
Diploid (2n = 14)		
Triticum boeticum	*AA*	Einkorn (wild)
Triticum monococcum	*AA*	Einkorn (cultivated)
Aegilops speltoides	*BB*	Goatgrass (wild)
Aegilops caudata	*CC*	Goatgrass (wild)
Triticum tauschii (= *Aegilops squarrosa*)	*DD*	Goatgrass (wild)
Secale cereale	*EE*	Rye (cultivated)
Tetraploid (2n = 28)		
Triticum dicoccoides	*AABB*	Emmer (wild)
Triticum dicoccon	*AABB*	Emmer (cultivated)
Triticum durum	*AABB*	Durum wheat (cultivated)
Triticum persicum	*AABB*	Persian wheat (cultivated)
Triticum polonicum	*AABB*	Polish wheat (cultivated)
Triticum turgidum	*AABB*	Poulard wheat (cultivated)
Triticum timopheevi	*AAGG*	Timopheevi (wild)
Aegilops cylindrica	*CCDD*	Goatgrass (wild)
Hexaploid (2n = 42)		
Triticum compactum	*AABBDD*	Club wheat (cultivated)
Triticum spelta	*AABBDD*	Spelt (cultivated)
Triticum aestivum	*AABBDD*	Bread wheat (cultivated)

chromosome sets is referred to as euploidy—that is, true ploidy. The following terms indicate the number of chromosome sets a plant contains:

No. of chromosome sets	Name of type
1	monoploid
2	diploid
3	triploid
4	tetraploid
5	pentaploid
6	hexaploid
7	septaploid
8	octaploid

In addition to variation of whole sets, there is variation in the number of chromosomes within a set (**aneuploidy**). Plants in which such variation occurs are usually low in fertility.

Polyploids, plants in which the number of chromosome sets is higher than the diploid number, may arise by the duplication of sets from a single species (**autoploidy**) or by the combination of sets from two or more species (**alloploidy**). Many of our crop plants are alloploids—for example, wheat, cotton, tobacco, and members of the cabbage family. The origin of their genomes can be traced by special genetic and cytological techniques. The complex genomic constitution of the wheats is shown in Table 19-1. Knowledge of the genomic constitution in polyploid crops enables the plant breeder to accomplish the transfer of genes from related wild species. This ability to juggle genes and chromosomes enormously enlarges the potential range of genetic variation available to the breeder. For example, it has been possible to reconstitute certain crop plants from their component species.

In referring to polyploid crops it is important to distinguish between the numbers of chromosomes in the somatic cells (the $2n$ number) and the basic number of chromosomes that make up each genome (the x number). The chromosome number in the somatic cells of the cultivated strawberry is 56 ($2n = 56$).

The cultivated strawberry, however, is octaploid, having eight sets of chromosomes. The number of chromosomes in each genome is seven ($x = 7$). The chromosome number in the gametes—the haploid number, or n number—is 28. The concept of n is halfness, whereas the concept of x is oneness. In diploid species, $n = x$.

Mutation has provided the variation that, through evolutionary processing involving recombination and selection, has given rise to the diversity of living things, each adapted to particular niches of existence. Mutations occur at very low rates and are usually deleterious. Any random change, however, that increases the fitness of an organism tends to be perpetuated and increased through natural selection. Although mutations can at present be induced by a number of techniques, only random mutations can be accomplished.

Inheritance

The transmission of genes is a consequence of the events that occur in meiosis (see p. 198). The transmission of genes from generation to generation can be followed in a cross between two individuals that differ in one trait conditioned by a single gene.

Let us assume the existence of a gene with two forms (alleles), A and a. The heterozygous plant Aa will produce two kinds of gametes (A and a) in equal proportions. Crossing Aa and Aa will produce three kinds of progeny in a predictable ratio if all gametic and zygotic types are equally viable. It can readily be seen that the possible types of zygotes resulting from all combinations of the two kinds of gametes A and a will be AA, Aa, aA, and aa, or $1AA$, $2Aa$ and $1aa$ (Fig. 19-1). If the AA plants are indistinguishable from Aa plants, the gene A is said to be **dominant** and a **recessive**. The phenotypic ratio becomes $3A-$ to $1aa$.

The genotypes AA and Aa may be distinguished by a genetic test. The genotype AA will produce a single type of gamete (A). The genotype Aa will produce two kinds of gametes (A and a). Thus by crossing plants of the dominant A phenotype with themselves, or with the double recessive aa, these two possible genotypes may be separated on the basis of their progeny:

Figure 19-1 Mendelian segregation, illustrating incomplete dominance in the four-o'clock (*Mirabilis jalapa*). Note that the phenotypic ratio of 1 red:2 pink:1 white is also 3 colored:1 noncolored. [Adapted from T. Dobzhansky.]

$$AA \text{ selfed } (AA \times AA) \rightarrow \text{all } AA$$
$$(AA \text{ plants "breed true"}),$$

in contrast with

$$Aa \text{ selfed } (Aa \times Aa) \rightarrow 3A-:1aa$$
$$(Aa \text{ plants segregate});$$

or,

$$AA \times aa \rightarrow \text{all } Aa,$$

in contrast with

$$Aa \times aa \rightarrow 1Aa:1aa.$$

Generations are designated by specialized terminology. The first cross (usually referring to homozygous genotypes that differ with respect to a particular character) is the P_1, or the **parental generation**. The progeny of such a cross is the first filial generation (called the F_1). Homozygous but genetically distinct parents will produce an F_1 that is heterozygous and does not segregate. The F_2 (second

Selfed seeds from F₁ plants represent F₂ generation.

The F₂ ratio will be:
9A–B–, smooth yellow ○
3A–bb, smooth green ●
3aaB–, wrinkled yellow ○
1aabb, wrinkled green ●

○ ← (F₁ seed) → ○
All seeds from cross-pollinated flowers are smooth (Aa) and yellow (Bb). [Seeds represent hybrid (F₁) generation.]

Seeds from selfed flower are smooth (AA) and yellow (BB).

Seeds from selfed flower are wrinkled (aa) and green (bb).

pollen (ab)

artificial crossing in one flower

pollen (AB)

smooth AA yellow BB seeds

wrinkled aa green bb seeds

Figure 19-2   Independent assortment of two genes in the pea. [After N. H. Horowitz.]

filial generation) results from intercrossing or selfing F₁ plants:

$$P_1 \quad AA \times aa$$
$$\downarrow$$
$$F_1 \quad Aa$$
$$\downarrow$$
$$F_2 \quad 1AA : 2Aa : 1aa.$$

The F₂ is typically the segregating generation, which, if large enough, could theoretically include every possible genotype. Further generations are known as F₃, F₄, F₅, and so on. A cross of the F₁ with one of its parents is known as a **backcross.** (Similarly, generations produced by selfing only are denoted as S₁, S₂, and so on.)

The inheritance of characters controlled by two or more genes follows from an expansion of the single-gene model. The assortment of one pair of chromosomes at meiosis has no effect on the assortment of the other pairs; consequently, genes on separate chromosomes segregate independently of each other. Thus a plant heterozygous for two gene pairs on different chromosomes (Aa and Bb) will produce four different types of gametes (AB, Ab, aB, and ab) in equal proportions. The self progeny of the double heterozygote AaBb should segregate into a 3:1 phenotype ratio for each factor, assuming complete dominance. The combined phenotypic ratio would then be 9A–B–:3A–bb:3aaB–:1aabb. This type of cross is diagrammed in Figure 19-2.

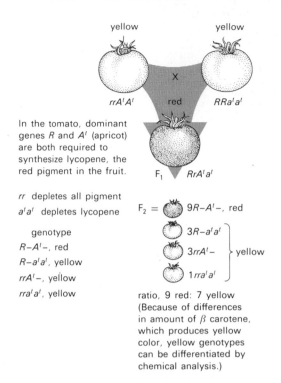

In the tomato, dominant genes R and A^l (apricot) are both required to synthesize lycopene, the red pigment in the fruit.

rr depletes all pigment

$a^l a^l$ depletes lycopene

genotype
$R-A^l-$, red
$R-a^l a^l$, yellow
rrA^l-, yellow
$rra^l a^l$, yellow

$F_2 =$ 9$R-A^l-$, red

3$R-a^l a^l$ ⎫
3rrA^l- ⎬ yellow
1$rra^l a^l$ ⎭

ratio, 9 red: 7 yellow
(Because of differences in amount of β carotene, which produces yellow color, yellow genotypes can be differentiated by chemical analysis.)

Figure 19-3 Gene interaction (epistasis) in the tomato.

When two genes affect the same biochemical pathway, the phenotypic ratios give some idea of the action of the genes. For example, the F_2 ratio, which involves a dihybrid in which two dominant genes interact to effect a single character, segregates into two classes in a ratio of 9:7 instead of four classes (in a 9:3:3:1 ratio) when two contrasting characters are involved. Either gene, when homozygous recessive, blocks some essential step, as shown in Figure 19-3. The phenomenon of gene interaction is called **epistasis.** Different types of epistasis can produce strange genetic ratios.

Linkage

Since a plant contains thousands of genes but only a limited number of chromosomes, each chromosome must contain many genes. In corn, for example, almost 1000 genes have been identified, but only 10 pairs of chromosomes. Because genes located on the same chromosome are not randomly assorted, they tend to be inherited together. This condition is referred to as **linkage.** There is a relationship between the physical closeness of genes and the intensity of strength of their linkage. The assortment of genes on the same chromosome is affected by crossing over, the actual exchange of chromosome material between chromosomes in meiosis.

By special techniques it is possible to determine on which chromosome a particular gene is located. Since the genes are in a linear sequence on the chromosomes, it is possible to determine their relation to each other by their linkage strength. In this way "road maps" of the chromosomes may be made, as shown in Figure 19-4 for the ten chromosomes of corn. The chromosome map is constructed only for recognizable genes—that is, genes for which alternate forms are known. The distance between the genes is related to the relative frequency of crossover events.

Linkages may help or handicap the plant breeder. The linkage of two desirable characters makes it easy to keep these characters together. In barley, for example, the genes for stem-rust resistance and loosesmut resistance are linked. But in order to free the plant of an undesirable gene, linkages between desirable and undesirable factors must be broken. If the genes are very close, extremely large populations of progeny may be required to obtain the exceptional but desired recombinants.

Quantitative inheritance

The plant characters, or traits, that have been discussed so far may be clearly separated into distinct, discontinuous classes, such as colored versus noncolored, or normal versus dwarf. The **qualitative characters** lend themselves to analysis because of the relative ease with which they can be classified. The greatest advances in genetic theory (beginning with Mendel) have been made using qualitative characters. Although many qualitative characters are of economic importance (for example, the red color of tomatoes and dwarfing in beans), most important characters, such as productivity, size, and quality, are continuous and range from one extreme to the other.

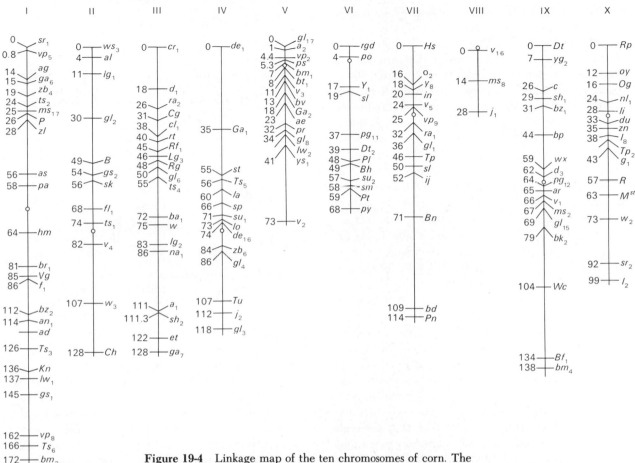

Figure 19-4 Linkage map of the ten chromosomes of corn. The numbers represent distance in the crossover units; the letters represent mutant genes. [Courtesy Dekalb Agricultural Association, Inc., after M. G. Neuffer.]

These are known as **quantitative characters.** For example, there is an almost imperceptible blending of color in wheat seed from very dark red to colorless. Similarly, corn plants vary continuously from very tall to almost dwarf. Such quantitative characters are difficult to investigate because of the problems encountered in excluding the effects of environment. Although the transmission of these characters was studied before the rediscovery of Mendel's paper, no general theory of inheritance was developed. It was clearly established, however, that this type of variation is inherited.

The present genetic theory holds that the inheritance of quantitative characters is essentially the same as for qualitative ones. The assumption is that there are many genes that affect these traits and that each contributes a small increment of effect to modify the character. These have been referred to as quantitative genes, polygenes, multiple factors, and modifiers.

The individual effect of each gene involved in a quantitative trait may be different, and the effect of environment may be large in relation to the small contribution of individual genes. As a result, it is difficult to determine the precise pattern of individual gene action. The special techniques used to analyze the inheritance of quantitative traits employ mathematical and statistical analysis of the variation encountered in crosses of plants that exhibit these characters. Some of these methods make it possible to estimate the number of genes involved and to assess their average contribution and average dominance. But because these methods rely on particular assumptions, the conclusions are difficult to prove precisely. For example, the genetic variability of quantitative characters can be divided into **additive, dominance,** and **epistatic components** that reflect the average action of the responsible genes. It is assumed that the additive component is responsible for the phenotypic difference between homozygotes of any single gene. All the genetic variability is additive when there are no intragenic (dominance) or intergenic (epistatic) interactions. The amount of additive genetic variance provides the best indication of the degree to which segregating populations will resemble their parents. An understanding of the components of variation that reflect gene action allows the breeder to predict results and thus make accurate decisions concerning various breeding procedures.

An example of the inheritance pattern for quantitative traits is shown in Figure 19-5. A cross between two homozygous lines of corn, one short-eared and the other long-eared, produced an F_1 population with a mean ear length intermediate between those of the parental lines. Presumably the parental and F_1 variation is entirely environmental. The distribution of the F_2 population includes the homozygous parental types. The variation of the F_2 can be shown to be greater than that of either the F_1 or the parental lines. This increase in genetic variation is due to gene segregation and recombination.

These results are similar to those that would be expected if a single partially dominant gene differentiated the parents, and if enough environmental variation were present to obscure the differences between the three expected F_2 classes—for example, *AA*, *Aa*, and *aa*. But if a single gene is involved, or if many are involved, markedly different consequences follow when F_2 plants are selfed to produce F_3 lines.

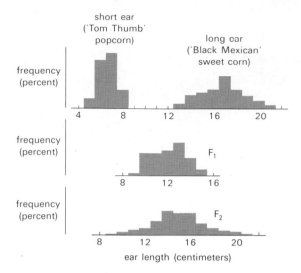

Figure 19-5 Distribution of ear length in a popcorn–sweet-corn cross. [Adapted from A. M. Srb, R. D. Owen, and R. S. Edgar, *General Genetics*, 2nd ed., W. H. Freeman and Company, copyright © 1965.]

If a single gene difference is assumed, only three F_2 genotypes in ratio of 1*AA*:2*Aa*:1*aa* would be expected. Therefore, one-fourth of the F_3 lines would resemble each parental type, and one-half would resemble the segregating F_2 distribution. In actuality, however, there are many more than three F_2 genotypes. It is possible to select F_2 genotypes that give an almost continuous range of ear length from that of the short-eared parent to that of the long-eared parent. This indicates that there are many genes controlling this character.

As the number of genes differentiating the parents with respect to any given trait (such as ear length) increases from one to n, the number of F_2 genotypes increases from 3 to 3^n. In contrast, the number of genotypes resembling each parental type decreases from one-fourth of the F_2 population to $1/4^n$. If only 10 genes differentiate a character, there are 3^{10}, or 59,049, different possible genotypes expected in the F_2 generation. To recover every genotype, a minimum population of 4^{10}, or 1,048,576, plants would have to be grown. In this population only one plant would be expected to duplicate each of the two parental genotypes.

The genetic value of a seed-propagated plant is the average performance of its progeny. Thus the basic problem in improving quantitatively inherited characters lies in distinguishing genetic from environmental effects. Those plants with the best appearance may not always be the most desirable genotypically. Because of the difficulties inherent in recognizing the effects of individual genes that govern quantitative characters, genetic analysis is heavily oriented toward the statistical approach. The great economic importance of quantitative traits demands that the breeder of crop plants be thoroughly familiar with the basic principles governing their inheritance.

Hybrid vigor

Hybrid vigor, or **heterosis,** is the increase in vigor, usually as compared to either parent, that occurs in certain crosses. The classic example of hybrid vigor is the mule, an interspecific cross of a mare and a jackass. The effect of heterosis in plants has been compared to the effects achieved by the addition of a balanced fertilizer. Hybrid vigor is often associated with an increased number of parts in indeterminate plants (whose main axes remain vegetative and whose flowers form on axillary buds; for example, cucumber) and with increased size of parts in determinate plants (whose main axes terminate in a floral bud; for example, maize). In perennial plants the vigor persists year after year.

Any organism is a hybrid if it results from a combination of genetically dissimilar gametes. (Although botanists often use the term hybrid to refer specifically to crosses made between species, geneticists use the term to refer to individuals that are heterozygous even for a single gene.) As used in the expression "hybrid corn" the term refers to a particular combination of inbred lines. To avoid confusion, the term hybrid should be qualified, such as single-gene hybrid, interspecific hybrid, or single-cross hybrid.

Heterosis is a phenomenon that is intimately associated with the decline in vigor brought about in some plants by continued crossing of closely related individuals (**inbreeding**). Selfing (crossing a bisexual

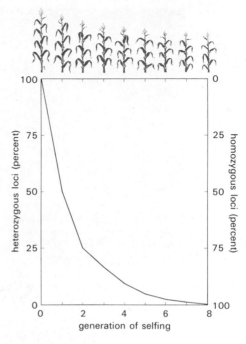

Figure 19-6 Self-pollination reduces the number of heterozygous genes by half in each generation. By the fifth generation more than 95 percent of these genes are expected to be homozygous. Most naturally cross-pollinated plants show a decrease in vigor as homozygosity is reached.

plant with itself) is an extreme example of inbreeding. Selfing has no effect on homozygous genes:

$$AA \times AA \to AA$$
$$aa \times aa \to aa$$

The selfing of a plant that is heterozygous for a single gene (*Aa*) produces progeny of which half are homozygous (*AA* and *aa*):

$$Aa \times Aa \to 1AA:2Aa:1aa$$

This increase in homozygosity may be generalized for any number of genes. With each generation of selfing, 50 percent of the heterozygous genes become homozygous (Fig. 19-6). Continued selfing in plants heterozygous for n genes will ultimately produce 2^n different lines, each of which is homozygous for all genes. Such plants are known as inbreds, or pure

Figure 19-7   Hybrid vigor in the onion. The F_1 hybrid in the center crate shows increased size and productivity compared with the parental inbreds on either side.

lines. Inbreeding can be defined as any breeding system that increases homozygosity.

The effect of inbreeding depends on the degree of heterozygosity and on the natural crossing mechanism in the plant. The self progeny of homozygous plants, of course, are unaffected. The self progeny of heterozygous plants increase in homozygosity. In some plants inbreeding often leads to a decline in vigor. This decline in vigor associated with an increase of homozygosity, or **inbreeding depression,** is usually high in those plants that are naturally cross-pollinated and therefore in a highly heterozygous state. Heterozygous individuals of self-pollinated plants can be obtained by crossing different homozygous lines. When these heterozygotes are selfed, the sharp characteristic decline in vigor associated with inbreeding in cross-pollinated plants is usually not found or is less pronounced.

The crossing of unrelated inbred lines of a cross-pollinated crop restores the vigor lost by inbreeding (Fig. 19-7). The progeny of such a cross are referred to as F_1 **hybrids.** In maize, some F_1 hybrids have shown yields as much as 25 percent higher than those of the original open-pollinated cultivars from which the inbred lines were derived. The increased vigor is due to the genotypic control made possible by careful selection of inbreds. An individual F_1 hybrid plant is not necessarily any better than the best plants of an open-pollinated cultivar. The fact that a vigorous, adapted F_1 hybrid line is composed of genetically homogeneous plants often guarantees increased performance over segregating open-pollinated cultivars. This lack of variation in F_1 hybrids, although disadvantageous under certain conditions, is a desirable feature in this age of mechanization.

This increase in vigor associated with the crossing of diverse selected (usually inbred) lines has been exploited by the commercial production of hybrids in such crops as petunia, zinnia, cabbage, cucumber, sugar beet, and sorghum, in addition to the well-known example of maize.

Crossing various cultivars or breeding lines of a self-pollinated crop (that in reality are inbred lines) increases vigor in isolated cases, but the increase is very much less than in cross-pollinated plants. An example of commercially produced hybrids in self-pollinated crops is the tomato. Hybrid wheats have now been introduced and are finding some success in the arid hard red winter wheat growing areas of the U.S. Great Plains.

Genetic interpretation of heterosis

Although heterosis is an important feature in plant improvement, its genetic basis is not completely established. Inbreeding depression and heterosis are related phenomena; when one occurs, the other can be expected with certainty. Consequently, any genetic explanation must reconcile both phenomena. The numerous explanations proposed for heterosis may be conveniently grouped under two general theories: dominance and overdominance.

The **dominance hypothesis,** first clearly stated by D. F. Jones in 1917, explains heterosis in terms of dominant or partially dominant growth factors. It is based on the assumption that cross-pollinated plants contain a large number of recessive genes concealed in the heterozygous condition. The recessive condition is due largely to the absence of some essential gene function, and homozygous recessives are therefore largely deleterious. As a result of inbreeding, concealed recessives become homozygous and cause loss of vigor. The restoration of vigor achieved by crossing inbred lines from diverse sources is due to the recessives of each inbred being mutually "covered up" by the other's dominant or partially dominant alleles, much as two moth-eaten bathing suits might cover up each other's holes. Thus, in successful combinations the two (or more) inbreds that make up a hybrid mutually complement each other's deficiencies. In its simplest case this situation may be expressed by the cross

$$aaBB \times AAbb \rightarrow AaBb$$

If each parent lacks vigor because of a deficiency in either dominant *A* or *B*, the hybrid *AaBb* would be more vigorous than either of its homozygous parents because of the presence of both dominant alleles.

According to the dominance hypothesis, then, the vigor of a hybrid is due to the large number of dominant genes that it contains or to the lack of particular homozygous recessive combinations that, like weak links, ultimately determine the strength of the chain of successful physiological reactions. The high heterozygosity of hybrids is merely a concomitant and incidental part of hybrid vigor. According to the theory, if it were possible to obtain each heterozygous gene in the homozygous dominant condition, the vigor of the resulting genotype would be as great as or greater than that of the heterozygous hybrid. The failure to obtain true-breeding homozygous lines that are as vigorous as F_1 hybrids is due to the large number of genes involved in traits such as yield. Thus the probability of obtaining one individual homozygous for each dominant gene is extremely low. In addition, linkages between deleterious and beneficial genes further reduce the expected frequency of recovery of homozygous dominant types.

The rate of decline of vigor caused by inbreeding depends upon the number of genes affecting a particular character and the frequency of deleterious genes in the population. The lack of a significant inbreeding depression and of heterosis in most naturally self-pollinating crops is explained as being due to the absence of deleterious recessives. Any deleterious recessives quickly become exposed by the natural inbreeding mechanism in self-pollinating crops and are rapidly eliminated by natural or artificial selection.

The dominance hypothesis has not been an entirely satisfactory explanation of heterosis to some geneticists. This is due in part to the high estimates of recessive alleles required for the phenomenon and in part to other statistical considerations. The **overdominance hypothesis** equates heterosis with heterozygosity. It assumes that heterozygosity, by itself, may be advantageous. If this is true, then the reduction in vigor upon inbreeding is a direct result of the attainment of homozygosity. According to the hypothesis, the heterozygous condition *Aa* is superior to either homozygous condition (*AA* or *aa*). This superiority has been termed overdominance.

There are a number of interpretations of overdominance at the gene level. If it is assumed that both alleles of a gene pair are physiologically active, then the combination of these alleles in the heterozygote may be superior to either alone. This may be due to supplementary action, alternative pathways, or the production of some hybrid substance. The combined activity of both alleles in the heterozygous condition produces heterosis. Another explanation proposes incomplete dominance and assumes that the amount of gene product produced by the heterozygote is optimum; the amount produced by the homozygous type is either deficient or inhibitory. There are experimental verifications, although admittedly rare, of these types of gene action.

It is difficult to distinguish in practice between the dominance and overdominance hypotheses, and, furthermore, they are not mutually exclusive. The essential features of the hybrid effect are explained by both. The correct explanation lies in determining the primary action of individual genes affecting heterotic traits. In general, however, the dominance hypothesis is thought to be the most likely explanation of hybrid vigor. This does not, of course, rule out the existence of overdominance, but it is felt by many that overdominance is at most only a contributory factor to hybrid vigor.

BREEDING METHODS

The goal of the plant breeder is to create superior crop cultivars. A **cultivated variety,** or **cultivar,** denotes a group of related plants within a species maintained either by sexual or asexual propagation, whose distinguishable characters are of agricultural significance (Box 19-1). The term "cultivar" (or "variety") has various meanings, however, depending on the mode of reproduction of the crop. With reference to asexually propagated crops, the term "cultivar" is used to mean any particular genotype that is considered of sufficient value to be graced with a name. With reference to sexually propagated crops the concept of cultivar depends on the method of pollination. The cultivar in self-pollinated crops is essentially a particular homozygous genotype—a pure line. In cross-pollinated crops the cultivar is not necessarily typified on the basis of any one plant but sometimes by a particular plant population, which at any one time is composed of genetically distinct individuals.

Methods of improvement have been developed for cultivars of all kinds. The improvement of vegetatively propagated crops (many fruit crops, woody ornamentals, potato, sugar cane) consists in producing a single desirable genotype. This particular

genotype may be of any degree of heterozygosity because elimination of the sexual process by vegetative propagation eliminates genetic segregation. There is no problem of genetic maintenance. Asexually propagated crops are usually cross-pollinating and consequently highly heterozygous; thus in developing populations from which to select a desirable genotype, inbreeding should be avoided. Unrelated plants are usually hybridized in order to obtain a vigorous population from which the most desirable individuals may be selected.

Breeding methods for the improvement of sexually propagated crops depend on the genetic structure of the cultivar, which is governed by the natural method of pollination. The amount of cross-pollination ranges from essentially none in such plants as soybean to 100 percent in dioecious and self-incompatible plants. For breeding purposes, however, two main groups are recognized: naturally self-pollinated plants, in which cross-pollination is less than 4 percent, and naturally cross-pollinated plants, such as maize, in which cross-pollination exceeds 40 percent. The intermediate types (cotton, sorghums) are usually considered with the cross-pollinated group.

Self-pollinated plants are ordinarily homozygous for practically all genes. The exceptions are the result of chance cross-pollination and mutation. Heterozygosity usually is quickly eliminated as a consequence of natural inbreeding. The basic problem in improving self-pollinated plants lies in producing and selecting the best homozygous genotype. Once this is accomplished, the problem of genetic maintenance is much smaller than it is with cross-pollinated species.

The genes in naturally cross-pollinated, seed-propagated plants are recombined constantly from generation to generation. The problems encountered in improving cross-pollinated plants include maintaining uniformity while avoiding the decline in vigor associated with homozygosity. Once a desirable population is achieved, there is still the perpetual problem of maintenance.

One method of maintaining both uniformity and heterozygosity is to produce hybrids through the crossing of selected diverse inbred lines. At present

473

Box 19-1 *International Code of Nomenclature for Cultivated Plants (1961)*

Article 5

The term cultivar, abbreviated cv., denotes an assemblage of cultivated individuals which is distinguished by any characters (morphological, physiological, cytological, chemical, or others) significant for the purposes of agriculture, forestry, or horticulture, and which, when reproduced (sexually or asexually), retains its distinguishing features.

Article 10

Cultivar is an international term for the category known in different languages by different names, for example, "variety" in English, "variété" in French, "variedad" in Spanish, "Sorte" in German, "sort" in the Scandinavian languages and in Russian, "ras" or "varieteit" in Dutch, and "razza" or "varietà" in Italian. Anyone is free to use the term cultivar or one of the equivalent terms. When the terms variety, variété and their variants are used in the sense of cultivar, confusion with the term "varietas" should be avoided by a suitable explanation. Varietas is a botanical category between species and forma. For the formation and use of cultivar (variety) names see Arts. 15-21.

Example: *Syringa vulgaris* 'Decaisne'. The first word (*Syringa*) is on the generic level being the name of the genus; the second word (*vulgaris*) is on the specific level and is the specific epithet; and the third word ('Decaisne') is the cultivar (variety) name.

Article 11

Kinds of Cultivars (Varieties). A cultivar (variety) may be any one of the following kinds of unit:

a. A *clone:* a genetically uniform assemblage of individuals (which may be chimaeral in nature), derived originally from a single individual by vegetative propagation, for example by cuttings, divisions, grafts, or apomixis. Individuals propagated from a bud-mutation form a cultivar (variety) distinct from the parent plant.

Examples: *Fraxinus excelsior* 'Westhof's Glorie'; potato 'Bintje'; *Cynodon dactylon* 'Coastal'; *Syringa vulgaris* 'Decaisne'; *Rubus nitidoides* 'Merton Early'.

Note. The word clone may be abbreviated cl. This abbreviation may be substituted for cv. if appropriate, e.g. *Fraxinus excelsior* cl. Westhof's Glorie.

b. A *line:* an assemblage of sexually reproducing individuals of uniform appearance, propagated by seeds or by spores, its stability maintained by selection to a standard.

Examples: *Petunia* 'Rosy Morn'; *Triticum aestivum* 'Marquis'.

c. An assemblage of individuals showing genetical differences but having one or more characteristics by which it can be differentiated from other cultivars (varieties).

Examples: *Cucurbita* 'Connecticut Field Pumpkin', a recognised cultivar (variety) known to vary markedly as to fruit size and shape, these variations not being sufficiently distinct for any group of individuals within the cultivar (variety) to warrant cultivar (variety) rank; *Medicago sativa* 'Ranger', the breeder seed of which is derived from intercrossing five seed-propagated lines each maintained under isolation; *Phlox drummondii* 'Sternenzauber', a mixture of different colour forms, all characterised by the same starlike shape of the corolla.

d. A uniform assemblage of individuals which is a first generation hybrid (F_1) reconstituted on each occasion by crossing two or more breeding stocks maintained either by inbreeding, or as clones, or recurrently made as F_1 hybrids.

Examples: Onion 'Granex', derived by crossing two other onion cultivars (varieties); maize 'US 13', a double-cross involving four inbred lines.

only technical problems prevent this from becoming a standard breeding technique for cross-pollinated, seed-propagated crops.

Introductions

An abundant source of variation is available to the plant breeder from cultivars or related wild species (Fig. 19-8). The first step in any crop-improvement program, therefore, is to assemble the natural variants available. From a study of their performance it may be possible to make an immediate improvement merely by choosing some cultivar not being grown. Because of the wide distribution of most crops, a representative collection should be worldwide in scope. It should include, if possible, old as well as new cultivars. In the United States the Plant Genetics and Germplasm Institute of the Agricultural Research Service (United States Department of Agriculture) carries out the exploration and introduction of genetic variability. Collections of germplasm or "germplasm banks" are currently available for many crops.

There are many examples of crop improvement by introduction from other parts of the world. 'Bartlett,' the main pear cultivar in the United States, was introduced (about 1797) from England and was originally called 'Williams' Bon Chrétien.' Similarly, the cultivar of pineapple grown in Hawaii is the 'Cayenne,' which was introduced from South America. In Brazil one of the valuable dry bean cultivars, 'Rico-23,' was introduced from Costa Rica. Some new selections turn out to be better adapted to areas other than the one in which they were developed. The Indiana wheat cultivar 'Redcoat' was slated for discard, but was subsequently named and released because it performed well in Pennsylvania. The Monterey pine, insignificant in its native California habitat, was introduced into Australia, where it is now one of the most important timber species grown.

With established crops and active breeding programs it becomes more and more difficult to make improvement through direct introduction. Usually, in spite of individually desirable attributes, no one introduction contains the proper combination of characters to be useful without further breeding. Often one or two desirable characters are found in plants that are otherwise unsuitable or unadaptable. Resistance to the apple-scab disease is found in species of apple with almost inedible fruit that is only a few centimeters in diameter. In such plants the approach is to recombine the characters by sexual crossing (**hybridization**) in the hope of obtaining plants that have a more favorable combination of desirable features. When these characters are distributed over many cultivars, or the desirable characters are associated with many undesirable ones, this procedure requires many crosses and many generations.

Genetic variability is the raw material of the plant breeder. The richest source of genetic variability for a particular cultivated species has been shown to be its geographical area of origin. The search for these areas of origin is not only intriguing and fascinating, but is important to the present-day improvement of crop plants. Many types of evidence—genetic, botanical, chemical, archeological, historical—can be brought to bear on individual crops.

Modern study in this field derives from Alphonse de Candolle's pioneering *Origin of Cultivated Plants* (1822). Drawing on many sources, de Candolle surmised that suitable climate and presence of wild relatives were good indications of original habitat, especially if supported by historical evidence. The Soviet plant scientist Nikoli Ivanovich Vavilov, with the benefit of new techniques and widespread collections, worked on the origins of cultivated plants for twenty years, from 1916 to 1936, and succeeded in expanding and revising de Candolle's conclusions. Vavilov was convinced that many crop plants do not occur outside of cultivation, and that wild relatives are often represented by narrow, isolated groups. He reasoned that centers of origin are marked by considerable variability and that, as plants spread from such centers, certain distinctive genotypes become successful and dominate the population. Dominant genes characterize dispersal centers, with recessive forms more likely to gain opportunity at the periphery of a plant's range.

Vavilov emphasized that the distribution of plant species is not uniform over the vegetated areas of the

Figure 19-8   Variation in corn. A collection of types grown in Guatemala. [Courtesy USDA.]

earth. For example, two small Central American countries, Costa Rica and El Salvador, possess as many native species as the United States and Canada combined, yet have only 1 percent of the land area. A number of regions especially rich in concentrations of different plants have been characterized as **centers of diversity** and are thought to coincide with the primary and secondary **centers of origin** of present-day cultivated plants (Fig. 19-9).

These areas, including only a small portion (2–3 percent) of the land area of the earth, are geographically distinct, being isolated by deserts or mountain ranges. The interaction of the rich local flora and the primitive human populations in these locations appears to have brought about independent pockets of agricultural development. Five-sixths of the 640 species listed by Vavilov are from the Old World and one-sixth from the New World.

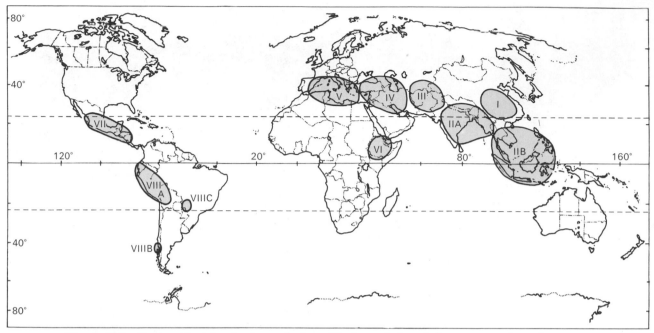

Figure 19-9   Centers of origin of cultivated plants. The key is below. [After N. V. Vavilov, *The Origin, Variation, Immunity and Breeding of Cultivated Plants*, Ronald Press, 1951.]

Key to the centers of origin of important cultivated plants

Old World

I   *China*—The mountainous and adjacent lowlands of central and western China represent the earliest and largest center for the origin of cultivated plants and world agriculture. Native plants include millet, buckwheat, soybean, many legumes, bamboo, crucifers, onion, lettuce, eggplant, cucurbits, pear, cherry, quince, citrus, persimmon, sugarcane, cinnamon, and tea.

II   *South Asia*—This area has two subcenters. The principal one, Hindustan (II A), is considered the center for rice, sugarcane, many legumes, and such tropical fruits as mango, orange, lemon, and tangerine. The other, Indo-Malaya (II B), is the center for banana, coconut, sugarcane, clove, nutmeg, black pepper, and manila hemp.

III   *Central Asia*—This region is most important as the center of origin of common wheat. Native plants include pea, bean, lentil, hemp, cotton, carrot, radish, garlic, spinach, pistachio, apricot, pear, and apple.

IV   *Asia Minor*—At least nine species of wheat, as well as rye, are native. This is the center for many subtropical and temperate fruits (cherry, pomegranate, walnut, quince, almond, and fig) and forages (alfalfa, Persian clover, and vetch).

V   *Mediterranean*—This is the home of the olive and many cultivated vegetables and forages. The effect of early and continuous civilizations is indicated in the improvement of crops that originated in Asian centers.

VI   *Abyssinia (Ethiopia)*—Wheats and barleys are especially rich; sesame, castor bean, coffee, and okra are indigenous.

New World

VII   *Central America*—Extremely varied native plants, including maize, bean, squash, chayote, sweetpotato, pepper, upland cotton, prickly pear, papaya, agave, and cacao.

VIII   *South America*—A number of South American centers are noted. The West-Central area (VIII A)—now Ecuador, Peru, and Bolivia—is the center of origin of many potato species, tomato, lima bean, pumpkin, pepper, coca, Egyptian cotton, and tobacco. The island of Chiloe in southern Chile (VIII B) is thought to be the source of the potato. The peanut and the pineapple originated in the semiarid region of Brazil (VIII C), and manioc and the rubber tree in the tropical Amazon region.

Selection

Selection, as a breeding term, refers to the differential reproduction of individuals in a population. Selection is achieved by preserving favorable variants and eliminating undesirable ones (Fig. 19-10). It is often a natural process, because, in the absence of human interference, those plants most adapted to survive and reproduce leave the most descendants. Almost all of the crops in cultivation today were domesticated before the advent of written history. Improvement of these crops has been continuous as a result of human selection. This process, carried out both consciously and unconsciously over the years, has been extremely effective; many of our cultivated plants are very different from their wild ancestors; some are known only in cultivation—for example, cabbage and maize. As a result of human selection over the centuries, they have changed so much that their wild ancestors have become obscure.

The efficient use of the selection process is one of the principal tools of the breeder. It is important to keep in mind, however, that selection does not create genetic variability but merely acts on the genetic variability already available. Thus the breeder must first create a variable population from which to select and then must recognize and propagate those individuals with superior characteristics. In order to obtain superior populations from which to select, a wise choice of parental varieties must be made. The efficient evaluation of potential parents is one of the main problems in modern plant breeding. The simplest technique used in choosing parent plants is to look for plants that possess complementing characters. More sophisticated methods involve the sampling of hybrid progeny to predict "prepotency," the ability of an individual to produce desirable offspring. The proper choice of parental materials is one of the crucial decisions in any breeding program.

The two general methods of selection used to sort desirable genotypes from segregating progeny are called **pedigree selection** and **mass selection** (Fig. 19-11). In pedigree selection, individual plants are selected on the basis of their ancestry or pedigree as well as their phenotype. It is assumed that the best

Figure 19-10 The effect of fifty generations of selection for high and low oil and protein content in corn. Solid lines indicate actual data; broken lines indicate trend. [After C. M. Woodworth, E. R. Leng, and R. W. Jugenheimer, reproduced from *Agronomy Journal* 44:60, 1952.]

genotypes will be derived from superior plants that are the progeny of superior parents. Therefore, the progeny of each plant are handled separately. Straight pedigree selection entails laborious and time-consuming effort. It involves keeping extensive records, and unless rigid selection is maintained the program soon becomes unwieldy. As a result, this method limits the amount of material that can be handled, but the method does allow detailed genetic studies to parallel the breeding program. In addition, self-pollinated crops are extremely responsive to efficient pedigree selection, and progress reflects to a great extent the individual breeder's skill.

In mass selection, individual desirable plants are

478

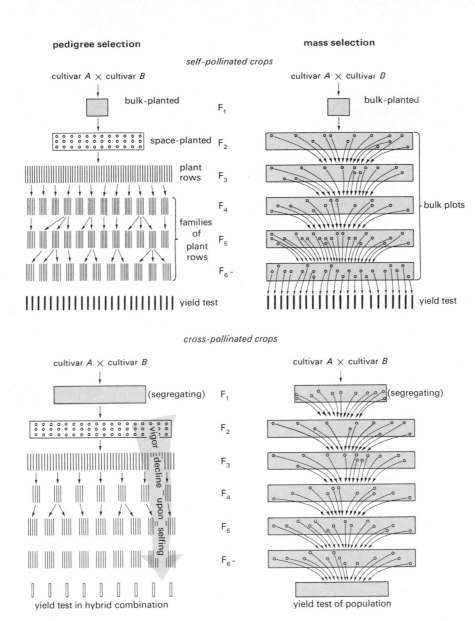

pedigree selection   **mass selection**

self-pollinated crops

cultivar *A* × cultivar *B*   cultivar *A* × cultivar *B*

bulk-planted F_1 bulk-planted

space-planted F_2

plant rows F_3

                    F_4

families
of
plant F_5 bulk plots
rows

                    F_6 -

yield test   yield test

cross-pollinated crops

cultivar *A* × cultivar *B*   cultivar *A* × cultivar *B*

(segregating) F_1 (segregating)

                    F_2

                    F_3

                    F_4

                    F_5

                    F_6 -

vigor decline upon selfing

yield test in hybrid combination   yield test of population

Figure 19-11 Pedigree selection and mass selection in self-pollinating and cross-pollinating crops. In pedigree selection, superior plants are selected from best rows and their progeny are kept separate. In mass selection, progeny of selected individuals are grown in mass. The consequences of the selection method depend on the natural method of pollination, as described in the text. Various modifications of these selection methods are possible.

selected solely on the basis of phenotype, and the resulting seed is pooled and the progenies are grown in mass. Individual progeny are not tested separately. Selection may be accomplished either by saving the best plants or by eliminating (**roguing out**) undesirable ones. Roguing out can be done mechanically for some crops, for example, by screening for seed size or harvesting at a particular date to select for season of maturity. Mass selection is based on the premise that the forces of natural selection will aid the breeder in eliminating undesirable types. The method permits a large pool of germplasm to be manipulated at one time and overcomes some of the limitations of pedigree selection. It is particularly suited to those crops that show decline in vigor under the forced inbreeding inherent in straight pedigree selection.

In actual practice, the processes of mass selection and pedigree selection may be combined. Pedigree selection may be utilized in the early segregating generations to exploit the major genetic differences and eliminate obvious undesirable types. Following this, mass-selection techniques may be used. Most modern breeding techniques involve a combination of these systems.

The particular method of selection used depends to a great extent on the mode of reproduction and the method of pollination of the crop. In general, pedigree selection is most useful for self-pollinated crops and mass selection for cross-pollinated crops.

Selection in self-pollinated crops

Two fundamentally different types of populations of self-pollinated crops exist. One is a mixture of different homozygous lines, as found in a collection of cultivars. Here selection consists of determining the best genotype by testing. The best genotype can be duplicated from its selfed seed. The other type of population is a mixture of different heterogeneous genotypes, as found in the F_2 generation of a cross between two homozygous cultivars. This population consists of many different genotypes with varying degrees of heterozygosity. In order to obtain an improved type, the best genotype must be selected and then transformed into a homozygous, true-breeding line without losing the essential characteristics of the selected individual.

Pedigree selection is widely used for self-pollinating crops. Individual plants are generally selected from a segregating population (typically, the F_2 generation of a cross between cultivars), and selection proceeds between and within progenies in each subsequent generation until genetic purity is reached. Since selfing brings about 50 percent more homozygosity in each generation, the variation between lines is halved with each generation. By choosing between lines rather than between plants, there is less chance of confusing genetic and environmental effects. This process is repeated each generation. By the fifth generation the lines derived from single plant selections will be homozygous for more than 95 percent of their genes. Such lines are for practical purposes considered to be true-breeding. If testing shows any one of these lines to be obviously superior to existing cultivars, it may be named to identify it as a new cultivar.

Mass selection may also be used in self-pollinated crops by planting segregating populations in large plots and harvesting in bulk. Selection may be practiced each generation by eliminating undesirable plants. After 5 to 10 generations the population will consist of a heterogeneous mixture of somewhat selected homozygous genotypes. The progeny of any plant can be expected to form the basis of a new cultivar. The problem at this stage is to determine the best genotype by testing. The method is successful if the better genotypes are retained. Mass selection, by virtue of its economy, permits a large pool of germ plasm to be manipulated and carried along.

The combination of mass selection and pedigree selection is well suited to self-pollinated crops. For example, F_2 plants can be pedigree-selected on the basis of the performance of the F_3 line. After the principal agriculturally desirable types are selected, the seed of the best F_3 (or F_4) lines may be pooled, and the process may then be carried on by mass selection until homozygosity is reached. At this time pedigree selection is resumed to isolate the superior genotypes. Some cultivars may intentionally consist of a mixture of pure lines to provide for some variability.

Selection in cross-pollinated crops

Pedigree selection depends on the evaluation of genotypes by inbreeding. Straight pedigree selection is therefore undesirable as a method of improving cross-pollinated plants when inbreeding leads to less vigor, unless some procedure is set up to restore vigor subsequently. Inbreeding is avoided in the mass-selection method by natural interpollination of selected individuals.

Mass selection often leads to an initial rapid improvement of cross-pollinated plants. Sometimes, however, it becomes difficult to increase genetic gain beyond a certain level. To achieve greater control of the genotypes making up the cross-pollinating population, pedigree and mass selection may be combined. This is accomplished by selfing individual plants and pedigree-selecting them on the basis of their progeny. The selected lines are then allowed to interpollinate to restore heterozygosity and vigor. The process may be repeated for a number of cycles. Such a system is called **recurrent selection.** There are a number of variations of this system. The benefit of the technique is that it allows steady progress without eliminating the variability brought about by cross-pollination. When the character evaluated is a fruit or a seed, selection must be made after pollination, in which case both pedigree selection and mass selection are carried out in the same generation. For example, the "female" parent, or seed-bearing plant, is pedigree-selected by planting its progeny separately, but the "male" parent, or pollen-producing plant, is mass-selected. This technique was first used on maize, in which connection it is known as "ear-to-row breeding."

In a very real sense, F_1 hybrids are an extreme product of recurrent selection. Here the inbreeding process is carried on to homozygosity, usually under a pedigree-selection system, and vigor is restored by crossing two (or more) inbred lines. The selection of inbred lines and the choice of inbred combinations results in uniform, genetically controlled hybrids.

In the past, mass-selection techniques have worked very well for highly heritable characters but have been disappointing for characters that have low heritability, such as yield. As a result, mass selection was generally abandoned in such crops as maize, and breeders switched to selection within and between inbred lines—that is, hybrid breeding. As more was learned about the genetics of populations, however, it became clear that what was lacking in the mass-selection technique were refined methods for distinguishing genetic from environmental variation. By using special selection techniques with careful replication and statistical control, steady gain in yield has been possible with mass selection. These techniques offer great promise for improving the yield potential of cultivars for use in locations where the uniformity of hybrids may not be desirable (as in mountainous areas) and for providing better populations from which to extract inbred lines.

Selection in asexually propagated crops

Selection is straightforward in asexually propagated crops, since any genotype may be perpetuated intact. The problem is one of testing to determine the most desirable genotype. Obtaining segregating populations from which superior genotypes may be found is the chief problem in breeding asexually propagated material. Unfortunately, the most desirable selections do not always make the best parents; consequently, potential parents are best selected on the basis of progeny performance.

Somatic mutations of spontaneous origin within vegetatively propagated material are commonly referred to as **sports.** Desirable mutations occurring in adapted, asexually propagated plants may result in an immediate improvement, such as the color sports in many apple varieties and superior tree types in coffee.

Mutations that occur in somatic tissue may be confined to only a sector of tissue, resulting in a **chimera** (named after a mythical beast with the head of a lion, the body of a goat, and the tail of a dragon). Such chimeras, when vegetatively propagated, tend to be unstable because buds may be formed from tissue with or without the mutation (Fig. 19-12). For

Figure 19-12 Plant chimeras contain mixtures of genetically different tissues. Various types may be derived from buds arising from a sectored chimeral stem. [After W. N. Jones, *Bot. Rev.* 3:545–562, 1937.]

example, the 'White Sim' carnation, a sport of the 'Red Sim' cultivar, is a chimera. Its white flowers tend to have islands of red tissue in them, and buds from internal tissue will produce red flowers. The selection of desirable sports is an important means of improving asexually propagated crops. This process is known as **bud selection.**

The backcross method

The backcross has been defined as a cross of an F_1 plant with one of its parents. The backcross method is a breeding technique for transferring single, readily identifiable characters from one cultivar or line to another. Characters controlled by single genes, such as growth habit and certain types of disease resistance, are easily transferred by this method. By repeated backcrossing of the hybrid with the parent variety that carries the most desirable characters (the **recurrent parent**), but selecting in each generation for a single character from the other (**nonrecurrent**) parent, a genotype will eventually be obtained that has all the genes of the recurrent parent plus the selected gene of the nonrecurrent parent.

Two sets of genes must be considered in the backcross method. One set consists of the particular gene (or genes) from the nonrecurrent parent that is to be transferred, and the other is the large group of genes from the recurrent parent that is to be retained. The genes of the recurrent parent are transferred by an increment of 50 percent in each generation of backcrossing. By selecting plants at random in each generation, the genes of the recurrent parent not only become incorporated but become homozygous. This can be shown easily with respect to a single gene.

Assume a cross involving a single gene pair, $AA \times aa$. The two types of backcrosses are:

$$Aa \times AA \rightarrow 1Aa{:}1AA$$
$$Aa \times aa \rightarrow 1Aa{:}1aa$$

Compare these with the self:

$$Aa \times Aa \rightarrow 1AA{:}2Aa{:}1aa$$

In backcrossing, as in selfing, half of the genes become homozygous. This can be generalized for any number of genes. The rate of return to homozygosity in backcrossing is the same as in selfing, but all of the homozygous plants resemble the recurrent parent. If backcrossing to the same homozygous type is continued for five or more generations, more than 95 percent of the genes of the hybrid will be identical to those of the recurrent parent and will be homozygous.

If the gene transferred is dominant, the procedure is straightforward. After the last backcross, the gene may be made homozygous by selfing. If the character transferred is recessive, special testing must be carried on in each generation to make sure that the plant selected for backcrossing is heterozygous and contains the desired recessive.

The backcross method has found its greatest usefulness in self-pollinated crops and in improving inbred lines of cross-pollinated crops. The reason for its limited usefulness in cross-pollinated crops is that in such crops backcrossing is equivalent to selfing and thus results in a decline in vigor. This may be overcome by using a large number of selections of the recurrent cultivar to avoid inbreeding (Fig. 19-13).

The main advantage of the backcross method is its predictability. The improved cultivar is often indistinguishable from the old, with the exception of the added character. As a result, extensive testing

A

B

C

D

E

Figure 19-13 Backcross breeding for scab resistance in the apple. (A) The fruits of *Malus floribunda* '821,' which is heterozygous for a dominant gene that confers resistance to apple scab. (B) A selection derived from a cross of *M. floribunda* '821' × 'Rome Beauty.' (C,D) Subsequent backcrosses to different large-fruited types produce resistant seedlings with increased fruit size and quality. (E) The fruits in the basket are the 'Prima' apple, the first release of the scab-resistant apple breeding program cooperatively conducted by Purdue University, Rutgers University, and the University of Illinois. 'Prima' has been successfully introduced in the United States as an early apple both for home gardens and for commercial use. Its use is expanding in eastern and western Europe, with the most extensive plantings in Yugoslavia. (F) Resistant seedlings are selected in the seedling stage. Susceptible seedlings are killed by the disease. [From J. Janick, Horticultural Science, 3rd ed., copyright © W. H. Freeman and Company, 1979.]

F

may be avoided. The backcross method cannot, however, be expected to improve the cultivar more than the increment achieved by the addition of the single selected character. Moreover, unless this character is inherited relatively simply, the method is difficult to use.

Hybrid breeding

Hybrid breeding refers to the production of heterozygous populations from crosses of homozygous lines. Hybrid breeding first came into prominence with maize. Although the feasibility of hybrid corn was suggested in 1909, the first commercial hybrids did not appear until the 1920s. From 1936 to 1945 the total amount of hybrid corn in the Corn Belt increased from less than 5 percent to more than 90 percent of the corn planted. The spectacular success of hybrid corn has given great impetus to this method of breeding. Because of the genetic nature of heterosis, hybrid breeding is most applicable to cross-pollinated crops, but it has limited usefulness for some self-pollinated crops.

Seed of F_1 hybrids must be remade each year. Consequently, for the success of this breeding method the extra advantage resulting from the use of hybrids must more than compensate for the additional cost of their production. Thus, whether hybrid breeding is used in a particular crop depends on the balance between the specific technical problems involved in producing and combining homozygous lines and the amount of the heterotic effect that is produced.

There are two technical steps in the production of hybrid seed for cross-pollinated crops: (1) the production of desirable homozygous lines; and (2) crossing of these lines to obtain hybrid seed. Self-pollinated crops are already naturally inbred, and therefore need only be crossed to produce hybrids.

Production of homozygous lines

The most common method of producing homozygous lines is through continued inbreeding. In self-pollinated crops homozygous lines occur naturally, but in cross-pollinated crops they must be produced by the breeder. The ease of producing inbred lines within a cross-pollinated species varies with the crop species. In perfect-flowered species, the inflorescence merely needs to be bagged to exclude foreign pollen. In monoecious crops, which have separate staminate and pistillate flowers, the pistillate flower must be protected to prevent random cross-pollination, and pollen must be collected and applied to the stigmal surface of the pistil. In maize this process is relatively simple because all the pollen is produced in one inflorescence, the tassel, and all the pistillate flowers are congregated in the ear.

The production of inbred lines is not straightforward in all crops. Dioecious species, such as spinach, consist of separate staminate and pistillate plants. Inbreeding in these crops is accomplished by sib (brother–sister) pollination. The rate of return to homozygosity in such crops is much lower than it is in self-pollinating crops.

A number of crops resist self-fertilization because of incompatibility mechanisms. Self-incompatibility is usually controlled by alleles at a single gene locus. When pollen contains the same incompatibility allele that is present in the style, tube growth is arrested. In a few crops, such as cabbage, incompatibility may be circumvented by pollinating before the flower opens (bud pollination). Self-incompatible lines may be increased asexually; this is being done with alfalfa.

Homozygous lines may be produced in a single step, avoiding the long process of inbreeding by means of some quite interesting and sophisticated techniques. Haploids are produced in certain crops, such as corn, with a very low but predictable frequency. They apparently result from a disruption in the double fertilization process. One male gamete combines with the two polar nuclei to form the triploid endosperm, but the fertilization of the egg nucleus does not occur. Haploids may be recognized in the seedling stage by the use of special marker genes. A doubling in the chromosome number of a haploid of a diploid species (= monoploid) produces an "instant" homozygous diploid. Haploids can also be produced by tissue-culture techniques from pollen. This has been achieved with tobacco, rice, tomato, and various other crops.

Another method of producing homozygous lines is

through a complex series of translocations. If a line of plants can be synthesized in which a portion of each chromosome is attached to another, it is possible to create inbred lines in one step. Such a series of translocations occurs naturally in evening primroses (*Oenothera* spp.). A similar complex series of translocations has been synthesized in maize and barley, but the incidence of sterility was too high for the method to be practical.

Combining inbred lines

The problem of producing crossed seed is the reverse of inbreeding. For maximum gain, selfing must be eliminated. The basic approach consists of transforming one parent of the cross into a "female" line (Fig. 19-13). The simplest method is to eliminate the pollen-bearing organ (emasculation). This is accomplished in corn by removing the emerging tassels. In perfect-flowered species (such as tomato), each flower must be emasculated. In dioecious species, staminate plants must be eliminated completely.

In perfect-flowered crops that produce small flowers or a small number of seeds per flower, the emasculation process is not feasible. In these crops the production of hybrids depends on genetic methods of eliminating selfing. One way to accomplish this is to induce male sterility. Chemicals that cause male sterility are known, but their use is not yet practical. At present, the most efficient way to eliminate selfing is to make use of the genetically determined male sterility that results from the interaction between a cytoplasmic factor and nuclear genes. The cytoplasmic factor is transmitted only through the pistillate parent. Cytoplasmic male sterility was first described in onions and subsequently found in many other crops. Hybrids have been produced commercially by means of cytoplasmic male sterility in onions, corn, and sorghum. This male sterility results from the interaction of the cytoplasmic factor with one or more recessive genes. In the dominant condition these genes are known as "restorers" of fertility. Cytoplasmic male sterility can be incorporated into inbred lines by backcrossing.

The use of cytoplasmic male sterility to eliminate detasseling in corn was so successful that by 1970

practically all hybrids had been produced by that method. The system worked so well that most hybrids contained sterile cytoplasm derived from a single plant discovered in Texas (T cytoplasm). However in the summer of 1970 an epidemic of Southern corn-leaf blight caused ruinous losses (valued at about 1 billion dollars, the greatest single plant-disease loss in history!) as a result of the appearance and spread of a new strain of the causal fungus, *Helminthosporium maydis*, that attacked *only* plants with the sterile (T) cytoplasm. The situation was totally unexpected because a linkage between susceptibility to a fungus and a cytoplasmic factor was all but unknown before this outbreak. As a result of the 1970 epidemic, a shift to normal cytoplasm and detasseling took place in hybrid-corn seed production. The epidemic was a striking demonstration of the way in which uniformity can result in the vulnerability of a major crop. Studies to evaluate new cytoplasms for male sterility are in progress, but it is unlikely that agriculture will ever again rely on a single source.

In cucumbers, hybrid seed can be produced by a gene that converts a normally monoecious species into a gynodioecious one, which gives rise to equal numbers of monoecious and completely pistillate plants upon selfing. When the monoecious plants are eliminated (rogued), they leave a completely pistillate line. It is also possible to produce a gynoecious, or "female," line. Such a line may be maintained by self-pollination after the induction of staminate flowers by means of gibberellins.

Self-incompatibility factors may be utilized to prevent selfing. This is done in the production of hybrid cabbage and alfalfa.

Where it is not possible to eliminate selfing completely, genetic markers may be utilized to distinguish cross-fertilized from self-fertilized plants. The "female" line is made homozygous for some recessive seedling character; the male line is homozygous dominant. All hybrid plants will carry the dominant marker gene; plants produced from selfing in the female line will show the recessive phenotype, and can be eliminated in seedbeds or in the field. Hybrid cotton has been produced by this method. A number of methods are used to increase the proportion of natural crossing and to reduce selfing, including chemically inducing male sterility and increasing insect activity.

The transfer of pollen in cross-pollinated crops is usually accomplished naturally by wind or insects. In self-pollinated crops the pollen must be transferred artificially. Nevertheless, even in naturally self-pollinated crops, cross-pollination can often be increased by encouraging insect activity. In male-sterile wheat, sufficient natural cross-pollination occurs by wind to make hybrids feasible.

Selection within and between inbred lines

The success of hybrid breeding depends on the efficient selection within and between inbred lines. Selection within lines refers to selection in the inbreeding process; selection between lines refers to the choice of combinations in the crossing of inbred lines.

The success of an inbred line is measured not by its phenotype or its self progeny but by its cross progeny. The ability of an inbred line to produce successful hybrids is known as **combining ability.** Combining ability may be partitioned into general and specific effects. **General combining ability** refers to the average performance of a particular inbred line in a series of hybrid combinations. **Specific combining ability** refers to the performance of a specific combination of inbred lines.

General combining ability is an inherited character. It can be estimated by crosses with a tester, usually an open-pollinated cultivar with a broad genetic base. Combining ability may be selected for early in the inbreeding process. Selected plants are crossed with the test cultivar at the same time they are selfed. The selfed seed is used only if the results of the general test of combining ability are favorable.

The genetic improvement of inbred lines is a specialized type of breeding. In general, inbred lines are handled as self-pollinated crops. Inbred lines can be improved, one character at a time (for example, disease resistance, or male sterility), by the backcross method.

One of the most difficult problems is to improve the combining ability of inbred lines. Present-day corn hybrids do not show great yielding superiority over the best of the early hybrids. Improvement has come by the incorporation of better agronomic characters, such as disease resistance and specific adaptation. In present-day corn breeding, emphasis is on improving populations in order to produce superior inbred lines. This is achieved by recurrent selection, with combining ability a principal criterion of selection.

Although maximum heterosis is obtained by crossing two diverse inbred lines, a number of other combinations produce hybrid vigor. The various kinds of crosses that are referred to as hybrids in the trade are designated as follows:

Type of cross	Name of hybrid
inbred × inbred	single-cross
F_1 hybrid × inbred	3-way-cross
F_1 hybrid × F_1 hybrid	double-cross
inbred × cultivar	top-cross
F_1 hybrid selfed	F_2 hybrid

The double cross, first suggested by D. F. Jones in 1918, was responsible for the commercial feasibility of hybrid corn. Inbred lines of corn are relatively weak, and their seed yields are low. By crossing certain single crosses, the cost of the seed was greatly reduced with little loss of heterosis. Double-cross yields are predicted from the yields of single crosses. Techniques have also been established for producing double-cross seed by means of cytoplasmic male sterility, which eliminates manual detasseling (Fig.

When the production of F_1 hybrids is impractical, heterosis may be obtained by mixing a number of inbred lines selected for their ability to combine well with each other. The subsequent population is referred to as a **synthetic variety.** Synthetics are not made anew each year but maintained from open-pollinated seed. Synthetic varieties have been most popular in forage crops whose floral structure makes the control of pollination difficult.

The success of hybrid breeding is due in large part to the inherent commercial possibilities. Control of

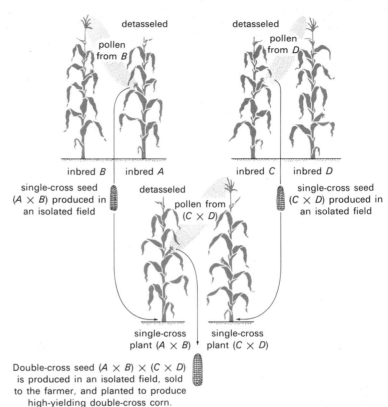

detasseling system

Figure 19-14 The production of single-cross and double-cross hybrid corn. The detasseling process (*right*) can be eliminated by manipulating male sterility (*opposite page*) determined by an interaction between a cytoplasmic factor $\boxed{S}$ and a double recessive nuclear gene *rf*. Only the combination $\boxed{S}$ *rf rf* is male-sterile. Plants with $\boxed{N}$ (normal) cytoplasm or with the pollen-restoring allele *Rf*, either homozygous or heterozygous, produce viable pollen. The cytoplasmic factor passes only through the egg. Appropriate genotypes are shown in shaded rectangles.

the unique sets of inbred lines effectively controls the hybrid, for hybrids do not breed true from seed. In hybrids between cultivars of self-pollinated crops, it is only necessary to know the two cultivars that are used to make the hybrid.

Polyploid breeding

Polyploidy is used in a number of important ways in breeding. The level of ploidy may in itself be important in conferring desirable characteristics. In addition, the manipulation of ploidy greatly increases the amount of variability available to the breeder, because many desirable characters are found in related species with different chromosome numbers.

Polyploids may occur spontaneously or be artifi-

cially induced by special treatment. For example, the formation of callus tissue by wounding is effective in inducing tetraploidy in the tomato. Colchicine, an alkaloid derived from the autumn crocus (*Colchicum autumnale*), has the remarkable property of doubling the chromosome number in a wide variety of plants when applied in concentrations of 0.1 to 0.3 percent. This drug interferes with spindle formation in mitosis; the chromosomes divide, but the cell does not. Tetraploids are produced routinely in many plants with the use of colchicine. Treatment of seeds with colchicine (usually for 24 hours) produces tetraploidy in about 5 percent of the surviving seedlings. This drug may also be applied to the growing points of seedlings or other vegetative parts by a variety of methods. Colchicine has not proven efficient with

cytoplasmic male-sterile system

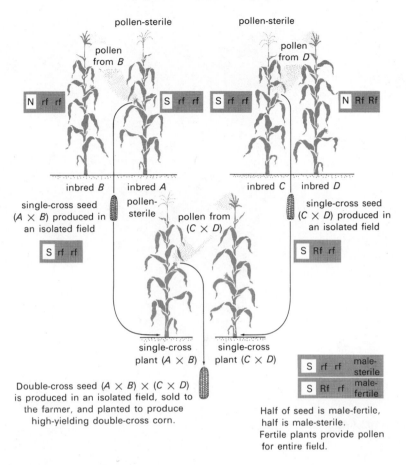

pollen-sterile   pollen-sterile

pollen from *B*   pollen from *D*

| N | rf | rf | | S | rf | rf | | S | rf | rf | | N | Rf | Rf |

inbred *B* | inbred *A*   inbred *C* | inbred *D*

single-cross seed (*A* × *B*) produced in an isolated field

pollen-sterile

| S | rf | rf |

pollen from (*C* × *D*)

single-cross seed (*C* × *D*) produced in an isolated field

| S | Rf | rf |

single-cross plant (*A* × *B*) | single-cross plant (*C* × *D*)

Double-cross seed (*A* × *B*) × (*C* × *D*) is produced in an isolated field, sold to the farmer, and planted to produce high-yielding double-cross corn.

| S | rf | rf | male-sterile |
| S | Rf | rf | male-fertile |

Half of seed is male-fertile, half is male-sterile. Fertile plants provide pollen for entire field.

many grasses. Tetraploids may be produced in corn by applying heat or by genetic methods involving the effects of *elongate*, a chromosome-doubling gene.

An artificially induced "autotetraploid" may usually be distinguished from a normal diploid by its larger, thicker leaves and organs (Fig. 19-15) and its somewhat slower growth. These **gigas** characteristics are a result of increased cell size. Large pollen size is thus a reliable indicator of tetraploidy, although actual chromosome counts are necessary for positive identification. Reduced fertility is another indication. Autotetraploids have four, rather than two, chromosomes of each type. Chromosomes that pair two by two (bivalent pairing) disjoin normally at

meiosis, but if all four chromosomes synapse (quadrivalent association), there is a possibility of a 3-to-1 distribution, which leads to reduced fertility because gametophytes having unbalanced numbers of chromosomes are either nonviable or at a great selective disadvantage. Because tetraploidy is so often accompanied by a reduction in fertility, it is used to greatest advantage in plants grown for vegetative or floral parts (forages, ornamentals, and nonseed food crops).

Inheritance in polyploids is more complex than in diploids. The single-gene ratios in diploid organisms are a result of the assortment of two alleles per gene. In tetraploids, however, there are four alleles per gene. There are two types of homozygotes (*AAAA*

Figure 19-15   Increased cluster size as a result of colchicine-induced polyploidy in 'Portland' grape: *left*, diploid; *right*, tetraploid. [Courtesy Haig Dermen.]

and *aaaa*) and three kinds of heterozygotes (*AAAa*, *AAaa*, and *Aaaa*). The genetic rations of tetraploids also differ from those of diploids. For example, the *AAaa* heterozygote will produce gametes in a ratio of 1*AA*:4*Aa*:1*aa*. If *A* is completely dominant (*Aaaa* producing the *A* effect), the progeny of *AAaa* selfed will segregate in a phenotypic ratio of 35*A*---:1*aaaa*. This ratio, however, is affected by the location of the gene on the chromosome. Thus, tetraploidy tends to decrease the frequency of recessive genotypes. Consequently, the rate of return to homozygosity upon selfing is very much slower in tetraploids than in diploids.

The promiscuous induction of tetraploidy seldom leads to changes of immediate value. Improvement,

if obtained, is usually found only as a result of breeding tetraploids. Examples of economic improvements include the development of large-flowered ornamentals (orchids, snapdragons, lilies, and daylilies) and higher yielding rye and forages (alsike and red clover).

In some plants, increases in vigor and in fruit size are associated with the triploid condition. Triploids can be produced by crossing tetraploids with diploids:

$$4n \ (2n \text{ gamete}) \times 2n \ (n \text{ gamete}) \rightarrow 3n$$

Triploids may also occur spontaneously as a result of the formation of unreduced egg cells.

Sterility in triploids is very high. In the assortment of three homologous chromosomes at meiosis, two move to one pole and the remaining chromosome moves to the other. Since the assortment of each group of three chromosomes is independent, gametes with chromosome numbers anywhere between n and $2n$ may be formed. The larger the chromosome number, the lower the frequency of balanced n and $2n$ gametes. The high incidence of sterility associated with triploidy has been exploited in watermelon. The seedless watermelon is a triploid produced by crossing a tetraploid and a diploid. Triploids can be propagated asexually (banana, apple), or produced anew each year from tetraploid–diploid crosses.

Almost a quarter of the apple and pear cultivars have turned out to be triploid, originating apparently from relatively rare unreduced gametes. They were selected because of larger fruit size and other desirable features. Triploid apples do not produce viable pollen. (An orchard with a triploid cultivar must have two additional diploid cultivars as pollinators in order for all of the cultivars to set fruit because of self-incompatibility factors.) Recently a number of promising triploid apple and pear cultivars have been produced from crosses of diploids with naturally occurring tetraploids. In sugar beets, triploidy is the optimum ploidy level: they have larger roots and more sugar than do either diploids or tetraploids.

There has been great interest in the use of polyploidy as a means of extending variability by creating new species or by transferring genes from other species. Induced polyploidy makes it possible to overcome sterility associated with some interspecific hybrids. This sterility is a result of the inability of chromosomes to pair at meiosis. Fertility is restored by doubling the chromosome number of the sterile hybrid or by crossing autotetraploids of each species. Using capital letters to designate genomes, we can represent this as follows:

$$AA \atop BB \searrow \atop \nearrow \times \rightarrow \quad AB \atop \text{sterile} \atop \text{hybrid} \quad \begin{array}{c} \text{— chromosome} \\ \text{doubling} \rightarrow AABB \\ \text{fertile} \\ \text{amphidiploid} \end{array}$$

$$\begin{array}{l} AA \text{ — chromosome} \\ \quad \text{doubling} \rightarrow AAAA \\ BB \text{ — chromosome} \\ \quad \text{doubling} \rightarrow BBBB \end{array} \searrow \atop \nearrow \times \rightarrow \begin{array}{c} AABB \\ \text{fertile} \\ \text{amphidiploid} \end{array}$$

The resulting amphidiploid is an allopolyploid made up of the entire somatic complement of each species. Fertility results from bivalent pairing within each genome.

These techniques have produced an interesting new crop called triticale, a hybrid of wheat and rye (*Triticum* and *Secale*). Selection within this amphidiploid is being carried out to combine the hardiness of rye with the quality of wheat. Other examples of natural and synthetic amphidiploids are found among the cole crops (Fig. 19-16).

The transfer of genes between species offers perhaps the greatest potential value of polyploid breeding. Tobacco-mosaic resistance from *Nicotiana glutinosa* ($n = 12$) has been transferred to the cultivated form of tobacco, *N. tabacum* ($n = 24$), which is a natural amphidiploid between *N. glutinosa* and *N. tomentosa*. Backcrossing and selection have resulted in the introduction of a whole chromosome from *N. glutinosa*, which contains the gene or genes conferring resistance to the disease.

The addition of whole chromosomes has certain disadvantages because of the possibility of the transfer of undesirable features. The transfer of small chromosome segments has been accomplished by irradiation techniques in wheat ($2n = 42$). A small chromosome segment containing resistance to rust has been transferred from a wild Mediterranean grass, *Aegilops umbellulata* ($2n = 14$), to wheat chromosomes by using amphidiploids with a related wheat, *Triticum dicoccoides* ($2n = 28$), as a genetic bridge.

Polyploidy has not proved to be a panacea in plant breeding. It is, however, another in the arsenal of weapons that the breeder can use to change plants to better suit human needs. Polyploid manipulation has greatly enlarged the genetic base from which the breeder may draw. These methods undoubtedly will contribute much to future crop improvement.

AA   CC   AACC

Brassica campestris   Brassica oleracea   rape
                                                        (Brassica napus var. olerifera)

turnip   kale   rutabaga
                                                        (Brassica napus var. rapifera)

Chinese cabbage   common cabbage   "00"
                                                        a synthesized fodder rape

Chinese cabbage   common cabbage   "hakuran"
                                                        a synthesized (head-forming)
                                                        leaf vegetable

Figure 19-16 Natural and artificial amphidiploids in *Brassica*. [From H. Kihara, *Seiken Ziho* 20:1–4, 1968.]

Mutation breeding

One of the limitations to crop improvement is the dependence upon naturally occurring variation. Mutations, the source of genetic variation, can be induced by radiation (x-rays, gamma rays, thermal neutrons) and by chemical means (ethyldimethyl sulfate), and can be increased by special genetic mechanisms. Mutation breeding refers to the use of artificially induced mutations in crop improvement. Mutagenic agents induce random changes, most of which are harmful, as are most naturally occurring mutations. Since such changes cannot be directed and are largely deleterious, the problem is one of detection and isolation of the rare desirable alteration. In order for a technique to be practical, an efficient screening procedure must be developed. For example, large populations of oats can be screened in the seedling stage for resistance to the Victoria blight disease, because susceptible plants will die soon after being inoculated with the causative fungus, *Helminthosporium victoriae*. Mutants that change self-incompatibility genes can be detected by using the stigma as a sorting device, for only mutated pollen grains carrying the desired mutation will germinate on the stigma.

Asexually propagated crops have been suggested as promising material for mutation breeding because adapted material may be screened in the search for a single desirable induced change. But because mutations in such material are largely chimeral, it is often as laborious to isolate and test for these mutations as it is to select for segregants from ordinary breeding programs. The saving in time has been more apparent than real.

A number of useful alterations have been accomplished by means of induced mutations. These include beneficial mutants of barley and higher-yielding strains of the mold *Penicillium* (for penicillin production) produced by exposing the mold to ultraviolet irradiation. A grape cultivar with too tight a cluster has been improved through the induction of reduced fertility. In general, however, progress with induced mutations has not been proportional to the effort expended. Because of the unpredictability of mutation breeding in a practical improvement program, it can be considered only as a last resort after other sources of variation have been exploited and found wanting.

Tissue-culture methods

Recently, great excitement has been generated about the possibility of using tissue-culture techniques for crop improvement. As we have seen in this chapter, crop improvement is based largely on the creation of variable populations from which individual genotypes are isolated. In traditional plant-breeding systems, genetic variation in the population is stored and released through gene assortment and recombination brought about by meiosis and fertilization. Because most crop improvement is based on the sexual cycle, genetic progress has been limited by the way the sexual cycle can be controlled and manipulated by the plant breeder.

It now appears that the sexual system for releasing variability may be short-circuited by imposing selection on populations of cells rather than on populations of whole plants. For the technique to be successful, it must be possible to regenerate plants from individual cells. This has been achieved with a great number of plants (see p. 217–221). The advantage of this system is that extremely large populations of cells (literally in the billions) can be grown in a very small space, such as a few culture tubes or flasks. The problem is to be able to select for variants at the cellular level that will be expressed as desirable characters in the growing plant.

Variation in cell populations occurs naturally as a result of spontaneous mutation, but it can also be induced by means of various mutagenic agents. Selection can be achieved by exposing cell or tissue cultures to environmental stresses, such as extremes in heat or cold, or to high concentrations of pathogenic toxins or herbicides: susceptible cells die while only resistant ones survive to grow and reproduce. Many research studies are now under way to test the practicality and usefulness of this system.

Tissue culture offers the possibility of restructuring the genetic organization of an organism through special techniques developed by molecular biologists. This so-called "genetic engineering," which includes

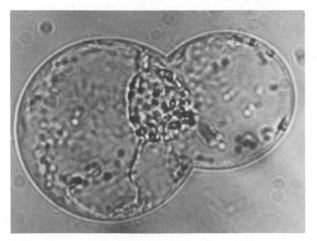

Figure 19-17 Initial stage of fusion between two protoplasts of tobacco. [From T. Murashige, *HortScience* 12:127–130, 1977.]

cell hybridization, organelle transfer, genetic transformation, and haploid production, offers potentially great opportunities for plant scientists.

Plant-cell hybridization has been shown to be feasible with protoplasts. Protoplasts can be isolated from plant cells by digesting cell walls with appropriate enzymes. A small but significant percentage of protoplasts grown in suspension culture show genetic fusion (Fig. 19-17). This means that, if protoplasts of two genetic backgrounds are mixed, some of the fused protoplasts will be genetic "hybrids." One method to isolate hybrid protoplasts from those that are not fused, and from those that are fused with protoplasts of the same genotype, is to use a medium in which only hybrid products will be able to grow. Mature plant hybrids have been produced from cellular hybridization in tobacco and petunia. The refinement of this process may make it possible to achieve wide crosses that are not possible with traditional methods of sexual hybridization. Various organelles that are capable of regeneration with their own DNA (such as chloroplasts) may also be transferred from one species to another in this way.

Gene transformation has been achieved in plants by extracting DNA from one plant and incorporating it into another through the use of bacterial viruses (bacteriophages) as transfer vectors. This technique is now only experimental, but it opens up the possibility of transferring to plant cells genes that have been created by isolating messenger RNA and producing DNA transcripts (gene copies).

Finally, pollen of various plants can be cultured to produce haploid cells, or even haploid "embryoids." Haploid cell lines of plants are extremely valuable for mutation breeding because, since dominance is eliminated, all mutated recessive genes will be immediately expressed. Further, doubling such haploids produces "instant" homozygous lines, a process that may require many years with ordinary sexual inbreeding. Haploid cells could also be used to produce "hybrids" directly by protoplast fusion.

Breeding techniques

Hybridization

Plant breeding relies on the control of the sexual process. Before two plants can be crossed they must be induced to flower. Various environmental factors, such as photoperiod, temperature, and nutrition, may have to be manipulated to achieve synchronization in flowering. For example, many cultivars of sweetpotato must be grafted to species of morningglory and then subjected to special photoperiods and temperatures to induce flowering. As a result, in many large breeding programs, crossing is often performed under controlled conditions in the greenhouse.

The basic rule in making artificial crosses is to avoid contamination before and after artificial pollination. In monoecious plants the pistillate flower must be protected before pollination from wind or insect pollination by such devices as paper or glassine bags or cheesecloth nets. In perfect-flowered plants, contamination by selfing is avoided by emasculating the flower before the anthers have begun to shed pollen. Selfing may also be circumvented by selective killing of the pollen. Immersion of the flower in hot water—45 to 48° C (113 to 118° F)—has been effective for sorghum and grasses, and immersion in 57 percent ethyl alcohol for ten minutes has been effective for alfalfa. The manipulation of genetic male sterility has been discussed.

Pollination can be achieved with fresh or stored

Figure 19-18 Techniques used in hybridizing western white pine: Male cones are collected (A) and washed (B) for pollen extraction (C). The extracted pollen is then squirted onto enclosed conelets from a hypodermic syringe equipped with a bulb (D). [Courtesy U.S. Forest Service.]

pollen. Often the anthers are collected the day before pollination and dried. Pollen is quite variable in its ability to remain viable. Wheat pollen remains viable only minutes at high temperatures, whereas apple pollen stored under low humidity and cool temperatures, −1 to 2° C (30 to 36° F), may retain its viability for years. Storage by freeze-dehydration techniques appear promising.

Pollen may be applied with a camel's-hair brush, the fingers, a blackened matchstick, or the dried flower itself. The pollen should be applied in large amounts. The pollination of onions is carried out by enclosing their flower heads in mesh cages containing pollinating insects (blowflies).

After pollination the flower should be protected from contamination by cross-pollination. This is usually accomplished by enclosing the flower in paper or glassine bags; if the petals have been left on, as is customary with many cucurbits and legumes, they may serve as protection if held closed with cellophane tape. When it is only necessary to keep away insects, cheesecloth is usually sufficient. Care must be used in selecting protecting devices that do not act as a heat trap and thus prevent successful fruit set. In citrus and apples, insects will not visit the flowers if the petals are removed during emasculation. Techniques used in hybridizing western white pine are shown in Figure 19-18.

Testing

The testing of genetically different material is a major undertaking in plant breeding. In order to avoid confusing genetic and enviromental effects, testing must be systematic. This requires that all plants be treated alike, or at least in an unbiased fashion. Soil differences are difficult to control; two plants cannot be grown in the same place at the same time. In order to minimize experimental error, or at least to be able to test its presence and to quantify it, various field-plot procedures have been established. These include randomization, replication, the use of controls, and the elimination of border effects. Wherever possible, quantitative measurements rather than qualitative evaluations are made, and the data are subjected to statistical analyses to determine the probability that differences are due to the variable being tested or to chance effects. Good field-plot technique and subsequent analysis require sound judgment and careful execution. It is axiomatic that accurate records be kept.

A number of artificial treatments are imposed to facilitate the selection process. These include extremes of heat and cold to test for genetic hardiness and the creation of disease epidemics and insect infestations for the evaluation of genetic pest resistance. Techniques for determining quality are often complex. The evaluation of wheat quality involves the testing of baking properties. Determining forage quality may require measuring the content of the toxic substance coumarin or checking grazing trails to gauge palatability.

Once a genetic improvement has been made, vigilance is still required to maintain the improvement. Mutation, natural crossing, contamination, and, in vegetatively propagated material, diseases, especially those caused by viruses, tend to cause the deterioration or "running out" of cultivars. Genetic deterioration may be reduced to a minimum by continued selection and careful propagation. The plant breeder's responsibilities do not end with the release of a superior cultivar.

Selected references

Allard, R. W., 1960. *Principles of Plant Breeding.* New York: Wiley. (An outstanding theoretical presentation of the genetic foundation of plant breeding.)

Ferwerda, F., and F. Wit (editors), 1969. *Outlines of Perennial Crop Breeding in the Tropics* (Miscellaneous Papers 4, Landbouwhogeschool). Wageningen: H. Veenman. (A review of tropical tree breeding, including breeding of many tropical fruits.)

Frankel, O. H., and E. Bennett (editors), 1970. *Genetic Resources in Plants: Their Exploration and Conservation.* Philadelphia: F. A. Davis. (An outline of methods for exploration and introduction in plant improvement.)

Frey, F. J. (editor), 1966. *Plant Breeding.* Ames, Iowa: Iowa State University Press. (A collection of articles on plant breeding with the emphasis on general methods rather than specific crops.)

Janick, J., and J. N. Moore (editors), 1975. *Advances in Fruit Breeding.* West Lafayette, Ind.: Purdue University Press. (A monograph on fruit breeding by crop, each chapter written by a distinguished fruit breeder.)

Poehlman, J. M., 1979. *Breeding Field Crops,* 2nd ed. Westport, Conn.: AVI. (Breeding of major field crops is discussed in detail.)

Reinert, J., and Y. P. S. Bajaj (editors), 1977. *Applied and Fundamental Aspects of Plant Cell, Tissue and Organ Culture.* New York: Springer-Verlag. (A collection of chapters by various authorities. Includes discussion of the uses of tissue culture for genetic engineering and crop improvement.)

Simmonds, N. W., 1979. *Principles of Crop Improvement.* London: Longmans. (An excellent new text in plant breeding; a broad view geared to principles.)

U.S. Department of Agriculture, 1936, 1937. *Better Plants and Animals,* vol. I and II (USDA Yearbooks, 1936 and 1937). (A valuable reference and a historical landmark of genetics and breeding of agricultural crops in the United States.)

Vavilov, N. I., 1951. *The Origin, Variation, Immunity and Breeding of Cultivated Plants,* translated by K. Starr Chester. New York: Ronald Press. (Contains the most important writings of the great Russian plant scientist.)

Wright, J. W., 1962. *Genetics of Forest Tree Improvement.* Rome: Food and Agriculture Organization of the United Nations. (A textbook of forest genetics and breeding with good coverage of world literature.)

Five
INDUSTRY OF
PLANT AGRICULTURE

20
Food Crops:
Cereals, Legumes, Forages

There is no problem confronting the human race more fundamental than that of feeding its expanding populations. A world in which the majority of the inhabitants go to bed hungry each night is inevitably rife with unrest and misery. Other chapters (especially 3, 4, and 15) have discussed this fundamental concern in some detail. This chapter focuses attention on the crop plants that supply the modern world with the major part of its food. It has been shown in Chapter 19 that crop improvement is a never-ending process (Table 20-1). Seldom do the cultivars of today satisfy the needs of tomorrow. Our purpose, then, is to discuss the worldwide importance of various species of cereals, legumes, and forages, not to make agronomic recommendations or to compare cultivars.

The greater proportion of our food is derived from relatively few plant species. Most of the staple crops are members of the grass family that yield grains. Rice, wheat, and corn are the "big three" of world cereal-grain production, which totals about 1.5 billion metric tons annually. Each of these grains accounts for nearly one-quarter of this total. Barley, oats, rye, millets, and sorghum together make up most of the remaining quarter.

Only one other food plant, the potato, a root crop, has nearly as much worldwide importance as the cereals. Total potato production almost equals that of the leading grains, but, since the potato is fleshy and consists of only 22 percent dry matter, its total food value is considerably less. Sweetpotatoes and yams, cassava, various seed legumes (inlcuding beans, peas, soybean, and peanut), and sugar cane are the other major noncereal food crops. For all practical purposes, people rely upon this handful of crop plants for their existence (Fig. 20-1).

Chapter 19 mentioned the centers of origin of most of the major crop plants. Lengthy lists of wild plants utilized by primitive cultures could be recited. The wild plants that made up part of the diet of the American Indian tribes were fairly well documented by the Old World settlers who invaded North America. Among the Indians of the Pacific Northwest, as Lewis and Clark found them, plant cultivation was unknown, except possibly for tobacco. Wild seeds and nuts made up a good part of their diet, but various roots, greens, and berries were also eaten. Piñon nuts, from *Pinus monophylla,* were something of a staple in southern areas, but seeds of wild grasses and of plants of the goosefoot and mustard families (Chenopodiaceae and Cruciferae) were also gathered in sizable quantities, as were acorns (*Quercus* spp.), hazelnuts (*Corylus cornuta*), and the seeds of the yellow pond lily (*Nuphar polysepalum*). Many species of *Vaccinium* (blueberries, huckleberries, cranberries) were eaten both fresh and dried, as were several

Table 20–1 Average and record yields. Improved cultivars and cropping practices make such astounding differences possible and offer hope for better feeding the world.

Food	Average, 1974	Record
Corn (bushel per acre)	72	307
Wheat (bushel per acre)	28	216
Soybeans (bushel per acre)	24	110
Sorghum (bushel per acre)	45	320
Oats (bushel per acre)	48	296
Barley (bushel per acre)	38	212
Potatoes (bushel per acre)	420	1400
Sugar Beets (ton per acre)	19	54
Milk production per cow (10^3 pounds)	10.3	50
Eggs per hen	230	365

Source: Adapted from S. H. Wittwer, *Science* 188 (4188):580.

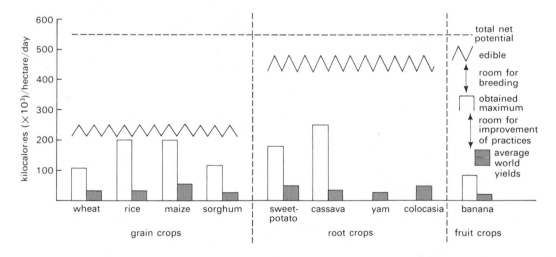

Figure 20-1 Average world yields, maximum yields obtained in selected tropical experiment stations, and estimated potential yields to be reached through breeding and research. As shown on the extreme right of the graph, the distance between the average yields and the maximum obtained indicates the scope for improved cultural practices. The distance between the obtained maximum and the edible yields indicates the scope for improvement in breeding. [Data courtesy *Netherlands Journal of Agricultural Science,* from T. Poleman, "World Food: A Perspective," *Science,* 188 (4188):510–518, fig. 6 (May 9, 1975), copyright © 1975 by the American Association for the Advancement of Science.]

Table 20-2   New World and Old World food plants before 1492.

New World		Old World		
maize	pumpkin	wheat	onion	wine grape
manioc	peanut	rye	garlic	apricot
lima bean	chayote	barley	spinach	peach
potato	papaya	oat	eggplant	olive
sweetpotato	avocado	millet	lettuce	fig
kidney and other beans	pineapple	sorghum	endive	almond
tomato	custard apple	rice	celery	quince
green pepper	soursop	buckwheat	asparagus	pomegranate
Jerusalem artichoke	cherimoya	turnip	peas, various	watermelon
sunflower	guava	cabbage	soybean	cucumber
quinoa	chocolate	rutabaga	broad bean	banana
squash	cashew	chard	dasheen	orange
	sapote	mustard	yam	lemon
		radish	apple	lime
		beet	pear	date
		parsnip	plum	mango
		carrot	cherry	clover
		breadfruit	alfalfa	
		mangosteen	bluegrass	

related plants of the heath family (Ericaceae). A number of fruits and berries of the rose family were gathered. A great many "roots" were dug, particularly those growing in marshy grounds, such as *Sagittaria latifolia*. Scores of other species could be named.

Perhaps it might be even more interesting to speculate on the wild plants that sustained the primordial civilizations, such as that of the Tehuacan Valley in Mexico (Chapter 1). Archeological findings, particularly human coprolites (desiccated feces), give definite evidence of primitive diet and the gradual changes that came about as the domestication of plants proceeded. A diet that depended at first upon wild game supplemented with wild beans, agaves, and cactus gradually changed to include foxtail millet (*Setaria*), squashes (*Cucurbita*), peppers (*Capsicum*), and eventually amaranths (*Amaranthus*) and maize (*Zea mays*). Archeological evidence shows that much the same sort of change took place in South America. Almost all important food plants had beginnings of this sort in prehistoric times, the ancestral type being so remote as to be extinct or unrecognized today.

The modern world would grind to a standstill if it could not depend upon the gradual improvement of these domesticates through new cultivars, which lend themselves so well to modern mechanized agriculture. Many thousands of wild plants were utilized by primitive cultures; some of them, of course, were used only in times of dire need. Table 20-2 categorizes a number of important crops according to whether they originated in the Old World or the New World. Since the fifteenth century, these species have been spread widely over the globe, wherever they proved adaptable to local climates. Areas of origin and major zones of today's production do not often coincide!

World food production more than doubled between 1950 and 1980. However, it is distressing to

Figure 20-2 Corn grains. Note in the longitudinal section that the embryo occupies only a small part of the seed; the rest consists mainly of endospermic food reserves. [Courtesy J. C. Allen & Son.]

note that there was little or no increase in food production per capita during this period; slight decreases occurred in Africa, parts of Asia, and Latin America. Furthermore, it is unlikely that the present overall rate of increase can be maintained for long, since most of the world's best lands are already under cultivation. It is ominous that the annual increase in cereal-grain yields, worldwide, fell from 2.9 percent in the decade 1960–1969 to only 1.9 percent in the decade 1970–1979, according to Environmental Fund statistics. If we are unable to produce more food, we will have to depend increasingly upon population control to avert widespread famine (Chapter 4).

Asia, with its huge land area, is the leading food-producing continent, of which the chief food plant is overwhelmingly rice (92 percent of world production). But because of its immense populations, Asia produces insufficient food for its needs, especially in India. Africa, in large part peopled by primitive societies, is in great measure self-sufficient on a subsistence level, if not by the standards of more sophisticated cultures. The mainstay crops of Africa—millets and yams, for example—are often rather quaint to the rest of the world. In Central America and South America, food production is generally inadequate (although in certain areas this may be more a reflection of poor dispersal than lack of capability). Maize, wheat, and potatoes (in the Andes) are the mainstay crops.

On the whole, Europe, in spite of its dense population, feeds itself adequately according to modern standards. Because of its industrialization, it is well equipped to "borrow" land elsewhere for food production, and to import foods that might be in short supply locally. Potatoes are perhaps the most important food plant in Europe, followed by wheat, barley, maize, and other grains. North America, with its high degree of mechanization and still relatively sparse population, has great potential for agricultural increase and export. The extent to which food production and export is carried out depends largely upon world economic conditions and the current political posture regarding foreign aid and foreign trade. More than military might, the responsive soils of the United States and Canada may be their real strength. Maize, wheat, and soybeans are the chief North American crops.

THE CEREALS

Cereals are the staff of life to civilization. More than 50 percent of the world's arable land of about 1.4 billion hectares (nearly 3.5 billion acres) is devoted to growing them. They are the direct source of half the food needs of the world, and the secondary source of much additional food when converted to meat, milk, eggs, and other animal products.

The true cereals are all members of the grass family (Gramineae). The fruit they yield is a **grain,** a

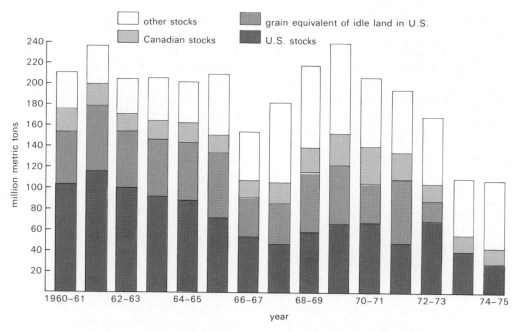

Figure 20-3 World grain stocks, including grain equivalent of idled land in the United States. [Adapted from T. Poleman, "World Food: A Perspective," *Science*, 188 (4188):510–518, fig. 2 (May 9, 1975), copyright © 1975 by the American Association for the Advancement of Science.]

type of fruit in which the ovary wall turns hard and durable, fusing with the single seed (Fig. 20-2). Buckwheat, amaranth, and a few similar seeds of other families ("pseudo-cereals") are sometimes regarded as grains because their seeds are quite similar to those of grasses. This section will be devoted to the true cereals, of which the major crops are rice (*Oryza*), wheat (*Triticum*), maize (*Zea*), barley (*Hordeum*), oats (*Avena*), rye (*Secale*), sorghum (*Sorghum*), various millets (*Panicum, Setaria,* and others), and a few other species of lesser world importance.

There are a number of reasons why cereals have evolved into our leading food source. Most of them are annuals, or are at least adapted to cultivation as annuals, permitting facility in cropping, whether on the midden of an early domestication center or as a component of today's highly mechanized farming. Grasses in general are quite versatile, too; they adapt well to a variety of soils, climates, and ways of handling. They are also relatively efficient in garnering the sun's energy and transforming it into usable food substance. They are generally hardy and recuper-

ative, and they are plagued by no more than their share of diseases and pests. But above all, the grain is a neat package of stored energy, conveniently harvested, easily cleaned and handled, amenable to storage without special drying.

It is no wonder that well-filled granaries have been the mark of security, even opulence, throughout the ages. The word *cereal* itself stems from the legendary Greek goddess Ceres, "giver-of-grain." She and other gods were often propitiated with gifts of grain. Because of overpopulation and the threat of starvation, efforts were initiated, during the 1960s and 1970s, to establish an international grain reserve. Unfortunately, it was a good idea that bore little fruit. In effect, the productivity of the United States and Canada has made them something of a grain reserve for the world (Fig. 20-3), with much of the production exported. However, the holding and financing of reserves to forestall world famine should be an international responsibility, not dependent upon the momentary technology and economic conditions of only a few countries.

As has been noted, rice, wheat, and corn are the

world's three major cereals. Rice is the prime source of sustenance for tropical populations; it is grown mostly on flood plains or where the land can be seasonally ponded. Wheat is chiefly grown on lands that were naturally prairies, too harsh, cold, and windy for maize. Maize is a crop that performs best with ample warmth and moisture; it is widely utilized as a summer annual in regions where general farming is practiced.

Rice, *Oryza sativa*

Rice, or paddy, widely grown in Asia from Afghanistan to the East Indies and north to Japan, is the principal food for 60 percent of the world's people. It is also a staple in parts of Africa, South America, and to some extent in the southern United States. World production exceeds 350 million metric tons annually, mostly from the Far East (China, India, Indonesia, Bangladesh, Japan, and Thailand are the major producing countries).

The research on rice genetics has been considerable, but details of domestication and discernment of species still remain unclear. The cultivated rices are generally regarded as belonging to the species *Oryza sativa*, a sort of catch-all complex. The various kinds of rices are generally regarded as varieties of *O. sativa*, although some botanists consider them to be separate species. The Asiatic forms are often called *O. sativa* var. *fatua*, *O. sativa* var. *nivara*, or *O. sativa* var. *rufipogon;* the American forms, *O. sativa* var. *perennis* (a name of doubtful validity); and the annual African forms, *O. sativa* var. *barthii* or *O. sativa* var. *glaberrima*. Some systematists have grouped rices into a *fatua* (or *rufipogon*) series and a *perennis* complex (of which the American forms are designated *O. sativa* var. *cubensis*, the Asiatic forms *O. sativa* var. *balunga*, and the African ones *O. sativa* var. *barthii* or *O. sativa* var. *longistaminata*). The wild forms of the *fatua* series generally grow in shallow swamps as annuals; the *perennis* forms (at least the Asiatic *O. sativa* var. *balunga* and the perennial African *O. sativa* var. *longistaminata*) generally grow in deep swamps. Floating rice can grow in water that is several meters deep, elongating as much as 30 centimeters (about 12 inches) per day. All rices

are customarily grown as annuals, although under tropical conditions they may be perennial, perpetuated by new tillers or even rhizomes. Indeed, "ratoon rice" growing from the crowns of harvested plants may prove to be quite profitable (considering the limited input required of the farmer), even though a ratoon crop seldom yields even half the grain of a primary planting. In Africa, wild forms often volunteer in cultivated plantings, and there is some harvest and marketing of them.

Rice has been cultivated in China for 5000 years, and in the Niger delta of Africa at least since 1500 B.C.E. The cultivated forms are thought likely to have originated from wild *O. sativa perennis* types in Southeast Asia. But so widely has rice been spread that it is entirely possible that wild rices from other tropical parts of the world have contributed to *O. sativa*. The diploid chromosome number of most rices is 24, but several forms (species?) are tetraploid ($2n = 48$). Triploid hybrids are often fertile. The International Rice Institute estimates that there are about 18 recognizable rice species, any of which may have made a contribution to *O. sativa*.

For practical purposes, rices are generally classified as long-grain, medium-grain, or short-grain. The long-grain types need a longer season to mature, and hence are grown mostly in tropical regions, such as Southeast Asia. Many people in the West prefer the long-grain type, which is nonglutinous and does not turn so soft and sticky when cooked. Medium-grain rices, somewhat softer, are most frequently grown in America. Short-grain types, even more sticky when cooked, are well adapted to more northerly climates, such as that of Japan.

Most rice is grown in paddy fields of standing water, the usual practice in Asia (Fig. 20-4). However, various forms of upland rice are grown in dry fields in much the same way as any small grain, the more common custom in Africa and Latin America. Yields of upland rice tend to be lower (0.5 to 1.5 metric tons per hectare in the Ivory Coast, for example) and its overall importance is not great; for example, in Indonesia and Bangladesh about 80 percent of the rice is grown on irrigated land. The best rice land is level, conveniently diked for alternating irrigation and drainage.

Figure 20-4 Paddy rice culture in the Philippines. [Photograph by Edwin G. Hoffman, courtesy World Bank.]

Only a little more than 1 percent of the world's rice is grown in the United States, but it is grown in a technologically advanced fashion. Rice was first introduced into the Virginia colony about 1609, and it was an established crop in the Carolinas by 1690. Major production has since shifted to the lower Mississippi Valley (principally Arkansas, eastern Texas and Louisiana) and to northern California. The more humid conditions of the Gulf Coast and Mississippi Valley areas are advantageous to rice production, requiring less irrigation water, which is becoming increasingly more difficult to provide in California. Immense fields are precisely flooded and drained to permit heavy equipment to prepare the land, plant the crop, and later harvest the grain in the field. Sometimes rice is sown, fertilized, and given pesticide protection by aircraft, especially in California. Elaborate research has elucidated soil behavior under irrigation and has determined optimum timing for fertilization under the anaerobic conditions that prevail in rice cultivation. Crop rotation is generally practiced, better to avoid physical deterioration of the soil and infestations of weeds and other pests.

Japan has the most progressive rice-growing practices in the Orient. Yields average more than 4 metric tons per hectare (about 2 tons per acre) and can be as high as 10 metric tons per hectare (about 4.5 tons per acre). Fertilization of rice fields in Japan is as intensive as on North American corn land (Fig. 20-5). A complete fertilizer is generally plowed down prior to flooding and hand planting of the seedling rice plants. Another topdressing is applied just before heading to increase yields. The fields are drained for harvesting and threshing, typically accomplished with small portable machines on the site. A green manure (vetch) is sometimes planted and plowed down. The paddy fields may also be used for a winter crop, such as barley, wheat, or rape.

Under more primitive circumstances, such as those that prevail in many tropical countries, rice growing is an exceedingly laborious and time-consuming hand operation. As noted in Figure 20-5, even fertili-

Figure 20-5   Rice yields related to fertilizer usage in several Asiatic production areas. [After Kemmler, *World Crops* 14:117, 1962.]

zation is quite limited in many places, although compost and sewage have been used since ancient times to fertilize paddy fields. Symbiotic nitrogen fixation by blue-green algae growing in association with the water fern *Azolla* commonly maintains fertility in Asiatic paddy fields.

Typically, after the paddy is flooded to soften the soil, the ground is worked, often by a water buffalo pulling a forked stick (small hand tractors may be used in more advanced countries, such as Japan). The rice seedlings are started elsewhere, sometimes under government auspices, and are transplanted by hand into flooded paddies, usually between November and January in Southeast Asia, to take advantage of the winter monsoon. Transplanting from a nursery allows for better weed control in the paddy, giving transplants an edge over paddy seedlings that could not be weeded. A few months later, depending upon length of season required by the cultivar, the paddies are drained and the stand is harvested—another laborious job, accomplished chiefly by women. The rice stalks are cut individually with a small, sharp sickle, banded together, and laid out in the sun to dry (Fig. 20-6). If there is sufficient water, either a second rice crop or some other crop may then be planted on the same land.

Various means are employed to clean and dehull newly harvested rice. In some countries the sun-dried heads are threshed in the field by treading, either by

people or by livestock (of course, with considerable loss of grain). In Afghanistan, green rice is mixed with heated sand to harden and "crisp" the hulls, which are then separated by crude milling on water-powered devices that pound the grain against the soil. Screening and winnowing follow (Fig. 20-7). In Ecuador, much of the rice is fermented by massing it on the floor while it is damp and covering it with a tarpaulin for a few days. Upon subsequent drying it cooks more quickly than unfermented rice in the high altitudes of the Andes. Fermentation may also provide healthful by-products.

The processing that rice undergoes in the so-called advanced countries removes no small part of its nutrients. After the hulls are removed, the grain is polished to remove the brown outer layers, which are rich in protein and B vitamins. The insistence on a pearly white grain carries with it the penalty of reduced nutritional value.

An outstanding crop breakthrough has occurred in rice cultivation, thereby extending the "green revolution" begun by Norman Borlaug's work with Mexican wheats (see p. 511) into the Oriental tropics. An extensive breeding program by the International Rice Institute in the Philippines has yielded improved cultivars, notably the double-recessive dwarf cultivar 'IR-8' and its pest-resistant successors, which are capable of markedly higher yields that can help to feed the dense populations of the Orient more

Figure 20-6 The way most rice is harvested throughout the world. [Photograph by Tomas Sennett, courtesy World Bank.]

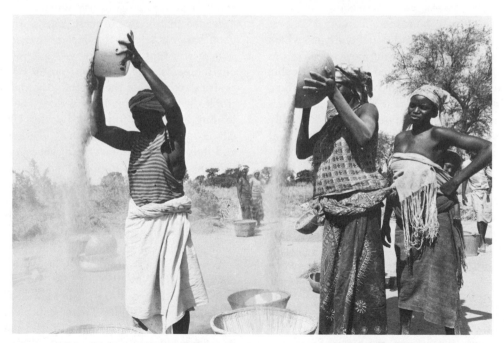

Figure 20-7 Winnowing rice (to separate the grain from chaff and dust) in Upper Volta. Upland rice is grown in much of Africa. [Photograph by Ray Witlin for IDA, courtesy World Bank.]

adequately. 'IR-8' is a short, stiff-stemmed cultivar with an ample panicle that can be traced back to the maternal parent 'Cina'. It responds well to nitrogen fertilization. The traditional taller cultivars would lodge if similarly fertilized for high yield. Although it is no longer widely planted, 'IR-8' is still very influential as an ancestor of most modern semidwarf rice cultivars. However, because so many modern cultivars bear 'Cina' maternal heredity, and thus have similar cytoplasmic inheritance, a potentially hazardous situation has been created that could conceivably lead to disease epidemics as devastating as the one that struck the U.S. corn crop in 1970 (see p. 514).

The introduction of high-yielding cultivars has not eliminated all problems attendant to rice-growing. Their use has even brought about a few new complications. In the Philippines, for example, where almost 70 percent of the plantings are of improved types, the peasant landholder is often unable to afford fertilizer. Thus the introduction of high-yielding rice has proved a boon to the advantaged but not necessarily to the masses, who can become caught up in a web of economic bondage. In many locations, the growing season of the traditional cultivars, which is often different from that of the new releases, was more suitable for the established custom of double cropping. In less developed parts of the world, then, improved methods that work efficiently in technically advanced countries (where research is readily accepted by a well-informed farming population) may not live up to expectations.

Indeed, imposition of "improved" growing techniques in technologically less advanced countries is fraught with difficulties and hazards. In the 1970s an extensive government-sponsored rice-spraying program was undertaken in Indonesia. Insecticide was applied from the air to hundreds of thousands of hectares to control a stem borer (the larva of a moth) that was said to cost the nation one-quarter of its crop. Aerial spraying was decided upon because it was deemed unlikely that peasant growers could be induced to practice preventive spraying for an "invisible" affliction (the borers are susceptible soon after the eggs are laid on the leaf surface and before the larvae hatch out, enter the stem, and cause damage). A spot check of moth abundance was used to determine the best time for the spraying, which could be done only by trained personnel available through a government contract with a European firm. Spraying was correlated with an "extension" program that included provision of the fertilizer in exchange for a sixth of the crop. Apparently, rice yields were increased by 25 percent or more, but it is difficult to assess whether the savings (primarily manifested as reduced need for rice imports, although acquainting the peasant populus with improved technology is an intangible gain) justified the drain on the public purse that resulted from contracting for such a tremendous program. The influence on the local ecosystem of massive spraying such as this is seldom considered, and after the foreign technologists leave, most peasants slip back to the traditional way of doing things.

Wheat, *Triticum aestivum*

World production of wheat, *Triticum aestivum* (*T. sativum* and *T. vulgare* are synonyms), has slightly exceeded that of rice in most years of the 1970s, and good harvests late in the decade exceeded 400 million metric tons. Wheat is well adapted to harsher environments, and is grown mostly on wind-swept prairies and steppes that are too dry and too cold for tropically inclined maize and rice. The USSR is the leading producer, followed by the United States, China, India, Canada, France, Turkey, Australia, and Italy. Wheat can be grown wherever the climate provides moderate moisture and cool weather for early growth, then bright summer months that gradually turn dry for harvesting. It is marginal in humid climates toward the tropics (where disease and debilitation are a hazard) and in the arctic North (where winterkill and short summers are risks).

Wheat is a gift from the Old World, a fair exchange for the New World maize. Wheat was apparently domesticated in the Near East, probably in the Taurus–Zagros foothills. There seem to have been a number of natural crossings with wild species during its early evolvement, as indicated in Table 19-1. As is true of many domesticated plants, a number of uncertainly related species constitute the genus, many of them growing in close association near the centers of origin. Actually, a number of domestications of wheat undoubtedly took place, depending upon which species is being considered. Wild and cultivated einkorn wheats (*T. boeoticum* and *T. monococcum*) each possess 14 chromosomes. A number of other wheats are tetraploid, including cultivated emmer (*T. dicoccon*), durum, or macaroni wheat (*T. durum*), and *T. timopheevi*. Bread wheat (*T. aestivum*) is hexaploid ($2n = 42$), as are club wheat (*T. compactum*) and spelt (*T. spelta*). Evidently, there have been natural crossing between several of these and wild species of *Aegilops* (and perhaps *Agropyron*).

A species newly recognized in Israel in 1979, named *T. searsii*, may have been the source of the B genome of *T. aestivum* (a hexaploid with three genomic complements, *AA*, *BB*, and *DD*). Genome *A* seems to have been derived from the diploid *T. monococcum*, which may have crossed with *T. searsii* to produce the tetraploid *T. turgidum* (*AABB*). This in turn probably hybridized with *T. tauschii* (= *Aegilops squarrosa*), *DD*, to yield the hexaploid *AABBDD*, perhaps as much as 8000 years ago. The taxonomic relations of the genomes have been substantiated by genetic, cytologic, and protein electrophoretic evidence.

Primitive einkorn and emmer wheats have been discovered in the Abu Hureya excavations in Syria that are more than 11,000 years old, and specimens from Jarmo (eastern Iraq) date from 9000 years ago. The cultivated and wild forms of einkorn (traditionally called *T. monococcum* and *T. boeoticum*) are scarcely distinguishable, although cultivated forms don't disarticulate so easily, the ears remaining intact until threshing. Almost solid stands of wild einkorn have colonized basaltic hillsides in southeastern Turkey, growing as thickly as they would if they had been sown. *T. (Aegilops) speltoides* occurs more sparingly, and these two are parents of wild emmer (usually called *T. dicoccoides*, but little different from cultivated emmer, except for the greater tendency to disarticulate). Wild emmer is not so "weedy" as its parents, and its presence may suggest some care (domestication). Other archeological findings show bread wheats to have been grown in the

Nile valley before 5000 B.C.E., and in India, China, and even England almost equally long ago. Apparently the species was widely disseminated by the civilizations that spread from the Near East, where Vavilov believed soft wheats to have originated in the mountains of Asia Minor and durum wheats in North Africa. Wheat was first grown in America in 1602, on an island off the Massachusetts coast.

At first, wheat was probably used hulls and all. The grain may have been boiled to make a porridge, or perhaps even fermented for preliminary digestion. It was soon discovered that the wheat grains could be cracked and ground, wetted to make a dough, and the dough baked or fried over an open fire. Thus originated the early unleavened flatbread, still a staple in the diet of many people of the Near East. There is evidence that raised or leavened bread was used in Egypt as early as 2500 B.C.E. Leavened bread results from the fermentation of dough by yeast, during which the carbon dioxide "bubbles" produced by the yeast are trapped in the glutenous dough as it bakes. Of the various cereals, only wheat is well suited to the making of leavened bread (rye is somewhat similar, although most rye bread is made from a mixture of rye and wheat flour).

There are two basic classes of bread wheat, hard and soft, a distinction recognized as early as Roman times. **Hard wheats** are more proteinaceous and make the best formed and longest keeping breads. **Soft wheats** are starchy and are primarily used for pastries or quickly consumed breads. Hard wheats come from growing areas having limited rainfall, such as the prairie zones of the United States and Canada, most of the USSR, China, and Argentina. Soft wheats are grown in the more humid general-farming areas—east of the Mississippi in the United States, and throughout Europe. Europe imports hard wheats from the United States to blend them with the soft wheats produced there in order to make better bread. Durum is a type of hard wheat particularly in demand for making macaroni and spaghetti. In the United States it is grown especially in eastern North Dakota, and there is considerable production in southern Europe and central-western India.

Wheats can also be grouped into winter and spring types. Ancestral wheats were probably of the "win-

Figure 20-8   Winter wheat planted near Aligarh, India, under irrigation. [Photograph by Mary N. Hill for IDA, courtesy World Bank.]

ter" type, and genetic research in Japan indicates that the winter habit was probably most powerfully reinforced by genes obtained from the *T. tauschii* (= *Aegilops squarrosa*) parent. **Winter wheat** (Fig. 20-8) is planted in autumn, gaining a head start from rooting and tillering before freezing temperatures set in. It is remarkably winter-hardy, standing temperatures well below freezing, especially with the presence of a snow blanket. The crop matures in early summer. Winter wheat is particularly well suited to the southern parts of the wheat belt in the United States, from Texas into South Dakota. North of this zone, **spring wheats** are usually preferred, and are planted as far north as the Arctic Circle (where winter wheats would be doubtfully hardy and autumn weather insufficiently protracted to give them a good start). Spring plantings of varieties adapted to a short season (as short as 90 days) still yield an adequate crop under the long summer days of these more northerly regions.

Wheat suffers seriously from the wheat rust disease, which is caused by the fungus *Puccinia graminis*. Rust is especially prevalent in the more southerly winter-wheat zones, where certain types of infective spores overwinter and rapidly spread northward on spring winds. Wheat rust can cut yields markedly, and it is a serious factor in discouraging wheat growing in subtropical latitudes or humid regions. Actually, the best wheat yields are obtained not in the famous wheat belts, but in the intensively managed general-farming areas, such as those of northern Europe and the northeastern United States. But since other, more remunerative crops can usually be grown in these regions, the emphasis on breeding better cultivars is usually reserved for wheats adapted to drier regions. The wheat breeder is especially concerned about rust-resistance. Developing rust-resistant wheat is a never-ending process, because mutation and natural selection of the rust fungus seem to occur almost as fast as new wheat varieties are produced, rendering new varieties susceptible to attack within a few years of introduction.

Some of the wheats that have probably figured in the origin of bread wheat are still grown locally. Einkorn (*T. monococcum*) yields a scanty harvest in the rocky, mountainous areas of southeastern Europe. It is cultivated nowadays chiefly in Turkish Thrace. Fossil evidence of einkorn has been found in

Figure 20-9   Combining wheat in Nebraska. [Courtesy USDA.]

the remains of Swiss lake dwellings that date back to the Stone Age. Emmer (*T. dicoccum*), which also has an ancient lineage in the Mediterranean area, is still cultivated to some extent in mountainous habitats in Yugoslavia and in western parts of the USSR, in southern India, and in Ethiopia. Spelt, Polish wheat, poulard (rivet or English) wheat, and club wheat are still grown in isolated areas, but in general cannot compete economically with modern bread-wheat varieties. The one exception is durum wheat (*T. durum*).

As noted in Chapter 15, wheat is a mainstay in areas where dry-land farming is practiced, where it is planted in alternate years. In more humid prairie habitats, wheat follows wheat year after year, often with relatively little fertilization on the naturally rich prairie soils. In humid climates, where diversified farming is practiced, wheat is often used in crop rotations. If soils are adequately fertilized, 3.4 metric tons per hectare (about 50 bushels per acre°) is not out of the question, and is about double the average yield in drier prairie environments where inbred varieties are grown. Yields of almost 7 metric tons per hectare (about 100 bushels per acre) are becoming rather commonplace in the Palouse wheatlands

of eastern Washington (the record is more than 13 metric tons per hectare). Yields of the 'Gaines' and 'Nugaines' cultivars, which are resistant to lodging under heavy fertilization, depend chiefly on the amount of accumulated soil moisture. Ten centimeters (4 inches) of water in the top 2 meters (about 6 feet) of soil is needed to grow the wheat plant, and every additional centimeter is worth an additional 170 to 175 kilograms per hectare of yield. Heavy fertilization is practiced, since 4.5 kilograms of nitrogen is needed for each 100 kilograms of grain, plus phosphorus, potassium, and sulfur.

Wheat lends itself especially well to mechanization, particularly on the prairie lands of the United States, where fields may run on for miles. Plowed soil is typically drill planted in the winter-wheat belt late enough in autumn to escape severe attacks by the Hessian fly. Autumn rains are usually favorable for sprouting and consolidative tillering, laying the groundwork for good growth and yields the following spring. If weeds are a problem, there are new herbicides to control most of them, although wheat growing is less able to sustain the expense of weed sprayings than other crops that bring higher returns per unit of land. As the golden heads of wheat mature—quite a sight across the wind-swept prairies— gigantic self-propelled combines complete the harvest in dawn-to-dusk operations (Fig. 20-9). In the

---

°One bushel of wheat amounts to about 60 pounds.

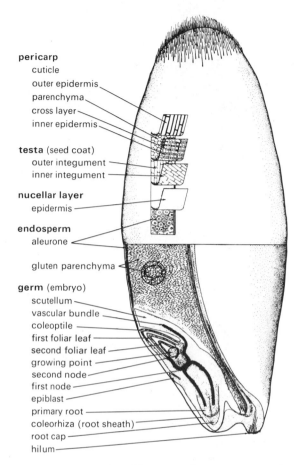

pericarp
  cuticle
  outer epidermis
  parenchyma
  cross layer
  inner epidermis

testa (seed coat)
  outer integument
  inner integument

nucellar layer
  epidermis

endosperm
  aleurone

gluten parenchyma

germ (embryo)
  scutellum
  vascular bundle
  coleoptile
  first foliar leaf
  second foliar leaf
  growing point
  second node
  first node
  epiblast
  primary root
  coleorhiza (root sheath)
  root cap
  hilum

Figure 20-10   The wheat grain. [Courtesy USDA.]

southerly parts of North America, harvesting begins in late spring and progresses with the season northward into Canada, where spring wheat may be harvested as late as September. In good wheat years there are not enough local elevators to store all the wheat at harvest time, and loose wheat is mounded in the open until transportation becomes available to move it to the urban grain centers.

The wheat grain (Fig. 20-10) is relatively proteinaceous; hard wheats usually contain 13 to 16 percent protein, soft wheats 8 to 11 percent protein. Glutenin and gliadin proteins make wheat flours "sticky" and suitable for bread-making. As with rice, the proteinaceous aleurone cells are situated toward the outer part of the grain, and may be removed with the bran during milling. Consequently, whole wheat is apt to be more nutritious than that further processed for white bread. Much of the hard wheat is used for human consumption, much of the soft wheat for livestock feeds. In flour milling, during which the grains are crushed between corrugated rollers, the oily embryos and the bran are screened off in shakers for livestock feed. The starchy endosperm is further mashed between smooth rollers to make flour. In primitive societies, wheat is hand crushed in such implements as hollowed-out stones to yield a coarse flour. A landmark of the American Colonial period was the water-powered flour mill, which utilized burred stone wheels to grind wheat.

Turkish wheat was first introduced into Kansas in 1873, though the state did not become a significant wheat-producing area until after 1900. In the last half century, well-adapted cultivars have been unceasingly bred in Kansas; the results include such improved varieties as 'Kanrad,' 'Tenmarq,' 'Pawnee,' 'Ponca,' 'Bison,' 'Kaw,' and 'Ottawa.' There has been a great deal of excitement about the potential use of hybrid wheat. Breeding problems have mostly been solved, but good yields of hybrid seed have been limited because the efficiency of pollination by interplanted male lines in the hybrid-seed production fields decreases with distance from the male plants.

In 1951 the Japanese scientist H. Kihara produced cytoplasmic male-sterility in a cross with a wild species of *Aegilops*. But because no fertility restorer could be found in American wheat cultivars for the Japanese male-sterile stock, similar crosses were attempted in America with *Aegilops* and primitive wheats to develop a male-sterile line. Most promising was a cross between *Triticum timopheevi* and 'Bison' bread wheat. Subsequently, fertility-restoring factors were also transferred from *T. timopheevi* to bread wheat, setting the stage for successful large-scale production of hybrid seed. A chromosomal sterility factor involving a recessive gene for male sterility has also been discovered; it is restorable through use of an alien chromosome bred into the line. Other research has indicated the possible usefulness of such growth regulators as ethephon—(2-(chloroethyl) phosphonic acid)—which often cause a planting upon which they are sprayed to become self-sterile.

Figure 20-11 Hand harvesting of wheat in Pakistan. CIMMYT wheats from Mexico being introduced into areas such as this are increasing production appreciably. [Photograph by Tomas Sennett, courtesy World Bank.]

Wheat is the principal crop underlying the "green revolution"—the development of major higher-yielding crop cultivars to nurture more adequately the burgeoning world population in the later decades of the twentieth century. Norman Borlaug received the Nobel Peace Prize in 1971 for his participation in the wheat-breeding program at the International Center for the Improvement of Maize and Wheat (CIMMYT) in Mexico. Borlaug, utilizing some of the 'Norin-10' germplasm brought from Japan at the end of World War II, created high-yielding varieties, well suited not only to Mexico, but also to areas of similar climate in other parts of the world, notably India (Fig. 20-11). Norin heredity has also been instrumental in the breeding of high-yielding wheats for the Pacific Northwest of the United States ('Gaines' and subsequent cultivars). Strangely, the Norin bloodline was observed in Japan by Americans as early as 1873, but was not "rediscovered" until 1946 when S. C. Salmon of the USDA brought 'Norin-10' (and other selections) to the United States for distribution to breeders. Borlaug's cultivars, hav-

ing strong, stiff stems, can accept heavy fertilization without lodging. (Prevention of lodging is sometimes accomplished nongenetically with cycocel sprays in Europe.) Fortunately, both in Mexico and in India, the new wheats were rather quickly accepted, and fertilizer was made available through various programs.

Even so successful an innovation as the Mexican wheats, which have eased the food crisis in many parts of the world, is not without its side effects, the seriousness of which remains to be determined. Like the new rices, cultivation of these high-yielding wheat cultivars requires more technically advanced operations, thereby placing small landholders and unskilled tenants at such a disadvantage that they may be forced to yield their own lands to the wealthier and more competent, with inevitable social consequences. Perhaps of even greater concern is the likelihood that the new cultivars will be so widely planted, even in remote areas, that old strains will be abandoned and lost. Experience has shown that new germplasm is needed continuously for the breeding of new cultivars that are able to resist rust and other afflictions. The only ready sources of such germplasm have been isolated ecotypes in remote corners of the world where modern cultivars are not planted.

Maize, *Zea mays*

Maize (Fig. 20-12), a more valuable find from New World exploration than all the golden tribute, is now an important crop throughout the world. But in no part of the world is maize cultivated so intensively as in the United States, where about half of the annual world production of 350 million metric tons of corn grain is grown; nor has the production of many other crops attained the degree of sophistication that has come to hybrid-corn production in North America. A distant second to the United States in maize production is China, with Brazil, the USSR, Romania, Yugoslavia, Mexico, South Africa, Argentina, and India following.

One of the best records of maize domestication comes from archeological findings in the Tehuacan Valley of Mexico. Pollen 80,000 years old, presumed to be from maize, has been collected from cores taken from old lake beds near Mexico City. Modern maize is presumed by many botanists to have arisen from teosinte, *Zea mexicana* (= *Euchlaena mexicana*), or a very similar ancestral wild plant, by a series of mutations and transformations under the guidance of domestication. Maize crosses readily with teosinte, and G. W. Beadle and his colleagues have created a convincing series of intermediates by making backcrosses. One of the pioneer investigators of maize ancestry, P. C. Mangelsdorf, long championed a complicated tripartite origin for cultivated maize, speculating that a now extinct wild form crossed perhaps first with gamagrass, *Tripsacum dactyloides* (which itself may have originated from a cross of primitive *Zea* with *Manisurus*), to yield teosinte, which then may have backcrossed with the primitive *Zea* to produce modern races of corn. Mangelsdorf has now abandoned much of this theory, and most experts lean to Beadle's simpler concept of direct transformation from wild teosinte (or at least a common ancestor). Garrison Wilkes has investigated three areas in Mexico and Guatemala

Figure 20-12   Well-filled ears of maize at the harvest stage. [Courtesy J. C. Allen & Son.]

where teosinte introgression with cultivated maize is still encouraged, with heterotic benefits. He sees little reason to doubt that this system has been operative for thousands of years.

Maize's place of origin has long been debated; if it truly were southern Mexico, which is an excellent possibility, maize soon spread throughout the Americas to become a mainstay food. Maize was an efficient source of sustenance for lowland and foothill settlements in northern South America (such as Loma Alta and other Valdivia sites in Ecuador) at least 5000 years ago. Even if it came from a single center of domestication in Mexico, maize might have been carried southward to the foothills of the Andes (where flint-corn genes seemed to have been acquired), eastward along the northern coast of South America, and thence northward from one Caribbean island to another and on to eastern North America. At the same time it most certainly spread overland northward from Mexico into what is now the western United States, and adapted forms spread northeastward as far as New England, where its famed rendezvous with the early English colonists took place.

Modern cultivars of maize resemble the primordial form only remotely. Moreover, modern maize is wholly a ward of the human species, unable to survive and perpetuate itself without human care in harvesting and planting. Selection through the millennia has yielded a specialized and very queer-looking crop, with the female (pistillate) flowers concentrated on one or a few lower cobs, or ears, and the male (staminate) flowers on separate, terminal tassels. The seeds (grains on the cob) shatter little, and, because they are enclosed in husks, they have scant chance for natural dispersal. Growing of maize today depends upon servicing this odd creation with a variety of contrivances to handle its soil preparation (cultivation, fertilization), planting (planter drills), care (pest control), harvesting (picker-shellers, combines), and cleaning and storing (drying equipment, grain elevators). Add to this all the special accouterments for hybrid seed production—inbreds, male-sterile lines, detasseling equipment, and so on. Here is a remarkable marriage of biology with technology—interlocking responses and mutual dependence between plants and people.

Early planting of maize by the American Indians was, of course, much more casual. But even by then selections had been made, and this subtropical species had been adapted to a wide variety of habitats, yielding such modern types as popcorn, northern flints, and southeastern dents, ancestral to most of today's Corn Belt cultivars. The Indians planted corn either in slash-and-burn forest clearings or on drier lands that were seasonally moist or irrigated. Planting was accomplished by such simple techniques as poking holes into the soil and inserting seeds. At least with the Mohawk Iroquois, the women soaked the corn seed for a few days before planting a few to a hill. In certain locations primitive fertilization was practiced, such as burying fish parts under the seed at planting time. Other crops, such as beans and squash, were often interplanted with the corn, and the weeds checked by hand pulling or crude hoeing. Vigil was kept to ward off freebooting birds and marauding animals.

In northern New Zealand, maize was introduced originally in 1772. The Maoris there have adopted only three introduced food plants, one of which is maize. Though they often prepare maize for eating in much the same way as Europeans do, the Maoris have a custom of rotting ears in water for six months or more prior to consumption. This is also done in parts of Peru! The slimy, odiferous kernels are crushed, boiled to a gruel, and eaten warm with milk and sugar, or are fried.

In Indonesia, maize is grown only on land unsuited to rice, the preferred cereal. The grain is ground to make puddings, as well as being consumed green as "roasting ears." It is also used as a feed grain. Stalks are utilized for fodder, and the leaves are even employed as cigarette wrappers (a practice common in South America as well). In India, flint corns are grown quite a bit in the northeast, usually sown during the period May to June for autumn harvest. With unselected varieties and primitive cultural techniques, yields are low. Dried grain is ground to a meal that is used like flour; green ears are roasted or else the kernels are removed from the ear and parched on beds of hot sand (thus enhancing its keeping qualities). The American Indian frequently parched or popped corn in much this same fashion for preservation in storage pits dug into the ground or into the sides of canyons. Corn was also preserved as hominy, by soaking it in lye made from wood ashes. But in dry climates less flinty corns have usually been grown; they keep well if merely allowed to mature naturally, and are consumed as meal made by grinding the kernels between stones, with mortar and pestle, or, in recent times, in mills. Human consumption in the United States today is chiefly as corn meal, or as grits, ubiquitous breakfast fare in the South prepared from ground hominy.

Crop breeding and mechanization have increased yields of corn fantastically in the United States. Moreover, now that corn borers and other pests can be controlled with pesticides, the trend is toward earlier planting. Corn is basically a long-season, moisture-loving crop. Starting corn with the first warm weather, rather than waiting until late spring, generally results in slightly higher yields from shorter plants that don't lodge so easily and that shade weeds better because of advanced growth (and thus conserve moisture too). They reach the pollinating and ear-forming stages earlier, at which time the extra sunlight of longer days helps make full, high-yielding ears. There is time enough then, too, to let the corn stand in the field for natural drying before harvest.

Noting the tremendous genetic flexibility of maize, which allowed transformation of a primitive ancestor quite unlike modern corn into the plethora of races that exist today, one wonders what can be expected next. Chapter 15 mentioned the trend toward denser plant populations in the field. Will this encourage breeders to develop still newer cultivars that are perhaps squatter, yield even more ears, and are better adapted to field combining? Will combining dictate changes in ear form from that developed for easy picking by hand, with grain attachment secure? Will kernels have to be developed that are less subject to damage by the combine? Of course, breeding for improved standing ability, disease and insect resistance, season-span to match climate, and yield rather than appearance (once important because of rural corn contests) can be taken for granted.

Probably the greatest changes to be wrought by breeding will come from an increase in the attention given to the chemical composition of the grain. It is possible through selection to produce corn high in protein or high in oil (see Fig. 19-10). More sophisticated qualitative changes are possible. For example, in normal corn the starch is composed of 27 percent amylose (straight chain) and 73 percent amylopectin (branching chain). Separation is expensive, and each kind of starch has particular industrial uses; for example, amylose is used for adhesives and thickenings in the food industry for such products as instant puddings, and amylopectin is used in the paper industry and as coatings or sizing in the fabrication of woven fiberglass. The *waxy* mutant produces 100 percent amylopectin; the *ae* gene (*amylose extender*) increases amylose to 60 percent of the total, and selection in these types has increased this to 80 percent. The discovery that certain mutants alter the amino acid composition of corn proteins has far-reaching consequences. Corn, like other grains, is not a good sole protein source for monogastric animals (pigs or humans) because it is low in lysine and tryptophan (see Chapter 3). By increasing the proportion of lysine and tryptophan, the *opaque-2* gene greatly increases the feeding value of corn. The combination of high protein and improved amino acid balance in corn and other grains offers tremendous potential in alleviating malnutrition in many parts of the world.

Hybrid seed corn is worthy of some special mention, since it has sparked the concept of (and has been the proving ground for) heterosis. Because modern corn is monoecious (ears and tassels separate on the same plant), it lends itself well to regulated crossing, which requires nothing more than clipping off the tassel to prevent self-pollination. Planting adjacent rows of a different inbred from which the tassels have not been removed assures a hybrid cross for seed collected from detasseled rows. In recent years, cytoplasmic male sterility had been introduced into inbred lines, which in effect accomplished the expensive detasseling procedure genetically. The Southern corn blight epidemic of 1970 in the United States—caused by the susceptibility of the sterile cytoplasmic factor to the "T" strain of the fungus responsible for the disease (see Chapter 19, page 484)—has resulted in a reversion to mechanical detasseling for the production of hybrid seed.

The mechanism of heterosis—the increase in yield

Figure 20-13   A 6-row combine, working after dark, is a sign of the high degree of mechanization and productivity associated with the American corn crop. [Courtesy USDA.]

due to hybrid vigor—was discussed in Chapter 19. It is widely recognized that select inbred lines can be hybridized to produce superior yields from first-generation progeny. Inbred suitability for such crosses has been determined by trial and error, and more recently through sophisticated mathematical techniques involving a cross to a tester line with a wide genetic base. The discoverer, having control of certain suitable inbred lines, has not only a valuable source but an exclusive source of that particular hybrid seed. The hybrid-corn seed industry began with a single cross of inbred strains, say A with B. But more remunerative yields of hybrid seed could be had if a seed were produced on a plant that exhibited hybrid vigor. Thus double-cross hybrids became commonplace, with $(A \times C)$ crossed with $(B \times D)$, or occasionally with $(A \times C)$ merely crossed with B as a pollinating parent (triple cross). The heterozygous seed so formed does not breed true in subsequent generations because of genetic reassortment;

thus the farmer must buy new seed the next year from reliable sources of supply.

In the United States much of the corn crop never leaves the farm. Most is used to feed cattle and hogs. There are three different field corns: (1) the **flint corns,** which have kernels of hard starch, were being used by the northeastern Indians when the Pilgrims arrived, and are probably related to certain South American genotypes (by way of dispersal through Central America and Mexico into the Southwest, and then northeastward); (2) the **dent corns,** which have kernels of hard starch capped by soft starch, were developed from stocks originally found in the southeastern United States, and are especially high yielding and much used in the Corn Belt; and (3) **flour corns,** which have kernels of soft starch and are amenable to hand grinding, a method much employed by Indians of the Southwest. In addition to field corns there are the **sweet corns,** the sugary kernels of which are much consumed in the green stage as human food ('Golden Bantam' and 'Country Gentle-man' are familiar cultivars), and the **popcorns,** whose kernels burst open upon heating, rendering them useful for primitive people lacking grinding mills. (Even teosinte will pop to some degree.) In addition to these standard types grown in the United States, there are hordes of other distinctive races grown throughout Latin America, often in isolated regions. Many are now being introduced into the United States as sources of germplasm for the breeding of future cultivars. Especially of interest in recent years is the rediscovery of perennial maizes in Mexico. The tetraploid *Zea perennis,* discovered by Hitchcock in Jalisco in 1910, but thought extinct since 1921, has been collected again; and a new diploid species, *Zea diploperennis,* was discovered in 1978. The implications for maize breeding are tremendous.

Corn is an important source of industrial products. More than 300 commercial items derived from the corn grain are shown on a Corn Refiners Association flow sheet. Perhaps the chief industrial derivative is corn starch, usually obtained by wet-milling. Starch may in turn be hydrolyzed or fermented to yield corn sugars, syrup, and special gumlike polymers such as phosphomannan, which is used for adhesives. The grain is nearly 10 percent protein, some of which is removed by initial steeping in weak acid. This valuable corn-steep liquor may be employed as a nutrient broth in the production of antibiotics and in animal feeds. The oily embryo, which makes up 6 to 13 percent of the grain, is next removed, usually by partly crushing the steeped grain, screening away the coarse hulls, and floating off the released "germ" (embryo). Expression or solvent extraction yields corn oil, which constitutes about 50 percent of the embryo. The residues are utilized, along with hulls and other by-products, in animal feeds. The corn oil is further refined and utilized principally as cooking and salad oils.

After removal of the embryos the remaining mass, chiefly endosperm, is crushed in burr mills, and any hulls are removed by screenings and siftings. Gluten is separated from starch either by differential sedimentation on long, inclined tables, over which the slurry is made to flow slowly, or by centrifugation. The gluten goes back into cattle feeds, and the starch is further washed, dried, and pulverized. The starch is either consumed directly as human food or is used by industry to make sizing, laundry starch, urethane plastics, and other products. Some is used as a raw material for further processing by digestion or hydrolysis with enzymes and weak acid catalysts to yield a glucose syrup or various mixtures of simple sugars and dextrins.

Barley, *Hordeum vulgare*

The annual world production of barley, *Hordeum vulgare* ($=$ *H. sativum*), the world's fourth most important cereal, may exceed 180 million metric tons (Fig. 20-14). Worldwide, the Soviet Union and China head the list of producers. France and the United Kingdom are the leading Western European producers. Canada, the United States, Turkey, India, and Morocco also produce significant amounts. The grain is much used for animal feed, little for direct human consumption. About a third of the grain goes for making malt, essential for the brewing of beer and liquor and used also in a number of confections and "health" foods.

A number of wild barleys have a diploid chromosome number of 14, including *H. spontaneum* and *H. maritimum.* The cultivated species all have 14 chromosomes, and are considered to be descended from *H. spontaneum,* which crosses readily with them, and is cytologically indistinguishable. Tetraploid barleys include *H. jubatum, H. marinum,* and *H. bulbosum,* the latter a source of male-sterile lines in crosses with *H. vulgare.* All cultivated barleys have been relegated to *H. vulgare* by many authorities, but others have reserved this appellation for the six-rowed types only. The two-rowed barleys often mingle with the similar indigenous wild barley in the Fertile Crescent area of the Near East. They are sometimes considered to be *H. distichon* (and the irregular in-between types are then referred to as *H. irregulare*).

Barley was a major grain crop in its apparent center of domestication (the Zagros–Taurus foothills)

production by kind

production by region

total 233 million metric tons

Figure 20-14 Relative world production of some small grains. [From *A Graphic Summary of World Agriculture*, USDA Misc. Publ. 705.]

more than 11,000 years ago, and it was probably one of the first cultivated cereals. Barley grains more than 10,000 years old, either cultivated or systematically collected from the wild, have been found in the Abu Hereyra diggings in the upper flood plain of the Euphrates River in Syria. Barley has even been discovered in late Paleolithic sites in the Nile Valley (near Aswan, Egypt) that have been dated between 17,000 and 18,300 years old, in a habitat where barley cannot survive today; it was probably utilized for food but not cultivated. Although annual, 2-rowed barley was no doubt the ancestral type, 6-rowed barley believed to be more than 8500 years old has also been excavated in the Near East.

As with wheat, there are spring and winter varieties of barley. **Spring barleys** mature in a short season of 2 to 3 months. In the United States they are planted north and west from middle Texas to southern New England. From the Ohio Valley southward **winter barleys** are generally planted, although spring barleys are winter-seeded near the Gulf for winter grazing, as well as for whatever grain eventuates when the cattle are removed in spring. As its wide distribution would imply, barley is tolerant of many soils and climates, but like wheat it is not well adapted to hot humid conditions.

Barley is grown in much the same fashion as wheat. It responds well to fertilization, and is generally planted in rotation rather than year after year in the same fields. As with wheat, shortness and straw

stiffness have been bred into barley from Japanese accessions, which makes it better adapted to generous fertilization. The trend is towards denser plant populations through closer spacing of rows. Attention has been given to production of hybrid barley seed. Male-sterile lines carrying an extra chromosome (trisomic) derived by the crossing of several species have led to some commercial hybrid-seed production. Cultivar yields have exceeded 4 metric tons per hectare in Denmark, although in the United States yields are seldom half as large. Yields in the rest of the world are considerably smaller.

In the United States, harvesting is done by field combining. In many of the less developed parts of the world, barley is still gathered and processed by hand (Fig. 20-15). Where humidity is relatively high, the grain may be windrowed for drying before threshing or combining. The moisture content of the grain should be less than 14 percent for storage.

Barley grain is usually crushed for incorporation into livestock feeds. Some is pearled for use in soups, baby gruels, and other foods. Pearling consists of rubbing the grain against abrasive disks to grind off the hulls and the outer layer of the kernel, much as in the polishing of rice. A minor portion of barley is milled into flour, particularly in Asiatic countries. It is used in India, for example, to make *chapatis*, a kind of flat bread similar to tortillas.

Much attention has been given to malting varieties of barley. Good malting qualities can be determined

Figure 20-15   Hand-harvested barley being threshed and winnowed in Ethiopia. [Photograph by Kay Muldoon, courtesy World Bank.]

in the laboratory by protein and diastatic-potential analysis. In contrast with wheat, of which proteinaceous varieties are most sought for bread, starchy varieties of barley are generally chosen for malt. Brewers demand uniform, carefully cleaned grain, for which a premium is offered. The making of malt is a relatively simple process. Grain is steeped in water and then germinated under controlled humidity, temperature, and atmosphere. Incipient sprouts develop a high content of amylase and other enzymes, valued for later digestion of starch adjuncts in brewing. Germination is stopped, when the enzyme content is maximal, by heating and drying the sprouting grain.

In brewing, malt is mixed with other grains to provide the mash; the enzymes of the malt reduce starch to simple sugars in the resulting wort, which is later fermented by yeast to make the beer. A limited supply of malt is also used for flavorings, breakfast cereals, icings, coffee substitutes, malted milk, infant food, malt flours, medicinal syrups, candies, and various industrial fermentations.

Oats, *Avena sativa*

In Samuel Johnson's *Dictionary of the English Language* (1755), oats are described as eaten by people in Scotland, but fit only for horses in England. The Scotman's retort to this is: That's why England has such good horses and Scotland such fine men. In spite of Samuel Johnson's scorn, oats are one of the world's important behind-the-scenes cereals, coming to market mainly in the form of livestock. The grain is nutritious and of high feed value.

Oats (Fig. 20-16) are chiefly a European and North American crop. World production is now little more than 50 million metric tons annually, on the decline ever since tractors replaced horses. The USSR, the United States, Canada, Poland, Germany,

Figure 20-16 An oat panicle with the grains separated. [Courtesy J. C. Allen & Son.]

France, and China are the leading producing countries. Like barley, oats are widely adapted.

The origin of oats is obscure. Recognized species represent a polyploid series with shared genomes. Classical Greek writers seemed to view oats as a weed, implying recognition but not domestication in that part of the world more than 2000 years ago. Some authorities believe that *A. sativa* developed as a mutation from wild oats, *A. fatua*, in Asia Minor or southeastern Europe not long before the Christian era. Most regard it as having entered domestication as an unavoidable weed in barley. Other experts suggest that *A. byzantina*, itself perhaps a subspecies of the wild red oat, *A. sterilis*, is more likely ancestral to at least certain forms of *A. sativa*. Some especially winter-hardy multiline cultivars (including 'Norline' and 'Pennlan') are considered to belong to the species *A. byzantina*. All of these are hexaploid ($2n = 42$). There are also a number of tetraploid and diploid species of little commercial importance. *A. barbata* is

a tetraploid widely represented in the Mediterranean area, with which the Ethiopian *A. abyssinica* is probably conspecific. Another tetraploid, *A. magna*, has close morphological resemblance to the hexaploids, to which it may have contributed. Diploids include *A. strigosa*, *A. ventricosa*, and *A. wiestii*.

Oats were brought to North America with other grain in 1602 to be planted on the Elizabeth Islands off the coast of Massachusetts. Today both spring and winter varieties are planted, the latter sown mostly in the South. Disease prevents oats from really prospering in the humid Southeast, however. In the United States most oats are grown in the Corn Belt, in rotation with corn. Continuous cropping of oats results in build-up of such pests as stem rust and crown rust, which in a short while make yields unremunerative. As with wheat, there is constant need for new disease-resistant cultivars. Planting and handling are as has been described for other small grains. Oats are harvested in summer (Fig. 20-17), and can be safely

Figure 20-17 Harvesting oats with a self-propelled field combine. [Courtesy International Harvester Co.]

stored when their moisture content is less than 14 percent. There is some processing of oats to flour, rolled oats, and oatmeal. Scotland, true to Dr. Johnson's allegation, is a heavy consumer of oats as human food. Several breakfast cereals made from oat flour are popular in the United States. Oat hulls, a by-product of milling, have served as a raw material for fermentation to furfural, a chemical solvent used in refining minerals and for making resins.

Rye, *Secale cereale*

Rye is chiefly a European grain, with the Soviet Union the leading producer, followed by Poland and Germany. Total world production is generally less than 30 million metric tons.

Rye is a diploid species ($2n = 14$). *Secale cereale* has not been authenticated in the wild, although it is familiar as a naturalized escape and it crosses with wild species. It may have been domesticated from wild ryes, *S. montanum* or *S. anatolicum*, of the Mediterranean region and Asia Minor. It is possible that domestication of rye may have come about by its appearance as a weed in barley or wheat crops. As with oats, domestication of rye may have come relatively recently, but wild-type rye grains (whether collected or cultivated) have been found in the Abu Hereyra excavations of Syria, indicating human usage dating back at least 10,000 years.

Rye prefers cool, dryish climates, and can stand winter weather well. It is widely adapted to temperate climates, and can often be successfully cultivated on soils too poor for remunerative growing of other grains. It is planted and harvested in the same way as other small grains.

The grain, usually about 13 percent protein, is somewhat glutinous like wheat. Thus rye can be made into a bread flour, and is much used in Germany, Poland, and the USSR for the making of a dark, bitter, "black" bread; in the United States "rye" bread is made largely of wheat flour with only a small proportion of rye flour. Most rye, however, is produced for hay and pasturage or for stock feed. Yields average more than 2 metric tons per hectare in Europe, and under intensive management they may be double this. The straw is often employed for stuffing, thatching, and paper-making, as well as for animal bedding. The crossing of rye (*Secale*) with wheat (*Triticum*) to yield × *Triticale* was reviewed in Chapter 19.

Sorghum, *Sorghum bicolor*

Sorghum is a vast and variable genus, mainly African, of which the cultivated grain types are grouped in the species *Sorghum bicolor* (= *S. vulgare*). The center for the domestication of grain sorghum was no doubt northern Africa, although secondary domestications seem to have been made in India after early domesticates were spread eastward. Grain sorghums are well adapted to sub-Saharan savanna climates, and the Sudan or Chad could have been the locale for the domestication of at least certain types (although N. V. Vavilov thought Ethiopia more likely). It has been hypothesized by J. M. J. de Wet and others that domesticated sorghums probably arose either from the "*aethiopicum*" race or from the "*verticilliflorum*" race (or from both) of the *arundinaceum* subspecies of *S. bicolor*. Both are locally abundant and commonly harvested from the wild for human consumption in parts of Africa today. Domestication may have been relatively recent—that is, within the

Figure 20-18   (A) Harvesting grain sorghum. (B) Grain sorghum. (C) 'Sordan,' one of the sorghum–Sudan grass hybrids. [A, courtesy Allis Chalmers; B and C, courtesy J. C. Allen & Sons.]

last 5000 years—and there may have been several independent domestications.

In any event, grain sorghum has become an increasingly important cultivated crop (Fig. 20-18), with formal production now centering in the United States and India more than in all of Africa. In excess of 55 million metric tons of grain are produced annually, and there is probably a good bit more produced in back-country African plantings that is not recorded in the statistics. The United States accounts for over a third of recorded world production, followed by India, Argentina, and the central African countries as a group. A tremendous germplasm collection of sorghum has been assembled under the aegis of the Rockefeller Foundation at CIMMYT in Mexico, and important breeding projects are underway in many locations, designed especially to make cultivars more reliably productive under erratic rainfall.

The genus *Sorghum* is taxonomically confusing, with a number of interfertile species, subspecies and innumerable cultivars. In the "sorghum" section of the genus fall the *S. bicolor* forms, all of which are annual diploids ($2n = 20$), and weedy Johnson grass, *S. halepense*, a tetraploid that nonetheless crosses readily with cultivated sorghums to yield various forms of Johnson grass and "sorghum almum". *S. halepense* is perennial, rhizomatous, a forage rather than a grain species, and so aggressive as to be classed a noxious weed in many states of the United States. Relationships in other sections of the genus are poorly understood, but they don't involve any of the grain sorghums. Wild representatives of grain sorghums are strictly African, falling within the *arundinaceum* subspecies. These races are found there: "*aethiopicum*," centered in the drier parts of northeastern Africa but extending to the western Sahel, where it can be locally abundant, occasionally invading cultivated fields and crossing readily with cultivated sorghums; "*arundinaceum*," in the tropical forest regions of central Africa along the Gulf of Guinea and through the Congo, where it is common on stream banks and occasionally invades fields; "*verticilliflorum*," quite similar to "*arundinaceum*," widespread through the savannas of eastern and southern Africa; "*drummondii*," including weedy types mimicking cultivated sorghums; "*virgatum*," limited to the central Nile Valley; and "*propinquum*," widespread in tropical Asia. All of these cross with the subspecies *bicolor*, repository for the cultivated sorghums (races "*bicolor*," "*caudatum*," "*durra*," "*guinea*," "*kafir*," and various combinations). The cultivated races are distinguished primarily by differing morphology of the inflorescence and grains. Sudan grass (*S. bicolor sudanensis* = *S. arundinaceum sudanense*, of the *drummondi* group) is grown as a forage plant rather than for grain. It has been utilized for hybridization with grain sorghums, to yield the so-called sorghum–sudan hybrids.

It is not surprising that, with so tremendous a range of types, grain sorghums can be found that are adapted to as little as 25 centimeters (about 10 inches) of rainfall or as much as 250 centimeters (about 100 inches). However, in severely dry habitats, pearl millet (*Pennisetum americanum*) is usually planted instead of sorghum, and in moist tropical habitats, root crops are preferred. Cultivated sorghums range in height from 1 to 5 meters (about 3 to 16 feet). Their grain color varies from white through yellow and brown to red and purple. The grain may be starchy or quite vitreous.

Hybrid grain sorghums exhibit strong heterosis. Yield increases of as much as 40 percent are obtained through hybridization. But because the flowers are perfect, with male and female parts together in a compact terminal head, production of hybrids had to await the development of male sterility. Only within recent years have practical male-sterile lines been developed, giving rise to a hybrid-sorghum seed industry that is especially important in the plains states from Texas northward into Nebraska. Since the climate there is too dry for successful maize growing, sorghum is an important alternative to wheat. Sorghum–sudan hybrids are also enjoying favor in the northeastern states as forage plants, especially where alfalfa has been hard hit by dry years and the alfalfa weevil. Grazing and green chop are the main uses of these hybrids in such normally humid areas.

Sorghum grain, although nutritious, is small in size and difficult to process. The planting of sorghum for grain rather than forage and silage is especially important in Africa and Asia, where it serves as human food (Fig. 20-19). In India the whole grain is eaten like rice; sorghum flour is used to make an unleavened bread. A red dye can be extracted from stem and leaf of sorghum, and the stems can be used as fuel and as a building material. Some sorghum is used as fodder. In the Mediterranean region grain sorghum is very important as a dry-land summer crop. Hybrid seed from the United States is used to advantage in Israel. To some extent it is grown under mechanized agriculture there, but in most of the Mediterranean region planting and harvesting are still done by primitive means.

Sorghum has an extensive root system for a plant with such a modest leaf area, equipping it well for survival in semiarid climates. Since rainfall is light where most sorghum is grown, yields are more de-

Figure 20-19 Hand-harvested sorghum being stacked for drying in a village in Upper Volta. [Photograph by Ray Witlin, courtesy World Bank.]

pendent upon moisture than fertility. Yields under mechanized agriculture in North America exceed 3 metric tons per hectare (about 2.7 tons per acre), but they are only half this in the Near East, and half again less in the Far East. In the United States, short-stalked combine cultivars are planted almost exclusively. The old "milo" and "kafir" varieties have almost disappeared. The grain is utilized mostly as a livestock and poultry feed, and to a limited extent for industrial processes, where it can substitute for corn. In Africa, sorghum grain is fermented to make a beer, which is a valuable dietary supplement in the

hinterland. About half of the land planted to sorghum is not harvested for grain, but is used instead for forage and silage. If Sudan grass used for hay and summer pasture is included, the forage value of sorghum outranks its grain value.

In some localities sorgo, or sweet sorghum ("cane") varieties are still grown, planted much like maize. Although these varieties serve as forage, sorghum syrup can be extracted from the juicy stems, much as with sugar cane. With broomcorn varieties the panicles are cut, dried, and bound to make brooms.

Millet

Various species of a number of different genera of grasses are termed millet. The millets are grown for grain principally in the Far East and in Africa. World production is well over 40 million metric tons; China, India, and a number of central African countries are the leading producers.

Archaeological evidence dating back 7000 years indicates that *Setaria* foxtail millets (probably both *S. geniculata* and *S. macrostachya*) were among the earliest domesticated crops in the primitive culture of the Tehuacan Valley of Mexico, but they were abandoned as maize (and later European introductions) proved more rewarding. By and large the small-seeded millets have remained a local grain of less developed areas. None has been able to compete with the major cereals in world commerce. In agriculturally advanced nations, millets are usually used as a component of feeds, especially chick and bird seed, or to make quick soil cover and pasture.

The major millet species of the world are foxtail millet, *Setaria italica;* pearl or cattail millet, *Pennisetum americanum;* finger millet, or ragi, *Eleusine coracana;* teff, *Eragrostis tef;* several species of *Panicum,* such as proso millet (*P. miliaceum*) and browntop millet (*P. ramosum*); several "finger millets," *Digitaria* spp., especially in Africa; west African finger millet, *Brachiaria deflexa;* kodo millet, *Paspalum scrobiculatum* of India; and a Japanese barnyard millet, *Echinochloa colona* (= *E. frumentacea*). They are generally grown as mixed crops in regions of light rainfall, in association with legumes. Pearl and finger millets are important crops in India, where they have been used in unleavened bread since prehistoric times. Teff is an important grain and forage in Ethiopia, and the flour is used in baking the ubiquitous *injara* "pancakes."

Minor cereals

Other seeds of grasses, of course, can serve as grain, and some are locally important as specialty items. Perhaps the most prominent in this respect is wild rice, *Zizania aquatica* (= *Z. palustris*), a rare example of an American cereal in the process of domestication during modern times. Wild rice was an important Indian food when the Northwest Territory of the United States was first explored. It grows in ponds and sluggish streams, and was usually harvested by flailing it into a canoe paddled into the wild stands. Today the wild stands are much decimated, but some 12 thousand hectares (about 30 thousand acres) in Minnesota are reserved for harvest by the Indians. Wild rice is a luxury food sold at prices generally more than ten times higher than those of other cereals. A considerable part of the harvest still comes from northern Minnesota, where the Chippewa Indians have a regulated "monopoly" on wild rice harvested on their reservations. It is still inefficiently gathered there by flailing the grain into canoes, although on private lands and in Canada mechanization has superseded hand gathering. As collected, the rice is full of wet foliage and lake debris, and it shrinks to about 40 percent of its green weight during processing. Wild rice is increasingly being planted in artificial paddies, which, as in *Oryza* rice growing, can be drained so that the crop can be combined. Seeding may be by plane or by boat, and yields are several times larger than those from wild stands, as much as a metric ton or more per hectare (about 0.45 ton per acre).

Zizania aquatica can be classified into three botanical varieties, of which the variety *interior* is that growing in northern Minnesota and southern Canada, the only one with large enough grain to make harvesting remunerative. Other varieties occur above tide level along the Atlantic Coast from Canada to Florida, where they are important as a wildlife food but are not directly utilized by people. The species is an annual, and it depends upon spontaneous reseeding each year. What with wild rice now being cultivated, shatter-resistant strains have been selected, and domestication is well under way, with pure-line cultivars being developed under isolation.

In nutritive quality, wild rice is on a par with oats, 14–18 percent of the grain being a nourishing form of protein. Although wild rice is still occasionally hulled by tredding and hand winnowing, even the Indians have begun to use such things as small fanning mills for processing the crop. In Canada, where the Indian methods are not protected as they are in Minnesota, harvest is often accomplished with scows fitted with paddle wheels; centralized processing includes drying, parching, mechanical hulling, and removal of chaff by suction. Wild-rice processing involves a week or so of fermentation of the green grain on the ground, during which the characteristic dark color and flavor develop.

A number of other grass species yield grain, especially in the tropics. Job's-tears, *Coix lacryma-jobi*, is a striking species native to southeastern Asia but widely introduced throughout the tropics. The large, shiny white grains are often used ornamentally, but they may be parched, boiled, or milled into flour for food. Seed of mannagrass (*Glyceria fluitans*), Guineagrass (*Panicum maximum*), fonio (*Paspalum longiflorum*), bergu (*Panicum vergii* and *Echinochloa stagnina*), common reed (*Phragmites communis*), and many others may be similarly utilized.

LEGUMES OR PULSES

If plant families can be rated according to importance, the family Leguminosae would rank not far behind the grasses. Indeed, the legumes may be more important fundamentally, since the family is noted for having evolved symbiotic associations with certain bacteria, such as *Rhizobium*. These symbionts, usually housed in nodules on the roots of the legume, fix gaseous nitrogen. The legumes are, by and large, the least dependent upon nitrogenous fertilizer of all crop plants, and thus are particularly an asset to

Box 20-1 *Symbiotic nitrogen fixation to extend the food supply*

The advancements in satisfying world food needs made possible by the "green revolution" (see the discussions on wheat and rice) depend heavily on fertilizer. However, under primitive circumstances fertilizer may not be available, and it is too costly for general use in typical subsistence farming. In some localities, its distribution may be erratic and means for its application may be inadequate. Even in technically advanced societies problems exist: in particular, there is concern that fertilization may overload the runoff and ground water with salts, causing a pollution problem.

None of these problems are caused by nitrogen secured through *Rhizobium*-type symbiosis, notably occurring in legume root nodules. Scientists have wondered whether strains of free-living bacteria, algae or fungi that are capable of fixing nitrogen can be made to thrive in the rhizosphere of monocotyledonous plants such as the various species of cereal grain, as well as in legume roots.

Nitrogen fixation by cyanophytes in association with *Azolla* fern is well documented for Asiatic rice fields, and other symbiotic relations may occur as well, for rice fields often remain productive without additions of nitrogenous fertilizer. Significant symbiotic nitrogen fixation associated with higher plants other than legumes occurs with *Alnus* in the Arctic, *Myrica* in the temperate coastal plains, *Casuarina* in the tropics, and other genera—even with certain grasses.

The enzyme nitrogenase has been shown to catalyze a conversion of nitrogen to ammonia in those biological systems in which it has been investigated. If nitrogenase is provided with the essential proteins, a high-energy source (ATP), and a strong electron donor, theoretically an effective nitrogen-fixing system might be developed with any number of unicellular organisms. Nitrogenase activity has been achieved in vitro by combining bacteria and cultured host tissues; why not, then, the "breeding" of broadly adapted microorganisms to live symbiotically in a wide range of host root tissues? Even partial success could do much to help delay worldwide famine.

primitive agricultural systems. By the same token, this degree of self-sufficiency in nitrogen makes legumes especially rich sources of protein, so often deficient in meager diets. Growing more legumes for food in less advanced parts of the world would materially reduce the prevalence of the dreaded kwashiorkor (protein deficiency) disease. For forages to feed livestock, legumes have no equal, especially in partnership with grass. When agronomists speak of "grasslands," they are thinking of legumes as well as grasses.

Table 20-3 compares the protein content of several food products. Note how much more protein legumes contain than cereals and major root crops. No wonder that edible legume seeds (or pulses) have had a history of domestication closely paralleling that of the grain cereals: the two complement each other for human nutrition, and are equally important to the evolution of civilization. Neither whole dried milk nor dried egg is richer in protein than are legumes. The protein content of dried meat or fish (generally 70 to 90 percent protein) is much greater, of course, but plant protein costs far less. No wonder legumes are so important a supplement to the diet of peoples whose subsistence depends primarily upon starchy cereals or root crops.

Legume seeds may be as important for their oil as for their protein. Soybeans are about 20 percent oil, peanuts around 50 percent. Indeed, in technologically advanced parts of the world they may be grown chiefly for their oils, with the protein a "by-product." In the United States about 70 percent of the soybean oil is utilized for derived foods, such as margarine, and the remainder mostly for industrial applications. Ninety-five percent of the protein becomes animal feed, especially preferred for poultry.

Table 20–3 Protein content of familiar foods and feeds.

Food or feed	Protein (percent)
Cereals	
whole rice	7.5–9.0
polished rice	5.2–7.6
wheat flour	9.8–13.5
corn meal	7.0–9.4
Legume seeds	
chick pea	22–28
soybean	33–42
peanut	25–30
Root crops	
potato (dry weight)	10–13
cassava	1.3
Miscellaneous	
walnut	15–21
dried alfalfa meal	18–23
dried *Chlorella* alga	23–44
dried *Torula* yeast	38–55

A

B

Figure 20-20 The soybean exemplifies legumes utilized for their seeds. (A) The growing plant, showing trifoliate, alternate leaves, and seeds borne in pods (legumes), a characteristic of the family Leguminosae. (B) The mature plant ready for combining. [A, courtesy USDA; B, courtesy J. C. Allen & Son.]

Various beans and peas have been eaten by people since time immemorial. But soybeans (Fig. 20-20)—the world's most abundantly grown seed legume today—have been esteemed only in the Orient. Even as an animal feed, the protein content of the soybean is not tops, presumably because certain minor proteins act as trypsin inhibitors and as mild hemagglutinins. The protein cake that remains after the oil has been extracted from soybeans is improved as a livestock feed by mild heating, which differentially alters these sensitive minor components. Soybean flour is used in human foods largely as an extender for other, more expensive forms of protein. In Israel, for example, government edict requires its inclusion as an "enrichment" in bread flour.

Although the legumes are only a little less important in world commerce than the grains, the relative importance of individual legume species is not easily determined. World-wide recorded production of edible beans and peas (pulses exclusive of soybeans) approaches 50 million metric tons annually. Soybeans account for almost twice this amount, with North America the leading producer. Around 18 million metric tons of peanuts are grown, chiefly in India. Dried beans come mainly from India, China, Latin America (Mexico especially), the United States, Japan, Italy, and Turkey. The leading producers of dried peas are the USSR and China. Broad beans are grown chiefly in China and North Africa. Chick peas are overwhelmingly an oriental crop, chiefly from India. Lentils are much grown in India and the Near East.

Soybean, *Glycine max*

The soybean, or soya, indigenous to eastern China, is one of the oldest of cultivated crops. It was probably domesticated 13 thousand years ago. Its cultivation in China was mentioned in a document written by Emperor Sheng Nung in 2838 B.C.E., at which time the soybean was regarded as one of the five "sacred grains" vital to Chinese civilization. Presumably the cultivated *G. max* was derived from *G. soja* (*G. ussuriensis*) of central China, with which it crosses readily. Both species have 40 chromosomes and are grouped in the subgenus *Soja*. In comparison, other species of the genus often have differing chromosome numbers (for example, species of the subgenus *Bracteata*, which are mostly African, have 22 or 44 chromosomes, and may be deserving of separate generic rank). Species of the subgenus *Glycine*, mostly native to Australia and the South Pacific, have diploid and tetraploid counts of 40 and 80.

Although the soybean was introduced into Japan by way of Korea before 200 B.C.E., it was little known outside of the Orient until the seventeenth century, when Engelbert Kaempfer, a German botanist, brought the soybean to Europe, and when soybean seeds were received at the Jardin des Plantes in Paris in 1739, sent there by missionaries. Benjamin Franklin, American ambassador to France from 1778 to 1785, probably received some soybeans as part of his seed exchanges, but the first official record of its introduction into North America was in 1804, when Dr. Mease grew it in Philadelphia. Thereafter the soybean was little grown in the United States until the twentieth century, and only recently has it been grown extensively. So fast has it been adopted in the Corn Belt in recent decades that it now rivals corn as an income crop. Breeding of improved cultivars, and their inherent adaptability to mechanization, have made soybeans a remunerative crop to grow; but the lack of indigenous races in this country and their gradual extinction in Asia creates difficulty in securing a broad germ-plasm base. Moreover, industry has found many ways to use soybean oil, creating a steady and increasing demand for the bean, an inducement to farmers of the Midwest.

Soybeans are well adapted to warm temperate climates as a summer annual. They are grown throughout the world—in middle and southern Europe, southern Russia, South Africa, Egypt, and South America. But the United States, Brazil, and China account for more than 90 percent of world production, which has reached more than 75 million metric tons annually in recent years. The United States alone produces nearly two-thirds of this amount.

Climates with warm, moist summers and at least a five-month growing season are ideal for most soybean cultivars. Soybeans are broadly tolerant of soils, but they do best on well-drained heavier types. In America, soybeans are generally row planted and often rotated with corn. Soybeans generally respond to phosphorus and potassium fertilization, but they are often erratic in response to nitrogen—applied nitrogen seems to substitute for some nitrogen fixation (Fig. 20-21). Strongly nodulating types make better use of nutrients, but for top yields nitrogen fertilizer may be used even with these cultivars to supplement the *Rhizobium* activity. Soybeans are self-fertile, exhibiting little outcrossing. Pre-emergence herbicides are utilized to control weeds in mechanized growing, especially with narrowly spaced rows (25 centimeters, or 10 inches) for high yields. Weed control is a big help for efficient combine harvesting.

Most of the soybean crop in the United States is processed for oil, which accounts for about three-fourths of the nation's production of vegetable oils—in the neighborhood of 5 million metric tons. The

Figure 20-21 Nitrogen fertilizer applied to soybeans often substitutes for a portion of the nitrogen normally fixed by symbiotic bacteria without greatly increasing yield. [Courtesy Potash and Phosphate Institute.]

bean is generally about 20 percent oil and 45 percent protein. Oil used to be expressed from the seed, but most is now extracted with hydrocarbon solvents (which cuts oil wastage from 4 or 5 percent to about 1 percent). About two-thirds of the vegetable oil is used for edible products—mainly shortenings, margarines, and salad oils—and about one-third is used industrially—in paints, varnishes, plastics, inks, caulking compounds, linoleum, and other products. The protein cake that remains after oil extraction may be processed into soybean flour or incorporated into animal feeds. Processing or fermentation is generally sufficient to counter the four antidigestive "trypsin inhibitors" found in the bean. Soybean protein balances some of the nutritional deficiencies of such grains as corn and wheat, which are low in the amino acids lysine and tryptophan. Lysine, generally present only in small amounts in feedstuffs, is present in reasonably adequate amounts in soybean meal. Lecithin, an important industrial soybean extract with antioxidant, emulsifying, and wetting properties, is used in foods, pharmaceuticals, printing inks, rubber, leather, textiles, cosmetics, and petroleum products.

As a garden plant, soybeans can be eaten green-shelled (though they are not easy to separate from the pods), salted and roasted, or dried. More commonly soybeans are used in human food as flour, incorporated with wheat flour to round out the protein

content in bread, cakes, cookies, and crackers; it is also used in soups and as an extender in potted meats, confections, and nut "butters." A small but thriving business has arisen in forming soybean protein to simulate meat, the so-called "textured vegetable protein" or TVP. In Japan much soybean is consumed as *miso* or *tofu*, of which an average of 30 grams total per day is consumed by the citizenry. Tofu is a bland, soft, cheeselike product made from soybean protein. Miso (which consists of fermented rice to which pulverized cooked soybeans are added, the fermentation being allowed to continue for several weeks before the material is ground into a paste) is used mainly in sauces and soups.

Peanut, *Arachis hypogaea*

The peanut, also called groundnut, pinder, and goober, is the strange fruit of *Arachis hypogaea*. The fruit begins as a fertilized, usually cleistogamous flower above the ground, but when the seed pod forms it is gradually pushed into the ground by elongation of the pedicel (the "peg"), which develops a meristem just below the ovary; the pod and seed then mature under the ground (Fig. 20-22). The species is indigenous to central South America south of Amazonia, but today it is known only in cultivation. It is grown the world over in the tropics and subtropics, and world production amounts to about 18 million metric tons, most of it in the Far East and Africa. India, China, the United States, Brazil, Senegal, and Sudan are leading producers. *A. hypogaea* is a tetraploid ($2n = 40$), while all of the fifteen or so wild species in South America except one (*A. monticola*, probably the closest relative of the cultivated peanut, with which it crosses readily) are diploids. Only three species, including *A. hypogaea*, are annual, the others being perennial. Peanuts are known to have been cultivated by South American Indians 3000 years ago, and they were taken to Africa by early traders and missionaries.

Peanuts fall into two distinct groups, usually considered subspecies. The larger-seeded Virginia peanuts have a distinctive alternately branching pattern, in which two nodes produce vegetative branches and

Figure 20-22   Peanuts as they appear on a newly unearthed plant. [Courtesy USDA.]

the next two produce inflorescences. Some Virginia cultivars are "runners," which spread along the ground; others have somewhat ascending branches and are then of the "bunch" type. Virginia peanuts require a lengthy season (120–180 days) and the fruit remains dormant in the ground. Oil is slightly less abundant than in the "Spanish–Valencia" subspecies.

By contrast the Spanish–Valencia cultivars are rather short-branched, and numerous nodes may bear inflorescences. They have an erect "bunch" growth habit, and they require a shorter season than Virginia types (90–110 days). The seeds do not remain dormant and will sprout if left in damp soil. The oil content of Spanish–Valencia peanuts is slightly higher than that of Virginia peanuts.

If they are not kept dry, peanuts are susceptible to infection by the mold *Aspergillus flavus*, which produces extremely toxic substances known as aflatoxins. Press cake made from infected peanuts has proved lethal to animals fed it. Peanuts used for human con-

sumption are generally given greater care and stored under dry conditions. The nuts contain nearly 50 percent oil and about 25 percent protein. Typically, some of the crop is used for feeding stock right on the farm. Unlike the soybean, however, peanuts are highly esteemed as human food. Peanut butter, which is made from pulverized peanuts, is a ubiquitous food in the United States, where it is a particular favorite of youngsters.

Peanuts do well in light, sandy soils, which facilitate "pegging" and crop digging. In the United States they are much grown along the southeastern coastal plain from Virginia through the Carolinas, Georgia, Alabama, and into Texas. The large-kerneled Virginia peanuts come chiefly from the Virginia–Carolina area, the "Spanish" nuts more from the deep South. In keeping with agricultural trends in the United States, cropping has become quite mechanized. Mechanized harvesting brings in nuts of varying maturity and creates the need for artificial drying, in contrast to the picturesque former method of hanging the vines on stackpoles to dry. One would suppose that interrow cultivation would provide a looser, more receptive soil for penetration of the maturing pedicels; but research has indicated that this does not increase yields, and that herbicidal control of weeds between the rows is often better for the plants because they are less disturbed than when they are cultivated. Peanuts are subject to a number of diseases, especially rusts indigenous to the American tropics, inoculae of which probably blow north to infect plantings in the United States. Selection of resistant cultivars and fungicidal sprayings are familiar controls.

In most of the Old World, planting, cultivation, and harvesting are still done by hand (Fig. 20-23). Considerable portions of the African and Indian peanut crops are exported to Europe for extraction of the oil. In the United States only a limited portion of the crop is grown for oil; extraction of peanut oil is often done in the same factories that crush and express cottonseed oil. Most of the peanut crop goes into peanut butter and confections, such as candy bars. Peanut oil finds much the same uses as soybean oil, principally in margarine, shortenings, and salad

Figure 20-23   Hand weeding peanuts in Malawi. [Courtesy World Bank.]

or cooking oils. The oil is high in oleic constituents, modest in linoleic acid, and low in certain other fatty acid components. It finds several industrial applications, and the protein-rich press cake is a valuable livestock feed.

Everyday pulses—beans and peas

Species of many different genera of the family Leguminosae provide seeds that are directly consumed for food. Included are dry beans, dry peas, dry broad beans, chick peas, and lentils, the recorded combined total for which comes to more than 45 million metric tons annually. Everyday pulses are especially important in the Far East, although much grown in Latin America, too. World production of dry beans is about 12 million metric tons annually, of which one-fourth are grown in South America and one-half in Asia. India, China, Brazil, and Mexico are the leading producing countries. Most dry peas are produced in the Far East, with Europe often a close second. Total annual world production is around 13 million metric tons, with the USSR and China leading by a wide margin. Some 6 million metric tons of dry broad beans are grown annually, mostly in China and northern Africa. Chick-pea production runs around 7 million metric tons annually, with India by far the leading source. Little more than 1 million metric tons of lentils are grown annually, mostly in India and the Near East.

Phaseolus

The genus *Phaseolus*, especially as represented by *P. vulgaris*, is the most important source of beans. Included in the genus are lima beans, scarlet runner beans, string beans, shell beans, white beans, black beans, pea beans, and kidney beans. The species now important commercially are widely spread throughout the New World, and they have been cultivated since time immemorial by the Indians. The early explorers found such beans an excellent protein-rich food for the long sea voyage between America and the home continent, and quickly introduced beans into most inhabited parts of the world. Today they are the main source of protein in the less affluent societies, where they substitute well for meat. The protein content of *Phaseolus* beans runs 17–31 percent, with an average of about 25 percent. They are not a rich source of oil, however.

All *Phaseolus* beans are warm-season diploid annuals or short-lived perennials that thrive on hot weather and ample moisture. They are a favorite garden plant, both the determinate bush type and the indeterminate climbing or pole type, and they are often interplanted with other crops in small subsistence gardens. Indeed, most of the world's bean crop is probably home grown and locally consumed. In the United States, commercial production is completely mechanized. Even green beans, marketed as "vegetables," can be mechanically harvested. The same species that yield green beans as a fresh vegetable also yield dry beans when allowed to mature.

The chief American beans of the genus *Phaseolus*, in addition to *P. vulgaris*, are *P. lunatus*, the lima or butter bean; *P. acutifolius*, the tepary bean; and *P. coccineus*, the scarlet runner bean. (A number of Asiatic species formerly included in the genus, all of which have slender cylindrical pods and small seeds, have recently been transferred to the genus *Vigna*, which is discussed on p. 532.) Most *Phaseolus* species cross with each other, and are thus potential sources of improved germplasm for bean breeding.

P. vulgaris is variously known as the kidney, field, garden, or haricot bean. It provides the familiar green, string, snap, or wax beans when the immature pod is eaten entire; and the kidney, pea, pinto, great northern, marrow, and other types of dry bean when matured and shelled. The exact origins and relations within the species have never been thoroughly worked out, but the various species may have had multiple domestications, one or more of them in Mexico where cultivated forms grade imperceptibly into wild types. Archeological bean finds in Mexico date back 7000 years. Cultivated *Phaseolus* beans at least 7000 years old have been discovered in cave diggings in Peru, but they may have come from the South American *P. aborigineus*, which is sometimes considered a subspecies of *P. vulgaris*.

P. lunatus is the lima or butter bean, another confusing assemblage of uncertain and equally ancient ancestry. From the culinary standpoint, lima beans are highly prized, whether fresh or dried. Fresh seeds, especially of the colored strains, may contain a cyanogenetic glucoside that renders them toxic until it is destroyed by cooking. Both pole and bush forms are grown in a wide selection of seed colors and shapes. Lima beans have had a long history of cultivation in tropical America, being found in some of the most ancient archeological sites in Peru and Brazil, which date back more than 7000 years. They are a bit more temperamental to grow than *P. vulgaris* beans, but they require essentially the same conditions.

P. acutifolius is the tepary bean, often grown in northern Mexico and in the southwestern United States. Tepary beans were much utilized by the Indians of that area, and they still have value for dryland subsistence farming. *P. coccineus* is the scarlet runner bean, perhaps more used nowadays as an ornamental than as a source of food seed.

Vicia

Vicia faba is the broad, Windsor, horse, or Scotch bean, probably native to the Mediterranean area or the Near East, where it has been under domestication for many millennia and widely cultivated since the Bronze Age. A vigorous annual, it was well known to ancient Greeks and Egyptians. The diploid chromosome number is 12, and the species crosses with a number of wild vetches, most of which are used for forage rather than seed beans. Historically, the broad bean is important, because it was the only widely used edible bean in the Old World before *Phaseolus* beans were introduced from America.

Bitter vetch, *V. ervillia*, is less palatable than the broad bean for human food, and it is mostly utilized by the very poor or for livestock feed. Both wild and cultivated forms are abundant in parts of Turkey, and it is well represented in Neolithic sites in Greece and the Balkans.

Pisum

Pisum sativum, the familiar green pea, has a diploid chromosome number of 14. It was probably domesticated about 7000 B.C.E. in western Asia in the same general area where wheat seems to have originated. Similar wild species with which it can be crossed are found there—*P. humile* and *P. elatius,* both of which might well be conspecific with *P. sativum. P. humile* is considered by some botanists to be a possible ancestor of *P. sativum.*

Pea remains have been found in ancient Swiss lake dwellings, and the plant was abundantly cultivated in Greek and Roman times. Apparently, peas were originally consumed exclusively in the dry or mature stage, but in recent centuries varieties grown as sweet or green peas have achieved great culinary status. Nonetheless, there is great world-wide production of dried peas (between 1 and 2 million metric tons most years), which in the United States are much used for "split-pea" soup. Most of the pea crop in the United States, however, is produced commercially for fresh or green peas, which are canned or frozen. Production exceeds half a million tons. The plants are well adapted to cool spring weather or climates with reasonably cool summers, and much of the crop is grown in such northerly locations as Wisconsin and Washington. There are both bush and vine forms of *Pisum*. In the United States, planting and harvesting are highly mechanized; even special "vining" machines are used to cut the plants and strip them of pods. A threshing device then removes the green peas from the pods. For machine harvesting, cultivars are bred for uniform rather than normally staggered maturation of the pods.

Cicer

Cicer arietinum is the chick pea, also called gram pea or garbanzo, a small, herbaceous annual with odd-pinnate leaves and normally cleistogamous flowers. It is native to western Asia, where the genus embraces a number of wild species. *C. arietinum* re-

mains have been found in Near Eastern excavations dated 5400 B.C.E. Chick-pea pods are short and contain only one or a few rather large seeds. This is one of the most important legumes grown in India, ranking fourth among Indian food sources (more than 5 million metric tons annually). The peas are often split to make one type of the ubiquitous *dhal* of northern India. In addition to consumption of the whole seed, or seed in young pods, chick peas are sometimes ground to flour and used in preparing unleavened bread or sweets. The seeds may be eaten raw, roasted, parched, or boiled. The chick pea is also an important source of livestock feed. Chick peas are drought-tolerant, and they are much planted as a cool-season crop in seasonally dry regions.

Vigna

Vigna unguiculata unguiculata (= *V. sinensis*) is the cowpea or black-eyed pea of the southern United States. It may have originated from the subspecies *dekindtiana* of Ethiopia. The cowpea is really more a "bean" than a "pea," and, like most food legumes, both vine and bush cultivars are available. The bean averages about 23 percent protein. The species is an aggressive, trifoliolate, usually cleistogamous vine, and an excellent contributor of soil nitrogen, which makes it especially helpful in subsistence agriculture. Its tropical background is reflected in its inability to withstand cold, although it grows passably on almost any soil. The cowpea is also an important forage crop. It is related to the "yard-long" or "asparagus" bean (*V. unguiculata* ssp. *sesquipedalis*), which has a pod that is 30–90 centimeters (1–3 feet) long.

There are several other useful species of *Vigna*, once included in the genus *Phaseolus. V. mungo*, the black gram or urd bean of the Orient, is probably native to India. Beans of this type are especially favored in India, where many of the people are vegetarians and subsist on *dhal* and rice. *V. radiata*, the mung, golden gram, or green gram bean, is also extensively cultivated in India and China. The germinating seeds of this bean are used as "bean sprouts" in oriental dishes. *V. aconitifolia*, the moth or mat bean, is another species cultivated in India and the Orient, both for its small seeds and its green pod.

V. angularis, the adzuki bean, is mostly grown in China and Japan, as is the rice bean, *V. umbellata.*

Lens

Lens culinaris (= *L. esculenta*) is the lentil, another of the legumes utilized since the dawn of agriculture some 10,000 years ago. It is believed to be derived from *L. orientalis*, one of four species that grow wild in southwestern Asia. *L. orientalis* is scarcely distinguishable from the cultivated lentil except for generally smaller seed size. Lentils were widely spread through the Mediterranean area before historical times. The plants are small annuals with pinnate leaves bearing a terminal bristle or tendril. The pods are short, and they contain only a couple of lens-shaped pealike seeds. These make a very tasty soup or porridge (*dhal*), the form in which they are usually consumed. Like the chick pea, lentils are chiefly a cool-season crop of drier areas. Almost the entire production in the United States, more than 200,000 metric tons of lentils and split peas combined, comes from the Palouse Hills of Washington and Idaho.

Cajanus

Cajanus cajan (= *C. indicus*) is the pigeon pea, cajan bean, or congo bean. It has been grown in the Nile Valley for thousands of years, and it is now widely grown in the Orient and in equatorial Africa. In India it is the second most important pulse, where it is much used as a cattle feed as well as for human food (*dhal*). India accounts for most of the world production of about 2 million metric tons. In Africa, the congo bean is often interplanted with cereals and root crops, and the foliage may be used as a forage. The species is notably free of pests, tolerant of heat and drought, and able to grow on poor soils with little attention.

Psophocarpus

Psophocarpus tetragonolobus is the winged bean or goa bean, a large, herbaceous tropical vine with tuberous roots resembling a sweetpotato. A handful of species of the genus are native to tropical Asia, but only *P. tetragonolobus* is cultivated, especially in Indonesia, both for its tubers and for its seeds. The seeds are said to contain 37 percent protein, including a highly desirable amino acid complement, and the tubers are reported to contain up to 24 percent crude protein. Because the species nodulates well with *Rhizobium*, it is also valued for soil improvement and green manure.

The winged bean is a perennial, but is most often cultivated as an annual. The plants are twining, trifoliolate vines, and the pods bear four prominent wings that run lengthwise (the source of the common name). Winged bean is being highly publicized as one of the "underexploited" tropical plants of great promise for helping to feed the growing world population. It is one of rather few protein-rich species well-adapted to the humid tropics and to culture by small farmers there. Winged bean has long been grown in highland Papua, New Guinea, where it is the second most important field crop after sweetpotato.

Immature pods of winged bean may be cooked much like green beans (*Phaseolus*) are in the United States, and the young leaves and shoots are eaten as leafy vegetables. Immature seeds can be utilized in soups, and mature ones can be roasted and eaten like peanuts. The seeds contain about 18 percent oil, more than half of which is constituted of unsaturated fatty acids. Seed yields of as much as 2.5 metric tons per hectare (a little more than 1 ton per acre) have been reported. Ample rainfall is required for good yields and abundant tuber production. Initial growth of the vines is slow, so that weeding of new plantings is generally required. As growth turns more vigorous, stakes or trellises are needed for the rampant vines. The bean crop takes four to five months from sowing; in Papua, New Guinea, most plantings are utilized primarily for the tubers.

Voandzeia

Voandzeia subterranea is the Bambarra groundnut, one of four pulses whose pods mature underground (the others are the peanut, *Arachis*; *Kerstingiella geocarpa* from the same general area of Africa as the Bambarra groundnut; and *Amphicarpaea bracteata* of North America). The Bambarra groundnut is still found wild in savanna areas of central Africa, where it was no doubt first domesticated. It is little known outside of Africa, but it is quite important there. In Ghana, for example, it is the second most important pulse, ranking behind the cowpea. The plants are trifoliolate, growing in dense clumps from a prominent taproot. Below each flower a swelling on the peduncle grows down into the soil after fertilization, dragging the ovary behind it. Usually only a single large seed matures, which is only 5 percent oil but up to 20 percent protein and 50–60 percent carbohydrate.

The species, one of the most drought-resistant pulses, is productive under high temperatures and low rainfall and does well on soils too poor for most other crops. In Africa it is often interplanted with cereals and root crops. Often individual pods are lifted as they ripen, minimizing their loss through deterioration in the ground without sacrificing the whole plant before it is fully mature. Immature seeds are consumed fresh or grilled, or they may be boiled when mature. In the Ivory Coast they are ground to flour and consumed as cakes. In many countries they are salted for snacks much as are peanuts. As with many legumes, the foliage is an excellent animal feed, although productivity is moderate.

Other legumes

The seed of almost any legume can be adapted to the needs of people or livestock. A few legumes that are somewhat less popular than the species discussed so far are the wild ground-bean, *Amphicarpaea bracteata*, of the northeastern United States and Canada, which has cleistogamous flowers that produce pods underground; the jack bean, *Canavalia ensiformis;* the sword bean, *Canavalia gladiata;* guar or cluster bean, *Cyanopsis tetragonaloba*, whose seeds are a promising source of gum as well as protein; lablab, *Lablab purpureus* (= *Dolichos lablab*); the geocarpa groundnut, *Kerstingiella geocarpa;* the kasari, *Lathyrus sativus;* the kudzu, *Pueraria lobata* (= *P. thunbergiana*); the velvet beans, *Mucuna* spp.; the pods of many tree legumes such as the carob, *Ceratonia siliqua;* the honeylocust, *Gleditsia triacanthos;*

locust beans *Parkia* spp.; rain tree, *Samanea saman;* and mesquite, *Prosopis* spp. Most of the herbaceous species are as important as forages as for pulses, although they are generally used for both.

FORAGES

Literally, **forages** are any plants consumed by livestock. Most generally the term is restricted to pasture and browse plants, hay, silage, immature cereals, and residues from grain harvest, such as straw. Livestock feed grain is normally not considered forage. There is no reason why a species from any plant family might not be considered a forage plant, but, practically speaking, only the grasses (Gramineae) and the legumes (Leguminosae) contribute much in the way of forage. Just as cereal grain and legume seeds make excellent complementary foods for human use, so does grass and legume vegetation provide excellent dietary complement for herbivorous animals (Fig. 20-24). In modern animal husbandry, cattle are generally fattened on grain before being sent to market in order to provide superior "finished" meat. But there is no reason why a grazing animal could not subsist indefinitely on grass–legume forage so far as its own well-being is concerned.

Thus forages supply additional human food through the intermediary of livestock. Cattle, sheep, hogs, poultry, and other domesticated animals become the instruments for transforming the proteins, carbohydrates, and fats of plant foliage that is indigestible to people into meat, eggs, and dairy products. At the same time that the forage plants sustain this important livestock industry, they also protect the land and build the soil, which cultivated row crops do not. Forages are especially suited to rolling or steeper lands that would erode readily if clean-cultivated, and to other lands that for one reason or another cannot be adapted to the growing of familiar cash crops. Because they are relegated to lands of secondary quality, forages have often seemed the neglected product of agriculture. However, cumulatively, their extent and value are as great as those of nonforage crops, and their response to management (fertilization, weed control, proper soil handling) is equally as rewarding. Forages are especially useful to

Figure 20-24 (A) The grass family furnishes many excellent forage species: Jersey cows graze Sudan grass on an Alabama pasture. (B) Fully as important as grass for forage are species of the family Leguminosae: cattle graze on the 'Hubam' cultivar of white clover in southern Florida. [Courtesy USDA.]

expanding civilization, because they provide a means of making good use of land that is not well suited to the conventional cultivated crops that are so much a part of the mechanized agriculture reviewed in Chapter 15.

Worldwide, 1.5 billion hectares (3.7 billion acres) of arable land is cropped; 3 billion hectares (about 7.5 billion acres) is permanent pasture and meadow. In North America, cultivated land and permanent pasture are about equal in extent, but in Latin Amer-ica, Africa, and Oceania pasture land far exceeds cultivated land. In more heavily populated Europe and most of Asia, cultivated cropland quite exceeds meadow and pasture. Of the livestock sustained on meadow and pasture, sheep, cattle, and pigs are most important in Europe; cattle, pigs, and sheep in North America; cattle, sheep, pigs, goats, and horses in South America; cattle, sheep, goats, buffalo, and pigs in Asia; cattle, goats, asses, and camels in Africa; and sheep in Oceania. Wordwide, there are about 1.2 bil-

Table 20–4   Nutrients withdrawn by three common forages grown under cultivation in the United States.

Crop production	Nutrients withdrawn by crop (kilograms/hectare)					
	Alfalfa		Clover–Grass		'Coastal' Bermudagrass	
	9 MT/ha	22.4 MT/ha	6.7 MT/ha	15.7 MT/ha	13.5 MT/ha	27 MT/ha
Nutrients						
N	252	673	168	392	336	673
P_2O_5	45	135	50	118	94	180
K_2O	224	673	202	471	282	565
Mg	22	59	17	39	30	61
S	22	57	17	39	30	61

Source: Data from the Potash and Phosphate Institute, *Better Crops with Plant Food,* 1979.

lion head of cattle, followed closely by sheep at a little over 1 billion, with about half as many pigs and a third as many goats. There are about 42 million head of horses and asses worldwide, 14 million camels, and 13 million water buffalo.

It is impossible to mention more than a few examples of the world's prominent forage plants. Almost all grasses (there are about 7000 species in the family) and most herbaceous legumes (there are about 12,000 species in the family) are useful as forages. The feed value of the species used for forage varies greatly with soil fertility, stage of harvest, and handling. Forages make better feed if grown according to advanced agricultural methods. Fertilizer, for example, increases both yields and nutrient value. The protein content of grasses is highest in lush, fresh foliage before the seed stalks form or the grass starts to turn dormant. Kentucky bluegrass in spring may be almost 20 percent protein, although this percentage decreases to little more than half that amount by summer. One mineral-content analysis of bromegrass, a much used forage in the Midwest, shows 2.2 percent nitrogen, 0.32 percent phosphorus, and 0.4 percent calcium. This analysis is for grass grown on good soil; on poor soil the mineral content is appreciably less. Table 20-4 indicates the nutrients utilized

by three familiar forages under cultivation in the United States. Alfalfa, the chief forage in the United States, is 3.5 percent nitrogen, 0.35 percent phosphorus, and 1.6 percent calcium—somewhat superior to the grass as far as protein and minerals are concerned. Mineral and protein analyses, however, provide only a rough indication of feed value. Balance in nutritional factors, including minor or trace elements, is essential to the health of livestock. Sometimes too much leguminous protein (as from heavy fertilization) along with insufficient grass carbohydrate leads to nitrogen poisoning, or livestock bloat. Contaminants sometimes poison livestock (for example, species of *Astragalus* may cause selenium poisoning on the western ranges).

In the United States, forages are estimated to have an annual worth of more than 10 billion dollars. They provide more than half the nutritional requirements of all livestock. About 75 kilograms (165 pounds) of meat is consumed yearly per person, and 315 kilograms (about 695 pounds) of milk; it is said of these that about 25 percent derives from grain (concentrates), and 75 percent from forage (roughage). About 23 cents of the agricultural dollar in the United States is said to come from beef. Where animal products are consumed in such quantities, obviously there is a need for improved forage plants, such as 'Coastal' bermudagrass, which is capable of yielding more than 2.2 metric tons of beef per hectare annually (about 1 ton of beef per acre) when it is fed in dried and pelleted form.

Though forage grasses have only recently begun to receive the attention that has been accorded the major cereal grasses, researchers are now exploring all facets of forage production. One need only scan agricultural advertising to note the tremendous diversity and widespread innovation in forage-harvesting machinery—cutters, choppers, pickups, blowers, balers, self-loaders, and so on. Almost everything can be done mechanically by a single operator. The plant breeder has been busy, too, and the future of forage plants lies particularly with select cultivars and hybrids. Already benefiting from heterosis are hybrid Sudan grasses (crosses made possible by deriving male sterility from sorghum), sorghum–Sudan crosses, hybrid bahiagrasses, sterile hybrid bermudagrasses ('Coastal' and 'Coastcross-1', such excellent yielders that they justify expensive vegetative propagation), and hybrid pearl millet (*Pennisetum typhoides*). There are also good prospects for white clover, trefoil, and other legumes.

Alfalfa, *Medicago sativa*

Alfalfa, or lucerne as it is called in Europe, is generally regarded as the "queen of forages." No forage is a better source of protein. Throughout the world more than 25 million hectares (about 62 million acres) are planted to alfalfa, more than 11 million hectares (about 27 million acres) in the United States alone, followed by Argentina, France, southern Europe, and parts of Asia. In the United States, about half of the crop comes from the Corn Belt, with Wisconsin, Iowa, Minnesota, Nebraska, Michigan, Kansas, and South Dakota the leading producers in the region. California, the largest producer outside of this area, generally produces only a little less than Wisconsin. Some 50,000 metric tons of alfalfa seed are produced annually in the United States. It is not uncommon for alfalfa or alfalfa–grass forage plantings to yield 13.5 metric tons per hectare (about 6 tons per acre) of hay each year (which amounts to a

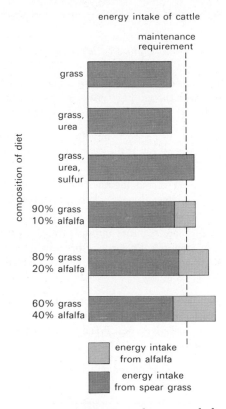

Figure 20-25   Contribution made by alfalfa when utilized with grass, in Australia. [Data from *Sulphur Institute Journal* 9 (2):3, 1973.]

dry nutrient content of 9–11 metric tons per hectare, of which nearly 3.5 metric tons is protein). On good lands giving such yields, alfalfa can be as rewarding as any other crop. The contribution alfalfa makes to grass pasture is graphed in Figure 20-25.

The origin of alfalfa is obscure, and its taxonomic relationships are uncertain. Some of its relatives, such as black medic, *Medicago lupulina*, are weeds in lawns. Wild alfalfa is scattered over central Asia, and the cultivated species is believed to have become domesticated in southwestern Asia, possibly in Iran. The chromosome number is $2n = 32$, the same as for several other species (*M. arborea*, *M. gaetula*, *M. glutinosa*, and *M. hemicycla*), indicating a tetraploid state as compared to black medic and other burclover species ($2n = 16$).

Box 20-2 *Why legumes with forage grasses?*

Combinations of grasses and legumes are planted in most pastures, even though high yields of grass alone could be obtained with fertilization. Some of the reasons are:

Legumes, owing to symbiotic nitrogen fixation, supply a good bit of the grass's nitrogen needs, an obvious economy.

The deeper-rooted legumes aid summer production at a time of year when grass growth often declines.

Proteinaceous legumes improve forage feed value and micronutrient balance (especially by providing additional magnesium, the lack of which may cause grass tetany).

Competition for space and survival occurs in the pasture ecosystem, as it does in nature. Grasses tend gradually to dominate legumes. Practices helping to favor the legume include:

Fairly close grazing (or mowing).

Rotational grazing, which provides recovery periods.

Mowing and weed control of grass patches that become undergrazed.

Making hay, if growth of the forage exceeds the capacity of the livestock to consume it.

Using fertilizers rich in phosphorus and potassium (legumes provide much of their own nitrogen).

In most climates the legume component may have to be replanted every few years, although, if liming and fertilization are ample, many clover stands remain adequate for a decade or two. As legumes disappear, pasture fertilization must be increased if the animal-carrying capacity is to be maintained.

Most grass–legume combinations will provide not only better animal nutrition than would grass alone, but greater daily gains per pound of TDN and more profitable cropping.

Roman records indicate that the ancient Medes introduced alfalfa into Greece from Persia about five centuries B.C.E. Even at this early date it was recognized as a forage without equal for chariot horses. The Spanish took alfalfa along with them as feed for their horses when they explored Mexico and South America. According to early records, the first alfalfa grown in the United States was in Georgia, in 1736. Its use spread slowly, however, until the Gold Rush of 1849, during which it was taken to California and was dispersed rapidly throughout the nation. An introduction from Germany by Wendelin Grimm paved the way for hardy cultivars that could be grown in the northerly states; the 'Grimm' cultivar was developed in Minnesota through continued selection of this introduction.

Alfalfa is well adapted to a wide range of climatic and soil conditions. It performs best on deep loam soils that are well drained, in climates where rainfall

is moderate, winters are cool, and summers are warm. Alfalfa does well in limed soils with ample potassium. In the northeastern United States it is very responsive to potassium and phosphorus fertilization, but on western lands where soils are rich in potassium, phosphorus is the main nutrient needed. As with legumes in general, alfalfa is symbiotically more or less self-sufficient in nitrogen. When alfalfa and other legumes are planted in association with forage grasses, nitrogenous fertilizer favors the grass and may encourage it eventually to dominate the legume. Similarly, close clipping favors the alfalfa, high clipping favors the grass. Meadow management is usually adjusted to favor the alfalfa, in order to hold it for a maximum number of years, after which fertilizer nitrogen can maintain grass yields as the alfalfa thins out. Favorably situated stands of alfalfa may persist for a decade or longer.

Because alfalfa has a deep taproot that can penetrate many feet into the subsoil on favorable ground, it is rather drought resistant. In the South it is generally autumn-seeded, and in the North spring-seeded, so that stands can become well established before cold winter weather. It is usually drill-planted, often with a companion crop (such as a small grain, which may be harvested the following spring). It is a good plant for improving the soil when used in rotation. Alfalfa gives the highest yields on the irrigated lands of the Southwest (Chapter 15), where as many as six cuttings a year can be made, compared with the more conventional three cuttings in the Corn Belt. Formerly alfalfa was swathed and windrowed for the making of hay, but artificial drying and compression into pellets are becoming increasingly popular.

Alfalfa is subject to a number of diseases, and in some of the northeastern states it is attacked by the alfalfa weevil. Where weevils are onerous and control costly, alfalfa is being abandoned in favor of other forages, such as sorghum-sudan. Needless to say, with so important a crop as alfalfa, there is constant endeavor to develop pest-resistant cultivars that are hardier and give better yields.

New breeding breakthroughs promise eventually to yield hybrid alfalfas. The genetics of alfalfa is somewhat confusing; some strains are self-pollinated, some are self-fertile, and others are almost obligately outcrossing. Because the flower must be pollinated by insects (leaf-cutting bees are usually used) and because it is perfect (with both male and female parts), practical techniques for producing hybrid seed have been slow to come. The technique that seems most promising is to breed self-incompatible plants (which produce little or no seed from their own pollen) from plants of desirable strains, and to propagate these by vegetative cuttings. If another self-sterile strain, also propagated vegetatively, is interplanted in alternate rows, this will ensure that all the seed produced will be hybrid. Although such a planting is expensive to make, it can be amortized over a period of several years. Hybrid seed generally increases yields 20 to 35 percent over yields from pure lines, and thus justifies a premium price.

Frequently interplanted with alfalfa for forage in the north-central and northeastern United States are orchardgrass (*Dactylis glomerata*), bromegrass (*Bromus inermis*), and timothy (*Phleum pratense*). About 11 kilograms per hectare (10 pounds per acre) of alfalfa seed and 7 to 11 kilograms per hectare (6 to 10 pounds per acre) of grass seed are usually used, sometimes with oats as a nurse crop (actually a hindrance to the establishment of the forage). The alfalfa seed is usually inoculated with an appropriate strain of nitrifying bacteria. Herbicides are being used increasingly to eliminate the less resistant annual grasses and weeds from alfalfa plantings. Yields from alfalfa–grass forage plantings are at least as high in protein, and often quite superior to, those obtainable from corn (about 33.5 metric tons of corn grain per hectare would be needed to match the protein from 18 metric tons per hectare of alfalfa hay).

Grazing or feeding of green alfalfa sometimes results in cattle bloat, which is often serious enough to be lethal. The exact cause of the trouble is not certain, but it may be due to saponins or other minor constituents of the "rich" alfalfa fare. Water-soluble phytotoxic substances have also been derived from the alfalfa leaf; these substances are at least partially repressive to grasses. Alfalfa foliage has served as a source of certain extractives, particularly the green pigment chlorophyll (not too long ago all the rage as an air purifier and deodorant, although it was useless for those purposes). Some extraction of protein and other fractionations have been undertaken with alfalfa in the hope of creating food suitable for direct human consumption (Table 20-5). The alfalfa is pulped, the juice expressed, the chloroplasts centrifuged out, and the protein precipitated by acidification and heating. So far, with only about half the protein recovered, this process has not proved particularly economical. Yet one day we may look for extracted leaf proteins to help feed a hungry world. In Wisconsin, an "on-farm" method of processing was developed for animal feeding, in which chopped forage is expressed, the solids relegated to silage, and the juice coagulated to yield a protein feed concentrate (with the remaining solubles used for fertilizer). About 11 percent more usable protein is obtained by this method, compared with baling the alfalfa for hay. However, most alfalfa protein still is converted into animal products, notably chick ration. The most efficient conversion is probably to milk, which, in

Table 20–5 The essential amino acids in leaves of alfalfa and other foods.

Food	Amino acid content (milligrams/gram of nitrogen)							
	Isoleucine	Leucine	Lysine	Methionine	Phenyl-alanine	Threonine	Tryptophan	Valine
Alfalfa leaf	357	596	441	91	400	360	100	460
Soybean leaf	362	723	522	82	421	381	120	485
Soybean	284	486	399	79	309	241	80	300
Yeast	239	475	300	69	538	338	150	238
Milk	399	782	450	156	434	278	—	463
Egg	393	551	436	210	358	320	—	428
Beef	301	507	556	170	275	287	—	313

Source: Adapted from J. E. Kinsella and A. Betschart, *New York's Food and Life Sciences Quarterly* 5:16, 1972.

well-run dairies, provides protein yields about equivalent to those derived from soybeans. Channeling alfalfa protein into meat is less efficient.

The clovers, *Trifolium*

There are a number of important forage species in the "true" clover genus, *Trifolium*. Most prominent are red clover (*T. pratense*), white clover (*T. repens*), alsike clover (*T. hybridum*), and crimson clover (*T. incarnatum*). Hop clovers, Persian clover, sub clover, strawberry clover, and others are somewhat less used. In the tropics, berseem or Egyptian clover (*T. alexandrinum*) is much planted, as is Kenya white clover (*T. semipilosum*) at high elevations.

Compared with alfalfa, the clovers are smallish plants, generally yielding a bit less forage, but they are excellent sources of protein and are very palatable to livestock. Clovers nodulate well and are perfect companions for grasses in mixed pasture. Indeed, Kentucky bluegrass (*Poa pratensis*) and white clover (*Trifolium repens*) are partners in world conquest, volunteering and supporting one another even where not planted. Together they were known as "English grass" in the American colonies. Though most clovers are perennial, the majority do not last long in managed pastures and meadows. The taproot

generally rots away by the second year (probably due to a complex of insects and fungi), and the plant must depend upon adventitious roots thereafter. Sometimes it is able to do so, but often it peters out after two or three years.

Red clover, *T. pratense*

Red clover probably originated in southwestern Asia, and it is a more recent domesticate than alfalfa. It was much used throughout Europe in the late Middle Ages, where it helped initiate crop rotation, which eventually superseded fallowing. Today it is planted and naturalized in almost all temperate climates. It thrives best in humid sections, and is probably the most important hay legume in the northeastern part of the United States. Red clover suffers from a number of diseases, and virus-resistant cultivars are especially sought. The species is largely self-sterile, so that outcrossing results, making it difficult to maintain a pure line for hybrid seed. Cultivars segregate wildly in the second year and thereafter. However, tetraploid red clovers are becoming available. Red clover seed production, although less than alfalfa seed production, amounts to a significant 17,000 metric tons annually in the United States. Red clover responds well to fertile, limed, phosphorus-rich, well-drained soils.

White clover, *T. repens*

White clover, which includes the robust ladino variety from Italy, behaves much like red clover, with which it may cross. Also like red clover, the species is generally self-incompatible. Genetic mechanisms have been discovered that permit self-pollination, giving hope for hybrid white clover. The species is probably a native of southern Europe or the Near East, and it undoubtedly served as a forage for the earliest European livestock. Today it is planted or found volunteering in temperate climates around the world, responding well to ample moisture, lime, and generous fertilization with phosphorus. White clover generally contains between 20 and 30 percent crude protein (on a dry-weight basis), depending upon the season. Unfortunately, plantings are frequently short-lived. But the plant regenerates and volunteers well, and it is undoubtedly one of the world's most important pasture plants (although it can't match alfalfa and red clover for hay and silage).

Alsike, *T. hybridum*

Alsike clover is much grown and well adapted in northern Europe and in high-altitude wet meadows in the United States. There are both diploid and tetraploid strains. The species persists poorly, and is often handled as a biennial. Its origin is obscure.

Crimson clover, *T. incarnatum*

Sometimes called scarlet, Italian, or incarnate clover, crimson clover is much utilized as a winter annual throughout the southeastern United States. It, too, seems to have originated in southeastern Europe or in the Near East, and was widespread in Europe by the Middle Ages. The species tolerates some acidity and some variation in soil types. It usually self-seeds sufficiently to re-establish itself a subsequent year. It is a good soil improver and excellent green manure. Crimson clover is often sown with annual ryegrass for winter pasture, but it can make a stand in established bermudagrass (which is semidormant through winter); its growth cycle and that of 'Coastal' bermudagrass complement one another nicely. It is

also useful for seeding between cotton rows to protect the soil while providing some winter forage.

Other prominent forage legumes

Bird's-foot trefoil, *Lotus corniculatus*

Bird's-foot trefoil is the most important of four *Lotus* species used for forage, two of which are annuals. It is a trailing perennial, probably first domesticated in the Mediterranean area. Both diploid ($2n = 12$) and tetraploid forms occur. The species is much used in Italy, both for hay and for pasture. In North America, it is planted principally east of the Great Plains (in Ohio and New York). Bird's-foot trefoil is tolerant of poor soils, but it is slow to develop a stand. Like other legumes, it is quite responsive to phosphorus. Its protein and mineral composition is not much different from that of alfalfa or clover. Once well established, it is said to persist in Kentucky bluegrass pasture better than do most legumes, and is then the equal of alfalfa–timothy or ladino–brome in beef-carrying capacity.

The vetches, *Vicia*

There are about 150 species of *Vicia*, most of them native to Europe and the Near East, of which hairy vetch (*V. villosa*), common vetch (*V. sativa*), purple vetch (*V. atropurpurea*), and Hungarian vetch (*V. pannonica*) are the most important. Other species may be useful as a source of germplasm (for example, *V. sativa* with 12 chromosomes can be crossed with *V. angustifolia* with 10 chromosomes to produce partially fertile F_1 hybrids). Most vetches are trailing vines with tendrils. Hairy vetch is winter-hardy in the North, but it finds its greatest use in the United States in the Cotton Belt, where it is widely naturalized. Common vetch is less winter-hardy, and purple vetch the least winter-hardy of all. The vetches are much used as winter-annual cover crops in the United States, especially in the Southeast and along the Pacific coast. They may be drill-planted or broadcast, and they are often mixed with rye or oats, especially when used for hay or silage.

Table 20-6 Some lesser-known leguminous forage and cover plants from around the world, mainly subtropical and tropical.

Botanical name	Common name	Comments
Herbs		
Aeschynomene americana		American tropics
Alysicarpus vaginalis	Alyce clover	Tropical Asia
Calopogonium mucunoides	Calopo	Guiana (green manure)
Centrosema pubescens, C. plumieri		Tropical America (green manure)
Clitoria ternata	Butterfly pea	World tropics
Cratylia floribunda		Brazil
Crotolaria spp.	Rattleboxes	World tropics
Cyanopsis tetragonoloba	Guar	India
Desmodium spp.	Beggarweeds or tick clovers	America and world tropics
Glycine javanica		Soybean relative from Oriental tropics
Indigofera spp.	Indigos	Tropical Africa and Asia
Lathyrus hirsutus	Rough pea	Mediterranean region
Lupinus spp.	Lupines	Europe and cooler habitats
Macroptilum atropurpureum		Central America
Medicago spp.	Bur clovers and other alfalfa relatives	Worldwide
Onobrychis viciifolia	Sainfoin	Near East
Pueraria lobata	Kudzu	Japan
Mucuna spp.	Velvet beans	World subtropics
Stylosanthes gracilis	Wild lucerne	Tropical America and Africa
Tephrosia candida	White hoary pea	Tropical Asia (green manure)
Trigonella foenum-graecum	Fenugreek	Southern Europe
Shrubs and trees (often for tropical plantation cover)		
Acacia albida	Winter thorn	Tropical Africa, browse in dry season
Albizzia spp.	Albizzias	Mainly Africa-Asia, plantation shade
Cassia sturtii	Australian shrub	Desert fodder
Erythrina spp.	Coralbean	Worldwide, plantation shade
Gliricidia sepium		Famous coffee shade, Central America
Prosopis tamarugo	Tamarugo	Saline habitat of Chile

Sweet clover, *Melilotus*

Species of the genus *Melilotus*, which are native to western Asia and temperate Europe, were introduced into North America in the early 1700s. Sweet clover has a deep taproot, and is prized as a drought-resistant forage plant and soil builder. Its preference is for well-limed soils. Two species, the yellow (*M. officinalis*) and the white (*M. alba*), are commonly grown in America, both of them as biennials. They are especially planted for forage from Texas northward in the eastern plains, and for soil improvement generally. Both are noteworthy honey plants. A relatively high coumarin content in sweet clover tends to make it less palatable to livestock than many other legumes.

The bush clovers, *Lespedeza*

The genus *Lespedeza* offers numerous useful forage species, of which the annual Korean lespedeza (*L. stipulacea*) and common lespedeza or Japanese bush clover (*L. striata*), and the perennial sericea (*L. cuneata*) are most important. All are native to eastern Asia. They were not introduced into the United States until the late nineteenth and early twentieth centuries, although they are now widely naturalized. Korean lespedeza makes an excellent catch crop for warm weather; it is adaptable to poor soils and even gravel, but it is slow to start in spring and is thus a poor competitor against weeds or companion crops. Korean and common lespedeza are most serviceable as a temporary cover while the land is out of other use (such as following early summer grain harvest). Sericea is much used as a roadside cover in the South, as well as for hay and pasture.

Various other forage legumes are listed in Table 20-6.

Forage grasses

Almost all grasses (and sedges) can serve as animal graze, for wildlife and for livestock. A number are converted to "haylage"—silage made from hay rather than from grain. Throughout much of the world, native grasslands serve as pasture and meadow to a greater extent than do cultivated lands. In the United States, tallgrass prairie once extended eastward from approximately 100° longitude west—that is, from the middle of the Dakotas, Nebraska, Kansas, Oklahoma, and Texas through much of Illinois and irregularly into Ohio. West from this longitude, the shortgrass prairie, with less abundant vegetation in response to less abundant annual rainfall, extended to the Rocky Mountains. Almost all of the original prairie has now been plowed for cropping, providing one of the most extensive sweeps of rich crop land anywhere in the world. Similar banding of vegetation types in response to rainfall occurs in all continents, creating such famed savanna and grazing habitats as the pampas of South America south of the Amazon, the Sahel–Sudan zone south of the Sahara and over most of East Africa, Asia Minor eastward into northern India, and the interior of Australia. Interestingly, there has been some speculation that modern control of the tsetse fly, reducing trypanosomiasis in tropical Africa, has permitted severe overstocking with cattle and consequent overgrazing pressure on fragile Sahel habitat that results in further desiccation and change of climate.

A few widely cultivated forage grasses are cited

Box 20–3 *Ecological change in rangeland*

Natural grasslands can be—indeed, usually are—degraded by the introduction of domesticated livestock. This has become a serious problem not only in "unsupervised" lands, such as the Sahel of Africa, but at home as well. Here, excerpted, is what Philip L. Fradkin has to say about it in *Audubon* magazine (January, 1979, "The eating of the West"):

Vegetative cover can be changed by overgrazing in the following manner: Cattle move onto an undisturbed range and first select those plants which are most accessible and palatable. Such species, called "dessert" or "ice cream plants," tend to have the highest nutritional value and are the sweetest. With prolonged grazing, these prime grasses and weeds—mostly perennials—are replaced by less desirable annuals. Shrubs, such as sagebrush and pinon–juniper, now invade and take over the weakened area. Less nutritious and even such poisonous species as larkspur and locoweed, which literally drive cattle crazy, appear. Exotic plant species—perhaps from Mediterranean Europe or the steppes of Asia—proliferate and crowd out the native species. The composition of vegetation has changed, and along with it the small mammals and invertebrates and their prey that are dependent upon it.

here. Some, such as sorghum–Sudan and pearl millets, were mentioned with the cereals. Others are native American grasses, such as bluegrama (*Bouteloua gracilis*), a shortgrass-prairie species now being planted on southwestern pastures; several lovegrasses (*Eragrostis*), such as the sand lovegrass from the southern Great Plains; and the famed bluestems (*Andropogon*), long a source of prairie hay in the cool dry plains regions. Rhodesgrass (*Chloris gayana*) is much used in Australia.

As has been pointed out several times, grasses go well with legumes in forage plantings. The legumes, which live in symbiosis with nitrogen-fixing bacteria, are the richer source of protein; but grass yields abundantly and persists under grazing or cutting. Where legumes falter, fertilizer nitrogen can substitute to a degree, not only making forage grasses more productive but also more proteinaceous. By and large, grasslands are very responsive to fertilization.

Sorghum–Sudan

Sorghum–Sudan crosses, bred for productivity, are especially high yielding. The forage may grow as tall as 4.5 meters (about 15 feet) without lodging, and under favorable conditions it can produce as much as 40 metric tons of dry matter per hectare (about 18 tons per acre). Although first looked upon as a catch crop in case of crop failure, sorghum–Sudan is now much planted as a planned pasture, hay, and haylage crop competitive with alfalfa. Harvesting is often possible within six weeks of planting. Kansas leads in production.

Pearl millet, *Pennisetum americanum (= P. typhoides)*

Pearl millet is more important for its grain in most of the world, in India and Africa especially, but it is increasingly being planted as a forage, especially in the southeastern United States. Discovery of male sterility has made possible the production of vigorous hybrid cultivars. The species is a freely tillering annual that may grow a few meters tall. Three or four cuttings per year are generally made for hay. Pearl millet, like sorghum, was apparently domesticated in northern Africa.

Kentucky bluegrass, *Poa pratensis*

Bluegrass probably originated in southeastern Europe, and was spread throughout that continent by the Middle Ages. It was introduced into America with early colonization, and it adapted so well that it soon became a prime pasture and forage species gaining the name "Kentucky" bluegrass in the early 1800s. Though not a high yielder, especially during hot summer weather, Kentucky bluegrass is nonetheless one of the richest and most palatable of grass forages. Because it spreads by rhizomes, it is also one of the world's finest sod formers and soil protectors. Kentucky bluegrass grows best on limy, phosphatic, well-drained loams, but it is adaptable and quite tenacious in almost any cool, humid climate. The nation's foremost lawngrass cultivars are Kentucky bluegrass.

The bromegrasses, *Bromus*

There are about 60 brome species, of which smooth bromegrass (*B. inermis*) is perhaps the most widely utilized. It is apparently native to central Europe, and it was introduced into the United States in the late 1800s. It does well on the northern plains, where it withstands moderate drought. Brome is much grown with alfalfa from Kansas to North Dakota and eastward to Michigan.

Timothy, *Phleum pratense*

Native to northern Europe and Asia, timothy was first grown in New England in about 1747, where it still holds sway as one of the major forage grasses. It is perhaps the best-yielding hay grass in Ohio and other parts of the eastern Corn Belt, where it is often interplanted with red clover.

Orchardgrass, *Dactylis glomerata*

Orchardgrass, which is native to western and central Europe, was introduced into the United States in colonial times. It has proved well adapted to cool, humid regions, where it frequently naturalizes, and today it is found worldwide in such climates. In the United States, orchardgrass has found its greatest use

Figure 20-26   Fine fescue being row-cropped in Oregon. Seed of most cultivars comes from that state. [Courtesy Union Pacific Railroad.]

slightly south of the timothy belt, from Virginia to Iowa. It is not as winter-hardy as Kentucky bluegrass, brome, or timothy, but its soil requirements are probably less exacting. In mixed plantings with legumes, it is a long-lived perennial, much used for hay or silage. Under close grazing it generally gives way to Kentucky bluegrass or other volunteer growth.

The fescues, *Festuca*

The fine or red fescues (*F. rubra*) are most esteemed for lawns (Fig. 20-26); the coarser meadow fescue (*F. elatior*) and tall fescue (*F. arundinacea*) are used as forages. The latter is especially important in the middle latitudes where summers turn too hot for Kentucky bluegrass, timothy, and other more northerly species. By the same token these fescues are doubtfully winter-hardy in northern states. Tall fescue, a vigorous grass with deep roots, adapts well and establishes easily on many types of soil, even in hot and difficult situations. It is much planted along waterways and roadsides as well as in pastures and meadows. In pastures it is usually sown with clover or lespedeza, which improve its nutrient value and palatability. Red fescue, naturalized widely in western Canada, serves as a forage there.

Bermudagrasses, *Cynodon*

Common bermudagrass, *C. dactylon*, is circumtropical, probably originating in Africa but possibly also in southern Asia. It was introduced into Georgia by at least 1751, and it has since spread entirely across the southern United States, where it is often derogatorily spoken of as "wiregrass" or "devilgrass" because of its proclivity for invading gardens, lawns, and fields with its vigorous runners. In many states it is classed a noxious weed. Bermudagrass has proved especially well adapted to the area around Yuma, Arizona, where most of the commercial seed is produced today. Mention was made of the cross between a local bermudagrass and an African introduction, to create 'Coastal' and, more recently, 'Coastcross-1.' These high-yielding hybrids must be vegetatively propagated through sprigs or stolons (live stems) because they are sterile.

Paspalums, *Paspalum*

Two species of Paspalum in particular are grown for forage in the southern United States, dallisgrass (*Paspalum dilatatum*) and bahiagrass (*P. notatum*), both native to South America. Dallisgrass is well adapted to the Cotton Belt, where it has volunteered widely and is often a pest in lawns; bahiagrass, in several strains, is much used as a seeded lawngrass in the deep Southeast, where it is also an important pasture species. Both are used more for pasture than for hay, and dallisgrass is often interplanted with lespedeza.

Other forage grasses

Reed canarygrass (*Phalaris arundinacea*) was introduced from northern Europe in the late 1800s and is especially planted to lowlands and stream banks across the northern United States. The ryegrasses, *Lolium multiflorum* (annual) and *L. perenne* (perennial), native to Asia Minor and the Mediterranean area, are used principally for winter pasture and cover crops in the southern United States and in New Zealand. Redtop and the bentgrasses (*Agrostis* spp.) are primarily adapted to damp locations, redtop often being interplanted with alsike clover, and volunteering on poorer soils in the southern Midwest. The wheatgrasses (*Agropyron* spp.) are well adapted to regions of limited rainfall, with introduced and native species both being of service in the northern Great Plains of North America. Johnsongrass (*Sorghum halapense*), native to southern Asia and North Africa, is valuable for hay pasture in the upper South, but it is so aggressive as to become a serious weed in middle latitudes. Carpetgrass (*Axonopus affinis*) is adapted to sandy, boggy soils near the Gulf Coast of the United States, where it may serve as pasture. Pangolagrass (*Digitaria decumbens*) is a relative of crabgrass from South Africa, used for pasture particularly in Florida. Green sprangletop (*Lepto-*

chloa dubia) is sometimes planted in the Southwest. Kleingrass (*Panicum coloratum*) is increasingly used in Texas. Limpograss (*Hemarthria altissima*) is a stoloniferous perennial adapted to damp, subtropical habitats. Additionally, many cereal species serve as forages, including oats, barley, and rye, which are grazed or used for hay. Corn and the sorghums are used primarily for silage. In the tropics, too, various species of *Brachiaria, Cenchrus, Chloris, Dicanthium, Echinochloa, Eragrostis, Hyparrhenia, Panicum, Pennisetum, Setaria,* and several other genera may be utilized for pasture or fodder.

Selected references

Abelson, P. H. (editor), 1975. *Food: Politics, Economics, Nutrition and Research.* Washington D.C.: American Association for the Advancement of Science. (A compilation of materials by recognized authorities that originally appeared in *Science.* An excellent resume of the global food situation, a potpourri of presentations with something of interest on all facets of food production.)

American Society of Agronomy, 1972. *Alfalfa Science and Technology* (ASA Monograph 15). Madison, Wisc.: American Society of Agronomy. (Consisting of 35 chapters by 75 authors, this work constitutes a complete review of the "queen of the forages" from its evolutionary background to modern alfalfa production.)

Barnard, C. (editor), 1964. *Grasses and Grasslands.* London: Macmillan. (A series of documented essays on the basic biology of grasses and pastures, by Australian authors adopting a worldwide approach; an authoritative over-all review.)

Beadle, G. W., 1980. The ancestry of corn. *Scientific American* 242(1):112–119. (A well-presented, concise overview of a long-debated subject, arguing for teosinte as the progenitor of maize.)

Bogdan, A. V., 1977. *Tropical Pasture and Fodder Plants.* New York: Longmans. (Bogdan gives worldwide emphasis to livestock management in the tropics with this wide-ranging presentation.)

Briggs, D. E., 1978. *Barley.* London: Chapman and Hall. (Covers all aspects of barley production, and is especially strong on biochemistry.)

Chandler, R. F., 1977. *Rice in the Tropics: A Guide to the Development of National Programs.* Boulder, Colo.: Westview Press. (A quick overview of the modern situation, under the auspices of the IRRI. There are many other IRRI publications, often issued as symposia from the IRRI at Los Baños, Laguna, Philippines.)

Chang, Te-tzu, 1976. *Manual on Genetic Conservation of Rice Germ Plasm for Evaluation and Utilization*. Los Baños, Philippines: International Rice Research Institute. (Brief coverage is ample to indicate the complexity of rice relationships and the difficulties facing an adequate germ-plasm bank. Extensive reference lists accompany each chapter, and a key to "valid" rice species is included as an appendix.)

Chheda, H. R., and L. V. Crowder, 1978. *Tropical Grassland Husbandry*. New York: Longmans. (Comprehensive coverage of a subject area just beginning to receive serious scientific attention.)

Cobley, L. S., 1977. *An Introduction to the Botany of Tropical Crops*, 3rd ed., revised by W. M. Steele. London: Longmans. (Meticulous exploration of the botanical background of major tropical crops, even those impinging only slightly on the tropics. Cereals, cane, legumes, root crops, vegetables, fruits and nuts, beverages, drug plants, spices, fibers, oils, essential oils, and rubber are all treated.)

Coffman, F. A. (editor), 1961. *Oats and Oat Improvement*. Madison, Wisc.: American Society of Agronomy. (A technical monograph.)

Coupland, R. T. (editor), 1978. *Grassland Ecosystems of the World*. New York: Cambridge University Press. (A diversified review of the development and maintenance of grasslands worldwide.)

Doggett, H., 1976. *Sorghum* (in microfiche). London: Longmans. (A thorough discussion of the subject typical of the "world crops" series.)

Duncan, P. D., 1978. *Tallgrass Prairie: The Inland Sea*. Kansas City, Mo.: Lowell Press. (A "popular" book made beautiful by outstanding photographs, by a lady native to the best remaining outpost of tallgrass prairie left, the Flint Hills of Kansas. Not a technical treatment, but rather a work of love about the prairie.)

Grist, D. H., 1965. *Rice*, 4th ed. London: Longmans. (Detailed monograph on rice, its genetics, and its cultivation by an author especially experienced in Malaya.)

Harrison, C. M. (editor), 1968. *Forage Economics—Quality* (ASA Special Publication 13). Madison, Wisc.: American Society of Agronomy. (Monographic treatment by a series of authorities having established credentials.)

Heath, M., D. Metcalfe, and R. Barnes (editors), 1973. *Forages: The Science of Grassland Agriculture*, 3rd ed.

Ames, Iowa: Iowa State University Press. (Long a standard reference about American grasslands, this book reviews changing trends in the production of forages, considers particularly grasses and legumes, and discusses management, harvesting, methods of utilization, and efficiencies for varying types of livestock.)

Hitchcock, A. S., 1950. *Manual of the Grasses of the United States*, 2nd ed., revised by Agnes Chase (USDA Miscellaneous Publication 200). (A manual of agrostology, the identification and life cycles of grasses.)

Inglett, G. E. (editor), 1974. *Wheat: Production and Utilization*. Westport, Conn.: AVI. (Fairly stereotyped coverage of breeding, production, harvesting, marketing, handling, and chemistry, with additional chapters on specialized usage of flour. Perhaps more of interest to the cereal technologists and chemists than to the general biologist.)

Mangelsdorf, P. C., 1974. *Corn: Its Origin, Evolution, and Improvement*. Cambridge, Mass.: Harvard University Press. (A well-written and beautifully illustrated volume by the dean of corn lineage research. A number of Mangelsdorf's conclusions are not shared by experts in the field, and Mangelsdorf himself has reversed his previous position regarding the tripartite origin of maize.)

Mueller, K. E., 1970. *Field Problems of Tropical Rice*. Los Baños, Philippines: International Rice Research Institute. (A book of more restricted scope than the title would indicate, devoted mostly to pest problems and nutritional disorders; pests of rice are described and illustrated. The book would seem of most use as a pocket-size reference for field use.)

Purseglove, J. W., 1968–1972. *Tropical Crops: Dicotyledons* (2 vols.). London: Longmans. (A thorough coverage of tropical crops in dicotyledonous families.)

Purseglove, J. W., 1972. *Tropical Crops: Monocotyledons* (2 vols.). London: Longmans. (A thorough discussion of tropical crops in monocotyledonous families, rounding out excellent earlier volumes on dicotyledons.)

Scott, W. O., and S. R. Aldrich, 1970. *Modern Soybean Production*. Cincinnati, Ohio: The Farm Quarterly. (Growth pattern, variety selection, planting, and caring for the soybean; the soybean's future for food and feed.)

Simmonds, N. W. (editor), 1976. *Evolution of Crop Plants*. London: Longmans. (One of the more authoritative, in-depth recitations on the origins of major domesticated species.)

Smartt, J., 1976. *Tropical Pulses*. London: Longmans. (Although not as comprehensive nor authoritative as one might wish, this book nevertheless reviews many aspects of tropical legume usage under one cover that are otherwise scattered through an immense literature.)

Sprague, G. F. (editor), 1977. *Corn and Corn Improvement*, 2nd ed. (ASA Monograph 18). Madison, Wisc.: American Society of Agronomy. (A comprehensive updating to 1977, about equally divided between consideration of the corn plant itself—origin, genetics, morphology, and so on—and the plant as a crop—requirements, pests, production, utilization, and the like. The authors of the various chapters are recognized experts in the field.)

Sprague, H. B. (editor), 1974. *Grasslands of the United States: Their Economic and Ecologic Importance*. Ames, Iowa: Iowa State University Press. (Twenty-four experts survey the nation's forage resources, considering livestock, soil and water conservation, wildlife interactions, and alternative uses.)

U.S. Department of Agriculture, 1966. *Rice in the United States: Varieties and Production* (USDA Agricultural Handbook 289). (Thorough coverage of the practicalities of rice growing as practiced under North American systems, including discussion of soil phenomenon for this "untypical" crop.)

Walden, D. B. (editor), 1978. *Maize Breeding and Genetics*. New York: Wiley. (An international symposium with participants from many countries. Six chapters by experts on the evolution of maize, including Nobel prize-winner Beadle.)

Wall, J. S., and W. M. Ross (editors), 1970. *Sorghum: Production and Utilization*. Westport, Conn.: AVI. (A major effort to compile in one volume information on the production and uses of sorghums; a worldwide overview by many authorities.)

Whyte, R. O., 1974. *Tropical Grazing Lands*. The Hague, Netherlands: W. Junk. (A survey of the vast grasslands of Australia, Asia, Africa, and America with an ecological assessment and consideration of management procedures to make these grasslands more productive. Individual chapters are devoted to the grasses and the legumes, which are often suggested as an addition to, or a replacement for, the natural vegetation.)

Woodroof, J. G., 1966. *Peanuts—Production, Processing, Products*. Westport, Conn.: AVI. (Cultivars, planting, care, and processing are given technical coverage.)

21
Food Crops:
Roots, Stems, Fruits, Nuts

Authorities indicate that nearly 60 percent of human sustenance (in terms of calories) derives from the cereal grains, 10 percent or less from roots and tubers, and another 10 percent from fruits, nuts, and vegetables. This, of course, is a quantitative rather than qualitative measure of value. Secondary foods are often the chief sources of necessary vitamins, minerals, and important proteins. This is not to say that the crop plants to be discussed in this chapter are minor, but they are in general less widely distributed, less easily preserved, and less conveniently transported than are cereal grains and legume pulses. Actually, the food plants of this chapter are somewhat more important than they appear to be statistically. In many parts of the world they are locally produced and locally consumed, and, because they are often bartered rather than sold for cash, the extent to which they are used is often not accurately indicated by recorded data. Some grow semiwild, and are simply gathered as needed in economically undeveloped places. Over much of the tropics and in the Orient, young shoots and foliage of innumerable species of various genera Westerners have seldom if ever heard of are regularly gathered and consumed as greens.

First discussed are the root crops. As was noted in Chapter 15, lush tropical ecosystems have tended to spawn vegetatively propagated crops rather than seeded annuals. The majority of the root crops are of tropical origin, even though some, such as the potato and sweetpotato, are now grown as annuals in temperate climates. Biennial root crops such as beets and onions, which are started from seed, are also adapted to annual cropping.

Root crops, along with the cereals and pulses, are the world's chief source of energy and protein; but the "vegetables" and fruits included in this chapter supply most of the nutritional supplements: the vitamins, the various trace substances, and—not least in importance—the appealing flavors dear to the gourmet. Meals would be far more dreary to most palates were it not for the wide choice of vegetables and fruits that advanced cultures, in particular, have come to take for granted as part of a balanced diet. In most parts of the world today, preservation techniques—drying, sugaring, canning, freezing—make these foods widely available and procurable at all times of the year. Except for the more massive root crops, then, the foods discussed in this chapter are "delicacies."

ROOT CROPS

"Root crops" is a convenient catch-all for a wide assortment of species scarcely less important as food plants than the groups discussed in the previous chapter. More than 500 million metric tons are produced annually for human food. The common feature of root crops is their fleshy underground storage organ, often a true root but sometimes a rhizome (tuber), corm, or bulb. Compared to cereal grains and legume seeds, food stored in fleshy roots and tubers is watery and less concentrated. Thus root crops are not generally as easily kept, transported, and marketed. Most are abundant in starch or sugar and low in protein and oil, and hence are useful "energy foods" but do not by themselves supply a balanced diet. Some consumed as fresh vegetables (such as carrots, garlic, parsnips, radish, and onions) are prized more for their flavor, vitamins, and subtle nutrient qualities than as sources of energy. Just the opposite is true of potatoes, sweetpotatoes, cassava, and sugar beets.

World gross tonnage of the major root crops is not much less than for the major cereals, although, as noted, the dry nutrient content of root crops is much less (a potato, for example, may be 80 percent water). Annual world production of potatoes is nearly 300 million metric tons, sweetpotatoes more than 140 million metric tons, manioc or cassava more than 100 million metric tons. Nearly 300 million metric tons of sugar beets are harvested annually. World production and consumption of fresh vegetables is difficult to ascertain, but some 18 million metric tons of onions and 7 million metric tons of carrots make up part of the total.

Numerous roots and rhizomes of wild species can be consumed for food in times of emergency, as many an early explorer learned perforce. Wildlife and primitive peoples have always made use of indigenous bulbs, roots, and tubers. Some species, changed little if at all from the ancestral form, are collected or planted in limited areas today. In the Peruvian Andes, for example, a great many root plants little known in other parts of the world (*Arracacia, Lepidium, Oxalis, Pachyrhizus, Tropaeolum, Ullucus*) are regularly eaten and marketed. The same is true of certain sedges (*Cyperus*), aroids (members of the family Araceae), and *Coleus* ("frafra potatoes") in Africa. In the hill country of India, soh-phlong (*Moghania vestita* of the Leguminosae) is a minor food source, in some respects more nutritious than conventional root crops (Table 21-1). But no

Table 21-1 Chemical analysis of soh-phlong, sweetpotato, and cassava.

Constituent	Approximate net weight (percent)		
	Soh-phlong	Sweetpotato	Cassava
Moisture	67.9	68.5	59.4
Protein	3.0	1.2	0.7
Carbohydrate	27.0	31.0	38.7
Mineral	1.0	1.0	1.0
Calcium	0.020	0.030	0.050
Phosphorus	0.064	0.049	0.040
Iron	0.003	0.001	0.001

Source: Adapted from "The Wealth of India," as reported by H. B. Singh and R. K. Arora in *Economic Botany* 27(3):336.

other root crop promises to rival the potato as a staple food in temperate climates, or the cassava in tropical climates.

Potato, *Solanum tuberosum* (Solanaceae)

The common potato is a member of another large and important plant family, the Solanaceae. Among its close relatives are eggplant and tomato, and in the same family is tobacco. The genus *Solanum* is a large one, comprising about 1700 species, many of them poorly described and incompletely understood. Various wild *Solanum* species produce small potatolike tubers, which from time immemorial have been grubbed from the ground by the Indians. One or more of these was domesticated in the Andes Mountains of Bolivia and Peru, certainly prior to 200 A.D., to become the cultivated *S. tuberosum*. The diploid chromosome complement in *Solanum* is 24, with *S. tuberosum* ($2n = 48$) being an autotetraploid. Triploid and pentaploid relatives are known as well. *S. tuberosum* spp. *andigena* is hypothesized as the ancestral subspecies of the cultivated potato, itself perhaps descended (at least in part) from *S. sparsipilum* or *S. stenotomum*. In typical Indian plantings in the American tropics, many "wild" potatoes flourish about the fields and between the planted rows. Doubtless some hybridization and introgression continues to take place today, as must have occurred abundantly in the evolution of *S. tuberosum*. Archaeologists have found representations of the potato incorporated in designs on early Andean pottery. The potato is still cultivated as a staple in its Andean homeland, where, better to preserve its food value and to build reserves against a poor harvest, potatoes are made into "chuño." Chuño is prepared by trampling and drying the potato during alternate freezing and thawing.

The potato was presumably first seen by a European in 1537, when the Spanish landed in what is now Columbia. New World explorers and monks became familiar with it in the decades that followed, and it was brought back to Europe by 1570. It was cultivated throughout the Continent before 1600, and in Ireland by 1663. The cultivated potato is said to have been first introduced into North America in 1621 (presumably via Bermuda). Not until 1700, however, was the potato extensively planted. One reason for its sudden prominence in Europe during the 1700s was that reigning sovereigns, recognizing its food potential, compelled the people by royal edict to plant it (Germany, 1744; Sweden, 1764). Especially in Ireland was the potato adopted as a mainstay food, and when the late-blight disease (caused by the fungus *Phytophthora infestans*) wiped out the crop two years in a row in the 1840s, famine forced large-scale emigration to America. One might speculate that the Irish introduction stemmed from one narrow genetic source, and that—had the great genetic variability of the potato in its South American homeland been available—famine might have been avoided and the course of history materially changed. Even before the Irish immigrants came to the United States, introductions of the potato into New England from Ireland had given it the appellation "Irish" potato. It is sometimes referred to as "white potato" to distinguish it from sweetpotato (a different family entirely), but the use of this adjective ignores the many colored varieties.

It may appear strange that a plant from the New World tropics should have become so important in temperate Europe. Keep in mind, however, that in the New World the potato was a highland crop, grown at elevations in the Andes that were too cold for corn. Basically it is adapted to a cool environment, and it will not yield well when temperatures average about 21°C (70°F). However, the many freely intercrossing species related to *S. tuberosum* afford great plasticity for selection, and types adapted even to the humid lowland tropics (certainly the dry highland tropics with irrigation) have proven to give yields as high as 15 metric tons per hectare in sixty days.

Today the potato is grown chiefly in Europe, with more than 200 million metric tons being harvested annually, if Soviet production is included. China produces more than 40 million metric tons, and North America about half that. The potato's ancestral homeland, South America, altogether produces only about 9 million metrics tons per year. The Soviet Union, Poland, and China are the leading potato countries. In the United States, some 8 million metric tons of potatoes are produced annually for the table, and another 6 million metric tons are processed (mostly for potato chips, frozen french fries, dehydrated mashed potatoes, and starch). Nearly 1 million metric tons each may be utilized for animal feed and propagation.

The food value of the potato varies, depending on variety, growth conditions, storage, and handling. Analyses have indicated its composition to be 70–81 percent water, 8–28 percent starch, and 1–19 percent protein (most cultivars have around 5 percent Kjeldahl nitrogen, perhaps half of it not in protein form, and with varying amino acid composition), with traces of minerals and other food elements. There seems ample opportunity to breed new cultivars nutritionally better balanced and as productive of protein as is the soybean. In comparison with other familiar foods, the potato is fairly economical, the cost per kilocalorie being roughly in the same range as that of bread and margarine, much cheaper than meat, but half again to three times as costly as sugar. In Europe much of the potato crop is fed to livestock, and an appreciable portion is used for fermentation or for starch for industrial purposes. In the United States, cull potatoes—those not used as food because of poor appearance—are utilized for animal feed and for industrial purposes. Potato starch is nowadays less important than corn starch, but potatoes mill more easily than corn, as they do not require preliminary soakings and separations, although final concentration by centrifugation or in starch tables is similar to the handling of corn starch. Potato starch goes chiefly into sizings for textiles and paper, and to some extent into confections and adhesives.

The potato tuber is anatomically a stem, with external buds (eyes) that are able to sprout into new growth (see Chapter 7). Sections of potato with a bud (eye) are conventionally used to vegetatively propagate and maintain potato varieties, since sexual reproduction through true seed would be slower to yield and would risk changes due to genetic segregation. In China and the tropics, however, newly inaugurated systems of growing potatoes from true seed are being undertaken. Potatoes saved for propagation, some 7 percent of the crop, are unfortunately termed "seed" potatoes, but of course they are not true seed.

Only since 1850 have there been serious attempts to improve the potato. A century ago plantings started "running out," gradually decreasing in yield, probably because of accumulation of tuber-transmitted virus diseases. Since then new introductions have been secured from throughout the Americas and entered into a breeding program. A large germplasm collection is maintained for breeding at the USDA Experimental Station at Sturgeon Bay, Wisconsin.

Potatoes are grown in all states of the United States, but Aroostook County, Maine, is responsible for about three times as much production as the next most productive counties (Kern County, California; Bingham County, Idaho; and Suffolk County, New York). Yields under irrigation in Idaho have been phenomenal—averaging some 33 metric tons per hectare (about 15 tons per acre), with some fields yielding 78 metric tons per hectare (about 35 tons per acre). (Some agronomists have predicted yields as high as 225 metric tons per hectare eventually. Few food plants can boast this productivity potential!) Winter and early-spring potatoes come from Florida and southern California, late-spring and early-summer potatoes from the southern and border states, and late-summer and fall potatoes from the northern half of the nation, where nearly three-fourths of the domestic crop is grown.

Sandy, well-drained loams, and soils high in organic content, are generally best for potato growing. In Idaho, the leading producing state in the United States, vital soil moisture is generally provided by irrigation. The potato is responsive to day length; in general, long days stimulate stem elongation, and short days stimulate tuberization. Under short days with low light intensity, foliage tends to be soft and more susceptible to late blight, whereas under long days and higher light intensities, foliage tends to be more resistant to late blight. Thus a variety grown in Florida in winter may have considerable blight trouble, but it may be relatively immune when grown in Maine in summer.

Potatoes are planted in well cultivated land, often after plowing down a legume or green manure. Ample organic matter keeps soil crumbly, which is beneficial to potato growth, and it facilitates digging. Moderately acid soils favor scab-free tubers. More than 2 metric tons of fertilizer per hectare (about 1 ton per acre) is used in commercial potato growing, bringing the average yields per hectare to more than 22 metric tons in the United States (triple the yield of only 30 years ago).

Seed tubers certified free of disease are the preferred planting stock. Fusarium wilt, wart, corky ringspot, and powdery scab are examples of diseases that might permanently infest the soil if uncertified tubers were used. It has been shown that the geographical source of "seed," as well as storage conditions, can have an influence on yields of subsequent plantings. Moreover, varieties differ in dormancy requirements, ability to sustain repeated desprouting in storage, and response to chemical dormancy treatment. It was once common practice to disinfect tubers before sectioning for planting, but with improved seed stock this is not so critical. In some areas whole (small size) potatoes may be planted as seed, both to avoid the spread of disease (on the sectioning knife) and the cost of sectioning. In the United States, both sectioning and planting are mostly done mechanically.

Choice of spacing, depth of planting, and mode of

Figure 21-1   Machine harvesting of potatoes from muck soil. [Courtesy Purdue University.]

cultivation are determined partly by variety and partly by local conditions. Spraying and dusting to control insects and diseases are common practices today, in keeping with intensified modern agriculture. Dozens of afflictions can affect tuber quality, and equally as many preventatives can be advocated. Weeds are controlled chiefly by cultivation, but pre-emergence herbicides are increasingly being used.

Potato tops must be eliminated to facilitate mechanical digging. This can be done chemically or physically. The digging machines then "sift" the soil from slightly below tuber depth, screening the tubers and depositing them on elevator lifts that carry them to adjacent vehicles for transportation from the field (Fig. 21-1). More than 90 percent of the crop is now harvested mechanically in the United States.

Considerable loss can arise from bruising potatoes during harvest. Care must be taken in handling and storage to avoid this injury as much as possible. Most potatoes are washed (with spray jets) to remove soil before marketing, and this, too, may encourage injury and rot. In northern producing areas, a goodly portion of the crop is stored on the farm in such facilities as pit houses, where low temperatures discourage sprouting. In larger storage facilities, special air circulation and ventilation may be needed. Potatoes are preferably kept at about 4°C (39°F). About 10°C (50°F), sprouting is stimulated within a few months, although this can be inhibited by chemical treatment with maleic hydrazide. The processing

quality of potatoes intended for french fries and potato chips is improved if they are held for some time (from a few days to a couple of weeks) at 15 to 21°C (60 to 70°F) before processing to avoid an undesirable brown discoloration caused by the formation of reducing sugars. This treatment is called reconditioning. During this period the concentration of reducing sugars (glucose and fructose) is decreased, partially by respiration and partially by conversion to starch.

Cassava, or manioc, *Manihot esculenta* (Euphorbiaceae)

Cassava, a member of the family Euphorbiaceae, is a species of the tropical lowlands, probably domesticated in northern Amazonia. It adapts to poor soil and casual cultivation, and has consequently become a staple food in many of the poor and developing parts of the world. Mandioca, yuca, tapioca, and sagu are other common names by which the cassava is known, and in Africa (where it was introduced by the Portuguese in 1558) more than a half-dozen vernacular names prevail. The Latin name, *Manihot esculenta*, represents a huge complex of cultivars that taxonomists have from time to time tried to separate into distinctive species with but little success. Thus *M. utilissima, M. dulcis, M. aipi, M. palmata*, and other proposed species are generally regarded as synonyms of *M. esculenta*. Scores of wild *Manihot* species grow in eastern South America, but it is not known whether any of these are ancestral to modern cassava cultivars; indeed, with widespread abandonment under primitive slash-and-burn agriculture, it is difficult to distinguish between occasional escapes and wild plants.

Although one of the world's most important food staples, cassava is known in North America and Europe almost solely as tapioca, an occasional dessert. But in remote tropical areas, such as the Amazon valley, it may be almost the sole cultivated plant. World production is reported to be about 110 million metric tons annually, but the figure may not reflect scattered local production in remote locations. Brazil is the leading producer of cassava, with Indonesia probably second (cassava is the third most important food staple in Indonesia, after rice and corn). Cassava is not as esteemed in Indonesia as are other foods, but it is often the only crop that can be economically grown on poor, exhausted land. Cassava is also an important staple in several tropical African countries, especially in Zaire and Nigeria.

The cassava plant is a shrubby, monoecious perennial that produces several swollen roots that resemble sweetpotatoes and develop from a central trunklike stem. The palmate leaves are quite characteristic of the genus, and they have been investigated as a tropical source of protein for livestock feed (the foliage is about 30 percent protein). The roots contain little protein (1.2–2.7 percent) or oil, but are about 30 percent starch, found chiefly in the voluminous pith. Yielding up to 40 metric tons per hectare with little attention, cassava is one of the more remunerative crops. Its cultivars are vaguely grouped as "bitter" or "sweet," having respectively relatively high or low concentrations of a cyanogenic glycoside. The poison is found mainly in the periderm and shallow cortex. Sweet cassavas require no special treatment before consumption, but bitter varieties are shredded, squeezed, cooked, and treated in other ways to rid them of the poisonous juice or to reduce its toxicity. There is no clear-cut demarcation between these two groups, as the glycoside content varies with individual plants and cultivars. There has been some speculation that low doses of cassava glycoside can be prophylactic, preventing several human ills!

Under the typical slash-and-burn agriculture of the tropics, cassava stem sections are hand planted in holes grubbed into the soil just before the rainy season. Extensive feeder roots grow quickly, making full use of the evanescent soil fertility. In most cases the crop is quite responsive to fertilizers, especially those high in potassium. If all goes well, within as little as seven months a number of starchy roots can be harvested from each plant, although best yields are not obtained until after 16 months. If allowed to grow for too long a time, however, the roots become rather woody and less edible. Cassava is a heliophile, performing best in full sun. Otherwise it is broadly tolerant, with local selections adapted to a wide variety of soils, pH values, and moisture conditions. In agriculturally advanced parts of the tropics, such as in Jamaica and parts of Central America, cassava fits well into rotation with corn and other cultivars. In some areas the cassava crop is undersown with a ground cover, such as *Stylosanthes,* to protect and improve the soil, and to increase top growth (important for livestock feeding, especially during the dry season). In Indonesia the "mukibat system" is sometimes employed, in which especially "good-eating" types of *M. esculenta* are grafted to stocks of *M. glaziovii,* which increase yield and facilitate crop growing.

Cassava is used in a number of ways. The whole root may be boiled; it has a sticky, "heavy" consistency, and of itself is rather tasteless. In Brazil it is usually shredded, then heated and dried, to make a meal known as *farinha,* eaten alone if need be, or mixed with other food and sauces. In Indonesia the roots are sectioned, dried in the sun, and later ground into meal. In the making of tapioca, an important export from Indonesia, the peeled roots are grated, the mass soaked in water for several days, kneaded, strained, dried, and then heated to partly hydrolize the starch to sugar and gel particles into "pearls" while being agitated on a grill. In Jamaica roots are ground into a mush called *bami,* or formed into cakes called *casabe.* In Africa boiled roots are often pounded into a thick paste called *fufu.* In many parts of the world cassava mush is fermented into a beer that is much like *chicha* (which is usually made from corn). The heat and pressure of grinding cassava serves to change or remove poisonous glycosides.

Juices expressed from bitter cassava are usually saved for fermentation to beer, or they may be variously treated to yield meat sauces such as "West Indian pepperpot."

Low-amylose starch from cassava is a minor item of commerce, being imported from the Far East into North America and Europe for foodstuffs, sizing, and adhesives. Leaf-protein extraction similar to that undertaken with alfalfa has been attempted. Regulated cassava fermentation can yield alcohol (important in Brazil as a back-up for fuel from sugarcane) or butanol and acetone. Or fungal protein can be secured by inoculation with *Aspergillus fumigatus* and the addition of traces of urea, potassium phosphate, and sulfuric acid (to obtain a pH of 3.5). Starch hydrolysis is a source of glucose.

Sweetpotato, *Ipomoea batatas* (Convolvulaceae)

In many respects the sweetpotato (Fig. 21-2) reminds one of the cassava root. It contains about 30 percent solids (mostly starch, but some sugar, and not inconsiderable minerals and such vitamins as ascorbic acid), concentrated especially in the storage parenchyma of the xylem. The protein content of most cultivars is about 5–7 percent of the dry weight. Yields as high as 40 metric tons per hectare (about 18 tons per acre) are obtained. The sweetpotato is a trailing vine adapted to tropical lowlands, and cultivars are propagated vegetatively. The cultivated sweetpotato is a hexaploid ($2n = 90$), and it may have been domesticated from the ill-defined *I. trifida*, a wild hexaploid species, or other tropical American entities, such as *I. tiliacea*, *I. gracilis*, and *I. trichocarpa*. *I. batatas* is a short-day plant, flowering with a photoperiod of eleven hours or less. It grows best where the average temperature is not less than 24°C (75°F).

The spread of the sweetpotato from its presumed ancestral home in Central America or South America has been gradual and unspectacular. Tubers 10,000 years old have been discovered in coastal Peru, so it has long been under cultivation. It was apparently introduced into southern Europe very early in the Colonial period, and was carried by navigators of the

Figure 21-2 The edible portion of the sweetpotato is the tuberous root. [Courtesy J. C. Allen & Son.]

sixteenth and seventeenth centuries to the Far East, where it found greater favor than in its homeland or Europe (but evidence of the sweetpotato in Polynesia dated to 500 A.D. is an enigma still to be explained). Especially in China and Japan, the sweetpotato has become an important food plant; it is very productive on the intensively managed uplands in rotation with other crops, such as grains. In many parts of the world the sweetpotato is used more as a nourishing livestock feed than for human consumption. Whether baked or boiled, the sweetpotato is more flavorful than most root crops, especially the moist, reddish cultivars favored in the southern United States, where they are popularly called "yams" (most northerners favor the drier, light-yellow cultivars).

World sweetpotato production runs about 140 million metric tons annually. China is by far the leading sweetpotato center, although the root is extensively grown in many other Asiatic countries, and in many Latin American and African ones, too. One-third of the United States' production comes from North Carolina. Some cultivars rarely flower, and the sweetpotato has proven temperamental in breeding programs. It is often self-incompatible due to three separate genes. Both incompatibility and sterility are influenced by environment. Commercially the sweetpotato is vegetatively propagated from numerous sprouts that originate at the head of a root when it is kept moist and warm. A certain percentage of the crop is saved to produce new planting "slips" (see p. 211). Although the plant is a perennial, slips grow fast enough to yield roots on an annual basis. Slips are generally planted into hills or rows by hand. In the Far East, whatever cultivation and weeding is needed is done by hand, as is digging of the roots at about the time of killing frost. In the southern United States the same type of mechanization as was developed for harvesting the potato is increasingly being used. Throughout much of the world the sweetpotato is a home-garden plant that never goes to market. Roots are generally dried for several days (to form a protective periderm) and then stored in a cool place. Because sweetpotatoes do not keep as well as potatoes, they are sometimes cut into slices and dried, especially when they are to be used for industrial purposes. In Japan and Korea the dried sweet potatoes may be ground into a meal, which in turn can be cooked for human consumption or fermented for alcohol.

Beets, *Beta vulgaris* (Chenopodiaceae)

There are two major groups of beet cultivars, the gross sugar beets and the delicate table beets, best adapted to temperate climates. The former are used somewhat as livestock feed, but have been selected principally for the extraction of their high sugar content. Both are biennial, although they are planted and harvested as annuals. Table beets are generally consumed when young, a few weeks after planting, when the swollen roots are sweet and tender. More than 0.5 million metric tons are grown annually in the United States. They are about 8 percent carbohy-

drate and 2 percent protein. Cultivated beets are presumed to be derived from the wild beet of northern Europe, *Beta maritima*. Close relatives are chard and the white-fleshed mangel-wurzels. The leaf of chard is eaten as a green, which seems to have been the case also with the original cultivated beet in pre-Christian times. In northern India, foliage of the annual *B. vulgaris* var. *orientalis*, the *palang* or *palanki* of the marketplace, is likewise consumed. The modern table beet was apparently not developed as a garden plant until about the sixteenth century.

Sugar beets have much more commercial value than table beets. Through selection and breeding, their sugar content has been increased from the 2–4 percent that is normal for table beets to as much as 20 percent. The annual harvest of sugar beets is about 300 million metric tons, and yields average about 50 metric tons per hectare (a little more than 22 tons per acre). Most of the world crop is grown in Europe. The USSR is the leading producer, followed by the United States, France, Germany, Poland, and Italy.

The sugar beet received its greatest impetus after 1747, when the German chemist Andreas Sigismund Marggraf proved that sugar from beets was identical with that from sugarcane. The King of Prussia became interested, and in 1802 financed the first beet-sugar factory. A few years later, Napoleon also became interested, for naval blockades had cut off the supply of imported sugar, and he hoped to supply his armies with sugar produced domestically. The French people never became enthusiastic about growing sugar beets, however, and after Waterloo the industry collapsed in France as the importation of cane sugar was resumed. In the United States, sugar beets were first grown in the Philadelphia area in 1830, and shortly thereafter in Massachusetts, Michigan, and finally, with most success, in California. California, Minnesota, Idaho, and North Dakota are prominent sugar beet producing states today.

Sugar beets are well adapted to a diversity of soil and climatic conditions, and today are grown in many states, drill planted in early spring. New, better yielding cultivars, and new monogerm seed (which produces a single plant rather than clusters and thus saves thinning) have given substantial boosts to the industry. Machines similar to potato lifters have mechanized harvesting, and the use of by-products (tops, pulp, and molasses for cattle feed) has given the crop further impetus.

Harvested sugar beets, weighing as much as 2.25 kilograms (about 5 pounds) each, are washed at the factory, cut with revolving knives into thin strips (cossettes), and soaked in hot water for diffusion of the sugar. Soluble contaminants are precipitated from the sugary solution with carbon dioxide and lime (clarification) and filtered out. The solution is then evaporated to a syrup, from which the sugar is crystallized by boiling under vacuum. The crystallized sugar is separated by centrifugation and further refined to make table sugar. The pulp may either be dried for stock feed or treated with dilute acid to yield commercial pectin.

Edible aroids (Araceae)

A number of genera of the family Araceae yield edible corms or tubers—food basic to many tropical areas. Cocoyam, dasheen, eddoe, tannia, taro, and yautia are a few of the more familiar common names. The two most prominent genera are *Colocasia*, the taros and dasheens; and *Xanthosoma*, the cocoyams, yautias, and tannias. Some species of these genera also serve as house plants ("elephant ears") in summer garden pools and greenhouses.

Colocasia esculenta, adapted to the rainy tropics, with some strains adapted even to subaquatic habitats, was apparently domesticated in southeastern Asia. It had been spread throughout the Pacific and as far east as the Mediterranean in prehistoric times. The corms, the source for the famed *poi* of Hawaii, are low in protein. *Xanthosoma sagittifolium* probably embraces most of the edible cocoyams, yautias, and tannias. They are 5–9 percent protein on a dry weight basis, and yields can run as high as 60 metric tons per hectare (nearly 27 tons per acre). The species is native to tropical America, but it was distributed worldwide in post-Columbian times. *Alocasia, Amorphophallus,* and *Cyrtosperma* are other genera prominently used for food, as are a number of others under emergency conditions.

The food value of aroid corms is similar to that of the potato, being about 20 percent starch with little protein and oil. Most of the corms are boiled like potatoes, and then crushed to make "cakes." Corms can also be roasted, steamed, or fried in oil to make "chips." In the making of *poi*, the crushed corm is allowed to undergo a natural fermentation before final table preparation.

Most edible aroids are simply cultivated in small patches by peasant farmers. They grow best in moist, lowland, frost-free locations, under light shade. Propagation is accomplished typically by means of satellite cormels or by sectioning of a corm. Yields as high as 45 metric tons per hectare (20 tons per acre) have been obtained experimentally, and under favorable conditions a crop can be harvested in as little as six months. Harvested corms can be stored for a few months if they are kept dry and ventilated. Cropping is mostly by hand in those tropical lands where aroids are an important crop. Accurate statistics on world production are not available, but several millions of metric tons of edible aroids are probably produced annually, which qualifies them as one of the world's basic food resources.

Other important root crops

Arracacha, *Arracacia esculenta* (Umbelliferae)

Also known as the Peruvian carrot, *Arracacia* is an herbaceous perennial of the Andes, mostly used locally because of short storage life. The starchy tubers (10–25 percent starch) are only half as costly as potatoes in many places, and the stems and foliage are eaten as well. Propagation is from offset tubers.

Carrot, *Daucus carota* (Umbelliferae)

The species is a wild weed of the family Umbelliferae. The cultivated variety (*D. carota* var. *sativus*) is believed to have been selected in the Near East long ago, but the root was generally accepted as human food only in recent centuries. Recorded world production is about 7 million metric tons, mostly from western Europe, China, and North America.

Figure 21-3 Rhizome of the Jerusalem artichoke, or sunchoke, *Helianthus tuberosus*, a rootcrop long enjoyed by the American Indians and now becoming more generally popular. These rhizomes have been washed free of soil. [Courtesy USDA.]

Figure 21-4 Sacking onion bulbs in Ecuador. [Photograph by Larry Daughters, courtesy World Bank.]

Horseradish, *Armoracia rusticana* (Cruciferae)

A crucifer native to Europe, this member of the mustard family has a white, carrot-shaped root that contains a pungent glucoside used mostly for flavoring.

Jerusalem artichoke, or sunchoke, *Helianthus tuberosus* (Compositae)

This rank sunflower has a potatolike rhizome (Fig. 21-3) that can be eaten boiled, pickled, or raw. The crisp flesh often substitutes for water-chestnut in oriental dishes, and is gaining favor in the United States.

Onion, *Allium cepa* (Liliaceae)

Along with its relatives—garlic, leeks, shallots, and chives, all members of the genus *Allium*—the onion is an important flavoring and food plant throughout the world (Fig. 21-4). The onion is a true bulb, consisting of condensed food-storage leaves that are rich in sugar and in the pungent compound allyl sulfide. It is a major crop in central and southern Europe and in the mucklands of the Great Lakes area of the United States (Fig. 21-5). Recorded world production is a sizable 18 million metric tons, mostly from China, India, the United States, Spain, and Japan.

Figure 21-5 Hybrid onions growing on muck soil.

Shallots, surprisingly, are a fairly important crop on the sandy soils of Ghana, in tropical Africa.

Parsnip, *Pastinaca sativa* (Umbelliferae)

Another relative of the carrot, the biennial parsnip has edible roots that turn quite sweetish after exposure to cold. A vigorous species mostly grown in Europe, the parsnip was presumably domesticated in the Mediterranean area; wild and escaped plants are common in many parts of the world.

Radish, *Raphanus sativus* (Cruciferae)

The radish, presumably native to China, is used chiefly in the Orient, where gigantic roots are grown and often pickled in brine.

Rutabaga, *Brassica napus,* Napobrassica group (Cruciferae)

Another of the cabbages (see p. 564), the rutabaga is also known as swede, or swede turnip, and seems to be a cross between the turnip and the cabbage in Europe during the Middle Ages.

Salsify, *Tragopogon porrifolius* (Compositae)

Also known as oyster plant, salsify has fleshy, parsniplike roots that are usually consumed boiled.

Turnip, *Brassica rapa* Rapifera group (Cruciferae)

Another biennial of the cabbage clan (see p. 564), the turnip has been long in cultivation and is probably native to Europe and central Asia. It is easily grown and is often used as stock feed.

Yam, *Dioscorea* (Dioscoreaceae)

Yams comprise several species of the genus *Dioscorea*, principally *D. cayenensis* and *D. rotundata*, which were domesticated in tropical Africa perhaps 5000 years ago, and *D. alata*, which is indigenous to southeast Asia. The true yam is not to be confused with the sweetpotato, sometimes popularly termed "yam." Mostly tropical vines with worldwide distribution, adapted to climates with abundant rainfall, yams can produce tremendous roots weighing as much as 40 kilograms each and yields of as much as 40 metric tons per hectare (nearly 18 tons per acre). Many yams resemble potatoes and are consumed in the same fashion; in western Africa, yams are pounded to make the prized dish *fufu*. Wild yams, of which there are more than 500 species, were much sought as a source of steroidal sapogenins in the development of cortisone, and yams have been the source of diosgenin for the manufacture of birth-control pills, the industry in Mexico first utilizing wild sources. Several species are cultivated in China and Africa for food (they contain about the same amount of starch as does the potato). Propagation is chiefly by replanting the top of a harvested root, and trellises are usually provided for the trailing vine. A growing season of 8 to 10 months is required to mature a crop, making yams unsuitable for any but tropical climates. World production must exceed 20 million metric tons, with Nigeria a major center for cultivation. Breeding of new cultivars has been undertaken there, and an assemblage of species and cultivars has been made at Mayaguez, Puerto Rico.

STEM AND LEAF CROPS

Except for a few tubers (underground stems), such as the potato, stems are more important as a source of wood and fiber than as a direct source of food. A few "vegetables," such as asparagus and kohlrabi, and such specialty items as bamboo shoots, are stems that are consumed directly, but none of these is very important as a world commodity. Sugarcane, however, is internationally important. Sugar, of course, is the carbohydrate extract from the cane stem, the stem itself being chewed only incidentally in the tropics. Sugar from cane is chemically identical with that from the sugar beet. But in international importance—and, one might add, in world romance, too—cane considerably outdistances the beet. Worldwide cane-sugar production is approaching 1 billion metric tons annually.

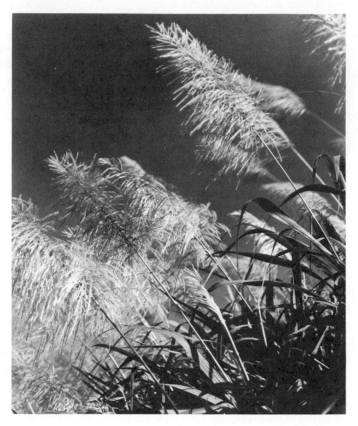

Figure 21-6 Sugarcane, a member of the grass family, as it appears in flower in Hawaii. [Courtesy J. C. Allen & Son.]

The tremendous use of plant leaves as forage for animals is, of course, an important indirect source of food for people. But some foliage, often modified (as in the onion), is much utilized directly for human food, typically as "fresh green vegetables" or in salads. Fresh foliage is quite perishable, and it is understandable that little international commerce exists in leafy vegetables and salad plants. In the United States a sizable and very specialized head lettuce industry has developed, in which the national market, at least early in the season, is serviced almost solely from intensively grown irrigated crops in southern Arizona and California. Celery, too, may be shipped widely. But on the whole, bulky and perishable leaf crops are best handled as local truck crops or preserved by canning (and sometimes freezing).

Sugarcane, *Saccharum officinarum* (Gramineae)

Sugarcane (Fig. 21-6) is a short-day crop of the warm tropical lowlands, adapted to a lengthy humid season followed (ideally) by a short dry season during which sucrose accumulates. It is one of the few species to store its carbohydrate reserves as sucrose. It has a great thirst, and it is said that about 2 metric tons of water is needed to produce each kilogram of sugar. (Drip irrigation has been introduced to the cane fields of Hawaii and elsewhere to conserve irrigation water.)

A little sugarcane is grown in southern Europe (Spain), a bit more in the southern continental United States (principally Florida and Louisiana), a fair amount in Africa (in a number of countries), and a great deal in South America (principally Brazil), the Caribbean (principally Cuba), Mexico, certain Pacific islands (especially Hawaii and the Philippines), and southern Asia (principally India and Pakistan). World production for 1977 was about 740 million metric tons.

Although sugarcane is identified botanically as *Saccharum officinarum,* most of the modern cultivars are derived from crosses of an original species (the *S. officinarum* "noble canes," probably native to New Guinea and possibly descended from *S. robustum*) with other species. *S. barberi* of northern India, itself perhaps a hybrid of *S. spontaneum* and *S. officinarum,* was long ago used for extraction and chewing, but today it is of no importance except as a germplasm source in breeding programs. Primitive chewing types were improved first by casual selection, then by intraspecific crop breeding, and finally by interspecific crossing that has yielded the prime clones so widely planted today. "Pure" *S. officinarum* has a diploid chromosome number of 80.

Breeding work with various sugarcanes began around 1885, particularly in Java. Crosses and backcrosses involving *S. spontaneum* and *S. sinense* (to confer disease resistance and vigor) yielded, in 1921, a markedly superior clone ('POJ 2878') that changed sugarcane production all over the world. Since then other improved cultivars have been bred, especially in Hawaii, with *S. robustum* "blood" entering the pedigree. The impact on sugar yields has been astounding. The average annual yield of crystal sugar in Java was 2 metric tons per hectare in 1840, 10 tons by 1910, and 20 tons by 1940. Of course, this phenomenal rate of increase is partly due to improved growing techniques, as well as breeding for higher yields and disease resistance. However, modern sugarcane is nonetheless one of the world's most productive crops, yielding 200 metric tons of canes per hectare (89 tons per acre) and 22 metric tons of sucrose per hectare (about 10 tons per acre) under good growing conditions.

Sugarcane was apparently introduced into the Western World from India via ancient trade routes to the Mediterranean, where primitive types are still grown to some extent for chewing. Sugarcane was grown in the Canary Islands before 1493, when, it is reputed, Columbus took it thence to the New World, where it was soon widely planted. In 1509, sugarcane was harvested in Santo Domingo, Hispaniola, and it had spread to Mexico by 1520, to Brazil by 1532, and to Peru by 1533. Cultivars from Java did not supplant this low-yielding sugarcane until late in the nineteenth century. The first sugarcane raised in what is now the continental United States was in the former French colony of Louisiana, in 1753.

In less developed lands, sugarcane has long been planted and tended by hand (Fig. 21-7), with sections of stem merely being stuck in the ground and the weeds kept reasonably in check. (Sections of stem 8–12 months old, called setts, sprout readily from axillary buds and adventitious roots.) However, few tropical crops have been as thoroughly acclimated to large-scale plantation agriculture and as extensively affected by advanced technology. Soil testing and customized field preparation, machine planting, efficient cultivation and herbicidal weed control, generous fertilization, preharvest flaming, and mechanized gathering are all commonplace. More and more the crop is sprayed with a "ripening agent," or a compound to prevent flowering, since the determinate inflorescence draws upon sucrose reserves and the stem becomes more fibrous. It is reported that nearly half the Hawaiian crop is treated with Polaris growth regulator, and elsewhere the herbicide diquat is used to repress flowering.

Twelve to eighteen months are generally required to complete a planting cycle. On rich lands, subsequent crops may be obtained from rhizomes left in the soil, but more often the crop is replanted like an annual to systematize mass production, or at most allowed to regenerate (ratoon crops) for only two or three years before replanting.

Whether sugarcane is harvested by gigantic cutting machines on a modern plantation or hand-harvested by machete in a distant jungle, sugar extraction follows much the same procedure. Basically, the sugarcane is brought to a centralized location where the juice is squeezed from it and concentrated. This is accomplished by crushing the stems between fluted rollers—hand or animal powered in small local operations, or part of a well-engineered system in

Figure 21-7 Planting sugarcane, Rharb Plain, Morocco. Women strip the cane before it is cut into pieces about a foot long and laid in trenches. [Photograph by Bill and Christina Graham, courtesy World Bank.]

plantation factories. The cane may be subjected to several successive expressions to extract the last bit of juice, the final one being undertaken after the nearly spent cane has been remoistened with water. The sugar content of sugarcane is about 12 percent. The fibrous residue of the sugar cane stem (bagasse) may be burned in the factory boilers, used for making fiberboard or paper, or composted to form an organic fertilizer.

The extracted sugarcane juice is handled in much the same way as the juice from the sugar beet. Successive boilings concentrate the sucrose, which under primitive circumstances may merely be dried to a sticky residue (the *rapadura* of Brazil) containing small percentages of other carbohydrates, proteins, and minerals. As such it is a nourishing staple in some countries where the refinements of technology are lacking. In technologically advanced societies the sugarcane juice is treated in modern mills with steam-heated chambers and vacuum pans, and is clarified by chemical precipitation of unwanted materials. The treated extract passes through filter presses to yield a solution that consists almost entirely of sucrose. When this is further concentrated, sugar crystallizes and is usually removed from the remaining syrup by centrifugation. The liquid portion, still containing a few brownish "impurities," is molasses. Molasses is used in foods (both human and animal) or is fermented to make rum, alcohol, or vinegar. The raw crystallized sugar, about 96 percent sucrose, is refined by repeated washings and recrystallizations to satisfy the demand for a pure white sugar, completely free from brown but nourishing "impurities."

Stem and leaf vegetables

In this category of convenience are pigeonholed the unrelated plants of relatively minor or specialized importance whose stems or foliage are eaten as human food. For example, asparagus is juvenile stem, broccoli an inflorescence of a cole plant, cabbage a leafy head of a cole plant, celery a petiole, spinach

Figure 21-8 Percentages of selected nutrients and calories contributed to the human diet by fruit and vegetables. [Courtesy USDA.]

an individual tender leaf. Compared with cereals, pulses, and root crops, they are not a rich source of energy, but they are excellent contributors of vitamins and minerals to the diet (Fig. 21-8). Most species listed are grown in temperate climates as summer annuals. Many others could be included: the bamboo shoots of the Orient and the many unusual "oriental vegetables" (including *Amaranthus, Arctium, Chrysanthemum*, and strange forms of "mustard"—see p. 564); the various potherbs of the tropics, such as Malabar spinach (*Basella alba*), and a host of other species yielding edible foliage (in such genera as *Alternanthera, Cnidoscolus, Cyrtosperma, Emilia, Hibiscus, Ipomoea, Monochoria, Moringa,* and *Sesbania*); New Zealand spinach (*Tetragonia expansa*), which is grown in southern temperate regions; and such "forgotten" temperate potherbs as roquette (*Eruca sativa*).

In technologically advanced countries, especially in North America, many commercial vegetables that don't hold their form well when preserved by canning—for example, broccoli, Brussels sprouts, cauliflower, and spinach—are now being marketed frozen. With a food freezer now in every supermarket, and a freezing compartment commonplace in the home refrigerator, preservation of perishable vegetables by freezing has become an important business. This requires cultivars that freeze well. Special han-

dling is also needed to preserve quality and flavor. Most frozen produce, after cleaning and sorting, is blanched (in commercial operations usually by steam) to inactivate enzymes that can cause deterioration or off-flavors. The blanched product may be completely packaged before quick freezing, after which it must be kept frozen until it is sold and used by the consumer.

Most vegetables have been known in substantially their present form for centuries. Broccoli, eggplant, kale, lettuce, and mustards (*Brassica*) are typical of species long utilized in the Old World. Spinach seems to have been introduced into China from Nepal in the seventh century. Specialized forms of cabbage (such as Brussels sprouts, kohlrabi, and head cabbage) were apparently selected out of more primitive forms (like kale and collards) during the Middle Ages. The stalk of celery was not eaten until the seventh century, although the seed was used as a flavoring prior to that time.

Modern vegetable breeders have had to consider a number of factors not previously of great importance. For the commercial market they have had to develop good keeping qualities (that allow long-distance shipping and rough handling), standardization (size and form convenient for canning or packaging), attractive appearance (to lure the consumer in some distant market), and acceptable processing qualities (to permit mechanical harvesting, handling, and canning or freezing). And, as with any crop, demand for better yields and improved disease resistance is unceasing. Cultivars of the following species are of fair commercial importance today as food plants.

Artichoke, *Cynara scolymus* (Compositae)

The artichoke was early domesticated in the Mediterranean region, originally for foliage (kept in the dark before harvest to make it tender). In recent centuries the young flowerhead (Fig. 21-9) has been considered a delicacy; the boiled fleshy bracts are peeled off and scraped clean of soft tissue by the teeth. The fleshy base, the artichoke "heart," is also pickled and preserved by canning. Only a little more

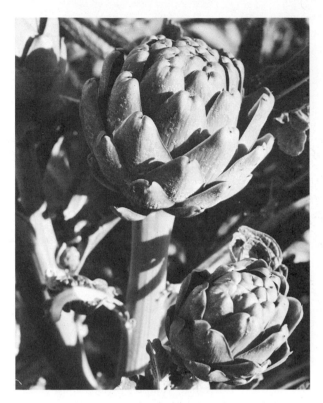

Figure 21-9 The fleshy flower head of the artichoke. [Courtesy J. C. Allen & Son.]

than 1 million metric tons of artichokes are grown annually, mostly in Italy and Spain.

Asparagus, *Asparagus officinalis* (Liliaceae)

Once regarded as a medicinal plant, and not greatly changed from the wild form, asparagus spears (young stems, cut just after they emerge from underground rhizomes in spring) are today considered a delicacy. Asparagus is a dioecious perennial, bearing for many years (if planted to fertile ground and if the spears are harvested in moderation). The plant has been cultivated from at least the time of Cato the Elder (234–149 B.C.E.), who gave directions for its growing in his *De Agri Cultura*. Asparagus is a heavy feeder and needs generous fertilization. Yields are only about 2.5 metric tons per hectare (a little more than 1 ton per acre), but the crop is highly prized and brings good prices.

Broccoli, *Brassica oleracea*, Italica group (Cruciferae)

The first selections of sprouting broccoli were probably made in Greece and Italy more than 2000 years ago; the modern heading types were developed mostly in northern Europe within the last three centuries. The fleshy terminal stems and young flowers and bracts are the parts consumed. Cultural requirements are similar to those for cabbage.

Brussels sprouts, *Brassica oleracea*, Gemmifera group (Cruciferae)

Brussels sprouts, another of the "cabbages," is a vegetable of relatively recent origin. Leafy buds in the axils of the main leaves become the "sprouts," which are consumed much as though they were miniature cabbages. They are grown in the same way as cabbage.

Cabbage, *Brassica oleracea*, Capitata group (Cruciferae)

Brassica is a large and taxonomically confusing genus. Wild *B. nigra*, black mustard of central and southern Europe ($n = 8$) is probably the source of the *b* genome contributed both to the oil-seed Indian mustard, *B. juncea* (*ab* genome, $n = 18$), and to Abyssinian mustard, *B. carinata* (*bc* genome, $n = 17$). *B. oleracea*, "cabbage" ($n = 9$) seems to be the source of the *c* genome; *B. rapa*, the turnip and Chinese cabbages ($n = 10$) is the sources of the *a* genome; and these two apparently hybridized to yield *B. napus* (*ac* genome, $n = 19$), which includes rape and rutabaga.

Ancestral cabbage (*B. oleracea*) was evidently brought into cultivation as much as 8000 years ago, in the low coastal areas of northen Europe, from which it was introduced into the Mediterranean area and the Near East before historical times. From the primitive cabbage have arisen not only the modern heading cabbage (Fig. 21-10), but cauliflower, Brussels sprouts, kale, kohlrabi, and some foliage mustards. In the second century B.C.E., Cato cited three forms of cabbage, and Pliny enumerated nine others in the first century A.D. Today heading forms of cab-

Figure 21-10 The cabbage head consists of large fleshy leaves attached to a compressed stem. [Courtesy J. C. Allen & Son.]

Figure 21-11 Celery before and after trimming for shipment. [Courtesy USDA.]

bage are among the most popular leaf vegetables in northern Europe and Japan; the loose-leaf "savoy" cabbages are more popular in southern Europe. Cabbages are often considered the "poor man's" vegetable. Although 90 percent water, cabbage is a good source of minerals and vitamins. Sauerkraut is a Teutonic dish, made by fermenting cut cabbage in brine until abundant lactic acid is produced. Some 30–40 percent of the water content of fresh cabbage then becomes waste, so that less succulent cultivars are sought for making sauerkraut.

Although biennial, cabbage is easily grown from seed, heads being harvested the first year (except when grown for seed). So many cultivars have been selected that head at different times that an almost continuous cabbage harvest is possible by proper selection of cultivars (a common practice in Japan). Typically, cabbage seed is started in the greenhouse or cold frame; the young plants are then transplanted to the fields, where they grow rapidly if protected from weeds. In recent years, techniques for direct seeding have been perfected for a number of crops in the cabbage group. Insecticidal spraying is generally needed to prevent damage from such pests as the cabbage looper (the voracious larva of the cabbage moth). Since harvest accounts for more than a quarter of production costs, the United States has increasingly turned to mechanized harvesters. World

production of cabbage exceeds 22 million metric tons, with China, Japan, the United States, and western Europe the leading producers.

Cauliflower, *Brassica oleracea*, Botrytis group (Cruciferae)

As with broccoli, the fleshy inflorescence of cauliflower is the part consumed; it is sometimes blanched to a lighter shade by tying the outer leaves over the head. Cultural requirements are similar to those for cabbage. More than 4 million metric tons of cauliflower are produced annually, especially in China, India, Italy, and France.

Celery, *Apium graveolens* (Umbelliferae)

Celery is believed to have been domesticated in the Mediterranean region. As a salad plant it is a comparative newcomer, with modern cultivars selected for a milder flavor and whiter stalk color (thus eliminating the need for mounding the stalks to blanch the petioles). In the United States, celery is grown mostly on moist soils in cool climates, as a summer crop in the North or a winter crop in the South (Fig. 21-11).

Figure 21-12 Swiss chard, a "beet" whose leaf and petiole are consumed as a potherb. [Courtesy USDA.]

Chard, *Beta vulgaris*, Cicla group, (Chenopodiaceae)

Chard (Fig. 21-12) is a "beet" without the enlarged root; its fleshy foliage is consumed instead. Like the table beet, it is easily grown in the home garden as a source of greens.

Chinese cabbage, or petsai, *Brassica rapa*, Pekinensis group (Cruciferae)

Petsai is primarily a crop of the Far East, where there is a bewildering array of cultivars. It is presumed to be native to eastern Asia, where it has been in cultivation since the dawn of history. It is a sister to the turnip, and a cousin to mustards and cabbages (see p. 564), and like them is grown as an annual. The central foliage, which forms a loose head, is the part consumed.

Collards and kale, *Brassica oleracea*, Acephala group (Cruciferae)

Collards and kale are primitive, leafy "cabbages," often grown for greens in the southern United States. They are very nutritious. In the tropics, they are grown as perennials, the leaves being harvested periodically or continually. Some types are extremely large leaved, and are used as forage.

Kohlrabi, *Brassica oleracea*, Gongylodes group (Cruciferae)

Kohlrabi, another member of the cabbage clan, is grown for its fleshy lower stem. Growing requirements are as for cabbage.

Endive, *Cichorium endivia* (Compositae)

Endive, a relative of chicory, is indigenous to the Mediterranean area. The lower frilled leaves are consumed in the same way as lettuce.

Lettuce, *Lactuca sativa* (Compositae)

Ancestral lettuce was likely a weedy trailside species much like its relatives, dandelion and chicory. It has since risen to become the prime salad plant, at least in America, where about 500 million dollars worth is produced annually.

Lettuce (Fig. 21-13) is known to have been grown by the Egyptians 4400 years ago, and it was generally utilized throughout the Old World by Christian times. It has long been a favorite home-garden plant, and in recent decades it has become a valuable commercial crop as well, particularly in the southwestern United States, where specially bred cultivars permit its being grown outdoors and shipped across the country (rather than being grown in greenhouses near population centers). Nearly 3 million metric

Figure 21-13   Greenhouse-grown 'Bibb' lettuce.

tons are produced annually in the United States, primarily in California and Arizona. Yields of 38 metric tons per hectare (17 tons per acre) are obtainable.

Lettuce cultivars have been developed in many varying forms. There are cold-tolerant varieties for midwinter production and slow-bolting or heat-tolerant ones for summer and autumn growing. Some form heads, and some are loose-leaved. To satisfy the present commercial market, disease-resistant varieties that form a firm head with wrapper leaves for protection are mandatory. To increase productivity and to effect other improvements, breeders have introduced genes from a wild lettuce, *L. virosa*, into *L. sativa* with the help of colchicine treatments. Some resistance to cabbage looper, from the wild lettuce *L. saligna*, has also been bred into *L. sativa*.

The southwestern crop is intensively managed on irrigated land, with fertilizer practices governed exactly to produce crisp solid heads and delayed bolting (flowering). Side dressings of nitrogen are usually applied. Special machines have been developed for planting, cultivating and harvesting. Modern equip-

ment simultaneously shapes up the planting bed for irrigation, plants the seed, and distributes starter fertilizer. Although much field harvesting is still done by hand, gigantic loading machines now enter the field for packaging the cut heads. Lettuce is trimmed (dead and discolored leaves removed) in automated sheds and is then carefully packed and iced for shipment (or vacuum-cooled before shipping in refrigerated cars, a more recent practice).

There is a variety of lettuce, (*asparagina* or *angustana*), known as celtuce, in which the fleshy stem rather than the leaf is the part esteemed. The name is coined from *cel*ery and let*tuce*, although the plant is in no way related to celery.

Parsley, *Petroselinum sativum* (Umbelliferae)

Parsley has been used as a garnish and flavoring since ancient times. Similar to its relative the carrot, it is a biennial. It is believed to have been domesticated in the Mediterranean area.

Rhubarb, *Rheum rhabarbarum* (Polygonaceae)

Edible rhubarb is believed to be native to Asia Minor, although a medicinal rhubarb has been used in China for several thousand years. Garden rhubarb is a perennial with immense leaves, the succulent petioles of which are cooked with sweetening. It grows best in cool, moist climates, and is more of a home-garden plant of rural areas than an item of horticultural commerce. It is typically propagated by dividing the clumps.

Spinach, *Spinacia oleracea* (Chenopodiaceae)

The crisp but fleshy foliage of spinach is much esteemed as a cooked green or as a salad leaf. It is a dioecious plant that grows quickly from seed, does well in cool spring weather, but bolts under long days. Spinach was domesticated long ago in southwestern Asia, but it was not introduced into Europe until historical times.

Figure 21-14 The muskmelon is a berry-type fruit called a pepo, typical of the family Cucurbitaceae. The rind is exocarp and the edible flesh is mesocarp (see Chapter 7). [Courtesy J. C. Allen & Son.]

FRUIT AND SEED VEGETABLES

In the grocery store you will find this group of foods among the "vegetables," although botanically they are truly fruits or seeds. Sweet corn, fresh peas, and green beans properly belong in this category but they have already been discussed among cereals and legumes in the previous chapter. "Vegetable" fruits are almost all grown as herbaceous annuals, although in the tropics several may be perennial. This is their chief distinction from typical grocery-store fruits— the seed-bearing structures of perennial trees or shrubs, such as apples, peaches, and brambles (bananas and strawberries are also regarded as "fruit," although the plants are herbaceous perennials). Obviously, this is a less than ideal categorization.

As with the leafy vegetables, most vegetable fruits are quite watery, and are not as concentrated a source of nutrients as are the grains, legume seeds, or even root crops. They do provide some vitamins and minerals, however, and above all add variety and fla-

vor to the diet. In the tropics a number of these vegetable fruits are grown only locally, either for home consumption or for sale in the village market. In the technically advanced societies, home gardens are less common today, and these "vegetables" are now important summer truck-gardening crops near metropolitan areas. In the United States, most are grown in the South for shipment to the North in winter, but there is some greenhouse production of them for sale in the North as high-priced winter delicacies.

Two plant families stand out as especially important in providing the world with fruit "vegetables." These are the potato family, Solanaceae, which includes the tomato, eggplant, and pepper among its 2000 or so species; and the gourd family, Cucurbitaceae, which includes various melons, squashes, cucumbers, and other pepo fruits (Fig. 21-14) among its 500 species. In both families the species grow rapidly under tropical conditions, but they are treated as summer annuals in temperate climates. The solanaceous plants are erect or sprawling herbs; the cucurbitaceous ones are trailing vines, usually with tendrils and rough, scratchy foliage.

Tomato, *Lycopersicon lycopersicum* (Solanaceae)

The tomato has been an unlikely success story among modern crops. Evidence is compelling that it was indigenous to the northern Andes, where the small-fruited cherry tomato, *L. lycopersicum* var. *cerasiforme*, still grows wild, and where crossing had apparently taken place naturally with other species, such as the currant tomato (*L. pimpinellifolium*) and possibly *L. hirsutum*. The diploid chromosome number is 24. Aboriginal peoples utilized the natural forms at least to some extent, and variants had spread northward into Mexico before the European conquest. The tomato was probably taken thence to Italy and other Mediterranean lands, where it was grown at first as an ornamental. During the 1600s, in France, it was known as the *pomme d'amour,* or love apple, and was considered poisonous.

Eventually, improved selections were reintroduced into North America from Europe, especially from Italy, and today the United States is by far the leading tomato-producing country. The USSR, China, Italy, Turkey, Egypt, and Spain also produce significant amounts. All told, some 50 million metric tons of tomatoes are grown worldwide each year (and this figure does not include home-garden production), which certainly qualifies the tomato as a major crop species. Rather little commercial production takes place in the tomato's original homeland, the Ecuadorian and Peruvian Andes, although the species is cultivated in almost all parts of the world to some extent. It is widely appreciated as a base for sauces and as a flavoring in cooking.

The tomato is a paradoxical plant. It is a perennial species that is grown mostly as an annual, and it yields a fruit that is popularly regarded as a "vegetable." It represents a genus that only a few centuries ago was unknown to the civilized world, but it is now a major world crop. When it first began to be cultivated outside the Americas, it was widely regarded as poisonous rather than edible. The tomato's popularity seems still to be growing; a half century ago, consumption per capita in the United States was less than 10 kilograms, but today it exceeds 25 kilograms.

Containing an abundance of water but relatively little in the way of solids, the tomato fruit is not rich nutritionally. Yet it provides reasonable quantities of vitamins and minerals, and, because of the tremendous tonnage consumed in the United States, the fruit is probably the leading contributor of these nutrients to the American diet—in this respect, it is ahead of oranges and such nutrient-rich vegetables as spinach and broccoli. Its popularity stems as much from its attractiveness as from its flavor and its nutritional qualities: What is a fresh salad without a garnishment of sliced ruby-red tomato?

The tomato has proven especially amenable to breeding for various shapes, sizes, and even colors. This plasticity is largely responsible for its economic success, because it makes the species adaptable to mechanized growing and harvesting (about which more will be mentioned in a moment). Its potential for improvement through breeding has been only slightly plumbed, and it seems possible that further genetic investigation will provide a wealth of "improved" characteristics, from greater salt tolerance (gained from *L. cheesmanii*, which is found on the shores of the Galapagos Islands) to greater pest resistance, and all sorts of variations that can arise from induced mutations. It has been relatively simple to change the originally open-pollinated tomato into self-pollinated true-breeding cultivars, simply by selecting for short style length so that the stigma is not exposed beyond the staminal column.

Stupendous advances have come from tomato-breeding programs. Yields of cannery tomatoes in California have increased four-fold within a few decades, to more than 50 metric tons per hectare (about 22 tons per acre), which is more than double the average yield worldwide. Hybrid F_1 cultivars are quite practicable, because of the ease in manipulating the macroscopic flowers and the abundance of seed from a single fruit. F_1 hybrids are often planted in home gardens in the United States, and in technically advanced lands such as Japan most of the commercial crop is now grown from hybrid seed.

Adaptation of the tomato to mechanization has been spectacular, especially in California and Florida. In California, where the crop is mostly processed, small, compact, determinate cultivars have been bred that are amenable to mechanical harvesting. Today almost all of the California crop is gathered by machines, from more than 100 thousand hectares (about 250 thousand acres). Even tomatoes for the fresh market, including much of the Florida winter crop, have been bred better to withstand handling, shipping, and distribution. (Many consumers lament, however, that these concerns seem to overshadow flavor and quality considerations, which are paramount in home-garden plantings.)

The tomato is temperamental about cold weather. Nor does it thrive on poorly drained soils, if it lacks adequate sunlight, or if it is grown under excessive nitrogen fertilization. In the tropics, plants may be left to grow indefinitely, but in temperate climates, where most of today's production takes place, tomatoes are treated as annuals. Traditionally seed was started in nurseries—usually in a greenhouse, in locations where the growing season is short (Fig. 21-15)—and the seedlings were later transplanted to the field. More recently, in southern latitudes at least, seed has been drilled directly into the soilbed, just as grains are planted. A well-prepared soil is necessary,

Figure 21-15 Greenhouse-grown tomato trained to string, showing fruit borne on trusses along the stem. [Courtesy L. Hafen.]

The cannery crop in California is produced on specially bred, determinate cultivars on which most of the fruit ripens concurrently. A self-propelled harvesting machine cuts off the entire plants with a blade that travels slightly below ground level, lifts and shakes the stems into a system of eccentrically moving horizontal rods from which the soil and debris is dropped to the ground and the fruit is diverted to canvas belts at the side of the machine where it can be sorted. In Florida, systems have been developed to direct-seed market tomatoes mechanically through a plastic mulch, giving more uniform growth and higher yields than with tomatoes seeded into unmulched beds. Plants are harvested by machine without disturbing the mulch, which allows for multiple cropping. Differences in tomato maturity, accounting for about 30 percent of the loss of potential yield, are said to be reduced by sowing pregerminated seeds through the plastic mulch.

Eggplant, *Solanum melongena* (Solanaceae)

The eggplant is believed to have been domesticated in northeastern India, and perhaps secondarily in China. It was brought by early traders to Europe, and by the Portuguese colonists to Brazil. The eggplant is an important food plant in Japan and other parts of the Orient, where it is often pickled. It is not as popular in western societies, but it is featured in French and Italian cookery. Oftentimes it is grown more as an ornamental; the oblong, round, or pear-shaped fruits are dark purple, white, tan, or green. The species is well adapted to "tropical" growing weather, requiring a long hot season to mature the fruit. In temperate climates, plants are often started in hotbeds and transplanted outdoors in late spring.

Peppers, *Capsicum* (Solanaceae)

The *Capsicum* peppers are not to be confused with spice pepper (*Piper nigrum*), the familiar condiment of the pepper shaker. Columbus found the New World *Capsicum* fruits quite as pungent as the pepper he sought in the Orient. Unfortunately, the misnomer "pepper" stuck for these fruits, although they

and special weed-control measures must be taken. Trellising is seldom practiced anymore, except with the limited growing of tomatoes in the greenhouse during winter, where space is at such a premium. Since fully ripe tomatoes are quite perishable, those meant for market are generally picked as they begin to turn pink, becoming red-ripe by the time they reach the point of sale or the processing plant. Ethephon sprays hasten ripening if such is needed.

are better termed *chilis* (from the Nahuatl word *chilli*) to distinguish them from true pepper.

After the discovery of America, *Capsicum* peppers were quickly introduced into other parts of the world. Some were grown as ornamentals, but the genus became best known for the pungent "spicy" types used in making "hot sauces." The less pungent "sweet" (green or red) peppers were grown but apparently not much esteemed as food, and are listed among the "medicinals" in many herbals. John Gerard's famous herbal of 1597 characterizes chili peppers as fruits that "warmeth the stomacke and helpeth greatly the digestion of meates."

As a food, green peppers are important for the generous amounts of vitamins A and C that they contain. They are eaten in salads, as hors d'oeuvres, or stuffed. The fruits of one type, when allowed to ripen until red, are known as "pimentos" (used chiefly for stuffing olives, mixing into cheeses, or in cooking and processing). Paprika, cayenne pepper, and chili powder all come from grinding the dried ripe fruit of certain *Capsicum* fruits. Paprika is used largely as a colorful garnish, and is produced mostly in Spain. Most chili powders include several other flavorings, such as oregano, cumin, garlic, and onion. *Capsicum* peppers still find their greatest use in Latin America. The highly pungent types are much favored for making the "hot" sauces so familiar south of the border. The pungent ingredient of *Capsicum* peppers is capaicin, a volatile phenol similar to vanillin; capaicin is so potent that it is said to be detectible to the taste in a dilution of one part in a million.

Some of the strongly pungent peppers, especially those grown in Africa, are still used medicinally. They are said to be useful for rheumatism, neuritis, and diarrhea. And the modern florist, at Christmas season, grows a dwarf form of *Capsicum annuum* as an ornamental.

Exact classification of the *Capsicum* peppers is quite confused. Three species, *C. annuum*, *C. frutescens*, and *C. chinense* probably account for most of the cultivated forms. But many "species" have been described on the basis of such varying characteristics as fruit shape and size. Remains of chilis 9000 years old have been recovered from Mexi-

can caves. The diploid number of chromosomes found in all cultivated species is 24, and sterility barriers between species appear to be well developed. *Capsicum* peppers require warm, moist weather and a long growing season to be productive, and considerable hand labor is required for their tending.

The Cucurbits (Cucurbitaceae)

The several genera and species of the family Cucurbitaceae that constitute the squashes, melons, pumpkins, and so on, can well be discussed together. They are not of great importance to world commerce, but they are mainstay foods in many primitive, tropical societies. Species of *Cucurbita* were major constituents of the diets of aboriginal peoples of Mexico and Central America. Squashes and pumpkins were not only eaten as fresh vegetables, but the matured fruits (which store reasonably well) could, like grain, be boiled or roasted after storage. The seeds of some were prized more than the fleshy parts, and were usually eaten roasted. It well may be that early domestications in the New World were grown more for the seed than for the fleshy pepo. Moreover, various gourds provided primitive household utensils.

Not all of the cucurbits of importance are of New World origin. Among the Old World contributions are the muskmelon (*Cucumis melo*), the gherkin (*C. anguria*), the teasel gourd (*C. dipsaceus*), the balsam pear, (*Momordica charantia*), the watermelon (*Citrullus lanatus*), and the sponge gourd (*Luffa aegyptiaca*) the vascular tissues of whose fruits may serve as a filter or sponge. All of these have been introduced into the New World, and some have naturalized widely throughout the tropics. The bottle gourd (*Lagenaria ciceraria*) is probably native to Africa, but it was spread throughout the world tropics many thousands of years ago, possibly by ocean currents.

Several cultivated cucurbits of the New World have been and continue to be important. *Cucurbita pepo* includes summer squashes and pumpkins; *C. mixta*, certain winter squashes, pumpkins and cymlings; *C. moschata*, other winter squashes and pumpkins; *C. maxima*, winter squashes and pumpkins; and *C. ficifolia*, the figleaf gourds. Except for the South American *C. maxima*, these were widely utilized in the primitive agriculture of the southern Mexican–Central American area. They may have arisen from or hybridized with several wild species also found in this region. Archeological evidence shows that *C. pepo* was in use in Mexico 9000 years ago and that *C. moschata* was used at least 7000 years ago.

Cucurbits of particular importance today include the ubiquitous balsam pear (*Momordica*); the tropical American chayote (*Sechium edule*), with its soft-spined, one-seeded fruit, which is cooked like a squash; the cucumber (*Cucumis sativus*), unknown in the wild but possibly originating in northern India; the gherkin, the small, prickly fruits of which are used mostly for pickles; the muskmelon (cantaloupe) and honeydews (*Cucumis melo*), available in various forms, such as casaba, snake, banana, and other types, possibly indigenous to Africa; the autumn pumpkin (*Cucurbita pepo*), generally consumed boiled or fried after mashing (in America mostly in pies); and the squashes. The more richly flavored squashes, such as *Cucurbita maxima*, generally substitute for the pumpkin in pumpkin pie; summer squashes (Fig. 21-16), which are similar to pumpkins, are usually eaten at once; winter squashes, which are allowed to develop a hard rind that aids in their keeping, are generally later baked or boiled ('Blue Hubbard,' 'White Cushaw,' 'Butternut,' and the like). The watermelon (*Citrullus lanatus*) presumably originated in Africa, where it grows wild and is often an important source of water during drought.

All of these are rank, trailing vines that are rough to the touch, although "bush" forms of squash and cucumber are being bred to facilitate harvest. Gynoecious strains of cucumber are often used in commercial plantings, and the fields are kept clean of weeds with herbicides. Over most of the world, however, cucurbits are planted to well-cleared and cultivated fields and then allowed to care for themselves. Cucurbits are warm-weather plants very susceptible to frost. Most are easily grown, thrifty, vigorous, and widely adapted. At one time, pumpkins were often interplanted with corn in the United States, but today most cucurbits are grown as specialized truck crops, and the melons and pumpkins are shipped widely to distant markets.

Several cucurbits grown in the tropics are apt to

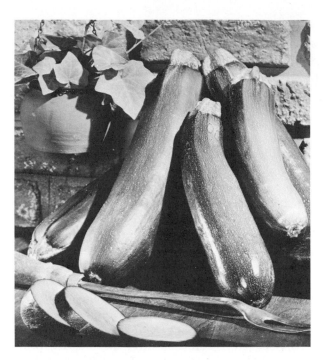

Figure 21-16 A popular summer squash, zucchini 'Aristocrat.'

assume greater importance in view of mushrooming world populations. Among these are the casabanana (*Sicana odorifera*), the fuzzy gourd or Chinese preserving melon (*Benincasa hispida*), the doan gwa or serpent gourd (*Tricosanthes anguina*), and the achocha (*Cyclanthera pedata*). Some of these are now being offered as garden curiosities in the United States, as "oriental vegetables." *Cucurbita ficifolia* is a high-altitude, short-day plant of Central America and the Andes, with large, watermelonlike pepos; it is now much grown in the Philippines.

Okra, *Abelmoschus esculentus* (Malvaceae)

Okra, or gumbo, is the young pod that quickly follows flowering of this plant, a close relative of *Hibiscus*. It is believed to have originated in northeastern Africa, and it is reportedly found wild in the upper Nile area. It has been used in northern Africa for centuries, and was introduced into the New World in the slave-trading days, where it found exceptional popularity in the French and Creole cookery of Louisiana. Most okra is still grown as a home-garden plant. The pod, which turns quite slimy on cooking, is most often used in stews, soups, and gravies. Handled much like okra is the pod of the unicorn plant (*Proboscidea louisianica* of the family Martyniaceae), which is often used for pickling in the southwestern United States and Mexico.

FRUIT CROPS

Fruits of perennial plants have been used as food since before the dawn of civilization, and they are perhaps more popular than their nutritional quality justifies. No doubt ancestral primates gathered wild fruits as a not inconsiderable part of their food supply. (A sampling of wild fruits native to North America is given in Table 21-2.) Symbolic of our early admiration of fruit is the Garden of Eden scene and its tempting fruit. Five of the Biblical seven species (Deuteronomy chapter 8, verse 8) are fruit trees. Fruit crops today, considered together and by weight, compare favorably with the world's staple agricultural crops.

In spite of their popularity, fruits in general are relatively unimportant to human sustenance (banana, coconut, and the date are exceptions). Most fruits are quite juicy (watery) and have low concentrations of nutrients, but some are excellent sources of vitamin C. Like many vegetables, however, they do add much appreciated variety and flavor to the diet. Indeed, most fruits have been selected to titillate the palate, and they commonly serve as a dessert or special treat.

A few fruits are more or less local staples. This is true of the banana and the coconut in parts of the tropics; avocados have been a prime source of nutrition in sections of the New World, and dates have been gathered in arid parts of the Old World since time immemorial. Indicative of the political importance attached to fruits, especially for sustenance in the tropics, is the voyage of the H.M.S. *Providence*, 1791–1793, under the command of the famed Captain William Bligh (whose first try with the *Bounty* resulted in failure because of the historic mutiny). Although organized principally for collecting breadfruit (*Artocarpus altilis*) from the South Pacific for

planting in the British colonies of the New World, records show that the *Providence* introduced scores of exotic species and was a veritable floating plant nursery.

Most fruits are preferred eaten raw. They become quite perishable once the climacteric (Chapter 8) is reached. Thus fruits have traditionally been preserved through a variety of treatments, such as drying in the sun, cooking in sugary syrups (candied fruits and jellies), "fumigating" with sulfur dioxide (prunes, maraschino cherries), and of course by canning and freezing. Technically advanced parts of the world now enjoy rapid transportation, and increasingly fruits can be marketed fresh even though they may be grown far from the market. A few, such as the coconut and citrus, with something of a protective rind, handle and ship reasonably well. Unfortunately, many of the most delicious fruits, such as the tropical cherimoya (*Anonna cherimola*) and mangosteen (*Garcinia mangostana*), are very delicate and can hardly be transported efficiently even by air.

In the constant search for better shipping cultivars, flavor is often sacrificed, as with the banana and the avocado, of which the best-tasting tree-ripened cultivars never reach temperate-zone markets. Even within a limited market area, poor-shipping cultivars, such as 'Belle of Georgia' peach (thought by many to be the best-tasting peach ever discovered), have had to give way to more marketable ones. Unfortunately, most fruits do not adapt well to freezing; they lose their form and substance upon thawing, and consequently this modern technique for providing distant markets with tasty perishables is not of great service as far as fruits are concerned.

Since a fruit is the normal reproductive structure of flowering plants, it is no cause for wonder that an almost unlimited selection of fruits occurs in the wild, from which people have chosen domesticates. Indeed, particularly in the tropics, a wide assortment of wild species is still important to local populations. For example, the wayfarer in Brazil escapes from the heat by stopping under the umbu tree (*Spondias* sp.), not only for its shade but to refresh himself on the tart fruit. In temperate climates the collecting of wild fruits and berries, once commonplace in rural societies, has now largely given way to mass marketing of specially grown cultivars. Even so, it is surpris-

Table 21–2   A sampling of North American wild plants, the fruits of which can be used for food.

Amelanchier spp., juneberry; shadbush—raw, cooked, or dried.

Annona glabra, pond apple—raw or made into jelly.

Arctostaphylos spp., manzanita; bearberry—raw, cooked, or dried and ground; also jelly, syrup, or cider.

Ardisia escallonioides, marlberry.

Asimina triloba, papaw—raw or cooked.

Chiogenes hispidula, creeping snowberry; birchberry.

Chrysobalanus icaco, cocoplum—raw or cooked.

Chrysobalanus pallidus, gopher apple.

Chrysophyllum oliviforme, satinleaf.

Coccoloba uvifera, seagrape—raw or as jelly.

Cornus spp., dogwood; bunchberry—raw or cooked.

Crataegus spp., haw—raw, cooked, or dried.

Diospyros virginiana, persimmon—raw, cooked, or dried and ground as meal.

Elaeagnus commutata, silverberry—raw or cooked.

Empetrum nigrum, crowberry—raw, cooked, or dried.

Ficus aurea and *F. laevigata*, wild fig.

Fragaria spp., strawberry—raw or cooked.

Gaylussacia spp., huckleberry—raw or cooked.

Gleditsia triacanthos, honey locust—fruit pulp raw; pods fermented for beer.

Lonicera spp., honeysuckle—raw, cooked, or dried.

Lycium spp., wolfberry—raw or boiled.

Malus spp., crabapple—raw, cooked, or preserved.

Mitchella repens, partridgeberry.

Morus rubra, red mulberry—raw, cooked, or preserved.

Nyssa spp., Ogeechee lime; tupelo—preserved.

Parthenocissus quinquefolia, Virginia creeper.

Passiflora incarnata, maypop—raw or as jelly.

Photinia salicifolia, Christmasberry—raw, roasted, or dried and ground for mush.

Physalis spp., groundcherry—raw or cooked.

Podophyllum peltatum, May apple—raw or cooked.

Prosopis spp., mesquite; raw, or ripe pods boiled for syrup, or dried and ground for meal, or used for beverage, fresh or fermented.

Prunus spp., cherry; plum—raw, cooked, or dried.

Reynosia septentrionalis, Darling-plum—raw or cooked.

Rhamnus spp., buckthorn—some species cathartic.

Rhus spp.; red-fruited are not poisonous; avoid *R. vernix* and other species with white fruit.

Ribes spp., currant; gooseberry—raw, cooked, or dried.

Rosa spp., rose—raw, stewed, or as jelly.

Roystonea elata, royal palm—raw, boiled for juice, sweetened.

Rubus spp., blackberry; dewberry; raspberry—raw, cooked, or preserved.

Sambucus spp., elderberry—raw, cooked, dried, or made into jelly or wine.

Shepherdia spp., buffaloberry—raw, cooked, or dried.

Smilacina spp., false Solomon's-seal.

Vaccinium spp., blueberry; cranberry; deerberry; whortleberry—raw, cooked, or dried.

Viburnum spp., black haw—raw, cooked, or dried.

Vitis spp., grape—raw, cooked, or dried.

Yucca spp., yucca; Spanish bayonet—unripe fruit cooked; ripe fruit raw, cooked, or made into paste.

Source: J. F. Morton, *Economic Botany* 17(4):319–330.

ing how relatively little most fruits have changed from the ancestral types, compared with the tremendous change that domestication has wrought on such food staples as the cereals. Over the ages new forms have arisen more from chance crossings or mutations than from genetic design. Breeding programs have dealt mostly with improving characteristics already pretty well established in earlier selections; the reasons are obvious if one considers the time necessary to undertake large-scale selective crossing with long-lived perennials.

Tropical fruits especially offer an exceedingly wide and diverse assortment. The numerically fewer temperate fruits are perhaps better known and more

Table 21-3 Food fruits familiar in village markets of Central Mexico.

Species	Common name	Family
Acrocomia mexicana	Palm fruit	Palmae
Ananas comosus	Pineapple	Bromeliaceae
Annona cherimola	Cherimoya	Annonaceae
Annona reticulata	Custard apple	Annonaceae
Bumelia laetevirens		Sapotaceae
Carica papaya	Papaya	Caricaceae
Casimiroa edulis	White sapote	Rutaceae
Citrus aurantiifolia	Lime	Rutaceae
Citrus paradisi	Grapefruit	Rutaceae
Citrus reticulata	Tangerine	Rutaceae
Citrus sinensis	Sweet orange	Rutaceae
Diospyros digyna	Black sapote	Ebenaceae
Hylocereus undatus	Cactus fruit	Cactaceae
Mangifera indica	Mango	Anacardiaceae
Manilkara zapota	Sapodilla	Sapotaceae
Musa spp.	Banana	Musaceae
Passiflora ligularis	Passion fruit	Passifloraceae
Persea americana	Avocado	Lauraceae
Physalis ixocarpa	Tomatillo	Solanaceae
Pouteria campechiano	Yellow sapote	Sapotaceae
Pouteria sapota	Mamey sapote	Sapotaceae
Psidium guajava	Guava	Myrtaceae
Punica granatum	Pomegranate	Punicaceae
Pyrus communis	Pear	Rosaceae
Malus pumila	Apple	Rosaceae
Tamarindus indica	Tamarind	Leguminosae

Source: T. W. Whitaker and H. C. Cutler, *Economic Botany* 20 (1):6–16.

thoroughly investigated as horticultural crops. However, very specialized industries have developed in the temperate zone to grow and market there such tropical fruits as the banana, pineapple, and citrus. The volume of citrus and of bananas entering world commerce exceeds apple production: that of citrus is some 50 million metric tons annually, that of bananas nearly 40 million metric tons. But a temperate fruit, the grape, leads all others in total production (more than 60 million metric tons annually), largely because of its use in making wine. The coconut, at more than 30 million tons, ranks just behind grapes, citrus, and bananas, but, because it is used mainly as a source of copra, it will be discussed later in a chapter dealing with extractives. Olives, another major subtropical fruit, are grown mainly for processing to oil, and will also be left for later discussion. Pears, peaches, apricots, plums, pineapples, dates, cherries, and figs rank next, essentially in this order of importance, among the fruits in world commerce. A tremendous number of fruits are found locally in tropical markets (Tables 21-3 and 21-4).

Table 21–4   Fruits commonly grown in India.

Species	Common name	Family
Ananas comosus	Pineapple	Bromeliaceae
Annona squamosa	Sweetsop	Annonaceae
Carica papaya	Papaya	Caricaceae
Citrullus lanatus	Watermelon	Cucurbitaceae
Citrus paradisi	Grapefruit	Rutaceae
Citrus reticulata	Mandarin	Rutaceae
Citrus sinensis	Sweet orange	Rutaceae
Cucumis melo	Muskmelon	Cucurbitaceae
Eriobotrya japonica	Loquat	Rosaceae
Ficus carica	Fig	Moraceae
Grewia asiatica	Phalsa	Tiliaceae
Litchi chinensis	Litchi	Sapindaceae
Malus pumila	Apple	Rosaceae
Mangifera indica	Mango	Anacardiaceae
Manilkara zapota	Sapodilla	Sapotaceae
Musa spp.	Banana	Musaceae
Phoenix dactylifera	Date	Palmae
Prunus armeniaca	Apricot	Rosaceae
Prunus domestica	Plum	Rosaceae
Prunus persica	Peach	Rosaceae
Psidium guajava	Guava	Myrtaceae
Punica granatum	Pomegranate	Punicaceae
Syzygium cumini	Jambolan	Myrtaceae
Vitis vinifera	Grape	Vitaceae

Source: P. Maheshwari and S. L. Tandon, *Economic Botany* 13 (3):205–242.

Grape, *Vitus vinifera* (Vitaceae)

Various species of the genus *Vitis* are known in the wild in many areas, both in the Old World and in the New World. Of chief importance today is the "bunch" or wine grape (*V. vinifera*) from the Old World, with 38 chromosomes. In many areas of the world, however, cultivars of this species must be grafted onto rootstocks of New World species to prevent devastation by the phylloxera root louse. Europe accounts for more than half of world grape production, which amounts to some 60 million metric tons annually (principally from Italy and France). As much as 90 percent of the European crop is used for wine. About a tenth of the crop is dried for raisins.

Vitis vinifera has long been domesticated. It apparently represents selection of bigger-berried, sweeter types of *V. silvestris*, a wild grape from Asia Minor and the Mediterranean basin northward along the Rhine and Danube valleys. *V. vinifera* and *V. silvestris* cross readily, and natural hybrids are often found near vineyards. Wild grapes, however, are dioecious (a trait determined by three alleles), a characteristic bred out of the domesticated species (which is hermaphroditic and self-fertile). Grape artifacts have been discovered with the earliest human remains in southcentral Europe and the Near East, with the earliest quite like *V. silvestris* and the later ones more like *V. vinifera*. Ancient Egyptian inscriptions indicate that the grape was widely grown by 2375 B.C.E. The Bible mentions the grape as having been planted by Noah; apparently it was introduced into Palestine from the north about 5000 B.C.E. Wine was a popular beverage in the Holy Land during Biblical times, and it is cited in practically every book of the Bible. Concoctions that include wine have also served as medicinals, pickling media (as for olives and vegetables), and even dyes for clothes. Grapes in summer and raisins in winter were familiar foods. In general the vineyard was a symbol of wealth and plenty.

The grape was equally as important in ancient Greece. Herodotus (484–425 B.C.E.) mentions the tremendous commerce in wine sent to Egypt. During the hegemony of Rome, terraced vineyards were widely established. Preserved records include instructions for the care of vineyards—clearing away stones, hoeing, manuring, pruning, primitive disease protection, and provisions for irrigation. Today *V. vinifera* remains important throughout Greece (and nearby Turkey), where grafted cuttings of select cultivars are generally planted 2 meters (about 6 feet) apart, about 3000 vines per hectare (about 1200 vines per acre), in holes punched in the soil. Most of southern Greece is phylloxera-free, but New World rootstocks are generally used as a precaution in newly planted vineyards. Vines are generally pruned to remain low, and are staked rather than trellised, or sometimes simply left to trail on the ground. Since hand labor is often inexpensive, girdling (removal of a ring of bark below a grape cluster into which food from the leaves above is to be directed) or its equivalent is not uncommon. Most cultivation is till done by hand.

In North America, centuries before Columbus, Indians were gathering fruit of such species as the fox grape (*V. labrusca*), the riverbank grape (*V. riparia*), and the scuppernong (*V. rotundifolia*). The Norse name *Vinland* lends credence to the claim that North America was discovered by Norsemen. Early settlers were impressed by the profusion of wild grapes, but for 200 years attempts to introduce and grow the more select *V. vinifera* from Europe failed. During these attempts there may have been some natural crossing between *V. vinifera* stock and the native American species. In 1852, E. W. Bull of Concord, Massachusetts, exhibited a seedling grape, which he named 'Concord,' that was to remake the grape-growing industry in the New World. 'Concord' came from a seed borne on a seedling of uncertain origin, but Bull speculated that it might have been a cross between a European grape gone wild and a *V. labrusca* 'Catawba' in the garden. In any event, it was far superior to cultivars then available, and even today 'Concord' is still the leading "juice" grape in the northeastern growing areas.

The greatest flowering of viticulture in America, however, awaited the settlement of California. The earliest grapes planted in California were apparently brought from Spain through Mexico by the Jesuit Fathers during the 1600s. The several California missions all had thriving vineyards before 1800. The 'Mission' grape, of obscure ancestry, was widely introduced and is still much grown in California. From 1850 on, with viticulture flourishing, many varieties of *V. vinifera* were introduced from France and other parts of Europe. During the 1960s "French hybrid" grapes (crosses of *V. vinifera* with such American species as *V. rupestris* and *V. lincecumii*, made by French hybridizers) were introduced into the midwestern United States with some success, reviving interest in grape growing there.

In the dry Central Valley of California, where phylloxera and mildew diseases do not have to be contended with, very efficient grape growing is possible on large irrigated ranches. Select varieties are planted at the rate of 1000 to 1500 vines per hectare (about 400 to 600 vines per acre) and trained to standardized trellises, with weeding and irrigation carefully controlled. Recently there has even been chemical management of fruiting: gibberellin sprays, properly timed, can substitute in large measure for tedious hand girdling to yield especially full clusters of large grapes for the fresh-fruit market. Although grapes must be hand pruned and harvested, irrigation and other growing procedures are increasingly being automated. Mechanical harvesting is past the experimental stage.

Figure 21-17   Abundant fruiting in the grape. [Courtesy Garth Cahoon.]

Grapes, perhaps more than many fruits, do their best only under intensive care. Although there are many varieties today, including crosses of *V. vinifera* with American species that provide suitable cultivars for most regions, the grape does require a long summer and winters mild enough—with temperatures generally above −18°C (0°F)—to avoid winterkill. Grapes reach best quality when the weather is bright and dry; the sugar content and the various subtle flavorings are then at their peak. Grapes are propagated chiefly by cuttings, and the vines perform best on well-drained soils. As has been noted, the tendency is to graft scions of *V. vinifera* onto phylloxera-resistant rootstocks of American origin. Most modern varieties are self-fertile, but some yield finer fruit when cross-pollinated with other varieties. Grapes flower and fruit on new wood (Fig. 21-17). In the United States, canes are generally pruned back to two or four paired branches trained to a wire trellis, on which about 30 buds are allowed to mature into new shoots, each of which may bear two or three bunches of grapes. Depending upon cultivar, use,

and market conditions, there may be thinning to provide especially select fruit. Wine making is discussed in Chapter 22.

Raisin production, of course, involves the drying of grapes, usually in the sun, to some 20 percent of fresh-fruit weight. Traditionally, raisin grapes ('Thompson Seedless' grapes are mainly used in the United States) were dipped in solutions of soda or oleate better to preserve the skins and to facilitate drying, but more recently a boiling-water bath followed by hot-air dehydration has proven preferable. Nearly 1 million metric tons of raisins (equivalent to about 5 million metric tons of fresh grapes) are produced annually, mostly in Turkey, Greece, and California.

Banana, *Musa* spp. (Musaceae)

The taxonomy of the banana is confused. Most cultivated bananas are referred either to *Musa acuminata* (= *M. cavendishii*), or to *M. paradisiaca* (= *M. sapientium*) which is apparently a hybrid between

M. acuminata and *M. balbisiana*. *M. acuminata* and *M. balbisiana* are diploid sources of genomes *a* and *b*, respectively, in crosses yielding triploid hybrids of genomic constitution *aab* or *abb*, such as 'French Plantain.' Triploid *M. acuminata* embraces the flavorful ladyfinger bananas and such Philippine cultivars as 'Lacatan.' A tetraploid from this source, with genomic constitution *aaaa*, is the cultivar 'Golden Beauty.' Diploid *M. acuminata* grows wild in the rain forests of Malaysia, and seems prone to production of autotriploids with bigger, more edible (parthenocarpic) fruits, which were probably the first cultivated bananas. Crosses with *M. balbisiana* occured in India to produce *M. paradisiaca*, from which other edible selections were made that, propagated vegetatively, spread throughout the world. The *b* genome, itself not a source of flavorful raw bananas, contributes hardiness, and today's cooking bananas or plantains have various combinations of the *a* and *b* genomes. A related species, abacá (*M. textilis*), is grown for the petiole fiber rather than the fruit.

Most of the fruit marketed until recent years has been the 'Gros Michel' cultivar of *M. acuminata*, also known as the "Poya banana" (so called for its discoverer, Jean François Pouyat, a Jamaican planter who came upon this triploid find in 1836 while strolling through a banana farm in Martinique). The 'Gros Michel,' being reasonably flavorful, of good size, and especially durable for shipping, was long the only banana available to temperate-zone people the world over. Since 1963 new triploid cultivars of *M. acuminata*, including 'Valery' and West Indian 'Lacatan,' have gradually replaced "Big Mike" (as the 'Gros Michel' is known) in importance, because of their better resistance to the dread Panama disease. The small, thin-skinned ladyfinger bananas, generally thought to be more flavorful, will not stand shipping to distant markets. The plantains are rather tasteless, large, green-skinned or red-skinned fruits that are usually baked, boiled, or fried rather than eaten raw, and they are mainly of local importance in the tropics.

As noted, the banana is believed to have been first domesticated in Southeast Asia millennia ago, although commercial banana production has now largely gravitated to the Western Hemisphere. The

banana had been introduced from the Far East into Africa by Arabian traders before historical times; it was encountered there when that continent was first explored in the fifteenth century. The name "banana" apparently became first applied to the fruit in Sierra Leone.

No doubt the banana was introduced into tropical America soon after Columbus' first voyage. Today Latin America accounts for more than half of the recorded world production of bananas, totaling approximately 37 million metric tons. In the Far East and Africa, some bananas are grown for export, but much of the crop is consumed locally (in some places the banana being a staple in the diet). Brazil leads the world in banana production, followed by India, Indonesia, and Central America.

Although some banana plants grow as tall as 9 meters (nearly 30 feet), botanically they are herbs rather than trees, being without woody tissue. The plant is propagated from a fleshy rhizome that produces a cluster of gigantic leaves, each nearly 3 meters (about 10 feet) long. Growth is rapid, and in less than a year a central, pendulous inflorescence forms, which, upon fruiting, terminates the plant's life (Fig. 21-18). About 14 months after planting, the stalk of bananas is ready for harvesting. Commercial bananas have sterile flowers, which develop unfertilized and yield fruit lacking seeds. Thus, perpetuation of the plant depends entirely upon suckers arising from the rhizome.

Most dwellings in the humid tropics have a few banana "trees" constantly in the dooryard, an aid to local subsistence. But the banana is also a remunerative cash crop in the steamy tropical lowlands. On good land it can yield more than 22 metric tons per hectare (about 10 tons per acre) annually, with reasonably little attention. Large-scale growing is generally of the shifting-agriculture pattern; forest is felled, the vegetation is burned or allowed to rot, and rhizome sections are planted by hand. Efficient plantations schedule plantings to permit cutting fruit the year around. Bananas are harvested green, since they are quite perishable when ripe (yellow). Experience has indicated just the proper stage for cutting, so that the green bananas, loaded onto refrigerated steamers for transport to temperate-zone markets, ripen in just

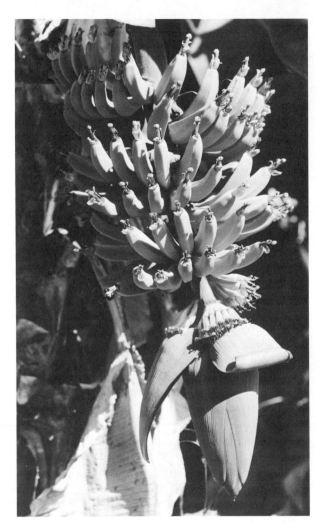

Figure 21-18 The fruiting stalk of the banana consists of many clusters called hands. Each female flower of the basal end of the stem develops into a fruit—the "finger." Male flowers are produced toward the apex of the stem. [Courtesy J. C. Allen & Son.]

cluding whole villages) and privately owned railroad and steamship lines for bringing the fruit to market. In recent years these "banana empires" have retrenched because of the growth of nationalism, and there is increasing tendency to buy fruit from native growers. Even the Indians living in isolated villages, such as in southern Panama and western Ecuador, may bring their green bananas by canoe to larger trading boats that ply the coast on regular schedules to carry produce to the larger ports.

Large land-clearing and leveling machinery may be used on well-capitalized plantations to prepare planting sites. Cultivation to restrain weeds until the bananas form a tight canopy is often a hand operation. On plantations, natural rainfall may be supplemented by irrigation, and fungicidal spraying is commonly practiced. Nowadays, on managed plantations, young stems of bananas are usually covered with a blue-tinted polyethylene plastic "sleeve"; this accelerates growth, resulting in a weight gain for the stem of about 2.5 kilograms (about 5.5 pounds) more than the stem of an uncovered plant, and it also affords some protection against insect and bird damage to developing fruit. Harvesting is by hand: a worker with a knife on a long pole lops the banana plant, settling the fruit cluster onto the shoulder of a second man, who is to carry it, and then severs the inflorescence from the rest of the plant. The carrier may take the bunch to a mule, small rail car, canoe, or other means of transportation for delivery to a major port. There the bananas may be washed, and are carefully loaded into air-conditioned ship holds by means of canvas conveyors or other sophisticated means to avoid bruising. There is even some individual packaging of clusters in cardboard cartons before shiping. The bananas may be treated with an antioxidant and a postharvest fungicide, too. During the long haul to market the temperature in the ship's hold is maintained at 14°C (57°F), the ideal temperature for preventing premature ripening. Where bananas are shipped in polyethylene bags, an atmosphere enriched in carbon dioxide and reduced in oxygen may be maintained, which prevents premature ripening and molding. A few days at 21°C (70°F)—or at 15.5–18.5°C (60–65°F) if ethylene is present in the atmosphere—causes the bananas to

a few days when brought to room temperature in the food store. Unfortunately, the flavor of such bananas is no match for that of tree-ripened fruit in the tropics.

Early in the twentieth century banana companies in the American tropics developed into very large, thoroughly organized corporate empires, with extensive plantation lands in diverse countries (often in-

Figure 21-19   Picking navel oranges from trees planted in 1913 in an orchard in Visalia, California. [Courtesy USDA.]

turn from green to yellow and the starch to hydrolyze to fruit sugars. Ripening with ethylene, from other fruit or from chemical sources, promotes shelf life and quality. The ripened banana is a wholesome and fairly well balanced source of nutrition, containing various minerals and vitamins, as well as a high complement of carbohydrate with some oil and a little protein.

Citrus, *Citrus* spp. (Rutaccac)

In the genus *Citrus* are the many familiar citrus fruits, of which the orange, tangerine, lemon, lime, and grapefruit are perhaps the most familiar. Sweet oranges are classed as *C. sinensis;* lemons, *C. limon;* limes, *C. aurantiifolia;* mandarin orange, or tangerine, *C. reticulata;* grapefruits, *C. paradisi;* citrons, *C. medica;* pomelos, pummelos, or "shaddocks" (after a Captain Shaddock), *C. maxima;* and sour oranges, *C. aurantium.* There are several other species that are seldom used for food. All except the grapefruit (which apparently arose from hybridization between the orange and the pomelo during the 1700s in the West Indies) seem native to Southeast Asia, where the more useful fruit types were selected by primitive people before historical times. Undoubtedly, a number of hybridizations are represented. All species have a diploid chromosome complement of 18, and cross with one another.

Most citrus is grown today in tropical and subtropical America, although there is considerable production in southern Europe, Japan, North Africa, and the Near East. Oranges and tangerines are the most popular of citrus fruits, totaling about 33 million metric tons annually, about twice the production of all other citrus fruits combined. The leading producing areas are the subtropical parts of the United States, Brazil, Japan, Spain, Italy, Argentina, Mexico, and Morocco. Grapefruits find their market almost solely in the United States. Lemons, limes, and other citrus fruits (such as sour orange) are grown mostly in Italy and the United States. Worldwide, recorded citrus production amounts to about 50 million metric tons annually.

Citrus fruits are borne on relatively small wiry trees, many of which apparently depend upon root mycorrhizae. Most are temperamental about soils, frequently suffering from trace-element deficiencies, excessive alkalinity (in Texas and California), nematode attacks, and similar afflictions. In the United States they do well on the sandy soils of south-central Florida, and in sunny southern California and southern Texas (Fig. 21-19). In these areas, highly mechanized orcharding is practiced, often with irrigation, pest control, protection against winter frost, and selection of improved cultivars. Intensive growing is also practiced in Japan, where the mandarin orange, or tangerine, is much esteemed. In less advanced parts of the tropics, however, citrus growing is rather casual, with occasional trees of doubtful origin planted in the home garden. In western Africa, where citrus trees should do well, fruit may be harvested only from chance trees that get little if any attention.

The citrus fruit, technically a **hesperidium,** is an excellent source of vitamin C and various fruit acids. It is approximately 40 to 45 percent juice, 20 to 40 percent rind, and 20 to 35 percent pulp and seeds. Chemically it is 86 to 92 percent water, 5 to 8 percent sugars, 1 to 2 percent pectin, with smaller quantities of acids, protein, essential oils, and minerals. The rind is full of oil glands, and it is the rind of the sour orange that is the part esteemed for production of an essential oil, flavorings, and marmalades.

The citron was well known in the Mediterranean area before Christian times, and it may have been the earliest of the *Citrus* domesticates, spread widely by the Jewish peoples (for whom it had religious significance), thus pointing the way for later *Citrus* plantings of various species throughout the Mediterranean basin. Some authorities have speculated that the citron was native to southern Arabia rather than Southeast Asia. The Western world became acquainted with the more esteemed types of citrus fruits only during the Middle Ages, when Arabic traders brought them from the East. The Crusades acquainted the northern Europeans with citrus fruits, and the Portuguese and Spanish explorers quickly introduced them into the New World. About this same time the sweet orange was introduced to Europe, apparently from India. Citrus was brought into Florida before 1565, and a number of "wild" orange groves (of relatively unpalatable types) were growing in Florida when it became a state in 1821. These wild groves were topworked with buds of superior strains, and Florida was thus started on the road toward leadership in the citrus industry. As with the grape, citrus trees were introduced into California through the Spanish missions.

The 'Washington Navel' orange sparked the growth of the orange industry in California. In 1869 a missionary, one Reverend Schneider, discovered the navel orange in Baia, Brazil, where it probably had originated as a mutation decades earlier. Budded trees were sent to the United States Department of Agriculture in 1870, and propagations from them were sent to Riverside, California. Two trees planted in the yard of Mrs. Luther Tibbetts started the 'Baia Navel' (later the 'Washington Navel') orange on its way.

In 1862, near Lakeland, Florida, William Hancock purchased a farm from one Mrs. Rushing, who had set out three seedling grapefruit trees. One of these happened to bear seedless fruit, and thus originated the 'Marsh' seedless strain of grapefruit, named after C. M. Marsh, a nurseryman who eventually purchased budwood from this cultivar. Today this variety and selections from it are widely grown in Florida, Texas, California, and in various other parts of the world. Also arising by chance in Florida were the pink-fleshed grapefruits. A bud sport was found on a 'Marsh' seedless tree near Oneco, Florida, in 1913; it became the 'Thompson Seedless' grapefruit, named after the Thompson groves where it was found. A new bud mutation from this stock was discovered at McAllen, Texas, in 1929, and became 'Ruby,' the first patented grapefruit cultivar, and the one most grown in Texas today.

Although the 'Temple' orange does not keep well, it is one of the more flavorful varieties. The 'Temple' originated in Winter Park, Florida, from trees grafted with budwood obtained in Jamaica. The cultivar may be the result of a cross between a tangerine and a sweet orange (a tangor). A "new" citrus was discovered in Puerto Rico in 1956, and was named the chironja. It is said to combine the better characteristics of the orange and the grapefruit, and to peel like a tangerine; it may have originated as a natural cross between the grapefruit and the orange, or between the orange and the tangelo (itself a cross between the tangerine and the grapefruit).

Select cultivars of citrus are usually budded on sour-orange stock that is started from seed. Trees are planted about 7.5 meters (25 feet) apart, often cover-cropped or green-manured, sprayed for insects and disease, and irrigated in an appropriate fashion. Precautions taken against soil problems and frost have already been mentioned. Occasional serious winter freezings have been disastrous in the southern growing areas of the United States.

Citrus fruits are much used for fresh consumption, since they ship rather well to distant markets. The fruit is generally allowed to ripen almost fully on the tree, except for lemons and limes, which are picked green. Many oranges are not deeply orange, and are artificially dyed or treated with ethylene before being marketed. The demand for preserved citrus juice is increasing. In the past most juice was preserved by canning, but frozen concentrate (prepared by boiling the juice in vacuum pans and then freezing the concentrate) has captured and revolutionized the industry in the United States. More than half the Florida orange and grapefruit crop is now processed. This had led to intensive study of possible uses for citrus by-products.

The need for disposing of peel, pulp, and seed after juice extraction has resulted in the use of dried citrus pulp for cattle feed and for production of a molasses that can also be used in stock feed. The processing residues (about 50 percent of the fruit) are shredded in hammermills, limed (which precipitates and coagulates the pectin, causing cells to break down further), cured, and expressed. The expressed liquor is evaporated to yield oils and a molasses liquor (nearly 50 percent sugar), and the press cake is dried separately (to a moisture content of 8 percent) for cattle feed. The essential oils from the expressed liquor, primarily terpenes, are used for flavoring, perfumery, pharmaceuticals, and soap. It is possible also to produce alcohol by fermenting the molasses or to grow yeast (for feed) on these residues. An oil that resembles olive oil can be extracted from the seeds. Sometimes pectin is extracted from the rinds, used chiefly in jellies and confectionary products. Of course, the fruit juices themselves can be fermented to vinegars and liqueurs or be utilized as a source of acetic acid.

Apple, *Malus pumila* (Rosaceae)

Malus pumila (= *M. communis*, *M. domestica*, and *Pyrus malus*) embraces most of the cultivated fruit apples. Additionally there are many wild species in the genus, including *M. sylvestris*, whose name has also been applied to the cultivated apples, and which may figure in the ancestry of certain cultivars. The apple was among the first plants sought in the wild; it was gradually adapted to orchards in the northern Mesopotamian center of domestication. Apparently the ancestral apple was indigenous to the Caucasus mountains of western Asia, where wild forms still grow. In the New World the genus is represented by a confusing group of indigenous crabapples, which, along with introduced forms, serve as attractive ornamentals.

Worldwide apple production comes to about 22 million metric tons annually. Europe is the leading apple-growing continent (much of the crop is consumed as a mild alcoholic cider), and North America is a rather distant second. France, Italy, and Germany are the leading European apple producers; the United States, Japan, and Argentina are all about equally as important in their respective parts of the world. Other than the grape, there is no temperate-zone fruit more important than the apple.

The apple fruit lends itself quite well to handling and keeping, in spite of having a thin skin. A ripe apple is approximately 84 percent water, 11 percent sugar, 1 percent fiber, with fractional percentages of protein, fat, mineral, and fruit acids. It is generally harvested fully ripe, because immature apples ripen poorly after removal from the tree. On the other hand, apples left too long on the tree turn somewhat mealy because of the incipient decomposition of the pectic portions of the cell wall. Ripe fruit kept in a cool, properly ventilated storage space will keep for months. Controlled atmosphere storage (see Chapter 3) permits year-round marketing of apples, which fruit only in summer. A large portion of the apple crop is converted into preserves, such as apple butter and jelly, canned apple sauce, cider, and cider vinegar.

Authorities estimate that there are nearly 2,000 named cultivars of the apple today. Not many have become important commercially; it has been estimated that in the United States only about two dozen cultivars have been much grown until modern times. Of the two dozen, only one, the 'Cortland,' is a product of directed plant breeding. Of the remaining 23,

Figure 21-20   Chemical thinning is a standard practice in apple growing. The left side of the tree was thinned; the right side was not. Overbearing may cause broken limbs, small, poorly colored fruit, and may lead to alternate bearing in some cultivars. Apples are thinned by materials that burn off the blossoms or cause abortion of the embryo. [Courtesy Purdue University.]

which include many of the best-liked cultivars, most originated as chance seedlings, and some from bud mutations; a few are of unknown origin. The 'Rhode Island Greening' apple, for example, is said to stem from a seedling tree raised by a Mr. Green at his tavern near Newport, Rhode Island, from which admiring travelers obtained scions after tasting the delicious fruit that this tree bore. Other well-known cultivars that seem to have been chance seedlings include 'Delicious,' 'Golden Delicious,' 'Jonathan,' and 'York Imperial.'

The apple is a long-lived, spreading tree, which may remain productive for a century, though becoming gnarled and broken. Apple trees for commercial orchards are propagated by grafting select cultivars onto seedling rootstocks or clonally propagated "dwarfing" rootstocks. Different varieties have distinctive fruiting habits, some flowering and bearing abundantly only in alternate years, and a number

requiring cross-fertilization from a different cultivar to set fruit adequately. Pollination is accomplished chiefly by bees. The apple fruit is botanically a **pome** (a fruit having an inner cartilaginous core surrounded by fleshy tissue).

Unfortunately, the apple is attacked by many insect and fungus pests, which is especially troublesome because, when this Old World species was introduced to the New World, natural predators of the pests were lacking. This may not have been of any great concern in rural areas, where most apples were probably consumed at home, or crushed for cider, applejack, and vinegar, but for modern marketing blemished fruits become unsalable. Thus the orchardist must undertake a complicated and unceasing series of pesticide sprayings, beginning before buds first appear and ending only just before the apple is harvested. With a half dozen or more such treatments required, apple growers have had to automate lavishly to keep production costs down. Elaborate mist blowers and efficient modern pesticides are commonplace in the orchards of the United States. Not only are such familiar pests as apple scab and the codling moth controlled by spraying, but chemicals are utilized to thin fruit when it is borne too abundantly (Fig. 21-20) or to hold ripening fruit on the tree longer (to prevent bruising due to preharvest drop). Fortunately, promising results are now forthcoming from the breeding of apples resistant to all of the major pests. For example, such cultivars as 'Prima,' 'Priscilla,' and 'Sir Prize' are immune to apple scab and resistant to several other afflictions.

Pruning of the orchard, to encourage vigorous bearing wood and to shape trees for ease in harvesting, does not lend itself well to mechanization. European growers are especially adept at training fruit trees to shapes that fit the limited area that can be accorded them. Fortunately, North American orchidists can handle pruning during winter, when the trees are dormant. If care is taken in pruning to avoid stubs and tearing of the bark, and if hygienic techniques are followed (including the painting of large wounds with a dressing), canker diseases are generally prevented.

One of the most laborious jobs in fruit production is harvesting, still done mostly by hand. Harvesting of apples for processing may eventually become mechanized. One portable harvesting machine has a special catching frame covered with foam plastic onto which apples fall as the tree is shaken by another mechanical device. Other labor-saving innovations are now used, including dwarfed trees for easier harvesting, direct field harvest into 30-bushel pallet boxes, and automatic grading and sorting devices.

The apple is adapted to cold climates, and is quite hardy even where winter temperatures are considerably below freezing. By the same token the apple requires winter-conditioning in order to fruit; it will not last in southern locations where winters are mild. The favorite apple-growing locations are to the leeward of bodies of water, or in rolling country where frost does not accumulate in low-lying pockets; thus early flowering is inhibited, and freezing that would ruin the year's crop is avoided. In the United States, Washington, Oregon, and California are the leading apple states of the West; New York, Michigan, and Virginia lead in the East. Nearly half the production is consumed as fresh fruit, and much of the remainder is processed for apple juice (sweet cider), which in turn may be fermented to produce an alcoholic product (hard cider) or, by controlled bacterial inoculation yield acetic acid (vinegar). Pectin is an important by-product from the cider mills.

Other fruits

The four fruits that have been reviewed serve as examples of the temperate and tropical types. Quite obviously, scores more could be discussed that are of equal interest, and many of them almost as important in world commerce. Brief treatments of a few of those that are fairly important follow.

Additional temperate fruits

Apricot, *Prunus armeniaca* (Rosaceae)

The apricot was probably native to China, but it was early brought to the Near East, and into California during the eighteenth century. It is a close relative of the peach, but more tender, and is cultivated in much the same way. Southern and central Europe and California account for most of the world production, which exceeds 1 million metric tons annually. Much of the crop is preserved by sun drying.

Blueberry, *Vaccinium* spp. (Ericaceae)

The blueberry (Fig. 21-21) occurs in "low-bush" and "high-bush" species, both wild and cultivated, in cultivars ranging up to the hexaploid. Considerable wild fruit is still harvested, especially in New England and eastern Canada. Commercial plantings are mostly on acid soils from Michigan into New England, and in Washington and Oregon. Blueberries have been domesticated from wild forms mostly within the last half century. The larger, high-bush types are used in orchards, now supporting a thriving small-fruit industry. The greatest cost is the hand picking of the berries, but machine harvesting has now gained a following. Wild fruit is harvested with a specially designed rake that combs the upright stems, gather-

Figure 21-21   The blueberry has become increasingly important, transforming many formerly worthless acid soils into valuable cropland. [Courtesy J. C. Allen & Son.]

ing many leaves and twigs as well as fruit. Most of the debris is removed by winnowing. Wild blueberries are generally used for processing. Total production, chiefly from the northern United States and northeastern Canada, probably does not exceed 50,000 metric tons annually.

Brambles, *Rubus* spp. (Rosaceae)

Grouped as "brambles" are the red and black raspberries, blackberries, dewberries, loganberries, and other popular types, each of these common names embracing perhaps a dozen species and botanical varieties. The plants are prickly perennial shrubs (hence the name), but individual canes are often biennial (bearing fruit the second year, then dying). The "berries" are an aggregation of separate pistils on a fleshy receptacle (aggregate fruits); in the raspberry the fruit separates readily from its core when picked (giving a "hollow" berry). As with blueberries, brambles are relatively recent domesticates, mostly since the Middle Ages. Appreciable amounts are still harvested from the wild, although cultivated forms are now widely grown, often as home-garden plants, but mainly in commercial orchards in Oregon and the northeastern states. In season, the fruit is consumed fresh locally, but most of the crop is preserved

(especially for jellies and sauces), canned, or frozen. Perhaps 100,000 metric tons are processed annually in the United States. World production of raspberries is about 150,000 metric tons (mostly from central Europe).

Cherry, *Prunus* spp. (Rosaceae)

Most grown are the diploid sweet cherries, *P. avium*, and the tetraploid sour or tart cherries, *P. cerasus*. Both are believed native to Asia Minor, where they were well known to the ancients. Wild sweet cherry often serves as a rootstock for grafted cultivars, as does *P. mahaleb*. Cherry orchards are handled in much the same was as apples. In the United States, sweet cherries are grown in several western states, sour cherries mostly in Michigan and New York. Production of cherries exceeds 250,000 metric tons annually in the United States. The most popular sour cherry in the United States is the 'Montmorency,' which originated in the Montmorency valley of France several centuries ago. World production of cherries is a sizable 1.5 million metric tons annually, chiefly from Europe. Italy, Germany, France, the United States, and Turkey are the leading producing countries. The cherry, plum, peach, and apricot, all species of *Prunus*, constitute the so-called **stone fruits.**

Cranberry, *Vaccinium macrocarpon* (Ericaceae)

The cranberry is a vine-like relative of the blueberry, grown in acid bogs of the northern United States. Another recent domesticate, it is still harvested from the wild in some locations, just as it was in colonial times. It is propagated from cuttings set by hand, an expensive procedure considering the difficult bog locations where these tart-fruited vines must grow. Regulated nitrogen fertilization increases yields. The fruit is gathered much like the wild blueberry, by combing with special rakes; or, where it is possible, the bogs are flooded to float the berries to the surface, where they can be churned with a mechanical beater (a large reel on the front of a motorized "garden tractor") and thus be more readily gathered. Cranberries are almost entirely a North American crop, utilized chiefly to make a tart red jelly or sauce that is consumed with fowl during the Thanksgiving and Christmas season. Production exceeds 100,000 metric tons annually.

Currant, *Ribes sativum* and other species (Saxifragaceae)

Currants and gooseberries are very hardy "bush fruits" grown largely in northern Europe (especially in Germany and Poland), but they are used to some extent for jellies and preserves in North America. Their planting has been discouraged in many parts of the United States because the plants are alternate hosts for the troublesome white-pine blister rust. World production is some 330,000 metric tons annually.

Peach, *Prunus persica* (Rosaceae)

The peach, in numerous cultivars, seems to have been first domesticated in China. It was brought to Persia before Christian times (hence the specific name), and then further spread throughout Europe during the era of Roman dominance. The peach is more tolerant of warmth and less tolerant of cold than is the apple; although it is hardly "subtropical," southern climates such as prevail in Georgia (the "peach state") are responsible for considerable production. The peach fruit, botanically a **drupe,** is much prized for its rich, "aromatic" flavor. Select cultivars are usually budded or grafted onto unselected peach or almond rootstocks (the almond is a close relative of peach in which the seed rather than the fleshy outer part of the fruit is consumed). Only in the nineteenth century were useful cultivars, such as the 'Elberta,' a freestone seedling from a Chinese cling peach, introduced into the United States to make peach orcharding successful. Active breeding programs have increased adaptation by incorporating winter hardiness, low chill requirements, and an extended harvest season. The world's peach crop exceeds 7 million metric tons annually, with Europe and North America about equally represented. The United States and Italy are the leading peach-growing countries, with no little production in Japan. The **nectarine** is simply a peach without fuzz, grown especially on the Pacific coast. There are a number of cultivars.

Pear, *Pyrus communis* (Rosaceae)

The pear is most important in Europe, where there are many cultivars. Total world production exceeds 7 million metric tons annually, mainly from Italy. The pear is believed indigenous to western Asia, where it has long been cultivated. Selection of modern cultivars has been accomplished only fairly recently, mostly in northern Europe. Varieties important to North America, such as the 'Bartlett,' were introduced from Europe (the 'Bartlett' is known as the 'Williams' in England; it was discovered by a schoolmaster as a wilding, and brought to America about 1797). The 'Seckel' pear "originated in America," having been discovered on a tract of land just south of Philadelphia that became the property of a Mr. Seckel, but the original stock had no doubt been brought from Europe. Growing requirements for the pear are much the same as for the apple, although it is less tolerant of extremes. (The pear and apple constitute the so-called **pome** fruits.) When mature, the pear fruit is quite perishable, the stage of edible ripeness coinciding with the peak of the climacteric. In the United States there is a sizable pear-canning industry based on the 'Bartlett' cultivar, and in Europe a good portion of the crop is crushed for pear cider (perry) and wine making. A limiting factor in the United States pear industry is the bacterial disease fireblight, which is caused by *Erwinia amylovora.*

Plum, *Prunus* spp. (Rosaceae)

Plums (Fig. 21-22) are the most extensively grown stone fruits. There are a number of species, some of them wild, and many of them cross freely. It is estimated that about 2000 cultivars are grown, selected from 15 different species; but cultivars of the European plum (*P. domestica*) are most important worldwide, followed by cultivars of the Japanese plum (*P. salicina*) and various North American species. Centers of origin are the Orient, Europe, and North America. With such diversity, it is no wonder that plum cultivars can be found suited to almost any climate or soil. Growing requirements are the same as for the peach. Some cultivars are self-sterile, and must be interplanted with other cultivars for pollination. But other than that, plums are perhaps the least

Figure 21-22   Harvesting plums, grown with the help of irrigation, on the Adana Plain of Turkey. [Photograph by Mary H. Hill, courtesy World Bank.]

temperamental of the temperate-zone orchard fruits, seeming to give at least some yield with only scant attention. The drupe fruits are quite perishable unless refrigerated. Plums are not so much eaten fresh as preserved (jellies and jams), canned, and dried (prunes). The prune industry is especially important on the Pacific coast, where "*domestica*" plums are dried, "sweated," and "glossed" with a steamy glycerine bath of sweetened juice that produces the sterile glossy skin we associate with prunes. Europe accounts for most of the total world production of plums, about 5 million metric tons annually; Yugoslavia, Romania, and Germany are the leading producers. In the United States, plums are grown for fresh fruit in many scattered locations, but prunes are produced chiefly in California and Oregon.

Strawberry, *Fragaria* spp. (Rosaceae)

One of the temperate zone's favorite fruits, and probably the most popular "berry," is the strawberry (Fig. 21-23). Yields of 90 metric tons per hectare (about 40 tons per acre) have been achieved with well-adapted cultivars. World production amounts to about 1.3 million metric tons, mostly from the United States, Japan, Italy, and Poland. Botanically, the strawberry is not a true berry, but an aggregate fruit, consisting of an assemblage of separate pistils on a fleshy receptacle. The modern octoploid cultivated strawberry has a rare genetic make-up. The male-sterile *Fragaria chiloensis*, native to the southern Andes and the Coast Ranges of western North America, and the eastern American species, *F. virginiana* (and possibly other European species) are thought to have been hybridized in Europe after introduction there, to produce the modern strawberry, often designated *F. × ananassa*. Cato the Elder (234–149 B.C.E.) grew strawberries of some type (perhaps *F. vesca*) at his villa in Rome. Of all fruits discussed, the cultivars of the strawberry stand out as the result of breeding programs rather than chance selections. Breeders are still making crosses of wild *F. chiloensis*, *F. virginiana*, and *F. ovalis*, which along with the cultivated *F. × ananassa* constitute the four octoploid species of strawberry. Selections have been bred that are adaptable to subtropical climates such as prevail in southern Florida, and for northern states as well. Some cultivars now yield tremendous fruits as big as a plum. Most commercial cultivars bear but once in the spring, but especially for the home garden there are everbearing sorts that yield almost continuously if the weather remains reasonably cool and moist through the summer.

Strawberries are herbaceous perennials that produce clusters of leaves from the condensed stem or crown. Under long days, trailing stems or stolons (runners) spread from the crowns, forming new plantlets. Plantings generally become too crowded after a few years, and the strawberries diminish in size and quality. When this happens, new daughter plants must be set in newly prepared beds, or the existing beds must be thinned and allowed to replenish by stolons from adjacent plants. Plants produced during summer will flower and bear fruit the following spring; in this sense the crop is a biennial, although the plants themselves are perennial. Many commercial plantings are replanted annually. In California, which is responsible for three-fourths of the United States production, improved cultivars are started annually under polyethylene mulch on fumi-

A

B

Figure 21-23   (A) The strawberry is one of the most widely adapted fruit crops. (B) Plantings in California are made on fumigated soil, and production is stimulated with polyethylene mulch. [Courtesy Victor Voth.]

gated soil. The practices are regulated to provide the plants with enough chilling for good fruit yields but not enough to encourage vegetative (runner) exuberance. Trickle irrigation is increasingly practiced, allowing increases in plant population of as much as 50 percent. Yields can be as high as 40 metric tons per hectare (about 18 tons per acre).

Several modern herbicides have been used to help control weeds in strawberry plantings, but picking has always been a laborious process. Near urban areas many growers invite the public in for "pick your own" harvesting, the customer paying for the quantity taken. Some new cultivars are nearly determinate, so that mechanical lifters can harvest the whole plant and enough ripe berries can be sorted from the trash to make the practice economical. Strawberries are soft and quite perishable, but so prized and delectable is this fruit that international air shipment has become a sizable business in recent years. Much of the commercial crop is processed into jellies, preserves, and sauces.

Additional tropical fruits

The richness of the tropical flora provides many fruits that are of direct importance to people and innumerable others that are important to the maintenance of tropical ecosystems. Many are little if at all known to inhabitants of temperate climates. Obvi-

ously, it is impossible even to list, much less discuss, the majority of these species, although a number of more major ones are touched upon in the few pages following. By way of example, Table 21-5 lists twenty-three fruits and vegetables grown in southern Florida, several of which have been discussed, but a number of which are unfamiliar to residents of the other forty-nine states (and even to most Floridians).

Avocado, *Persea americana* (Lauraceae)

The avocado, alligator pear, or aguacate is a rainforest tree native to the American tropics, but it has now been introduced to many areas with suitable climate worldwide. It was among the wild plants first utilized by the early Central American civilizations; in some it was a major source of nutrition. Archeological finds show it to have been important in Mexico 9000 years ago. The fruit was strikingly increased in size by prehistorical selection, as archeological remains show. The avocado is still widely grown, esteemed as a nutritious food, although more a delicacy than a staple in the modern diet. The large greenish or purplish fruits contain a single large seed surrounded by rich flesh containing as much as 30 percent digestible fat and 4 percent protein. Compared to most fruits it thus packs quite a nutritional wallop. Many local selections (some of them appreciably more tasty than the commercial avocado) are thin-skinned and ship poorly; they are consumed lo-

590

Table 21-5   Some tropical fruits and vegetables grown in Dade County, Florida.

Scientific name	Common name	Origin	Uses
Aegle marmelos	Bael	India, Sri Lanka	Yellow pulp is eaten fresh.
Antidesma bunius	Bignau	Philippines	Fruit is made into jelly and wine.
Artocarpus heterophyllus	Jackfruit	Southeast Asia	Pulp is eaten fresh or cooked; seeds are eaten cooked.
Averrhoa carambola	Carambola	Southeast Asia	Tart to sweet pulp is eaten fresh or made into preserves.
Blighia sapida	Akee	Africa	Yellowish-white aril is edible when fruit is mature, but very poisonous when immature.
Carissa carandas	Karanda	India	Pulp is eaten fresh or made into jellies or preserves.
Carissa grandiflora	Natal plum	South Africa	Pulp is eaten fresh or made into jellies or preserves.
Citrus aurantiifolia 'Tahiti'	Tahiti lime (Persian lime)	Unknown	Juice is used in beverages, marinades, seasonings. Oil is used as flavoring and in cosmetics.
Citrus aurantiifolia	West Indian lime (Key lime)	Southeast Asia	Same as Tahiti lime.
Citrus maxima	Pummelo (Shaddock)	Southeast Asia	Pulp is eaten fresh.
Cocos nucifera	Coconut	Tropics	Important source of food, oil, and fiber.
Cucurbita moschata	Calabaza (Cuban squash)	Asia	Fruits are boiled, baked, or used to make pie.
Dioscorea alata	Yame (yam)	West Africa	Tubers are boiled; used to manufacture flour and flakes.
Diospyros diayna	Black sapote	Central America	Sweet brown pulp is eaten fresh.
Ipomoea batatas	Boniato (Cuban sweetpotato)	American Tropics	Tubers are boiled, baked, fried, or used to make preserves.
Manihot esculenta	Cassava (yuca, manioc)	Brazil and Mexico	Roots are boiled; used to manufacture starch and other products.
Manilkara zapota	Sapodilla (nispero)	American Tropics	Pulp is eaten fresh; tree is source of chiclé.
Monstera deliciosa	Ceriman	Mexico	Soft, sweet pulp is eaten fresh.
Persea americana	Avocado	American Tropics	Oily pulp is eaten fresh, usually in salads or dips
Pouteria sapota	Mamey sapote	American Tropics	Sweet red pulp is eaten fresh or made into ice cream or sherbet.
Psidium guajava	Guava	American Tropics	Tart to sweet pulp is eaten fresh or made into purees, jellies, and paste.
Spondias cytherea	Ambarella	South Pacific	Sweet yellow pulp is eaten fresh or made into jelly.
Xanthosoma sagittifolium	Malanga (yautia, cocoyam)	South America	Tubers are boiled, baked, fried, or added to soup.

Source: Agricultural Research and Education Center, Homestead, Florida. [Courtesy Florida Cooperative Extension Service, University of Florida.]

cally, and do not reach the large temperate-zone markets. Cultivars grown as a cash crop in the southern United States (California and Florida) are chosen especially for their keeping and shipping qualities. Even so, once ripe the fruit is quickly perishable, and therefore it is expensive to market. World production well exceeds 1 million metric tons annually, with Mexico the leading producing country.

Breadfruit, *Artocarpus altilis* (Moraceae)

The breadfruit is a handsome East Indian tree, and its large fruit is recognized as an excellent source of food. The species does well in hot, humid climates and was widely dispersed by Britain during her colonial era. Captain Bligh, against whom the famed mutiny took place on the *Bounty,* was on a mission to gather young breadfruit trees in Tahiti and transport them to the West Indies. Ordinarily, the meaty aggregate fruit develops parthenocarpically, and it is about 20 percent starch. It is usually eaten cooked in much the same fashion as potatoes. When cultivars do set seed (then called breadnut), little edible flesh develops, but the seeds can be cooked and eaten. Propagation is by root cuttings and offshoots. A similar species, the jackfruit, *A. heterophyllus,* also yields large, edible fruits, but it is less esteemed than the breadfruit.

Cashew, *Anacardium occidentale* (Anacardiaceae)

This small Brazilian tree is better known for the seed (nut) than for the fleshy receptacle (pear) that bears it (Fig. 21-24), and is used in its homeland as a fresh or preserved fruit. The cashew is discussed further under Nuts, later in this chapter.

Coconut, *Cocos nucifera* (Palmae)

With world production of some 30 billion coconuts annually, this is hardly a fruit that can be overlooked. Chief sources of the coconut are the Philippines, Indonesia, Sri Lanka, Malaysia, and Mexico. Most of the crop is used to produce copra, a source of oil that will be discussed in the next chapter. Both volunteer and planted coconuts provide food and other essen-

Figure 21-24 Fresh cashews as harvested in Guatemala. The true fruit at the tip contains the cashew nut kernel; the fleshy pear-shaped pedicel and receptacle can be eaten fresh or preserved. [Courtesy USDA.]

tials for simple maritime societies throughout the tropics.

Custard apple, *Annona* spp. (Annonaceae)

Several annonas, including the cherimoyas or pinhas as well as the custard apple, are flavorful tropical fruits little known in temperate countries. They are relatives of the wild papaw (*Asimina*). Delicate and quickly perishable, they ship poorly even by air. The delightfully "fruity" sweet types are consumed raw, the pulp (with seeds) being spooned from the inedible rind. Other, more acid annonas are used mostly to make drinks or flavor frozen desserts. There are relatively few selected cultivars, and often volunteer or wild trees are the source of this delicacy, which is consumed locally.

Date, *Phoenix dactylifera* (Palmae)

Dates appear to have been in cultivation in arid parts of the Near East since 4000 B.C.E., and they were much utilized by the Sumerian and Assyrian civilizations. Puny-fruited wild and feral palms are still found in the area. *P. dactylifera* is interfertile with other wild species, including the east African *P. reclinata* and the northwest African *P. atlantica,* as

Figure 21-25   The fruit of the date palm. [Courtesy J. C. Allen & Son.]

well as with island and Indian species. Today the annual production exceeds 2 million metric tons, most being grown in Asia Minor and northern Africa. In the Arabic world the date has long been a prime source of subsistence for nomadic tribes, being a rich source of nutrients when dried (about 70 percent carbohydrate, 2 percent protein, and 2.5 percent fat).

Dates may be eaten fresh, pounded into a paste, mixed with milk products (bolstering the protein content), or fermented to arrak (once described as "the strongest and most dreadful drink every invented"). The date palm is usually propagated by "suckers" (side shoots from the base of the trunk). Flowering begins about the fourth year, and the flowers of the pistillate (female) tree are dusted by hand with inflorescences obtained from a relatively few staminate (male) trees grown apart from the female trees. In modernized date orchards the developing fruit may be protected from birds or other

damage by being "bagged" in netting. Date palms require hot, dry weather for flowering, but they are ordinarily planted in oases, or where irrigation is possible, to supply necessary ground water. A single tree when mature may yield more than 45 kilograms (100 pounds) of fruit yearly, the fruit usually ripening in autumn (Fig. 21-25). The fruit must be hand picked, after which it is usually heated or treated chemically to aid in its preservation, and then allowed to ripen further to develop sugar content and to precipitate astringent components. Date palm fronds are widely used for thatching, braiding, and basketry, and the stems are in many places almost the only wood available.

Fig, *Ficus carica* (Moraceae)

The fig's relationship with people has been long and involved. *F. carica* is native to the eastern Mediterranean region, and either wildings or escapes are found scattered throughout the Near East, as are many native species of *Ficus*. Certainly the fig has been cultivated in the Holy Land for more than 5000 years, and escapes are found in rocky crevices all along the Jordan River and around the Dead Sea. The fig is cited in Egyptian documents of the fourth dynasty (about 2700 B.C.E.), and it is mentioned repeatedly in both the Old and New Testaments of the Bible. Most figs are self-pollinated and parthenogenetic, producing a nearly continuous sequence of fruits. But the larger, more luscious Smyrna figs (*F. carica* var. *smyrniaca*) bear only pistillate flowers, and so are dependent for fertilization upon the pollen of other figs. For this the male forms of the wild or semicultivated caprifigs (*F. carica* var. *sylvestris*), possibly ancestral to many of the select cultivars, are used as a source of pollen. The caprifigs are the host for a small fig wasp that carries the pollen from the caprifig to the Smyrna; the process is known as caprification. Caprification seems to have been discovered in Greece, several centuries B.C.E. In ancient Greece the biological facts of caprification were not clear, but it was realized that to assure good yields caprifig stems had to be hung in the elite fig trees, often with suitable ritual and tribute to the gods. In California the Smyrna figs, termed 'Calimyrna' there, were not a commercial success until caprifigs and fig wasps were introduced in 1899. Most of to-day's fig cultivars develop fruit parthenocarpically, so they do not require caprification.

The fig fruit is a curious structure called a **syconium,** in which unisexual flowers are borne inside a pear-shaped, nearly closed receptacle. In figs that bear only pistillate flowers and require pollination, the fig wasp enters the syconium, pollinates the fruit, lays its eggs, and dies. Wasp and eggs are absorbed by the fig. The early fruits are usually larger and are consumed fresh, subsequent "second crop" figs often being dried or preserved. Fig trees vary widely in size and shape, ranging from frail climbers to sturdy orchard trees about the size and shape of an apple tree. The latter, propagated by cuttings, are planted commercially. The fruit is generally gathered by hand when fully ripe and is often sun-dried (figs are typically grown in semiarid climates, where orchards must be irrigated). World production of fresh figs totals about 1.5 million metric tons, with about 200,000 metric tons dried. Portugal, Italy, and Turkey are the major fig-producing countries. In the United States, production is chiefly in California, with some importation from the Near East. In the Near East figs are utilized for stock feed as well as for human food. The sycamore figs of the Nile Valley, *F. sycomorus,* have long been important in Egypt, where both the fruits and the timber have been extensively used.

Guava, *Psidium guajava* (Myrtaceae)

Now widely distributed and naturalized, the guava is believed native to tropical Brazil, domesticated more than 2000 years ago. The fruits are somewhat gritty, and are more often used to make a dessert paste or jelly than eaten fresh. The fruit is a rich source of minerals and vitamins. Most of the harvesting is done casually and by hand from relatively unselected dooryard trees, although some modern orcharding is practiced, especially in Florida. A smaller tree, *P. littorale,* the strawberry guava, is cultivated on a small scale in the West Indies.

Litchi, *Litchi chinensis* (Sapindaceae)

Well known in the Far East, the litchi has been in cultivation since time immemorial. It is a favorite fruit in China, where the species is believed to be indigenous.

Maracujá, *Passiflora* spp. (Passifloraceae)

Known also as the passion flower, or granadilla, species of *Passiflora* yield tart, aromatic fruits about the shape of a small gourd; the fruit is usually used to make drinks or sherbetlike desserts.

Mango, *Mangifera indica* (Anacardiaceae)

The mango is a large, handsome tropical tree that bears an oblique, one-seeded drupe with a distinctively aromatic outer pulp. It is apparently native to Southeast Asia, where it has been cultivated for thousands of years. Wild forms are found from the Malay Archipelago to northeast India (it is mentioned in ancient Hindu mythology). There are about forty species in the genus *Mangifera,* and more than a thousand cultivars. Many are delectable, tasting like something between a peach and a pineapple; others are quite rich, and have a turpentine taste that must be cultivated to be enjoyed. The diploid chromosome number is 40, and some cultivars are apparently polyploid. Today the mango grows throughout southern Asia, the Pacific islands, and the lowlands of Central America and South America. The Portuguese brought the mango from India to Brazil, and it is said that Captain Cook found the fruit abundant in Rio de Janeiro in 1768. Seeds reportedly brought from Rio de Janeiro introduced the mango into the West Indies in 1742. Introductions into Florida, both from Mexico and from India, occurred early in the nineteenth century. An unusually diversified mango germplasm bank has been assembled in southern Florida, where better cultivars yield 30 metric tons per hectare (about 13 tons per acre) on a regular basis (unselected mangos tend to bear heavily in alternate years, and in irregular cycles). World production of mangos exceeds 13 million metric tons annually, two-thirds of the crop coming from India.

The mango thrives best in climates having a humid rainy season alternating with a dry season (during which the plant flowers and fruits). The species adapts well to a wide variety of soils, but it should have adequate drainage. Most mango trees are seedlings, and, although apomixis often occurs, variability is high. Inflorescences produce voluminous flowers, but often as few as 1 percent are hermaphroditic (the remainder being staminate), depending upon cultivar and conditions. Selections having superior fruit, made mostly in India, are vegetatively propagated by grafting, budding, marcottage (stem layering), or inarch grafting (bringing into contact partially debarked stock and scion until they grow together—a method widely practiced in India). Plantations are fertilized, often planted to a cover crop or an intercrop until the mango trees are of bearing age, and provided with irrigation when necessary (especially for young trees being established). Harvesting generally involves shaking the fruit from the trees; the fruit is then put into baskets and taken to the market. The fruits are quite astringent until fully ripe; eating mangos gives many people a rash on the lips, much like that caused by poison ivy (to which the mango is related). Where modern facilities are available, mangos keep well for many days in cold storage. Unfortunately, up to this time the mango has not been sufficiently important in technologically advanced markets to have been accorded the necessary experimentation on how best to keep and ripen the fruit (such as has been given the banana); consequently, much of the market fruit is inferior in taste to tree-ripened mangos locally consumed. The fruit contains as much as 20 percent sugar, plus fruit acids, minerals, and a slight bit of fat and protein. As with other fruits, surplus mangos can be fermented to vinegar or wines and brandy.

Mangosteen, *Garcinia mangostana* (Guttiferae)

This small tree from Southeast Asia yields a fruit of exquisite flavor, often termed the "queen of tropical fruits." Production in the Far East is mostly from dooryard plants, and transplantings to the New World have seldom succeeded. The mangosteen is a lowland species of the humid tropics, and quite susceptible to cold. Protective shade is helpful to young trees.

Olive, *Olea europaea* (Oleaceae)

Like the fig, the olive was intimately involved in early civilizations in the Mediterranean area, and it is frequently mentioned in the Bible. Carbonized remains of olives that are 5700 years old have been unearthed in archeological diggings near the Dead Sea. Wild olives (frequently considered *O. europaea* var. *oleaster*) are found throughout the Mediterranean basin, and they cross readily with *O. europaea*. The wild and cultivated forms together probably represent one species complex. World production exceeds 8 million metric tons annually, mainly from the Mediterranean area (Spain, Italy, Greece, Tunisia, and Turkey). Perhaps 90 percent of these olives are crushed to yield oil, which is discussed in the following chapter on extractives. Olives for table use were first introduced into California through the San Diego Mission in 1769, and the 'Mission' cultivar is still a leading one. The olives are carefully handpicked—when straw-colored for green olives, and when black for ripe olives. The fruit is kept for several months in a weak salt solution, where a lactic acid fermentation takes place. The pickled fruit is then treated with sodium hydroxide, washed, and stored in dilute brine. The lactic acid fermentation is omitted for green olives.

There are hundreds of cultivars of the olive tree, a small evergreen native to western Asia. The species has rather exacting climatic requirements, since it is killed outright at temperatures below $-9°C$ ($15°F$), though a certain amount of winter chilling is required for floral induction. Olives, typically grown in semiarid regions, are apparently tolerant of alkaline soils. For steady and certain production irrigation is required. Propagation is vegetative, in the Near East mainly from knobs ("ovuli") developing at the base of the trunk, which root easily. Select cultivars are budded onto seedling olive trees or propagated from cuttings or by layering. In California, softwood cuttings are made under mist, and hardwood cuttings may be set vertically in the nursery row rather than as layers. Several decades are required for olive trees to reach full productivity. Well-established trees in Spain are estimated to yield nearly 750 kilograms of olives per hectare (about 670 pounds of olives per acre). In California, olives are more intensively fertilized, and yields are better. Olives, like mangos and many other fruits, are prone to **alternate bearing**—a heavy fruit set one year so exhausts the tree that only a light crop is borne the next. In California, this tendency is overcome to some extent by hormonal thinning. Harvesting is increasingly being done by mechanical shakers, especially for olives from which oil is to be extracted.

Papaya, *Carica papaya* (Caricaceae)

The papaya or mamão is a fast-growing, short-lived dioecious tree native to tropical America (Fig. 21-26). It was introduced into India before the seventeenth century, and it now grows throughout the Pacific islands, where it is more highly esteemed than in its homeland (many Americans think of it as a Hawaiian fruit). Recorded world production is well over 1 million metric tons annually, mostly from India and Brazil. The tree propagates readily from seed, and it may bear ripe fruit resembling the honeydew melon in as little as a year. The fruit is not very tasty unless it is eaten at just the proper stage of ripeness, one of the factors contributing to its limited usefulness outside of the area of its immediate production. It is usually consumed fresh in the same fashion as a cantaloupe, but it can be squeezed for juices or be pickled, candied, made into jellies, or even cooked like squash. The fruit is 90 percent water, but it is rich in vitamins A and C, and it contains some carbohydrates and minerals.

A constituent called papain, which acts as a proteolytic enzyme, is sometimes used in meat tenderizers; consumption of the fruit is believed to be an aid to digestion and perhaps also a vermicide. Papain is extracted commercially by scratching the green, immature fruits, from which a latex exudes and drips to a collecting platform. Most papain is produced in Tanzania, with imports to the United States chiefly for use in "chill-proof" beer (to prevent cloudiness due to precipitation of proteins). The papaya tree volunteers readily in warm, humid climates, performing best in full sunlight. It is weak, subject to

Figure 21-26 Papaya is a popular tropical fruit and the source of the enzyme papain, which is used as a meat tenderizer. [Courtesy J. C. Allen & Son.]

damage where not protected from strong winds, but it tolerates a wide variety of soils. Plantations are established by clearing the land of all vegetation, planting seeds in a nursery, transplanting the young plants to the field, and keeping weeds in check (usually by hand labor, in the tropics). As with dioecious cultivars, all but a few staminate (male) trees are destroyed; some cultivars have perfect (hermaphroditic) flowers and sometimes sex reversal occurs (male to female, but never the reverse). Since the life of the papaya tree is only a few years, rotation planting is often practiced to ensure a continuous supply of bearing trees. Fertilizing and mulching are well rewarded by increased yields.

Pineapple, *Ananas comosus*
(= *A. sativus*) (Bromeliaceae)

The pineapple, also kown as ananas, piña, and abacaxí, is one of the outstanding examples of the domestication and plantation production of a tropical fruit. So popular has it become that more than 6 million metric tons of pineapples are produced annually, mainly in Hawaii, but also in Brazil, Thailand, Taiwan, Malaysia, Mexico, the Philippines, and South Africa. Yields as high as 67 metric tons per hectare (about 30 tons per acre) have been obtained in Hawaii. However, production is shifting from Hawaii to other locations because of the high costs and expensive land capitalization.

The pineapple is a native American species, first discovered by Europeans when Columbus landed on Guadaloupe Island in 1493. At that time the pineapple had spread throughout most of tropical America, where it was cultivated here and there by the Indians. Several essentially seedless types were noted, although the wild forms were quite seedy. This suggest that primitive Americans had already made considerable progress in selection of parthenocarpic types, possibly derived from such wild South American species as *A. ananassoides*, *A. bracteatus*, and *A. erectifolius*.

Early introductions to Europe were grown in the greenhouse, mostly as a curiosity and the plaything of monarchs. But some selection and improvement was achieved, although it is not clear how much of this was transferred back to the pineapple-growing regions. The pineapple was taken by the Portuguese to India in 1548, and it was grown in the East Indies soon after. Breeding work was undertaken in southern Florida about 1900, and shortly after in Hawaii. Little came of this, however; most commercially produced cultivars, bearing such varietal names as 'Cayenne,' 'Queen,' 'Spanish,' and 'Pernambuco,' have been obtained in various other parts of the world. Pigeon-holing of cultivars has never been very satisfactory, but the more recent studies suggest these groupings:

'Spanish'—A spicy-acid flavor, very good for shipping, worldwide.

Figure 21-27 Pineapple may be propagated from slips (leafy shoots originating from axillary buds borne on the base of the fruit stalk), from the crown of the fruit, or from suckers lower down on the stem. [Courtesy Dole Corporation.]

'Queen'—Sweeter, less acid, rather small fruits, mainly African and Australian.

'Abacaxi'—Sweet, tender, juicy, best suited to local use, mostly South American.

'Cayenne'—Large, tender, low fiber, excellent for canning (most of the Hawaiian crop), worldwide.

'Maipure'—A newer spineless form, sweeter than Cayenne but more fibrous, mainly northern South American.

The pineapple fruit grows on a central stalk from a rosette of leaves, which are usually prickly. The fruit is multiple, consisting of fused individual flowers. Strains are self-incompatible, but cultivars will cross and set seed, as will crosses with wild types. The fruit contains up to 16 percent sugar, various acids, vitamins, and minerals, and a supposed digestive enzyme (bromelin). Plants are propagated by basal suckers or "slips" that arise from beneath the fruit (Fig. 21-27). Under the careful, automated culture accorded plantation growing in Hawaii, a sucker will bear a new pineapple head in about one year. After the first fruit is cut, the plant branches to yield a second crop of two fruits, and when these are cut, a further branching yields four small pineapples, which are used for cut pineapple or for juice. After the third crop, the field is replanted to avoid any further reduction in fruit size. Planting is typically through a paper or plastic mulch to conserve water and to control weeds. In Hawaii, extensive research has determined just what fertilization, trace elements, and pesticide schedules are needed to give the highest yields of high-quality fruit. The whole operation, including field harvesting, is almost completely automated. Although the individual pineapple heads are generally still cut by hand, portable conveyor belts carry the fruits to central loading facilities.

The pineapple has been suggested as an efficient source of biomass production, being about as productive as sugarcane or cassava on far less water. Pineapple has crassulacean acid metabolism, providing good carbon dioxide assimilation with maximum water conservation. One study has indicated that pineapple, sugarcane, and cassava can yield enough sugar to produce 964, 921, and 611 liters of ethanol per hectare per month, respectively, with pineapple requiring less than half the water of sugarcane and only a little more than half that of cassava. The crop is suggested for seasonally dry regions, such as central and northeastern Brazil.

Pomegranate, *Punica granatum* (Punicaceae)

The pomegranate is a small shrubby tree found wild in northeastern Turkey and the Caspian region. The fruit contains many pulpy seeds, each with a red, juicy aril—an additional covering surrounding the seed. Pomegranate domestication goes back at least to the Bronze Age, and the species is often represented in ancient Egyptian tombs. Pomegranates are more eaten in the Old World than in the New World, but they are grown somewhat in California and Latin America. There are few select cultivars. Plants can be propagated from seed or cuttings, or by layering.

NUTS

A number of woody plants (trees and shrubs) of which the seed is the important food constituent are grouped here as "nuts." Nuts are not of great commercial importance, although nutritionally they are quite a concentrated source of food. Unlike most fruits of the previous section, they are generally rich in oil and fat, and contain moderate amounts of carbohydrate and protein. They ship and keep rather well, and they can be used in much the same fashion as cereal grains and seed legumes. Nuts provide relatively little of the food needed for the world's burgeoning populations. Perhaps they have been too much overlooked and will increase in importance, especially on those tropical soils that cannot stand continuous tillage and annual cropping. Tropical forestlands, yielding little in the way of food directly, would seem to be a logical habitat for more extensive nut cropping. Certainly one would suppose that the seeds of trees could be as important a source of food in their adapted habitat as are the seeds of grasses and annual legumes.

Most nuts are something of a luxury; they are expensive, and mostly consumed in small amounts as sweetmeats or delicacies. But in certain times and places, nuts have been used as staple foods, as the piñon nut was by the American Indian in sections of the Southwest. Indeed, the piñon nut crop is still collected from wild stands of the piñon pine rather than from cultivated orchards. In fact, nuts are more often gathered from the wild than any of the foods discussed in this chapter. Wild trees are the source of most Brazil nuts, cashews, black walnuts, certain hickory nuts and pecans, and of course coconuts, if you prefer to consider it here rather than with fruits or oil sources. Fortunately, others, such as the almond, filbert (hazelnut), and domesticated pecans have been brought into cultivation, and cultivars have been selected that yield vastly improved nuts. Usually these are grown in well-tended orchards with the aid of many of the techniques employed in fruit orchards. There would still seem to be tremendous opportunities ahead for domestication and improvement of nut crops; a handicap, of course, is the long generation time, a problem encountered with all perennial tree crops.

Almond, *Prunus dulcis* (Rosaceae)

The almond, *Prunus dulcis* (= *P. amygdalus* and *P. communis*), is a species closely related to *P. persica*, the peach. In the almond, however, the seed of the drupe is utilized and the flesh is discarded. The outer flesh is astringent and tough, unlike that of its near relative the peach. The two species may have had a common Near Eastern ancestor, and some believe that the peach was developed from an almondlike progenitor through selection of types having edible flesh. The almond has been under cultivation for thousands of years in the eastern Mediterranean region, and it is repeatedly referred to in the Bible. There are two varieties of almond, the sweet (*P. dulcis* var. *dulcis*), and the bitter (*P. dulcis* var. *amara*). The former is the source of edible almond nuts used in confectionery; the latter (poisonous to eat because it contains prussic acid) is grown chiefly for the extraction of bitter almond oil. Almonds contain 18-20 percent protein; the "skin" surrounding the nutmeat is rich in minerals and should be consumed with the nut for its food value.

The cultivation of almond trees is similar to that of peaches. They require a subtropical climate that is nearly frost-free, similar to that of their native Mediterranean region. In the United States the center of almond production is in semiarid California, where almonds were introduced in 1843 by Franciscan Fathers; the orchards are irrigated, and harvesting is mechanized. The trees begin bearing within seven years. Planted 7.5 to 9 meters (about 25 to 30 feet) apart, they yield slightly more than 1 metric ton of nuts per hectare (about 0.5 ton per acre) at maturity.

The pit or seed separates readily from the split flesh as the fruit ripens and cracks. The pit is mechanically cracked, freeing the seed, which may be bleached, but is more commonly roasted for consumption in candy, mixed nuts, and confectionery.

Almond production in the United States (mainly California) runs more than 200,000 metric tons annually. World production approaches 1 million metric tons in some years, with Italy and Spain almost as productive as the United States.

Brazil nut, *Bertholletia excelsa* (Lecythidaceae)

Known also as the cream nut, or castanha do Pará, the Brazil nut is an important indigenous food plant of northern South America. Closely related is the Brazilian sapucaia, or paradise nut, *Lecythis* spp. Both are magnificent forest trees that grow in the Amazon Valley and bear clusters of seeds (nuts) in horny, urnlike fruits. The tip end of the fruit dehisces, and within it, clustered much like segments of an orange, are the individual Brazil nuts that have become so familiar a component of mixed nuts at temperate-zone cocktail parties. The nuts contain as much as 70 percent oil. Most nuts are still collected from fallen pods in the forest. Production amounts to approximately 30,000 metric tons annually, most of which is exported to the United States and to Europe.

Cashew, *Anacardium occidentale* (Anacardiaceae)

The cashew, known also as the marañón, or cajuil (Spanish), cajou (French), and cajú (Portuguese), is a remarkable plant of many uses (see Fig. 21-24). In a previous section it was noted that the fleshy, pear-shaped stalk or receptacle, which bears the single-seeded fruit, is often utilized fresh or preserved, especially in South America (in India, the chief commercial source of the cashew, the receptacle is usually discarded). In addition to the receptacle and the nut, the species yields a phenolic nut-shell "oil," cardol (90 percent anacardic acid) that is used as a medicinal, an irritant, a wood protectant, and a source of resin; the young leaves are sometimes used as a flavoring and as a medicinal; a yellow gum obtained as an exudate from incisions made in the bark is used mainly as an adhesive; the wood is utilized in several ways; and the bark is used to some extent for tanning and to make a yellow dye.

The cashew is native to coastal parts of northeastern Brazil, but for 400 years has been grown in many parts of the world. It is a sprawling broad-leafed evergreen, well adapted to poor soils and dry, sandy locations. Various parts of the tree are quite astringent and toxic, especially the shell that surrounds the nut, which may cause painful blistering, especially before the nut is fully ripe. The tree was first introduced by the Portuguese into India as a shore tree to bind the soil; it flourished, and it now grows as an escape throughout India, Sri Lanka, and Southeast Asia. The Portuguese also introduced the cashew into Africa, where there is considerable nut production (the unshelled nuts are generally shipped to India for processing). Nut production in India may reach 150,000 metric tons annually, with an additional 350,000 metric tons from Africa (mostly Mozambique) and lesser amounts from South America and various oceanic islands.

The cashew has been propagated chiefly by seed. Typically, three nuts are planted into holes spaced 6 meters (about 20 feet) apart; only one of the three seedlings (the most vigorous) is allowed to persist. In recent years cultivars have been selected in India that are propagated by inarching, budding, and other vegetative techniques. The trees may begin bearing in as little as three years, but they seldom yield profitably until they are seven years old. Some mature trees have been reported to yield as much as 90 kilograms (about 200 pounds) of nuts. Adapted as the cashew is to the sterile, sandy *taboleiros* of Brazil, the trees seldom receive the benefits of any fertilizer.

Nuts are generally harvested from the ground, where they fall with the receptacle when ripe. They are dried in the sun briefly until their moisture content is reduced to about 7 percent (marked by a rattling sound when the nut is shaken). At this stage they may be stored for prolonged intervals. Occasionally in India the kernels are eaten in this state, but more frequently the nut is roasted. In India this is customarily done in a shallow, open pan. The shell oil that is exuded by the roasting nuts must be absorbed by shaking the nuts in sand to avoid blistering the hands of handlers and shellers. A few factories in India have mechanized the roasting of cashew nuts and specialized in the recovery of by-product cashew-shell oil. The actual shelling is done by hand, typically by women utilizing a mallet. In Africa a machine was developed for cutting into the shell to allow its easy separation, giving a more perfect, undamaged nut. The nuts are chiefly exported to North America and Europe, where they are consumed as table nuts and in confectionery. The kernel is nearly half oil (of which about 74 percent is the oleic type) and about 18 percent protein.

Hazelnut, Filbert, or Cobnut
Corylus spp. (Betulaceae)

Species of *Corylus* are sometimes also called Turkish nuts or Barcelona nuts, with good reason; Turkey produces as much as 300,000 metric tons annually, and Spain about 30,000 metric tons. Italy, too, is a major producer, with the crop often approaching 100,000 metric tons per year (compared with 10,000 metric tons per year in the United States, mostly from Oregon).

There are native species of *Corylus* in North America and in Eurasia. They make an excellent understory in open woodlands in temperate climates. *C. americana* furnished wild hazelnuts to many a school boy when America was still rural, and several cultivars have been domesticated. But of greater economic importance is the larger European tree, *C. avellana*, a vegetational dominant bordering peat bogs in northern Europe that enjoys climates with mild winters and cool summers. Select varieties are propagated vegetatively, often by budding or grafting onto seedlings, which are then transplanted to the orchards. Trees first bear when about four years old, and they eventually provide nut yields of about 1 metric ton per hectare (nearly 0.5 ton per acre). Nuts contain about 60 percent oil and 13 percent protein. The species bears biennially, and cultivars are self-incompatible (so about 10 percent of the trees in an orchard must be pollinator trees of a different cultivar).

Macadamia, *Macadamia ternifolia* (Proteaceae)

The macadamia nut, also known as the Queensland, Australian, gympie, bush, or bopple nut, is native to Australia, whence it was introduced into Hawaii less than a century ago. In Hawaii it has gained favor so rapidly that it has become one of the most important orchard crops of the islands. Cultivars have been selected both in Australia and in Hawaii. Of little consequence in the world food picture, macadamia nonetheless serves as an interesting example of a tropical nut industry in the making.

Macadamia trees are large, broad-leafed evergreens that reach heights of 18 meters (about 60 feet) or more. The fruit is a follicle with a fleshy husk that splits to reveal a spherical seed about an inch in diameter. This seed, or nut, consists of a meaty kernel in a relatively thin enclosing shell that must be cracked to get at the kernel. Machines have been devised to remove the outer husk, after which the unshelled nuts can be kept satisfactorily for several months in dry storage. For final processing, the unshelled nuts are dried to approximately 2 percent moisture in a warm draft of air. The shell is difficult to crack by hand, but, in commercial production, powered nut-cracking machines are used quite effectively. The shelled nuts are "french-fried" in oil for several minutes, and after being drained are salted and vacuum-sealed in glass containers for export.

Best yields of macadamia are obtained on well-drained lands where rainfall is generous. Harvesting consists of gathering nuts as they fall, a chore that lasts for several months. Annual yields of 45 kilograms (100 pounds) or more per tree are obtained after the trees are 15 years of age.

Pecan, *Carya illinoensis* (Juglandaceae)

The pecan was highly regarded as a wild nut tree by settlers in what are now the border states and the upper South of the United States. Several species of hickory (the genus *Carya*) yield edible nuts, but *C. illinoensis* is responsible for those pecan cultivars

that today support a sizable orchard industry. These thin-shelled ("paper shell") varieties account for about half the domestic production (which in most years well exceeds 100,000 metric tons), and are grown chiefly in Georgia and Alabama. The harvest from wild and seedling trees exceeds that from select cultivars in Louisiana, Texas, and Oklahoma. There is little growing of the pecan outside of the United States, although successful introductions have been made into Australia and China.

Typical of hickory fruits, the pecan has an outer leathery husk that splits open when mature to reveal an inner shell that surrounds the two kernels. Harvest is by picking the nuts from fallen fruits. Sometimes the fruits are knocked from the trees with long poles, but, since a mature pecan tree is of large forest stature, mechanical harvesting aids are not frequently feasible. The nuts are dried, or "cured," for a few weeks, and they are then ready for eating without roasting or other treatment. Cracking and shelling machines handle most of today's crop, which is usually marketed shelled.

Pistachio, *Pistacia vera* (Anacardiaceae)

The pistachio (or pistache), native to Western Asia, has been widely cultivated since ancient times in the Middle East. It is in the same family as the mango, the cashew, and poison ivy. There are a dozen or more species of *Pistacia*, some serving as rootstock for grafted *P. vera*, but only *P. vera* has the dehiscent fruit that yields commercially acceptable edible nuts. In India and Afghanistan, nuts are still obtained from wild trees. In general, the orchard cultivars of Iran, Turkey, and Italy differ little from wild pistachios, having been selected mainly for larger nut size. An increasing number of pistachio orchards have been established in the interior valleys of California, where the locally developed 'Kerman' cultivar yields larger, better-flavored nuts than can be imported. Also, modern California techniques free the nut from their outer hulls quickly, thus preventing the "moldy" discoloration that traditionally required imported shells to be dyed red. Most commercial production comes from Iran and Turkey, but California production is exceeding 5000 metric tons. World production approaches 70,000 metric tons annually.

Since *P. vera* does not root readily from cuttings, most cultivars are propagated by budding onto seedling rootstocks. Choice of a suitable species as rootstock can result in a dwarf tree more easily reached for harvesting. Budded seedlings should be planted to the field very early, since they tend to develop a taproot that suffers in transplating. Transplanting is usually accomplished during the dormant season (December–January). Mature trees are usually spaced about 9 meters (30 feet) apart; trees planted closer together for earlier yield can be thinned. The trees seem to prefer well-drained soils having adequate calcium; the tree thrives on sandy loams, and is quite tolerant of drought. The pistachio is relatively slow-growing and needs little pruning, at least with budded stock. It is one of the few dicotyledonous trees known to attain an age of more than 1000 years.

The species is dioecious. As low a ratio as one staminate (male) tree to twenty pistillate (female) trees is considered adequate in some orchards. The diploid chromosome number for *P. vera* has been reported both as 30 and as 32. Fruit buds are initiated by winter conditioning. Like the apple, the pistachio tends to bear heavily in alternate years. Fruiting begins about the fifth year, but it is not really remunerative until the tree is at least 10 years old. Trees 20 years old are reported to yield 68 kilograms (about 150 pounds) or more of marketable nuts per tree. The fruit is a small drupe that matures in autumn, at which time the outer "pulp" slips easily from the inner shell. The latter splits longitudinally, making it relatively easy to free the edible seed within. In the Near East, picking is done largely by hand. The husks are laid out on a cloth spread under the tree, and the nuts are removed from the husks by hand. These are dried in the sun. In Turkey the dried nuts are often soaked in water, then hulled on stone rollers, accompanied by winnowing.

The nuts are typically consumed roasted and salted, or they are absorbed by the confectionery trade. When harvested, the nuts contain about 40 percent moisture. Dried kernels are approximately 20 percent protein and nearly 60 percent oil. The oil is approximately 70 percent oleic, 20 percent linoleic, and 8 percent palmitic.

Walnuts, *Juglans* spp. (Juglandaceae)

There are a number of kinds of walnuts, ranging from the wild black walnut (*J. nigra*) and the butternut or white walnut (*J. cineria*) of the United States, of which there has been only limited domestication, to the English or Persian walnut (*J. regia*), which is extensively orcharded throughout southern Europe, in China, and more recently on the Pacific coast of the United States.

The black walnut is one of the outstanding timber trees of the eastern deciduous hardwood belts in the United States. It has been much decimated for timber and veneer, but some nuts are still collected from wild trees for home consumption and, in certain rural areas, for commercial sale. Walnut fruits are drupes. The husk of the black walnut does not readily separate from the shell; it is usually allowed to rot away. Moreover, the husk strongly stains hands and clothing. Add to this the hard-to-crack inner shell, and it is no wonder that the black walnut nut industry has not made great commercial strides. Production of the meat is usually a sideline activity in rural areas for "pin money." The nut fragments are mostly used as confectionery flavoring, for it is difficult to remove the kernels entire.

World production of the English walnut, however, comes to nearly 1 million metric tons annually. California and Oregon produce 200,000 metric tons, Italy more than 60,000 metric tons, and Europe as a whole about 250,000 metric tons. Significant additional production occurs in Turkey and China. *J. regia* is native to Persia. It is a handsome ornamental tree in warmer climates. In California orchards it is customarily grafted onto American walnut rootstocks. The husks fall away much more readily than those of the black walnut, leaving an attractive shell in which the English walnut is customarily marketed (Fig. 21-28). Harvesting is by collection of the fallen nuts under

Figure 21-28   Fruit of the English walnut, showing the outer husk and the shell-covered nut. [Courtesy USDA.]

the trees in the orchard. Walnut meats contain about 15 percent protein as well as abundant oil.

Other nuts

The eight kinds of nuts reviewed so far by no means exhaust the possibilities. These are the chief nuts in commerce today, and they serve to exemplify this group of food plants, but others, such as the chestnuts (of the family Fagaceae) are equally significant. Chestnuts—*Castanea sativa* of southern Europe, *C. mollissima* of China, and *C. crenata* of Japan—are all much appreciated for their edible seeds, world

production of which exceeds 500,000 metric tons (about 170,000 metric tons in Europe, 165,000 in China, and 67,000 in Japan). Many other nuts are of local interest only, especially those harvested from the wild in agriculturally less advanced parts of the world. One of the few temperate-zone nuts that still comes largely from wild trees is the piñon, *Pinus edulis* and related species of the family Pinaceae, of the American Southwest. An October ritual with the New Mexican Indians was the gathering of the "Pinones," mostly picked up from the ground (but occasionally gained by robbing rodent caches, or shaking the seeds from their cones into a box or onto a blanket, if collecting has been timed to coincide with the few days of nut drop). Bumper crops occur in a given locality erratically over a period of years, so the market supply is generally opportunistic. Piñon nuts will store for years, and they are generally roasted before consumption.

As noted in the opening paragraphs, a number of nuts would seem to have great potential as food plants for the future, especially in tropical climates. Some of the nuts that are well known and were used at least in earlier days are: acorns (fruit of the oak tree, *Quercus*); African locust beans (*Parkia*); beech-nuts (*Fagus*); sal-nuts (*Ginkgo*); hackberry (*Celtis*); several hickories (*Carya*); oysternut (*Telfairia*); pignolia (seeds of certain European pines, *Pinus*); pili (*Canarium*, of the Far East); souari (*Caryocar*, of South America, also known as butter, paradise, or guiana-nut); *Terminalia;* and wingnut (*Pterocarya*).

Selected references

Barnes, A. C., 1974. *The Sugar Cane*, 2nd ed. New York: Wiley. (A readable but erratic presentation, at its best for discussing practical production and harvesting in far parts of the World, but often lacking scientific rigor.)

Blackburn, F. H. B., 1977. *Sugar Cane*. New York: Longmans. (One of the more recent treatments in the abundant literature about sugarcane, comprehensive and thorough.)

Brooks, R. M., and H. P. Olmo, 1973. *Register of New Fruit and Nut Varieties*, 2nd ed. Berkeley and Los Angeles: University of California Press. (A handy listing of cultivars—3897 of them—for scores of familiar and not-so-familiar fruit species; primarily for the professional.)

Childers, N. F., 1976. *Modern Fruit Science*, 7th ed. New Brunswick, N.J.: Horticultural Publications. (An encyclopedic treatment in a teaching text.)

Cobley, L. S., 1977. *An introduction to the Botany of Tropical Crops*, 3rd ed., revised by W. M. Steele. London: Longmans. (A thorough discussion of many important crops, both of this chapter and of others, strong on general characteristics and morphology, with a selected bibliography for "further reading.")

Cock, J., R. MacIntyre, and M. Graham (editors), 1977. *Proceedings of the Fourth Symposium of the International Society for Tropical Root Crops*. New York: AID, Unipub. (Diversified treatment by numerous authors on the origin, dispersal, evolution, basic productivity, harvest problems, and utilization of important root crops.)

Collins, J. L., 1960. *The Pineapple: Botany, Cultivation and Utilization*. New York: Wiley. (The history, taxonomy, genetics, and cultural importance of this tropical fruit.)

Condit, I. J., 1969. *Ficus: The Exotic Species*. Berkeley: University of California, Division Agricultural Sciences. (A more recent treatment updating the author's classic *The Fig*, of 1947, itself a thorough-going technical review of an age-old crop.)

Coursey, D. G., 1977. *Yams*, 2nd ed. New York: Humanities Press. (One of the principal food sources for tropical forest people of West Africa is discussed in detail. Non-food uses, such as extraction of steroids—often from wild species—for use in such products as birth-control pills, are also reviewed.)

Darrow, G. M., and others, 1966. *The Strawberry: History, Breeding and Physiology*. New York: Holt. (A readable review of this small fruit, involving history, breeding, and culture.)

Duckworth, R. B., 1966. *Fruit and Vegetables*. Elmsford, N.Y.: Pergammon Press. (Well-expressed, dwelling more upon chemical composition, structure, physiology, production, and nutrition than upon the botany of the plants.)

Eck, P., and N. F. Childers (editors), 1967. *Blueberry Culture*. New Brunswick, N.J.: Rutgers University Press. (A broad compilation of information on this special crop, including its botany, breeding, and growing.)

Edmond, J. B., and G. R. Ammerman, 1971. *Sweet Potatoes: Production, Processing, Marketing*. Westport, Conn.: AVI. (A remarkably comprehensive, factual review distilling, in understandable language, the sum total of knowledge about the sweetpotato to the time of publication.)

Gould, W. A., 1974. *Tomato Production, Processing and Quality Evaluation*. Westport, Conn.: AVI. (Complete coverage is given in three parts—production, processing, and technology. The first of these has the greatest botanical interest, reviewing breeding and harvesting.)

Harris, P. M. (editor), 1978. *The Potato Crop*. New York: Wiley. (Comprehensive treatment of this important crop, with opening chapters on history and biosystematics by J. G. Hawkes.)

Hawkes, J. G., R. N. Lester, and A. D. Skelding (editors), 1979. *The Biology and Taxonomy of the Solanaceae* (Linnaean Society Symposium 7). New York: Academic Press. (Covers a wide range of species in this economically important family, including potatoes, tomatoes, peppers, tobacco, and several medicinals. There may be more detail than the general reader wants, especially on chemical qualities and biosystematics, but the volume provides an excellent overview of the family.)

Herklots, G. A. C., 1972. *Vegetables in South-East Asia*. London: George Allen and Unwin. (Encyclopedic coverage about a subject on which there is little reference literature in the western world; the opening chapters on "principles" are followed by successive chapters on greens, cabbages, beans and peas, cucurbits, other fruits, root crops, and market vegetables.)

Janick, J., and J. N. Moore (editors), 1975. *Advances in Fruit Breeding*. West Lafayette, Ind.: Purdue University Press. (Experienced horticulturists offer up-to-date accounts of their fields of specialization.)

Jaynes, R. A. (editor), 1979. *Nut Tree Culture in North America*. Hamden, Conn.: Northern Nut Growers Association. (An updating, with revised format, of the 1969 publication, utilizing the talents of twenty-five authorities conversant with the important nut trees. The physiological factors that determine nut-tree habits and performance are mentioned. The book is well illustrated, and should be of interest both to practical nut growers and to those with an academic interest.)

Jones, H. A., and L. K. Mann, 1963. *Onions and Their Allies: Botany, Cultivation, and Utilization*. New York: Wiley Interscience. (A "World Crop" series monograph on the cultivated species of *Allium*.)

Martin, F. W., 1974–1975. *Tropical Yams and Their Potential:* I, *Dioscorea esculenta;* II, *D. bulbifera;* III, *D. alata* (USDA Agricultural Handbooks 457, 466, and 494). (Short accounts of promising yam species determined from research on the collection assembled at Mayaguez, Puerto Rico.)

McCann, T., 1976. *An American Company.* New York: Crown. (An entertaining account of the promise and the pitfalls of being in the banana business, with respect to the ups and downs of United Fruit.)

Menninger, E. A., 1978. *Edible Nuts of the World.* Stuart, Fla.: Horticultural Books. (A more popularized approach than Woodroof's, in a field where there is little comprehensive literature.)

Mortensen, E., and E. T. Bullard, 1964. *Handbook of Tropical and Subtropical Horticulture.* Washington, D.C.: U.S. Department of State, Agency for International Development. (A manual written for the nonspecialist.)

Nagy, S., P. E. Shaw, and M. K. Veldhuis (editors), 1977. *Citrus Science and Technology* (2 vols.). Westport, Conn.: AVI. (A wide-ranging review of this important subtropical industry.)

Nieuwhof, M., 1969. *Cole Crops.* London: Leonard Hill. (A "World Crop" series monograph by a Dutch authority on the cultivated species of *Brassica,* detailing history, botany, growing, and handling.)

Onwueme, I. C., 1978. *The Tropical Tuber Crops.* New York: Wiley. (A didactic approach, with emphasis on culture and care of yams, cassava, sweetpotato, and cocoyams.)

Plucknett, D. L. (editor), 1979. *Small-Scale Processing and Storage of Tropical Rootcrops.* Boulder, Colo.: Westview Press. (This is an attempt to give modern consideration to methods of handling and storage of perishable but basic staples for millions of people, especially taro, yam, and sweetpotato.)

Reuther, W., H. J. Webber, and L. D. Batchelor (editors), 1967, 1968. *The Citrus Industry* (2 vols.). Berkeley: University of California, Division of Agricultural Sciences. (A monumental work on citrus.)

Rogers, D. J., and S. G. Appan, 1973. *Manihot and Manihotoides* ("Flora Neotropica" Monograph 13). New York: Macmillan. (The authors explore the cultivated species of cassava, especially abundant in the American tropics where taxonomy of the genus *Manihot* has received little attention.)

Ryder, E. J., 1979. *Leafy Salad Vegetables.* Westport, Conn.: AVI. (A review of the production of familiar herbaceous salad crops.)

Salaman, R. N., 1949. *The History and Social Influence of the Potato.* Cambridge: Cambridge University Press. (The potato throughout its history; a delightful treatment.)

Samson, J. A., 1978. *Tropical Fruits.* London: Longmans. (Businesslike examination of a subject too long given little attention.)

Shoemaker, J. S., 1978. *Small Fruit Culture,* 5th ed. Westport, Conn.: AVI. (A revised and updated printing of this landmark treatment, which, in spite of some deficiencies, is almost the only source of information about the whole field.)

Simmonds, N. W., 1966. *Bananas,* 2nd ed. London: Longmans. (The definitive monograph about bananas by a recognized authority.)

Singh, L. B., 1960. *The Mango: Botany, Cultivation, and Utilization.* New York: Wiley. (A comprehensive treatise on one of the most popular and luscious of tropical fruits by an Indian authority.)

Smith, O., 1977. *Potatoes: Production, Storing, Processing,* 2nd ed. Westport, Conn.: AVI. (A comprehensive review of the potato, with emphasis on the crop and rather little concerning its origin and ancestral relations. One chapter is devoted to the processing of potatoes, the most important usage for the crop.)

Teskey, B. J. E., and J. S. Shoemaker, 1978. *Tree Fruit Production,* 3rd ed. Westport, Conn.: AVI. (Comprehensive treatment brought up to date on the production of commercially important deciduous tree fruits—apple, pear, peach, cherry, plum, apricot, and nectarine. A good reference and guide for orchard practices.)

Viegas, A. P., 1976. *Estudos sobre a mandioca.* Instituto Agronomico Estado Sao Paulo. Brascan Nordeste, Brasil, 1976. (Even though in Portuguese, Viegas's monograph, abundantly illustrated, deserves reference concerning this important crop species, which he believes to have originated in Brazil and for which he utilizes the name *Manihot utilissima.* Viegas is an accomplished botanist, and certainly Brazil is the heart of *Manihot* diversification. Morphology is well covered, and there are chapters devoted to the practicalities of cultivation and marketing, with an extensive multilingual glossary and a bibliography included.)

Weaver, R. J., 1976. *Grape Growing.* New York: Wiley. (Covers the history and classification of grapes briefly, with emphasis on the strategy of growing grapes, especially in California.)

Westwood, M. N., 1978. *Temperate-Zone Pomology.* San Francisco: W. H. Freeman and Company. (An excellent treatment of fruit and nut production in temperate climates.)

Whitaker, T. W., and G. N. Davis, 1962. *Cucurbits*. London: Leonard Hill. (Part of the "World Crops" series, this is one of the definitive modern reviews of the important members of the family Cucurbitaceae, useful as background for more recent literature.)

Woodroof, J. G., 1979. *Tree Nuts: Production, Processing, Products*, 2nd ed. (2 vols.). Westport, Conn.: AVI. (A thorough review, with the commercial flavor characteristic of AVI publications, covering almonds, brazil nuts, cashews, chestnuts, filberts, pecans, walnuts, pinenuts, and pistachios.)

Yen, D. E., 1974. *The Sweet Potato and Oceania*. Honolulu: Bishop Museum Press. (Thorough research on the sweetpotato, its origin, distribution, and importance to susbsistence farming communities, especially in the Orient, mark this excellent study. Yen confirms the American origin of the sweetpotato, but also its presence in Polynesia by 500 A.D., with later spreading by the Spanish and Portuguese in the fifteenth and sixteenth centuries. The sweetpotato's place in ten contrasting cropping systems is of special interest.)

22
Plant Extractives and Derivatives

This chapter is an overview of the ways in which many plants support civilization in addition to providing sustenance. Direct consumption as food is, of course, the most obvious and important use for plants. But after food wants are satisfied, people turn to somewhat more sophisticated uses of plant products. Extractives and derivatives from plants supply us with the raw materials to make other products, many of them of a complicated industrial nature requiring much technical skill to fabricate—paints, automobile tires, and plastics, for example. Other derived products are little different from some of the foods reviewed in the previous chapter, and they are rather arbitrarily discussed here rather than there—beverages, seed oils, and spices in particular.

Yet many of these foodlike products are processed to such an extent that they bear no resemblance to the natural plant structures from which they are derived. Many others are specialized components that are isolated and often concentrated—perhaps useful for flavoring, but with little real nutritive value. Thus the grape was included in Chapter 21, but wine, derived by fermenting grape juice, is treated here. There is no compelling reason for having lumped soybeans and sugars among the food plants rather than in this chapter, except that they are both commonly thought of as foods (even though in technically advanced countries the soybean is mostly

processed for oil and meal, and only secondarily made into foods; while sugar is extracted from the sugarcane stem or the beet root, which are consumed directly only in minor quantities).

Organic chemists have learned to synthesize many of the products once derived solely from plants, and they have provided excellent substitutes for many others, particularly those from limited or obscure wild plant sources. This has been very much the case with medicinals, for example. Thus a change in emphasis has been occurring in recent decades, away from wide exploratory efforts with a diversity of limited derivatives and extractives and toward mass production of a few crops yielding steady, economical supplies of substances the organic chemist can use to mold what modern civilization demands. Consequently, there have arisen gigantic enterprises based upon huge volumes of such things as soybean and palm oil, processed starch, and resins or essential oils obtained as by-products of pulping. The chemical industry depends even more upon the residues from ancient plants—petroleum and natural gas—than upon living plant derivatives. But as these fossil raw materials become exhausted, civilization must turn ever more to raw materials from renewable crops or biomass.

Because of obvious similarities to the food plants just discussed, beverage plants will be the first group treated. Then will come oils, fats, and waxes from vegetational sources; next, spices and other products important for essential oils; then medicinals, including psychoactive drugs; and then fermentation products. Next we will examine fumatories and masticatories, such as tobacco and betel. It is only a step further to those extractives employed almost entirely for industrial purposes (that is, those that are not taken internally), such as latex products, resins, tannins, and dyes.

BEVERAGE CROPS

Even stronger than the need to satisfy hunger is the need to satisfy thirst. Water has always been and will continue to be the prime thirst-quencher. But since the beginning of civilization, inquiring people have sought additives for water that might make it more tasty and zestful, or have turned to various juices and fermentations for their enjoyment as well as to satisfy their thirst.

The earliest "beverage" was probably juice sucked or squeezed from fruit. What with spores of wild yeast blowing everywhere about, it was only a small step further to the discovery that natural fermentation could convert these sweet juices into wines. Nor would it take a great deal of insight to discover that hard, dry grains, which must have been difficult and unpleasant to consume under primitive cooking conditions, could be softened by soaking, and that the resulting liquor would ferment to a pleasant and nutritious beer. (Sumerian tablets more than 5000 years old describe in detail the making of several varieties of beer.) These are still the basic techniques by which people make their alcoholic beverages, the products of fermentation; wines and beers are among the oldest and most cherished of beverages.

Unfermented fruit juices, of course, contain no alcohol. Modern "soft drinks" are, essentially, synthetic fruit juices, compounded of sugar, fruit acids, and other flavorings. But more universally utilized than these are coffee and tea, in which tasteful substances (including stimulating caffeine and related alkaloids) are diffused from plant parts steeped in water. Cocoa, or chocolate, is perhaps consumed less as a beverage these days than it is as chocolate candy. But in its ancestral American homeland, it was highly esteemed as a ceremonial beverage, and it is still much used in hot chocolate or chocolate sodas.

Almost any sugary or starchy substance can be fermented. Primitive tribes in various parts of the world have long used locally available substances for fermentation to alcoholic beverages—the potato in the Andes (for *chicha*), cassava in the tropical lowlands (manioc beers), palm toddy in the Eastern tropics, and various obscure plant materials "secretly" chosen by the shaman for ceremonial purposes. Some of these products are hallucinatory and even semipoisonous. A few secondary beverage plants can be listed, but coffee, tea, chocolate, beers, and wines must serve here as examples of beverages derived from plant sources.

Coffee, *Coffea arabica* (Rubiaceae)

Unlike most of the crop species discussed in this chapter, coffee (Fig. 22-1) is a relatively recent domesticate, having become popular as a beverage only since the eighteenth century (although the berry may have served as a stimulant and "medicinal" much earlier in tribal cultures of the Near East). So quickly and widely has coffee become popular that today it ranks among the leading international agricultural commodities. Total world production of dry "beans" reaches about 4 million metric tons annually, nearly one-fourth of which comes from Brazil. Colombia is the next most important coffee-raising country, followed by Mexico, Central America, Indonesia, and several African nations (principally Ivory Coast, Uganda, and Ethiopia).

The genus *Coffea* is a large one, numbering about 40 species. Only two of these species have attained commercial importance for the fruit or berry: *C. arabica* (accounting for about 75 percent of the coffee crop), and caffeine-rich *C. canephora* ("robusta" coffee, mainly produced in Africa, and especially suitable for making "instant" coffees). The species *C. arabica* is tetraploid with 44 chromosomes and is self-compatible; all other species are diploid with 22 chromosomes and are apparently self-incompatible.

Recent investigations leave little doubt that *C. arabica* is native to and still grows wild in the moun-

Figure 22-1 Woman stripping coffee berries in the highlands of Kenya. [Photograph by Per Gunvall, courtesy World Bank.]

tainous sections of Ethiopia. It is difficult to tell the escapes from indigenous plants, so little has domestication altered the species. Perhaps during the Middle Ages the first coffee plants were brought from Ethiopia to Yemen, where cultivation may have begun, and these yielded the planting stock taken to Java by the Dutch in the seventeenth century and to the New World early in the eighteenth century. Brazil gained highly restricted live coffee seeds when serving in 1727 as arbitrator of the boundary dispute between French Guiana and Dutch Guiana. Plantings were first made in lower Amazonia, but by the end of the century had extended to southern Brazil, the center of production today. Most Brazilian coffee is from the 'Mundo Novo' cultivar, said to be a hybrid between *C. arabica* var. *arabica* and a mutant from Réunion Island, *C. arabica* var. *bourbon*.

A serious leaf-rust disease, caused by the fungus *Hemileia*, has accompanied coffee wherever it is grown, and native breeding stock capable of resisting the disease has been sought in the Ethiopian homeland. The rust does not seem too troublesome on the robusta coffees in Africa, provided the trees grow in the shade of an overstory of such species as *Albizzia*. Because coffee has attained such great commercial importance within the last century, a number of elite clones have been selected in various parts of the world for vigor, yield, and degrees of resistance against disease.

Coffee thrives best in a rainforest habitat at moderate elevations. The southwestern Ethiopian homeland has an average annual rainfall of nearly 160 centimeters (63 inches), with the months from April to September having the heaviest rainfall. Coffee cannot stand freezing, and it performs best where seasonal temperature variation is small and the average temperature is about 20°C (68°F). It grows best on deep soils, which in its native African homeland are generally volcanic. When first introduced into Brazil it was grown on low valley land, and was only later found to prosper far better at the higher elevations in São Paulo, on deep soils rich in iron and potassium.

In Africa the coffee trees are often grown as dooryard plants for local use. But a number of large coffee plantations have been established, in which crop production is more efficient. Yields have gradually been increased to nearly 700 kilograms per hectare (about 620 pounds per acre), but under experimental conditions yields nearly twice as high have been obtained. Coffee yields are greatly increased by mulching (usually alternate rows are mulched each year). In Africa, coffee plants are usually started from seed in nurseries and transplanted to the field. In Brazil, direct field planting is common, with 3 or 4 seeds sown to a hole to provide several "stems." Recommended spacing for mature multiple-stemmed trees is about 3 meters (10 feet). In Africa, the tendency is to shade the plantation (thus reducing bearing, which would otherwise have to be controlled by pruning, and also helping to inhibit rust); in Brazil, plantings are generally in the open. In most tropical lands, weeds are controlled by hand and fertilization is limited. The desirable flavor in coffee seems to result more from the location in which the coffee is grown than the kind or amount of care. Colombian, Blue Mountain Jamaican, and highland Central American coffees are usually considered superior to African coffees, even though the same cultivars may be planted.

Coffee berries are harvested by hand. In much of Africa and the Near East "dry" harvest is practiced: individual berries are dried in the open long enough to prevent deterioration of the pericarp. At a later time the hard seeds, commonly called beans, are separated mechanically from the pulp, often by special mobile machines circulated from village to village. "Wet" processing is more common in the Americas, the berries first being floated to separate the defective berries from the good ones. The good berries settle to the bottom of the flotation tanks, and pulping machines remove most of the fleshy outer parts. The remaining fragments of pulp are fermented for approximately a day, and the beans are then washed again. The beans are then dried in open paved areas, either on mats or in trays, and polished in special mills. After the beans have been polished they are ready for roasting and grinding. They are about 14 percent protein, 10–13 percent oils (including volatile caffeol), and 1.5 percent caffeine.

The roasting process develops the coffee aroma; the intensity of roast varies from location to location according to the preference of the market. Roasted and ground coffee gradually loses its flavor, and eventually turns rancid, so that "fresh-roasted" coffees have particular appeal to the connoisseur. When the ground coffee bean is steeped in hot water, it loses about 25 percent of its weight as soluble constituents, including caffeine, the stimulant for which coffee is noted. To varying degrees, caffeine relieves fatigue and promotes a feeling of well-being, hence the modern "coffee break."

The coffee industry has been beset by cycles of glut and scarcity. Most scarcities have been caused by serious freezes in the chief Brazilian producing regions, such as the one in July, 1975; freezes may kill all above-ground parts and necessitate a three- or four-year recovery period for the new sprouts to mature. This can have serious consequences on prices and can mean economic hardship for countries like Brazil, which depend upon coffee for a large share of their foreign exchange. Encouraged by high coffee prices, growers often make extensive plantings that in a few years oversupply the market and depress the price. Brazil has attempted to regulate production by destroying huge surpluses of coffee from time to time or by subsidizing removal of coffee trees. With most of the good coffee land in Brazil now already planted, overproduction may not be so voluminous as in the past; but there are still ample opportunities for increasing coffee production in Africa and the Far East.

Tea, *Camellia sinensis* (Theaceae)

Tea (Fig. 22-2), like coffee, is a beverage prized for the stimulating effects of the caffeine it contains. About half the people of the world drink it regularly. Tea is made from young leaves of the tea plant, a tropical broadleafed evergreen that may reach a height of 12 meters (40 feet) if not pruned. It grows best in equable, moist environments where the rainfall is at least 150 centimeters (about 59 inches) annually and the temperature varies from 21° to 32°C (70° to 90°F). The best tea-growing locations are Assam (northeastern India) and Sri Lanka, which

Figure 22-2   Women picking tea leaves from an unshaded planting in Kenya. [Courtesy World Bank.]

together account for half of the world production. Other important tea-producing countries are China, Japan, Indonesia, and Kenya. World production comes to nearly 2 million metric tons annually.

The exact origin of the tea plant is unknown, but the art of growing tea has been practiced in China for 4000 years, and presumably the ancestral tea plants originated there. Tea drinking was not prevalent in Europe and England (where perhaps it is now most common) until the Dutch, and eventually the English East India Company, initiated the tea trade during the 1600s. How important it became as an item of international trade is known to every American school child who has studied about the "Boston Tea Party" and the American Revolution. Tea grows well in the United States, too, some commercial plantings having been made in South Carolina, but labor costs there compared with those in the Orient have not been conducive to the development of a domestic tea industry.

Tea plants are usually started in nurseries as seed-lings and then transplanted to the plantation fields. Clonal selections are increasingly being propagated vegetatively. In Sri Lanka, average yields of 895 kilograms per hectare (nearly 800 pounds per acre) are said to increase to 1680 kilograms per hectare (nearly 1500 pounds per acre) where improved cultivars are planted. Tea seeds don't preserve well, and after ripening in October and November they are usually sown immediately. Fields are generally planted 1.25 meters (about 4 feet) apart in locations with ample drainage. Taller trees are sometimes maintained over the plantings for shade (shade conserves moisture to about two-thirds that needed in unshaded plantings, and the trees contribute nutritive litter to the soil). The tea plants are top-pruned on about a three-year cycle to give the plants a low, spreading structure that facilitates leaf picking. The tea plant is adapted to acid soils. It is a "heavy feeder," and, in the normal course of harvesting, 55 kilograms or more of nitrogen is lost per hectare (about 50 pounds per acre). For good production, nitrogen and other nutrients must be restored through mulches, shade-tree drop-pings, manures, and fertilizers. Foliar sprays of urea are often employed, sometimes containing trace elements (especially zinc). Weeds have generally been controlled by hand cultivation, but nowadays herbi-cides (such as paraquat) are being more and more used. Insecticide application is carefully regulated to avoid affecting the salability of the tea leaves.

Women circulating through the plantation pluck the leaves by hand (Fig. 22-3), removing the termi-nal bud and the two or three leaves immediately be-low it. A tea bush generally yields nearly 1 kilogram (about 2 pounds) of plucked green shoots per year. The basketfuls of green leaves are brought to a re-ceiving station and subjected to a "withering" (fer-mentation) that develops taste and aroma. Leaves rich in tannin are considered superior. The withered leaves are rolled to rupture the cells, freeing juices and enzymes that play a part in further fermenta-tion, for which the leaves are spread in moist rooms at a temperature of 26–27°C (79–81°F) for a few hours. Here the brown color and pleasing aroma de-velop, after which the leaves are dried by warm air until their moisture content is reduced to about 3 percent and the leaf turns black. Orange Pekoe is a tea of this type, highly esteemed in the western world. Green tea may be preferred locally, but it is a minor export item. Green tea is not withered.

As with the coffee berry, the quality of the tea leaves varies with the area of production (and also with the care taken in its cultivation and processing). Expert tea tasters can almost pinpoint the location from which a sample comes. Obviously much hand labor is involved in tea growing, and the industry remains largely confined to countries with an abun-dant source of inexpensive labor. In China and throughout Southeast Asia, isolated dooryard trees yield appreciable tea for home consumption. On the large tea plantations, waste leaf and dust fragments serve as a source of commercial caffeine. Tea leaf ready for marketing contains up to 45 percent solu-ble materials, including dextrins, pectins, and essen-tial oils as well as 1.5–5.0 percent caffeine (theine). Tea accumulates several parts per million of fluorine, an element with a reputation for preventing dental caries.

Figure 22-3 Close-up of tea plucking in Indonesia. [Courtesy International Development Association.]

Cacao, *Theobroma cacao* (Sterculiaceae)

Cacao (Fig. 22-4), the source of cocoa and chocolate, had already been domesticated when the first Euro-peans reached the New World. There is speculation that the species may be native to the South American lowlands just east of the Andes, but certainly it was well known (and most famed) as the prized *xocolatl* of Montezuma and his Aztec empire of Mexico. Cacao was soon introduced by the Portuguese into Africa, chief seat of modern-day production, which each year totals about 1.5 million metric tons of cacao beans worldwide. About one-fourth of the world production comes from tiny Ghana, and one-sixth each from Nigeria, Ivory Coast, and Brazil.

INDUSTRY OF PLANT AGRICULTURE

Figure 22-4 Cacao fruits.

The cultivated cacaos have been divided into three groups: the Criollo (with the finest flavor), the Forastero (the South American cacaos that were introduced into Africa and now account for most production), and the Trinitario (grown in Trinidad, possibly a hybrid of Criollo and Forastero). Although there is some vegetative propagation of cultivars, most of the cacaos are grown from seeds and exhibit typical seedling variability. Selection for yield and disease resistance has, however, provided superior populations to those found in the wild. The trees are believed to be partially self-pollinated and partially cross-fertilized, with at least the widely planted Amelonado populations reputed to be uniformly self-compatible.

Cacao is generally planted as an understory, beneath the high shade of larger trees. Plantings in full sun usually suffer more from weed competition and pests and often grow poorly, although yields may be higher if good husbandry is practiced. Fertilization

improves yields. Virus and "black pod" diseases are the main hazards, although the pests attacking cacao are legion. Resistant cultivars are constantly being sought.

The cacao trees bear small flowers along the trunk, which, when fertilized, develop into pendulous, football-shaped fruits that hang from the larger branches. The fruits have a tough outer husk and a soft pulp that contains several seeds. A single prime tree may bear 70 or more fruits in the course of a year, providing annual yields of cacao beans that average 335 kilograms per hectare (about 300 pounds per acre) but that, in select locations, may be as high as 2.25 metric tons per hectare (about 1 ton per acre). Cacao fruits are carefully removed by hand to avoid injuring the flowering "cushions" along the trunk. Usually the pods are split by hand, and the seeds are hand-scraped from the pulp.

The seeds are cured by subjecting them to a period of fermentation (during which unpleasant odors develop as the temperature mounts in the mounds of seeds); it is this step that helps develop the characteristic chocolate aroma. No longer viable after being cured, the seeds are then cleaned or polished further, and sent on to market. Final processing includes roasting, mechanical cracking of the seed shell, separation of the kernels, and expression of the meats to yield the oils and fats that constitute the "cocoa butter" of the trade. Typically, 50–57 percent of a bean is cocoa butter. This is one of the most valuable edible fats known, and it is a constituent of the finest chocolate candy. The "cake" that remains after expression is a source of the caffeinelike alkaloid theobromine. The cake may also be employed as flavoring for various products of the grocery and confectionery trade.

Other nonalcoholic beverage crops

As noted earlier, a great many plant species yield beverage flavorings. These range from the "tonics" of folklore to stimulants and even hallucinatory substances that often play a part in primitive (and present-day) rituals. Some are as important in limited regions as coffee, tea, and chocolate are the world over—for example, maté (Paraguayan tea), in southern South America. Others are only occasionally used and enter little (if at all) into commerce.

Cola, *Cola nitida* and *C. acuminata* (Sterculiaceae)

Cola trees, native to western Africa and now widely introduced throughout the world tropics, gave their name to several of today's most popular soft drinks. Most commercial production is from Africa (especially Nigeria, where more than 100,000 metric tons are produced annually) and Jamaica. Cola seeds are relatively rich in essential oils, caffeine, and theobromine. They may be chewed by tribesmen or they may be pulverized and boiled to make a beverage that is said to inhibit fatigue and to forestall hunger. The seeds are borne in star-shaped fruits, from which they are removed by hand. They are sun-dried before shipment to commercial markets. Propagation is from seed.

Guaraná, *Paullinia cupana* (Sapindaceae)

Guaraná is the "cola" of Brazil. The seeds yield a paste considerably richer in caffeine than either coffee or tea, and small quantities of guaraná are reputedly sufficient to counteract fatigue. *Paullinia* is a trailing shrub found in the wild and casually cultivated in Brazil. The fruits are hand picked, pulverized, and dried for sale to make the guaraná beverage especially popular in northern Brazil.

Maté, *Ilex paraguariensis* (Aquifoliaceae)

Known also as yerba maté, or Paraguayan tea, maté was apparently long used by the Guarani Indians of Paraguay before South America became settled. The maté tree, a relative of holly, is common in the broken forests of southern Brazil, Paraguay, and northern Argentina. Harvest is from both wild and planted trees. The Jesuit missionaries of the seventeenth century encouraged maté planting, and are probably responsible for having spread the species widely in this part of South America. At harvest time the smaller branches are cut by hand and are usually dried over an open fire. When dry, the caffeine-rich leaves are stripped from the branches and crushed, after which they are ready for use in the same fashion as tea leaves. Travelers in this part of South America generally carry a pouch of maté tea leaves and a strawlike *bombilla* with a perforated sievelike base. From time to time, when crossing a stream, someone makes an infusion in a hollowed gourd, and the beverage is passed around to all members of the party, as each in turn sip from the *bombilla*. A national institute to supervise the production and trade of maté was created in Brazil in 1938.

Beer

Beer, next to water, is the world's most popular beverage—perhaps 7.5 billion liters (nearly 2 billion gallons) are consumed annually. Beer making was known in Babylon 6000 years ago, and beer was a favorite beverage among the ancient Egyptians, for

Box 22–1 *Fruit beverages and ciders*

From the earliest of times, juices have been squeezed from various plant parts, particularly from fruits, to provide tasty and healthful beverages. Rapid natural fermentation of such sweet extractives, principally by yeasts and bacteria, yields a variety of end products ranging from sought-after alcoholic beverages to "spoiled" vinegarlike accumulations of acids, aldehydes, and ketones. Modern technology has devised means for preserving juices in essentially their natural state: by using preservatives, by freezing, by pasteurizing, by concentrating and then reconstituting, and so on. Tremendous juice industries have arisen, especially for citrus fruits (orange, grapefruit), pineapple, concord grapes, tomatoes, even "vegetable" mixtures that include root crops. A good opportunity for making wider use of tropical fruits that are difficult to ship would seem to be juice extraction and preservation. Some papaya juice already reaches temperate-zone markets, and certainly in the tropics a number of fruits such as passion vine and soursop are widely used for *refrescos*.

The dividing line between nonalcoholic juice and alcoholic beverages is ill-defined. Some alcoholic fermentation begins almost immediately, because of inoculation by air-borne yeast, when any fruit juice is allowed to stand. Controlled fermentations have been highly perfected in the making of wines and beers, which are discussed on subsequent pages. Among fruit drinks, grape juice, and the wine made from it, dominate wherever grapes grow well. But elsewhere, especially in more northerly latitudes, ciders have long been produced. Strictly speaking, cider is derived from apples, but the name has also been extended to expressions from pears (perry) and other fruits. "Sweet" ciders are freshly extracted, having developed little or no alcoholic content. If fermentation is allowed to continue, a "hard" cider develops that contains up to 8 percent alcohol, usually within about 6 weeks. Ciders are popular in northern Europe, where fermentation has been perfected through the centuries to a sophisticated state, as in wine making. One directed fermentation of cider yields vinegar, the production of which constitutes a considerable industry. In France, hard apple cider is distilled to yield a brandy called Calvados; and applejack was made in colonial days by letting hard cider freeze and then removing the watery ice from the concentrate. The production of about 450 million liters (about 120 million gallons) of cider is recorded annually, and doubtless a lot more consumed "on the farm."

In producing bottled ciders, bisulfites and sulfur dioxide may be added to the juice to check fermentation, after which the juice is inoculated with a sulfur-tolerant yeast. Sometimes a secondary fermentation is encouraged by the addition of sugar, which produces a bubbly beverage similar to champagne. A great advantage of producing cider fruit rather than dessert-quality fruit is that "culls"—misshapen, bruised, and defective fruits—can be utilized. Moreover, because the appearance of the fruit is not important, spraying and other orchard labors may be minimal. Most cider orchards are sodded, and the fruit is simply picked up from the ground after it falls, often mechanically. It is washed and sorted, and the trash is discarded.

whom it served as a "liquid bread." In some parts of the world, beer is still a major source of nutrition: the caloric intake of Tibetans, for example, is said to be one-third to one-half from beer. Tibetan beers are made from the millet *Eleusine* and buckwheat as well as from barley, from both of which can be distilled a potent drink known as *arak*. Beer is a healthful dietary adjunct, especially where sanitary conditions are primitive and milk and water are suspect. The mild alcoholic content of beer acts as a sanitizing agent and as a preservative, in addition to engendering a feeling of pleasantness in the drinker. Writing about the beers of old New England, Sanborn Brown (*American Scientist* 66:460, 1978) noted that "the nutritional and medicinal values of spent yeast (left over from the brewing process) had been recognized for centuries. Since colonial beers were not clarified but were murky and muddy with yeast, they supplied many of the vitamins and minerals needed for good health."

Beer is not the product of a single plant species, but rather the end result of fermentation by yeast (*Saccharomyces*) of any of a number of carbohydrate sources; modern beers are usually flavored by hops (*Humulus lupulus*). There are a number of local "beers," such as chicha, fermented from potatoes in the South American Andes; sake, fermented from rice in Japan; pulque, fermented from *Agave* leaves in Mexico; and mead, fermented from honey in Africa and Scandinavia. Early Americans, lacking malting grain, fermented maple and birch sap, honey, immature corn ears, and just about any garden plant. Even such natural fermentations as hard cider and palm toddy might loosely be considered beers. Beer brewed from malt, as is typical in North America and Europe today, will serve as our example here. The various steps and the several plant products involved in making beer are shown in Figure 22-5.

The integration of the various products and procedures to make the flavorful beverage bought as bottled and draft beer is as much an art as a science. Brewers still zealously guard their particular methods and their prized yeast strains. Regionally there are wide differences in public taste for beers: the Irish like it dark, Berliners light; the British drink it warm, the Americans cold; the Swedes view it as intoxicating, the Russians as an antidote for drunkenness. Bavaria can be considered the center for brewing, both on the basis of consumption per capita (said to be about 1.3 liters per day) and the number of breweries (reported to be 1400, one-fourth of those in existence). Munich, Germany, is responsible for one of the two great types of beer, the Münchener—a golden lager, heavy-bodied and mild. The other great type is Pilsner (after Pilsen, Czechoslovakia), which is lighter in color and body, more tart and alcoholic, with an especially creamy head.

Conventional beer is brewed from malt. Malt is derived by germinating certain cereal seeds, especially barley. During sprouting, enzymes are produced in the grain that are capable of hydrolyzing not only the remaining carbohydrates of the barley seed, but also other starchy products added to the malt before fermentation. In the United States, barley used for malting is primarily of the two-row type,

90 percent of which is produced in the Midwest. After being thoroughly cleaned, it is washed and then steeped in huge vats of water for 2 days. It is then germinated in rotating drums, where temperature and humidity are carefully controlled. At about the stage when the primary root has emerged, but before the coleoptile has ruptured, germination is halted by drying the malted barley in kilns at about 82°C (180°F). For bock beers, the drying temperature is a bit above this, caramelizing the malt to give a darker, sweeter beer.

In the typical brewing procedure an adjunct—rice, corn, wheat, potatoes, or even cassava—is added to the malt before brewing, which is more economical than using malt alone. Munich, however, prohibits by law the use of adjuncts in beer brewed there, it is said. Many brewers use about 65 percent malt, and 35 percent adjunct. This combination becomes the mash, which is soaked in hot water to produce the wort. During mashing many soluble substances are released from the malt and adjunct under the influence of the malt enzymes. After the wort is drawn off for fermentation, the spent grains are retrieved, dried, and usually utilized for stock feed. The wort is boiled with hops for two or three hours to provide the characteristic bitter flavor of beer and to help clarify the liquor.

Hops, *Humulus lupulus* (family Cannabaceae) have long been used as a medicinal and a salad green. There is no evidence that hops were used to flavor beer until the eighth century, though the species was in cultivation before that time. During the Middle Ages many monasteries became famous for the hopped beers they produced, and doubtless the hops helped to preserve the brew against deterioration in storage, even as it does today. Annual world production of hops is about 115,000 metric tons, chiefly from Germany, the United States (Pacific Coast states), and Czechoslovakia.

The hop plant is a dioecious vine, with female plants bearing spikelike clusters of bracts called "cones," which become the commercial hop. The vines are usually propagated from rhizome cuttings and trained onto specially constructed trellises. In Europe the cones are hand-harvested; in the United States they are frequently machine picked (the entire

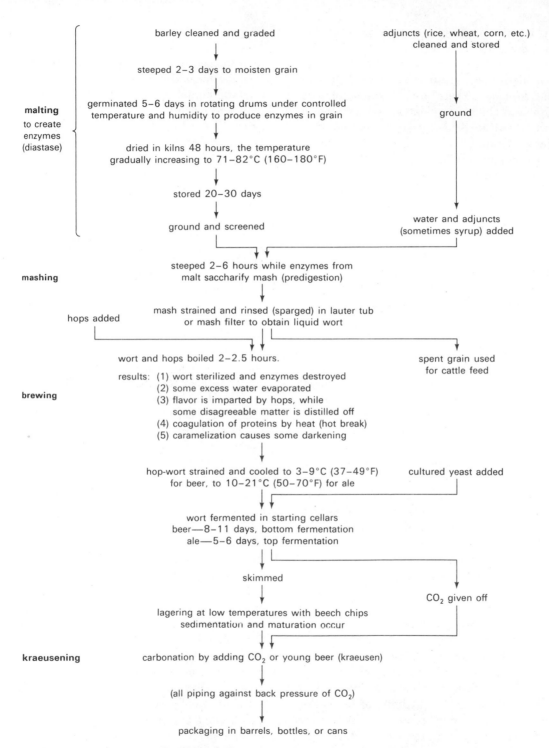

Figure 22-5 Brewing flowsheet. [After R. W. Schery, *Plants for Man*, Prentice-Hall, 1972.]

cone-bearing branches being fed into the picking machine by rotary drums). Freshly picked hops contain as much as 80 percent moisture, and must be dried until their moisture content is around 12 percent. Usually this is done in kilns or by blowing warm air through the hops. After being dried the hops are cured for about two weeks, during which time the bracts become tough and pliable and develop a characteristic aroma. In the United States each 31-gallon barrel of beer utilizes less than a pound of hops, but European beers may contain four times as much, which gives a more bitter brew. Hops without seeds (that is, female plants grown in the absence of male plants) bring a premium, for it is thought that the seeds sometimes contribute objectionable odors, and in any event they add to the mass of spent material that must be discarded after the final brewing of the wort.

The cooled wort is piped under antiseptic conditions to the starting cellars, where in huge tanks similar to swimming pools it is inoculated with a select strain of yeast (*Saccharomyces cerevisiae*), which is sometimes obtained from the sediment of a previous brew. The yeast ferments the wort to produce carbon dioxide, alcohol, and minor amounts of organic compounds that impart the characteristic flavors to the beer. After a day or two of fermentation in the starting cellar, the beer is transferred to lagering vats (*lager* means "to store"), where a secondary fermentation takes place for a number of days and various substances settle out. In traditional brewing practice the beer is then aged in closed vats for weeks or even months, where it becomes naturally carbonated, since the carbon dioxide is unable to escape. In modern, faster brewing, the beer may be artificially carbonated; in any event, before it is bottled or kegged, a bit of new "green" beer is added to give a final fillip of fermentation. The beer is then filtered and usually pasteurized (draft beer is unpasteurized, and is best kept refrigerated if it is to be stored for long). Finished beer contains about 90 percent water, 5 percent ethyl alcohol, and minor quantities of maltose, gum, dextrins, and nitrogenous substances.

There are a number of other fermented malt beverages. **Ale,** originally a strongly alcoholic English brew made from malt alone, later became an unhopped, cloudy beer fermented by floating colonies of English yeast (rather than the German yeast strains, which "work" from the bottom, giving a clearer brew). Today, however, ales are often indistinguishable from beers, at least as bottled in the United States.

Kvass, or **quass,** is a Russian beer made from barley and rye with peppermint flavoring; **pombe,** or **bousa,** is an African beer made from millet; **porter,** a dark, sweetish ale similar to bock beer, is usually made from a mixture of dark and light malts; **stout** is a porter of higher alcoholic content and strong hop flavor; and **weiss** is a light malty ale made mostly from wheat.

Wine

Grape growing was reviewed in Chapter 21. It was mentioned there that most of the grape crop is not consumed fresh but is fermented to wine. Wine making is nearly as ancient as is civilization itself in the Near East, and wine consumption has long been a cherished tradition in western culture. Wine making has evolved into a special art in Europe, and more recently in California and southern South America. A geographical band that extends from central France eastward to Prague, Vienna, and Budapest yields some of the world's finest wines—"fruity" and naturally acid, but not highly alcoholic, with a subtle blending of flavors from nicely adapted grape strains that grow in special climatic pockets. South of this band, through the Mediterranean basin, blander, "softer," less acid and more alcoholic wines of less local distinctiveness are produced. Wine making consumes about 40 million metric tons of grapes annually, largely in France and Italy, but appreciably in Spain, the USSR, Argentina, and the United States. World wine production amounts to about 30 million metric tons annually.

The principle of wine making is relatively simple: the juice of crushed grapes (called **must**), rich in fermentable sugars, is inoculated with the proper strain of yeast (or, under primitive circumstances, is allowed to ferment naturally), until the alcohol content reaches 10 to 14 percent (when the yeast is automatically killed by the alcohol). Such wines are called beverage wines, or table wines, and are consumed with meals. Fortified wines are those to which additional alcohol is added to raise their alcohol content to about 20 percent. They are generally termed apertif, or dessert, wines. Sherry and port are examples of fortified wines.

Wines are further classified as **red** (made from highly colored grapes, usually with the skins included in the must when the juice is expressed) or **white** (made from lightly colored grapes, or from highly colored grapes but with the skins excluded); **sweet** (in which not all of the sugar is fermented) or **dry** (in which almost all of the sugar is fermented); **sparkling** (if final fermentation is in an enclosed container so that carbonation results) or **still** (fermented with escape of the carbon dioxide); and so on. Subtle differences in flavor can be attributed to grape cultivar, strain of fermenting yeast, soil and growing conditions, and suitability of weather (summers with dry bright weather yield vintage wine-grape crops).

The yeast responsible for wine fermentation is generally considered to be *Saccharomyces ellipsoideus*. As in beer fermentation, a poor strain of yeast or the invasion of unwanted bacteria and molds will lower the quality of the wine, producing musty and undesirable flavors. Grape juice for wine is generally mechanically expressed. The must may be pasteurized or perhaps treated with sulfur dioxide to control unwanted organisms, and is then inoculated with a pure culture of yeast. Fermentation continues for many days, accompanied by settling out of solids, tannins, and pigments. Final fermentation usually takes place in closed vats, slowly (perhaps for a period of years), at reduced temperatures, to produce suitably mature wines. Final fermentation of champagne and sparkling burgundy takes place in sealed bottles from which the carbon dioxide cannot escape, making a carbonated wine. Some wines, such as vermouth, have bitter and aromatic flavors blended in to provide a particular taste.

Broadly considered, wine can be fermented from fruits other than the grape, to make, for example, cherry, blackberry, or elderberry wines. All of these are 70 percent or more water, about 10 to 14 percent alcohol, with highly variable amounts of sugar and trace organic substances. (Most wines have up to 1 percent acids, around 0.1 percent tannin, and a pH of 3.0–3.5.)

Distilled spirits

As noted in the discussion of wine, natural fermentation can continue only until the alcohol content is about 10 to 14 percent. To make beverages with a higher alcohol content, beers and wines are distilled to yield highly concentrated potions or even pure ethyl alcohol (industrial alcohol). Distillation involves heating the beer or wine, usually in a copper still, with fractionation of the alcohol and other volatile substances in suitable condensers. Depending upon the type of beer or wine distilled, characteristic components accompany the alcohol, to yield whiskeys, rums, brandies, vodka, and so on. The process sounds simple enough, as indeed it is in backwoods distilleries; but for choice spirits the process requires great skill and experience. Proper blending of minute quantities of ingredients imparts the characteristic flavor esteemed in a distilled beverage.

Brandy is made by distilling wine or fermented fruit mash. The distillate is generally aged in wooden casks to develop its own characteristic flavor, and it usually contains 40 to 50 percent alcohol. Legend has it that brandy was first invented by a Dutch shipmaster attempting to devise a wine concentrate to lower shipment costs. French brandies are named for the region in which they are produced (Cognac, Armagnac), and are often said to be impossible to duplicate because of distinctive characteristics of the fruit of the region and characteristic yeast strains in the casks that have developed through the centuries, to say nothing of "secret" formulations.

Whiskey is to cereal beers what brandy is to wine. Fermented grain is distilled, and the distillate is customarily aged in charred white-oak casks. Most whiskeys are about 50 percent alcohol, the "blended" types being brought up to approximately this alcoholic strength by the addition of ethyl alcohol (grain neutral spirits). Scotch whiskey is made from a barley mash in which the malt is cured in the smoke of peat fires to impart the characteristic flavor. Irish whiskey is similar, but the malt is kiln-dried. Bourbon is a corn whiskey (not less than 51 percent corn grain in the mash), and rye is a rye whiskey (not less than 51 percent rye in the mash).

Gin and **vodka** are made by distilling the mash to obtain nearly pure alcohol. In making gin, aromatics, such as juniper "berries" or sloe (*Prunus spinosa*) fruits, are added to impart the flavor. The type of grain fermented is relatively immaterial, for it will have little affect on the flavor (ethyl alcohol is ethyl alcohol no matter what the source of the mash).

Rum is the distillate from fermented sugarcane juices or molasses, suitably aged in wood. Rum distillation began in the Orient, but it was quickly picked up in the Caribbean area during the days of the Yankee Clipper trade, and most rum is now produced mainly in the West Indies and on the mainlands around the Caribbean sea. Characteristic flavorings (even a bit of brandy) may be added. Raw rums are often locally fermented and distilled in primitive areas. Broadly considered, rums may be made from other carbohydrates besides sugar or molasses.

Liqueurs, or **cordials,** were originally devised as elixirs, to which mystical properties were attributed. They are usually produced by combining brandy with a variety of flavorings. Often dried fruit of a particular type is soaked in brandy to produce such things as "cherry liqueur" or "apricot cordial." For example, creme de cacao is flavored with chocolate; creme de menthe is flavored with mint. Absinthe is made from a strong brandy flavored with wormwood (*Artemisia absinthium*) and other aromatics. Tequila is the distillate of the Mexican pulque (the fermented juice of *Agave* spp.).

VEGETABLE OILS, FATS, AND WAXES

Vegetable oils, fats, and waxes, are derivable from a great many sources; economics—the price of the oil in its market—usually governs which crops will dominate the vegetable-oil industry. In recent years, soybean oil has been a leading product—both for edible and industrial purposes—to the extent that more than 10 million metric tons of it is produced each year in the United States alone. Ponderous production, rather than glamour, characterizes the vegetable-oil business, with only moderate exploration

and search for new sources and no wars fought to command the trade. Alternate supplies of standard types of oil are too readily available. Indeed, today's vegetable-oil chemist is so adept at changing oils (as by hydrogenation) and substituting one for another that price differentials often overshadow inherent qualities. And not only can one vegetable oil be made to substitute for another, but petroleum or mineral derivatives often take over markets from botanicals, and vice versa. An example is provided by the paint industry, in which synthetic resins and latices have appreciably substituted for the once dominant drying oils, linseed and tung.

About a dozen crop plants make up the main vegetable oil-seed market. Most are traditional sources, for which an established industry has been built, which gives these plants considerable commercial advantage. It takes a while to work out details with new crop introductions and to create grower and market acceptance; but government agencies such as the USDA are on the lookout for newer specialty oils, and *Crambe* (Cruciferae), safflower (*Carthamus,* Compositae), jojoba (*Simmondsia,* Buxaceae), and others have spawned small but promising industries based upon the unique chemical qualities of the oils they produce. Since oils are relatively abundant in almost any seed, it is no wonder that researchers have looked at such an assortment as ironweed (*Vernonia*), *Cuphea,* wild carrot (*Daucus*), *Limnanthes,* and *Comandra,* and even spruce (*Picea*), *Tribulus,* and grape seed (*Vitis*). A table 50 pages long lists the oil content and other characteristics of seeds and surrounding tissues for 1253 species investigated by the USDA (Barclay and Earle, *Economic Botany* 28(2), 1974).

Vegetable oils, fats, and waxes serve many uses. About two-thirds are used for foods and one-third for industrial purposes. Several, such as corn, cottonseed, peanut, olive, and soybean oil, are used for cooking oils, margarine, and salad dressings; some are incorporated into many food products and animal feeds. But equally important are the industrial uses of these and other oils, including coconut and palm oils, but especially linseed oil, castor oil, tung oil, grape seed oil, and tall oil (a by-product of papermaking). The major industrial uses are for paints, coatings, plasticizers (especially for vinyl), polyamide resins for plastics, and surface-active agents (including soaps and detergents), and as components in linoleums and oilcloths. Other industrial uses are for lubricants, additives, hot-dip tinning of metal products, hydraulic fluids, glycerol, synthetic fibers, lecithin, food coatings, cosmetics, medicinals, printing inks, plastic foams, and fatty acid raw materials.

Of course, many plant parts contain oils, fats, or waxes, which are chemically quite similar. Oils occur in small droplets within the plant cells, and are typically most abundant in seeds. The globules are removed from the oil-bearing tissue by crushing (expression) or a combination of crushing and solvent extraction. Vegetable fats differ only slightly from oils, having acidic constituents that are more or less solid at ordinary temperatures rather than liquid. They, too, are usually most remuneratively garnered from seeds. Waxes are the fatty acid esters of monohydroxy alcohols; oils, by contrast, are fatty acid esters of trihydroxy glycerol. Waxes are found mostly as protective coatings on the leaves and stems of plants, where they function to retard water loss. Waxes are a less voluminous (and typically more expensive) commodity than the vegetable oils and fats.

The oil seeds utilized commercially generally have an oil content of 40 percent or more, some as high as 70 percent. With more than 75 million metric tons of soybeans, 27 million metric tons of cottonseed, and 18 million metric tons of peanuts produced annually, most of which is subjected to oil extraction, it is apparent that we are talking about sizable quantities of vegetable oils. In addition, about 12 million metric tons of sunflower seed, 8 million metric tons of rapeseed, 5 million metric tons of copra (dried coconut meat, used for oil extraction), about 4 million metric tons of palm-husk oil, 1.5 million metric tons of palm-kernel oil, 3 million metric tons of linseed, 2 million metric tons of sesame seed, and 1.6 million tons of olive oil are produced annually.

The production of vegetable oil is not concentrated in any single part of the world; rather, production of vegetable oil is important to less advanced agricultural economies as well as to countries that are agriculturally mechanized and industrialized.

Vegetable oils are fundamental to the functioning

Table 22–1 Percentages of polyunsaturates in major edible oils.

Oil	Polyunsaturates (percent)
Safflower	75.5
Sunflower	72.5
Soybean	60.0
Corn	55.5
Cotton	51.5
Peanut	26.0
Coconut	1.5
Lard	11.0
Olive	9.5
Butter	5.0
Linseed	70.5

Source: After C. Moses, 1977, Proceedings of *Twenty-third Annual Farm Seed Conference.* Washington, D.C.: American Seed Trade Association.

of the modern industrialized world. Thus they will be in continuing, steady demand, and their importance can be expected to increase as nonrenewable mineral resources become exhausted and petrochemicals become more costly. In this respect the crop plants we are reviewing here are equally as basic as are food plants.

There would seem to be little cause to dwell on the uses of vegetable oils as foods. Their main use, as has been noted, is for making cooking fats and margarines. There has been a great deal of commotion in recent years about substituting unsaturated oils (Table 22-1) for saturated fats in the diet, because of the concern about cholesterol in the blood. Thus a moderately unsaturated oil, such as corn oil (rich in oleic and linoleic components), carries something of a premium for the manufacture of margarine. But almost any edible oil can be converted to a cooking fat or margarine by chemical manipulation, such as changing fluid oils into fats by elevating them to a higher degree of saturation through **hydrogenation.** This is accomplished by introducing hydrogen under pressure into the heated oil in the presence of catalysts, such as pulverized nickel compounds. Margarine is formulated through such techniques, with the addition of small percentages of milk products, vitamins, and other ingredients, to make a spread that is nutritionally the equal of butter and scarcely different in taste. As a matter of fact, margarines are generally advertised as being less fattening than butter, particularly when formulated from unsaturated vegetable oils.

Industrially, many things can be made from vegetable oils (Fig. 22-6). The oils, being fatty acid esters of glycerol, can readily be broken down to fatty acids and glycerol. In fact, most fats and oils are naturally unstable: when kept for considerable lengths of time, they become rancid due to breakdown of the glycerides into various lesser products. When oils are largely glycerides of saturated acids, such as the oleics characteristic of palm, peanut, olive, and grape oils, they are said to be **nondrying** oils. They remain in a liquid state when exposed to air. If the oil is high in glycerides of the unsaturated type, particularly linoleic and linolenic with little oleic and lauric, the oil is said to be a **drying** oil (capable of forming an elastic film upon absorption of oxygen from the air). Linseed, safflower, soybean, and tung are oils of this type. Drying oils are more important commercially, and they are used in such products as paint and varnish. The nondrying oils (and an intermediate group termed "**semidrying,**" such as cottonseed, corn, and sunflower oils) are used in the manufacture of soaps and detergents and as raw materials for other industrial purposes. The classification as to drying type is not as meaningful as once it was, what with many specialty oils prized for particular chemical characteristics rather than degree of acid-chain saturation.

Saponification is the process whereby soap is made. Vegetable oils treated with alkalis, such as potassium or sodium hydroxide, split into three molecules of the metal salt of the particular fatty acid and one molecule of glycerin. Early civilizations mostly utilized animal fats boiled with wood ashes from the fire to make soap. (Wood ashes are rich in potassium carbonate.) Gradually vegetable oils, such as olive oil, were substituted for animal fat, and sodium hydroxide (which makes a harder soap than does its potassium equivalent) was substituted for wood ashes. The by-product, glycerin, has found many industrial uses, including the making of explosives (trinitroglycerin and TNT).

Figure 22-6   One of the major sources of vegetable oil is the soybean. Represented here are the proportions of drying oil and protein meal resulting from milling. A bushel of soybeans produces about 11 pounds of oil and 48 pounds of meal. The price of beans depends largely on the value of oil and meal in the marketplace.

Linoleum was invented about a century ago, when it was discovered that boiled drying oil made an excellent binder for pulverized cork (or other particles). Today many plastic materials substitute for the original linoleum (the name derived from *linum* for flax, and *oleum* for oil). Likewise, drying oils made an excellent vehicle for pigments of various sorts, which hardened into a resinous protective coating, or paint. Varnishes and enamels are basically similar to paint.

Compared to oils, vegetable waxes are of relatively minor importance. The waxes are used chiefly for polishes and carbon paper, and to a minor extent for such diverse products as candles and chewing gum.

Nondrying oils

Oil from palm fruits (Palmae)

Because oils from palm fruits of many species are remarkably similar, they are often interchangeable and mixed in commerce. Palmitic, lauric, and oleic fatty acids tend to predominate. M. J. Balick listed 23 palm species in 13 genera in Amazonia alone as sources of "oil palms of promise" (*Economic Botany* 33(1), 1979). Most are still not cultivated and are therefore only of incidental commercial importance. As commercial successes, two palms stand out, the coconut (*Cocos nucifera*) and the African oil palm (*Elaeis guineensis*). But throughout the tropics the palm family is as ubiquitous as the grass family is in temperate climates—there are 1250 species indigenous to the Western Hemisphere—and it is no

Figure 22-7   Three-month-old oil palms in a government-sponsored nursery, Ivory Coast. They will be planted in plantations when about eight months old. [Photograph by Hilda Bijur, courtesy World Bank.]

wonder many other species of palm contribute oil-bearing fruits of greater or lesser importance.

In the lower Amazon Valley, there is considerable traffic in murumuru palmnut (*Astrocaryum murumuru*). In northern Brazil (especially Maranhão), hundreds of thousands of tons of fruits of the babassú (*Orbignya speciosa*) are gathered from extensive wild stands for use in the soap industry. The ouricuri (*Syagrus coronata*) of eastern Brazil yields not only an oil from the fruit, but a wax that is scraped from the leaves, much like the wax of the carnauba palm. Extensive stands of the mbocayá or gru-gru (*Acrocomia totai*) grow in Paraguay, supplying many materials for rural living. One palm abundant in Venezuela, the bataua (*Jessenia bataua*), contains as much as 24 percent oil in the pulp (rather than the nut); the oil is chemically very similar to olive oil (about 7 percent palmitic, 9 percent stearic, 73 percent oleic, and 5 percent linoleic acids). Usually, the transportation, marketing, and processing facilities

are primitive or lacking in such tropical locales, which prevents much greater production from wild sources.

In contrast are the oil palm and coconut (copra). Nearly 4 million metric tons of palm-husk oil reach commerce annually, chiefly from west-central Africa, but significantly also from Indonesia and Malaysia and from scattered parts of the New World where *Elaeis guineensis* has been introduced or naturalized. World production of copra is about 5 million metric tons, chiefly from the South Pacific region. More than 30 billion coconut fruits are harvested annually, again mostly from the tropical Far East but also from Mexico and Brazil.

The African oil palm (Fig. 22-7) is one of the world's most useful oil-yielding species. The "palm oil" of commerce comes from the outer husk or pericarp of the fruit; a less prized kernel oil is obtained from the nut or seed. The "pulp oil" from the husk is used for making soap, in metal plating (to protect the metal surface from oxidation before plating), and in foods (margarine). The kernel oil is used mostly in the making of soap and as a source of glycerin. Under favorable circumstances, fantastic annual yields that exceed 4.5 metric tons of oil per hectare (about 2 tons per acre) can be achieved.

The African oil palm is a heliophile, growing naturally in a coastal belt about 150 miles deep from Angola to Senegal, where it is often adventive on disturbed land. A similar species, *Elaeis oleifera*, is found in the New World, and is even better adapted to wet habitats. The two species can be cross-bred. The palm fruit is still much harvested from wild stands, but some extensive plantations have been developed, particularly in Nigeria and the Congo Basin, where attempts at mechanization have been made. In Nigeria, palm oil is both an important source of food and a leading export source of foreign exchange. Several groups of cultivars have been selected there. In 1910 the African oil palm was introduced into Sumatra and Malaya, where it has been widely interplanted with rubber trees. A great many cultivars were selected, and Dutch agriculturists sent expeditions to Africa in search of potentially more valuable breeding stock. As a result, in the Far East, and to a limited extent in the New World, there has been considerable plantation growing of the African oil palm.

Under primitive circumstances, the palm fruit is boiled, the free oil is skimmed off the cauldron, and residual oil is squeezed from the pulp. The nuts are marketed or processed separately. On managed plantations, the fruit is subjected to a preliminary digestion, after which the pulp oil is centrifuged or pressed free and purified; the nuts are cleaned, dried, and cracked, and the freed kernels are further dried.

The exact home of the coconut is uncertain, for it has been widely scattered throughout the tropical coastal environments of the world. These picturesque, graceful trees of the sea coasts are often sources of food, thatch, and income for island people. Coconut fruits are among the largest in the plant kingdom, and as many as 400 may be produced annually from a single tree under favorable circumstances. Production is chiefly from the Philippines,

Kampuchea, and Indonesia. The husk of the coconut is usually removed and discarded (although there is a limited market for it as a fiber, coir). Copra oil, from the coconut, is of course a kernel oil, being derived from the endosperm of the coconut seed. In making copra, the inner shell is cracked and the "meat" is removed to be smoked or sun-dried before being sent to the extraction plants.Most of the harvesting and splitting is done by hand. The copra is gathered to a central location (or even imported into industrialized countries without processing), where it is usually expressed while heated to yield coconut oil. The press cake is sold as stock feed.

Attempts to cross coconut lines to yield "hybrid" planting stock were made as early as 1926 in Fiji, but the enterprise proved to be too cumbersome and expensive to be practical. A system of bringing select pollen to an isolated field of high-yielding mother palms (emasculated) is now producing F_1 hybrid coconut planting stock in Ivory Coast and Jamaica. But much of the world's production still comes from unselected, chance trees.

Other important nondrying oils

The **castor bean,** *Ricinus communis* of the family Euphorbiaceae (Fig. 22-8), is potentially an important industrial oil, but is of uncertain supply and poses difficulties in adapting it to mechanization. The chief fatty acid component is ricinoleic (which is amenable to dehydrogenation). Presumably native to northeastern Africa, castor bean is grown today throughout the world tropics, and it frequently "goes wild." It was introduced into India and China before recorded history. It is adaptable as a summer annual in temperate climates, although it can be a short-lived perennial in the tropics. Efforts have been made to breed cultivars that will not shatter seeds so readily, making them more suitable for mechanical harvesting. Yet, world production is only about 700,000 metric tons annually, mainly from Brazil and India. The seeds, borne in spiny capsules, have an oil content of as much as 50 percent. The oil is removed mainly by expression; whatever remains in the press cake is extracted with solvents. Because the castor bean contains an extremely toxic alkaloid,

Figure 22-8   An Indian farmer examines a new castor bean bred for quick maturation, near Hyderabad, India. [Courtesy Agency for International Development.]

ricin, the press cake is unsuitable for stock feed. The oil is used mostly as a medicinal (purgative) and high-grade lubricant, and to some extent in soaps, linoleum, plastics, inks, and various finishes. Through chemical alteration castor bean oil is often made to serve other purposes, such as substituting for drying oils and in the plastics industry.

The **olive** was discussed at some length as a fruit in the previous chapter. More than 8 million metric tons of olives are produced annually, and perhaps 90 percent of the crop is crushed to produce oil. Annual production of olive oil amounts to about 1.6 million metric tons, most of it from Spain, Italy, and Greece. Many of the cultivars have been selected with oil production in mind rather than edible fruit. Olives grown for oil may contain as much as 40 percent oil. The fruit, usually gathered by hand on small, private holdings, is crushed to yield the oil. Several expressions are made, the first giving the more valuable "virgin" oil suitable for use without further refining. The press cake is generally treated with solvents to

get the final measure of oil, and the press cake may then serve as cattle feed or as a fertilizer for the orchards. Olive oil finds its greatest use as a salad and cooking oil.

Also discussed at some length in Chapter 20 was the **peanut,** grown for its edible seed as well as its oil. Much of the African and Indian production is imported into Europe, where the oil is extracted. The nuts contain as much as 45 percent oil, with oleic and linoleic constituents predominating. It may be recovered by expression or extraction with solvents. The oil is used mostly in edible products, and the press cake, rich in protein, is a valuable livestock feed.

Semidrying oils

Major semidrying oils

The major semidrying oils of commerce originate as by-products from crops grown primarily for other purposes. Especially important are **corn** (mainly a feed cereal, yielding corn oil) and **cotton** (mainly grown for the fiber, but yielding cottonseed oil). Cottonseed production generally exceeds 27 million metric tons annually, chiefly from the United States, the USSR, China and India. Most cottonseed contains 30 to 40 percent oil, rich in linoleic glycerides, and the remaining press cake contains approximately 10 percent nitrogen.

Corn-oil production is the "other half" of wet-milling the grain for starch (see the discussion of corn in Chapter 20). The oil-rich "germ," or embryo, constitutes 6 to 13 percent of the grain. In wet milling this is freed by "cracking," which follows the initial steeping (which yields "corn-steep liquor," used in industrial fermentations). The germ is removed by flotation, and is then subjected to expression or solvent extraction. Refined corn oil, which is used principally as a cooking oil and in margarine, is rich in linoleic fatty acids.

Huge quantities of cottonseed are left after the ginning of cotton fiber. In oil production, the hulls are first removed from the seed, and they may be incorporated into cattle feeds as roughage. The kernels are then crushed between rollers, heated (to expel moisture, to make the oil fluid, and to coagulate albumenoids), and subjected to hydraulic expression or solvent extraction. After refining, the oil may be hydrogenated for margarine or used to some extent as a liquid cooking oil and in the manufacture of soap.

Other semidrying oils

Rapeseed, from *Brassica napus* and other members of the cabbage group, enters commerce to the tune of more than 8 million metric tons annually. Canada, India, and China are the chief producing nations. Rapeseed is used chiefly to produce an edible oil. The seeds contain as much as 40 percent oil, largely of erucic, oleic, and linoleic types, and about 24 percent protein. Rape was grown in India and Japan at least 4000 years ago.

Sesame or benniseed (*Sesamum indicum,* of the family Pedaliaceae) yields about 2 million metric tons of commercial seed annually, mostly from India, China, and Sudan. It is one of the most ancient of oil seeds, originating in the Near East at least 4000 years ago. The seed, which is often used as a garnish on bakery products, contains as much as 55 percent oil, largely of the oleic and linoleic types. The oil, which is extracted by expression, is used to a great extent as a salad or cooking oil. The crop is adapted to semiarid conditions, and most production comes from areas that are not agriculturally mechanized. Harvesting is often done by hand, by flailing the pods with a stick. It is said that, under mechanized agriculture, this oil seed can yield more per acre than any other annual oil crop; efforts have been made to develop nonshattering cultivars suitable for mechanized growing in the United States. Sesame oil is said to keep for several years without turning rancid, and the press cake is an excellent livestock feed.

The **sunflower** (*Helianthus annuus,* of the family Compositae) is native to North America, but the oil-crop cultivars are highly modified from the wild plants of the plains. It produces an oily seed that may be toasted for direct consumption, but chiefly the oil is expressed to make margarine and other foods, or to be used for industrial purposes. More than 12 million metric tons of sunflower seeds are produced annually, chiefly in the USSR, the United States, Argentina, and southeastern Europe. The oil, which contains large amounts of linoleic acid, is resistant to rancidity, and it has attracted attention as an extender for more expensive oils. The press cake contains more than 50 percent protein. Plant breeding in the USSR is chiefly responsible for transforming this "cinderella" species (25 percent oil, 50 percent hulls) to a modern oil crop (50 percent oil, 25 percent hulls). The oil is advocated for low-cholesterol diets.

Drying oils

Major drying oils

Linseed or flax (*Linum usitatissimum,* of the family Linaceae) is a source both of fiber and of a valuable oil (Fig. 22-9). However, cultivars selected for the stem fiber generally differ from those used for production of oil seed (from which flax fiber is the "by-product"). Total world production of linseed is approximately 3 million metric tons annually, chiefly

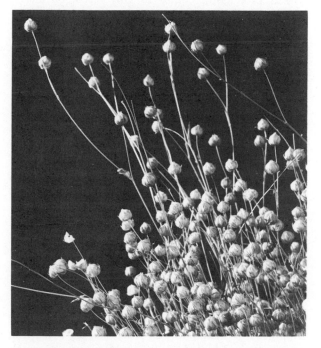

Figure 22-9 Flax plants with maturing seed pods ready for harvest and extraction of linseed oil. [Courtesy J. C. Allen & Son.]

from Argentina, Canada, India, and the United States. Linseed oil has long been the chief ingredient in oil paints, but its importance there has been declining as various latex paints and plasticlike substitutes have been developed. Linseed contains as much as 43 percent oil, which is extracted by expression. When refined, linseed oil is said to keep almost indefinitely in sealed containers. Typical of drying oils, however, when it is exposed to air (oxygen) it turns into the tough, impervious coating that makes it so valuable for paints, oilcloths, and linoleum.

Tung (*Aleurites fordii*, of the family Euphorbiaceae) once showed great promise as the source of one of the world's best drying oils. *A. fordii* is native to central and western China. *A. montana* of southern China and Malaya yields an oil of equally high quality. Tung is a small monoecious tree, with fruits that contain 2 to 5 heavy-shelled seeds with white kernels that are about 65 percent oil, chiefly of the eleostearic type. The oil is not edible because of its laxative effect, and it is used mainly in the manufacture of paints, linoleum, water-proofing, and printers ink. World production comes to about 100,000 metric tons annually, two-thirds of it from China, but Paraguay also produces a significant amount. In the 1940s a significant tung-oil industry developed along the Gulf Coast of the United States, but rising costs have now made tung growing uneconomical there.

A. fordii is highly susceptible to frost kill once it buds out; but it can stand mild winter frosts, and indeed some cold is needed to initiate full flowering and fruit production. The species does best on well-drained, fertile soils in climates with ample rainfall. The tree starts bearing at about three years of age, and it has a productive life of about 30 years. It tends to bear more heavily in alternate years. Tung responds well to fertilization, especially with nitrogen fertilizer. Since the press cake that remains after oil expression is toxic as a feed, it is usually returned to the fields for fertilizer, often accompanied by applications of ammonium sulfate.

The tung fruit is collected after it has fallen to the ground. The seed should be collected immediately and dried to prevent any hydrolysis of the oils. Often the fruits are hulled in the field; transporting the husks is eliminated by the use of portable hulling machines. Nut shells are removed mechanically at centralized locations, after which the kernels are heated, ground, and expressed, usually in a screw expeller. Solvent extraction is also feasible. Except for filtering, no refining is necessary. Tung oil dries rapidly into a hard, waterproof coating, resistant both to acids and to alkalis. In this respect, it is said to be superior to linseed oil.

Other drying oils

Scores of other species yield oils that are either minor items of commerce or are potentially useful as drying oils. By and large, only safflower, among the new entries, and soybean, among the established agricultural crops, have been of much significance. **Safflower** (*Carthamus tinctorius*, of the family Compositae) yields an oil that is rich in oleic and linoleic constituents. Production, chiefly from Mexico and India, amounts to about 1 million metric tons each year. Originally the species was grown for the yellow dye that the flowers yield, a substitute for saffron. Safflower is apparently indigenous to the Near East, and it may have arisen from *C. oxycanthus* or *C. lanatus*.

Safflower is a slow-growing thistlelike plant that produces a rosette of leaves close to the ground and orange-yellow flowers on stems that may be nearly 1 meter tall. It is tolerant of heat and drought. Safflower is planted and grown in about the same way as some of the small grains. Combines like those used for barley have also been used to harvest safflower seed in the United States. Higher-yielding lines have been developed from Sudanese and Egyptian germplasm, some producing seed with an oil content as high as 38 percent. The oil is either extracted with solvents or expressed in a screw expeller.

Safflower oil has long been used for cooking and illumination in India and as an edible oil in Europe. In the United States, safflower oil has been used for margarine, but also to make synthetic resins. The press cake is quite suitable as a livestock feed, but is not very palatable when used alone.

The **soybean** (*Glycine max*, of the family Leguminosae) has been discussed in Chapter 20. It is noted here because one of its chief products is the oil extracted from the seeds. The oil, which is high in lineolic acid, is expressed either by expeller or by hydraulic press, or it is extracted with solvents. In the United States, the most common method is continuous expelling, in which the beans are crushed or ground, steam heated to about 66°C (150°F), then fed into the expeller. The oil is chiefly used for shortening, margarine, and salad oil, but it is also used to some extent for making soap, paint, ink, linoleum, and oilcloth. The protein cake remaining after oil expulsion is especially valuable.

Nigerseed (*Guizotia abyssinica*, of the family Compositae) has long been cultivated in Africa and India for its seed, which yields an oil that is about 70 percent linoleic acid. Only recently has it gained much attention in the United States, primarily as "thistle" birdseed, which is especially esteemed by goldfinches. Imports are chiefly from Ethiopia, where production is about 200,000 metric tons annually. India produces less than 100,000 metric tons.

Buffalo gourd (*Cucurbita foetidissima*, of the family Cucurbitaceae), and other members of its genus, yield a seed oil that can run as high as 65 percent linoleic acid. Because buffalo gourd is native to the arid southwestern United States and northern Mexico, it has been suggested as a possible crop plant for this region, where few productive species are adapted. Other cucurbit seeds have an even greater content of unsaturated oil (the oil from *Lagenaria* seeds, for example, contains as much as 78 percent linoleic acid), and would seem to offer promising opportunities for the production of highly digestible drying oils as well as useful protein cake.

Specialty vegetable oils

A number of species have been proposed as prospective oil sources because their oils contain particularly useful chemical compounds. Seeds of *Euphorbia lagascae*, of the family Euphorbiaceae, are rich in epoxy acids, which are valuable for making plastics and protective coatings. Also rich in epoxy acids are the seed oils of *Cephalaria setosa* (Dipsacaceae) and *Vernonia anthelmintica* (Compositae). *Crepis alpina* (Compositae) yields crepennynic (acetylenic) acids, which are useful for resins and coatings. Various members of the family Cruciferae, such as species of *Crambe* and *Eruca*, can be good agronomic sources of erucic acid, which is useful as a lubricant in continuous casting of steel, as a rubber additive, and as a raw material in the production of various chemicals. Many other potentially valuable oil species are listed by White et al. (*Economic Botany* 25(1), 1971).

Waxes

Vegetable waxes differ from the oils in being esters of a long-chain alcohol and a fatty acid, rather than triglycerides of three fatty acid molecules. Most are solid rather than liquid at room temperature.

Carnauba palm (*Copernicia prunifera*, of the family Palmae) is the world's premier wax plant. Both wild and planted carnauba palms grow in arid northeastern Brazil, the chief area for the production of carnauba wax. Several other palms of the tropics yield waxy substances as well as oil seeds. For example, the ouricuri palm (*Syagrus coronata*) of eastern Brazil and the caranday palm (*Copernicia australis*) of Paraguay yield leaf wax that approaches carnauba wax in quality. Wax from various desert shrubs or even from sugarcane stems may substitute for carnauba wax when prices are high or the supply is short.

C. prunifera is a slow-growing fan palm that develops thick secretions of wax on the leaves in response to the seasonal drought and scorching winds of its native habitat. But the tree prefers to root in moist soil, and it is commonly found in low areas that, in certain seasons, have a high water table or even standing water. Harvest of the leaves is regulated, to prevent overcutting and exhaustion of the plants. The leaves of young trees that have been planted are cut by machete; those from wild trees, which are older and taller, are cut with the aid of long pole-saws and ladders. The leaves are taken to a drying shed where they are fed into shredding machines. The shredded leaves are left to dry for several days, during which they wither and the waxy coat becomes loose, falling to the floor as a whitish powder. Laborers further flail the shredded leaf parts to remove adhering wax. The loose wax is swept up and put into containers for further processing. Spent leaf is then returned to the field as a mulch.

The dustlike wax is melted, strained, and poured into molds, where it hardens for shipping to market. Most crude wax goes to the United States, where it is the most prized natural wax for polishes and floor coatings; but it is also much used for plastics, cosmetics, films, and other products that require a very hard, high-melting wax.

Candelilla (*Euphorbia antisyphilitica*, of the family Euphorbiaceae) is a native desert plant of the southwestern United States and Mexico. The leafless candelilla secretes an epidermal wax in adaptation to the arid environment. For years candelilla has been gathered in rural Mexico, in some places to the point of extermination. Plants are pulled by hand and transported on burros to an extraction camp, where they are boiled in water in sunken tanks. About 8 kilograms of sulfuric acid is added for each 100 kilograms of candelilla in the brew. The wax floats to the surface as a light-colored foam, which is skimmed into other containers for further refining. It eventually solidifies as a cake, which can be cut to convenient size for shipping to market. This crude wax, containing much debris, must be further refined before being put to commercial use.

In Mexico, candelilla wax is often mixed with paraffin to make candles. Although of limited and erratic industrial supply in the United States, it often substitutes for more expensive waxes, such as beeswax and carnauba wax. Candelilla finds occasional use in coatings and adhesives, automobile polishes, printing inks, chewing gum, cosmetics, explosives, and various dressings or polishes. Since only wild plants are harvested, candelilla production is destined to be a declining industry.

Jojoba (*Simmondsia chinensis*, of the family Buxaceae) is native to the southwestern United States and Mexico, in spite of the name *chinensis*. It bears a seed that contains about 50 percent liquid wax, which is prized as a lubricant. Jojoba substitutes for sperm whale oil, and its adoption as an agronomic crop has received much impetus from various restrictions on whaling, including the ban on sperm whale imports into the United States under endangered-species legislation. By hydrogenation, the liquid wax

of jojoba can be transformed into a substitute for carnauba wax and the other hard waxes of commerce. It seems the equal of hard waxes for furniture polishes and as an extender for other waxes in a variety of uses.

Wild jojoba is found from southern California and Arizona south into northwestern Mexico, usually at elevations between 600 and 1200 meters (roughly 2000 to 4000 feet). The plants are dioecious. As much as 4.5 kilograms (about 10 pounds) of seeds, which contain about 50 percent wax, can be harvested from a single, well-nurtured crop plant. Although mainly gathered from the wild by American Indian families, some plantings have been made in Israel, Africa, and Australia, as well as in the species' native range in the southwestern United States and Mexico. Jojoba has been experimentally planted directly to the field, but it is usually transplanted from nurseries. In cultivation, it responds to proper spacing, pruning, and irrigation, but mechanical harvest of the seed has proven difficult.

SPICES, FLAVORINGS, PERFUMES, AND OTHER ESSENTIAL OILS

Extractives important for their essential-oil content obviously interrelate with other extractives discussed in this chapter (such as medicinals, resins, and masticatories), and of course with foods, the subtle flavors of which are frequently due to essential oils.

Essential oils are highly aromatic substances, chemically simpler than the oils of the preceding section (see Chapter 8). Mostly they are benzene or terpene derivatives or straight-chain hydrocarbon compounds of intermediate molecular length (seldom more than 20 carbon atoms long). The highly aromatic nature of this diverse group of substances provides the "zing" of spices, the taste of flavorings, the fragrance of perfumes, and the "clean" smell of antiseptics and medicinals. Usually only very small quantities of an essential oil are needed to produce a noticeable effect.

The political importance of essential oils, especially as represented by the spices, has been out of all proportion to their limited usefulness and present minor economic value. Historically, spices have been responsible for the making and breaking of empires, the exploration of far corners of the globe, and, of course, the discovery of the New World by Columbus. But even before this heyday of Western interest in spices and aromas, which reached its zenith between the fifteenth and nineteenth centuries, essential oils have had a compelling appeal to almost all civilizations. Aromas and flavorings have been part of the magical rites of the shaman; they were used in the earliest temples, and have been in demand for various "purification" ceremonies throughout the centuries. Essential oils were used by the early Egyptians for mummification, and were thus important even in afterlife. A number of primitive cultures believed that spices of one kind or another were aphrodisiacs, and many of the ancient wines were flavored with essential oils. The early healers used spices and fragrances in the first medicinals, and the Romans demanded aromatics for the bath. One could recite at length the ways in which essential oils have had a bearing upon the rituals and development of various cultures. But they have seldom been necessities, in the same sense that food and drink are necessities, though they may have motivated people to a greater extent (having been the cause for the first extensive trade between East and West, and their procurement having instigated numerous wars, both great and small).

Sesame, already mentioned as an oil seed, seems also to have been the source of one of the earliest essential oils. Hieroglyphics deciphered from ancient Egyptian tombs mention sesame. But certainly almost as early in use were such prized spices of the pre-Christian world as ginger, cardamon, cassia, and cinnamon. These were the mainstay of early Arabian commerce with the Malabar Coast of India, and they were vaguely thought by the European and Mediterranean world to be produced somewhere in Arabia ("the spices of Araby"). Also highly prized by the Mediterranean and Near Eastern civilizations, and treated with much respect in the Old Testament of the Bible, are such essences as anise, basil, balm, coriander, cumin, fennel, frankincense, mace, marjoram, mint, myrrh, mustard, nutmeg, and thyme.

In Assyria and the Fertile Crescent, about 5000 years ago, many plants were grown for their essential oils, as in the gardens of the kings of Babylon. In an era having little concern for sanitary procedures, their fragrance would no doubt be thought a counteraction against the foul smells of the developing cities. The streets were often fumigated with spices before royal visits, and essences were used to "counteract" the plague for centuries. Herodotus mentioned their use in embalming: the bodies of the deceased were filled with "the purest myrrh, cassia, and every other sort of spicery except frankincense."

The spice trade has few parallels for wild romance. The Biblical story of the sale of Joseph to the spice merchants indicates that the spice trade was well established very early. So far as the Western World was concerned, the spice trade was chiefly in the hands of Arabian peoples as late as the fourteenth century. The sources of these spices lay in the Far East, with centers in southern China, Malaysia, and India. Apparently there was commerce by sea from the Red Sea and the Persian Gulf to the Malabar Coast of India, to which intermediate point the Chinese and other East Indian merchants delivered their spices, often procured in the mysterious East Indian islands. The Arabs, who controlled the traffic in these spices, intentionally spread false rumors of their sources and of the "insurmountable" difficulties in procuring them (requiring death-defying ventures against war-like tribes and hostile environments). During the centuries of Arabian hegemony there was, of course, no little overland caravan trade from India and the Persian Gulf, but early records indicate that much of the Egyptian trade, at least, was consummated along the Red Sea. In later centuries the spice traffic extended throughout southern Europe, with the Arabian peoples of the Near East serving as intermediaries.

Most of the spices used today were already items of commerce at the time of the Greek and Roman empires. Indeed, taxes and tribute on their transshipment sustained the royal coffers of many a sheikdom and kingdom. Around the time of Christ, the Egyptians apparently tried to break the Arabian monopoly of the spice trade, and sent expeditions from Alexandria to India. But it was not until the fifteenth century that the Arabians really lost their dominance. By then, information about the Far East had filtered back to the West, especially after Marco Polo's travels to China. Polo spoke of cloves in the Nicobar Islands, and sesame, pepper, ginger, and cinnamon in Ceylon and along the Malabar Coast. He spoke of ships arriving at the port of Aden from India, "loaded with spices which they sell to great advantage." In the East Indies this spice trade was of considerable moment too, with Javanese and Chinese merchants doing a thriving business in cloves, nutmeg, pepper, cassia, and other spices destined for shipment to southern India. The Indian way-station for Far Eastern spices was to become gradually less important, as mariners from Europe reached the eastern spice lands. Even in the fifteenth century, Venice still lay astride the spice routes through the Mediterranean, and her merchants and bankers gained their profits from the transshipment of spices from the Near East. This was indeed a thriving business, for spices were becoming more and more in demand in the West for such uses as the preservation and flavoring of food on shipboard during increasingly long voyages and for seasoning the favorite dishes of the European nobility.

The control of Venice and Constantinople over the spice trade was broken when the dauntless Portuguese went to the Far East themselves. At first they crossed the Mediterranean and re-embarked on Moslem ships in the Arabian Sea. But between 1497 and 1499 Vasco da Gama skirted Africa, reaching India to find "Calicut the centre of a region producing cinnamon, ginger and pepper—and learned that many other aromatic and pungent spices came to this port from more distant spice lands." The handwriting was on the wall for the princes of Venice and the sultans of Egypt.

The voyage of Columbus, which failed to find the sources of Eastern spices but resulted in the discovery of the New World, is well known. With Spain looking West, Portugal soon dominated the sea routes from Borneo to Europe. Tiny Portugal now had a stranglehold on the spice trade, to be exercised even more zealously than it was by the Arabians. The monopoly was maintained by subjugating coastal lands in the East, and by curtailing trade by native islanders in that area. No spices were to leave the East Indies except in Portuguese ships!

The reaction of other European powers to the Portuguese monopoly could be predicted. By the 1500s, English mariners, and then Dutch, were challenging the Portuguese. In 1574 there was an uprising against the Portuguese in the Moluccas, and in 1579 Sir Francis Drake reached the East Indies by sailing westward around South America. Soon England was in control of India and the nearby spice islands, establishing the famed East India Company. Meanwhile, the Dutch had broken Portuguese power in the East Indies, and were in control of the spice trade there. A shipload of cloves and pepper loaded in the spice islands would bring twelve times its purchase price a few months later in northern Europe.

The Dutch East India Company, patterned after the English company, eventually gained control of the producing regions for pepper, most cinnamon, cloves, ginger, mace, and nutmeg. In an effort to maintain the monopoly, spice trees were extirpated in all except certain proscribed areas. For a short time the Dutch were the undisputed masters of the East Indian spice trade. But such avarice was bound to backfire, for economic pressure to break the Dutch monopoly steadily mounted. Before long both the French and the English succeeded in smuggling spice species from the Dutch islands into those that they controlled, including many in the New World. After nearly 200 years of operation, the Dutch East India Company was dissolved. British ascendancy over the Far East was established by the early 1800s, and the spice trade was a monopoly no more.

Thus did essential oils influence the ebb and flow of empire, and the search for spices create such giants as Dias, da Gama, Columbus, Magellan, Drake, Cavendish, and Van Houtman. While all this history-making was taking place, it is strange how little information accumulated about the spice plants themselves. The intentional confusion spread by the Arabian traders was understandable, but even until the recent century much uncertainty existed about the exact identity and growing requirements of the major spices from the East. Table 22-2 indicates the families, species, and common names of many oriental spices.

Table 22–2   Botanical sources of major oriental spices.

Species	Common name	Family
Indian center A (Himalayan foothills)		
Cinnamomum tamala	Indian cassia	Lauraceae
Piper longum	Long pepper	Piperaceae
Murraya koenigii	Curry-leaf tree	Rutaceae
Amomum aromaticum	Bengal cardamom	Zingiberaceae
Amomum subulatum	Greater or Nepal cardamom	Zingiberaceae
Curcuma zedoaria	Zedoary	Zingiberaceae
Zingiber officinale	Ginger	Zingiberaceae
Zingiber zerumbet		Zingiberaceae
Indian center B (Malabar Coast and Sri Lanka)		
Cinnamomum cassia	Cassia	Lauraceae
Cinnamomum zeylanicum	Cinnamon	Lauraceae
Myristica fragrans	Nutmeg and mace	Myristicaceae
Myristica malabarica		Myristicaceae
Areca catechu	Betel-nut palm	Palmae
Piper betle	Betel pepper	Piperaceae
Piper cubeba	Cubeb pepper	Piperaceae
Piper longum	Long pepper	Piperaceae
Piper nigrum	Black pepper	Piperaceae
Curcuma longa	Turmeric	Zingiberaceae
Curcuma zedoaria	Zedoary	Zingiberaceae
Elettaria cardamomum	Cardamom	Zingiberaceae
Zingiber officinale	Ginger	Zingiberaceae
Zingiber zerumbet		Zingiberaceae
Malaysian center		
Cinnamomum cassia	Cassia	Lauraceae
Cinnamomum loureirii	Saigon cinnamon	Lauraceae
Cinnamomum zeylanicum	Cinnamon	Lauraceae
Myristica fragrans	Nutmeg and mace	Myristicaceae
Syzygium aromaticum	Clove	Myristicaceae
Areca catechu	Betel-nut palm	Palmae
Piper betle	Betel pepper	Piperaceae
Piper cubeba	Cubeb pepper	Piperaceae
Piper longum	Long pepper	Piperaceae
Piper nigrum	Black pepper	Piperaceae
Piper retrofractum	Javanese long pepper	Piperaceae

(continued)

Table 22–2 (continued)

Species	Common name	Family
Malaysian center		
Amomum compactum	Round cardamom	Zingiberaceae
Amomum xanthioides	False cardamom	Zingiberaceae
Curcuma longa	Turmeric	Zingiberaceae
Curcuma zedoaria	Zeodary	Zingiberaceae
Elettaria cardamomum	Cardamom	Zingiberaceae
Zingiber officinale	Ginger	Zingiberaceae
Zingiber zerumbet		Zingiberaceae
South China center		
Cinnamomum cassia	Cassia	Lauraceae
Illicium anisatum	Chinese anise	Illiciaceae
Illicium verum	Star anise	Illiciaceae
Amomum zanthioides	False cardamom	Zingiberaceae
Curcuma longa	Turmeric	Zingiberaceae
Curcuma zedoaria	Zeodary	Zingiberaceae
Zingiber officinale	Ginger	Zingiberaceae
Zingiber zerumbet		Zingiberaceae

Source: R. M. Newcomb, *Economic Botany* 17:127–132, 1963.

Some noteworthy spices

Black pepper, *Piper nigrum* (Piperaceae)

Commercial pepper is obtained from the fruits of *Piper nigrum*, a perennial climbing vine adapted to humid tropical climates. It is indigenous to Sri Lanka and southern India, but it is now extensively cultivated throughout much of Southeast Asia and, more recently, it has been much encouraged in Amazonian Brazil. It was already well known in pre-Christian times, but it became highly commercialized with rise of the spice trade from the Dutch East Indies. Clusters of the hard berries, which are known as peppercorns (Fig. 22-10), are hand picked, piled in heaps, and left in the sun for several days to dry. Natural fermentation causes the fruits to turn black. Occasionally the clusters may be dipped in boiling water to accelerate the blackening and drying. If the weather turns rainy, artificial drying may be used. The dry berries are detached from the stalks by beating with sticks and by rubbing with the hands. They are then screened to remove impurities and are readied for shipment. The black pepper for temperate zone pepper shakers is derived by grinding these peppercorns. White pepper is produced by immersing ripe pepper fruits in running water for a few days to soften the outer skins, after which they may be removed easily by hand or by treading. The smooth white kernels are washed and dried on mats in the sun (or sometimes artificially in small smokehouses). The pungency of pepper is due to various "resins" and the alkaloid piperine.

Figure 22-10 Developing peppercorns, and foliage of the pepper vine blighted by a local infestation, in Cameroon. [Courtesy Food and Agriculture Organization of the United Nations.]

Cinnamon, *Cinnamomum zeylanicum* (Lauraceae)

As we have seen, cinnamon is one of the classic spices in the Eastern spice trade. It comes from the bark of cinnamon trees native to Sri Lanka and India. Cassia, also a famed name in the early spice trade, is from another species, *C. cassia*. In Biblical stories the two may have been confused under the name of cinnamon.

The bark of the small cinnamon tree contains as much as 1 percent essential oil, which, in turn, is nearly three-fourths cinnamic aldehyde. The bark is stripped from the trees just after the monsoon season (from May to August), preferably from second-year wood. An active cambium in the twigs makes peeling relatively easy then. This is a hand operation, accomplished with the aid of a knife. The best grade of bark forms hollow tubes, which after thorough drying are scraped free of epidermal particles to become the cinnamon "quills" or "sticks" of commerce. Fragments and less-than-perfect quills are ground to yield powdered cinnamon.

Most cinnamon is produced on small native-owned plantations. The trees may be propagated by seed or by cuttings; seedlings are started in the nursery and are later planted at spacings determined by land conditions.

Clove, *Syzygium aromaticum* (Myrtaceae)

Cloves are another of the exotic tropical spices once under complete control of the Dutch monopoly. Their production was confined then to a single island of the Moluccas (the Spice Islands), where the tree is native and was grown under close supervision. Clove trees had been eliminated from all other East Indian territory.

The highly aromatic clove is the dried unopened flower bud and twig tip of *Syzygium aromaticum* (= *Eugenia caryophyllata*). Cloves were used long ago in China, where it was customary for court officials to perfume their breath with them before addressing the sovereign. Clove trees are small, and the buds are generally harvested by women who pick

In the producing countries, the pepper plant is usually propagated by making cuttings that are stuck in the ground around the base of a shade tree, which later supports the vines. Pepper is often interplanted with coffee, utilizing the coffee shade as support and the coffee crop as a source of income until the pepper vines mature. Layering, marcottage, and separation of rhizomes are occasionally practiced as more certain means of propagation where nurseries are available. Apparently, the seeds germinate poorly. Production is chiefly a home operation by small landowners; the total crop probably exceeds 100,000 metric tons annually, mostly from Malaya, Indonesia, and India.

The long pepper is generally considered to be *Piper longum*, although in a loose sense the name refers to a number of species. Burrlike clusters of fruits constitute the item of commerce. The long pepper is considered more a medicinal than a condiment, since it contains an alkaloid that induces salivation and numbness. In India it is often cultivated along with *P. nigrum*.

them as they begin to color. The plucked twig tips are sun-dried and are then ready for marketing as the familiar spice. Most of the world's cloves are grown in Tanzania (on the islands of Pemba and Zanzibar) and in Madagascar, but there is some production in Indonesia, where "oil of cloves" is extracted by distillation. Clove oil consists largely of eugenol, which in turn can be used as raw material for the synthesis of vanillin. Clove oil finds considerable use in disinfectants, as a flavor in dentifrices, and in perfumery blends.

Vanilla, *Vanilla planifolia* (Orchidaceae)

This climbing orchid (Fig. 22-11) and other related species yield the world-famous "vanilla bean," the fully grown but unripened and fermented seed pod of the species. Fermentation develops the characteristic vanilla odor, and the essential oil is extracted with alcohol to yield the natural vanilla extract of commerce. The active fragrance, vanillin (see p. 168) can be synthesized inexpensively, but it lacks the overtones of other aromatic compounds found in natural vanilla that enhance its aroma.

Bernal Diaz del Castillo, an officer under Cortés, was perhaps the first European to make note of vanilla when he observed its use by Montezuma in compounding the chocolate drink, *xocolatl*. In the Aztec empire, vanilla beans were valued as tribute. The Spanish were quick to import vanilla beans, and even today cultivation and production of vanilla beans is a reasonably important minor industry in Mexico. During the 1800s vanilla was introduced into the Far East and into Madagascar, where plantation growing of vanilla is still part of the agricultural economy. Most of the present-day production comes from islands of the Western Indian Ocean.

The horticulture of vanilla growing is reasonably complicated. Land must be cleared, except possibly for trees that are to serve as supports for the climbing vanilla plants, and mechanical supports provided. Vanilla prefers shade, so suitable shade trees and windbreaks are necessary. After cultivating the soil, cuttings of the vanilla vine are planted, and the vines are trained to climb over the supports. Since vanilla

Figure 22-11   Commercial vanilla extract is made from the pods of the orchid *Vanilla planifolia*. [Courtesy USDA.]

roots grow mostly near the surface, it is sound practice to maintain a layer of humus about the plant. A certain amount of pruning is necessary, and a watch must be maintained for parasites and diseases. The flowers are often hand-pollinated and the fruits thinned to the proper number to allow full development. With this much attention required, vanilla tends to be grown in areas with inexpensive labor.

The mints, *Mentha* spp. (Labiatae)

Distillation of oil from peppermint (Fig. 22-12), *Mentha piperita*, was known to the Egyptian pharaohs. The species may be a natural hybrid, and it seldom sets seed. Other important mints are spearmint (*M. spicata*) and *M. arvensis*, a source of menthol. Commercial production of mint oil is not great, totaling only somewhat more than 1000 metric tons annually in the United States. But mint plants are often utilized as garden plants and herbs around the home. Commercial mint production is primarily on muck soils, today principally in the Columbia River basin of Oregon and Washington.

Both spearmint and peppermint are propagated vegetatively from rhizomes or young plants. Mint is usually planted in rows, which are cultivated to con-

Figure 22-12 The foliage of peppermint (*Mentha piperita*) contains an essential oil. The oil, produced under long days, is extracted for flavors and medicinals.

trol weeds during the first year. Typically the mint is "plowed down" before frost, and it spreads from the buried parts into a complete field of "meadow mint" the second year. This mint is cut with a mowing machine, and left to dry in the field until its moisture content is reduced to approximately 35 percent. It is then taken to a still for extraction of the oil by steam distillation. Distillation usually takes about 45 minutes, and the spent hay is spread back on the fields.

Peppermint oil, consisting largely of menthol, is used for antiseptics, lotions, dentifrices, medicinals, and culinary flavors. Yields are reportedly not much over 34 kilograms of distilled oil per hectare (about 30 pounds per acre).

Oregano (Labiatae)

Oregano can serve as our example of savory herbs—those used in cookery. Several plants turn up as commercial oregano, among them *Origanum vulgare*, various species of *Lippia* (of the related family Verbenaceae), and about a dozen other genera in several families. *Origanum vulgare*, often called wild marjoram, was brought to America by the early colonists from the Mediterranean area. American oregano seems to lack the rich flavor typical of Old

World oregano. A. O. Tucker, of Rutgers University, has determined that most commercial samples he was able to obtain in 1973 were principally *O. heracleoticum*. "Origanum oil" has been field-distilled from *Coridothymus capitatus*, which grows wild in Andalusia, Spain. Syria has also been a source of commercial oil.

Other spices

Allspice, which is derived from *Pimenta officinalis* (Myrtaceae), is produced mainly in Jamaica. Angelica oil is distilled from the roots of *Angelica* spp., of the family Umbelliferae, and anise is derived from the seed of *Pimpinella anisum,* which belongs to the same family. Balm comes from distillation of the leaves of *Melissa officinalis* (Labiatae), which is native to the Mediterranean area. Cardamom is the aromatic dried fruit of *Elettaria cardamomum*, of the family Zingiberaceae. Caraway, celery, coriander, cumin, and fennel are all seed spices from species of various genera of the family Umbelliferae. Ginger is obtained from the rhizome of a large perennial herb, *Zingiber officinale* (Zingiberaceae), which is native to tropical Asia. Mustard is a spice obtained from the seeds of various species of *Brassica* (Cruciferae), an important genus among food plants. Nutmeg is the seed of *Myristica fragrans* (Myristicaceae), which is native to the Molucca islands, and mace is derived from the aril surrounding the nutmeg kernel. Parsley is from the foliage of *Petroselinum hortense* (Umbelliferae). Sage is obtained from the foliage of *Salvia officinalis* and related species of the family Labiatae. Turmeric, which serves as a yellow dye and a kitchen flavoring, is obtained from rhizomes of *Curcuma longa* (Zingiberaceae). There are many other spices and savory herbs, too many to mention here.

Essential oils for perfumes and cosmetics

Essences used in perfumery are most frequently obtained from flowers; often those used in soaps and cosmetics are derived from vegetative parts. Most flowers have nectar glands that secrete essences attractive to insects—an aid in pollination. Foliage oils generally occur in epidermal glands or hairs of uncertain usefulness to the plant.

Oils cherished for their scent are often extracted by absorption into cold fat, a method termed **enfleurage.** Enfleurage is typically used for delicate essential oils when the harsher distillation process may have deleterious effects upon them. Enfleurage is the chief means of extracting floral odors in the famous perfume centers of Provence, France. There huge volumes of flowers are grown for the perfume industry, which, when harvested, are applied to plates covered with cold fat that absorbs the floral essences. Later the saturated fat is subjected to alcoholic extraction of the essential oil.

More commonly, and especially with fragrances derived from vegetative portions of the plant, distillation, solvent extraction, or expression are practiced, much as they are with edible oils. In distillation, the plant is generally boiled in water, the distillate separating into an immiscible oil layer that floats on top of the condensed water. Sometimes steam treatment substitutes for the boiling. Direct extraction with solvents is used occasionally with flowers and commonly with other aromatic plant parts. It is usually the simplest and least expensive extraction method; the solvent dissolves the fragrance and is then evaporated itself, leaving the essential oil behind. Expression, which involves the squeezing out of the essential oils, is usually reserved for gross plant parts, such as rinds of citrus or whole seeds and fruits of various kinds.

Most essential oils deteriorate or polymerize to resins rather readily. Delicate essential oils like those used for perfumery are consequently stored in hermetically sealed bottles kept in dark, cool cellars. These essences are blended according to secret recipes to yield the famed perfumes of commerce. Fragrances used to mask foul odors have been in demand since the earliest days of civilization, and they have often been involved in the religious rituals of the times. The art of perfumery has discovered through the ages various additives that extend and make more permanent the expensive floral essential oils. Sometimes these are animals oils, or the fatty oils and resins from plants. (Ambergris, a fatty substance from the intestine of the sperm whale, is a highly valued ingredient.) Synthetic materials are coming more and more into use.

Rose oil, *Rosa* spp. (Rosaceae)

Rose oil, or attar of roses, is chosen here as an example of an expensive floral perfume oil. It is obtained from flowers of various species of *Rosa*. Shrub roses, such as *R. damascena*, are much grown in southern Europe and Asia Minor. They bloom in late spring, at which time the flowers are collected for distillation of the essential oil (Bulgaria) or for enfleurage and solvent extraction (France). Less than 0.5 gram of oil is obtained from each kilogram of flowers. The principal constituent of rose oil is citronellol; lesser amounts of such essential oils as geraniol, nerol, linalol, and other organic compounds are also present. Because pure rose oil is so expensive, it is generally extended with less expensive essential oils from other sources. Flowers are generally harvested by women and children when the rose is in the late bud stage.

Patchouli, *Pogostemon cablin* (Labiatae)

Patchouli, a mint native to the Philippines, has been widely grown in the Orient since time immemorial. In India it has served as an insect repellent and as a constituent of wardrobe cachets. Patchouli was once prized as a fine perfume, but it later came to be associated with women of low repute and fell into disfavor. It is still regarded as one of the finest fixatives for heavy perfumes, and it is used for scenting soaps and cosmetics. Patchouli is much grown in the Seychelles and in various parts of Southeast Asia. Because it seldom flowers, it is propagated vegetatively by cuttings. It is grown under light shade and is generally mulched to preserve moisture and to provide some fertilizer. Harvesting takes place about six months after planting, and is continued every three or four months; a planting is said to last for several years before replanting is necessary. The plucked leaves are kept overnight in small heaps to induce a fermentation that is reputed to increase the yield of oil. They are then spread out to dry in sheds or barns. Steam distillation is utilized to retrieve the oil, which amounts to 3 to 5 percent of the dry weight of the leaf. The oil consists mainly of sesquiterpenes, which upon aging develop a cedarlike odor. Probably no more than 100 metric tons of oil is produced annually throughout the world.

Other essential oils of importance to perfumery

Bergamot is expressed from the peel of the bergamot orange, *Citrus aurantium* ssp. *bergamia,* which is cultivated in southern Italy. About 150 metric tons of oil are produced annually. Camphor oil is distilled from the heartwood of *Cinnamomum camphora,* the camphor tree, which grows wild and under cultivation from Japan to southern China. Cananga or ylang-ylang is the famed essence of *Cananga odorata* (Annonaceae), which is prized for fine perfumes and is much grown in Java; the essence is derived from the flowers. Cassia is the essence derived from *Cinnamomum cassia* (Lauraceae), a tree that grows in China and was especially important in the early spice trade. Cassia oil, mostly cinnamic aldehyde, is distilled from the terminal foliage. Citronellol is obtained mainly by distillation of the citronella grass, *Cymbopogon nardus,* which is much grown in Sri Lanka and the Far East. Citronellol was once used as an insect repellent, but it has been displaced by modern synthetics. It can also be fractionated from other essential oils, such as menthol. There is considerable production of citronellol in Taiwan and Guatemala. *Eucalyptus* trees (Myrtaceae) provide several medicinal and perfumery oils from the distillation of their foliage, chiefly from *E. citriodora,* which is native to southern Australia. The oil is produced in Spain, Portugal, and Brazil, as well as Australia, and similar eucalyptus oils are extracted from other species in Southeast Asia.

Geranium oil is distilled from the flowers and foliage of various species of *Pelargonium* (Geraniaceae), mostly on the island of Réunion and in Algeria. Lavender oil comes from *Lavandula officinalis* (Labiatae), a fragrant herb much cultivated in southern France. More than 100 metric tons of lavender

oil is produced in some years. Malabar lemongrass, is another of the *Cymbopogon* species (*C. flexuosus*), is grown and distilled in India, where annual production has sometimes exceeded 1 million metric tons. Along with West Indian lemongrass (*C. citratus*), it is grown in many parts of the tropics and is used as a source of citral. Palmarosa (*C. martini* var. *motia*) is a source of geraniol. Petitgrain is the oil distilled from the foliage of the bittersweet orange, *Citrus aurantium* ssp. *amara.* Rosemary oil, produced chiefly in Spain and Tunisia, comes from *Rosmarinus officinalis* (Labiatae). Thyme oil is distilled from the flowering tops of several species of *Thymus* (Labiatae), chiefly in Spain. Tuberose, *Polianthes tuberosa* (Agavaceae), yields an essence obtained by enfleurage or solvent extraction of its flowers.

Industrial essential oils

A number of aromatic oils serve industrial rather than culinary and esthetic purposes. Turpentine and camphor stand out as two of the most consistently useful, and they will serve here to exemplify essential oils used industrially.

Camphor, *Cinnamomum camphora* (Lauraceae)

The camphor tree is a close relative of the cinnamon tree (*C. zeylanicum*), but it yields its essence from the wood rather than from the bark. The camphor molecule is not uncommon, and it is also obtained from various species in the families Labiatae, Compositae, and Dipterocarpaceae. It has long been known that camphor can be synthesized by oxidation of borneol, a fairly common ingredient of many essential oils.

The camphor tree is native to Japan and eastern China. Originally the chipped wood was crudely distilled locally. Such destruction of the trees of course threatened the industry, and it is perhaps good that other sources of camphor have been found. It is said that as much as 3 metric tons of camphor could be derived from a single tree, and that the wood chips contained as much as 5 percent of the essence. Camphor finds use in medicinals, liniments, and insecticides, but it was especially important for making celluloid from nitrocellulose (an industrial use that once accounted for most of the camphor production). Celluloid has now been supplanted by a wide variety of plastic materials; consequently, the demand for camphor has diminished.

Turpentine, *Pinus* spp. (Pinaceae)

Turpentine is a mixture of resins and essential oils that was originally derived by injuring the sapwood of coniferous trees, especially the longleaf pine (*Pinus palustris*) and the slash pine (*P. elliotti*). It is the basis of the **naval stores** industry, so named because in the days of wooden sailing ships the resinous residue left after the volatile turpentine removal (called rosin) was vital for the caulking and treatment of hulls and rigging. Naval stores were first collected by digging a pit and filling it with pine logs. They were then set afire and immediately covered with soil. The heat from the smoldering fire caused oleoresins to drip from the wood and accumulate in the bottom of the pit. When cool, the dirt and ashes were removed and the dirty rosin dug out of the waste. Men working in these pits got it on their feet, hence the name "tar heels," which is used for North Carolinians even today. The old pits used in this manner are still to be seen throughout the slash pine and longleaf pine region of the southeastern United States.

In the early years of American colonization, when forests seemed limitless, turpentine was collected in the southeastern pine belt by "boxing" the tree—that is, cutting a deep hollow in the base, into which the exudate drained from chipping the bark above. This is still the practice in developing countries such as Belize. Less wasteful collection has evolved in the United States. Trees must be at least 25 centimeters in diameter (more than 9 inches) for tapping. The bark near the base of the tree is shaved off, and into a shallow cut made a few inches above the ground an apron or gutter is fitted. Below this is placed a collecting vessel, into which the gutter drips. A shallow wound is made in the bark above the gutter, from which drips the oleoresin containing the volatile turpentine (Fig. 22-13). During the turpentining season

Figure 22-13   Gutter (apron) and cup for collecting turpentine from pine. [U.S. Forest Service.]

(spring and summer), the "chipper" makes approximately weekly rounds to each tree and renews the wound by slicing away a thin section of bark or "streak" above the last one. If the streak is treated with sulfuric acid, or with herbicides such as 2,4-D or paraquat, the duration of turpentine flow is increased. Chipping may then be needed only one-half or one-third as often.

Formerly, the oleoresin was locally distilled in wood-fired iron kettles. Liquid turpentine was condensed in iron coils, and the rosin that remained behind was independently barreled. In later years water came to be added to the oleoresin, and distillation was accomplished in copper stills, in effect becoming steam distillation. Turpentine is immiscible with water, and can be separated quite readily. More recently, collected oleoresin has been taken to processing centers, where modern stainless-steel batch stills and flash stills effect the separation. The rosin remaining after extraction of the turpentine is used in preparing paints and varnishes, as a sizing for paper, in polishes and waxes, and for many other purposes. It is several times more valuable than the turpentine. Annual turpentine production in the United States runs to more than 11,000 metric tons (nearly 500,000 barrels containing 50 pounds each); U.S. rosin production is more than 250,000 metric tons annually. Only 5 percent of the turpentine is "gum turpentine," which is obtained by streaking pine bark as described above; about 10 percent is "wood turpentine," which is obtained from the distillation of wood (primarily old pine stumps and roots); and about 85 percent is "sulfate turpentine," a by-product of sulfate paper processing, which utilizes southeastern pines for pulp. Spain, Greece, Russia, India, Yugoslavia, Austria, Indonesia, and Mexico also produce turpentine oleoresin from various species of pine.

Typical wood turpentine contains about 80 percent α-pinene, a terpene hydrocarbon. The pine-oil fraction of the wood distillate (a secondary product of the distillation process) is a complex mixture of terpene hydrocarbons, aldehydes, alcohols, ethers, ketones, and phenols. The α-pinene fraction is used chiefly as a solvent; the pine-oil fraction is used in the textile industry as a wetting agent and as an inhibitor of bacterial growth. It is also used in paper making as a wetting agent during the coating stages. Sulfate turpentine is condensed from the vapor that arises during the cooking of pine chips to make wood pulp for paper making. Depending upon species and the condition of the pine wood, as much as 18 liters of crude turpentine may be obtained for each metric ton of air-dried pulp produced (about 4.3 gallons per ton). Sulfate turpentine must usually be further washed or treated to remove the sulfate odors. During the sulfate processing the resinous constituents are saponified, and the solution is known as **tall oil;** this oil is composed essentially of oleic and lineoleic acids, and it serves such purposes as were described in the section on vegetable oils.

Other volatile oils used industrially

Volatile oils of relatively minor importance, usually recovered by steam distillation from tree parts, include: cedarwood oil, from *Juniperus virginiana*, used in soaps, polishes, and perfumes; conifer leaf oil, from the foliage of various spruces, hemlocks, and junipers, used mainly for ointments, soaps, and shoe polishes; sweet birch oil, from the bark of *Betula lenta*, used in drugs, candies, and for flavorings; sassafras oil, from various parts of *Sassafras albidum*, used in dentifrices, carbonated beverages, and medicinals; eucalyptus oils, previously mentioned; and sandalwood, from the heartwood of *Santalum album*, a tree of southern India, used chiefly in cosmetics and soaps.

MEDICINALS AND RELATED DERIVATIVES

Among medicinals, too, there is much overlap with other sections of this chapter. For example, castor oil has been a purgative since the earliest times; gum arabic, primarily an adhesive, serves also as a demulcent; olive oil has traditionally been used in ointments and as a lotion; cloves and clove oil serve as stomachics and are used in a wide variety of remedies; camphor has long been used as an antiseptic and as a stimulant; betel (*Areca*) is an anthelmintic; nicotine (from tobacco) makes an effective natural insecticide; and always has there been but the scantiest of distinctions between the use of essences for esthetic purposes and their use for hygienic ones.

Perhaps the earliest extractive used medicinally was gum arabic, derived from *Acacia* spp. (Leguminosae), which grow wild in northern Africa; it is recorded by Herodotus as being of medicinal importance before the fifth century B.C.E. It is interesting that gum arabic is still listed in some pharmacopeias, although today it is used only as a vehicle or as a demulcent (a soothing substance) in such remedies as cough syrups. Gum tragacanth, which is derived from *Astragalus* (Leguminosae), a spiny shrub of the Mediterranean area, has been used medicinally since time immemorial.

People have been especially concerned with and awed by curatives, which until recent years were almost exclusively derived from plant sources, often from species found only in the wild. The fear, mysteriousness, and highly personal nature of disease has led people to try almost any remedy offering hope. It is no wonder that much superstition surrounds the choice of medicinals, especially under primitive conditions where symptoms rather than causes must be attacked. And who is to say that, if the patient is convinced of medicinal efficacy, this of itself is not therapeutic? In modern experiments a placebo has proved beneficial in many cases. Civilized people, of course, no longer regard disease as "divine punishment" or "visitation of the devil"; but they are still as intensely interested as ever in finding more effective controls for disease. The search has involved systematic screening of many plant species. Among other things, the remedies discovered for various diseases have helped increase life expectancy from an average of about 20 years at the time of the Roman Empire, to somewhat more than 70 years in medically advanced parts of the world today.

The fall of Troy, Hannibal's failure against Rome, the defeats suffered by Lee during the Civil War, and the French abandonment of their Panama Canal project have all been attributed to malaria; had quinine been known earlier, it might have changed the course of history! Morphine, derived from opium, has long been used to ease pain—and it has also led to drug addiction. Hippocrates was well acquainted with oil of wintergreen (which consists largely of methyl salicylate), and he recommended it for many maladies; its investigation led eventually to the synthesis of acetylsalicylic acid, or aspirin, in 1853, of which today more than 40 metric tons are consumed daily in the United States alone! Lysergic acid diethylamide, LSD, was accidentally discovered years ago in the dust from an obstetrical drug made from the fungus ergot (*Claviceps purpurea*, an ascomycete); today its use as a hallucinogenic agent creates social problems, even though it may sometimes serve in the treatment of mental illness. Domesticated animals frequently die from toxic principles (mycotoxins) consumed in graze.

It is more and more recognized that the folklore about medicinal plants and the home remedies of an earlier era are not all based on sheer superstition. The biological influence of extracts ranges from the extreme toxicity of such substances as strychnine or potassium fluoroacetate (in *Dichapetalum cymosum*)

to the relative inertness of oils and sugars. The search continues for unusual substances that have profound physiological effect when administered in small doses. Important discoveries have yielded the antibiotics and the plant growth regulators, many of them from molds. Useful plant sources of the rather commonplace medicinals, such as emollients and cathartics, and even of the more specialized medicinals, such as cardiac glycosides (digitalis), have been known for many years. The alkaloids, a class of nitrogenous bases, are notable for the profound physiological responses they produce in animal systems. Alkaloids occur commonly in such plant families as Solanaceae and Apocynaceae, especially in tropical genera. Many of the medicinals we list are important for the alkaloids they yield; for example, quinine (from *Cinchona*, Rubiaceae) and reserpine (from *Rauvolfia serpentina*, Apocynaceae) (see Table 8-3).

The search for healing botanicals even in primitive and circumscribed cultures was, indeed, extensive. De Montellano investigated twenty-five medicinal herbs of the Aztecs, and found most of them to be effective for the purposes to which the Aztecs put them (*Science* 188(4185):215–220, 1975). A recent list of medicinal plants used by the Maoris of New Zealand runs to well over 100 species, and 177 species are known to have been used by the Hawaiians. Table 22-3 compares plants included in the first edition (1820) of the U.S.P. (*The Pharmacopeia of the United States of America*), and the sixteenth revision (1960). Note that many of the old folk remedies have been dropped over the years. It cannot be assumed that all these older remedies were without merit; possibly the modern chemist should analyze more carefully those species that were once thought to be effective, isolating components that might be medicinally useful. Compared with the many scores of species dropped, there have been only 20 new additions to the most recent revision.

Primitive people accorded hallucinogenic plants the highest respect: Psychic more than physiological effects were the tools of the shaman in his medico-religio-magical rites. The use of hallucinogens (Fig. 22-14) by primitive peoples makes fascinating lore, and shows that there is really nothing new about "drug culture." Today's experiences with psycho-

Table 22–3 A comparison sampling of botanicals in the 1820 and 1960 editions of *The Pharmacopeia of the United States of America.*

Official title	Listing in 1820 edition	Listing in 1960 edition	Family
Acacia	*Acacia vera*	*A. senegal*	Leguminosae
Aconite	*Aconitum neomontanum*	NLO°	Ranunculaceae
Allium	*Allium sativum*	NLO	Liliaceae
Aloe	*Aloe spicata*	*A. perryi*	Liliaceae
	A. barbadensis	*A. barbadensis*	Liliaceae
		A. ferox	Liliaceae
		A. africana	Liliaceae
		A. spicata	Liliaceae
Ammoniacum	*Heracleum gummiferum*	NLO	Umbelliferae
Almond	*Amygdalus communis*	NLO	Rosaceae
Angustura	*Bonplandia trifoliata*	NLO	Polemoniaceae
Anise	*Pimpinella anisum*	*P. anisum*	Umbelliferae
		Illicium verum	Illiciaceae
Anthemis	*Anthemis nobilis*	NLO	Compositae
Armoracia	*Cochlearia armoracia*	NLO	Cruciferae
Asafetida	*Ferula asafoetida*	NLO	Umbelliferae
Aurantii cortex	*Citrus aurantium*	*C. sinensis*	Rutaceae
Avenae farina	*Avena sativa*	NLO	Gramineae
Belladonna	*Atropa belladonna*	*A. belladonna*	Solanaceae
Benzoin	*Styrax benzoin*	*S. benzoin*	Styracaceae
		S. parralleloneurus	Styracaceae
Cajeput	*Melaleuca cajuputi*	NLO	Myrtaceae
Camphor	*Laurus camphora*	*Cinnamomum Camphora*	Lauraceae
	Dryobalanops camphora		Dipterocarpaceae
Canella	*Canella alba*	NLO	Canellaceae
Capsicum	*Capsicum annuum*	NLO	Solanaceae
Caraway	*Carum carvi*	NLO	Umbelliferae
Cardamom	*Elettaria cardamomum*	NLO	Zingiberaceae
Caryophylli	*Eugenia caryophyllata*	*Eugenia Caryophyllus*	Myrtaceae
Cascarilla	*Croton eleutheria*	NLO	Euphorbiaceae
Cassia fistula	*Cassia fistula*	NLO	Leguminosae
Cassia marilandica	*Cassia marilandica*	NLO	Leguminosae
Catechu	*Acacia catechu*	NLO	Leguminosae
Chenopodium	*Chenopodium anthelminticum*	NLO	Chenopodiaceae
Cinchona	*Cinchona lancifolia*	*Cinchona* spp.	Rubiaceae
	C. oblongifolia		Rubiaceae
	C. cordifolia		Rubiaceae
Cinnamomum	*Laurus cinnamomum*	*L. cinnamomum*	Lauraceae
Colchicum	*Colchicum autumnale*	*C. autumnale*	Liliaceae
Colocynthis	*Cucumis colocynthis*	NLO	Cucurbitaceae

°NLO = no longer official.

(continued)

Table 22–3 *(continued)*

Official title	Listing in 1820 edition	Listing in 1960 edition	Family
Conium	*Conium maculatum*	NLO	Umbelliferae
Copaiba	*Copaifera officinalis*	NLO	Leguminosae
Coriandrum	*Coriandrum sativum*	NLO	Umbelliferae
Cornus florida	*Cornus florida*	NLO	Cornaceae
Crocus	*Crocus sativus*	NLO	Liliaceae
Cubeba	*Piper cubeba*	NLO	Piperaceae
Digitalis	*Digitalis purpurea*	*D. purpurea*	Scrophulariaceae
Dolichos	*Dolichos pruriens*	NLO	Leguminosae
	Pubes leguminis	NLO	Leguminosae
Dracontium	*Dracontium foetidum*	NLO	Araceae
	Ictodes foetidus	NLO	Araceae
	Symplocarpus foetidus	NLO	Araceae
Dulcamara	*Solanum dulcamara*	NLO	Solanaceae

Source: B. R. Hershenson, *Economic Botany* 18:342–356, 1964.

active drugs were anticipated by witch doctors and practitioners long ago, in innumerable parts of the world, utilizing mushrooms of a half-dozen genera, extracts of *Rivea* and other "morning glories," strange trees and vines of the mysterious tropical forest (including curare or blow-dart poisons from such genera as *Banisteriopsis*, *Psychotria*, and *Virola*, and various genera of the families Apocynaceae, Leguminosae, and Solanaceae), as well as the more familiar peyote (*Lophophora*, of the family Cactaceae), marijuana, and the opiates (which will be briefly discussed later in this chapter). The historical and medical value of their study can be of considerable significance.

Plant extracts that have dramatic influences on animal systems do not come only from rare exotics. Yams (*Dioscorea* spp.), which are important food crops, are also an important source of steroidal sapogenins, which are used in the synthesis of cortisone and birth-control drugs. The fruit and seed of *Balanites aegyptiaca* (Zygophyllaceae), of semiarid northern Africa, also yield a sapogenin, diosgenin, perhaps in commercial quantities. Thujone ($C_{10}H_{16}O$) is the primary constituent of the essential oil from wormwood, the flavoring for absinthe, an often illicit alcoholic beverage with the popular reputation of being "mind destroying." Thujone can also be derived from *Salvia*, *Tanacetum*, and *Thuja*, among other readily available sources. One unusual source of an extract is the serendipity berry, *Dioscoreophyllum cumminsii* (Menispermaceae), which yields a protein of low molecular weight called monellin, the sweetest known natural substance (up to 3000 times sweeter than sucrose).

For want of a better place to pigeonhole them, insecticides, growth regulators, and similar biologically active agents obtained from plant sources are mentioned in this section too. Like medicines, they have profound influence on biological systems in small doses. They are found in a wide variety of plants and plant parts, from simple molds to the flowers of evolutionarily advanced flowering plants. Some discoveries have resulted directly from biochemical research, and, in turn, the isolation of natural regulators has helped elucidate the functioning of biological systems. Fundamental work on auxins and other growth hormones led directly to the immense modern pesticide industry, in which today most herbicides, insecticides, and fungicides are synthesized rather than obtained from plant materials.

Figure 22-14 The main psychoactive constituents of hallucinogenic plants.

In general, many medicinals and related products are today chemically synthesized rather than obtained from higher plants (almost the only source even a few generations ago). A sophisticated chemical industry has made possible economical substitution of even more potent products patterned after the natural ones.

Aloe, *Aloe* spp. (Liliaceae)

Various *Aloe* species have been used medicinally since antiquity. Juice from the fleshy leaves is used as a laxative, well known to the pre-Christian Greeks. In the West Indies, juice of *A. barbadensis* is used as an antiseptic applied to minor cuts and skin injuries. (Sometimes, mixed with rum and honey, it is employed to treat colds.) The pulp has at times been popular for treatment of x-ray burns. There has been considerable confusion about the identity of medicinal aloes. *A. perryi*, native to islands in the Arabian Sea, is one of the most valued. *A. barbadensis* is probably of Old World origin, but it is widely naturalized in the West Indies (and often cited as *A. vera* or *A. vulgaris*). *A. ferox*, a distinctive arborescent species of southern Africa, has been considered as a source of cortisone.

A. ferox was brought into cultivation in England as early as the middle 1700s, and "gum aloe" was offered by London pharmaceutical houses before the nineteenth century. Leaves are cut to release the juice, which drips from the butt end; the juice is collected and the bitter principle is concentrated by boiling. In the West Indies, cut leaves are stacked in V-shaped troughs while the juice drains. Production comes almost entirely from wild plants in Africa; in the West Indies the naturalized *E. barbadensis* is frequently propagated by offsets, especially in Aruba, where the industry has been an important source of livelihood.

Antibiotics

Natural antibiotics have generally held their own in competition with synthetics. True, it is possible to synthesize such well known "active ingredients" as tetracycline. But so efficient are modern methods of growing various microorganisms on prepared media under controlled conditions that economics has usually favored this means of production even where alternatives exist. This is especially so when high-yielding strains of organisms are developed.

The Chinese realized 4000 years ago that green mold applied to skin ulcers aided in their cure and stopped the festering. Pasteur, too, realized that his disease-causing bacterial cultures might be destroyed by other microorganisms. But practical application of this ecological fact of life among microorganisms had to await Sir Alexander Fleming's observation that a green mold (*Penicillium*) contaminating his bacterial cultures destroyed the bacteria. From this chance observation, made in a British hospital in 1928, arose the modern concept of pitting one microorganism against another to prevent disease. Penicillin is the name given the active ingredient produced by the ascomycete *Penicillium notatum* (and other related species), which started the wave of antibiotic discoveries. Since then useful molds and microorganisms of many genera have been unearthed, to yield such well known antibiotics as aureomycin, bacitracin, chloromycetin, neomycin, polymixin, streptomycin, and terramycin.

Many technological problems had to be worked out in order to make the production of antibiotics practical. Growing media (generally utilizing corn-steep liquor) had to be devised and kept sterile yet properly aerated to support the mass growth of what are essentially surface-living molds. Suitable methods of solvent extraction, purification, and crystallization into stable form had to be found. To increase yields of antibiotics, new species or strains of microorganisms had to be sought continously, or new mutants had to be created by irradiating microorganisms with ultraviolet light. As disease organisms became resistant, through mutation, to antibiotics to which they were originally susceptible, new antibiotics had to be found. The original penicillin was simply the filtrate of the nutrient broth in which *Penicillium notatum* was grown. Eventually it was proven that such solvents as ether were suitable for extracting the active penicillin, and that through subsequent precipitations a purified product could be produced that was far more potent than Fleming's original mold.

Modern production of antibiotics is not unlike the fermentation of beer, discussed in an earlier section. Highly pedigreed molds (analogous to the special strains of yeast used in brewing) are kept in a sterile environment for starting cultures in small flasks. These are used to inoculate larger "seed tanks," which in turn start the mold in huge vats containing thousands of gallons of nutrient broth. At a determined point of maximum growth, the solution is filtered, subjected to solvent extraction, and the purified antibiotic is precipitated, all under aseptic conditions. Sources of the most useful antibiotics have been fungus molds and actinomycetes, yielding substances mainly effective for controlling bacteria. Since not all disease-producing bacteria are equally affected, a great deal of empirical screening goes into selecting and proving an antibiotic. Under modern conditions, millions of dollars may be invested before any income derives from the medicinal. And for every useful antibiotic discovered, thousands of others have been dropped in the screening because they are insufficiently effective, competitively impractical, or because they produce undesirable side effects.

Belladonna, *Atropa belladonna* (Solanaceae)

Also called deadly nightshade and death's herb, belladonna was well known to the herbalists, who wrote much of medicinal plants (both real and fanciful) during the Middle Ages. The juice of belladonna was first used in salve preparations, and apparently not until the nineteenth century were extracts and tinctures of the plant much used. Belladonna alkaloids are most abundant in physiologically active tissues; atropine, scopolamine, and hyoscyamine are medicinally the most important. In preparation of alkaloidal extracts, the plant parts are typically dried for several weeks, after which solvent extraction (with ether or ethyl acetate) and crystallization of the alkaloids is undertaken. These serve as stimulants to the sympathetic nervous system (and hence as antidotes for

opium alkaloids), for dilation of the pupil of the eye in optical treatment, in ointments, for asthmatic paroxysms, as a circulatory stimulant, and in many other ways. Shortages of belladonna alkaloids from *Atropa* have led to their derivation from species of *Duboisia* and other genera of the family Solanaceae, especially in India.

Cascara, *Rhamnus cathartica* (Rhamnaceae)

Both *R. cathartica* and *R. purshiana* are sources of the cathartic usually referred to as cascara. The former was used in the Middle Ages, but the latter is the official drug of the U.S.P.; it is called "cascara sagrada." The bark of these and perhaps other buckthorns is pulverized to yield the medicinal. *R. purshiana* is native to the Pacific Northwest of the United States, where it was discovered by the Lewis and Clark expedition.

Cocaine, *Erythroxylum coca* (Erythroxylaceae)

The stimulant cocaine is almost as renowned as opium. It instills a feeling of self-confidence and intense exhilaration and has been used as a local anesthetic. But perhaps its greatest fame rests upon its use by porters traversing the rigorous Andean mountain trails. It is said that workers who chew coca can carry on with superhuman endurance for days, even with little food. The coca leaf is typically chewed with powdered lime, which is said to be required to achieve the full effect. It is perhaps significant that lime is also used in commercial extraction of the alkaloid.

In the Andean countries, coca is one of the secondarily cultivated crops. The world's commercial supply comes chiefly from Java, where cocaine was introduced. Once utilized for "cola" flavorings, after removal of much of the alkaloid, coca leaf is today chiefly solvent-extracted for the 1 percent cocaine contained, used medicinally as an analgesic. Since *E. coca* is a small tree, the foliage can be readily collected by hand, after which it is fermented briefly and dried. The cocaine alkaloid is usually extracted with solvents. Chemists have learned to synthesize novocaine, which is very similar to cocaine and is used to a large extent by dentists and doctors as a substitute for the natural product.

Digitalis, *Digitalis purpurea* (Scrophulariaceae)

Digitalis, or foxglove, was recorded by the herbalists as useful for treating epilepsy, wounds, and other ailments. William Withering, an Englishman, first introduced it to modern medicine in the eighteenth century, when he obtained samples from the garden of "the Witch Woman of Shropshire." Withering used digitalis successfully in treating dropsy, and published a treatise on his results. His findings were soon recognized and put into practice in Europe. Used for many purposes during the nineteenth century, it is used today almost exclusively as a heart stimulant and to regulate the pulse, for which digitalis ranks as one of the world's most important drugs. Several glycosides, all quite similar, are the active ingredients.

Although a common garden plant, digitalis grown for drug extraction is usually started under glass and transplanted to the fields. It is a biennial, but the leaves (which yield the glycosides) can be gathered at any time. These are dried in the shade, sometimes pulverized for direct usage, or more commonly subjected to solvent extraction.

Ipecac, *Cephaelis ipecacuanha* (Rubiaceae)

The rhizomes and roots of *C. ipecacuanha*, an herbaceous perennial of Latin America, yield an alkaloidal extract (emetine) that has been used as an emetic and to treat intestinal amoebas. Wild plants, gathered from the rainforest, are a minor item of commerce in South America, but India produces modest commercial supplies. The related *C. acuminata*, also from South America, is also used as a source of the alkaloid.

Marijuana, *Cannabis sativa* (Cannabaceae)

Marijuana (the word is said to be a corruption of the Portuguese *mariguango*, signifying "intoxicant") has been in use for thousands of years. Chinese writings nearly 5000 years old indicate medicinal and psychic uses of marijuana, and it figured importantly in Indian religion as early as 1000 B.C.E. It had spread through the Near East and into Europe before 500 A.D. A hashish cult was fashionable in Paris during the middle nineteenth century, at about which time a British study in India determined that it was not harmful (Christian missionaries had objected to it). Marijuana became a focus of contention in America, after its widespread adoption as a social amenity during the 1960s.

Cannabis sativa is also the source of hemp fiber, for which it can legally be cultivated in the United States, but only under government permit. Hemp oil and oil cake are legal commodities. The species grows readily in temperate regions, and is often found naturalized as an escape along roadsides and in old fields. All soft tissues contain active principles, although these principles are most concentrated in the young foliage and especially in the bracts surrounding the female inflorescence. Subtropical strains of marijuana are said to have higher potency than northern-grown plants. Hashish is the concentrated resin scraped from the pistillate inflorescences; it is several times as potent as typical marijuana "leaf." *Hashish* is the Arabic name for hemp. From the name *Hashashin*—a Mohammedan sect of the Middle Ages ready to eliminate enemies by assassination, presumably while under the influence of marijuana—comes the English word *assassin*. Other designations for marijuana are bhang (the cut-off tops), ganja (a purified Indian bhang), and kif (the term for all forms of marijuana in northwest Africa).

Among other ingredients, marijuana contains tetrahydrocannabinol, or THC (Fig. 22-14), which is hallucinatory under certain circumstances, producing illusions, dreaminess, and euphoria. Marijuana smokers report enhanced perceptivity and sensitivity as the chief pleasures from social use of the drug. Marijuana smoking is estimated to be 50 percent effi-

Box 22-2 *The modern "drug culture"*

On a per capita basis, the addictive "drug problem" was probably as serious some generations ago as it is today, although it caused relatively little stir then. Beginning in the middle 1960s, considerable interest in psychedelics or "mind-expanding" drugs developed, particularly among younger people. Curiosity centered not only on the age-old natural hallucinogens such as marijuana and peyote, but on synthetics such as LSD (lysergic acid diethylamide). Even the sniffing of glue or other aromatics having a physiological influence was undertaken by the unsophisticated.

Emotions, both pro and con, have been intense, forcing political action that was often based on rather little sound information. Highly restrictive laws have been passed in some localities, leading to such incongruities as an elderly judge (at liberty to enjoy martinis at the end of his day) perhaps sentencing some youth to decades of imprisonment merely for possession of a marijuana cigarette. All evidence indicates that marijuana is less habituating and no more addictive than are tobacco and alcohol.

However, serious human degradation has resulted from use of depressants such as the narcotic heroin, even though used only by a small minority (perhaps one in a thousand). Whether the harsh measures taken to limit the availability and use of addictive drugs are wise is a matter for debate. When availability is limited, prices rise, and more serious crime is required for the addict to gain funds enough to satisfy his craving. Great Britain has tried regulated availability, treating the matter as a medical rather than a criminal problem, thus eliminating the potential for large profit that draws organized crime.

Whatever the answers may be, drugs have had an extraordinary influence on contemporary "life styles." In America it is said that nine out of ten adults drink alcohol, and that at least one of them becomes a "problem drinker." More than 10 percent of the people are thought to be smoking marijuana. Equally as many take tranquilizers; a like percentage, barbiturates almost to the point of addiction (ostensibly to relieve insomnia). Many have tried amphetamines (to prevent drowsiness or to lose weight). Almost everyone drinks the stimulant caffeine,

in coffee, tea, and colas. Half smoke tobacco in spite of medical hazards. "Pill taking" without medical supervision is widely accepted; some sort of pill seems to be offered for every problem!

Psychoactive drugs have been part of human culture since the beginning of time; it is unrealistic to suppose that people will suddenly forsake their euphoric and relaxing qualities. Society's problem may well be to learn how best to live with a modicum of "drug culture," avoiding police-state repressive tactics that could be even more injurious to individual well-being than are the drugs.

The main types of drugs in use are these:

Depressants: habituation very strong; addiction, moderate to strong (except essentially nonaddictive inhalants).

  alcohol—natural fermentation products, such as beer, wine, and whiskey.

  sedatives—phenobarbital and other barbiturates, chloral hydrate, etc.

  inhalants—synthetic aerosols and solvents.

  narcotics—mostly natural opiates (heroin, morphine).

  tranquilizers—synthetics such as Miltown, Librium, etc.

Psychedelics: habituation moderate to low; not addictive.

  marijuana—in various forms (hashish, etc.).

  hallucinogens—various natural products, such as scopolamine or mescaline, and such synthetics as DMT, LSD, and STP.

Stimulants: habituation strong (except the essentially nonhabituating antidepressants); not addictive.

  amphetamines—synthetics such as Benzedrine and Methedrine, for depression and fatigue.

  antidepressants—synthetics such as Elavil, for relief of depression.

  caffeine—natural alkaloid in coffee, tea, cola.

  cocaine—alkaloid from coca foliage, produces feeling of exhilaration.

  nicotine—alkaloid from tobacco (extremely toxic).

cient in picking up the quantity of THC that selected marijuana leaf contains (which is about 2–3 percent). Equally effective is a beverage made from the powdered upper leaves and seedheads. The oil may also be inhaled after vaporization.

Marijuana is not considered addictive, although it may be mildly habituating. Efforts have been made to curb marijuana under the supposition that its use may lead to the use of such addictive drugs as heroin; studies show, however, that only a small proportion of marijuana users ever progress to opiates, and that there is probably little impairment to health even from long-term use.

Natural plant growth regulators

Many extractives have profound hormonal effects on plant growth. It has been realized for some time that leachates from plants may inhibit seed germination and the growth of nearby plants (allelopathy); this is especially true of desert plants. But the widespread development and use of herbicides had to await the results of basic research on the hormones (auxins) present in growing shoot tips. Chemical analysis showed that such substances as indolebutyric acid had profound effects upon cell growth, a line of investigation that led eventually to the discovery of the remarkable growth-influencing compound 2,4-dichlorophenoxyacetic acid (2,4-D). Here was a growth regulator that could be synthesized economically and was highly effective in minute amounts; very small dosages stimulated plant growth, but slightly larger ones caused plants, especially dicots, to twist, curl, and finally die, their physiology disrupted. Here was the basis for selectivity; certain plants, such as grasses, are not injured at doses that kill broad-leafed weeds. Intensive research on the way in which such herbicides affect plant physiology, and extensive empirical screening, have since yielded scores of specific, selective, and highly effective general herbicides, to the lasting benefit of modern agriculture. Unfortunately, this also ushered in a new type of chemical warfare—a means of killing the crops that support the population of an enemy, by aerial application of a potent herbicide.

Equally of interest are some of the other growth regulators, a number of them derived from botanical sources. One that has received much scientific attention in recent years, though its practical applications are limited, is gibberellin. It is obtained by methods similar to those used for producing antibiotics from the fungus *Gibberella fujikoroi*, a parasite that attacks rice plants in the Orient. Mild attack by *Gibberella* causes rice to grow abnormally tall. Japanese scientists isolated the biologically active ingredients responsible and named them, collectively, gibberellin. Very small quantities of gibberellin applied to growing plants (usually in solution) markedly increase their growth rate and, in some plants, their size, or differentially alter their growth pattern. The response differs with species; sometimes there is no response. Perhaps the most significant results from commercial application have been obtained in grape production in California, where spraying with gibberellin has produced larger, fuller clusters of grapes on shorter stems.

Other growth-controlling chemicals have been discovered, and botanical science may be on the threshold of controlling certain kinds of plant growth that are now controlled largely by means of expensive breeding programs. The prospects for the development of medicinals for plants as well as for animals seem promising for agriculture's future.

Opium, *Papaver somniferum* (Papaveraceae)

Opium in its crude form contains about 11 percent morphine and 1 percent codeine (as well as many other alkaloids, resins, and oils); it is perhaps the most renowned of all botanicals used medicinally. Morphine is world famous for relieving pain. Smoking or eating opium leads to addiction, which in turn leads not only to mental and physical deterioration but often forces the addict to turn to crime to support his habit. The opium poppy is believed to have originated in Asia Minor, where it often grows as an escape; what its wild ancestral form might be is not known, so long has the species been in cultivation. Opium taking appears to be as ancient as written history, and the opium poppy seems to have had a fearful fascination for all civilizations. The drug is prepared in the Balkans, the Near East, and Southeast Asia, as well as in Japan and China. Commercial supplies for medicinal use come mostly from the Near East, under careful regulation.

The opium poppy is a spiny-leafed annual. In commercial plantings several successive sowings are generally made in spring to ensure that a crop will result, since the small seeds and seedlings are temperamental about unfavorable weather. By early summer, about two weeks after the petals have fallen, the large, urn-shaped seed capsules of the poppy are ready for opium production. The capsules are incised by hand (dictating that opium production be confined to areas with inexpensive labor), usually in late evening. The next morning, the coagulated drops of milky exudate are scraped from the capsule. The exudate is dried and is eventually kneaded into balls of crude opium. Yields as high as 45 kilograms per hectare (about 40 pounds per acre) are possible under good growing conditions. Yields this size may require nearly 50,000 poppy plants per hectare (about 20,000 per acre) and untold hours of human labor.

In Turkey, where opium growing was banned in 1971 under pressure from the United States, its production was reinstated in 1974, but only from unincised capsules let grow to maturity and marketed dry to governmental extraction plants. In Western countries opium can be sold legally only for medical purposes. *P. bracteatum*, which yields thebaine rather than opium, has been suggested as a substitute for the opium poppy. Thebaine is readily converted to codeine for medicinal use, but it is difficult to convert it to morphine (from which heroin is made).

Pyrethrum, *Chrysanthemum* spp. (Compositae)

The daisylike flowers of several species of *Chrysanthemum*, notably *C. cinerariifolium* and *C. coccineum*, yield a natural insecticide, pyrethrum, especially useful because it is nontoxic to warm-blooded animals. The plants (Fig. 22-15) are perennial herbs; the active ingredient, pyrethrum, is most abundant in the mature ovaries. The most important species

Figure 22-15 The dried flowers of *Chrysanthemum cinerariifolium* are the source of the natural insecticide pyrethrum. [Courtesy E. R. Honeywell.]

seems to be *C. cinerariifolium*, native to southeastern Europe, now widely cultivated as a source of insecticide in Japan, Africa (the Rift Valley of Kenya and Tanzania), and New Guinea (Fig. 22-16). Pyrethrum plantings may be started from seed in nursery beds, but the chief means of propagation is through crown separation. Planted 60 centimeters (about 2 feet) apart, as much as 560 kilograms of flower heads can be harvested annually per hectare (about 500 pounds per acre), usually by hand, in the inexpensive-labor areas where most pyrethrum is grown. The dried flowers yield about 3 percent crude pyrethrum, the active ingredients of which are several pyrethrins and volatile oils. Pyrethrum is much used in aerosol sprays for use around the house, in dusts for garden produce, and in protection of foodstuffs, such as grain stored in commercial elevators.

Figure 22-16 Women harvest a root crop in front of a pyrethrum planting in the Kenya highlands. [Photograph by Per Gunvall, courtesy World Bank.]

Quinine, *Cinchona* spp. (Rubiaceae)

Quinine, today principally derived from the bark of such species as *C. calisaya* (and its "variety" *ledgeriana*), *C. officinalis,* and a few other species, was probably in use by the South American Indians before colonization. It contains four alkaloids collectively termed totaquina. Jesuit missionaries are believed to have first taken the bark to Europe. Legend has it that the name "chinchona" derives from the successful treatment of the wife of the Spanish Viceroy of Peru, the Countess Ana of Chinchón, who suffered a severe attack of malaria there in 1638. By the middle 1600s, "peruvian bark," as quinine was called, was utilized for making a variety of extracts, infusions, and tinctures. In addition to quinine's usefulness in treating malaria, the component alkaloid quinidine is a cardiac depressant. Synthesis of quininelike substances has been successful, particularly in Germany, where quinachrine hydrochloride (atabrine) was developed in 1932. Still later, primaquine, chloroquine, and plasmoquine were syn-

thesized. Malaria seemed pretty well under control, until the war in Vietnam brought to the attention of military doctors various forms of the disease more or less resistant to all existing malarial drugs.

Quinine species are native to the Andean highlands of South America. But after *Chinchona* was introduced into Java for plantation growing, Java accounted for about 95 percent of the world's commercial supply until it was captured by the Japanese during World War II. The several associated alkaloids constituting crude quinine occur in the bark of native trees in amounts up to about 7 percent (Fig. 22-17). Selection of superior stock on the plantations increased yields to nearly 20 percent with high-yielding clones budded onto seedling root stock (Fig. 22-18). The trees are planted under shade, rather closely spaced; harvesting gradually thins the stand until the entire planting is consumed after a number of years. The trees are uprooted to provide bark from the roots, as well as from trunks and branches. The bark is first beaten with a mallet to loosen it, and it is then peeled by hand and quickly dried (in drying

Figure 22-17 Bark from wild *Cinchona calisaya,* harvested in Bolivia. [Courtesy USDA.]

Figure 22-18 Seven-year-old plantation of *Cinchona* in Guatemala. [Courtesy USDA.]

ovens on the plantations). It is usually sent in crude form to the consuming country for solvent extraction. Powdered bark is typically mixed with lime, and the alkaloids are dissolved in amyl alcohol or ether. The alkaloids are removed from the solvent by acidified water, and then precipitated by increasing the alkalinity.

Reserpine, *Rauvolfia serpentina* (Apocynaceae)

Derived from *R. serpentina*, reserpine has long been used in India as a medicinal for treatment of the mentally ill. Today it is widely employed against hypertension (high blood pressure). It is effective as an antipsychotic tranquilizer, and it has even been used to counteract insect bites, fevers, and dysentery. Reserpine decreases serotonin levels in the brain. One hypothesis suggests that an imbalance in the metabolism of serotonin is responsible for schizophrenia.

Demand for *Rauvolfia* roots increased so much during the 1950s that most of the wild stands were exterminated in the Indian and East Indian forests. Most production comes today from field plantings. *Rauvolfia* is a perennial shrub about 1 meter tall, capable of growing under a wide range of climatic conditions. It is propagated by seeds, cuttings, or root separations. Plants started from seed, usually in nurseries, are transplanted to the field. Satisfactory growth requires ample rainfall, or, in lieu of this, irrigation. Roots are ready for harvest after 2 or 3 years of growth. As much as 9 metric tons of fresh root may be harvested per hectare (about 4 tons per acre). The alkaloid content of fresh roots is somewhat less than 2 percent. The alkaloid content is higher in the bark than in the wood. India, the main producer, exports the root to Europe and America.

Rotenone, *Derris* spp. and *Lonchocarpus* spp. (Leguminosae)

Rotenone, along with pyrethrum and nicotine, is one of the most widely used natural insecticides. It is primarily employed where there is fear of chemical contamination of edibles or of human habitations. Long known as a fish poison in the jungles of South America, it was recognized that this "barbasco" in no way poisoned the fish for human consumption, or in any way caused ill effects to warm-blooded animals. *Lonchocarpus nicou*, in tetraploid form ($2n = 44$), was primitively cultivated and vegetatively propagated by South American Indians. In the Far East, *Derris*, very similar to *Lonchocarpus*, served as a source of rotenone. In recent decades there has been interchange of planting stock, usually propagated by cuttings, often in small plantations. Planting methods may amount to little more than sticking a live stem into a hole. About half the stems will grow into sprawling shrubs that are uprooted 2 or 3 years later. Hand-weeding must be carried on in the interim; sometimes an intercrop is planted in the first year. For harvest, the tops are cut away and the roots are dug, sectioned, and dried. Rotenone extract is a mixture of aromatic compounds that are soluble in oils but not water. Concentration in the dry roots ranges from almost nothing to as much as 20 percent. The root may be pulverized for direct use as an insecticide (usually mixed with talc or clay), or an extraction may be made with solvents.

Strychnine, *Strychnos nux-vomica* (Loganiaceae)

Strychnine, like opium, is another of the awesome names among plant extractives. Originally it was a component of arrow poisons (curare) made by primitive tribes in the South American jungles. Until even more potent synthetics were developed, it was best known as a "poison," even in modern societies. Yet properly used it is a tonic and a stimulant to the central nervous system, and it can relieve some kinds of paralysis. Actually, many species of *Strychnos* are used; the one cited here, however, is the source of most commercial production, now centering in the Orient. The alkaloid strychnine and a less poisonous one, brucine, are the active components; they are extracted mainly from the seed, where they are most concentrated. Production comes chiefly from uncultivated plants. The large orangelike fruit is collected by local people in their off hours, and the seeds are removed, washed, and dried. These are bought by local merchants and eventually end up in industrialized centers, where they are pulverized and the alkaloids are extracted. The exact procedure for compounding curare to tip arrows and blowgun darts is uncertain; plants other than *Strychnos* are included in the brew, and probably there is no exact formula.

Other botanicals

The foregoing are random examples from thousands of species that have been, or are, used in analogous ways.° Many included in the latest edition of the U.S.P. have not been listed, such as benzoin (*Styrax*, Styracaceae), colchicine (*Colchicum*, Liliaceae), licorice (*Glycyrrhiza*, Leguminosae), mandrake (*Podophyllum*, Berberidaceae), and tolu balsam (*Myroxylon*, Leguminosae). One could search the herbals for the hordes of remedies that were a part of early folklore. These are fairly well known for Europe and the New World, but information on the Far East is less accessible.

INDUSTRIAL FERMENTATIONS

Fermentation,[†] broadly considered, is the degrading of energy-rich molecules by microorganisms. As such, it is a pervading biological fact of life. It is responsible for the decomposition of organic wastes and residues (detritus) and the return to the soil and the atmosphere of elements that will be used again in plant growth. A myriad of saprophytic micro-

organisms live by fermenting; they are capable of utilizing almost any energy source, yielding end products of tremendous variability. By means of a specific enzyme the microorganism catalyzes the decomposition of a complex molecule, splitting it into simpler end products with the release of energy that the microorganism uses for its own growth. One of the most familiar examples is the fermentation of a sugar solution to alcohol and carbon dioxide:

$$C_6H_{12}O_6 \xrightarrow{\text{yeast enzymes}} 2C_2H_5OH + 2CO_2$$

In a nontechnical sense, "fermentation" refers to the bubbling off of gas during such microorganismal decompositions.

Several of agriculture's most important fermentations have already been mentioned in this chapter. The production of alcoholic beverages by the action of yeast is especially important. The production of antibiotics is another example. Mention has also been made of the several "curing" processes in the development of flavors and aromas in such things as cacao, tea, and several essential oils. In the next chapter we will see that microorganisms are useful in **retting**—freeing fibers from surrounding tissues. They are important in producing acids during the fermentation of fodder to ensilage, which keeps the silage preserved. Fermentation makes bread dough rise, converts milk curd to cheese, cabbage to sauerkraut, and cider to vinegar. Many plant diseases (such as rots) involve fermentation, as does food spoilage and the decay of structural timber. Fermentation is used in sewage disposal; gardeners prepare compost by fermenting plant residues.

All of this, of course, is not "industrial." But these examples do serve to illustrate the ubiquitous activity of microorganisms. Because these applications relate so closely to the discussions in other sections and chapters, industrial fermentations need only brief coverage here. As a direct source of materials usable in a modern economy, fermentation industries are relatively unimportant (compared to agricultural crop growing or chemical synthesis from such raw materials as natural gas and petroleum). Indeed, some of the major industrial fermentations of a for-

°More than 100 species are discussed briefly as being of some drug value in Schery, *Plants for Man*, Prentice-Hall, 1972, including some as well recognized as *Arnica* (Compositae); Chaulmoogra, from various genera of the family Flacourtiaceae, which is used to treat leprosy; ephedrine, from *Ephedra* (Gnetaceae); ergot, from *Claviceps* (an ascomycete); ginseng, from *Panax* (Araliaceae), a mystical Chinese remedy; peyote, or mescal buttons, *Lophophora* (Cactaceae), containing a hallucinatory alkaloid similar to LSD; quassia, from genera of the family Simarubaceae, both an anthelmintic and insecticide.

[†]Fermentation to the physiologist refers more specifically to the anaerobic degradation of carbohydrates, often to CO_2 and an alcohol or an organic acid; the degradation of protein is called putrefaction.

mer era, such as the production of ethanol, acetone, and butyl alcohol, have lost much of their importance in competition with chemical synthesis from petrochemicals. This, of course, is a matter of economics: when the cost of molasses, grain, and other raw materials suitable for fermentation rose, petrochemical production generally proved less expensive—that is, until the energy crises of the late 1970s. Since petroleum and natural gas are not renewable resources, industrial fermentations of renewable crop materials would seem to have time on their side, and already the tide is turning in that direction in the 1980s. The world's greatest reserve of carbohydrate, cellulose, which is abundant in old newspapers, awaits being called upon as an energy source, for digestion to glucose and other foods (a process already practiced on hay and browse by ruminants).

Production of industrial ethanol, or ethyl alcohol, is one of the more important industrial fermentations. Up until World War II most alcohol was obtained by fermentation, molasses being the chief raw material. Production exceeded about 2 billion liters (over 500 million gallons) in 1945, a year of unusual demand because of World War II. Since then, total demand has exceeded 760 million liters (about 200 million gallons) annually, but only about 20 percent of this has been produced by fermentation, the rest having being synthesized from ethylene. The story is much the same for acetone and butyl alcohol. During World War II about two-thirds of the butyl alcohol production was by fermentation, and around 1940 nearly half the acetone produced was a fermentation product. As for ethanol, the raw material was cheap blackstrap molasses. Since World War II, synthetic production of these materials has supplied most of the market. The reason is that the cost of molasses has increased from about one cent per pound of sugar content before World War II to several times this price since.

A number of organic acids are also produced by fermentation, and these have fared better economically. Citric acid and lactic acid are both produced almost entirely by fermentation. The demand for citric acid is about 50,000 metric tons annually in the United States. Acetic acid (vinegar) is another fermentation product that has held its own. Among the less well-known but important acids manufactured by fermentation are gluconic, fumaric, itaconic, and tartaric, whose esters find some use in the plastics industry.

Vitamins are also produced largely by fermentation. Production of ascorbic acid alone amounts to thousands of tons in the United States, and riboflavin and vitamin B_{12} are produced in lesser amounts. Similar growth factors are produced by yeast cultivation, which is often undertaken to make use of by-products from other industries. For example, *Torula* yeast, which may be grown on pulp liquors from the paper industry, constitutes a nourishing source of proteins and vitamins, which are utilized to some extent in animal food.

Of course, enzyme preparations other than malt are frequently the by-products of fermentation. And the production of antibiotics, discussed in the previous section, is a billion-dollar industry. Dextran has been prepared by fermentation; it is used in food products, as an extender for blood plasma, and as an additive to well-drilling muds. It can be produced by fermenting almost any sugar source with *Leuconostoc mesenteroides*. This is an example of a potentially important new field—the production of polymers by microbial fermentation.

Not commercially of much consequence, although also potentially useful, is the production of resin substitutes from microorganisms: the same form of polyisoprene that occurs in natural rubber has been recovered in small amounts from such fungi as *Lactarius* and *Peziza*. Other fungi are capable of converting carbohydrate into fats that may possibly be useful in special diets (although it is unlikely that they will ever compete with such inexpensive products as soybean oil). Microorganisms can also aid in the production of amino acids, or even proteins: lysine has been commercially produced by fermentation. Bacteria and yeasts double their cell mass in as little as ten minutes, a rate hundreds of times faster than the rate for forage crops. Since about half the dry weight of bacteria, molds, and yeasts is protein, 1 ton of microorganism can become 12 tons in 24 hours, yielding 6 tons of protein (assuming no limitations on food supply or growth restraints)!

These examples merely indicate the industrial

possibilities of fermentation: new strains of micro-organisms capable of executing specific fermentations are discovered almost daily. As in the improvement of antibiotics, the selection of new strains of microorganisms should increase the efficiency of even the most well-known fermentation processes. All that seems necessary is to find fermentable raw materials that are less expensive than those now being used. For the present, however, industrial fermentations are confined chiefly to nonfood specialties and to the utilization of waste and by-products.

FUMATORIES AND MASTICATORIES

Plant products smoked or chewed for pleasure are not uncommon. Coca leaves are chewed in the Andean highlands for the stimulating effect that they have; marijuana and opium are smoked for the feeling of euphoria they induce. We have considered the cola nut as a flavoring for beverages, but it is also used as a stimulating masticatory in western Africa. One plant in particular is smoked more than any other—tobacco. From time to time tobacco has been considered to have medicinal value, and it would be difficult to make a hard and fast distinction between it and the many plants whose medicinal value is established.

Tobacco, in the form of chewing tobacco, also serves as a masticatory, although this unappealing custom is less and less practiced in sophisticated societies. In the Orient, perhaps more people chew the betel nut, a locally produced stimulant, than smoke tobacco. Betel will serve as our example of a much used masticatory, and tobacco as an example of a fumatory that is smoked throughout the world. Chewing gum, a popular masticatory in the United States, is reserved for discussion along with latex products.

Other than tobacco, fumatories and masticatories are not important items of commerce. Many, such as coca and betel, although intensively used regionally, enter commerce almost solely as local items of trade. But tobacco has for centuries been an important international commodity, produced in significant quantities on all continents. Nearly 6 million metric tons of tobacco leaf are grown annually, almost all of which is used for smoking.

Tobacco, *Nicotiana tabacum* (Solanaceae)

Tobacco, grown and used moderately by the American Indians, was discovered by the explorers of the New World. It was first introduced into Europe in 1558, where it attracted attention as a curiosity and was claimed to possess medicinal values. Throughout the seventeenth and eighteenth centuries, tobacco was recommended as a sort of "cure-all" for treating many common afflictions ranging from the toothache and rheumatism to indigestion and blood poisoning. For a while the use of tobacco in the form of snuff was socially accepted. Gradually this use gave way to smoking. Tobacco was used almost exclusively by men until the advent of the ready-made cigarette in the twentieth century, when it came to be used by women as well. The use of tobacco continues rather steadily in spite of evidence that it is hazardous to health. Certainly it is habituating, more so than such drugs as marijuana. The chief centers of production are China and the United States, in each of which nearly 1 million metric tons of leaf is grown annually, with India and Brazil following at some little distance. Most parts of the world grow some tobacco.

Tobacco is one of the most pampered of crops. It must be started in sterilized nurseries and tended intensively from planting until harvest. A glance at the technical literature about tobacco suggests that it is prone to more ailments than even the human hypochondriac; the tobacco breeder endlessly searches for genetic resistance to anthracnose, black root-rot, black shank, blue mold, brown spot, cyst nematodes, frog-eye leaf spot, fusarium wilt, Granville wilt, powdery mildew, root-knot nematodes, rattle virus, tobacco etch, tobacco-mosaic virus, tobacco-streak virus, wildfire, and several others. Since the tobacco genus is, fortunately, very diversified and genetically variable, potential sources of disease resistance are always at hand, even though the situation is complicated by the general sterility of F_1 interspecific hybrids.

Figure 22-19 Tobacco culture: (A) Fields of broad-leaf burley growing in terraced and contoured fields in Kentucky. (B) Shade houses for cigar wrapper production in Connecticut. [A, courtesy USDA; B, courtesy Connecticut Agricultural Experiment Station.]

Nicotiana contains in the neighborhood of 70 species. *N. tabacum* is the specific name applied to most cultivated tobaccos. In its many cultivars, it is a widely variable and polymorphic species. A review of the genus by Thomas Goodspeed indicates that *N. tabacum* may have originated as a chance cross between ancestral forms of *N. sylvestris* and some distant relative of *N. otophora.* There is no definite record of *N. tabacum* in the wild state, and it appears that the species may be of relatively recent origin, having been brought quickly into domestica-

tion after formation as an amphidiploid, possibly in central South America. *N. tabacum* is a tetraploid, with 48 chromosomes, as is *N. rustica,* a second species utilized by the North American Indians (and perhaps the first to be introduced into Europe; it is still grown in Africa and the Near East).

The tobacco plant is grown as a perennial in tropical environments, but it is customarily grown as an annual in temperate localities (Fig. 22-19). Although some cultivars of tobacco are grown ornamentally for flowers, the inflorescence is removed as it develops

Figure 22-20   Picking tobacco leaves in Malawi. [Photograph by James Pickwell for International Development Association, courtesy World Bank.]

in commercial leaf production. The cut leaf is fermented ("cured") to develop the desired aroma and taste, which is raw and harsh in fresh tobacco. The alkaloid nicotine is perhaps the most distinctive constituent of tobacco leaf. It volatilizes readily when the leaf burns and is responsible in part for the irritating vapor in tobacco smoke or nicotinic insecticides. There are, of course many other organic components in tobacco smoke; in recent years these have been analyzed repeatedly in the search for carcinogenic factors. Tobacco "tar," which includes resins, essential oils, organic acids, and phenols, is usually implicated as the health hazard.

Nicotiana is named after the French Ambassador to Portugal during the late 1500s, Jean Nicot, who was responsible for introducing tobacco to the court of France. Although tobacco had been sent to Europe from the West Indian colonies of Spain for half a century, it remained for John Rolfe to initiate the first significant commercial planting, in the Virginia colonies of England, in 1612. Ever since, tobacco growing and export have remained an important agricultural enterprise along the Atlantic seaboard. In the early years, the crop was sent to Europe for ci-

gars (the Spanish custom), pipe tobacco (the English custom), snuff (the French custom), and eventually cigarettes (the American custom, adopted from the Turkish). It is perhaps historically significant that cigarette manufacturers pride themselves upon tobacco blends that include small-leaved, untopped, aromatic Turkish cultivars.

The growing of tobacco (Fig. 22-20) varies, in various parts of the world, in the intensity of care given to the plants, but it follows a similar pattern wherever the species is cultivated. Tobacco growing in Africa is typical of the intricate procedures practiced in major growing regions. Seedbeds—usually about 9 square meters (100 square feet), sufficient to supply seedlings for several acres—are meticulously cultivated and fumigated (to kill nematodes, weed seeds, and soil-borne diseases). Fertilizer rich in phosphate is liberally mixed into the top 5 to 8 centimeters (2 or 3 inches) of soil. Tobacco seed is extremely small; it is generally extended with an inert material to help achieve uniform distribution, or it is mixed with water and applied with a sprinkling can. Only 25 grams of seed will plant more than 100 square meters (about 1 ounce plants 1200 square feet), with a density of nearly 650 seedlings per square meter (about 60 seedlings per square foot). Hybrid seed, although very expensive, is mostly used in the burley belt of the United States, and is produced through hand pollination of a male-sterile line. After sowing, a mulch is applied and the bed is sprinkled several times daily until the seed is sprouted and the seedlings are well established. In the United States some growers cover the seedbed with polyethylene plastic to give the seedlings more adequate protection and to maintain better control of environmental changes than is possible with the more general practice of shading seedlings with cloth. The nursery is sprayed regularly with fungicides to control disease, and no smoking is allowed nearby for fear of introducing diseases. (Tobacco-mosaic virus, for example, may be transmitted to the plants by the hands of smokers.) As the seedlings become large enough for transplanting to the fields, watering is withheld to harden the transplants.

In the meantime, the growing fields have been thoroughly worked, fertilized, and perhaps fumigated. The seedlings are typically planted in ridges, the rows roughly 1 meter (about 3 feet) apart, the plants spaced about 60 centimeters (2 feet) apart in the row. The plant population runs approximately 17,300 plants to the hectare (about 7000 to the acre). Transplanting is usually a hand operation in Africa (there is some mechanization in the United States), and the transplant is given a measured amount of water as it is set. This is sufficient to carry it until the rains take over. Such weeds as develop are normally controlled by cultivation. As the plants begin to flower, the inflorescence is topped by hand. This stimulates side branching or "suckering," which must be checked repeatedly (either by hand or by applying sprays) if the leaves are to achieve their full potential.

The leaves are harvested progressively from the bottom upward on each stalk, two or three at a time, as they reach the proper degree of ripeness. They are gently collected, covered for protection from the sun, and taken to a "tying shed" for fastening to sticks with twine. They will be suspended by these sticks in the curing barn, in a framework that accommodates many thousands of leaves at a time. The curing barns are fired with wood (or occasionally coal), and the heat is radiated from a series of flues. The barns are ventilated at roof level. Curing procedures vary with the kind of tobacco and with the region of production, but they may start with a temperature of 29°C (85°F) for one day, after which the temperature is raised to about 32°C (90°F) until the leaf is well colored, and then increased to 38°C (100°F) until no more green shows in the leaf. Subsequently the temperature may be raised even higher, with a final drying at about 71°C (160°F). Curing takes several days, after which the dried tobacco is allowed to absorb sufficient moisture to make it pliable for convenience in handling. Grading, baling, and eventually auctioning follow. Oriental and Turkish tobaccos may not be so elaborately handled. Moreover, they are generally cured in the sun. The curing process reduces the leaf weight about 80 percent. After auctioning, tobacco is generally further aged, typically in huge casks, for a period that may extend into years.

The tobacco industry has developed many individualized customs, some of which have even hindered acceptance of new disease-resistant cultivars. The classification of tobacco as "burley," "fire-cured," "piedmont," "filler," "cigar wrapper," and so on, is confusing to those not conversant with the tobacco industry. These designations may be based as much upon the region in which the tobacco is produced, techniques of culture, and the method of curing as on the genetic characteristics of the tobacco. The blending of different tobacco types constitutes the professional secret of the manufacturing company, and it provides such distinctiveness as there is among the different brands. Cigarette manufacture, the greatest use of tobacco, is especially competitive. Manufacturers vie for public acceptance with tremendous advertising programs and varying types of cigarettes (with or without filter, menthol, and so on). The basic operations involved in cigarette manufacturing are: conditioning the fully aged tobacco to a uniform moisture content, removing coarse parts (petioles and large veins), and chopping the leaf into "strip." The strip may be further aged, then blended with other tobaccos (such as Turkish). Humectants are applied to keep the tobacco moist (often glycerine, glycols, or apple juice), and, depending upon the type of cigarette, flavorings may be added (sugars, coumarin, rum, and so on). The different tobaccos are thoroughly blended in rotating drums, and the mix is sent to shredding machines. After remixing it is routed to the "making" machine, which mechanically rolls the tobacco in cigarette paper. The slender strands of rolled tobacco are automatically cut into proper lengths, affixed with filters if required, and packaged by elaborate machines capable of handling thousands of cigarettes per minute.

Betel, *Areca catechu* (Palmae)

Among the dense populations of Southeast Asia, betel chewing is as commonplace as is the chewing of chewing gum and the smoking of tobacco in the Western world. It is said that one-tenth of the world population chews betel. The chief component of betel is the seed, or "nut," of the betel palm, *Areca catechu*. The betel nut is typically chewed along with a fresh or processed leaf of a native pepper, *Piper betle*, of the family Piperaceae. The combination of betel nut, betel leaf, and a small quid of lime gives the mouth and saliva a wine-red coloration. The mass is worked in the mouth without swallowing, which of course stimulates a copious flow of saliva that is continuously expectorated. After a time the teeth of a betel addict turns blackish. Betel contains a narcotic that provides mild stimulation. Although its use seems disagreeable to westerners, it is comparable to smoking and chewing tobacco in other parts of the world. In India, Malaysia, Indonesia, and eastern Africa the practice is socially quite acceptable.

The betel palm has been domesticated for so long that betel is mentioned even in the earliest Hindu mythology. Early writings vaunted the medicinal virtues of betel, claiming that it could be used to "expel wind, remove phlegm, kill germs, subdue bad odors, beautify the mouth, remove impurities, and induce love," among other things. Arabian visitors to India in the Middle Ages reported, "the betel brightens the mind and drives away the cares." Marco Polo commented upon the Oriental habit of betel chewing in his writings of 1298. Exactly where

the betel palm may have originated is uncertain— possibly the Malay peninsula or the Philippines. The trees are now widely distributed from India to the Philippines and Japan, all seeming to be the cultivated form. *Areca* has also been introduced into the Mediterranean area and into Florida as an ornamental. A number of botanical varieties of *A. catechu* have been described in the Far East, and about 50 other species are recognized in the genus (only one of which is occasionally chewed). The various forms of betel nut found in the Eastern markets represent seedling races or local cultivars.

Areca palms reach heights of 30 meters (about 100 feet) and bear large pinnate leaves. They are generally planted in gardens or small plantations, flourishing best where rainfall is abundant and temperature rather unvarying. Since *Areca* tolerates shade well, the palms are often planted mixed with fruit trees such as mango, citrus, and coconut. Propagation is by seed, for which the seed nuts are allowed to ripen completely on the tree before drying and sowing. The seeds are generally sown in nurseries, and the seedlings then transplanted to the permanent site. The palm flowers in about its seventh year, and it reaches full bearing potential in somewhat more than a decade. The fruits, which form in large bunches, are hand harvested (by climbing the tree) when ripe (bright red). A tree that yields well will supply 500 to 1000 nuts a year. Like other crops, betel palms suffer from various diseases and insect pests, although steps are seldom taken to control them.

Betel nuts may be consumed raw or cured, following hand removal of the outer husk. Surplus nuts may be stored in pits in the soil, or under water in earthenware jars. Sometimes nuts are dried in the sun and then marketed. Sliced nuts are often boiled for several hours before drying. The endosperm contains several alkaloids, as well as tannins, fats, carbohydrates, and proteins. At least three different alkaloids, including arecoline, are present, constituting as much as 0.5 percent of the endosperm. The tannin content may be as much as 18 percent, which contributes to the astringency during mastication. As is typical of palms, the betel palm has much oil in its endosperm, and the husks constitute a potential by-product.

LATEX PRODUCTS

Latex is a milky, colloidal secretion found especially in species of the families Apocynaceae, Asclepiadaceae, Compositae, Euphorbiaceae, Moraceae, and Sapotaceae. Latex occurs in specialized cells, or cavities, that drain when ruptured, thus permitting its flow when a plant is tapped (usually by gouging or slicing the bark). Lactiferous plants are found in all parts of the world, in many sizes and shapes, from the humble dandelion (*Taraxacum*) and milkweed (*Asclepias*) of temperate lands to the impressive lactiferous trees of the tropical forests. You may recall that several of the species reviewed earlier yield a latex, particularly the opium poppy (drug) and the papaya (meat tenderizer).

The colloidal particles in latex consist mainly of hydrocarbons and resins, but latex also contains oils, essential oils, various carbohydrates, proteins, enzymes, organic acids, minerals, salts, alkaloids, sterols, tannins, and other substances. It has been theorized that latex functions in the plant to seal wounds and that it serves as a disinfectant and a food reserve. It may, however, consist of nothing but waste materials. The coagulated latex (coagulum that results when the colloidal particles are precipitated) is of two general types: **rubber,** which contains a high percentage of long-chain *cis*-polyisoprene, a highly elastic polymer $[(C_5H_8)n]$, or **balata,** which contains high proportions of shorter-chain *trans*-polyisoprene and resin, yielding an inelastic coagulum. Rubber is, of course, the most important. It is derived chiefly from the Pará rubber tree (*Hevea brasiliensis*). To a lesser extent *Castilla elastica*, species of *Parthenium*, various figs (*Ficus*), *Manihot*, *Hancornia*, *Cryptostegia*, *Funtumia*, *Landolphia*, and many others have served as sources of rubber. The principle balata gums are chiclé, or chewing gum (*Manilkara zapota*), chilte (*Cnidoscolus*), guttagum (*Couma* and *Manilkara*), gutta-percha (*Palaquium*), and jelutong (*Dyera*).

Latex extraction procedures vary. Most commonly these days the bark of growing trees is tapped, as with *Hevea*. Sometimes the tree is felled and the latex is freed by girdling, as with wild *Castilla* and *Manilkara*. Solvents can be used to extract balata from foliage, as with *Palaquium*. Or maceration can be used as a physical means of freeing the polymers, as with *Parthenium*. Only with *Hevea* has tapping been well investigated and an efficient technique worked out. This has been possible because of concentrated plantation cultivation. Most other latices are gathered from wild sources in remote areas by uneducated tappers lacking supervision.

Before a latex can be used the solids must be separated from the liquid supernatant "serum." **Coagulation** can be induced in various ways. If the latex is merely left standing for a day or two, coagulation will sometimes occur naturally, due to enzymatic changes in the proteins or to acidification through fermentation. Usually, however, the latex is heated or boiled, often with the addition of salts or acids (in primitive locations, fruit juices or sea water serve). Such treatments break the repelling forces between colloidal particles, causing them to coagulate. The fresh coagulum is whitish, soft, and pliable. As it ages or is further treated, rubber becomes tough and elastic; balata is usually dark, hard, and brittle. The many uses of rubber are well known, principally for pneumatic automobile tires; balatas are used mainly for insulating materials, for making molded products, and for chewing gum.

Rubber

Rubber, *Hevea brasiliensis* (Euphorbiaceae)

In intrigue and romance the rise (and partial fall) of the natural-rubber industry challenges that of the spice trade. The first European notice of rubber was by Columbus in the Caribbean and by Cortez in Mexico, where crude rubber balls were noted as playthings of the American Indians. More likely they were of *Castilla* rather than *Hevea* rubber. Rubber gained its appelation from the discovery that "bouncy" tree-gum balls brought back to England from the New World could be used to "rub out" pencil marks on paper. Rubber became more than a curiosity in 1823, when Charles MacIntosh learned to waterproof fabric with it, and especially when

Charles Goodyear learned to vulcanize it (heat it with sulfur to eliminate stickiness or "tackiness") in 1839. Interest first centered on *Landolphia* and other apocynaceous vines of equatorial Africa. The latex was obtained by cutting the vines at ground level, a practice that soon depleted supplies and, of course, wrote a finish to the African rubber boom. While the boom lasted, however, the exploitation of native peoples became a worldwide scandal. Peaceful tribes were taxed ever increasing quotas of rubber by their colonial overlords. If the quotas were not met, women were seized and men put in chains—even mutilated and left to die a lingering death as new sources of the rapidly disappearing rubber trees in the Congo jungles were sought. This was, unfortunately, a pattern again to be repeated in the upper Amazon at the height of the wild rubber boom there.

Soon the boom spread to the New World, where the species destined to become the world's main source of rubber, *Hevea brasiliensis*, grew wild in the Amazon Valley. *Castilla*, which grew in Central America and northern South America, had its moment of glory, and later *Funtumia* was introduced into the West Indies from Africa, but only to become another of the West Indian casualties, as *Hevea*, the genus most amenable to repeated tapping, came into its own. With only wild rubber to be had, and demand soaring because of the newly burgeoning automobile industry, rubber prices increased astronomically. In 1910 the price of rubber reached the all-time high of $3.06 per pound, an especially heady price considering the value of the dollar in those days. No wonder those were tumultuous times.

As early as 1875 the competent British Colonial Service had engaged Henry A. Wickham to procure seeds of *Hevea brasiliensis* in Brazil, and Robert Cross to bring back seeds and plants of *Castilla* from Central America (Cross later went to Brazil for young trees of *Hevea*). An ocean liner was chartered to rush the perishable seeds to Kew in England, where a modest proportion were germinated and the seedlings shipped carefully to Ceylon and Singapore. There they were received initially with little enthusiasm. These few seedlings, however, were to spawn the tremendous rubber plantation industry of the Far East. But in the early 1900s the world still relied upon wild rubber from the Amazon. At that time the rubber capital of the world was Manaos, situated in the sweltering jungles thousands of miles up the Amazon River. At the height of the rubber boom, so great was the wealth flowing into Manaos that a tremendous opera house was erected there from construction materials imported from Italy. The boom was to be short-lived, however, for by the end of World War I the once-scorned rubber trees of the British colonies in the Far East had been widely planted and were of tappable age. These plantations yielded more rubber than the market could then consume. The price broke to 14 cents per pound, rose again as the "Stevenson plan" cartel restricted rubber production, then fell again to an all-time low of 3 cents per pound in the depression years. With synthetic rubber now available, a ceiling of sorts exists on rubber prices, which are not likely again to become so chaotic and so susceptible to political maneuvering as in the early days of the industry.

A small amount of wild rubber is still harvested in the Amazon Valley, most of it by lonely *seringueros*, living in thatched huts and crudely tapping distant trees that may still bear the tapping scars of the boom years. There was quite a revival of interest in wild rubber during World War II, when Japan gained control of the Far Eastern plantations, and a frantic search was made by the Allies for alternate plant sources for latex. Chemists, however, synthesized fairly short-chain rubberlike polymers from isoprene and butadiene in 1945, and things have never been the same with wild rubber since. However, these synthetic elastomers, although great for certain purposes, have never been so tough for hard-use tires as natural rubber; a generous proportion of the natural product is preferred for radial tires, for example. In competition with synthetics, natural rubber fell to about 20 percent of annual world production during the epoch of petroleum abundance, but demand exceeded supplies by 1980 and the proportion climbed back nearer to equality. Of the nearly 4 million metric tons of natural rubber that are produced annually, 90 percent comes from plantations and small holdings in the Orient (principally Sri Lanka, Malaysia, and Indonesia).

Tapping of wild rubber trees in the Amazon Valley has always been a wasteful and expensive procedure. The trees are not easily accessible, and much time and effort are spent in collecting the latex. Gouging of the bark with crude tools mutilates the trees and interferes with efficient continuing production. There are few processing centers, a necessity for economical mass handling of latex; instead, individual tappers have traditionally smoked the latex over open fires on paddles, forming it into huge balls, which in turn must be transported tremendous distances to market. Such a system can hardly compete with the highly efficient plantation growing of rubber in the Far East, where abundant, skilled, and inexpensive labor can tap and tend concentrated plantings. Moreover, trained botanists early initiated breeding and selection programs to develop high-yielding clones of *Hevea* that are especially amenable to tapping; these clones are much more productive than the wild trees of the New World. So much has tree quality been improved that during and after World War II, in an effort to re-establish rubber growing in Central America and South America, planting stock was zealously sought in the Far East. In less than a century that little-appreciated introduction from the New World had been thoroughly domesticated, one of the few documented cases of domestication of an important crop species.

The **jebong system,** the most efficient system of tapping, was developed on the Far Eastern plantations. A "panel" is opened half way round the tree, from the upper left to the lower right (since the latex canals in the bark spiral to the right, more canals are opened by a cut made in this way). Thin slices of bark are removed with each new tapping from the lower edge of the panel, to renew the flow of latex. More recently it has been demonstrated that tapping upward on the panel yields as much latex on a once-in-four-day tapping schedule as does tapping downward on a once-in-three-day schedule. Laborers are trained not to cut too deeply into the tree, so that the cambium will not be injured and will be able to regenerate new bark (Fig. 22-21). By the time tapping reaches ground level on one side, the opposite side is worked, after which the bark will have healed over the original panel and be ready for tapping again. Tapping is done early in the morning, when the flow

of latex is greatest and showers are unlikely. A small spout is inserted at the lower end of the panel, which guides latex into an affixed cup. With high-yielding trees, as much as 1 metric ton of latex per year can be obtained per tree, although the intensity and season of tapping varies somewhat with local conditions. Good trees on well-tended plantations can endure tapping as frequently as every second day, except for a rest period during change of foliage.

The latex is carefully kept free of bark and dirt. At each tapping, latex coagulated on the cut is removed by hand and kept separate as "scrap." After tapping his quota of trees, the tapper retraces his route and carefully empties each cup of latex into a clean, covered pail, which is then taken to a central processing plant. The treatment of the latex varies with locality. In some places, latex is treated with an anticoagulant (usually ammonium hydroxide) and shipped in bulk to the location where it will be processed; more commonly, however, it is coagulated locally by the addition of acid. The soft coagulum is squeezed free of "serum," and the resulting thin sheet is typically "smoked" in a special smoke house to dry and preserve it. Properly smoked rubber will not deteriorate, and it can be shipped to consuming markets as needs and prices dictate. This is a far cry from grimy smoking over an open fire, or coagulation by fermentation in ill-provided jungle camps. Coagulation formulas specified by the purchaser are often undertaken nowadays, and much raw rubber is shipped as "heveacrumb," (compressed granular unsmoked crepe) rather than smoked sheet.

High-yielding *Hevea* clones are perpetuated vegetatively, especially by budding. So technically specialized has propagation become that "composite" trees are often created from separate, exceptionally good root stocks, high-yielding trunks, and vigorous tops. A trunk from a high-yielding clone is budded to a proven root stock and grafted with a vigorous top that is especially efficient photosynthetically. A plantation tree can be tapped when 5 years old, but peak yields don't come until it is at least 12 years of age. Usually the tree is abandoned when it is about 25 years old, mainly because better types will have been bred by then. Research has indicated just which are the best fertilizer regimens and cover crops for the

Figure 22-21 Rubber tapper using special jebong knife that scores the bark deeply enough to cause latex to flow without damaging the tree. [Courtesy USDA.]

particular locality. Usually, in addition to the major fertility elements, iron and manganese are required for vigorous growth, and occasionally other minor elements. Thus, in contrast to what is generally true of tropical tree crops, as much is known about the requirements of plantation-grown rubber trees as is known about the requirements of almost any agricultural annual. It has even been found that very light applications of 2,4-D or such growth regulators as ethephon just below the tapping cut can increase the yield as much as 30 percent without damaging the tree.

Fortunately for the Far Eastern plantations, *Hevea* diseases inherent in the Brazilian homeland were not transported with Wickham's seeds to England, and thence on to Ceylon in 1876. As improved stock is reintroduced into the New World, it often suffers from leaf blight and other diseases, which prevent the establishment of economic plantings. Thus for the New World rubber industry, additional research is necessary to develop clones adapted to New World conditions. Because it is expensive and difficult to control diseases with fungicidal sprays, the most practical approach has been to develop disease-resistant clones through selection and breeding. Two

other *Hevea* species yielding latex in commercial quantity, *H. benthamiana* and *H. guianensis*, are used in crosses, and they contribute to a much broader germplasm base than was represented by Wickham's original collections.

Panama rubber, *Castilla* (Moraceae)

Several species of *Castilla* have been utilized for rubber. Indeed, the first reports of rubber from the New World, by the French geographer Charles Marie de la Condamine, who explored the Amazon in 1734, probably referred to *Castilla* rather than *Hevea*. La Condamine called the extract *caoutchouc*, but reported the Guiana Indian name of *heve*, which later led the French botanist Jean Baptiste Christophe Fusée Aublet to name a different genus *Hevea*, mistakenly believing it to be the species reported by la Condamine.

In the heyday of the natural-rubber boom, high hopes were held for *Castilla*. Royal Botanic Garden botanists, testing rubber species in Trinidad, felt that the rapid growth and wide tolerance of *Castilla* indicated that it might be of greater promise than *Hevea*. But they were proved wrong. Following enthusiastic (and often black-market) distribution of planting stock in the West Indies, the combination of inexperience in growing and tapping, the lack of knowledge about handling latex, the vagaries in the price of rubber, and insufficient research, led to complete demoralization of not only the *Castilla* rubber industry, but of all plantation rubber growing. *Castilla elastica* still grows as an escape in the West Indies, and it grows wild throughout Central America and northern South America. In booms or times of emergency, there is some tapping. Usually, however, the wild trees are felled and the latex is drained. The latex canals in *Castilla*, unlike those in *Hevea*, are continuous, and not amenable to frequent tappings as with the jebong system. A large tree can yield several gallons of latex, which can be processed into as much as 32 kilograms (70 pounds) of rubber.

Manicoba, *Manihot* spp. (Euphorbiaceae)

Several species of this genus, especially *M. dichotoma* and *M. piauhyensis,* grow wild in the arid sections of northeastern Brazil and yield a good rubber latex, but only in small quantities. The trees have been exploited only in times of emergency and when rubber prices are high. There have been only desultory attempts to cultivate the tree, so that tapping has largely been of an exploitative unscientific nature, involving a series of gashes from top to bottom of the tree in a fashion that causes the latex to drain progressively downward into a receptacle at the base. The bark is hard, not suitable to jebong tapping; and, in the desertlike *caatinga* of Brazil, growth is slow and replenishment of tree stands is poor.

Mangabeira, *Hancornia* (Apocynaceae)

Like manicoba, mangabeira (especially *H. speciosa,* native to the sandy coastal belts from eastern Brazil south into Paraguay) has served as a source of a rubber latex during times of emergency. The quality is not as good as that of other rubbers, and the small trees of the scrub forest provide relatively small yields for the effort expended in gathering the latex.

Palay rubber, *Cryptostegia grandiflora* (Apocynaceae)

This and other species from Madagascar have been introduced into Central America as a promising source of rubber. The plants are sprawling vines, and the latex is typically gathered by cutting off the tips of the stems and hand-gathering the small drops of exudate. Attempts made during World War II to grow the plant extensively and to develop mechanized harvesting proved a failure; consequently, interest in this widely naturalized "weed" as a source of rubber has died.

Guayule, *Parthenium argentatum* (Compositae)

Guayule, native to Mexico and the southwestern United States, had been a limited source of rubber before World War II, and it contributed a little (but vital) supply to the war effort. It was hoped that extensive stands could be cultivated eventually as a field crop in semiarid regions, and the rubber could be retrieved by maceration of the harvested foliage. Although progress has been made in developing strains of acceptable yield, many practical difficulties have prevented guayule production from becoming competitive with other sources. But there is still much interest in developing the species as a field crop that might make industrial nations less dependent on distant tropical sources of rubber.

Other rubber sources

A great many other genera have been tried from time to time as possible sources of rubber. In the family Apocynaceae, *Alstonia, Apocynum, Forsteronia, Funtumia, Landolphia, Mascarenhasia, Odontadenia,* and *Tabernaemontana* have all been tried. So have various milkweeds, *Asclepias* spp. (Asclepiadaceae), and figs, *Ficus* spp. (Moraceae). In the family Compositae, Russian dandelion (*Taraxacum kok-saghyz*), various goldenrods (*Solidago* spp.), and plants of a number of other genera have been investigated. *Euphorbia, Sapium,* and other genera of the family Euphorbiaceae have at times shown promise, too. The USDA's Northern Regional Research Center at Peoria, Illinois, has investigated scores of rubber-bearing species as potential crop plants, with the thought of multiple (solvent) extractions better to utilize the whole biomass: *Monarda* spp. (Labiatae) and other mints have ranked surprisingly high, and even the pokeweeds, *Phytolacca* spp. (Phytolaccaceae), seem quite usuable.

Balata gums

Chiclé, *Manilkara zapota* (Sapotaceae)

Chiclé, the latex from the sapodilla tree of Central America, provided the original chewing gum. It is still of minor importance, but such synthetic elastomers as polyvinyl acetate mostly substitute for (or extend) chiclé. Chiclé has always been harvested from wild trees of *M. zapota*, which are found scattered throughout the dense forests of Latin America, especially in southern Mexico and Guatemala. Because of this remoteness, the harvest of chiclé follows much the same pattern as the harvest of wild rubber in the Amazon valley. Because the tree cannot be repeatedly tapped, but must recover for many months before yielding appreciable quantities of latex again, the balata is collected by itinerant individuals or small groups who scour the jungles for untapped trees. Tapping is done by cutting a series of gashes from treetop to base. It is practiced principally during the rainy season.

Chilte, *Cnidoscolus* (Euphorbiaceae)

Species of *Cnidoscolus* are characteristic of semiarid environments like those of northern Mexico and northeastern Brazil. Only two species, the highland (or red) chilte, *C. elasticus,* and the lowland (or white) chilte, *C. tepiquensis,* have served in a minor way as commercial sources of latex, the latter since remote times in southern Mexico, where it has been used for making ornamental articles. The latex from both species is high in resin and unsuitable for rubber. All production is from wild trees, red chilte

being found in the mountains of west-central Mexico, and white chilte a little farther south along the coast into the state of Jalisco. The small trees are typically tapped by gouging the soft bark in a series of channels that form a sort of herring-bone pattern; several tappings are made over a period of three months, after which the trees are given a nine-month resting period. The latex coagulates naturally if left standing for a day or so, and it is then pressed into rectangular blocks for export. The dried coagulum is nearly half resin, half hydrocarbon. Early peoples in southern Mexico used white chilte to mold various ornamental articles, but today chiltes are used chiefly as extenders for more important elastomers, such as guttapercha.

Gutta-gum, *Couma* spp. (Apocynaceae)

Couma is found throughout northern South America and Central America. *C. rigida,* known as mucugé in Brazil, and *C. macrocarpa,* leche-caspi, or sorva, have been tapped for latex from time to time; during rubber booms it has been used to a certain extent as an adulterant for rubber latices. The smooth bark of *Couma* makes it possible to tap the tree with the same tools used for *Hevea* in the Amazon area. *Couma* is not, however, believed to be amenable to repeated tapping, as with the jebong system. Often the trees are felled and completely drained by girdling, as with *Castilla.* The latex is usually coagulated by boiling, being only partially coagulated by most acids and alum. Coagulated latex is often stored immersed in water until it is ready for sale; exposure to the air oxidizes it and makes it friable. The dry coagulum has a very high resin content.

Gutta-percha, *Palaquium* spp. (Sapotaceae)

Gutta-percha is a well known, traditional balata, supplied principally from *P. gutta. Palaquium* grows naturally in India and throughout the South and Central Pacific. The balata becomes pliable when heated, turns hard, and darkens when exposed to air. It is an excellent nonconductor of heat and electricity, and it is impervious to water; thus it has found widespread use as an insulating material, particularly for oceanic cables, but also for such miscellaneous purposes as golf-ball centers, acid-resistant receptacles, denture plates, and certain types of transmission belts. In many of these uses it has been superseded by modern synthetic and plastic materials.

Long before it was known to Europeans, gutta-percha was used as a molding material by the native peoples of Malacca, Borneo, and Sumatra. British colonials in the Far East found it useful in the humid tropics for handles, splints, and other articles, since it is inert and not subject to deterioration. Not until the middle 1800s was its botanical identity recognized, about which time a German officer discovered its excellent insulating qualities for submarine telegraph cables. This sparked quite a market demand—a demand that resulted in extending the tapping of wild *Palaquium* trees into the remoter sections of Borneo, the Philippines, and other parts of the southeastern Pacific. The original gutta-percha came from felling *P. gutta* trees on the island of Singapore.

The felling of *Palaquium* trees has since been prohibited, and tapping has been substituted. Tapping consists in cutting a series of channels in the bark to conduct the latex to some receptacle at the base of the tree. Several tappings a year are possible, but the tree is usually so injured in the process that it may die. Much of the latex coagulates directly in the gashes in the bark, and is gathered as scrap. At a central camp the latex is heated to complete its coagulation. A newer method of extracting the balata from leaves has also been developed. Twigs with foliage are ground, washed, and treated with cold water, causing small threads of balata to adhere to each other and float to the surface. This mass is skimmed off, purified, softened in hot water, and formed into compact blocks of nearly pure balata. This method of extraction is now preferred commercially. Nevertheless, the gradual extermination of trees has prevented any sizable capital investment in facilities for extraction from leaves. Gutta-percha is generally preserved under water, to avoid oxidation of the surface. The balata softens on heating, and then may be molded into any shape, or even drawn into threads by hand.

Juletong, *Dyera* spp. (Apocynaceae)

Juletong is obtained from several species of *Dyera*, tropical forest trees of Malaysia and the East Indies, especially *D. costulata*. Indiscriminate tapping causes high mortality, although the trees are now protected in many areas. Improved tapping schemes have now been developed, one of which is similar to the jebong panel system utilized for *Hevea*. Nearly 0.5 kilogram (about 1 pound) of latex may be obtained per tapping from top-yielding trees. Coagulation occurs naturally, but it is more frequently stimulated with acid or other chemical agents. As with gutta-percha, juletong is plastic when warm, but it hardens upon cooling and exposure to air. It is generally stored under water to prevent brittleness and crumbling.

Other balata sources

Many other latex-bearing species, particularly tropical trees, yield balata. Their tapping can be locally important if the market warrants. Almost any can substitute for those named. Genera of the family Sapotaceae are especially important, including *Manilkara* and *Mimusops*, as well as *Pouteria*, *Sideroxylon*, *Dipholis*, *Bumelia*, and *Chrysophyllum*. In the family Apocynaceae are *Stemmadenia*, *Tabernaemontana*, *Thevetia*, *Plumeria*, *Cameraria*, and others. *Brosimum* and *Pseudolmedia*, of the family Moraceae, are sometimes also used.

RESINS, GUMS, AND RELATED EXUDATES

This is something of a catch-all section. Some of the products discussed here are related to those discussed in various other sections. For example, resins are an important component of latex, especially of the balata types. Many medicinals owe their effects to their resin content. Rosin, the important industrial resin obtained as a by-product of turpentining, was mentioned among essential oils.

When civilization was young, natural resins and gums were much more important than they are now. They were often the only source of materials for waterproofing, caulking, and incense, and they were used in emollients, paints, medicinals, and embalming compounds. As with spices, they were often procured in distant lands at considerable effort. Almost all were extracted from wild plants, and some still are today. But natural resins and gums have largely given way to synthetics in the modern economy.

The resins constitute a more diversified and complicated group than the gums. Basically they are polymerized from carbohydrates that combine with various organic substances. They are amorphous, brittle, generally clear or translucent, and insoluble in water (but often soluble in alcohols or other organic reagents). Most burn with a smoky flame (one of their primitive uses was for all-weather torches). As with latex, their usefulness to the plant is obscure. They occur as secretions in special ducts in a wide variety of plants and plant parts, often mixed with other materials, such as latex or essential oil. Many resins seem to be oxidative products of an essential oil. They may be antiseptic and prevent desiccation, thus aiding in healing plant wounds. Commercial supplies are typically derived by bruising or cutting the bark of stems and roots. Because there is no truly satisfactory way to classify resins, they are usually identified by source. The more important natural resins come principally from the families Pinaceae, Leguminosae, and Dipterocarpaceae. Those from the family Pinaceae are often "oleoresins," like crude turpentine; resins from legumes include not only the "copals," but those that contain benzoic or cinnamic acid, loosely termed "balsams"; the "damars" are Far Eastern resins from members of the family Dipterocarpaceae.

One of the more interesting and valuable resins is amber, a catch-all group of fossilized secretions protected from oxidation by having been buried. Baltic amber, which is the copious exudate of an extinct pine (*Pinus succinifer*), is 3–8 percent succinic acid: more than 450 metric tons have been gathered in some years. Miocene amber from tropical America probably comes from *Hymenaea* of the Leguminosae. Manila or East Indian copal derives from a conifer, *Agathis dammara*. Amber occurs in subterranean deposits, particularly along the northern coast of Europe, where it frequently washes up on the beaches. In ancient times it was quite a mysterious substance, with many wild surmises as to its origin. It has always been prized for ornaments and charms. Within recent centuries, better pieces have been considered too valuable to be used in varnishes and have been reserved for carved jewelry. Other amber of lower quality was used for chemical processing, as are other resins. Clean fragments may be melted for reconstituted amber. That amber was once fluid is attested to by the fact that fossil insects found trapped in fragments of amber are preserved just as if embedded in a transparent plastic: paleontologists have described new species and "missing links" known only from specimens discovered in amber.

Today's dominant synthetic resins are derived from such inexpensive base chemicals as phenol, formaldehyde, and glycerin. They substitute for natural resins in adhesives, plastic coatings, paints, varnishes, safety glass, waterproofing compounds, and so on. But natural resins still find some market in many of these products, as well as in paper and textile sizings, in some pharmaceuticals, coatings, inks, and incenses. The compounding of resins with an oil or solvent yields a **varnish.** Varnish leaves a hard and essentially waterproof coating as the oil oxidizes or the solvent volatilizes.

Gumlike hydrocolloids (including pectin) are chemically simpler than the resins. By the same token they are much more widespread in plants, and thus more easily substitutable. Although widely used, they have generally commanded neither the price nor the human interest that resins have. Gums and pectins occur naturally as cementing substances in plant tissues, showing up especially upon injury, disease, or tissue disintegration. Most gums are amorphous colloids without any definite melting, freezing, or boiling points. They consist largely of polymerized carbohydrate constituents and, in contrast to resins, are hydrophillic, being soluble or dispersable in water (but not in familiar organic solvents and oils). Their greatest use stems from their capacity to form viscous colloidal gels: they are used industrially as sizings, thickeners, and stabilizers. Gelation is due to chemical cross-linkage of adjacent polymers to give a three-dimensional network; many food gels are firm when they are no more concentrated than 0.1 percent. Mucilaginous types are mainly polysaccharides, with little resin content.

One of the best-known gums is agar, an alginate obtained from seaweed, mostly in Japan. Seaweeds such as *Gracilaria* (Rhodophyceae) are boiled, and the agar gum is flaked out by alternate freezing and thawing. Attempts have been made to discover economical substitutes for this among crops adaptable to North American farming. More than 300 species were reviewed by H. L. Tookey and Quentin Jones, who concluded that four plant families in particular contain promising genera—Leguminosae, Plantaginaceae, Cruciferae, and Convolvulaceae. In all of these families there are species in which the seed endosperm contains more than 18 percent gum (36 species in the family Leguminosae alone). Leguminous gums are mostly classed as galactomannans. Extraction usually involves no more than separation of the endosperm and its grinding to the fineness of flour. Galactomannans are important as paper additives, in the waterproofing of explosives (their swelling prevents further penetration of water), as textile sizings, as flocculants in sewage treatment, or as thickeners and stabilizers in cosmetics and drugs.

Pectins, which bind adjacent cell walls in the plant, occur with greater or lesser frequency in all parts. Commercially they are extracted chiefly from fruit by-products, such as citrus rind and apple pomace from cider milling (vinegar fermentation). The most familiar use of the methoxyl pectins is to insure solidification of jellies. But they are also widely used in pharmaceuticals, cosmetics, medicinals, candies, ice creams, and various food products. Industrially, pectins are incorporated into adhesives and sizings and are used for the "creaming" of rubber latex, in the quench-hardening of steel, and in various other ways. Apple pomace and citrus residues contain as much as 6 percent pectin. This is extracted by mild acid hydrolysis, followed by precipitation with alcohol or other hydroxides.

It is evident that resins and gums are derivable from hordes of botanical sources; price and competition with synthetics determine which will enjoy economic success. By-products are especially inexpensive raw materials for the production of resins and gums, and many promising ones have not yet been exploited for that purpose. A useful resin can be derived from cashew shells, for example, and a gum that is used in water paints and mucilages can be derived from solvent-extracted flax seed meal. Waste from western larch (*Larix occidentalis*) can be made to yield an arabogalactan gum that can be substituted for gum arabic. Resinous substances are derived from tall oil, mainly a by-product in kraft paper-pulp production. Of course, in less affluent regions, many little-used wild plants may possibly prove to be economical sources of gums, such as certain trees of tropical West Africa (*Anogeissus* and *Kahya*). Gum tragacanth, a classical gum-resin, is obtained from wild sources (*Astragalus* spp.), and guar (*Cyamopsis tetragonolobus*) is a potential new crop source for mechanized agriculture.

Gum tragacanth, *Astragalus* spp. (Leguminosae)

Tragacanth is a traditional gum, produced chiefly in Iran. On occasion production has reached nearly 2000 metric tons annually, although production has declined in recent years because of the exhaustion of wild stands, the only gum source. The several species of *Astragalus* that yield gum tragacanth grow in the arid highlands, where they are often used for badly needed firewood in spite of a national conservation law designed to protect them. Even the larger species grow only to moderate shrub size in the inhospitable environment from the Zagros Mountains northward, where the gum is collected seasonally by nomadic peasants.

Perhaps as an adaptation to the arid environment, such species as *A. gossypinus* and *A. echidenaeformis* develop an amorphous cylinder of gum in the center of the stem and root, where pith would normally be expected. This gum is under pressure in summer, and will exude if the stem or root is cut. Twenty-three different species of the genus are known to produce such gum. Originally the bark alone was tapped; eventually it was discovered that incisions to the central gum cylinder provided better yield. Tapping is accomplished by digging the earth away from the taproot and by incising the upper part of the root deeply enough to reach the gum cylinder.

"Ribbons" of gum exude, progressively more slowly as the days go by, and are collected every few days for a week or two, after which a new tapping may be initiated. Tapping starts in June, and tapers off before the autumn rains come. A large, healthy bush may yield as much as 10 grams (about 0.35 ounce) of gum on first tapping, although the average is nearer 3 grams (0.1 ounce).

As was mentioned in the discussion of medicinals, gum tragacanth has been collected since before historical times. Initially it may have served as a food during times of stress, and later as a medicinal because of its bitterness. Greek physicians recommended it as early as the seventh century B.C.E. In the Zagros Mountains, small insects puncture the branches of *Astragalus ascendens* in late summer, causing exudation of flakes of high-fructose (34 percent) manna.

Guar, or cluster bean, *Cyanopsis tetragonolobus* (Leguminosae)

Guar, an ancient crop of India, is also grown in Africa, Australia, Brazil, and to a limited extent in the southwestern United States. In general appearance it is very similar to its relative the soybean. The seed, a source of the colloidal guar gum, is consumed in India as human food, the young pods being eaten like string beans. The mature seed contains 34 percent protein and 40 percent oil, as well as 23 percent gum. The gum, a polymer of galactose and mannose, is extracted from the endosperm. Guar gum exhibits high viscosity at low concentration, swells well in cold water, is useful over a wide pH range, and has excellent film-forming and stabilizing properties. It is used as a flocculent and filter aid in mining; it is also used in cosmetics, drugs, paper manufacture, and foods.

Guar seems to be the kind of plant that would adapt well to mechanized agriculture. Moreover, it is drought resistant, and should grow well in the arid southwestern United States, where new crops are constantly sought and where, indeed, it was grown to a limited extent decades ago for soil improvement and forage.

Other resins and gums

A list of plants would be almost unending were it to include most past and potential sources of gums and resins. Pectins, for example, can come from almost any fruit pulp from almost any part of the world. The costs of gathering, processing, and shipping determine which ones are economical.

Gum arabic is another of the traditional gums from wild species of *Acacia* (Leguminosae), native principally to north Africa and the Near East; gum arabic was mentioned among the medicinals. **Karaya gum** is the dried exudate from *Sterculia urens* (Sterculiaceae) of India. **Locust bean gum,** or **carob gum,** is derived from the fruit of *Ceratonia siliqua* (Leguminosae) of the Mediterranean area.

Other resins are **Balsam of Peru,** from *Myroxylon balsamum* (Leguminosae); **Canada balsam,** from *Abies balsamea* (Pinaceae); **copaiba balsam,** from *Copaifera* spp. (Leguminosae) of South America and Central America; various **copals,** from genera of the family Leguminosae, such as *Hymenaea* in eastern Africa and South America; various **damars** from dipterocarpaceous genera, such as *Balanocarpus, Hopea, Shorea,* and *Vateria;* **elemi,** from genera of the family Burseraceae, in various parts of the world; **frankincense,** from *Boswellia* (Burseraceae), of much interest to the ancients in the Near East; **kauri resin,** from *Agathis australis* (Araucariaceae) of New Zealand; **lacquer resins** from species of the family Anacardiaceae, such as *Rhus verniciflua,* of China; **mastic,** from a shrubby pistachio, *Pistacia lentiscus* (Anacardiaceae); **myrrh,** from *Commiphora* (Burseraceae), mainly from Africa, companion to frankincense in the olden days; **sandarac,** a varnish resin from *Tetraclinus* (Cupressaceae) of North Africa; and **storax,** from Oriental and North American species of *Liquidambar* (Hamamelidaceae).

TANNINS AND DYES

As with several of the natural products discussed in previous sections, the heyday of trade in tannins and (especially) dyes lies in the past. Products from chemical synthesis have almost entirely substituted for natural dyes, and more and more leather is being tanned with chemical agents or being substituted for by synthetic leather or rubber and composition shoe soles.

Tannins are astringent substances basically of carbohydrate origin. They characteristically produce a dark coloration when combinated with iron salts; they precipitate gelatins, proteins, and alkaloids; and they combine with skin proteins to form leather. Natural tannins have been divided into two groups, catechols and pyrogallols, largely on the basis of their origin rather than on the basis of their rather obscure chemical structure. Some tannins are hydrolized by acids and enzymes, and others are not; this difference forms the basis for still another grouping.

Tannins occur in small amounts in most plant tissues. You will recall that tannin is relatively quite abundant in tea leaves, and that tea of high quality generally has a high content of tannin. Tannins are also relatively abundant in tree bark, certain types of wood, most foliage, and many young fruits, and they often are in relatively high concentration in insect galls. As with rubber and resin, their usefulness to the plant is obscure; tannin may be merely a by-product of plant metabolism, or it may serve some antiseptic purpose.

Bark and wood have been the chief commercial sources of tannin. In North America the supply long came from the wood of the American chestnut (*Castanea dentata*), now a thing of the past, since chestnut blight largely eliminated the species. Hemlock, oak, and Douglas-fir bark have all been of some importance. But since domestic production has seldom met demand, there has been much importation of tannin—quebracho especially, derived from the wood of a South American tree; mangrove tannin, from the Far East; and wattle tannin, from *Acacia* spp. in South Africa.

Tanning may have been discovered when pelts were soaked for softening in bogs charged with leaves and twigs. In any event, since time immemorial hides have been treated with tannin to yield an especially durable product, leather. The Romans used various barks, berries, galls, and leaves as sources of tanning agents. Since tannins leach readily, they can be extracted and concentrated, to give more certain and uniform tanning results. In tan-

ning, hides are first soaked in lime solution to loosen the hair, which is removed by special machines. Then it is immersed in tanning solution, and afterward washed, bleached, oiled (and in some cases rolled or polished and coated with various substances).

In gathering bark for tannin, trees are typically girdled and the lowermost bark removed while the tree is still standing. The trees are then felled, and additional sections of bark are removed. Dry bark is shipped to processing centers, where it is shredded (if wood is used, it is chipped), and the tannin is extracted by various methods of water diffusion or percolation (often under pressure in autoclaves, especially with wood chips). Usually the extraction liquor is heated by steam and circulated progressively from exhausted bark to the newest charge. Chemicals are added to some extraction liquors; sodium sulfite is added in quebracho tannin extraction.

Tanning liquor from the initial extraction is concentrated by vacuum evaporation. Liquid concentrates contain up to 35 percent tannin, solid ones up to 70 percent. Tannins are shipped in coated tanks, to prevent the tannin from reacting with the metal. Natural tannin is primarily used for heavy leathers, but an appreciable quantity has been used to control viscosity in oil-well drilling mud. It was formerly used to produce ink. The mining industry uses small quantities for flotation of ores, and a little is used as an astringent in medicinals.

The importance of natural dyes in the ancient world is evidenced by their use in decorating mummy cerements in Egypt; untold toil went into the collection and extraction of yellow from saffron flowers, blues from indigo, and reds from madder. The aboriginal peoples of Britain even colored their skin with the blue dye woad. Everywhere, until the middle 1800s, dyes from vegetable sources were coveted nearly as much as spices. Almost every region had its favorite dyes, for plants capable of yielding dyes are legion. But in 1856, when an Englishman named William Perkin developed a coloring material much like indigo from formerly "useless" coal-tar substances, the death knell sounded for natural dyes. Particularly in Germany the synthesis of coal-tar dyes was pursued, until the civilized world eventually came to rely almost entirely on aniline derivatives for its coloring materials. Not only are the synthetics less expensive, but they are in more certain supply, truer, and more lasting.

In early times, most dyes were not very permanent; they washed out of the fabrics rather readily and faded or discolored easily. It is understood now that dyes applied in soluble form must be made "fast" to prevent their running. This is accomplished with chemical **mordants,** which combine with the dye and fiber molecules and fix the color on the fabric. Most mordants are salts of metals, and often they are effective in helping to develop the characteristic color of the dye. In fact, different mordants can cause a particular dye to yield different colors.

Quebracho, *Schinopsis lorentzii* (Anacardiaceae)

Quebracho is probably the most important tannin derived from wood, now that the American chestnut is nearly extinct in eastern North America. The tree is native to southern South America, where it grows in scattered stands on the savannas and lowlands of Argentina and Paraguay. The hard wood has long been used for posts and cabinet wood, preserving well since it contains about 22 percent tannin. For tannin production, logs are felled in the interior and transported to local extraction plants along the rivers after trimming them of bark and sapwood (which contain relatively little tannin). The heartwood, about 30 percent tannin, is chipped on special machines, and the tannin is extracted by steam under pressure to yield a heavy dark liquor that is evaporated to produce tannin cake. Quebracho tannin is well thought of, acting quickly to give a strong, tough leather.

Mangrove, *Rhizophora mangle* (Rhizophoraceae)

This abundant small tree grows worldwide along tropical seacoasts. In the Far East, where abundant and inexpensive labor is available, cutting the trees has often proved profitable. The tannin is extracted from the bark, which is stripped from the small stems

and transported to local extraction factories. After drying, it is pulverized and boiled to yield the commercial extract.

Wattle, *Acacia* spp. (Leguminosae)

Several species of *Acacia,* especially species native to Australia (but planted in various parts of the world), yield wattle bark, from which tannin is extracted. The bark contains as much as 50 percent tannin. Pulverized bark may be used directly, or concentrates can be made as previously described. Wattle production is chiefly from South Africa and Brazil.

Canaigre, *Rumex hymenosepalus* (Polygonaceae)

Canaigre, looking much like dock, the familiar weed of lawn and garden, has a fleshy taproot abundant in tannin. It is mentioned here because it should be adaptable to mechanized agriculture were economics to favor its growing. The root resembles a parsnip and contains up to 35 percent tannin. The tannin is extracted by pulverizing the root and eliminating soluble sugars by fermentation.

Sumac, *Rhus* spp. (Anacardiaceae)

Various sumacs, in the same family with quebracho, yield tannin rather abundantly from the foliage. *R. coriaria* is said to contain 35 percent tannin in the dry leaves, which have been a minor source of tannin in world trade.

Hemlock, *Tsuga canadensis* (Pinaceae)

Hemlock bark was once wastefully stripped from the trees by the North American colonists as an important source of tannin. The bark contains up to 50 percent tannin, which is easily extracted and useful for tanning heavy leathers such as sheepskins. With supplies depleted, colonists soon turned to chestnut wood, and with that now gone too, some bark is harvested from the western hemlock, *T. heterophylla,* mostly as a sideline to logging operations in the Pacific Northwest.

Oaks (Fagaceae)

Several oaks provide both tannins and dyes. Examples are the chestnut oak, *Quercus prinus*, which grows in the eastern United States, and the tanbark oak, *Lithocarpus densiflorus*, which grows in the Pacific forests. Acorn cups of the Turkish oak, *Quercus macrolepis*, abundant in Asia Minor, yield as much as 45 percent tannin, known in the trade as valonia.

Indigo, *Indigofera tinctoria* (Leguminosae)

Indigo was once one of the world's most important dyes, possessing unusual fastness. It does not achieve its characteristic color until undergoing aerobic fermentation. The precursor, the glucoside indican, is found in a number of species, but has been principally derived from *Indigofera*. Preparation involves holding chopped indigo plants in water for some hours, followed by aeration of the solution (by agitation) until the blue color develops. After a precipitate settles out, further fermentation is stopped by heating. The filtered sludge, indigo, is formed into cakes for shipment. Indigo was especially important to the commerce of India in the days of the spice trade. It is reported that in 1631 three Dutch ships brought to Holland a cargo of indigo worth more than half a million dollars. Some indigo was grown in the English colonies in North America during the 1700s, but the industry was rather short-lived in competition with less expensive production in the Orient.

Madder, *Rubia tinctorum* (Rubiaceae)

Another of the classical dyes was madder, a brilliant red ("turkey red") obtained from the root of *Rubia tinctorum*. The plant is a weak herbaceous shrub, the dye found principally in the cortex of the slender roots. Typically, two-year-old plants are dug, and the roots are washed, dried, and pulverized. An infusion was made from the powder as needed. Different mordants give decidedly different colors with madder. Madder was once grown in India, but introduction into Europe in the late Middle Ages made Holland a center both of dye production and of textile fabrication.

Other tannins and dyes

Gambier, *Uncaria gambier* (Rubiaceae), native to the Far East, yields a resinous extraction from the leaves that contains as much as 40 percent tannin; myrobalan, the unripe fruit of *Terminalia bellirica* (Combretaceae), is a tropical source of tannin somewhat used in India; divi-divi is the fruit of *Caesalpinia coriaria* (Leguminosae), a small tree of lowland tropical America; the pods of this plant yield up to 50 percent tannin and are collected from wild trees for export, principally to Europe.

Many dye plants have been in use since ancient times. Woad, *Isatis tinctoria* (Cruciferae), yields from its leaves the same colorless glucoside as indigo; the species was generally cultivated in Europe and Russia before the advent of aniline dyes, and for a while it was protected in England by prohibition of the importation of indigo. Extraction involved fermentation, and only after oxidation of the dyed fabric did the blue color develop. Logwood dye is obtained from *Hematoxylon campechianum* (Leguminosae), a small thorny tree of tropical America. The red heartwood is a source of the dye, still somewhat used in the staining of laboratory slides. Annatto, the seed pulp of *Bixa orellana* (Bixaceae), a plant native to Brazil but widely naturalized throughout the tropics, yields a yellow dye that was once much used as a food coloring (mainly in margarine). Saffron, the principal yellow dye of the ancients, is derived from the stigmas of the saffron crocus, *Crocus sativus* (Iridaceae). It is said that about 140 stigmas are needed to yield 1 gram of the dye (nearly 4000 stigmas per ounce); no wonder saffron was supplanted by synthetic substitutes! Safflower, *Carthamus tinctorium* (Compositae), is known today as an increasingly important source of edible oil. Originally, however, a yellow dye obtained from the flowers was used as a substitute for saffron.

Selected references

Akehurst, B. C., 1969. *Tobacco*. New York: Humanities Press. (An in-depth review of "the weed" and its uses.)

Amerine, M. A., and E. B. Roessler, 1976. *Wines: Their Sensory Evaluation*. San Francisco: W. H. Freeman and Company. (The book draws upon the experiences of a veteran wine expert who has been concerned with wines and their technology for many years.)

Appelqvist, L. A., and R. Ohlson (editors), 1972. *Rapeseed: Cultivation, Composition, Processing and Utilization*. Amsterdam: Elsevier. (The first monograph on rapeseed, by eminent European editors and authors. Deals with seeds of several *Brassica* species, discussing history, production, and various aspects of the crop.)

Arctander, W., 1960. *Perfumes and Flavor Materials of Natural Origin*. Westport, Conn.: AVI. (Comprehensive. Part one discusses processing and evaluation; part two, 537 raw materials sources.)

Baum, V., 1971. *The Weeping Wood*. New York: Greenwood Press. (A republishing of that delightful mid-century novel about the history of the rubber industry, accurate and a pleasure to read.)

Burgess, A. H., 1965. *Hops*. London: Leonard Hill. (Another of the authoritative treatments in the Leonard Hill series, comprehensive to the date of publication.)

Casida, J. E. (editor), 1973. *Pyrethrum, The Natural Insecticide*. New York: Academic Press. (The book, which grew out of a conference sponsored by the Pyrethrum Board of Kenya, dwells upon natural pyrethrum without mentioning the new synthetics. The book provides a comprehensive review of all aspects of natural pyrethroids, perhaps giving more detail than most people want about a single product.)

Child, R., 1974. *Coconuts*, 2nd ed. Atlantic Highlands, N.J.: Humanities Press. (A worldwide look at this important tropical crop.)

Eden, T., 1958. *Tea*, 2nd ed. London: Longmans. (Discussion of history, botany, planting, care, harvest, and processing of this important beverage, with references, by the Director of the Tea Research Institute in Sri Lanka and Africa.)

Furst, P. T., 1976. *Hallucinogens and Culture*. San Francisco: Chandler and Sharp. (A small volume that surveys major hallucinogens, in which Dr. Furst stresses the interplay between the nature of the hallucinogens and social settings. An extensive bibliography is included.)

Guenther, E., 1948. *The Essential Oils*. Princeton, N.J.: Van Nostrand. (A classical reference.)

Haarer, A. E., 1962. *Modern Coffee Production*. London: Leonard Hill. (On the history, genetics, physiology, and practical growing of coffee by an English expert with wide practical experience in East Africa.)

Hardman, M., 1978. *Beer, Naturally*. London: Bergstrom and Boyle. (A paperbound album replete with handsome photographs and brief but detailed text sums up the brewing industry, especially as it relates to England.)

Hartley, C. W. S., 1977. *The Oil Palm*, 2nd ed. New York: Longmans. (An updating of this standard text.)

Heiser, C. B., 1978. *The Sunflower*. Norman, Okla.: University of Oklahoma Press. (A leading economic botanist looks at his specialty, the source of an increasingly important oil crop.)

Howes, F. N., 1949. *Vegetable Gums and Resins*. Waltham, Mass.: Chronica Botanica. (A definitive discussion.)

Jensen, A., and J. R. Stein (editors), 1979. *IXth International Seaweed Symposium*. Princeton, N.J.: Science Press. (A review of possibilities from the sea, as land resources become fully exploited.)

Jutkowitz, J. M., and K. Kerner (editors), 1979. *Drug Crops: Public Policy and International Control* (AAAS Symposium). Boulder, Colo.: Westview Press. (An attempt to achieve some understanding in the kaleidoscopic area of modern drug usage.)

Krochmal, A., and C. Krochmal, 1974. *The Complete Illustrated Book of Dyes From Natural Sources*. New York: Doubleday. (A husband-and-wife team noted for popular books writes about a subject once of high botanical interest.)

Parry, J. W., 1969. *Spices* (2 vols.). New York: Chemical Publishing. (Chiefly directed to food chemists, this work emphasizes the morphology, histology, and chemistry of the familiar spice plants.)

Pirie, N. W., 1978. *Leaf Protein and Other Aspects of Fodder Fractionation*. New York: Cambridge University Press. (Another of the explorations of "far-out" ideas about how to help nourish a hungry world.)

Polhamus, L. G., 1962. *Rubber: Botany, Production, and Utilization*. New York: Wiley. (An excellent monograph on rubber and rubber-bearing plants.)

Porter, J., 1975. *All About Beer*. New York: Doubleday. (A readable account of beer and the brewing industry.)

Poucher, W. A., 1974. *Perfumes, Cosmetics and Soaps*, vol. 1, 7th ed., revised by G. M. Howard. New York: Halsted Press. (Volume 1 has the greatest botanical interest; volumes 2 and 3 are eighth editions. The three volumes together represent a voluminous updating of what has been the standard reference work for the perfume and cosmetic trade for half a century.)

Purseglove, J. W., C. L. Green, D. G. Brown, and S. R. J. Robbins, 1978. *Spices.* New York: Longmans. (A commendable effort to assemble wide-ranging information on a category of plants still attracting high interest.)

Reed, G., and H. J. Peppler, 1973. *Yeast Technology.* Westport, Conn.: AVI. (Yeasts are extremely important for industrial fermentations as well as for baking and brewing, the featured topics of this book. A final chapter discusses yeast-derived products.)

Robertson, S. M., 1976. *Dyes From Plants.* New York: Van Nostrand. (One of the infrequent reviews of a subject now largely obsolete due to development of dyes from chemical sources.)

Rose, A. H. (editor), 1977. *Alcoholic Beverages.* New York: Academic Press. (A most successful beginning to a projected series of six volumes dealing with economic microbiology; comprehensive, up-to-date, a readable blend of the scientific and the practical.)

Rubin, V., and L. Comitas, 1975. *Ganja in Jamaica: A Medical Anthropological Study of Chronic Marijuana Use.* The Hague: Mouton. (Of special interest because of the publicity given marijuana, this is one of the few unprejudiced reviews of habitual use of the drug. In Jamaica, however, marijuana is smoked along with tobacco, in tremendously larger quantities than is typical in the United States, and it is an accepted custom. Contrary to popular belief, users worked harder and displayed more energy than nonusers. So far as was determinable, chronic use of marijuana was not toxic to the mind or body.)

Schery, R. W., 1972. *Plants For Man,* 2nd ed. Englewood Cliffs, N.J.: Prentice-Hall. (A comprehensive work on economic botany by one of the authors of *Plant Science.*)

Smith, J. E., and D. R. Berry (editors), 1975. *The Filamentous Fungi,* vol. 1: *Industrial Mycology.* New York: Halsted Press. (The wide-ranging perspectives of industrial mycology are brought out in this book—organic acids, enzymes, antibiotics, mushrooms, waste disposal, and so on. Fungus metabolism is reviewed.)

Snyder, S. H., 1971. *Uses of Marijuana.* New York: Oxford University Press. (A concise, readable book, which, after reviewing marijuana's history and pharmacology, comes to the conclusion that "moderate social use of marijuana . . . is unlikely to have grave harmful consequences.")

Trainor, F. R., 1978. *Introductory Phycology.* New York: Wiley. (With the likelihood that greater emphasis will be put on resources from the sea in the coming years, a nodding acquaintance with phycology would seem rewarding.)

Tso, T. C., 1972. *Physiology and Biochemistry of Tobacco Plants.* Stroudsburg, Pa.: Dowden, Hutchinson & Ross. (An authoritative survey of worldwide scope by a leading tobacco scientist from the USDA; discusses various types of tobacco, their growth and handling, metabolic factors, and regulation of the tobacco industry.)

Weiss, E. A., 1971. *Castor, Sesame and Safflower.* New York: Barnes and Noble. (A superbly published text, but unfortunately at a very high price. The history, production, consumption, and general characteristics of these crops, their breeding and cultivars, and recommendations for their culture are given in detail. Complete with excellent plates, figures, and tables.)

Wellman, F. L., 1961. *Coffee: Botany, Cultivation and Utilization.* London: Leonard Hill. (Another of the thorough "world crops" series, dealing with all aspects of knowledge about the crop up to the time of publication.)

Whistler, R. L. (editor), 1973. *Industrial Gums: Polysaccharides and Their Derivatives,* 2nd ed. New York: Academic Press. (Full, updated coverage with chapters devoted to natural gums, seaweed extracts, plant exudates, seed gums, and so on. The definitive treatment of a subject about which there is little comprehensive literature.)

Wood, G. A. R., 1974. *Cocoa,* 3rd ed. London: Longmans. (An updating of the classic treatment on this important beverage crop.)

Wright, L. N., and J. R. Wilcox (editors), 1974. *Oil Seed Crops For Today's Society.* Madison, Wisc.: American Society of Agronomy. (A review by experts of a fundamental class of crops for modern economies.)

23
Fiber, Forest, and Ornamental Crops

Chapter 22 is primarily concerned with the plant extracts that are important for their chemical and combining properties. In this chapter, emphasis is on direct employment of relatively permanent plant substances for their mechanical, structural, or esthetic qualities. Strength and rigidity are implied; the cell wall and the cell-to-cell cementing substances are important here, rather than the cell content or living protoplast. The cell wall is durable and chemically rather inert. It consists mainly of a polymerized carbohydrate, cellulose (Fig. 23-1). It is deposited by the protoplast, which is typically dead and long gone by the time the plant tissue is utilized by people.

Before petroleum was exploited, cellulose served as the chief source of energy, and the present trend seems to be a return to cellulose as an energy source. Estimates indicate that up to 50 billion metric tons of cellulose are produced by land plants annually, of which much goes to yield wood, paper, fiber, and fuel as the world's biomass is recycled (wood is about 40 percent cellulose). The chemical feedstocks of the future could very well be mostly cellulose (and starch) from the biomass.

In the making of paper, the fibers may be separated by dissolving the cementing lamellae between cells with chemical solutions, a chemical transformation. The substances dissolved from the wood in such operations may be hydrolyzed to simpler carbohy-

drates and concentrated for certain uses. In the preparation of fiberboard, for example, solutions used to soften and separate the fiber can yield a "wood molasses" that is as useful for feeding cattle (or for industrial purposes) as is the traditional molasses obtained from sugarcane. Decay of woody tissues is itself a natural chemical recycling of elements essential to a functioning ecosystem.

Especially noteworthy is the changing foresty picture. In the modern slaughterhouse, it is said, "everything is used but the squeal"; forestry, once wasteful of a considerable portion of the tree, now takes a similar tack. Where formerly only wood of lumber grade was used, the industry now integrates operations to utilize more fully all parts of a tree. Chapter 15 pointed out how forestry is fast changing into a planned cropping system, albeit on a long-term cycle as compared to annual cropping on the farm. With the exhaustion of virgin timber in sight, forest management must simulate farming by making trees a renewable crop. The economy already views wood substance as a convertible raw material, to be made into the necessities of modern life by whatever processing is most economical. If trees cannot supply enough good lumber, perhaps even scrawny stock will make good plastic. If the fiber needed for paper pulp cannot be had economically from trees, perhaps perennial bamboos or fibrous stems of annual crops can be substituted. Already technology has shown

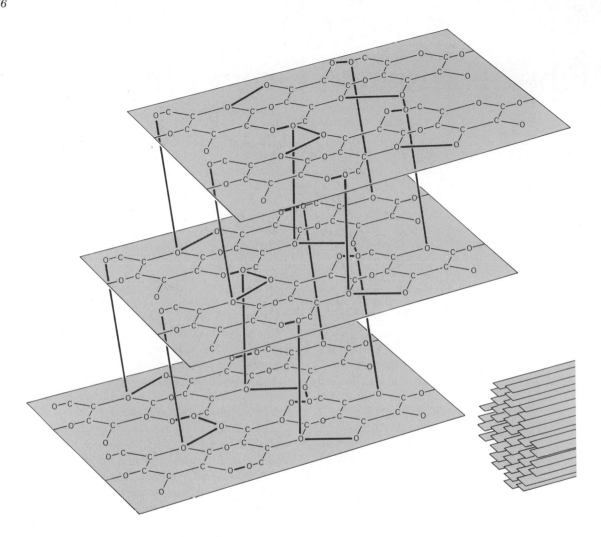

Figure 23-1 Cellulose is made up entirely of glucose units linked together in long chains. In a cellulose fiber there are about 40 of these chains, called glucan, in an orderly array (diagram at lower right). The glucan chains are straight, and they line up next to one another to form sheets. The chains also line up one above the other, but in this dimension they are staggered at half the length of a single glucose unit. Within a glucan chain the units are joined by glycosidic bonds, and each unit is upside down with respect to the preceding one. The chains are held together by hydrogen bonds (heavy lines) between oxygen atoms and the hydrogen atoms of hydroxyl groups. (In this diagram the hydrogen atoms are not shown; the bonds are indicated as if they extended from one oxygen atom to another. In addition the lengths and angles of the bonds are distorted because the sugar units have been flattened and the chains separated for clarity.) The large number of hydrogen bonds give cellulose its strength and resistance to chemical degradation. This model of the cellulose fiber is based on an analysis performed by Anatole Sarko and Reto Muggli of the State University of New York at Syracuse. Several of its features, including the pattern of hydrogen bonding, remain tentative. [From P. Albersheim, "The Walls of Growing Plant Cells." Copyright © 1975 by Scientific American, Inc. All rights reserved.]

Box 23–1   *The once "limitless" forests: another nonrenewable resource?*

People are exhausting the world's forest bulk (biomass) as relentlessly as they are using up accumulated fossil fuels (petroleum, coal). With a new highway in Amazonia, the end of virgin forest tracts of any great size is in sight. On a crowded earth, good land able to produce forest fairly quickly and abundantly is being planted to cash crops with an annual or short cycle, not left to bulky, slow-maturing trees (which, for the most part, are now only allowed to regenerate on, or are planted on, rocky slopes, these often degraded by erosion after earlier forest felling).

The situation seems especially critical for the tropical rainforest, the ecology and succession of which are poorly known. Since the rainforest has been traditionally subjected to no greater pressure than the limited clearing inherent in shifting agriculture, regeneration seems to have occurred without mass extinctions. Now, as part of the tropical Green Revolution, much land is being brought into permanent use for food crops. What regeneration is allowed to occur is mostly of the secondary heliophiles ("weedy" species), and succession by the lordly primary species of the forest never occurs.

As the primary species continue to be felled on neighboring virgin land, even seed sources are eliminated. And because of repeated land disturbance, these species face increasing competition with more weedy growth and, consequently, extinction. Indeed, in much of the virgin rainforest, the soil and surroundings become permanently unsuitable for tall forest, at least for forest containing very many of those species that have been most prized for wood and fiber. Again, the human proclivity for exploitation seems to be killing the goose that lays the golden egg! This seems more applicable to the tropical rainforest (which has a fragile ecology poorly adapted to serious disturbance) than to temperate forests (which are less rich in species).

ways to utilize what were formerly "useless" species and waste tree parts, for the making of pulp and other products, such as charcoal briquets and soil conditioners.

The beauty and "warmth" of wood can hardly be duplicated by synthetics. Thus, while it is hard to imagine the abandonment of wood for esthetic uses, one can visualize the day when houses, for example, might be constructed mostly with plastic materials, perhaps molded to exact size and shape—"plastic lumber" as it were. Whether the raw material for such plastics would come from trees, from annual crops, or from mineral sources will be determined by economics. No plant substance is these days entirely secure in its market. We can be fairly confident that certain plants now managed as crops will yield the raw materials for future civilizations, since exhaustable mineral resources must diminish and thus become more expensive. But we can't be sure which species will be used, nor to what extent. The products discussed in this chapter are particularly amenable to competition from synthetics and other structural materials, such as metals; pronounced change seems in store for many of the traditional uses.

Plants used as ornamentals are also reviewed in this chapter. Although people have tried to beautify their surroundings in one way or another since time immemorial, only in recent decades has the affluence provided by industrialization created a mass market for ornamental plants. We don't often think of ornamentals as a "crop," ranking in dollar value in the United States on a par with a number of important food and industrial species such as rice and citrus. But market research indicates that the "outdoor living" market amounts to many billions of dollars annually. Of course, this includes swimming pools, patio furniture, and maintenance equipment, as well as lawns, shrubs, trees, and products for their care. Nonetheless, the market is based on esthetic employment of plants. Where once agronomy and horticulture dealt largely with farm crops, today many specialists in the United States give full attention to lawn and home garden problems. With only some 3 percent of today's population on the farm, it is to be expected that increasing interest will center on urban and suburban horticulture.

FIBERS

The word "fiber" has many meanings. To the anatomist a fiber is a slender, tapered cell with an unusually thick wall and simple pits. Stringlike masses or entrainments of cells are also called fibers, at least commercially. Single-cell fibers are most familiar as constituents of hardwood xylem. Fibrous bundles are more often found in the pericycle, phloem, or cortex of the stems of such plants as flax, hemp, and ramie. Cotton, the world's most important natural fiber, consists of individual, elongated cells borne on the seed. A wide assortment of plant substance is sometimes used as stuffing material along with wool, silk, feathers, and similar products that are loosely termed "fiber" in the trade. Examples are dried Spanish moss (*Tillandsia*), coconut husks (coir), broomcorn inflorescences, frayed bark (especially of palms), and leaf thatch.

Chemically, the fiber cell wall is chiefly cellulose. Most cell walls contain an appreciable percentage of lignin, the dark residuum evident in humus as plant tissues decompose. Other substances, too, are contained in minor quantities, including tannins, dyes, and various soluble or readily converted components. But the better fibers are nearly pure cellulose, a complex polymer of linked glucose residues (Fig. 23-1) of great molecular length. Although we have had no reason to synthesize cellulose, so amply and inexpensively do plants provide it, we have nevertheless created similar polymers from a variety of raw materials (mostly from petrochemical sources) to yield such synthetic fiber materials as rayon, nylon, acrylics, polyesters, and polypropylenes. So successful have these tailor-made synthetics been that they have greatly supplanted natural fibers in the modern market.

Rayon deserves special mention, both because it was the first important synthetic fiber, and because it is a chemurgic derivative of plant cellulose. Antedating nylon, grandaddy of the "new" synthetics, rayon

at one time accounted for most of the synthetic fiber market. Nowadays rayon, like many natural fibers, has relinquished its position to nylon, acrylics, polyesters, and polypropylenes, but it is still important in many markets.

Rayon traditionally has been made either from wood cellulose or from lint removed from cottonseed after ginning of the cotton fiber. The cellulose is dissolved in any of several solvents to make a viscous solution, which is then drawn out into long strands by forcing it through apertures or dies. The strands are hardened again into a cellulosic ester by being plunged into a chemical hardening bath or by evaporating the solvent. A number of filaments are twisted together to form thread. A process was patented as early as 1855 for making thread from dissolved nitrocellulose, but commercial production of synthetic fiber did not begin until 1891 in France, and not until 1911 in the United States. Production of rayon exceeded that of silk by 1923.

The viscose process of making rayon is the most commonly employed. Clean cellulose is steeped in a solution of sodium hydroxide, shredded, and aged under controlled conditions. It is then treated with carbon disulfide to make cellulose xanthate. This is dissolved in dilute sodium hydroxide to provide the spinning solution, to which pigments may be added. The solution is forced through a spinneret, made from a precious metal, each orifice yielding a thin fiber filament. The filaments emerge into a hardening bath of sulfuric acid, various sulfates, and other materials. A rotating wheel twists the several filaments into a yarn. The yarn is washed, bleached, and then spun and woven in the same way as any other fiber.

Natural fibers are often grouped according to their botanical origin. **Surface fibers** are obtained from the surfaces of leaves, fruits, or seeds; cotton and kapok are the two chief examples. **Soft, or bast, fibers** develop in the bark tissues (i.e., external to the xylem), and are typically found in clusters that make strong, durable strands, as do those from flax, ramie, hemp, and jute. **Hard fibers** are strands of veinlike supportive cells (typically fibrovascular bundles consisting of both xylem and phloem plus ensheathing cells) found primarily in the leaves of monocotyledonous plants; they are somewhat more lignified, rougher, and weaker than soft fibers. **Wood fiber** will be left for discussion in the section on forest products.

Fibers are also classified according to use. Those used for stuffing (upholstering, mattresses, and so on), for caulking, or for reinforcement and insulation (as in wall board) are designated as **filling fibers.** Those used to weave mats and baskets or to make brooms are termed **brush, broom,** or **braiding fibers.** Fibers used to make rope or twine are called **cordage fibers,** and those used to weave cloth are called **textile fibers.**

Surface fibers are usually separated mechanically, by the process called ginning. The principle behind the cotton gin, invented by Eli Whitney in 1793, is to catch the fibers onto toothed discs or combs and pull them through slots too narrow for the seed to follow, thus tearing the fiber free for subsequent spinning into yarn. Hard fibers frequently, and soft fibers rarely, are separated mechanically, too, by machines that scrape away soft surrounding tissues; this work is usually done by hand where labor is cheap. Hand scraping is often the practice with abacá, sisal, and New Zealand hemp, although mechanical cleaning may be combined with biological retting.

Retting is a fermentation process that rots away weaker cells and leaves the strong fiber strands; it is employed mainly with soft fibers. In some parts of the world retting is accomplished merely by immersing sheaves of stems in ponds or rivers; this method is frequently used in India in the retting of jute. Flax fiber is often separated by dew retting, which involves nothing more than leaving the stems in the field, where dew and other natural moisture allows microbes to do the job. In some parts of the world elaborate retting ponds are built, in which the stems undergo a more precisely controlled retting process that yields a more uniform fiber.

Fibers are today very much taken for granted, but primitive people did not have things quite so easy. Thongs of hide may first have served in bows, snares, nets, and skin garments. Long before recorded history, however, primitive people had turned to plant fibers, not only for tying and thatching, but eventually for weaving more suitable and lighter fabrics

than animal skin could provide. The invention of the knot was almost as monumental as the invention of the wheel! Fragments of woven flax fiber have been found in Swiss lake dwellings 10,000 years old. Egyptian cerements almost as old document the use of both flax and papyrus. Hemp was grown in southeastern Asia before 2000 B.C.E., and the pre-Christian armies of the Near East were clothed in cotton cloth. Palm leaf fibers are found among 12,000-year-old artifacts in the Tehuacan diggings of Mexico. *Typha, Agave,* and *Yucca* fibers almost as old are found in archaeological sites in western North America. A beautifully crafted net of extracted *Apocynum* bast fiber, found in Danger Cave, Utah, dates back to about 5000 B.C.E.

Before fabric can be woven, fiber must be spun into yarn. Until just a few centuries ago, this was done entirely by hand, by straightening and thinning the fibers between the fingers and then twisting them into yarn. The yarn becomes a continuous thread of twisted, interlocking fiber strands. The distaff and the spindle, among the first hand tools used to spin yarn, gradually gave way to the familiar spinning wheel, so much a part of the cottage textile industry of the Middle Ages. The spinning wheel consisted of a hand- or foot-powered wheel that rotated a spindle and twisted the fiber into yarn—a foretaste of eventual automation. The principle behind the immense powered spinning machinery utilized by today's textile industry remains the same.

Since fiber cells are found in most plants, and in all parts of the world, it is evident that there is no lack of candidate species for exploitation as sources of fiber. One could gather bast fibers from milkweed stems as well as from flax, surface fiber from dandelion seeds as well as from cotton, hard fibers from grass as well as from abacá. Plants from 44 different families have yielded fiber of greater or lesser commercial value. Which fiber is to prove most fruitful for commerce depends not only upon its natural qualities, but upon cropping and mechanization possibilities—including suitable machinery for cleaning and spinning—labor, market familiarity, and economical mass production (which may call for genetic improvement leading to higher yield and better quality). Only a favored few natural fibers have attained world stature by satisfying these requirements, and even these are hard pressed by synthetics.

We will concentrate here chiefly upon cotton, still the world's single most important fiber. Flax, hemp, jute, ramie, and kenaf among the soft fibers will be mentioned; abacá and sisal among the hard fibers. Nor can the brush fibers be entirely overlooked. This is not to ignore the fact that dozens of other plants have been and still are grown for fiber: some have consistently served specialty purposes. The quality of most of these fibers is satisfactory; they are not generally marketable, however, because they cannot compete economically with other major fibers. Everything considered, fiber is an abundant and low-cost commodity, but the expense involved in harvesting, cleaning, shipping, spinning, and weaving some fibers often overshadows their basic virtues. Add a touch of politics, such as government subsidies, tariffs, and the export duties entailed in worldwide commerce, and it is apparent that distantly related events may often determine the success or failure of any natural fiber in the marketplace. The possibilities for substitution and interchangeability among fibers are sufficient to preclude runaway prices for any single one.

Surface and soft (bast) fibers

Cotton, *Gossypium* spp. (Malvaceae)

The cotton genus, *Gossypium* (Fig. 23-2), is represented by approximately 30 tropical and subtropical species of the Old World and the New World, many of the wild forms being lintless or essentially so. There has been considerable crossing, both directed and natural; five diploid genome groups ($2n = 26$) and one tetraploid combination (of the A and D groups, $2n = 52$) can be distinguished. Of the more familiar species, *G. arboreum* and *G. herbaceum* are Old World diploids. *G. arboreum* was cultivated in India as much as 5000 years ago; while *G. herbaceum* of Africa, which may have been domesticated in eastern Africa, seems to be ancestral to the Indian species as well as to the African domesticates. *G. hir-*

sutum and *G. barbadense* are New World tetraploids, about which more in a moment. Collectively these species account for by far the greatest world production of any natural fiber, more than 14 million metric tons of lint annually. The crop is grown worldwide, but the United States and the USSR are the leading producers, followed by China, India, Brazil, Pakistan, Egypt, and Mexico. Although *Gossypium* is tropical and perennial by nature, it can be cultivated as an annual as far north as 47° north latitude in the Ukraine, and, in the United States, into southern Illinois in the Mississippi Valley. Cotton is a heliophile and relishes warm weather, growing best at temperatures around 32°C (90°F). Wild cottons are semixerophytic. Cultivated cottons have become so pampered that they are today beset by an increasing array of ills and pests that require both skilled and persistent attending. Cotton is one of the more temperamental crops with respect both to weather and to local ecological conditions.

The world importance of cotton is relatively recent. It was singularly little mentioned by the Greeks or the Romans; both of these cultures depended mostly upon flax among botanical fibers, and wool and silk. Cotton had been grown in Africa, however, for many centuries, and, when the Moors invaded Spain at the beginning of the tenth century, they introduced an already highly developed cotton industry. Barcelona became a center of textile manufacturing, especially of sailcloth for the Spanish fleet.

By the fourteenth century some cotton was being shipped from Asia Minor into northern Europe for looming. The establishment of trade with the Far East brought cotton cloth manufactured in India to Europe. During the 1600s a textile industry sprang up in England, first based upon a mixed fabric using linen as the warp and cotton as the woof. Weaving of cloth was still largely a cottage industry, until the invention of several successful spinning and looming devices in the middle of the nineteenth century brought about mechanization. With machine weaving, demand for cotton fiber rose astronomically. Cotton growing was pursued assiduously in the United States, eventually causing the virtual ruination of much land in the South. The longer, finer

A   B

Figure 23-2   Cotton. (A) Commercial cotton as it is grown in the southern United States. (B) Open cotton boll ready for picking. [A, courtesy USDA; B, courtesy J. C. Allen & Son.]

fiber from the American species *G. barbadense* and *G. hirsutum* became the main source upon which the dominant British textile industry depended. The political and social ramifications of the cotton industry in the United States, involving slavery and the Civil War, attest to the influence such a plant product can have on human affairs. During the American Civil War, when cotton supplies were largely cut off, the British introduced the famed Sea Island cotton (*G. barbadense*) and better strains of upland cotton (*G. hirsutum*) into their colonies throughout the world.

New World cottons seem to have contained Old World genomes long before the crop achieved world importance. How this may have occurred is quite obscure. Archaeological findings suggest that *Gossypium* is of considerable antiquity in the New World. Remains of cultivated upland cotton from

Tehuacan, Mexico, date from 3500 B.C.E., and bolls from 500 B.C.E. are scarcely distinguishable from modern cottons. Although the plant was probably cultivated by the Indians there, as were both *G. barbadense* and *G. hirsutum* in the Caribbean even before Columbus, there is no certainty that the fiber was used for more than twisting into simple strings, more flexible than cord made of hard fibers. Twine fabrics may have been produced shortly thereafter, and certainly cotton textiles are known from Peru before 2500 B.C.E., where primitive forms of *G. barbadense* have been unearthed in caves. Cotton fibers from *G. arboreum* were woven into cloth in Pakistan equally as long ago.

In some parts of the world cotton is still grown and handled in a relatively unmechanized fashion (Fig. 23-3). In Ethiopia, for example, cotton still supports a

Figure 23-3 In Malawi, a woman removes the seeds from cotton bolls by hand and grades the cotton lint. [Photograph by James Pickerell for International Development Association, courtesy World Bank.]

cottage industry. American types have been introduced and are gaining favor. These have crossed with Old World cottons. All are grown as short-lived perennials, the planting lasting for about five years. The fibers are harvested mostly by hand, and the spinning and weaving are largely done at home. Such hand crafting permits the use of variable cotton fiber, which can be separated for staple length and selected for different spinning qualities. It is interesting that machine-ginned cotton is rejected by spinners as inferior to cotton separated by hand. The preferred hand "ginning" is accomplished simply by pulling the fiber from the seed with the fingers (often with aid of a smooth stone or wooden block). Blankets and garments are woven from the handspun yarn. The cotton plant is also used for other

than its fiber. Leaves, flowers, seeds, and roots are all reputed to have medicinal value, used both internally and externally. A fermented cake that contains cottonseed is eaten in certain parts of Ethiopia.

Quite in contrast is the highly automated and businesslike production of cotton in parts of the southern United States (Fig. 23-4), where the plant is grown as an annual and the land is clean-cultivated. The most remunerative growing has shifted from the traditional southeastern producing region to the irrigated lands of the Southwest. Remunerative cotton growing demands a long frost-free season, moderate moisture, and either ample labor or well-capitalized mechanization. Without extensive mechanization it is estimated that cotton requires 370 hours of labor per hectare (about 150 hours per acre), a greater labor input than for any other major crop.

Top growers in such locations as the Imperial Valley of California average between 12 and 15 bales of cotton per hectare (about 5 to 6 bales per acre), more than twice the yields on good cotton land in the Southeast. Throughout the Southwest, narrow-row planting has supplanted conventional 100-centimeter (40-inch) row seeding, with plant populations approaching 150,000 to 200,000 per hectare (about 60,000 to 80,000 per acre) in 25- to 35-centimeter (10- to 14-inch) rows. Environmental factors such as proper timing of planting and pest control are extremely important in determining cotton yields.

In the Imperial Valley of California, and in southern Arizona, cotton planting begins about the first week in March, and it is often staggered somewhat to avoid overtaxing of facilities at harvest time. About 55 kilograms of delinted seed of a modern cultivar (such as 'Deltapine'), treated with fungicide, is mechanically planted to each hectare (about 50 pounds of seed per acre), many rows at a pass. After the sprouts appear the rows are mechanically thinned to approximately one plant every 7 or 8 centimeters. About 225 kilograms of phosphate and 100 kilograms of nitrogen (as anhydrous ammonia) per hectare are worked into the seedbed ahead of planting. About six weeks after planting, supplementary fertilizer side dressings are made, and later fertilizer is applied in the irrigation water about every two weeks. Up to 670 kilograms of nitrogen per hectare is used to bring

Figure 23-4   Cotton harvesters munch their way through a field, each picking as much in an hour as one person could pick in 72 hours. [Photograph by Larry Rana, courtesy USDA.]

home a crop. Herbicides are increasingly employed to save labor. Insecticides are sprayed as needed (Fig. 23-5). They arc usually applied by airplane at night, when application seems to be most effective. Efficient sled cultivators weed within an inch of the row, a half-dozen rows at a time, for weed control and soil loosening in the early stages of growth. Harvesting, which is completely mechanized, usually consists of going over a field twice with a cotton picker, a machine with rotating spindles onto which the cotton fiber sticks and is drawn from the boll. This is generally done after spraying with a defoliant to cause leaf drop. "Scrapper" machines usually follow the pickers once or twice over, to salvage an additional fraction of a bale of fiber per acre that escaped the picker. Integration of all these complex procedures requires constant programing and businesslike management of what a century ago was little more than a hand operation tended by slaves.

Research shows that heterosis in cotton can increase fiber yields greatly. Unfortunately, cotton does not lend itself well to regulated crossing; in the field it exhibits quite a mixture of self-pollination and cross-pollination (the degree of cross-pollination often depending upon the bee population). There are indications that suitably irradiated cottonseed grows into plants that show some pollen sterility; it has been suggested that a fair degree of hybridization could be obtained either by this technique or by chemical manipulation, providing hybrid seed economically, and more uniform F_1 populations.

The seed hair that constitutes a cotton fiber is but a single cell, from 1000 to 6000 times as long as it is wide in those species grown for fiber. In 1 kilogram of cotton there are nearly 200 million seed hairs! As the hair grows, its wall thickness increases daily; growth is partly influenced by the weather and other momentary environmental influences. By the time

Figure 23-5 The boll weevil, shown here on a forming cotton fruit, was a pest even on prehistoric cottons. Boll weevils are now being fought with modern techniques and sprays. [Courtesy USDA.]

the fiber is mature, many secondary wall layers have formed, and very little lumen (cavity) remains. As the seed ripens, the hair collapses into a twisted filament of nearly pure (90 percent) cellulose. Its unadulterated cellulosic nature makes for a fiber of surprising durability and strength; and its twisted form makes it well suited for spinning.

After picking, ginning, and grading, cotton goes to a mill, where the fiber is further fluffed and cleaned. It is then carded, combed, and drawn—a sequence of operations accomplished by feeding the fiber onto a series of barbed cylinders that progressively draw it out and thin it. One calculation estimates that during this process an inch of the product, after carding and combing, is drawn to a length of miles. The final combing insures parallel orientation of the fibers, which in pedigreed cotton are nearly of uniform staple length. These are twisted (spun) into yarn or thread, which in turn is used for everything from cordage to the weaving of fabrics. Depending upon its use, the fiber may be chemically treated, dyed, or coated. Most fiber, of course, goes for the weaving of cotton cloth. A considerable amount of cotton is used

for cordage; cotton was formerly used to make automobile tire cord and carpeting.

A by-product of cotton-fiber production is cottonseed, a source of a semidrying oil mentioned in Chapter 22. For each kilogram of lint produced, 1.7 kilograms of seeds result. Cottonseed from gossypol-free cultivars is useful as a source of protein as well as a source of oil.

Flax, *Linum usitatissimum* (Linaceae)

Flax (Fig. 23-6) is perhaps the world's finest bast fiber. For centuries it was Europe's chief fiber, and in some places the only textile fiber. Egyptian mummies dating back four millennia are wrapped in cerements made of linen, attesting to flax's antiquity in the Near East. Its discovery among artifacts from the ancient Swiss lake dwellings was noted earlier. Today flax growing is confined mostly to temperate parts of Europe. The leading flax country is Russia, accounting for more than half of the world production of considerably less than 1 million metric tons annually. Northern and central European nations account for most of the rest. Growing flax for fiber has practically ceased in North America.

The once pre-eminent position of linen, the weaving of which centered in Belgium, has gradually given way under competition, first with cotton and wool and more recently with synthetic fibers. Even so, flax fiber still has few equals for quality. It is often used for fabrics that must be especially tough yet flexible, such as the fiber for making fire hose. Different flax cultivars are used for fiber than for linseed oil (Chapter 22). Growing techniques differ too, including field spacing. Some of the early investigations on cytoplasmic male sterility were done with flax, and there seems to be little doubt that acceptable hybrid seed could be produced. Normally flax is mostly self-pollinated. With cytoplasmic male sterility and proper vectors to facilitate crossing, hybrid seed should be more readily procurable than with many other crops.

Flax does best in cool climates with rather moist summers. The crop is handled as an annual, its planting and care being much the same as with small

Figure 23-6 Branched and unbranched flax. The phloem fibers of the stem, which are used to make linen, are most easily extracted from stems that have few branches. (See also Fig. 22-9.)

grains. Like grain, it lends itself well to mechanization up to the point of retting, for which considerable labor is needed. In much of Europe, plants are still harvested by hand-pulling rather than cutting, to take advantage of full stem length. The harvested stems are generally left in the field to dry until the foliage shrivels and drops, after which retting takes place. In dew retting, the stems are simply left in the field for several weeks until the soft tissues decay, freeing the bast. Where ponds (or tanks, often with controlled water temperature) are available, cold-water retting is practiced; tied bundles of flax stems are completely immersed. It is said that controlled retting in tanks produces a superior fiber in just a few days, as compared to the several weeks required for dew retting. After retting, the fiber is subjected to scutching. This consists of feeding the newly dried stems between fluted rollers that crush and break the central tissues and allow the peripheral bast fibers to be worked free by various forms of shaking (as in revolving drums).

The flax fiber as it comes to market is actually a bundle or group of fiber cells cemented together. The bundles may be up to three feet long in well-grown flax. They are remarkably flexible, tough, lustrous, strong, and durable. Moreover, they are stronger wet than dry, which makes linen an excellent material for towels and fabrics subjected to weather. Just as with other fibers, linen goes through an elaborate series of cardings and combings to clean and draw out the fiber before spinning it into a yarn. Naturally, the quality of the fibers ("slivers") determines the quality of the thread; over-retted fiber tends to be of an inferior blackish appearance. Spinning can be done either by a wet or a dry process. In the former the fiber is first drawn through warm water to soften it and is then further attenuated into particularly delicate strands. Spun yarn may be bleached or dyed before weaving. Today, most weaving is done on elaborate power looms, but there is still some traditional craft weaving in Ireland and other countries. Because of its strength, linen thread is still in demand for shoe sewing and for making fine embroidery or lace. The stems of oil flax find some use in paper making, but its fiber is not of sufficient quality for weaving. Cigarette papers are often made from flax because the nearly pure cellulose (only 2 percent lignin) burns evenly with no objectionable taste or odor.

Hemp, *Cannabis sativa* (Cannabaceae)

At least in China, hemp was domesticated for its fiber as long ago as flax was in other parts of the world. For many purposes these two fibers are interchangeable, although linen is of higher quality for fabrics. Cloth is made from hemp almost nowhere but in the Far East, where a good bit of flax fiber is imported from Europe. The world hemp crop is only about half that of flax, amounting to no more than 250,000 metric tons annually. Most hemp is grown in regions slightly to the south of those in which flax is grown; the leading producing countries are the USSR and India. Very little hemp is grown either in the Americas or in Africa.

Hemp is of course the plant from which marijuana is obtained (see Chapter 22); it also yields a seed oil of minor importance. Apparently a native of Asia, it had been introduced into the Near East and Europe by the Middle Ages. It is grown and harvested in much the same way as flax. Occasionally, as in Korea, the fiber is separated by steam retting, a process taking only a few hours. The quality of steam-retted fiber, however, is not as good as that of dew-retted or pond-retted fiber. Hemp fibers occur in bast bundles, which may be nearly 2 meters (more than 6 feet) long. The thick wall of the cells consists largely of cellulose, but not as pure a cellulose as that in flax. By and large, hemp has been displaced by other fibers in most of its traditional uses, although hemp twines are still much used, as are hemp fabrics for summer clothing and shrouds in parts of Asia. Gibberellin treatment of growing hemp not only produces a greater number of fibers per plant, but it produces individual cells that are larger in diameter and up to ten times as long as normal.

There has been some selection of higher-yielding hemp cultivars, and these have been crossed to give simple hybrids showing some heterosis. But in less advanced areas where much of the hemp is grown, there has been little genetic improvement; a few plants are simply left to mature in a corner of the field for seed. Hemp yields from primitive growing in the East are about 620 kilograms per hectare (about 550 pounds per acre), some 20 to 50 percent less than where more advanced methods are used, as in parts of southern Europe. About 110 days are required to mature hemp, which may cause some of it to be harvested prematurely in the Far East to permit planting of food crops. In the Orient, hemp is often alternated with rice, both in paddy fields and in upland locations. Seed is generally broadcast at a rate of about 50 kilograms per hectare (about 45 pounds per acre), even though tests indicate that row planting and proper spacing provide higher yields and better fiber. In recent years some fertilization has been practiced, nitrogen markedly increasing yields. Weeding is generally done by hand with a hoe. Insecticidal sprays may be used to control borers.

Hemp is usually harvested by hand with sickles. Steam-, dew-, or pond-retting follows. After retting, the hemp stems are dried and then "broken" (partly crushed), either by hand or in primitive machines. The woody core and bark fragments are typically shaken away by hand. The freed fibers are twisted into yarn on hand spindles and woven into a rather coarse fabric at home. The fabric may be "degummed" by soaking in sodium hydroxide. Hemp-cloth garments are traditional for mourners in the East, and for wrapping corpses.

Jute, *Corchorus* spp. (Tiliaceae)

Jute is the world's foremost bast fiber—foremost in quantity, not quality. From the standpoint of quality, it is among the poorest; it is rough and weak, and its cellulose is of relatively low purity. Jute is derived from two domesticated species of *Corchorus*, *C. capsularis* and *C. olitorius*, both of which have a diploid chromosome number of 14. The genus embraces about forty species. Considering the tremendous consumption of a fiber that has so little to recommend it, jute must claim its markets because of low cost. It is produced in regions where labor is inexpensive:

Bangladesh, India, and China account for 90 percent of the world's supply, which amounts to more than 4 million metric tons annually.

C. olitorius is believed to be native to Africa; *C. capsularis* is probably native to eastern Asia. Fiber from *C. olitorius* is frequently finer and softer but more yellowish than that from *C. capsularis*. In India it is designated *tossa*, while the more abundantly grown *C. capsularis* is known as *white* jute. But there is much variation among the many cultivars of both species. Both are chiefly self-pollinating. Jute is grown as an annual; the plants mature within 3 to 5 months, by which time they have developed stems as tall as 4.5 meters (about 15 feet). The fibers occur in long wedge-shaped bundles in the bark, the individual fiber cells seldom being longer than 2 or 3 millimeters. The cell walls are considerably more lignified than those of flax or hemp.

Jute is grown during the rainy season, and it performs best on deep alluvial soils in tropical climates with about 150 centimeters (about 60 inches) of annual rainfall. Seed is planted to well-cultivated, preferably fertilized (manured) soil, and it is usually broadcast at a rate of 7 to 11 kilograms per hectare (about 6 to 10 pounds per acre). Whether broadcast or row planted, stands are generally thinned down (by hand) to populations of less than 500,000 plants per hectare (roughly 200,000 plants per acre) as growth progresses. At the same time, weeds are hoed. Land used for raising jute is generally planted once or even twice during the year to some other crop, such as rice.

As seed pods form, the stems are cut close to the ground with a sickle. In flooded areas this may entail working under water. The stems are dried to rid them of foliage and then tied into bundles for retting. The bundles are submerged in pools or ditches until the tissues binding the fibers disintegrate. Depending upon the temperature, this may take from 1 to 4 weeks. A number of bacteria and some of the fungi most active in retting have been isolated. Experiments have shown that these microorganisms are capable of hydrolyzing the pectins (thus loosening the fiber) in as little as five days. Workers separate the fiber strands from the retted stalk by hand while standing in the pools or ditches, so that the freed fiber can be floated on the surface while fragments of the adhering bark are swished away and picked off. Fiber yield averages about 6 percent of the green weight of the stem; chemically, the fiber is about 75 percent cellulose, 11 percent lignin, and 12 percent xylan. It is stronger when dry than wet; prolonged bleaching reduces its rather poor strength even further. After retting, the fiber is dried and usually sold to small buyers circulating within the community. Most of the crop is eventually transported to Calcutta, the chief weaving and export center for jute. Mills purchasing jute characteristically test for fiber strength as an indication of quality. Careless retting is the chief cause of poor fiber.

Jute fiber, used in India since ancient times, became important as sackcloth during the late sixteenth century. Yarns were spun by peasants and generally woven into cloth at home on hand looms. During the 1800s the commerce of British merchants in India created a great demand for gunny sacks made of jute. Jute mills sprang up near Calcutta, and hand looming declined. Much raw jute fiber was exported to Scotland for spinning and looming there. The earliest American mill (at Ludlow, Massachusetts) is still making jute fabric for such specialty purposes as roadside ditch matting. Burlap sacks still accommodate much grain seed, and produce (even protecting bales of a rival fiber, cotton); burlap is favored because it is not torn even when perforated by sharp hooks or other implements in handling. Jute fiber wastes are sometimes used as stuffing material, and as an extender or raw material for plastics.

Kenaf and roselle, *Hibiscus* spp. (Malvaceae)

Kenaf, *Hibiscus cannabinus,* and the related roselle, *H. sabdariffa,* are bast fibers of minor commercial importance. Kenaf has shown promise as a jute substitute in the Western Hemisphere, and roselle has substituted somewhat for jute in Southeast Asia. Crash research was undertaken on kenaf during World War II, when jute and other fiber imports were cut off from the United States; the species proved well adapted to a variety of growing conditions, yielding a crop quickly (90 to 120 days) with reasonably little care. Roselle requires 180 days.

Kenaf is a diploid with 36 chromosomes, roselle a tetraploid. Kenaf appears to have been domesticated in western Africa, where wild plants and escapes are found. It has been considered a suitable annual crop for paper pulp in America and Australia. Roselle in particular serves other uses in addition to yielding fiber. The seeds may be roasted or ground into meal; the leaves and shoots serve as a vegetable; the flowers are used in jellies and confections; the fleshy calyx makes the ubiquitous "tea" of Egypt and the Sudan, *cacody;* and a fermented drink is made from the juice. Both kenaf and roselle are also thought to have medicinal virtues.

Ramie, *Boehmeria nivea* (Urticaceae)

Ramie, or Chinagrass, has always shown much promise but little fulfillment as a fiber crop. The bast fiber is strong, very durable, and has many of the attributes of flax. Growing and fiber extraction have unfortunately proved difficult and expensive. Consequently the crop has remained a casual one; it is grown in scattered parts of the Orient, where it sometimes substitutes for hemp, jute, and other fibers.

The biggest difficulty with ramie is that the cementing gums binding the bast bundles are not easily destroyed by retting. Separation of the fiber has generally had to be done at least in part mechanically, by pounding, scraping, and other means. Even then the freed fiber is "gummy" and must be chemically treated to meet flexibility standards. The fiber cells themselves are among the longest and strongest of natural fibers, often being 30 centimeters (about 1 foot) long. But they are quite smooth, which is a disadvantage, because spinning machines for other fibers are ill-fitted to handle a smooth yarn.

Ramie, like jute, which it much resembles, is adapted to moist tropical climates and deep soils. Unlike jute, however, it is a perennial, and thus must occupy the land the year around, preventing the growing of a second crop. Per unit of land, it is thus generally less remunerative than either hemp or jute.

Where grown in the Orient it is typically harvested by hand as the canes mature. In areas of inexpensive labor the bast fiber is still removed by hand and generally cleaned by repeated pulling across a knife edge. Most fiber is locally spun and woven into coarse cloth, often without degumming. Degumming is accomplished by boiling in lye or acid, which frequently weakens the fiber, but the processing is necessary to provide yarn adaptable to machine spinning. In spite of the inherent excellence of ramie fiber, it finds only a limited market.

Hard fibers, monocotyledonous genera

A preliminary word is in order about hard fibers as a group. They are really no "harder" than the bast fibers of the several genera just discussed, but they are thicker and stiffer because they constitute the entire fibrovascular bundle derived from leaves of tropical and subtropical monocotyledons. Of course they are not as pure in cellulose as the bast fibers, and they are generally employed for rough use, such as making rope. *Agave* (principally sisal and henequen) accounts for some four-fifths of the approximately 1 million metric tons of hard fiber produced worldwide annually, with abacá accounting for most of the remainder.

Abacá, or Manila hemp, *Musa textilis* (Musaceae)

Musa textilis is closely related to the banana, which it resembles. As the alternative name would suggest, it is primarily a crop of the Philippine Islands, where it is native. Ninety-five percent of world production of somewhat less than 100,000 metric tons comes from the Philippines, where the fiber has even been used for weaving textiles.

Abacá was not known to the western world until 1686, and it did not significantly enter commerce until the nineteenth century. Since then it has gained an excellent reputation, especially for cordage. The fiber strands range in length from 2 to 3.5 meters (about 6 to 12 feet) and have a high tensile strength. Moreover, they are light in weight and resistant to appears especially useful in maintaining the crop. Harvesting consists in cutting the outer stems as soon as the inflorescence begins to emerge, care being taken not to damage young shoots and developing suckers. The internal leaf sheaths yield a fiber of higher quality. The sheaths are cut to approximately 1.5-meter (5-foot) lengths for transporting to decorticating centers. Under primitive circumstances the leaf sheaths are simply sectioned longitudinally into strips and the gross pulp is discarded. The strips are run between a wood block and a knife to scrape the fibrovascular bundles clean. Powered decorticators are used on plantations. A mature abacá rootstock may produce as many as twenty separate stems rising to nearly 9 meters (30 feet). About one-third reach flowering stage each year, at which point they are harvestable.

In addition to its main use for making rope, abacá is much used for fishing nets in the Philippines. There is also some use of the fiber there for weaving table "linens," hats, purses, and other handicraft. Waste abacá, such as from old rope, is suitable for making paper. In fact, the term "manilla," as used in "manilla envelope" or "manilla folder," originated from abacá being used to make a heavy paper when fibers were in short supply during the American Civil War.

Agave spp. (Agavaceae)

Several species of *Agave* are important sources of hard fiber, their production amounting to well over 500,000 metric tons annually. Sisal, from *A. sisalana,* is the most important, accounting for two-thirds of this. Sisal comes mainly from Brazil and East Africa (Tanzania, Kenya, and the Malagasy Republic). Henequen, from *A. fourcroydes,* is produced primarily in Mexico and Cuba, and accounts for less than one-fourth of the world's total *Agave* fiber production. Other *Agave* fibers of lesser importance are cantala, *A. cantala;* lecheguilla, *A. lophantha poselgaeri;* Mexican henequen, *A. lurida;* and letona, *A. letona.* Most of the commercially important *Agave* species are polyploids and are sterile. The basic chromosome number appears to be 30, but various investigations of sisal and henequen show chro-

Figure 23-7 Abacá harvest in Panama. The stalk is cut at the base after the plant has flowered, and the leaves are cut from the top.

water, especially salt water; hence their demand for rope and ship caulking. Originally, the fiber was produced from wild plants as a home industry. In recent centuries, however, plantation growing similar to banana growing in the American tropics, has accounted for most of the harvest (Fig. 23-7). Abacá is truly tropical, performing best where annual rainfall exceeds 175 centimeters (about 70 inches) annually and mean annual temperatures are around 27°C (80°F).

Although abacá, unlike banana, produces viable seed, it can be propagated more quickly from rhizomes or from "suckers" (side shoots of a clump). Abacá is usually planted in rows, individual holes often being dug among fallen brush and trees after land clearing. Later the fields may be cultivated, even interplanted with food plants. As is typical of slash-and-burn agriculture, the soil is generally exhausted within a decade (unless fertilized). Potassium mosome numbers ranging from 137 to 149. Probably a good deal of ancient crossing is represented in modern cultivars.

Most species of *Agave* are native to warm arid environments, but they perform best as fiber sources where rain is reasonably abundant. Soil type is not too important, but the soil should be well drained. Sisal is generally grown in equatorial plantations. It is propagated by hand-planting bulbils that form on the inflorescence or basal "suckers" that are separated from mature plants. Sisal, native to Central America, was first introduced into Florida in 1836. It did not thrive, and success awaited introduction into the West Indies and Africa. African production has dominated the sisal industry in the twentieth century.

During the life of an *Agave* plant, about 300 leaves are produced. Upon flowering, the plant dies; in favorable environments, plants may flower in as little as five years; in less favorable environments, they may need as much as fifty years. When a leaf is old enough to have attained an almost horizontal position, it is ready for cutting. First cutting of sisal in Africa is generally about $2\frac{1}{2}$ to 3 years after planting; with henequen in Yucatán, the first cutting is made about five years after planting. Cutting is done by hand with a machete. With spiny-leaved *Agave*, the spines are removed before bringing the leaves to a central decorticating center. As with abacá the pulpy tissues are scraped away mechanically. On plantations this is done with special machines after the leaf has been crushed. After initial separation, the mass is washed free of loose tissue, dried, and further cleaned by shaking, rubbing, and brushing, often on special machines. The fibers consist chiefly of sheathing cells around the vascular strands; the internal xylem and phloem are generally destroyed in the processing. There will be differences in size of the fiber strand and in its degree of lignification, depending upon the part of the leaf from which it is taken, as well as upon the species and the conditions under which it was grown. Waste material from decortication is a valuable source of fertilizer nutrients, and it is generally returned to the fields. Short and off-grade fibers unsuitable for twine may be made into paper or fiberboard.

Other hard fibers

There are numerous additional sources of hard fiber similar to that produced from *Agave*, world production amounting to about 50,000 metric tons annually. Perhaps most encountered as minor items of world commerce are palma istlé, from *Samuela carnerosana* (Liliaceae), a relative of yucca, and often popularly so-called, produced chiefly in Mexico; formio, or New Zealand flax, from *Phormium tenax* (Liliaceae), produced in Argentina, New Zealand, Chile, and some of the oceanic islands near Africa; fiqué, from *Furcraea macrophylla* (Agavaceae), produced in Colombia; Mauritius fiber, *Furcraea gigantea*, mainly from the islands of Mauritius and Réunion; and caroá, *Neoglaziovia variegata* (Bromeliaceae), from Brazil.

Brush and filling fibers

As noted earlier, a miscellaneous assortment of materials, ranging from the delicate seed hairs of kapok to coarse twigs or fragments such as split bamboo, are also marketed as fibers. Most of these are used as stuffing materials or for making brooms.

One filling fiber that has found a successful specialty market is kapok, or silk-cotton, the seed hair from fruits of a giant tropical tree native to Central America, *Ceiba pentandra*, of the family Bombacaceae. The fiber is of high quality; like cotton, it is nearly pure cellulose, but because of its smooth, untwisted form it does not spin well. The fiber is very resistant to wetting, however, and has long filled a need for a resilient stuffing material in life preservers, outdoor cushioning, sleeping bags, and so on. In East Africa there is much demand for kapok fiber to stuff mattresses, for which a demand has erupted as the local standard of living has advanced. Many years ago kapok was introduced into the Far East and East Africa, where it is rather extensively cultivated, today supplying most of the kapok fiber entering commerce. Fruits are gathered by hand, often being cut with knives on long poles, the fiber typically separated with the fingers.

Another tropical fiber long used as filling material is coir, the fibrous outer husk of the coconut, *Cocos nucifera* (Palmae). Coir is generally loosened by retting in salt water for prolonged periods, after which it is beaten and manipulated to separate the fibers. The stiff fibers are used mostly as an inexpensive stuffing or mat-making material.

Strips of the palmlike leaf of *Carludovica palmata* (Cyclanthaceae) are woven into the famed Panama hats (which, incidentally, are produced mostly in Ecuador).

Broomcorn, *Sorghum bicolor* (Gramineae), which was mentioned among grains and forages, is grown for a limited fiber market. Cultivars used for making brooms have a dense, bristly inflorescence which, when cut, dried, and assembled, makes the typical household and whisk brooms.

Several tropical palms have a fibrous accumulation at the base of the petiole, which is collected by hand and marketed as piassava fiber. This is something of an export from eastern Brazil, gathered chiefly from wild *Attalea* palms; such fiber from various other palms is locally used for stuffing and cordage in oases in the Sahara Desert.

FOREST CROPS

Because different tree species are adapted to different climates and soil types, many kinds of forests occupy the earth today. This has not always been the case. Primitive forests of several hundred million years ago consisted of fewer kinds of trees. In fact, the earliest "trees" (psilophytes), which grew nearly 500 million years ago, were like giant clubmosses. They lacked true roots and consisted of a tangle of specialized branches that clambered over rocky ground. Fifty million years later came the dense forests of tree ferns that prevailed in tropical climates of that era. The forerunners of modern conifers were on the scene 300 million years ago, when plant life abundantly colonized marshy land, building the tremendous coal and oil reserves so important today. By 200 million years ago, seed plants had evolved, and many trees assumed an appearance quite similar to those of today. By the time the dinosaurs roamed the earth some 180 million years ago during the Cretaceous period, seed-bearing trees had evolved that

shed their leaves in winter; from these have sprung the angiosperms and our present deciduous forests.

The ability to produce seeds that can survive unfavorable seasonal cycles was a significant evolutionary advance that occurred at about the end of the Paleozoic era. These primitive plants (related to the seed ferns) bore seeds not entirely unlike those of the modern conifers, which are gymnosperms. Consider the familiar pine tree as an example of the gymnosperms. Two ovules are borne on the upper side of each pine-cone scale, fertilized by pollen from separate, smaller male cones, carried by the wind directly to the exposed ovule. The ovule matures into a free seed, which drops to the ground when ripe. The fundamental distinction between such gymnosperms and the later-evolving angiosperms is that the angiosperm ovule is surrounded by a fleshy protective tissue through which a pollen tube must grow, and which becomes a fruit that encloses the seed. Typical of the angiosperms are such deciduous trees as the oaks (*Quercus*), maples (*Acer*), ashes (*Fraxinus*), and palms (*Cocos* and others).

Gymnosperms and angiosperms are distinguished by the lumbermen as well as the botanist; the gymnosperms are the "softwoods" of the lumber industry, the dicotyledonous angiosperms the "hardwoods." In temperate climes, hardwoods are readily distinguishable in the forest from softwoods; they usually have flat, broad leaves rather than needles or scales, and they shed their leaves annually. There are discernible differences in wood anatomy, too (see p. 145). The wood of gymnosperms has a less complicated structure, lacking certain distinctive cells (vessels) characteristic of angiosperms and often having resin canals that hardwoods lack. (Some hardwoods develop resin canals when injured, however.) Gymnosperms comprise the bulk of commercial temperate-zone forests; angiosperms generally predominate in tropical forests. There are fewer gymnosperm species than angiosperms, and the commercially important trees are confined almost entirely to the order Coniferales (the needle-bearing gymnosperms, or conifers). In the United States about two-thirds of the commercial forest land is in softwoods, accounting for nearly 80 percent of the timber cut. Softwoods also predominate at high elevations in the tropics and at high

Figure 23-8   The Manistee National Forest in Michigan as it appeared around 1900 when it was first being logged. [Courtesy U.S. Forest Service.]

latitudes in northern Europe, Siberia, and Japan. In winter-wet temperate climates like those of the Pacific Northwest, from which most old-growth timber is harvested, conifers (being evergreen) enjoy the advantage of better utilizing the large pulses of nutrients leached in winter compared to deciduous trees that have shed their leaves by then.

In temperate regions, hardwoods such as birch (*Betula*) and aspen (*Populus*) mingle with softwoods in forests that consist mainly of gymnosperms (Fig. 23-8). Hardwoods predominate only at middle and subtropical latitudes. The great diversity of angiosperms, especially in tropical forests, leads to some practical difficulties in managing and marketing. Total yields per unit of land are frequently great, but production consists of small lots of a myriad of species, and for some there is not a ready market. Temperate hardwood forests predominate in a central belt of the United States, roughly between Tennessee and the Great Lakes, and in central and southern Europe eastward to southern Siberia and Korea;

tropical hardwood forests are circumtropical wherever there is sufficient rainfall. Tropical hardwood forests are perhaps the greatest remaining timber reserve in the world, and no doubt they will become increasingly more important as a source of forest products.

In the humid tropics, luxuriant forests (Fig. 23-9) represent, as the ecological climax, the world's most complex biotic community. There are, worldwide, nearly 1 billion hectares (about 2.5 billion acres) of such forests. The virgin forests of southeastern Asia are especially rich. In one hectare of Malayan rainforest, 227 species with a girth of 10 centimeters or greater were counted. Note the tremendous productivity of the tropical forest compared with other types (Table 23-1). It is difficult and usually uneconomic (particularly in view of the limited human and material resources) to reject the wealth of the tropical forest in favor of annual cropping systems that are frequently inappropriate for the soil and the climate. Yet that is an all too familiar approach.

Figure 23-9 Logs being harvested from a tropical rainforest in Gabon. Unfortunately, much of this harvest of virgin timber is not replaced by new growth under sustained-yield practices. [Photograph by Alain Prott/CIRIC, courtesy World Bank.]

Indeed, tree crops are the logical use for most marginal lands, a use that has been much neglected. Since primeval times the tropical forest has been a vast storehouse of natural wealth, not only for wood, but for extractives. A genus such as *Manilkara* yields not only excellent timber, but a latex (balata) of export value, and fruit useful at least to wildlife. Rubber, resins, waxes, oils, nuts, and medicinal or insecticidal alkaloids long have been important forest products. Among timber trees, West Indies mahogany (*Swietenia mahagoni*) has for centuries been a prized cabinet wood; rosewood (*Dalbergia*), Spanish cedar (*Cedrela*), balsa (*Ochroma*), African mahogany (*Khaya*), and several others likewise have been selectively much harvested but seldom planted. Only the vastness of the forest has generally prevented their complete exhaustion.

Nonetheless, it is encouraging that in many places

Table 23-1 The productivity of forest types and other forms of vegetation.

Vegetation	Net primary production (grams of carbon per square meter per year)	
	Range	Mean
Forests		
Tropical (wet)	450–1600	990
Temperate	270–1125	560
Boreal	180–900	360
Savannahs	90–900	400
Grasslands	90–675	270
Cultivated land	45–2800	290

Source: Data from G. M. Woodwell, *Scientific American* 238(1), 1978.

Box 23–2   *Forest ecology*

The influences that shape forest regeneration are only beginning to be understood. Forest change is often a continuing process, and foresters attempt preservation of what is a successional stage. This, of course, is not possible without some form of management. In many instances there may be no stable climax, since the overall environment is being altered by human activity and perhaps by climatic change.

Fire as an influence on forest succession has been recognized as having both beneficial and negative effects. It has always been an integral part of the forest environment, and it has contributed to a mosaic of successional subclimaxes within an overall species climax. For example, fire in a redwood forest may promote the growth of grassland, which is only slowly reconverted to redwood, depending upon the frequency of subsequent fires.

The succession that occurs in the northern Rocky Mountain forests is influenced by fire. Lodgepole pine is the offspring of fire, occurring in even-aged stands after burning. If fire does not reoccur, species of *Picea* and *Abies* come to dominate, and, if fire control is exercised, they come to constitute the climax. The fact that lodgepole pine will not reproduce under its own canopy helps to explain the even-aged stand.

On the other hand, wildfire can alter soil conditions so much that it prevents succession. This has been demonstrated in stands of ponderosa pine in the Southwest, where no new ponderosa catch occurs after burning even though reproduction takes place on land not burned. The soil constituency following fire—or for that matter after clear-cutting and the exposure this procedure brings—may materially depauperate the biological composition and influence forest quality far into the future.

In north-central Colorado, an outbreak of spruce beetles caused death of large-diameter spruce trees in patches, and altered the species composition of the forest in favor of fir and lodgepole pine. The outbreak developed in logging slash (and was thus an upset caused by people), but it remained epidemic for only two years, until natural competitors (for example, such parasites as nematodes, and woodpeckers, which lent great assistance) responded sufficiently to control the beetles.

As people assume greater influence in the forest through harvesting and attempted management, the balance and buffering characteristic of virgin forest can be expected to diminish. The rapidity with which natural forests are being cut allows little time for the research necessary to avoid habitat degradation.

tropical authorities are beginning to adopt a more far-sighted approach. After centuries of cutting and depletion, reforestation with select species is being undertaken. Teak (*Tectona grandis*) is often used. In salubrious tropical climates the cycle from planting to harvest is probably much less than the 50 or so years deemed necessary for maturing a forest in temperate localities. For the teak plantations, scrub forest is first cleared (the wood often made into charcoal) and the land cropped with food plants for a year before teak trees from a nursery are planted. Food cropping may be continued between the trees until they are tall enough to shade the crop. Thinnings useful for poles and fencing are had in as little as four years, and small lumber stock not long after. In much the same way, quick-covering *Euca-*

lyptus and *Leucaena* have been successfully introduced into many subtropical parts of the world, and Monterey pine (*Pinus radiata*) is a reforestation mainstay from New Zealand to southern Africa. Sometimes introductions succeed so well that they become pests, as with *Casuarina*, *Melaleuca*, and *Schinus* in the Florida everglades.

As nations have developed, the regard in which they hold their forests seems to pass through three stages. First, the forests are obstacles to colonization—hurdles to be surmounted. Land is cleared, and the trees are piled and burned in order to provide necessary living space and agricultural land. During the next stage, they become storehouses to be exploited. They are merely sources of energy and construction materials. In the last stage, forests are fi-

Figure 23-10 A productive, planted forest of introduced species in southern Africa, yielding timber products where none were before. [Photograph by John Moss for International Development Association, courtesy World Bank.]

nally recognized as a renewable resource and are husbanded, protected, and increased (Fig. 23-10).

Forests around the world

Before examining North American forests, which are of greatest concern to most readers of this book, a glance at forest conditions around the world is in order.

In heavily-populated Europe, most virgin forest has been cut and replaced by planted species, especially Scotch pine (*Pinus sylvestris*) and Norway spruce (*Picea excelsa*). Birches, as an important hardwood, mingle with the conifers in the Scandinavian countries, which are still heavily forested (Finland, for example is two-thirds "woods"). The forests of central Europe are very well managed. In parts of southern Europe, such as Yugoslavia, forests are mostly state-owned; about three-fourths of the timber is hardwood, although conifers are being intro-

duced. In the Mediterranean basin, most of the original forest has long been cut, although small stands persist in the mountains. Wood is very much at a premium in this area, as it is down through the Sahara to the rainforests of central Africa (which are now becoming badly overcut). Only in southern Africa is sensitive forestry practiced sufficiently to assure wood for the future.

The Near East is lightly timbered, and once abundant forests (such as the famous "Cedars of Lebanon") have been almost obliterated. However the USSR has more than 1 billion hectares in forest (about 2.5 billion acres), of which about one-quarter is under management. The forests of Siberia are only partly exploited, and often poorly known. Three-quarters of the Siberian forests are coniferous, accounting for almost half of the world's total. Much of the timber is mature, but it is dwarfed in this rigorous environment. In European Russia, with good soils and ample rainfall, forests flourish. But the country there is heavily populated, and demands for other land uses are great, so forest reserves are not extensive. Across the continent, in southeastern Siberia adjacent to Korea, forests become rather luxuriant—far more productive than in the cold, frozen-soil taiga belts to the north. As in Europe, pine and spruce mingle with birch in Siberia, while larch, fir, and "cedar" are also found. Southward, oaks (*Quercus*), maples (*Acer*), lindens (*Tilia*), and hazelnut (*Corylus avellana*) begin to dominate, and many familiar genera provide a rich botanical assortment. Overall, *P. sylvestris* is probably the most important commercial species in northern Eurasia.

In Southeast Asia, dense populations practicing shifting cultivation, combined with extensive warfare, have materially degraded what in the virgin state were some of the world's most magnificent forests. From China south through the Malay Peninsula and on into the East Indies, luxuriant tropical hardwoods once prevailed. In China, most forest has long been exhausted, although there is some government-sponsored replanting, especially for shelter belts. A few of the East Indian islands, such as Borneo, are still generously forested, and not all of the timber trees have even been identified. The most important family is Dipterocarpaceae, which supplies about 90 percent of the export timber (reaching the United States as "Philippine mahogany"). There are at least 150 species in the family Dipterocarpaceae, which yields the chief timber trees from the East Indies westward along the southern flank of the Himalayas (especially species of *Shorea*).

By modern forestry standards, such immensely diverse and little-explored forests as those of Borneo are underutilized. Transportation and human labor are both limited in the jungle, and careless procedures are employed in harvesting the trees. Undoubtedly many fine timber trees are wasted, simply because the wood is not known in outside markets and because familiar species (such as those of the family Dipterocarpaceae) continue to be called for in specifications. Facilities for local processing are crude, and in spite of an abundance of some of the world's finest woods, quality lumber (as is needed for making furniture) must be imported from traditional trade sources.

Under these primitive circumstances it is no wonder that very little is known about regeneration of cutover land or the quality that can be expected from second-growth forest. The prevailing custom is to girdle and kill all large trees (those more than 2 meters in girth) left after selective logging in the belief that they will be overmature by the next cycle of rotational harvesting. It is not known which species will regenerate under prevailing ecology, what the shade requirements of desirable species might be, nor what rotational cycle is best fitted to these particular forests (the forest authorities presume an 80-year cycle). In Malaysia, where forestry is more advanced than in Borneo, and the virgin forest is similar, it has been found that smaller trees of the families Lauraceae, Myristicaceae, Rosaceae, and Leguminosae, as well as certain oaks, once considered useless, are of considerable value in providing cover that allows a new stand of dipterocarps to emerge. When the forest composition has been changed on cacao lands, there have been serious consequences, such as lowering of the water table and slow debilitation of the plantation crop.

Aside from teak (*Tectona grandis*, Verbenaceae) timber trees from the tropical hardwood forests of Southeast Asia are largely unknown in the Western world. Useful and commercially quite important to Borneo are *Shorea, Parashorea, Dryobalanops, Dipterocarpus, Koompassia, Cratoxylum, Anthocephalus*, and several other genera, including species of the families Leguminosae and Sapotaceae, as well as Dipterocarpaceae.

Australia is quite limited in the extent of its forests because of the inhospitable climate of the interior. The chief indigenous forest, along the moister, eastern coast, is dominated by *Eucalyptus*. New Zealand has abundant beech (*Nothofagus*) forest on the South Island, of local usefulness but rather little commercial value; and the magnificent kauri (*Agathis australis*) forests on the North Island, of which logging of natural stands was forbidden in 1958, are now decimated and of little commercial importance. A few snatches of virgin kauri forest have been preserved, which show the indigenous trees to grow 50 meters (more than 160 feet) tall and 12 meters (nearly 40 feet) in girth. With exhaustion of her original forests, New Zealand has turned to tree planting, successfully utilizing such species as California's Monterey pine, *Pinus radiata*.

Across the Pacific, South America has seen exhaustion of many of its forests, particularly the paraná pine (*Araucaria*) of southern Brazil and the limited stands of mixed hardwoods in northern Argentina. In recent years massive efforts to settle and exploit the Amazon basin have caused great devastation there, arousing fears that alteration of this, the most extensive tropical rainforest in the world, may have serious consequences not only locally but for world climate (Box 23-3). As in Amazonia, clearing and cutting continues northward through Central America into Mexico, so that very little of what was once a vast strip of virgin tropical forest remains.

Forests of North America

Forests in North America exhibit general latitudinal banding as do those of Asia. The northernmost belt is chiefly coniferous, extending from Alaska and western Canada through northern and eastern Canada into the Great Lakes and New England sections of the United States. It also extends southward at the high elevations of the Appalachian Mountains as far

Box 23–3 *Fuelwood and related problems*

In many parts of the world, insufficient fuelwood has long been a pressing problem, and the problem becomes more serious as population pressures increase. In the Himalayas the original forest has been almost completely felled, even on steep and nearly inaccessible slopes. Inevitably, much of the topsoil washes away during torrential monsoon rains. In the semiarid Near East and northern Africa, once luxuriant forest land is now desert. In Tierra del Fuego, slow-growing *Nothofagus* is being wantonly exploited for export, and most merchantable timber from northern Argentina to southern Brazil has already been harvested. The seemingly limitless virgin forests of North America and the tropical rainforests of central Africa and Amazonia are now suffering significant degradation. Matters are especially critical in the developing countries, which utilize 86 percent of their wood for fuel (and when faggots of fuelwood are not available, as in many parts of India, dried animal dung must substitute, which might better serve to fertilize impoverished soils).

The consequences are staggering. Stands of virgin timber are about to play out, and trees soon will have to be grown as a crop. The need for productive biomass, as fossil fuels become more expensive and are eventually used up, is a brand-new concern. The problem is apparent, and its solution, if one is possible, will no doubt be costly. Less apparent is the possibility that world climates may be altered as forests disappear. The tropical rainforest, now being decimated at an accelerating pace, may be especially essential for maintaining normal climates and stable ecosystems. Many valuable species have already been exterminated because of habitat change. If

this continues, might our rich biosphere retrogress and become impoverished? And what of the release of enormous quantities of carbon dioxide when wood and fossil fuels are burned? Is the resulting "greenhouse" effect likely to raise temperatures as speculated, thereby melting polar ice masses and raising sea levels catastrophically? The average carbon dioxide concentration in the atmosphere, measured in Hawaii, has risen from 315 parts per million to 330 parts per million in less than two decades! There would seem to be only one practical way to tie up excessive carbon dioxide: to rebuild carbon into the semipermanent biomass, particularly the trunks of large trees.

Modern civilization, in its thirst for energy and a continuously growing standard of living (and in its widespread refusal to institute effective population control), ignores a simple ecological tenet—that many natural resources are finite, and that excessive exploitation hastens their exhaustion. Civilization seems to have ignored the fact that people, too, are part of, and are dependent upon, the environment. Once all of the cheap energy from fossil fuels is exhausted and most of the world's forest lands are changed to grasslands, the consequences cited will have to be fully faced, even though opportunities for their amelioration will have passed. Inflation is already taking its toll on standards of living throughout the world, as resources become increasingly scarce and dear. People cannot begin too soon to recognize their dependence upon the other organisms in the biosphere. Conservation of environmental systems, including forest ecosystems worldwide, must be undertaken immediately.

---

as North Carolina. South of this belt, between the Great Plains and the Atlantic Coast, is hardwood deciduous forest; it stretches roughly from east-central Texas through northern Georgia and northward along the piedmont into Maryland. South of this is a belt of southern coniferous forest that stretches into middle Florida and is dominated by several so-called southern yellow pines (except in the bottomlands, where hardwoods predominate).

Subtropical and tropical forests are of limited importance in the continental United States, existing only in southern Florida and in a narrow band along the Gulf Coast from Florida to Texas; southward in the lower-lying lands of Mexico they become fairly extensive, but even there they are of limited economic usefulness. In the extreme western part of North America are found the most majestic forests, especially as a coastal band on the moist Pacific slopes from central California north to Alaska—magnificent stands of redwoods, coastal Douglas-fir and other species. Inland is a drier Rocky Mountain forest belt, extending from western Canada through Colorado into central Mexico. Important and extensive forests occur, although confined chiefly to mountainous outcroppings that punctuate otherwise desertlike country.

The forest belts in the United States are depicted in Figure 23-11. They consist of 200 million hectares (nearly 500 million acres) of potentially harvestable trees, about three-fifths from private holdings (farms and nonindustrial lands), a little over one-fifth from federal and state properties, and a little less than one-fifth from timber-company holdings. By and large the national forests contain the best stands and most of the country's remaining virgin timber, but some say they are being overlogged at "give away" prices to timber companies. There is also much questioning of policies that permit the sale of whole trees to Japan while a hue and cry is heard about long range inadequacy of the timber reserves of the United States.

In contrast to most tropical forests, and to some extent the Siberian forests, the forests of North America are well known, the tree species well identified. Outstanding in the eastern quarter of the northern coniferous forests, at least as the early colonists found it, was the eastern white pine, *Pinus strobus*. White pine alone supported almost the entire lumbering industry in the northeastern United States for nearly two centuries; regrowth of white pine makes it still of importance. However, lumbering has now shifted to the virgin forests of the West and the managed forests of the Southeast. When the United States was still a colony of Great Britain, so important was the eastern white pine that an act was passed for its preservation on crown lands, withholding the finest trees for masts of the sailing ships of the Royal Navy. This was another of the irritations that eventually led to the American Revolution.

In addition to eastern white pine, there are several other species of *Pinus* in the eastern portion of the

Figure 23-11   Forest types of the United States. [Courtesy St. Regis Paper Co.]

Northern Forest covers a region of abundant lakes, river valleys, and hills. Some areas are predominantly coniferous (mainly pine, spruce, balsam, fir, hemlock), some hardwood (mainly maple, birch, beech), and some mixed, as shown here.

Central Forest is characterized by hardwood species. Much of this region was long ago converted to farmland. Important species include oak, hickory, yellow poplar, as well as such conifers as yellow pine and red cedar.

West Coast Forest is characterized by tall, straight conifers like the Douglas-fir and, most notably, the redwoods (shown here). Their growth is favored by the abundant moisture that is blown in from the Pacific Ocean.

 Mountain Forest of the West has less rainfall because moisture from the Pacific is barred by the Cascades and Coast Ranges. Prominent species are ponderosa pine (front), western white pine, sugar pine, Englemann spruce, and Douglas fir.

 Southern Forest is dominated by southern yellow pines, notably loblolly, slash, and longleaf pine (shown here). However, many hardwoods are found here too, including sweetgum, tupelo, and a large number of oak species.

 Tropical Forests of the United States are found only in the southern tip of Florida and a small area in southeastern Texas. Mangrove (shown here in foreground) and mahogany are the principle trees of our tropical forests.

northern forests, of which the red pine, *P. resinosa*, is the only large one of great commercial interest. There are several important spruces (*Picea rubens, P. glauca, P. mariana*), hemlock (*Tsuga canadensis*), fir (*Abies balsamea*), arborvitae, or northern white-cedar (*Thuja occidentalis*), and a scattered interspersion of hardwoods, including principally the aspen (*Populus tremuloides*) and birch (*B. alleghaniensis*).

The arctic portion of the northern forest belt is composed largely of *Abies balsamea* and *Picea glauca* with the *A. balsamea* disappearing westward. The northern forest merges westward with the forests of the Pacific slopes, which are similar in general composition but marked by different species. The white pine there is western white pine (*P. monticola*) and sugar pine (*P. lambertiana*); western hemlock is *T. heterophylla;* spruce is mainly Sitka spruce (*P. sitchensis*); the arborvitae is *T. plicata;* and the firs are represented by an important group of different species (*A. grandis, A. nobilis, A. magnifica,* and *A. amabilis*). But the Pacific coast forests are primarily noted for genera having no close counterparts in the East. In recent decades the most important timber tree has been the Douglas-fir (*Pseudotsuga menziesii*), and in northern California the most dramatic forests in the world consist of almost pure stands of redwood (*Sequoia sempervirens*) containing the tallest trees in the world, some more than 113 meters (about 370 feet) tall! Some stands of redwood contain nearly 500,000 board feet to the acre, but unfortunately in the felling of such gigantic trees much breakage and waste occurs. Although most of the finest stands of coastal Douglas-fir and redwood have already been cut, significant old-growth Douglas-fir still exists in remote sections of California, Oregon, Washington, and British Columbia. This forest belt is the richest timber reserve in North America today.

A number of other distinctive trees of no little economic importance are found in the Pacific forest belt, too. Incense-cedar (*Libocedrus decurrens*) is found in Oregon and California, where it is often a companion to the giant-sequoia, or big-tree (*Sequoiadendron giganteum*), the largest tree in the world, some specimens of which are 4000 years old and weigh 6000 metric tons (Fig. 23-12). The few giant-sequoias remaining in the Sierra Nevada of Califor-

Figure 23-12 The big-tree, or giant-sequoia (*Sequoiadendron giganteum*), in Sequoia National Park, California. This mountain species, which is confined to a few small groves in the Sierra Nevada, is a relative of the equally famous coast redwood (*Sequoia sempervirens*), which grows abundantly in the Coast Ranges of northern California. [Courtesy U.S. Forest Service.]

nia are today protected, and in fact have never been of any great commercial importance because of their limited number, weak wood, and tremendous size, which makes cutting and handling difficult. There are several "cedars," too, in the Pacific forest belt, of the genus *Chamaecyparis*, although they are of secondary importance compared with Douglas-fir.

The Pacific forest belt merges with that of the Rocky Mountains in western Canada; in the United

Figure 23-13 Second-growth deciduous forest in Indiana, composed mainly of oak and hickory. [Courtesy Purdue University.]

States these two zones are separated by the arid Great Basin. An inland subspecies of Douglas-fir grows in the Rocky Mountain forests as far south as middle Mexico, although seldom so luxuriantly as in the coastal forest belt. The western white pine grows in the northern Rockies, but the pine species of chief importance as a Rocky Mountain timber tree is the ponderosa, or western yellow, pine (*P. ponderosa*), which is also well represented in the Pacific belt. On the more arid southerly plateaus, piñon pine (*P. edulis*) assumes minor importance; it is too small for general timber use, but the seeds (piñon nuts), which have long been important to the Indian tribes as a staple, are marketed as a modern-day confection. Lodgepole pine (*P. contorta*), another small pine, is as widespread as ponderosa pine but not as useful for lumber. The western larch (*Larix occidentalis*) is found abundantly in the northern Rocky Mountains, and two spruces (*Picea engelmannii* and *P. pungens*) grow at higher elevations throughout this forest region. Several firs also grow in the Rocky Mountain forests, of which the most important species is probably the white, or Colorado, fir (*Abies concolor*).

As is typical of hardwood forests (Fig. 23-13), great diversity is found in the forest belt just south of the coniferous zone, from southern New England and the middle Great Lakes states (where hardwoods mingle with the conifers) southward to central Texas and the northern parts of the states to the south and east of Tennessee. The most humid eastern portion of this forest belt is characterized especially by yellow birch, beech, sugar maple, and hemlock in the more northerly portions and by several oaks, yellow-poplar (*Liriodendron*), and sweetgum (*Liquidambar*) in the more southerly portions. Oaks, hickories, and maples dominate the drier westerly and southeasterly portions. Although these genera characterize the forest, diverse other hardwoods that grow along with the dominants are often of even greater economic value (black walnut, for example, or old-growth black cherry). There is almost no virgin timber left.

Chestnut (*Castanea dentata*) was once an important component of the eastern hardwood forest, but, except for short-lived stump sprouts, it is nearly extinct, owing to introduced chestnut blight. Today, the oaks in the northern and western portion of this belt and yellow-poplar in the southeastern portion are probably the most important trees. There are many species of oak, but the white oak (*Quercus alba*) is the most prized, not only for lumber but for such specialty purposes as barrel staves. Black walnut (*Juglans nigra*) and black cherry (*Prunus serotina*)

Figure 23-14 Old-field loblolly pines, 50 years old, growing in Arkansas. [Courtesy U.S. Forest Service.]

are perhaps the most valuable individual species; they serve primarily for making cabinet and furniture veneers. Large trees of these species are now very scarce. Among the other hardwood trees of commercial importance are ashes (*Fraxinus* spp.), basswood (*Tilia americana*), beech (*Fagus grandfolia*), the hickories (especially the shagbark hickory, *Carya ovata*, and the pecan, *C. illinoensis*), maples (especially the hard or sugar maple, *Acer saccharum*), black locust (*Robinia pseudoacacia*), honeylocust (*Gleditsia triacanthos*), sycamore (*Platanus occidentalis*), the tupelos (species of *Nyssa*), sweetgum (*Liquidambar styraciflua*), yellow-poplar, or tulip tree (*Liriodendron tulipifera*), many species of oak (*Quercus*), walnuts (the butternut, *Juglans cinerea*, as well as the black walnut), and such commercially unimportant (but abundant) trees as willows (*Salix*), poplars (*Populus*), and elms (*Ulmus*).

The final forest belt is that of the Southeast, characterized especially by longleaf, slash, and loblolly pines (Fig. 23-14). This coniferous forest has assumed major economic significance in the eastern United States, and it is destined to become the "number one" wood-producing region within the United States within the next decade or so. Although forest land in the Southeast is mainly planted to pine, some elite,

specially-bred hardwoods are being planted, too. Major holdings by large integrated companies are being built up in this zone rather than in northern regions because the trees grow fast and the human population density is relatively low. Much cut-over land is being reforested solely for the production of pulp stock and small timber. In some areas second growth of sufficient size for lumber is available for harvest, but only in very inaccessible locations are there any virgin forests.

The longleaf pine (*P. palustris*) is probably the most famous of the southern yellow pines. It is a major source of lumber and pulp, and was once a major source of turpentine. Regeneration of longleaf pine is made difficult because seedlings are slow growing and do not survive well when planted. Slash pine (*P. elliottii*) was originally more restricted in its southeastern range than the longleaf pine, but it has been established successfully in forest stands over the entire deep South; it is probably the chief source of natural naval stores today. Loblolly pine (*P. taeda*) is the most important yellow pine species in the middle South. It colonizes abandoned agricultural land quickly, and it is the principal commercial pine of the piedmont region today. Shortleaf pine (*P. echinata*) is the widest-ranging of the southern yellow pines, occurring as far north as Missouri and Connecticut. Another important timber species of the Southeast is the baldcypress (*Taxodium distichum*), typically found in swamps and bayous along the coast and river bottoms from southern Virginia to Texas.

In the eastern United States, most forest land is under private ownership, whereas most western forests are in federal hands (mostly as national forests or as holdings of the Bureau of Land Management). In volume of growing stock the New England forests contain around 1 billion cubic meters (about 35 billion cubic feet), of which slightly more than half is softwoods. The central hardwood belt contains nearly 3 billion cubic meters (about 100 billion cubic feet) of growing stock, almost entirely hardwoods. The southeastern forests contain about 4.5 billion cubic meters (about 160 billion cubic feet), about evenly divided between softwoods and hardwoods. The Pacific forests contain about 6.8 billion cubic meters (240 billion cubic feet), almost entirely softwoods; and the Rocky Mountain forests nearly 3 billion cubic meters (about 100 billion cubic feet), also almost entirely softwoods. The West accounts for one-half of the country's total of about 18 billion cubic meters (roughly 650 billion cubic feet), the Southeast for one-fourth, and the central and northern forests together for about one-fourth. In terms of growth potential, the South is far ahead of the rest of the nation.

In total reserves, Douglas-fir is by far the leading species, with about 600 billion board feet of sawtimber. Next come western hemlock, with about 250 billion board feet, and ponderosa pine, with about 190 billion board feet. All firs together total about 218 billion board feet, all spruces about 130 billion board feet. In the East, shortleaf and loblolly pine together constitute the biggest reserve, amounting to about 200 billion board feet. The once abundant eastern white pine runs to perhaps 25 billion board feet. Among the hardwoods, the red oaks as a group provide the greatest reserve, about 100 billion board feet, followed by the white oaks as a group at about 80 billion board feet. Hickory, maple, sweetgum, black tupelo, and yellow-poplar follow in about that order.

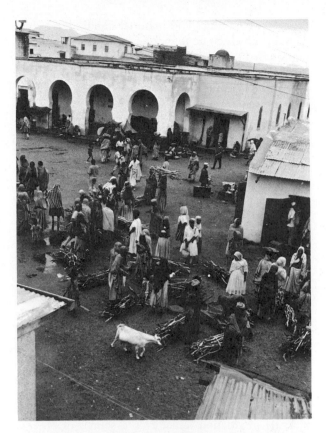

Figure 23-15   A marketplace in Harar, Ethiopia, is typical of conditions worldwide, in less developed lands. Faggots for fuel are brought from the distant countryside for sale. [Photograph by Kay Muldoon, for IDA, courtesy World Bank.]

Forest products

Forest trees have traditionally been the most abundant and economical source of cellulose. Cellulose is a staple of the economy, being used for many varied purposes. Even throw-away clothing has been marketed, made of especially attractive but nonetheless disposable paper. World production of wood products exceeds 2 billion cubic meters (about 70 billion cubic feet) annually, and more than half is consumed for fuel (Fig. 23-15); about one-third becomes lumber, and approximately one-eighth becomes pulp. Cellulose consumption is an indicator of a nation's economic conditions and material standard of living. So many goods come to market packaged in paper or paperboard that pulpwood consumption usually foretells shipments to market. Similarly, consumption of newsprint is a rough indication of the adequacy of a country's communication and of its level of literacy. Paper and paperboard are used at the rate of about 275 kilograms (roughly 600 pounds) per person per year in the United States.

Forest products represent nearly 5 percent of world trade. Forest products are especially important to the Scandinavian countries (accounting for nearly half the trade of Finland, for example). Canada exports nearly as much in the way of forest products as all of Scandinavia (about half the world's newspapers are printed on paper from Canadian pulp). Although much of South America is forested, it exports only modest amounts of forest products, and Asia exports relatively little. The absolute value of forest products from tropical Africa is not great, but proportional to the total trade it is quite appreciable. The greatest importers of forest products are Japan, the United States, the United Kingdom, and Germany. As we enter the final decades of the twentieth century, a new strategy of forest use is emerging—that of biomass production, chiefly as a source of energy. Coppicing (frequent harvesting of sprouts and small stems) may in part replace tree growing as biomass harvest becomes more important. But forest products of traditional types remain of greatest importance to a modern economy. A few are reviewed in the following paragraphs.

Wood

While forests serve many useful purposes, they are mainly noted for the wood products that they yield. Wood was once so abundant, when populations were not so voluminous, that it was profligately wasted. Modern demand for wood is now increasing beyond the ability of forests to provide it. In the United States, demand for wood increased 65 percent in the three decades following World War II, and it is projected to double again by the year 2000. What is this remarkably economic substance, wood?

As a twig elongates, the apical meristem lays down a train of primary tissues with relatively little bulk. One of these is a thin cylinder of secondary meristem, the cambium, located between the phloem and other outer tissues (the bark), and the xylem (the wood). Throughout the life of the tree, the cambium continues to produce more xylem on the inside, more phloem on the outside. The xylem produced by the cambium is secondary xylem, or wood, and makes up the great mass of the tree. Xylem cells nearest the cambium remain physiologically active for some time as **sapwood;** when the sapwood grows old it is infiltrated with additional lignin, dyes, tannins, and other substances to become **heartwood.**

Activity of the cambium varies primarily with the season. Even in tropical forests there is apt to be seasonality. In temperate climates there is a very pronounced seasonal activity. It is responsible for xylem growth rings, which give the characteristic banded appearance to wood. New xylem is usually produced very abundantly as growth resumes in spring after winter dormancy. The cells produced then are relatively large and have comparatively thin walls. As spring turns into summer, xylem formation slows, and cells have comparatively thicker cell walls. The abrupt change from the thick-walled cells to the new xylem of the next year marks an annual growth ring. The relative thickness of a tree's annual ring (that is, the abundance of xylem produced) is a reliable indication of yearly growth, and in turn a reflection of whether environmental conditions are favorable. For example, trees shaded by an overstory grow very slowly and have extremely narrow annual rings; but when overstory trees are removed or die, **release** occurs, and wider annual rings reflect a higher growth rate.

Visualized as a whole, the cambium of a tree forms a narrow cone that ensheathes the wood of each stem. The base of the trunk usually contains all of the growth rings produced since the tree was a seedling. When the tree was a sapling, it bore branches low to the ground; as it matured, these were shaded and died. The branches leave scars, which become knots. Large trees have deposited many additional layers of xylem outside of the old sapling branch scars, and

Debarking the log is essential to its full utilization because bark cannot be used for papermaking, and therefore any piece dropped in the chipper has to be free of bark. But the bark can be used for fuel and soil mulch.

The rounded sides of the log, called "slabs," are the first pieces sent to the chipper as the log goes through the sawmill. This idealized picture shows the entire log being used for lumber, except for the slabs. Actually, as cutting continues, other pieces go to the chipper, including edgings, trim ends, and other parts of the log not usable as lumber. Each log presents different problems and can be handled differently.

The outer portions of the log have the fewest knots. This "clear" lumber is usually made into boards or planks varying in thickness from one to three inches.

Toward the center of the log, knots increase and the wood is less suitable for boards. Heavier planks and square or rectangular beams are normally sawed from this section. The center of the log is used primarily for structural beams, which are of a large enough size that knots do not weaken them. Knots are most numerous here because this is the oldest section of the tree. Branches that were removed during the early years of the tree's life left knots that were covered over as the tree grew outward.

Plywood is, in effect, a sandwich of thin wooden veneers. Veneer is made by "peeling," that is, holding a long blade against a rotating log. The wood is continuously peeled off, down to an eight-inch core. The core is then treated as though it were a small log. It can be made into lumber and the rounded portions sent to the chipper.

Figure 23-16 Disposition of wood in a sawlog. [Courtesy St. Regis Paper Co.]

hence lumber from large trees is relatively knot-free (**clear timber**) toward the outer portion of the trunk. Typically the rings deposited in youth and middle age are wider then those deposited in maturity. Lumber of the highest quality comes from steady, moderate growth, in which all annual rings are the same width. Wide rings from temporarily lush growth may be weak, tending to shrink more and to split.

Drawings of blocks of wood are given in Figure 7-6, representing both a softwood and a hardwood. Figure 23-16 indicates how a log of wood might be utilized. The disposition of the cells and their appearance not only delineate the annual rings, but are responsible for cell pattern, or **figure.** The way a log is

sawed to expose transverse, tangential, and radial faces will determine its figure; the type, size, and arrangement of the cells, however, will determine the inherent characteristics of the wood. Growing conditions are reflected in variations of annual rings, knots, scars or blemishes, and so on. The diagrams of softwood and hardwood illustrate some of the distinguishing differences between the two types. Large vessel cells characterize some hardwoods, softwoods consisting entirely of tracheid cells with bordered pits. The horizontally aligned cells are rays, running from the outside of the xylem to the center. The scattering of large vessels through the wood varies according to species. For example, in sugar maple the vessels are small and rather uniformly distributed between spring and summer xylem, so that this type of wood is said to be **diffuse-porous.** In contrast, the wood of red oak has its large vessels confined to the spring xylem, and is thus said to be **ring-porous.**

Characteristics and quality of wood

The character of wood is determined by the nature of its cell walls and the structural configurations of its cells. It burns well, so it has been our most important fuel since time immemorial, only in recent centuries giving way to fossil fuel and today to atomic energy. Wood is strong, yet it is easily cut, nailed, and formed—qualities that make it highly suitable as a construction material. Esthetically, wood is highly pleasing; it polishes well to exhibit beautiful design, hence its great popularity for furniture, interior trim, and other decorative purposes. Consisting as it does of individual cellular air pockets, wood is a good insulator, comfortable to the touch and efficient in resisting temperature changes. It is a poor conductor of electricity, and it is more flexible and more resilient than metals. Along with these virtues it does have drawbacks; it decays when certain conditions of moisture and temperature prevail (although it can be treated to prevent this). It shrinks and swells appreciably with changes in temperature and humidity, and it occasionally twists or cracks.

The physical qualities of wood are reflected in certain measurable ways. Some woods are so dense that they will not float (chiefly such tropical hardwoods as ironwood, lignum vitae, and quebracho). But most woods, when dry, are less dense than water. They float well, having a specific gravity generally between 0.3 and 0.7. But if soaked in water long enough so that all air space in the cells is replaced by water (**waterlogging**), then wood will sink. Cell-wall substance itself is about half again as dense as water. A piece of wood compacted to eliminate all intracellular space would sink in water readily.

Durability is a feature of wood that varies greatly. Natural durability depends mainly upon resistance to decay, which in turn depends upon the kind of deposits in the cell walls of the heartwood. Where tannins, dyes, and other organic substances are prevalent, the fungi and bacteria that cause decay grow poorly. Woods that absorb little water may remain too dry for most decay organisms to flourish. Most woods preserve almost indefinitely if kept dry; witness the wooden artifacts taken from Egyptian tombs. When wood is waterlogged, it is not likely to decay, since most decay fungi must have air to grow. Wood decays most rapidly when it is moist and exposed, as in contact with soil or in locations where humidity is consistently high. The peculiar impregnations of the heartwood of members of the family Taxodiaceae, such as redwood (*Sequoia*) and baldcypress (*Taxodium*), make these species especially resistant to decay; to a greater or lesser extent so are various "cedars" and junipers (*Juniperus*), black locust (*Robinia*), Osage-orange (*Maclura*), black walnut (*Juglans*), chestnut (*Castanea*), certain mulberries (*Morus*), and catalpa (*Catalpa*). At the other extreme, such weak, soft woods as poplar, willow, and basswood decay very readily. Of course almost any wood can be made more durable by impregnating it with preservatives such as creosote, pentacholorophenol, or sodium arsenate.

The hardness of wood is a reflection of the sturdiness of the individual cell walls, the amount of lignin they contain, and other factors. The harder the wood, the better it will resist wear. Some species, such as black locust (*Robinia*) and oak, are extremely hard. Other woods are so soft they can be dented even with a thumbnail. Balsa (*Ochroma*), perhaps the most familiar of the unusually soft woods, has been imported into the United States for fashioning lightweight model airplanes. Tensile strength and resistance to shear are other fairly important characteristics of wood. In some species the cells separate much more readily than they do in others; easy cleavage may be an advantage in splitting firewood, but a disadvantage where toughness is needed, which is what was needed for wagon-wheel spokes, and still is for handles, athletic equipment (skis, bats, hockey sticks), and so on. Hickory, elm, maple, ash, and persimmon are strong woods, with considerable bending strength and shock resistance.

A characteristic of wood that gives certain types particular appeal is the grain (orientation of cells), which determines the figure (design). As already mentioned, this is determined not only by the inherent arrangement of the cells in the wood, but by the direction of cutting. If a log is sawed along its radius (quarter-sawn), the rays are revealed as broad patches, but if it is cut tangentially (plane-sawn), the rays, like the annual rings, will be cut across and be seen in end view as narrow bands. Figure is chiefly of importance for cabinet woods; other characteristics of esthetic importance are color, luster, polishability, and ability to hold glue.

The porosity of wood may be of importance in its ability to take paint, or even to resist decay. White oak (*Quercus alba*), esteemed for making barrels in which to age whiskey, has low porosity in spite of being a hardwood with large vessel cells; it develops balloonlike **tyloses** that fill the cell lumen, making the wood essentially impermeable. Such tyloses help make black locust resistant to decay.

Shrinkage occurs as cell walls dry. This is especially manifest in sapwood, which may shrink as much as 20 percent in certain dimensions. Greatest shrinkage is across the grain (at right angles to the long dimension of the cells). Such shrinkage causes twisting or warping due to differences in orientation of cells or differences in types of cells in different parts of a board. Quality lumber is seasoned before being used, to dry it out slowly and uniformly, holding shrinkage and strains to a minimum. Since seasoning in the open air requires a considerable length of time, much lumber is seasoned in kilns (**kiln-dried**) under controlled temperature and humidity. In most instances, the moisture content of the wood is reduced to around 6 to 10 percent; there is no point in drying wood more than this, since it will absorb at least this much moisture from the air in most localities.

Lignin

The importance of cellulose (which constitutes 40 to 50 percent of most wood) has been emphasized, but **lignin,** the second most abundant component of wood (constituting as much as 25 to 30 percent of the wood) is often dismissed as a humuslike residue left after wood decomposes. Considering the volume of wood consumed, the amount of lignin wasted annually in pulping operations is tremendous. There have been many effots to decipher the chemical nature of lignin, and to find worthwhile uses for it. To a

limited extent it has been broken down to yield vanillinlike intermediates of minor value, and itself has served as a plasticlike binder for road-construction soil and in fertilizers. But the cost of handling and transportation is such that it is unprofitable to move lignin very far to market; moreover, as a residue from pulping operations, it is dirty, sticky, and foul-smelling. Hence, lignin continues to be more a problem than an asset; means have to be found for disposing of it without polluting streams.

There are at least three, possibly more, lignins of slightly differing nature, with chemical structures almost as complicated as those of proteins. It is generally agreed that lignins are polymers of phenylpropanoid units. Studies with radioactively labeled carbon dioxide show that they are slow to form in developing wood, and are thus considered to be an end-product in cell-wall formation. Apparently, cellulose forms the basic structural framework around which the lignin infiltrates and cements the lattice. Very little lignin is found in young cell walls, but perhaps 20 to 35 percent is found in old heartwood. Lignin is chemically bonded to cellulose, and serves in strengthening and stiffening the cell wall.

Lumber, plywood, and veneer

Worldwide, the greatest use for wood (next to fuel) is for lumber (Fig. 23-17). Nearly 1 billion cubic meters (about 35 billion cubic feet) of sawlogs for lumber and veneer are cut annually, most of it from the United States and the USSR, and about 80 percent of it softwood! Home construction provides the greatest outlet for lumber. Even houses with brick or other exteriors utilize appreciable lumber for studding, floors, doors, windows, and trim. Properly constructed to resist rotting and termite damage, a frame house can last indefinitely. In New England, frame houses centuries old are still in use. Techniques are at hand for bonding stock of small dimensions that is light in weight and conservative of wood, yet just as effective for many purposes in house construction as traditional lumber. The middle of the twentieth century witnessed appreciable change from solid planks to plywood, especially for subflooring and sheeting. Wooden crating, too, once consumed sizable quantities of rough lumber, but

Figure 23-17   Cedar and pine logs are harvested from vestigial stands in the mountains of eastern Algeria. [Photograph by H. Null, courtesy World Food Program.]

this is being replaced by fiberboard and fiber (cardboard) containers. Never have there been widely accepted substitutes for wood in making furniture, although veneer is now accepted as being superior to solid wood for most cabinetry because it will not check and split so readily and because it is usually less expensive.

Veneers provide an economical use of fine logs, since a thin layer of wood is peeled and cemented to the surface of a less expensive, more abundant core wood. Cut fairly thickly and bonded several odd-number plies together, veneers become plywood, an increasingly important construction material. Consumption of timber for veneer is nearly 20 percent of that for lumber. Fine veneers for furniture are made from expensive and often scarce woods, frequently imported. Examples are walnut, bird's-eye maple, cherry, mahogany, teak, and rosewood. Plywood, however, utilizes abundant supplies of inexpensive wood, chiefly Douglas-fir, although southern pines also compete for the plywood market.

Both veneers and plywood achieved public acceptance only when suitable bonding agents (phenolic resin glues) had been developed that would solidly and lastingly cement the thin veneer. Today's excellent bonding agents set stronger than the wood itself, and are suitable for exterior as well as interior use. Successive plys are positioned with the grain at right angles, thus reinforcing one another and making the finished panel far stronger and much more dimensionally stable than an equivalent panel of solid wood. It also comes in far larger sizes (normally 4' × 8') than is obtainable in lumber (and in various thicknesses and numbers of plies). Construction is speeded up greatly because fewer individual pieces are handled. For balance, the center ply is made twice as thick as the adjacent ones. Inner plies are generally of less select grades of wood; those containing no visible defects are reserved for facing. However, special machines cut out such blemishes as knots from a plywood veneer and replace them with a duplicate insert that is scarcely noticeable. Plywood is assembled by elaborate machines that automatically apply the adhesive and affix adjacent plies. With the familiar thermosetting adhesives, the plywood can be heated to a specified temperature for curing in a few seconds.

Veneers may be produced in many ways, such as by sawing, slicing, or peeling. The most frequently used and most economical method is rotary peeling.

Figure 23-18   Poles ready to go into pressure chambers for treatment with preservatives. [Courtesy U.S. Forest Service.]

The log (**bolt**) is softened by boiling or steaming for a number of hours. It is then fastened securely to a rotating shaft, which works against a large stationary knife much like a wood lathe works against a chisel or the reel of a lawn mower works against the bed-knife. The knife is set to adjust continuously inward and maintain a uniform width of veneer, which peels off the bolt much as a roll of paper unwinds. Virtually all of the bolt becomes veneer, except for the very center, which is used for pulp. Rotary cutting is employed for most inexpensive veneers and all plywood, although sawing and slicing to obtain a special figure may be practiced with certain expensive cabinetry veneers.

There has been a tremendous rush into plywood production. In the Pacific Northwest, the bounteous supplies of old-growth Douglas-fir have assured economical raw materials for some years to come. Vertical integration has been inaugurated, in most plants, including chip-handling facilities to dispose of by-product wood profitably. Technological developments have also made it possible to use peeler logs of smaller dimension ("shallow-clear" logs), thus reducing dependence on butt sections of old-growth timber ("deep-clear" logs), which can be reserved for top-quality lumber and timbers. With modern steaming methods, even the sapwood makes acceptable veneer. This in turn has made greater care necessary in the handling of logs, to prevent, for example, jabbing them with spiked poles, the traditional way of moving them about the millpond. Plywood is increasingly made from southern pines, too; there are scores of plywood plants in Arkansas, Mississippi, and Louisiana, for example. Newer techniques permit plywood to be formed into graceful curves, used much like molded plastics. Modern glues, which permit the use of plywood outdoors, are said to make it stronger than steel, pound for pound.

Posts, poles, pilings, and ties

In spite of the diminishing availability of large timber in many regions, wood poles are still used for stringing utility lines, for pilings and ramps, and in many parts of the world for fence posts and mine timbers. Worldwide this accounts for about 5 percent of all lumber and veneer. In technically advanced parts of the world, both fenceposts and utility poles are treated with preservatives to improve their durability. At one time poles were merely soaked in a preservative, which provides a peripheral barrier against fungi but does not protect the center wood of the pole. More thorough impregnation of preservative is achieved by pressure and vacuum treatments in sealed chambers, in which the preservative is forced into all cells (Fig. 23-18). Creosote is still a frequently used preservative, but such chemicals as pentachlorophenol are equally useful.

Although railroads are scarcely expanding in settled parts of the world, the continuous replacement of old railroad ties causes a minor drain on timber. So far no metal or synthetic tie has been developed that is as useful and inexpensive as wood for undergirding the rails. Wood is resilient, easily fashioned, and amenable to the driving of spikes to hold the rails. Ties were once hand hewn, but today they are mostly sawn. The modern railroad tie, treated with preservatives, has a potential life of nearly a half century. The kind of wood used for ties depends partly upon local availability, and in the United States red oaks, yellow pines, gums, and Douglas-fir have all found favor.

Cooperage

Cooperage accounts for another minor use of wood. Although such commodities as cement and flour are no longer transported in wooden barrels or tubs, as they once were, and though most industrial liquids are shipped in steel drums, wooden barrels still find some specialty use, especially for the storage and aging of liquors. **Tight cooperage** for holding beer, whiskey, wine, syrups, and corrosive chemicals must be carefully constructed of woods chosen for their impermeability. Hardwoods have been favorites, especially white oak, the heartwood of which is made impermeable by tyloses. White oak imparts no disagreeable odor or taste, and it is required by statute that barrels made from white oak be used for the aging of bourbon whiskey. The manufacture of white oak barrels is of local importance in the midwestern hardwood forest belt. Staves cut to the proper curvature by special saws are made flexible by steam heating for the "raising" of the barrel. After the barrel is assembled, it is flame-charred within; the charred wood gives whiskey its amber color. Redwood and baldcypress have also been much used for tight cooperage, both species being noted for their natural resistance to decay.

Miscellaneous wood products

Limited quantities of excelsior are still made from weak, soft woods that shred well, such as poplar. Excelsior has declined in importance as a packing material, but there is some interest in it as a mulch. Modest interest continues in wooden shingles and shakes for roofing, mainly for esthetic reasons, even though synthetic roofing materials are less expensive and more practical. Wood is used for toys, novelties, picture frames, gunstocks, and a multitude of other purposes. Wood chips and wood flour are compressed into fiberboard for insulation and construction materials. They are also compressed into fireplace "logs" and made into charcoal. Wood flour is used as an abrasive for cleaning oily floors, as a bulking material in the manufacture of linoleum, in explosives, and as a diluent or raw material for plastics. The use of wood cellulose to make acetate and viscose rayon was discussed in the previous section on fibers. Charcoal has many uses, from the purification of sugar to fuel for barbecue fires. In Brazil tremendous amounts are used as a source of carbon in the manufacture of steel.

Pulp and paper

Wood is the chief raw material for conversion to pulp, which is principally used for making paper (Fig. 23-19). In technologically advanced parts of the world, pulpwood consumption is nearly half that of lumber. In the United States about 75 million metric tons of pulpwood are consumed annually, about two-thirds of which is domestically produced. It is estimated that in the United States nearly 275 kilograms (roughly 600 pounds) of paper and fiberboard are consumed annually per capita. Some two-thirds of this is derived from softwoods. At one time only select species such as spruce were thought fit for making paper, but these species now account for less than one-fifth of production. Technological improvements have made many short-fibered hardwoods acceptable. The resinous yellow pines are used for heavy kraft and linerboard papers. Synthetic sponges, honeycomb insulation, and disposble linens, draperies, and clothing are all being made from pulp these days. In mild climates, we may one day see paper houses!

Papermaking is an ancient art. The modern concept of paper—a felt or mat of fibers—appears to have originated among the Chinese during the first century A.D. Stems, bark, and silk were pulverized, the fiber dispersed in water, then sieved out onto a porous mold of reeds. Bamboo and the inner bark of the paper mulberry (*Broussonetia papyrifera*) were early sources of fiber. The principle behind modern

Figure 23-19   Wood ready for conversion to pulp. [Courtesy U.S. Forest Service.]

papermaking is still the same: fibers are separated and dispersed and then matted into a thin layer that is filled, coated, and compressed on elaborate machines (Fig. 23-20).

The name "paper" is derived from the word *papyrus*, a sheet made by pasting together thin sections of an Egyptian reed (*Cyperus papyrus*). But almost any sort of fiber can be used for making paper. In Europe, textile fibers were used during the Middle Ages, principally flax derived from old linen cloth, but cotton and hemp rags were also used. *Agave* or *Ficus* bark were used in the early Spanish colonies of the New World. With Gutenberg's invention of the printing press in the middle of the fifteenth century, the demand for paper increased greatly, and the supply of old rags could no longer satisfy the demand. Straw and wood were tried in England and France as early as 1800, but it was not until about the middle of the nineteenth century that practical techniques were developed for converting wood into pulp. In Germany, a wood grinding mill was built at about

this time, and in 1854 a patent was issued in the United States for chemically reducing wood to pulp (the soda process). Sulfite chemical pulping was introduced right after the American Civil War, and some years later the sulfate process now so much employed for making kraft paper was brought out. These processes have made wood the most economical source for paper pulp, which accounts today for more than 90 percent of all paper made. A few special papers, such as those used for cigarettes and banknotes, are still made from flax fiber, but no other fiber is as abundant, consistently available, and economical as wood fiber. Moreover, wood supplies have generally been near water, which is needed in abundance for pulp processing.

In the **mechanical,** or **groundwood,** method of preparing pulp, short debarked logs are pressed sideways against a revolving grindstone sprayed with water to produce a slurry of short fibers and fragments. The quality of groundwood pulp varies with the kind of wood used and with the care taken in

Figure 23-20 One of the gigantic calendering machines in a modern paper mill. The machine is guided electronically for precisely controlled thickness by the computerized device in the foreground. [Courtesy Accuray Corp.]

grinding. But inasmuch as the lignin and other noncellulosic components are not removed, groundwood paper "ages" and discolors easily. Groundwood pulp is chiefly employed for making newsprint or inexpensive sheeting of secondary quality. The groundwood method is the most economical of the pulping methods, and it produces the highest yields. A ton of air-dry pulp can be derived from little more than one cord of wood, nearly double the yield from chemical methods. Although power costs are high, the number of hours of labor required is generally only two-thirds that for chemical methods, and less skilled supervision is needed.

Paper of better quality is made from pulp pro-duced by chemical methods. Logs with the bark removed are mechanically chopped into small chips that are screened for size. The chips are chemically treated to dissolve most of the lignin, leaving nearly pure cellulose fiber. This in turn may be bleached to produce white papers. In the **sulfite process** the chips are digested in a solution of bisulfites at a temperature of 130–150°C (266–302°F). The amount of cooking determines the degree of lignin removal; prolonged cooking may cause some degradation of the cellulose. Highly resinous woods cannot be converted in this process. In the **sulfate process** the chips are digested in a solution of sodium hydroxide and sodium sulfide at a temperature of 170–180°C

(338–356°F). The process is particularly suitable for pulping resinous pine wood, and it is used to make kraft wrapping papers and paperboard. Tall oil, sulfate turpentine, and alkaline lignin are by-products of sulfate pulping. The **soda process** uses sodium hydroxide alone as the cooking agent; it finds limited use for making book and magazine papers from softer hardwoods. A so-called **semi-chemical process** removes only a portion of the lignin from the wood by chemical cooking, after which the softened chips are mechanically pulped.

Tree species used for pulping vary widely around the world. In the coniferous belt of the southeastern United States, the yellow pines are much used; in the northern coniferous zones, various spruces, firs, and hemlocks are used; in the West, hemlocks, firs, Douglas-fir, larches, and several pines and spruces are employed. Canada, the world's largest producer of groundwood pulp for newsprint, processes principally softwoods, such as spruces, pines, hemlocks, and firs. In Latin America, sugarcane bagasse is often used instead of wood, but the use of tropical hardwoods and conifers for pulp is increasing. Paraná pine (*Araucaria*) from southern South America still helps to supply the Brazilian mills. In northern Europe, pulp is made chiefly from softwoods, much as it is in Canada. In Australia, Chile, and Spain, *Eucalyptus* is used, as well as herbaceous fibers. Japan utilizes spruce, fir, and pine pulp, as well as several textile fibers. China depends more upon bamboo, rice straw, cotton stalks, and bagasse. In South Africa, introduced *Eucalyptus* and pines, as well as native *Acacia*, are utilized. It is apparent that almost any locally abundant and inexpensive fiber may serve as a source of pulp. In India, bamboo is grown as a crop and is an important pulping species. Should wood sources become scarce or too costly, annual crops or high yielding "shrubs" could substitute as a source of pulp. In the United States, many species have been studied as possible future sources of pulp. Sunn, or sann hemp (*Crotalaria juncea*, Leguminosae), has specifically been suggested as a potential annual pulp crop, and kenaf fiber (*Hibiscus cannabinus*) has been used in India.

In the making of printing papers, fillers (such as clay, chalk, calcium sulfate, and talc) are added to the bleached and processed pulp to give weight, opacity, and "body" to the finished product. The forerunner of the modern paper-making machine was invented in France by Nicolas-Louis Robert in 1799. In the early 1800s improvements were made by Henry Fourdrinier, whose name is still applied to the endless belt of wire cloth onto which the pulp is discharged—the **Fourdrinier screen.** This screen vibrates in two directions, dispersing the pulp fibers and intermeshing them as the pulp is moved continuously from the discharge box to the drying and calendering rolls at the other end. During the interval on the Fourdrinier screen, most of the water drains from the pulp; a series of suction boxes near the terminal position remove all remaining free water. The incipient paper so formed is picked up by another endless belt, on which it is further dried and consolidated by heated rollers. **Sizing,** the addition of rosin or some other substance, is done to give paper resistance to the penetration of water. **Coating,** the application of materials consisting of some combination of mineral pigments and adhesives, is done to increase opacity, surface smoothness, and whiteness. Coatings give paper a smooth surface that is impervious to ink. Clay is the most common coating agent; others are chalk, titanium dioxide, and zinc oxide. The more common adhesives are animal glue, casein, starch, soybean protein, various synthetic resins, and cellulose derivatives. The final step is an intense squeezing between huge rollers of the calender, which **finishes** the paper. Finished paper emerges from the calender as a continuous sheet, usually rolled, although it may be cut to a desired size, stacked, and warehoused.

Associated forest industries

There are many forest products and industries of specialized interest. The production of tannins and naval stores (turpentine, rosin, and pine oils) were discussed in the preceding chapter. A number of food crops, particularly nuts, but also such things as maple syrup and sugar are significant local industries in various parts of the world. Typical small industries linked to the forest are those that produce cork and Christmas trees.

Cork

Commercial cork comes from the bark of the cork oak, *Quercus suber* (Fagaceae), a subtropical evergreen. The cork oak is native to the western Mediterranean region, and it is important principally to Portugal, Spain, and northwestern Africa, although it is utilized to some extent in France, Italy, and Corsica, and in other islands in the Mediterranean. The bark is stripped, in a way that does not damage the tree, from trees that are sufficiently mature; new bark grows from the uninjured cambium, and renews sufficiently within a decade to be stripped again. Cork has been harvested in the Mediterranean region for nearly two millennia. Forests containing cork oak cover more than 2 million hectares (about 5 million acres); most of these forests are natural rather than planted. Only recently have management steps been taken to control fire in these forests and to reforest select areas. Government regulation now extends to stripping, to ensure that the operation be done carefully, in a fashion to protect and perpetuate this important resource.

Commercial cork continues to find a steady world market for both century-old and new-found uses. Whether it can continue to be used for seals, gaskets, flooring, and so on will depend upon its ability to compete economically with synthetic materials, such as plastics, which are not of great importance in the rural areas where cork is produced. So far, no synthetic has been substituted completely for cork, the isodiametric cells of which are like so many resilient, air-filled balloons, each coated with a wax that makes them impervious to liquids. Cork is resilient, light of weight, easy to hold when wet, and imparts no odor or flavor to substances with which it comes in contact. It is slow to catch fire, lasts almost indefinitely, is an excellent insulator, and is a nonconductor of electricity. An outstanding attribute is its chemical inertness; cork not only makes an excellent closure, but does not deteriorate in contact with most liquids, although strong alkalis or halogens do affect it.

Cork was known to the early Greek and Roman civilizations; there are references, for example, to its use as a float when soldiers traversed rivers. It has also been used as a seal for jars and casks since early times, although it remained for the invention of the glass bottle in the fifteenth century to extend this use throughout the world. There were some beginnings of cork-oak cultivation in the eighteenth century, as its commercial importance became increasingly apparent. The thick bark of the cork oak adapts it well to the arid climate and desiccating winds of the western Mediterranean.

The cork oak grows about 18 meters (roughly 60 feet) tall and has a bole circumference of 3 meters (10 feet) or even greater. Generally the trees have not been pruned for a straight bole, and consequently they exhibit a spreading, branching habit. Careful examination of cork bark shows that the cells are laid down in annual layers like those of wood, although cork does not have a layered appearance. The outer part of the bark, which is quite rough and fissured, must be scraped away to provide the smooth cork used for stoppers.

Cork bark is stripped in summer, when the cork is most easily separated from the trunk. The dead cork cells contain no sap. Circular cuts are made at the base of the tree and just below the first branches, and then a vertical slit is made the length of this section. Using special hatchets, skilled workmen pry away the cylinder of bark. They climb into the tree to remove sleeves of bark from the larger branches in the same fashion. Bark must be at least 2 centimeters (almost 1 inch) thick to be worth stripping. A skilled stripper generally collects more than 100 kilograms of bark per day; exceptionally large trees may yield as much as a ton of cork. Stripped bark is stacked and left to season for a few weeks before being boiled in huge tanks to soften it (incidentally removing much tannin and extraneous matter), after which the rough outer surface is scraped away. It is then dried, trimmed, graded, and baled, coming to market in flat sheets several square feet in size.

In our modern economy, cork is used for gaskets, seals, floats, nonslip walkways, handles, corkboard, floor tile, and linoleum. The manufacture of composition cork is today one of the more important phases of the industry. Pulverized cork is mixed with a suitable adhesive, pressed into large molds, and heated until cured. Corkboard made in this fashion is lightweight and nonconducting, making it an excellent insulation for coldrooms and refrigerator compartments. The original linoleum was made by bonding a mixture of ground cork, resins, and linseed oil to a burlap backing. Today many less expensive filling materials, such as wood flour, are used, with a minimum of cork.

Portugal accounts for about one-half of the world's cork commerce of about 335,000 metric tons annually. Spain and Spanish Morocco account for most of the remainder. The United States frequently imports more than 100,000 metric tons of cork in a year, and attempts have been made to establish the cork industry in California, where the tree grows well. It seems unlikely, however, that cork can flourish as a crop in the United States, requiring as it does so lengthy a production cycle and so much hand labor in its harvest. If inexpensive imported cork is not available, it appears likely that such materials as plastic foams and rubbers will substitute for it.

Christmas trees

The custom of buying a small evergreen tree for ornamentation and display in the home at Christmas season has gained widespread acceptance in Christian societies. The idea was initiated during the Middle Ages in Germany, where individual families would select and harvest trees from the forest. From that simple beginning the production of Christmas trees has grown into an important industry, for which trees are planted and pruned ("sheared") to make dense growth. Many species are used, preferences varying with region. Spruces are generally not favored, since they shed needles rather readily. Scotch pine, an increasingly favorite plantation-grown species usually started from seed imported from Europe, is much used in the eastern United States. Large numbers of balsam firs are imported from Canada, 3 to 4 million harvested mostly from the wild (although shearing and shaping may have been practiced). Eastern red-cedar (*Juniperus virginiana*) is used somewhat in the Southeast. Various pines and firs are used in the West, the noble fir being among the most esteemed species.

Land that is to yield a Christmas-tree crop is generally earmarked for that purpose, and is not meant to be used for future timber production. One- or two-year-old purchased seedlings are generally planted with a mechanical tree planter, and weed control is assiduously practiced for the next few years. As stature is attained, trees are sheared or pruned each spring–summer season to achieve a pyramidal shape and good density. Measures for the control of pests, ice and snow, and fire are undertaken as necessary, and some species (such as Douglas-fir) are typically fertilized. Harvest is usually a hand operation since it is necessary to practice some selectivity as well to avoid damaging the foliage. Pines are usually harvested the seventh or eighth year; spruces and firs, the eleventh or twelfth year. After a tree is cut, several side branches remaining on the stump will develop, of which one of the more promising can be trained for a future harvest four or five years later. Many conservation departments offer conifer stock for reclaiming waste land, and permit the stock to be harvested for Christmas trees. In some areas, where reforestation is desired, thick initial plantings will actually benefit from thinning for Christmas trees. Harvest of the Christmas trees in the early years can subsidize the cost of reforestation.

In addition to Christmas trees, greenery, often obtained from the forest, is also marketed at Christmas. Holly (*Ilex*), a traditional decorative greenery, is grown in favorable climates, such as western Oregon, where orchards are established primarily to supply decorative material for sale during the Christmas season.

Recreational public benefits

Public uses of forest lands must be considered along with forestry. Recreation (Fig. 23-21), hunting, wildlife habitat, and such are increasingly important to growing civilizations, even if they cannot easily be assigned a dollar value. Nor can indirect benefits from the forest, such as protection of watersheds, control of flooding, and the contribution to the atmosphere, be accorded a monetary value. These may well be the most important uses of the future!

Figure 23-21 Winter sports in Bitterroot National Forest, Montana. Forests have become important recreation centers.

The business of forestry

The tremendous reserves of wood in tree trunks throughout the world are a capital asset that has long seemed inexhaustible. Yet in the twentieth century we are witnessing the final stages of exploitation of the forests of the temperate zone. This has been brought about by extensive colonization and ever-increasing demands for timber and pulp. Forest land has become cropland, the tree crops cut and planted (Fig. 23-22) or replenished every 15 to 100 years. Consequently, there are few virgin forests left except in isolated reserves.

In Chapter 15 the impact of recent changes in the forest industries was discussed. Whereas forests were once thought of as an "enemy" of civilization, to be felled and disposed of as rapidly as possible in order to permit field cropping, they are today regarded as an important raw-material source to be managed and husbanded. Today even secondary species and residues formerly wasted are now utilized. Scrub land and cut-over holdings are potential sources of biomass and cellulose. Chemurgy has come to the forest, and we can expect to witness economic battles over whether trees or some other crop can serve most eco-nomically as the source of cellulosic raw materials. Land suitable for agriculture will likely be pre-empted for raising food crops as populations increase, with less productive lands managed to yield forest crops.

Forest cropping is still an unsophisticated business in many parts of the world. Half of the trees felled in the world are used for fuel rather than timber, and many prime logs are wasted in slash-and-burn preparation of agricultural lands, especially in the tropics. The selective cutting and high-grading that are practiced in tropical forests leaves only cull trees and species of lesser value to perpetuate the forest. Remoteness, lack of ready market, and a lack of appreciation of the principles of conservation perpetuate this wastefulness. Even in technically advanced parts of the world, forest resources are not always efficiently employed. Witness what is happening in the midwestern United States, where the few remaining stands of mature white oak are often cut for firewood and where sizable trees are girdled in local woodlots to increase graze, while lumber is imported from the western states for repair and construction on the farm.

But the trend is unmistakable: the forest products

Figure 23-22   Planting ponderosa pine seedlings with a Lowther tree planter pulled by a tractor in Shasta National Forest, California. [Courtesy U.S. Forest Service.]

industry is becoming more businesslike and more conscious of overall values, especially in technically advanced parts of the world. Research, lavish by former standards, is sponsored by the large and well-capitalized firms in the forest-products industry, as well as by public agencies. Projects range from practical investigations on tree breeding, forest fertilization, pest control, fire prevention, and selective weeding to fundamental studies on the influence of the forest on the watershed, weather and climate, snow melt and water yield, energy requirements of various species, and so on. Forest management increasingly leans upon research, and its practices are tempered by overall economic considerations.

The integrated forest-products industry

Nowhere is the change from a "cut and move on" forest philosophy to a firmly established business base more manifest than in the integrated, business-managed paper and timber corporations that today dominate the forest-products industry in North America. Massive investment demands long-range planning for sustained yield.

Much company forest land is acquired outright, inventoried, managed, and improved by modern techniques. Generally, even more timber is bought from government holdings. Complete utilization of the harvest is stressed (peeler logs for veneer, saw logs for lumber, small logs and veneer cores for studs, mill residues such as chips, sawdust, and planer shavings for pulping, and bark for garden mulch), all processed at a single location. Tremendous automated log-handling equipment and storage yards are maintained. Professional foresters manage the forests on a sustained-yield basis. For public-relations reasons, public use of some of its timberland for recreation is allowed. Research-and-development centers look into lower-cost higher-yielding pulping processes, better uses for chemicals derived from the noncellulosic components of the tree, and better

Box 23–4   *Impediments to high-yield forestry*

As the demand for prime timber continues to mount, the forest-products industry dips more and more into its remaining capital of virgin or old-growth timber. Foresters assume this depletion can be compensated for by utilizing highly productive management techniques and genetically superior trees. Draw-down of existing forest is predicted upon the assumption that an era of high-yield forestry lies ahead. Studies are made of the potential productivity of new inbred and clonal lines, requirements for quick regeneration in all sorts of habitats, and the stimulation that can be expected from fertilization, thinning, and efficient harvesting under foreseeable economic and technological conditions. All these variables are fed into the computer, which suggests an acceptable harvesting rate.

Immense regional "programs," such as "third forest" expectations for the Southeast, have been launched: The first forest was the virgin wilderness that yielded a trillion board feet of lumber; the second forest was that which regenerated naturally after the virgin forest was felled, and which has supplied most of the region's needs for the past century; the "third forest" is to consist of tailor-made plantings for enhanced croplike yields in this productive climate. In such projections, proponents visualize millions of superior trees, planted and grown by revolutionary means, eventually logged by gigantic machines that mop up the slash after harvest and set out thousands of new seedlings per hour (these perhaps even protected by "minihothouses," which soon disintegrate). Fertilization, stand density, pest control, and so on are all optimized.

An intensive program, yes, but what of the problems resulting from such a completely regulated ecology? A switch to intensive tree monoculture should have consequences similar to those encountered in the cultivation of annual crops (Chapter 15).

One conflict that has already surfaced has been initiated by the conservation-minded, who question the wisdom of converting most natural forest to a cropping system that is likely to diminish esthetic enjoyment, wildlife habitat, and recreational values. No matter how well the trees are growing, a "walk in the woods" loses its savor beneath unvarying trees all in a row. Gordon Robinson, staff forester for the Sierra Club, states, "Biologically the forest that is most secure against fire, insect and disease is one with a variety of ages and species." He prefers selective cutting of individual trees as they reach their prime, and contends that the cutting of old trees in anticipation of a new harvest that may never result is an exercise in brinksmanship.

manufacturing and finishing processes. Computers are used to monitor and analyze data and provide accurate operating information. By-products carry over into the food field (yeast production), and into chemicals for plastic coatings and bondings (for waterproof wrappers and all-weather plywood). This is a far cry from small operations on family farms.

The changing forestry picture

In many areas forestry has begun to change from being concerned mostly with husbanding and protecting (for example, fire prevention) to being more concerned about the future, when all forests, like today's farm plantings, will be intensively managed. In Europe and in the United States, this change is already manifest. Elsewhere, especially in the vast, largely unexploited tropical forests, enlightened forest management is still many years away. But more and more, attention is being given to fitting the forest to overall land-use patterns.

Clear-cutting, the most efficient means of harvest for intensive forest management, is not necessarily best for the ecosystem and for multiple-use forestry. During the 1970s much controversy arose over a judicial decision that in Appalachia only selective cutting, not clear-cutting, was legal (Congress then passed new legislation relaxing the old prohibition). There is much interest in practical management techniques, of course, to keep forest production efficient and wood competitive. Progress and exciting change are evident in many fields, including tree improvement, fertilization, and pest control.

Tree improvement

Because of the prolonged cycle for tree generations, 25 years or more for some species, it is no wonder that tree breeding has generally lagged behind breeding of annual crops. There have been some notable exceptions, as with rubber and some fruit crops (peaches, for example). Here the emphasis has been upon selection of naturally occurring superior genotypes that are then perpetuated vegetatively. In many countries (the United States, for example) the finest forest trees have been harvested, leaving the culls and unwanted types behind to reproduce and dominate. Much genetic degradation occurs as this process is repeated over time. In Europe a general practice has evolved of saving a few of the "best" trees to become seed sources for the regenerating forest.

Unlike fruit trees, which are usually propagated vegetatively, most forest trees are propagated by seed. Where efficient means of vegetative propagation can be achieved for forest trees, improvement programs will be geared to isolating unique genotypes, as is the practice for most fruit-tree breeding. This is now true for improvement programs in *Populus* species (aspen, cottonwood) in the United States and Europe and for *Cryptomeria* in Japan. Where seed propagation should prove to be the only practical alternative, the improvement program must be geared to the identification and production of superior seed trees. This is slow work because progeny testing should proceed until maturity, or at least until evidence can be obtained that the performance of young trees can be used to predict the productivity at normal harvestable age. Thus the extent to which hybridizing, backcrossing, inbreeding, ploidy manipulation, and so on, can be practiced usefully with forest trees remains to be determined for most species. There is no doubt that in dealing with anything so bulky and slow to reproduce as forest trees, there are formidable obstacles to merely making available a sufficient amount of the proper habitat to permit a comprehensive breeding program.

Figure 23-23 Hybrid vigor in pine. The large tree on the left is a Jeffrey pine–Coulter pine backcross, *Pinus jeffreyi* × (*P. jeffreyi* × *P. coulteri*); the tree on right is *P. jeffreyi*. Both trees are 13 years old. [Courtesy U.S. Forest Service.]

Nevertheless, some of the sophisticated techniques developed for annual crops (see Chapter 19) may well prove adaptable to forest-tree crops. Improvement similar in magnitude to those achieved by hybrid corn can be expected (Fig. 23-23), but speedy progress seems unlikely unless new seedling evaluation techniques are developed.

Mapping the world's soils and climates has progressed far enough that we have a fairly good idea of where to look in the world for potential stock adaptable to a given location. The Chinese chestnut, *Castanea mollisima*, has already been introduced into the eastern United States as a blight-resistant species, and its resistance can be transmitted to offspring of crosses with the American chestnut, *C. dentata*. It may be possible to introduce a disease-resistant clone in time to permit natural crossing, with the hope that some progeny may exhibit disease resistance. More needs to be known about interspecific heterosis in trees. This should be a continuously rewarding field of investigation, in spite of the awkwardness of handling tree populations.

Already some seed is produced in orchards developed by forest geneticists. When seed of a high, uniform quality is produced in quantity, much of it will go into the national and international market. A guarantee of quality is therefore necessary to the buyer. For field crops, certification programs have proved to be of great value in keeping the sale of undesirable seed to a minimum (see Chapter 9). The same device, certification of seed by a public agency, is being used in some states in the United States to insure tree seed quality. It seems likely that all forest-tree seed in international commerce may eventually be certified as to source (progeny-tested parents, seed orchards, isolated open-pollinated trees, or merely selected trees).

Seed collected from superior parents for seeding nurseries and for restocking is open-pollinated, and the progeny are therefore variable. However, off-type seedlings can be eliminated in the nursery. Such factors as rapidity of seedling growth, suitability to climate and cultivation, even resistance to disease, can often be noted in a relatively few years, making it possible to avoid the long wait until maturity to judge these characteristics. Species that do mature sufficiently fast to produce seed in a few years offer the opportunity for controlled crossing analogous to that undertaken for herbaceous crops. Lending urgency to programs for improving tree stock is the need to incorporate disease resistance in once-important species now sadly decimated by introduced diseases, such as the American chestnut (chestnut blight, caused by *Endothia parasitica*), white pine (blister rust, caused by *Cronartium ribicola*), and the American elm (Dutch elm disease, caused by *Ceratostomella ulmi*, and phloem necrosis, caused by a virus). Poplars, oaks, maples, locusts, sycamores, and many other genera, too, are today threatened by

serious disease or insect afflictions, both native and introduced. It is frequently impossible to control epidemics in the forest, and even if possible the cost may be exorbitant. The advantage of breeding resistant stock for reforestation is obvious.

Although the tree breeder experiences certain disadvantages compared to the breeder of annual farm crops, there may be compensations. An exactly uniform plant population may not be so important in the forest as it is with crops grown in more restricted areas and that must be all harvested simultaneously. Thus the advantages of heterogeneity and variability can be utilized, lessening the chance of a forest population being suddenly wiped out by a new affliction.

Systematic screening of tree introductions has been nowhere so intensive as the screening of herbaceous plants. There would seem to be considerable possibility of using exotic trees in many parts of the world, in locations where they are well adapted and perhaps resistant to common insects or pathogens.

Forest fertilization

Until recently there has been little reason to encourage more rapid growth of forest trees through fertilization. Moreover, timber has been traditionally a bulky material, and it has been doubted whether the returns from fertilization would justify the cost. Other than in orchards and plantations, relatively little tree fertilization is undertaken. But as concentrated fertilizers and mass means for their dispersal are developed, forest fertilization may in some instances become remunerative. On some soils, lack of nitrogen may prove the chief limitation to growth, but elsewhere phosphorus, potassium, or trace minerals are limiting.

In the United States, forest-tree fertilization has generally been limited to nurseries and occasionally new field plantings, although some fertilization with urea has been practiced from the air. In Scandinavia, the practice of fertilizing the forest from the air seems to be growing. With forest products so important to the economies of Sweden, Norway, and Finland, all are feeling pressure to increase the output of wood products in order to remain fully competitive. If Sweden is to realize its goal of raising pulp production to 250 million metric tons, forest fertilization

Figure 23-24 Spraying a stand of young western white pines with an antibiotic fungicide to arrest damage caused by white-pine blister rust. [Courtesy U.S. Forest Service.]

will certainly be necessary. It is recommended that the forest be fertilized every 5 years. Swedish foresters estimate a profitable annual return on the outlay in fertilizing the forest, a practice apt to become more commonplace in other parts of the world where mineral nutrition is a limiting factor.

Forest pest control

Control of pests in the forest has traditionally been slow and expensive, because of the immense area to be covered and the difficulty of reaching inaccessible parts. Now with slow, stable, low-flying aircraft available, spraying of the forest is becoming commonplace. In the United States, large-scale spraying programs carried out by airplane or helicopter have been limited to extensive acreages away from croplands (Fig. 23-24); herbicides have been used for such purposes as killing the hardwood overstory in the southeastern coniferous belt to encourage growth of young pines, and pesticides have been used to stop insect epidemics in remote northern or western forests.

A large business has been developed around "brush killing" along roadsides and powerline rights-of-way with phenoxy herbicides and similar compounds. In general, the intention has been to eliminate all vegetation rather than to remove certain species selectively. Proximity of farmland often makes air application unfeasible, and drift, contamination of runoff, and possible air pollution are becoming matters for concern. Ground-spraying with low-volatility materials permits better pinpointing of the treatment. Soil applications of herbicides are often used, which are lethal to many weed species in dosages as low as a few ounces per acre. Other sterilants and nonselective herbicides are often used along fence rows, around buildings, under guardrails, and so on. Sometimes sufficient labor is available for frilling (partially girdling the tree and applying the herbicide to the girdle), basal treatment (using a portable sprayer to apply the herbicide, usually mixed with an oil carrier, around the tree base or stump, typically to prevent resprouting), or even injection of herbicide into the tree for a systemic effect.

ORNAMENTAL AND HORTICULTURAL SPECIALTY CROPS

Many crops are employed chiefly for esthetic purposes—for beautification rather than for food, fiber, or timber. It is difficult to decide, sometimes, whether a nut- or fruit-bearing tree, for example, is more important for its yield or for its esthetic contribution to the landscape. But, in general, any noncommercial plantings, especially around the home, are regarded as ornamental (Fig. 23-25).

Economically, ornamental-plant industries "sit at the second table," compared with the food-crop industry; ornamental plants are one of many luxuries enjoyed by affluent societies, in which basic needs are easily fulfilled and in which there is leisure time as well as surplus money to make possible the enjoyment of plants. Today, North America and Europe constitute the main commercial market for horticultural specialties. Individual properties throughout the world are graced by ornamental plants; in many parts of the world, home plantings also have a practical purpose, yielding fruit, for example, along with beauty. Japan is an outstanding example of a country in which the use of ornamental plants has become highly specialized and where landscaping is meticulously kept and often designed for miniature gardens. Throughout the tropics, ornamentals are more casually handled, often being locally propagated and passed from party to party rather than supporting a significant commercial industry.

In the United States the sale of ornamentals has achieved a high order of importance, linked with the extensive population shift from city and farm to the suburbs. The value of all horticultural specialties has been estimated to be 1 percent or more of all farm crops; about 1 percent of the gross national product is spent in the "outdoor living" market. The lion's share of the "outdoor living" market goes to tools, power equipment, outdoor furniture, and recreational devices associated with ornamental landscaping. In some regions the retail value of horticultural specialties averages 0.7 percent of the total retail trade. Every year, more than 50 million home gardeners, tending some 4 million hectares (about 10 million

A

B

Figure 23-25 (A) In the garden of a suburban home native trees, introduced ornamentals, and a well-tended lawn are combined for a pleasing landscaping effect. (B) Ornamental plantings are also considered vital for cities. [A, courtesy The Lawn Institute; B, courtesy American Association of Nurserymen.]

acres), directly consume perhaps 68,000 metric tons of lawnseed, and more than a billion dollars worth of "green goods" (bedding plants, bulbs, deciduous shrubs, evergreens, perennials, shade trees, and so on). Related sales of fertilizers, soil amendments, and pesticides amount to as much.

Ornamentals

Horticultural specialties used for home landscaping and plant materials used for outdoor display (such as along highways) are here grouped with ornamentals.

This is primarily a "discretionary spending" market, which is likely to vary appreciably with the general state of the economy. Since the end of World War II, the use of ornamentals has expanded rapidly. The industries dealing in ornamentals are consumer-oriented, with high markups, promotional advertising, appeal to fashion and status, and impulse buying. Competition is keen. Most industries have specialized either in production or in merchandising, but competitive pressures have brought both specialization and integration into large operations, better to enable them to capitalize the extensive and complicated activities required.

The seed industry

During the twentieth century the farm seed industry has evolved into two segments, one primarily supplying the home gardener, the other supplying the farmer. The home-garden seed industry has divided into two specialties, one supplying seeds for flowers and vegetables, the other supplying seed for lawns and fine turf. The market for home-garden seeds has not grown greatly, since the move to suburbia brought less rather than more need for home food production; even annual flowers are often purchased from the local nurseryman as seedlings rather than started at home. The lawnseed market, however, has grown well. A lawn is fundamental to the landscaping of many a suburban home, whether the seed is planted in situ or purchased indirectly as sod. A sizable proportion of the lawnseed consumed is used for rejuvenating deteriorated turfs. New demands have been created for seed; for example, plans for the extensive interstate highway system call for nearly 30 acres of landscaped right-of-way per linear mile.

The lawnseed industry

Grass seed was once supplied by the field-seed industry, the main concern of which is forage grasses. Right after World War II, demand arose for new quality standards and special turf cultivars solely for lawns. Today, highly specialized farmers, utilizing up-to-the-minute production techniques (for example, weed control with new herbicides, and mainte-

nance of genetic identity in select cultivars), produce lawnseed essentially free of weed and contaminating "crop" seed.

The most important lawn grasses grown from seed in the United States are the Kentucky bluegrasses, *Poa pratensis;* the perennial ryegrasses, *Lolium perenne;* the fine or red fescues, *Festuca rubra;* bentgrasses, *Agrostis*, in several species; and bermudagrass, *Cynodon dactylon*. The first four are so-called cool-season (C_3) grasses, well adapted to the northern two-thirds of the United States. Bermudagrass (C_4) is adapted from Tennessee southward, and in the deep South, near the Gulf of Mexico, bahiagrass (*Paspalum notatum*), carpetgrass (*Axonopus*), and centipedegrass (*Eremochloa ophiuroides*) are also planted from seed. A limited amount of *Zoysia* seed is available, genetically mixed because of cross-pollination (many grasses are apomictic). Special cultivars and hybrids of *Zoysia* and bermudagrass are propagated vegetatively, with plugs, sprigs, and sod sold through nursery-type outlets. Another lawn grass that is widely used in the South is st. augustinegrass, *Stenotaphrum secundatum;* because it produces few seeds, it is marketed mostly as sod and occasionally as sod and sprigs.

Not very many years ago most lawnseed was secured from pastures and meadows left ungrazed and unmowed through the seeding season. To a great extent, bahiagrass is still so produced in Florida and Texas; limited amounts of natural Kentucky bluegrass may be produced in this way in the Midwest, as are fine fescues in the Peace River country of Canada. Mostly, however, select cultivars are agronomically grown in fields free of genetically unlike strains. Once sufficiently large and steady markets for lawnseed became available, such intensive growing proved more profitable than gathering seed from pasture and meadow. The sizable plantings in the Pacific Northwest, heavily fertilized for high yields and carefully tended to avoid weeds and competing vegetation, justified their high capitalization once the demand for select cultivars of lawnseed was large enough. The transition from the harvesting of "natural" Kentucky bluegrass in Kentucky and other sections of the Midwest to agricultural production in the Northwest occurred in less than a decade begin-

Figure 23-26   Harvesting natural Kentucky bluegrass seed in Kentucky. These seed-stripping machines are pulled by a tractor. [Courtesy The Lawn Institute.]

ning in the late 1950s. A résumé may be of interest, for the economic pressures that have affected bluegrass mark a characteristic trend.

Kentucky bluegrass seems to have been introduced from Europe by the first North American colonists, perhaps as chance seed in hay and cattle bedding aboard ship. Wherever the soil was turned, newly arrived Kentucky bluegrass volunteered. Evidently it was spread as far west as Illinois by the early French missionaries, and it was well established in the Ohio Valley by the 1800s. Early in the nineteenth century it gained the name "Kentucky" bluegrass because it flourished so well on the rich phosphatic soils of north-central Kentucky. By the 1900s it had spread westward into Missouri and northward into Canada, being utilized chiefly for pasture and hay.

It was from such fields that Kentucky bluegrass lawnseed was harvested. There is much to recommend this grass; natural selection should have proven out the diverse ecotypes found in every field, which when blended for lawnseed provide a widely adaptable genetic base. In a sense, such a seed crop is subsidized by the livestock industry, since after seed har-

vest in June or early July the bluegrass is grazed or cut for hay until the next May. But savings in grass maintenance are now more than offset by increased harvesting and handling costs. Instead of harvesting bluegrass from scattered pastures and meadows, it is more economical to grow it strictly for seed under intensive management that gives yields of as much as 1125 kilograms of seed per hectare (about 1000 pounds per acre), compared with an "excellent" yield of 225 kilograms per hectare (about 200 pounds per acre) from pasture or meadow. Moreover, the soils of Oregon and Washington, virgin to bluegrass, permit easy maintenance of genetic purity now that proprietary cultivars have become important.

Special stripping machines harvested natural Kentucky bluegrass in the Midwest (Fig. 23-26). These had to be capitalized for but a few weeks of activity during the harvest season. In the "western district," from Missouri northward into Canada, portable stripping machines were used, moved (at considerable expense) from location to location as the seed ripened and contracts for its gathering were negotiated. Little wonder that sedentary production in the Pacific

Figure 23-27   Combining Highland bentgrass in Oregon for lawnseed. [Highland Bentgrass Commission.]

Northwest came to be favored, where conventional farm machinery can be used and can be capitalized for diversified and year-round activity rather than sitting idle for most of the year. Today, most lawnseed is forced to maximum yields by heavy fertilization, and even if lodging occurs it is lifted, cut, windrowed, and field combined at appropriate seasons (Fig. 23-27).

Where fine lawnseed is grown agronomically, it is sometimes feasible to establish local cleaning plants to remove chaff and contaminating seed and thus avoid the expense of shipping rough seed. The larger farms even build their own seed-processsing equipment. The end result is an integrated operation for the particular area, from planting to finished seed. In a few favored regions, established channels of trade and experienced farmers and handlers make lawnseed growing exceedingly efficient. This small specialty crop has become established in these regions not only because growing conditions are favorable (weather, soil, rainfall), but because of human and economic factors. It has become, on the whole, less expensive to produce quality seed in Oregon and

Washington and to ship it all the way across the country to eastern markets than to harvest seed from eastern fields.

The garden-seed industry

Although many large seed firms have both garden-seed and lawnseed divisions, the respective operations are not alike, and the products are usually marketed separately. Garden seed is produced mainly in the western states—most flower seeds in California, and much vegetable seed in Idaho. The main market, as with lawnseed, is in the more populous eastern states. More so than with lawnseed, cultivars of garden plants must be grown in isolation to maintain their distinctive characteristics. The production of hybrid garden seed has increased rapidly in recent years, partly for the same reason that hybrid corn proved more profitable to the seedsman than did open-pollinated seed (that is, exclusive control through inbred lines). Generally, F_1 hybrids are superior because of heterosis, and the control of parent lines assures premium prices for several years. The industry has initiated a number of merchandising flairs, such as the annual "All-America" selections of new cultivars. It is standard practice to allocate a percentage of the profits from new All-America originations for promotion; competing firms must pay a royalty for the privilege of also offering these new cultivars.

Seed production requires a great deal of technical competence, as well as meticulous attention to planting time, controlled pollination, and customized harvesting techniques for each crop. The flower-seed industry centers in the interior valleys of California, where a mild climate with little rain during the growing and harvesting seasons permits fairly certain production and harvest under irrigation. Not until World War I, however, did garden-seed production in the western United States gain much momentum; prior to that time much seed was imported from Europe, some of it poorly adapted to the United States. Today, domestic garden seed is produced by a relatively small number of firms, most of which are rather large; each has its own research department for breeding new cultivars, widespread trial grounds, and a corps of experts. Once a new cultivar is considered to be ready for market, it may be grown under contract by independent farmers of the area. By the time a new cultivar is ready to be marketed, the investment in research, in building up supplies of foundation seed, in storage, handling, and so on, is appreciable. Moreover, the investment is staked on presumed but uncertain market acceptance. The risk is apparent, should public taste suddenly change, or some unforeseen circumstance endanger a lengthy commercial life for the new cultivar.

Most of the home-garden seed traditionally has been sold as packet seed, from catalogs. (This, however, is a small portion of the vegetable-seed business.) Several large firms specialize in this activity, one vying with another to see who can produce the most attractive, superbly illustrated catalog each year. The highlight of the winter season for many gardeners is to receive the attractive catalogs, which are increasingly expensive for the seed houses to produce and account for much of their overhead. Seed houses operate under two economic disadvantages. One is that seed unsold at the end of the season is generally destroyed, since by law there is a time limit to the germination guarantee (it is obviously impractical to rework small lots of seed in sealed packets). The other is that the sales season lasts but a few months at the most, usually oriented to spring planting.

Since World War II the garden seed industry has increased the use of glass and screened structures, particularly for production of high-value items such as F_1 hybrid petunias. Controlled crossing can be better regulated under technically precise conditions. It is obviously necessary to screen out insects that may contaminate a cross with outside pollen, and precise emasculation of perfect flowers is more easily handled on benches than in open fields.

The processing, storage, and testing of both lawn and garden seeds is an exacting and highly technical procedure. Seed-testing laboratories manned by registered seed technologists operate under precise rules governing each species in order to provide the guarantees for germination and purity that are the basis for trade. In general, seed remains viable longest if kept cool and dry; this is often best accomplished by keeping the seed in sealed containers in a cold room. Government control officials constantly sample seed supplies on dealers' shelves to determine whether the seed meets the purity and germination claims on the label.

Growers typically overplant somewhat, to assure ample supply regardless of weather. This is not entirely unselfish, since markets once sacrificed are regained with difficulty. Large firms generally hedge against the weather by growing seed in several locations. The entire genetic stock of prized cultivars is seldom risked in a single planting, in a single year. In fact so valuable is the germplasm, the real wealth of the industry, that seed stocks are usually stored in fireproof vaults, to be withdrawn from time to time only as a field planting demands. Touchy items, such as seed of asters, onions, and parsnips, are generally packaged in special foils, plastics, or cans to control humidity.

Vegetative materials and the nursery industry

Living ornamental stock is fundamental to landscaping. Seasonal demand for corms, bulbs, and rhizomes, and for herbaceous perennials, has been steady if modest. Even more important to home landscaping is shrubbery; evergreens (principally yews, junipers, and arborvitaes among needle types; azaleas, hollies, *Mahonia,* boxwood, and so on among broad-leafed types) constitute the chief segment of the nursery industry. Production of select forms of deciduous shrubs and shade trees has enjoyed rapid expansion in recent years, what with the native forest almost gone and needing replacement in the mushrooming subdivisions, and with roadside-beautification programs in full force. Because of their tremendous popularity, the production and sale of roses is almost equally as important, and it has been settled upon as a specialty by several national firms. The offering of vines, ground covers, bramble and tree fruits, and perennial "vegetables" (such as asparagus and rhubarb) rounds out the ornamental nursery trade.

Bulbs

In the horticultural trade the word "bulb" is used for corms, tubers, rhizomes, and swollen roots as well as for true bulbs. Thus tulips, hyacinths, narcissus, and even iris, daylily, and dahlia fall in this category. There is limited trade in nonhardy bulbs for planting indoors or outdoors in summer only. Included are amaryllis (*Hippeastrum*), *Anemone,* various tuberous begonias, *Caladium, Canna, Clivia, Dahlia, Freesia, Gladiolus, Hymenocallis, Montbretia, Ranunculus, Tigrida,* tuberose (*Polianthes*), *Zephyranthes,* and a few others. Hardy bulbs generally left in the soil outdoors include various crocuses (*Colchicum* and *Crocus*), *Fritillaria,* snowdrops (*Galanthus*), hyacinths, lilies, daffodils (*Narcissus*), *Lycoris, Scilla,* tulips, iris, and several others.

Many species of these genera are of Old World origin, introduced into horticulture long ago, subjected to unrecorded selection and crossing through the years to yield the many modern cultivars. One of the most popular is the tulip (*Tulipa*). Tulips are widely grown in gardens as botanical species, but they are especially prized as select forms of the garden tulip (which arose from crosses between thousands of cultivars representing several species). Garden tulips are roughly grouped as early tulips, breeder's tulips, cottage tulips, Darwin tulips, lily-flowered tulips, triumph tulips, Mendel tulips, parrot

tulips, and so on. The garden tulips seem to have been developed first in Turkey, but they were spread throughout Europe and were adopted enthusiastically by the Dutch. Holland has been the center of tulip breeding ever since the eighteenth century, when interest in the tulip was so intense that single bulbs of a select type were sometimes valued at thousands of dollars. Holland remains today the chief source of tulip bulbs planted in the United States; hundreds of thousands are imported annually. Holland has also specialized in the production of related bulbs in the lily family, and it provides tens of thousands of hyacinths, narcissuses, crocuses, and others to the United States market annually. The Dutch finance extensive promotion of their bulbs in the United States to support their market.

Years of meticulous growing are required to yield a commercial tulip bulb from seed. Thorough soil preparation, high fertility, constant weeding, and careful record-keeping are part of the business of intensive growing in Holland, which also employs much hand labor. Bulbs sent to market meet specifications as to size and quality, which assure at least one year's bloom even if the bulb (having been subjected to cold induction) is supplied with nothing more than warmth and moisture. The inflorescence is already initiated, and the necessary food is stored in the bulb. Under less favorable maintenance than prevails in Holland, a subsequent year's bloom may be smaller and less reliable; it is not surprising that tulip-bulb merchants suggest discarding bulbs after one year's performance and replanting with new bulbs. For featured display, such as at botanical gardens, this is the general practice, although there is no reason why the homeowner cannot derive several years of reasonable satisfaction from a single bulb planting if he allows maturation of the tulip foliage after flowering.

Bulbs generally grow best when planted fairly deeply (several times the vertical dimension of the bulb itself) in friable, loamy soil supplied with phosphatic fertilizer. Autumn, a few weeks before freezing weather, is the recommended planting time for tulips, hyacinths, daffodils, and crocuses. The bulbs have a chance to root and become established, and they will flower in spring as soon as the soil warms. After the foliage has died back it is often recommended that the bulbs be dug up and reset, a practice that is of some value for rearranging the planting pattern and determining which bulbs are in good condition, if nothing else. Some bulbs, such as daffodils and hyacinths, proliferate; digging them up provides additional stock.

Herbaceous perennials

Garden perennials include a number of herbaceous species grown for their flowers, or occasionally used as vegetative groundcovers. Under adequate growing conditions the plants of course persist and increase year after year. The biggest drawback to perennials as compared with annuals is that they must be maintained throughout the growing season, but they usually have only a limited flowering period. Typical perennials are hollyhock (*Althea rosea*), columbine (*Aquilegia*), bellflower (*Campanula*), *Chrysanthemum*, *Delphinium*, pinks (*Dianthus*), coralbells (*Heuchera sanguinea*), *Phlox*, poppies (*Papaver*), primroses (*Primula*), many speedwells (*Veronica*), and scores of others.

Perennials are often produced and sold as a sideline to other nursery activities; some are sold through seed dealers. Perennial production could be undertaken on a massive scale, with attendant economies, but the market is neither large enough nor predictable enough (except for the greenhouse growing of such cut flowers as chrysanthemums and carnations) to interest most growers. The homeowner himself can provide his long-term needs by dividing a few plants and perhaps exchanging cultivars with friends and neighbors. Perennials can of course be started from seed, but flowering is more immediate through vegetative propagation. Most perennials are planted in spring, kept properly weeded, fertilized, and watered. In northern climates, many varieties must be mulched in autumn to protect the plant from winter kill.

Shrubs

Production of ornamental shrubs is the backbone of the nursery trade, and it amounts to approximately half the dollar value of florist crops. The nursery industry is especially important in the Great Lakes and Middle Atlantic states and in Florida and California. In some of these areas, nursery sales (including sales of roses, shade trees, and small fruits for the home) account for nearly 40 percent of the retail trade in horticultural specialties. The nursery business is about equally divided among the production of (1) coniferous evergreens, such as yew, juniper, spruce, and pine; (2) broad-leafed evergreens, such as rhododendron, camellia, holly, and boxwood; (3) deciduous plants, such as *Forsythia*, *Viburnum*, barberries, privets, lilacs, and flowering vines; and (4) roses, which will be discussed separately.

Fields of specialization have evolved within the ornamental-shrub industry. Some firms confine their activity mostly to the production of "lining out" stock—seedlings and rooted cuttings for sale to growers for field plantings that must be tended for several years before the plants reach salable size. Such firms will generally offer not only the bare-root liners themselves, but small plants started in plant bands or slightly larger ones in containers. They may also offer small grafted plants of expensive or rare cultivars. Lining out stock of a typical evergreen, such as the Pfitzer juniper (*Juniperus chinensis* 'Pfitzerana'), is relatively inexpensive; it is more economical for a grower to purchase these ready to plant in the field than to undertake the specialized propagation himself.

The field grower may, in turn, specialize in mass growing for the wholesale trade only. The field plantings are tended until they attain marketable size, having been shaped and sheared as necessary, provided with pest control, irrigated, culled, and so on. Because of the time required to produce a marketable crop, and because of growing labor costs, this phase of the nursery industry can easily be caught in a profits squeeze, especially if landscaping preferences should change and there were no quick market for a field of mature shrubs. But wholesale growing escapes the high overhead of retail marketing in urban areas, and although many growers do sell retail at the nursery they generally avoid the expensive merchandising required of the typical urban-area garden center. Growers are especially interested in new technology that will lessen labor requirements, and they are turning to herbicidal control of weeds (as with dichlobenil) and short-cut methods for transplanting (to substitute for the traditional hand digging and burlapping).

Most shrubs are usually marketed **balled and burlapped** ("B and B"). The plants are dug with considerable soil left around the roots, which are wrapped with burlap or other suitable material. Recently a trend has developed toward **container-grown stock**—nursery stock grown in the container in which it is sold. This allows year-round sales of plant material. Deciduous shrubs may be marketed bare-root. There are obvious economies in not having to dig and wrap the plant carefully and to handle added weight and soil volume. Fairly satisfactory techniques have been developed for holding bare-root ornamental shrubs in viable condition until marketed. Usually this involves digging after leaf fall in autumn and storing the plant through winter in cold cellars with regulated humidity for spring sale. Unfortunately, the plants are often handled carelessly. The chief cause of loss during storage is desiccation, but respiration also plays a part when temperatures rise. But even if the plants are properly cared for during cold storage, they often suffer from warming and drying at the sales outlet. Bare-root plants ordered by mail often arrive in poor condition because of warming and drying in transit. It is apparent that the great effort spent to control humidity and temperature during winter storage can be defeated if there are no cold-storage facilities at retail outlets.

In recent years a breed of "nurseryman" that is more merchandiser than plantsman has taken over much of the retail trade. Exemplifying this change are the numerous diversified garden centers that sell all sorts of horticultural merchandise and maintain display yards of shrubs, trees, and roses purchased wholesale from large growers. Many such firms have associated landscaping servies, and offer the homeowner counsel, product, and service.

Roses

The production of roses is even more specialized than that of other shrubs; the rose grower often deals solely in rose plants. Whereas other shrubs are typically started from seed or cuttings on their own roots, most roses are bud-grafted onto rootstocks (typically of *Rosa multiflora*). This is the only way to achieve rapid and economical increase of a new selection to meet market demands. Large-scale production of roses has tended to center in the southwest (from Texas to southern California), where the nearly year-long growing season makes rapid production possible. Rose production is also carried on near the important consuming areas, such as northern Ohio.

Partly because of widespread promotion, and partly because of the great popular interest in the rose, the marketing of roses is nearly as important a part of the nursery industry as is the production of other kinds of shrubs. Mail order sales from attractive catalogs that picture the roses in color account for an appreciable part of the market. Enthusiasm for new selections is heightened by an elaborate scheme for testing and selecting "All-America" roses (only a few roses receive the coveted award each year) and by a strong national rose society with headquarters in Shreveport, Louisiana, that encourages rose clubs and sponsors rose shows throughout the nation. Each year accredited judges compile and issue a rating sheet that ranks all available named cultivars on a numerical performance scale. Thus even the neophyte can determine readily which have been the most successful named roses through the years, and purchase his 'Peace' or 'Tropicana' on the basis of authoritative testing rather than just catalog hearsay.

Because the budding operation calls for skilled hand labor, and because field maintenance is expensive, few economies can be practiced in the production of roses. But techniques have been developed for the distribution of roses that do offer certain economies. In the mail-order handling of bare-root roses, coated paper or plastic bags are now used to retain humidity, in contrast to the former practice of wrapping the roots in damp moss. For mass marketing through chain outlets, the stems of dormant stock are coated with wax to prevent desiccation during prolonged storage and handling, and the root system is "potted" in a moss-filled carton on which the name and a picture of the rose appear. Although much stock held dormant for long periods is not of first quality, this method of penetrating the mass market is well established in the nursery trade today.

Trees

Ornamental shade trees are usually grown and marketed in conjunction with shrubs. Deciduous trees and shrubs together have a somewhat greater value than the rose crop. While the value of fruit trees around the home has been recognized since colonial times, only in recent decades has there been much of a market for nursery-grown shade trees. Again, as part of the migration to the suburbs in the middle of the twentieth century and the construction of houses on cleared land, shade trees have become an increasingly important part of the nursery picture. The growing of shade trees was especially stimulated by the interstate highway program, which, for the first time, undertook the landscaping of highway rights-of-way rather than merely protecting them from erosion. As interest in shade and ornamental trees increased, creation of improved cultivars followed. There is still some activity in transplanting native trees from the woodlot, and some trees are still grown from genetically unselected seed or cuttings. But more and more trees, like roses and shrubs before them, are being offered as named cultivars and as patented horticultural varieties when grafted, budded, or propagated vegetatively.

There are some obvious drawbacks to the tree-growing part of the nursery industry. The crop takes a fairly long time to reach marketable size, and trees not saleable within a year or two become too large for convenient handling. The investment in a tree is high, and consequently so is the price. Risks are entailed, too, in trying to anticipate what the market will be for a crop planted today but not usable for five or six years. Yet so indispensable is an ornamental tree to attractive landscaping that sales in suburban communities have been quite rewarding. Some mortgage lenders even require landscaping, including shade trees, for new homes (Fig. 23-28). There has been a notable trend toward the use of small trees in small yards and under utility lines, especially trees that bear colorful flowers or fruits, such as crab apples, redbuds, dogwoods, hawthorns, and magnolias. Larger shade trees have been selected for shape and form, color of foliage, and other characteristics, so that there are available both spreading and upright cultivars of hard maple, ash, linden, and other favorites. Thornless honeylocust, of a type whose new foliage is yellow rather than green; dense-growing Scotch and Austrian pines; and once rare introductions, such as *Ginkgo* and dawn redwood (*Meta-*

sequoia glyptostroboides), can readily be had today, although only a generation ago selection was limited and trees came as often as not from some rural woodlot.

The digging, handling, and transporting of trees have always been difficult. Except in the sapling stage, trees are seldom sold bare-root. A good deal of experienced labor is needed to dig a tree and wrap its root system in burlap, the traditional way of moving larger trees. Through the years, however, special machines with slings and hoists have been developed for handling large trees. A powered tree spade is now widely used, first to dig a hole where the tree is to be planted, then to dig the tree so that the root ball is the exact size of the previously dug hole, and finally to transport the tree to the hole. Thus there is no need for wrapping or separate handling. Such a machine makes it possible to move relatively large trees easily, but of course the tree must be contracted for in the field rather than selected from a nursery display.

Whether indigenous or transplanted from the nursery, trees about the modern home usually encounter an unnatural environment. Indigenous trees usually suffer damage to the roots from grading and compaction during house construction. The root systems of indigenous trees are often buried by soil from basement excavation. In order to save such trees, dry wells are built around the trunks, but this is by no means an ideal solution. The leaves of deciduous trees are raked rather than allowed to decompose in situ to recycle nutrients, although in most yards there is compensatory fertilization from the lawn. Because of size, lack of familiarity, and less understanding of the physiology of trees than of herbaceous plants, trees are often poorly tended. For example, when trees are pruned, instead of merely removing dead, moribund, or interfering branches, they are often indiscriminately pollarded or topped, regardless of their natural shape or growth habit. When branches are removed, the base is often left as a stub, preventing healing of the scar. In urban and industrial areas, air pollution may be serious enough to damage trees, especially many of the needle evergreens, some of which simply cannot survive in cities with serious air pollution. Shade trees seem ever more susceptible to

Figure 23-28   Landscaping has become an integral part of the home-construction business. [Courtesy Ford Motor Co.]

"decline" from uncertain causes, such as the wilt of oaks and the debilitation of hard maples, and to complete loss to devastating diseases, such as the Dutch elm disease and chestnut blight. Because of their size, trees are especially difficult to treat for disease; spraying is difficult with the usual home equipment, and it is expensive to hire tree specialists to do the job.

Floriculture

The growing of plants in the greenhouse for cut flowers, potted plants, and greenery has long been an important industry (especially in Europe), and even today it accounts for about half of the horticultural-specialties market. It is a competitive and sophisticated business, requiring both technical skill and appreciable capitalization. Maintenance of a greenhouse is itself a demanding occupation. Even with the best of greenhouses, the grower must choose his specialty crop in light of local demand, provide efficient operating conditions, and be constantly alert against infestations of pests, heat failure, or other inadvertencies that could doom his crop overnight. Exact scheduling must be arranged, because many floral crops are only seasonally in demand (such as poinsettias at Christmas, lilies at Easter, and roses and carnations on Mother's Day and other occasions). Because the keeping quality of the crop is limited, established trade channels or auctions must function smoothly for quick disposal.

The chief cut flowers are roses, carnations, chrysanthemums, gladioluses, orchids, and gardenias. Some of these, such as gladioluses and orchids, are produced outdoors in the south and in Hawaii, and are shipped to market by air freight. Potted flowering plants include hydrangeas, poinsettias, lilies, African violets, cyclamens, and others. Foliage plants include ferns, palms, ivy, *Philodendron*, and various succulents. Bedding plants, such as begonias, geraniums, *Vinca*, and so on, may be started in pots or plant bands and sold to homeowners for planting out. Seasonal annuals such as petunias, marigolds, salvias, and home garden vegetables may be started

Figure 23-29 Substances that reduce the height of ornamental flowering plants without reducing flowering time or flower size are finding promise in the production of florist crops. The chrysanthemums on the left were dwarfed with foliar applications of succinic acid 2,2-dimethylhydrazide (SADH, or daminozide). [Courtesy Uniroyal.]

indoors for sale to the homeowner as seedlings. In most instances the grower will concentrate on but a few such items, arranging his seasonal schedule in such a way that valuable greenhouse space will not stand idle.

As in the nursery business, the floriculture trade involves both the grower, who mass-produces flowers for the wholesale market, and the retail florist, who markets to the public and includes such services as delivery, floral arrangements, and custom orders for weddings, funerals, and various other occasions. As is characteristic of other industries in the horticultural-specialties field, the retail florist is perhaps more an expert in business than in floriculture. Purchased cut flowers are the most important florist's product, since they have about three times the value of potted plants. Among potted plants, the foliage types rank first. There is also some greenhouse growing of vegetables (especially tomatoes), and mushrooms in specialized habitats.

While it would seem to be impossible to automate greenhouse operations completely, the production of plants under glass is constantly becoming more complex technically. Soil sterilization and proper hygiene in the greenhouse (insect and disease control with fumigants and pesticides) are now taken for granted. Optimum fertilization, pH, temperature, water regimen, and illumination have been determined for most crops, and they must be understood by the grower. Where appropriate, hydroponic culture can be practiced, or at least subirrigation of the growing bed. Automatically timed misting is commonplace today in beds for rooting cuttings.

The modern floriculturist must have training and practical experience in chemistry, plant physiology, and related subjects; the occupation today demands a trained craftsman, not an amateur. In the era that lies ahead, the grower will probably be capable of controlling plant growth even more precisely than is possible now. Already chemical growth regulators determine shape and size (Fig. 23-29), rate of production, and even disbudding (in chrysanthemums). Some growers already enrich the greenhouse atmosphere with carbon dioxide to increase luxuriance of growth. Greenhouse floriculture is in many ways the most sophisticated part of crop agriculture.

Selected references

Anderson, J. F., and H. K. Kaya (editors), 1976. *Perspectives in Forest Entomology*. New York: Academic Press. (Diverse presentations are brought together concerning subject matter widely scattered in the literature.)

Bailey Hortorium staff, 1976. *Hortus Third*. New York: Macmillan. (A comprehensive encyclopedia, the "bible" of horticulturists for reference to genera, species, and certain subject areas; a modern update of the landmark publications by L. H. Bailey.)

Bernatzky, A., 1978. *Tree Ecology and Preservation*. Amsterdam: Elsevier. (After a brief account of the interrelations of trees and civilization, morphology, physiology, and environmental considerations are covered. Later chapters deal chiefly with tree performance in altered environments, with reviews of research from Germany, the United States, and Great Britain.)

Bormann, F. H., and G. E. Likens, 1979. *Pattern and Process in a Forested Ecosystem*. New York: Springer-Verlag. (Based upon the extensive studies at Hubbard Brook, New Hampshire, this presentation integrates various ecological concepts and their influence upon woodland productivity.)

Braun, E. L., 1967. *Deciduous Forests of Eastern North America*. New York: Hafner. (A reprinting of Lucy Braun's definitive ecological examination of the eastern North American forests, first published in 1950.)

Burley, J., and B. T. Styles (editors), 1976. *Tropical Trees: Variation, Breeding and Conservation*. New York: Academic Press. (Wide-ranging discussions from a symposium held at Oxford University, concerned with the heavy demand being placed on forests throughout the world. Interdisciplinary collaboration is stressed, and a case-study approach used. Emphasis is on breeding systems and genetics.)

Chapman, A. G., and R. D. Wray, 1979. *Christmas Trees for Pleasure and Profit*, 2nd ed. New Brunswick, N.J.: Rutgers University Press. (A look at this specialized field, a sideline of conventional forestry.)

Cooke, G. B., 1961. *Cork and the Cork Tree*. London: Pergamon Press. (Perhaps more than you will want to know about all aspects of cork and its production in the Mediterranean area.)

Dempsey, J. M., 1975. *Fiber Crops*. Gainesville, Fla.: University of Florida Press. (One of the few domestic discussions relating to this vast subject.)

Duerr, W. A., D. E. Teeguarden, N. B. Christiansen, and S. Gutterberg, 1979. *Forest Resource Management*. Philadelphia: Saunders. (What is almost a case-study approach is used in weighing economical considerations of forest management.)

Eckholm, E., 1979. *Planting For The Future: Forestry For Human Needs*. Washington, D.C.: Worldwatch Institute. (A stimulating booklet exhorting the world to pay more attention to forestry in the Third World, and citing the general dire consequence to be faced if improvement does not come.)

Edlin, H. L., and M. Nimmo, 1978. *The Illustrated Encyclopedia of Trees: Timbers and Forests of the World*. New York: Harmony Books. (A "quick look" at the world of trees.)

Edlin, H. L., 1976. *Trees and Man*. New York: Columbia University Press. (A book for the layman, discussing the basics of tree growth and forestry practices. A well-written introduction by an experienced British forester.)

Forbes, R. D., 1976. *Woodlands for Profit and Pleasure*, 2nd ed. Washington, D.C.: American Forestry Association. (A rather popularly presented review of forest basics designed for the small property owner.)

Fritts, H. C., 1977. *Tree Rings and Climate*. New York: Academic Press. (A complete and authentic summarization of dendroclimatology, from the center for such investigations at the University of Arizona.)

Haden-Guest, S., J. K. Wright, and E. M. Teclaff (editors), 1956. *A World Geography of Forest Resources*. New York: Ronald Press. (A partial replacement for the classical Zon and Sparhawk, *Forest Resources of the World*, now outdated; 35 authorities discuss species, ecology, and products, providing a general picture of world forests.)

Hannebaum, L., 1980. *Landscape Operations*. Reston, Va.: Reston. (A discussion of the landscape-nursery business, with emphasis on its management, including business aspects.)

Harlow, W. M., E. S. Harrar, and F. M. White, 1979. *Textbook of Dendrology*, 6th ed. New York: McGraw-Hill. (An authoritative work on American tree species.)

Harlow, W. M., 1975. *Inside Wood*. Washington, D.C.: American Forestry Association. (A close look at the basic substance that trees have provided since time immemorial.)

Hocker, H. W., Jr., 1979. *Introduction to Forest Biology*. New York: Wiley. (The book discusses forest trees at length, their reproduction, evolution, growth, succession, and classification. It touches on basic background, biological influences, and the importance of site.)

Kaul, R. N. (editor), 1970. *Afforestation in Arid Zones*. The Hague: Junk. (With arid or semiarid climates covering about one-third of the world's land surface, this publication has widespread pertinency. Coverage includes the Mediterranean and northern Africa, the Near East, Central Asia and parts of India, western parts of the Americas, and Australia.)

Kirby, R. H., 1963. *Vegetable Fibres*. London: Leonard Hill. (This technical monograph in the "World Crops" series deals in detail with a large number of important fiber crops excepting cotton.)

Little, E. L., Jr., 1975, 1976, and 1977. *Atlas of United States Trees*: vol. I, *Conifers and Important Hardwoods* (USDA Misc. Pub. 1146), vol. II, *Alaska Trees and Common Shrubs* (USDA Misc. Pub. 1293); vol. III, *Minor Western Hardwoods* (USDA Misc. Pub. 1314); vol. IV, *Minor Eastern Hardwoods* (USDA Misc. Pub. 1342). Washington, D.C.: U.S. Department of Agriculture, Forest Service. (This series provides an atlas with little text and many maps, showing the geographical distribution of important North American trees and shrubs. Indices of both common and scientific names are included.)

Lock, G. W., 1969. *Sisal*, 2nd ed. New York: Longmans. (An important, comprehensive treatment.)

Loewus, F. (editor), 1973. *Biogenesis of Plant Cell Wall Polysaccharides*. New York: Academic Press. (Not elegantly printed and with a number of errors, this book is nonetheless useful in exploring the macromolecules of plant cell walls. The chapters are uneven, some being highly technical, others presenting an overview. The results of a symposium held by the American Chemical Society.)

Longman, K. A., and J. Jenik, 1974. *Tropical Forest and Its Environment*. London: Longmans. (Another of the thorough-going treatments in the "World Crops" series, this book discusses tropical forests and the stresses to which they have been subjected in recent times.)

Luckwill, L. C., and C. V. Cutting (editors), 1970. *Physiology of Tree Crops*. New York: Academic Press. (Proceedings from a symposium held in England, whose opening presentations deal with several aspects of forestry, and whose later ones treat mainly with such crop species as apple, coffee, oil, palm, tea, rubber, and grape.)

McDaniel, A. R., 1979. *Ornamental Horticulture*. Reston, Va.: Reston. (Explores floriculture, nursery, turfgrass and landscaping opportunities, including greenhouse operation, crop production, maintenance, and design.)

Mirov, N. T., 1967. *The Genus Pinus*. New York: Ronald Press. (An "in-depth" treatment of one of the most important forest genera: paleo-record, geography, genetics, morphology, physiology, ecology, chemistry, and taxonomy.)

Panshin, A. J., Carl De Zeeuw, and H. P. Brown, 1964. *Textbook of Wood Technology*, 2nd ed., vol. I. New York: McGraw-Hill. (The first volume of this two-volume work deals with structure identification, uses, and properties of the commercial woods of the United States; the second volume, which was written by H. P. Brown, A. J. Panshin, and C. C. Forsaith and published in 1952, deals with the physical, mechanical, and chemical properties.)

Poulsen, G., 1978. *Man and Tree in Tropical Africa*. Ottawa: International Development Research Center. (Impoverishment of lands in Africa due to deforestation is extensive, with overuse of trees for charcoal, firewood, and cattle feed causing deterioration of ecosystems. This book effectively calls attention to the plight.)

Prentice, A. N., 1972. *Cotton*. New York: Longmans. (A definitive review of the world's most important fiber crop.)

Preston, R. J., 1978. *North American Trees*, 3rd ed. Cambridge, Mass.: MIT Press. (One of the standard reviews of familiar forest trees.)

Reader, J., and H. Croze, 1978. *Pyramids of Life*. Philadelphia: Lippincott. (A nonspecialist's book about East African forests, nicely balancing excellent illustrations with summarizing text to provide an eminently readable book.)

Richards, P. W., 1979. *The Tropical Rainforest*. New York: Cambridge University Press. (An important republication of an earlier treatment, in an economical paperback edition.)

Sarkanen, K. V., and C. H. Ludwig (editors), 1971. *Lignins: Occurrence, Formation, Structure and Reactions*. New York: Wiley. (A definitive text for the first time sifting widely disbursed literature on the subject to provide a good overview.)

Sharpe, G. W., S. W. Allen, and C. W. Hendee, 1976. *An Introduction to Forestry*, 4th ed. New York: McGraw-Hill. (A teaching text, useful for broad coverage and background.)

Stainton, J. D. A., 1972. *Forests of Nepal*. New York: Hafner. (Chiefly of interest for examining a "corner of the world" where forests are so fast disappearing that most will be exhausted in the next decade.)

Tomlinson, P. B., and M. H. Zimmermann (editors), 1978. *Tropical Trees as Living Systems*. New York: Cambridge University Press. (This book, a product of the fourth Cabot Symposium held at Harvard Forest, is important especially in view of the degradation tropical forests are now undergoing. The botany of tropical forests is emphasized, but ecological influences are not neglected.)

Whitmore, T. C., 1975. *Tropical Rain Forests of the Far East*. Oxford: Clarendon Press. (A useful presentation about an area unfamiliar to most Westerners.)

Wilkins, A. H., 1978. *Ten Million Acres of Timber: Forest Production in the Maine Forestry District (1909–1972)*. Woolwich, Maine: TBW Books. (After a lifetime of service in the Maine Forest Service, the author is well equipped to trace the history of this forest district, with emphasis on the practical working relationships.)

Vankat, J. L., 1979. *The Natural Vegetation of North America*. New York: Wiley. (Up-to-date background on vegetational geography of North America.)

Zacharin, R., 1978. *Emigrant Eucalypts: Gum Trees as Exotics*. Melbourne: Melbourne University Press. (*Eucalyptus*, one of the most productive tree genera, now widely dispersed from its Australian homeland, is reviewed for its impact around the world.)

Zohary, M., 1973. *Geobotanical Foundations of the Middle East*. Stuttgart: Gustav Fischer. (Professor Zohary of the Hebrew University, Jerusalem, an expert on the Middle East flora, provides transects and describes various tree-and-shrub associations.)

Six
THE MARKETPLACE

24

The Economics of Crop Production and Distribution

The food and fiber industry encompasses a wide range of activities and interactions, including those involved in manufacturing and distributing the industrial inputs used in farming; the farm production of crops, animals, and animal products; the processing of these materials into finished products; the provision of products at a time and place demanded by consumers; and an extensive set of interactions with government programs and agencies (Fig. 24-1).

In the low-income developing countries a large share of the total value added to the national product by the food and fiber industry is contributed by farm production. Farmers in poor countries use limited quantities of industrial inputs, such as fertilizer, pesticides, and machinery. The value added in marketing, by processing and distribution, is low. A high percentage of farm output is consumed at the farm or in villages, near the source of production. Only a small share of the final output reaches urban consumers in a packaged or processed form.

In the high-income countries, farm-level production makes up a much smaller share of the value added to national output by the food and fiber industry. The purchased industrial inputs represent a larger share of the expenses incurred in farm output. Farm produce itself constitutes a relatively small share of the total value of food and fiber production. Consumers want a wide variety of fresh and processed foods throughout the year. And they demand that these products be packaged, frozen or dehydrated, and processed for easy preparation in the home. Both the farm and the household supply a smaller proportion of the value of the total output of the food and fiber industry than in the past, or than in today's less developed countries.

In 1978, sales of final products by the food and fiber industry amounted to $448 billion, or about one-fourth of the United States gross national product. Farm-level production, the gross value of farm output, amounted to $118 billion, or just slightly more than 26 percent of the value of output by the entire food and fiber industry (Fig. 24-2).

Economic decisions are involved at every phase of production, marketing, and utilization in the food and fiber industry. Decisions by producers concerning what to produce, how much to produce, and what inputs to use depend both on technical input–output ratios and on the prices of inputs and products. When consumers decide what to consume and how much to consume, they are influenced not only by their tastes and preferences but also by their incomes and the relative prices of the products avail-

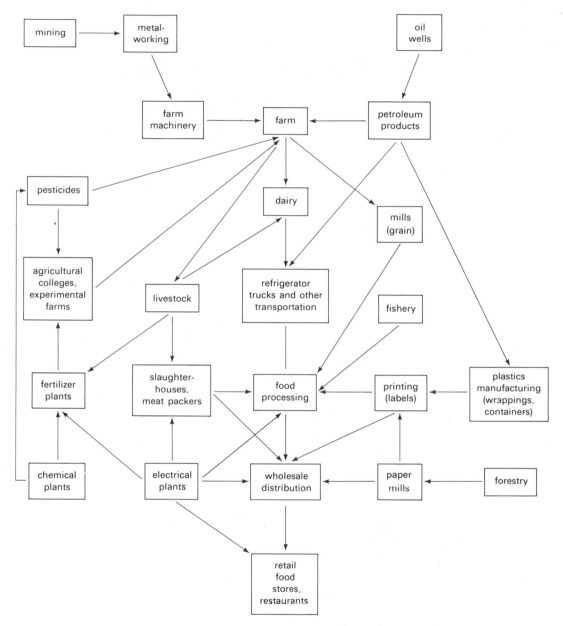

Figure 24-1 "Agribusiness," the entire food industry of the United States, is intricately interwoven with many sectors of the economy and encompasses far more than farming. The flow chart gives a visual impression of the interaction of farming and other activities. On the input side of the farm are industries that supply such items as farm machinery and fertilizer; on the output side of the farm is the food-processing industry. [From E. O. Heady, "The Agriculture of the U.S.," copyright © 1976 by Scientific American, Inc. All rights reserved.]

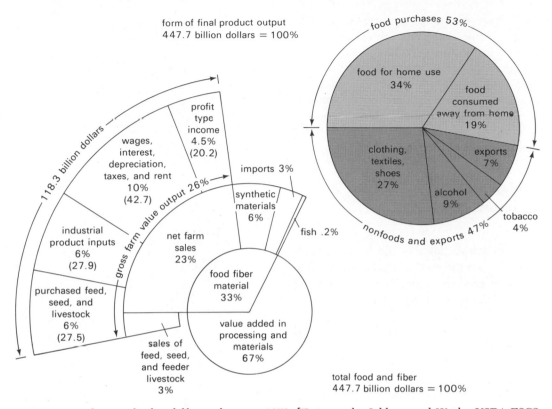

Figure 24-2   The U.S. food and fiber industry in 1978. [Estimates by Schluter and Weeks, USDA ESCS, September 1979.]

able in the market. Decisions by the marketing, processing, and distribution firms are influenced by the decisions of the crop producer and the consumer and by the prices associated with these decisions.

ECONOMICS OF CROP PRODUCTION

Crop-production decisions fall into three general categories:°

1. What combination of inputs should be used in producing a particular crop? For exam-

    ple, should a straight nitrogen fertilizer or some nitrogen-phosphate-potash formulation be used on corn? What combination of intensity of land preparation and cultivation, herbicide use, and hand labor should be used to control weeds?

2. How much of each input should be used? For example, should 120 or 160 kilograms of nitrogen per hectare be applied on corn? How many applications of irrigation water should be used on tomatoes?

3. What combination of crops, or of crops and livestock, should be produced? How many hectares should be planted to corn? How many hectares to soybeans? Should feed grain be sold in the market or fed to livestock?

°The authors are indebted to Randolph Barker of the International Rice Research Institute (Philippines) and to Martin Christiansen, Willard Cochrane, and Earl Fuller of the University of Minnesota for assistance in the preparation of this section.

To answer these questions, two types of information or data are needed. Data are needed, either from records or experiments, on the effects of decisions to change input combinations and/or the product mix. It is also necessary to know as accurately as possible the prices of the inputs and of the products that are affected by such decisions. Failure to evaluate the technical or economic data accurately leads to inefficient production decisions.

Partial budgeting of costs and returns

When these data are available or can be obtained, it is possible to calculate the costs, the gross returns, and the net gain from production decisions. The simplest procedure for making such calculations is a technique called **partial budgeting.** Partial budgeting focuses attention only on those inputs, products, and prices that are expected to change as a result of a particular decision. This is in contrast to the more detailed procedure of **whole-farm budgeting,** which is frequently used to summarize costs and returns to the total farm operation over a particular time period, such as a year. In the generalized partial budget format presented below, any change in use of inputs or in the amount or mix of products can be classified under one of the four headings.

Partial budget of estimated change
in annual net cash farm income
from a change in resource use

---

1. Added costs _____   3. Added returns _____

   _____   _____

2. Reduced returns   4. Reduced costs

   _____   _____

   _____   _____

Subtotal A _____   Subtotal B _____

Estimated change (B − A) _____

The simplicity of partial budgeting has made it a valuable planning tool for agricultural-research, mar-

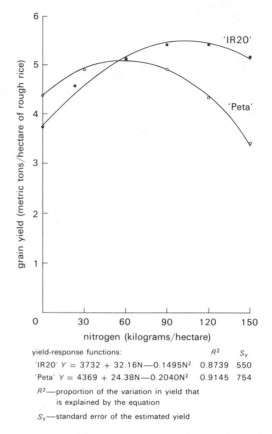

yield-response functions:
R^2 S_y

'IR20' $Y = 3732 + 32.16N - 0.1495N^2$ 0.8739 550

'Peta' $Y = 4369 + 24.38N - 0.2040N^2$ 0.9145 754

R^2—proportion of the variation in yield that is explained by the equation

S_y—standard error of the estimated yield

Figure 24-3 Yield response of two cultivars of rough rice to nitrogen at the Maligaya Experiment Station (Philippines) during the 1970 dry season. [Data from IRRI, March 1972.]

keting, and extension organizations, as well as for producers. The applied research organization needs to select, from among the preliminary results obtained by its research staff, those lines of research that should be developed and tested more completely. To conserve time and effort, certain alternatives must be eliminated because they lack scientific, technical, or economic interest. After completing more intensive development and testing, private research organizations must decide which results should be incorporated into new products and placed on the market. Public research agencies need to decide which results should be recommended to farmers or to agencies working with farmers. Even after

gross returns (subtotal *B*)

total	$0.00	$58.44	$98.44	$118.88	$120.88	$103.88
change		$58.44	$40.00	$20.44	$2.00	$ −17.00

costs (subtotal *A*)

total	$0.00	$16.48	$29.89	$40.04	$47.11	$53.85
change		$16.48	$13.41	$10.15	$7.07	$6.74

net return (*B* − *A*)

total	$0.00	$41.96	$68.55	$78.84	$73.77	$50.03
change		$41.96	$26.59	$10.29	$ −5.07	$ −23.74

Figure 24-4   Cumulative costs and returns from fertilizing the rice cultivar 'IR20' during the dry season (based on fertilizer response reported in Fig. 24-3). X = optimum fertilization level. [Data from IRRI, March 1972.]

rather thorough screening by researchers, extension workers or distributors may frequently find it desirable to conduct trials and to budget the costs and returns in their own country or market area.

Determining how much fertilizer to use

Use of partial budgeting to determine how much fertilizer to use is illustrated in Figures 24-3, 24-4, and 24-5 and Tables 24-1 and 24-2. Figure 24-3 presents data on the relations between inputs of nitrogen and yield of rough rice from experiments conducted at the Maligaya Experiment Station in the Philippines by the Philippines Bureau of Plant Industry and The International Rice Research Institute. Table 24-1 presents partial budgets showing the change in costs,

Figure 24-5   Cumulative net return from fertilization at alternative prices for rough rice (based on rice 'IR20' fertilizer response reported in Fig. 24-3). X = optimum fertilization level. [Data from IRRI, March 1972.]

gross returns, and net returns from increasing the level of nitrogen from 30 to 60 kilograms per hectare (27 to 54 pounds per acre).[*] Table 24-2 shows similar changes when the level of nitrogen is increased from 60 to 90 kilograms per hectare.

The results of a larger series of partial budgets are presented in Figure 24-4. Note that the net return curve (*B* − *A*) reaches its maximum while both the gross-return curve (*B*) and the cost curve (*A*) are still rising. (Geometrically, the tangent to the net-return curve is horizontal at the point where the tangent to the gross-return curve and the cost curve are parallel.)

---

[*]A useful rule of thumb is that one metric ton (1000 kilograms) per hectare is roughly equivalent to 1000 pounds per acre. One kilogram is equivalent to 2.2046 pounds; one hectare is equivalent to 2.471 acres. One kilogram per hectare is, therefore, equivalent to approximately 0.89 pounds per acre. See appendix for more detailed conversion factors.

Table 24–1 Application of 60 kilograms of nitrogen versus 30 kilograms of nitrogen to rice "IR20" during the 1970 dry season, Maligaya experimental station, Philippines.

1.	Added costs		3.	Added returns	
	Materials[1]	$ 6.00		Change in rough rice produced[6]	$40.00
	Interest[2]	0.36			
	Equipment[3]				
	Application[4]	0.38			
	Harvesting[5]	6.65			
2.	Reduced returns		4.	Reduced costs	
	Change in rough rice produced	———			———
	Subtotal A	$13.39		Subtotal B	$40.00
				Estimated change $(B - A)$	$26.61

[1]30 kg of nitrogen at $0.20/kg ($4.53/50 kg bag of urea containing 45% N).
[2]6% for 6 months.
[3]Broadcast, no equipment cost.
[4]Broadcasting: 4 additional hours of labor/hectare at $0.094/hour = $0.38.
[5]Harvest cost (at 1/6 of 571 kg), 95 kg valued at $0.07 per kg = $6.65 ($1.00 = $6.40 pesos).
[6]Increased production of 571 kg valued at $0.07 per kg ($3.12/cavan of 44 kg = $40.00.

Table 24–2 Application of 90 kilograms of nitrogen versus 60 kilograms of nitrogen to rice "IR20" during the 1970 dry season, Maligaya experimental station, Philippines.

1.	Added costs		3.	Added returns	
	Materials[1]	$ 6.00		Change in rough rice produced[6]	$20.44
	Interest[2]	0.36			
	Equipment[3]				
	Application[4]	0.38			
	Harvesting[5]	3.43			
2.	Reduced returns		4.	Reduced returns	
	Change in rough rice produced	———			———
	Subtotal A	$10.17		Subtotal B	$20.44
				Estimated change $(B - A)$	$10.27

[1]30 kg of nitrogen at $0.20/kg ($4.53/50 kg bag of urea containing 45% N).
[2]6% for 6 months.
[3]Broadcast, no equipment cost.
[4]Broadcasting: 4 additional hours of labor/hectare at $0.094/hour = $0.38.
[5]Harvest cost (at 1/6 of 292 kg), 49 kg valued at $0.07 per kg = $3.43 ($1.00 = $6.40 pesos).
[6]Increased production of 292 kg valued at $0.07 per kg ($3.12/cavan of 44 kg = $20.44.

Net returns are, in this example, quite sensitive to the price of rough rice. Figure 24-5 presents net-return curves obtained by budgeting the yield-response data of Figure 24-3 at three alternative price levels for rough rice. Note also that the most profitable amount of fertilizer to apply per hectare rises as the price of rough rice rises relative to the price of nitrogen fertilizer.

Net returns are, in this example, also highly responsive to the tenure arrangements under which the crop is produced. Figure 24-6 compares the farmer's net return from the use of fertilizer under different tenure arrangements. When the landlord shares the cost of the fertilizer and the increased production with the tenant, the rate of return on the tenant's fertilizer investment is the same as that of the owner-operator, but the absolute size of his return is reduced. When the tenant pays the full cost of the fertilizer and shares the increase in output with the landlord, the tenant's rate of return is reduced, and the absolute size of his return is lowered even further. The effect is to further reduce the tenant's incentive to use fertilizer.

Other elements also enter into even a simple decision, such as how much fertilizer to use. Some uncertainty regarding the yield response to fertilizer always exists. Even under carefully controlled experimental conditions, some of the observations lie well off the calculated response curve (see Fig. 24-3). Under field conditions, variations in weather or attacks by insects or pathogens may result in significant departure from expected response levels. Under these circumstances, a farmer having a strong aversion to risk, or whose access to credit is inadequate, may use less fertilizer than the calculated "optimum." Decisions that affect the use of farm resources over the course of a long period of time, such as the acquisition of additional land or investment in equipment and buildings, are even more difficult because of the greater degree of risk and uncertainty. In recent years, economists have developed and implemented rather elaborate computer programs for use by farmers in these more complex decision processes. The basic principles employed and the information requirements are, however, essentially similar to those

Figure 24-6   Implications of tenure on net returns and optimum level of fertilization of rough rice (based on fertilization response curve for rice 'IR20' reported in Fig. 24-3 and a price of $70/metric ton). X = optimum fertilization level. [Data from IRRI, March 1972.]

that have been illustrated in the partial-budgeting analysis of decisions on fertilizer use (Figs. 24-3, 24-4, and 24-5; Tables 24-1 and 24-2).

MARKETS AND PRICES

Prices play an important part in guiding farmers' decisions about how much of each input to use, what combination of inputs to use, and how much of each product to produce.

If production is to be organized so that farm commodities are produced in the proportions desired by consumers, some mechanism must be available for transmitting information about consumer demand to producers. Similarly, some mechanism is needed to make consumers aware of differences in the costs incurred by farmers in producing different commodities, so that consumption choices can be made in a manner that enables individual consumers, and society as a whole, to obtain the greatest satisfaction from the use of individual income and from the use of national resources.

The role of prices
in an exchange economy

In an exchange economy, this function is performed by market prices. When prices are not regulated, either by government intervention or by private economic power, they function as a thermostatic or feedback device to achieve a balance between production and consumption.

If production of a product such as soybeans fails to expand as rapidly as the demand for soybean meal and oil, the price of soybeans will rise. As a result, farmers will find it more profitable to use their land and other inputs to produce soybeans rather than other crops, such as corn. At the same time the rise in the price of soybeans will cause users of soybeans to substitute other sources of protein feed and vegetable oil for soybean meal and oil. The effect of these two simultaneous responses to the initial rise in the price of soybeans—the stimulation of production and the curtailment of consumption—acts to bring about a new balance between production and consumption.

Demand and supply curves

The functioning of the supply–demand thermostat, with its system of information feedback through prices, is illustrated graphically in Figure 24-7. In this figure, curve SS is referred to as a supply curve, or supply function. It describes the quantities of a particular commodity, such as soybeans, that farmers as a group are willing to produce and offer for sale at different prices during a particular time period, such as a crop year. It may shift to the left or to the right in response to changes in the availability of resources or the level of technology in the region or country to which the supply curve applies. It rises to the right because, in order to produce and market larger quantities of a particular commodity, it is necessary to bring inferior land into production, or to compete with other crops such as corn for labor, capital equipment, credit, and other inputs, thus forcing production costs upward.

The curve DD is referred to as a demand curve, or

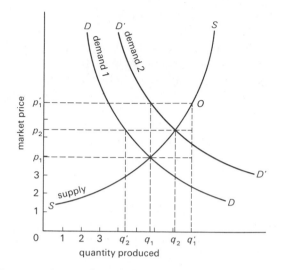

Figure 24-7 Supply and demand functions illustrating the effect of a shift in demand on price and output.

demand function. It describes the quantities of a commodity, such as soybeans, that buyers as a whole are willing to purchase at different prices during a particular time period. Its position on the chart is determined by such factors as the level of income and the size of the population of the region or country to which the curve applies. For many commodities, international demands are also important. The steepness with which the demand curve slopes downward to the right depends on the availability of the commodity and the willingness of users to substitute other commodities for the particular commodity in question as prices change. If there are no close substitutes, or if users are unwilling to use substitutes because of rigid preferences, the curve will have a steep slope: small changes in the amount supplied will be accompanied by large changes in price. If close substitutes are readily available, and are acceptable to users, the curve will be relatively flat: relatively large changes in the amount supplied will be accompanied by small changes in price. With supply and demand represented by DD and SS, the quantity supplied and the price that would just clear the market—that is, the price that would balance farmers willingness to produce and the consumers willingness to buy—can be represented by q_1 and p_1.

The market can be forced out of adjustment if, as a result of some combination of foreign demand, population growth, and increase in income, the demand curve shifts to $D'D'$. It could be said to be out of adjustment, or in a disequilibrium state, because, while producers would be willing to produce the quantity q_1 for a price of p_1, consumers would be willing to pay a price of p'_1 if only the amount q_1 were supplied. On the other hand, if consumers would be willing to pay a price of p'_1, producers would be willing to produce and market the amount q'_1.

This is the point at which prices begin to function as "feedback" devices. As prices begin to rise toward p'_1, producers begin to expand production. If the price system functions with great precision, the system will rapidly shift to a new "equilibrium" that just balances producer and consumer interest with an amount q_2 produced and marketed at a price p_2. With supply and demand represented by $D'D'$ and SS, the quantity supplied and the price that would just clear the market—the price that would balance the farmers' willingness to produce and the consumers' willingness to buy—can be represented by q_2 and p_2. The day-to-day fluctuations that are observed in cash and futures markets reflect the efforts of market participants to "discover" the longer-run equilibrium prices. If the demand–supply feedback system is not highly sensitive or efficient—if it works like an unreliable thermostat—prices and quantities may fluctuate around the equilibrium values of q_2 and p_2 to produce a "cobweb" pattern of price and output.

Spillover effects, externalities, and market failure

The importance of an effective supply–demand feedback mechanism for efficient market behavior can not be overemphasized. When the value that society places on the inputs used in production or on final products is not accurately reflected in the market, the production decision that is "efficient" for the producer—that maximizes his return—may become socially "inefficient," and thereby suboptimal for the entire economy.

The failure of society to reform the market system so that it accurately reflects social values is one of the causes of the excessive environmental impact of many agricultural production practices. Technical externalities or "spillover" effects that result in a divergence between private and social cost are an important cause of decisions that, although rational from the point of view of an individual producer, may result in excessive soil erosion, destruction of valuable wildlife habitat, or unacceptable nutrient pollution of watercourses when viewed from a social perspective. The problem of excessive or imprudent use of pesticides has come to be of particular concern because of the health hazards to agricultural workers and consumers.

Farmers have typically made their decisions about levels of pesticide use in the same manner as their decisions about seed, fertilizer, and labor use. But most pesticide prices have not accurately reflected the "spillover" effects of their use: they have not included the cost of potential damage to neighbors' crops, the health hazards to farm workers and consumers, and the damage to wildlife. The effect has been to encourage the farmer to use larger quantities of pesticides per unit of production, and to prefer chemical methods of pest control to other methods.

The sensitivity of pesticide use to price is illustrated in Figure 24-8. The yield-response curve for pesticides on most crops typically rises sharply and then flattens out. This is reflected in the shape of curve R in the figure. The lines p_1 and p_2 reflect the price ratio of the pesticide and the crop per unit of output. Line p_1 reflects a relatively high pesticide price per unit and line p_2 reflects a lower price. The optimum level of pesticide use is the point at which the price line is tangent to the gross-returns line. There, the increment to cost is the same as the increment to returns. Because the upper range of the response curve is quite flat, a small reduction in the pesticide price induces a large increase in the optimum level of application. And the farmer may be tempted to apply amounts much larger than the opti-

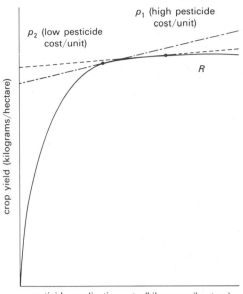

Figure 24-8   Sensitivity of pesticide use to price. The slopes of p_1 and p_2 represent alternative ratios of the price of pesticide per unit to the price of crop per unit. [After J. C. Headley and J. N. Lewis, *The Pesticide Problem: An Economic Approach to Public Policy*, The Johns Hopkins University Press, 1967, p. 19.]

mum to insure against uncertainty, since typical pesticide costs per unit of output are low, relative to the price of the commodity.

Market reforms that could lead to greater consistency between farm-production decisions and efficient use of resources for the entire economy would include a combination of changes in property rights, tax incentives and subsidies, regulation of pesticide use, and public management of integrated pest-control programs. In Michigan, for example, the introduction of an experimental pest-management program for apples has permitted a reduction in the level of pesticide application by as much as three-quarters. Under this program, complex insect-detection, monitoring, and scouting methods have been developed, and the cooperation of the Michigan State University Cooperative Extension Service, the Michigan Department of Agriculture, and the U.S. Department of Agriculture has been enlisted. The Michigan program is publicly supported. An alter-

native would be to establish a pest-management district and pay for the service by an assessment or a direct charge to the producers who benefit from the program.

MARKETS

Markets are the institutions through which the price-making forces, which have been summarized in terms of demand and supply schedules, are channeled and interpreted. Markets perform this role by converting (1) information on changes in yield and changes in the area planted, and (2) information affecting the quantities of a commodity desired by consumers or other users into price signals that trigger the production–consumption feedback system described above. Markets are cultural artifacts. They were invented by human beings. Their effectiveness depends on human inventiveness and ingenuity in market organization and development.

Marketing institutions

Markets in developing countries tend to be localized. Buying, selling, and market-information activities are typically concentrated at a single place, such as a village market or a transportation terminal (Fig. 24-9). At these locations, buyers and sellers meet each other face to face, and sales are consummated by the physical transfer of commodities and money. In a modern economy with a sophisticated information system, the concept of a specific market "place," where producers and consumers meet to bargain over prices and quality, is largely obsolete. Instead, producers and consumers are linked by a large number of intermediaries, each of whom may perform one or several market services or functions.

In addition to the buying, selling, and market-information services, modern market institutions perform a number of other services, including storage, transportation, and whatever services are necessary to facilitate buying and selling, such as establishing standards, arranging financing, and assuming risks. These services are provided by retailers, wholesalers, brokers, and commission agents, credit agencies, transportation firms, and others. The organizations

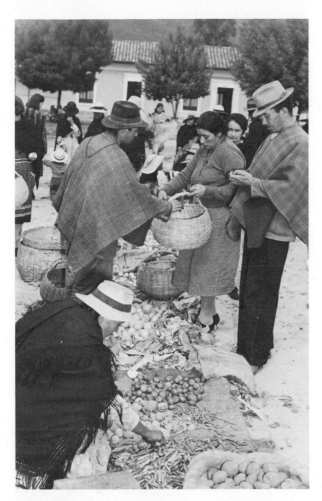

Figure 24-9   The village market concentrates the buying, selling, and market-information functions at a single location. [Courtesy USDA.]

that provide these services are also frequently engaged in other activities, such as the processing or manufacturing of basic commodities into consumer goods. The processing or manufacturing phase is sometimes done by separate firms. For many commodities, more resources are used in the marketing and processing activities than in crop production itself. In the conversion of wood to paper, the value added to the original forest product represents perhaps 90 percent or more of the final price to the consumer.

Marketing efficiency

There are two ways in which to evaluate the efficiency of the marketing system. The first is in terms of pricing efficiency. This is measured by the speed and precision with which the equilibrium or market-clearing process is discovered and made known to consumers, producers, and processors. The second is in terms of technical efficiency. This is measured by the resources utilized in marketing and processing agricultural commodities into finished consumer goods.

Technical efficiency should not, however, be confused with the percentage share of the total consumer dollar that goes to pay for marketing services and processing. The fact that farmers in the United States receive 60 percent of each dollar the consumer spends for eggs and only 17 percent of each dollar the consumer spends for bread says nothing about the efficiency of the marketing system. Bread is a much more highly processed product than eggs. Technical efficiency can only be determined by a careful examination of whether the marketing and processing functions that are preferred by the consumer could be performed at a lower cost.

Market structure

Markets are, of course, not the only method of transmitting information about consumer preferences and production alternatives between producers and consumers. Markets are in some ways very inefficient devices for transmitting information. In the process of translating information about the forces that affect demand and supply into market prices, much relevant information about producers' plans, market stocks, and quality preferences, for example, may be lost.

When large numbers of buyers and sellers are involved, when the commodity is not perishable, and when quality differentials can be accurately specified and measured, organized commodity markets, such as the grain exchanges in Chicago, Minneapolis, and Kansas City, appear to be relatively efficient.

When the commodity is perishable or if technical efficiency is increased by agreements between producers and processors about quality and timing of delivery, as in the canning industry, open markets have frequently been replaced by contractual arrangements. Farmers agree to delivery schedules and quality specifications in return for price and quantity guarantees by processors. In some arrangements, the terms of sale are established by bargaining between producer organizations and processor organizations.

One of the most important factors leading to new patterns of marketing and distribution has been the rapid development of transportation facilities and communication media. The development of national railway systems led to the establishment of great concentrations of marketing organizations at such rail centers as Kansas City, Minneapolis-St. Paul, and Chicago. Central wholesale markets, futures markets, and processing facilities were established at these centers because the produce from large regions was forced to pass through these centers on the way to population centers and port cities.

With the development of more efficient long-distance communication systems, the construction of a national highway system, and the development of modern motor-freight equipment, the necessity to move goods through a few central locations has been reduced. Many of the old established central markets for fruits and vegetables, livestock, and dairy and poultry products that developed during the railway era have declined in importance. Many products are now being processed closer to the point of production. The major food-distribution organizations now send their own buyers directly to the producing areas. Many go directly to growers for such fresh products as fruits and vegetables.

GOVERNMENT AND MARKETING

In addition to private marketing institutions, government plays an important role in the marketing of crops in most countries. Governments attempt to improve pricing efficiency—the sensitivity and precision with which the marketing system reflects producer and consumer interests and behavior. And governments attempt to overcome inequities in the distribution of income that result from the effects of prices on the allocation of resources.

Programs to improve pricing efficiency

A number of conditions must be met if marketing procedures are efficiently to perform their function in establishing prices and if prices are to act as efficient guides for determining the future quantities to be marketed. Effective standards for weights and measures and for quality must be established and adhered to. Information on prices and quantities must be equally available, as nearly as possible, to all market participants—producers, marketing agencies, processors, and consumers. In all advanced economies, government plays an important role in establishing effective market standards and in providing market information (Fig. 24-10).

Weights and measures

The standardization of weights and measures is one of the oldest recognized functions of government. Without such standardization internal trade within countries could hardly be conducted on any extensive scale. As international trade has expanded, the need for greater standardization of weights and measures has been increasingly recognized, although discrepancies still exist, even between countries employing more or less the same systems. For example, liquid and dry volumetric measures are not the same in the United States and Great Britain. The Imperial gallon used in Canada and Great Britain to measure liquid volume is about one-fifth larger than the United States gallon. And in both systems the dry volume measures differ from the liquid—the dry pint is larger than the liquid pint.

The English system of weights and measures gained wide use throughout many parts of the world because of the British dominance of trade during the last half of the nineteenth century and the early part of the twentieth century. The metric system, developed by a commission of French scientists and adopted as the legal system of weights and measures in France in 1799, has gained increasing acceptance because of its great simplicity and logical structure. It is widely used for scientific work even in countries that continue to use the English system for commercial and industrial use, and it is continuously

maintain
currency

inspect weights
and measures

control
quality

established grades
and standards

provide market
information

enforce
contracts

Figure 24-10   Governmental services essential to marketing. [After A. T. Mosher, *Getting Agriculture Moving: Essentials for Development and Modernization*, Praeger, 1966.]

achieving even wider commercial and industrial acceptance. As of 1980, the United States is one of the few major trading nations that has not officially adopted the metric system.

Grades and standards

Establishment of effective quality standards for agricultural commodities represents a much newer function of governments. Standardization of volume measurement has little meaning if the content of the container varies in quality. With the increasing commercialization of agriculture, commercial standards tended to evolve. At a later stage, government efforts were often exerted to achieve uniformity or in-

creased precision of such standards. In the early days, for example, each grain market had its own grades and grading methods. Different early standards for No. 2 corn required that the corn "be dry," "reasonably dry," or "have not more than 16 percent moisture." Lack of accepted fruit and vegetable standards resulted in especially chaotic trade conditions. In the United States the passage of the Cotton Futures Act of 1914 and the Grain Standards Act of 1916 initiated a series of laws that have gradually resulted in broadening federal and state responsibilities in the establishment of uniform grades and standards for agricultural commodities.

It is clear, however, that much of the grading system that has evolved from early commercial practices and has been given broader legitimization by government is no longer as relevant as it was in the past. Improvements are being made in many areas. Mechanical and chemical tests are being used to replace sensory tests, which often vary with the sensory ability of the individual grader. Less progress has been made in developing criteria that consumers can effectively use to distinguish between different qualities for which they are willing to pay different prices.

The establishment of standards for the quality and performance of industrial inputs used in agricultural production has become increasingly important during the last decade. Early efforts in this area were primarily confined to the establishment of accurate labeling of such inputs as fertilizer, livestock feeds, and pesticides. The objective was to make certain that the farmer knew what he was buying and that, for example, the plant nutrients in the fertilizer bag were as specified on the label. More recently, increased attention has been focused on the registration and use of pesticides. Regulation has been intended to ensure the health and safety of the industrial workers in the factories that produce pesticides and of the farm workers who apply them. Increasingly stringent efforts are also being made by the Environmental Protection Agency and the Food and Drug Administration to remove from use those pesticides that have measurable impact on environmental quality or the health of the consumer. Requirements for the registration of new pesticides have become so stringent that some observers believe that they are inhibiting research leading to the development of new pesticide materials. There is little doubt, however, that inadequate regulation in the past led to the marketing of pesticides that created excessive health hazards for workers and consumers alike.

Market information

The establishment of prices will clearly be inefficient or distorted if information on the quantity produced and the prices at which commodities are sold is unavailable or available only to some of the participants in the marketing process. In recognition of the importance of such information, most governments have sharply increased their efforts to provide producers, consumers, and marketing agencies with increasingly adequate data on the production, storage, movement, and prices of commodities. The USDA publishes information on production stock and prices for approximately 150 crop products and 50 livestock and related products.° The following list illustrates the type and extent of the information available from the Market News Service of the United States Department of Agriculture.

Cotton
1. Prices of lint cotton, cotton linters, and cottonseed.
2. Price quotations from principal markets.
3. Quality data during the ginning season.

Fruits and vegetables
1. Prices and supply movements for the large city markets.
2. Shipping-area data on prices, volume of loadings, quality, and market trends.
3. Cold-storage holdings in principal centers.

Grain
1. Grain prices and supply movements (cash and futures) for the principal markets.
2. Prices, market movements, and production data for feedstuffs.
3. Prices at country shipping points.

Similar information is provided by governments in an increasing number of countries. In the developing countries, such data are more frequently provided for export crops, on which the countries' foreign-trade earnings depend, than for the food crops used for domestic consumption.

As the food economies of the world have become more interdependent, major information gaps have been recognized. The seriousness of these gaps was dramatized in 1973 when a combination of drought in the Sahel region of Africa and exceptionally large

---

°For a complete listing of these reports and how to obtain them, see U.S. Department of Agriculture, *Crop Reporting Board Catalog*, Economics, Statistics and Cooperatives Service, January 1979.

grain purchases in the United States by the USSR resulted in dramatic increases in grain prices and a disruption of trading patterns. At the 1974 World Food Conference, the inadequacies in the world food-information system were reviewed. Following the conference, the Food and Agricultural Organization of the United Nations established a Global Information and Early Warning System in Food and Agriculture. The system includes:

1. A monthly food-outlook report that provides an up-to-date review of world food stocks and short-term projections of production and stocks.
2. A monthly report on food-crop shortages, which identifies those countries in which crop conditions give cause for concern or in which there are acute food shortages.
3. A periodic report on cereal-import requirements of food-deficit developing countries and food-aid allocations and shipments by developed countries.
4. Special reports on the food situation in countries or regions in which deterioration of the food supply is threatened.

Government agencies are increasingly undertaking the responsibility of analyzing and interpreting the information they collect. Major efforts are made by such agencies as the Council of Economic Advisers, the World Outlook Situation Board, and the Economics, Statistics, and Cooperatives Service of the United States Department of Agriculture to analyze the economic outlook and its implications for the several sectors of the economy and for individual commodities over the years ahead. At the state level, the agricultural extension services hold area and county "outlook" meetings to acquaint farmers with the economic forces that will affect the outlook for prices during the coming year.

At one time it was hoped that such economic information activities would, in themselves, help to stabilize commodity output and prices. By anticipating the forces that would affect prices, farmers were expected to restrict production if low prices were anticipated and to expand production in anticipation of rising prices.

Many students of commodity-price behavior have been convinced, however, that even with adequate information the price system could never become a completely reliable guide upon which to base decisions governing the allocation of resources. Most analysts agree, however, that the system does do an effective job of determining the prices that will effectively clear the market of whatever is produced. The major criticism that has been made is that the price fluctuations that result from unregulated market processes are too great and that farmers need more definite guides in making their production decisions. An implication that has frequently been drawn from this criticism is that the government should actively intervene in the market to stabilize prices and to protect commodity producers from the price fluctuations resulting from the free play of market forces.

Programs to overcome inequities in income distribution

When the pricing system is functioning efficiently, it is completely impersonal in its operation. When consumer tastes shift away from a product for which demand was formerly strong, or when technological change creates industrial substitutes for products formerly produced by agriculture, or when transportation or natural-resource development result in the opening up of new regions, supply and demand functions shift and the prices of products or inputs are affected. The adjustments in production that are induced by the price change also result in income gains and losses for individual families, for whole producing regions, and for the workers and owners of resources in the industrial and commercial sectors that serve agriculture.

The income adjustments forced upon families and regions as a result of such changes represent a second reason why farmers and their representatives frequently demand that governments intervene in the market process to prevent production, consumption, or prices from freely responding to changes in supply or demand.

Dissatisfaction with the supply–demand feedback system, which relies on price as a signal for changes in production or consumption activities, therefore rests on two perceptions. The first is that the system is inefficient—that it does not operate with sufficient speed and precision to avoid distortions in resource use. The second perception is that it is inequitable—that when it works efficiently some families and regions are unfairly penalized by income losses and other families and regions receive unearned "windfall" gains. Both sources of dissatisfaction were present in the demands for "fair" or parity prices for agriculture and the associated acreage-control and price-support programs that were instituted in the United States following passage of the Agricultural Adjustment Act of 1933. Parity price, as defined by the Agricultural Adjustment Act of 1933, was the price that would ensure that an agricultural commodity (say, one bushel of wheat) would have the same purchasing power that it had during the base period, which, for most commodities, was from August 1909 to July 1914. The 1909–1914 base was chosen because of the relatively favorable prices for agricultural commodities that prevailed at that time. Technological and other changes that have modified the relation between production costs and the parity price have made this earlier base obsolete. Today, there is no clear relation between "parity price" and "parity income" for farm and nonfarm families.

The problems posed by the inefficiency and inequity criticisms of the supply–demand feedback system is that efforts to improve the pricing efficiency of the system are frequently inconsistent with efforts to use the price system to achieve income-distribution objectives. Figure 24-7 can be used to illustrate this problem. Suppose that the demand for a crop, such as potatoes, is initially represented by $D'D'$ and the supply by SS. Price is equal to p_2 and the production is q_2. Suppose, as incomes rise, that consumers' tastes shift away from potatoes and can, after a time, be represented by DD. In order to protect the incomes of potato producers, the government could attempt to hold the price at p_2. At p_2 prices, consumers will now be willing to consume only the amount q'_2, whereas producers will be willing to produce the amount q_2. A surplus equal to $q_2 - q'_2$ will appear unless the government also takes action to restrict the amount of potatoes that can be pro-

duced through acreage restrictions or marketing quotas.

THE EVOLUTION OF FARM COMMODITY POLICY

The conflicts that are inherent in efforts to utilize the supply–demand feedback system to achieve pricing efficiency and equity in income distribution simultaneously have been of continuing concern in farm-policy debates since the 1920s. Table 24-3 provides a schematic outline of major agricultural legislation—from the Agricultural Adjustment Act of 1933 to the Food and Agriculture Act of 1977—in the context of related economic and political developments.

During the 1950s, the major controversy in debates over farm-price supports was the issue of high, fixed price supports versus flexible price supports. The policies that prevailed combined relatively high fixed prices with weak production controls. The result was the accumulation of large agricultural surpluses. Throughout this period, a major program instrument used to transfer income to farmers was the nonrecourse commodity loan. The nonrecourse loan was first employed to support the price of cotton, at 10 cents per pound, in 1933. Under the nonrecourse arrangement, the government loaned the farmer an amount equal to the support price (loan level) on the amount of the farmer's crop that was eligible for support. If the market price remained below the loan level, the government stood ready to accept the commodity in full satisfaction of the loan. If the market price rose above the loan level, the farmer could choose to redeem his crop by repaying the loan, plus interest and a service charge, and sell the crop at the market price. The nonrecourse feature allowed farmers to gain from any price rise with no risk of loss.

During the 1930s and again after the mid-1950s, efforts were made to limit the accumulation of surpluses by diverting land from crop production. In the 1930s, acreage allotments, based on historic land use, were established for each major crop. The farmer was required to divert a portion of his base acreage from production to be eligible for a price-support loan. Under the "Soil Bank" program that was initiated in 1957, payments were made to farmers (1) to reduce the land planted to specific crops, such as wheat, cotton, corn, tobacco, peanuts, and rice, and to convert the land to soil-conserving uses (the acreage reserve); and (2) to designate a certain portion of their cropland to be devoted to conservation use (the conservation reserve).

The agricultural commodity programs of the 1960s and 1970s have tended to place increased reliance on the use of direct payments to farmers as incentives to adjust production to demand and less reliance on price supports to transfer income to farmers. Price supports have been more closely related to estimates of long-run market price levels than to the parity concept. Farmers have been given greater freedom in deciding how to adjust their crop mix to their particular resource and economic conditions.

Under the Food and Agriculture Act of 1965, which culminated in a set of reforms of commodity policies during 1961–1965, farmers were given the option of whether to participate in the programs. Farmers who participated in the programs by holding 20 percent of their historical "base" acreage out of feed-grain production were eligible for a payment of 30 cents per bushel on the "projected" yield on half of their base acreage. Additional "diversion" payments could be obtained by diverting more than 20 percent of the base acreage (but not more than 50 percent). These diversion payments were calculated as half of the "projected yield" multiplied by the support level ($1.00 + 0.30) for any acres diverted from production in excess of the 20 percent (but not more than 50 percent) necessary to participate in the program. The principle of providing the Secretary of Agriculture with a flexible set of incentives to induce farmers to adjust production to demand has been established.

Additional flexibility was introduced by the Agricultural Act of 1970. Receipt of price support payments for feed grains, for example, was not tied directly to planting restrictions on feed grains. Farmers were eligible for price-support loans on feed grains if they diverted a specified percentage of their total cropland from production. The diverted acreage could come out of any crop, not just feed grains. In 1971 the only requirement for diversion payments

and price-support loans was that the farmer leave idle a quantity of cropland equal to 20 percent of the base acreage. In 1972 the minimum acreage to be set aside was increased to 25 percent. The loan rate was set at $1.05 and the diversion payment was fixed at 80 cents. Additional payments were also made available as an inducement to divert acreage beyond the mandatory minimum. To some extent, precision of output control was traded for greater freedom to plant by the 1970 act.

The Agricultural and Consumer Protection Act of 1973 was passed during a period of world food shortage. World demand, spurred by devaluation of the dollar, export subsidies, and crop failures in the USSR and in parts of Africa and Asia, had liquidated the stocks that had accumulated under previous price-support programs. Commodity prices had risen dramatically to exceed loan rates and target prices by substantial margins (Figs. 24-11, 24-12, 24-13, and 24-14). The new act placed increased reliance on market arrangements to facilitate the movement of agricultural commodities into world markets. Several actions were taken to meet this objective:

1. The price-support loan rates were set at levels that were intended to be below market prices in order to encourage production for domestic consumption or export rather than for accumulation in government hands.
2. Payments to farmers would be made only when market prices fell below target prices. Target prices were to be adjusted annually to reflect changes in productivity and prices paid by farmers for inputs. The payments, termed deficiency payments, were equal to the difference between the target price and the market price, but they could not exceed the difference between the price-support loan and the target price.

The Food and Agricultural Act of 1977 reflected the less buoyant market conditions that prevailed at that time. Prices of food and feed had receded from the 1973–1975 highs. Producers and their friends in the Congress were again concerned with problems of "oversupply and depressed prices." The administra-

Table 24-3 Chronology of agricultural legislation, 1933–1980.

	President	Secretary	Major farm legislation		Other important events
1933	Roosevelt	H. A. Wallace	Agricultural Adjustment	Depression	Worldwide protectionism
34	(Democrat)		Act, 1933		
35					
36			Soil Conservation and Domestic Allotment Act, 1936		
37			Agricultural Marketing Agreement Act, 1937		
38			Agricultural Adjustment Act, 1938		
39					
1940					
1		Wickard		World War II	Wartime price and wage controls
2			Stabilization Act, 1942		
3					
4					
5					
6	Truman	Anderson		Cold War	
7	(Democrat)				Marshall Aid Plan
8			Agricultural Adjustment Act, 1948		
9		Brannan			Brannan Plan rejected
1950				Korean War	
1					
2					Emergence of U.S. and USSR as nuclear powers
3	Eisenhower	Benson	Agricultural Act, 1954	Cold War	
4	(Republican)		Agricultural Trade Development and Assistance Act, Public Law 480, Food for Peace, 1954		
5					

Source: A. D. O'Rourke, *The Changing Dimensions of U.S. Agricultural Policy,* © 1978, pp. 96–97. Reprinted by permission of Prentice-Hall, Inc., Englewood Cliffs, N.J.

	President	Secretary	Major farm legislation		Other important events
1956			Agricultural Act, 1956		Japanese-German "economic miracles"
7					EEC founded
					Treaty of Rome
					Sputnik launched
8			Agricultural Act, 1958		
9					
1960					
1	Kennedy	Freeman	Feed Grain Act, 1961		Cochrane Bill rejected
2	(Democrat)		Food and Agriculture Act, 1962		
3				Kennedy assassination	Wheat referendum on mandatory controls, 1963
4	Johnson		Wheat-Cotton Act of 1964		
5	(Democrat)		Food and Agricultural Act, 1965	Vietnam War	
6					World food crisis
7					Domestic poverty problems
8					
9	Nixon	Hardin			
1970	(Republican)		Agriculture Act, 1970		
1		Butz			Peacetime wage and price controls
2					First Russian grain deal
3			Food and Consumer Protection Act, 1973	Cold War ends	Energy crisis
4					World food crisis
5	Ford		Emergency Farm Bill, 1975 (vetoed)		
6	(Republican)				
7	Carter	Bergland	Food and Agricultural Act, 1977		Recovery from energy and food crises
8	(Democrat)				
9					
1980					
1	Reagan (Republican)				

* total support level, 1963–1973; target prices, 1974–1978

△ estimated

Figure 24-11   Wheat support levels and farm prices.
[Courtesy USDA/ESCS.]

* total support level, 1963–1973; target prices, 1974–1978

△ estimated

Figure 24-12   Corn support levels and farm prices.
[Courtesy USDA/ESCS.]

* no loan rate for 1975–1976

△ estimated

Figure 24-13   Soybean loan levels and farm prices.
[Courtesy USDA/ESCS.]

△ average, August–December

Figure 24-14   Cotton support levels and farm prices.
[Courtesy USDA/ESCS.]

tion was eager to avoid what it viewed as excessively high program costs. There was agreement, however, on the continued use of target prices and the use of loans at lower levels than target prices to allow crops to move freely in international trade. Discussions dealt more often with the issues of target price, loan level, and deficiency payment than with general principles. Several new principles were introduced, however.

1. The historical base for acreage allotments was dropped. Rather, 1977 plantings of row crops were designated as the normal crop acreage (NCA) base. Beginning with the 1978 program, national program acreages for wheat, feed grains, and upland cotton to meet estimated domestic and export needs were to be determined by the Department of Agriculture. A farmer's

acreage eligibility for deficiency payments was determined by multiplying his current acreage planted for harvest by an allocation factor based on the ratio of planted acreage in the United States to an estimate of needed acreage made by the Secretary of Agriculture. But if farmers voluntarily reduced their plantings of wheat, corn, soybeans, and barley below their 1977 level of plantings, all their acres would be eligible for deficiency payments. A set-aside program was again authorized in the event that the Secretary of Agriculture might determine a need for it. The set-aside was used in 1978 and 1979 but not in 1980.

2. A farmer-owned grain-reserve program was mandated for wheat, and the program could be expanded, at the Secretary's discretion, to include feed grains. The program involved the use of price-support loans of 3 to 5 years duration, payment of annual storge costs of the grain, and waiver or adjustment of interest rates on the loan. Under the 1978 program, farmers were not to release the wheat to the market until the farm price exceeded $3.29 a bushel (40 percent above the loan level). The farmers' reserve could be called on for domestic consumption or export whenever the market price reached 175 percent of the current loan rate.

During the 1960s, such minor crops as rice, peanuts, and tobacco were exempted from the program reforms described here. During the 1970s, however, the program provisions were gradually modified to bring them into line with the reforms in the other commodity program.

It seems apparent that the dominant trend in the development of policy governing agricultural commodities in the United States is toward utilizing the supply–demand feedback system to achieve greater precision and efficiency in allocating resources. Programs to achieve equity in income distribution between farmers and nonfarm families, and among farm families, are more likely to rely on use of direct

payments and extension of such government services as Social Security, Medicare, family assistance, and other programs that do not directly affect commodity prices.

PERSPECTIVE

This chapter has presented a framework for understanding the role of economic forces in crop production and marketing decisions. Particular emphasis has been placed on (1) the role of supply–demand feedback systems in influencing production and marketing decisions, (2) the role of markets and marketing institutions in the functioning of this system, and (3) the contribution that government makes in increasing marketing efficiency and in balancing efficiency with equity among the many individuals and organizations that contribute to the production and marketing of agricultural commodities.

The general principles developed in this chapter are treated in greater depth in the next four chapters. Chapter 25 discusses some of the domestic-market and international-trade policies that nations have developed to regulate prices and market activity. Chapter 26 discusses the organization and management of agricultural research systems. Chapter 27 examines the economics of technical advances in crop science and in crop production. In Chapter 28, specific attention is given to the dynamic role that technological change in agriculture plays in national economic development.

Selected references

Barker, R., 1971. *The Economics of Rice Production (A/D/C Teaching Forum*, Farm Operations and Management, No. 5). New York: Agricultural Development Council. (An excellent discussion of farm-management decision processes and information needs and simpler analytical methods used in making farm-management decisions.)

Cochrane, W. W., 1958. *Farm Prices: Myth and Reality.* Minneapolis: University of Minnesota Press. (Price-income problems of commercial agriculture in the United States. Section one deals with farm price-income behavior, in particular the myth of an automatically adjusting agriculture; section two presents an analysis of farm-price behavior; and section three discusses certain aspects of economics and policy, such as the economist as a policy adviser, blind alleys, and hard policy choices.)

Cochrane, W. W., and M. E. Ryan, 1976. *American Farm Policy, 1948–1973*. Minneapolis: University of Minnesota Press. (A detailed treatment of the evolution of farm policy between 1948 and 1973. Part one describes the evolution and general content of the programs and explores the policy formulation process; part two documents the farm policy legislation enacted during that period and describes the program operations and costs; and part three contains an interpretation and appraisal of the programs of the 1948–1973 period.)

Dahl, D. C., and J. W. Hammond, 1977. *Market and Price Analysis: The Agricultural Industries*. New York: McGraw-Hill. (A text intended for use in an introductory course in marketing. Its approach to problems of market organization and agricultural price behavior is more analytical than descriptive or institutional. It is particularly strong in its presentation of market-price determination over space and time.)

Greig, W. S., 1971. *Economics of Food Processing*. Westport, Conn.: Avi. (Discusses the characteristics of the food-processing industry in the United States, including its structure, growth regulation, technological base, locational characteristics, vertical integration, and future trends.)

Kohls, R. L., and W. D. Downey, 1972. *Marketing of Agricultural Products*, 4th ed. New York: Macmillan. (A marketing text written for the beginner in economics. Its approach is mixed—partly functional, partly institutional, and partly by commodities. Part one develops the framework of the marketing problem; part two discusses some functional problems—that is, agricultural prices, market organization, and market functions; and part three presents commodity-marketing problems on a commodity-by-commodity basis.)

O'Rourke, A. D., 1978. *The Changing Dimensions of U.S. Agricultural Policy*. Englewood Cliffs, N.J.: Prentice-Hall. (An elementary text that gives special attention to the role of the conflicting pressure groups that have influenced the evolution of U.S. agricultural policy. It describes the important role of U.S. agriculture in the world economy, and is critical of the failures of U.S. agricultural policy to respond to the economic and social problems of small farmers and rural communities.)

Rasmussen, W. D., and G. L. Baker, 1979. *Price-Support and Adjustment Programs from 1933 through 1978: A Short History* (Agricultural Information Bulletin 424). Washington, D.C.: USDA, Economics, Statistics and Cooperatives Service. (Many USDA programs, particularly those concerned with farm price-support and adjustment legislation, resulted from a series of interrelated laws passed by Congress from 1933 to 1978. This bulletin provides a history of how congressional legislation and programs have been modified for changing economic situations.)

25
World Commodity
Markets and Policies

During the years since World War II, a fundamental restructuring of the world market for agricultural products has occurred. The general direction of agricultural trade has shifted from a North–South direction to an East–West direction. The developing countries' share of agricultural trade has declined. These have, as a group, shifted from being net exporters to net importers of basic food crops. Their exports of agricultural raw materials to the developed countries have also lagged. Among the developed countries, feed and oil crops have replaced food grains as the dynamic factor in agricultural trade. The socialist countries, particularly the USSR and the People's Republic of China, have become major importers of food and feed grains.

In discussing the factors that have brought about these shifts in market patterns, attention will be focused on the food and feed crops, tropical beverages, agricultural raw materials, and forestry products (Table 25-1). The extent to which these crops are traded among countries or are consumed domestically varies widely. Only about 2 percent of world

rice production enters international trade, whereas 88 percent of rubber production is traded internationally.

WORLD MARKET TRENDS

In the international economic system that emerged during the nineteenth century and persisted until World War II, raw materials, particularly food and fiber, were exported from the most recently settled countries of the temperate region and from the tropical colonial areas to the developed countries. Industrial products were exported by the industrial countries to the developing countries. It was believed that it would be to the economic advantage of both the developed and the developing countries if each country pursued its comparative "natural advantages." Furthermore, by capitalizing on their natural advantages in the production and export of raw materials, the developing countries could earn the foreign exchange, or attract an inflow of foreign capital, that would permit their economies to evolve first into developed agricultural economies and thence into diversified agricultural and industrial economies.

This was the pattern that was actually followed by the United States, Canada, Australia, and a few other countries during the last half of the nineteenth cen-

The authors are indebted to James P. Houck and Jimmye S. Hillman for having reviewed and made comments on a draft of this chapter.

Table 25-1 Importance of world trade in terms of percentage of world production of selected crop classes, 1952–1956 and 1975–1977.

Category	1952–1956	1975–1977
Food and feed		
Wheat	14	17
Rice	2	2
Coarse grains	4	11
Sugar	37	28
Fats and oils[1]	26	32
Citrus fruit	16	14
Tropical beverages		
Coffee	80	76
Cocoa	83	74
Tea	58	49
Wine[2]	13	14
Tobacco	20	23
Agricultural raw materials		
Cotton	21	31
Rubber	95	88
Jute and allied fibers	40	21
Hard fibers	78	49
Forestry products[3]	12	12

Sources: Commodity Review, FAO, 1961; *Commodity Review and Outlook 1977–1979,* FAO, 1978; *Commodity Review and Outlook 1978–1980,* FAO, 1979.
[1]Including oil content of oilseeds traded as such.
[2]Wine statistic shown as 1955–1957 average: *Commodity Review and Outlook 1968–1969,* FAO, 1969.
[3]Excludes fuelwood. Calculated from production and export data.

tury and the early half of the twentieth century. Before 1890, the United States, Russia, and the Danube countries of Eastern Europe were the major grain exporters. After 1890, grain exports from Australia, Argentina, and Canada expanded rapidly. India was a major exporter of wheat between 1880 and 1900. Burma and Thailand became major exporters of rice following the opening of the Suez Canal. Exports of such tropical products as coffee, tea, cocoa, and rubber expanded rapidly. At the beginning of the twentieth century, the system of comparative "natural" advantages appeared to be working well.

This system gradually broke down after World War I and was essentially reversed after World War II, with many raw materials and foods being exported from the developed to the developing countries. Three major factors brought about these changes. The first was the erection of trade barriers by most of the industrial nations on the imports from developing countries that might compete with the products of their own farmers. The United States, for example, found it necessary to erect barriers to the importation of wheat, cotton, corn, tobacco, and other products as farm-price programs were developed that held prices above the world-market level. The temperate-zone countries of Europe and North America erected trade barriers against the imports of raw sugar from tropical areas in order to encourage domestic production of sugar beets.

The result was a shift away from a system of free trade based on agricultural commodities to one that placed severe limitations on imports of tropical agricultural products, except for those beverages (such as coffee, tea, and cocoa), raw materials (such as rubber), and food crops (such as bananas) that did not compete directly with the agricultural products grown by farmers in the developed countries. This in turn depressed the prices and the growth of export earnings by the countries that traditionally depended on agricultural exports for foreign-exchange earnings (Fig. 25-1). The lag in prices and export earnings forced the exporting countries to reduce their own development efforts.

The second factor, the rapid growth of population in the developing countries (see Chapter 4), resulted in a sharp reduction in the exports of basic food crops, particularly food grains, from the developing to the developed countries. The rise in the rate of population growth reduced the internal supply of food available for export at the same time that trade barriers were reducing the demand for food exports from the developing countries.

The third factor, technological change in the developed industrial countries, has contributed to a reversal in the direction of trade (see Chapter 27). A developed industrial economy is capable of making fertilizer, farm machinery, and various mechanical,

Figure 25-1 Prices of food and nonfood agricultural commodities in developed and developing areas. [After *U.N. Statistical Yearbook*, 1979 and earlier issues.]

biological, and chemical inputs available to agriculture in large quantities and on favorable terms. The relatively high prices enjoyed by farmers in the industrial countries as a result of trade restrictions added further incentives to the adoption of the new technological inputs. The strong technological and economic incentives to expand production, plus the slow growth in demand, particularly in the United States, combined to produce agricultural surpluses in the developed countries. Policies were then adopted that subsidized the export of these surpluses to the major commercial markets and to the countries that

had rising food deficits. Some subsidized exports from the developed countries have been directly competitive with the products produced for commercial export by the developing countries. Subsidized grain exports from the United States to India and Indonesia have competed with commercial exports from Thailand and Burma.

The net effect has been a fundamental restructuring of the world trade pattern for food crops. This is particularly true of food grains (Fig. 25-2). Before World War II, from 1934 to 1938, the developing countries exported 11 million metric tons of grain annually. Between 1948 and 1952 the developing countries became a net importer of grains. Their grain imports continued to rise and exceeded 25 million metric tons in the mid-1960s. By the early 1970s, there was some hope that the long-term trend might be reversed. The introduction of new fertilizer-responsive varieties of wheat and rice had resulted in rapid growth in yield per hectare in some developing countries. By 1970, net imports of food grains by these nations had declined to approximately 15 million metric tons. This decline in exports from developed countries to the developing countries was at least partially offset by rising exports to the socialist-block countries.

During the time that the developing countries shifted from being net exporters to net importers of food grains, they continued to expand their production and exports of tropical beverages and agricultural raw materials. The developing countries have continued to rely heavily on exports of agricultural products and raw materials in order to earn the income they need to purchase imports from the developed countries. Yet exports of their agricultural commodities have tended to be smaller than their imports of the commodities produced by the developed countries. In part, this lag has been due to relatively slow growth in the demand for tropical beverages (tea, coffee, cocoa) and raw materials (rubber, cotton, jute, sisal). The slow growth in exports has also reflected, in some nations, limits in supplies, since the production of agricultural raw materials, beverage crops, and food crops for export has become increasingly competitive with the use of the

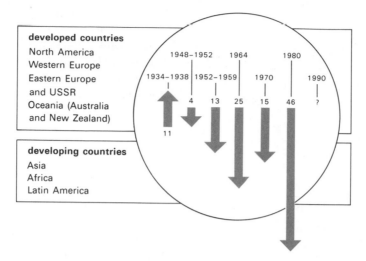

developed countries
North America
Western Europe
Eastern Europe
and USSR
Oceania (Australia
and New Zealand)

developing countries
Asia
Africa
Latin America

Figure 25-2   Net flow of world grain in millions of metric tons. [After *Food for Freedom*, U.S. Departments of State and Agriculture, 1966; *World Grain Trade Statistics, 169/70,* FAO, 1971.]

same resources, particularly land, for domestic food production.

WORLD FOOD CRISES IN THE 1960s AND 1970s

During the 1950s and 1960s, the grain reserves that had been accumulated in North America (as a result of the agricultural programs that were described in Chapter 24) provided a remarkable degree of price stability for food and feed grains. This stability was reinforced by the relatively fixed exchange rates that prevailed among national currencies since the framework of the "old" international economic order had been formulated at the Breton Woods Conference of 1944.

Stability in international commodity markets was briefly disturbed in the early and middle 1960s. Following a poor grain crop in 1960, China entered the world grain market as a major importer of wheat. The USSR, which had exported about 5 million metric tons of grain annually in the early 1960s, became a net grain importer of about the same magnitude

following poor grain harvests in 1963 and 1965. Drought resulted in disastrous grain harvests in India in 1964 and 1965. Massive shipments of grain were required to prevent starvation. The total stocks of wheat and feed grains held by the major exporting countries declined from 139 million metric tons in 1961 to 70 million metric tons in 1967 (Table 25-2).

The food crisis of the mid-1960s was of short duration. The new high-yielding varieties of wheat and rice were rapidly adopted in a number of developing countries. Grain production and grain stocks in the developed grain-exporting countries rose again in 1967 and 1968. Grain prices declined. Forecasters who had predicted a permanent food crisis in the mid-1960s were, by the late 1960s, predicting a new agricultural era of permanent food surpluses. By the late 1960s, the grain exporting countries shifted from attempting to meet world food needs to restraining the growth of food stocks. Acreage devoted to grains was restricted and production was reduced.

In the early 1970s, the U.S. Secretary of Agriculture was still trying to remove the burden of surpluses on farm prices when a new and more severe world food crisis emerged. In the United States

Table 25-2 Wheat and feed-grain stocks in major exporting countries, world grain production, and total stocks as a percentage of production.

Year	Stocks or production (millions of metric tons)				
	Wheat stocks[1]	Feed-grain stocks[1]	Total stocks	World grain production[2]	Stocks as percentage production
1950–1954	30	33	63	[4]	
1955–1959	54	54	108	[4]	
1960	55	73	128	885	14.5
1961	57	82	139	857	16.2
1962	48	69	117	906	12.9
1963	47	63	110	902	12.2
1964	40	69	109	949	11.5
1965	40	54	94	946	9.9
1966	27	43	70	1013	6.9
1967	31	39	70	1053	6.6
1968	36	48	84	1096	7.7
1969	53	52	105	1099	9.6
1970	60	52	112	1115	10.0
1971	45	38	83	1109	7.5
1972	41	52	93	1083	8.6
1973	27	40[3]	67	1250	5.4
1974	22	32	54	1201	4.5
1975	22	25	47	1226	3.8
1976	30	24	54	1353	4.0
1977	47	36	83	1336	6.2
1978	45	48	93	1445	6.4
1979	45	63	108	1383	7.8
1980					

Sources: D. G. Johnson, *World Food Problems and Prospects* (Foreign Affairs Studies 20), American Enterprise Institute for Public Policy Research, June 1975, pp. 54–55; *World Agricultural Situation* (WAS-14), USDA/ERS, October 1977, pp. 31–32; *Commodity Review and Outlook, 1977–79*, FAO, 1979, p. 52.
[1]Stocks at beginning of crop year for Australia, Canada, Argentina, and the United States.
[2]Includes rice in milled form; stock estimates do not include rice.
[3]Includes Thailand.
[4]Data not available on a comparable basis.

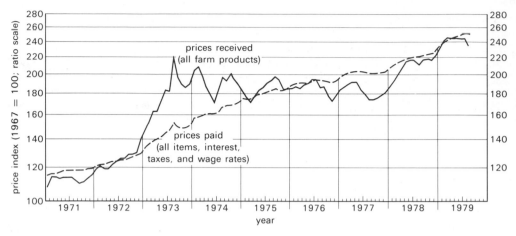

Figure 25-3 Prices received and paid by U.S. farmers, 1971–1979. [After U.S. Council of Economic Advisors, *Economic Indicators*, September 1979, p. 25.]

the most visible symbol of the new food crisis was a dramatic increase in the prices of agricultural commodities between the third quarter of 1972 and the same period a year later (Fig. 25-3). The prices of some grains and oilseeds not only doubled but went on to triple and even quadruple (Figs. 24-11, 24-12, 24-13, and 24-14).

The explanations that have been suggested for the dramatic changes in commodity prices since 1972 reflect a remarkable convergence of natural, technical, and economic forces:

1. A decline in world grain production due to adverse weather over large areas of the world. World grain production was above the long-run trend in 1971 and below the trend in 1972. But the actual decline was only about 26 million metric tons—3 percent of the world grain production (Table 25-2).

2. A drastic reduction in the Peruvian anchovy catch in 1972. Although this was small when considered in the context of total world consumption of fats and oils, equivalent to less than 10 million metric tons of soybean meal, its impact was mag-

nified because of its coincidence with lower grain production.

3. Large purchases of grain by the USSR. In 1972–73, the USSR imported 20.5 million metric tons of wheat and feed grains. Soviet imports declined in 1974 and fluctuated widely from 1975 to 1978 (Fig. 25-4). The USSR has clearly shifted from its earlier position as a grain exporter to that of a major grain importer.

4. Slowing of the growth in agricultural productivity. The recovery from the food crisis of the mid-1960s was greatly aided by the rapid adoption of new high-yielding varieties of food grains in developing and developed countries alike. By the early 1970s, the easy gains from the new varieties had been realized. The rate of increase in crop yield per hectare began to decline, both in the developed countries and in the developing countries.

5. Increases in fuel and fertilizer prices. Between 1972 and 1974, the Organization of Petroleum Exporting Countries (OPEC) increased oil export prices from approxi-

Figure 25-4   U.S. agricultural exports to the USSR, 1970–1978. (1977–1978 is partially estimated.) [After U.S. Department of Agriculture, *1978 Handbook of Agricultural Charts*, USDA/ESCS, November 1978.]

mately $2.00 to $12.00 per barrel. Oil prices have continued to rise. During the food crisis period of 1973–1975, prices of fertilizer also rose sharply. Many observers attributed the increase to an emerging energy crisis. In retrospect, the crunch on fertilizer supplies and the dramatic increase in prices reflected the effects of a temporary shortage of production capacity in the fertilizer industry at a time when the increases in the prices of agricultural commodities were resulting in a rapid increase in fertilizer demand.

6. Rising affluence in the developed countries. The growth of income per capita in North America, Western Europe, Japan, and the higher-income socialist countries has resulted in a significant shift in consumption habits. When income increases, consumers shift from the direct consumption of cereal products to indirect consumption in the form of meat. It takes 3–4 kilocalories in the form of feed grains to produce 1 kilocalorie in the form of poultry, meat, and eggs, and 6–7 feed-grain kilocalories to produce 1 kilocalorie in the form of beef.

7. Devaluation of the U.S. dollar. During most of the 1950s and 1960s, the fixed exchange rates between major currencies resulted in undervaluation of the U.S. dollar. This made it expensive for other countries to purchase U.S. agricultural commodities. When the dollar was devalued and allowed to "float" in March 1973, the price of U.S. agricultural commodities became cheaper in terms of other currencies. A ton of wheat cost less, valued in terms of German marks, Japanese yen, USSR rubles, or such monetary commodities as gold or silver.

As in the case of the food crisis of the mid-1960s, disagreement again arose as to whether the food crisis of the mid-1970s was the beginning of a long-run trend of increasing food scarcity or the result of an unfortunate combination of short-run events. In retrospect, it appears that short-run factors were dominant in 1972 and 1973. In its anxiety to dispose of surpluses, the U.S. Department of Agriculture encouraged excessive depletion of grain stocks. As a result, a modest decline in world output, combined with the Russian grain purchases and strong demand for food and feed grains elsewhere, resulted in an explosive increase in grain prices. Because of the importance of feed grains in livestock production, this set off an inflationary feed–livestock price cycle that diffused the effect of the higher grain prices throughout the entire agricultural economy in the United States and the world.

It is also clear that during the 1970s a number of longer-run factors have emerged that may imply greater instability in agricultural-commodity markets in the future than in the 1950s and 1960s. These include the impact of growth in affluence in the developed world and the unstable relations among national currencies.

PRINCIPLES OF COMMODITY TRADE*

One objective of commodity-trade policy is to achieve greater efficiency in production. A second objective is to achieve stabilization in commodity markets. These two objectives often conflict. Consumers in importing countries are anxious to take advantage of the expanded trade and lower prices that result from more liberal trading arrangements. Producers in exporting countries are anxious to avoid the downward adjustments in prices, production, and employment that occur as a result either of long-term shifts in comparative advantage or of short-term shocks such as those associated with the world food crises of the 1960s and 1970s. This section examines some of the principles of trade expansions and stabilization.

Protection from trade

Despite strong arguments about the benefits of freer trade, the economic changes it generates can impose hardship on some industries and individuals. Those affected often argue successfully for policies that provide protection from freer trade. Trade protection occurs when any group of producers or consumers is insulated from the full force of international competition by deliberate economic policy.

The classic method of import protection is the tariff, a system of duties or taxes. The tax can be either a fixed charge per unit imported or a fixed percentage of the value of each shipment. The former is called a specific tariff, and the latter is an *ad valorem* tariff. Other nontariff protection devices include import quotas (direct quantity controls), mixing regulations, complex packaging and labeling requirements, health and sanitary regulations, foreign-exchange restrictions, and variable import levies. All have one thing in common. They make it more difficult (or even, in some cases, impossible) for foreign sellers to compete with domestic sellers. Their effects are similar to tariffs. They drive a wedge between international prices and domestic prices.

Who bears the burden of a tariff or a quota? The European Economic Community (EEC) uses a variable levy—a tariff that is adjusted to take into account the differences in import prices and domestic prices—to regulate the import of wheat and feed grains (Fig. 25-5). The effect is to increase the price European farmers receive for wheat and feed grains. This increases the price European consumers pay for bread and livestock products. The European livestock-feed industry has responded by using more materials that are not subject to the variable levy in their feed formulations—more soybean meal from the United States and Brazil, more copra meal from the Philippines, and cassava chips from Thailand. Producers of these products gain and producers of feed grains lose. In the United States, feed-grain producers have shifted land from the production of feed grains to the production of soybeans in order to take advantage of the different treatment of feed grains and oilseeds and oil meal by the EEC.

The gains from foreign trade

Individual traders generally use current prices to guide their decisions. They buy commodities (and services) wherever they can obtain them most cheaply. They sell to the buyers offering the highest prices. This behavior is observed across national borders as well as within them. A country exports what it can produce more cheaply than others and imports those items that others can produce more cheaply. A country has an "absolute advantage" in the production of an item if its production costs are lower than those of other countries at prevailing prices and exchange rates.

The principle of absolute advantage appeals to common sense. But as our perspective expands from

---

*This section was drawn directly from James P. Houck and Peter K. Pollak, "Basic Concepts of Trade," in Martin K. Christiansen (editor), *Speaking of Trade: Its Effect on Agriculture* (University of Minnesota Agricultural Extension Service Special Report 72), pp. 23–24, 1978.

common market minimum import prices

Figure 25-5   Common Market minimum import prices
in 1977. Threshold price is the minimum import price at
which the EEC allows grain to enter a member country.
[After U.S. Department of Agriculture, *1978 Handbook
of Agricultural Charts,* USDA/ESCS, November 1978.]

single products or industries to whole nations, then
the principle of absolute advantage becomes less
clear. What happens if a nation has an absolute advantage in all its products because of very efficient
labor, abundant resources, and highly advanced
technology? History indicates that, even under such
circumstances, trade still occurs. This leads to the development and use of the principle of "comparative
advantage" to analyze trade flows. (The principle of
comparative advantage is illustrated in Box 25-1.)

In the context of comparative advantage, mutually
advantageous trade can occur among nations as long
as there is a difference in the rates at which one
commodity can be substituted for another in production. And there is a difference when the climate, natural resources, labor productivity, and technology of
the countries are different. The opening up of new
land resources and the adoption of higher-yielding
crop varieties, as in the case of soybeans in Brazil,
can change a nation's comparative advantage in producing and exporting a particular commodity.

The actual size and composition of trade flows and
the international price level also depend on demand

conditions in trading nations. Demand conditions are
shaped by national income and its distribution, by
population levels and growth rates, and by tastes and
preferences. Changes in demand can alter international trade just as much as changes in comparative
advantage. The increased demand for animal protein
that has accompanied rapid increases in income per
capita in Japan has transformed Japan into the
world's major importer of feed grains.

Commodity-market stabilization

The problem of how to achieve an effective balance
between the objectives of trade expansion and protection and the uncertainty of weather and politics
has led to a search for international institutions that
offer a possibility of greater stability in international
commodity markets. Achievement of a satisfactory
answer to the problem of price instability of internationally traded products has come to represent a major policy objective of many of the developing countries. The United States has also committed itself at a
number of international conferences and conventions
to work toward policies that "should contribute toward stabilization of agricultural prices at levels fair
to producers and consumers alike." Interest by the
developing countries has been stimulated by the success of the Organization of Petroleum Exporting
Countries (OPEC) in achieving dramatic increases in
petroleum prices in the early 1970s.

The rhetoric that surrounds the discussion of stabilization policies tends to reflect the different interests
of exporting and importing nations. When representatives of exporting nations discuss stabilization, there
is usually the implication that prices should be stabilized at levels above the long-run market-clearing
levels. When representatives of importing nations
discuss stabilization policies, they usually have somewhat lower price levels in mind.

There are three general policies that can be
adopted to stabilize market prices, or prices paid to
producers, whether in the domestic market or in the
international market. These are policies that would
(1) control or influence supply, (2) control or influence demand, or (3) require the payment of subsidies
(incentives) or the collection of taxes (penalties).

Box 25-1 *The key to comparative advantage in trade*

Think of two countries, Alphaland and Betaland. Each can use its resources and people to produce two major products—agricultural goods (*F*) and manufactured goods (*M*). Panel *A* of Figure 1 shows Alphaland's possible outputs of *F* and *M* based on its particular combination of natural resources, capital, and people.

If Alphaland used all its resources in agriculture, it could generate 50 units of *F* per year and no units of *M*. This is point *a* in Figure 1. If only manufactured goods were produced, 50 units of *M* could be had per year and no units of *F*, point *c*. By shifting resources between farms and factories, many output combinations of *F* and *M* are possible along the line *abc*. Points inside *abc*, within the shaded area, are possible too. But they are inefficient, reflecting resource unemployment or underemployment. From any shaded area point inside *abc*, more *M* or *F* (or both) can be obtained without sacrificing any other output. Along *abc*, more *M* can be obtained only at the expense of some *F*, and vice versa. The slope or steepness of *abc* reflects the rate at which *F* and *M* can be substituted for each other *in production* by rearranging fully used resources inside Alphaland. In this particular case, that rate of substitution is 1.0 *F* for 1.0 *M*. No output combination outside of *abc* is possible for Alphaland, given its resources.

Panel *B* of Figure 1 shows the same thing for Betaland. But there are some differences. First, Betaland is a smaller economy than Alphaland. No matter what it does, Betaland cannot match the potential production of Alphaland in either *F* or *M*. Betaland could possibly produce 40 units of *F* (and no *M*) at point *d*, or 20 units of *M* (and no *F*) at point *j*. Any of the points along or inside *dej* are feasible, but only the points along *dej* are efficient. The rate at which *F* and *M* can be substituted for each other in production is different in Betaland than in Alphaland. For Betaland it is 2.0 units of *F* for 1.0 unit of *M*. This country-to-country difference in the rate of substitution of one output for another is the key to the concept of comparative advantage used in international trade analysis.

To grasp this concept, imagine that Alphaland is now producing and consuming the combination of 20*F* and 30*M* denoted by point *b* in Figure 1. Assume similarly that Betaland is at point *e*, which is 32*F* and 4*M*. Now

visualize taking panel *A* in your hand, flipping it over, and placing it upsidedown on panel *B* so that points *b* and *e* lie exactly on top of each other. This is point *e(b)* of Figure 2. The size of the rectangle in Figure 2 formed by this maneuver is the total amount of *F* and *M* produced in Alphaland and Betaland together. This "world" output is 34*M* and 52*F*. Point *e(b)* shows how this "world" production is shared between the two. Betaland produces 4 units of manufactured goods and Alphaland contributes 30 units, the total being 34. On the other hand, Betaland produces 32 units of agricultural products while Alphaland grows 20 units, totaling 52.

Up to now, these two nations were isolated from each other. Now suppose that they look into possible international trades. Why might they wish to do this? For one thing, they could, via trade, separate the combination of *F* and *M* that each produces from the combination that each consumes. In Figure 2, the two nations could possibly trade away from *e(b)* to any point inside the large rectangle by exchanging *F* and *M* with each other.

Alphaland would not be interested in any trade that would deliver it to a shaded-area point inside *abc*. Those points are available to Alphaland without trade and are inefficient besides. Similarly, Betaland would disdain trades leading to shaded-area points inside *dej*. However, there are points in the rectangle that are outside the capacity of each nation to achieve independently yet are available through trade. These are inside the unshaded area of Figure 2. This unshaded area exists because the rate of substitution of *F* for *M* differs between Alphaland and Betaland. The greater this difference, the larger this unshaded area of potential exchange.

If these two nations are jointly producing *F* and *M* at point *e(b)*, demand analysis will show that, in general, the people of each nation will be better off if they trade away from *e(b)* down into the unshaded area. Determining a precise point of mutually agreeable exchange in that area is beyond this discussion, but it exists. Suppose for instance that it is point *g* in Figure 2. At *g*, Betaland would have 18 units of *M* and 14 units of *F* available for use, while Alphaland would have 16 units of *M* and 38 units of *F*. Naturally, this distribution also uses up the total "world" output of 34*M* and 52*F*.

Figure 3 is a close-up view of part of Figure 2. In order

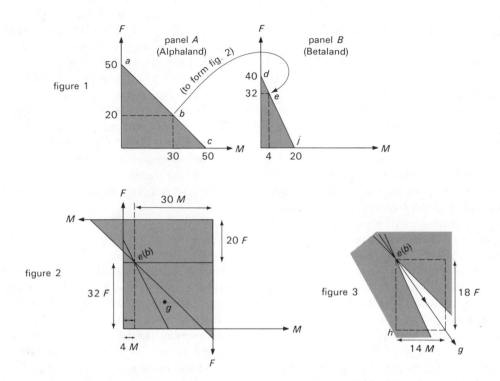

figure 1

figure 2

figure 3

for the two nations to get from point $e(b)$ to point g via trade, Betaland would need to import 14 units of M and export 18 units of F. On the other hand, Alphaland would export $14M$ in exchange for $18F$.

Notice that this is a better trade-off of F for M than either Alphaland or Betaland could make by rearranging its own resources internally. Alphaland could obtain only 14 more units of F internally by shifting resources and giving up 14 units of M; but on the "world" market it can get 18 units of F. Similarly Betaland could get only 9 more units of M internally by releasing resources from 18 units of F; through international trade it can gain 14 units of M.

In this example, Betaland has a "comparative advantage" in agriculture relative to Alphaland. This is because, within its own resource structure, it can generate 2.0 units of F for each 1.0 unit of M it gives up, and Alphaland can get only 1.0 unit of F for each 1.0 unit of M it gives up. The reverse argument shows that Alpha-

land has a "comparative advantage" in manufactures relative to Betaland. The existence of comparative advantage produces an area of potential trade (an unshaded area) within which *each* nation can make better deals for itself by international exchange than by adjusting its own resources internally.

Further analysis shows the validity of the common-sense notion that trading nations can capture even further trading gains by specializing, at least to some extent, in the products for which they have comparative advantage. As times goes by, however, nations' resources and abilities may change. Such changes can drastically alter the worldwide patterns of comparative advantage.

Source: James P. Houck and Peter K. Pollak, "Basic Concepts of Trade," in Martin K. Christiansen (editor), *Speaking of Trade: Its Effect on Agriculture* (University of Minnesota Agricultural Extension Service Special Report 72), pp. 23–24, 1978.

Within each of these general categories there are limited alternatives.

Control of supply may be achieved through control of the amount produced or marketed, destruction of output, restrictions on imports or exports, and storage.

Control of demand may be undertaken by instituting income supplements to consumers, such as the Food Stamp Plan; multiple-price systems, which permit some individuals or countries to purchase at lower prices than others; and forced utilization of products, such as mixing requirements that specify a certain percentage of flour from domestic wheat to be mixed with flour milled from imported wheat.

Subsidies and taxes can be used as incentives or penalties to induce desired production or consumption behavior. Subsidies can be paid to induce desired changes in land use—for example, to induce Malay farmers to plant rubber trees or to induce U.S. farmers to remove crop land from production. In the United States, taxes can be imposed as a penalty, to prevent planting of wheat and feed grains beyond acreage allotments.

Two institutions that have received increased attention since the food crisis of the early 1970s have been commodity agreements and grain reserves. Food aid has become a less important factor in agricultural trade since the early 1970s. International trade negotiations hold promise for reducing some of the nontariff barriers that nations have erected to protect their less efficient agricultural sectors. These institutional arrangements are discussed in the next sections.

INTERNATIONAL COMMODITY AGREEMENTS*

The negotiation of international commodity agreements capable of turning the terms of trade in favor of the developing countries is a major objective of the

---

*This section was drawn directly from Jimmye S. Hillman, Anne E. Peck, and Andrew Schmitz, "International Trade Agreements," in Martin K. Christiansen (editor), *Speaking of Trade: Its Effect on Agriculture* (University of Minnesota Agricultural Extension Service Special Report 72), 1978.

proponents of a new international economic order. Commodity agreements have, however, been difficult to negotiate and even more difficult to operate. After many years of trial and error, there existed in the late 1970s only five commodity agreements—for coffee, cocoa, sugar, wheat, and tin. Other agreements have existed for olive oil, skim-milk powder, and butterfat. International consultative organizations have been set up for wine, cotton, rubber, wool, and lead. There are also associations of producing countries built around particular commodity interests, such as raisins, copper, coconut, pepper, and oil.

The principal features of the five major agricultural-commodity agreements are presented in Table 25-3 and are discussed briefly in this section.

Wheat

International discussion aimed at stabilizing wheat prices began in the early 1930s. There have been seven agreements since 1949. Up to 1968, the International Wheat Agreement (IWA) was a "multilateral contract" under which exporters and importers agreed to buy or sell wheat among themselves at prices within an agreed-upon range. This mechanism provided a useful umbrella under which the world's two major exporters—the United States and Canada—prevented prices from falling below the floor level. The IWA was renegotiated in 1967, as part of a broader set of trade negotiations known as the General Agreement on Tariff and Trade (GATT)—Kennedy round. The agreement broke down in 1968 under the pressure of surpluses. The IWA also included a food-aid provision, partly to spread the financial burden more evenly among the surplus-grain producing countries who were donors of food aid and partly to remove some of the surplus grain from commercial markets. This worked relatively well. In 1971, the negotiators were unable to agree on a new price range. Since then, the IWA has contained no commercial-trade provision. Efforts were begun in 1977 to work out a new type of IWA based on an internationally coordinated system of national stocks. The negotiations collapsed in February of 1979.

Sugar

Various attempts have been made to stabilize the world free market for sugar beginning as early as 1937. Between 1969 and 1973, an arrangement was in effect that relied mainly on export quotas to ensure the sharing of markets and to keep the prices within a preset range. Provisions existed for use of stock maintenance to stabilize prices as well as for supply guarantees to traditional importers. After operating satisfactorily in its first three years, the agreement became ineffective after shortages developed and prices could no longer be controlled. The price range was narrow and the upper limit not high enough to encourage sufficient investment at the right time. The market regulation scheme lapsed between 1973 and 1977. The United Nations negotiation conference on sugar convened in April 1977 and once again ended inconclusively, mainly because of disagreements over whether the agreement should be basically a minimum-price scheme (quotas) or establish minimum and maximum prices (combined quotas and stocks). After the conference convened again in the summer of 1977 in Geneva, an agreement was reached. However, as of early 1980, the new agreement had not yet come into effect because the U.S. Congress had not ratified the proposed sugar arrangement.

Tin

The international marketing of tin has been controlled since 1921. For the first 35 years, the marketing agreements were solely among the producing nations. The first International Tin Agreement, which included both consumers and producers, came into effect in July 1956 and lasted for a period of five years. There have been several agreements since that time. Most major producing and consuming countries have been parties to the agreements, except the People's Republic of China (the fourth largest exporter) and the United States (the largest consumer). One reason that the agreeements have worked to the reasonable satisfaction of producers and consumers alike is that an attempt has been made to stabilize prices at near market levels rather than to maintain prices well above market levels.

Coffee

Two long-term coffee agreements have been negotiated, one in 1962 and the other in 1968. However, groups of exporters, primarily in Latin America, have been involved off and on in intercountry coffee arrangements since 1902. Several one-year agreements were in effect between 1957 and 1962. Supplies entering world trade were regulated by export quotas in order to maintain prices within an agreed-upon range. An innovative plan was adopted later—a diversification fund set up to help coffee producers to shift to the production of other crops, which took pressure off coffee supplies. The agreements worked reasonably well in stabilizing prices at relatively low levels, and this, coupled with a series of disastrous frosts in the late 1960s in Brazil, ultimately eliminated the burdensome surpluses. The 1972–1975 agreement eliminated the diversification fund. The current agreement, which began in 1976, contains provisions for export quotas that can be invoked if prices fall below a given level. There are no contingencies for shortages.

Cocoa

An agreement on cocoa was a long time in gestation—from the mid-1950s until 1972. The agreement attempted to establish both a price floor and a price ceiling along with a system of adjustable export quotas and a buffer stock. The buffer stock and a price ceiling were not implemented because of the poor harvests and high prices of 1972–73 and 1973–74. A new agreement was negotiated in 1976 that contains some more flexible features, but it still relies essentially on the same mechanisms. The producing and consuming nations have found it difficult to agree upon a price range.

Experience with commodity agreements has revealed a number of serious problems that have

Table 25-3   Principal features of international commodity agreements.

Title	Entry into force	Participating countries	Price targets
International Wheat Agreement	1971. Preceding agreements 1949, 1953, 1956, 1959, 1962, 1965, 1966, 1967.	Major exporting and importing countries with exceptions during certain periods.[1]	Not at present. Earlier agreements included a range between maximum and minimum prices.
International Sugar Agreement	1969. Preceding agreements 1954 and 1959.	All major exporting and importing countries trading on free-world market.	Not at present. A range of trigger prices applied from 1969 to 1973, when economic provisions of the agreement lapsed.
International Tin Agreement	1976. Preceding agreements 1956, 1961, 1966, 1971, and 1976.	All major producing and importing countries.	Yes. Range between maximum and minimum prices.
International Coffee Agreement	1976.[3] Preceding agreements 1962 and 1968.	All major exporting and importing countries.	Yes. Range between minimum and maximum prices to be established by the ICO.
International Cocoa Agreement	1976.[4] Preceding agreement 1973.	All major exporting and importing countries, except U.S.A.	Yes. Range between minimum and maximum prices.

Source: J. S. Hillman, A. E. Peck, and A. Schmitz, "International Trade Arrangements," in Martin K. Christiansen (editor), *Speaking of Trade: Its Effect on Agriculture* (University of Minnesota Agricultural Extension Service Special Report 72) p. 68, 1978.

limited their effectiveness. First, an effective price-raising agreement creates a strong incentive by suppliers to "cheat." Second, entry into the industry becomes more profitable for additional suppliers. And third, product substitution by consumers is inevitable as prices go higher relative to other products. In addition, negotiating commodity agreements to stabilize—or to raise—prices has presented major difficulties. Exporters and importers have, in practice, had trouble agreeing upon an appropriate price range. For example, there was a great controversy between

the United States and the producing countries on the price range in the first Coffee Agreement. Exporters have had differences regarding the proper basis for establishing export quotas. And financing any stock acquisitions has been a formidable obstacle when any serious talks get under way.

In spite of these difficulties, the search for solutions to commodity problems through some type of commodity policy, commodity agreements, or broader economic arrangements by governments has intensified since efforts began in the 1920s.

Formal export or production quotas	Buffer stocks	Other policy measures	Institutional support
No.	No.	From 1959 until 1971 importers undertook to buy a minimum share of their imports from member countries and exporters to sell the equivalent of their average exports over the preceding 4 years to member countries at prices within the agreed range.[2]	Administered by the International Wheat Council.
Not at present. Export quotas applied to free market from 1961 to 1973.	Not at present. Minimum and maximum stocks to be held in producing countries were stipulated from 1969 to 1973.	Not at present. Trigger mechanism linking quota adjustments and stock releases to movements in market prices operated from 1969 to 1973.	Administered by the International Sugar Organization.
Yes. Export quotas.	Yes. Producing and consuming countries contribute to the buffer stock.		Administered by the International Tin Council.
Yes. Export quotas consisting of a fixed part and a variable part, linked to the price range established by the ICO, or in its absence to average indicator prices for "other milds" and Robusta coffees in 1975.	No.	No.	Administered by the International Coffee Organization.
Yes. Export quotas. The agreement provides for different combinations of export quotas and buffer stock operations at different price levels.	Yes.	No.	Administered by the International Cocoa Organization.

[1]For example, U.K. in 1953 and 1956, Argentina in 1949 and 1953, and USSR in 1949, 1953, 1956, 1959, and 1967.
[2]The 1967 and 1971 agreements include a Food Aid Convention involving a commitment on the part of each member to supply a fixed quantity of cereals in the form of aid to the developing countries.
[3]Not yet operative.
[4]Not yet operative.

GRAIN RESERVES AND FOOD SECURITY

From the end of the Korean War in 1953 until 1972, the grain reserves of North America served as the world's grain reserve. These reserves were large enough to give remarkable stability to world grain prices. They were adequate to meet most of the shortfalls in world production that occurred during those two decades (Table 25-2, Figure 25-6). The reserves were not the consequence of deliberate policy decisions by the American and Canadian governments but were the unintended consequences of domestic agricultural price policies (see Chapter 24).

The existence of the reserves in North America did, however, make it possible for most of the world

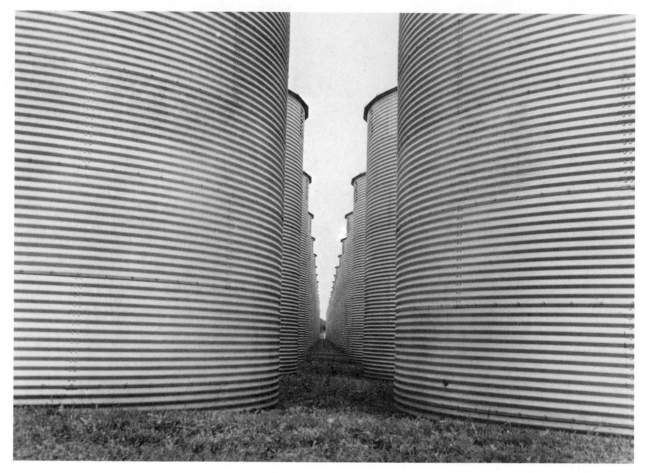

Figure 25-6   U.S. grain stocks have had a major stabilizing influence on prices.

to avoid the cost of holding substantial grain reserves. By the early 1970s, it was clear, however, that neither the private grain trade nor the governments of the major exporting nations were willing to bear the cost of holding the reserves that would be necessary to prevent the kind of price instability that has characterized the 1970s.

World wheat and feed grain stocks had fallen to about 5 percent of annual consumption in 1973 and 1974. When the World Food Council met in 1974, it recommended the establishment of a World Food Security program. The Committee on World Food Security of the Food and Agriculture Organization (FAO) of the United Nations has estimated that a

reasonable degree of security could be assured if world grain stocks were in the neighborhood of 17 to 18 percent of annual consumption. Of this amount, 5 or 6 percent should constitute a "reserve" component that would be available only under emergency conditions. The rest would be "working" stocks, which would be made available to meet normal marketing and trade requirements.

Much smaller reserves would be required under a free-trade regime. During the crisis of 1973 and 1974, the increase in grain prices relative to livestock prices in the United States resulted in a diversion of some 25 million metric tons of grain from use as animal feed to exports. When governments intervene in market relations to protect their consumers from price increases, as did the European Economic Community, the USSR, and China during the crisis of the early 1970s, there is little incentive for such adjustments. As a result, larger reserve stocks are required to avoid greater instability in grain prices. Trade liberalization can reduce the cost of achieving food security by reducing grain-reserve requirements.

In the late 1970s, under the auspices of the United Nations Council on Trade and Development (UNCTAD) and the International Wheat Council (IWC), considerable effort was devoted to the negotiation of an internationally coordinated system of national wheat reserves as part of a new International Wheat Agreement. After several years, these negotiations collapsed in February 1979. The collapse of the wheat negotiations was a major setback in efforts to design a multilateral approach to greater food security.

International efforts to achieve greater food security continue to receive attention. One proposal is that the United States modify its food-aid program to guarantee to any developing country that, in any year when its grain production falls more than a given percentage below trend, the shortfall would be supplied out of U.S. stocks.

Another suggestion is that international financial arrangements be established, under the International Monetary Fund, for example, to protect members against undue fluctuations in the cost of imports. Drawing rights on the fund could depend, for example, on the difference between the cost of imports and export earnings. Simulations based on the food-crisis experience of 1962–1965 indicate that such schemes for "compensatory financing" would be less expensive than the establishment of international buffer stocks necessary to meet food-security objectives.

FOOD AID: FROM DISPOSAL TO DEVELOPMENT

The United States has utilized programs designed to facilitate the noncommercial disposal of surpluses to poor countries in order to reduce the surpluses of wheat and other commodities that accumulated during the period of high domestic prices and ineffective control programs of the 1950s and early 1960s. This program is carried out primarily through the Agricultural Trade and Development Act of 1954, generally known as Public Law (PL) 480.

The original PL 480 consisted of three titles: Title I, local currency sales; Title II, government-to-government emergency food aid; and Title III, domestic food distribution, foreign food distribution, and barter. It was first developed as a temporary program to eliminate surplus stocks in storage. It was hoped that U.S. farm prices would increase as sales and disposals came into closer balance with production and that this would permit the government to retire from intervention in agriculture. As it became clearer to Congress and to the Administration that the problem of excess capacity could not be solved with the control programs then in force, PL 480 was periodically extended, but it continued to be considered a short-term, or temporary, program.

Over time, other goals began to supplement the surplus-disposal objectives of PL 480. Increased attention was given to the possibilities of using food as a means of supporting economic development in low-income countries. In 1960 and 1961, greater efforts were made to implement this concern through long-term sales contracts to such countries as India, and partial wage payments in the form of food to workers on development projects. Furthermore, the emphasis of the program, particularly the sales for foreign currencies, was shifted toward those countries with heavy or rapidly growing populations and limited foreign-exchange earnings.

In addition, the humanitarian objectives of the program have been expressed in the form of (1) assistance programs after earthquakes, floods, drought, and other natural disasters; (2) support of efforts by CARE (Cooperative for American Relief Everywhere) and UNICEF (United Nations Children's Fund) to provide food and other aid; and (3) efforts by religious organizations to improve economic well-being.

Food aid to developing countries has been a substantial portion of U.S. agricultural exports since the program was initiated in 1954. During most of the

1960s, annual program expenditures were roughly between $1.5 billion and $1.6 billion. This amounted to 25–30 percent of the total of U.S. agricultural exports. In the case of wheat, PL 480 shipments accounted for more than 60 percent of total exports in the mid-1960s. Program expenditures in the late 1960s and early 1970s fell to the $1.0 billion range and accounted for approximately 15 percent of agricultural exports. By the late 1970s, food-aid expenditures were running in the neighborhood of $1.5 billion. This was less than 6 percent of the value of agricultural exports. Exports of wheat under PL 480 auspices amounted to 10–12 percent of wheat exports. PL 480 shipments were also substantially smaller in physical volume than the amounts being exported in the form of food aid a decade earlier. Because of the shift in food aid toward reinforcing national development efforts and toward meeting the basic food needs of the poorest segments of society in the developing countries, its positive impact may actually have been greater than when larger volumes were being shipped. In spite of its declining absolute volume, food aid still accounts for 20–25 percent of the U.S. foreign-assistance effort.

Two factors were important in shifting the emphasis of United States agricultural-trade and food-aid programs away from a price-stabilization and commodity-disposal focus and toward a development focus in the mid-1960s. One important element was the increasing effectiveness of the commodity programs of the early and middle 1960s in balancing U.S. food and fiber production with the growth of domestic and foreign demand. Farm surpluses were reduced to near the levels required for long-term price-stabilization policy. This permitted a shift in emphasis from simply disposing of accumulated surpluses to programming the use of U.S. agriculture's capacity for production to meet the demands of the domestic market, the requirements for profitable commercial export, and the need for food aid.

A second factor was the growing realization that the food needs of the developing countries would have to be met, for the most part, by the countries themselves by expanding their own capacity to produce. This change in perspective was reflected in the extension of PL 480 by Congress in 1966. The

Act directed the President to place particular emphasis on assistance to countries that were taking appropriate steps to improve their own productivity. Consistent with this objective, Congress authorized the use of local currencies generated by PL 480 sales to promote food production in the developing countries.

This change in perspective by Congress, from disposal to development, was instrumental in enabling the U.S. Agency for International Development (AID) to focus a much more significant part of its total effort on assistance in the improvement of food-production capacity in the developing countries. This trend was reinforced by the 1975 Foreign Assistance Act, which mandated that a larger share of food aid be used for development purposes. The 1977 act amended the earlier legislation to place greater emphasis on meeting the food needs of the poorest people in developing countries.

At the same time that the food-aid efforts of the United States were becoming focused on development and nutrition, changes were taking place in some developing countries that would, by the late 1960s, begin to reduce their dependence on food aid. By the late 1960s, substantial progress was being made in the development of new high-yielding, fertilizer-responsive varieties of wheat, corn, and rice suitable for the tropics (see Chapters 26 and 27).

The views of the policy makers of many of these countries regarding the potential role of agriculture in national economic development also changed. In India, for example, the drought of 1965 and 1966 impressed on the Indian government its vulnerability to external political pressures as a consequence of its heavy reliance on food aid. As a result, the Government of India embarked on a "new agricultural strategy" designed to take full advantage of the new cereals-production technology. The achievement of self-sufficiency in food-grain production in India was delayed by a drought in 1972. However, by the mid-1970s, India, once the world's largest recipient of food aid, had been able to eliminate the need for food aid.

The net effect of all of these changes, then, both in the United States and in some of the developing countries, was, by the mid-1970s, to reduce substantially the absolute volume of both bilateral and multilateral food aid, and to focus the remaining food aid more directly on the support of developmental and nutritional goals.

The reform of food-aid objectives does not, however, assure that the objectives will be achieved. If the food supplied by aid programs is simply released into the market and allowed to depress local food prices, it can have a negative impact on the incentives needed to expand domestic food production. A major challenge in making more effective use of food aid is to design programs that enable the poor to be fed while at the same time improving the capacity of the country to meet its own food needs. One method to achieve this objective is to use the food aid to improve the health and education of the young—to invest in the development of human resources. This will require the design of food-aid programs that focus on the nutritional needs of children, pregnant women, and lactating mothers and that support school feeding programs that enhance nutrition and nutrition education.

MULTILATERAL TRADE NEGOTIATIONS

International trade relations were disrupted during the depression of the 1930s and World War II. After the war, efforts were made to establish a new basis for a worldwide system of liberal trading relations. After a series of negotiations, a General Agreement on Tariffs and Trade (GATT) was established as an international agency with responsibility for implementing a code of conduct for international trade.

The objective of GATT was to expand trade through negotiated reductions in the obstacles to trade. Four basic principles were adopted.

1. Trade should be nondiscriminatory. All members of GATT should receive equal treatment regarding import and export duties and charges.

2. Domestic industries should be protected mainly by tariffs. In order to get agreement, agriculture was exempted from the provision. Import quotas on agricultural products were permitted if domestic production restrictions (such as those employed by the United States) are in force or if a country is experiencing severe balance-of-payment problems.

3. Agreed-upon tariff levels bind each country. Changes should be arrived at through consultation in order to avoid excessive damage to the interests of other member countries. If tariffs are raised above the agreed-upon levels, the injured countries must be compensated. Consultation procedures to settle disputes were established.

4. GATT provides the appropriate institutional framework within which to hold negotiations for the reduction of tariff and nontariff trade barriers.

Since its inception, seven major "rounds" of negotiations have taken place under GATT auspices. A major accomplishment has been a substantial reduction of industrial tariffs. Little progress was made in freeing agricultural trade. In the 1950s, the United States requested a waiver from the general rules to legalize its use of import quotas to protect domestic dairy producers. This waiver then became an obstacle to the reduction of nontariff trade barriers in agriculture. Other nations have cited the U.S. waiver when they have wanted to resist attempts to improve access to their protected markets.

The most recent round of GATT negotiations (the Tokyo/Geneva Round) began in Tokyo in 1973 with nearly 100 participating nations. A major effort was made by the United States to obtain greater access to markets for U.S. agricultural commodities. Attention was shifted from tariffs to nontariff barriers—variable levies, import quotas, export subsidies, packaging and labeling standards, government procurement practices, custom valuation methods, import licensing, and sanitary regulations.

As tariff protection has been reduced, there has been increasing use of nontariff barriers to protect domestic producers from foreign competition. An example has been the use, by the United States, of grading standards to prevent the importation of

Mexican tomatoes in order to protect winter tomato growers in Florida.

When the Tokyo/Geneva Round was completed in April 1979, a number of modest gains had been achieved and principles established that should lead to further expansion of agricultural trade, particularly with Japan and the European Economic Community (EEC). The gains to agriculture from this latest round of negotiations fall into three categories:

1. Tariff bindings. The Japanese, for example, bound themselves not to increase tariffs on soybeans. A similar "binding" agreement with the EEC in the 1963 round has been partially responsible for the rapid growth of soybean exports to the EEC.

2. Tariff cuts. Tariff cuts were negotiated with Japan, the EEC, Canada, and several developing countries, on a number of products, including vegetable oils, soybean meal, orange juice, tobacco, fruits, vegetables, pork, and poultry.

3. Quota increases. A new precedent—that quotas are a legitimate subject of multilateral negotiations—was established in the Tokyo/Geneva Round. Substantial relaxation of quotas on beef, oranges, and orange and grapefruit juice was negotiated with Japan. Relaxation of nontariff barriers on beef was negotiated with the EEC, Austria, and Switzerland.

In the process of negotiating freer access to foreign markets for U.S. agricultural products, the United States agreed to enlarge its quotas on cheese imports and to reduce tariffs on a number of products, including coconut oil, canned pineapple, barley, lamb, and wool.

In the language of trade negotiations, the reduction in trade barriers by an importing country is referred to as a concession. From an economic perspective, it should be counted as a gain. Consumers in the importing countries and producers in the exporting countries gain from the more efficient organization of commodity production and consumption. Profes-

Figure 25-7   The U.S. agricultural trade balance. (1978–1979 is a forecast.) [Courtesy USDA/ESCS.]

sor James P. Houck of the University of Minnesota has estimated that the annual net gains in agricultural trade for the United States arising from the Tokyo/Geneva Round will amount to somewhere in the neighborhood of $500 million in terms of 1980 price levels.

AGRICULTURAL TRADE IN THE WORLD AND U.S. ECONOMIES

Since the early 1950s, the volume of production of all goods, agricultural and nonagricultural, has more than tripled. Total world trade has increased fivefold. During this same period, agricultural production has approximately doubled and the volume of agricultural trade has nearly tripled.

Agricultural trade has fallen from about 30 percent of the total world trade value in the period 1950–1954 to approximately 15 percent in the late 1970s. There are three reasons for this decline. First, agricultural production has expanded less rapidly than industrial production. Second, agricultural prices have fallen relative to the prices of other primary products, such as minerals and fuels, while maintaining a rough parity with the prices of manufactured goods. Third, the barriers to trade discussed above have restricted the growth of agricultural trade more than industrial trade. In spite of this lag, the nations of the world have become more dependent upon one another for food, fiber, and agricultural raw materials than at any time in history.

As world markets for agricultural products have expanded, the United States has tended to occupy an increasingly important role in world agricultural trade. The United States now accounts for more than 16 percent of all world agricultural exports and for almost 10 percent of world agricultural imports. Agricultural trade is an important contributor to the nation's balance of payments (Fig. 25-7).

The continued importance of agricultural exports in an economy in which agriculture used only about 4 percent of the labor force and contributed 3 percent of the national output is due to a very significant extent to the rapid gains in productivity that have resulted from technical change in American agriculture. The shift from a resource-based to a science-based agriculture has enabled agriculture in the United States to compete more effectively with other sectors of the economy for labor and capital. The growth of U.S. agricultural exports has reflected also the rapid growth of income and the changing composition of demand in Japan, Western Europe, Eastern Europe, and the USSR. As their incomes have risen, consumers in these nations have increased their consumption of meat and poultry. This increase has created a rising demand for imports of feed grains, and the developed countries of Western Europe and Japan now account for a very large share of U.S. exports (Figs. 25-7, 25-8, and 25-9). The agricultural imports into the United States, however, come primarily from the less developed countries (Figs. 25-9, 25-10, and 25-11). But the growth in demand for these commodities—cocoa, coffee, tea, bananas, sugar, and natural rubber—is slow.

The shifts in demand toward livestock products and feed grains that have occurred since World War II have tended to reinforce the long-term disequilibrium in trade patterns between the developed and developing countries. This disequilibrium has also been reinforced by the more rapid rates of technical change and productivity growth in the developed than in the developing countries.

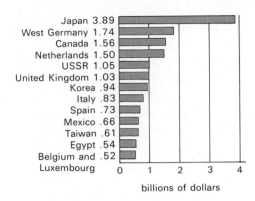

Figure 25-8 Where we shipped our agricultural exports in 1977. [Courtesy USDA/ESCS.]

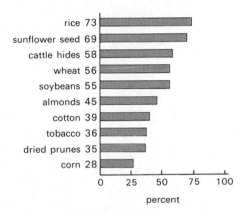

Figure 25-9 Percentage of U.S. farm products exported during the year ending September 30, 1978. The data were partially estimated. [Courtesy USDA/ESCS.]

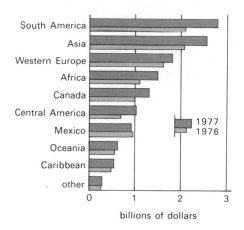

Figure 25-10 Where we got our agricultural imports in 1976 and 1977. [Courtesy USDA/ESCS.]

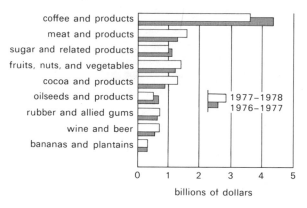

Figure 25-11 Leading U.S. agricultural imports by value in 1976–77 and 1977–78 (October–September years). The data for 1977–78 were partially estimated. [Courtesy USDA/ESCS.]

FOOD PRODUCTION AND FOOD SECURITY IN THE 1980s

The 1970s have not been kind to the reputations of the prophets who have attempted to plot the course of world food production. Perspectives have shifted from a sense of impending catastrophe engendered by the world food crisis of the 1960s, to the euphoria of the new potentials opened up by the "green revolution," to the crunch on world grain stocks and the explosion of grain prices in 1970–1974.

The resurgence of world grain production and stocks in the late 1970s appears to be leading to a new round of complacency about the prospects for world food supplies and the stability of agricultural markets. Yet the simple mathematics of economic growth, population increase, uncertain weather, inelastic food demand, and improvident food-stock policies suggests the possibility of another food crisis of global significance before the end of the 1980s.

Most recent projections suggest a widening gap between food production and demand in the developing countries over the next decade. The International Food Policy Research Institute (IFPRI), for example, has projected an increase in the staple food deficit of the developing market economies from approximately 12 million metric tons in the mid-1970s to 75–80 million metric tons in 1990 (see Fig. 4-5). The IFPRI projections are not predictions of what will happen. They are extrapolations of the implications of recent trends.

Deficits of the magnitude projected by the IFPRI would, if compensated for by commercial imports, represent an annual drain on foreign exchange earnings of the developing countries of more than $20 billion at 1980 prices. It seems clear that the developing countries would face great difficulty in financing food imports of this magnitude, in addition to the burden they face from rising petroleum prices, without compromising other development objectives. The projected deficits will have to be reduced through some combination of slower growth in consumption, more rapid growth in domestic production, and expanded food aid.

The next three chapters are devoted to an exploration of some of the possibilities of increasing commodity production and enhancing the development of rural areas.

Selected references

Christiansen, M. K. (editor), 1978. *Speaking of Trade: Its Effect on Agriculture* (Special Report 72). St. Paul, Minn.: University of Minnesota Agricultural Extension Service. (A useful introduction to agricultural-trade policy issues designed for undergraduate classroom or public-affairs extension use. Includes a historical perspective on the development of international trading relations; an exposition of the theory of trade and protection; discussion of the relation between international trade and domestic agricultural policies; the strengths and weaknesses of international commodity agreements; and an overview of major commodity flows.)

Food and Agriculture Organization, 1978. *The State of Food and Agriculture, 1978.* Rome: Food and Agriculture Organization of the United Nations. (An annual report published by the FAO on the world food and agricultural situation. Each annual issue presents an in-depth review in a specific area of food and agricultural development or policy, in addition to a review of current developments in agricultural commodity markets.)

Houck, J. P., 1979. Agricultural trade: protectionism, policy and the Tokyo/Geneva Negotiating Round. *American Journal of Agricultural Economics*, 61(5). (A careful analysis of the implications of the latest round of GATT negotiations for agricultural trade.)

Johnson, D. G., 1975. *World Food Problems and Prospects.* Washington, D.C.: American Enterprise Institute for Public Policy Research. (An analysis of the sources of the 1972–1974 world food crisis and of the policies that have been proposed to stabilize international commodity markets.)

Schuh, G. E., 1976. The new macroeconomics of agriculture. *American Journal of Agricultural Economics*, 29(5):802–811. (A useful exposition of the implications of international monetary policy, particularly the shift from fixed to floating exchange rates, for international trade in agricultural commodities.)

26

The Organization and Management of Agricultural Research Systems

A belief that the application of science to the solution of practical problems is a sure foundation for human progress has been a persistent theme in American intellectual and economic history. During the two decades after World War II, this belief was seemingly confirmed by the dramatic association between the progress of science and technology and rapid economic growth. The technological revolution in American agriculture (see Chapter 27), the growth of industrial productivity, and the contributions of science to military and space technology seemed to reinforce this perspective.

By the late 1970s, however, the formula that had permitted the United States to move into a position of scientific, economic, and political leadership in the world community was being seriously challenged. The opposing view that has emerged is that the potential consequences of the power created by modern science and technology—as reflected in the cataclysm of war, the degradation of the environment, and the psychological cost of rapid social change—are obviously dangerous to the modern world and to the future of civilization. Science, including agricul-

tural science, is regarded by increasing numbers of people as a new vested interest exerting an irrational claim on financial support, rather than as an efficient source of national growth. The result has been to question seriously the significance of scientific progress, technical change, and economic growth for human welfare.

This change in perspective has also posed a new challenge to science administrators in universities, government, and business, and consequently the impetus to redirect funds to assure their use in ways that are scientifically valid, economically efficient, and socially viable is stronger than ever before. In agriculture, then, as well as in other disciplines, this shift is leading to a search for new allocative methodologies and for wider definitions of the objectives of scientific and technical effort.

A HISTORICAL PERSPECTIVE

Public support for education and research as an instrument of economic progress is a relatively recent institutional innovation. In the United Kingdom, the basis for modern agricultural science was established in the later part of the eighteenth and the first half of the nineteenth centuries, the responsibility for agricultural improvement being carried out by the owners of the great estates. For example, the famous Rothamsted Experimental Station was established in 1843 and financed personally by Sir John Bennet Lawes for the rest of the nineteenth century. The Edinburgh Laboratory (founded in 1842) was supported by the Agricultural Chemistry Association of Scotland, a voluntary agricultural society.

In contrast, a publicly supported agricultural experiment station was successfully established in Germany (at Möckern, Saxony) in 1852. A group of Saxon farmers drafted a charter for the station, which the Saxon government legalized by statute, and secured an annual appropriation from the government to finance the operations of the experiment station. Although the German system of agricultural research evolved later than the British, it provided a more effective environment for the "enlargement" of new scientific and technical knowledge. The development of publicly supported agricultural research institutions in Germany reflected the emergence of a social and political climate that fostered a perception of science and technology as instruments of economic growth, and their advancement as a major responsibility of the state.

In the United States, the institutionalization of public-sector responsibility for research in the agricultural sciences and technology can be dated from the 1860s. The Act of May 15, 1862, "establishing the United States Department of Agriculture," and the Act of July 2, 1862, "donating public lands to the several states and territories which may provide colleges for the benefit of American agriculture and the mechanic arts," became the first federal legal authority under which a nationwide agricultural-research system was to develop.

The institutional pattern that emerged for the organization of agricultural research in the United States drew heavily on the German experience. Samuel W. Johnson of the Yale Analytical Laboratory, who had studied agricultural chemistry in Germany (under the great organic chemist Baron Justus von Liebig), was familiar with the German experiment-station movement.* He became America's first advocate of state agricultural experiment stations. The first one, the Connecticut State Agricultural Experiment Station, was not established, however, until 1877. When the Hatch Act, which provided federal funding for the support of the land-grant college experiment stations, was passed in 1887, only six state experiment stations had been established.

In institutionalizing agricultural research, the United States created a dual federal–state system. The federal system, administered by the United States Department of Agriculture, developed more rapidly than the state system. Yet it was not until the later years of the nineteenth century that the Department of Agriculture achieved any significant capacity to provide the scientific knowledge needed to deal with the urgent problems of agricultural development. The ability of the land-grant colleges to produce new scientific and technical knowledge for agricultural development was even more limited than that of the Department of Agriculture. It was well after the turn of the century before the new state experiment stations could be regarded as productive sources of new knowledge or significant contributors to the growth of productivity in U.S. agriculture. In 1900, after 50 years of "institution building," expenditures for agricultural research by the state experiment stations and by the Department of Agriculture were less than $2.0 million.

The U.S. federal–state agricultural research system developed rapidly after 1900. In 1914, upon the passage of the Smith-Lever Act, a firm institutional basis, in the form of a cooperative federal-state extension service, was established for the educational functions of the Department of Agriculture and the

---

*The German experience also played a seminal role in the development of agricultural science in Japan. Japan initially (1870) invited British and U.S. scientists and educators to assist in the development of Japanese agricultural colleges and experiment stations. The failure of these efforts led to their replacement by German agricultural scientists and an emphasis on German agricultural chemistry and soil science rather than on mechanized agriculture of the Anglo-American type. The result was a direction of scientific effort to achieve increases in output per hectare rather than output per worker.

A

B

Figure 26-1 Installations of the United States Department of Agriculture. (A) Agricultural Research Center, Beltsville, Maryland. (B) Federal Experiment Station in Puerto Rico, Mayaguez.

land-grant colleges. By the early 1920s, a national agricultural research and extension system had been effectively institutionalized at both the federal and the state levels. The institutional innovations in the organization of public sector agricultural research, development, and extension between 1860 and 1920 proved capable, during the 1925–1965 period, of absorbing new resources at a relatively rapid rate and of generating rapid growth in new scientific and technical knowledge. Figure 26-1 shows two USDA installations that are currently in operation.

The rapid growth in public resources devoted to agricultural science and technology after 1900 was clearly a response to economic demand. Throughout the nineteenth century, U.S. agriculture had experienced rapid growth of output and labor productivity as a result of expansion in the amount of land under cultivation and the "horse mechanization" of land preparation and harvesting operations. Industry was able to supply the necessary advances in mechanical technology with little help from public-sector research. Between 1900 and 1920, however, the constraints on agricultural growth imposed by an increasingly inelastic supply of land became more severe. One index of this problem was that corn prices more than doubled, relative to the general price level, between 1895 and 1920. The economic conditions had been established for the development of a new biological and chemical technology to complement the increasingly effective mechanical technology.

The successful "socialization" of agricultural research necessitated the establishment of viable organizational patterns, both federal and state, that were capable of generating effective financial support. In the U.S. Department of Agriculture, the result was the organization of scientific bureaus, such as the Bureau of Plant Industry and the Bureau of Animal Industry, focusing on a particular set of problems or commodities. These innovations served to generate long-range visibility for the research effort and support by client groups. In the states, experiment-station research efforts were oriented to meet local and regional needs. Experiment-station directors were able to generate financial support from state legisla-

tures by directing their research in such a way that it had a direct impact on state economic development. Minnesota farmers were provided with corn and soybean varieties adapted to shorter growing seasons, for example. Consequently, there evolved an effective feedback relation between the effectiveness of the applied research effort and the economic viability of the state experiment station. And continuous competition, as well as cooperation, developed between the state and federal research systems.

This system did not lead to uniform growth rates in the public resources devoted to agricultural research. In fact, a marked disparity in the funding of research has existed among states and among commodities. In general such funding lagged in the South, the Appalachian region experiencing the greatest retardation. North Carolina, Louisiana, and Texas are important exceptions to this general pattern, however. In the Pacific states and the Midwest, agricultural research has generally been well supported. In the East, support for agricultural research has generally not been strong, except in New York and Pennsylvania. Among commodities, horticultural crops and cotton have received the highest levels of support, and field crops have received the lowest levels, relative to the value of production (Table 26-1, Fig. 26-2).

As the agricultural-research system has grown, the source of funding has become increasingly diverse and the management of the system has become increasingly complex. In the late 1970s, industry was conducting approximately half of the agricultural research; the U.S. Department of Agriculture, 20 percent (Fig. 26-3) and the state agricultural experiment stations, 26 percent. Of the research conducted at the state stations, approximately two-thirds was funded by state appropriations.

The returns to U.S. farmers and consumers from the investment in agricultural research have been high by any standard (Table 26-2). During the 1960s and the 1970s, however, the system came under considerable criticism. Support for agricultural research, particularly at the federal level, was weakened (Fig. 26-4).

The sources of criticism and the adjustment in re-

786

Table 26-1 Expenditures on research by state stations and the USDA per thousand 1967 dollars of product.

Year	Field crops	Horticultural crops	Livestock
1969	2.66	12.57	3.01
1975	1.76	10.68	3.50

Source: R. E. Evenson, P. E. Waggoner, and V. W. Ruttan, "Economic Benefits from Research: An Example from Agriculture," *Science* 205 (4411):1106, 1979. Copyright © 1979 by the American Association for the Advancement of Science.

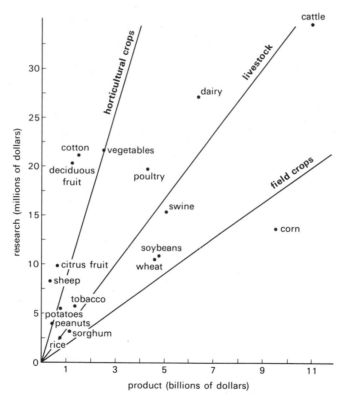

Figure 26-2 Expenditures (in 1967 dollars) of state stations and the USDA on research on 16 products as a function of the gross income for the product in 1975. [From R. E. Evenson, P. E. Waggoner, and V. W. Ruttan, "Economic Benefits from Research: An Example from Agriculture," *Science* 205(4411):1104, 1979. Copyright © 1979 by The American Association for the Advancement of Science.]

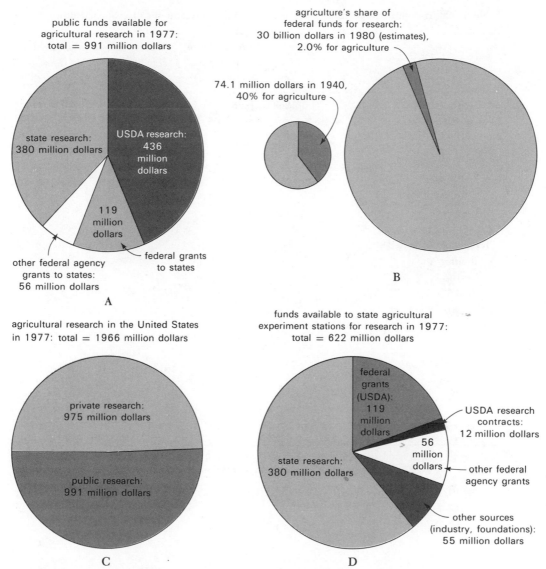

Figure 26-3 Agricultural and forestry research in the United States. (A) The sources, by amount and proportion, of public funds available for agricultural research in 1977. (B) Agriculture's share of federal funds for research for the years 1940 and 1980. (C) The amount of funds devoted to agricultural research by the public and private sectors in 1977; figures for private expenditures are estimates. (D) The amounts and sources of funds available to state agricultural experiment stations for research in 1977. [After USDA Science and Education Administration, Current Research Information System (CRIS) printout, 1979.]

Table 26–2 Estimated impacts of research and extension investments in U.S. agriculture.

Subject	Annual rate of return (percent)	Percentage of productivity change realized in the state undertaking the research
1868 to 1926		
All agricultural research	65	[1]
1927 to 1950		
Agricultural research		
Technology-oriented	95	55
Science-oriented	110	33
1948 to 1971		
Agricultural research		
Technology-oriented		
South	130	67
North	93	43
West	95	67
Science-oriented	45	32
Farm management and agricultural extension	110	100
1969		
Commodity groups		
Cash grains	36	[1]
Poultry	37	[1]
Dairy	43	[1]
Livestock	47	[1]

Source: R. E. Evenson, P. E. Waggoner, and V. W. Ruttan, "Economic Benefits from Research: An Example from Agriculture," *Science* 205 (4411):1105, 1979. Copyright © 1979 by the American Association for the Advancement of Science.
[1]Not estimated.

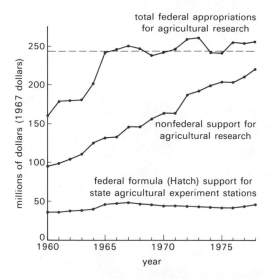

Figure 26-4 Purchasing power of federal appropriations and nonfederal support of agricultural research programs in the United States for fiscal years 1960–1978. [After B. J. Liska and J. Havlicek, Jr., "Statement on Experiment Station Committee on Organization and Policy, Fiscal Year 1981 Budget Recommendations," USDA Science and Education Administration, September 27, 1979.]

search priorities and in the organization of the agricultural-research system will be discussed later in this chapter.

THE AGRICULTURAL EXPERIMENT STATION

The basic functional unit in any national agricultural research system is the experiment station, research center, or laboratory.° The productivity of the human and material resources that are devoted to the development and operation of an agricultural research station, whether operated as part of the federal–state system or in the private sector, must be

---

°This section is adapted from Melvin G. Blase and Arnold Paulson, "The Agricultural Experiment Station: An Institutional Development Perspective," *Agricultural Science Review* 10:11–16, 1972.

evaluated in terms of the station's contribution to social progress. The experiment station can be viewed as a system for transforming raw inputs (human and physical capital) into outputs (new knowledge). This new knowledge becomes available in the form of research papers, books, bulletins, information releases, and consultation with other scientists and producers. It is frequently embodied in blueprints and formulas. And its social and economic impact is ultimately realized in the form of technical or institutional change.

Research system output

One useful depiction of an institution, such as a research station, is shown in Fig. 26-5. Stock and flow resources are combined into intermediate products. These products are, in turn transformed into final outputs. The three final outputs can be categorized as information, capacity and influence.

The most important and visible output of an experiment station or a research laboratory is the **information,** or new knowledge, that is generated and released. In the long run, the use of resources for agricultural research must be justified in terms of the economic value of the output of the new knowledge that it produces.

If a research institution is to remain a valuable social asset, it must also devote resources to the enlargement of its own **capacity.** This means diverting some resources from the production of information that has immediate application. It means expanding the capacity of its scientific staff by allowing more time to be designated for graduate education, study leaves, and "basic research." It also includes modernization of facilities, administrative structure, and the ideology that serves as the rationale for a research program. The frequent arguments about the amount of emphasis that should be placed on "basic" research, and the amount on "applied" research, often reflect a more fundamental disagreement about how much emphasis should be given to the production of current services relative to investment in future capacity.

All successful research institutions devote significant amounts of resources to increasing their **influ-**

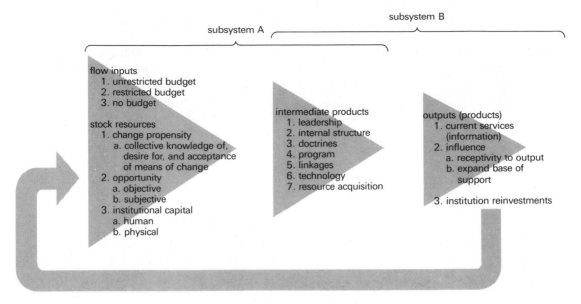

Figure 26-5   A systems model of experiment-station performance and development. [After Blase and Paulson, *Agricultural Science Review* 10(2):13, 1972.]

ence. In order to establish a successful claim on current and future resources, a research organization usually finds it necessary to maintain effective relations with funding agencies—legislative bodies, operating bureaus and divisions, and private foundations. The organization may also find it useful to devote effort to building a public image as a valuable institutional resource. Although the resources devoted to the production of influence may have little direct value to society, such activities are essential to the maintenance and continuity of both public and private research institutions. Institutions that have outlived their social function as producers of information frequently devote excessive effort to organizational maintenance activities.

Intermediate products

The intermediate products or services identified in Figure 26-5 are for internal use. They have little direct value to society. But they are indispensable to the research institution itself. They represent the "en-

gine" that determines the efficiency with which the research system makes use of the resources available to it. By and large, these intermediate services must be produced by the research institution itself rather than be purchased from external sources.

The **linkage services** include the establishment and maintenance of contacts and relations with individuals and institutions outside the research station or laboratory, with other scientists, with the clients who use the services of the station, and with sources of support. The linkages carry messages in both directions. It is through the linkages with the outside that a research institution influences technical and institutional change and is in turn influenced by the external changes in society and in science and technology. The state agricultural experiment stations have an exceedingly complex set of linkages with the external environment—with state crop-improvement associations, community and regional development councils, producer and consumer interest groups, the extension service, and the entire hierarchy of local, state, and federal administrative and political institutions. The problem of maintaining effective communication with outside groups without becoming either a captive or an adversary is exceedingly difficult.

Leadership is an extremely important intermediate product. The myth that all a research director needs to do is hire good people and let them "do their thing" has only minimal support at a time when the solution to significant technical and social problems requires concerted multidisciplinary research effort. Leadership must be sensitive to changing social goals and effectively transmit their implications to the scientific staff. Leadership also includes **resource acquisition.** It must be able to visualize and communicate effectively the potential contribution of the experiment station or research laboratory to the solution of emerging problems. Then, it must be capable of mobilizing and allocating both financial and scientific resources to produce the high returns from scientific and technological effort that society has come to expect. To do so, not only must leadership acquire the necessary staffing and finances, but it must also create an environment in which they can be used productively. The **technology,** or the methodology, of research is in continuous flux. The research program must be organized in such a way that the research staff is aware of the advances in its own and closely related fields and contributes to them.

Doctrine is reflected in the articulation of institutional goals and philosophy and in the operating style of an institution. The traditional production-oriented doctrine of the state and federal agricultural-research system has experienced severe stress as a result of the increasing pressure to give more emphasis to the environmental "spillover" effects of technical change in agriculture and to the problems of personal and community development in the formulation of research priorities. The operational manifestation of a shift in doctrine is the reformulation of a research station's **program.** Modifications in doctrine are sterile unless accompanied by the allocation of research resources in such a way that they contribute to the realization of the revised priorities. Many modifications in program imply drastic revisions in the **internal structure** of a research station. Problem-oriented interdisciplinary "centers" erode the decision-making authority of discipline-oriented departments. During periods of tension, the reformulation of doctrine, the redirection of program, and the reorganization of internal structure may absorb substantial resources and seriously compete with the production of information. These efforts must be justified primarily in terms of their impact on the future productivity and viability of the experiment station.

THE RESEARCH-AND-DEVELOPMENT PROCESS

What are the processes by which an experiment station or a research institute generates new knowledge or information? The generation of information and its embodiment in the form of a new crop variety, a synthetic fiber, or a new communication device is a complicated process that requires the cooperation of large numbers of individuals and organizations. Educated people—including scientists, engineers, technicians, farmers, and laborers—are critical to the process that produces technological change and economic growth.

A striking feature of modern research and development is the extent to which it has become institutionalized. In applied research and development, the industrial laboratory and the research institute have increasingly replaced the individual inventor or the academic scientist who did research in the free time he could divert from teaching.

The process of innovation

The design of effective institutions for research and development requires a clear understanding of the processes by which new things—**innovations**—emerge in science and technology. This subject has received much attention in applied technology, sociology, and history. In economics it has occupied only a peripheral, though expanding, role.

The distinction between the role of insight and the role of skill in the emergence of a new concept or technique has posed major difficulties in developing a theory of innovation. A. P. Usher proposed the following distinction:

Acts of skill include all learned activities whether the process of learning is an achievement of an isolated adult individual or a response to instructions by other individuals . . . acts of insight are unlearned activities that result in new organizations of prior knowledge and experience. . . . Such acts of insight frequently emerge in the course of performing acts of skill, though characteristically the act of insight is induced by the conscious perception of an unsatisfactory gap in knowledge or mode of action.

Usher distinguished among the several concepts of how innovation occurs on the basis of the relative prominence given to insight and skill.

At one extreme is the **transcendentalist approach,** which attributes the emergence of invention to the inspiration of the occasional genius who from time to time achieves new insight through intuition. This theory lies behind the great-inventor concept—the assertion that Fulton invented the steamship or that Marconi invented the radio. At the other extreme is the **mechanistic-process approach,** which assumes that "necessity is the mother of invention." In this, the individual inventor is one of a number of individuals with the appropriate skills who makes the final modification that is acclaimed as an invention.

A more comprehensive view is to regard major innovations as emerging from the cumulative synthesis of relatively simple innovations, each of which requires an individual "act of insight" by a person who also possesses a high degree of skill, acquired through training and experience in the particular area of investigation. In this view, the individual invention involves four steps:

1. *Perception of the problem,* in which an incomplete or unsatisfactory pattern or method of satisfying a want is perceived.

2. *Setting the stage,* in which the elements or data necessary for a solution are brought together through some particular configuration of events or thought. Among the elements of the solution is an individual who possesses sufficient skill in manipulating the other elements.

3. *The act of insight,* in which the essential solution of the problem is found. Large elements of uncertainty surround the act of insight. It is this uncertainty that makes it impossible to predict the timing or the precise configuration of an innovation.

4. *Critical revision,* in which the newly perceived relations become more fully understood and effectively worked into the entire context to which they belong. This may again call for new acts of insight.

A major or strategic invention represents the cumulative synthesis of many individual inventions. It will usually involve all the separate steps that may be found in the case of the individual invention. Many of the individual inventions do no more than set the stage for the major invention, and new acts of insight are again essential when the major invention requires substantial critical revision to adapt it to particular uses. A schematic presentation of the elements of the individual act of insight and the cumulative synthesis as visualized by Usher are presented in Figures 26-6 and 26-7.

A major advantage of the cumulative-synthesis approach is that it clarifies the points at which conscious efforts to speed the rate or alter the direction of innovation can be effective. The possibility of affecting the rate or direction of innovation is obscured by the transcendentalist approach, with its dependence on the emergence of "the great inventor," and it is denied by the mechanistic-process approach, with its dependence on broad historical trends or forces.

The focus of much conscious effort to affect the speed or direction of innovations centers on the second and fourth steps in the process—in "setting the stage" and in "critical revision." By consciously bringing together the elements of a solution—by creating the appropriate research environment—the stage can be set in such a manner that fewer elements are left to chance. It would be inaccurate to suppose that it is yet possible to set the stage in such a manner as to guarantee a breakthrough in any particular area, although the probability of a breakthrough may be increased as more resources are devoted to research on a given problem. As more is learned about the effectiveness of various research environments, the probability that breakthroughs will be achieved should increase. For the present, it can only be emphasized how little is known about the administration of basic research.

At the level of "critical revision," considerable progress has already been made in bringing economic and administrative resources to bear. Many of the elements of critical revision require "acts of skill" in contrast to "acts of insight." The effectiveness of modern research procedures in shortening the time span from the test tube to the production line testifies to our ability to exert conscious direction at the level of applied research.

The role of skill and insight in a biological innovation: the rice-blast nursery case*

The four stages in the process of innovation—(1) *perception of the problem;* (2) *setting the stage,* or bringing together the elements of a solution; (3) *the act of insight,* which permits solution of the problem; and (4) *critical revision,* including refinement and improvement of the original solution—are ideally illustrated in the development of a recent biological innovation, the rice-blast nursery. The example also illustrates the complex interactions that frequently accompany technological change.

Rice blast is an airborne fungus disease caused by *Piricularia oryzae,* which attacks the leaves and stem of the rice plant and substantially reduces yield. It has emerged as an important disease in Southeast Asia during the last several decades. In the past, a combination of natural selection and farmer selection resulted in the use of rice varieties that were relatively resistant to local strains of the rice-blast fungus. Most of these varieties were also adapted to conditions of low fertility, but unfortunately their yields were usually low and they were not responsive to

Figure 26-6 The emergence of novelty in the act of insight. Synthesis of familiar items: (1) perception of an incomplete pattern; (2) setting the stage; (3) the act of insight; and (4) critical revision and full mastery of the new pattern. [After A. P. Usher, *A History of Mechanical Inventions,* Harvard University Press, 1954.]

Figure 26-7 The process of cumulative synthesis: a full cycle of strategic invention and part of a second cycle. Roman numerals *I–IV* represent steps in the development of a strategic invention; small figures represent individual elements of novelty; arrows represent familiar elements included in the new synthesis. [After A. P. Usher, *A History of Mechanical Inventions,* Harvard University Press, 1954.]

*This section was developed with the assistance of Dr. S. H. Ou, a plant pathologist working at The International Rice Research Institute. For a description of the blast nursery, see S. H. Ou, "Some Aspects of Rice Blast and Varietal Resistance in Thailand," *International Rice Commission Newsletter* 10 (3):11–17, 1961. See also, "Report of the Committee on the Establishment of Uniform Blast Nurseries," *International Rice Commission Newsletter* 11(1):23–32, 1962.

improved soil fertility. Attempts to increase yields by the use of nitrogen fertilizer and by introducing new varieties have been accompanied by increased virulence of the rice-blast fungus in those areas where farmers had adopted new varieties and cultural practices most extensively.

The first step, perception of the problem, was the recognition of the rising incidence of the rice-blast disease following the introduction of new rice cultivars. Increased recognition that rice-blast disease was an important factor in limiting attempts to achieve higher rice yields in tropical areas led to the search for an economically and biologically efficient technique for screening rice cultivars to be used in rice-breeding programs and the new cultivars produced by such programs.

The object of the screening was to select cultivars of rice that were highly resistant to infection by the fungus. Initial screening efforts involved inefficient laboratory techniques or inexact field observations. To complicate matters further, the fungus spores were found to be present in the atmosphere in concentrations that exceeded those that could be effectively produced under laboratory conditions.

The second step, setting the stage, involved a high degree of technical skill but relatively little new knowledge. The problem was to create an optimum environment for the growth of spores of the rice-blast fungus that provided, at the same time, favorable conditions for the mass production of rice seedlings. The components of such an environment, well known at the time, are:

1. *Adequate moisture in the environment.* The fungus spores multiplied most rapidly and attacked the rice plant most successfully under conditions of adequate soil moisture and high humidity. The importance of direct contact between the plant and the water meant that the optimum time for water application was in the evening. The influence of humidity and the requisite skills required for controlling humidity were well known by many competent researchers.

2. *High levels of nutrition.* A given cultivar of rice tends to be more susceptible to rice-blast disease when grown under high levels of nutrition, particularly nitrogen nutrition, than when grown on poor soils. This was also well known to competent research workers.

3. *Upland nursery conditions.* The rice blast develops more rapidly when grown under upland (that is, nonflooded) conditions because of the slower accumulation of silica in the tissues of the rice plant. Susceptibility to the rice-blast disease is inversely proportional to the amount of silica in the plant. This information was obtained through experimental work.

4. *Tropical temperature.* The blast spores multiply and attack the plant more effectively in the range 20–30°C (68–86°F). As tropical temperatures are typically within this range, special temperature controls are not necessary to create a favorable environment for spread of the disease.

The third step, the act of insight, was facilitated by the increasing concern with the damage caused by the rice-blast disease in Thailand during the mid-1950s, and by the presence of a pathologist in Thailand who possessed the professional skill necessary to combine the elements of the solution into the design for a blast-disease nursery that would permit effective identification of blast-resistant cultivars and identification of the races of the blast pathogen.

The fourth step, critical revision of the solution, involved the addition of refinements aimed at improving the effectiveness of the solution—for example, the addition of border rows of susceptible cultivars and the installation of automatic moisture-control devices to increase the concentration of spores and to maintain the required moisture conditions.

RESOURCE ALLOCATION IN AGRICULTURAL RESEARCH

As the agricultural-research system has grown increasingly complex, the problem of the use of research resources—research laboratories, field plots, and time spent by scientists—has also become increasingly complex. Yet the fundamental issues can be condensed into two simple questions.

1. What is it technically or scientifically feasible to accomplish?
2. What value does society place on the results?

The first question can be answered only by the most imaginative and skillful research scientists in each field. There is no alternative source of information on potential outcomes (in terms of new knowledge or new technology) from the allocation of research resources to a particular project or problem area. Furthermore, only the most productive research scientists, who are most intimately involved in an area of research, are able to supply the estimates of the resources and time required to achieve the objectives of a particular research project or program. Finally, even under the best of circumstances, the precision of these estimates is subject to great uncertainty.

Although research scientists are in a strong position to outline what can be done, they are in a weak position to evaluate the priority that society places on research in a particular field or on a particular scientific or technological innovation. Evaluations of the scientific significance and the social or economic significance of new knowledge may not be the same. A low-income developing country, such as India, may place a relatively high value on research leading to increases in crop yields or more efficient water use, whereas a high-income country, such as the United States, may place more emphasis on solving the medical problems of the aged (cancer, heart disease, and arthritis). These kinds of choices affect not only the allocation of funds for applied research, but also the selection of those areas of basic research in which research effort may have substantial application in applied research.

Both the efficient selection of research alternatives and the efficient design of a national agricultural-research strategy require that both sets of considerations—the resources needed to accomplish feasible alternatives and the value of the research product to society—be taken into account. This decision-making system—in which the individual scientist, the re-search team, the academic department, the experiment station, the government research bureau, the corporate research division, the planning and budgeting offices of the university, the federal government, or the corporation, and the state and federal legislatures all participate—has become so complex that more systematic, more formalized procedures are now being sought. This search is also a response to the new skepticism regarding the role of science in society.

The demand for more efficient decision-making in the allocation of research resources has emerged simultaneously with the evolution of a set of "third generation" planning-oriented budgeting techniques. The first generation of techniques consisted of efforts to develop increasingly effective systems of expenditure control (1920–1935). The second generation was management or performance oriented, with its major thrust devoted to (1) casting budget categories in functional terms and (2) providing work-cost measurements to facilitate the efficient performance of prescribed activities. The thrust of the third generation program budgeting approach is planning oriented. As such it has shifted the planning perspective toward a more careful evaluation of the purposes to be fulfilled by the allocation of resources to economic activity—toward systems efficiency in the place of the more limited objectives of process efficiency.

The elements of any systematic research planning and resource allocation system, regardless of its level of formalization, can be outlined in a systems model such as Figure 26-8. The steps outlined in the figure are implicit even in the more intuitive traditional decision systems.

Since the mid-1960s, there have been a number of attempts to design and implement research-evaluation and research-planning procedures that would be more sensitive both to the perceptions of research scientists regarding the opportunities for productive research effort and to changing social values and priorities.

The pressure for more effective planning of agricultural research was reinforced by a series of critical reviews and populist challenges to the effectiveness and orientation of the U.S. agricultural-research system. The system was criticized by its detractors as being a servant of the "vested interests" in agriculspecification

Figure 26-8   A modular form of research-resource allocation. [After W. L. Fishel, *Resource Allocation in Agricultural Research*, University of Minnesota Press, 1971.]

ture and agribusiness. The agricultural-science establishment was viewed by its critics as unresponsive to environmental, distributional, and humanitarian concerns. New clientele groups insisted that such concerns as nutrition, rural development, environmental impact, and the welfare of hired farm workers should receive a higher priority in the allocation of research resources.° Even the defenders of the agricultural-research system criticized it as an "aca-

demic ghetto" out of touch with current scientific and intellectual developments.

In view of the high rates of return and the rapid growth in productivity that have resulted from agricultural research, many of the criticisms were excessive. Nevertheless, they have resulted in the reorientation of research priorities and in the implementation of more coordinated research-management systems.

The national program of research for agricultural projections

One of the most ambitious research-planning attempts was the joint effort of the U.S. Department of Agriculture and the state experiment stations to project future needs for publically supported re-

---

° Among the many items in the literature of criticism of the U.S. agricultural-research system, two have been particularly significant: Rachel Carson, *Silent Spring*, Houghton Mifflin, 1962; and Jim Hightower, *Hard Tomatoes, Hard Times: The Failure of the Land Grant College Complex*, Agribusiness Accountability Project, 1972.

search. The impetus for the study originated within the U.S. Senate Appropriations Committee. It was conducted by a joint USDA–State Experiment Station Agricultural Research Planning Committee. It was carried out by a series of task forces, each composed of a broadly representative group of research administrators and scientists.

The 1977 projections that emerged from the study are summarized, in terms of scientist-years, in Figure 26-9. In general, the 1977 goals implied both a substantial increase in scientific effort and a modest shift in priorities from the commodity-production, protection, and product-marketing categories toward a consumer and community orientation.

The projections developed in the National Program study have not been closely followed. Federal support for agricultural research remained almost unchanged, in real (corrected for inflation) dollars (Fig. 26-4). Total scientist-years in the USDA, the state agricultural experiment stations, and related institutions increased by only about 500 between 1966 and 1976, from 10,500 to 11,000. This overall increase conceals an actual decline in the number of USDA scientists.

There has been, however, a shift in resources that is roughly in line with the National Program recommendations. Public sector research in product development and marketing has declined. This reflects a judgment that the research in these areas is most appropriately conducted in the private sector. There has been a substantial reallocation of resources to the areas of consumer health and nutrition, rural development and consumer services, and the environment.

Reorganization of the research system

Another significant result of the study was the establishment of the computer-based Current Research Information System (CRIS) for the storage and retrieval of USDA and state research information. The establishment of this system, and its gradual improvement, are producing an information base for improved agricultural-research planning at both the federal and the state levels.

In 1977, the U.S. Congress authorized the estab-

lishment of a Science and Education Administration to coordinate all USDA support for agricultural research and extension work (Fig. 26-10). An effort was also made to assure more effective coordination of public- and private-sector agricultural research through the establishment of two advisory bodies— the Joint Council on Food and Agricultural Sciences and the National Research and Extension Users Advisory Board.

An attempt was also made to design new research-support instruments to increase the responsiveness of the research system to the new priorities in both basic and applied research. Support for research at the predominately black agricultural colleges in the South, referred to as "1890 colleges" because of the year they were authorized by Congress, was provided through direct grants rather than through the state experiment stations located at the land-grant universities. A new competitive research-grant system was initiated to complement the traditional formula-support system. Under the Hatch Act, federal funds are allocated to the state agricultural experiment stations by a formula based on the number of farms and the size of the rural population in each state. The competitive grant funds are not restricted to scientists connected with the state experiment stations but are awarded on a competitive basis, for research in priority areas.

Underinvestment in agricultural research

Why has investment in agricultural research grown so slowly since the mid-1960s? The data presented in Table 26-2, as well as a large body of other research data, suggest that the social rates of return realized from agricultural research are high compared to almost any other investment available to the U.S. economy. Agricultural research is like an undervalued stock with a low price–earnings ratio. Yet the data in Table 26-1 indicate that the investment in research to improve productivity in grain production decreased from $2.66 to $1.76 per $1000 of grain produced between 1969 and 1975. In most industries, a high rate of return is a signal that returns could be increased by increasing investment. Why, despite the

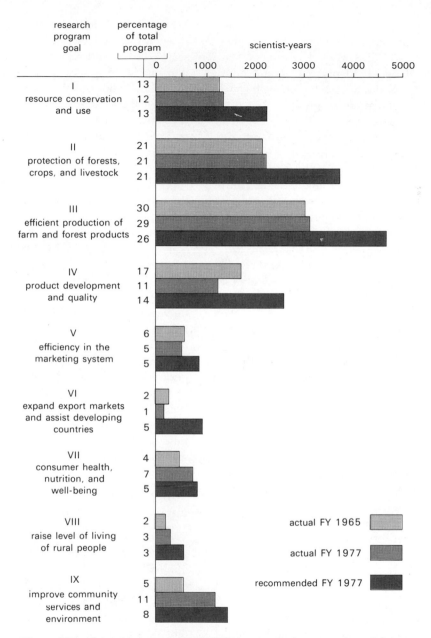

Figure 26-9 Scientist-years in SAES-USDA program by research goal for fiscal years 1965 and 1977 compared to recommendation for 1977. [After U.S. Department of Agriculture, *A National Program of Research for Agriculture*, October 1966; U.S. Department of Agriculture, *Inventory of Agricultural Research*, USDA Science and Education Administration, March 1979.]

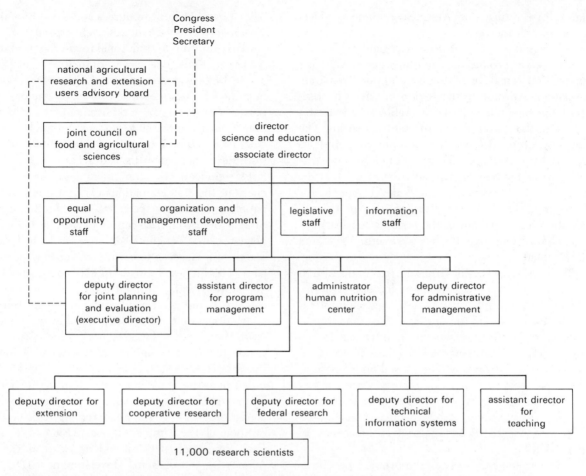

Figure 26-10 The organization of research supported by the U.S. Department of Agriculture. In cooperation with federal departments and agencies, the states, state agricultural experiment stations, colleges and universities, and user groups, the Science and Education Administration conducts and supports research, extension, and teaching programs in the food and agricultural sciences responsive to local, state, regional, national and world needs. These needs include human nutrition, home economics, consumer services, family life, rural and community development, agricultural energy, agricultural economics, environmental quality, natural and renewable resources, forestry, range management, animal and plant production and protection, aquaculture, and the processing distribution, marketing, and utilization of food and agricultural products. [Courtesy USDA/ARS.]

high rates of return, does America continue to undervalue agricultural research?

One reason is spillover. Research paid for by one state increases productivity in other states. Although agricultural research, particularly applied research, is relatively specific to the region in which it originates, the research reported in Table 26-2 indicates that other states also receive substantial benefits. The state legislature in Iowa, for example, may limit its support for research on the genetics of corn because only part of the benefits will accrue to Iowa farmers and consumers. Although the federal support for state agricultural research represents a partial compensation for the spillover, the formula for distributing federal funds has still left agricultural research undervalued.

The division of benefits from agricultural research between producers and consumers has also been important in limiting the support for agricultural research. If the demand for agricultural commodities is inelastic, or growing slowly (as it has been in the United States during most of the last 50 years), a large share of the gains will be passed on to consumers in the form of lower commodity prices. If demand is elastic, or growing rapidly (as when overseas markets are expanding), producers may retain a relatively large share of the gains from increases in productivity.

Since consumers have been the primary beneficiaries of agricultural research, one would think that they would insist on increased investment in research. The individual consumer, however, receives only a small share of the benefit from higher productivity. As a result, consumers have tended to support agricultural research actively only during periods of rapid inflation in food prices. This support has been episodic. Agricultural research remains undervalued by consumers as well as producers.

INTERNATIONAL SPECIALIZATION AND COOPERATION IN AGRICULTURAL RESEARCH

A major reason for low productivity in agriculture in the developing countries of the tropics has been the lack of capacity to produce an agricultural technology adapted to local resources and the economic environment. This lack of research capacity was particularly limiting to traditional food-crop production.

Consequently, one of the dramatic developments of the 1960s was the growth of agricultural sciences in some of these countries. A major factor in this growth has been the establishment of an international research, training, and communications network centered on a new set of international research institutes. These new institutes have been particularly significant in strengthening agricultural sciences in the smaller nations of the tropics. A small nation with either a specialized or a limited agricultural or industrial base cannot capture as high a proportion of the benefits of its investment in basic research as a large nation with a diversified economic base. The highest returns from research investment in countries with a limited economic base will typically be found in applied problem-solving research, which draws on the knowledge produced by research institutions with a broader base. This is the major reason for the development of a new set of international research institutes by such philanthropic organizations as the Ford Foundation and the Rockefeller Foundation and for the support of international research programs by the Food and Agriculture Organization of the United Nations, (FAO), the World Bank, and such national aid agencies as the U.S. Agency for International Development (AID).

The series of steps that led to the establishment of the new international institutes began in 1943 with the establishment of the Office of Special Studies (Oficina de Estudios Especiales) in the Mexican Ministry of Agriculture with the assistance of the Rockefeller Foundation. Field-research programs were first initiated with wheat and corn. The program was later expanded to include field beans, potatoes, sorghum, vegetable crops, and animal sciences. Norman Borlaug (Fig. 26-11), who received a Nobel Prize in 1970 for his research on wheat, was stationed in Mexico by the Rockefeller Foundation in 1944 and continues as leader of the wheat program.

By 1980, thirteen international research centers had been established. To gain knowledge of how these institutes function and how they are organized, it is useful to examine one of them in detail. The International Rice Research Institute (IRRI) in the

Figure 26-11   Norman E. Borlaug, 1970 recipient of the Nobel Prize for peace, working with hybrid wheats at the CIMMYT Toluca station, Mexico.

Philippines will be our example. (There are, of course, substantial differences in the organization and in the research policies of the several institutes. The International Potato Center, for example, conducts more of its research program through contractual relations with other research institutions, in both the developed and the developing countries, than do most other institutes in the international system.)

Because of its importance throughout Southeast Asia, research on rice was particularly well adapted to an international cooperative effort. Basic work at the IRRI has found broad application by researchers throughout the rice-growing regions of the tropics— in Asia, Africa, and Latin America. The interchange of knowledge fostered by the Institute's fellowship program and international conferences could be used to disseminate basic research results, applied research methods, and research skills rapidly throughout the region.

Because of these considerations, the Ford and Rockefeller foundations, in cooperation with the government of the Philippines, established in the Philippines a world center for the study and improvement of rice (Fig. 26-12). The Ford Foundation contributed a capital grant of $7,150,000. The foundations together agreed to provide annual operating funds in the neighborhood of $1,250,000.

The Institute, dedicated in February 1962, assembled an international staff of scientists from Australia, Sri Lanka, Taiwan, India, Japan, the Philippines, and the United States. In addition to its research program, the IRRI conducts a training program for rice technicians and scientists from the rice-producing regions of the world in cooperation with the University of the Philippines. Participants in the training program include students who are interested in pursuing graduate work at the University of the Philippines while conducting their research at the IRRI, plus

Figure 26-12 The International Rice Research Institute (IRRI), established at the University of the Philippines, Los Baños, Laguna.

more mature scholars who are primarily interested in expanding their research experience. Scholars from most of the countries of Southeast and South Asia, as well as many countries in Latin America and Africa, study at the IRRI.

A major aim of the research program at the International Rice Research Institute has been to breed new high-yielding rice cultivars that are adapted to tropical conditions. By 1966 the first new IRRI cultivars were available for field tests under actual farm conditions. The yields of the new cultivar 'IR8' were typically almost double the yields of a number of widely used local cultivars grown under essentially the same conditions.

The stage had been set for the development of the new IRRI cultivars by the successful development of similar cultivars by Japan in Taiwan between 1900 and 1930. The much more rapid development of the new cultivars at the International Rice Research Institute was the direct result of an IRRI breeder's insight, which led to the identification of a specific plant type as a breeding objective. The new cultivars are responsive to fertilizer, resistant to lodging, and insensitive to photoperiod. The resulting biologically efficient rice plant permitted the IRRI plant breeders to design a whole group of new higher-yielding cultivars with much greater precision and in much less time than was believed possible by many rice breeders.

By 1980, the new rice cultivars developed at the IRRI, or similar cultivars developed in other countries as a result of the stimulus of the Institute, were planted on more than 25 million hectares in Asia. Similar cultivars were being developed and adopted in Africa and Latin America. At the IRRI itself, increasing attention was being given to problems of pest management, and to broad-spectrum resistance to fungal, bacterial, and viral diseases, such as rice blast, bacterial blight, bacterial streak, and tungro-virus disease. Increased attention was also being given to nutritional value and cooking quality, to the thermodynamics of flooded rice soils, to the physiology of plant nutrition and growth, to the economics of rice production, marketing, and trade, and to the economics of the rural communities in the rice-producing areas of Southeast Asia.

The new international institutes do not, of course, represent completely new concepts of research organization. Commodity-centered research institutes established in the tropics under British, French, Belgian, and Dutch colonial auspices have been responsible for substantial gains in productivity in the production of tropical export crops, such as rubber and sugar. The Rubber Research Institute in Malaysia, an industry-supported research institute, is one of the world's outstanding research organizations. And the sugar station established in West Java under the Dutch (Proefstation Oost Java) was the major source of the advances in sugarcane breeding that led to dramatically higher yields in almost all major sugar-producing countries.

The significance of the new international institutes does, however, extend well beyond their direct impact on the availability of new wheat, corn, and rice technologies. The most important contribution has been the evolution of an institutional pattern for the organization of scientific resources that can be replicated for work with a wide variety of crops, in different localities, with a reasonable probability of success. It is now possible to organize a multidisciplinary team of biological, physical, and social scientists capable of developing or adapting a new biological and chemical technology for crop production to local growing conditions and to make that technology available to farmers in a form they are capable of accepting within a period of five to ten years.

Another important contribution of the new international centers has been the evolution of a technique for establishing a set of linkages with national and local education and research centers. The technique includes such activities as exchanges of staff, professional conferences, support of graduate and postgraduate training, personal consultations, and exchanges of genetic materials. This communications function of the international institutes is particularly important for the experiment stations located in the smaller countries where the feasibility of the development of a broad-based national agricultural-research system is limited.

Stages in the development of national agricultural research systems

If the world is to be able to meet its future food and fiber requirements, agricultural research in the developing countries will have to be strengthened. It is estimated that world expenditures on agricultural research amounted to about $8 billion in 1980. Of this total, less than $2 billion was spent in the developing countries of Latin America, Africa, and Asia. The development of national research capacity is a slow process.

Among the presently developed countries of Western Europe, North America, and Japan, agricultural research has typically progressed through a series of relatively distinct stages. The newer national systems have drawn heavily on the experience of the older national systems. But the pattern of evolution of the newer systems has been conditioned in some countries by whether the initial stages evolved under colonial auspices and how they have been influenced, in later stages, by particular bilateral or multilateral assistance programs.

Box 26–1 *Present structure of the international agricultural research network*

Center	Location	Research	Coverage	Date of initiation	Budget for 1979 (millions of dollars)
IRRI (International Rice Research Institute)	Los Baños, Philippines	Rice under irrigation, multiple cropping systems; upland rice	Worldwide, special emphasis on Asia	1959	13.803
CIMMYT (International Centre for the Improvement of Maize and Wheat)	El Batan, Mexico	Wheat (also triticale, barley); corn (also high-altitude sorghum)	Worldwide	1963	14.533
IITA (International Institute of Tropical Agriculture)	Ibadan, Nigeria	Farming systems; cereals (rice and corn as regional relay stations for IRRI and CIMMYT); grain legume (cow-peas, soybeans, lima beans, pigeon peas); root and tuber crops (cassava, sweetpotatoes, yams)	Worldwide in lowland tropics, special emphasis on Africa	1965	14.880
CIAT (International Centre for Tropical Agriculture)	Palmira, Colombia	Beef; cassava; field beans; swine (minor); corn and rice (regional relay stations to CIMMYT and IRRI)	Worldwide in lowland tropics, special emphasis on Latin America	1968	13.557
WARDA (West African Rice Development Association)	Monrovia, Liberia	Regional cooperative effort in adaptive rice research among 13 nations with IITA and IRRI support	West Africa	1971	2.253
CIP (International Potato Centre)	Lima, Peru	Potatoes (for both tropical and temperate regions)	Worldwide, including linkages with developed countries	1972	7.267
ICRISAT (International Crops Research Institute for the Semi-Arid Tropics)	Hyderabad, India	Sorghum; pearl millet; pigeon peas; chickpeas; farming systems; groundnuts	Worldwide, special emphasis on dry semiarid tropics, non-irrigated farming. Special relay stations in Africa under negotiation	1972	11.207

Source: The Secretariat of the Consultative Group on International Agricultural Research, World Bank, Washington, D.C.

Center	Location	Research	Coverage	Date of initiation	Budget for 1979 (millions of dollars)
IBPGR (International Board for Plant Genetic Resources)	FAO, Rome, Italy	Conservation of plant genetic material with special reference to crops of economic importance	Worldwide	1973	2.772
ILRAD (International Laboratory for Research on Animal Diseases)	Nairobi, Africa	Trypanosoiasis; theileriasis	Mainly Africa	1974	8.510
ILCA (International Livestock for Africa)	Addis Ababa, Ethiopia	Livestock production system	Major ecological regions in tropical zones of Africa	1974	9.265
ICARDA (International Centre for Agricultural Research in Dry Areas)	Lebanon Syria Iran	Crop and mixed farming systems research, with focus on sheep, barley, wheat, broad beans, and lentils	Worldwide, emphasis on the semiarid winter precipitation zone	1976	10.369
IFPRI (International Food Policy Research Institute)	Washington, D.C., United States	Food policy	Worldwide	1975	1.938
ISNAR (International Service for National Agricultural Research)	The Hague, Netherlands	Strengthening the capacity of national agricultural research programs	Worldwide	1979	1.199

Associate Centers

Center	Location	Research	Coverage	Date of initiation	Budget for 1979 (millions of dollars)
AVRDC (Asian Vegetable Research and Development Centre)	Shanhua, Taiwan	Vegetable improvement (Mung beans, soybean, tomato, sweetpotato, Chinese cabbage, potato); cropping systems	South and Southeast Asia	1971	2.000
IFDC (International Fertilizer Development Centre)	Muscle Shoals, United States	Development of new and improvement of existing fertilizer materials and processes	Worldwide	1975	5.200

The first stage

The first stage in the progress of agricultural science and technology is based on the innovative activities of individual farmers and inventors. In the United States, Jefferson's experiments with soil fertility, Washington's efforts to introduce new crops, and the invention of new land-preparation and harvesting equipment by Deere and McCormick come immediately to mind. In the United Kingdom, Arthur Young dramatized the efforts of progressive landowners in the development of improved breeds of animals and farming practices. In Japan, veteran farmers (*rōnō*) and farming landlords (*gōnō*) were the source of new technology, such as improved methods for seed selection, and were responsible for the diffusion of the best traditional practices.

The second stage

The second stage is characterized by the organization of agricultural experiment stations staffed by research workers with professional specialization in agricultural science or technology. The distinguishing feature of this second phase is the research laboratory or experiment station. The laboratory and the station represent the necessary infrastructure to attract professionally trained scientists to agricultural research, as well as the physical and institutional environment needed to search for "methods of applying science to agriculture." It permits both the framework for assembling the resources needed to support agricultural research and for organizing the diverse scientific and technical skills necessary to advance knowledge. During the second stage, support for research leading to advances in biological technology is, in most countries, supported primarily by public revenues (or by granting organized groups of producers the right to tax themselves); advances in chemical technology are produced primarily in the laboratories of the chemical companies; and advances in mechanical technology continued to be due primarily to farmer inventions but with engineering research by the machinery companies contributing to their commercial success.

The third stage

The third stage is characterized by the evolution of integrated national agricultural-research systems. The characteristic feature of the third stage is the establishment of an agricultural-research planning capacity that is capable of relating research priorities to the allocation of professional and financial resources. The third stage may evolve out of efforts by private scientific bodies, or by such quasi-public advisory councils as national academies of sciences, to draw attention to gaps in the research effort—between research in crop production and in animal or human nutrition, for example, or between private-sector research in plant breeding and public-sector research in genetics. As the level of public-sector research support rises, the national agency that is primarily responsible for the funding of agricultural research typically develops substantial capacity for "in house" planning and resource allocation. One result of the development of a national research capacity is an attempt to rationalize the organization of individual commodity-research institutes or experiment stations, to reexamine the location of branch stations, and to attempt to achieve more effective integration of national, state or provincial, and private agricultural research.

Selected references

Arndt, T. M., D. G. Dalrymple, and V. W. Ruttan, 1977. *Resource Allocation and Productivity in National and International Agricultural Research.* Minneapolis: University of Minnesota Press. (A comprehensive analysis of the implications of agricultural science and technology for agricultural development. The authors examine evidence on the returns to investment in national and international agricultural-research systems and explore the relevance of natural, economic, and social factors for the organization and management of research systems. The chapters are based on papers presented at a conference sponsored by the Agricultural Development Council, the U.S. Agency for International Development, and the World Bank.)

Arnon, I., 1975. *The Planning and Programming of Agricultural Research.* Rome: Food and Agricultural Organization of the United Nations. (A useful survey of methods for planning and evaluating agricultural research programs.)

Dupree, A. H., 1957. *Science in the Federal Government: A History of Policies and Activities to 1940*. Cambridge, Mass.: Harvard University Press. (This book contains an excellent discussion of the emergence of agricultural research in the U.S. Department of Agriculture. Particular attention is given to the search for a method or organizing scientific effort in agriculture to achieve scientific and financial viability.)

Fishel, W. L. (editor), 1971. *Resource Allocation in Agricultural Research*. Minneapolis: University of Minnesota Press. (This book presents the views and findings of a number of experts concerned with the problems, issues, and procedures involved in the allocation of resources for agricultural research. The twenty-one contributors include agricultural and general economists, public-administration officials, scientists, and other specialists. The chapters are based on papers given at a symposium sponsored jointly by the Minnesota Agricultural Experiment Station and the Cooperative States Research Service of the U.S. Department of Agriculture.)

Knoblauch, H. C., E. M. Law, and W. P. Meyer, 1962. *State Agricultural Experiment Stations: A History of Research Policy and Procedure* (USDA Misc. Publ. 904). (A history of the development and growth of the U.S. Department of Agriculture–land-grant college system of agricultural research in the United States.)

Moseman, A. H., 1970. *Building Agricultural Research Systems in the Developing Nations*. New York: Agricultural Development Council. (An excellent introduction to the strategy and practice of developing agricultural-research capacity in new nations.)

Ruttan, Vernon W., 1981. *Agricultural Research Policy*. Minneapolis: University of Minnesota Press. (A comprehensive treatment of agricultural research policy. Includes chapters on the development of national research systems, the evolution of the international research system, allocation of research resources, and social responsibility in agricultural research.)

27
Technical Change and Agricultural Development

Economic growth is generally thought of in terms of rising consumption or in terms of an increase in real income per person. Economic growth can occur as a result of advances in the techniques of production (in technological progress) that result in a greater output with the expenditure of a constant total quantity of resources. It can also occur as a result of an increase in the quantity of other factors of production, such as capital equipment and raw materials per unit of labor, in such a manner that real income per person rises even though the ratio of output to the total quantity of resources used in production remains unchanged or even declines.

A crucial element in the rise in material well-being in most rapidly developing countries has been the progressive use of a growing store of knowledge. The fundamental significance of **technological change** for the growth of the developing countries is that it permits the substitution of knowledge and skill for resources. The effect is to permit the substitution of less expensive or more abundant resources for more expensive or relatively scarce resources, and to release the constraints on growth imposed by inelastic resource supplies.

To be sure, technological change is only one of a number of interrelated elements that contribute to economic growth. The stock of capital and the availability of raw materials have already been men-tioned. Other important elements are (1) the quality of the labor force, including its general education, health, occupational skills, incentives, and motivations; and (2) the way that economic activity is organized—(a) the ability of the government to provide facilities and services that individual firms find difficult to provide themselves; (b) the efficiency with which the markets for factors of production and products are organized; and (c) the manner in which economic and social institutions, such as the land tenure system, the tax structure, and the monetary system, reinforce or dampen personal incentives.

It is clear, however, that technical change is an essential element in the growth of agricultural production and productivity from the very beginning of the process of agricultural development. It is also evident that the process of technical change in agriculture occurs, to a substantial degree, as a dynamic response to the resource endowments of a country and to the economic environment in which a country finds itself at each stage in the history of its development. The design of a successful strategy for agricultural development is based on a unique pattern of technical change and growth in productivity consistent with the resource endowments in each country or major region. It may also require that social and political systems be modernized in order to bring about an economic and social environment conducive to an effective response by firms and individuals to the new technical opportunities.

The previous chapter describes the organization and management of scientific and technological resources for the generation of new scientific knowledge and the progress of agricultural technology. This chapter discusses the various aspects of technical change: its role in agricultural development; its diffusion from farm to farm, from region to region, and in the course of time; and its implications for the development of market institutions. Linkages between technical change and the supply–demand feedback system described in Chapter 24 are critical to producers' decisions to adopt new technology. Also, the "spillover effects" of technical change on environmental quality must be carefully considered in evaluating the implications of a particular change for resource utilization.

ALTERNATIVE PATHS
OF TECHNICAL CHANGE*

In any attempt to develop a meaningful perspective on technical change in agricultural development, the first step is to abandon the naive view of agriculture in premodern or traditional societies as essentially static. Even in premodern times, agriculture underwent a continuous, though relatively slow, development—in agricultural tools and machines, in cultivars and animal breeds, and in crop and animal husbandry practices (see Chapter 1).

The frontier model

Throughout most of history, expansion of the area cultivated or grazed has been the dominant reason for increases in agricultural production. The most dramatic example in Western history was the opening up of the new continents—North America, South America, and Australia—to European settlement

---

*This section draws heavily on Yujiro Hayami and Vernon W. Ruttan, *Agricultural Development: An International Perspective*. The Johns Hopkins University Press, 1971.

during the eighteenth and nineteenth centuries. With the advent of cheap transportation during the latter half of the nineteenth century, the countries of the new continents became increasingly important sources of food and agricultural raw materials for the metropolitan countries of Western Europe.

Throughout history, similar processes had been at work, though at a less dramatic pace, in the peasant and village economies of Europe, Asia, and Africa. The first millennium A.D. saw the agricultural colonization of Europe north of the Alps, the Chinese settlement of the lands south of the Yangtze River, and the Bantu occupation of Africa south of the tropical forest belts. Intensification of land use in existing villages was followed by pioneer settlement, the establishment of new villages, and the opening up of forest or jungle land to cultivation. In Western Europe there was a series of successive changes from neolithic forest fallow to systems of shifting cultivation on bush and grass land, followed first by short fallow systems and in recent years by annual cropping.

Where soil conditions were favorable, as in the great river basins and plains, the new villages gradually intensified their system of cultivation. Where soil resources were poor, as in many of the hill and upland areas, new areas were opened up to shifting cultivation or to nomadic grazing. Under conditions of rapid population growth, the limits to the frontier model were often quickly realized. Crop yields were typically low—measured in terms of output per unit of seed rather than per unit of crop area. Output per hectare and per hour of labor tended to decline—except in the delta areas, such as in Egypt, South Asia, and the wet-rice areas of East Asia. In many areas, the result was increasing misery for the peasantry.

There are relatively few remaining areas of the world where development along the lines of the frontier model will represent an efficient source of growth during the last quarter of the twentieth century. The 1970s saw the "closing of the frontier" in most areas of Southeast Asia. In Latin America and Africa, the opening up of new lands awaits the development of technologies for the control of pests and diseases (such as the tsetse fly in Africa) or for the release and maintenance of productivity of problem soils.

The twentieth century can be viewed as the transition from a period when most of the increases in world agricultural production were a result of expansion of the area under cultivation to a period when most of the increase in crop and animal production will come from increases in the frequency and intensity of cultivation—from changes in land use that make it possible to crop a given area of land more frequently and more intensively and hence to increase the output per unit of area and per unit of time.

The conservation model

The emergence of modern agricultural science itself was in part generated by an attempt of the founders of agricultural science, such as Arthur Young (1741–1820) in England and Albrecht Thaer (1752–1825) in Germany, to interpret and extend the English agricultural revolution of the late eighteenth and early nineteenth centuries. The English agricultural revolution consisted of the evolution of an intensive, integrated system of crop and livestock husbandry. More intensive crop-rotation systems evolved to replace the older system in which arable land had been allocated between permanent cropland and permanent pasture. Crop rotation, in turn, led to the introduction and heightened use of new forage and green-manure crops and an increase in the availability and use of animal manures to maintain soil fertility. This "new husbandry" practice induced an expansion in the cultivation of new crop and livestock production by the recycling of plant nutrients, in the form of animal manures, to maintain soil fertility. The advances in technology were accompanied by the enclosure of the "common" pastures, by the consolidations of fields, and by investment in drainage, fencing, and other forms of land development. The net effect was substantial growth, both in total agricultural output and in output per unit of land. The inputs used in this conservation system of farming were supplied largely by the agricultural sector itself, with little dependence on the industrial sector. Power, in the form of labor and animals, was produced primarily within the community in which it was used. The organic fertilizers, which enhanced output per unit of land, were produced as green manures or animal manures, primarily on the farm on which they were used. Even the capital improvements were, for the most part, produced by the labor available on the farm or within the local community.

The conservation model of agricultural development—based on the development of increasingly complex land- and labor-intensive cropping systems, the production and use of organic manures, and investment in physical facilities to utilize land and water resources more effectively—has been clearly capable in England, and in many other areas of the world, of sustaining a rate of growth in agricultural production of about 1 percent per year for relatively long periods of time. Agricultural growth rates that are this low, however, are inadequate to meet the requirements of developing countries today, with their prevailing conditions of explosive population growth and rising per capita incomes.

The organic sources of plant nutrients that served to increase productivity under the conservation model of agricultural development continue to make an important contribution to agricultural production. In areas of the world where industrial development has lagged and where transportation is inadequate and expensive, there are few alternative sources of plant nutrients available to farmers; and, in the industrial countries, the rising cost of energy is directing the attention of scientists to the possibilities of developing more efficient biological sources of plant nutrients.

The induced-innovation model

Modern agricultural growth is heavily dependent on the use of new technical inputs that are the products of research in agricultural science and technology and that reach the farmer in the form of items that must be purchased from the industrial sector of the economy. Three types of investment are essential if society is to achieve sustained growth in agricultural productivity: (1) in the capacity of agricultural experiment stations to produce new scientific and technical knowledge; (2) in the capacity of the industrial sector to develop, produce, and market new technical inputs; and (3) in the capacity of farmers to use modern agricultural inputs effectively.

This does not imply, however, that there is only one path to the achievement of a modern agricultural economy undergoing rapid technical change and rapid growth in productivity. On the contrary, multiple paths of technological development are available to a nation. Technology can be developed to relieve the constraints on growth imposed by a scarcity of certain factors of production. In some societies this constraint may be a limited supply of labor. In others it may be a scarcity of land. In countries having high population densities and a scarcity of land, as in eastern and southern Asia, increases in output depend on the development of an agricultural technology, including fertilizer-responsive crops, that can release the constraints on growth imposed by an inelastic supply of land. Likewise, in economies with a scarcity of labor, such as those of the United States and other countries of recent settlement, substitution of land and capital for labor is made possible primarily by improving agricultural implements and machinery. The constraints imposed on agricultural development by a scarcity of land may be offset by advances in biological technology. Those imposed by a limited supply of labor may be offset by advances in mechanical technology. The capacity of a country to achieve rapid growth in agricultural output and productivity seems to hinge on its ability to make an efficient choice among the alternative paths. Failure to choose a path that effectively relaxes the constraints imposed by a shortage of resources can depress the whole process of economic development.

Two alternative patterns of technical change, influenced by shortages of two diverse resources, are illustrated in the recent histories of the United States and Japan. In the United States, it was primarily the progress of mechanization that facilitated the expansion of agricultural production and productivity by increasing the amount of land that could be worked by a single worker, thus compensating for the labor shortage. In Japan, it was primarily the progress of biological technology, resulting in seed improvements, that increased the yield response of plants to larger quantities of fertilizer, thus permitting the rapid growth of agricultural output in spite of severe limits to the supply of land. In Japan, the impact of biological technology has exerted a dramatic influence on the growth of production and productivity throughout much of the period since 1870. U.S. agriculture has undergone significant biological innovations since the 1930s, and farm mechanization in Japan has been progressing at an accelerating pace since the 1950s.

The factors of production include natural-resource inputs (land), physical inputs (capital equipment), and human resources (labor and management). The contrasting patterns of productivity growth and factor use in U.S. and Japanese agriculture can best be understood if they are thought of as processes of dynamic adjustment to changing relative factor prices—dynamic in the sense that production technology changed in response to the alterations in relative factor prices, which in turn reflected changes in the availability of resources in the two economies. In the United States, the long-term decline in the prices of machinery relative to wages (Fig. 27-1) encouraged the substitution of land and power for labor. This substitution was dependent on continued progress in the development of larger-scale mechanical technology.

For example, an optimum factor combination with the reaper (such as the McCormick or Hussey), assuming two weeks for harvesting and two shifts of horses, was approximately five workers, four horses (two horses for the original models), and 140 acres of wheat. Only when a new technology, in the form of the binder, was introduced was it possible for the farmer to change this proportion to two workers, one reaper (binder), four horses, and 140 acres. This process has continued. Introduction of the tractor permitted the replacement of human and animal power by mechanical power. This greatly raised the possibilities of substituting power for labor by making it much easier to command more power per worker. Substitution of more highly powered tractors for low-powered tractors has had a similar effect.

In Japan, the supply of land was limited and the price of land rose relative to wages. It was not, therefore, profitable to substitute land and power for labor. Instead, the new opportunities arising from

Figure 27-1 Farm-machinery price relative to farm wage, and fertilizer price relative to arable-land price (1880 = 100), in logarithms, in the United States and Japan, 1880–1960. Prices are the averages for the five years preceding the year shown. [After Y. Hayami and V. W. Ruttan, *Agricultural Development: An International Perspective,* The Johns Hopkins University Press, 1971, p. 123.]

continuous declines in the price of fertilizer relative to the price of land were exploited by advances in biological technology. Seed improvements were directed to the selection of more fertilizer-responsive cultivars. The traditional cultivars had yields that were equal to or higher than those of the improved cultivars at low levels of fertilization, but they would not respond to the application of greater quantities of fertilizer. The enormous changes in fertilizer input per hectare that have taken place in Japan since 1880 and in the United States since the 1930s reflect not only the effect of a decline in the price of fertilizer but the effect of crop-improvement efforts designed to achieve a yield response to larger amounts of fertilizer in order to take advantage of the decline in the real price of fertilizer.

In Japan, then, the decline of fertilizer prices relative to land prices was a strong inducement for farmers and experiment-station workers to develop such biological innovations as high-yielding, fertilizer-responsive cultivars. It is significant that, in the United States, the biological innovations, the first important one being hybrid corn, began to appear about ten years after the rate of increase in arable land area per worker decelerated (around 1920), and that biological innovations and fertilizer application were accelerated after acreage restrictions were imposed by the government. The changes in the land supply coupled with a dramatic decline in the price of fertilizer induced a more rapid rate of biological innovation in the United States after the 1930s.

Adjustments in factor proportions in response to changes in relative factor prices can be thought of as movements along the isoproduct surface of an innovation-possibility curve or a metaproduction function. This is illustrated in Figure 27-2. The curve U in Figure 27-2 (*left*) represents the land-labor isoquant, or constant-output curve, of the innovation-possibility curve, which is the envelope of less elastic isoquants such as u_0 and u_1 corresponding to different types of machinery or technology. A certain technology represented by u_0 (for example, the reaper) is created when a price ratio, p_0, prevails for a certain length of time. When the price ratio changes from p_0 to p_1, another technology represented by u_1 (for example, the combine) is induced.

The new technology represented by u_1, which

Figure 27-2   Graphs illustrating the ways in which factor prices induce technological change. [After Y. Hayami and V. W. Ruttan, *Agricultural Development: An International Perspective*, The Johns Hopkins University Press, 1971, p. 126.]

makes it possible for one worker to work a larger area of land, generally corresponds to a higher intensity of power per worker. This implies a complementary relation between land and power, which may be drawn as a line representing a certain combination of land and power $[A, M]$. In this simplified presentation, mechanical innovation is conceived as the substitution of a combination of land and power $[A, M]$ for labor (L) in response to a change in wage relative to an index of land and machinery prices—though, of course, in actual practice, land and power are substitutable to some extent.

In the same context, the relation between the fertilizer–land price ratio and biological and chemical innovations, such as the development of crops that are more responsive to application of fertilizers, is illustrated in Figure 27-2 (*right*). The curve V represents the land-fertilizer isoquant of the innovation-possibility curve, which is the envelope of less elastic isoquants such as v_0 and v_1, corresponding to cultivars with different degrees of responsiveness to fertilization. A decline in the price of fertilizer relative to the

price of land from r_0 to r_1 makes it more profitable for farmers to search for cultivars or alternative crops described by isoquants to the right of v_0. The farmers also press public research institutions to develop new cultivars. Through a kind of dialectic process of interaction between farmers and experiment-station workers, a new cultivar, such as that represented by v_1, is developed.

Such movements along the innovation-possibility curve are consistent with U.S. and Japanese data on the relation between fertilizer use per hectare of arable land and the fertilizer–land price ratio as plotted in Figure 27-3. Despite the enormous differences in climate and other environmental conditions, the relation between these variables is almost identical in both countries. As the price of fertilizer has declined relative to the price of land, both Japanese and U.S. agricultural scientists have developed more fertilizer-responsive crops. Farmers in both countries have responded similarly to the increased availability of fertilizers and of fertilizer-responsive crops.

It should be recognized, of course, that the identi-

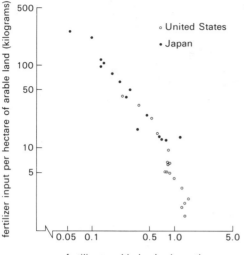

Figure 27-3 The relation between fertilizer input per hectare of arable land in the fertilizer–arable land price ratio, in logarithms, for the United States and Japan, based on five-year observations taken from 1880 to 1960. This relation is expressed in terms of hectares of arable land that can be purchased by 1 metric ton of $N + P_2O_5 + K_2O$ contained in commercial fertilizers. [After Y. Hayami and V. W. Ruttan, *Agricultural Development: An International Perspective*, The Johns Hopkins University Press, 1971, p. 127.]

fication of mechanical technology as "labor saving" and of biological technology as "land saving," as we have done in this section, can be oversimplified. All mechanical innovations are not necessarily motivated by incentives to save labor, nor are all biological innovations motivated by incentives to save land. For example, horse plowing was developed in Japan as a device to cultivate more deeply, to increase yield per unit of land rather than primarily to save labor. In the United States in recent years attempts have been made to develop crops more suitable for mechanical harvesting. For example, in order to facilitate mechanical harvesting, tomatoes have been developed that have a sturdier skin and nearly synchronous ripening. Actually, the most sophisticated technological

progress may depend on simultaneous advances in both biological and mechanical technology. Thus in the mechanization of tomato harvesting, the plant-breeding research and the engineering research were conducted cooperatively in order to invent new machines capable of harvesting the tomatoes that were specifically bred for mechanical handling.

DIFFUSION OF NEW TECHNOLOGY

Once a technological innovation has been developed and introduced by a research organization or a commercial firm and adopted by one or a few farms, how long does it take another farm to adopt the innovation? What factors determine the speed with which other farms imitate the innovator? By what process is technical change diffused among farms, regions, and nations?

Diffusion among farms

The speed with which innovations are adopted affects both the private and the social return to investment in research and development. In economies with high interest rates, the passage of only a few years before a new practice or a new product is adopted may sharply reduce the return on investment in research and development.

Even under relatively favorable conditions, the diffusion of new technology is a fairly slow process. The introduction of such a relatively simple innovation as hybrid corn into the relatively sophisticated farming community of Iowa took eight years from the time commercial hybrids were offered for sale until hybrid corn was adopted by 90 percent of the Iowa farmers (see Fig. 27-4). The same pattern, with even longer delays, has been characteristic of the more complex agricultural and industrial innovations (Fig. 27-5). There are, of course, substantial differences in the rates of adoption of the several innovations identified in Figures 27-4 and 27-5.

Studies of the diffusion process identify five important generalizations that explain many of the differences in rate of adoption of innovations.

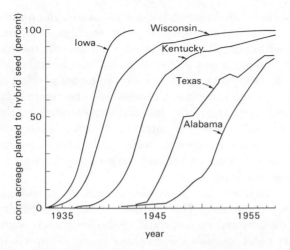

Figure 27-4 Percentage of all corn acreage planted to hybrid seed. [After Z. Griliches, "Hybrid Corn and the Economics of Innovation," *Science* 132(3422):275–280, 1960. Copyright © 1960 by the American Association for the Advancement of Science.]

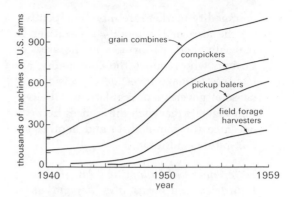

Figure 27-5 Machines in use on farms in the United States, 1940–1959. [After Z. Griliches, "Hybrid Corn and the Economics of Innovation," *Science* 132(3422):257–280, 1960. Copyright © 1960 by the American Association for the Advancement of Science.]

1. *As the number of farms in an area adopting an innovation increases, the probability of its adoption by a nonuser increases.* As experience and information about a particular innovation accumulate, the element of risk associated with its introduction diminishes. Competitive pressures mount, and general adoption of the innovation is encouraged. Where the profitability of an innovation is difficult to assess, the mere fact that a large proportion of neighboring farms have adopted the innovation may prompt those who have not adopted it to consider its adoption all the more seriously.

2. *The expected profitability of an innovation influences the probability of its adoption.* The more profitable the investment in an innovation promises to be, relative to other available investments, the greater will be the probability that its potential profitability will compensate for the risk involved in its adoption.

3. *For equally profitable innovations, the probability of adoption tends to be smaller for innovations requiring relatively large investments.* Firms will be more cautious in committing themselves to large, expensive changes, and they will have more difficulty in financing such changes. The same principle applies to relatively complicated innovations. Adoption of innovations that require a complete reorganization of the farm layout and equipment or that impose heavy demands on managerial ability are adopted more slowly than innovations that just replace one piece of equipment or one crop cultivar with another.

4. *The rate of adoption of an innovation is influenced by the geographic region or type of farm enterprise in which the innovation is introduced.* For equally profitable innovations requiring the same investment, the rate of adoption in one region or in one commodity sector might be higher than in another because (a) farmers in that region, or farmers producing a particular commodity, are more inclined to experiment and take risks, (b) the area or

commodity markets are more competitive, (c) the area or commodity is healthier financially, and (d) farmers in the area, or farmers who produce the commodity, are less free to make investment decisions because of government regulations. Observation suggests that such differences between one area (or commodity) and another affect rates of diffusion.

5. *Differences in adoption rates among industries and firms also depend on a number of social and personal factors.* Sociologists, in particular, have studied the factors that cause some individuals to be classified as (a) innovators, (b) early adopters, (c) early majority, (d) late majority, and (e) laggards in the adoption of new ideas or techniques. By and large, the early adopters are more venturesome. They tend to have a wide range of social relations; and their circle of acquaintances and contacts tends to have a wide geographic radius. They also tend to be younger and to have higher social status than those who are slow to adopt.

International diffusion

The diffusion of technical change among regions and among countries depends not only on the personal and economic factors identified previously but on the speed with which the innovation is adopted and made available to farmers by experiment stations, extension services, and agribusiness firms. For example, when hybrid corn was developed, the lag in the initial adoption by farmers in Texas and Alabama (Fig. 27-4) was not due to a resistance to new technology, but rather to the fact that the agricultural experiment stations and the seed producers did not produce hybrids suitable for the South as soon as they did hybrids for the Midwest. Hybrid corn cultivars are quite "location specific" and must be individually tailored to a particular environment. The primary factors limiting geographic adaptability are sensitivity to temperature (see Chapter 11) and to photoperiod (see Chapter 10). These factors limit adaptability in a

north–south direction but not necessarily in an east–west direction. However, other ecological factors, such as rainfall and pathogens, combine with the first two to make most hybrid-corn cultivars location specific from east to west as well. The technical effort of the public experiment stations and the commercial seed companies was initially concentrated in the Midwest where the intensity of corn production promised the greatest social and private returns from the development and marketing of the higher--yielding hybrids.

The international diffusion of agricultural technology is not new. The diffusion of better husbandry practices and of crops and livestock between countries and continents was a major source of growth in productivity even in prehistory. It is well known that transfer of new crops (potatoes, corn, tobacco, and so on) from the new continents to Europe, after the discovery of America, had a dramatic impact on European agriculture. Before agricultural research and extension were institutionalized, this diffusion took place as a by-product of travel and communication undertaken for other purposes. In the course of a long gestation period (several decades or centuries) exotic plants and techniques were gradually adapted to local conditions. These relatively slow processes are clearly not consistent with the requirements for technical change in agriculture in today's developing countries where the rate of growth in demand for agricultural output is typically from 3 to 6 percent per year.

It appears useful to distinguish the three phases of transfer of international technology as (1) **material transfer,** (2) **design transfer,** and (3) **capacity transfer.** The first phase is the simple transfer or import of new materials, such as seeds, plants, animals, and machines, as well as the techniques needed for using these materials. Local adaptation does not take place in an orderly and systematic fashion. The naturalization of plants and animals tends to occur primarily as a result of "trial and error" by farmers.

The second phase is the transfer of certain designs (blue-prints, formulas, journals, books, and so on). During this period, exotic plant materials and foreign equipment are imported by the developing country in order to obtain new plant-breeding materials or to copy equipment designs. New plants and animals are subject to orderly tests and are propagated by means of systematic multiplication. Domestic production of the machines imported in the previous phase is initiated. This phase usually corresponds to an early stage of the evolution of publicly supported agricultural research, in that experiment stations are established to conduct, for the most part, simple tests and demonstrations.

In the third phase, the scientific knowledge of technology and the capacity to implement it are transferred, thus enabling the production of locally adaptable technology, modeled on a "prototype" technology already existing in another country. Increasingly, plants and animals are bred locally to adapt them to local ecological conditions. The imported machinery designs are modified in order to meet climatic and soil requirements and factor endowments of the economy.

An important element in the process of capacity transfer is the travel, and at least temporary migration, of agricultural scientists. In spite of advancements in communications, diffusion of the ideas and craft of agricultural science depends heavily on extended personal contact. The participation of scientists in international research conferences can be of critical importance in counteracting the limited supply of scientific and technical people in the developing countries, and it is essential if these countries are to enter into the capacity-transfer phase (see Chapter 26) as soon as possible.

The development of sugarcane

The development of improved sugarcane cultivars is a useful example of the international diffusion and transfer of biological technology in agriculture.* It is of particular interest because the process has evolved from a simple transfer of biological materials to the capacity-transfer phase. There have been four stages in the development of sugarcane.

Stage I: diffusion of indigenous cultivars

Sugarcane was cultivated in India as early as 400 B.C.E. Sugarcane and the art of sugar making were diffused from India to China, to Arabia, and to the Mediterranean region very early. Shortly after 1400 A.D., sugarcane was introduced on Madeira and in the Azores. Columbus took it with him to Hispaniola on his second voyage to the New World in 1493, and soon thereafter it was carried to Cuba and Puerto Rico, and later to Mexico, Peru, and Brazil. Throughout this period, the sugarcane cultivars used in commercial cultivation were of two closely related species indigenous to India, *Saccharum sinense* and *S. barberi*. In 1791 Captain Bligh introduced the thick-stemmed *S. officinarum* from Tahiti to Jamaica, and it rapidly replaced the thin-stemmed forms within a few years after it was introduced. The most prominent cultivars (clones) were 'Bourbon,' 'Lahaina,' 'Cana Blananca,' and 'Otaheito.' They diffused widely throughout the world and were the basis for commercial production for nearly a century.

Stage II: sexual reproduction

In nature the sugarcane plant reproduces only asexually. Until methods of sexual reproduction were discovered, this limited the selection of superior clones from indigenous cultivated or wild forms. Procedures for the sexual reproduction of sugarcane were discovered independently in Java (1887) and in Barbados (1888). The sugarcane plant can be induced to flower and produce seedlings under conditions of appropriate temperature and light. Each new seedling is then a potential new cultivar, since it can be produced asexually. The Java station (Proefstaten Oost Java) was the first to develop a new cultivar of commercial significance ('POJ 100'). It later introduced many more important cultivars. Important commercial varieties were also developed by 1900 at experiment stations in Hawaii, Barbados, and elsewhere. The Coimbatore station in India released the first of its improved cultivars in 1912. By 1930 POJ

---

*The material in this discussion is based on the following sources: J. J. Ochse, M. J. Soule, Jr., M. J. Dijkman, and C. Wehlburg, *Tropical and Sub Tropical Agriculture* (2 vols.) (New York: Macmillan, 1961), pp. 1197–1251; R. E. Evenson, J. P. Houck, Jr., and V. W. Ruttan, "Technical Change and Agricultural Trade: Three Examples—Sugar Cane, Bananas and Rice" in Raymond Vernon (editor), *The Technology Factor in International Trade,* (New York: Columbia University Press, 1970), pp. 415–480.

and Coimbatore cultivars were planted commercially in every major sugarcane-producing area of the world. Only simple tests and demonstrations, if any, were required for recipient countries to propagate these cultivars. Many of them, however, were susceptible to local diseases, and their initial yield advantages were lost.

Stage III: interspecific hybridization

Breeding for disease resistance became a dominant concern, since many of the new cultivars were found to be susceptible to local diseases and pests. The Java station played a leading role in the development of disease-resistant varieties. In 1921 the variety 'POJ 2878' was produced. This was achieved by crossing a 118-chromosome disease-resistant thin-stemmed wild sugarcane, S. *spontaneum,* with one of the 80-chromosome S. *officinarum* sugarcanes. By a series of crosses and back crosses, new interspecific hybrids were developed that incorporated the hardiness and disease resistance of the noncommerical species. Later, the Coimbatore station in India developed a series of tri-hybrid sugarcanes by introducing a third species, S. *barberi.* Use of S. *barberi* resulted in new cultivars adapted to local climate, soil, and disease conditions. The high-yielding, disease-resistant stage III cultivars were transferred to every producing country in the world. Although this international transmission was widespread, it was much more dependent on local experiment-station capacity than was the transmission of the stage II cultivars.

Stage IV: location-specific breeding

The Coimbatore station set the stage for modern breeding activity directed toward the development of cultivars suited to the specific soil, climate, disease conditions, and management practices of smaller regions. For the most part, this breeding is conducted by experiment stations located in the specific producing region. More than 100 sugarcane experiment stations exist in the world. Very little international transfer of cultivars is taking place. Almost every important sugarcane-producing country is now using locally adapted stage IV cultivars. Yet the cultivars

developed at the Java station and Coimbatore are important sources of genetic materials in these locally adapted canes.

It seems possible to allocate sugarcane cultivar transfers of stage I to the material-transfer phase, and those of stage IV to the capacity-transfer phase. Stage II appears to be a transition from the material-transfer phase to the design-tranfer phase, and stage III, a transition from the design-transfer phase to the capacity-transfer phase. This sequence indicates the important function that public research has had, not only in producing and "naturalizing" sugarcane cultivars but also in developing increasingly effective modes of transfer, from the initial international diffusion of superior sugarcanes to the international diffusion of the capacity to "invent" location-specific cultivars superior to the "naturalized" cultivars.

The green revolution

The most dramatic example of the international transfer and diffusion of agricultural technology during the last several decades has been the development and rapid diffusion of new high-yielding varieties (HYVs) of rice and wheat in the tropics (Fig. 27-6). The development and diffusion of the new cereals technology was made possible by a series of institutional innovations in the organization, financing, and management of agricultural research in a number of developing tropical countries (see Chapter 26).

The development of the HYVs illustrates the process of transfering agricultural technology from the temperate zone to tropical and subtropical zones by means of the transfer of scientific knowledge and capacity. Within the tropics, the diffusion of the new cereals technology from Mexico (wheat) and the Philippines (rice) was that of an initial materials-transfer phase. The initial impact of this diffusion on grain production in Pakistan, India, Malaysia, Turkey, and certain other countries resulted from the direct transfer of seed from Mexico and the Philippines, and from transfer of fertilizer, insecticides, and fungicides from Japan, the United States, and Western Europe. In other countries, such as Thailand, the impact was delayed until the design-

¹Bangladesh, Burma, India, Indonesia, Korea (South), Malaysia (West), Nepal, Pakistan, Philippines, Sri Lanka, and Thailand.
²Bangladesh, India, Nepal, and Pakistan.

Figure 27-6   Estimated area planted to high-yielding cultivars of wheat and rice in eastern and southern Asia. [After Dalrymple, *Development and Spread of High-Yielding Varieties of Wheat and Rice in Less Developed Nations*, USDA, in cooperation with U.S. Agency for International Development, 1978, p. 117.]

transfer and capacity-transfer phases could be achieved, in order to maintain the quality characteristics of the Thai rices, which are important in the export market. In the countries that benefited initially from capacity transfer (Mexico and the Philippines), there has been a rapid movement to develop the capacity of the local experiment station so that it will permit them to move to the design-transfer and capacity-transfer phases in the development of ecologically adapted cultivars. There is also, in many countries, a move toward the development of a domestic fertilizer and agricultural-chemical industry based primarily on models existing in developed countries.

The diffusion of the new "green revolution" rice and wheat cultivars has generated a great deal of controversy among students of rural development. Critics have argued that the new technology has been adopted primarily by the larger farmers and that the smaller farmers and the landless laborers in peasant communities have actually been made worse off. The argument of the critics typically runs somewhat as follows: The green-revolution technology tends to be monopolized by the large commercial farmers who have better access to information and credit. Small farmers are unable to obtain the credit needed to purchase the fertilizer and chemicals needed to obtain high yields. The profit realized by the large farmers from adopting the new cultivars encourages them to enlarge their farming operations by evicting their tenants or purchasing land from their neighbors. The result is a polarization of rural communities into large commercial farmers and a landless proletariat.

These criticisms often represent a failure to distinguish the effects of the biological-chemical technology of the green revolution from the effects of mechanical technology. It was pointed out earlier in this chapter, in the discussion of induced innovation, that the biological-chemical technology of the green revolution tends to be much more divisible, and hence more suitable for small farms than mechanical technology. This view is supported by a study of the adoption patterns for modern rice cultivars and tractors on small and large farms in 30 villages in five countries in Asia (Figs. 27-7 and 27-8). The tractor was adopted more rapidly by the larger farmers, but modern rice cultivars were adopted more rapidly by the smaller farmers. Other studies indicate that, in most countries in Asia, labor requirements per hectare also rose sharply when the high-yielding modern cultivars were introduced. One reason why the new cultivars have been adopted more rapidly by the smaller farmers is that they permit the more effective use of family labor.

TECHNOLOGICAL CHANGE AND PRODUCTIVITY

A basic characteristic of technological change is that it permits the use of a given set of resources in a way that better satisfies human wants. It includes the application of new organizational and managerial concepts as well as advances in science and technology.

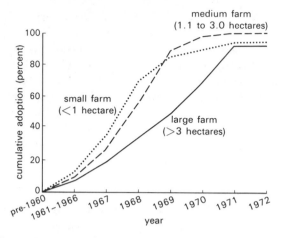

Figure 27-7   Cumulative percentage of farms in three size classes adopting modern high-yielding rice cultivars in 30 villages in South and Southeast Asia. [After International Rice Research Institute, *Interpretive Analysis of Selected Papers from Changes in Rice Farming in Selected Areas of Asia,* 1978, p. 91.]

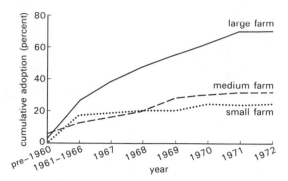

Figure 27-8   Cumulative percentage of farms in three size classes adopting tractors in 30 villages in South and Southeast Asia. [After International Rice Research Institute, *Interpretive Analysis of Selected Papers from Changes in Rice Farming in Selected Areas of Asia,* 1978, p. 91.]

It is useful to distinguish between technological changes that reduce the cost of producing existing private and public consumer goods and those that create new or substantially improved products that enlarge our choices in production or consumption. Radio, penicillin, nylon, and the automobile are examples of new products from technological changes. The development of transistor radios and the automatic transmission represent qualitative improvements in existing technology. Improved rices and chemical fertilizer are the result of technological advances that reduce the cost of producing goods or services already available in the market. By and large, technological change in agriculture reduces the cost or improves the quality of existing commodities.

Because of the complex ways in which technological change interacts with the other components of economic growth, its effects have been described or measured in many ways. Individual inventions or innovations may be described in terms of changes in the blueprints or the specifications for individual items of capital equipment. A new cultivar may be described in terms of the genetic stock from which it was derived or in terms of its responses to different levels of fertilizer application (Chapter 24). For economic purposes, however, it is most useful to measure technological change in terms of input–output, or productivity, relations—that is, in terms of such ratios as yield per hectare, output per hour of labor, and kilograms of gain in weight per kilogram of feed.

A program to increase output at the national level involves a hierarchy of potential changes in products, inputs, and techniques of production. Such a system for achieving increased food output for a tropical rice-producing economy is illustrated schematically in Figure 27-9. The changes on the right side of the chart are made by individuals and by organizations engaged primarily in research, development, and education. The changes in the center of the chart fall within the area of farm production and management decisions. The changes on the left are made primarily in the distribution, or marketing, sector.

Each arrow represents a technical input–output linkage. Changes in technology introduced at these points change the factor-product, or output–input, ratios that produce the total and partial productivity changes for the entire food-producing sector of the economy.

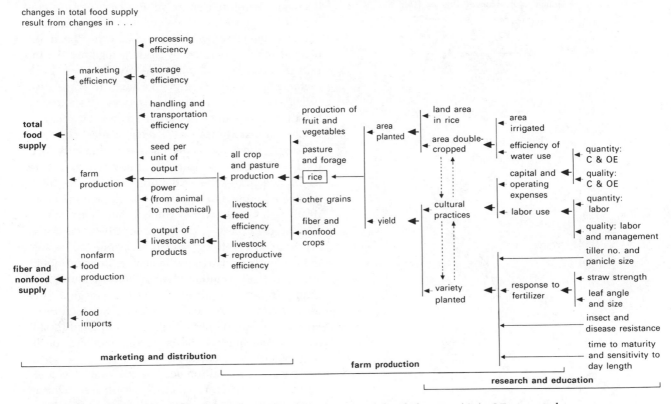

Figure 27-9 Sources of potential change in food supply, with special details for rice. ($C \& OE$ = capital and operating expenses.)

Measures of partial and total productivity

The adoption of innovations by individual farms and industrial firms results in more efficient use of resources throughout the entire economy. These industry-wide and economy-wide gains in efficiency are frequently measured by partial-productivity ratios or indexes.

Changes in partial-productivity ratios, such as output per hour of labor or yield per hectare, are the most frequently employed measures of technological change. Partial-productivity measures, such as yield per hectare or output per hour of labor, have several advantages. They are relatively easy to compute and they summarize a great deal of information in a single measure.

Although useful for some purposes, partial-productivity measures have some limitations. They are not precise indexes of technological change. A rise in output per hour of labor may occur as a result of mechanization of operations that were formerly performed manually—for example, the substitution of mechanical cotton pickers for hand picking. If this is the case, the resource savings resulting from the rise in output per unit of labor input are at least partially offset by an increase in the use of capital, which results in a decline in output per unit of capital.

Dissatisfaction with the precision of partial-productivity measures as indexes of technological change has led to the development of total-productivity measures. Total-productivity indexes measure the change in output per unit of total input.

They are therefore a more valid measure of technological change than partial-productivity measures. In constructing total-productivity measures, the index of total input is derived by adding together the inputs of labor, land, capital, and raw material. Each factor is given a weight proportional to its relative importance in the production process. As a result, substitution of capital for labor has less tendency to result in a biased measure of technological change.

The relations between partial-productivity and total-productivity measures is illustrated in Figure 27-10 (see also Tables 28-4 and 28-6). The total-productivity measure indicates the growth in output relative to the growth in total input. This can be compared to the two partial-productivity measures: (1) **labor productivity,** which measures the change in farm output relative to the change in labor inputs, and (2) **land productivity,** which measures the change in farm output relative to the change in land use. In addition to land and labor, the total-input measure includes inputs of capital equipment (buildings and machinery) and operating expenses (fertilizer, insecticides, and fuel).

Note the wide differences in the rate of change in the two partial-productivity measures. During recent years use of either the land-productivity or the labor-productivity measure, instead of the total-productivity measure, would overstate the true rate of technological change. This is because the total-productivity measure takes into account the substitution of other resources—in the form of capital equipment and operating expenses—for part of the land and labor that has been withdrawn from production.

MARKET IMPLICATIONS OF TECHNOLOGICAL CHANGE

How does technological change influence, and how is it influenced by, the supply and demand relations that underlie price determination and behavior, which were studied in Chapter 24 (see Fig. 24-6)? The effect of technological change is to introduce a dynamic element into the supply relations studied earlier. A rise in output per unit of total input results in higher net returns to farm operations at one of the

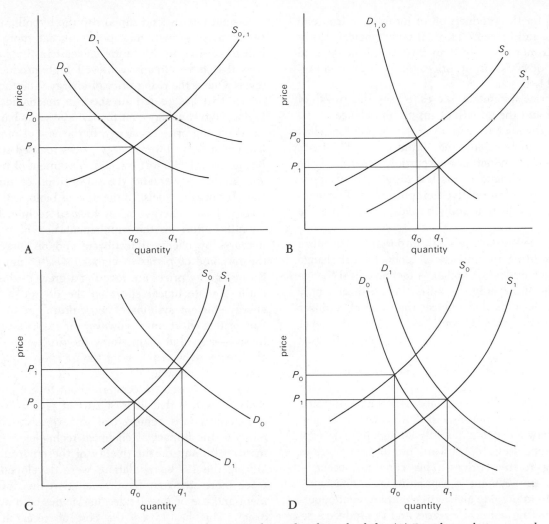

Figure 27-11 Price effects of four alternative demand and supply shifts. (A) Rapid population growth, income growth, or high income elasticity results in rising demand; resource availability and technology remain unchanged. Prices rise. (B) Resource availability or advancing technology shifts product-supply curves rapidly to the right; unchanged population and income holds demand unchanged. Prices fall. (C) Resource availability or advancing technology shifts product curves only moderately to the right, whereas rapid population and income growth (or high income elasticity) result in rapid growth in demand. Prices rise. (D) Resource availability or advancing technology shifts product-supply curves rapidly to the right, whereas slow population growth or income growth (or low income elasticity) results in slow growth in demand. Prices fall.

input–output linkage points of Figure 27-11—for example, increased returns to fertilization—and shifts the supply curve to the right, as in Figure 27-11, B. Changes in population and income also introduce dynamic elements into the demand relations. A rise in

population with a constant or rising income per capita shifts the demand curve to the right, as in Figure 27-11, A. The net impact of these two dynamic forces depends on their relative growth rates. If technological change occurs more slowly than growth of demand for the products of an industry, prices tend to rise, as in Figure 27-11, C. If technological change proceeds more rapidly than the growth of demand for an industry's output, prices tend to fall, as in Figure 27-11, D.

The price system also serves to keep the growth of demand and the growth of supply in balance. If demand grows more rapidly than supply, product prices rise more rapidly than input prices. This tends to stimulate production and curtail demand for the product. If demand grows less rapidly than supply, product prices fall relative to input prices. This tends to curtail production and stimulate demand for the product.

The price system is also an important mechanism in transmitting the impact of technological change from one industry to another. A technological change in the fertilizer industry shifts the fertilizer-supply curve to the right and reduces the price of fertilizer relative to the price of crops. This increases the use of fertilizer by producers and, in turn, reduces the cost of producing crops.

The most favorable economy for rapid adoption of new technology in agriculture is one that is growing rapidly enough to create labor shortages or rising wage rates in agriculture and that is capable of generating new technology rapidly enough to make capital and operating inputs available to agriculture at declining relative prices. This can only occur if nonfarm employment is expanding at a sufficiently rapid rate to provide alternative employment opportunities for agricultural workers and to produce a labor "shortage" in the agricultural sector. Indeed, a labor shortage in agriculture is one of the most reliable indicators of economic development.

THE FUTURE SOURCES OF PRODUCTIVITY GROWTH

Will technical change continue to represent a dynamic source of productivity growth in the future? By the late 1960s there were disturbing signs that productivity growth in the United States was slowing down—both in agriculture and in the total economy (see Fig. 27-10).

Several factors lend support to the hypothesis that productivity growth may be somewhat more difficult to achieve in the future, at least in those countries that have already achieved high productivity levels. One is the rising price of energy. The focusing device that has guided advances in mechanical and biological technology over the last 100 years has been the declining price of energy. In the case of mechanical technology, lower energy prices focused scientific and technical effort on the replacement of human and animal power and the substitution of mineral fuels for organic fuels. In the case of biological technology, lower energy prices focused technical and scientific effort on the supplementation of organic manures by chemical fertilizers and on increasing the response of plants to higher levels of nutrition. Rising energy prices are focusing a great deal of scientific and technical effort on the design of more energy-efficient systems of agricultural production. Part of this effort must, however, be regarded as defensive—aimed at maintaining the productivity levels already achieved—rather than as new sources of productivity growth.

A second reason for concern about future productivity growth is that at least part of past growth in productivity was achieved at an excessive cost in terms of the impact of chemical technology on human health and on the quality of the environment. During the 1970s, regulations were developed that place more stringent requirements on the development and use of insecticides, herbicides, and animal drugs. This has raised the cost of research and reduced the availability of new innovations for commercial use.

A recent USDA study of opinions by leading scientists identified areas that could emerge as significant sources of productivity growth in the period 1980–2000 (Box 27-1). Most of these technologies are on the technical or scientific frontier. Most will not be available to be adopted by farmers until well into the 1990s. The authors believe that there may be an extended period through at least the 1980s during which the rate of productivity growth in U.S. agriculture will fall below the long-term trend that began in the late 1930s.

Box 27-1 *Potential sources of growth in productivity*

1. *Enhancement of photosynthetic efficiency:* Improving the process by which living plants form carbohydrates, through genetic selection, physical modification, and chemical modification; enhancing the biological capacity of living plants to absorb nitrogen for protein synthesis; and enhancing plant growth through elevating atmospheric levels of carbon dioxide.

2. *Water and fertilizer management:* Increasing efficiency of inputs used in production through combined water-management and fertilizer-management systems, such as the one developed for potatoes in Washington, expanded trickle or drip irrigation, new subirrigation techniques, and foliar application of fertilizer.

3. *Crop pest-control strategies:* Adopting total pest-management systems that incorporate resistant varieties, sex attractants, juvenile-hormone analogs, and other biological controls that reduce energy inputs, environmental hazards, and pest-control costs.

4. *Controlled environment or greenhouse agriculture:* Using plastic or glass covers over plants with or without the addition of heat and carbon dioxide—likely restricted to high-value and specialty crops.

5. *Multiple and intensive cropping:* Double cropping and intensive cropping to increase yields per acre.

6. *Reduced tillage:* Expanding use of minimum-tillage or reduced-tillage techniques, a process minimizing the number of times farmers must cultivate a green field.

7. *Bioregulators:* Using certain natural and synthetic compounds to regulate the ripening and senescence of horticultural products. When applied at the preharvest stage, they can enhance ripening and facilitate mechanical harvesting. When applied after harvest they can slow down processes and prolong shelf life of some fruits and vegetables and reduce cooling costs.

8. *New crops:* Developing new and improved hybrids and searching for alternate food crops.

9. *Bioprocessing:* Extending traditional agricultural production so that unpalatable or indigestible raw products, such as cellulose and petroleum materials, can be converted into edible protein, carbohydrates, and fats to provide additional animal feed.

10. *Antitranspirants:* Inhibiting the tendency of plants to lose water through evaporation.

11. *Developing plants to withstand drought and salinity:* Developing plants that genetically resist drought better and thrive on saline water.

12. *Twinning:* Enhancing multiple births in beef cattle through breeding and selection of livestock for genetic twinning, multiple ovulation through hormonal control, and embryo transfer.

Source: Yao-chi Lu and Leroy Quance, *Agricultural Productivity: Expanding the Limits,* (USDA/ESCS AI Bulletin 431), 1979, pp. 9–10.

Selected references

Harwood, R. R., 1979. *Small Farm Development.* Boulder, Colo.: Westview Press. (This book is intended to help agriculturalists and others understand farming systems in the humid tropics and how those systems can be improved. It stresses that higher productivity will result not only from cultivars that have a higher yield potential but also from economic crops capturing the sunlight and water that strike the land and from the fostering of subtle natural interactions between animals and crops, and between the crops themselves, that favor higher and more stable yields.)

Hayami, Y., and V. W. Ruttan, 1971. *Agricultural Development: An International Perspective.* Baltimore: The Johns Hopkins University Press. (Identifies the capacity to develop a technology that is consistent with environmental and economic conditions as the single most important variable that explains the growth of agricultural productivity. Analyzes the historical and economic factors in the develoment of agriculture in the United States and Japan, and draws the implications of the U.S. and Japanese experience for less developed countries.)

Lappé, F., and J. Collins, 1977. *Food First: Beyond the Myth of Scarcity.* Boston: Houghton Mifflin. (A radical criticism of current agricultural technology and food policies in the developed and the developing countries alike. The main message is that all countries could and should become self-sufficient in basic foods and provide an adequate diet to all of their citizens.)

Lu, Yao-chi, P. Cline, and L. Quance, 1979. *Prospects for Productivity Growth in U.S. Agriculture* (USDA/ ESCS Agricultural Economic Report 435). Washington, D.C.: U.S. Department of Agriculture. (An assessment of the implications of alternative levels of research support for the development of new technologies and for productivity growth in U.S. agriculture.)

Rogers, E. M., and F. F. Shoemaker, 1971. *Communication of Innovations: A Cross Cultural Approach.* New York: Macmillan. (The definitive sociological treatment of the diffusion of technical and institutional innovations. Includes a very thorough review of the theoretical and empirical work on diffusion of innovations.)

Spicer, E. H. (editor), 1952. *Human Problems in Technological Change.* New York: Russell Sage Foundation. (Case studies of the introduction of new ideas in agriculture, industry, and medicine to areas deficient in these technologies. Particular emphasis is placed on the social and cultural dynamics involved in the adoption of new technologies.)

Wortman, S., and R. W. Cummings, Jr., 1978. *To Feed This World: The Challenge and the Strategy.* Baltimore: The Johns Hopkins University Press. (The authors argue that the technical means needed to expand world food production are now emerging. They argue that a sense of urgency on the part of political leaders in implementing food-production programs is needed if widespread hunger is to be avoided.)

28
Agriculture and Economic Development

Economic growth was previously defined as a "sustained rise in the level of consumption or real income per person." Economic development, a broader concept, relates to progress along a wide spectrum of technological, economic, social, and cultural dimensions. Thus it is much more difficult to characterize a country in terms of development than it is to characterize it in terms of growth. In many respects, India is a highly developed low-income country, whereas the United States is an underdeveloped high-income country.

One index of development is the extent to which an economy can free its resources from the production of basic food and fiber requirements. In a society in which a high proportion of resources must be devoted to meeting basic subsistence requirements, most people are engaged in the production of food and fiber for themselves and for their neighbors. They do not have the time, the ability, or the tools to produce other goods and services. In a highly developed economy, only a small proportion of the population is engaged in agriculture, agricultural productivity is relatively high, and most of the working population is engaged in the production of other goods and services.

The processes by which a nation achieves the transition from a low to a high level of productivity and income are still poorly understood. It is clear that rapid industrial development is difficult to achieve without the support of a progressive rural economy. And it is equally true that a highly productive agricultural sector is easier to achieve when it is part of a rapidly growing urban-industrial economy. Most countries of the world continue to be characterized by low productivity and income in rural and urban areas alike. The significance of the problem of agricultural development was dramatized when Professor Theodore W. Schultz of the University of Chicago was awarded the Nobel Prize in economics in 1979 for his contributions to a better understanding of agricultural and economic development (Fig. 28-1).

There are four ways in which the agricultural sector can contribute to national economic development. First, and most essential, it can provide the food and fiber necessary for an expanding population that is growing in income and wealth. Second, it can release workers needed for the production of nonagricultural goods and services. Third, it can serve as a market for nonfarm goods and services, thus providing a stimulus for expansion of employment and output in the nonfarm sector. Finally, it can provide a source of capital that can be invested in improved productive facilities in the rest of the economy.

A progressive urban-industrial economy contributes, in turn, to the rapid development of agriculture

Figure 28-1 In 1979 Professor Theodore W. Schultz of the University of Chicago received the Nobel Prize in economics for his contributions to a better understanding of agricultural and economic development. Research by Professor Schultz demonstrated the high personal and social returns on investment in the education and health of rural people and the importance of providing farmers with new and more productive inputs and technologies.

by expanding the markets for agricultural products; by supplying the farm machinery, chemical fertilizers, and so on, that raise the level of agricultural technology; by expanding productive employment opportunities for workers released from agriculture by technological change; and by making possible improvements in the quality of rural life by raising standards of consumption both in urban and in rural areas.

The net effect of this dynamic interaction between the agricultural and nonagricultural sectors in a developing economy is the complete transformation of the agricultural sector. Indeed, the term agriculture itself may have little meaning for the modern food and fiber industries that are in the process of emerging from the economic and technological revolution now occurring in the United States.

AGRICULTURAL DEVELOPMENT UNDER LABOR SCARCITY: THE UNITED STATES

From the time of the Plymouth and Jamestown settlements until the closing years of the nineteenth century, the encounter with the frontier represented a dominant theme in American agricultural development. This encounter created an opportunity for the evolution of an agriculture based on an abundance of land and a relative scarcity of labor. This, in turn, stimulated the development of an agricultural technology that was primarily directed toward achieving gains in labor productivity rather than gains in land productivity.

Since the closing of the frontier, in the last quarter of the nineteenth century, the encounter with an increasingly dominant urban-industrial sector has emerged as a major theme in American agricultural development. By 1880, nonagricultural employment exceeded agricultural employment. By 1929, manufacturing employment alone exceeded agricultural employment. By 1970, agricultural employment was less than total unemployment in the United States (Table 28-1).

We can best visualize the interactions between the farm and the nonfarm sectors that led to this fundamental restructuring of the American economy by looking in turn at each of three sets of market relations: the **product market,** through which the output of the agricultural sector is transmitted to the nonfarm sector and through which incomes are generated in the farm sector; the **input markets,** through which move the manufactured inputs, equipment, and capital used in agricultural production; and the **labor market,** through which labor is allocated between the agricultural and nonagricultural sectors and among firms in each sector.

The product market

Throughout most of American economic history, the product market—the market for things farmers sell—represented the primary link between the farm and the nonfarm sectors of the economy. It was the dominant channel through which international shifts in the terms of trade, national fluctuations in non-

Table 28–1 Employment by sector in the United States, 1880–1978, and projections to 1985 (in thousands of workers)[1]

Year	Agriculture	Total nonagriculture	Manufacturing	Unemployment
1880	8,585	8,807	NA	NA
1929	10,450	37,180	10,534	1,550
1950	7,160	51,760	15,241	3,288
1970	3,462	75,165	19,369	4,088
1978	3,380	87,763	20,332	6,012
1985 (est)	2,300	100,504	22,597	5,309

Sources: 1880: United States Bureau of the Census, *Historical Statistics of the United States, Colonial Times to 1957,* 1960.
1929 to 1971: United States Department of Labor, Bureau of Labor Statistics, *Employment and Earnings* 18 (9):21, 49, 1972.
1978: United States Department of Labor, Bureau of Labor Statistics, *Employment and Earnings* 26 (3):50, 1979.
1985 Projections: N. C. Saunders, *Monthly Labor Review* 101 (12):44, 1978.
NA = Not available.
[1]Before 1950, data include people of age 14 and older; after 1950, data include people of age 16 and older.

farm income, and local variations in nonfarm demand were channeled into the agricultural sector.

In most low-income countries, in which a substantial share of increases in income per capita are devoted to dietary improvement, the product market is still the main link between the peasant and the urban-industrial sector of the economy. As income per person rises, consumption of agricultural products expands less rapidly (Table 28-2). At the highest income levels, there may be almost no additional food consumed as income continues to rise. In the United States, the declining response in consumption of food and fiber to increases in nonfarm income has almost eliminated the commodity-market effects of growth in economic activity in the nonfarm sector.

In addition, agricultural trade and commodity policies have been designed that partially insulate the prices of agricultural commodities, particularly crop prices, from normal trade and market fluctuations (Chapters 24 and 25).

As a result of these changes, the rate of growth of domestic demand for food and fiber products in the U.S. economy now barely exceeds the rate of population growth. Opportunities for growth in U.S. agricultural output have become increasingly dependent on the growth of export demand and on food aid to the developing countries (Chapter 25).

Technological change in agricultural production has been accompanied by institutional changes in the product markets that link the agricultural sector to the urban-industrial sector. The production of some products has become essentially industrialized. Broiler "factories" have almost entirely replaced farm production of poultry meat. Commercial production of fruits and vegetables has become highly concentrated. As a result of technological and organizational changes in processing, transportation, and distribution, regional specialization in the production of fruit, vegetables, and animal products has reduced the impact of local urban-industrial development on the demand for locally produced farm products. Milk, protected by a series of local-market trade barriers, remains a major exception to this generalization.

These developments in the farm sector have been accompanied by the replacement of small-scale retail stores served by local wholesalers with large chain-store distribution systems served by integrated wholesale and retail operations. There is an increasing tendency for these large distributing organizations to establish direct linkage with the large specialized fruit and vegetable producers, poultry producers, and livestock feeders, and to bypass traditional marketing channels.

This type of development has not yet had any substantial impact on the major agricultural commod-

Table 28–2 Comparison of growth of demand for agricultural commodities, at different stages of development (hypothetical cases).

Levels of development	Percentage of population in agriculture	Annual rate of population growth	Annual rate of income growth	Income elasticity of demand	Annual rate of growth in demand[1]
Very low income	70	2.0	.5	1.0	2.5
Low income	60	3.0	1.0	.9	3.9
Medium income	50	3.0	3.0	.6	4.8
High income	25	2.0	4.0	.5	4.0
Very high income	10	1.0	3.0	.2	1.6
Highest income	5	0.1	2.0	.1	0.3

Source: Adapted from J. W. Mellor, *The Economics of Agricultural Development*, Cornell University Press, Copyright 1966. Used by permission of Cornell University Press.

[1]The total annual rate of growth in demand is the sum of the population effect and the income effect. The income effect is calculated by multiplying the income elasticity of demand by the rate of income growth.

ities, such as wheat, cotton, corn, and soybeans, nor on most hog and beef production. But the emergence of highway truck transportation as a major competitor with the railways in the hauling of bulk agricultural commodities has eliminated many of the older terminal markets.

The markets for purchased inputs

The markets for manufactured capital equipment and current inputs have become increasingly important in transmitting the effects of changes in the nonfarm sector to the agricultural sector. Much of the new agricultural technology is embodied in the form of new capital equipment or more efficient fertilizers, insecticides, and other manufactured inputs. In 1870, the typical American farm was still a subsistence unit, with inputs purchased from the nonfarm sector amounting to less than 3 percent of the value of farm production. By 1900, nonfarm inputs still amounted to only 7 percent. But by the late 1970s, they accounted for approximately 40 percent of the value of farm production.

The use of purchased inputs has been closely related to developments in the labor market. The demand for labor, resulting from rapid urban-industrial development, reinforced the economic pressure for substitution of capital equipment for labor in American agriculture at precisely that period when the frontier was disappearing as a major factor in agricultural development. It was during this period that the tractor, first functioning as the source of motive power for formerly horse-drawn equipment, and later as the power unit around which new equipment was designed, became the symbol of technological change, both in U.S. agriculture and around the world.

The rapid growth of labor productivity in U.S. agriculture was not, at first, accompanied by parallel changes in land productivity. Grain yield per hectare in American agriculture remained essentially unchanged from the end of the Civil War in the 1860s until well into the twentieth century (Table 28-3). By the mid-1920s, however, the production of fertilizer and other agricultural chemicals was beginning to be reflected in higher yields. Higher yield potentials were also emerging as a result of the application of advances in genetics to plant breeding by scientists in the Land Grant College system and the U.S. Department of Agriculture.

This emergence of a new chemical and biological technology in U.S. agriculture in the 1920s was itself the product of a long sequence of institutional developments. As early as 1862, the U.S. Congress passed legislation granting public lands to the states, to be used for the "support and maintenance of at least one

Table 28–3 Acreages, yields, and production of grains in the United States.

Period	Acreage (millions)	Production (millions of metric tons)	Yield (metric tons per acre)	Indexes (1880–1889 = 100)		
				Acreage	Production	Yield
1880–1889	135.0	72.2	0.535	100	100	100
1890–1899	162.8	86.8	0.533	121	120	100
1900–1909	185.2	103.5	0.559	137	143	104
1910–1919	208.1	111.2	0.534	154	154	100
1920–1929	218.2	120.2	0.551	160	169	103
1930–1939	219.6	102.8	0.468	163	142	87
1940–1949	212.5	134.7	0.634	157	187	119
1950–1959	187.9	143.3	0.762	139	198	142
1960–1964	160.7	176.7	1.099	119	244	205
1965–1969	152.9	215.3	1.408	113	298	263
1970–1974	155.0	223.9	1.445	115	310	270
1975–1979	173.0	282.4	1.632	128	391	307

Sources: From USDA, *Agricultural Statistics*, 1938, 1952, 1957, and 1964; and *Crop Production, 1964 Annual Summary, 1971 Annual Summary*, and *1978 Annual Summary*. (The crops included are wheat, rye, rice (from 1900), buckwheat, corn, oats, barley, and grain sorghums (from 1920). In 1900 the acreage of rice was 0.36 million, while in 1919 the grain sorghum acreage was 3.63 million. Acreage are in terms of harvested area.)

college where the leading object shall be . . . to teach such branches of learning as are related to agriculture and the mechanic arts." It was not until 1887, however, that an act was passed that established agricultural experiment stations in each state, usually at the "land grant" colleges.

By the mid-1920s, substantial progress in agricultural research, embodied in crops, production practices, and other forms of new knowledge, was beginning to flow from the agricultural-education and research system to the farmer via the newly organized federal–state extension service. Hybrid corn was a particularly dramatic product of this research.

The combination of rapid advance in biological research plus the high volume of relatively inexpensive agricultural chemicals created a new dimension in agricultural productivity in U.S. agriculture. Land productivity, which had experienced no real growth between 1900 and 1925, rose by 1.4 percent per year for the period 1925–1950, by 2.5 percent per year between 1950 and 1965, and by 0.8 percent per year between 1965 and 1979. This higher output per hectare combined with continued mechanization to produce a rate of growth in labor productivity of 6.6 percent per year between 1950 and 1965 and by 4.8 percent per year between 1965 and 1979 (Table 28-4).

The labor market

The labor market has become an increasingly important channel of interaction between the farm and the nonfarm sectors. Technical and economic developments have made it increasingly profitable to substitute inputs purchased from the industrial sector for farm labor. Between 1950 and 1965, the slow growth in domestic demand for farm products, the insulation of domestic markets from changes in demand in other countries through import restrictions and export subsidies, and the rapid growth in labor productivity combined to place most of the burden of balancing the rate of growth in agricultural output with the rate of growth in demand for agricultural products on the labor market.

Table 28–4   Annual average rates of change (percent per year) in total outputs, inputs, and productivity in United States agriculture, 1870–1979.

Item	1870–1900	1900–1925	1925–1950	1950–1965	1965–1979
Farm output	2.9	0.9	1.6	1.7	2.1
Total inputs	1.9	1.1	0.2	−0.4	0.3
Total productivity	1.0	−0.2	1.3	2.2	1.3
Labor inputs[1]	1.6	0.5	−1.7	−4.8	−3.8
Labor productivity	1.3	0.4	3.3	6.6	6.0
Land inputs[2]	3.1	0.8	0.1	−0.9	0.9
Land productivity	−0.2	0.0	1.4	2.6	1.2

Sources: USDA, *Changes in Farm Production and Efficiency*, 1979; and D. D. Durost and G. T. Barton, *Changing Sources of Farm Output* (USDA Production Research Report 36), 1960.
[1]Number of workers, 1870–1910; hours of labor, 1910–1971.
[2]Cropland used for crops, including crop failure and cultivated summer fallow.

With the demand for agricultural output expanding at less than 2 percent per year and labor productivity rising by about 6 percent per year, the burden of adjustment in the labor market was extremely heavy. It was particularly difficult in the low-income agricultural regions in which local nonfarm employment did not expand fast enough to absorb both the excess agricultural labor force and the new entrants to the labor force from rural areas.

Between 1940 and 1960, the annual net emigration of farm people averaged close to 1 million people (Fig. 28-2). During this same period, annual average employment on farms declined from 11 million to 7 million workers. Between 1960 and 1980, farm employment declined by another 3 million workers—to just under 4 million. There is today, however, a better balance between the size of the farm population and the availability of farm and nonfarm employment in rural areas than in the past. The incomes of farm families, which averaged less than 60 percent of the incomes of nonfarm families in the 1950s, are, when cost-of-living differences are considered, close to parity with nonfarm incomes (Table 28-5).

One of the more surprising developments in the agricultural-labor market is the rise in the number of hired farm workers since 1970. The average annual employment of farm workers rose from 1.17 million in 1970 to approximately 1.25 million in 1980, while employment of operator and family labor declined from 4.5 million to less than 4 million.

Hired farm workers remain the most disadvantaged group of American workers.° They have less protection than other workers from arbitrary action by employers and they receive fewer employment-related benefits when they are out of work. A fundamental limitation in achieving legislation to improve the position of farm workers is that they are not organized to reflect their own economic interests. Organization of farm workers is more difficult than organization of industrial workers because of the seasonal nature of much farm work and the casual attachment of many farm workers to the agricultural labor force. It is also difficult because of the weak legal protection that is given to organization efforts on the part of farm workers. Farm-operator interests are represented through general farm organizations, through commodity organizations, and through cooperatives. The neglect of the problems of hired farm workers in America is in particularly striking contrast to the attention that has been given to the economic problems of farm operators.

°See Stephen H. Sosnick, *Hired Hands: Seasonal Farm Workers in the United Stated* (Santa Barbara: McNally and Loftin, 1978).

1920-1925	666,000
1925-1930	593,000
1930-1935	58,000
1935-1940	708,000
1940-1945	1,602,000
1945-1950	677,000
1950-1955	1,115,000
1955-1960	910,000
1960-1965	794,000
1965-1970	594,000
1970-1975	199,000
1975-1978	312,000

Figure 28-2   Annual net emigration from the farm population. [From *1979 Handbook of Agricultural Charts* (USDA/ESCS Agriculture Handbook 561), 1979.]

Table 28-5   Mean income of farm and nonfarm families by race and origin, United States, 1977.

Mean income	All families	White	Black	Spanish origin
Total	16,100	16,729	10,791	12,565
Nonfarm	16,168	16,811	10,846	12,579
Farm	14,111	14,496	6,531	NA

Source: Bureau of the Census, *Money Income in 1977 of Households in the United States,* Series P-60, No. 117, December 1978.
NA = Not available.

Structural change

The agricultural-development policies of the United States have been uniquely successful in meeting national farm-output and productivity objectives. These policies have been less successful in meeting the income objectives of all of the families who are engaged in the production of agricultural commodities. One observer has pointed out that "behavior of rural people, their representatives and their institutions implies a materialistic bias in favor of plants, land, and animals and against people." Although this is perhaps overdrawn, it is true that the policies of the past were designed primarily to solve technological and commodity problems rather than to solve the income problems of rural people and the development problems of rural areas.

It can be argued that this was a valid choice at the time these policies were established. It was important for U.S. economic development that the agricultural sector achieve sufficiently high rates of output and productivity growth to meet national requirements for food and fiber and at the same time to release substantial numbers of farm workers for nonfarm employment.

In spite of the productivity of American farms and the prosperity of the families that account for most of the nation's food and fiber production, there has been a growing unease about the role of rural people and of rural communities. In March 1979, in a

speech at a Farmers Union convention in Kansas City, Secretary of Agriculture Bob Bergland suggested that the United States engage in a national dialogue about the future structure of American agriculture.°

Three issues have been of particular concern to American farm people and to rural communities:

1. Who will control agricultural production in the future?

2. Is there a role for the small farm?

3. What is the future of rural communities?

In the following sections we discuss these three issues.

Who will control agricultural production?

Agricultural production has been increasingly concentrated. The number of farms has declined from a peak of nearly 7 million in the 1930s to 2.7 million in the late 1970s. These numbers are somewhat mis-

---

°The USDA has produced a series of reports to serve as background for the discussion of structural issues. See U.S. Department of Agriculture, *Structural Issues of American Agriculture* (Washington, D.C.: USDA/ESCS Agricultural Economic Report 438, 1979).

leading. A farm is defined somewhat unrealistically in the official statistics as any place with annual sales of $1000 or more.

In 1978, 82 percent of farm sales were accounted for by the 577,000 farms with sales of more than $40,000. The share of sales accounted for by the 50,000 largest farms rose from 23 percent in 1967 to 36 percent in 1977. It is not difficult to anticipate that by the year 2000 close to two-thirds of farm sales could be accounted for by the 50,000 largest farms and 80 percent by the 100,000 largest farms.

Who will own the 100,000 farms that will account for most of U.S. farm output in the future? Will they be family farms or will they be owned by agribusiness firms that integrate production, processing, and marketing activities? It seems quite clear that, unless remedial action is taken, an increasing share of the production on specialized farms engaged, for example, in cattle feeding and poultry production, fruit and vegetable production, nursery and greenhouse plant production, and sugarcane production will be corporate farms. Family farms will probably continue to dominate in the more diversified farming systems and in the production of soybeans, food and feed grains, dairy products, and pork. Even in these areas, however, many family farms may incorporate in order to facilitate the intergeneration transfer of resources.

The rapid inflation of land prices in the 1970s is another factor that has led to a concentration of land ownership. Prices for agricultural land more than doubled in the decade between 1970 and 1980. The combination of larger farm size and higher prices for land and other assets has made it more difficult for young people to get started in farming without substantial family backing. This has led to a concern that the ownership of farmland may become increasingly concentrated in the hands of the wealthy—that rural America will be owned by a class of hereditary landed proprietors and by integrated corporate enterprises.

An important issue for the future will be whether changes should be made in the tax laws and the inheritance laws that would reduce the incentives for farmland ownership by those who are not operating farmers.

Is there a role for the small farm?

Small farms have been regarded by most agricultural scientists, farm-program administrators, and farm-organization and political leaders as outside the mainstream of American agriculture. Small farms have been viewed as subsistence operations for those with no alternative employment opportunities; as retirement residences for those who are no longer active farmers or who have retired to rural areas; and as a source of part-time employment or income supplementation for persons employed in nonagricultural occupations. The information presented in Figure 28-3 supports the view that, for small farmers as a group, income from farming accounts for a relatively small share of total income.

There has been a growing commitment to the idea that small farms ought to have a larger role in contemporary agriculture. The Rural Development Act of 1972 and the Food and Agriculture Act of 1977 included provisions for research and extension directed to the agricultural-production problems of small farmers. A number of pilot programs, such as the Texas Intensified Farm Planning Program and the Missouri Small Farm Program, have been started by the state extension services.

Others have been started with the help of private philanthropic foundations. In Mississippi and Alabama, two programs modeled on the Israeli *moshav* farm-cooperative concept have been started. Each farm involves 1200 acres and 40 farm families. The projects, which are initially heavily subsidized both from public and from private sources, are managed by a Small Farm Development Corporation. Such programs tend to be rather controversial. One newspaper that objected to the assistance provided by the USDA ran the following headline: "Government Will Establish Collective Farms under the Guise of Family Farm Development."

The most active support for the strengthening of small-farm agriculture comes from what is often referred to as the "alternative-agriculture" movement. The alternative-agriculture movement emphasizes both technological and social alternatives to the dominant trends in U.S. agricultural and rural development over the last fifty years. In the area of tech-sales classes

all figures in thousands of dollars

off-farm income   net income from farming

sales class	off-farm income	net income from farming
all farms	12.8	10.0
below 2.5	17.2	1.7
2.5–4.9	16.2	1.9
5–9.9	13.6	3.3
10–19.9	10.1	5.9
20–39.9	7.8	11.7
40–99.9	6.8	21.6
100 and over	10.9	52.3

Figure 28-3   Farm and off-farm income per farm-operator family by value of farm products sold in 1978. [From *1979 Handbook of Agricultural Charts* (USDA/ESCS Agriculture Handbook 561), 1979.]

nology, it emphasizes organic approaches to maintaining and improving soil fertility. Alternatives to the intensive use of pesticides are stressed. Less capital-intensive systems of crop production are employed. At the social level, smallness, self-sufficiency, decentralization, and the development of the rural community and culture are emphasized.°

What is the future of the rural community?

During the 1970s, there has clearly been a reversal of the long-term trends in population in many rural areas (Fig. 28-4). The nonmetropolitan areas of the southern Appalachians, the Ozark Uplands, and the upper Midwest that had lost population in the 1950s and 1960s were among the regions with the most rapid population growth in the 1970s. Improvements in the health and education services in rural communities have made them more attractive places to live. Decentralization of employment opportunities have made them more attractive places to work. And the new preference for rural living has resulted in a more positive view of the advantages of rural life.

As the population of rural areas has expanded, the nonagricultural character of the rural economy and the rural community has become more pronounced. As a result, economic-development and community-development programs must address the needs of the entire rural and small-town population if they are to have any chance of success. The need to respond to expanding needs for improvements in the quality of education, to meet the health and housing needs of the elderly, and to provide for water, sanitation, and fire-protection services is placing increasing pressure on the capacities of voluntary community agencies and local governments.

A major challenge that faces rural communities is how to respond to the opportunities and pressures for growth without losing the very qualities that have led to a preference for rural life by farm people and other rural residents.

THE DEVELOPMENT OF A LABOR-INTENSIVE AGRICULTURAL SYSTEM: JAPAN

Japan was the first Asian country to succeed in bringing about a striking transformation in the productivity of its agriculture. In contrast to the United States, agricultural development in Japan occurred within a

---

°For a useful introduction to the philosophy of alternative agriculture, see Wendell Berry, *The Unsettling of America: Culture and Agriculture* (New York: Avon Books, 1977).

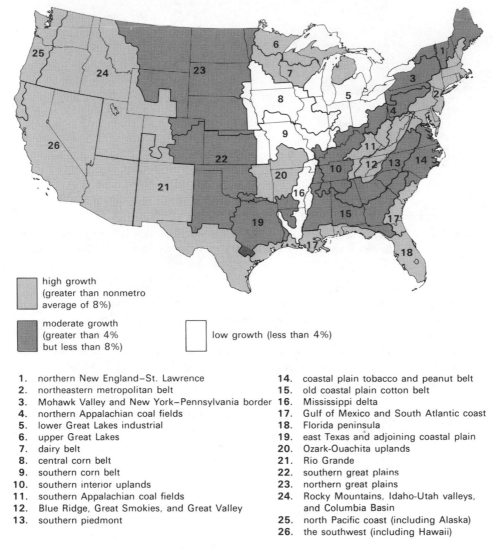

high growth
(greater than nonmetro
average of 8%)

moderate growth
(greater than 4%
but less than 8%)

low growth (less than 4%)

1. northern New England–St. Lawrence
2. northeastern metropolitan belt
3. Mohawk Valley and New York–Pennsylvania border
4. northern Appalachian coal fields
5. lower Great Lakes industrial
6. upper Great Lakes
7. dairy belt
8. central corn belt
9. southern corn belt
10. southern interior uplands
11. southern Appalachian coal fields
12. Blue Ridge, Great Smokies, and Great Valley
13. southern piedmont

14. coastal plain tobacco and peanut belt
15. old coastal plain cotton belt
16. Mississippi delta
17. Gulf of Mexico and South Atlantic coast
18. Florida peninsula
19. east Texas and adjoining coastal plain
20. Ozark-Ouachita uplands
21. Rio Grande
22. southern great plains
23. northern great plains
24. Rocky Mountains, Idaho-Utah valleys, and Columbia Basin
25. north Pacific coast (including Alaska)
26. the southwest (including Hawaii)

Figure 28-4 Nonmetropolitan population change by economic subregion, 1970–1976. [From *1979 Handbook of Agricultural Charts* (USDA/ESCS Agriculture Handbook 561), 1979.]

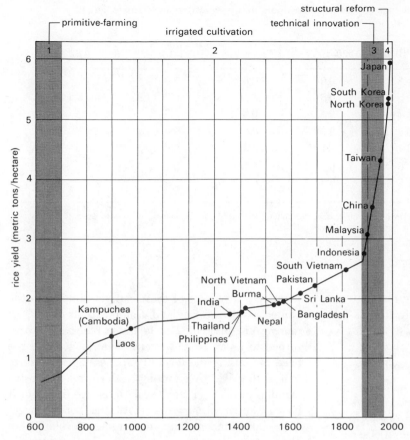

Figure 28-5   Current rice yields (rough rice) in selected countries related to Japan's historical trend. Intensification of farming on land now being farmed is the other way to grow more food. This means moving farming in developing regions to higher stages of development, in effect recapitulating the historic progression exemplified here by the case of Japan. Typical rice yields in Japan increased (*black curve*) as Japanese agriculture moved from the traditional stage through the advent of irrigation to scientific agriculture and finally to structural transformation. Current yields in most Asian countries, where less than 50 percent of the rice land is cultivated, place them still in second stage, as plotted. [From W. D. Hopper, "The Development of Agriculture in Developing Countries." Copyright © 1976 by Scientific American, Inc. All rights reserved.]

framework of increasing labor intensity. The average size of the Japanese farm was approximately 1 hectare (2.2 acres) in 1878 and 0.8 hectare (1.7 acres) in 1962.

In many respects, the Japanese experience was the outcome of factors peculiar to the Japanese environment. Japan is an island economy, and social institutions had been cultivated and refined over generations to permit an efficient balance between population and a limited resource base.

Despite other differences, at the time of the "take-off," Japanese agricultural development was similar in many respects to that of other countries in Asia—a traditional agriculture characterized by small-scale subsistence farms dominated by a hierarchical social structure. The level of productivity in Japanese agri-

culture prior to the Meiji Restoration (1868), as measured by yields per hectare or per person, were probably only slightly higher than the levels of productivity that persist today throughout the rest of Asia (Fig. 28-5).

The remarkable feature of the Japanese agricultural transformation, and the feature that makes it of special interest in the rest of Asia today, is that it took place within a traditional framework of small-scale agriculture. Although the average farm was declining in size, the average rice yield rose from 1.8 metric tons per hectare (1868 to 1882) to 4.0 metric tons in the late 1950s and 5.8 metric tons in the mid-1970s.

During this development, Japanese agriculture went through four phases (Table 28-6):

Table 28-6 Annual average change in total outputs, inputs, and productivity in Japanese agriculture, 1880–1975.

Item	1880–1920	1920–1935	1935–1955	1955–1965	1965–1975
Farm output	1.8	0.9	0.6	3.6	1.4
Total inputs	0.5	0.5	1.2	0.7	—
Total productivity	1.3	0.4	−0.6	2.9	—
Labor inputs	−0.3	−0.2	0.6	−3.0	−3.6
Labor productivity	2.1	1.1	0.0	6.6	5.0
Land inputs	0.6	0.1	−0.1	0.1	−0.7
Land productivity	1.2	0.8	0.7	3.5	2.1

Source: S. Yamada and Y. Hayami, "Growth Rates of Japanese Agriculture, 1880–1965," in *Agricultural Growth in Japan, Korea, Taiwan, and the Philippines,* University Press of Hawaii, 1979.

1. A period of rapid growth lasting from approximately the time of the Meiji Restoration (1868) until the end of World War I.

2. A period of slower growth lasting from the end of World War I through World War II.

3. A new period of rapid growth, starting almost immediately after the end of World War II and continuing to the mid-1960s.

4. A period of slower growth since the mid-1960s.

Productivity growth from intensification of traditional agriculture, 1870–1920

The rapid increase in Japanese agricultural productivity between 1870 and 1920 was due to the diffusion of the superior practices already in use in Japanese traditional agriculture combined with the limited adoption of Western methods, primarily soil and fertilizer science adopted from Germany.

The basic agricultural policy of the new Meiji Government, established in 1868, developed around the ideas of Lord Iwakura, who visited North America and Europe in 1871 and 1873. Iwakura's visits were followed by tours made by other Japanese officials, and western agricultural experts were invited to

Japan. At first, most of these experts came from the United States and England. Later, experts from Germany were invited. The Japanese officials who returned to Japan from western countries stressed the need for raising the level of Japanese agriculture to that of the West by adopting western-style extensive farming in place of small-scale intensive farming. It became apparent, however, that instead of the large-scale farm-management techniques developed in North America and England, the knowledge and techniques of Germany, particularly the new knowledge of soils and fertilizers, would become more realistic and practical in application to the small-scale Japanese agriculture. Only in the northern island of Hokkaido did extensive farming based on livestock production and western-style "horse mechanization" find a permanent home.

In 1887, a new agricultural policy was adopted that shifted emphasis away from introducing western farming methods to bolstering traditional farming methods. There were three major elements in this new policy: (1) selection and diffusion of high-yielding cultivars of rice; (2) the establishment of a fertilizer-consuming agriculture; (3) introduction and diffusion of new cultural practices and implements.

These three policies were closely interrelated. The new rice cultivars selected were those that responded well to nitrogen fertilizer. Such cultural practices as deep plowing, double cropping, straight-row planting, revision of field layouts, irrigation and drainage, and others were also oriented toward obtaining a favorable response from increased fertilization. Table 28-7 traces the evolution of the use of fertilizer in one Japanese prefecture from 1877 to 1957.

These policies were effective for several reasons. Over the preceding three hundred years of the Tokugawa period, agricultural techniques had been slowly improving, but the restraints of the feudal system had suppressed the diffusion of new techniques. Under the feudal system, peasants were bound to their land and not allowed to leave their villages except for religious pilgrimages. Nor were they free to choose which crops to plant or which cultivars to sow. The feudal lords were anxious to raise agricultural productivity in their own territories, but they frequently prohibited the transfer of techniques or cultivars outside their bounds. It is even recorded that one village placed a guard at its border to prevent a variety of seed selected in the village from being taken out.

A second factor in the success of the Meiji agricultural policy was the pattern of investment in agricultural education by the government. Agricultural schools were organized as early as 1876. An agricultural experiment station and seed breeding station was established in 1877—ten years before the U.S. Congress passed the Hatch Act, which established experiment stations in each state. By 1900, Japan had developed a number of national research institutions as well as a network of experiment stations at the prefectural level.

Initially, the most successful "veteran farmers" were used to carry improved techniques to other farmers in their own and in other prefectures. By 1893, eleven years before the establishment of the federal–state extension service in the United States, the prefectural experiment stations were given responsibility for formal extension activity.

Thus the prefectural experiment stations became extension centers for the dissemination of new knowledge to associations of village farmers. Failure to adopt improved technologies was frequently punished by fines or arrest. This obviously created a somewhat different level of receptivity than the county agents in the United States faced when they began their work in 1914.

Stagnation of traditional agriculture, 1920–1946

Shortly after World War I, Japan appeared to have reached the limit of agricultural development that could be attained by using traditional methods. The rate of growth of agricultural output declined. Food shortages developed in the face of a growing urban population. Whereas agriculture had been a major source of support for Japan's industrial revolution during the Meiji era, it now became a depressed area in the economy. The land-tenure system placed increased burdens on the tenants and dampened incentives to produce.

A number of significant changes took place during this period. Agricultural land development was pushed into marginal areas. Farmers began to shift their production away from rice into the even more labor-intensive livestock production, silkworm raising (sericulture), and fruit and vegetable production. The shift away from rice was partly in response to the lowering of prices that resulted from the importation of rice from Formosa (Taiwan) and Korea.

The rapid shift of farm workers from agricultural to nonagricultural employment during the period was not sufficient to reduce the size of the agricultural labor force. Farm employment did decline, however, from roughly 40 percent of the labor force at the beginning of the period to less than 20 percent by the beginning of World War II.

By the beginning of World War II, Japanese agriculture was again beginning to experience rapid economic growth. During the war, however, agriculture suffered from shortages of labor and materials. Availability of commercial fertilizer dropped sharply. The yield of rice fell in 1945 to only about 70 percent of the prewar level.

The modernization of Japanese agriculture

Following World War II Japanese agriculture experienced a new burst of productivity. Two factors appeared to have been particularly important. One was the increased incentive to produce, which resulted from the land reform of 1947–1950. A second was the backlog of modern technology resulting from the

Table 28–7 Changes in application of fertilizers in Kanaya-Machi, Shizuoka prefecture, Japan, 1877–1977.

Year	Major fertilizer	Rice		Barley		Tea	
		Plains	Mountains	Plains	Mountains	Plains	Mountains
1877	Natural organics						
1887		Grass	Grass	Night soil	Unknown	Soybean meal, shrimp	Unknown
1897	Green manure and organic by-products (fish, soybean, rapeseed meal)	Grass, compost, soybean and fish meal	Grass, compost, soybean and fish meal	Night soil, compost, soybean meal, superphosphate	Night soil, compost	Soybean meal, shrimp, organic mixes	
1907		Grass, compost, soybean and rapeseed meal, spent distillery mash, green soybean	Grass, soybean and rapeseed meal, seaweed	Night soil, compost, soybean meal, superphosphate	Night soil, compost, superphosphate	Night soil, shrimp, organic mixes, soybean meal, green soybean	
1917		Grass, compost, soybean and rapeseed meal, sodium nitrate, ammoniated superphosphate	Grass, soybean and rapeseed meal, ammonium sulfate, superphosphate	Night soil, compost, superphosphate, soybean meal, ammonium sulfate	Night soil, compost, superphosphate	Night soil, meal, organic mixes, shrimp	Soybean and fish meal
1927	Chemical	Grass, compost, fish meal, ammonium sulfate, sodium nitrate, mixed	Grass, compost, night soil, organic mixed fertilizers, superphosphate, ammonium sulfate, sodium nitrate	Night soil, compost, meal, ammonium sulfate, super phosphate, mixed	Night soil, compost, meal, ammonium sulfate, super phosphate, mixed	Night soil, meal, green soybean, organic mixes	Grass, organic mixes, super phosphate, ammonium sulfate

Source: From T. Ogura (ed.), *Agricultural Development in Japan,* Japan FAO Association, 1963, pp. 372–373 (after Y. Ienaga); unpublished data for 1957–1977 from Y. Ienaga.
[1]The distinction between plains and mountains has tended to disappear as a result of continued modernization.

Year	Major fertilizer	Rice		Barley		Tea	
		Plains	Mountains	Plains	Mountains	Plains	Mountains
1937	Chemical	Compost, fish meal, ammonium sulfate, superphosphate, potassium chloride, calcium cyanamide, ammonium phosphate, mixed	Grass, compost, wood ash, night soil, ammonium sulfate, calcium cyanamide, super phosphate, potassium chloride, mixed	Night soil, compost, ammonium sulfate, super phosphate	Night soil, compost, ammonium sulfate, super phosphate, mixed	Superphosphate, ammonium sulfate, mixed	Rapeseed meal, grass, ammonium sulfate, mixed
1947		Rapeseed meal, ammonium sulfate, calcium cyanamide, potassium chloride, fused phosphate, mixed	Night soil, compost, ammonium sulfate, potassium chloride, urea, fused phosphate, mixed	Night soil, compost, potassium chloride, ammonium sulfate, calcium cyanamide, mixed	Night soil, compost, ammonium sulfate, super phosphate, mixed	Grass, ammonium sulfate, calcium cyanamide, mixed	Fish and rapeseed meal, ammonium sulfate, calcium cyanamide, urea, mixed
1957		High synthetic, low synthetic compound, calcium cyanamide (granular, dust), calcium cyanamide compound, potassium phosphate ammonium mixed, special compound for rice seedling bed, silic calcium, ammonium chloride, magnesium lime, mixed, urea intermixed compound, ammonium potassium compound[1]		Low synthetic compound	None grown	Superphosphate of lime, fused phosphate, potassium chloride, potassium sulfate, low synthetic compound, high synthetic compound, organic matter intermixed compound, multiphosphate, phosphate potassium ammonium mixed, phosphate ammonium nitrate, potassium mixed, magnesium lime, fish meal, seed cake, chicken droppings, ground bonemeal, ground composts[1]	
1977							

increased sophistication of experiment-station research and the increased industrial capacity that emerged from World War II.

The postwar land reform represented the culmination of land improvement that began during the Meiji restoration rather than any sharp break with the past. All farmland owned by absentee landlords and all farmland leased by resident landlords in excess of 4 hectares was appropriated by the government and sold to the actual tenants. About 2 million hectares, approximately 80 percent of the tenant-cultivated land, was acquired by tenant farmers. The result was a major improvement in incentives to adopt new technology and increase production (Table 28-6).

The change in agricultural technology since World War II can be illustrated by the following. The new rice cultivars being planted in Japan today are the result of experiment-station breeding programs designed to produce high-yielding, disease-resistant varieties that respond well to fertilization. This is in contrast to the basis of selection that prevailed during the Meiji period, when the best cultivars were selected from the many already in existence. The new breeding programs, however, did not depart in their objective from that of the older selection programs. The varietal improvements and associated cultural practices continued to be directed toward the development of a "fertilizer-consuming rice culture."

The increased capacity of Japanese industry to produce fertilizers and other agricultural chemicals has complemented this traditional objective. There has, however, been one major change. As a result of Japan's continued rapid industrial development, the agricultural labor force began to decline in the 1950s. Agricultural wage rates have risen. Small-scale mechanization, including the use of power sprayers and dusters and the use of electric motors and internal-combustion engines for threshing and pumping irrigation water, has expanded rapidly. Most striking of all has been the rapid small-scale mechanization of plowing and other field operations. The use of mechanized equipment for field operations was trivial before World War II, but by 1960 there were half a million small tractors or cultivators in use. By the 1970s, this small-scale equipment was being replaced by larger land-preparation and tillage equipment.

A major problem of structural reform in Japanese agriculture is how to increase the size of the farm operating unit in an equitable manner. Programs of assistance for land consolidation and cooperative use of farm equipment have been implemented. The Farmers' Pension Fund is authorized to purchase agricultural land from older farmers and sell it to farmers who want to enlarge their farms. The fund can also assist in financing the purchase of land from retiring farmers.

The significance of Japanese agricultural development

This history of Japanese agricultural development illustrates how the agricultural sector of the economy was able to fulfill its traditional role in the strategy of overall development. Japanese agriculture, in the course of its transformation, was able to earn foreign exchange, to provide savings and investment for a developing urban-industrial sector, and to supply raw materials and foodstuffs for the rest of the economy.

Most significant of all, this was achieved within a system of small-scale, labor-intensive farming made possible by placing greater emphasis upon the "biological revolution" than upon the "mechanical revolution." In the next several decades, as the Japanese agricultural labor force declines, it seems likely that Japanese agriculture will successfully complement the "biological revolution" of the last 100 years with a "mechanical revolution," leading to a fully modern system of agriculture. The Japanese example, with its initial stress on the "biological revolution," represents a more valid model for many of the developing economies than the United States model.

THE TAKE-OFF THAT FAILED: ARGENTINA

Agricultural development in Argentina presents an interesting contrast to both the United States and Japan. Like the United States, Canada, and Australia, Argentina developed, during the nineteenth century, an agricultural system based on an abundance of land and a relative scarcity of labor. By the late 1920s, Argentina ranked among the leading countries

Table 28-8 Share of Argentine exports in world exports of selected commodities (expressed as percentages).

Commodity	1934–1938	1959–1962	1968–1970	1975–1977
Wheat and wheat flour	19.3	5.5	4.6	5.2
Corn	64.0	18.5	14.2	7.3
Greasy wool	} 11.7	9.9	5.8	5.6
Degreased wool		13.9	13.8	11.6
Linseed and linseed oil[1]	67.6	49.3	29.3	23.5
Linseed, cake and meal	NA	71.6	59.5	55.2
Sunflower seed and its oil[1]	NA	14.1	5.3	4.9
Sunflower seed, cake and meal	NA	95.2	75.0	74.4
All fresh, chilled, frozen meat	39.7	17.5	11.8	5.4

Sources: From *Trade Yearbook, 1963* and *Trade Yearbook 1977*, FAO, and *Analisis y Proyecciones del Desarrollo Economico, V: El Desarrollo Economico de la Argentina*, United Nations, part 2, p. 48, 1959.
[1]It has been assumed that three tons of linseed and sunflower seed are equivalent to one ton of linseed oil and sunflower-seed oil.
NA = Not available.

in the world with respect to income per capita. Its prosperity was based on an economic system that involved (1) exportation of agricultural raw materials, particularly food and feed grains, oil crops, and meat and wool, and (2) importation of manufactured capital equipment and consumer goods.

The disorganization of international markets during the world depression of the 1930s and during World War II impressed upon Argentina's policymakers the desirability of developing a more diversified economy. Industrialization became a major policy goal.

In an effort to increase the foreign exchange earnings needed to finance industrialization, Argentina established a government monopoly to handle the export of primary commodities. The monopoly paid low prices to Argentina's agricultural producers and attempted to bargain for the highest possible prices for the export commodities on world markets. The prices received by Argentine farmers were kept far below world market prices. This was equivalent to a heavy tax on agricultural production.

The effect was to dampen incentives either to produce or to export agricultural commodities. Agricultural production increased by less than 10 percent between the periods 1945–1949 and 1959–1961. Argentina's share of world exports of a number of major agricultural commodities declined sharply from prewar levels (Table 28-8). This weakened Argentina's capacity to import. By the mid-1950s, imports of capital goods needed for industrialization had dropped below prewar levels.

The lack of price incentives might have been partially offset if Argentina had given support to technological change in agriculture by building agricultural research and extension programs in a manner comparable to the United States and Japan. Even today, however, Argentina has only a relatively weak agricultural-research and extension establishment. Corn yields remain relatively low compared to those of the United States and a number of other countries.

Thus the heavy taxation of Argentina's agricultural technology killed the goose that laid the golden eggs that were expected to pay for the program of industrialization. Today Argentina's income per capita ranks her either as one of the less prosperous developed nations or one of the more prosperous developing nations, rather than in the position among world leaders that she held in the 1920s.

Argentina's heavy taxation of agricultural exports, in such a way as to reduce directly the prices received by farmers, was in sharp contrast to the policies adopted by the Japanese during the last third of the nineteenth century. Japan successfully utilized the export earnings from tea and silk to finance industrialization. Distortion of production incentives was avoided by using a heavy land tax to raise revenue from agriculture instead of directly interfering with prices. The favorable prices for silk and tea were allowed to act as incentives to produce supplies for export, and the land tax, which was not affected by the level of production, was used to finance industrialization.

During the 1970s, there were some indications that Argentina was beginning to adopt policies that would provide greater incentives for agricultural production. The response of agricultural production and exports to these greater incentives has, however, been obscured by the extremely high rates of inflation in Argentina.

DEVELOPMENT IMPERATIVES

There are numerous examples of situations in which rapid progress in agricultural development did not lead to sustained economic growth because of the absence of other elements critical to the process of development. Many examples of this type can be drawn from the former colonial economies. The Dutch experience in Java is particularly striking. The Dutch were remarkably successful in their efforts to develop improved crop varieties and cultural practices for a number of tropical export crops. The Agricultural Research Institutes that they developed at Bogor were among the best in the world.

The colonial system was not, however, successful in developing the capacity of the Javanese population. The result was a dual economy that provided high incomes for the expatriate owners and employees who ran the plantation and for the firms that handled the export commodities. Production of food for local consumption continued to be produced by traditional methods by the local population. Gains in productivity per hectare were absorbed by population growth with no rise in the standard of living of the indigenous population.

The same pattern was repeated, although perhaps not as strikingly, in India, in the Philippines, and in many other former colonial economies in Asia and Africa.

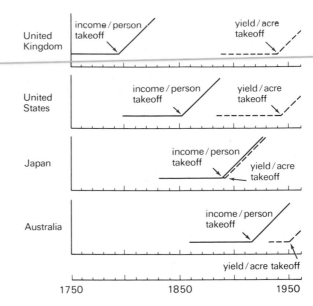

Figure 28-6 Take-off dates for income per person and yield per acre in selected countries. [After Brown, *Increasing World Food Output* (USDA Foreign Agricultural Economics Report 25), 1965.]

Successful economic development involves a complex of technological, economic, and cultural changes. Rapid productivity changes in agriculture are becoming an increasingly essential component in this complex. In the past, countries like the United Kingdom, the United States, and Australia were able to achieve rising incomes in advance of growth and land productivity (Fig. 28-6). The United Kingdom was able to postpone the yield takeoff because it was more profitable to capitalize on its early technological lead by exporting manufactured goods and importing foodstuffs and raw materials. The United States and Australia, because of a relatively favorable ratio of land resources to labor, were able to concentrate on achieving rapid growth of labor productivity prior to emphasis on land productivity.

Japan's experience in generating an income takeoff and a yield takeoff at the same time is unique. This is clearly the optimum pattern for most of the presently developing countries, particularly those that already have a high population relative to land resources. For many of the developing countries, particularly those that are unable to limit the rate of

Markets for
farm products
provide one spoke.

New farm technology
adds a second.

Local availability of
farm supplies and
equipment is a third.

Adequate incentives
for farmers
provide the fourth.

Transportation facilities
complete the wheel.

Figure 28-7 The five essentials for agricultural development are like the parts of a wheel. [After A. T. Mosher, *Getting Agriculture Moving: Essentials for Development and Modernization*. Praeger, 1966.]

population growth, it may be necessary for a yield takeoff to precede an income takeoff.

Important as yield increases are, they are but one of a number of closely interrelated factors involved in agricultural development. The development of the countries studied in this chapter—the United States, Japan, and Argentina—illustrates the workings of five "essentials" and five "accelerators."

The five essentials are: (1) new farm technology, (2) availability of purchasable inputs, (3) markets for products, (4) transportation, and (5) incentives for agricultural producers. The adequacy of these five determine the possibilities of agricultural development. They are like the parts of a wheel (Fig. 28-7); none is useful without the others.

The accelerators are those factors that, although not absolutely essential for agricultural growth, can contribute to speeding up the rate of growth once the essentials are met. The five accelerators are: (1) education, (2) production credit, (3) effective farm organizations or associations, (4) improving or expanding the land base, and (5) effective agricultural planning.

There can and will be some growth in agricultural

productivity wherever all of the essentials are present, but without all of them there will be none. The situation is different with the accelerators. Each of them is important but none are indispensable. But in many countries it is necessary that agricultural development proceed as rapidly as possible. Recent studies have documented that investment in agricultural research often gives returns 5 to 10 times higher than conventional returns on capital investment. By placing emphasis on the accelerators, as well as on the essentials, the pace of progress can be accelerated and the return on the investment in the essentials can be raised.

DEVELOPMENT CONSTRAINTS

It is conventional to emphasize the social and economic gains brought about by agricultural development. These gains have been substantial. In short, productivity growth in agriculture has been an essential component of the total development process.

The harmful effects of agricultural development are frequently ignored. They are primarily of two types. First, there is the environmental stress that occurs as a by-product of productivity growth. Second, there is the social stress resulting from inequities in the partitioning of the costs and benefits of agricultural development.

Environmental stress

The environmental damages caused by productivity growth were discussed in Chapters 5, 24, and 27. They are largely the by-products of the increases in intensity of cropping systems associated with the use of insecticides and herbicides, and of nitrogen and phosphate fertilizers, and of the processing of agricultural and forest products. Concern has also been expressed about the loss of genetic diversity in crops that have been bred for high yields and about the effects of deforestation on climate and on oxygen production.

These social costs have not been adequately reflected in most calculations of the economic returns from investment in research or in trends in output per unit of input (Tables 28-4 and 28-6). There can be no question that the problems of environmental congestion and pollution, both from agricultural and from industrial sources, have reached serious dimensions in specific localities and regions. The casual use and diffusion of certain materials, such as chlorinated hydrocarbons, pose a serious threat to environmental stability, public health, and economic activity.

Recent debate on environmental policy has tended to polarize around two alternatives. One is the antigrowth movement. It is founded on the view that the relation between people and the environment is so delicate that the effects of economic growth on the natural world may seriously impair the capacity of the earth to support life. The capacity of the ecosystem to sustain production and to absorb the byproducts of productive activity is regarded as finite; the limits to growth are being approached exponentially; and the conclusion is that, if present growth trends continue, the world will face ecological disaster within a matter of decades.

An alternative view is that scientific and technical effort can be redirected to permit the reduction of environmental stress and the continued acceleration of the performance of the ecosystem.

In the past, the capacity of the environment to absorb pollutants or residues seemed to be infinite. Consequently the tendency was to bias the direction of technical change toward excessive production of residues. This process has been clearly apparent in agriculture. Although one effect of the agricultural commodity programs has been to make the use of land in agricultural production "expensive" by restricting its use, at the same time, the ability of the environment to absorb the residues from crop and livestock production has been assumed to be limitless. As a result, in the development of agricultural innovations in the United States, too much emphasis has been placed on the development of land substitutes—plant nutrients, plant-protection chemicals, crop cultivars, and management systems that reflected the overvaluation of land and the undervaluation of the social consequences of the absorption or disposal of the residues from agricultural production processes. It also seems apparent that these biases in resource pricing have led to underinvestment in sci-

entific and technological efforts directed toward those pest-management and soil-management systems that would be more compatible with current efforts to preserve the quality of the environment for the people who must live in it.

The implication of this perspective is that the appropriate response to the environmental crises is the redirection of scientific and technical effort to reduce the environmental stress caused by the production of agricultural commodities. This move will also require the innovation of social institutions having the authority to establish and regulate both private and public property rights in environmental resources. One possibility might be the establishment of private firms or public authorities with appropriate incentives to manage environmental resources. Another alternative is to design market or marketlike mechanisms to direct the production and use of environmental commodities and services.

The basic limitation of the first, or "crisis," approach to environmental stress is that too much preoccupation with the threat of ultimate disaster diverts attention from the efforts needed for solution of the problems that are of immediate concern. The skills attained in scientific, technical, and institutional innovation required to solve the more immediate, if less dramatic, problems will add to our capacity to solve the more distant, through more dramatic, problems as they emerge.

Social stress

Agricultural and economic development places great stress on the social systems as well as on the ecosystem. Technological change widens the options available to a society. It makes available to society new income streams that can be used to support a wide range of individual and social objectives. Economic units, such as firms, households, and public agencies, are engaged in a continuous struggle to capture or internalize the new income streams resulting from economic growth and to avoid the costs associated with growth. The broader society is simultaneously engaged in a struggle to force the economic units to bear the costs of growth and to diffuse or externalize the benefits.

In agriculture, the socialization of much of agricultural research, particularly the research leading to advances in biological technology, is an example of public-sector institutional innovation designed to realize for society the potential gains from advances in agricultural technology. The political and legislative history of farm-price programs, from the mid-1920s to the present, can be viewed as a struggle between agricultural producers and society in general regarding the partitioning of the new income streams between agricultural producers and consumers.

The gains from agricultural development are also distributed unequally among the several social classes engaged in agricultural production. For example, when the benefits of technical change are computed, usually little consideration is given to the losses undergone by displaced workers. In a study of the returns from the invention, development, and use of the tomato harvester in the United States, it was found that the social returns, assuming that workers would be compensated for their loss of jobs, were several orders of magnitude below the level of returns when such losses are not included.° Effective social institutions to facilitate such compensation are not available. Therefore, the adoption of the tomato harvester provided a net gain to society but also imposed substantial uncompensated losses on workers. Part of this gain, then, was actually a subsidy to the producers by the workers.

In the United States, the effects of such displacement are often of relatively short duration because of the capacity of economic growth in other sectors of the economy to absorb the labor displaced by technological change in the agricultural sector. Where the capacity to absorb displaced workers does not exist, the intensity of the sociopolitical struggle over the partitioning of gains—the new income streams—resulting from agricultural development may be quite intense.

Some of the conflict is between land owners, ten-

ants, and landless laborers over the distribution of the increase in farm income. Some is between the rural and urban sectors regarding how much of the gains will be transferred to urban workers in the form of lower prices. There is a struggle between the private and public sectors concerning control of the uses of the growth "dividends." Also, within the public sector, the "military" and the "development" bureaucracies frequently struggle over the uses of new resources available to the public sector. The effects of agricultural development, and of economic development generally, may be to generate more social tension than the political systems of many countries seem able to absorb.

A viable sociopolitical environment is essential for the successful solution of the problems of environmental stress accompanying economic growth. The evolution of a viable sociopolitical system, capable of introducing institutional innovations that effectively allocate the gains of growth among classes and sectors, is fundamental to the development process.

SCIENCE, TECHNOLOGY, AND AGRICULTURAL PROGRESS

This book has given major attention to the scientific and technological foundations that underlie the creation of new agricultural practices. New technology, particularly new crop-production technology resulting in higher yields, is one of the five essentials for agricultural progress.

The foundation of biological science on which this new crop-production technology rests is becoming increasingly sophisticated. It requires greater depth in the training of biological and agricultural scientists in their respective disciplines. If the new agricultural technology is to exert an impact on agricultural production, increasing sophistication is also required in the agribusiness and public sectors of the economy, since these sectors organize and make available to agricultural producers the other four essentials—purchasable inputs, markets, transportation, and incentives.

It is also true that the biological and agricultural scientists who develop the new science and technol-

°Andrew Schmitz and David Seckler, "Mechanized Agriculture and Social Welfare: The Case of the Tomato Harvester," *American Journal of Agricultural Economics*, vol. 52, no. 4, pp. 569–577, 1970.

ogy can make the greatest contribution to agricultural progress if they can achieve a broad understanding of the process of agricultural development itself. Agricultural progress depends not only on the specialized skills of a wide variety of persons engaged in many occupations and activities in each country attempting to develop its agriculture but also on an understanding of the nature of agricultural development. Such understanding is needed not only by agricultural scientists, educators, planners, and administrators but by legislators, editors, merchants, bankers, and many others.

For this reason the scope of this book is considerably wider than the typical introduction to a field of science. It provides a broad international and interdisciplinary orientation to the problems of crop science and production. There are, however, many chapters that remain unwritten or incomplete. Their completion will depend on the potential scientists, educators, farmers, and citizens who will be organizing and carrying out the agricultural development efforts of the future.

Selected references

Brubaker, S., 1975. *In Command of Tomorrow: Resource and Environmental Strategies for Americans*. Baltimore: The Johns Hopkins University Press. (The author outlines potential resource-policy choices and suggests ways to make a transition from our present technology to one based on less perishable resources. Particular attention is given to problems of land use, energy, pollution control, and population growth.)

Cochrane, W. W., 1979. *The Development of American Agriculture: A Historical Analysis*. Minneapolis: The University of Minnesota Press. (A historical analysis of the development of American agriculture written from the perspective of economic development. A particularly useful discussion of the technical and institutional sources of U.S. agricultural development.)

Hayami, Y., and V. W. Ruttan, 1971. *Agricultural Development: An International Perspective*. Baltimore: The Johns Hopkins University Press. (A theoretical and empirical analysis of the role of technical and institutional change in the agricultural development of the United States and Japan. Implications of the U.S. and Japanese models of agricultural development for the developing countries and for agricultural-development policy are examined.)

Hayami, Y. (with M. Akino, M. Shintani and S. Yamada), 1975. *A Century of Agricultural Growth in Japan: Its Relevance to Asian Development*. Minneapolis: The University of Minnesota Press. (This book reveals how agriculture has contributed to, and benefits from, the transition of the Japanese economy into one of the most highly developed economies of the world. A unique feature of this growth is that it was achieved against the strong constraint of limited land resources.)

Mosher, A. T., 1966. *Getting Agriculture Moving*. New York: Praeger. (This book deals with the needs and problems at early stages of agricultural development and its principal purpose is to serve as a framework for systematic discussion of the development process in agriculture in in-service programs.)

Owens, E., and R. Shaw, 1972. *Development Reconsidered: Bridging the Gap Between Government and People*. Lexington, Mass.: D. C. Heath. (This book sets forth a strategy of development in which participation by the mass of small producers—farmers, artisans, entrepreneurs—is both the means of and the end of development. This approach is contrasted with the capital-intensive dualistic strategy based on capital-intensive development in the big cities and on big farms.)

Schultz, T. W., 1964. *Transforming Traditional Agriculture*. New Haven, Conn.: Yale University Press. (A classic discussion of the sources of agricultural development by the recipient of the 1979 Nobel Prize in economics. Major emphasis is placed on the farmer in developing countries as a rational decision maker who utilizes the resources available to him in an efficient manner. Stress is placed on the importance of research and development in making new and more profitable inputs available to farmers and on the importance of the education of farm people.)

Appendix
Conversion Tables

TEMPERATURE

To convert a temperature in either Celsius or Fahrenheit to the other scale, find that temperature in the center column, and then find the equivalent temperature in the other scale either in the Celsius column to the left or in the Fahrenheit column to the right. For example, if the Celsius temperature is 44°, the equivalent Fahrenheit temperature (in the right-hand column) is 111.2°. If the Fahrenheit temperature is 44°, the equivalent Celsius temperature (in the left-hand column) is 6.67°.

On the Celsius scale the temperature of melting ice is 0° and that of boiling water is 100° at normal atmospheric pressure. On the Fahrenheit scale the equivalent temperatures are 32° and 212°, respectively. The formula for converting Celsius to Fahrenheit is $°F = 9/5 \ °C + 32$, and the formula for converting Fahrenheit to Celsius is $°C = 5/9(°F - 32)$.

850

C	C or F	F	C	C or F	F	C	C or F	F	C	C or F	F
−73.33	−100	−148.0	− 6.67	20	68.0	15.6	60	140.0	43	110	230
−70.56	− 95	−139.0	− 6.11	21	69.8	16.1	61	141.8	49	120	248
−67.78	− 90	−130.0	− 5.56	22	71.6	16.7	62	143.6	54	130	266
−65.00	− 85	−121.0	− 5.00	23	73.4	17.2	63	145.4	60	140	284
−62.22	− 80	−112.0	− 4.44	24	75.2	17.8	64	147.2	66	150	302
−59.45	− 75	−103.0	− 3.89	25	77.0	18.3	65	149.0	71	160	320
−56.67	− 70	− 94.0	− 3.33	26	78.8	18.9	66	150.8	77	170	338
−53.89	− 65	− 85.0	− 2.78	27	80.6	19.4	67	152.6	82	180	356
−51.11	− 60	− 76.0	− 2.22	28	82.4	20.0	68	154.4	88	190	374
−48.34	− 55	− 67.0	− 1.67	29	84.2	20.6	69	156.2	93	200	392
−45.56	− 50	− 58.0	− 1.11	30	86.0	21.1	70	158.0	99	210	410
−42.78	− 45	− 49.0	− 0.56	31	87.8	21.7	71	159.8	100	212	414
−40.0	− 40	− 40.0	0	32	89.6	22.2	72	161.6	104	220	428
−37.23	− 35	− 31.0				22.8	73	163.4	110	230	446
−34.44	− 30	− 22.0	0.56	33	91.4	23.3	74	165.2	116	240	464
−31.67	− 25	− 13.0	1.11	34	93.2	23.9	75	167.0	121	250	482
−28.89	− 20	− 4.0	1.67	35	95.0	24.4	76	168.8	127	260	500
−26.12	− 15	5.0	2.22	36	96.8	25.0	77	170.6	132	270	518
−23.33	− 10	14.0	2.78	37	98.6	25.6	78	172.4	138	280	536
−20.56	− 5	23.0	3.33	38	100.4	26.1	79	174.2	143	290	554
−17.8	0	32.0	3.89	39	102.2	26.7	80	176.0	149	300	572
			4.44	40	104.0	27.2	81	177.8	154	310	590
−17.2	1	33.8	5.00	41	105.8	27.8	82	179.6	160	320	608
−16.7	2	35.6	5.56	42	107.6	28.3	83	181.4	166	330	626
−16.1	3	37.4	6.11	43	109.4	28.9	84	183.2	171	340	644
−15.6	4	39.2	6.67	44	111.2	29.4	85	185.0	177	350	662
−15.0	5	41.0	7.22	45	113.0	30.0	86	186.8	182	360	680
−14.4	6	42.8	7.78	46	114.8	30.6	87	188.6	188	370	698
−13.9	7	44.6	8.33	47	116.6	31.1	88	190.4	193	380	716
−13.3	8	46.4	8.89	48	118.4	31.7	89	192.2	199	390	734
−12.8	9	48.2	9.44	49	120.2	32.2	90	194.0	204	400	752
−12.2	10	50.0	10.0	50	122.0	32.8	91	195.8	210	410	770
−11.7	11	51.8	10.6	51	123.8	33.3	92	197.6	216	420	788
−11.1	12	53.6	11.1	52	125.6	33.9	93	199.4	221	430	806
−10.6	13	55.4	11.7	53	127.4	34.4	94	201.2	227	440	824
−10.0	14	57.2	12.2	54	129.2	35.0	95	203.0	232	450	842
− 9.44	15	59.0	12.8	55	131.0	35.6	96	204.8	238	460	860
− 8.89	16	60.8	13.3	56	132.8	36.1	97	206.6	243	470	878
− 8.33	17	62.6	13.9	57	134.6	36.7	98	208.4	249	480	896
− 7.78	18	64.4	14.4	58	136.4	37.2	99	210.2	254	490	914
− 7.22	19	66.2	15.0	59	138.2	37.8	100	212.0	260	500	932

AREA

Metric

1 square centimeter	=	0.155 sq inch
	=	100 sq millimeters
1 square meter	=	1,550 sq inches
	=	10.764 sq feet
	=	1.196 sq yards
	=	10,000 sq centimeters
1 square kilometer	=	0.3861 sq mile
	=	1,000,000 sq meters
1 hectare	=	2.471 acres
	=	10,000 sq meters

Imperial

1 square inch	=	6.452 sq centimeters
	=	1/144 sq foot
	=	1/1296 sq yard
1 square foot	=	929.088 sq centimeters
	=	0.0929 sq meter
1 square yard	=	8,361.3 sq centimeters
	=	0.8361 sq meter
	=	1,296 sq inches
	=	9 sq feet
1 square mile	=	2.59 sq kilometers
	=	640 acres
1 acre	=	0.4047 hectare
	=	43,560 sq feet
	=	4,840 sq yards
	=	4,046.87 sq meters

LENGTH

Metric

1 nanometer	=	0.001 micrometer
1 micrometer	=	0.001 millimeter
1 millimeter	=	0.001 meter
	=	0.0394 inch
1 centimeter	=	10 millimeters
	=	0.3937 inch
	=	0.01 meter
1 meter	=	39.37 inches
	=	3.281 feet
	=	1,000 millimeters
	=	100 centimeters
1 kilometer	=	3,281 feet
	=	1,094 yards
	=	0.621 mile
	=	1,000 meters

Imperial

1 inch	=	25.4 millimeters
	=	2.54 centimeters
1 foot	=	30.48 centimeters
	=	0.3048 meter
	=	12 inches
1 yard	=	0.9144 meter
	=	91.44 centimeters
	=	3 feet
1 mile	=	1,609.347 meters
	=	1.609 kilometers
	=	5,280 feet
	=	1,760 yards

WEIGHT

Metric

1 milligram	=	0.001 gram
	=	0.0154 grain
1 gram	=	0.0353 avoirdupois ounce
	=	15.4324 grains
1 kilogram	=	1,000 grams
	=	353 avoirdupois ounces
	=	2.2046 avoirdupois pounds
1 metric ton	=	1,000 kilograms
	=	2,204.6 pounds
	=	1.102 short tons
	=	0.984 long ton

Imperial

1 grain	=	1/7000 avoirdupois pound
	=	0.064799 gram
1 ounce (avoirdupois)	=	28.3496 grams
	=	437.5 grains
	=	1/16 pound
1 pound (avoirdupois)	=	453.593 grams
	=	0.45369 kilograms
	=	16 ounces
1 short ton	=	907.184 kilograms
	=	0.9072 metric tons
	=	2,000 pounds

YIELD

Metric

1 kilogram per hectare	=	0.89 pound per acre
1 cubic meter per hectare	=	14.2916 cubic feet per acre

Imperial

1 pound per acre	=	1.121 kilograms per hectare
1 ton (2000 lb) per acre	=	2.42 metric tons per hectare
1 cubic foot per acre	=	0.0699 cubic meter per hectare
1 bushel (60 lb) per acre	=	67.26 kilograms per hectare

VOLUME

Metric

1 liter	=	1.057 U.S. quarts (liquid)
	=	0.9081 quart (dry)
	=	0.2642 U.S. gallon
	=	0.221 Imperial gallon
	=	1,000 milliliters
	=	0.0353 cubic foot
	=	61.02 cubic inches
	=	0.001 cubic meter
1 cubic meter	=	61,023.38 cubic inches
	=	35.314 cubic feet
	=	1.308 cubic yards
	=	264.17 U.S. gallons
	=	1,000 liters
	=	28.38 U.S. bushels
	=	1,000,000 cubic centimeters
	=	1,000,000,000 cubic millimeters

Imperial

1 fluid ounce	=	1/128 gallon
	=	29.57 cubic centimeters
	=	29.562 milliliters
	=	1.805 cubic inches
	=	0.0625 U.S. pint (liquid)

1 U.S. quart (liquid)	=	946.3 milliliters
	=	57.75 cubic inches
	=	32 fluid ounces
	=	4 cups
	=	1/4 gallon
	=	2 U.S. pints (liquid)
	=	0.946 liter
1 quart (dry)	=	1.1012 liters
	=	67.20 cubic inches
	=	2 pints (dry)
	=	0.125 peck
	=	1/32 bushel
1 cubic inch	=	16.387 cubic centimeters
1 cubic foot	=	28,317 cubic centimeters
	=	0.0283 cubic meter
	=	28.316 liters
	=	7.481 U.S. gallons
	=	1,728 cubic inches
1 U.S. gallon	=	16 cups
	=	3,785 liters
	=	231 cubic inches
	=	4 U.S. quarts (liquid)
	=	8 U.S. pints (liquid)
	=	8.3453 pounds of water
	=	128 fluid ounces
	=	0.8327 British Imperial gallon
1 British Imperial gallon	=	4.546 liters
	=	1.201 U.S. gallons
	=	277.42 cubic inches
1 U.S. bushel	=	35.24 liters
	=	2,150.42 cubic inches
	=	1.2444 cubic feet
	=	0.03524 cubic meter
	=	2 pecks
	=	32 quarts (dry)
	=	64 pints (dry)

Index